McCoy's RCRA Unraveled

2015 Edition

McCoy and Associates, Inc.
Lakewood, Colorado
www.understandrcra.com

McCoy's RCRA Unraveled (ISBN: 978-0-930469-70-2) is published one time per year (January) by McCoy and Associates, Inc., 12596 West Bayaud Avenue, Suite 210, Lakewood, Colorado 80228-2035. www.understandrcra.com. Copyright 2015 by McCoy and Associates, Inc. Manufactured in the USA.

No part of this publication may be stored in a retrieval system, transmitted, or reproduced in any way, including but not limited to photocopy, photograph, magnetic or other record, without the prior agreement and written permission of the publisher.

Considerable care has been exercised in preparing this document; however, McCoy and Associates, Inc. makes no representation, warranty, or guarantee in connection with the publication of this guidance manual. McCoy and Associates, Inc. expressly disclaims any liability or responsibility for loss or damage resulting from its use or for the violation of any federal, state, or municipal law or regulation with which this guidance manual may conflict. McCoy and Associates, Inc. does not undertake any duty to ensure the continued accuracy of this publication.

This guidance manual addresses problems of a general nature related to the federal RCRA regulations. Persons evaluating specific circumstances dealing with the RCRA regulations should review state and local laws and regulations, which may be more stringent than federal requirements. In addition, the assistance of a qualified professional should be enlisted to address any site-specific circumstances.

How to reach McCoy and Associates. Web: www.understandrcra.com; Email: info@understandrcra.com; Phone: 303-526-2674; Fax: 303-526-5471; Physical address: McCoy and Associates, Inc., 12596 West Bayaud Avenue, Suite 210, Lakewood, Colorado 80228-2035.

Technical Editors
Paul Gallagher, P.E., President
Eric Weber, VP of Product Development
Rodger Goffredi, P.E., Senior Environmental Engineer
Brian Lindman, Senior Environmental Scientist

Design Editor
Robyn Weber, VP of Marketing and Strategy

Production Designers
Allan Shearer
Peter Hancik
Kristen Weber

Office Manager
Erin Beaver

Senior Administrative Assistant
Brian Ciccarone

Administrative Assistant
Molly Garvey

Surviving an avalanche.

We live in a complex, interdependent, hyper-competitive business world. And the environmental managers who must make accurate, rapid-fire operational decisions on behalf of their organizations—deal with an avalanche of pressure. Most of the time, the impact of the myriad of RCRA decisions they make are benign. But when unintentional missteps and errors happen, things can break loose in a blink of an eye.

Because you cannot predict where the next problem will arise, environmental managers must prepare like pragmatic backcountry skiers. You must fully understand the risks, obtain the necessary knowledge and skills through experience, reinforce concepts through training, and be acutely aware of human dynamics when making difficult decisions.

Those who deal with RCRA get comfortable with the risks and complexities. However, a wild card can appear when making your "compliance case" to coworkers, senior management, regulators and others, because people can behave in unpredictable ways. And complex social interactions can trump your RCRA experience and planning.

In 2004, avalanche researcher Ian McCammon, released a seminal study, *Heuristics Traps in Recreational Avalanche Accidents: Evidence and Implications*. He identified six human, social, and psychological factors present in more than 95 percent of the fatal accidents. They are:

Familiarity: I've done this many times before and never had problems.

Acceptance: I need to please my coworkers and boss.

Consistency: I'll stick with what I know and minimize new information.

The Expert Halo: The most experienced people know what they're doing.

Tracks/Scarcity: Let's get the goods before someone else does.

Social Facilitation/Proof: Everyone else is doing it.

At McCoy, our goal is to support good decision making through our educational process. We encourage customers to be careful how they deliberate. There are no silver bullets or short cuts. Check and recheck assumptions. Listen carefully to others. Encourage discussion and vigorous debate with your team and regulators. Pour over the best available information. Then check it again. All to ensure that compliance, and risks, are managed well. Visit understandrcra.com to learn more.

Preface

EPA's regulations implementing the Resource Conservation and Recovery Act (RCRA) are extremely complex. As its title implies, *McCoy's RCRA Unraveled* unties the knots often created by the RCRA regulations and clarifies a number of difficult issues. This publication takes a unique look at the areas in RCRA that result in the most confusion, concern, and enforcement actions.

The 2015 edition of *McCoy's RCRA Unraveled* includes guidance/clarification on most of the troublesome areas in Parts 261–266, 268, 273, and 279 of the RCRA regs contained in Title 40 of the *Code of Federal Regulations* (*CFR*). As a companion publication, *McCoy's RCRA Reference* contains the up-to-date federal hazardous waste regulations in a user-friendly format along with an extensive keyword index. The regulatory text in *McCoy's RCRA Reference* is cross-referenced to the appropriate section numbers in this book so that the reader can read the text of the regulations and then review any interpretations/guidance pertaining to them. Both of these publications are included in our electronic offering: *McCoy's RCRA Compliance CD*. Our goal is to provide an electronic product that is powerful and easy to use. We reformatted the content with large type so on-screen reading is comfortable. We've included over 45,000 hyperlinks, 1,450 RCRA Online documents, and 360 *Federal Register* notices (1995 and later) to lighten your research load. And there are three ways to find the specific information you need: hyperlinked table of contents, knowledge-based indexes, and Find and Search capability. Take a look at a few sample pages at www.mccoyseminars.com.

The material presented in this book is used extensively in McCoy's RCRA seminars—call 303-526-2674 for information. However, this publication is designed for standalone use by readers of all experience levels.

Significant Rule Changes in Last Three Years

Final rules issued by EPA that have significantly changed the RCRA Subtitle C regulatory program in the past three years (2012–2014) are summarized in the table on the following page and are discussed briefly below. Please note that some of these rules made the federal regulations less stringent and are not applicable in RCRA-authorized states until they are adopted by the state.

Electronic manifest rule—EPA issued its e-manifest rule on February 7, 2014. [79 *FR* 7517] The rule implements a portion of the Hazardous

PREFACE

Rule	Issue date	*Federal Register* reference	Federal effective date
Electronic manifest rule	February 7, 2014	79 *FR* 7517	August 6, 2014
Exclusion for CO_2 sequestration	January 3, 2014	79 *FR* 350	March 4, 2014
Solvent-contaminated wipes rule	July 31, 2013	78 *FR* 46448	January 31, 2014

Waste Electronic Manifest Establishment Act (e-Manifest Act), which was signed into law on October 5, 2012. The act authorizes EPA to 1) develop an electronic hazardous waste manifest system; and 2) impose user fees as necessary to recover costs incurred in developing, operating, maintaining, and upgrading the system. EPA was given a one-year deadline to promulgate regulations to carry out the new law and a three-year deadline to have the system fully operational.

The primary purpose of the new rule is to clarify that e-manifests, completed and signed as specified in the rule, are legally valid for all RCRA purposes. When the e-manifest system is operational, e-manifests will be used for shipments of hazardous waste, state-only hazardous waste, and any other waste that a state requires be accompanied by a hazardous waste manifest. The benefits of an e-manifest include greater access by emergency responders to information about a waste shipment, higher quality and more timely waste shipment data, lower cost, and fewer burdens on the regulated community.

Until the e-manifest system is ready, the paper manifest will continue to be used. Ultimately, operation of the e-manifest system will be paid for by fees collected from users. EPA plans to issue a separate rule to establish the fee structure in fiscal year 2015.

Waste handlers will be able to opt out of the electronic system and continue using paper manifests if they choose. If a paper manifest is used, the facility that receives the waste shipment and terminates the manifest will be required to send a copy to the operator of the e-manifest system so that the data from the manifest can be entered into the system. Thus, the electronic system will be a comprehensive repository for all information on hazardous waste shipments.

Newly codified §262.25 says that the electronic signature methods for the e-manifest must be: 1) legally valid and enforceable signatures under applicable EPA and other federal requirements pertaining to electronic signatures, and 2) designed so as to be cost-effective and practical. What constitutes a valid and enforceable signature is governed by EPA's Cross-Media Electronic Reporting (CROMERR) regulations, which are codified at 40 *CFR* Part 3. EPA believes that the first-generation system should support a PIN/password signature method and/or a digitized handwritten signature using a signature pad and stylus. Based on preamble language, EPA is recommending the PIN/password approach, with security questions, but the agency also expects to include the digitized handwritten signature method pending the outcome of studies to demonstrate its forensic reliability.

To satisfy DOT requirements, a generator originating an e-manifest must also provide the initial transporter with one printed copy of the manifest. After the initial transporter has signed the e-manifest to take custody of the waste shipment, if the e-manifest system goes down or if the e-manifest cannot be completed electronically for any reason, the transporter must make a copy of the printed manifest for each waste handler and two additional copies for the designated facility. These paper copies will then become the manifest for that shipment and must be signed and handled accordingly by each waste handler.

Under the e-manifest rule, the regulated community can no longer assert that information entered on paper manifests, or on an e-manifest when the system is available, is confidential business information (CBI). EPA gave two reasons for this policy:

> "First...as manifests are shared with several commercial entities while they are being processed and used, a business concerned with protecting its commercial information would find it exceedingly difficult to protect its individual manifest records from disclosure by all the other persons who come into contact with its manifests.... Second, we explained that much of the information that might be claimed by industry commenters to be CBI is already available to the public from a number of government and other legitimate sources, because a large number of states now require the submission of generator and/or TSDF copies of manifests to state data systems, and the data from these manifests are often made publicly available through state Web sites or reported and disclosed freely in federal and state information systems." [79 FR 7540]

The e-manifest rule was effective on August 6, 2014. However, the implementation and compliance date for these regulations will be delayed until the e-manifest system is ready for operation and the schedule of fees for manifest-related services has been issued. EPA will publish a further document subsequent to this rule's effective date to announce the user fee schedule—this document will also announce the date upon which compliance with this regulation will be required. Similar to HSWA regulations, the e-manifest rule will be effective in all states at the same time and will be implemented by EPA in RCRA-authorized states until the state adopts the rule.

Exclusion for CO sequestration—On January 3, 2014 [79 FR 350], EPA excluded waste carbon dioxide (CO_2) streams injected underground for geologic sequestration from the definition of hazardous waste. Generators and well owners/operators must meet the following conditions to claim the new exclusion: these CO_2 streams must be 1) captured from emission sources (e.g., power plants or other industrial sources), 2) transported via DOT-regulated (or state equivalently-regulated) pipelines or vehicles, and 3) injected into SDWA Class VI wells for purposes of geologic sequestration.

Generators of these excluded CO_2 streams and Class VI well owners/operators must separately certify that no hazardous wastes have been mixed, or otherwise co-injected, with the CO_2 stream. These certifications must be updated annually, kept onsite for three years, and be accessible on the facilities' publicly-available websites.

EPA noted in the preamble to the rule that the exclusion does not affect the regulatory status of CO_2 streams that are injected into wells other than SDWA Class VI wells for geologic sequestration. The agency specifically noted that "should CO_2 be used for its intended purpose as it is injected into UIC Class II wells for the purpose of [enhanced oil or gas recovery], it is EPA's expectation that such an injection process would not generally be a waste management activity." [79 FR 355]

The exclusion has been added as a new section, §261.4(h), effective March 4, 2014 in the federal RCRA regs. Because this exclusion makes the federal program less stringent, states are not required to adopt the new provision (although EPA encourages all states to do so). "EPA notes that in situations involving the interstate transportation of conditionally-excluded waste, the exclusion must be authorized in the state where the waste is generated, any states through which the waste passes, and the state where the UIC Class VI injection well is located, in order for that conditionally-excluded waste to be managed as excluded from Subtitle C from point of generation to injection in a UIC Class VI well. A state that has not adopted the conditional exclusion may impose state requirements, including the uniform hazardous waste manifest requirement (where applicable) if characteristically-hazardous CO_2 streams are being transported through that state." [79 FR 360]

Solvent-contaminated wipes rule—In an attempt to provide national consistency for the management of solvent-contaminated rags and wipes, EPA has issued exclusions from the definitions of solid and hazardous waste for:

1. Wipes that contain one or more of the F-listed spent solvents or the corresponding P- or U-listed unused solvents (e.g., that were spilled and wiped up),
2. Wipes that exhibit a characteristic if that characteristic results from a solvent listed in Part 261 (e.g., ignitability due to acetone, toxicity due to methyl ethyl ketone), and
3. Wipes that exhibit the characteristic of ignitability due to the presence of one or more solvents that are not listed in Part 261.

Solvent-contaminated wipes that contain listed hazardous waste other than solvents, or exhibit a characteristic of corrosivity, reactivity, or toxicity due to contaminants other than solvents are not eligible for the new exclusions. For example, wipes that exhibit the toxicity characteristic because they're contaminated with RCRA metals are not excluded. The two new exclusions for solvent-contaminated rags and wipes are:

1. *Exclusion from the definition of hazardous waste*—To be excluded from the definition of hazardous waste, solvent-contaminated wipes that will be sent for disposal (to a landfill or combustion facility) must meet the following requirements:
   - The wipes cannot be hazardous due to trichloroethylene.
   - Containers used for accumulation and transportation must be 1) non-leaking and closed [defined in new §261.4(b)(18)(i)], 2) able to contain free liquids, and 3) labeled "Excluded Solvent-Contaminated Wipes."
   - Generator accumulation time cannot exceed 180 days (regardless of generator status) from the start date of accumulation for each container (i.e., the date the first solvent-contaminated wipe is placed in the container).
   - At the point of being transported for disposal, the solvent-contaminated wipes must not contain free liquids as determined by the paint filter test (SW–846 Method 9095B), and there cannot be any free liquid in the container holding the wipes. Free liquids removed from solvent-contaminated wipes or from containers holding the wipes 1) are not excluded, 2) must be characterized, and 3) must be managed appropriately based on the hazardous waste characterization.
   - Documentation per new §261.4(b)(18)(v) must be maintained at the generator's site.
   - The disposal facility must be one of the following: 1) a municipal solid waste landfill regulated under Part 258, including §258.40 (design criteria for new landfills); 2) a hazardous waste landfill regulated under Part 264 or 265; 3) a municipal waste combustor or other combustion facility regulated under Section 129 of the Clean Air Act; or 4) a hazardous waste combustor, boiler, or industrial furnace regulated under Part 264, 265, or Subpart H of Part 266.

2. *Exclusion from the definition of solid waste*—EPA is excluding solvent-contaminated wipes that are sent for cleaning and reuse from the definition of solid waste because the agency believes they are more commodity-like than waste-like. To be excluded from the definition of solid waste, solvent-contaminated wipes that will be cleaned and reused must meet the following requirements:
   - The exclusion is extended to wipes contaminated with any of the F- or U-listed solvents, including trichloroethylene.
   - Containers used for accumulation and transportation must be 1) non-leaking and closed

- [defined in new §261.4(a)(26)(i)], 2) able to contain free liquids, and 3) labeled "Excluded Solvent-Contaminated Wipes."
- Generator accumulation time cannot exceed 180 days (regardless of generator status) from the start date of accumulation for each container.
- At the point of being sent for cleaning, either onsite or offsite, the wipes and the container they are in must not contain free liquids.
- Documentation per new §261.4(a)(26)(v) must be maintained at the generator's site.
- The wipes must be sent to a laundry or dry cleaner whose discharge, if any, is regulated under Sections 301 and 402 or Section 307 of the Clean Water Act.

The effective date of the new rule was January 31, 2014 at the federal level. Because, the conditional exclusions discussed above are not HSWA regulations, they will not become effective in authorized states until those states adopt the rule. The rule includes requirements and conditions that are less stringent than those required under the base RCRA hazardous waste program. Thus, states, except as described below, are not required to adopt the conditional exclusions (although EPA encourages such adoption to reduce regulatory burden on businesses and maximize national consistency). Some conditions required in the final rule may be more stringent than certain existing state programs. As a result, authorized states whose programs include requirements that are less stringent than the final rule are required to modify their programs to maintain consistency with the federal program.

Obtaining Guidance Used in This Book

In writing this book, we made a significant effort to find and read every bit of publicly available guidance that EPA has issued for each topic included. We subsequently reviewed and distilled all of that guidance into an easy-to-read article—one article for each subject. However, we don't claim to have reviewed and incorporated all of the guidance that EPA has issued, in part because we don't have access to internal agency documents. The thousands of pages of guidance that we did gather, however, are specifically referenced in this book, giving the reader access to pertinent documents. Having the applicable guidance in hand will give the reader the ammunition he/she needs to argue their case with co-workers or regulators. These references and their means of acquisition fall into four primary groups:

1. *Federal Register* preamble language—When EPA publishes a rule affecting the RCRA regulations, it does so in the *Federal Register*, a document published every working day by the Office of the Federal Register in Washington, DC. The agency includes a discussion of its rationale for the rule and response to public comments in what's called the "rule preamble." The preamble doesn't get codified into the *CFR*, but it is a great source of guidance on EPA's intent and interpretation of its regulations. Every time we used such preamble language as the source for our discussion in this book, we have listed the date, volume number, and page number of the applicable *Federal Register*.

   All 1995 and later *Federal Register* rules and preambles used as a source of guidance in this book are available from a Government Printing Office website at http://www.gpo.gov/fdsys/browse/collection.action?collectionCode=FR. Pre-1995 rules and preambles specific to major RCRA rules can be obtained from an EPA website: http://www.epa.gov/osw/hazard/dsw/fedregnotices.htm. Rules not covered by that EPA site can be obtained through a number of subscription-based services, or by calling us at 303-526-2674. All 1995 and later *Federal Register* rules and preambles used as a source of guidance in this book are loaded and hyperlinked on *McCoy's RCRA Compliance CD*, 2015 Edition.

2. RCRA Online (RO) documents—Many of the thousands of letters and other guidance documents (including questions and answers from the RCRA Hotline) that the agency has issued over the last 30 plus years are available online. The agency has assigned each letter/guidance document a specific five-digit RO number (previously called FAXBACK number). You will see those numbers listed as the reference for a significant amount of the discussion and many of the examples included in this book.

   The most convenient method of getting your hands on these RO documents is to use EPA's online document retrieval service. This service, called RCRA Online, is an excellent, free database developed by EPA. Go to http://www.epa.gov/epawaste/inforesources/online/index.htm, and click on "Advanced Search" on the right side of the opening page. The second data-entry box from the top on the right-hand side of the Advanced Search page asks for the RCRA Online Number. Enter the five-digit RO code you are interested in and click on the Search button. A "Search Results" page will come up giving you an abstract of the document and a hyperlink. Clicking on the hyperlink will retrieve the "Record Detail" of the document, providing more information and (finally) a page icon to click to download a text or PDF file of the document itself.

   The 2015 edition of *McCoy's RCRA Compliance CD* also has all of the applicable RO documents loaded and hyperlinked for easy reference and printing.

3. EPA reports—A number of EPA reports were used in the preparation of this book. These reports all have unique EPA numbers associated with them (e.g., EPA/530/K-01/004), and the availability of each of these reports from various EPA websites is included in each reference.

4. OSWER Directives—Finally, there are a few guidance documents that are referenced in this publication that are not available through RCRA Online or other websites. These documents are referenced as EPA's Office of Solid Waste and Emergency Response (OSWER) Directives and are identified by unique numbers assigned by OSWER. They can typically be obtained by calling the RCRA Docket at 202-566-0270 or by calling us at 303-526-2674.

Although we have relied on EPA guidance whenever possible, we have also included in this book insights imparted to us by the thousands of people who have attended our RCRA seminars over the years. Additionally, in areas where EPA's guidance was inconsistent or not to be found at all, we have tried to bring this to your attention. Although at times we are unable to be definitive in our discussions about specific topics, the information presented in this book reflects our best professional judgment on the intent of the regulations.

Comments

If you have any comments or questions regarding this publication, please don't hesitate to call us at 303-526-2674 or email us at info@mccoyseminars.com. For up-to-date information regarding our seminars and publications, visit our website at www.mccoyseminars.com.

Table of Contents

Table of Contents ··· xi

Chapter 1—Solid Wastes ·· 1

1.1	**Introduction to solid wastes** ········ 1	1.2.3	Recycling activities subject to RCRA control ··························· 10
1.1.1	Discarded materials ············· 1	1.2.4	Putting it all together ··········· 11
1.1.1.1	Products ························ 2	1.2.5	Classifying your secondary material ··· 12
1.1.1.2	Garbage, refuse, and sludge ········ 3	1.2.5.1	Spent materials ················ 12
1.1.1.3	Materials that are thrown away, abandoned, or destroyed ··········· 4	1.2.5.2	Sludges ······················ 14
1.1.1.3.1	Wastewater ····················· 4	1.2.5.3	By-products ··················· 15
1.1.1.3.2	Process wastes ·················· 5	1.2.5.3.1	By-products vs. co-products ······· 15
1.1.1.3.3	Gaseous emissions from manufacturing operations ······················ 5	1.2.5.4	Commercial chemical products ····· 16
		1.2.5.4.1	Two definitions of "commercial chemical product" ··············· 17
1.1.1.3.4	Obsolete chemicals ················ 5	1.2.5.4.2	Use constituting disposal ········· 19
1.1.1.4	Spent materials ·················· 5	1.2.5.4.3	Burning for energy recovery ······· 19
1.1.1.5	Incidental residues ················ 5	1.2.5.4.4	Reclamation ··················· 20
1.1.1.6	"Discarded materials" is a very broad net ····························· 5	1.2.5.5	Scrap metal ··················· 22
		1.2.5.5.1	Excluded scrap metal ············ 23
1.1.2	Solid wastes are a subset of discarded materials ······················ 6	1.2.6	Determining your recycling activity ··· 24
		1.2.6.1	Use constituting disposal ········· 24
1.1.3	Hazardous wastes are a subset of solid wastes ························ 6	1.2.6.1.1	Fertilizer production ············· 24
		1.2.6.1.2	Cement or aggregate production ···· 25
1.1.4	EPA defines "solid wastes" to capture recycling activities ··············· 7	1.2.6.1.3	Materials applied to the land when that is their ordinary manner of use ·· 27
1.1.4.1	Abandoned materials ············· 7	1.2.6.1.4	Additional examples involving use constituting disposal ············· 27
1.1.4.1.1	Materials that are disposed ········ 7	1.2.6.2	Burning for energy recovery ······· 28
1.1.4.1.2	Materials that are burned or incinerated ····················· 7	1.2.6.2.1	Fuel blending ·················· 30
		1.2.6.3	Reclamation ··················· 31
1.1.4.1.3	Materials that are stored in lieu of being abandoned ················ 8	1.2.6.3.1	DSW rule establishes exclusions for materials being reclaimed ········ 33
1.1.4.2	Materials that are inherently waste-like ··· 8	1.2.6.3.2	Secondary materials from mineral processing ···················· 36
1.2	**Materials that are solid wastes when recycled** ······················ 9	1.2.6.4	Speculative accumulation ········· 36
1.2.1	Background ···················· 9	1.2.7	The big caveat ················· 37
1.2.2	Materials that may be RCRA-regulated when recycled ·················· 9		

Chapter 2—Characteristic Wastes ···································· 39

2.1	**Introduction to hazardous wastes** ···· 40	2.2.2.1	Ignitability of solvent-contaminated rags/wipes ···················· 49
2.1.1	Four types of hazardous wastes ····· 40	2.2.3	Ignitable compressed gases ········ 49
2.1.1.1	Declared hazardous wastes ········ 40	2.2.4	Oxidizers ····················· 49
2.1.1.1.1	Declaring wastes as hazardous may save analytical costs ············ 41	2.3	**Corrosivity** ·················· 51
2.2	**Ignitability** ·················· 42	2.3.1	Aqueous vs. liquid wastes ········ 51
2.2.1	Liquids with low flash points ······ 42	2.3.2	Aqueous wastes with low or high pH ·· 52
2.2.1.1	Determining if a waste is a liquid ··· 43	2.3.2.1	State definition of corrosivity may be more stringent ················ 52
2.2.1.2	Flash point testing ·············· 43	2.3.2.2	Derivation of pH levels ··········· 52
2.2.1.3	The alcohol-content exclusion ····· 46	2.3.3	Liquid wastes that corrode carbon steel ·· 53
2.2.2	Ignitable solids ················ 46		

TABLE OF CONTENTS

2.3.4	Do both D002 criteria have to be evaluated?	53
2.3.5	What about corrosive solids?	54
2.4	**Reactivity**	**54**
2.4.1	No test methods available for determining reactivity	55
2.4.2	Reactive cyanides and sulfides	56
2.4.2.1	Release thresholds for cyanide and sulfide wastes no longer valid	56
2.4.2.2	Knowledge-based determinations for reactive cyanide and sulfide wastes	56
2.4.3	Explosives	57
2.4.3.1	Off-specification small arms ammunition	57
2.4.3.2	Explosives-contaminated soil/sediment	57
2.4.3.3	Claiming the immediate response exemption	57
2.4.4	Management of reactive wastes in lagoons	58
2.5	**Aerosol cans**	**58**
2.5.1	Regulation of cans depends on management method	58
2.5.1.1	Punctured aerosol cans are not regulated when recycled	58
2.5.1.2	Aerosol cans destined for disposal may be hazardous waste	59
2.5.2	Liquid material from aerosol cans may be hazardous waste	60
2.5.3	What about the propellant?	60
2.5.3.1	Combusting aerosol can propellant for energy recovery	60
2.5.4	Accumulating aerosol cans prior to puncturing/shipment	61
2.5.5	Household waste exclusion	61
2.5.6	State programs may be different	61
2.6	**Toxicity**	**62**
2.6.1	Background	62
2.6.1.1	EP toxicity test initially used	64
2.6.1.2	TC revised in 1990	64
2.6.2	TCLP-based toxicity determinations	65
2.6.2.1	Cost	65
2.6.2.2	Reproducibility	65
2.6.2.3	Analyzing the extract	66
2.6.2.4	Oily wastes and organic liquids pose problems	66
2.6.2.5	Sample holding times	66
2.6.2.6	TCLP not valid for manufactured gas plant wastes	67
2.6.3	Knowledge-based toxicity determinations	67
2.6.3.1	Total waste analyses in lieu of TCLP results	67
2.6.4	Exemptions from the TC	69
2.6.5	Other TC examples	73
2.6.6	Masking of lead-bearing wastes prior to disposal is prohibited	74
2.6.6.1	Lead-based paint waste	74
2.6.7	Delisting petitions	75
2.7	**Electronic wastes (e-wastes)**	**75**
2.7.1	E-wastes often exhibit the toxicity characteristic	76
2.7.2	Management of e-wastes	76
2.7.3	Cathode ray tubes (CRTs)	77
2.7.3.1	Current regulatory status of CRTs	78
2.7.3.1.1	Processed CRT glass sent to smelters	79
2.7.3.2	Exporting CRTs	79
2.7.4	Circuit boards in electronic components	80
2.8	**Hazardous waste lamps**	**80**
2.8.1	History	80
2.8.2	Waste lamps managed as hazardous waste	81
2.8.2.1	Accumulation	81
2.8.2.2	Crushing	81
2.8.2.3	Transportation	82
2.8.3	Making a hazardous waste determination for lamps	82
2.8.4	Fluorescent light ballasts	83
2.8.4.1	PCB-containing light ballasts	83
2.8.4.2	DEHP-containing light ballasts	83
2.8.5	CERCLA reporting requirements	84

Chapter 3—Listed Wastes ... 85

3.1	**Introduction to listed wastes**	**85**
3.1.1	Four lists of hazardous wastes	86
3.1.2	ICR-only listed wastes	87
3.1.3	State-listed wastes	88
3.2	**Spent solvents**	**88**
3.2.1	Description of the spent solvents	88
3.2.1.1	F001	88
3.2.1.2	F002	88
3.2.1.3	Differences between F001 and F002 listings	88
3.2.1.4	F003	89
3.2.1.5	How nonignitable F003 spent solvents are regulated	89
3.2.1.6	F004	89
3.2.1.7	F005	89
3.2.2	Two criteria must be met for F001–F005 listings to apply	89
3.2.2.1	Manufacturing operations	90
3.2.2.2	Painting operations	91
3.2.2.3	Laboratory activities	91
3.2.2.4	Miscellaneous	92
3.2.3	Spent solvent listings often do not apply	92
3.2.3.1	Process wastes and products do not fall under the F001–F005 listings	92
3.2.3.1.1	Solvents used as reactants	92

3.2.3.1.2	Solvents used as ingredients · · · · · · 93	3.4.2.5.1	Zinc plating · · · · · · · · · · · · · · · 113
3.2.3.2	Other solvent-contaminated process wastes are not F001–F005 · · · · · · · 95	3.4.2.5.2	Chemical conversion coating · · · · · 114
		3.4.2.5.3	Printed circuit board manufacturing · · 114
3.2.4	Spent solvent mixtures · · · · · · · · · · 96	3.4.2.6	F006 storage · · · · · · · · · · · · · · · 114
3.2.4.1	Mixtures containing F001, F002, F004, and F005 solvents · · · · · · · · · · · · 96	3.4.2.7	F006 recycling · · · · · · · · · · · · · · 115
		3.4.2.7.1	F006 exception from the derived-from rule · 116
3.2.4.2	Mixtures containing F003 solvents · · · 97		
3.2.4.3	Other oddities with F003 mixtures · · · 100	3.4.3	F007 wastes · · · · · · · · · · · · · · · 116
3.2.4.3.1	Mixtures and residues of F003 wastes · · 100	3.4.4	F008 wastes · · · · · · · · · · · · · · · 116
3.2.4.3.2	F003-contaminated soil and the contained-in policy · · · · · · · · · · · 101	3.4.5	F009 wastes · · · · · · · · · · · · · · · 117
		3.4.6	F019 wastes · · · · · · · · · · · · · · · 117
3.2.5	Reclaimed solvents · · · · · · · · · · · · 101	3.4.6.1	Exemptions from the F019 listing · · · 118
3.2.6	Miscellaneous spent solvent issues · · · 102	3.4.7	The big disclaimer · · · · · · · · · · · · 118
3.2.6.1	Spill cleanup residues · · · · · · · · · · 102	**3.5**	**P- and U-listed hazardous wastes · · · 118**
3.2.6.2	Used oil · · · · · · · · · · · · · · · · · 103	3.5.1	Regulatory background · · · · · · · · · 119
3.2.6.3	Still bottoms · · · · · · · · · · · · · · · 103	3.5.2	Practical hints on using the P- and U-list · · · · · · · · · · · · · · · 119
3.3	**Solvent-contaminated rags/wipes · · · 103**		
3.3.1	Background · · · · · · · · · · · · · · · 103	3.5.2.1	CAS numbers, variants, and isomers · · 119
3.3.2	Rags regulated on a case-by-case basis from 1991–2013 · · · · · · · · · 104	3.5.2.2	Hazard codes · · · · · · · · · · · · · · 120
		3.5.2.3	Tip for identifying commercial chemical products · · · · · · · · · · · 121
3.3.3	Exclusions for solvent-contaminated rags/wipes · · · · · · · · · · · · · · · 104		
		3.5.3	What is a "commercial chemical product"? · · · · · · · · · · · · · · · 122
3.3.3.1	Exclusion from the definition of hazardous waste · · · · · · · · · · · · 106		
		3.5.3.1	Pure, unused chemicals · · · · · · · · · 122
3.3.3.2	Exclusion from the definition of solid waste · · · · · · · · · · · · · · · 106	3.5.3.1.1	Chemicals used for their intended purpose · · · · · · · · · · · · · · · · · 123
3.3.3.3	Recordkeeping · · · · · · · · · · · · · 107	3.5.3.1.2	Manufactured articles aren't P- or U-wastes · · · · · · · · · · · · · 125
3.3.3.4	State implementation · · · · · · · · · · 107		
3.3.4	Ignitable rags · · · · · · · · · · · · · · 107	3.5.3.2	Manufacturing chemical intermediates · · · · · · · · · · · · · · 125
3.3.5	Rags contaminated with used oil · · · 107		
3.4	**F006, F007, F008, F009, and F019 wastes · · · · · · · · · · · · · · · · · · 107**	3.5.3.3	Off-specification commercial chemical products · · · · · · · · · · · 125
		3.5.3.4	Container residues · · · · · · · · · · · 126
3.4.1	The plating process · · · · · · · · · · · 108	3.5.3.4.1	Partially used containers · · · · · · · · 127
3.4.2	F006 wastes · · · · · · · · · · · · · · · 108	3.5.3.4.2	Used vials and syringes · · · · · · · · · 127
3.4.2.1	Baths and rinsewaters · · · · · · · · · 109	3.5.3.5	Cleanup residue and debris · · · · · · 127
3.4.2.1.1	Chemical etching and milling · · · · · 109	3.5.3.6	What does "technical grades of the chemical" mean? · · · · · · · · · · · 128
3.4.2.1.2	Cleaning and stripping · · · · · · · · · 110		
3.4.2.1.3	Confusion with K062—spent pickle-liquor wastes · · · · · · · · · · 110	3.5.3.7	What is a "sole active ingredient"? · · · 128
		3.5.3.8	Manufacturing process wastes · · · · · 131
3.4.2.1.4	Filtrate and supernatant · · · · · · · · 111	3.5.3.8.1	Manufacturing or product handling? · · 131
3.4.2.2	Sludge · · · · · · · · · · · · · · · · · · 111	3.5.3.8.2	Unreacted reagent from reactors · · · 133
3.4.2.2.1	Sludge from sequential treatment of electroplating wastewaters · · · · · · 111	3.5.4	Equipment regulated under the contained-in policy · · · · · · · · · · 134
3.4.2.2.2	Surprise! You've got F006 · · · · · · · 112	3.5.5	Generator issues · · · · · · · · · · · · 134
3.4.2.2.3	Court decision creates confusion · · · 113	3.5.6	Recycling issues · · · · · · · · · · · · · 135
3.4.2.3	Examples of F006 wastes · · · · · · · 113	3.5.6.1	Limitations on certain recycling practices · · · · · · · · · · · · · · · · 135
3.4.2.4	Methods to avoid F006 generation · · · 113		
3.4.2.5	Exemptions from the F006 listing · · · 113		

Chapter 4—Exclusions and Exemptions · 137

4.1	**Domestic sewage exclusion · · · · · · · 140**	4.1.3	At what point does the exclusion apply? · · · · · · · · · · · · · · · · · 143
4.1.1	Background · · · · · · · · · · · · · · · 140		
4.1.2	How do I qualify for the exclusion? · · 141	4.1.3.1	Notification for discharging hazardous waste to a POTW · · · · · · · · · · · 144
4.1.2.1	"Domestic sewage" · · · · · · · · · · · 142		
4.1.2.2	"POTW" · · · · · · · · · · · · · · · · · 142	4.1.4	What if I mix hazardous waste into my facility's domestic sewage? · · · · 144
4.1.2.3	CWA pretreatment program · · · · · 142		

Section	Title	Page
4.1.4.1	Does the exclusion apply to material removed from the sewer line?	145
4.1.5	What about the sludge from the POTW?	145
4.1.6	Can federally owned treatment works qualify for the exclusion?	146
4.1.7	Are there any LDR implications?	146
4.2	**NPDES discharge exclusion**	**147**
4.2.1	Background	147
4.2.2	Where does the exclusion apply?	147
4.2.3	Discharges to ground water that has a direct connection to surface water	149
4.2.4	Dredged sediments	150
4.2.4.1	Dredged sediments now exempt	150
4.3	**Irrigation return flows exclusion**	**150**
4.4	**Radioactive materials and mixed waste**	**151**
4.4.1	Two types of radioactive materials	151
4.4.1.1	Source, special nuclear, and by-product materials	151
4.4.1.2	NARM	151
4.4.2	Mixed waste	152
4.4.2.1	Commercially generated (non-DOE) mixed waste	152
4.4.2.2	DOE mixed waste	152
4.4.2.3	Hazardous NARM is not mixed waste	153
4.4.3	The RCRA exclusion for radioactive materials	153
4.4.3.1	DOE acquiesces	153
4.4.3.2	AEA takes precedence over RCRA if conflicts occur	154
4.4.3.3	Examples	154
4.4.4	Mixed waste management under RCRA	155
4.4.4.1	Storage/accumulation	155
4.4.4.1.1	RCRA storage provisions may be inapplicable	156
4.4.4.2	Transportation	156
4.4.4.3	Land disposal restrictions	157
4.4.4.3.1	LDR storage prohibition	157
4.4.4.4	Corrective action	158
4.4.5	2001 rule reduces dual regulation of LLMW	158
4.4.5.1	The storage and treatment exemption at non-DOE facilities	159
4.4.5.2	The transportation and disposal exemption for all facilities	160
4.4.6	Mixed waste resources	161
4.5	**In situ mining exclusion**	**161**
4.6	**Pulping liquors exclusion**	**162**
4.6.1	Kraft mill pulping process	162
4.6.2	RCRA implications of pulping liquor recovery	162
4.7	**Oil-bearing secondary materials and recovered oil exclusions**	**164**
4.7.1	Oil-bearing secondary materials exclusion	164
4.7.2	Recovered oil exclusions	166
4.7.2.1	Recovered oil from petroleum operations	166
4.7.2.2	Recovered oil from an associated organic chemical manufacturing facility	168
4.8	**Comparable fuels exclusion**	**169**
4.9	**Exclusion for spent caustic solutions from petroleum refining**	**169**
4.9.1	Spent cresylic, naphthenic, or phenolic caustic	169
4.9.2	Spent sulfidic caustic	171
4.10	**Household hazardous waste exclusion**	**171**
4.10.1	Types of waste excluded	172
4.10.2	Exclusion may be lost if mixed with other wastes	173
4.10.3	Management/treatment of household hazardous wastes	173
4.10.3.1	Household hazardous wastes burned in resource recovery facilities	174
4.10.4	Importing household hazardous wastes	174
4.11	**Exclusion for agricultural waste/manure returned to the soil as fertilizer**	**174**
4.12	**Exclusion for mining overburden returned to the mine site**	**175**
4.13	**Fossil fuel combustion wastes exclusion**	**175**
4.13.1	Four large-volume fossil fuel combustion wastes	175
4.13.2	Fossil fuel combustion wastes from other sources	177
4.13.3	Other, low-volume fossil fuel combustion wastes	178
4.13.3.1	Boiler chemical cleaning wastes	180
4.13.4	Effect of co-burning wastes in boilers	181
4.13.5	Beneficial uses of FFC waste	181
4.13.6	Importing fossil fuel combustion wastes	181
4.14	**Oil and gas exploration and production exclusion**	**182**
4.14.1	Background	182
4.14.2	Exploration, development, and production processes	183
4.14.2.1	Oil and gas processes	183
4.14.2.1.1	Exploration	183
4.14.2.1.2	Development	183
4.14.2.1.3	Production	183
4.14.2.1.4	Workover	184
4.14.2.2	Geothermal energy processes	185
4.14.3	Determining what wastes are excluded	185
4.14.3.1	"Uniquely associated" wastes	185
4.14.3.1.1	Primary field operations	185
4.14.3.1.2	Rule of thumb for determining if wastes are "uniquely associated"	186

Section	Title	Page
4.14.3.1.3	Materials must be used in order to be "uniquely associated"	186
4.14.3.2	Examples of wastes that are excluded	187
4.14.3.3	Examples of wastes that are not excluded	187
4.14.4	Additional guidance for determining if the exclusion applies	191
4.14.4.1	Transportation wastes	191
4.14.4.1.1	Crude oil transportation	191
4.14.4.1.2	Natural gas transportation	191
4.14.4.1.3	Status of excluded wastes shipped offsite	193
4.14.4.2	Manufacturing wastes	194
4.14.4.3	Crude oil reclamation	194
4.14.4.4	Service companies	195
4.14.4.5	Natural gas wastes	195
4.14.4.5.1	Natural gas condensate	195
4.14.4.5.2	Gas-plant cooling tower wastes	196
4.14.4.5.3	Natural gas regulators	196
4.14.4.5.4	Underground natural gas storage fields	196
4.15	**Exclusion for mining and mineral processing wastes**	**196**
4.15.1	Extraction and beneficiation wastes	197
4.15.1.1	Extraction wastes	197
4.15.1.2	Beneficiation wastes	198
4.15.1.3	"Uniquely associated" wastes	201
4.15.2	Mineral processing wastes	204
4.15.2.1	Reclaiming mineral processing wastes	207
4.15.3	Effect of co-processing hazardous wastes in furnaces	207
4.16	**De minimis wastewater exemptions**	**208**
4.16.1	Listed spent solvents	209
4.16.2	Refinery wastes	210
4.16.3	De minimis losses of commercial chemical products and F- and K-wastes	211
4.16.4	Laboratory wastes	213
4.16.5	Carbamate production wastes	215
4.16.6	Miscellaneous issues	215
4.17	**Active manufacturing process unit exemption**	**216**
4.17.1	Background	216
4.17.2	What qualifies as a manufacturing process unit?	217
4.17.2.1	What doesn't qualify?	217
4.17.2.2	What about pipelines?	217
4.17.2.3	And ships?	220
4.17.3	Wastes from manufacturing process units	220
4.17.3.1	Wastes removed from the unit	220
4.17.3.2	Wastes remaining in the MPU for more than 90 days	221
4.17.3.2.1	Temporary or permanent removal from service	222
4.17.4	MPUs as SWMUs	223
4.18	**Analytical sample/treatability study exemptions**	**224**
4.18.1	Analytical sample exemption	224
4.18.1.1	Scope of the exemption	224
4.18.1.2	Is the sample collector the "generator"?	225
4.18.1.3	How should analytical samples be shipped?	226
4.18.1.4	What are the implications for laboratories?	226
4.18.1.4.1	Sample storage	226
4.18.1.4.2	Sample analysis	227
4.18.1.4.3	Disposal of excess sample	227
4.18.1.5	Does the exemption include product samples?	228
4.18.2	Treatability study exemption	228
4.18.2.1	What is a treatability study?	228
4.18.2.2	Scope of the exemption	229
4.18.2.3	Treatability study sample quantity and time limits	229
4.18.2.4	Sample import/export for treatability studies	231
4.18.2.5	Laboratory/testing facility requirements	231
4.18.2.5.1	Mobile treatment units	231
4.18.2.6	Disposal of excess sample/analytical residues	231
4.18.2.7	Reporting and recordkeeping requirements	232
4.19	**Precious metals exemption**	**232**
4.19.1	Extent of the exemption	233
4.19.2	Recovery of precious metals	233
4.19.2.1	Precious metals recovery in industrial furnaces	233
4.19.2.2	Legitimate vs. sham recycling	235
4.19.3	Importing and exporting precious metals	235
4.19.3.1	Special requirements for OECD countries	236
4.20	**Lead-acid battery exemption**	**236**
4.20.1	The Part 266, Subpart G exemption	236
4.20.1.1	Battery generators/collectors are exempt from most RCRA standards	236
4.20.1.2	Battery regeneration facilities are also exempt	239
4.20.1.3	Battery reclamation facilities are not exempt	239
4.20.2	Exporting spent lead-acid batteries	239
4.21	**Scrap metal exemption**	**240**
4.21.1	What is scrap metal?	240
4.21.2	"Excluded scrap metal"	241
4.21.3	Other recycled scrap metal is exempt	241

4.21.4	Scrap metal that is not exempt from RCRA · · · · · · · · · · · · · · · · · ·	242
4.21.4.1	Scrap metal that is mixed with something else · · · · · · · · · · ·	243
4.21.4.2	Scrap metal managed improperly loses the exemption · · · · · · · ·	243
4.21.5	Printed circuit boards · · · · · · · · · ·	243
4.21.5.1	Whole circuit boards · · · · · · · · · ·	243
4.21.5.2	Shredded circuit boards · · · · · · · · ·	244
4.21.6	Used oil and fuel filters · · · · · · · · ·	244
4.21.7	Punctured aerosol cans · · · · · · · · ·	245
4.21.8	Lead shot reclaimed from shooting ranges · · · · · · · · · · · · · · · · · ·	245
4.22	**Exclusion for geologic sequestration of CO_2** · · · · · · · · · · · · · · · · · ·	**246**

Chapter 5—Rules for Management · 247

5.1	**The mixture rule** · · · · · · · · · · · · · ·	**247**
5.1.1	Background · · · · · · · · · · · · · · · · ·	248
5.1.2	Three elements to the mixture rule · ·	248
5.1.2.1	Mixtures of characteristic wastes · · ·	250
5.1.2.2	Mixtures of ICR-only listed wastes · · ·	250
5.1.2.3	Mixtures of other listed wastes · · · · ·	251
5.1.2.4	Special mixture rule for Bevill wastes · · · · · · · · · · · · · · · · · · ·	252
5.1.2.5	Special mixture rule for oil and gas wastes · · · · · · · · · · · · · · · · · · ·	253
5.1.3	LDR ramifications · · · · · · · · · · · · ·	254
5.1.3.1	Dilution prohibition · · · · · · · · · · · ·	254
5.1.3.2	Complying with LDR treatment standards · · · · · · · · · · · · · · · · · ·	254
5.1.3.2.1	Characteristic wastes · · · · · · · · · · ·	255
5.1.3.2.2	ICR-only listed wastes · · · · · · · · · ·	255
5.1.3.2.3	Other listed wastes · · · · · · · · · · · ·	256
5.1.4	Mixing is treatment · · · · · · · · · · · ·	256
5.1.5	Has mixing occurred? · · · · · · · · · ·	256
5.1.5.1	Petroleum refinery wastewaters in contact with listed sludges · · · · · · ·	257
5.1.6	Miscellaneous issues · · · · · · · · · · ·	257
5.1.6.1	Precipitation run-off · · · · · · · · · · ·	257
5.1.6.2	Rinsates from product tanks and containers · · · · · · · · · · · · · · · · · ·	258
5.1.6.2.1	Rinsate from P-listed containers · · ·	258
5.1.6.3	Sludges formed from mixtures of wastewaters · · · · · · · · · · · · · · · · ·	258
5.1.6.4	Household and CESQG wastes · · · · ·	259
5.1.6.5	Used oil ·	259
5.1.6.6	Mixtures of product and wastewater · ·	259
5.1.6.7	Mixtures of excluded hazardous secondary materials and hazardous waste ·	260
5.2	**The derived-from rule** · · · · · · · · · ·	**260**
5.2.1	Background · · · · · · · · · · · · · · · · ·	261
5.2.2	Three elements of the derived-from rule ·	261
5.2.2.1	Residues from treating characteristic wastes ·	262
5.2.2.1.1	Watch out for wastewater treatment sludges ·	263
5.2.2.2	Residues from treating ICR-only listed wastes · · · · · · · · · · · · · · · ·	263
5.2.2.3	Residues from treating other listed wastes ·	263
5.2.2.3.1	Residues from incineration/burning · ·	263
5.2.2.3.2	Residues from reclamation · · · · · · ·	264
5.2.3	Exceptions to the derived-from rule · ·	265
5.2.3.1	Exempt in, exempt out · · · · · · · · · ·	265
5.2.3.2	Exception for residues from treating petroleum refinery wastes · · · · · · ·	266
5.2.3.3	Exception for supernatant/filtrate from F006/F019 · · · · · · · · · · · · · ·	267
5.2.3.4	Regulatory exceptions · · · · · · · · · ·	268
5.2.3.5	Cement produced from hazardous waste fuels · · · · · · · · · · · · · · · · · ·	270
5.2.3.6	Exception for hazardous debris · · · · ·	270
5.2.3.7	No exception for residues from wastewater treatment units · · · · · · ·	271
5.2.3.8	Residues derived from recycling excluded materials · · · · · · · · · · · ·	271
5.2.4	Regulatory status of reclaimed materials · · · · · · · · · · · · · · · · · ·	272
5.2.5	LDR ramifications · · · · · · · · · · · · ·	274
5.2.5.1	Characteristic wastes · · · · · · · · · · ·	274
5.2.5.2	ICR-only listed wastes · · · · · · · · · ·	274
5.2.5.3	Other listed wastes · · · · · · · · · · · ·	274
5.2.6	Miscellaneous issues · · · · · · · · · · ·	274
5.2.6.1	Landfill leachate · · · · · · · · · · · · · ·	274
5.2.6.2	Precipitation run-off · · · · · · · · · · ·	275
5.3	**Contained-in policy** · · · · · · · · · · · ·	**275**
5.3.1	Contaminated environmental media ·	276
5.3.1.1	Contained-in policy for ICR-only listed wastes · · · · · · · · · · · · · · · ·	278
5.3.1.2	Contaminated sediments · · · · · · · ·	278
5.3.1.3	Contaminated rainwater · · · · · · · · ·	278
5.3.1.4	Applicability of the land disposal regulations · · · · · · · · · · · · · · · · · ·	278
5.3.2	Contaminated debris · · · · · · · · · · ·	279
5.3.3	Policy creep · · · · · · · · · · · · · · · · ·	281

Chapter 6—Generator Issues · 283

6.1	**Waste counting** · · · · · · · · · · · · · · ·	**283**
6.1.1	Three classes of generators · · · · · · ·	284
6.1.2	Wastes must be counted monthly · · ·	284
6.1.2.1	Episodic generators · · · · · · · · · · · ·	285
6.1.3	Wastes that must be counted · · · · · ·	286
6.1.3.1	Containerized waste · · · · · · · · · · · ·	287

6.1.3.2	Spent filter cartridges and paint filters	287
6.1.3.3	Batteries	287
6.1.3.4	Diluted P-listed chemicals	287
6.1.3.5	Wastes generated from cleaning out tanks and containers	288
6.1.3.5.1	Product tanks/manufacturing process units	288
6.1.3.5.2	Hazardous waste tanks/containers	288
6.1.3.6	Conditionally exempt small quantity generator wastes	288
6.1.3.7	Imported waste	289
6.1.4	Wastes exempt from counting	289
6.1.5	Waste counting examples	289
6.1.5.1	Solvents in manufacturing process units	289
6.1.5.2	Analytical samples	291
6.1.5.3	Waste managed in elementary neutralization units	291
6.1.5.4	Onsite solvent recycling	291
6.1.5.5	Offsite solvent recycling	292
6.1.5.6	Wastes managed in totally enclosed treatment units	292
6.1.5.7	Use wet weight for counting	292
6.1.6	Don't double count your waste	293
6.1.7	Acute hazardous waste counted separately	293
6.2	**Satellite accumulation**	**293**
6.2.1	Satellite accumulation "areas"	294
6.2.2	At or near the point of generation	296
6.2.3	Physical limitations	296
6.2.4	Dating requirement	297
6.2.5	Miscellaneous issues	299
6.3	**90- and 180-day accumulation**	**300**
6.3.1	Regulatory history	300
6.3.2	Summary of accumulation standards	300
6.3.3	Allowable facilities	300
6.3.4	Allowable units	302
6.3.4.1	Tanks	302
6.3.4.2	Containers	303
6.3.4.3	Drip pads	303
6.3.4.4	Containment buildings	303
6.3.5	Allowable activities—accumulation and treatment	303
6.3.5.1	Treatment in sequential units	304
6.3.5.2	Thermal treatment	304
6.3.6	The 90-day clock	304
6.3.6.1	Unknown wastes	304
6.3.6.2	When does the clock start?	304
6.3.6.3	When does the clock stop?	305
6.3.6.3.1	Containers	306
6.3.6.3.2	Tanks	306
6.3.6.3.3	Drip pads	307
6.3.6.3.4	Containment buildings	307
6.3.6.4	Accumulation time extensions	307
6.3.6.5	Generators get a new 90- or 180-day clock for returned shipments	307
6.3.7	Preparedness and prevention	308
6.3.7.1	Design and operation of facility	308
6.3.7.2	Internal communication or alarm system	308
6.3.7.3	External communication system	308
6.3.7.4	Testing/maintenance of equipment	309
6.3.7.5	Aisle space	309
6.3.7.6	Arrangements with local authorities	309
6.3.8	Contingency planning	310
6.3.9	Waste minimization program	311
6.3.9.1	What constitutes a waste minimization program?	312
6.3.10	Training requirements	314
6.3.11	Air emission standards	314
6.3.12	Corrective action	314
6.3.13	Closure standards	314
6.3.14	Financial assurance	315
6.3.15	Converting permitted/interim status units to 90-day units	315
6.3.16	SQG standards	315
6.3.16.1	Accumulation units	316
6.3.16.2	6,000-kg limit	316
6.3.16.3	270-day limit	316
6.3.16.4	No contingency plan	316
6.3.16.5	No buffer zone	316
6.3.17	Conditionally exempt small quantity generators	316
6.3.17.1	Hazardous waste identification	316
6.3.17.1.1	Hazardous waste mixture rule for CESQGs	318
6.3.17.1.2	Used oil mixture rule for CESQGs	318
6.3.17.1.3	Hazardous waste derived-from rule for CESQGs	318
6.3.17.2	Onsite management	318
6.3.17.2.1	Generation rate and accumulation quantity limits	319
6.3.17.2.2	Onsite treatment	319
6.3.17.3	Offsite treatment or disposal	319
6.3.17.3.1	EPA ID number exemption	319
6.3.17.3.2	Manifesting exemption	320
6.3.17.3.3	Land disposal restrictions exemption	320
6.3.17.3.4	Offsite transportation	320
6.3.17.4	Generator issues	320
6.3.17.4.1	Container accumulation, labeling, and inspection	320
6.3.17.4.2	Training	320
6.3.17.4.3	Recordkeeping and biennial reporting	321
6.3.17.4.4	Universal wastes	321
6.3.17.4.5	Used oil	321
6.3.17.4.6	Radioactive mixed waste	321
6.3.18	Signage requirements	321
6.4	**Training**	**322**
6.4.1	Large quantity generators	322
6.4.1.1	Who must be trained?	323
6.4.1.1.1	Training exception for satellite accumulation areas	324

TABLE OF CONTENTS

6.4.1.2	When must training be completed?	324
6.4.1.3	What are the basic RCRA training requirements?	324
6.4.1.3.1	Hazardous waste management procedures training	324
6.4.1.3.2	Emergency response training	325
6.4.1.3.3	Three training options available	327
6.4.1.4	Who may perform the training?	328
6.4.1.5	What are the recordkeeping requirements?	328
6.4.2	Small quantity generators	328
6.4.3	Conditionally exempt small quantity generators	330
6.4.4	TSD facilities	330
6.4.4.1	Training not required in post-closure permits	330
6.4.5	Universal waste handlers	330
6.4.5.1	Large quantity handler training standards	330
6.4.5.2	Small quantity handler training standards	331
6.5	**Contractor relations**	**331**
6.5.1	Cogenerator agreements	331
6.5.2	Manifesting and recordkeeping	332
6.5.3	Waste counting for generator status	332
6.5.4	Contractor training	332
6.5.5	Accumulation requirements	332
6.5.6	Examples of contractors/cogenerators	332
6.5.6.1	Painters	332
6.5.6.2	Sample collectors	332
6.5.6.3	Household contractors	332
6.5.6.4	Mobile solvent recyclers	333
6.5.6.5	Equipment maintenance contractors	333
6.5.6.6	Contractors dealing with shipboard wastes	333
6.5.7	The states' role	334
6.6	**Manifesting**	**334**
6.6.1	Background	334
6.6.1.1	Uniform hazardous waste manifest	335
6.6.2	Exemptions from manifesting	335
6.6.2.1	Manifesting exemptions for CESQGs and SQGs	335
6.6.2.2	Manifesting exemption for onsite shipments	337
6.6.2.2.1	DOT requirements may still apply	338
6.6.2.3	Manifesting exemption for immediate responses	338
6.6.2.4	Manifesting exemption for noncharacteristic ICR-only listed wastes	338
6.6.2.5	Manifesting exemption for wastes shipped by pipeline	339
6.6.3	Filling out the manifest	339
6.6.3.1	Emergency response information	339
6.6.3.2	Generator's name and EPA ID number	340
6.6.3.3	Transporters' information	341
6.6.3.3.1	Transfer facilities	341
6.6.3.4	Designated facilities' information	344
6.6.3.5	DOT shipping description	344
6.6.3.6	Container types	346
6.6.3.7	Weight designation	346
6.6.3.8	Units of measurement	347
6.6.3.9	Waste codes	347
6.6.3.10	Special handling instructions and additional information	348
6.6.3.11	Generator's/offeror's certification	348
6.6.3.12	Discrepancies	350
6.6.3.12.1	Standardized procedures for rejected loads	351
6.6.3.12.2	Generators may receive manifested shipments	352
6.6.3.13	Management method codes	353
6.6.3.14	TSD facility certification of receipt	353
6.6.3.15	Continuation sheets	353
6.6.3.16	Making changes to the manifest	353
6.6.4	Exception reports	353
6.6.4.1	Exception report for rejected shipment	354
6.6.5	Manifest retention	354
6.6.6	Rail shipments	355
6.6.7	Universal wastes	355
6.6.8	Imports	356
6.6.9	Exports	356
6.6.10	Electronic distribution of manifests	359
6.7	**Recordkeeping**	**360**
6.7.1	Manifests/exception reports	360
6.7.2	LDR forms	362
6.7.3	Hazardous waste characterization records	362
6.7.3.1	Test results and waste analyses	363
6.7.3.2	"Other determinations" (documentation of knowledge)	363
6.7.4	Waste analysis plans	364
6.7.5	Inspection records	364
6.7.6	Training records	364
6.7.7	Emergency responder agreements	364
6.7.8	Contingency plans	365
6.7.9	Biennial reports	365
6.7.10	Export reports	366
6.7.11	Miscellaneous recordkeeping	366
6.8	**Biennial reports**	**366**
6.8.1	Wastes to be included	368
6.8.2	Weight designation	369
6.8.3	Waste minimization efforts must be included	369
6.8.4	Generator certification	369
6.8.5	State-specific requirements	369
6.9	**EPA ID numbers**	**370**
6.9.1	Obtaining an EPA ID number	370
6.9.1.1	One EPA ID number for one site	370
6.9.1.1.1	Multiple generators on a single contiguous property	371

6.9.1.1.2	Hazardous waste generated on ships and vessels during maintenance activities · · · · · · · · · · · · · · 375		6.9.1.3.1	Episodic generators · · · · · · · · · · 377	
6.9.1.2	Obtaining a temporary, emergency, or other one-time-use EPA ID number · · · · · · · · · · · · · · 375		6.9.2	Exemptions from having an EPA ID number · · · · · · · · · · · · · · 377	
			6.9.2.1	Conditionally exempt small quantity generators · · · · · · · · · · · · · · · · 377	
6.9.1.3	When renotification of hazardous waste activities is required · · · · · · 376		6.9.2.2	Managing hazardous waste during an immediate response · · · · · · 377	
			6.10	**Academic labs/Subpart K** · · · · · · · · **377**	

Chapter 7—Hazardous Waste Treatment · 379

7.1	**The definition of "treatment"** · · · · · · **380**		7.2.3	Miscellaneous WWTU exemption issues · 396	
7.1.1	Does the activity change the hazardous waste? · · · · · · · · · · · · 380		7.2.3.1	Eligibility decisions are unit-specific · · 396	
7.1.1.1	Point of generation issues · · · · · · 380		7.2.3.2	WWTUs that leak are not WWTUs · · 396	
7.1.1.2	Container-related issues · · · · · · · · 381		7.2.3.3	WWTUs are SWMUs · · · · · · · · · · 396	
7.1.2	Why was the hazardous waste changed? · · · · · · · · · · · · · · · · · · 382		7.2.3.4	LDR notices · · · · · · · · · · · · · · · · 396	
			7.3	**Treatment in 90/180/270-day accumulation units** · · · · · · · · · · · · **397**	
7.1.2.1	Waste bulking and containerization · · 382		7.3.1	History of the accumulation unit treatment exemption · · · · · · · · · · 397	
7.1.2.2	Mixing · 383				
7.1.2.3	Fuel blending · · · · · · · · · · · · · · · 383		7.3.2	Exemption qualifying criteria · · · · · 398	
7.1.2.4	Recycling · · · · · · · · · · · · · · · · · · 384		7.3.2.1	Allowable units and associated management standards · · · · · · · · · 398	
7.1.2.5	Miscellaneous processes · · · · · · · · 384				
7.2	**Wastewater treatment units** · · · · · · **384**		7.3.2.1.1	Upward mobility · · · · · · · · · · · · · 398	
7.2.1	The key definition · · · · · · · · · · · · 385		7.3.2.2	Prohibited treatment operations · · · 399	
7.2.1.1	Part 1—The CWA must apply to the wastewater treatment system · · · · · 386		7.3.2.3	Onsite generation versus offsite receipt · 400	
7.2.1.2	Part 2—The unit must be managing hazardous wastewater or hazardous wastewater treatment sludge · · · · · 387		7.3.2.4	Treatment in multiple vessels · · · · 400	
			7.3.2.5	LDRs and WAPs · · · · · · · · · · · · · · 400	
			7.3.3	Check with your state · · · · · · · · · · 400	
7.2.1.2.1	Wastewater · · · · · · · · · · · · · · · · 387		7.3.4	Examples · · · · · · · · · · · · · · · · · · 400	
7.2.1.2.2	Wastewater treatment sludge · · · · 387		**7.4**	**Recycling is treatment** · · · · · · · · · · · **401**	
7.2.1.3	Part 3—Tanks and tank systems only · 387		7.4.1	Scope and applicability · · · · · · · · · 403	
			7.4.1.1	Shipment of hazardous waste to a recycling facility · · · · · · · · · · · · · 403	
7.2.1.3.1	Evaporators, filter presses, dryers, etc. · 388				
			7.4.1.2	Storage before recycling · · · · · · · · 403	
7.2.1.3.2	Sumps · 388		7.4.1.2.1	Storage at generator facilities · · · · · 405	
7.2.1.3.3	Sinks and sewers · · · · · · · · · · · · · 389		7.4.1.3	Recycling units are exempt except for air emission standards · · · · · · · 405	
7.2.1.3.4	Ancillary equipment · · · · · · · · · · · 389				
7.2.1.3.5	Incinerators, lagoons, and surface impoundments · · · · · · · · · · · · · · · 390		7.4.1.3.1	Mobile recycling units · · · · · · · · · 405	
			7.4.1.3.2	Partial recycling by the generator is still exempt · · · · · · · · · · · · · · · 406	
7.2.2	Scope of the WWTU exemption · · · 390				
7.2.2.1	WWTU exemption applies to equipment, not to wastes · · · · · · · 390		7.4.1.4	Storage after recycling · · · · · · · · · 406	
			7.4.1.5	Some processes can never be exempt recycling · · · · · · · · · · · · · · · · · · · 406	
7.2.2.2	What are WWTUs exempt from? · · · 390				
7.2.2.3	WWTUs must be dedicated to onsite wastewater treatment systems · · · · 391		7.4.2	Examples · · · · · · · · · · · · · · · · · · 406	
			7.4.2.1	Qualifying processes · · · · · · · · · · 406	
7.2.2.4	Onsite handling with subsequent shipment for offsite treatment/ disposal · · · · · · · · · · · · · · · · · · · 391		7.4.2.2	Other examples · · · · · · · · · · · · · · 407	
			7.5	**Other permit-exempt treatment options** · **408**	
7.2.2.4.1	Offsite shipment of sludge for treatment/disposal · · · · · · · · · · · · 392		7.5.1	Elementary neutralization units · · · 408	
			7.5.1.1	Implementation hinges on ENU definition · · · · · · · · · · · · · · · · · · 408	
7.2.2.5	Common wastewater treatment facilities · · · · · · · · · · · · · · · · · · · 393				
			7.5.1.1.1	Neutralization and ENU eligibility · · 408	
7.2.2.6	Treatment of hazardous wastewater/sludge generated offsite · · 395		7.5.1.1.2	Tanks, containers, etc. · · · · · · · · · 408	
			7.5.1.2	Scope of the ENU exemption · · · · · 410	
7.2.2.7	Portable tanks · · · · · · · · · · · · · · · 395		7.5.2	Totally enclosed treatment facilities · · 411	

7.5.2.1	Two characteristics of a TETF	411
7.5.2.1.1	Directly connected	412
7.5.2.1.2	No release of hazardous constituents	412
7.5.2.1.3	What does qualify as a TETF?	413
7.5.2.2	A final word on TETFs	414
7.5.3	Adding absorbents to wastes	414
7.5.3.1	Implementation	415
7.5.3.1.1	Containers are key	415
7.5.3.1.2	Order does not matter, but timing does	415
7.5.3.1.3	Other implementation issues	416
7.5.3.2	Examples	416
7.5.4	Immediate responses	417
7.5.4.1	Scope of the exemption	417
7.5.4.1.1	Sumps	418
7.5.4.2	Examples	418
7.5.5	Burning small quantities of hazardous waste in onsite units	419
7.5.5.1	Waste must be burned at the site of generation	420
7.5.5.2	Quantity/specification limitations on hazardous waste burned	420
7.5.5.3	Accumulation before burning	420
7.5.5.4	Notification and recordkeeping requirements	421

Chapter 8—Universal Wastes ... 423

8.1	**What are universal wastes?**	**424**
8.1.1	State universal wastes	426
8.1.2	Other issues	426
8.2	**Universal waste program structure**	**426**
8.3	**Universal waste handler requirements**	**428**
8.3.1	Specific handler requirements	429
8.3.1.1	Waste management	430
8.3.1.2	No treatment allowed	432
8.3.1.3	Accumulation location	432
8.3.1.4	Exporting universal waste	432
8.4	**Universal waste transporter and destination facility requirements**	**433**
8.4.1	Transporters	433
8.4.1.1	DOT requirements	433
8.4.2	Destination facilities	435
8.5	**State authorization issues**	**436**
8.5.1	Current state status	436
8.5.2	Interstate transport	436
8.6	**Mercury-Containing and Rechargeable Battery Management Act**	**438**

Chapter 9—Container Issues ... 439

9.1	**What is a container?**	**440**
9.1.1	Regulation of container-like equipment	440
9.1.1.1	Containers versus debris—LDR implications	441
9.1.2	Container-related issues	441
9.1.2.1	DOT standards	441
9.1.2.2	Counting containerized waste	441
9.1.2.3	Containers are not ancillary equipment to tanks	442
9.1.2.4	Containers and the liquids-in-landfills ban	442
9.1.3	Container management 101	442
9.2	**Containers must be closed**	**442**
9.2.1	Satellite accumulation containers	444
9.2.2	90/180/270-day containers	444
9.2.3	Preventing spills	445
9.2.4	Miscellaneous container closure guidance	445
9.3	**Container labeling requirements**	**445**
9.3.1	Labeling requirements for satellite accumulation containers	446
9.4	**Secondary containment requirements**	**447**
9.4.1	Technical details	447
9.4.1.1	Secondary containment exemption	448
9.5	**RCRA-empty containers**	**448**
9.5.1	Containers that held nonacute hazardous waste	449
9.5.1.1	Commonly employed practices	449
9.5.1.2	The one-inch requirement	449
9.5.1.3	The weight-limit alternatives	450
9.5.1.3.1	Empty aerosol cans may still be hazardous	451
9.5.2	Containers that held acute hazardous waste	451
9.5.2.1	Triple rinsing	451
9.5.2.2	Alternative cleaning methods	452
9.5.2.3	Regulatory status of warfarin residues in containers	452
9.5.3	Containers that held compressed gas	453
9.5.3.1	Cylinders returned to gas suppliers and removal of residues	453
9.5.4	Beneficial reuse of residues	454
9.5.5	Regulatory status of residues removed from empty containers	454
9.5.5.1	Discharging rinsates into wastewater treatment systems	455
9.5.5.2	Miscellaneous empty-container examples	456
9.5.6	RCRA and DOT definitions of "empty" are different	456
9.5.6.1	Manifesting nonempty shipments back to a generator	456

9.5.7	Waste counting issues · · · · · · · · · 458		9.6.1	Satellite containers do not require weekly inspection · · · · · · · · · · · · · 459
9.5.8	CERCLA issues · · · · · · · · · · · · · · 458		**9.7**	**Lab packs · · · · · · · · · · · · · · · · 459**
9.5.9	EPCRA reporting · · · · · · · · · · · · · 458		**9.8**	**Segregation of incompatible wastes in containers · · · · · · · · · · · · · · 460**
9.6	**Container inspection requirements · · 458**			

Chapter 10—Tank Issues · 463

10.1	**What is a hazardous waste tank? · · · 464**	10.2.4	Examples · · · · · · · · · · · · · · · · · 478
10.1.1	Hazardous waste tanks and tank systems · · · · · · · · · · · · · · · · · · 464	**10.3**	**Secondary containment and release detection requirements · · · · 479**
10.1.2	Exempt/unregulated tanks and tank systems · · · · · · · · · · · · · · · · · · 466	10.3.1	Applicability · · · · · · · · · · · · · · · 480
		10.3.1.1	Avoiding an endless loop · · · · · · · · 480
10.1.2.1	Permit-exempt treatment can exempt tanks as well · · · · · · · · · · · · · · · 466	10.3.2	Technical details · · · · · · · · · · · · · 480
		10.3.2.1	More on required containment system capacity · · · · · · · · · · · · · 480
10.1.2.2	Waste exclusions/exemptions can apply to tanks · · · · · · · · · · · · · · 466	10.3.2.2	Containment structures must be impermeable · · · · · · · · · · · · · · · 483
10.1.2.3	Subtitle I USTs · · · · · · · · · · · · · · 468	10.3.2.3	USTs · · · · · · · · · · · · · · · · · · · 483
10.1.2.4	Temporary tanks · · · · · · · · · · · · · 469	10.3.2.4	Ancillary equipment · · · · · · · · · · · 483
10.1.2.5	Closed-loop recycling tanks · · · · · · · 471	10.3.3	Release detection · · · · · · · · · · · · 484
10.1.3	Equipment that isn't a tank or tank system · · · · · · · · · · · · · · · · · · · 471	10.3.3.1	Responses to releases · · · · · · · · · · 486
		10.3.3.1.1	Reporting · · · · · · · · · · · · · · · · · 486
10.1.3.1	Waste piles · · · · · · · · · · · · · · · · 471	10.3.4	Implementation timing for secondary containment · · · · · · · · · · · · · · · 486
10.1.3.2	Surface impoundments · · · · · · · · · 472		
10.1.3.3	Vaults · · · · · · · · · · · · · · · · · · · 473	10.3.4.1	Interim measures for existing tanks without secondary containment · · · 487
10.1.3.4	Drip pads · · · · · · · · · · · · · · · · · 473		
10.1.4	Tank systems include ancillary equipment · · · · · · · · · · · · · · · · · 473	10.3.5	Variances may be requested · · · · · · 487
		10.4	**RCRA-empty tanks · · · · · · · · · · · · 488**
10.1.5	Hazardous waste tank design and management standards · · · · · · · · · 474	10.4.1	Emptying a 90-day accumulation tank · 488
10.1.5.1	New vs. existing hazardous waste tanks · · · · · · · · · · · · · · · · · · · 474	10.4.2	What about heels? · · · · · · · · · · · · 488
		10.4.3	Check with your state · · · · · · · · · · 489
10.1.5.2	Tank integrity assessments · · · · · · · 476	**10.5**	**Tank inspection requirements · · · · · 489**
10.2	**Are sumps tanks? · · · · · · · · · · · · 477**	10.5.1	Weekly tank inspections allowed under certain conditions · · · · · · · · 491
10.2.1	Temporary sumps · · · · · · · · · · · · 477		
10.2.2	Secondary containment sumps · · · · · 478		
10.2.3	Primary containment sumps · · · · · · 478		

Chapter 11—Recycling · 493

11.1	**Overview of the five-step recycling determination process · · · · · · · · · · 494**	11.2.3.1.2	Materials generated and reclaimed at different facilities · · · · · · · · · · · 502
11.1.1	What is a hazardous secondary material? · · · · · · · · · · · · · · · · · 494	11.2.3.1.3	Materials generated and reclaimed per a tolling agreement · · · · · · · · 503
11.1.2	What are the five steps? · · · · · · · · 494	11.2.3.1.4	Notification requirement for hazardous secondary materials · · · · 503
11.1.3	Putting it all together · · · · · · · · · · 494		
11.2	**Step 1—Check the table with the asterisks · · · · · · · · · · · · · · · · · · 494**	11.2.3.2	The transfer-based exclusion · · · · · · 504
		11.2.3.2.1	Materials exported for reclamation · · 508
11.2.1	Classifying hazardous secondary materials · · · · · · · · · · · · · · · · · 497	11.2.3.2.2	Notification required · · · · · · · · · · 508
		11.2.3.2.3	State authorization for DSW rule · · · 509
11.2.2	Categorizing recycling activities · · · 498	11.2.4	Examples of Step 1 recycling scenarios · · · · · · · · · · · · · · · · · 509
11.2.3	Exceptions to the asterisks · · · · · · · 498		
11.2.3.1	Materials reclaimed "under the control of the generator" · · · · · · · 499	11.2.5	Why "—" results *may* be exempt · · · 511
		11.2.5.1	Managing exempt materials · · · · · · 511
11.2.3.1.1	Materials generated and reclaimed at the generating facility · · · · · · · · 500	11.2.6	Don't quit if you hit an "*" · · · · · · · 511
		11.3	**Step 2—The four use/reuse recycling exclusions · · · · · · · · · · · · · · · · · 511**

TABLE OF CONTENTS

11.3.1	Use/reuse as an ingredient or feedstock	512
11.3.1.1	Regulatory nuts and bolts	512
11.3.1.2	Reclamation vs. incidental processing	512
11.3.1.2.1	The reclamation process	514
11.3.1.2.2	Some materials can be reclaimed without being solid wastes	516
11.3.1.3	POG considerations	516
11.3.1.4	Examples	517
11.3.1.5	Special provision for recycling spent sulfuric acid	520
11.3.2	Use as an effective substitute for a commercial product	521
11.3.2.1	No reclamation please	521
11.3.2.2	Other limitations/considerations	522
11.3.2.3	Examples	522
11.3.3	Closed-loop recycling with no reclamation	524
11.3.3.1	Condition 1—Reuse must be in the original process	524
11.3.3.2	Condition 2—No reclamation can be occurring	525
11.3.3.3	Condition 3—Reuse must be as a feedstock	525
11.3.3.4	Other eligibility issues	525
11.3.3.4.1	No placement on the land	525
11.3.3.4.2	Primary and secondary processes on equal footing	526
11.3.3.5	Examples	526
11.3.4	Closed-loop recycling with reclamation	526
11.3.4.1	Condition 1—Reuse must be in original process	527
11.3.4.2	Condition 2—Storage and conveyance limitations	528
11.3.4.3	Condition 3—No flame-based reclamation	528
11.3.4.4	Condition 4—No production of fuels or products applied to the land	528
11.3.4.5	Additional details	529
11.3.4.6	Examples	529
11.4	**Step 3—Additional qualifying criteria**	**529**
11.4.1	General qualifying criteria	530
11.4.1.1	Use constituting disposal	531
11.4.1.1.1	Use as an ingredient to produce fertilizer	531
11.4.1.1.2	Use as an ingredient to produce animal feed supplements	533
11.4.1.1.3	Use as an ingredient to produce cement/aggregate	533
11.4.1.1.4	Use as an ingredient to produce pesticides	534
11.4.1.1.5	Use as an ingredient to produce absorbent material	534
11.4.1.1.6	Spent material reuse as wastewater conditioner	534
11.4.1.2	Burning for energy recovery/use to produce a fuel	534
11.4.2	Specific qualifying criteria	535
11.5	**Step 4—Recycling must be legitimate**	**536**
11.5.1	Recycling legitimacy factors codified	537
11.5.1.1	Mandatory recycling legitimacy factors	537
11.5.1.2	Nonmandatory recycling legitimacy factors	538
11.5.1.3	Economics to be considered when determining recycling legitimacy	540
11.5.2	Previous legitimacy guidance	541
11.5.2.1	Questions to ask to determine if sham recycling is occurring	542
11.5.2.2	Some activities are under close scrutiny	542
11.5.2.3	Economics—not the key factor in determining legitimate recycling	544
11.5.3	Legitimacy evaluation may involve regulator input	544
11.6	**Step 5—Documentation requirements**	**544**

Chapter 12—Used Oil · · · 547

12.1	**Applicability of the used oil management standards**	**547**
12.1.1	Definition of used oil	548
12.1.1.1	Examples of used oil	548
12.1.1.1.1	Synthetic oil is used oil	548
12.1.1.1.2	Consolidated used oil is just used oil	548
12.1.1.1.3	Is spent grease used oil?	551
12.1.1.2	Materials that are not used oil	551
12.1.2	All used oil is recyclable until a decision is made to dispose	552
12.1.3	Mixing used oil with hazardous waste	553
12.1.3.1	Used oil mixed with characteristic hazardous waste	553
12.1.3.1.1	EPA trying to close a loophole	554
12.1.3.2	Used oil mixed with listed hazardous waste	554
12.1.4	Rebuttable presumption	555
12.1.4.1	How to make a rebuttal	556
12.1.4.2	Exemptions from the rebuttable presumption	557
12.1.5	Used oil filters	558
12.1.5.1	Exclusion for non-terne-plated filters	559
12.1.5.2	Scrap metal exemption also available	559
12.1.5.3	Nonexcluded oil filters	559
12.1.6	Other mixtures of used oil and solid wastes/other nonhazardous materials	560

Section	Title	Page
12.1.6.1	Used oil mixed with fuels	560
12.1.6.2	Used oil-contaminated materials	560
12.1.6.2.1	Contaminated materials are used oil when burned for energy recovery	561
12.1.6.3	Residues from managing used oil	561
12.1.6.4	Used oil-contaminated wastewater	562
12.1.6.4.1	Used oil-contaminated wastewater at petroleum refineries	562
12.1.6.5	Used oil mixed with nonhazardous waste	562
12.1.7	PCBs in used oil	562
12.1.8	Used oil managed in the petroleum industry	564
12.1.8.1	Used oil inserted into crude oil pipelines	564
12.1.8.1.1	Storage/transportation of used oil/crude oil mixtures before insertion into crude oil pipelines	566
12.1.8.2	Used oil inserted into petroleum refineries	566
12.1.8.2.1	Used oil inserted prior to crude distillation or catalytic cracking	567
12.1.8.2.2	Used oil inserted after crude distillation or catalytic cracking	567
12.1.8.2.3	Used oil captured in hydrocarbon recovery/wastewater treatment system	569
12.2	**Used oil management standards for generators**	**569**
12.2.1	Used oil generator standards	570
12.2.1.1	Used oil storage	570
12.2.1.1.1	Containers and aboveground tanks	570
12.2.1.1.2	USTs	573
12.2.1.1.3	Other units	574
12.2.1.1.4	SPCC plans	574
12.2.1.2	Offsite shipments	574
12.2.1.3	Other generator operations	575
12.2.1.4	No recordkeeping required	576
12.2.2	Generators: don't process the oil you generate	576
12.2.3	Used oil disposal	578
12.3	**Burning used oil**	**578**
12.3.1	The used oil specification	578
12.3.2	Burning on-specification used oil	581
12.3.2.1	For CAA purposes, on-spec used oil is not solid waste when combusted	582
12.3.3	Burning off-specification used oil	582
12.3.3.1	Allowable units	582
12.3.3.2	The space heater exception	583
12.3.3.3	Off-spec used oil burner requirements	583
12.3.3.4	For CAA purposes, off-spec used oil is solid waste when combusted	584
12.3.4	Burning used oil-contaminated and derived materials	584
12.3.5	Used oil fuel marketer requirements	585

Chapter 13—Land Disposal Restrictions · 587

13.1	**Overview of the LDR program**	**587**
13.1.1	Does the LDR program apply to you?	587
13.1.2	Summary of requirements under the LDR program	588
13.1.3	What triggers the land disposal requirements?	589
13.1.3.1	Avoiding LDR requirements	591
13.1.4	What is the objective of the LDR program?	592
13.1.5	How were treatment standards established?	593
13.1.5.1	Treatability groups established	594
13.2	**Requirements apply at the point of generation**	**594**
13.3	**Underlying hazardous constituents**	**598**
13.3.1	Wastes for which UHCs have to be identified	601
13.4	**Treatment standards for characteristic wastes**	**602**
13.4.1	Understanding the table of treatment standards	602
13.4.2	D001 ignitable wastes	605
13.4.3	D002 corrosive wastes	610
13.4.4	D003 reactive wastes	610
13.4.5	TC wastes (heavy metals, pesticides, and organics)	611
13.5	**Treatment standards for listed wastes**	**615**
13.6	**Waste coding**	**619**
13.7	**Contaminated soil standards**	**621**
13.7.1	Soil contaminated with characteristic wastes	624
13.7.2	Soil contaminated with listed wastes	624
13.7.3	Dealing with residues from soil treatment	626
13.8	**Contaminated debris standards**	**626**
13.8.1	Three options for managing hazardous debris	628
13.8.2	Interpreting the alternative treatment standards table for debris	629
13.8.3	Constituents subject to treatment	631
13.8.4	Is debris treatment subject to permitting?	631
13.8.4.1	Containment buildings	632
13.8.4.1.1	Generator sites	632
13.8.4.1.2	Interim status facilities	632
13.8.4.1.3	Permitted facilities	632
13.8.5	Miscellaneous debris issues	632
13.9	**Mixed waste standards**	**634**
13.9.1	Mixed wastes at DOE facilities	634
13.9.2	Mixed wastes at commercial sites	634
13.10	**The LDR storage prohibition**	**636**

TABLE OF CONTENTS

13.10.1	Mercury Export Ban Act	637
13.11	**The LDR dilution prohibition**	**637**
13.11.1	Aggregation for centralized treatment	638
13.11.2	Dilution as a consequence of treatment	639
13.11.3	McCoy's dilution diagram	640
13.12	**LDR paperwork requirements**	**640**
13.12.1	Figuring out what forms to use	643
13.12.1.1	The §268.7(a)(7) one-time notice	643
13.12.2	Forms that are not available from someone else	649
13.12.3	Common paperwork examples	649
13.12.3.1	The §268.7(a)(7) one-time notice	649
13.12.3.2	Wastes sent to fuel blenders	652
13.12.3.3	Wastes sent to storage facilities	653
13.12.3.4	Wastes sent to TSD facilities	653
13.12.3.5	Wastes sent to recycling facilities	653
13.12.3.6	Wastes sent to Subtitle D facilities	653
13.12.3.7	Decharacterized wastes sent to Subtitle C facilities	654
13.12.3.8	Wastes treated in 90/180/270-day units	654
13.12.3.9	Wastes used in a manner constituting disposal	654
13.12.3.10	Wastes sent to Canadian facilities	654
13.12.3.11	Miscellaneous paperwork issues	655
13.13	**Miscellaneous issues**	**655**
13.13.1	Lab packs	655
13.13.2	Used oil	656
13.13.3	Universal wastes	657
13.13.4	Variances, extensions, and exemptions	657
13.13.4.1	Case-by-case extensions to the effective date	657
13.13.4.2	No-migration exemptions	657
13.13.4.3	Alternative treatment methods/determinations of equivalent technology	658
13.13.4.4	Treatability variances	659

Chapter 14—Point of Generation 661

14.1	**Point of generation issues**	**661**
14.1.1	The fundamentals: when and where	661
14.1.1.1	When is a hazardous waste generated?	662
14.1.1.2	Where is the POG of a hazardous waste?	662
14.1.2	POG for waste coding	663
14.1.3	POG for waste counting	668
14.1.4	POG for residues from 90-day units	668
14.1.5	POG issues associated with the LDR program	669
14.1.5.1	The change-in-treatability-group principle	669
14.1.5.2	Examples of POG under the LDR program	670
14.1.5.3	Intermediate-step treatment residues	671
14.1.6	POG for spills	672
14.1.6.1	POG for spills of characteristic wastes/products	672
14.1.6.2	POG for spills of listed wastes/products	673
14.2	**Unknown wastes**	**673**
14.2.1	Determining if unknown wastes exhibit a characteristic	674
14.2.2	Determining if unknown wastes are listed	674
14.2.3	Accumulation time for unknown wastes	675
14.3	**Spills and spill residues**	**675**
14.3.1	Do we have to clean up spills?	676
14.3.1.1	Spills of hazardous waste versus spills of products	676
14.3.2	Spill response	677
14.3.2.1	Generator requirements	677
14.3.2.2	TSD facility requirements	677
14.3.2.3	Transporter requirements	677
14.3.2.4	Cleanup levels	678
14.3.2.5	Exemption for immediate response treatment and containment activities	678
14.3.2.6	Point of generation issues	679
14.3.3	Spills of characteristic wastes/products	679
14.3.4	Spills of listed wastes/products	680
14.3.4.1	Spills of F- and K-wastes	680
14.3.4.1.1	Leachate from spill-contaminated soil	680
14.3.4.2	Spills of commercial chemical products (P- and U-wastes)	680
14.3.4.3	Did we spill a listed waste or not?	681
14.3.5	Spills of oil or used oil	682
14.3.5.1	Oil spills	682
14.3.5.2	Used oil spills	683
14.3.6	Used absorbents	683
14.3.7	Recycling spilled products	684
14.3.7.1	Leaking petroleum underground storage tanks	684
14.3.8	Corrective action requirements	685
14.3.9	Land disposal restrictions	685
14.3.10	Spill reporting	685
14.3.10.1	RCRA spill reporting	686
14.3.10.2	CERCLA spill reporting	686
14.3.10.3	EPCRA spill reporting	687
14.3.10.4	SPCC spill reporting	687
14.3.10.5	DOT spill reporting	688
14.3.10.6	Release "to the environment"	688
14.3.11	Spill response training	689

14.3.11.1	Generator requirements · · · · · · · 689		14.5	**Point of generation for waste military munitions · · · · · · · · · · · · 696**	
14.3.11.2	TSD facility requirements · · · · · · 689		14.5.1	What are "military munitions"? · · · · 697	
14.3.11.3	Transporter requirements · · · · · · 689		14.5.2	Determining if military munitions are solid wastes · · · · · · · · · · · · · 697	
14.4	**Gases · 689**		14.5.2.1	Unused munitions that are designated for disposal · · · · · · · · 697	
14.4.1	EPA's stance on gases has changed over time · · · · · · · · · · · · · · · · · 689		14.5.2.1.1	Munitions that are disposed, burned, or otherwise treated · · · · · · · · · 697	
14.4.2	Gaseous emissions from manufacturing operations · · · · · · 691		14.5.2.1.2	Munitions removed from storage and then disposed, burned, or otherwise treated · · · · · · · · · 699	
14.4.2.1	Gases vented from compressed gas cylinders · · · · · · · · · · · · · · · · · 691		14.5.2.1.3	Munitions that are leaking or deteriorated · · · · · · · · · · · · · 699	
14.4.2.2	Activated-carbon/filtered control of manufacturing emissions · · · · · · · 691		14.5.2.1.4	Munitions determined by an authorized military official to be a solid waste · · · · · · · · · · · · · · 699	
14.4.2.2.1	Spent carbon/filters from product handling areas · · · · · · · · · · · · · 693		14.5.2.2	Unused munitions that are disassembled, repaired, or otherwise recovered · · · · · · · · 699	
14.4.2.3	Fume incinerators are regulated under the CAA—not RCRA · · · · · · 693		14.5.2.3	Munitions used for training, research and development, or evaluation · · · 699	
14.4.3	Gaseous emissions from hazardous waste management activities · · · · · 694		14.5.2.3.1	Military training exercises · · · · · · · 700	
14.4.3.1	Vent streams from hazardous waste management units · · · · · · · · · · · 694		14.5.2.3.2	Weapons testing, research and development, or evaluation · · · · · · 700	
14.4.3.1.1	Boilers and process heaters used to destroy organic vapors · · · · · · · · 694		14.5.2.4	Range clearance operations · · · · · · 701	
14.4.3.1.2	Activated-carbon control of organic vapors · · · · · · · · · · · · · · · · · · · 695		14.5.3	Military munitions that are solid waste may also be hazardous waste · · 701	
14.4.3.2	Gases vented from treating hazardous ground water · · · · · · · 695				
14.4.3.3	Landfill gas condensate · · · · · · · · 695				
14.4.3.4	Synthesis gas · · · · · · · · · · · · · · · 696				

Chapter 15—Air Emission Standards · 703

15.1	**Air emission standards for process vents—Subpart AA · · · · · · · · · · · · 704**		15.2.1.2	Only specific types of equipment are subject to controls · · · · · · · · · · 714	
15.1.1	Applicability of the Subpart AA regulations · · · · · · · · · · · · · · · · · 705		15.2.1.3	Organic concentration of 10 percent by weight triggers requirements · · · 716	
15.1.1.1	Units handling hazardous waste · · · 705		15.2.1.3.1	Knowledge-based determinations · · · 716	
15.1.1.2	Only specific separation processes are subject to controls · · · · · · · · · · · · 705		15.2.1.4	A 300-hr/year contact time is required · · · · · · · · · · · · · · · · · · 717	
15.1.1.3	Organic concentration of 10 ppmw triggers requirements · · · · · · · · · · 707		15.2.1.5	Only equipment associated with certain categories of units is regulated · · · · · · · · · · · · · · · · · · 717	
15.1.1.3.1	Knowledge-based determinations · · · 707		15.2.1.6	Equipment in compliance with CAA requirements is exempt · · · · · · · · · 719	
15.1.1.4	Only certain categories of units are regulated · · · · · · · · · · · · · · · · · · 708		15.2.2	Complying with Subpart BB · · · · · · 720	
15.1.1.5	Only certain process vents are regulated · · · · · · · · · · · · · · · · · · 709		15.2.2.1	Light liquid service · · · · · · · · · · · 720	
15.1.1.6	Process vents in compliance with CAA requirements are exempt · · · · 710		15.2.2.2	The LDAR program · · · · · · · · · · · 720	
15.1.2	Complying with Subpart AA · · · · · · 711		15.2.2.2.1	Identification of Subpart BB-regulated equipment · · · · · · · · · · · · · · · · · 721	
15.1.2.1	Subpart AA emission limits · · · · · · 711		15.2.2.3	Recordkeeping and reporting · · · · · 721	
15.1.2.2	Subpart AA control devices · · · · · · 712		**15.3**	**Air emission controls for tanks, containers, and surface impoundments—Subpart CC · · · · · · 721**	
15.1.2.3	Inspection and monitoring · · · · · · · 712				
15.1.2.4	Recordkeeping and reporting · · · · · 712		15.3.1	Applicability of the Subpart CC regulations · · · · · · · · · · · · · · · · · 723	
15.2	**Air emission controls for equipment leaks—Subpart BB · · · · · · · · · · · · 713**		15.3.1.1	Units managing hazardous waste · · · 724	
15.2.1	Applicability of the Subpart BB regulations · · · · · · · · · · · · · · · · · 713		15.3.1.2	Only specific types of units are subject to controls · · · · · · · · · · · · 724	
15.2.1.1	Equipment managing hazardous waste · 714		15.3.1.2.1	Exemptions · · · · · · · · · · · · · · · · · 724	

15.3.1.2.2	Equipment in compliance with CAA requirements is exempt	725		15.3.4.2.2	Other requirements	740
15.3.1.2.3	Examples	726		15.3.4.2.3	Recordkeeping and reporting	740
15.3.1.3	Volatile organic concentration of 500 ppmw triggers requirements	727		15.3.4.3	Tank examples	741
				15.3.5	Control standards for containers	742
15.3.1.3.1	Volatile organics	727		15.3.5.1	Level 1 containers	743
15.3.1.3.2	Point of waste origination	727		15.3.5.1.1	Compliance options	743
15.3.2	Basic intent of Subpart CC	728		15.3.5.1.2	Container loading/opening	744
15.3.3	Compliance options for Subpart CC	729		15.3.5.1.3	Inspection and monitoring	744
15.3.3.1	Sampling and analysis	729		15.3.5.1.4	Recordkeeping and reporting	744
15.3.3.2	Knowledge-based determinations	730		15.3.5.2	Level 2 containers	744
15.3.3.3	Process changes	730		15.3.5.2.1	Compliance options	744
15.3.3.4	Rendering the waste nonhazardous	730		15.3.5.2.2	Container loading/opening	745
15.3.3.5	Treating wastes for VO reduction	731		15.3.5.2.3	Inspection and monitoring	745
15.3.3.6	Meeting emission control standards for all potentially affected units	735		15.3.5.2.4	Recordkeeping and reporting	745
				15.3.5.3	Level 3 containers	745
15.3.4	Control standards for tanks	735		15.3.5.3.1	Container loading/opening	746
15.3.4.1	Level 1 tanks	735		15.3.5.3.2	Inspection and monitoring	746
15.3.4.1.1	Prerequisites for Level 1 tanks	736		15.3.5.3.3	Recordkeeping and reporting	746
15.3.4.1.2	Specifications for the fixed roof	737		15.3.5.4	Container examples	746
15.3.4.1.3	Recordkeeping and reporting	738		15.3.6	Control standards for surface impoundments	748
15.3.4.2	Level 2 tanks	738				
15.3.4.2.1	Compliance options	738		15.3.7	Recordkeeping and reporting	748

Chapter 16—Inspections and Enforcement · 751

16.1	**Common RCRA violations**	**751**		**16.4**	**ECHO—EPA's enforcement database**	**768**
16.1.1	Training violations	752				
16.1.2	Contingency plan violations	752		**16.5**	**EPA's self-disclosure policy**	**769**
16.1.3	Labeling violations	753		**16.6**	**Preparing for a RCRA Inspection**	**769**
16.1.4	Failure to make hazardous waste determinations	754		16.6.1	State's/EPA's right to inspect	769
				16.6.2	Types of inspections	770
16.1.5	Open containers	754		16.6.3	State/EPA inspections involve three stages	770
16.1.6	Land disposal restrictions violations	754				
16.1.7	Satellite accumulation violations	755		16.6.3.1	Preparing for the inspection	770
16.1.8	Improper tank management	755		16.6.3.1.1	Review past inspection reports/compliance history	771
16.1.9	Preparedness and prevention violations	756		16.6.3.1.2	Review hazardous waste management equipment/areas	771
16.1.10	Used oil violations	756				
16.1.11	Inadequate emergency response provisions	756		16.6.3.1.3	Review process areas	771
				16.6.3.1.4	Review product and raw material warehouses/storage areas	772
16.1.12	Recordkeeping problems	757				
16.1.13	Failure to perform weekly container inspections	757		16.6.3.1.5	Review contractor areas and waste generation activities	772
16.2	**RCRA enforcement**	**757**		16.6.3.1.6	Review shared services with other companies	773
16.2.1	RCRA civil penalty policy	757				
16.2.1.1	Step 1—Determine gravity-based penalty	757		16.6.3.1.7	Review applicable RCRA records	773
				16.6.3.1.8	Prepare facility personnel	775
16.2.1.2	Step 2—Calculate multiday penalty	758		16.6.3.2	The inspection itself	775
16.2.1.3	Step 3—Determine any adjustment factors	759		16.6.3.2.1	Denying access	776
				16.6.3.2.2	Pre-inspection (opening) conference	776
16.2.1.4	Step 4—Adjust for economic benefits of noncompliance	760		16.6.3.2.3	Confidential business information	777
				16.6.3.2.4	Records/physical inspection	777
16.2.2	RCRA enforcement cases	760		16.6.3.2.5	Post-inspection (closing) conference	779
16.2.2.1	EPA's right to overfile	760		16.6.3.3	Inspection follow-up	780
16.3	**EPA's national RCRA enforcement initiatives**	**768**		16.6.4	Resources for RCRA inspection checklists	781

16.7	**"Cradle-to-grave" RCRA liability—What does that mean?** **781**		16.7.2	RCRA regulatory liability for generators	783
16.7.1	RCRA statutory liability for generators	781	16.7.3	EPA guidance on generator liability	783
			16.7.4	What about CERCLA liability ...	784
16.7.1.1	Civil penalties	782	16.7.5	Ways to minimize RCRA/CERCLA liability for hazardous waste shipments	784
16.7.1.2	Criminal penalties	782			
16.7.1.3	Imminent and substantial endangerment	782	16.7.6	Conclusion	784

Chapter 17—Corrective Action · **785**

17.1	**Few regulations codified** **786**		17.8.5.1	Media cleanup levels	801
17.2	**Solid waste management units** **787**		17.8.5.2	Points of compliance	801
17.2.1	Routine and systematic releases ...	788	17.8.5.3	Compliance time frames	802
17.2.2	Manufacturing and product storage areas	789	17.8.5.4	Natural attenuation	802
			17.8.5.5	Summary of the remedy selection process	802
17.2.3	EPA discourages arguing about "SWMUs"	789	17.8.6	Financial assurance	802
			17.8.7	Corrective measures implementation	803
17.2.4	Areas of concern	789	**17.9**	**Completion determinations** **803**
17.3	**Hazardous waste and constituents** · · **789**		**17.10**	**Environmental indicators** **804**
17.4	**Releases** **790**		**17.11**	**Corrective action management units** · · **805**
17.5	**Facility** **790**		17.11.1	Background	805
17.6	**Applicability issues** **791**		17.11.2	CAMU-eligible wastes	805
17.6.1	Corrective action at interim status facilities	792	17.11.3	CAMU designations	806
			17.11.4	Liners and caps	808
17.6.2	Additional applicability examples ...	792	17.11.5	Treatment requirements	808
17.7	**RCRA corrective action vs. Superfund** **793**		17.11.6	Treatment/storage CAMUs	808
			17.11.7	Offsite disposal of CAMU-eligible wastes	808
17.8	**Corrective action process** **794**				
17.8.1	RCRA facility assessment	794	17.11.8	State authorization issues	809
17.8.2	RCRA facility investigation	796	**17.12**	**Temporary units** **809**
17.8.2.1	Action levels	797	**17.13**	**Staging piles** **809**
17.8.2.2	Determination of no further action ...	798	**17.14**	**The area of contamination policy** ... **811**
17.8.3	Interim corrective measures	798	**17.15**	**Remedial action plans** **813**
17.8.4	Corrective measures study	799	**17.16**	**Useful references** **815**
17.8.4.1	Performance-based approach	799			
17.8.5	Remedy selection	800			

Chapter 18—Waste Characterization/Sampling · **817**

18.1	**Waste characterization** **817**		18.2.2.5	Knowledge often included	826
18.1.1	Is it a solid waste?	818	18.2.3	Waste analysis plans	827
18.1.2	Is it exempt?	818	18.2.3.1	WAPs for generators	827
18.1.3	Is it listed?	818	18.2.3.2	WAPs for TSD facilities	828
18.1.4	Is it characteristic?	819	18.2.3.2.1	Characterization of each movement of waste	829
18.1.5	Who makes the determination? ...	819			
18.1.6	How often should I recharacterize? ...	820	**18.3**	**Knowledge-based determinations** · · **830**
18.2	**Analysis-based determinations** **821**		18.3.1	What is acceptable knowledge?	831
18.2.1	General analytical considerations ...	822	18.3.1.1	Process knowledge	831
18.2.1.1	SW–846	822	18.3.1.1.1	Safety data sheets and other manufacturers' data	832
18.2.1.2	Performance-based measurement ...	823			
18.2.1.3	Method-defined parameters	823	18.3.1.2	Analytical data from other facilities ...	833
18.2.2	Analyzing for characteristics	824	18.3.1.3	Old analytical data	833
18.2.2.1	Ignitability	824	18.3.1.4	"Acceptable knowledge" looks like waste analysis data	834
18.2.2.2	Corrosivity	825			
18.2.2.3	Reactivity	825	18.3.2	Knowledge for listings	834
18.2.2.4	Toxicity	825	18.3.2.1	F- and K-wastes	834

18.3.2.1.1	F-wastes	834	18.5.1.2	Representative samples for LDR compliance	852
18.3.2.1.2	K-wastes	835	18.5.2	Sampling and analysis plan/DQOs	852
18.3.2.2	P- and U-wastes	835	18.5.2.1	Where should I take a sample?	853
18.3.2.3	Analysis can be included	836	18.5.2.1.1	Probability sampling	857
18.3.2.3.1	Wipe/chip sampling for contaminated debris	836	18.5.2.1.2	Authoritative sampling	860
18.3.3	Knowledge for characteristics	836	18.5.2.1.3	Sampling heterogeneous wastes	861
18.3.3.1	Ignitability	837	18.5.2.2	How many samples do I need?	863
18.3.3.2	Corrosivity	838	18.5.2.3	How should I take a sample?	865
18.3.3.3	Reactivity	838	18.5.2.3.1	Containers	870
18.3.3.4	Toxicity	839	18.5.2.3.2	Tanks	871
18.3.3.5	Analysis may be included	841	18.5.2.3.3	Surface impoundments/lagoons	872
18.3.4	Use knowledge for mixed wastes	842	18.5.2.3.4	Piles	873
18.4	**LDR program characterization**	**843**	18.5.2.4	"Hot spots"	873
18.4.1	Meeting LDR standards	843	18.5.2.4.1	Hot spots in containerized waste	874
18.4.2	Generator LDR testing/knowledge	844	18.5.2.4.2	Compositing to identify hot spots	875
18.4.2.1	Identifying underlying hazardous constituents	845	18.5.2.5	Field sampling techniques	875
			18.5.2.6	Sample holding times	875
18.4.3	Treatment facility LDR testing	846	18.5.2.7	Minimizing sampling error	877
18.4.4	Disposal facility LDR testing	847	18.5.2.8	Choosing a sampling strategy	877
18.4.5	Documentation for LDR program characterization	848	18.5.2.8.1	Preliminary information	878
			18.5.3	Statistical analyses	878
18.5	**Sampling issues**	**848**	18.5.3.1	An 80% confidence interval of the mean	879
18.5.1	"Representative sample"	848			
18.5.1.1	How do I get a "representative sample"?	850	18.5.3.2	Using a proportion or percentile	881
			18.5.3.3	Nonnormal distributions	881
18.5.1.1.1	A single sample	850	18.5.3.4	Handling nondetects	883
18.5.1.1.2	A set of samples	850	18.5.3.5	Outliers	883
18.5.1.1.3	One or more composite samples	850			

Chapter 19—Remediation and Demolition · 887

19.1	**Remediation**	**887**	19.1.3.2.2	Remediation wastes contaminated with ICR-only listed wastes	899
19.1.1	Site investigation	888			
19.1.1.1	Systematic planning using the DQO process	888	19.1.4	Hazardous remediation waste management options	899
19.1.1.2	Other methods of site investigation	888	19.1.4.1	Onsite waste management options	899
19.1.2	Point of generation for remediation wastes	890	19.1.4.1.1	Areas of contamination	899
			19.1.4.1.2	Satellite accumulation units	903
19.1.2.1	Retroactivity of RCRA	890	19.1.4.1.3	90/180-day units	903
19.1.2.2	"Active management"	890	19.1.4.1.4	Temporary units	903
19.1.2.2.1	Exception for plant construction activities	891	19.1.4.1.5	Staging piles	903
			19.1.4.1.6	Corrective action management units (CAMUs)	904
19.1.2.2.2	Leachate derived from previously disposed wastes	892			
19.1.2.2.3	Wastes in surface impoundments	892	19.1.4.1.7	Remediation waste management sites	904
19.1.2.2.4	Corrective action and CERCLA provisions still apply	893	19.1.4.1.8	RCRA-permitted units	905
			19.1.4.1.9	Is soil consolidation considered dilution?	906
19.1.2.3	Movement of wastes within a unit or AOC	893			
			19.1.4.2	Offsite waste management options	906
19.1.2.4	Summary of POG for remediation wastes	894	19.1.4.2.1	Remediation wastes to Subtitle C TSD facility	906
19.1.3	Are remediation wastes hazardous?	894			
19.1.3.1	Are remediation wastes characteristic?	895	19.1.4.2.2	CAMU-eligible wastes to offsite Subtitle C landfill	906
19.1.3.2	Are remediation wastes listed?	896	19.1.4.2.3	Hazardous remediation waste sent offsite triggers LDR program	907
19.1.3.2.1	Soil contaminated with pesticides	897			

19.1.5	RCRA/CERCLA remediation interface 907		19.2.2	Point of generation for demolition wastes 923	
19.1.5.1	How RCRA sites become subject to CERCLA 907		19.2.2.1	POG for components/equipment removal 923	
19.1.5.1.1	Imminent and substantial endangerment 907		19.2.2.2	POG for residues from cleaning/ decontaminating 923	
19.1.5.1.2	RCRA-permitted facilities 908		19.2.2.3	POG for demolition debris 924	
19.1.5.1.3	Federal facilities 909		19.2.3	Are demolition wastes hazardous? 924	
19.1.5.2	How CERCLA sites become subject to RCRA 909		19.2.3.1	Are demolition wastes characteristic? 924	
19.1.5.2.1	ARARs drive the cleanup process 909		19.2.3.1.1	Components/equipment removed from the building/structure 925	
19.1.5.2.2	Substantive vs. administrative requirements 912		19.2.3.1.2	Residues generated from cleaning/ decontaminating the building/ structure 925	
19.1.5.2.3	"To be considered" guidance 914		19.2.3.1.3	Building/structure demolition debris 925	
19.1.6	Applicability of the G5 MACT During Site Remediation 915		19.2.3.2	Are demolition wastes listed? 926	
19.2	**Demolition 915**		19.2.4	Hazardous demolition waste management options 927	
19.2.1	Building/structure assessment and pre-demolition activities 915		19.2.4.1	Onsite waste management options 927	
19.2.1.1	Lead-based paint assessment/ removal 917		19.2.4.1.1	Scrap metal containers/yards 927	
19.2.1.2	Mercury-containing equipment assessment/removal 919		19.2.4.1.2	Satellite accumulation units 927	
19.2.1.3	CFCs assessment/removal 919		19.2.4.1.3	90/180-day units 927	
19.2.1.4	PCBs assessment/removal 919		19.2.4.1.4	Universal waste units 928	
19.2.1.5	Asbestos assessment/removal 920		19.2.4.1.5	RCRA-permitted units 930	
19.2.1.6	Process equipment within building/ structure 920		19.2.4.2	Offsite waste management options 930	
19.2.1.7	Utilities within building/structure 920		19.2.4.2.1	Metallic demolition wastes to scrap metal recycling facility 930	
19.2.1.8	Hazardous waste tanks/container accumulation areas within building/ structure 921		19.2.4.2.2	Universal wastes to destination facilities 930	
19.2.1.9	Spills/stains on walls, concrete 922		19.2.4.2.3	Nonmetallic demolition wastes to Subtitle C TSD facility 930	
19.2.1.10	Assessment of contaminated soil/ ground water adjacent to or underlying building/structure 922		19.2.4.2.4	Hazardous demolition waste sent offsite triggers LDR program 930	

Appendix 1—List of Case Studies 931
Appendix 2—List of Figures 933
Appendix 3—List of Tables 937
Index 939
Acronyms 983

TABLE OF CONTENTS

Solid Wastes

Answering your question: is my material a solid waste?

This book is a tool for complying with the Resource Conservation and Recovery Act's (RCRA's) hazardous waste regulations—those regs mandated by Subtitle C of the RCRA statute. When Congress and EPA structured this program, they decided that, before a material can be a hazardous waste, it must first meet the definition of a "solid waste." Thus, determining if a material is a hazardous waste must necessarily begin with a discussion of what constitutes a solid waste.

The goal of this chapter is to provide enough information to answer the question: "Is my material a solid waste?" (Chapters 2 and 3 include detailed discussions on when solid wastes are regulated as hazardous wastes.)

1.1 Introduction to solid wastes

When Congress passed the RCRA statute in 1976, they included a definition of "solid waste":

"[A]ny garbage, refuse, sludge from a waste treatment plant, water supply treatment plant, or air pollution control facility, *and other discarded material*, including solid, liquid, semisolid, or contained gaseous material resulting from industrial, commercial, mining, and agricultural operations, and from community activities...." [Emphasis added.] [RCRA Section 1004(27)]

From this definition, it appeared to EPA that Congress intended solid wastes to be discarded materials. Therefore, when determining what constituted a solid waste, the agency first had to consider the universe of "discarded materials." As EPA put it when it first issued the RCRA regulations:

"A review of both RCRA and its legislative history indicate that Congress intended to regulate four broad categories of materials as solid wastes under RCRA, and particularly Subtitle C, irrespective of their ultimate disposition. The common thread linking all these materials is that they are 'sometimes discarded.' Because they are 'sometimes discarded,' they not only fall within the general rubric 'waste,' but also may become part of the 'discarded materials disposal problem'...which Congress sought to remedy under RCRA." [May 19, 1980; 45 *FR* 33093]

1.1.1 Discarded materials

The four categories of "discarded materials" identified by the agency are [45 *FR* 33093]:

1. Garbage, refuse, and sludge—Congress apparently regarded these materials as waste, regardless of their disposition.

2. Materials that are thrown away, abandoned, or destroyed—This category includes those materials that are intended to be thrown away, abandoned, or destroyed as well as those that actually are so managed.

3. Spent materials—EPA referred to these materials as "wastes which have served their intended purpose…. While acknowledging that some of these post-consumer wastes might be recycled…, Congress also recognized that they were sometimes discarded and therefore were wastes." Included in this category are waste solvents, waste acids, used drums, and waste oil.

4. Tars, residues, slags, and other materials that are incidentally generated as part of a manufacturing or mining process—These incidental residues, as we call them, consist of the "waste by-products [from] the nation's manufacturing processes."

To illustrate EPA's concepts of what constitutes a discarded material, look at the hypothetical manufacturing plant in Figure 1-1. This facility produces products; they could be automobiles, chemicals, pharmaceuticals, fuels, electricity, airplanes, computers, nuclear weapons, etc. It also produces a number of waste streams. We will classify all of the materials that come out of the plant with respect to the four categories of discarded materials.

1.1.1.1 Products

The RCRA regulations have nothing to do with products that are used for their intended purpose. Raw materials are purchased and stored in warehouses, and the plant then uses these materials in its manufacturing processes to make products that it sells. Both the raw materials and products are being used for their intended purpose, and RCRA has no say in the transportation, storage, use, or other management of those materials. [RO 11398, 11501]

In another example, suppose the plant buys a pesticide for pest control. Although the pesticide is used in accordance with the directions on the label, some pesticide residue remains in the soil of the plant yard. As long as the product was used for its intended purpose, RCRA doesn't regulate that activity or resulting contamination. [January 4, 1985; 50 FR 628, RO 11291, 12357] Some other environmental statute might apply to the pesticide's use (such as the Federal Insecticide, Fungicide, and Rodenticide Act—FIFRA), but not RCRA.

Where we do potentially have to worry about products under RCRA is when they aren't used for their intended purpose. For example, no matter how careful plant personnel are, some raw materials and/or products are bound to be spilled from time to time. Raw materials and/or products are also discarded occasionally if they get too old, are no longer needed, etc. If a product is spilled (and cannot be reclaimed) or discarded, it is considered to be discarded material, and RCRA Subtitle C potentially applies.

The RCRA statute actually consists of two parts that are of interest here: Subtitle C and Subtitle D. In Subtitle C, Congress specified the requirements for managing hazardous waste. Spills and discarded product are potentially regulated under the Subtitle C hazardous waste provisions. If hazardous, they will have to be managed in Subtitle C units (e.g., a hazardous waste incinerator or hazardous waste landfill). All of the materials that exit the plant that are potentially subject to RCRA Subtitle C are indicated by solid lines in Figure 1-1.

RCRA Subtitle D governs nonhazardous waste. For example, trash will also be generated in the manufacturing plant. These office, packaging, and cafeteria wastes (e.g., waste plastic, cardboard, paper, and orange peels) are regulated under RCRA, but it is the Subtitle D, nonhazardous provisions that apply. Typically, these wastes will be placed in a dumpster and will eventually be managed in a Subtitle D, nonhazardous waste combustor or landfill. Materials that are not subject to RCRA at all or that are subject only to the Subtitle D provisions are indicated by dashed lines in Figure 1-1.

Figure 1-1: Regulatory Status of Discarded Materials From a Manufacturing Plant

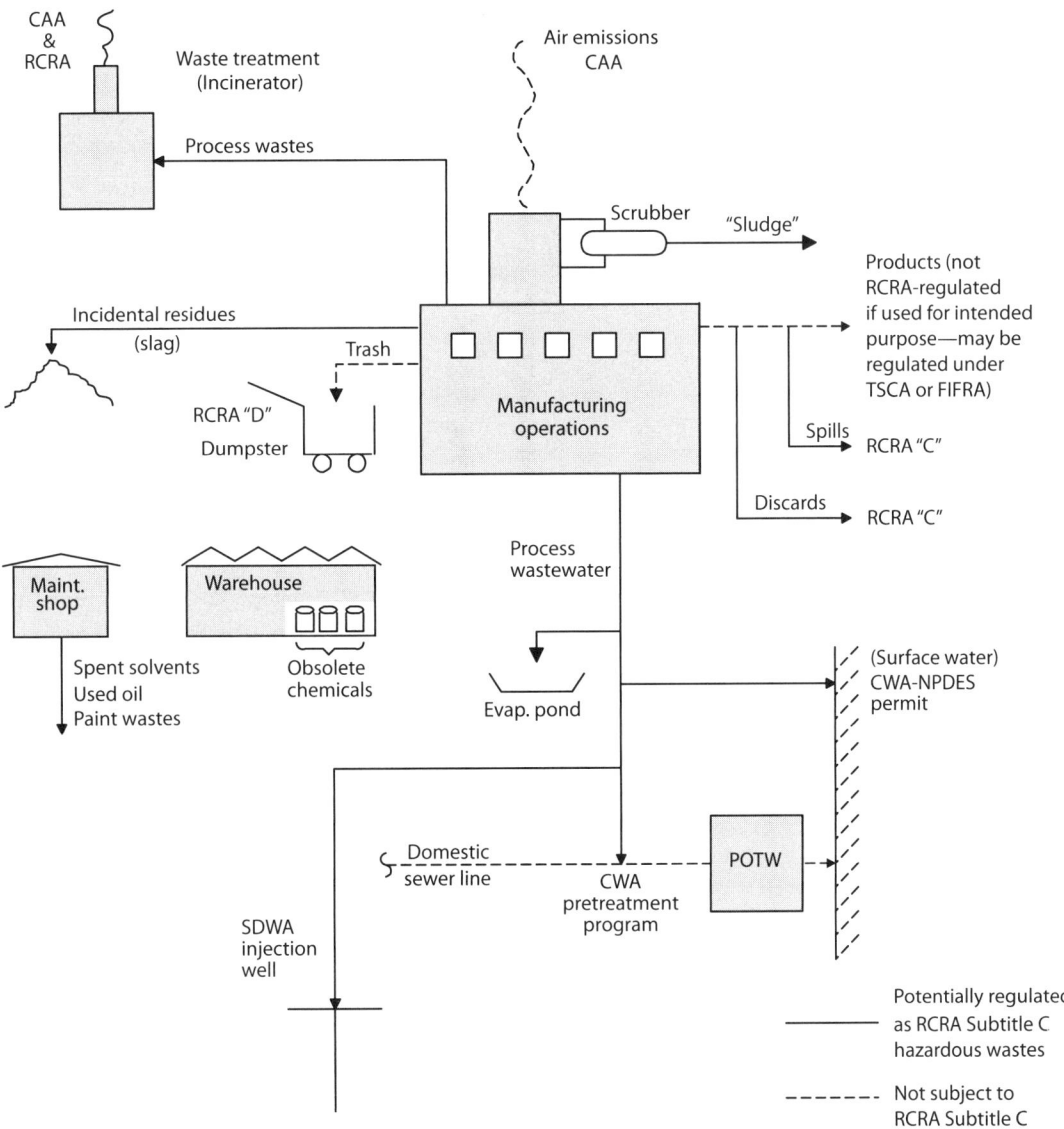

Source: McCoy and Associates, Inc.

1.1.1.2 Garbage, refuse, and sludge

Garbage and refuse make up the facility's trash stream that we mentioned previously; these discarded materials will be disposed in a Subtitle D facility.

A "sludge" is actually a defined term in the RCRA regulations (that's why it's in quotation marks in Figure 1-1). In essence, "sludges" are residues from either water or air pollution control devices. [RO 11879]

In the manufacturing plant, scrubbers may be installed on the stacks from the boilers, process heaters, and furnaces that are used in manufacturing operations. Air emissions from manufacturing operations are generally subject to the Clean Air Act (CAA)—not RCRA. RCRA is limited to regulating gaseous emissions from waste management activities. However, residues from the scrubbers (e.g., fly ash, slurries, scrubber water) are usually disposed and are discarded materials.

The residues from the scrubber, which is an air pollution control device, are considered "sludges" and are potentially subject to RCRA Subtitle C. Sludges often contain toxic substances, so EPA wants to regulate their management under RCRA (when discarded or recycled).

1.1.1.3 Materials that are thrown away, abandoned, or destroyed

A number of the waste streams that are generated in the manufacturing plant fall into this category of discarded materials. If spilled or discarded product is hazardous and cannot be reclaimed, it will be thrown away or abandoned in a hazardous waste landfill or possibly destroyed in a hazardous waste incinerator.

1.1.1.3.1 *Wastewater*

Most manufacturing facilities produce wastewater. As shown in Figure 1-1, process wastewater can be managed in a number of different ways. But no matter how it is managed, the plant is trying to get rid of it; it is being thrown away or abandoned. Several possible wastewater management scenarios are as follows:

- Wastewater can be discharged to surface water of the United States (e.g., a lake or river). The discharge is regulated under a Clean Water Act (CWA) program called the National Pollutant Discharge Elimination System (NPDES) program. It is the NPDES permit that limits or controls what is discharged from the facility into the waters of the United States. RCRA Subtitle C potentially applies to the management of the wastewater from its point of generation (where it comes out of the manufacturing process) to the point where it is actually discharged. At the point of discharge, however, the wastewater is subject to the CWA. Thus, an interface or handoff exists between RCRA and the CWA at the discharge point. The details of this interface are discussed in Section 4.2.

- Wastewater is sometimes abandoned or thrown away by being discharged into a domestic sewer line which leads to a publicly owned treatment works (POTW). In this case, the wastewater is simply mixed with domestic sewage for subsequent treatment in a public sewage treatment plant. Again, RCRA Subtitle C potentially applies to the wastewater from its point of generation until it is discharged into the sewer line. At that point, RCRA stops, and the CWA regulates the wastewater flowing to the POTW. Standards established under the CWA pretreatment program limit or control what can be put into the sewer line. So, there is another RCRA/CWA interface associated with this wastewater management option; this interface is discussed in Section 4.1.

- A third management option for wastewater generated at the plant is to get rid of it by simply injecting it into an underground well. This practice is common in certain parts of the U.S., such as Texas, Louisiana, Indiana, and Ohio. Underground injection wells are not regulated under RCRA, but under the Safe Drinking Water Act (SDWA). In this situation, management of the wastewater is potentially subject to RCRA Subtitle C from its point of generation until it is injected; at that point, the SDWA takes over, delineating another interface between RCRA and a separate environmental program. Even after injection, however, the wastewater may continue to be a hazardous waste.

- Finally, wastewater can be managed in onsite evaporation ponds. In this option, which isn't used very often for hazardous wastes, no discharge of wastewater actually occurs—it's just evaporated. One of the reasons that this management practice is utilized infrequently for hazardous wastewater is that the evaporation ponds would be hazardous waste surface impoundments (which are regulatory nightmares). The wastewater remains potentially subject to RCRA Subtitle C from its point of generation through and including its storage/treatment in the evaporation ponds.

1.1.1.3.2 Process wastes

Process wastes are frequently generated at manufacturing facilities. These could be liquid, semisolid, or solid waste streams. At this hypothetical plant, some high-organic process waste is treated onsite. The process waste is potentially subject to Subtitle C regulation from the point it exits the manufacturing unit all the way through the treatment process. If the process waste meets the definition of a hazardous waste, the facility is conducting hazardous waste treatment, which usually requires a RCRA permit. Because of the high organic concentration, the treatment selected at this plant is onsite incineration. Incinerating the waste constitutes destroying it, making it a discarded material.

Gaseous emissions from the incinerator stack are subject to regulation under RCRA as well as the CAA (creating another interface between environmental statutes/regulations). These emissions will be treated to meet certain contaminant concentrations and then will be discharged to atmosphere; that is, the gaseous emissions will be thrown away (abandoned).

1.1.1.3.3 Gaseous emissions from manufacturing operations

We mentioned previously that gaseous emissions will be generated from manufacturing operations. Like those from the waste incinerator, these emissions will be discarded by being thrown away or abandoned. The only regulations that apply to these gaseous emissions, however, are those promulgated under the CAA.

1.1.1.3.4 Obsolete chemicals

The last waste stream that fits under this category of discarded materials is obsolete chemicals. There may be some old or unwanted raw materials or products sitting in plant warehouses. Because the facility knows it'll never use or sell them, they really are wastes that the plant hasn't yet sent for disposal. These obsolete chemicals are being abandoned and/or they are materials that are intended to be thrown away. In either case, they are considered discarded materials.

1.1.1.4 Spent materials

A maintenance shop at the facility generates spent solvents, used oil, and paint-related wastes. These wastes fit into the spent materials category of discarded materials and are thus very likely to be subject to Subtitle C regulation, even if the facility usually recycles them. Residues from maintenance activities are common RCRA hazardous wastes.

1.1.1.5 Incidental residues

Finally, a slag is generated at the facility. These wastes are common at steel plants or metal refineries. Such incidental residues are simply stored in a slag pile that extends for many acres across the plant site. These slags often contain leachable heavy metals, and EPA is concerned about their management. Incidental residues like these are discarded materials because they are being abandoned.

1.1.1.6 "Discarded materials" is a very broad net

By looking at all of the materials generated at a hypothetical manufacturing plant (Figure 1-1), we have determined that, from a practical standpoint, "discarded materials" are potentially anything that comes out of a facility other than products used for their intended purpose. EPA is concerned about discarded materials that have the potential to enter or contaminate the environment. This includes materials that are intended for discard but are being stored instead (potentially to avoid disposal costs). As we noted, this might include raw materials or products that are no longer usable and that cannot be reclaimed or recycled.

From EPA's perspective, a discarded material can also be something that will be sent for recycling, not disposal. For example, the spent solvents and used oil coming out of a maintenance shop will more than likely be recycled; but, they are still considered discarded materials, because these materials are "sometimes discarded." [45 *FR* 33093]

1.1.2 Solid wastes are a subset of discarded materials

Now that we have a grasp on the concept of a "discarded material," we can continue to think about "solid wastes." When EPA wrote the hazardous waste regulations to implement the RCRA statute, the agency knew that not all discarded materials could be solid wastes. For example, Congress explicitly excluded uncontained gaseous materials from the definition of solid waste. Therefore, these gaseous emissions, while discarded materials, are not solid wastes.

When EPA published the RCRA regulations in 1980, it excluded a number of other discarded materials from the definition of solid waste. Most of these materials are also specifically excluded from EPA's jurisdiction in the RCRA statute. The exclusions for materials that are not solid wastes are found in §261.4(a) of the hazardous waste regulations.

Figure 1-2 summarizes the concepts we have discussed thus far. Discarded materials (which are encompassed by the outer-most circle) include anything produced at a plant other than products used for their intended purpose, and they include materials that will be recycled. However, gaseous emissions and discarded materials listed in §261.4(a) are not included in the definition of "solid wastes" (i.e., are not within the middle circle in Figure 1-2). Saying it another way, solid wastes are a subset of discarded materials.

1.1.3 Hazardous wastes are a subset of solid wastes

Hazardous wastes are represented by the innermost circle in Figure 1-2—the bull's-eye. As we stated at the very beginning of this chapter, before a material can be a hazardous waste, it must first meet the definition of a solid waste. In other words, hazardous wastes are a subset of solid

Figure 1-2: Relationships Among Discarded Materials, Solid Wastes, and Hazardous Wastes

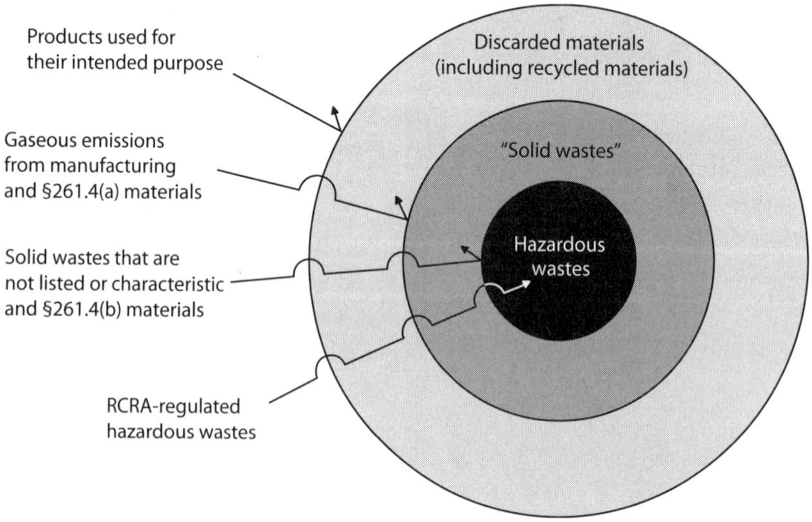

- Products used for their intended purpose
- Gaseous emissions from manufacturing and §261.4(a) materials
- Solid wastes that are not listed or characteristic and §261.4(b) materials
- RCRA-regulated hazardous wastes
- Discarded materials (including recycled materials)
- "Solid wastes"
- Hazardous wastes

Source: McCoy and Associates, Inc.

show that "during the calendar year (commencing on January 1), the amount of material that is recycled, or transferred to a different site for recycling, equals at least 75 percent by weight or volume of the amount of that material accumulated at the beginning of the period." [§261.1(c)(8)] If such a demonstration cannot be made, the material is being speculatively accumulated and is a solid waste.

1.2.4 Putting it all together

As indicated in Section 1.2.2, EPA identified five categories of materials that may be RCRA-regulated when recycled. The agency also identified four recycling activities (summarized in Section 1.2.3) that are of particular concern from an environmental standpoint. The key to understanding the relationships between these materials and activities is Table 1 in §261.2(c), reproduced in this book as Table 1-1.

This table is used to determine if recycled materials are solid wastes. The first step is to classify the material that you want to recycle as a spent material, sludge, by-product, commercial chemical product, or scrap metal. This tends to be the most difficult part of the solid waste determination for materials that are to be recycled. Note in Table 1-1 that the regulatory status of sludges and by-products depends, in part, on whether the material exhibits a characteristic of hazardous waste or is listed in §261.31 or 261.32.

After you have classified the material, determine if the manner in which the material will be recycled is addressed in the four column headings in Table 1-1. If it is, read across from the type of material to the column corresponding to the type of recycling activity, and, if there is an asterisk at their intersection, the material is a solid waste when recycled in that manner. For example, if your secondary material is a spent solvent, it is a spent material. If your intention is to burn the spent solvent as fuel, the recycling activity is energy recovery/fuel. Reading across in Table 1-1 from "Spent materials," you find an asterisk in the column with the heading "Energy recovery/fuel." The asterisk indicates that your spent solvent is a solid waste when you burn it.

That's the quick overview. There are lots of details and exemptions, however, that can affect the outcome of a solid waste determination for materials

Table 1-1: Determining If Recycled Materials Are Solid Wastes

Secondary material	Use constituting disposal [§261.2(c)(1)]	Energy recovery/fuel [§261.2(c)(2)]	Reclamation [§261.2(c)(3)] [except as provided in §§261.2(a)(2)(ii), 261.4(a)(17), 261.4(a)(23), 261.4(a)(24), or 261.4(a)(25)]	Speculative accumulation [§261.2(c)(4)]
Spent materials	(*)	(*)	(*)	(*)
Sludges (listed in 40 *CFR* 261.31 or 261.32)	(*)	(*)	(*)	(*)
Sludges exhibiting a characteristic of hazardous waste	(*)	(*)	—	(*)
By-products (listed in 40 *CFR* 261.31 or 261.32)	(*)	(*)	(*)	(*)
By-products exhibiting a characteristic of hazardous waste	(*)	(*)	—	(*)
Commercial chemical products listed in 40 *CFR* 261.33	(*)	(*)	—	—
Scrap metal that is not excluded under §261.4(a)(13)	(*)	(*)	(*)	(*)

Note: The terms "spent materials," "sludges," "by-products," and "scrap metal" are defined in §261.1.
Source: §261.2(c).

that will be recycled. Many of those details are discussed in the following two subsections. First, we give additional guidance on classifying your secondary material into the right category, followed by a more in-depth review of the four regulated recycling activities. Note that there are several exceptions in the "Reclamation" column heading in Table 1-1. Most of these exceptions were added in an October 30, 2008 rule [73 FR 64668], and these exceptions are discussed in Sections 1.2.6.3.1 and 1.2.6.3.2.

Our focus in this section is determining whether a material is a solid waste when recycled; you will note in the following examples, however, that if a material is determined to be a solid waste, we often go one step further by noting if the solid waste is also hazardous. Although we're getting ahead of ourselves (whether a solid waste meets the definition of a hazardous waste is discussed in Chapters 2 and 3), we think going the extra step is helpful for bringing closure to the examples.

1.2.5 Classifying your secondary material

As mentioned above, it is sometimes difficult to determine into which of the five categories your secondary material fits. EPA guidance in this area is summarized in the next several pages.

1.2.5.1 Spent materials

Spent materials are materials that we bought for use, then we used them, and now we are taking them out of service. They include spent pickle liquor, used batteries, spent solvents, used mercury switches, spent catalysts and etchants, and spent acids and caustics. [January 4, 1985; 50 FR 650, RO 11822] EPA noted that wastewater is a spent material in EPA/530/SW-86/015, available at http://nepis.epa.gov/EPA/html/Pubs/pubtitleOSWER.html by downloading the report numbered 530SW86015. Two commonly asked questions about the spent material definition in §261.1(c)(1) are presented below.

Q *Must a material be spent as a result of contamination in order to meet the definition of "spent material"?*

A No. EPA has consistently interpreted the definition of "spent material" as applying to materials that have been used and, as a result of contamination, are no longer fit for use without being regenerated. However, the agency considers "contamination" to be "any impurity, factor, or circumstance that causes the material to be taken out of service for reprocessing." [RO 11822; see also RO 11419]

For example, consider lead-acid batteries that are taken out of service and sent to a lead reclaimer. EPA's position is that the batteries pose the same risks and are handled the same way no matter how many or how few physical and chemical impurities they contain, and no matter how much or how little the presence of impurities contributes to the decision to stop using the battery. The batteries are spent materials regardless of the reason they were taken out of service. [RO 11822]

Q *To meet the definition of "spent material," must a material be nonfunctional in the sense that it could not continue to be used for its original purpose?*

A No. EPA considers the part of the definition stating that a spent material "can no longer serve the purpose for which it was produced" as being satisfied when the material is no longer serving its original purpose and is being reprocessed instead. As an example, when a generator removes mercury-bearing thermostats from buildings as part of an upgrade to the building's heating system, the thermostats could continue to be used for their intended purpose. However, assuming the generator intends to ship these thermostats to a reclamation facility for mercury recovery, these thermostats would be considered to be spent materials irrespective of the reason for their removal and despite the fact that the thermostats are potentially capable of being used as thermostats in another building. [RO 11822, 11876, 14762]

The agency provided additional guidance as follows:

"Regarding whether a material must be nonfunctional to meet the definition of spent material, the fact that a material can continue to be used for its original purpose is not relevant to the issue of whether or not it is a spent material when it is clear from the facts that the material will not be used but instead will be treated by reclamation. The mere potential for continued original use does not preclude a material from being defined as spent.... [T]he fact that it is actually removed from service establishes, as to this generator, that it can no longer serve its original purpose.

"If all that were required to avoid RCRA Subtitle C regulation would be a showing that a secondary material could continue to be used, then generators would be able to circumvent RCRA simply through changing their operating practices to remove secondary materials just prior to that material being unfit for its original use. Thus, spent solvents that are heavily contaminated but might still be fit for metal degreasing (even though they were being sent to be regenerated into new solvents), spent lead-acid batteries that still hold a charge (or were capable of holding a charge), and mercury-bearing thermostats removed from buildings sent for reclamation would not be subject to RCRA regulation in spite of the fact that the generator was no longer using the material but instead was sending it to be treated by reclamation.

"Clearly, this result is not consistent with the cradle-to-grave purpose of RCRA Subtitle C regulation. Used materials taken out of service and sent for reclamation pose the same risks and are handled in the same manner regardless of the reason they are taken out of service. For this reason, EPA has consistently interpreted spent materials as including materials which could continue to be used for their original purpose but are, in fact, being taken out of service for reclamation, showing that for this generator, they can no longer serve the purpose for which they were produced.

"Because spent materials being reclaimed (or to be reclaimed) are within the definition of solid waste, it is important to be able to distinguish among spent materials, other categories of solid wastes such as sludges, and products which are still in use that have not been discarded. Spent materials are distinguished from products and other categories of solid wastes in that they have been used previously and have been taken out of service and are going to be treated by reclamation." [RO 11822]

Conversely, if a product such as a battery, thermostat, or solvent is taken out of service from one situation but continues to be used for its intended purpose in another (e.g., a used battery is reused as a battery in a different application without reclamation), then the material never became a waste subject to RCRA. [January 4, 1985; 50 *FR* 624, RO 11258, 11541, 11868, 14281, 14677]

Q *A castings manufacturer uses potassium hydroxide solution to clean casting parts. When the solution becomes so contaminated that it no longer effectively cleans the parts, it is sent as a commodity—not a waste—to a fertilizer producer who incorporates the solution into its fertilizer product to control pH and provide a source of potassium. The castings manufacturer claims that since the used potassium hydroxide solution is taken out of service from one situation and continues to be used in another situation without reclamation, the material never became a spent material subject to RCRA. Is the castings manufacturer correct?*

A No. EPA's position is that a used material is not considered a spent material only if the continued use is "sufficiently similar to or consistent with the material's *initial* use." Since the castings manufacturer was taking a material that had been used as a solvent and using it for a completely different purpose (an ingredient in fertilizer), it clearly met the definition of a spent material. [*Howmet Corporation v. EPA*, D.C. Circuit Court of Appeals, Docket Number 09-5360, decided August 6, 2010]

Note that leftover, unreacted raw materials from a process are not spent materials, since they have never been used. [January 4, 1985; 50 *FR* 624] Similarly, leftover fuel (e.g., fuel removed from aircraft) is not classified as a spent material, since fuel, by definition, is spent only when it is destroyed to produce energy. [OSWER Directive 9441.00-2]

Q *Parts washers sometimes consist of some sort of cleaning apparatus attached to the top of a drum of solvent. Solvent is drawn up into the cleaning apparatus for use and is discharged back into the drum afterward. Following a period of use, the solvent in the drum becomes too contaminated to clean effectively. Periodically, facility personnel move the cleaning apparatus from the contaminated drum and place it atop a fresh drum of solvent. After several drums of spent solvent are accumulated in this manner, the supplier arrives to replace the spent solvent with fresh material. When, if ever, is waste that is generated in a parts washer regulated under RCRA?*

A When the solvent can no longer be used effectively, it is classified as spent material. A spent material sent for reclamation is a solid waste per Table 1-1 (i.e., there is an asterisk at their intersection), unless one of the exclusions noted in Section 1.2.6.3.1 applies. If it's a solid waste, and if the solvent exhibits any of the hazardous characteristics or meets a listing description, it is regulated as hazardous waste when the facility owner/operator decides it has become too contaminated for further use. The facility would thus become a generator of hazardous waste when the cleaning apparatus is removed from the drum. [RO 12790]

Q *Fixer solution used in a photographic film development process is removed from the working baths once its concentration of ammonium thiosulfate decreases to a certain level. At that point in time, the fixer solution could still function effectively as a fixer. The used fixer is transported to an offsite facility where it is treated, filtered, and "refortified." The refortified fixer will then be sold for reuse in developing film. Is the used fixer a solid waste prior to being refortified?*

A Yes. The treatment, filtering, and refortification are reclamation steps. Even though it could still be used for its intended purpose, the fixer is considered a spent material because it is removed from service and sent for reclamation.

From Table 1-1, it is a solid waste [RO 11541], unless one of the exclusions noted in Section 1.2.6.3.1 applies.

1.2.5.2 Sludges

A sludge is any residue from a water or air pollution control device, including flue and baghouse dust and wastewater treatment sludge. The following examples illustrate EPA's guidance regarding sludges that are recycled.

Q *Is spent carbon considered a spent material or a sludge?*

A "[S]pent carbon would normally be considered a spent material, unless it results from pollution control [activities], in which case it is considered a sludge." [RO 11143; see also RO 11089]

Q *According to Table 1 in §261.2(c) (reproduced in this book as Table 1-1), sludges exhibiting a characteristic of hazardous waste are not solid wastes when reclaimed (i.e., there is no asterisk at their intersection). An aircraft maintenance facility generates acrylic plastic dust while blasting paint and coatings off of aircraft. The dust is collected in the facility's air filtration system and is subsequently sent offsite for reclamation. Would the dust, which exhibits a characteristic, be considered a "sludge"?*

A Since the primary purpose of the air filtration system is not air pollution control per se but rather collection of acrylic plastic dust for further processing, the filtration system would not be considered an air pollution control device. Therefore, the dust would not be considered a sludge. The dust more clearly fits within the meaning of "spent material" and would, therefore, be regulated as a solid waste [RO 11937], unless one of the exclusions noted in Section 1.2.6.3.1 applies.

Q *Silver-recovery units are used to treat wastewater from photoprocessing operations. Would residues from these units be classified as "sludges," even if the primary purpose of the units is silver recovery rather than meeting specific wastewater discharge limits?*

A   Yes. The definition of "sludge" is not limited to materials generated from wastewater treatment undertaken specifically to meet federal, state, or local discharge or pretreatment requirements. Instead, the term applies to materials generated from wastewater treatment regardless of whether such treatment is required by law or regulation. [RO 11879]

Q   In an electroplating process, ion-exchange units are located upstream in the process rather than prior to discharge. For example, resins are used to take metals out of rinsewater from rinse tanks, and the treated water is reused in the process rather than disposed. Would the resin be considered a "sludge" under these circumstances?

A   The definition of "sludge" is tied to the type of unit that generated the waste, not the disposition of treated effluent or the intent of processing. Even if treated rinsewater is reinserted into the process rather than disposed, the act of removing contaminants from the water is still considered to be pollution control. The used resins are pollution control residues and meet the definition of "sludge." Specifically, the used resins are regulated as F006 wastewater treatment sludge. [RO 11857]

1.2.5.3 By-products

The term "by-product" is defined in §261.1(c)(3). By-products are residual materials that are not produced intentionally and that are unfit for end use without substantial processing. EPA has viewed by-products as a "catch-all" category that includes materials that are not spent materials or sludges. [April 4, 1983; 48 *FR* 14476] These materials result from a production process but are not the intended product. They tend to be residual in nature, rather than a highly processed material intentionally produced for sale to the public. [RO 11617, 11793] By-products may have some economic value, but they must be processed substantially before use. [RO 11342, 11750, 11793]

Listed by-products are solid wastes when recycled in any of the ways addressed by Table 1-1 (i.e., there is an asterisk associated with use constituting disposal, burning for energy recovery, reclamation, or speculative accumulation), although they would not be solid waste if one of the exclusions noted in Section 1.2.6.3.1 applies when they are reclaimed. There is also an exclusion for characteristic by-products that are reclaimed (there is no asterisk at their intersection).

Q   In a zinc galvanizing process, metal parts are first placed in a kettle of molten zinc and then placed in a chromic acid quenching bath for chrome passivation. During this process, zinc and charcoal residues are carried over from the molten-zinc bath into the chromic acid bath (the passivation solution). As a result of this continuous process, fine particles of chrome-coated zinc and charcoal accumulate in the passivation solution. These particles settle out of the solution and are then partially dewatered and reintroduced to the molten zinc kettle as a substitute for raw material feedstock for the process. These chrome-coated zinc and charcoal particles would be considered hazardous wastes if disposed because they exhibit the toxicity characteristic for chromium. What regulations apply?

A   The particles of chrome-coated zinc and charcoal meet the definition of a by-product found in §261.1(c)(3). The process of dewatering the accumulated by-product is defined as reclamation. Thus, the chrome-coated particles are reclaimed from the liquid portion of the by-product to make those particles available for use in the zinc kettle or more amenable for reintroduction into the process. From Table 1-1, a by-product that is hazardous solely because it exhibits a characteristic is not a solid waste if it is reclaimed. Therefore, the particles of chrome-coated zinc and charcoal are not subject to RCRA regulation when managed as described. [RO 11415]

1.2.5.3.1 By-products vs. co-products

A by-product is not the same thing as a co-product. Co-products are products that are intentionally produced, can be used as commodities in trade by the general public, and can be used "as is" without additional processing (or require minimal processing to become usable). [January 4, 1985; 50 FR 625, RO 11750, 11767] These materials are considered products because they are manufactured to specifications. [RO 11677, 11750, 11767, 11936, 14589] Another feature of a co-product is that it appears to have a legitimate or guaranteed market. [RO 11750, 11793, 14589] For example, lead recovered from primary copper smelting operations is a co-product. Gasoline, kerosene, fuel oil, and asphalt are co-products from petroleum refining. These co-products are regulated the same as any commercial product; that is, they are not subject to RCRA regulation if they are used for their intended purpose. However, petroleum product tank bottoms are a by-product. [50 FR 630]

Co-product vs. by-product determinations are supposed to be made by authorized states or EPA regions based on site- and process-specific information. [RO 11767, 11793, 11795, 13671] One criterion evaluated is whether the material is managed as a valuable commodity (e.g., managed to prevent releases and inventory records are maintained). Another is whether the material contains any Part 261, Appendix VIII constituents in concentrations higher than those found in comparable virgin products. Also, the state/EPA region will consider whether the residual material is as effective in its identified use (e.g., used in roughly similar amounts such that the displaced virgin product is not also used) as the alternative virgin product. [RO 11750, 11767, 11936, 14589] Table 1-2 gives the state of Texas' criteria for determining whether a material is a co-product fuel or a by-product hazardous waste. Based on these criteria, a co-product fuel determination is given as Case Study 1-1.

Q The F021 hazardous waste listing includes wastes (except wastewater and spent carbon from hydrogen chloride purification) from the production or manufacturing use of pentachlorophenol. A company produces hydrochloric acid from the hydrogen chloride gas obtained as a co-product during the manufacture of pentachlorophenol. The raw acid contains traces of phenol and is passed through activated carbon to reduce the phenol concentration in the finished product to less than or equal to 3 ppm. Spent carbon that was used to scrub the hydrochloric acid is specifically exempted from the F021 classification. How is the hydrochloric acid product regulated?

A Materials produced intentionally and which in their existing state are ordinarily used as commodities in trade by the general public are defined as co-products—not by-products. These co-products are not solid wastes and hence are not hazardous wastes. The hydrochloric acid produced as a co-product from pentachlorophenol manufacture is, therefore, not an F021 waste. [RO 11260]

Q A company generates distillation-column bottoms from the production of chlorobenzenes (listed waste K085). The company claims the material is not a waste at all because it is processed to recover muriatic acid, chlorobenzenes, and/or benzene. How are the distillation-column bottoms regulated?

A EPA considers this material a by-product, not a co-product. The bottoms, although they may have some economic value, must be processed before use. "[B]y-products are materials, generally of a residual character, that are not produced intentionally or separately, and that are unfit for end use without substantial processing. Examples are still bottoms…." [January 4, 1985; 50 FR 625] Per Table 1-1, listed by-products that are reclaimed are solid wastes, unless one of the exclusions noted in Section 1.2.6.3.1 applies. If solid wastes, the distillation-column bottoms (which would be solid wastes that meet a listing description) are hazardous wastes subject to regulation under RCRA Subtitle C. [RO 11297, 11342]

Table 1-3 gives some examples of co-products vs. by-products from EPA guidance.

1.2.5.4 Commercial chemical products

Normally the term "commercial chemical products" refers to unused, essentially pure, chemical products listed in §261.33 (the P- and U-lists). However, there are actually two meanings to this term, as discussed in the next subsection.

Table 1-2: Questions Used to Determine If a Material Is a Co-Product Fuel

Intent of producing the material
- Can it be demonstrated that the material is produced more or less intentionally?
- Are there meaningful chemical or physical specifications that the material has to meet? Is a legitimate safety data sheet (SDS) available?
- Is there evidence of reasonably good quality control for the material? Is there evidence that the material has a stable and relatively invariant composition over time?
- Do the physical properties and chemical composition of the material meet commonly recognized industry standards for fuels or fuel additives?

Value of the material
- Can it be demonstrated that the material has a significant intrinsic economic value?

Likelihood of a broad-based and stable market for the material
- Are there contracts or comparable documents in place?
- Are there copies of written requests from legitimate customers for the material?
- Has a market feasibility study been prepared?

Degree of processing required
- Can the material be used as a fuel or fuel additive as is?
- Does the material have an energy value of at least 5,000 Btu/lb?
- Would the physical properties and chemical composition of the material meet the comparable fuel specifications in §261.38(a)(1) if burned in a stationary unit?

Contaminants contained in the material ("toxics along for the ride")
- Does the material contain halogenated or inorganic constituents found in Part 261, Appendix VIII?
- Does the material contain Part 261, Appendix VIII constituents above those levels found in §261.38(a)(1)?

Performance of the material
- When used as a fuel additive (including as a cutter stock), can it be shown that the material does not adversely affect the performance, physical properties, and chemical composition of the fuel to which it is added?

Handling of the material
- Is the material managed in an environmentally responsible manner such that it is protected from loss?
- Does the generator have adequate and sufficient documentation to show that the criteria above are met?

Source: Adapted from Texas Commission on Environmental Quality, "Factors to Consider When Determining Whether a Secondary Material is Either a Legitimate Fuel or a Legitimate Fuel Additive," December 2003.

1.2.5.4.1 Two definitions of "commercial chemical product"

The term "commercial chemical product" has one meaning when determining whether a material is a solid waste when recycled, but a different meaning when deciding if a material is a hazardous waste. As applied to §261.2, the definition of solid waste, EPA interprets the category of commercial chemical products to include all types of unused commercial products (e.g., circuit boards, batteries, gasoline, or paint), whether or not they would commonly be considered chemicals and whether or not they would be included on the P- and U-lists. Although Table 1 in §261.2(c) (our Table 1-1) specifies "commercial chemical products listed in 40 *CFR* 261.33," EPA interprets the term to also include those products that

> ### Case Study 1-1: Co-Product Fuel Determination
>
> EPA has frequently been asked to distinguish between co-product fuels, which are not RCRA-regulated when burned for energy recovery, and by-product hazardous waste fuels, which are. The following example is typical.
>
> A material known as LX-830 results from reaction of petroleum and/or coal-tar naphtha feedstocks used in a resin manufacturing process. LX-830 is not the principal product of the process, and most of it is burned onsite as a substitute for conventional petroleum-based fuels. However, the generator also markets this material as a fuel or fuel additive. Is LX-830 regulated under RCRA when used as a fuel or fuel additive?
>
> It would appear that LX-830 better meets the definition of a co-product than a by-product and, hence, is not a solid or hazardous waste unless discarded. LX-830 has market value as a fuel product or fuel additive (comparable to conventional fuels), a conclusion based on its Btu value, product specifications, and market history. There is no evidence that the material was burned, either onsite or offsite, with the intent to discard it (e.g., burning amounts in excess of what was needed as a fuel source).
>
> Another factor supporting a determination that LX-830 is better classified as a co-product is that the material contains no hazardous constituents that are not typically found in conventional fuels. Thus, burning LX-830 does not constitute the discard of hazardous constituents and does not raise any greater environmental concerns than those raised by burning conventional fuels.
>
> Therefore, since the chemical makeup and subsequent handling and use of LX-830 is essentially similar to that of a commercially available fuel, the agency believes it should be considered a co-product. If, however, the LX-830 is mixed with any nonfuel materials and then burned, the agency would be concerned, not only about the other materials being burned, but about whether LX-830 is truly a co-product rather than a by-product. Such mixing would be an indication that LX-830 is not truly managed as a product. In other words, to the extent that LX-830 is produced to product specifications and handled in a manner consistent with a valuable product, the agency considers LX-830 to be a co-product. However, to the extent that LX-830 appears to be simply a process residue that happens to have high Btu content and is handled as a waste stream with little concern for product integrity, the agency would consider it to be a by-product. [RO 11936]
>
> Similar EPA guidance on distinguishing between co-product fuels and by-products being burned for energy recovery is contained in RO 11677, 11793, 14589, and 14609. In all of these documents, however, the agency emphasizes that this type of determination is best made on a site-specific basis by EPA regional offices or state environmental agencies. Also, note that there is another option available to producers of by-product fuels today that was not available in the past. These by-product fuels may qualify as "comparable fuels" and thus be excluded from the definition of solid waste under §261.38. (See Section 4.8.)

are not listed in §261.33, but exhibit one or more characteristics of hazardous waste. [April 11, 1985; 50 *FR* 14219, RO 11713, 11726, 13356, 13490]

Conversely, for the purposes of determining if a product is a hazardous waste on the P- and U-lists, EPA intended to include *only* those commercial chemical products, manufacturing chemical intermediates, and off-specification products known by the generic chemical name listed in §261.33. For example, EPA considers the P- and U-list definition of "commercial chemical products" to exclude manufactured articles such as thermometers or fluorescent lamps. Therefore, manufactured articles that contain a P- or U-listed chemical are not considered listed wastes when discarded in an unused form. This concept is discussed in detail in Section 3.5.3.1.2.

Table 1-3: EPA Examples of Co-Products vs. By-Products

Material description	Co-product or by-product	RO reference
Distillation-column bottoms	By-products	11101, 11297, 11342[1]
Hydrochloric acid produced from pentachlorophenol manufacture	Co-product	11260
Slags, drosses	By-products	11101, 11395
Coal-tar distillates produced to meet fuel specs and sold as fuels	Co-products	11677, 11936
Disulfide oil burned in a furnace to produce sulfuric acid (provides both material and energy value)	Maybe co-product but probably by-product	11750, 14086
Tank bottoms	By-products	11101, 11561
Natural gas pipeline condensate (with high energy value) used as a fuel additive	Co-product (off-spec fuel)	11831, 13049
Hydrochloric acid produced in an air pollution control system	Co-product	13671
Purge monomer from polystyrene manufacturing burned for energy recovery or used as a feedstock for a lesser grade of polystyrene	Co-product	14589

[1] Also see January 4, 1985; 50 *FR* 629.

Source: McCoy and Associates, Inc.

One question that often arises under RCRA is whether commercial chemical products automatically become solid wastes when their shelf life has been exceeded (i.e., the expiration date of the label has passed). EPA addressed this issue in RO 11606, 12996, and 14163. The agency's position is that expired commercial chemical products do not become solid wastes (i.e., they remain commercial chemical products) until a determination is made that the material will be discarded.

Table 1-1 shows the regulatory status of commercial chemical products when they are recycled in four different ways. Determining whether such products are solid wastes when so recycled is expanded upon as follows.

1.2.5.4.2 *Use constituting disposal*

First, because there is an asterisk at the intersection of "Commercial chemical products" and "Use constituting disposal" in Table 1-1, such products are solid wastes if recycled by being applied to the land. Solid wastes that exhibit a characteristic or meet a listing description are hazardous wastes. The introductory paragraph of §261.33 reiterates this conclusion by limiting certain recycling practices for P- and U-chemicals that might otherwise allow these chemicals to escape regulation; if these chemicals are mixed with used oil or other materials and then used for dust suppression (a form of use constituting disposal), they would be listed hazardous wastes. This prevents someone from claiming that the chemical is an ingredient in a product (dust suppressant), which might meet the recycling exemption at §261.2(e)(1)(i).

Section 261.33 also states that, if the chemicals are recycled by being applied to the land, and that is *not* how they are intended to be used, they are hazardous wastes. For example, it would be acceptable to recycle a P- or U-listed herbicide into a product and apply it to the land to kill weeds, because land application is an intended use for herbicides. [§261.2(c)(1)(ii)] On the other hand, if unused benzene is applied to the land (ostensibly to kill weeds), the benzene would be a hazardous waste (U019). Using benzene in this manner would be considered use constituting disposal.

1.2.5.4.3 *Burning for energy recovery*

If commercial chemical products are recycled by being burned for energy recovery or used as a component of fuels, and if this is not their original intended use, they are solid wastes (note the asterisk at their intersection in Table 1-1). If the product exhibits a characteristic or is P- or U-listed, it would also be a hazardous waste when recycled in this manner. For example, blending excess creosote (U051) into fuel would make the fuel a hazardous waste.

Conversely, unused commercial chemical products that are normal components of fuels (e.g., benzene, toluene, and xylene) are not solid wastes if they are burned for energy recovery. [§261.2(c)(2)(ii), RO 12505] EPA has indicated in guidance that this exclusion applies to off-specification commercial chemical products, as well as to products that are not listed in §261.33 but that exhibit a characteristic (e.g., gasoline, jet fuel, kerosene, diesel, and other waste petroleum products with a flash point <140°F). [RO 11138, 11449, 11713, 11848, 12825, 14503] Propane or butane propellants from full or partially full aerosol cans would also be considered commercial products that are not solid wastes when burned for energy recovery. [RO 11717]

The agency noted that, during an audit by a regulatory agency, a facility would likely be asked to show that the material for which this exclusion is claimed is "suitable for, and actually is used as fuel, or to make a fuel." [RO 11713]

Other examples of materials that could qualify for the §261.2(c)(2)(ii) exclusion include recovered free product from spills or leaks (e.g., recovered gasoline) and off-specification solvents, if the solvent is itself a fuel (e.g., methanol—some dragsters use methanol for fuel). However, an unused acetone-based solvent wouldn't qualify, because acetone is not normally a fuel. Similarly, burning a pesticide for energy recovery would not qualify under this exclusion. Waste-derived fuels (e.g., solvent-based fuel recovered from spent solvents) are also not exempt. [RO 11713, 13208]

Sometimes personnel drain fuel out of aircraft or other equipment, and they want to know whether such drained fuel is a spent material or still a commercial chemical product not listed in §261.33. EPA clarified that "leftover fuel is not classified as a spent material, since fuel, by definition, is spent only when it is destroyed to produce energy." [EPA/530/SW-86/015, cited previously]

1.2.5.4.4 Reclamation

There is a dash at the intersection of "Commercial chemical products" and "Reclamation" in Table 1-1, indicating that products may be reclaimed outside of the RCRA regulatory program. This includes manufacturing chemical intermediates, off-spec commercial chemical products, and residues from cleaning up spills of P- or U-listed commercial chemical products. Nonlisted commercial chemical products that exhibit a characteristic of hazardous waste also are not solid wastes when reclaimed. [April 11, 1985; 50 *FR* 14219, RO 11713, 11726, 13356, 13490] In fact, EPA uses the same approach to determine the regulatory status of any unused product that will be reclaimed, even if the product is not a chemical at all. The following examples and Case Study 1-2 illustrate how commercial chemical products are evaluated when reclaimed.

Q *A distributor returns pharmaceutical products to the manufacturer, who credits the distributor for the products and determines whether the products will be reused, reclaimed, or disposed. The products are returned to the manufacturer for many reasons, including: 1) an oversupply at the distributor, 2) expiration of the recommended shelf life, 3) a product recall initiated by the manufacturer, or 4) product damage. Are the pharmaceutical products solid waste when returned?*

A It is the distributor's obligation to ask the manufacturer what it will do with the returned products. Returned products are unused commercial products and are not solid wastes if there is a reasonable expectation that they will be recycled when returned to the manufacturer. There are no provisions in the RCRA regulations restricting shipments of commercial chemical products. Such products are not solid wastes until a decision has been made to discard them. However, in many cases, the manufacturer will not know whether the

Case Study 1-2: Mercury and Thermometers

An instrument manufacturer has off-specification (but unused) mercury thermometers that it wishes to either reclaim or discard. Would the thermometers be subject to hazardous waste regulation if reclaimed? What if they were discarded instead? What about contaminated mercury that has not been put into thermometers yet? How about used thermometers? Finally, what is the status of mercury removed from old thermometers and manometers?

If the unused thermometers in question are to be reclaimed, they would be considered commercial chemical products being reclaimed and, from Table 1-1, would not be solid wastes (there is no asterisk at the intersection of "Commercial chemical products" and "Reclamation"). Since a material must be a solid waste in order to be considered a hazardous waste, the thermometers destined for reclamation could not be regulated as hazardous waste. If the thermometers are to be discarded, on the other hand, they would be solid wastes [because they are being disposed per §261.2(b)(1)], and the manufacturer must then consider whether the off-specification thermometers are listed or characteristic hazardous waste. The thermometers would not meet the P- or U-listing criteria because they are considered manufactured articles, not commercial chemical products. As a result, discarded, unused thermometers would not be regulated as U151, and would only be subject to regulation as hazardous waste if they exhibit a characteristic (such as toxicity for mercury). [RO 13310, 14012]

If unused, contaminated mercury (an off-specification product) is sent to a reclamation facility to be purified, Table 1-1 indicates that commercial chemical products being reclaimed or stored speculatively are not solid wastes; hence, the contaminated mercury is not a hazardous waste when sent to a reclamation facility. Conversely, commercial chemical products identified on the P- and U-lists are solid and hazardous wastes when discarded or intended to be discarded. Thus, if the unused, contaminated mercury is shipped offsite for disposal, it would be listed waste U151. [RO 11378]

A thermometer removed from service does not meet the definition of a commercial chemical product because it has been used. Instead, it would be considered a spent material. Spent materials sent for reclamation are solid wastes per Table 1-1 (unless one of the exclusions noted in Section 1.2.6.3.1 applies); if solid waste, the used thermometers would be hazardous waste if they exhibit a characteristic for mercury (D009). If the used thermometers are to be disposed, they would also be solid wastes, and the hazardous determination would be the same as if reclaimed (i.e., they would be hazardous waste if they fail the TCLP for mercury). [RO 11123, 11378]

Finally, mercury recovered from used thermometers or manometers generally would be considered a spent material. However, EPA has provided (in guidance) an exemption for reclaimed metals that are suitable for direct reuse or that only have to be refined to be usable; such reclaimed metals are not wastes at all, but products. [January 4, 1985; 50 *FR* 634] Thus, such mercury that is removed from used instruments (assuming that it is pure enough to be reused as is or after refining) is not subject to the hazardous waste regulations. [RO 11123, 11159] If the removed mercury will not be reused but will be disposed, it is a solid waste that would undoubtedly exhibit the characteristic of toxicity for mercury and would be regulated as D009 (not U151, because it has been used).

Note that as of August 5, 2005, thermometers, barometers, and manometers may be managed as universal waste at the federal level. [70 *FR* 45508] Authorized states must adopt such mercury-containing equipment into their state universal waste programs before generators in those states may use this rule. (See Chapter 8.)

material can be reclaimed until a sample is analyzed. In this situation, the decision to reclaim or discard the material is made by the manufacturer, and the returned products would not be solid wastes (e.g., they would not need to be manifested to the manufacturer as a hazardous waste, although they would still be shipped pursuant to applicable DOT regulations). [RO 11606, 12996]

If, on the other hand, the returned products are simply being collected and accumulated by the manufacturer prior to disposal, they meet the definition of "abandoned material" in §261.2(b)(3) and are solid wastes at the distributor's facility. [RO 11492] If they exhibit a characteristic or are P- or U-listed, they would have to be manifested to the manufacturer. It is therefore incumbent upon the distributor to ask the manufacturer what it is going to do with the material upon receipt.

Q: *A mixture consisting of two or more unused chemicals is shipped to an offsite facility for reclamation. If the mixture is ignitable, must it be shipped as hazardous waste?*

A: No. When a commercial chemical product (or a mixture of commercial chemical products) is reclaimed, it is not a solid waste and, therefore, cannot be a hazardous waste. [RO 11147, 11321]

Q: *A paint manufacturer produces a batch of off-spec paint that he/she wants to reclaim by distillation to recover solvents. Assuming that it would exhibit a characteristic, is the paint a solid waste?*

A: No. The paint is considered to be a nonlisted commercial chemical product that, when reclaimed, is not a solid waste per §261.2(c)(3). [RO 11726]

Q: *Circuit boards from manufacturing are periodically rejected for not meeting specifications during the quality control (QC) inspection process. These units are dismantled, and the materials are reclaimed for use in the production of new circuit boards. Assuming the circuit boards would exhibit a characteristic of hazardous waste, would recycling the boards be subject to the hazardous waste regulations?*

A: No. Circuit boards that are rejected during the QC inspection are considered off-specification commercial chemical products. EPA interprets this category to include all types of unused commercial products that exhibit characteristics, whether or not they would commonly be considered chemicals. Reclamation of the off-spec circuit boards would not be subject to the hazardous waste regulations because the products are not solid wastes when reclaimed. [RO 13490]

1.2.5.5 Scrap metal

Other than "excluded scrap metal," which will be discussed in Section 1.2.5.5.1, scrap metal is a solid waste when reclaimed per Table 1-1. This sometimes surprises people. EPA explained its position as follows: "Scrap metal is waste-like in that it is a used material that is no longer fit for use and must be reclaimed before it can be used again, or is a process residue that must be recovered in a different operation from the one in which it was generated." [January 4, 1985; 50 *FR* 624] Thus, scrap metal that is reclaimed is a solid waste. However, hazardous scrap metal (scrap metal that exhibits a characteristic or is contaminated with listed hazardous waste) *that is recycled* is exempt from the hazardous waste regulations under §261.6(a)(3)(ii). This exemption is discussed more thoroughly in Section 4.21.

Given that recycled scrap metal is exempt, it is important to be able to distinguish scrap metal from other metal-containing wastes that are subject to RCRA control. Scrap metal includes pieces of metal as well as metal parts or equipment that may be bolted or soldered. Put another way, scrap metal includes products made of metal that become worn or are off-specification and are recycled to recover their metal content. Metal pieces that are generated from machining operations (i.e., turnings, stampings, etc.) are also considered scrap metal.

The term "scrap metal" does not apply to residues generated from smelting and refining operations (i.e., drosses, slags, and sludges). [RO 11083, 11446] It also does not include liquid wastes containing metals (i.e., spent acids, spent caustics, or other liquid wastes with metals in solution), liquid metal wastes (i.e., liquid mercury), or metal-containing wastes with a significant liquid component, such as spent batteries. [January 4, 1985; 50 FR 624] Furthermore, "the regulatory definition of scrap metal is not based on either the value of the material or the process by which it is to be reclaimed. Rather, the definition of scrap metal is based more on a material's physical appearance and previous use." [RO 11446]

Q *Do natural gas regulators that contain liquid mercury meet the definition of scrap metal when sent for reclamation?*

A No. The natural gas regulators, which contain mercury, cannot be managed as scrap metal. In general, any quantity of liquid mercury other than trace amounts attached to or contained in a material precludes that material from being scrap metal. Although these mercury-bearing natural gas regulators cannot be regulated as scrap metal, they may meet the definition of scrap metal and be exempt from regulation once the mercury is removed. [RO 11860] On August 5, 2005 [70 FR 45508], EPA added mercury-containing equipment as a type of universal waste subject to the Part 273 provisions. This category of universal waste includes regulators that contain elemental mercury integral to their function. [§273.9] Thus, generators of natural gas regulators that contain elemental mercury integral to their function could manage them as universal waste, if not scrap metal. See Sections 8.1 to 8.3.

Q *Are solder skimmings considered to be scrap metal?*

A No. Scrap metal is defined [§261.1(c)(6)] as "bits and pieces of metal parts (e.g., bars, turnings, rods, sheets, wire) or metal pieces that may be combined together with bolts or soldering (e.g., radiators, scrap automobiles, railroad box cars), which when worn or superfluous can be recycled." Solder skimmings or dross do not meet the physical description of scrap metal. [RO 11446]

Q *If a treatment, storage, or disposal (TSD) facility is storing scrap metal destined for recycling, does the scrap metal have to be managed as hazardous waste during storage if it exhibits a characteristic?*

A No. Scrap metal is a solid waste if destined for recycling, but it is exempt from the hazardous waste regulations found in Parts 262–266, 268, 270, and 124 if it is recycled. [§261.6(a)(3)(ii)] The recycling activity is viewed prospectively; that is, provided that the generator intends to recycle his/her scrap metal at some point in the future, the scrap metal is exempt from the hazardous waste regulations at the point of generation. [RO 14277] Note that respondents in enforcement actions who claim that a material is excluded or exempt based on recycling (e.g., the generator, recycler, or owner/operator of the TSD facility conducting storage) must be able to document a claim of legitimate recycling per §261.2(f). [RO 11877]

Q *A treatment facility is applying for a RCRA operating permit and, therefore, must address releases of hazardous waste or constituents from any solid waste management unit at the facility pursuant to RCRA Section 3004(u)—the corrective action program. Among other things, the facility currently manages scrap metal destined for reclamation. Must this facility address releases from the scrap metal management area (which is not subject to RCRA regulation) under corrective action?*

A Yes. Even though scrap metal destined for recycling is not regulated under Subtitle C, it is considered a solid waste. Thus, a release of a hazardous waste or constituent from the scrap metal management area (solid waste management unit) must be addressed during RCRA corrective action. [RO 12415]

1.2.5.5.1 Excluded scrap metal

The provision in §261.2(c) designating scrap metal as a solid waste when recycled does not apply to "excluded scrap metal" as defined in §261.1(c)(9). "Excluded scrap metal" is "processed scrap metal, unprocessed home scrap metal, and unprocessed prompt scrap metal [as these materials are defined in §261.1(c)(10–12)]." [§261.1(c)(9)] Excluded scrap metal being recycled is excluded from the definition of solid waste because EPA believes it is more commodity-like than waste-like. See §261.4(a)(13) and Section 4.21.2.

1.2.6 Determining your recycling activity

Once you have classified your secondary material into its proper category, you must determine if the recycling activity you want to employ is one of the four listed at the top of Table 1-1. The following pages review EPA's guidance clarifying these activities.

1.2.6.1 Use constituting disposal

Materials noted with an asterisk in Column 1 of Table 1-1 are solid wastes when they are applied to or placed on the land in a manner that constitutes disposal, or used to produce products that are applied to the land (in which case the product itself remains a solid waste). However, §266.20(b) in Part 266, Subpart C exempts such products from RCRA regulation if 1) they are produced for the general public's use, 2) the recycled materials undergo a chemical reaction so as to be inseparable from the product by physical means, and 3) the product meets land disposal restrictions treatment standards for all hazardous wastes used to produce it.

As you can see from Table 1-1, any secondary material (i.e., spent material, sludge, by-product, or commercial chemical product) that is applied to or placed on the land in a manner constituting disposal is a solid waste. The only exception to this general rule is that commercial chemical products are not solid wastes if they are applied to the land and that is their ordinary manner of use. (More on that in Section 1.2.6.1.3.)

Examples of use constituting disposal include using a waste-like material as fill or cover material, use for structural support, a soil conditioner or a dust suppressant, or as an ingredient in fertilizers or construction materials (e.g., asphalt, road base, concrete) that are applied to the land. [50 FR 628]

When EPA revamped its solid waste definition in 1985 to address recycling, the agency wasn't clear if recycling of a waste-like material into a pesticide and applying the pesticide to the land would be regulated the same way:

> "We note that we are *not* asserting RCRA jurisdiction over pesticides or pesticide applications. Use of a pesticide involves use of a product, not recycling of a waste. Thus, if a pesticide (including off-specification pesticide, pesticide rinsewaters, or unused dip solution applied in accord with label instructions) is applied to the land for beneficial use, the practice is not viewed as use constituting disposal." [Emphasis in original.] [50 FR 628]

Although EPA's language above says nothing about adding a hazardous waste as an ingredient into a pesticide before it is applied, the regulatory status of wastes recycled into pesticides was unclear. In May 2000 guidance, however, the agency has apparently decided that a hazardous waste recycled into a pesticide should be managed in the same way as such a waste that is used as an ingredient in fertilizer. In that guidance [RO 14437], EPA discussed the regulatory status of fertilizers and pesticides containing hazardous waste on the same footing. Several common examples of use constituting disposal follow.

1.2.6.1.1 Fertilizer production

Q *Sulfuric acid is used in a chemical manufacturing process and becomes contaminated with water, such that it can no longer be used for its primary purpose. The contaminated acid is subsequently used to make fertilizer. How is the contaminated sulfuric acid regulated?*

A If the contaminated sulfuric acid can no longer be used for its intended purpose in the chemical manufacturing process, it is a spent material. If the spent acid is then used to make fertilizer (i.e., use constituting disposal), it is a solid waste. If the spent acid exhibits a characteristic (e.g., corrosivity), it will also be a hazardous waste. [RO 11361]

Q Sulfuric acid is generated in an acid plant at a primary lead smelter. If the acid is used to manufacture fertilizer, is this use constituting disposal?

A In this situation, the acid plant is an air pollution control device and the sulfuric acid, which is a residue from the device, meets the definition of "sludge." Characteristic sludges used to produce a product applied to the land (e.g., fertilizer) are considered to be used in a manner that constitutes disposal and are, therefore, solid wastes. [RO 14437]

Q Waste pickle liquor from steel finishing (K062) is used to produce fertilizer. How is the waste pickle liquor regulated?

A The waste pickle liquor is a solid waste because it is a spent material being used in a manner constituting disposal. A solid waste that meets a listing description is a hazardous waste. Additionally, the fertilizer product is subject to the provisions of Part 266, Subpart C discussed above. [RO 11112]

Q A copper plating operation produces a spent copper sulfate bath that exhibits the corrosivity characteristic. However, the bath is not disposed but is shipped to a recycling facility that incorporates it into a commercial fertilizer; the fertilizer, which is not corrosive, is sold to farmers. What is the regulatory status of the bath generated at the copper plating facility?

A The bath is D002 and must be manifested to the recycling facility. Although the material is recycled, all secondary materials that are placed on the land for beneficial use or are incorporated into products that are placed on the land for beneficial use are solid wastes per Table 1 in §261.2(c). [RO 11124]

Q A characteristically hazardous flue dust is used to produce a commercial fertilizer. As such, the dust is a sludge used in a manner constituting disposal and, under §261.2(c), is classified as a solid waste. Would the classification of the flue dust be the same if 1) the resulting fertilizer does not exhibit any characteristic, 2) the flue dust reacts to generate a different compound in the process of producing fertilizer, and 3) payment to the generator exceeds the cost of delivery?

A The answer in each of these cases is that the classification remains the same. The solid waste determination for a recycled material is made at the point the waste is generated and takes into account the entire waste recycling process. The determination is based on the type of material (in this case sludge) and the type of recycling activity (in this case use constituting disposal). None of the three factors noted would change either the type of material or the type of recycling activity, so they won't change the classification of the flue dust as a solid waste. [RO 11774]

Note that on July 24, 2002 [67 FR 48393], EPA finalized conditional exclusions from the definition of solid waste for: 1) hazardous secondary materials used to make zinc fertilizers, and 2) zinc fertilizers made from hazardous wastes. See Case Study 1-3.

1.2.6.1.2 Cement or aggregate production

Q Spent abrasive from sandblasting is used as a substitute for silica, aluminum, and iron in the manufacture of Portland cement. How will it be regulated?

A In this case, the spent material most likely is being used in a manner constituting disposal (i.e., the Portland cement produced from these materials will be, or is likely to be, placed on the land, although this is a rebuttable presumption). Therefore, the spent abrasive material is a solid waste. If it exhibits the toxicity characteristic (e.g., for lead), it must be managed as a hazardous waste, including manifesting requirements. [RO 11433]

Q K061 electric arc furnace (EAF) dust is used as an ingredient to make cement. How is the EAF dust regulated?

A As noted in the previous example, cement is considered to be a product that is typically applied to the land. The EAF dust is a listed sludge

> ### Case Study 1-3: Zinc Fertilizer Production From Waste Materials
>
> A generator produces baghouse dust containing high concentrations of zinc and other heavy metals. The baghouse dust is not a listed waste but fails the TCLP for lead. The generator would like to send the dust to a manufacturer that could use it as an ingredient in micronutrient fertilizer due to its zinc content. What regulations apply?
>
> Prior to July 24, 2002, the baghouse dust would have been a solid waste (from Table 1-1) because it was a characteristic sludge recycled to produce a product that is applied to or placed on the land (i.e., used in a manner constituting disposal). Since the dust exhibited the toxicity characteristic, it was also a hazardous waste. [RO 11644]
>
> Since EPA published its zinc fertilizer rule on July 24, 2002 [67 *FR* 48393], the baghouse dust is excluded from the definition of solid waste [by §261.4(a)(20)]. Therefore, this material is not a solid or hazardous waste when used to produce zinc fertilizer, and so it is not subject to any of the RCRA storage, transportation, or paperwork requirements.
>
> The baghouse dust generator must comply with the following requirements to maintain this exclusion from the definition of solid waste: 1) no speculative accumulation is allowed; 2) storage of the excluded material must be in engineered buildings (with nonearthen floor), covered tanks, or closed containers (Super Sacks® for outdoor storage are allowed if stored on concrete pads with built-in containment systems); and 3) a one-time notice claiming this exclusion must be sent to the state or EPA, and a notice must be included with each offsite shipment of material noting the exclusion (copies of such notices must be maintained by the generator for three years).
>
> Zinc fertilizer manufacturers receiving shipments of this excluded material must comply with the following to maintain the exclusion: 1) store the material in the same manner as generators noted above; 2) submit a one-time notice claiming the exclusion to the state or EPA, followed by annual reports summarizing the quantities of excluded material used to manufacture zinc fertilizer; and 3) maintain for three years shipping records for all receipts of the excluded material.
>
> Zinc fertilizer produced from this excluded material is also excluded from RCRA regulation if it either 1) meets maximum contaminant concentrations for five heavy metals and dioxins, or 2) meets land disposal restrictions (LDR) treatment standards for each hazardous waste code that would have applied to the secondary material (in this example, the baghouse dust) had the exclusion not applied. [§§261.4(a)(21), 266.20(d)]

used in a manner constituting disposal. From Table 1-1, the material is a solid waste and, therefore, it is a listed hazardous waste. [RO 11491]

Q *F006 (wastewater treatment sludge from electroplating operations) is being recycled by being used as an ingredient in the manufacture of 1) aggregate, or 2) cement. Under these circumstances, would the wastewater treatment sludge be considered a solid waste?*

A If F006 waste is used as an ingredient to produce aggregate that is applied to the land (e.g., road-base material), it is a listed sludge being used in a manner constituting disposal. Therefore, it is a solid and hazardous waste and subject to regulation from the point of generation to the point of recycling. Such produced aggregate would remain a solid waste if so used. The aggregate product is, however, entitled to the exemption under §266.20(b). Conversely, if the aggregate is not used on the land, then the F006 used to produce it might not be a solid waste at all, and in such case, neither those materials nor the aggregate would be regulated. The same results would apply to cement produced using F006 as an ingredient. [RO 11426]

1.2.6.1.3 Materials applied to the land when that is their ordinary manner of use

Commercial chemical products are not solid wastes if they are applied to the land and that is their ordinary manner of use. For example, in a situation where soil was contaminated with chlordane in the normal course of agricultural use, EPA provided the following guidance:

> "The pesticide chlordane is a listed commercial chemical product (U036) that becomes a hazardous waste when discarded or intended to be discarded. The agency did not intend to cover those cases when the chemical is released into the environment as a result of use…. In addition, 40 *CFR* 261.2(c)(1)(ii) specifically states that commercial chemical products listed in §261.33 are not solid wastes (and, thus, not hazardous wastes) if they are applied to the land and that is their ordinary manner of use. Therefore, the contaminated soil would be treated as a hazardous waste (if it is dug up) only if it exhibits one or more of the four RCRA hazardous waste characteristics defined in 40 *CFR* 261.21 through 261.24." [RO 11182; see also RO 12357]

EPA provided the following guidance for petroleum-contaminated soil that is used as an ingredient in asphalt batching:

> "In general, a hazardous secondary material (and soils contaminated with a hazardous secondary material) used to produce a product used in a manner constituting disposal is a solid waste, unless it is a commercial chemical product that is normally used in this manner, such as a petroleum product normally used as an ingredient in asphalt batching [see §261.2(c)(1)(ii)—although the commercial product may not be listed in §261.33, the same regulatory approach applies]. The regulatory status of soils contaminated with crude oil would be determined using the same approach. The crude oil, while not a secondary material, would be a solid waste because it is being discarded by use in a manner constituting disposal, unless crude oil is a normal ingredient in asphalt batching. We expect that most petroleum-contaminated soils are not contaminated with a petroleum product that normally is used in asphalt production and would, therefore, be solid wastes. (For example, if gasoline is not normally used in asphalt production, then gasoline-contaminated soil is a solid waste when used in asphalt production.) However, there may be specific cases where the soil is contaminated with a petroleum product normally used to make asphalt, in which case the contaminated soil would not be a solid waste when used in asphalt batching.

> "Also, you should note that any media (including soil) or debris resulting from remediation of an underground storage tank cleanup under Part 280 is excluded from regulation as hazardous waste (for the D018–D043 constituents) regardless of the intended disposition, so these soils could be used in asphalt production.

> "In summary, with the exceptions of soils contaminated with petroleum materials normally used in asphalt production and soils resulting from underground storage tank cleanups, soils contaminated with petroleum materials that are listed waste or exhibit one of the characteristics would be hazardous and solid waste." [RO 11616]

1.2.6.1.4 Additional examples involving use constituting disposal

Q *Spent pickle liquor is used as a substitute for ferric chloride to condition wastewater. Is this use constituting disposal?*

A No. When secondary materials are used as wastewater conditioners, EPA does not consider this to be use constituting disposal. The activity is not similar to land application because the secondary material is chemically combined as part of a conditioning process and is subsumed as an ingredient in the conditioned wastewater. [January 4, 1985; 50 *FR* 628, RO 11081]

Q Lead containers are used to dispose radioactive wastes because the lead functions as a radiation shield. When the filled containers are landfilled, are they solid wastes that are hazardous for lead?

A No. The containers are products performing their intended function and would not be considered to be a discarded material. Hence, they are not solid wastes. Note that even though the containers are not regulated as hazardous wastes, they should be disposed in an environmentally safe manner (e.g., they could be macroencapsulated to prevent the shielding from leaching). [RO 12956, 13468, 13538]

Q If hazardous wastes are recycled as nutritional supplements in animal feed, is this use constituting disposal?

A In general, EPA does not believe that recycling hazardous wastes as nutritional supplements in animal feed is use constituting disposal. In contrast to crop fertilization, many animal feed preparations are not applied to the ground directly (although there may be some exceptions to this that would need to be determined on a case-by-case basis). [RO 11932]

Q For training purposes, a fire department sprays virgin diesel fuel on the ground. The fuel is set ablaze and then extinguished. The resultant residues are collected and disposed as RCRA hazardous wastes. Does the act of spraying the virgin diesel fuel meet the use constituting disposal classification?

A No. Spraying virgin fuel on the ground for fire-fighting practice does not meet the use constituting disposal classification. In this case, the fuel is a primary material and not a waste. Had the fuel been spent or a secondary material, such usage could be considered use constituting disposal. Additionally, burning fuel is how it is intended to be used. [RO 12488] In contradictory guidance, however, EPA noted that when commercial fuels are used in this manner, a soil contamination problem may develop. The release of a material that is either listed in §261.33 or that exhibits one of the characteristics (e.g., ignitability) onto or into land or water is hazardous waste disposal. Anyone wishing to conduct these exercises with ignitable or listed material would be advised to conduct the burn in a tank or lined pit to prevent illegal disposal. [RO 11267]

1.2.6.2 Burning for energy recovery

In the 1980s, EPA's position was that when wastes containing less than 5,000 Btu/lb are burned, they are being burned for destruction, not energy recovery, and hence, are solid (and potentially hazardous) wastes. Today, materials (other than commercial fuels) that are burned for energy recovery are solid and potentially hazardous wastes regardless of their energy content.

Simply stated, any secondary material (i.e., spent material, sludge, by-product, or commercial chemical product) that is burned for energy recovery or used to produce a fuel is a solid waste. The only exception to this general rule is that commercial chemical products that are themselves fuels, or normal components of fuels, are not solid wastes when burned for energy recovery. [§261.2(c)(2)(ii)] For example, benzene (listed as U019 when disposed) is a normal component of gasoline, and can be burned for energy recovery without being considered a solid or hazardous waste. [RO 12773] Here are some common examples of how burning for energy recovery is regulated.

Q As noted, commercial chemical product benzene is not a solid waste when burned for energy recovery. Would off-spec unused benzene, therefore, be acceptable as a start-up fuel in an incinerator?

A No. Boilers and industrial furnaces burn wastes for energy recovery. The primary purpose of an incinerator, conversely, is to burn for destruction. Therefore, materials burned in incinerators are always considered to be solid wastes. [§261.2(b)(2)] Solid wastes that meet a listing description (in this case U019) are hazardous wastes. The operating requirements for hazardous waste incinerators in §§264.345 and 265.345 state that hazardous wastes must not be fed to an incinerator during start-up or shutdown unless the incinerator is operating within steady-state conditions or conditions specified in the permit. Since off-spec benzene burned in an incinerator is regulated as a hazardous waste even if it is used as a start-up fuel, this would not be allowable. [RO 12773]

Q. *A toluene stream is used to extract product in a manufacturing process until it becomes so contaminated with benzene (roughly 15%) that it is no longer suitable for use as an extractant. Can the facility owner/operator burn the resulting toluene/benzene stream as a fuel outside of RCRA per §261.2(c)(2)(ii), since it contains materials that are normal components of fuel?*

A. No. Any manufacturing process waste that contains benzene, toluene, etc. that is burned for energy recovery or used to produce a fuel is a solid and potentially hazardous waste. [§261.2(c)(2)(i)] According to EPA, the used toluene is a spent material per the definition in §261.1(c)(1)—not a commercial chemical product. Thus, §261.2(c)(2)(ii) does not apply; instead, spent materials burned for energy recovery are solid and potentially hazardous wastes. [RO 14814]

Q. *Still bottoms from the recovery of an F-listed spent solvent are blended with oil and then sold as fuel. Is the fuel subject to RCRA regulation?*

A. Yes. When a spent material is burned for energy recovery, it is a solid waste per Table 1-1. Because the still bottoms are derived from a listed waste, they remain a listed waste, and the fuel mixture is also listed via the mixture rule. Therefore, the waste-derived fuel is considered a hazardous waste per §261.2(c)(2). [RO 12650] Solvent-based fuels recovered from spent solvents are also subject to RCRA regulation. [RO 13208]

Q. *Gasoline, jet fuel, kerosene, diesel, etc. are not listed commercial chemical products (i.e., they do not appear in §261.33). Are off-specification variants of these materials (including absorbent pads and other residues from cleaning up fuel spills) still exempt from the definition of solid waste under §261.2(c)(2)(ii) when burned for energy recovery?*

A. Yes. These nonlisted commercial chemical products are not solid wastes when burned for energy recovery. [RO 11138, 11449, 11713, 11848, 12825, 14503]

Q. *When determining if an off-specification aviation fuel is a solid waste, does EPA care how the fuel becomes off spec or how it is subsequently upgraded for use as fuel?*

A. The manner in which fuels become off specification does not determine how they are regulated, with one exception: If the fuels are mixed with hazardous waste, they are regulated as hazardous waste when burned for energy recovery. Also, EPA does not distinguish between different types of burning for energy recovery. For example, the fuel may be blended with other types of fuel (e.g., gasoline, diesel, etc.) and then used to power aircraft or burned in industrial furnaces and boilers. [RO 11938]

Q. *If a generator blends unused product xylene (U239) into used oil, does he/she then have a hazardous waste fuel?*

A. No. Xylene is a normal component of fuel. Therefore, the unused xylene is neither a solid nor hazardous waste when burned for energy recovery. The generator's used oil has not been mixed with hazardous waste and is not a hazardous waste fuel. [RO 12505]

Q. *Is natural gas pipeline condensate that exhibits the characteristic of ignitability and that is going to be burned for energy recovery a hazardous waste?*

A. According to the January 4, 1985 *Federal Register* [50 *FR* 630], off-spec fuels burned for energy recovery are not by-products and thus are not considered wastes. This includes natural gas pipeline condensate. The condensate contains many of the same hydrocarbons found in liquified natural gas and certain higher hydrocarbons that have energy value. It is generated in the pipeline transmission of natural gas and is not considered to be a by-product or waste when burned for energy recovery. [RO 13049]

EPA added additional guidance for natural gas pipeline condensate in a later interpretation:

"A more precise example [of "sham" burning for energy recovery], and one that the agency has found to have occurred, would be the sale or use of contaminated, low-energy-value 'natural gas pipeline condensate' as a motor fuel or fuel additive (such that additional octane enhancers also had to be added).

"In general, the January 4, 1985 preamble discussion (50 FR 630) still applies. Accordingly, use of unadulterated natural gas pipeline condensate with high Btu content as a fuel or fuel additive could, in fact, constitute a legitimate type of burning for energy recovery. It is important to note, however, that energy value is not the sole determinant of whether the natural gas condensate is being legitimately burned as a fuel. Additional sham recycling criteria...are equally relevant to a regulatory status determination. Of particular relevance to a determination regarding natural gas pipeline condensate is whether the condensate contains toxic constituents not found in normal fuels and, if so, whether these constituents contribute to the recycling objective or are simply being destroyed. Assuming that burning is for legitimate energy recovery, high-energy-value natural gas condensate would not be considered a RCRA solid or hazardous waste when used as a fuel or fuel additive. However, the determination would have to be made case-by-case based on the facts relevant to both the specific material and the manner in which it is being burned." [RO 11831]

Q Vapor-recovery units (VRUs—carbon-containing filters) are used to capture offgases and liquid condensate from petroleum product storage tanks. The contaminated VRUs are removed and heated, releasing hydrocarbons that are returned to the petroleum refinery for reuse. The spent carbon is then tested for characteristics and land disposed. Are the VRUs subject to RCRA regulation when removed from service?

A No. The captured hydrocarbons in the VRUs are not secondary materials; they are products rather than wastes and are not subject to RCRA regulation. The VRUs containing the product would be considered commercial chemical products being reclaimed, which are excluded from the definition of solid waste per Table 1-1. [RO 14555]

1.2.6.2.1 Fuel blending

Hazardous wastes are often blended with other waste or nonwaste materials to produce a hazardous waste fuel that will meet specifications established by the burner. Although not defined in the regulations, the term "fuel blending" is generally used to describe the combination of hazardous wastes and other materials to create a material that is amenable to burning for energy recovery. Since processes that fit this description change the properties of a hazardous waste in order to meet a fuel specification and recover energy, both parts of the treatment definition are met. Hence, fuel blending is considered treatment by EPA. [RO 11497, 11881, 13577, 13764] See Section 7.1.2.3.

Facilities conducting fuel blending have tried to argue that their hazardous waste management practices are a form of recycling (i.e., recovery of energy from a waste stream) and should be exempt from RCRA permitting or management standards under §261.6(c)(1). EPA's response is that "[t]here may be some recycling operations at a fuel blending facility that are exempt from permitting, even though the fuel blending process itself is not exempt." [RO 11881] See also April 13, 1987; 52 FR 11820, RO 11238, 11411, 11497, 13512.

Generators and fuel blenders frequently have questions about Btu-content restrictions that apply to materials blended into fuels. EPA provides the following guidance:

"[U]nlike the situation prior to adoption of the boiler and industrial furnace (BIF) rules in 1991, the 5,000 Btu/lb (as-generated) heating value criterion no longer determines the regulatory status of the boiler or industrial furnace (see 56 *FR* 7134; February 21, 1991). Currently, a fuel blender can blend wastes of any Btu value for burning in a BIF; however, there are consequences in doing so for industrial furnaces that use such blended wastes.

"Specifically, the 5,000 Btu/lb value, as-generated, is used as a reasonable yardstick to distinguish between waste fuels being burned for energy recovery versus those burned for destruction or, potentially, as an ingredient, unless the facility can demonstrate that the waste with less than 5,000 Btu/lb is being burned for legitimate energy recovery. Thus, if an industrial furnace produces a product that is used in a manner constituting disposal (e.g., a cement or light-weight aggregate kiln) and uses a blended fuel with a portion that has an as-generated heating value of less than 5,000 Btu/lb, the agency will generally assume that the waste is being burned for destruction. In such cases, the resulting product will be considered waste-derived (i.e., subject to regulation as hazardous waste), unless the facility can document that the low-heating-value waste is being burned for legitimate energy recovery." [RO 11885]

"[I]f an industrial furnace burns a listed hazardous waste with an as-generated heating value less than 5,000 Btu/lb and the facility does not document that the waste is burned for legitimate energy recovery, then any product applied to or placed on the land in a manner that constitutes disposal (e.g., cement) would be a waste-derived product subject to regulation as a hazardous waste…. [An] industrial furnace may burn a waste with an as-generated heating value less than 5,000 Btu/lb and avoid waste-derived product implications only if the facility documents that the lower-heating-value waste contributes substantial, usable energy to the furnace. Documentation could be provided by, for example, empirical data showing that substitution of a lower-heating-value waste results in a substantial reduction in fuel (e.g., coal) usage that would otherwise be consumed. Other approaches may also be used to demonstrate that low-heating-value waste contributed significant energy input to the furnace. However, facilities should discuss their approach(es) to document that lower-heating-value wastes are being burned for legitimate energy recovery with the appropriate permitting agency to be sure that it is acceptable." [RO 11883]

1.2.6.3 Reclamation

A material is reclaimed if it is processed to recover a usable product. Examples include recovery of lead from spent lead-acid batteries, recovery of oil from petroleum refining sludges or by-products, recovery of silver from spent photoprocessing solutions, dewatering, ion exchange, and distillation. Metal smelters are considered reclamation units. [50 *FR* 633, RO 11338, 11426, 11910, 14026]

Reclamation also occurs when a secondary material is regenerated. Materials are regenerated when they are processed to remove contaminants in a way that restores the material to its original condition. Common examples are regeneration of spent carbon and catalysts. [RO 11089, 14176]

Incidental processing operations such as briquetting, sintering, agglomerating, crushing, or grinding that make secondary materials easier to handle typically are not considered to be reclamation. [RO 11201] A complete discussion of reclamation vs. incidental processing is given in Section 11.3.1.2.

It is important to distinguish between situations where material values in a spent material, sludge, or by-product are recovered as an end product of a process (as in metal recovery from secondary materials) as opposed to situations where these secondary materials are used as ingredients to make new products. The former situation is reclamation; the latter is a type of direct use that usually is not considered to constitute waste management. When a secondary material is first reclaimed and then put to direct use, the material is still a solid waste until reclamation has been completed. [January 4, 1985; 50 *FR* 633]

A material can have monetary value and still be a solid waste when sent for reclamation. For example, a lead smelting company that reclaims ingots from lead plates of spent automobile batteries must manage the batteries as solid waste (there is an asterisk at the intersection of "Spent materials" and "Reclamation" in Table 1-1). The fact that these materials are solid waste under RCRA does not change "just because a reclaimer has purchased or finds value in the components." [*U.S. v. ILCO*, 996 F.2d 1126 (1993)]

EPA has recently revised its regulations with respect to how spent materials, listed sludges, and listed by-products are regulated when sent for reclamation. (See Section 1.2.6.3.1.) Even before that rule was promulgated, the agency had begun to reevaluate its position in interpretive guidance memos.

One of the agency's first reevaluations in this area involved the foundry industry. As part of casting metal components, molds are made using sand. After the castings cool, the molds are broken to remove the castings, which are sent for cleaning. The used sand from the broken molds exhibits the toxicity characteristic for lead and is processed to remove contaminants (e.g., pieces of metal) using vibratory screens and conveyors. The cleaned sand is then used to make new molds.

Consistently throughout the 1990s, EPA held the position that the contaminated sand is a spent material being reclaimed. From Table 1-1, it is a solid waste and, because it exhibits a characteristic, a hazardous waste (D008). See RO 11900.

In March 2001, however, the agency revisited this issue and changed its position:

"We believe that reusing sand onsite within the sand loop for mold-making, including the separation step described…above, is part of a continuous industrial production process by the generating industry. Consequently, the sands being reused onsite in the primary production process on a continuous basis in the sand loop are not solid wastes.

We also believe that sand stored indoors that is reused in the sand loop is part of the ongoing production process, as long as it is handled as a valuable material for the manufacture of castings. On the other hand, if sand is handled in a way that would indicate the material is not valuable (e.g., a sand pile exposed to the elements), it would raise questions of whether the sand is being legitimately reused as part of the continuous production process." [RO 14534]

Based on Table 1-1, sludges and by-products that exhibit a characteristic, but are not listed, are not solid wastes if reclaimed. EPA structured the regulations in this way so as to avoid regulating nonlisted sludges and by-products that are routinely processed to recover usable products as part of ongoing production operations.

The determination of whether a material being reclaimed is a solid waste is made at the point of generation; that is, the recycling activity is viewed prospectively. The following examples illustrate these concepts.

Q *A manufacturer generates a characteristic by-product that is excluded from the definition of solid waste if reclaimed. The generator wants to recycle the material but cannot afford the transportation costs to send the material to the reclaimer. Instead, he/she sends the material to a nearby TSD facility that, in turn, ships it to the reclaimer. Is the material subject to RCRA while in storage at the TSD facility?*

A The recycling activity is viewed prospectively; that is, the status of certain secondary materials is determined by knowing how the material is going to be recycled. The term "when" as it is used in §261.2(c) for recycling activities (e.g., when reclaimed) is not meant to refer only to the moment in time reclamation occurs. If the generator intends to reclaim his/her characteristic by-product at some point in the future, he/she would not be considered to be managing a solid or hazardous waste. Of course, when secondary materials are excluded or exempt based on a claim of recycling, the material will no longer be excluded or exempt if it is accumulated speculatively prior to recycling. Also, respondents in enforcement actions who make such a claim (e.g., the generator, recycler, or owner/operator of the TSD facility conducting storage) must be able to document a claim of legitimate recycling per §261.2(f). [RO 11747, 11877]

Q *A facility operates a photoprocessing laboratory that produces waste fixer solution containing silver at more than 5.0 mg/L TCLP. If the fixer is sent offsite for silver recovery, does it have to be managed as a hazardous waste? Is the answer the same if the fixer is passed through an onsite ion-exchange column, and the spent resin is sent offsite for silver recovery?*

A Used fixer that contains silver is a spent material. When sent for reclamation, the spent fixer is a solid waste per Table 1-1, unless one of the exclusions noted in Section 1.2.6.3.1 applies. If it is a solid waste that exhibits the characteristic for silver, it would be subject to the hazardous waste management rules as D011. However, because silver is a precious metal, the reduced requirements of Part 266, Subpart F would apply (i.e., the fixer would not have to be stored in units meeting hazardous waste standards, but would have to be counted and manifested, and no speculative accumulation could occur). [RO 11541, 11914, 14092] The precious metal exemption is discussed in Section 4.19.

If the fixer is treated in an onsite ion-exchange column (from which the treated water is discharged to a wastewater treatment system), the spent resin from the column (which now contains most of the silver) would be considered a sludge. That is, the resin is a residue from a water pollution control device, so it meets RCRA's definition of a "sludge." Such characteristic sludges, when reclaimed, are not solid wastes and therefore are not hazardous wastes. [RO 11857, 11879, 14331]

Q *Wastewater from electroplating nickel and chrome bumpers is treated in tanks. Periodically, a metal hydroxide sludge is removed from the tanks and sent to a copper smelter for metal recovery. How is the sludge regulated?*

A Sludge from treating wastewater from electroplating operations is F006 hazardous waste. Metal smelters are reclamation units. Listed sludges sent for reclamation are solid wastes; therefore, the sludge is a listed hazardous waste [RO 11910], unless one of the exclusions noted in Section 1.2.6.3.1 applies.

1.2.6.3.1 DSW rule establishes exclusions for materials being reclaimed

In 2008, EPA amended the regulations that apply to hazardous secondary materials destined for reclamation. Partly due to a series of decisions by the U.S. Court of Appeals for the DC Circuit [including *Association of Battery Recyclers, Inc. et al. v. EPA*, 208 F.3d 1047 (2000)], EPA had to reevaluate whether all materials noted with an asterisk in the third column in Table 1-1 really are solid (and therefore potentially hazardous) wastes when reclaimed.

EPA revised the definition of solid waste (DSW) for certain hazardous secondary materials being reclaimed on October 30, 2008. [73 *FR* 64668] The hazardous secondary materials to which that rule applies are spent materials (that are hazardous by either listing or characteristic), listed sludges, and listed by-products that will be reclaimed. Examples are metal-bearing sludges and spent catalysts sent for metals recovery and organic-bearing liquids sent for recovery of solvents. Such materials are not solid or hazardous wastes if reclaimed under one of the exclusions codified in the October 30, 2008 final rule. Under the pre-existing RCRA regulations, these hazardous secondary materials typically were solid and hazardous wastes when reclaimed via §261.2(c)(3) and associated Table 1. To add these exclusions, EPA revised this table by providing several exceptions in the column heading under "Reclamation" (see Table 1-1). The five exceptions are:

1. §261.2(a)(2)(ii)—Reclamation "under the control of the generator" within the U.S. or its territories, where the materials are managed in non-land-based units (discussed below);

2. §261.4(a)(17)—Reclamation of spent materials generated in the primary mineral processing industry (discussed in Section 1.2.6.3.2);

3. §261.4(a)(23)—Reclamation "under the control of the generator" within the U.S. or its territories, where the materials are managed in land-based units (discussed below);

4. §261.4(a)(24)—Materials are transferred to another company for reclamation within the U.S. or its territories and certain conditions are met (discussed below); and

5. §261.4(a)(25)—Materials are exported for reclamation in a foreign country and certain conditions are met (discussed below).

Basically, EPA finalized two self-implementing exclusions from the definition of solid waste. The first exclusion is for materials that are generated and reclaimed under the control of the generator. The rule requires that these materials, when managed in either non-land-based units (e.g., tanks, containers, and containment buildings) [see §261.2(a)(2)(ii)] or land-based units (e.g., piles and surface impoundments) [see §261.4(a)(23)], be "contained" in such units (basically, this means that the material is not released into the environment) before recycling. "Under the control of the generator" means:

1. The material is generated and reclaimed at the generating facility (located in the U.S. or its territories); or

2. The material is generated and reclaimed at different facilities (both of which are located in the U.S. or its territories), both facilities are controlled by the same "person," and the controlling "person" must certify responsibility for safe management of the recycled material; or

3. The material is generated pursuant to a written contract between a tolling contractor and a toll manufacturer (both of which are located in the U.S. or its territories) and is reclaimed by the tolling contractor (who certifies ownership and responsibility for the secondary materials, even while they are managed at the toll manufacturer's facility).

A more detailed review of the exclusion for reclamation under the control of the generator is found in Section 11.2.3.1.

The second exclusion (the "transfer-based" exclusion) is for materials that are transferred to another person or company for reclamation, but only if the following conditions are met [see §261.4(a)(24)]:

1. The material is not handled by any person or facility other than the generator, the transporter, an intermediate facility, and one or more reclaimers. An intermediate facility is allowed to store the material for more than 10 days, but the generator/intermediate facility must ensure that the material is sent to the generator-designated reclamation facility.

2. The generator must make "reasonable efforts," repeated, documented, and certified every three years, to ensure that each reclaimer will legitimately recycle the material and that any intermediate facility and reclaimer will manage the material in a manner protective of human health and the environment (e.g., the generator could audit these facilities). "Reasonable efforts" are only required for reclaimers and any intermediate facilities where management of the material is not addressed under a RCRA permit or interim status standards.

3. The generator, any intermediate facility, and each reclaimer must keep records of each shipment of hazardous secondary material for at least three years. Additionally, generators must receive confirmations of receipt from each intermediate facility and reclaimer and maintain them for at least three years.

4. The reclaimer and any intermediate facility must manage the material in a manner that is at least as protective as that employed for the analogous raw material and the secondary material must be contained.

5. Any residues generated from the reclamation process must be managed in a manner that is protective of human health and the environment (e.g., any hazardous wastes must be managed in accordance with RCRA Subtitle C).

6. The reclaimer and any intermediate facility must obtain financial assurance (similar to that required for permitted TSD facilities) per Subpart H of Part 261, thereby demonstrating that they will not abandon the secondary material.

EPA also excluded from the definition of solid waste hazardous secondary materials that are exported from the U.S. and reclaimed in a foreign country. To use this exclusion, the following conditions must be met [see §261.4(a)(25)]:

- The generator must notify EPA of its intent to export the material to a foreign reclamation facility and obtain, through EPA, written consent from the receiving country (Acknowledgment of Consent). The Acknowledgment of Consent must accompany each shipment.

- The material cannot be handled by any person or facility other than the generator, the transporter, an intermediate facility, and one or more reclaimers, and the material must be contained. The generator must ensure that the material is sent to the generator-designated reclamation facility.

- The generator must make "reasonable efforts," repeated and documented every three years, to ensure that each reclaimer will legitimately recycle the material and that each reclaimer and any intermediate facilities will manage the material in a manner protective of human health and the environment.

- The generator must keep records of each shipment of hazardous secondary material and all Acknowledgments of Consent for at least three years and must file an annual report summarizing all such exports for reclamation.

A more thorough review of the transfer-based reclamation exclusion is given in Section 11.2.3.2.

These exclusions are not applicable to K171/K172 spent catalyst from petroleum refineries. EPA is planning to address whether these materials are solid wastes when reclaimed in a future rulemaking, where the agency will address the potential pyrophoric properties of these materials, particularly during transportation and storage prior to reclamation. EPA also wants spent lead-acid batteries to continue to be managed under either Part 266, Subpart G or Part 273 (see Section 4.20). That is because these are actually hazardous waste regulations, although regulations with less-stringent management standards, that are appropriate due to the unique nature of these batteries.

The exclusions do not apply to materials that are otherwise subject to material-specific management conditions under §261.4(a) when reclaimed; such materials must continue to meet the existing conditions/requirements in that section of the regs to be excluded from the definition of solid waste. For example, the exclusion for used, broken CRTs at §261.4(a)(22)(iii), that requires these CRTs to be managed in accordance with §261.39, is not changed or superseded.

The exclusions also do not apply to materials that are speculatively accumulated, used in a manner constituting disposal, or burned for energy recovery or used to produce a fuel.

Generators, reclaimers, and any intermediate facilities, tolling contractors, and/or toll manufacturers that manage hazardous secondary materials that have previously been subject to regulation as hazardous waste, but which are excluded from regulation under the 2008 rule, must send notification to EPA or the state. The notification, which must be sent prior to managing the material under one of the exclusions and by March 1 of each even-numbered year thereafter, must identify: 1) the name, address, EPA ID number, and NAICS code of the facility; 2) the name and phone number of a contact person; 3) the exclusion [e.g., §261.2(a)(2)(ii), 261.4(a)(23), (24), and/or (25)] under which the material will be managed and whether the material will be managed in any land-based units; 4) the materials that will be excluded, when the materials will begin to be managed in accordance with the exclusion(s), and the quantities managed annually; and 5) documentation that reclaimers and any intermediate facilities have the required financial assurance [applicable for materials managed under §261.4(a)(24) or (25) only]. The notification must be submitted using EPA Form 8700-12, and the certification included on that form must be signed by an authorized representative of the facility.

Since the 2008 DSW rule excludes from regulation materials that were previously subject to regulation as hazardous waste, the rule results in a less-stringent RCRA program. Therefore, states are not required to adopt the rule, although EPA strongly encouraged them to do so. Moreover, a state may choose to adopt only parts of the rule.

As of August 2014, the exclusions from the definition of solid waste promulgated on October 30, 2008 were in effect in only six states: Alaska and Iowa, where EPA administers RCRA, and Idaho, Illinois, New Jersey, and Pennsylvania, which have adopted the rule. Puerto Rico has passed a resolution allowing application of the rule.

1.2.6.3.2 Secondary materials from mineral processing

In the land disposal restrictions Phase IV rule issued on May 26, 1998 [63 FR 28556], EPA amended §261.2(c)(3) and associated Table 1 by adding parenthetical references to §261.4(a)(17) under the "Reclamation" column. That section, as amended by a March 13, 2002 final rule [67 FR 11251] resulting from settlement of a lawsuit [*Association of Battery Recyclers, Inc. et al. v. EPA*, 208 F.3d 1047 (2000)], provides a conditional exclusion from the definition of solid waste for spent materials from mineral processing operations that will be reclaimed. If the conditions in §261.4(a)(17) are met, spent materials from mineral processing are not solid wastes. If the conditions are not met, the spent materials are regulated as solid and potentially hazardous wastes. The most significant of the conditions is that the spent materials cannot be stored in surface impoundments or piles.

1.2.6.4 Speculative accumulation

Secondary materials that are being accumulated before being recycled are presumed to be speculatively accumulated, unless the person accumulating the material can show that it is "potentially recyclable and has a feasible means of being recycled." [§261.1(c)(8)] Furthermore, the person accumulating the material must be able to show that during the calendar year (commencing on January 1), the amount of material that is recycled, or transferred to a different site for recycling, equals at least 75% by weight or volume of the amount of that material on-hand on January 1.

EPA created the speculative accumulation provision to mitigate the risk posed by facilities that overaccumulate hazardous secondary materials prior to recycling. The provision serves as a safety net, preventing recyclable materials that are not otherwise regulated under RCRA from being stored indefinitely and potentially causing environmental damage. EPA subjects facilities that accumulate speculatively (i.e., facilities that fail to recycle at least 75% of a recyclable material during the calendar year or fail to demonstrate that a feasible means of recycling exists) to immediate regulation as hazardous waste generators or storage facilities.

For example, if in December 2013, a facility determines that it has speculatively accumulated a secondary material (e.g., it couldn't recycle 75% of what it had onsite on January 1, 2013), that material becomes subject to regulation as a solid (and potentially hazardous) waste on January 1, 2014. If the material is hazardous, the facility becomes a hazardous waste generator and would be able to store the hazardous waste onsite for either 90 days or 180/270 days without a storage permit, depending on its monthly hazardous waste generation rate. [RO 14199]

The speculative accumulation provisions generally apply to secondary materials that are not solid wastes when recycled, such as characteristic sludges and by-products that will be reclaimed. The speculative accumulation provisions also apply to spent materials, listed sludges, and listed by-products that will be reclaimed under the control of the generator, since these materials are not solid wastes per §261.2(a)(2)(ii). [Note that prior to the definition of solid waste rule in 2008, EPA's position was that the speculative accumulation provisions did not apply to spent materials, listed sludges, and listed by-products being reclaimed, because under the rules in effect at that time these materials were already considered solid wastes when awaiting recycling.] The speculative accumulation provisions do not apply to commercial chemical products that are stored prior to reclamation, because, by definition, these materials are not regulated as solid wastes until they are abandoned or intended for discard. [RO 13528, 13755]

Q: *In March 1991, a facility generated 200 kg of sludge that exhibits the toxicity characteristic for lead. The operator of the facility placed these materials in storage to await reclamation of lead. Since the sludge will be reclaimed, it is not considered a solid waste while stored prior to reclamation. On December 31, 1991, the facility still had not recycled any of this material. Was the sludge speculatively accumulated, since 75% was not recycled during the year?*

A: No. The sludge was not speculatively accumulated. By December 31 of a given year, a facility owner/operator must show that he/she has recycled 75% of the material that was in storage on January 1 of that year. For the above facility, the amount of lead-bearing material in storage on January 1, 1991 was zero; so, on December 31, 1991, the operator does not have to show that any amount was recycled during the calendar year. On January 1, 1992, however, 200 kg of characteristic lead sludge are in storage. Thus, the facility must be able to show that 75% of this material, or 150 kg, has been recycled or sent for recycling by December 31, 1992.

If the operator cannot demonstrate this 75% recycling rate, the sludge remaining in storage is speculatively accumulated and becomes subject to regulation as a solid waste on January 1, 1993. Because it exhibits a characteristic, the generator would have to begin to handle the material as a hazardous waste on that date. EPA noted that "this approach could allow essentially a free year to accumulate where a generator starts a year with little or no waste." [April 4, 1983; 48 *FR* 14490] The period of one calendar year starting on January 1 was selected, however, to facilitate enforcement and achieve uniformity. In making the calculation, the 75% requirement applies to all materials of the same class being recycled in the same way. If this facility also generated a by-product that exhibited the toxicity characteristic for chromium and reclaimed it, the owner/operator would make a separate speculative accumulation calculation for this by-product. [RO 13528]

1.2.7 The big caveat

When EPA is asked to clarify how recycled materials are regulated under RCRA, it typically responds that it is providing hypothetical answers as to the general applicability of RCRA. The agency always goes on to state something like this:

"For each individual facility, the appropriate region or RCRA-authorized state will have to make the final determination as to the applicability of RCRA regulations based on an analysis of the actual facilities and processes. Regardless of their RCRA authorization status, states may impose regulations more stringent or broader in scope than those in 40 *CFR* Parts 260–270 as a matter of state law." [RO 11385]

Characteristic Wastes

Does my solid waste exhibit a characteristic?

If you have determined that a material you are managing is a solid waste per the discussions in Chapter 1, the next step is to determine if it is hazardous. In fact, the RCRA regulations in §262.11 state that the generator of a solid waste "must determine if that waste is a hazardous waste…." A systematic procedure for making such a determination is given in Section 2.1.

Table 2-1 gives the big picture regarding hazardous waste categorization. As shown in that table, EPA developed two primary approaches for designating a solid waste as hazardous. First, the agency identified four generic physical/chemical properties that, if exhibited by a solid waste, make it a hazard to human health or the environment. In other words, this first set of wastes are hazardous because of their physical or chemical character. Such wastes are known as the "characteristic" hazardous wastes. The regulations [§261.3(a)(2)(i)] specify that if a solid waste exhibits a characteristic, it is a hazardous waste. The four hazardous characteristics are ignitability, corrosivity, reactivity, and toxicity. Each of these is examined below in Sections 2.2 through 2.8.

Table 2-1: Hazardous Waste Categories

Hazardous waste type	EPA hazardous waste code(s)	Applicable regulations	Text reference
Ignitable characteristic waste	D001	§261.21	Section 2.2
Corrosive characteristic waste	D002	§261.22	Section 2.3
Reactive characteristic waste	D003	§261.23	Sections 2.4–2.5
Toxic characteristic wastes	D004–D043	§261.24	Sections 2.6–2.8
F-listed wastes from nonspecific industrial sources	F001–F039	§261.31	Sections 3.1.1, 3.2–3.4
K-listed wastes from specific industrial sources	K001–K181	§261.32	Section 3.1.1
P-listed acutely hazardous, unused commercial chemical products, manufacturing chemical intermediates, and off-spec variants	P001–P205	§261.33(e)	Section 3.5
U-listed toxic, unused commercial chemical products, manufacturing chemical intermediates, and off-spec variants	U001–U411	§261.33(f)	Section 3.5

Source: McCoy and Associates, Inc.

EPA has developed an excellent guidance document for the four characteristic hazardous wastes: *Hazardous Waste Characteristics—A User-Friendly Reference Document*, October 2009, available at http://www.epa.gov/waste/hazard/wastetypes/wasteid/char/hw-char.pdf. The document is designed to be web-based, so its usefulness is maximized when viewed on a device that is online. It includes hyperlinks to *Federal Register* notices, letters and memoranda issued by EPA, and other relevant documents that provide clarification of the hazardous waste characteristics.

The second approach used by the agency to regulate a solid waste as hazardous was to make lists of waste streams or unused chemicals that EPA knew, based on its experience, presented a threat to human health or the environment when disposed. So, EPA put together four lists of hazardous wastes; any waste or discarded chemical on one of those lists became known as a "listed" hazardous waste. Listed wastes are discussed in Chapter 3.

2.1 Introduction to hazardous wastes

Whenever facilities are trying to make a hazardous waste determination for a particular material, they should ask themselves the following four questions *in order*:

1. Is it a solid waste?
2. Is it exempt?
3. Is it listed?
4. Is it characteristic?

Asking these four questions (which come right out of §262.11) will hopefully keep mistakes in identifying hazardous wastes to a minimum.

If the answer to the first question is "No," you can stop. Unless a material is a solid waste, it cannot be a hazardous waste. See, for example, RO 11138, 11398, 11631, and 12641.

Even if the material is a solid waste, it may be excluded from management under RCRA Subtitle C based on several sets of exclusions and exemptions in §261.4 of the regulations. A number of these exclusions and exemptions are examined in Chapter 4.

A solid waste that is not exempt from RCRA regulation will have to be evaluated to determine if it is hazardous. Section 262.11 specifies that the generator should first determine whether the solid waste is a listed hazardous waste. This third step is accomplished by ascertaining if the solid waste meets any of the listing descriptions in §§261.31–261.33.

After the generator has determined if the solid waste is or is not listed, the waste must still be evaluated to determine if it exhibits any of the four hazardous waste characteristics. This fourth question must be answered to comply with the land disposal restrictions (LDR) program, even if the answer to the third question is "Yes." [§262.11(c)]

2.1.1 Four types of hazardous wastes

Although we have alluded in our opening comments that there are only two types of hazardous wastes, there are actually four:

1. Declared hazardous wastes,
2. Characteristic hazardous wastes,
3. Listed hazardous wastes, and
4. Mixtures of hazardous waste with solid waste (such mixtures are often hazardous).

This book provides detailed discussions of the last three types of hazardous wastes in Chapter 2, Chapter 3, and Chapter 5 (Section 5.1), respectively. But before moving on to those discussions, we want to introduce the concept of a "declared" hazardous waste.

2.1.1.1 Declared hazardous wastes

A declared hazardous waste is a solid waste that does not meet the definition of a hazardous waste, but the generator says it does. There are a number of reasons generators sometimes declare a waste to be hazardous:

1. They make a mistake.
2. They want to be conservative (or maybe they're not really sure), so they'll overclassify

the waste as hazardous to minimize their risk of getting enforced against. For example, a facility may have a policy that certain nonaqueous wastes will be declared D002 because they might absorb sufficient water from the atmosphere to become corrosive between their point of generation and the TSD facility.

3. They know the waste is not hazardous, but they want to manage it as hazardous because it contains some very toxic chemicals.

As noted at the beginning of this chapter, if a person generates a solid waste, he or she must determine if it is hazardous. The regulations allow that determination to be made either by testing a representative sample of the waste or using knowledge of the waste or the process that generated the waste. If the generator says the waste is hazardous (even if it really isn't), that is considered a knowledge-based determination, and EPA and the state agency will consider that waste to be hazardous just like any other. So, from the point of generation, the waste has to be managed in compliance with all of the hazardous waste regulations.

We generally don't recommend declaring nonhazardous wastes to be hazardous. First, we don't like to see people make mistakes (hopefully, this book will help out in that regard). There are also significant costs associated with hazardous waste management, including storage, transportation, treatment, disposal, and recordkeeping costs, as well as potential hazardous waste taxes and fees levied by some states.

Additionally, we think that being overly conservative may actually result in increased liability. Additional regulatory oversight and scrutiny are associated with hazardous waste management. As a result, a significant chance exists of being found out of compliance with one or more of the numerous applicable RCRA regulations while managing a declared hazardous waste.

Sometimes a waste is legally nonhazardous even though it may contain one or more very toxic constituents. Generators of such wastes will sometimes declare them to be hazardous and then ship them to a commercial hazardous waste TSD facility. Because they will be treated or destroyed at such a facility under all of the Subtitle C management standards, generators think they are minimizing any future liability. However, this may not be the case because manifesting a declared waste leaves a paper trail that may actually increase the generator's potential liability under CERCLA. A suggested alternative to declaring a nonhazardous waste to be hazardous is to inquire if the TSD facility will accept and manage your nonhazardous waste in the same manner as they would if it were hazardous. Such facilities usually will accept such wastes (perhaps without a manifest since they are not hazardous), but they may charge you the same as if it were hazardous. Once at the facility, the waste will be managed with the high degree of care you are seeking but without having to declare it as hazardous.

2.1.1.1.1 Declaring wastes as hazardous may save analytical costs

Although we don't like declared wastes, it is sometimes more cost-effective to declare a small quantity of solid waste hazardous than it is to properly evaluate it. For example, a laboratory generates a 5-gallon bucket of waste that may contain a small amount of benzene. Instead of conducting a TCLP analysis for the benzene in the waste (which might cost $200 or more), the facility declares it to be hazardous and ships it to a TSD facility with all of its other hazardous waste. "If a person believes his waste to be hazardous, he may also simply declare it to be so without any reference to [the regs] or to scientific literature." [December 18, 1978; 43 *FR* 58969]

A "declared" hazardous waste is not a term that you'll find in the RCRA regulations, but it is a valid type of hazardous waste. "The regulations allow a generator to characterize its waste based on process knowledge, and it is understood that generators may at times characterize their wastes conservatively, rather than incur the costs of testing every batch or stream." [RO 11918; see also RO 11608]

2.2 Ignitability

From §261.21(a), a solid waste exhibits the characteristic of ignitability (and is a D001 hazardous waste) if a representative sample of the waste meets any of the following criteria:

- A liquid (other than an aqueous solution containing <24% alcohol by volume) with a flash point <140°F. [§261.21(a)(1)]

- A nonliquid that 1) can ignite under standard temperature and pressure (68°F and 1 atm) through friction, moisture absorption, or spontaneous chemical changes; and 2) burns so vigorously and persistently after ignition that it creates a hazard. [§261.21(a)(2)]

- An ignitable compressed gas defined as a compressed gas that 1) is flammable when in a mixture of 13% or less with air (i.e., its lower flammability limit is 13% or less), or 2) has a flammable range with air of more than 12% (i.e., the difference between its upper and lower flammability limit is greater than 12 percentage points) regardless of the lower flammability limit. [§261.21(a)(3)]

- An oxidizer, which is a substance that yields oxygen readily to stimulate the combustion of organic matter. [§261.21(a)(4)]

EPA's objective with the ignitability characteristic is to require hazardous waste management for those wastes "capable of causing fires during routine transportation, storage, and disposal and wastes capable of severely exacerbating a fire once started." [May 19, 1980; 45 *FR* 33108] Each of these four categories of ignitable wastes is explored in more detail below.

2.2.1 Liquids with low flash points

Liquids with a flash point <140°F make up the majority of ignitable wastes. Almost all of the nonchlorinated solvents have flash points below 140°F and so exhibit the ignitability characteristic when they become wastes. Spent paint thinners, contaminated fuels, various waste alcohols and other liquids, and some adhesives are also

Table 2-2: Closed-Cup Flash Points for Common Materials

Material	Flash point (°F)
Propane	–156
Gasoline	–36
Acetone	–4
Cyclohexane	–1
Benzene	12
Methyl ethyl ketone	19
Toluene	39
Methanol	54
Isopropyl alcohol	54
Ethanol	55
Ethylbenzene	59
Isobutyl alcohol	82
o-Xylene	90
n-Butyl alcohol	95
Kerosene	100
Mineral spirits	105–140
No. 4 fuel oil	130
No. 6 fuel oil	150
Phenol	174
Vegetable oil	400–600

Source: McCoy and Associates, Inc.; adapted from safety data sheets from ChemBioFinder.com (http://chembiofinder.cambridgesoft.com).

typically ignitable. Flash points of some common materials are listed in Table 2-2.

EPA's definition of an "ignitable" liquid is a little different from DOT's definition of a "flammable" liquid. A DOT flammable liquid is a liquid with 1) a flash point ≤140°F, or 2) a flash point ≥100°F that is intentionally heated and offered for transportation at or above its flash point. [49 *CFR* 173.120(a)] The primary reason EPA selected the flash point level of <140°F for D001 wastes was that these temperatures can be encountered during routine storage and transportation conditions (e.g., liquid waste in a black drum sitting in the sun on a summer day). EPA specifically chose the term "ignitable" to try to eliminate confusion between a RCRA ignitable liquid and a DOT flammable liquid.

2.2.1.1 Determining if a waste is a liquid

Since the criteria for ignitable waste determinations are different for liquids versus nonliquids, the generator must first determine if his/her waste is a liquid. There are a number of ways this can be accomplished. A knowledge-based determination may be made: "EPA believes that, for purposes of the characteristics of ignitability and corrosivity, it will generally be obvious whether or not the waste is a liquid." [April 30, 1985; 50 *FR* 18372] For instance, a generator can use knowledge to determine that waste paint thinner or solvent is a liquid.

For wastes with both a liquid and solid phase (semisolids), however, it can be more difficult to make such a determination. For these materials, generators commonly run the Paint Filter Liquids Test (Method 9095B in SW–846). This test, which was added to SW–846 in 1985 [50 *FR* 18370], is used to determine the presence of free liquids in a representative sample of a waste. [RO 13601] The sample is placed in a conical paint filter and allowed to drain for 5 minutes; if even one drop of liquid passes through the filter in that time frame, the sample contains a free liquid. Although the paint filter test was added to the agency's test methods to ensure no free liquids were disposed into landfills, EPA noted that "this test provides a practical method of testing ignitable and corrosive materials to determine the presence of liquids, and assists the regulated community in complying with the Part 261 requirements until further evaluation is done." [50 *FR* 18372]

Chapter 7 of SW–846 (which discusses the four hazardous waste characteristics and how they should be evaluated) currently includes the following statement: "Use Method 9095…to determine free liquid." EPA proposed in 1993 to modify that language by replacing the paint filter test with the pressure filtration technique specified in Method 1311 (the toxicity characteristic leaching procedure or TCLP). [August 31, 1993; 58 *FR* 46052] Although the agency never finalized that proposal, the pressure filtration technique remains its preferred method:

"The definitive procedure for determining if a waste contains a liquid for the purposes of the ignitability and corrosivity characteristics is the pressure filtration technique specified in Method 1311. However, if one obtains a free liquid phase using Method 9095, then that liquid may instead be used for purposes of determining ignitability and corrosivity. However, wastes that do not yield a free liquid phase using Method 9095 should then be assessed for the presence of an ignitable or corrosive liquid using the pressure filtration technique specified in Method 1311." [January 13, 1995; 60 *FR* 3092]

The pressure filtration technique specified in Method 1311 defines a "liquid" as the material that passes through a 0.6- to 0.8-micron filter when a representative sample of the waste is subjected to a 50-psi differential pressure. Clearly, this is a more stringent procedure than the gravity-based paint filter test, but EPA has never codified this requirement in the regulations or in SW–846—it is just guidance. In fact, two weeks *after* the agency issued the above-noted preamble language preferring the pressure filtration technique over the paint filter test, it released guidance noting that, until it finalizes the proposed change in SW–846, "the paint filter test is the method to use to determine if a free liquid is present for ignitability determination." [RO 11935] As such, it is still acceptable to use the paint filter test to determine if a material is a liquid, and most generators are employing this approach. Table 2-3 summarizes the "liquid" definition for the ignitability characteristic.

2.2.1.2 Flash point testing

Whether a generator uses knowledge, the paint filter test, or the pressure filtration technique, the ignitability of liquids must be evaluated using knowledge or flash point testing. If testing is used, flash point must be determined using a Pensky-Martens closed-cup tester (SW–846 Method 1010A) or a Setaflash closed-cup tester (SW–846 Method 1020B). Only one test method needs to be used. The choice of the method should be based on the applicability of the

Table 2-3: RCRA Definitions for "Aqueous," "Liquid," and "Wastewater"

Term	Application	RCRA definition
Aqueous	Alcohol-content exclusion to the ignitability characteristic [§261.21(a)(1)]	Aqueous solution means a solution containing at least 50% water by weight [RO 11060, 13548]
	Corrosivity characteristic definition [§261.22(a)(1)]	Aqueous phase constitutes at least 20% of the total volume of the waste [SW–846, Method 9040C]; at least 20% free water by volume [RO 11738]
Liquid	Ignitability characteristic definition [§261.21(a)(1)]	Free liquid phase obtained using paint filter test (SW–846, Method 9095B) [April 30, 1985; 50 *FR* 18372, Chapter 7 of SW–846]; or liquid obtained using the pressure filtration technique in the TCLP (SW–846, Method 1311) [January 13, 1995; 60 *FR* 3092]
	Corrosivity characteristic definition [§261.22(a)(2)]	Free liquid phase obtained using paint filter test (SW–846, Method 9095B) [April 30, 1985; 50 *FR* 18372, Chapter 7 of SW–846]; or liquid obtained using the pressure filtration technique in the TCLP (SW–846, Method 1311) [January 13, 1995; 60 *FR* 3092]
	Free liquid in containerized or bulk waste [§§264/265.314, 264.552(a)(3)]	Free liquid phase obtained using paint filter test (SW–846, Method 9095B) [§§264.314(b), 265.314(c)]
Wastewater	For purposes of the wastewater treatment unit exemption [§§264.1(g)(6), 265.1(c)(10), 270.1(c)(2)(v)]	Wastes that are substantially water with contaminants amounting to a few percent at most [RO 11020, 14472]
	For land disposal restrictions treatment standards [§268.40]	Wastes that contain less than 1% by weight total organic carbon *and* less than 1% by weight total suspended solids [§268.2(f)]

Source: McCoy and Associates, Inc.

method to the waste being tested. [RO 11594] For example, the Setaflash method is not applicable to liquids with nonfilterable, suspended solids and viscosities above 150 stokes at 25°C. The Pensky-Martens test can be used for liquids that contain nonfilterable, suspended solids.

Both of the closed-cup flash point testers specified in the regulations were originally developed by ASTM and work about the same way. A sample of liquid is put into a metal cup, which is then placed in the test apparatus. The lab technician toggles a switch on the apparatus, which directs a test flame into the headspace of the cup, and he/she watches for a flash due to ignition of the fumes or vapors coming off the liquid. If the sample flashes while at room temperature (e.g., 70°F), then the flash point of the liquid is 70°F or less. If no flash is detected, the technician increases the sample temperature and toggles the switch again. If the sample still has not flashed when its temperature is raised all the way to 140°F, then the generator can conclude that the waste is not an ignitable liquid. [RO 12909]

When testing is used for determining compliance with the <140°F criterion, closed-cup flash point testing must be employed. Note that flash point can also be obtained through open-cup testing, but those results are valid only if the measured flash point is <140°F. "Ordinarily, open-cup tests…will produce higher flash points than the closed-cup tests required by EPA." [RO 12296]

Q *A flash was observed when testing a solvent mixture that should not (based on composition) be ignitable. For example, a flash was observed* *at 68°F when evaluating a spent solvent containing 99% Freon 113 and trace levels of cyclohexane. Should this spent solvent be considered ignitable under RCRA?*

 Yes. [RO 12909]

Some confusion has arisen over the use of flash point testing for wastes that are essentially all solids (e.g., solvent-contaminated rags or gasoline-contaminated soil) or for semisolid wastes containing both a liquid and solid phase. According to EPA, "[n]either test, however, is approved by ASTM for use in evaluating the flash point of solids or sludges." [OSWER Directive 9443.00-1A] "If your samples contain filterable solids, they are not amenable to the Pensky-Martens flash point test. Flash point testing is only appropriate for liquid samples. It should not be applied to solids." [RO 13759] See also RO 12909, 13550, and 14669.

Based on this guidance, wastes that are essentially all solids or are semisolids should be evaluated using the paint filter test (or pressure filtration technique if you want to be very conservative). If no free liquid comes through the filter, the waste is not a liquid and will not be D001 via the §261.21(a)(1) criteria. [RO 11619, 11787, 13328] However, the waste may still be an ignitable solid or a DOT oxidizer (which are D001 wastes, as discussed in Sections 2.2.2 and 2.2.4, respectively). We've heard of people running flash point tests on solvent-contaminated rags or gasoline-contaminated soil that would pass the paint filter test (i.e., no free liquid would pass through the filter); if a flash from one of these materials is recorded by the lab at a sample temperature <140°F, the generator may think he/she has to apply the D001 code to the waste when they really don't have to.

"If a solid flashes using some modification of the flash point test, this may indicate there is a potential problem with the sample, such as contamination with ignitable volatiles, and further investigation may be in order. The flash point test alone is not definitive for determining the ignitability of solids, but may be used with other evidence [e.g., the criteria in §261.21(a)(2)] to build a case for a waste being classified as an ignitable hazard." [RO 13761]

It is also our understanding, however, that some waste haulers and TSD facilities may require flash point data on certain waste solids that do not contain free liquids as a condition of transporting or accepting the waste. If running the flash point test is a condition of doing business with such an entity, then the generator will either have to comply or find another supplier. But, even if a flash is detected from such solid materials at <140°F, the D001 code still will not have to be applied to such materials unless they meet the §261.21(a)(2) or (a)(4) criteria discussed in Section 2.2.2 or 2.2.4. "You are correct in asserting that the absence of free liquids precludes the application of the ignitability characteristic as defined in §261.21(a)(1)." [RO 11787]

If the paint filter test produces a free-liquid phase from wastes that are solid or semisolid, EPA says you "should separate the solid/liquid phases of your samples and test each phase separately; liquids by flash point and solids by the DOT procedure [Method 1030 in SW–846 (discussed in Section 2.2.2)]." [RO 13759; see also 50 *FR* 18372] Then what? "A waste liquid or mixture containing a free-liquid phase (as defined by our paint filter liquids test) is ignitable under RCRA if the waste (*or liquid phase*) has a flash point <140°F." [Emphasis added.] [RO 11619] "All such wastes which contain or consist of liquids which have a flash point below [140°F] are to be considered as ignitable wastes." [RO 12034]

Our discussions with EPA's Methods Information Communications Exchange (MICE—see Section 2.6.2.3 for contact info) corroborate the agency's previous guidance. MICE indicated that, if either of the phases fails their respective ignitability test, the entire mixture should be considered D001 hazardous since it will be disposed as a mixture of solids and liquids.

That's the theory. The reality is a lot fuzzier. Most people we talked to about this matter are using the paint filter test to determine if a waste is a liquid. But there is no apparent consensus as to what is subsequently tested for flash point: just the free-liquid phase generated from the filter test or the entire waste matrix (liquid plus solids). Testing just the free-liquid phase is more conservative, because it is almost certainly going to have a lower flash point than the entire waste matrix. If the free-liquid phase is what is tested and has a flash point <140°F, these facilities would designate the entire waste as D001. Based on EPA's guidance regarding the applicability of the flash point tests, this appears to be the more defensible approach. Additionally, due to the physical limitations associated with the two flash point testers, it is often difficult to introduce much of the solids into the test cup anyway. (Of course, facilities that have a waste analysis plan must conduct ignitability testing as prescribed in their plan.)

2.2.1.3 The alcohol-content exclusion

The characteristic of ignitability does not apply to an aqueous solution that contains <24% alcohol by volume, even if it has a flash point <140°F. This exclusion, set forth in §261.21(a)(1), was originally intended to exclude waste alcoholic beverages and some types of latex paints from hazardous waste management. These materials exhibit low flash points due to their alcohol content but do not sustain combustion because of their high water content. [45 *FR* 33108] The waste must have some alcohol in it (but <24%) to qualify for the exclusion. [RO 12274]

The alcohol exclusion is a carryover from DOT's old regulatory definition of combustible liquids. In those DOT regulations, the term "aqueous" was defined as no less than 50% water. To remain consistent with DOT's approach, EPA has interpreted "aqueous solution" for the purpose of the alcohol exclusion to mean a solution containing at least 50% water by weight. [RO 11060, 13548] Note that this definition of "aqueous" associated with the ignitability characteristic is considerably different from the one used for determining corrosivity. (See Table 2-3.)

EPA has also clarified that the exclusion applies to wastes that contain a nonalcoholic component (e.g., a mixture of 77% water, 13% alcohol, and 10% nonalcoholic liquid). In other words, the presence of a nonalcoholic constituent will not require the waste to be regulated as D001 if the mixture has a flash point <140°F. [RO 13548]

EPA originally intended that the alcohol exclusion apply only for solutions containing <24% ethanol. When they codified the regulatory language, however, the agency said "alcohol." EPA clarified on June 1, 1990 that the term "alcohol" in the §261.21(a)(1) exclusion refers to any alcohol or combination of alcohols. [55 *FR* 22543] Additionally, "all aqueous wastes which are ignitable only because they contain alcohols (here using the term alcohol to mean any chemical containing the hydroxyl [–OH] functional group) are excluded from regulation." [RO 11060] However, "if the alcohol is one of those alcohols specified in EPA hazardous waste codes F001–F005 and has been used for its solvent properties, the waste must be evaluated to determine if it should be classified as an F-listed spent solvent waste." [RO 13548]

2.2.2 Ignitable solids

The second category of wastes exhibiting the ignitability characteristic is nonliquids that ignite due to friction, moisture absorption, or spontaneous chemical change and, once ignited, burn vigorously and persistently so as to create a hazard. [§261.21(a)(2)] The regulatory definition for ignitable solids sometimes causes confusion because the language is unclear. (We think it's easier to call this group "solids" instead of "nonliquids.") EPA has tried to clarify the definition by noting that two distinct criteria must *both* be met before a solid will be considered an ignitable waste: 1) it must be capable of ignition through friction, moisture absorption, or spontaneous ignition (e.g., a pyrophoric material); *and* 2) once ignited, it must burn vigorously and persistently so as to present a hazard. For example, titanium swarf may burn vigorously after ignition such that it poses a hazard, but since it is difficult to ignite, the material is *not* a D001 ignitable waste. Both criteria must be met. [RO 12089]

In the preamble to the May 19, 1980 rule establishing the ignitability characteristic, the agency tried to clarify the type of materials it wants to regulate under this category: "[EPA] has no intention of designating such things as wastepaper and sawdust to be hazardous and is only interested in capturing the small class of thermally unstable solids which are liable to cause fires through friction, absorption of moisture, or spontaneous chemical changes." [45 *FR* 33108]

In a June 1, 1990 preamble [55 *FR* 22545], EPA provided additional guidance on this category of ignitable wastes:

"These wastes are typically generated on a sporadic basis in low volumes and are characterized as primarily inorganic solids or wastes containing reactive materials. Ignitable reactive materials include reactive alkali metals or metalloids (such as sodium and potassium) and calcium carbide slags. Most of these are very reactive with water and will generate gases that can ignite as the result of heat generated from the reaction with water. Other reactive ignitable solids in this subcategory include metals such as magnesium and aluminum that, when finely divided, can vigorously react with the oxygen in the air when ignited.

"There appears to be an overlap between wastes in this D001 subcategory and certain D003 (characteristic of reactivity) wastes. A close examination of the definitions in §261.21(a)(2) for ignitable wastes and §261.23(a)(2), (3), and [(4)] for reactive wastes reveals the distinction between these two groups. The key difference is in the definition of ignitable wastes, which states: 'when ignited, burns vigorously and persistently.' This phrase implies that the hazard is due primarily to the ignition potential rather than to the extreme reactivity."

The regulations do not specify any test methods for determining the first criterion: if a waste is capable of ignition through friction, moisture absorption, or spontaneous ignition. Instead, the generator must evaluate this criterion based on "best engineering judgment" [RO 14285] and on "operational experience." [RO 14176] The agency delineated the following issues that a generator can evaluate when applying best engineering judgment and operational experience as to the ignitability of solids; however, only the first one given below would be sufficient by itself for definitive classification of a solid as a D001 waste under §261.21(a)(2) [October 20, 2003; 68 *FR* 59940]:

- Have there been landfill or other fires attributable to disposal of the solids?
- Have the solids been observed emitting smoke during any phase of waste management?
- Have the solids been packaged or transported with a DOT designation of pyrophoric or self-heating material?
- Have the solids given a positive result in the DOT test for self-heating materials, as discussed in 49 *CFR* 173.124(b)(2) and 173.125(c)?
- Is there any information on a safety data sheet (SDS) indicating the possibility of ignition due to friction, moisture absorption, or spontaneous ignition?
- Have the solids ever been stored in special containers or under inert gas such as nitrogen?
- Have the solids ever been stored in any other way so as to limit their exposure to the air, such as coating with oil or wetting with water?

The DOT regulations discuss spontaneously combustible and self-heating materials in 49 *CFR* 173.124(b) and reference test methods for materials liable to spontaneous combustion or self-heating in the UN Manual of Tests and Criteria. [RO 11935]

On January 3, 2008 [73 *FR* 486], EPA added Method 1050—Substances Likely to Spontaneously Combust to SW–846. The test procedures in Method 1050 are intended to identify two types of wastes with spontaneous combustion properties:

> "Wastes (including mixtures and solutions, liquid or solid) which, even in small quantities, ignite within five minutes of coming in contact with air. These wastes are the most likely to spontaneously combust and are considered to have pyrophoric properties.
>
> "Other solid wastes which, in contact with air and without an energy supply, are susceptible to self-heating. These wastes will ignite only when in large amounts (kilograms) and after long periods of time (hours or days) and are considered to have self-heating properties."

This test method may be helpful to generators in determining whether a solid waste is capable of ignition through spontaneous chemical change.

EPA has approved Method 1030 in SW–846 for determining the second criterion: whether a material "burns so vigorously and persistently that it creates a hazard." [June 13, 1997; 62 *FR* 32452] This method, which is appropriate for pastes, granular materials, solids that can be cut into strips, and powdery substances, is based on DOT's burn-rate test. A 1,000°C flame from a Bunsen burner is used to ignite a 100-mm test strip or powder train of the material; if the material burns at a rate faster than 2.2 mm/sec (0.17 mm/sec for metals), the material is considered to have a positive result for ignitability *according to the DOT regulations*. However, the method is guidance only; its use is not required under §261.21(a)(2). As an alternative to using Method 1030, any valid, documented knowledge may be used to make this determination. [RO 14259]

It is important to note that, even if a solid tests positive using Method 1030, it is not automatically an ignitable hazardous waste under RCRA. Method 1030 only evaluates the second criterion of whether a material will burn vigorously and persistently; the first criterion, its capability of ignition through friction, moisture absorption, or spontaneous chemical change, must also be satisfied in order for the waste to be a D001 ignitable waste. However, if a generator's waste does not test positive when using Method 1030 (i.e., an ignited test strip or powder train burns slower than the thresholds given in the test method), the generator should be able to use this knowledge to conclude that the waste solid is not ignitable per §261.21(a)(2).

EPA has noted that ASTM D4982-89 (Standard Test Method for Flammability Potential Screening Analysis of Waste) is not an appropriate method to determine if a solid is ignitable. [RO 14405]

Q *Iron "sponge" is used to remove hydrogen sulfide from natural gas. It sometimes consists of redwood chips coated with hydrated ferric oxide. When spent, the sponge is removed from the absorption towers and placed on the ground. Due to contact with oxygen in the air, it begins to smolder; the smoldering continues until the sponge is reduced to ashes. Is the waste iron sponge ignitable at its point of generation (removal from the absorption towers)?*

A Possibly. The spent iron sponge is clearly not a liquid, so the §261.21(a)(2) criteria apply. Since the material will ignite due to moisture absorption and/or spontaneous ignition, the first criterion is satisfied. The generator must then use knowledge (e.g., the use of special fire-fighting techniques to extinguish a fire) or Method 1030 to determine if the smoldering material will burn vigorously and persistently so as to create a hazard. [RO 12115] Note, however, that even if ignitable, this waste may be excluded from hazardous waste regulation under the oil and gas exploration and production exemption in §261.4(b)(5). [July 6, 1988; 53 *FR* 25454]

Q *How should waste nitrocellulose filter fabric be prepared prior to testing using Method 1030? Should it be chopped into small pieces?*

A EPA recommends that materials such as nitrocellulose filter fabric be simply cut into long strips prior to Method 1030 testing. The agency does not recommend grinding, shredding, or chopping the sample into small pieces, because that would alter the form of the waste and greatly increase its surface area. [RO 14259]

2.2.2.1 Ignitability of solvent-contaminated rags/wipes

EPA's long-standing policy (since 1991) was that the regulatory status of solvent-contaminated rags and wipes should be determined by authorized states or EPA regions on a case-by-case basis. [RO 11576, 11813, 11935, 14405] However, the agency finalized a rule on July 31, 2013 [78 *FR* 46448] that established national consistency for the management of ignitable solvent-contaminated rags/wipes. This rule is discussed in Section 3.3.4.

2.2.3 Ignitable compressed gases

To define this category of D001 wastes, EPA simply adopted DOT's definition of "flammable" compressed gas. Until July 2006, the RCRA definition of an ignitable compressed gas in §261.21(a)(3) cited DOT's regulations at 49 *CFR* 173.300, which has been an obsolete DOT citation since 1990. To fix that problem, EPA issued a technical correction on July 14, 2006. [71 *FR* 40254] However, instead of replacing the obsolete reference in §261.21(a)(3) with correct references to DOT's current regs, the agency replaced the out-of-date reference with the complete text of the definition from the 1980 version of DOT's regulations (i.e., the DOT definition that was in effect at the time the RCRA regs were first issued).

So today, the RCRA regs define an ignitable compressed gas as a compressed gas that 1) is flammable when in a mixture of 13% or less with air (i.e., its lower flammability limit is 13% or less), or 2) has a flammable range with air of more than 12% (i.e., the difference between its upper and lower flammability limit is greater than 12 percentage points) regardless of the lower flammability limit. All percentages are based on volume at 68°F and 1 atm absolute pressure.

That 1980 definition of a flammable compressed gas is basically the same as DOT's current one in 49 *CFR* 173.115. However, the old DOT regulations that are now in the RCRA regs include Bureau of Explosives test methods to determine flammability, while DOT's new regulations specify ASTM E681-85 (Standard Test Method for Concentration Limits of Flammability of Chemicals).

Examples of D001 ignitable compressed gases are cylinders of waste propane or acetylene.

2.2.4 Oxidizers

As with ignitable compressed gases, EPA based its regulatory definition of a D001 oxidizer on DOT's definition. However, the RCRA definition of an oxidizer in §261.21(a)(4) referenced DOT's regulations at 49 *CFR* 173.151, which has been an obsolete DOT citation since 1990. When EPA corrected this problem on July 14, 2006 [71 *FR* 40254], the agency replaced the out-of-date reference with the complete text of the oxidizer definition from the 1980 version of DOT's regulations (i.e., the DOT definition that was in effect at the time the RCRA regs were first issued).

That 1980 definition includes a material "that yields oxygen readily to stimulate the combustion of organic matter." The current DOT definition in 49 *CFR* 173.127 is similar: a Hazard Class 5.1 oxidizer is "a material that may, generally by yielding oxygen, cause or enhance the combustion of other materials." However, DOT's current regulatory language includes test methods for quantifying such a designation—the old DOT regulations that are now in the RCRA regs give none.

Chlorates, permanganates, inorganic peroxides, and nitrate compounds are all examples of oxidizers that are hazardous for ignitability when they become wastes, because they meet DOT's definition of an oxidizer.

On January 3, 2008 [73 *FR* 486], EPA added Method 1040—Test Method for Oxidizing Solids to SW–846. The test procedures in Method 1040 are intended to evaluate the relative oxidizing hazard posed by wastes that are physically solid, including granular and other materials that can be formed into a conical pile. The method is based on a conical pile-type burning test method adapted from the UN regulations and classification procedures, and it provides a qualitative means to measure the potential of a physically solid waste to increase the burning rate or burning intensity of a combustible substance. This test method was specifically developed to illustrate the oxidizer properties of materials and thus may be helpful to generators in determining whether a solid waste is capable of stimulating the combustion of organic matter per §261.21(a)(4).

The July 14, 2006 rule also clarified in a Note 4 to §261.21 that organic peroxides (which are currently defined by DOT as Hazard Class 5.2 in 49 *CFR* 173.128) are a type of oxidizer. Therefore, these materials will also be identified as D001 when they become wastes. The organic peroxide table [found in 49 *CFR* 173.225(c)] may be used for the classification and packaging of organic peroxides. This table, although not exhaustive, identifies the common organic peroxides found in industry. By referencing the organic peroxide table, one can determine if a waste is an organic peroxide, thus warranting the D001 waste code. For example, in the organic peroxide table, methyl ethyl ketone peroxide is listed three times. The listings are concentration- and diluent-dependent. All three listings identify the material as an organic peroxide, so no matter which description is selected by the generator, it will be classified as 5.2, which warrants the D001 waste code when this material is intended for disposal.

There is a possible overlap between ignitable oxidizers (which would be D001) and reactive oxidizers (which would be D003). EPA noted in RO 51434 that an oxidizer that reacts violently should be considered a D003 reactive waste; on the other hand, an oxidizer which reacts in a milder manner should be considered a D001 ignitable oxidizer.

Probably the easiest way to determine if a waste is a D001 oxidizer is to look up the material on DOT's hazardous materials table at 49 *CFR* 172.101. Column (3) on this table identifies the Hazard Class of each material; a Hazard Class of 5.1 or 5.2 indicates that the material is a DOT oxidizer. We don't profess to be knowledgeable about all DOT requirements; therefore, our bias is to consider all 5.1 and 5.2 materials (when they become wastes) to be D001 wastes, unless advised otherwise by an authorized regulatory agency or a DOT expert. However, if a generator stops here, the classification is not complete, for Column (3) in the hazardous materials table identifies only the primary hazard associated with a given hazardous material. For a complete classification, the generator must also reference Column (6), which identifies the primary and subsidiary hazards associated with a given hazardous material.

For example, should spent nitric acid be coded D001 due to its oxidizer potential? EPA has given peripheral guidance noting that the D001 code may apply for nitric acid in RO 11480. The answer depends on the concentration of the nitric acid being disposed. By referencing the DOT hazardous materials table, if the nitric acid concentration is 65–70 percent, the subsidiary hazard found in Column (6) is 5.1, thus warranting the D001 waste code when disposed. If the concentration of the nitric acid is less than or equal to 65 percent, the oxidizer hazard is not present and the D001 waste code would not be justified.

Q *A bath containing inorganic peroxides is used to clean brass and other copper alloys prior to plating, lacquering, or other finishing activities. Could the spent baths be D001 ignitable wastes?*

A Yes. Because the baths contain inorganic peroxides, they could meet the definition of a DOT oxidizer. However, once the hydrogen peroxide has been chemically decomposed, the wastes would no longer exhibit the ignitability characteristic due to the presence of an oxidizer. [RO 11854]

Q *Potassium permanganate is sometimes used as a bleach in the garment industry. When spent, can the material be disposed in a nonhazardous waste landfill?*

A Yes. "Since spent permanganate materials are already reduced, they are not likely to be powerful oxidizers and, therefore, are not likely to exhibit a hazardous waste characteristic." [RO 11628]

2.3 Corrosivity

Per §261.22, a solid waste exhibits the corrosivity characteristic (and is a D002 hazardous waste) if a representative sample of the waste meets either of the following criteria:

- It is aqueous with a pH ≤2.0 or ≥12.5 as measured by a pH meter using Method 9040C in SW–846. [§261.22(a)(1)]
- It is a liquid that corrodes SAE 1020 carbon steel at a rate >0.25 inch per year as measured by Method 1110A in SW–846. [§261.22(a)(2)]

Concerning this characteristic, the agency wanted to require Subtitle C management of wastes that would 1) mobilize toxic metals if discharged into a landfill; 2) corrode waste handling, storage, transportation, and management equipment; or 3) damage human or animal tissue in the event of inadvertent contact. Acidic and basic wastewaters are the primary D002 hazardous wastes. To evaluate potential corrosivity in solid wastes, EPA:

"chose pH as one barometer of corrosivity because wastes exhibiting low or high pH can cause harm to human tissue, promote the migration of toxic contaminants from other wastes, react dangerously with other wastes, and harm aquatic life. EPA chose metal corrosion rate as its other barometer of corrosivity because wastes capable of corroding metal can escape from the containers in which they are segregated and liberate other wastes." [May 19, 1980; 45 *FR* 33109]

Both of these corrosive waste criteria are examined below after a general discussion of the difference between "aqueous" and "liquid."

2.3.1 Aqueous vs. liquid wastes

As noted above, one corrosivity criterion applies to "aqueous" wastes and the other to "liquid" wastes. What's the difference? The RCRA regulations do not define "aqueous." However, EPA has concluded that an aqueous waste for purposes of the corrosivity characteristic is a waste for which pH is measurable (i.e., measurable dissociated hydrogen ions are present) via Method 9040C. The scope and application section of Method 9040C specifies that it should only be used when "the aqueous phase constitutes at least 20% of the total volume of the 'waste.' Therefore, any waste for which this method is applicable must contain at least 20% free water by volume." [RO 11738]

According to EPA's Methods Information Communication Exchange (MICE—see Section 2.6.2.3 for contact info), the 20% provision was intended to represent the amount of free water present in a multiphase waste. This could be estimated after allowing the sample container contents to settle and ensuring that the aqueous phase represents at least 1/5 of the total sample volume. It is recognized that water-miscible solvents would not liberate a discernable separate phase, and a provision should be available to determine the water content. In these situations, the MICE suggests that the water content be measured using either a Karl Fisher titration or calcium hydride reaction according to SW–846 Methods 9000 or 9001, respectively. The latter technique can be completed either in the field or in the lab using commercially available test kits.

Note that this definition of "aqueous" associated with corrosivity is considerably different from the one used for determining the applicability of the alcohol-content exclusion in the ignitability characteristic. Table 2-3 summarizes the differences in this term, depending on the characteristic being evaluated.

Determining whether a waste is a liquid for purposes of the corrosivity characteristic is identical to that used for ignitable wastes (see Section 2.2.1.1). If the paint filter test is used to determine whether free liquids are present, it "may also be used to obtain the liquid portion of the waste for subsequent flash point evaluation (in the case of an ignitable waste) or for corrosivity evaluation (in the case of a corrosive waste)." [April 30, 1985; 50 *FR* 18372] See also RO 12095 and 13328.

2.3.2 Aqueous wastes with low or high pH

Aqueous materials with a pH ≤2.0 or ≥12.5 are considered to be D002 wastes when discarded. [§261.22(a)(1)] Most corrosive wastes fit into this category. As specified in the regulations, corrosivity determinations for these types of wastes should be made with a pH meter using Method 9040C in SW–846. If the pH of the waste is above 12.0, pH measurement for the corrosivity characteristic must be taken when the sample is at 25±1°C. This requirement was added to Method 9040C because sample temperature affects pH results for highly alkaline wastes. Measurements for wastes with pH levels less than 12 may be made at other sample temperatures. [April 4, 1995; 60 FR 17003]

Although the regulations at §261.22(a)(1) specifically give the low-end pH as "less than or equal to 2," guidance from EPA notes that it is actually "less than or equal to 2.0." [RO 51432] In this same guidance, the agency notes that pH values should be reported to the nearest 0.1 pH unit.

Sometimes generators make pH determinations using other analytical methods. For example, pH paper can be used to make rough acid-base measurements; however, such a corrosivity determination would fall under the category of generator knowledge because the prescribed method (Method 9040C) was not used.

Q *Is sludge from the bottom of a steel pickling tank D002 if its pH is ≤2.0?*

A Yes, but only if the sludge meets the definition of aqueous as noted above. Or, if the sludge meets the definition of a liquid, it could also be hazardous if it corrodes carbon steel at a rate >0.25 inch per year. [RO 11346]

Q *A food processing plant generates an aqueous waste stream from a caustic peeling process. The waste sometimes has a pH of ≥12.5, but it is always neutralized to well below that level before it is discharged from the plant. Is the waste D002 at its point of generation?*

A Yes. Even though the waste is neutralized, it meets the definition of a corrosive hazardous waste at its point of generation. Depending on how the waste is managed before and during neutralization, it may not have to be counted against the generator's monthly hazardous waste quantity determination (see Section 6.1). However, the land disposal restrictions may apply since it is hazardous at its point of generation. [RO 12005]

2.3.2.1 State definition of corrosivity may be more stringent

Even though some wastes would intuitively have a pH ≤2.0 or ≥12.5, they may not actually be D002 hazardous wastes. For instance, solid sodium hydroxide pellets and concentrated sulfuric acid are not corrosive wastes under the federal pH criteria because they are not "aqueous" (i.e., they contain less than 20% free water and their pH is not measurable). Again referring to Method 9040C, "[t]he corrosivity of concentrated acids and bases, or of concentrated acids and bases mixed with inert substances, cannot be measured. The pH measurement requires some water content." (However, in the case of concentrated acids or bases, they still may be D002 if they corrode carbon steel at a rate >0.25 inch per year.)

The above discussion reflects current federal guidance. Some states, however, have adopted a more stringent approach for determining corrosivity. In those states, a waste that might have a pH ≤2.0 or ≥12.5 if in the presence of water must be evaluated for corrosivity as follows: a sample of the waste is mixed with a specified amount of water (usually 1:1 on a mass basis), and the pH of the resulting solution is measured using Method 9040C. If the pH of the *mixture* is ≤2.0 or ≥12.5, the original waste is considered corrosive at the state level. Make sure you know how the states you do business in define corrosivity for these types of wastes.

2.3.2.2 Derivation of pH levels

How did EPA come up with pH levels for the corrosivity characteristic of ≤2.0 or ≥12.5? Rather than being scientifically derived, these pH levels are based primarily on the types of wastes EPA wanted to capture under the D002 characteristic. To keep lime-stabilized wastes and sludges out of hazardous waste management, EPA set the high-end pH level at 12.5. [May 19, 1980; 45 *FR* 33109] (The typical pH of lime slurries is 11.5–12.2.)

The agency set the low-end pH level so as not to include "a number of substances generally thought to be innocuous and many industrial wastewaters prior to neutralization." [45 *FR* 33109] Although EPA didn't specify what "substances" it was trying to keep out of RCRA regulation, we understand anecdotally that it was waste soft drinks and fruit juices. The pH of many of these beverages ranges from 2.5–3.0.

Finally, the pH ranges selected for defining this aspect of corrosivity address "the problem of tissue damage more realistically." [45 *FR* 33109] While some entities argued for skin corrosion tests to be part of the corrosivity definition, EPA decided that there was sufficient correlation between pH and tissue damage to use pH alone.

2.3.3 Liquid wastes that corrode carbon steel

If a waste is not aqueous with a low or high pH, it can still be a corrosive hazardous waste if it is a liquid that corrodes carbon steel at a rate >0.25 inch per year. [§261.22(a)(2)] A slightly revised version of the National Association of Corrosion Engineers Standard TM–01–69 is to be used to measure such corrosion rates. This method has been standardized in SW–846 as Method 1110A, and the regs require generators to utilize this standardized version when conducting this test. [June 14, 2005; 70 *FR* 34549] RO 13389 provides some guidance on running the test.

We're not aware of very many liquids that carry the D002 code because of this second corrosivity criterion. One such material is ferric chloride, which is sometimes used to condition wastewater in wastewater treatment systems. Ferric chloride will not fail the pH thresholds in the first criterion, but it can corrode a carbon steel tank or drum at a rate >0.25 inch per year. Such a material, when it becomes a waste, would be D002.

2.3.4 Do both D002 criteria have to be evaluated?

In many cases, high- or low-pH wastes will also corrode carbon steel. For such wastes, do both the pH and steel corrosion evaluations have to be conducted to determine if they are hazardous? EPA addressed this issue in September 1980:

"Strictly speaking, in order to be absolutely certain that a waste is not [D002] hazardous, both corrosivity criteria should be evaluated. However, from a more pragmatic standpoint, both tests may not always be necessary. For instance, if a given waste has a pH of 5 and is known not to contain any oxidizing materials, then the corrosivity toward steel test may not be warranted.

"The generator of the waste should decide which tests need to be performed on a case-by-case basis. Available engineering data are usually quite useful in determining which tests are most appropriate for a particular waste.

"I might add that even though certain tests may not be warranted, this does not relieve the generator of liability in the event that the waste is subsequently determined to be hazardous." [OSWER Directive 9443.01(80)]

So, aqueous liquids must be evaluated for both pH and steel corrosivity. Even if an aqueous liquid has a pH >2.0 but <12.5, it will exhibit the corrosivity characteristic if it corrodes carbon steel at a rate >0.25 inch per year. Aqueous liquids with a pH between 2.0 and 4.0, for example, are not corrosive by the pH criteria but still may significantly attack carbon steel surfaces. [RO 13561] While pH is usually measured using Method 9040C, knowledge is often used to determine that the waste will or will not corrode steel (as implied in the above statements from EPA).

Under EPA's interpretation for aqueous corrosives, they can be in nonliquid form (e.g., suspensions, colloidal dispersions, and gels that would pass the paint filter test). Such aqueous nonliquids would only be subject to pH testing; the steel corrosion test would not be necessary since it applies only to liquids. Conversely, nonaqueous liquids (those that contain <20% aqueous phase) need only be tested using the steel corrosion test. [RO 11719, 13561]

2.3.5 What about corrosive solids?

For a waste to be D002, it must be either aqueous or liquid—there are no corrosive solids at the federal level. [RO 11278] When EPA was developing this characteristic, it received a number of comments suggesting the inclusion of solids in the corrosivity definition, but no one "described situations where the improper disposal of such wastes would be likely to cause damage. EPA has concluded that, inasmuch as the great majority of [D002] wastes are presumed to be in liquid or semiliquid form, there is no demonstrated need to address corrosive solids at this time." [45 FR 33109]

The agency has added a test method to SW–846 to determine the pH of solids and soil; it is Method 9045. The test is applicable to solids, sludges, or nonaqueous liquids that contain <20% water by volume. Basically, water is added to the solids or soil, and the pH of the resulting solution is measured. Although this test is available to generators, EPA has clarified that "Method 9045 is not [to be] used for corrosivity characteristic determinations." [April 4, 1995; 60 FR 17003] That is EPA's position at the federal level; thus, lye, solid acids, and some baghouse dusts are not D002 when they become wastes. [RO 13533] As noted in Section 2.3.2.1, some states have chosen to have a more stringent corrosivity definition and require use of this test for potentially corrosive solids.

Q *Are granules or pellets of sodium or potassium hydroxide characteristically corrosive wastes if disposed?*

A Not at the federal level. [RO 12337] However, they may be in the states that have expanded their definition of corrosivity to include solids.

2.4 Reactivity

According to §261.23(a), a solid waste exhibits the characteristic of reactivity (and is a D003 hazardous waste) if a representative sample of the waste:

1. Is normally unstable and readily undergoes violent change without detonating;
2. Reacts violently with water;
3. Forms potentially explosive mixtures with water;
4. When mixed with water, generates toxic gases, vapors, or fumes in a quantity sufficient to present a danger to human health or the environment;
5. Is a cyanide- or sulfide-bearing waste which, when exposed to pH conditions between 2 and 12.5, generates toxic gases, vapors, or fumes in a quantity sufficient to present a danger to human health or the environment;
6. Is capable of detonation or explosive reaction if subjected to a strong initiating source or if heated under confinement;
7. Is readily capable of detonation or explosive decomposition or reaction at standard temperature and pressure; or
8. Is a forbidden explosive as defined in 49 CFR 173.54, or is a Division 1.1, 1.2, or 1.3 explosive as defined in 49 CFR 173.50 and 173.53.

EPA intended that this characteristic include wastes that are unstable, tend to react violently or explode, and/or give off toxic gases. The narrative standards in §261.23(a) are, to a large extent, a paraphrase of the reactivity definition employed by the National Fire Protection Association. The following materials are typically D003 reactives when they become wastes: aluminum alkyls, cyanides, lithium- and sodium-containing materials, and sulfides. Examples of other specific materials that exhibit the characteristic of

Table 2-4: Examples of Materials That EPA Has Determined Are Reactive

Material	Reason for exhibiting reactivity	RO reference
Aluminum chaff roving bundles	Release hydrogen gas when exposed to moisture	11810
Charged lithium-sulfur dioxide batteries that are spent and/or discarded[1]	Release hydrogen gas when exposed to moisture; release sulfur dioxide and hydrogen cyanide under certain conditions; batteries are capable of violent rupture if subjected to a strong initiating source or if heated under confinement	11033, 11229, 14756
Out-dated blasting caps	Capable of detonation or explosive reaction if subjected to a strong initiating source or if heated under confinement	12308
Unused flameless ration heaters for the U.S. Army's Meals Ready-to-Eat[2]	React violently with water; form potentially explosive mixtures with water, including production of hydrogen gas	14567, 14774

[1]Lithium-sulfur dioxide batteries that have been fully discharged no longer exhibit the characteristic of reactivity. "Fully discharged" means that each cell within each battery has been discharged to a voltage of one volt or less.
[2]EPA generally considers multiple unused flameless ration heaters that are not packaged with Meals Ready-to-Eat to be D003 when disposed.

Source: McCoy and Associates, Inc.

reactivity are provided in Table 2-4. Aerosol cans are sometimes considered reactive hazardous waste; a separate discussion addressing aerosol cans is presented in Section 2.5.

2.4.1 No test methods available for determining reactivity

Probably the most difficult aspect in identifying D003 wastes is that the regulatory language in §261.23(a), as paraphrased above, does not specify any test methods for determining if a waste meets the reactivity characteristic criteria. The agency has found no appropriate test methods for this purpose, noting that existing methods: 1) are not general enough to measure the various reactivity criteria, 2) do not take enough factors (e.g., mass, surface area) into account to reflect the reactivity of the whole waste (as opposed to the reactivity of the sample itself), and 3) require subjective interpretation of the results instead of providing pass/fail results. [May 19, 1980; 45 *FR* 33110]

The lack of test methods to determine reactivity isn't an obstacle in EPA's mind: "The unavailability of suitable test methods for measuring reactivity should not cause problems. Most generators of reactive wastes are aware that their wastes possess this property and require special handling. This is because such wastes are dangerous to the generators' own operations and are rarely generated from unreactive feedstocks. Consequently, the prose definition [found in §261.23(a)] should provide generators with sufficient guidance to enable them to determine whether their wastes are reactive." [May 19, 1980; 45 *FR* 33110] Similarly, RO 14176 notes that EPA has "given reasonable deference to the operational experience of the waste generator or facility" to determine the potential reactivity of a solid waste.

For wastes that might be reactive due to their explosive nature, EPA noted that "[s]pecific methods used by agencies such as the Bureau of Alcohol, Tobacco, and Firearms to determine if a substance is an explosive could, however, be applied to determine whether a waste is reactive according to §261.23(a)(6) or (7)." [RO 13735] Method 8095—Explosives by Gas Chromatography (added to SW–846 in January 2008 [73 *FR* 486]) may be used to determine the concentration of nitroaromatics, nitramines, and nitrate ester explosives in water and soil. This method uses capillary column gas chromatography with an electron capture detector (GC/ECD). The results of this analysis can help a generator decide whether a particular waste meets the narrative description for a reactive waste.

For advice on possible tests for determining if a waste is reactive or not, interested parties can contact EPA's National Enforcement Investigation Center (NEIC) in Denver, Colorado [(303) 462-9000]. Staff at the NEIC are the agency's experts on this topic. [RO 14259]

2.4.2 Reactive cyanides and sulfides

Most of the volume of D003 wastes generated nationwide consist of reactive cyanides. [§261.23(a)(5)] They are typically generated by the electroplating and metal finishing industries. Examples of reactive cyanide wastes are mixed cyanide salts, cyanide solutions, and cyanide-bearing sludges. Treatment technologies appropriate for these types of waste include electrolytic oxidation, alkaline chlorination, and wet-air oxidation. [June 1, 1990; 55 *FR* 22550]

The narrative definition of reactive cyanide and sulfide wastes says that such wastes release toxic gases *when exposed to pH conditions between 2 and 12.5.* "The pH range chosen is that which is considered nonhazardous by the corrosivity characteristic (2 < pH <12.5). This range was chosen because any liquid outside the range is hazardous and requires management within the Subtitle C regulations. Only liquid wastes inside this range can be landfilled without regard to the strictures on compatibility imposed by the Subtitle C regulations and co-disposed with wastes containing soluble cyanides or sulfides. These are then the most stringent pH conditions which a waste could be subjected to outside of a Subtitle C facility." [RO 51435]

2.4.2.1 Release thresholds for cyanide and sulfide wastes no longer valid

EPA received numerous inquiries into what the agency meant in §261.23(a)(5) by "a quantity sufficient to present a danger to human health or the environment" for cyanide- and sulfide-bearing wastes.

As a result, they issued thresholds for releases of gases from these wastes in July 1985. The agency determined that wastes releasing more than 250 mg of hydrogen cyanide (HCN) gas/kg waste or 500 mg of hydrogen sulfide (H_2S) gas/kg waste should be regulated as reactive hazardous wastes. [RO 11091, 12436] Test methods for determining the amount of gases released were also provided and incorporated into Chapter 7 of SW–846.

However, EPA later discovered that "critical errors" were made in developing the release thresholds and test methods discussed above. Therefore, the use of these thresholds for making reactive waste determinations and the test methods were withdrawn from SW–846. [June 14, 2005; 70 *FR* 34548] Revised guidance for determining reactivity of these wastes is under development. [RO 14177, 14407] This action has reestablished the uncertainty associated with the regulatory status of cyanide- and sulfide-bearing wastes. EPA remains convinced, however, that generators can rely on knowledge of their wastes to make a proper waste classification: "generators and other persons can use other appropriate methods or process knowledge in determining whether a particular waste is hazardous due to its reactivity." [70 *FR* 34549]

Even though these thresholds and test methods have been withdrawn at the federal level, some states may still use the above cyanide and sulfide release threshold levels for determining the characteristic of reactivity. We recommend that you check with your state to determine how it defines reactivity for cyanide- and sulfide-bearing wastes.

2.4.2.2 Knowledge-based determinations for reactive cyanide and sulfide wastes

As noted in Section 2.4.2.1, EPA requires generators to use their knowledge to make D003 determinations per §261.23(a)(5) for cyanide- and sulfide-bearing wastes. In old guidance documents (none of which are in EPA's RCRA Online database), the agency noted that, if the answer to any of the four questions below is "Yes," the waste should be considered a D003 reactive hazardous waste:

1. Has the waste ever caused injury to a worker because of HCN or H_2S generation?
2. Have the OSHA workplace air concentration limits for either HCN or H_2S been exceeded in areas where the waste is generated, stored, or otherwise handled?
3. Have air concentrations of HCN or H_2S above a few parts per million been encountered in areas where the waste is generated, stored, or otherwise handled?
4. Would a chemist with knowledge of the waste believe that one or more of the above might occur if the waste was subject to acidic conditions?

[Four EPA Headquarters memos, the most recent of which is a letter from William Collins, Jr. to Donald Searles, U.S. Attorney for the Eastern District of California, December 17, 1996]

2.4.3 Explosives

The reactivity definition includes a number of references to wastes that will explode under certain conditions. Guidance on waste explosives follows.

2.4.3.1 Off-specification small arms ammunition

Off-specification small arms ammunition (ball or sporting ammunition of calibers up to and including 0.50) contains an ignition source that may be shock and heat sensitive and is designed to generate high pressure during use. As a result, EPA originally believed that this type of ammunition should be regulated as a reactive waste when disposed. However, based on test data provided by the U.S. Army and one of the ammunition manufacturers, EPA now believes that small arms ammunition is not capable of detonation or explosive reaction if it is subjected to a strong initiating source or if heated under confinement [the §261.23(a)(6) reactivity criteria]. Therefore, the agency has concluded that waste ammunition up to and including 0.50 caliber is not reactive; however, nonreactive wastes exhibiting another characteristic (e.g., toxicity for lead) would be regulated under Subtitle C. [RO 12339, 13712]

2.4.3.2 Explosives-contaminated soil/sediment

Cleanup activities at DOD and commercial explosives manufacturing facilities often include management of explosives-contaminated soil/sediment. Whether such soil/sediment is reactive under EPA's regulatory definition is often an issue. To answer that question, the U.S. Army tested soil/sediment containing varying concentrations of research department explosive (RDX—an explosive nitroamine) and trinitrotoluene (TNT). They concluded that explosives-contaminated soil/sediment containing 12% explosives or less will not propagate a detonation or explode when heated under confinement. [*Testing to Determine Relationship Between Explosive-Contaminated Sludge Components and Reactivity*, AMXTH-TE-CR-86096, January 1987] Appendix A of SW–846 Method 8330B says the same thing.

The U.S. Army Environmental Center considers all soils containing more than 10% by weight secondary explosives to be susceptible to initiation and propagation. [*Handbook: Approaches for the Remediation of Federal Facility Sites Contaminated With Explosive or Radioactive Wastes*, EPA/625/R-93/013, September 1993, available at http://nepis.epa.gov/EPA/html/Pubs/pubtitleORD.html by downloading the report numbered 625R93013]

2.4.3.3 Claiming the immediate response exemption

EPA has provided a possible exemption for personnel managing potentially explosive wastes. For example, consider a facility that is cleaning out old laboratory chemicals, including shock-sensitive and explosive materials such as picric acid. Is the transportation and disposal of these materials regulated under RCRA?

"Under EPA's RCRA regulations [§270.1(c)(3)], all activities taken in immediate response to a discharge of hazardous waste, or an imminent and substantial threat of discharge of a hazardous waste, are exempt from the RCRA permitting and substantive requirements. Since the chemicals in question would be hazardous by virtue of their reactivity, any actions you take to eliminate the imminent and substantial danger would qualify under this exemption. If the response action involves transportation to a remote site for destruction, then the transportation as well as the destruction would be exempt. However, the transportation is exempt only to the extent necessary to respond to the immediate threat. Hence, we expect the [exempt] transportation would normally cover a relatively short distance and would occur in special transportation equipment such as bomb trailers." [RO 13574; see also RO 11363]

2.4.4 Management of reactive wastes in lagoons

As a final note to this section, the following example gives EPA's perspective on managing a reactive waste in a lagoon. A wastewater, which is not a listed hazardous waste, is placed in a surface impoundment for dewatering. A constituent of the wastewater is reactive, but only when dry. If the surface impoundment never dries out during its active life, is it subject to the hazardous waste regulations?

"The surface impoundment becomes subject to RCRA when it dries out after receiving its final volume of waste. If the waste is immediately removed as it becomes reactive (dewatered), then the operator would be a generator for that waste. If the waste dries out first and becomes reactive [before it is removed], then the surface impoundment would be subject to the regulations." [RO 12085]

2.5 Aerosol cans

Aerosol cans pose a unique hazardous waste determination problem because they consist of three different types of materials, each of which may be classified as hazardous waste: 1) the can itself, 2) the liquid product contained in the can, and 3) the gaseous propellant. Depending on how they are managed, these wastes may be excluded from RCRA regulation or be subject to the full hazardous waste requirements.

2.5.1 Regulation of cans depends on management method

When the RCRA regulations were first promulgated in 1980, EPA's position was that the hazardous waste regulations apply to the contents of an aerosol can, but not to the cans themselves. [RO 12020] This interpretation has been modified over the years. Notably, in 1987, the agency revised its position, saying that even empty aerosol cans could be hazardous due to reactivity [i.e., an empty can could detonate or explode when subjected to a strong initiating source or if heated under confinement per §261.23(a)(6)]. [RO 13027, 13435]

To date, the agency has not been able to make a categorical determination on whether or not various types of aerosol cans that may have contained a wide range of products are reactive. As a result, it is the responsibility of the generator to determine if an aerosol can is hazardous in accordance with §262.11. [RO 11806, 14235] If such cans are determined to be hazardous and are destined for disposal, they are regulated as hazardous wastes; however, as discussed below, punctured cans that are recycled are not regulated due to the scrap metal exemption.

2.5.1.1 Punctured aerosol cans are not regulated when recycled

According to EPA's current regulatory interpretation, a steel aerosol can that does not contain a "significant" amount of liquid would meet the definition of "scrap metal" in §261.1(c)(6). Consequently, if the can were recycled as scrap metal, it would be exempt from RCRA regulation under §261.6(a)(3)(ii), which excludes recycled scrap metal from RCRA regulation. As such, a determination of reactivity or any other characteristic would not be necessary. [RO 11782, 11806] According to EPA, aerosol cans do not contain "significant" liquids if they "have been punctured so that most of any liquid remaining in the can may flow from the can…and drain (e.g., with punctured end down)…." However, it is very important to note that this exemption applies only to the punctured can itself;

any liquids and propellants removed from the aerosol cans are subject to regulation as hazardous wastes if they are listed or if they exhibit a characteristic. [RO 11782]

Since emptying aerosol cans that are to be recycled is part of a recycling process (i.e., scrap steel recycling), this activity is exempt from RCRA regulation under §261.6(c), except as specified in §261.6(d); specifically, the puncturing step would *not* be considered hazardous waste treatment requiring a permit. Even so, puncturing and draining aerosol cans should be performed in a safe and environmentally protective fashion, and any gaseous or liquid materials removed from the cans should be managed appropriately (as discussed below). [RO 11782]

A number of different aerosol can puncturing devices are available; many attach to standard 55-gallon drums. In such units, the liquid removed from punctured cans drains into the drum, while the propellant exhausts through some kind of filter (typically screwed into one of the bung openings). Sometimes the question comes up as to whether the drained liquid has to be shipped offsite or transferred to a permitted storage area every 90 days, even if the drum isn't full yet. An argument can possibly be made that the 55-gallon drum is part of the aerosol can recycling process and therefore is exempt from RCRA regulation (as noted above) until the day the puncturing device is removed. That would be Day 1 of the 90-day accumulation period, assuming that the drained liquids are hazardous. Another argument could be made that the drum is a satellite accumulation unit, since the drained liquid in the drum could be considered a newly generated waste. Satellite accumulation units don't have a specified accumulation period (see Section 6.2). State policy will dictate whether an aerosol can puncturing device mounted on a drum can be considered a 1) recycling unit, 2) 90-day unit, or 3) satellite accumulation unit.

Some aerosol cans containing pesticides (e.g., wasp killer) are labeled "Do not puncture." When asked about this situation, EPA's Office of Prevention, Pesticides, and Toxic Substances wrote a letter noting the following: "EPA has determined that the puncturing of disposed aerosol pesticide containers for recycling is consistent with the purposes of FIFRA, provided that: 1) the puncturing of the container is performed by a person who, as a general part of his or her profession, performs recycling and/or disposal activities; 2) the puncturing is conducted using a device specifically designed to safely puncture aerosol cans and contain the residual contents; and 3) the puncturing, waste collection, and disposal are conducted in compliance with all applicable federal, state, and local waste (solid and hazardous waste) and occupational safety and health laws and regulations." [Letter, Lois Rossi and William Diamond to John Wildie, April 30, 2004]

2.5.1.2 Aerosol cans destined for disposal may be hazardous waste

If an aerosol can is to be disposed rather than recycled, a hazardous waste determination must be made (for both the can itself and its contents). [RO 11782, 11806] According to EPA, a discarded aerosol can is a hazardous waste if:

- The can is not empty as defined in §261.7 and contains a commercial chemical product that is listed in §261.33(e) or (f) or contains a material that exhibits one or more of the hazardous waste characteristics; and/or

- The can itself exhibits any of the characteristics of hazardous waste (i.e., reactivity). [RO 13225, 14656]

Consequently, if a generator wants to dispose an aerosol can as nonhazardous waste, he/she must determine that 1) the can is empty according to §261.7 (or that the product it contains is not hazardous), *and* 2) the can itself is not reactive. [RO 11782]

If the generator determines that the can is hazardous, two options exist: 1) the hazardous can may be manifested as is to a TSD facility, or 2) the can may be punctured, with the resulting residues disposed separately. Conflicting interpretations have been given by EPA on the topic of whether puncturing a hazardous can to release the liquid/propellant prior to disposal constitutes treatment and whether a RCRA permit is required. Originally, in 1980, EPA determined that the puncturing, crushing, or shredding of nonempty aerosol cans that contain hazardous wastes was not considered treatment. [RO 12020] However, this stance has changed; according to current EPA policy, it is now up to EPA regional offices to make this determination on a case-by-case basis. [RO 13225] It appears that most regions consider the puncturing/venting of aerosol cans prior to disposal to be treatment. [RO 11414]

Since hazardous waste treatment can be performed in 90-day units without a permit (see Section 7.3), such puncturing prior to disposal might be accomplished in a 90-day container. (We say "might" because one of the requirements for treating wastes in 90-day containers without a permit is that the container must be closed per §265.173(a). Because propellants must be vented from the 90-day container during puncturing, the authorized agency might have to make a site-specific determination on whether the provisions of §265.173(a) are met.) Once the can has been punctured and drained, the can itself may be disposed as nonhazardous waste if it meets §261.7 requirements, but the liquid and propellant must be managed as discussed below.

2.5.2 Liquid material from aerosol cans may be hazardous waste

The liquid material removed from aerosol cans that are punctured, crushed, or shredded may be subject to RCRA hazardous waste regulation if 1) it is a commercial chemical product, such as a solvent, listed in §261.33(e) or (f), or 2) it exhibits any hazardous waste characteristic. [RO 14656] However, such liquids are only regulated if they are discarded or intended to be discarded. If the materials are recovered for 1) repackaging and beneficial use, or 2) other legitimate recycling or reclamation, they are not solid wastes, and consequently are not subject to the RCRA regulations. [RO 12020]

Liquid removed from RCRA-empty containers (i.e., those that meet §261.7 standards) may or may not be subject to RCRA regulation. EPA headquarters [RO 14708] and some states (e.g., Alabama) have stated that residues removed from RCRA-empty containers *are* hazardous wastes if they exhibit a characteristic. From a practical standpoint, however, since most aerosol can puncturing operations drain liquids from both empty and nonempty cans into the same drum, this issue may be moot.

2.5.3 What about the propellant?

Our experience indicates that the propellant released during aerosol can puncturing operations is typically vented from the unit through a coalescing/activated-carbon filter. Because such propellant releases are no longer "contained gases," they are not subject to RCRA hazardous waste regulation; Clean Air Act standards (e.g., state air regulations or a facility's air permit) may apply instead. The activated-carbon filter will be hazardous when disposed if it contains any P- or U-listed waste or exhibits a characteristic. If the propellant is captured in a bottle or cylinder instead of being vented, it is a contained gas subject to possible regulation under RCRA.

A word of caution is in order for facilities that puncture aerosol cans. We are aware of a situation where several fatalities occurred during a can puncturing operation. Cans were punctured in a hood that vented propellant gases to a control device. The liquids drained into a nonsealed container in the workroom. Because the liquids were supersaturated with propellant, they continued to evolve propellant vapors. When vapors reached flammable levels in the workroom, an ignition source caused a fire that proved fatal to personnel in that room.

2.5.3.1 Combusting aerosol can propellant for energy recovery

During recovery of aerosol cans, recycling facilities often segregate the resulting materials into three streams: 1) scrap metal (punctured, crushed, or shredded cans), 2) liquid material removed from the cans, and 3) gaseous propellant. In situations where the propellant has significant Btu content (e.g., butane and propane), it may be burned for energy recovery in onsite boilers. [RO 11717]

Questions have arisen over whether the combustion of such propellant mixtures constitutes burning of hazardous waste (which would consequently be subject to the RCRA boiler and industrial furnace or BIF regulations). According to EPA, such combustion of propellants would constitute burning for energy recovery. However, if the butane and propane propellants, which are normally used as fuels, are classified as commercial chemical products, the propellant mixture is not a solid waste (and subsequently not a hazardous waste) under §261.2(c)(2)(ii). (This section exempts commercial chemical products that are listed or exhibit RCRA hazardous waste characteristics from classification as solid wastes when they are burned for energy recovery if they are themselves fuel.) As EPA explains:

> "[I]f the aerosol cans are full (not used), or partially full (in which case they would be considered off-specification with the remaining propellants in the cans also being unused), then the butane and propane propellants would be classified as commercial chemical products. Since these products are fuels and are being burned for energy recovery, they would not fall within the definition of solid waste and would consequently not be considered a hazardous waste."

As a result, the BIF regulations would not apply. However, this determination assumes that hazardous constituents have been removed from the propellant mixture; the exemption does not apply if the propellant mixture contains hazardous constituents and is being burned to destroy the hazardous constituents instead of for energy recovery purposes. [RO 11717]

2.5.4 Accumulating aerosol cans prior to puncturing/shipment

Whether spent aerosol cans will ultimately be recycled or disposed, they generally are accumulated by the generator for some period of time before they are punctured and/or shipped offsite. Such cans will likely be considered hazardous waste (either due to their contents or reactivity of the cans themselves) before the puncturing and/or shipping step. Remember that EPA's interpretation of when an aerosol can meets the definition of scrap metal is *after* it has been punctured and drained. As such, accumulation before puncturing is typically accomplished in satellite accumulation units. Either 90-day or satellite accumulation would be appropriate for unpunctured cans that will be shipped offsite directly. We are aware of a facility that received an NOV when state inspectors found an unpunctured can of WD-40 in a scrap metal bin.

2.5.5 Household waste exclusion

Under the federal RCRA regulations, household waste (including full and empty aerosol cans) is excluded from the definition of hazardous waste. [§261.4(b)(1)] Consequently, any aerosol cans generated by households (including military housing) are not regulated as hazardous waste. This also applies to aerosol cans collected from households and managed by municipal recycling programs. [RO 11782] However, if waste aerosol cans from households are mixed with aerosol cans from industrial applications (e.g., waste aerosol cans from a maintenance shop on a military base), such mixtures will be subject to all Subtitle C regulations (see Section 4.10.2).

2.5.6 State programs may be different

The guidance outlined above pertains only to the federal regulation of aerosol cans. Some authorized states have taken a different approach to managing these wastes. For example, Colorado has added aerosol cans as a universal waste. Additionally, some states (e.g., California) define an empty aerosol can to be nonhazardous. Consequently, it is important to contact the appropriate state agency to find out if they have different regulations or policies that apply to these wastes.

2.6 Toxicity

A solid waste exhibits the toxicity characteristic (TC) if the extract (obtained using the toxicity characteristic leaching procedure or TCLP) from a representative sample of the waste contains any of the TC constituents identified in §261.24 Table 1 at a concentration greater than or equal to the applicable regulatory level (also given in that section). Forty TC constituents are listed in §261.24, including eight heavy metals, six pesticides, and 26 solvents and other organics. Because many of the TC constituents are commonly found in industrial wastes (e.g., benzene, chromium, lead, mercury, methyl ethyl ketone, silver, vinyl chloride), large quantities of solid wastes generated in this country exhibit the characteristic of toxicity. Although there may be many chemicals other than the 40 included in §261.24 that are toxic to human health and the environment, only those 40 make up the TC in the federal RCRA regulations. State programs may be more stringent, however, and some states (e.g., California) have added other constituents to their state toxicity definition.

Two waste streams that often exhibit the TC and that are problematic for almost all industries are electronic wastes (e-wastes), including cathode ray tubes (CRTs) associated with outdated computers, and burned-out fluorescent light tubes and other spent lamps. These two wastes are discussed separately in Sections 2.7 and 2.8, respectively.

2.6.1 Background

When EPA was working on the TC in the late 1970s, the agency knew it needed to 1) establish concentration-based regulatory levels for toxic contaminants in industrial wastes, above which the wastes would be considered hazardous; and 2) develop a test that industry could use to quickly and inexpensively evaluate their wastes for this characteristic.

To achieve these goals, EPA first came up with a worst-case hazardous waste mismanagement scenario: the co-disposal of industrial waste containing toxic constituents with municipal waste in an unlined, municipal solid waste (MSW) landfill. This scenario is depicted in Figure 2-1. Based on the agency's research and modeling, an actively decomposing municipal landfill creates acidic conditions, and the weak organic acids produced will begin to work on and eventually leach the toxic constituents from the industrial waste. The resulting acidic, toxic leachate will migrate downward into the underlying soil, and if there is an aquifer underlying the landfill, the leachate will eventually contaminate the ground water.

The agency included in this scenario the possibility that homes near the landfill would use the aquifer as a source of drinking water. The resulting exposure pathway was EPA's primary concern when developing the TC. At the time the agency was finalizing the initial RCRA regulations, the best federal ground water protection standards available were the maximum contaminant levels identified in the National Interim Primary Drinking Water Standards (NIPDWS), which had previously been promulgated under the Safe Drinking Water Act (SDWA). For example, the SDWA level for lead was 0.05 mg/L in the late 1970s. If the concentration of lead in the contaminated aquifer exceeded that level, it would not be safe for people to drink the water.

But EPA also knew that the toxic constituents leached from the industrial waste would be subject to dilution and attenuation as they migrated through the subsurface. For example, some of the lead would likely adsorb to the soil underlying the landfill as the leachate passed through it; additionally, the leachate would be diluted in the aquifer. The agency evaluated and modeled these dilution and attenuation processes and concluded in 1980 that the hazardous constituents would be diluted and/or attenuated by a factor of 100 in the subsurface.

Continuing our example, applying the dilution and attenuation factor (DAF) of 100 to the health-based SDWA level for lead of 0.05 mg/L resulted in a TC regulatory level for lead of 5.0 mg/L. What does that tell us? If lead leaches out of an industrial waste and

Figure 2-1: Regulatory Scenario for the Toxicity Characteristic

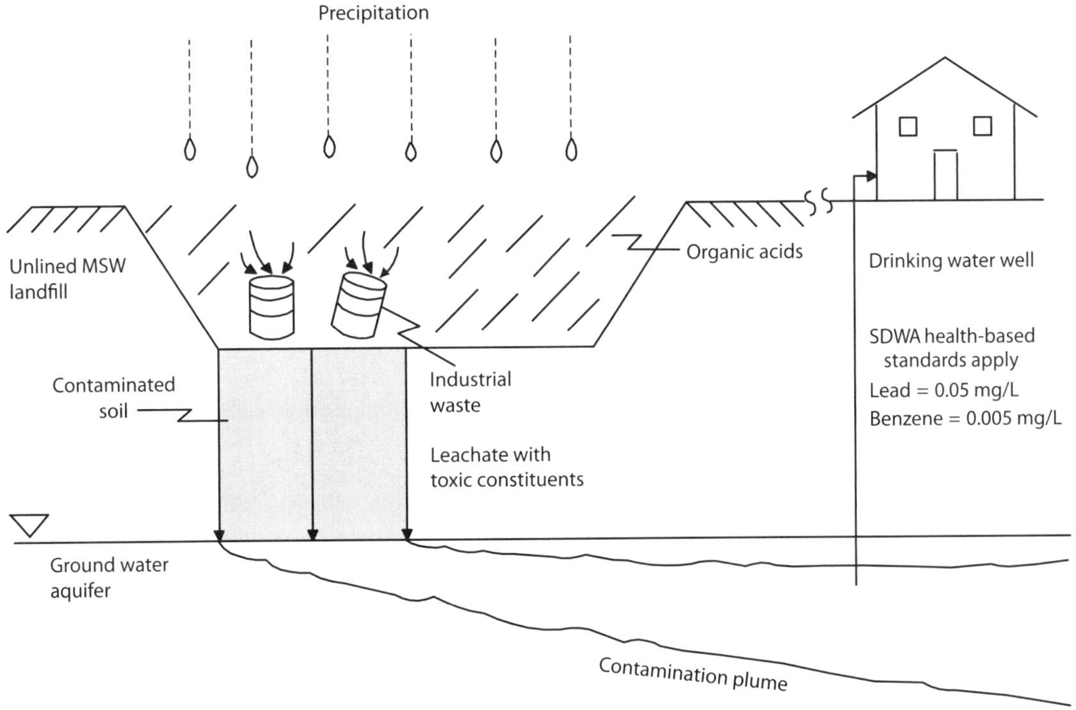

SDWA health-based standard	Dilution-attenuation factor	TC regulatory level (§261.24 Table 1)
Lead = 0.05 mg/L	100	5.0 mg/L
Benzene = 0.005 mg/L	100	0.5 mg/L

Source: McCoy and Associates, Inc.

its concentration in the leachate is greater than or equal to 5.0 mg/L, the concentration of lead in the underlying aquifer will exceed the SDWA standard (even factoring in the dilution and attenuation effects of the subsurface as the lead passes through it), and it will not be safe for people to drink the water. Therefore, such industrial waste cannot be placed into a municipal waste landfill; it must be disposed in a hazardous waste landfill, which will be much more protective.

Based on this approach of multiplying the NIPDWS by 100, EPA established TC regulatory levels for 14 toxic constituents in 1980. The constituents included 8 heavy metals (arsenic, barium, cadmium, chromium, lead, mercury, selenium, and silver), 4 insecticides (endrin, lindane, methoxychlor, and toxaphene), and 2 herbicides (2,4-D and 2,4,5-TP). These were the constituents included in the NIPDWS at that time.

Obviously, there are numerous additional constituents that might be toxic to human health and the environment. EPA noted in RO 51433 "that although the EP Toxicity Characteristic regulates only those wastes containing toxic constituents for which National Interim Primary Drinking Water Standards have been established, other wastes may still be regulated as toxic via the listing mechanisms under §§261.31 through 261.33 of the regulations." Thus, although the TC is limited to a very small list of toxic constituents, the F-, K-, P-, and U-listings in §§261.31–261.33 will require hazardous waste management of many other toxic chemicals.

2.6.1.1 EP toxicity test initially used

After EPA established regulatory levels for 14 toxic contaminants often found in industrial wastes, the second step was to develop a test industry could use to determine if their wastes exceeded those levels. When the TC regulations were first promulgated in 1980, they were based on use of the extraction procedure (EP) toxicity test. To determine if their wastes were hazardous due to toxicity, generators could test their wastes using this method.

The EP toxicity test was designed to simulate the acidic leaching processes occurring in the municipal waste landfill discussed above. An acetic acid solution was added to a sample of the waste to simulate the leaching of toxic constituents from the waste. (Acetic acid was chosen because it is the most prevalent acid found in municipal landfill leachate. [RO 51433]) The extract (leachate) was then analyzed for the 14 TC constituents, and if it contained any of the contaminants at concentrations greater than or equal to regulatory levels, the waste was considered hazardous for toxicity.

2.6.1.2 TC revised in 1990

EPA received many complaints about the EP toxicity test due to concerns over its reproducibility and ability to accurately measure a waste's toxicity. Therefore, when the Hazardous and Solid Waste Amendments of 1984 (HSWA) were enacted, Congress directed EPA to revise the EP toxicity test and expand the TC constituent list.

In compliance with the Congressional mandate, the agency developed the TCLP, which was promulgated on March 29, 1990 [55 *FR* 11798], to replace the EP toxicity test. On that date, EPA also added 26 organic chemicals to the list of toxic constituents regulated under the TC. Based on that action, the TC now consists of 40 constituents, all of which are included in Table 1 in §261.24. The waste codes associated with the TC are D004–D043, depending on which chemical or chemicals in Table 1 leach out at concentrations greater than or equal to regulatory levels. If none of the 40 constituents in Table 1 leach out of a solid waste at or above regulatory levels, it does not exhibit the toxicity characteristic at the federal level.

The chemicals added to the list in 1990 were Part 261, Appendix VIII constituents for which EPA had adequate chronic toxicity reference levels and fate and transport data. (Chronic toxicity reference levels are levels below which chronic exposure for individual toxicants in drinking water is considered safe or pose minimal risk. For the 26 new TC chemicals, these levels were based on SDWA maximum contaminant levels, where available, or reference or risk-specific doses.) As before, EPA calculated regulatory levels for the 26 new chemicals by multiplying the chronic toxicity reference level of a constituent times the DAF (which again was set at 100).

For purposes of determining toxicity of wastes containing chromium, the TC level is based on total chromium. The two major valences of chromium are tri- and hexavalent. Trivalent chromium is much less toxic than the hexavalent form. Because trivalent chromium can be oxidized to the hexavalent form and because the NIPDWS was developed for total chromium, EPA believes that the regulatory level for chromium in the TC should be based on total levels. [March 29, 1990; 55 *FR* 11812] See also RO 12587.

Sometimes questions arise as to the relevance of Footnote 3, which applies to three constituents in Table 1 of §261.24 that were added in 1990. For those three compounds (2,4-dinitrotoluene, hexachlorobenzene, and pyridine), the calculated regulatory levels were below the levels measurable using available analytical methods. Consequently, EPA set the TC regulatory levels for these compounds at the "quantitation limit," which is the lowest concentration that can be reliably measured under routine laboratory conditions. For example, the concentration of 0.13 mg/L for hexachlorobenzene was set by EPA in 1990 because it was the lowest concentration that could be reliably measured under routine laboratory conditions that existed in 1990. Even if today's labs can measure lower than that, the regulatory level is still 0.13 mg/L today. So the regulatory level under the TC for hexachlorobenzene is 0.13 mg/L, not the quantitation limit at the lab running your TCLP test.

Although the revised RCRA regulations in §261.24 base the definition of the TC on the TCLP, EPA does *not* specifically require that generators utilize the TCLP to determine if a waste is toxic. True, generators must make a determination as to whether or not their solid waste is hazardous (§262.11), but such a determination may be made by either testing a representative sample of the waste or applying knowledge of the waste or the processes that generated the waste. Therefore, determination of the potential toxicity of a solid waste can be made by either running the TCLP or applying knowledge. These options are discussed in the next two subsections.

2.6.2 TCLP-based toxicity determinations

The TCLP remains the test that we use today to determine toxicity and is included as Method 1311 in SW–846. Briefly, for wastes containing greater than or equal to 0.5% filterable solids, the lab performing the TCLP separates the liquid (if any) from the solids in the representative sample. The solids are then extracted (leached) by adding acetic acid in an amount equal to 20 times as much, by weight, as the weight of the solids portion of the sample. The extraction vessel holding the solids and acid is tumbled for 18 hours, after which the acidic extract (leachate) is separated from the solids by filtration. If compatible, the liquid initially separated from the solids is combined with the acidic extract, and the mixture is then analyzed for total concentrations of the 40 constituents in §261.24. If the two liquid streams are incompatible, they are analyzed separately, the results are mathematically combined, and the resulting volume-weighted total concentrations are compared to the regulatory levels.

For wastes that contain less than 0.5% filterable solids (e.g., wastewater), the waste itself (after filtering using the procedures outlined in the TCLP) is considered the extract and directly analyzed for total concentration of toxic constituents—no extraction (i.e., leaching with acetic acid) is necessary.

After the TCLP was promulgated to replace the EP toxicity test, several concerns cropped up as industry began using the new method. These problems and EPA guidance on these topics are discussed below.

2.6.2.1 Cost

One of the most significant obstacles to using the TCLP is its cost. Our experience indicates that a lab will charge $700–$1,500 to run the TCLP when testing for all 40 constituents. The cost goes down considerably if you ask the lab to evaluate a sample for only a subset of the 40 TC constituents (such as the metals). But the test, in general, is relatively expensive.

2.6.2.2 Reproducibility

Another concern with the TCLP is its reproducibility. From a scientific standpoint, the method is not very precise. While EPA was in the process of modifying the TC to incorporate the TCLP, an intergovernmental cooperative program was undertaken to evaluate the reproducibility of the test. Three different laboratories were given samples from the same batch of five discrete waste types. Within a given laboratory, the TCLP was performed on two to six replicate samples of each waste.

Data from the three laboratories indicated that 67% of the TCLP results for metals were reproducible and 13% were marginally reproducible. For the nonmetals, 60% of the results were considered reproducible. Some of the results for certain contaminants were reproducible within a given laboratory but not among the three laboratories.

These results were reported by personnel from the U.S. Army Engineer Waterways Experiment Station in 1990. The paper, "Reproducibility of Toxicity Characteristic Leaching Procedure" by Teresa T. Holmes et al., appeared in *Proceedings of the 7th*

National Conference on Hazardous Wastes and Hazardous Materials.

2.6.2.3 Analyzing the extract

When the extract produced from the TCLP is analyzed for the 40 constituents listed in Table 1 of §261.24, what test methods have to be used? According to EPA, "the extract obtained from the TCLP may be analyzed by any method as long as that method has documented quality control and the method is sensitive enough to meet the regulatory limit. In other words, the lab does not have to use SW–846 methods [to analyze the extract for TC constituents] because these methods are intended to serve only as guidance for the regulated community." [RO 11568; see also RO 11579, 11649]

However, the agency noted that the following EPA test methods could be used to analyze the extract [EPA/530/R-94/024]:

- Methods 3010 and 6010 for arsenic, barium, cadmium, chromium, lead, silver, and selenium;
- Method 7470 for mercury;
- Methods 3510 and 8081 for pesticides;
- Method 8151 for herbicides;
- Method 8260 for volatile organics; and
- Methods 3510 and 8270 for semivolatile organics.

For assistance on determining appropriate analytical methods, contact EPA's Methods Information Communications Exchange (MICE) in Washington, DC at (703) 818-3238 or mice@techlawinc.com.

2.6.2.4 Oily wastes and organic liquids pose problems

Many problems have been reported when laboratories have attempted to run the TCLP on oily wastes and organic liquids. In some situations, analysis of TCLP extracts has resulted in detection limits above regulatory levels. This typically happens when an organic liquid waste passes through the initial filtering stage and is, by EPA definition, a liquid and therefore its own extract (i.e., no acid extraction is required). The analysis of this liquid extract for organics entails diluting it before injection into a gas chromatograph or mass spectrometer. The dilution often results in detection limits being higher than regulatory thresholds. [RO 11579, 11592, 11627] In these circumstances, it is not possible to determine conclusively whether the waste is hazardous or not.

In addition to the detection limit problems, oily wastes have also caused premature filter clogging, emulsion formation, and difficulty in estimating percent solids. [RO 11721] EPA acknowledges that the TCLP was not intended to be applied to certain matrices, such as oils or neat solvents, and is investigating ways to solve these problems. [RO 11579, 11592]

2.6.2.5 Sample holding times

In order to ensure that accurate results are obtained from the TCLP, EPA specifies quality control measures (including sample holding times) in Method 1311. Sample holding time is the storage time allowed between field collection of a sample and completion of the laboratory analysis. [RO 11306, 13589] It is an important parameter because certain constituents, such as volatile organic compounds, can degrade or volatilize over time.

The sample holding times that EPA specifies for TCLP extraction and analysis are given in Section 8.5 of Method 1311. When these sample holding times are exceeded, the analytical results may be invalid or inconclusive because constituent levels of volatiles in expired samples may be lower than if the samples were fresh.

Analytical results obtained if sample holding times are exceeded may still have some use. For instance, if TCLP results from samples that exceeded specified holding times indicate that the concentrations of one or more TC constituents are above the applicable regulatory threshold, then those concentrations can be treated as minimum values, and the waste can be determined to be hazardous via toxicity. No additional testing is necessary. However, if the results from an expired sample reveal concentrations that are below the regulatory thresholds, further testing may be necessary to demonstrate that the waste is not toxic. [RO 13612]

2.6.2.6 TCLP not valid for manufactured gas plant wastes

EPA's application of the TCLP to manufactured gas plant (MGP) wastes was successfully challenged in court [*Association of Battery Recyclers, Inc. et al. v. EPA*, 208 F.3d 1047 (2000)]. The basis of the court challenge was that MGP wastes are not managed in municipal solid waste landfills, which was the disposal scenario that serves as the basis for the TCLP. EPA has determined that, under the court's opinion, the TCLP cannot be applied to MGP wastes; therefore, these wastes (which are not listed) are hazardous only if they exhibit any of the ignitability, corrosivity, or reactivity characteristics. Because MGP remediation wastes are unlikely to exhibit any of these characteristics, they are typically not RCRA hazardous wastes unless otherwise regulated under state authority. [RO 14491, 14492]

2.6.3 Knowledge-based toxicity determinations

If a generator does not run the TCLP to determine toxicity, the other option is to use knowledge. For example, the generator can use material balances and/or knowledge of the raw materials and the processes that generated the waste to make a toxicity determination. If a generator has good knowledge that a certain waste would pass or fail the TCLP, no testing is necessary. An example is the U.S. Army managing waste munitions as toxic for lead and/or 2,4-dinitrotoluene; the service didn't want to perform the TCLP because of the inherent safety problems associated with the method's particle-size reduction criteria. The Army could simply declare the waste to be hazardous for toxicity and manage it as such. [RO 11608, 13472] Where no information or knowledge is available to assist a generator, or when TCLP results are not available or are inconclusive, the agency noted that it might be prudent for the generator to manage the waste as hazardous. [RO 11579, 11592]

Another source of knowledge, especially for products that will be discarded, is chemical composition data on safety data sheets (SDSs). However, be advised that SDSs are not always an appropriate reference for determining if a material exhibits the toxicity characteristic. The OSHA regs, as modified to reflect the provisions of the United Nations Globally Harmonized System of Classification and Labeling of Chemicals, generally require manufacturers to identify constituents present in the material at concentrations ≥1% (10,000 ppm) for noncarcinogens or ≥0.1% (1,000 ppm) for carcinogenic constituents. [Appendix A to 29 *CFR* 1910.1200] Therefore, the product might contain toxicity characteristic constituents above RCRA regulatory levels even though they are not identified on the SDS.

Sometimes, a generator will use a combination of the two approaches; that is, if the generator has knowledge that certain TC constituents could not be present in the waste, the TCLP will have to be performed only for the other constituents that could possibly be in the material. [EPA/530/R-93/007, RO 11603, 14695]

A test method other than the TCLP can be used that may be more appropriate for determining the toxicity of the waste matrix. For example, the oily waste extraction procedure (OWEP—Method 1330 in SW–846) can be used to evaluate metal concentrations in oily sludges, slop oil emulsions, and other oily wastes (although this method is typically used only in support of delisting petitions). This method was developed for wastes containing oil or grease in concentrations of 1% or greater. [RO 11522, 12450]

The generator can also use total waste analyses as an alternative to TCLP results. This option is discussed in detail below.

2.6.3.1 Total waste analyses in lieu of TCLP results

Section 1.2 of the TCLP states that "[i]f a total analysis of the waste demonstrates that individual analytes are not present in the waste, or that they are present but at such low concentrations that the appropriate regulatory levels could not possibly be exceeded, the TCLP need not be run." Therefore, while the TCLP is typically performed to make a TC determination, a generator can alternatively use total waste analyses to determine that a waste does not exhibit the TC. A total waste analysis is also a convenient and cost-effective screening tool for determining if a TCLP is needed.

The methodology for using total waste analyses to make a TC determination varies depending on the type of waste [RO 11721, 13563, 13647, 14533, 14695]:

- *Liquids*—Liquids (i.e., wastes that contain less than 0.5% filterable solids) do not require extraction. Instead, per the last sentence in §261.24(a), a generator can characterize such a liquid waste by filtering it, analyzing the total constituent concentrations in the resulting filtrate, and comparing those concentrations directly to regulatory levels.

- *Solids*—For wastes that are 100% physical solids (i.e., they contain no filterable liquid), the total concentrations of the 40 TC constituents are determined, and then these total levels are converted to the maximum theoretical leachate concentrations that could possibly result from performing the TCLP. This is accomplished by dividing each total constituent concentration by 20 (reflecting the 20 to 1 weight ratio of extraction fluid to solid in the TCLP) and then comparing the resulting maximum theoretical leachate concentration to the applicable regulatory level. "If no maximum theoretical leachate concentration equals or exceeds the appropriate regulatory limit, the solid cannot exhibit the toxicity characteristic and the TCLP need not be run." [RO 13647]

For example, a facility wants to know if chromium-containing soil that is going to be excavated and sent offsite is hazardous for toxicity. A representative sample of the soil is sent to a lab, but instead of asking the lab to perform a TCLP for metals (which would cost a couple hundred dollars), the facility asks the lab to run a total chromium analysis (which costs about $30). The lab would probably use either atomic absorption (AA) or inductively-coupled plasma (ICP) methods to determine total chromium concentration in the soil, which is measured at 50 mg/kg. Dividing that concentration by 20 gives a maximum theoretical leachate concentration for chromium in the soil of 2.5 mg/L. (That is, if 100% of the total chromium leaches via the TCLP, it would have been measured in the extract at 2.5 mg/L.) The facility can conclude that the soil cannot exhibit the TC for chromium (chromium's regulatory level is 5.0 mg/L), because even if all of the chromium leaches during a TCLP analysis, not enough chromium is present to make the soil hazardous.

What if the total chromium concentration measured by the lab was 140 mg/kg? Again dividing by 20 gives a new maximum theoretical leachate concentration of 7.0 mg/L. Since this value is greater than the regulatory level, does the facility conclude that all of the soil is hazardous? No; the facility concludes that now it is worth spending the money to have the lab run the TCLP. [RO 14533] Our experience indicates that significant percentages of chromium and other heavy metals in soil will not leach in the TCLP.

This is why some generators send samples of wastes to a lab with the following instructions: Run a total analysis first. If the resulting total concentrations are less than 20 times the TC regulatory levels, stop. Conversely, if total concentrations are greater than or equal to 20 times the TC regulatory levels, run the TCLP. Sometimes people ask: "If total concentrations divided by 20 are greater than or equal to TC regulatory levels for a 100% physical solid, why would you ask the lab to run the TCLP—couldn't you just use the total concentration results and classify this waste as hazardous?" You could do that, but it would be very conservative. For example, depending on the contaminant and waste matrix, it is often the case that less than 20% of the total concentration of a contaminant will leach out during the TCLP test. The "Rule of Twenty" can be used to prove that a material is not a TC hazardous waste, or it can also be used as a screen to determine when the TCLP needs to be run. Either way, it is a useful tool for generators, enabling them to save time and money when making TC determinations.

- *Dual-phase wastes*—The generator of a dual-phase waste (a waste that has both a solid and a filterable-liquid component) can perform a total waste analysis on both the solid and liquid portions and calculate maximum theoretical leachate concentrations for the waste as a whole. This is accomplished by combining results mathematically through use of the following formula:

$$M = \frac{(A \times B) + (C \times D)}{B + (20\ L/kg \times D)}$$

where:

- A = Total concentration of the analyte in the liquid portion of the sample (mg/L),
- B = Volume of the liquid portion of the sample (L),
- C = Total concentration of the analyte in the solid portion of the sample (mg/kg),
- D = Weight of the solid portion of the sample (kg), and
- M = Maximum theoretical leachate concentration (mg/L).

An example of the use of this equation is given in Case Study 2-1.

This method may also be used for nonaqueous wastes (e.g., oily wastes). If this approach is used for such wastes, the concentration of constituents in the liquid portion of the waste (A) may be expressed in mg/kg instead of mg/L; if that is true, the liquid volume (B) would have to be converted to kg, and the final leachate concentration would be expressed in mg/kg. [RO 11721]

2.6.4 Exemptions from the TC

Even though a waste fails the TCLP, it may not be subject to hazardous waste regulation. Examples of some of the most significant exclusions and exemptions that may apply to wastes exhibiting the TC are summarized below.

Case Study 2-1: Total Waste Analyses for Toxicity Determination

A generator of a dual-phase waste wants to use total waste analysis data to determine if the waste is hazardous for lead. Analysis shows the waste contains lead at a total concentration of 0.023 mg/L in the liquid phase and 85 mg/kg in the solid phase. The liquid-phase volume of the sample is 0.025 L, and the weight of the solid-phase portion of the sample is 0.075 kg. Does the waste exhibit the TC for lead?

The generator can calculate the waste's maximum theoretical leachate concentration using EPA's recommended formula as follows:

$$\frac{(0.023\ mg/L \times 0.025\ L) + (85\ mg/kg \times 0.075\ kg)}{0.025\ L + (20\ L/kg \times 0.075\ kg)}$$

$$= 4.18\ mg/L$$

Since the 4.18-mg/L maximum theoretical leachate concentration is below the regulatory threshold for lead of 5.0 mg/L, the generator definitively concludes that the waste cannot exhibit the TC for lead. [RO 11721, 13647]

For all types of wastes, if the maximum theoretical leachate concentration is less than the §261.24 applicable regulatory threshold, the waste does not exhibit the TC and the TCLP does not need to be performed. "If, on the other hand, total waste analysis data yield a maximum theoretical leachate concentration that equals or exceeds the toxicity characteristic threshold, the data cannot be used to conclusively demonstrate that the waste does not exhibit the toxicity characteristic. The generator may have to conduct further testing to make a definitive toxicity characteristic determination." [RO 13647]

- *Lead-based paint waste from residences*—Residential lead-based paint waste is defined as waste "generated as a result of activities such as abatement, rehabilitation, renovation and remodeling in homes and other residences. The term residential lead-based paint waste includes, but is not limited to, lead-based paint debris, chips, dust, and sludges." [40 *CFR* 258.2] These wastes may be disposed in municipal solid waste landfills and construction and demolition waste landfills under the household waste exclusion of §261.4(b)(1). The exclusion is available to both contractors and do-it-yourselfers that generate and dispose lead-based paint wastes. [October 23, 2001; 66 *FR* 53537, RO 14459, 14673]

- *Trivalent chromium wastes*—As mentioned in Section 2.6.1.2, for purposes of determining toxicity of wastes containing chromium (D007), the TC level is based on total chromium. However, EPA acknowledged there is a significant difference in the relative hazard of the two major valences of chromium, trivalent chromium and hexavalent chromium. Trivalent chromium is much less toxic than hexavalent chromium; but, trivalent chromium can be oxidized to the more toxic hexavalent form. With this in mind, EPA developed a conditional exclusion in §261.4(b)(6) for wastes that are exclusively (or nearly exclusively) hazardous due to trivalent chromium only. The exclusion is for wastes that are characteristic for chromium only, or wastes that are listed due to the presence of chromium only. [October 30, 1980; 45 *FR* 72035] The exclusion applies to wastes from the leather tanning and finishing industry and to wastewater treatment sludge from a titanium dioxide (TiO$_2$) pigment manufacturing process. These wastes are generated in processes that exclusively (or nearly exclusively) use trivalent chromium. The wastes are routinely managed in non-oxidizing environments (e.g., landfills, deep wells) which preclude the formation of the more toxic hexavalent chromium in the waste. Only the wastes identified in §261.4(b)(6)(ii)(A–H) are clearly eligible for this exclusion and the provisions of §261.4(b)(6)(i)(A–C) must be satisfied. [RO 11319, 14655] However, EPA guidance in RO 14733 indicates the exclusion may apply to other wastes that aren't specifically identified in §261.4(b)(6)(ii)]. See Case Study 2-2.

- *Petroleum-contaminated media and debris*—Media (soil and ground water) and debris resulting from the cleanup of petroleum underground storage tanks (USTs) are excluded from the definition of hazardous waste, even if they exhibit the TC. [§261.4(b)(10)] This exclusion applies if the media and debris 1) exhibit the TC for D018–D043 only, and 2) are subject to the corrective action requirements in 40 *CFR* Part 280 of the UST regulations. The exclusion was promulgated to prevent the large quantities of benzene-contaminated soil at old gas stations from overwhelming the commercial hazardous waste management capacity in the United States. The exclusion does not apply if the media or debris exhibit any characteristic other than D018–D043 or have become contaminated with a listed hazardous waste. [RO 11569]

- *Used chlorofluorocarbon refrigerants*—When chlorofluorocarbons (CFCs or Freon) are removed from refrigeration or air conditioning systems, they are usually considered spent materials. Such materials are solid wastes when they are reclaimed [see Table 1 in §261.2(c)], unless an exclusion under EPA's October 30, 2008 definition of solid waste rule is applicable. [73 *FR* 64668] Spent CFCs can exhibit the TC due to the small amounts of carbon tetrachloride (D019) or chloroform (D022) contaminants often found in them. Since spent materials that exhibit the TC may require hazardous waste management even if they are reclaimed, EPA was concerned that requiring RCRA management (e.g., manifesting, etc.) would provide a disincentive for facilities to recycle the CFCs—instead, they would just vent them to the atmosphere. To remove such a disincentive, an exclusion in §261.4(b)(12) allows nonhazardous management of these materials, but it applies only if the CFCs will be reclaimed for reuse. This exclusion also applies to hydrochlorofluorocarbons

Case Study 2-2: Toxicity Characteristic Determination for Leather Shoes and Gloves

New leather shoes and gloves may exhibit the TC for chromium, as a result of the tanning process. Thus, when such articles are to be thrown away after use, they often still exhibit this characteristic. Would such wastes be considered hazardous, requiring Subtitle C management?

There is an exemption from the definition of hazardous waste for seven specific types of leather tanning and finishing wastes at §261.4(b)(6)(ii). Per §261.4(b)(6)(i), this exemption was promulgated for wastes that 1) contain exclusively trivalent chromium, 2) are generated in processes that use only this form of chromium, and 3) are not typically managed in oxidizing environments. Discarded leather shoes and gloves are not among the wastes specified in §261.4(b)(6)(ii), however, and so they apparently have no regulatory exemption.

A generator recently asked its state regulators if leather shoes and gloves, when discarded after use in an industrial setting, would be considered hazardous waste. The facility argued that, since these materials were characteristic only because of the chromium which they contained before use, they were similar to the wastes specified in §261.4(b)(6)(ii) and thus should be exempt.

The state issued a determination that, even though discarded leather shoes and gloves are not specifically included in §261.4(b)(6)(ii), they are not hazardous waste. The state's decision is based on the generator's 1) assertion that the shoes and gloves are not contaminated with any hazardous waste, and 2) compliance with the conditions of §261.4(b)(6)(i).

To cover all its bases, the generator subsequently asked for a similar determination from EPA headquarters. In RO 14733, the agency noted that it has not seen any data which show that discarded leather shoes and gloves exhibit the TC for chromium as a result of the tanning process. However, EPA supported the state determination. In the same letter, EPA suggested that the generator contact the state agency in any other state where the generator has facilities to obtain their position on the issue.

We had previously thought that a generator would have to submit a rulemaking petition to EPA to receive an exemption for wastes that are not included in §261.4(b)(6)(ii). (See RO 11319, 14655, and October 30, 1980; 45 *FR* 72036.) However, the current guidance in RO 14733 indicates that a letter from the state agency is all that is needed. Keep in mind, though, that even if these wastes are not hazardous in one state, they may be considered hazardous waste if the shipment crosses state lines. An interpretation is needed from any state in which the discarded shoes and gloves (which exhibit the TC for chromium) will be managed or transported as nonhazardous.

For example, the states of Ohio and Washington have recently determined that such discarded shoes and gloves are not excluded per §261.4(b)(6). EPA Region 10 noted in an April 28, 2009 letter that waste leather shoes and gloves do not meet the §261.4(b)(6) criteria but that generators may petition EPA to modify §261.4(b)(6)(ii)(G) to include an exclusion for these waste leather products.

(HCFCs) but not to hydrofluorocarbons (HFCs). [February 13, 1991; 56 *FR* 5911–3, RO 14323] CFCs used as refrigerants are typically not subject to the F001–F005 spent solvent listings, because, as refrigerants, the CFCs are not used as solvents. [54 *FR* 31336]

- *Scrap metal that is recycled*—Scrap metal that is recycled is exempt from hazardous waste regulation at its point of generation. [§261.6(a)(3)(ii)] For example, scrap metal that exhibits the TC does not need to be managed as a hazardous waste before it reaches the reclamation facility (e.g., manifests are not required). However, if the scrap metal is *not* reclaimed (e.g., if it is disposed), the exemption does not apply, and the scrap metal "remains subject to all applicable hazardous waste regulations from the point of generation. Only if the facility ensures that the material will be reclaimed will the hazardous waste regulations not apply." [RO 14277] According to EPA, unprocessed, spent (i.e., used) printed circuit boards qualify for the scrap metal exemption. [RO 11689] A more complete discussion of the scrap metal exemption is in Section 4.21.

- *PCB dielectric fluids and the electrical equipment in which they are contained*—Polychlorinated biphenyls (PCBs) will not, by themselves, fail the TCLP. However, chlorobenzene, which is a degradation by-product contained within these materials, will sometimes cause them to exhibit the TC as D021. Thus, such materials, when they become wastes, would be regulated as both a toxic material under the Toxic Substances Control Act (TSCA) and a hazardous waste under RCRA. To avoid such dual regulation of PCB wastes, EPA established an exemption at §261.8 from RCRA regulation for PCB fluids themselves and the electrical equipment in which they are contained. The exemption is based on compliance with two criteria: 1) the waste must be regulated under TSCA (40 *CFR* Part 761) in order to escape RCRA, and 2) the waste must be hazardous only because it exhibits the TC for D018–D043. If the waste exhibits any other characteristic or contains a listed hazardous waste, it would be regulated under both TSCA and RCRA. [RO 13324, 14014]

Examples that illustrate how some of the above exemptions apply are given below.

Q A facility generates a chromium-containing waste that fails the TCLP for this metal. However, analysis of the waste indicates that the chromium is essentially all trivalent in form. The waste does not fail the TCLP for any other constituent and does not exhibit any other characteristic. The waste is not one of the eight specific trivalent chromium wastes listed in §261.4(b)(6)(ii). Can the generator take advantage of the §261.4(b)(6) exemption in the regulations and manage the waste as nonhazardous?

A No. A facility that generates a waste that meets the criteria in §261.4(b)(6)(i) but that is not one of the eight wastes specifically excluded in §261.4(b)(6)(ii) cannot take advantage of the exclusion without first submitting a rulemaking petition to EPA or the states involved. If the regulatory agency agrees with the petition, it will amend the federal or state regulations or otherwise grant a variance to exclude the facility's waste. [October 30, 1980; 45 *FR* 72035, RO 11319, 14477, 14655] See Case Study 2-2.

Q During the closure and corrective action of a petroleum UST, sludge is removed from the UST that exhibits the TC for benzene (D018). Is the sludge excluded from regulation as a hazardous waste under §261.4(b)(10)? What about sand that was added during closure and now exhibits the TC?

A Neither the sludge nor the sand qualify for the §261.4(b)(10) exemption. According to EPA, the exemption "was intended to cover contaminated debris and media such as soil, ground water, surface water, and air that have become contaminated with petroleum substances as a result of a release from an UST." It does not apply to wastes (such as sludges) generated in an UST or to inert materials (such as sand) that have been introduced into an UST for the purposes of closure. Consequently, in these situations, both the sludge and the sand would have to be managed as hazardous waste if they exhibit the TC. [RO 13409, 14316] In further guidance, EPA noted that media (soil or ground water) that migrated into the UST and became contaminated from the petroleum within the UST would also not qualify for the exemption. [RO 14738]

Q During the servicing of air conditioners, a facility generates spent CFC-11, which exhibits the TC for carbon tetrachloride (D019). The used refrigerant is reclaimed for future reuse and therefore is excluded from the definition of hazardous waste under §261.4(b)(12). During the reclamation process, contaminated filters that also exhibit the TC for carbon tetrachloride are generated. Are the spent filters also *excluded from RCRA regulation under §261.4(b)(12) when they are discarded because they are generated from the reclamation of an excluded waste?*

A No. The §261.4(b)(12) exclusion applies only to used CFC refrigerants that are reclaimed for further use. Wastes derived from the refrigerant reclamation process itself (e.g., the spent filters) are not excluded and, therefore, must be managed as hazardous waste if they exhibit a hazardous waste characteristic. [RO 13560]

Q *Can natural gas regulators that contain mercury (and therefore would exhibit the TC for mercury) qualify for the scrap metal exemption?*

A No. EPA has determined that the scrap metal exemption does not apply in this situation. The agency noted in January 4, 1985 preamble language that liquid metal wastes (e.g., liquid mercury) could not be included in the term "scrap metal." [50 *FR* 624] The agency expanded this concept in 1994 guidance: "In general, any quantity of liquid mercury other than trace amounts attached to or contained in a spent material precludes that material from being a scrap metal." Only when the mercury (and any other liquid) is removed from the regulator can it be considered scrap metal. [RO 11860] On August 5, 2005 [70 *FR* 45508], EPA designated mercury-containing equipment as universal waste subject to the Part 273 provisions. This category of universal waste includes regulators that contain elemental mercury integral to their function. [§273.9] Thus, generators of natural gas regulators that contain elemental mercury integral to their function could manage them as universal waste, if not scrap metal. See Sections 8.1 to 8.3.

Q *Steel beams are coated with lead-based paint in sufficient quantity to make them toxic for lead. The facility wants to send the beams to a scrap metal reclamation facility. How should this material be managed?*

A Even though the scrap metal exhibits the TC, as long as it will be recycled and meets the scrap metal recycler's acceptance criteria, it is exempt from RCRA regulation per §261.6(a)(3)(ii). [RO 11769]

2.6.5 Other TC examples

The following examples provide additional EPA insight into application of the TC.

Q *Lead containers or container liners are sometimes used as shielding when low-level radioactive wastes are landfilled. A representative sample of the container plus contents would likely fail the TCLP. Therefore, would these containers have to be considered mixed waste when land disposed?*

A No. The lead containers or container liners are not regulated as hazardous waste because they are products fulfilling their intended purpose. Assuming that the container contents do not meet the definition of a hazardous waste, such containers of low-level radioactive waste could be disposed in a radioactive waste landfill without concern for the lead in the container. While the lead shielding is not a solid waste, EPA recommends that the outside be macroencapsulated (e.g., with a polymer coating) to prevent the shielding from leaching lead into the environment. [RO 12956, 13468, 13538]

Conversely, if the radioactive contents are removed from the container prior to disposal, and the empty lead containers or container liners are now to be landfilled, they would be solid wastes (no longer products) being discarded. Solid wastes that exhibit the TC are hazardous when disposed.

Q *If lead-lined pipe or lead-shielded phone cable is abandoned in place, are any hazardous waste requirements triggered?*

A Yes. Abandonment constitutes disposal of a solid waste. If hazardous for lead (D008), the material would have to meet land disposal restrictions. [RO 13468]

Q *Is used antifreeze removed from radiators a hazardous waste?*

Used antifreeze is typically not ignitable, corrosive, or reactive. Ethylene glycol, the primary ingredient in antifreeze, is not one of the 40 TC constituents; therefore, it will normally be hazardous only due to lead or other heavy metal content (picked up from the radiator). According to one reference, up to 40% of antifreeze drained from radiators can exhibit the TC for lead. [RO 11554, 14003] Each generator should conduct TCLP testing on representative samples of its used antifreeze to make a facility-specific determination of whether its used antifreeze is toxic or not. [RO 13521] If it is, it should be managed in a satellite accumulation or 90/180-day container prior to hazardous waste disposal or recycling. If the drained coolant does not exhibit any hazardous waste characteristic, however, it may be recycled to, or disposed in, a nonhazardous waste facility.

We sometimes hear of used antifreeze that, upon TCLP testing, contains high levels of arsenic and/or selenium; however, knowledge indicates that neither arsenic nor selenium are present in the materials of construction of the radiator and neither are contained in any process fluids which could leak into the antifreeze. An EPA-funded study found that arsenic and/or selenium may appear as false positives during the analytical process. Although the study was not able to definitively determine the cause of the analytical discrepancy, it did indicate that false positives were not observed when using Methods 6020 or 7060, instead of the more common Method 6010, to test for heavy metals in TCLP leachates from used antifreeze. Note that Method 7060 was deleted from SW–846 in 2007. A complete discussion of these results is contained in *Waste Analyses Project for Auto Dealerships—Waste Antifreeze Summary*, September 2006, available online at http://www.understandrcra.com/rccd/WasteAntifreeze.pdf.

2.6.6 Masking of lead-bearing wastes prior to disposal is prohibited

It was common practice in the 1990s for some industries to add iron fines, filings, or dust to wastes characteristically hazardous for lead. This practice occurred frequently at foundries that would mix iron filings with D008 foundry sand. The iron filings interfered with the TCLP, resulting in treated sand that appeared to meet LDR treatment standards and therefore could be land disposed. Based on EPA's research into this practice, the agency concluded that the "masking" effect is only temporary:

> "[T]he addition of iron metal is not a permanent treatment because the iron inevitably oxidizes and loses its adsorptivity for soluble lead ions. After oxidation of the iron surfaces, surface adsorption of lead ions ceases and the lead-bearing waste returns to its original state; all pretext of treatment is lost. Since iron addition is not effective, it cannot be allowed for lead-containing hazardous wastes that are to be land disposed, regardless of their origin (i.e., all lead-bearing wastes, not just foundry sands). The agency concludes that addition of iron metal, in the form of fines, filings, or dust, fails to provide long-term treatment for lead-containing hazardous wastes. EPA is codifying this determination by calling this practice impermissible dilution, and so invalidating it as a means of treating lead in lead-containing hazardous wastes." [May 26, 1998; 63 *FR* 28568]

The dilution prohibition is codified in §268.3(d).

2.6.6.1 Lead-based paint waste

Some contractors involved in removal of lead-based paint (LBP) use iron-containing additives in the abrasives used to blast off the paint. Other products are applied to the paint, causing it to separate from the equipment or structure for subsequent removal. These practices have the same effect as discussed above—the removed LBP waste (which includes the abrasive or paint remover) typically passes the TCLP because of the masking effect of the iron. When the dilution prohibition noted in Section 2.6.6 was approaching promulgation, these entities asked EPA if the new regulation would prohibit this practice. EPA responded that "the dilution prohibition does not apply to processes which generate a waste, only to processes that treat a waste which already has been generated." [63 *FR* 28568]

Based on this position, the contractors' practice discussed above continues to be acceptable. What is not allowed under §268.3(d) is the addition of iron filings or shot to LBP waste that has already been generated. EPA defines the point of generation for such wastes as "once the paint has been removed from the surface of the structure." [RO 14069]

Although the use of iron-containing abrasives for removing LBP is not prohibited, EPA noted that, because the iron in the abrasive only temporarily prevents lead from leaching from the waste, two factors may come back to haunt the generators of such waste:

1. If the LBP waste passes the TCLP initially but then fails later prior to disposal, the waste is hazardous, subject to all applicable Subtitle C regulations.
2. CERCLA liability is independent of any previous hazardous waste determination. LBP waste that passes the TCLP, is disposed in a Subtitle D landfill, and then subsequently causes environmental damage from leachable lead will subject responsible parties to CERCLA liability. [63 *FR* 28568, RO 11624, 14069]

2.6.7 Delisting petitions

EPA requires the use of the TCLP for all testing used to support delisting petitions submitted under §§260.20 and 260.22. For all delisting demonstrations, the agency requires that the TCLP be used to predict the leaching potential of any inorganic and organic constituents listed in Appendix VIII to Part 261. As such, TCLP extracts should be analyzed for any inorganic or organic constituents that may be present in the waste. Analysis of total constituent concentrations of metals, cyanides, sulfides, and any organic constituents will also be required.

The leaching step of the TCLP is not performed on liquid wastes (i.e., those with less than 0.5% filterable solids). As a result, the TCLP will not be required for delisting demonstrations pertaining to liquid wastes. Instead, total constituent data will be used to determine if a liquid waste should remain hazardous.

As mentioned in Section 2.6.3, the TCLP is often not appropriate for evaluating oily wastes. Consequently, for wastes that contain more than 1% total oil and grease and that are difficult to filter using the TCLP (e.g., tars), EPA requires the use of the OWEP in place of the TCLP to determine the leaching potential of inorganic constituents. [RO 11522, 12450]

2.7 Electronic wastes (e-wastes)

Electronic products (computers, printers, cell phones, tablets, pagers, two-way radios, televisions, H_2S monitors, etc.) are integral to industry operations. The rapid development of new technology has brought with it huge amounts of obsolete electronics destined for storage, disposal, or recycling, which are typically know as e-wastes.

The acceleration of e-waste generation has caused several states to begin looking at ways to recycle and reuse the equipment and keep it out of landfills. Many states have enacted some form of e-waste management law. Although the goal of each law is similar—to avoid landfill disposal or incineration of certain types of e-waste—approaches taken to achieve that goal differ significantly.

At the federal level, EPA is trying to develop national policies/regulations to avoid patchwork implementation by the states. To that end, the agency has been working with stakeholders to improve awareness of the need for recovery of electronics and access to safe reuse and recycling options. In 2011, EPA released a study of e-waste generation and management patterns: *Electronics Waste Management in the United States Through 2009*,

EPA/530/R-11/002, May 2011, available at http://www.epa.gov/osw/conserve/materials/ecycling/docs/fullbaselinereport2011.pdf.

The report notes the following for U.S. electronic products management in 2009:

- An estimated 438 million electronic products were sold.

- About 5 million tons of used electronic products were in storage. Residential households store 5 times more used computer products than commercial establishments.

- Approximately 2.4 million tons of used electronics were sent for disposal or recycling. Cathode ray tube (CRT) TVs and CRT monitors comprised nearly half, by weight, of the electronics that entered this waste stream. About 141 million mobile devices, including cell phones, smart phones, personal digital assistants, and pagers, were sent for disposal or recycling, more than any other type of product included in the analysis; yet, they comprise less than 1% of waste electronics by weight.

- About 25 percent of the above 2.4 million tons of used electronics were sent for recycling.

2.7.1 E-wastes often exhibit the toxicity characteristic

Lead, mercury, and cadmium are among the substances of concern in electronics when they become wastes. These substances are included in products for important performance characteristics but can cause resulting e-wastes to exhibit the toxicity characteristic. Lead was used in glass in television and computer CRTs as well as solder and interconnects; thus, many CRTs leach lead at higher than RCRA regulatory levels. [RO 14723] Mercury is used in small amounts in bulbs in flat panel computer monitors and notebook computers. Cadmium has been widely used in Ni-Cd rechargeable batteries for laptops and other portables. Newer batteries (nickel-metal hydride and lithium ion) do not contain cadmium.

Although traditionally thought to be nonhazardous [RO 14723], computer components other than the monitor may also fail the TCLP for lead, resulting in wastes exhibiting the characteristic of toxicity. Generators must acknowledge this potential hazardous characteristic when their equipment becomes a solid waste. In one study, 1 of 23 computer central processing units (CPUs) failed the standard TCLP, although 21 of 41 (51%) exceeded the regulatory value for lead when tested in a modified large-scale version of the TCLP (where the units were disassembled and leached in their entirety). In the same report, 6 out of 6 laptop computers, 0 out of 3 computer keyboards, 15 out of 15 computer mice, and 33 out of 43 cell phones failed the standard TCLP for lead. [T.G. Townsend et al., "RCRA Toxicity Characterization of Computer CPUs and Other Discarded Electronic Devices," July 2004, available at http://www.epa.gov/region5/waste/solidwaste/ecycling/pdfs/uf_ewaste_final.pdf]

2.7.2 Management of e-wastes

Electronic devices, including certain computer equipment, telephones, radios, calculators, etc. continuing to be used for their intended purpose are products—not wastes. Materials used and taken out of service by one person are not considered wastes if a second person reuses them in the same manner without reclaiming them. [§261.2(e)(1)(ii)] Many businesses that take electronic devices out of service do not know if they can be reused or have to be sent for reclamation/disposal. Therefore, such entities sometimes send them (under a contract) to a reseller who evaluates the equipment, may make minor repairs to it, and ultimately either sells it for reuse or sends it for reclamation/disposal.

EPA has clarified that a user sending electronic equipment to a reseller for potential reuse is not a RCRA generator. Electronics undergoing repairs before resale or distribution are not being reclaimed and are considered to be products "in use" rather than solid wastes. Therefore, used electronics from a business are not considered solid wastes when sent to resellers and would not be subject to

RCRA requirements. [July 28, 2006; 71 *FR* 42930, June 12, 2002; 67 *FR* 40511, RO 14668] If the reseller decides after evaluation to send some of the used devices for reclamation/disposal, the reseller is the generator of a solid waste and must make a hazardous waste determination for that material.

In some situations, used electronic equipment cannot be reused and must be reclaimed or disposed. A number of states (e.g., California, Colorado) have added certain e-wastes to their universal waste program, which eases the regulatory burden on facilities sending such equipment for reclamation or disposal.

EPA's guidance on the regulatory status of used electronic devices destined for reclamation is as follows:

> "While used electronics sent to a reseller are not solid wastes, used electronics sent to a recycler could, under certain circumstances, be considered spent materials undergoing reclamation and could therefore be solid wastes. However, EPA believes that in some instances, electronics sent for recycling do not resemble spent materials. To determine how electronics must be managed in particular situations, users and recyclers of electronics should check with their implementing agencies to see which, if any, RCRA Subtitle C requirements apply when used electronics are sent toward specific recycling pathways." [RO 14668]

Any unused (usually off-specification) electronics that are sent for reclamation are not solid wastes per Table 1 in §261.2(c), because they are considered commercial products since they haven't been used. [RO 11689] Table 1 also indicates that such unused electronic devices are not subject to speculative accumulation provisions when recycled. Of course, unused devices sent for disposal are solid (and potentially hazardous) wastes.

Finally, used electronics from households or conditionally exempt small quantity generators are exempt from regulation under RCRA per §§261.4(b)(1) and 261.5, respectively.

In addition to the above general guidance, EPA has provided rules and/or guidance for the management of specific e-wastes: cathode ray tubes (CRTs) and printed circuit boards. The management of these two wastes is discussed separately in Sections 2.7.3 and 2.7.4, respectively.

2.7.3 Cathode ray tubes (CRTs)

Cathode ray tubes (CRTs) are a major component of older-style computer monitors and television screens. These units contain glass, which may contain lead to protect the user from x-rays present inside the CRT. Older CRTs typically contain an average of 4 pounds of lead, while newer CRTs contain closer to 2 pounds.

Research has been conducted at the Florida Center for Solid and Hazardous Waste Management (Gainesville, Florida) to determine if color and monochrome CRTs fail the TCLP for lead. The results indicate that color CRTs typically exceed the 5-mg/L TCLP limit for lead and consequently may be considered hazardous wastes when recycled or discarded. This has stymied CRT recycling efforts and has caused a large number of CRTs to be put into indefinite and indeterminate storage.

Of the 30 color CRTs tested, 21 (70%) exceeded the TCLP for lead. (The 9 color CRT samples that did not fail the TCLP limit were not believed to be representative samples due to an absence of leaded frit.) None of the six monochrome CRTs that were tested failed the TCLP. (Color CRTs typically use a leaded frit seal between the funnel and face; no such seal is present in monochrome CRTs. In addition, color CRTs commonly contain more lead in the funnel section.) [T.G. Townsend et al., "Characterization of Lead Leachability From Cathode Ray Tubes Using the Toxicity Characteristic Leaching Procedure," December 1999; see also RO 14723]

As a result of fast-breaking technological advances in electronic equipment (e.g., increasing computer speed and memory, high-definition television, and flat-panel computer monitors), the number of CRTs that are projected to be discarded in the future is quite high. Because so many CRTs are projected to be discarded in the future, EPA and a number of states have investigated ways to encourage recycling of these wastes instead of land disposal.

2.7.3.1 Current regulatory status of CRTs

The RCRA regs define a CRT as follows: "a vacuum tube, composed primarily of glass, which is the visual or video display component of an electronic device." [§260.10] Based on a final rule issued July 28, 2006 [71 *FR* 42928], EPA regulates CRTs as follows:

- Used CRTs that continue to be used as a computer or television, as is or after minor repairs, are not solid wastes. One option for getting rid of used CRTs that you don't want is to give them away to schools or other entities; such CRTs are not wastes if they continue to be used for their intended purpose.

- Used, intact CRTs (defined as CRTs whose vacuum has not been released) are excluded from the definition of solid waste if they are sent for recycling within the United States—including glass processing, glass manufacturing, or smelting—in lieu of disposal. [§261.4(a)(22)(i)] Such units are not subject to any RCRA requirements at the generator facility. "[W]e are not imposing speculative accumulation requirements on persons who use computers or televisions and then send the intact CRTs to collectors and glass processors." [71 *FR* 42932] Thus, they could be held indefinitely at the generator's facility without becoming solid waste, provided there is a reasonable expectation they will be recycled at some point. However, CRT collectors (defined as persons who receive used, intact CRTs for recycling, repair, resale, or donation) and CRT glass processing facilities (defined as persons receiving intact or broken CRTs, breaking intact CRTs and/or further processing broken CRTs, and sorting or otherwise managing glass removed from CRTs) are subject to the speculative accumulation provisions in §261.1(c)(8).

- Used, broken CRTs (defined as glass removed from its housing or casing whose vacuum has been released) are conditionally excluded from the definition of solid waste if they 1) are recycled within the United States, and 2) meet certain conditions. [§§261.4(a)(22)(iii), 261.39] The conditions include storage prior to recycling in a labeled, closed container or a building with roof, floor, and walls; compliance with speculative accumulation provisions; compliance with use constituting disposal requirements in Part 266, Subpart C; transportation in a labeled, closed container; and processing conducted in a building with roof, floor, and walls and at temperatures below that capable of volatilizing lead from the CRTs.

- Used CRTs sent for disposal are clearly solid wastes and hazardous wastes if they exhibit a characteristic.

- Unused CRTs would be considered off-spec commercial chemical products if they were found to be inoperable after manufacture. If these unused CRTs are sent for reclamation/recycling, they are not solid wastes per Table 1 in §261.2(c). They also are not subject to speculative accumulation provisions. Therefore, as long as the unused CRTs or glass from the unused CRTs are kept segregated from used CRTs or glass from used CRTs, they would not be regulated under RCRA. If these materials are mixed with regulated materials, the combined materials would be subject to the applicable provisions of the July 28, 2006 CRT rule. [RO 14757]

- Unused CRTs sent for disposal are solid wastes and hazardous wastes if they exhibit a characteristic.

- Used CRTs from households are exempt from regulation under RCRA.

- CRT glass cullet (that contains lead) used as a raw material in the video display manufacturing process is a product (commodity) and not a waste. [RO 14734]

Another option for managing used computer monitors is to send them to a reseller as noted in Section 2.7.2.

2.7.3.1.1 Processed CRT glass sent to smelters

To further encourage the recycling of CRTs, when EPA added the CRT recycling exclusion in July 2006, the agency added an exclusion from the definition of solid waste for processed CRT glass sent to lead smelters for recycling. [§261.39(c)]

In order to understand this exclusion better, let's take a closer look at CRT recycling. When CRTs are recycled, they contain two types of glass. The first type, panel glass, is basically the screen at which we are used to looking. The second type, funnel glass, is the back side of the CRT which is usually underneath a plastic housing (i.e., the back of the unit). In addition, processed glass (i.e., glass chunks) is generally called "cullet." Thus, the terms "panel cullet" and "funnel cullet" are commonly used in CRT recycling to refer to processed panel glass and processed funnel glass.

Because funnel cullet generally contains a lot more lead than panel cullet, it is more difficult to recycle. However, it makes a good substitute for virgin silica that is used as a fluxing agent in lead smelters. Thus, funnel cullet is the beneficiary of the §261.39(c) exclusion. Be aware that the exclusion is lost if the processed CRT glass is speculatively accumulated.

Q: *Will the §261.39(c) exclusion apply if processed CRT glass is used as a fluxing agent for copper smelting?*

A: No. While the degree of processing that is required for use in copper smelting appears to be the same as that required for lead smelting, lead is not recovered when CRT glass is used for flux at a copper smelter. For this reason, EPA did not include copper smelters in the §261.39(c) exclusion.

Q: *Emerging technologies have made it possible for CRT recyclers to process CRT glass into lead and silica sand. This has the potential to expand CRT glass recycling because there is a greater market for these individual components compared to the market for leaded CRT glass. Does the §261.39(c) exclusion apply if CRT glass is processed into the primary components of lead and silica sand, which are then sold for a variety of commercial uses?*

A: No. While EPA supports the development of technologies that can recycle CRT glass in an environmentally protective fashion, the agency noted that the exclusion "only applies to processed CRT glass sent for recycling to a CRT glass manufacturer or a lead smelter." [RO 14839] This is not to say that CRT glass cannot be recycled using the new technologies—only that the exclusion would not apply, and the CRT glass would be a solid and potentially hazardous waste during these activities.

However, the agency has noted in recent guidance [RO 14835] that processed CRT glass used, without reclamation, as an effective substitute for virgin fluxing agent at copper smelters would be excluded from RCRA regulation as an effective substitute for a commercial product under §261.2(e)(1)(ii). Furthermore, EPA noted that this interpretation is in line with a similar finding for foundry sands used as a fluxing agent at copper smelters. [RO 11900]

2.7.3.2 Exporting CRTs

Export provisions were also included in the July 28, 2006 rule to address concerns about CRTs that are exported and recycled under unsafe and/or inappropriate conditions in other countries. CRTs that are exported for recycling are not solid wastes, but only if they meet the following requirements:

- Used, intact CRTs exported for reuse (i.e., not for processing or other reclamation) are not solid wastes if the exporter sends a one-time notice to the applicable EPA regional administrator and keeps specific documentation. [§261.41]

- Used, intact CRTs exported for recycling (i.e., for processing or other reclamation) are not solid wastes if the exporter notifies EPA of the export, receives written consent from the receiving country, and complies with speculative accumulation provisions. [§§261.4(a)(22)(ii), 261.40]

- Used, broken CRTs exported for recycling (i.e., for processing or other reclamation) are not solid wastes if the exporter notifies EPA of the export; receives written consent from the receiving country; complies with speculative accumulation provisions; and meets the same storage, labeling, and transportation requirements noted in Section 2.7.3.1 for used, broken CRTs. [§§261.4(a)(22)(iii), 261.39(a)(5)]

- Processed CRT glass being exported to a CRT glass manufacturer or a lead smelter is not a solid waste and is not subject to any export requirements unless it is speculatively accumulated. [§261.39(c)] "Processed CRT glass" is defined as glass that has been broken, separated, and sorted, or otherwise managed after it has been removed from CRT monitors. To meet this definition, the glass does not have to be cleaned; that is, the coatings do not have to be removed. Under these conditions, EPA considers the processed CRT glass to be a commodity rather than a waste. [RO 14805]

2.7.4 Circuit boards in electronic components

Printed circuit boards are a major component in computers and other electronic products. Sometimes these components are removed from electronic devices for recycling/reclamation. EPA has given guidance specific to these materials. In a nutshell:

- Whole, used circuit boards meet the definition of scrap metal and so are exempt from RCRA under §261.6(a)(3)(ii) if recycled [May 12, 1997; 62 *FR* 26013, RO 11689, 14155];

- Whole, unused circuit boards are off-spec commercial products and so are not solid wastes if sent for recycling/reclamation [§261.2(c), Table 1, RO 11689]; and

- Shredded used or unused circuit boards are not solid wastes if they are 1) free of mercury switches, mercury relays, nickel-cadmium batteries, and lithium batteries; 2) stored in containers prior to recovery; and 3) sent for recycling. [§261.4(a)(14), RO 14692]

Details on the above are contained in Section 4.21.5.

2.8 Hazardous waste lamps

Many fluorescent (including compact fluorescent), high-pressure sodium, mercury vapor, and metal halide lamps exhibit the toxicity characteristic for mercury. Some high-intensity discharge (HID) and incandescent lamps may contain lead solder, causing them to be characteristic for lead. When these lamps burn out, how should they be managed?

2.8.1 History

During the 1980s, no one was worried about burned-out light bulbs under RCRA, so they were routinely thrown in a nonhazardous waste dumpster. Shortly after the toxicity characteristic leaching procedure (TCLP) became the required test to determine the characteristic of toxicity (effective September 1990), a federal facility had a few spent fluorescent bulbs evaluated using the new procedure. When the results showed that the bulbs failed the TCLP for mercury, the regulated community and EPA became aware that burned-out bulbs, especially fluorescent light tubes, could be characteristic hazardous waste. [RO 11715]

What if the bulbs will be recycled rather than disposed? From a regulatory standpoint, burned-out bulbs most closely fall into the category of a spent material—something that was bought for use, was used, and can't be used again without reprocessing. [RO 14468] From Table 1 in §261.2(c), spent materials sent for reclamation are solid wastes, unless the exclusions under EPA's October 30, 2008 definition of solid waste rule are applicable. [73 *FR* 64668] So, even spent lamps (that fail the TCLP) sent for recycling may be hazardous waste at the federal level.

In the 1990s, some states developed creative ways to allow generators to circumvent having to manage their spent bulbs as hazardous waste. At least one state considered burned-out bulbs sent for reclamation to be characteristic by-products in lieu of spent materials. Table 1 in §261.2(c) thus exempted such bulbs from solid or hazardous waste management. Other states set a de minimis number of bulbs generated per month, below which a facility would not have to manage them as hazardous.

In most states, however, generators of spent lamps that exhibit a hazardous waste characteristic were required to manage them under the full RCRA Subtitle C hazardous waste regulations (Parts 260 through 272), unless they qualified as conditionally exempt small quantity generators (CESQGs—generators of less than or equal to 100 kg of nonacute hazardous waste in a calendar month). Spent lamps generated by consumers in their homes are not regulated under RCRA when discarded because they are excluded from the definition of hazardous waste under the household waste exclusion. [§261.4(b)(1)]

In order to encourage better management of spent hazardous waste lamps (i.e., more recycling and less illegal management as nonhazardous waste), EPA added them to the list of federal universal wastes on July 6, 1999, with an effective date of January 6, 2000. [64 *FR* 36466] The rule, which is codified in Part 273, allows handlers of hazardous waste lamps to comply with less-stringent standards for storing and transporting these wastes.

Therefore, generators of spent *hazardous* waste lamps today can choose to either comply with the universal waste standards for spent lamps (if their state has adopted them as universal wastes) or continue managing them as hazardous waste under the full Subtitle C regulatory program. Chapter 8 discusses how hazardous waste lamps are managed under the universal waste program of Part 273, while Section 2.8.2 describes how these lamps are managed under the conventional hazardous waste program.

2.8.2 Waste lamps managed as hazardous waste

Generators may handle hazardous waste lamps under the conventional Subtitle C standards of Parts 260–272. Some facilities have chosen this approach for a couple of reasons: 1) based on the relatively small number of hazardous bulbs generated, it is easier to manage them under the existing, smooth-running hazardous waste management system compared to setting up a parallel universal waste management system; and/or 2) facilities want to crush their bulbs.

2.8.2.1 Accumulation

If a generator chooses to manage its burned-out bulbs as hazardous waste, they must be accumulated in satellite accumulation areas, 90- or 180-day accumulation areas, or permitted areas. In any case, the containers in which the bulbs are accumulated must be closed at all times, except when adding or removing waste. Some facilities have received NOVs for having the flaps of the spent bulb boxes hanging open.

Additionally, satellite accumulation containers of burned-out bulbs must be marked with the words "Hazardous Waste," "Spent Bulbs," or something equivalent, while 90- or 180-day containers must have the words "Hazardous Waste" on them.

2.8.2.2 Crushing

Some generators have decided to manage their bulbs as hazardous so they can crush them. (However, some states allow crushing of bulbs under their universal waste program—see Section 8.3.1.2.) You can get approximately 500 4-ft fluorescent light tubes in a 55-gallon drum if you crush them, significantly reducing transportation costs. But, crushing hazardous lamps is a potential RCRA problem, because it meets the definition of hazardous waste treatment [64 *FR* 36477] and, therefore, potentially requires a permit. Two options are available for dealing with the treatment issue:

- Bulbs may be crushed in a 90- or 180-day tank or container without a permit (see Section 7.3). If crushing takes place in containers, §265.173(a) must be addressed (i.e., the container must be closed except when necessary to add or remove waste). This could be accomplished by crushing the bulbs simultaneously as they are added to a container.

- If the bulbs are to be recycled, on a case-by-case basis, the crushing step may be interpreted as part of the recycling process and therefore may not be subject to permitting per §261.6(c)(1). [RO 11759, 11906]

Another problem with crushing bulbs has nothing to do with RCRA but is more of an OSHA issue. That is, does crushing spent lamps expose plant personnel to airborne mercury emissions? A number of facilities that started out crushing bulbs have since stopped the practice because of personnel exposure concerns.

If you are crushing waste lamps or considering it for the future, you may want to read a study on the performance of drum-top crushers (DTCs) which was released by EPA in August 2006. [*Mercury Lamp Drum-Top Crusher Study*, EPA/530/R-06/002, available at http://www.epa.gov/epawaste/hazard/wastetypes/universal/drumtop/index.htm] The agency evaluated three DTCs (a fourth DTC was dropped midstudy due to high mercury emissions) to assess their ability to contain the mercury released from crushed lamps. While not endorsing or discouraging the use of DTCs, the EPA study concluded the following:

- Measurable concentrations of mercury were detected in the air, but all three of the devices usually maintained mercury levels below the OSHA permissible exposure level (PEL) within the structure and in the operator breathing zone.

- DTCs must be operated optimally to achieve low exposures. Emissions were higher with even minor mistakes in DTC assembly or operation. An external mercury monitor and alarm were usually needed to determine when mercury was not being contained.

- The risk of mercury exposure is increased when full drums are being replaced with empty ones and the full drums are being fitted with shipping lids.

- Performance of DTCs may change over the lifetime of the device.

EPA also developed a guidance document entitled *Fluorescent Lamps Recycling* (EPA/530/R-09/001, February 2009) that describes best management practices for DTCs. The document is available at http://www.epa.gov/osw/hazard/wastetypes/universal/lamps/lamp-recycling2-09.pdf.

2.8.2.3 Transportation

When bulbs that are managed as hazardous waste are shipped offsite, a manifest and appropriate LDR notification form must be used.

2.8.3 Making a hazardous waste determination for lamps

Lamps that are removed from service but are not spent (burned out) can be reused; such lamps, if reused, are not wastes at all.

How do you know if your burned-out bulbs are hazardous? Generators of spent lamps that are solid wastes (i.e., spent materials being reclaimed or disposed) are responsible for making a hazardous waste determination either by testing or knowledge. If testing is selected, the TCLP should be used.

In lieu of testing, it may be difficult to use generator knowledge in making a hazardous waste determination for spent lamps because the generator typically has little process knowledge on which to make a judgment. However, the generator could make a knowledge-based determination using data obtained from the manufacturer. A number of lighting manufacturers have developed low-mercury fluorescent light tubes that they claim will pass the TCLP. The information that these manufacturers provide about the TCLP results may be used for making a knowledge-based nonhazardous waste determination. Our experience is that most states will accept manufacturer's data as generators' knowledge-based evaluations, although some states require generators to test their lamps to determine that they are nonhazardous. EPA added the following final thought: "Under both federal and state laws, the ultimate responsibility for determining whether a lamp is hazardous lies with the waste generator, not the lamp manufacturer." [*Frequent Questions about Regulations that Affect the Management and Disposal of Mercury-Containing Light Bulbs (Lamps)*, available from http://www.epa.gov/epawaste/hazard/wastetypes/universal/lamps/faqs.htm]

Spent bulbs that the generator determines are not hazardous at their point of generation may be managed as nonhazardous waste in Subtitle D facilities. Making a hazardous waste determination for lamps is discussed in Case Study 2-3.

2.8.4 Fluorescent light ballasts

In addition to the hazardous constituents contained in the lamps themselves, fluorescent light fixtures may contain hazardous constituents in their ballasts—the small metal box-shaped devices that control the flow of electricity to the light tubes. Two constituents that may affect how such ballasts are managed when they become wastes include polychlorinated biphenyls (PCBs) and diethylhexyl phthalate (DEHP).

2.8.4.1 PCB-containing light ballasts

Most fluorescent light ballasts manufactured prior to 1979 contain PCBs. Ballasts manufactured after 1979 that do not contain PCBs are labeled "No PCBs." According to EPA, if the ballast is not labeled, the generator should 1) assume that it contains PCBs, or 2) contact the manufacturer to determine whether the ballast contains PCBs (if the manufacturer is unsure, assume the ballast contains PCBs).

PCB-containing light ballasts are regulated under the Toxic Substances Control Act (TSCA). Under the TSCA regulations, entities that are not PCB ballast manufacturers may dispose small, *nonleaking* PCB ballasts (containing less than 3 lb of dielectric fluid) as municipal solid waste as long as they have approval from the solid waste landfill. [§761.60(b)(2)(ii)] However, EPA recommends that these ballasts be managed per the guidance available on the agency's PCBs website: http://www.epa.gov/epawaste/hazard/tsd/pcbs/pubs/guidance.htm#ballasts.

> **Case Study 2-3: Making a Hazardous Waste Determination for Lamps**
>
> In order to determine if its spent fluorescent lamps were hazardous, a large manufacturing facility had one spent fluorescent tube tested using the TCLP. The results came back showing the tube was nonhazardous (i.e., it passed the TCLP). Based on this testing, can the facility manage all of its burned-out bulbs as nonhazardous?
>
> No. EPA requires that a generator take a representative sample when making a hazardous waste determination using the TCLP. As defined in §260.10, a "representative sample" is a sample of a universe or whole (e.g., waste pile, lagoon, ground water) that can be expected to exhibit the average properties of the universe or whole. Testing only one spent fluorescent tube and concluding that all of the facility's spent tubes are nonhazardous does not qualify as a representative sample because of the potential variability among fluorescent lamps (e.g., new vs. old, different manufacturers, different wattages, etc.). "A representative selection of lamps randomly chosen should be analyzed to make this determination." Guidance on obtaining a representative sample is given in Chapter 9 of EPA's *Test Methods for Evaluating Solid Waste, Physical/Chemical Methods* (SW–846). [RO 11907]

2.8.4.2 DEHP-containing light ballasts

Beginning in 1979, some manufacturers started to use DEHP instead of PCBs in certain light ballasts. DEHP has been found in ballasts designed for the following lighting fixtures: 4-ft fluorescent fixtures manufactured between 1979 and 1985; 8-ft fluorescent fixtures manufactured between 1979 and 1991; and HID fixtures manufactured between 1979 and 1991. [EPA/430/B-95/004]

DEHP, as a pure, unused commercial chemical product, is listed hazardous waste U028 if disposed.

However, once it has been used in a light ballast, it no longer meets the listing description. Ballasts containing DEHP that will be discarded are only hazardous if they exhibit a characteristic. [RO 11842]

2.8.5 CERCLA reporting requirements

Under CERCLA Section 103(a), a person in charge of a facility from which a hazardous substance has been released in a quantity that equals or exceeds its reportable quantity (RQ) must immediately notify the National Response Center (NRC). As defined in CERCLA Section 101(22), a "release" includes (among other things) "the abandonment or discarding of barrels, containers, and other closed receptacles containing any hazardous substances."

Since spent light bulbs and light ballasts may contain mercury, lead, PCBs, or other CERCLA hazardous substances, their abandonment or discard could trigger the CERCLA notification requirements. According to EPA, "[i]f light ballasts or lamp bulbs are abandoned or discarded into the environment…and they contain a hazardous substance such as PCBs (in the case of ballasts) or mercury (in the case of lamps), then a release has occurred. This release must be reported to the NRC if the amount [of hazardous substance] released exceeds the applicable RQ [in a 24-hour period]. If ballasts, bulbs, or both are moved without being abandoned, discarded, or otherwise released, no reportable event has occurred." [RO 11878]

Key RQs to keep in mind when dealing with waste lamps and fixtures are [EPA/430/B-95/004]:

- 1 lb for mercury,
- 1 lb for PCBs (roughly 12 to 16 PCB-containing fluorescent light ballasts), and
- 100 lb for DEHP (approximately 1,600 DEHP-containing fluorescent light ballasts).

Disposal of a hazardous substance at a RCRA-permitted facility does not require CERCLA notification. [April 4, 1985; 50 *FR* 13461] However, disposal of PCB-containing ballasts at a TSCA-permitted landfill would require CERCLA reporting if the RQ was exceeded; such disposal is not yet exempt from CERCLA notification. Finally, disposal of nonhazardous lamps in a municipal waste landfill would trigger CERCLA reporting if the RQ is exceeded. [RO 11878]

Listed Wastes

Does my solid waste meet a listed waste description?

As mentioned in the introduction to Chapter 2, EPA used two approaches for designating wastes as hazardous. Those solid wastes that exhibit one or more of four generic, hazardous properties are known as characteristic wastes. Chapter 2 explored the four hazardous characteristics of ignitability, corrosivity, reactivity, and toxicity.

The second approach used by the agency for defining a hazardous waste was to make lists of specific waste streams or chemicals that EPA knew from experience presented a threat to human health or the environment when disposed. EPA ended up with four such lists, each covering different types of hazardous wastes: they are the F-, K-, P-, and U-lists. The letter designation for each list is incorporated into the waste code; for example, hazardous wastes on the F-list all have waste codes that begin with the letter F.

Section 261.3(a)(2)(ii) specifies that if a solid waste appears on any of these four lists and has not otherwise been delisted, it is a hazardous waste.

3.1 Introduction to listed wastes

As described in Chapter 2, generators of a solid waste determine if it exhibits a characteristic primarily by testing or otherwise evaluating the properties of the waste. Generators determine if they have listed wastes, conversely, principally based on knowledge of the source of the waste. A listed waste is hazardous, not because of the concentration of any contained constituents, but because it meets a listing description on one of the four lists of hazardous wastes in the regulations. The primary criterion for applying a listed code to a waste is that you know the source of the waste (i.e., the process that generated the waste).

Appendix VII of Part 261 identifies the hazardous constituents contained within each listed waste that EPA was worried about when it listed the waste (i.e., these constituents form the basis for the listing). However, if a solid waste meets a listing description, it is that listed hazardous waste, even if it does not contain any of the Appendix VII constituents (unless the generator gets a delisting petition approved). [RO 13586, 14103, 14482, 14691, 14699]

EPA has recently started to take a new approach with regard to listed wastes: conditional listings. Under this approach, a waste is not listed on the condition that it will be managed in a particular manner that EPA considers to be protective of the environment. For example, K174 wastewater treatment sludges are not listed hazardous wastes if they are managed in Subtitle C landfills or nonhazardous waste landfills permitted by the state or federal government, and provided the sludges are not placed on the land prior to final disposal.

This chapter begins with a brief description of the four lists of hazardous wastes and where they can be found in the regulations. Following that, we examine in detail the four types of listed hazardous wastes that we find cause the greatest confusion in the regulated community: spent solvents, solvent-contaminated rags, electroplating and heat treating wastes, and the P- and U-chemicals.

3.1.1 Four lists of hazardous wastes

The F-wastes are found in §261.31—Hazardous wastes from nonspecific sources. Twenty-eight F-wastes are currently identified having waste codes ranging from F001 through F039 (obviously, some gaps exist in the numbering system). The concept behind this set of listed wastes is that they are manufacturing process wastes produced by a wide variety of industrial operations; that is, they are process wastes generated from nonspecific sources. The 28 different wastes can be grouped as follows:

- F001–F005—These are the spent solvents; an in-depth discussion of these wastes follows in Section 3.2. A separate discussion in Section 3.3 addresses solvent-contaminated rags.

- F006–F019—Specific wastes from electroplating or heat treating operations carry these listed codes. EPA is primarily worried about their heavy metal and cyanide content. A review of these wastes is provided in Section 3.4.

- F020–F023 and F026–F028—These wastes all contain dioxins or dioxin precursors. Although EPA considers them to be acutely hazardous, hardly anyone manages significant quantities of these wastes.

- F024 and F025—Process wastes from the production of chlorinated aliphatic hydrocarbons by free-radical catalyzed processes are assigned these listed waste codes. Here's where EPA's system of assigning F-codes to wastes from nonspecific sources starts to break down; these wastes are not generated outside of the organic chemical industry.

- F032, F034, and F035—These three wastes are all associated with the wood preserving industry. Here again, these three wastes don't seem to belong on the F-list because they are only generated by one industry.

- F037 and F038—Primary and secondary sludges produced primarily in wastewater treatment systems at petroleum refineries are F037 or F038. These waste codes were developed to capture wastes not otherwise covered by the K048–K052 listings.

- F039—This waste code is for multisource leachate produced at facilities (typically landfills) conducting land disposal of more than one type of listed hazardous waste.

Manufacturing process wastes from specific industries/sources make up the K-list in §261.32. This list is subdivided into groups of wastes generated from several specific industrial categories: inorganic pigments, organic chemicals, pesticides, etc. In order to use the K-list, a generator first determines if his/her operations fit within any of the industrial categories (e.g., wood preservation or inorganic pigment manufacturing). Once the industrial category is identified, the generator checks to see if his/her wastes meet any of the K-waste listing descriptions in that category. The generator should not be concerned about the K-wastes associated with the other industrial categories. For example, if a generator is in the petroleum refining industry, the nine K-wastes included under that industrial category may be applicable; the K-wastes in all the other industry categories will not be applicable. Obviously, if a facility fits into several industrial categories (e.g., it is a petroleum refinery and an organic chemical manufacturing facility), the K-wastes in each industrial category could be applicable.

The listing descriptions associated with the K-wastes are generally very specific and clear. When questions are put to EPA as to the applicability of a particular K-waste code, the agency notes in general that "[o]ur interpretations on the applicability of RCRA [K-] codes are based on the consideration of 1) the descriptive regulatory language, 2) the regulatory intent of the original listing, and 3) facts specific to the waste stream at issue." [RO 13679]

When people aren't sure if a particular K-waste listing applies to their waste, we usually suggest that they obtain and read EPA's listing background document that the agency prepared for every F- and K-waste. These documents are quite specific about the particular waste streams the agency wanted to capture via the listings. Copies of these listing background documents can be obtained online from http://www.regulations.gov by entering RCRA-2004-0016 for F-listed wastes or RCRA-2004-0017 for K-listed wastes in the search field. Alternatively, you can request this information by emailing rcra-docket@epa.gov or phoning the EPA Docket Center at (202) 566-0270.

The last two lists of hazardous wastes, the P- and U-wastes, are both found in §261.33. These two lists identify unused commercial chemical products that, when discarded, pose a threat to human health or the environment, usually due to their toxicity. A complete discussion of these listed wastes is presented in Section 3.5.

EPA has developed an excellent guidance document for the listed hazardous wastes: *Hazardous Waste Listings—A User-Friendly Reference Document*, Draft, March 2008, available at http://www.epa.gov/waste/hazard/wastetypes/pdfs/listing-ref.pdf. The document is designed to be web-based, so its usefulness is maximized when viewed on a device that is online. It includes hyperlinks to *Federal Register* notices, letters and memoranda issued by EPA, and other relevant documents that provide clarification of the hazardous waste listings.

3.1.2 ICR-only listed wastes

There are 29 listed wastes that were listed solely because they exhibit the characteristic of ignitability, corrosivity, and/or reactivity (what we call ICR-only listed wastes—see Table 5-1 in Chapter 5). At the federal level, these 29 F-, K-, P-, and U-wastes are not hazardous if, at the point of generation, they do not exhibit any characteristics. [§261.3(g)(1), May 16, 2001; 66 *FR* 27266, RO 14638, 14654] EPA clarified this requirement as follows: "if a listed hazardous waste is listed solely because it exhibits the characteristics of ignitability, corrosivity, and/or reactivity, and the waste does not exhibit the characteristic for which it was listed, then it is not a hazardous waste...." [December 2, 2008; 73 *FR* 73524] Because this rule, issued in May 2001, is less stringent than the pre-existing regulations, states were not required to adopt it. To see if your state has adopted this rule, download this document: *Authorization Status of all RCRA/HSWA Rules*, at http://www.epa.gov/osw/laws-regs/state/stats/authall.pdf, and scroll down to page 282.

Q: *A generator in a state that has adopted the May 2001 ICR-only listed waste rule manages spent acetone solvent that does not exhibit a characteristic as nonhazardous waste. What happens if the generator ships the waste to a TSD facility located in a state that has not adopted the rule?*

A: While traveling through any state that has not adopted the May 2001 rule, the spent acetone solvent meets the F003 listing description and must be managed as hazardous waste. Thus, a hazardous waste transporter and manifest must be used, and all other Part 263 requirements must be met. At the TSD facility, the waste must be managed as any other F003 hazardous waste. [RO 14716]

Q: *A generator has spent acetone solvent. The spent acetone solvent is not ignitable at the point of generation; however, vinyl chloride is present at a concentration >0.2 mg/L by the TCLP. Because the waste exhibits a characteristic at the point of generation, is the waste required to carry D043 only, or D043 and F003?*

A: Only the D043 code would apply. Because the waste does not exhibit the characteristic of ignitability when generated, the F003 code would not apply. [Email from EPA to McCoy and Associates, Inc., dated August 28, 2008]

3.1.3 State-listed wastes

Because the RCRA statute allows states to have broader, more-stringent regulations than the federal program, some states regulate what we call "state-listed wastes." These are wastes that are not included on any of EPA's F-, K-, P-, or U-lists, but they are considered listed hazardous wastes within that state's boundary. For example, some states have added PCBs as listed waste, and there would be a special code for this state-listed waste. A couple of states have listed waste codes for chemical warfare agents that are being destroyed at facilities within these states.

3.2 Spent solvents

Because spent solvents were involved in incidences of environmental damage more often than any other waste type, the initial RCRA regulations included five spent solvent listings. [May 19, 1980; 45 *FR* 33123] The concept is simple: 1) a solvent is used for its solvent properties, 2) it becomes too contaminated to use any more (i.e., it becomes spent), and 3) the spent solvent chemical is identified by name in one of the F001–F005 listing descriptions.

Those initial F001–F005 listings were for spent solvents that resulted from using pure or technical-grade solvents only. Spent solvents resulting from mixtures of the listed solvents were not regulated until January 1986. [December 31, 1985; 50 *FR* 53315]

3.2.1 Description of the spent solvents

The five spent solvent listings are found in §261.31. They each include specific solvent chemicals, but all of the listing descriptions read about the same. We start our discussion of these wastes by providing a short description of each of the five listings.

3.2.1.1 F001

This listing covers spent halogenated solvents used in large-scale industrial degreasing operations. It includes five specific solvent chemicals (carbon tetrachloride, methylene chloride, tetrachloroethylene, 1,1,1-trichloroethane, and trichloroethylene) and one class of compounds [chlorinated fluorocarbons (CFCs)]. CFCs used as refrigerants are not typically subject to the spent solvent listings, because, as refrigerants, they are not being used as solvents. [July 28, 1989; 54 *FR* 31336] CFCs and hydrochlorofluorocarbons (HCFCs) used in degreasing operations are included in the F001 listing. [RO 13580] Conversely, hydrofluorocarbons (HFCs) would not be considered F001 spent solvents if used for degreasing. [RO 14323] The five specific spent solvent chemicals were listed due to their toxicity [i.e., they have a hazard code of (T)], while CFCs were listed because they may deplete the earth's stratospheric ozone layer if released.

3.2.1.2 F002

Spent halogenated solvents used for purposes other than large-scale industrial degreasing operations comprise the F002 listing. Nine specific solvent chemicals (chlorobenzene, *o*-dichlorobenzene, methylene chloride, tetrachloroethylene, 1,1,1-trichloroethane, 1,1,2-trichloroethane, 1,1,2-trichloro-1,2,2-trifluoroethane, trichloroethylene, and trichlorofluoromethane) are identified in the listing. When they become spent, they are included in the F002 listing. As with F001, these spent solvents were listed due to their toxicity.

3.2.1.3 Differences between F001 and F002 listings

You will notice that four solvents (methylene chloride, tetrachloroethylene, trichloroethylene, and 1,1,1-trichloroethane) appear in both the F001 and F002 listings. There has been some confusion over which listing is appropriate when one of these solvents becomes spent. EPA stipulated in guidance that the F001 listing applies when one of these solvents has been used in large-scale industrial degreasing operations, such as cold cleaning or vapor degreasing (open top and conveyorized). On the other hand, the F002 listing applies when the solvent has been used for equipment cleaning or smaller scale degreasing operations, such as industrial maintenance and repair. [RO 13469]

3.2.1.4 F003

The following nonhalogenated solvents, when spent, are F003 listed hazardous wastes: acetone, *n*-butyl alcohol, cyclohexanone, ethyl acetate, ethyl benzene, ethyl ether, methanol, methyl isobutyl ketone, and xylene. The F003 spent solvent listing has created a great deal of confusion because these nine solvents were listed only because they are ignitable [i.e., they have a hazard code of (I)]. In the past, some parties have asked EPA why it needed to retain the F003 spent solvent listing, suggesting that these wastes would be adequately regulated under the ignitability characteristic. The agency defended the F003 listing on the basis that "these solvents are likely to contain other toxic contaminants. In fact, solvents become spent when they have been contaminated with other materials (e.g., toxic heavy metals or toxic organic compounds) and must be disposed, reprocessed, or reclaimed." [50 *FR* 53317] Therefore, EPA believes the F003 listing is necessary.

3.2.1.5 How nonignitable F003 spent solvents are regulated

Historically, EPA's position was that a spent solvent did not have to be ignitable to be an F003 waste. If a solid waste met the listing description for F003 in §261.31, it was an F003 spent solvent. Period. Whether or not the F003 spent solvent was ignitable was irrelevant. [RO 11707] This continues to be the position of most states that have not adopted the rule described in the next paragraph.

On May 16, 2001 [66 *FR* 27266], EPA published a final rule that changed the regulatory status of F003 spent solvents (and other ICR-only listed wastes). The primary focus of this rule was described as follows:

"[U]nder today's final rules, all wastes listed solely for an ignitability, reactivity and/or corrosivity characteristic (including mixtures, derived-from and as-generated wastes) are excluded once they no longer exhibit a characteristic." [66 *FR* 27268]

The final rule removed §261.3(a)(2)(iii), which dealt with mixtures of ICR-only listed wastes, and added a new §261.3(g) that begins:

"(g)(1) A hazardous waste that is listed in subpart D of this part solely because it exhibits one or more characteristics of ignitability as defined under §261.21, corrosivity as defined under §261.22, or reactivity as defined under §261.23 is not a hazardous waste, if the waste no longer exhibits any characteristic of hazardous waste identified in subpart C of this part."

Hence, at the federal level, wastes meeting the listing description for F003 spent solvents are not F003 wastes if they do not exhibit the ignitability characteristic. (The full scope of this rulemaking is described in Sections 3.1.2, 5.1.2.2, and 5.2.2.2.)

3.2.1.6 F004

Wastes that result from using cresols, cresylic acid, or nitrobenzene for their solvent properties are F004. These wastes were listed due to their toxicity.

3.2.1.7 F005

Finally, the F005 listing encompasses the following nonhalogenated spent solvents: benzene, carbon disulfide, 2-ethoxyethanol, isobutanol, methyl ethyl ketone, 2-nitropropane, pyridine, and toluene. These spent solvents were listed for both toxicity and ignitability.

3.2.2 Two criteria must be met for F001–F005 listings to apply

For all listed wastes, including the F-wastes, you should be able to read the listing description in the text of the regs and figure out from that language whether you are managing that hazardous waste or not. Unfortunately, the F001–F005 listing descriptions in the regulations do not include all of the information that you need to know. EPA provided significant clarifying guidance as to what it wanted to cover in these listings in the December 31, 1985 *Federal Register* [50 *FR* 53315]; however, this guidance did not get added to the regulatory language. Instead, it is summarized in the following subsections.

The F001–F005 spent solvent codes apply to the specific solvents contained in the listing descriptions only if *both* of the following criteria are met [December 31, 1985; 50 *FR* 53316, RO 11384, 14656]:

1. The solvent was used for its solvent properties (i.e., to solubilize or mobilize other constituents). Specifically, "solubilizing or mobilizing" include degreasing, cleaning, or fabric scouring, and use as carriers, diluents, extractants, or reaction and synthesis media.

2. The solvent is spent (i.e., the solvent has been used and is no longer fit for use without being regenerated, reclaimed, or otherwise reprocessed).

Although application of these two criteria may seem simple on the surface, it is not. To help clarify the F001–F005 listings, we provide specific examples below of solvent applications that meet both of these criteria.

3.2.2.1 Manufacturing operations

Q *A solvent, which contains 80% toluene prior to use, is used to wash out ink mixing tubs. What is the regulatory status of the resulting washes and sludges?*

A Due to its use as a solvent (cleaning), the spent solvent and resulting sludges are F005 wastes. (These wastes may also carry the K086 code.) [RO 11249, 13041, 13605]

Q *Petroleum-based solvents are used to clean oily machine parts. Are the spent solvents (that contain oily residues) regulated as used oil under Part 279?*

A No. The solvents were used for their solvent properties (degreasing or cleaning), and the residues from cleaning are spent. Therefore, the residues become a solid waste subject to a hazardous waste determination. If the solvent is one of the chemicals specified in the F001–F005 listings, the residues will qualify as F001–F005 wastes. The used oil regulations do not apply. [RO 14396]

Q *Methylene chloride is used as a reaction medium in a manufacturing process. When spent, the medium is removed, stored in containers, reclaimed, and then returned to the production unit. What is the status of the spent medium?*

A Since it is used as a reaction medium, the methylene chloride has been used for its solvent properties. Consequently, when the spent material exits the production process unit, it becomes a listed F002 spent solvent. [RO 12384] Note that the closed-loop recycling with reclamation exemption [§261.4(a)(8)] is not applicable because the spent medium is stored in containers—not tanks that are hard-piped. Also note that under the federal regulations promulgated on October 30, 2008 [73 *FR* 64668], the spent methylene chloride solvent would likely not be a solid waste per §261.2(a)(2)(ii). However, many states have not adopted the October 30, 2008 rule.

Q *A parts washer is mounted on top of a drum of clean solvent. Solvent is drawn up into the washer while it is in use and discharged back into the drum afterward. Eventually, the solvent in the drum becomes too contaminated to clean effectively. At that time, the washer is moved to a new drum, and the spent solvent is sent offsite for reclamation. What is the status of the spent solvent?*

A In December 1986, EPA indicated that the type of parts washer described above is not a manufacturing process unit. Therefore, the solvent from the parts washer is a spent material as soon as the operator decides that it has become too contaminated for further use. In general, spent materials that will be reclaimed are solid wastes [see Table 1 in §261.2(c)]. If the solvent is one of the chemicals in the F001–F005 listings, it will be a listed hazardous waste when spent. [RO 12790]

EPA changed the federal regulations applicable to spent materials being reclaimed on October 30, 2008. [73 *FR* 64668] Under the new rules, spent materials that will be transferred to another person for the purpose of reclamation are not solid wastes if certain conditions specified in §261.4(a)(24) are met. If the spent solvent is excluded from the definition of solid waste under §261.4(a)(24), it cannot be a hazardous waste. In states that have not adopted the October 30, 2008 rule, the spent solvent is still solid and potentially hazardous waste.

Q: *A solvent (50% acetone and 50% methanol, before use) is used to remove water from products in a manufacturing unit. The solvent stream that exits the unit has picked up enough water that it is not ignitable. What is the status of the solvent waste?*

A: Use as a drying agent meets the definition of solvent use because the solvent is used to extract water (i.e., it is an extractant). Under the May 16, 2001 rule discussed in Section 3.2.1.5, if the spent solvent is not ignitable at the point of generation, it is not F003 hazardous waste. (Note that this is the federal interpretation. In states that have not yet adopted the May 16, 2001 rule, the solvent mixture is an F003 waste, regardless of whether it is ignitable. [RO 11447])

3.2.2.2 Painting operations

Solvents are often used to solubilize or mobilize during painting operations.

Q: *Paint is stripped from aircraft prior to repainting. If the stripping compound used is one of the solvents identified in the F001–F005 listings, what is the regulatory status of the stripped paint waste?*

A: Use as a paint stripper is considered to be use as a solvent because the material is being used to solubilize or mobilize the paint. Thus, the stripped paint waste is F001–F005. [RO 11340, 11787, 12906] Note, however, that if the stripper consists solely of F003 constituents, and the spent stripper does not exhibit the ignitability characteristic at the point of generation, it is not F003. [§261.3(g)(1)]

Q: *Xylene is used to clean paint spray guns, and the xylene/paint mixture is collected on the water wall of a paint spray booth. What is the status of the resulting sludges skimmed off the paint booth sump?*

A: If the sludges are ignitable at the point of generation, they are F003 hazardous wastes. The xylene is being used for cleaning (i.e., to solubilize the paint), and the resulting spent solvent waste is regulated. [RO 12925]

3.2.2.3 Laboratory activities

Q: *Scintillation fluids or cocktails are used in laboratories to count the amount of radioactivity in samples. Xylene or toluene is used in the cocktail to suspend the radioactive sample so that sampling may occur. When disposed, how are the scintillation vials regulated?*

A: The xylene or toluene in the scintillation fluid is used for its solvent properties to mobilize other constituents in solution and to act as a reaction medium. Therefore, disposed (spent) liquid scintillation vials are F003 wastes (for xylene, assuming it is ignitable at the point of generation) or F005 wastes (for toluene). [RO 11639, 13258]

Q: *Solvents in the F001–F005 listings are often used for analytical and research work in laboratories (e.g., in liquid chromatography, rinsing paraffin off tissue culture slides, in ion-exchange columns, in layer separation, as the final step of organic synthesis). Are the wastes generated from these activities regulated as listed hazardous waste?*

A: If the solvents are used to solubilize or mobilize other constituents, the wastes produced are listed spent solvents. Note that for F003 solvents, the wastes would have to be ignitable at the point of generation to carry this code. [RO 11249, 12917]

Q: *Laboratory quality assurance (QA) standards are sometimes created by dissolving P- and U-listed commercial chemical products in listed solvents (e.g., methanol). What is the regulatory status of excess or expired solutions that are to be disposed?*

A: When a solvent is used to formulate a compound or product, neither the solvent nor the formulated product meets the listing description for spent solvents. The QA standards may be P- or U-listed if the commercial chemical product is on the P- or U-list and is the sole active ingredient in the standard. [RO 11523]

3.2.2.4 Miscellaneous

Q *Dry cleaning (fabric scouring) operations generate spent solvents and spent cartridge filters containing perchloroethylene (tetrachloroethylene) or 1,1,2-trichloro-1,2,2-trifluoroethane. Are they F002 when disposed?*

A Yes. [RO 12201, 12818, 13565] Note that dry cleaning wastes generated from hotels would *not* be excluded under the household waste exclusion. "Even though generated on premises of a temporary residence (i.e., a hotel), dry cleaning waste is not household waste because the spent solvents from the dry cleaning operations are not similar to wastes typically produced by a consumer in the home." [RO 13736]

Q *Tetrachloroethylene is used to remove PCBs from an electrical transformer. How is the resulting PCB-contaminated material regulated?*

A Tetrachloroethylene was used in this situation to remove (i.e., to solubilize or mobilize) PCBs from the transformer. When spent, therefore, it is an F002 spent solvent. (TSCA considerations would also apply to the PCB portion of the waste.) [RO 11470]

3.2.3 Spent solvent listings often do not apply

As noted in the above examples, the F001–F005 listings apply in many situations. However, we have seen people assign these codes to wastes when they really don't apply. A typical overclassification scenario occurs when a drum of waste material is found in the back of a maintenance shop at a large facility. Plant personnel don't know what the material is or where it came from, so they have it analyzed at their lab to find out what's in it. The analysis indicates that it contains methylene chloride. The facility wants to get rid of this stuff, but is concerned it may be a RCRA-regulated waste. The first place methylene chloride shows up in the RCRA waste identification (Part 261) regulations is in the F001 spent solvent listing. So, the facility puts the F001 code on the drum and manifests it offsite to a TSD facility. In other words, the facility declared the waste to be F001. And they might be right....

But EPA's position is that to really know that a solvent-containing waste is an F001–F005 listed waste, a generator would have to know how the solvent was used. The solvent must have been used to solubilize or mobilize something, and, after it is used for that purpose, it must be spent. As the following subsections demonstrate, there are numerous situations in which we will not have to apply the spent solvent codes to wastes that contain listed solvent chemicals.

3.2.3.1 Process wastes and products do not fall under the F001–F005 listings

In the December 31, 1985 spent solvent guidance referenced previously, EPA provided direction as to when the F001–F005 listings would not apply. The agency clearly stated that wastes from processes in which solvents were used as reactants or ingredients in the formulation of commercial chemical products (i.e., nonsolvent uses) are *not* covered by the listings nor are the products themselves. [50 FR 53316] This guidance has also not made it into the §261.31 listing descriptions.

As a result, wastes that contain solvents that were used in an industrial process as reactants or ingredients are not included within the scope of the spent solvent listings. In addition, these waste streams are not hazardous by virtue of the mixture rule since a spent solvent is not being mixed with another solid waste. [RO 12417] Examples of situations where the F001–F005 waste codes do not apply to process wastes and products are given below.

3.2.3.1.1 *Solvents used as reactants*

When an F001–F005 listed solvent is used as a reactant in a manufacturing process to produce a product, any wastes generated from that process that contain the solvent will not meet the spent solvent listing description. [RO 11384] "The same substances may also be used in a manufacturing process as chemical reactants or process intermediates and, when so used, are not considered to be spent solvents." [November 17, 1981; 46 *FR* 56584] Such solvent-containing manufacturing process wastes still may be hazardous if they carry a K-code or if they exhibit a characteristic, but they will not be F001–F005 listed wastes. This concept is illustrated in the following examples.

Q *Listed solvent chemicals (i.e., m-cresol, methanol, and toluene) are used as reactants during pesticide production. For process reasons, the solvents are used in such excess that large amounts of these materials do not react and are removed from the process as waste streams. What is the regulatory status of these wastes?*

A Pesticide production wastes that contain the unreacted materials are not listed spent solvents. Use as a feedstock is considered to be a nonsolvent use. Because the solvents are used in significant excess quantities, however, the company was concerned that the solvents could be considered reaction and synthesis media. EPA clarified that a reaction medium "refers to a substance that is capable of dissolving another substance (i.e., the solute) to form a uniformly dispersed mixture or solution thereby enhancing the ability of the solute to undergo a chemical reaction with other soluble substances." [RO 11079] In this situation, the solvents are reactants—not reaction media.

Q *Toluene is used to convey reactants into a reactor during pesticide production. After the product is formed, the toluene (which is a reaction medium) is separated from the product-bearing stream and returned to the process to convey more reactants. The product-bearing stream is distilled to remove any impurities; the separated impurities are sent offsite for disposal. Is the impurities stream an F005 listed hazardous waste?*

A No. The toluene that is removed from the product-bearing stream is not spent because it is still in use as a reaction medium. Since it is not F005, the product-bearing stream remaining after the toluene is removed would not be derived from an F005 waste. Finally, the residue remaining after product distillation would be considered a solid waste, but not an F005 hazardous waste or a waste derived from the treatment of an F005 waste. [RO 11703]

Q *Xylene is used as a reactant in the production of sodium xylenesulfonate. Excess xylene from the reactor is separated from the product and distilled before reuse. Are the still bottoms from the recovery of the xylene F003?*

A No. The xylene is used as a chemical reactant in the production process, and, therefore, the excess xylene stream is not an F003 spent solvent. Hence, the still bottoms generated from the recovery of the xylene are also not spent solvent wastes (although they may exhibit a characteristic). [RO 12809] An almost identical interpretation was rendered by the agency for still bottoms generated during purification of an unreacted toluene stream. [RO 11326]

3.2.3.1.2 Solvents used as ingredients

Many commercial products, including paints, inks, and adhesives, contain solvents as ingredients. Formulations of these products may contain a number of the solvents in the F001–F005 listings, such as acetone, methyl ethyl ketone, toluene, and/or xylene. When excess or old paints, inks, and adhesives are thrown out, the F001–F005 listings will not apply per EPA's guidance. [50 *FR* 53316, RO 11220, 11340, 11349, 12334, 12906, 13273]

These discarded, unused products that contain solvents may still be hazardous, however, if they exhibit a characteristic (e.g., D001). In addition, even though such products will not carry an F001–F005 code, they may be regulated under §261.33 as P- or U-wastes if they contain only one active ingredient and that ingredient is on the P- or U-list. [RO 11787] EPA published similar guidance for waste aerosol cans. Unused solvent residues in non-empty aerosol cans will not be F001–F005 but will be P- or U-listed if the product meets a listing description. [RO 14656]

The F001–F005 listings also do not apply to unused off-specification solvents, which may be regulated under §261.33 as P- or U-wastes. Take, for instance, a company that recycles spent 1,1,1-trichloroethane solvent in a distillation process. After distillation occurs, the recovered solvent may sometimes fail to meet market specifications and, therefore, must be disposed. In this case, the discarded 1,1,1-trichloroethane would be a U226 off-specification commercial chemical product. [RO 12276]

Addition of a solvent to modify a commercial chemical product is also not solvent use that will result in an F-listed spent solvent waste. For example, commercially purchased products (such as paints) to which solvents have been added by the end user to adjust the viscosity (e.g., toluene or xylene added to thin the paint) are not regulated as F001–F005 spent solvents when an unused portion of the modified product is later discarded. [RO 11220, 14005, 14109] "The agency does not recognize a distinction between paints that contain solvents and paint where solvents have been added." [RO 11249, 12925] However, if spent solvent that does meet a listing description (e.g., from cleaning out paint spray apparatus) is mixed with one of the unlisted products to be discarded, the entire mixture would be subject to the hazardous waste mixture rule. Examples where solvents are used as ingredients are given below.

Q *A manufacturer of "pre-coat" uses 2-ethoxyethanol as an ingredient in the process. A waste from the process contains 2-ethoxyethanol—is the waste a spent solvent?*

A No. Chemicals used as ingredients in the formulation of commercial chemical products are not being used as solvents. The process waste is, therefore, not a listed spent solvent. Note, however, that 2-ethoxyethanol is toxic in the environment and should be managed properly. [RO 11169]

Q *Are activated-carbon canisters used to collect solvent vapors (e.g., Freon 113; 1,1,1-trichloroethane; and methylene chloride), that are generated during the application of paint products, considered to be hazardous wastes?*

A Since the incorporation of solvents into paint formulations does not constitute solvent use, the solvent vapors collected from paint application are not spent solvents. [RO 11151] The canisters could be hazardous due to a characteristic, however.

Q *During painting operations, a base coat and a top coat are sprayed onto a product. The base coat consists of a resin mixture containing toluene, and the top coat consists of a different toluene-containing resin. In order to clean out the spray gun between coats, the spray gun is directed toward a drum and the top coat is used to clear out base coat or vice versa. Is the waste in the drum a listed spent solvent?*

A The definition of spent solvent does not extend to cases in which the solvents are strictly reactants or ingredients in a commercial chemical product formulation. "The coating materials merely push the residue of the previous coating out of the nozzle so that pure top or bottom coat can be applied to the products. The toluene is there as part of the manufacturing process itself. It is therefore part of the formulation of the commercial chemical product and not covered by the listing." [RO 11685, 13585] (Although EPA did not say it in this guidance, we think the waste base or top coat that is contaminated with the other is an off-specification commercial chemical product. Off-spec products are listed only if they contain a sole active ingredient on the P- or U-list. Since the paints will likely not have a sole active ingredient on the P- or U-list, they will only be hazardous via characteristic.) Of course, if pure toluene is used to clean the spray gun, the waste would be considered F005 spent solvent.

Q *A solvent-based product is used as a machining coolant during metal machining, drilling, etc. The coolant contains 80% 1,1,1-trichloroethane and 20% lubricating oil. What is the status of coolant wastes containing this solvent?*

A The 1,1,1-trichloroethane is being used as an ingredient in product cutting oil—not as a solvent. Assuming that it has not been mixed with a spent solvent, the cutting-oil waste is not an F001 or F002 waste. [RO 11212, 13257]

3.2.3.2 Other solvent-contaminated process wastes are not F001–F005

EPA has determined that process wastes that become contaminated with small amounts of listed solvents during manufacturing operations are also not within the scope of the spent solvent listings. For instance, rinsewaters that pick up small amounts of solvents during product or equipment rinsing are not considered spent solvents—they are process streams that may have become contaminated with solvents. [RO 11384] In addition, if these solvent-contaminated rinsewaters then contaminate the plant's process wastewater and wastewater treatment tank sludges, those streams are also not listed spent solvent wastes. [RO 11447]

In general, solvent-contaminated process wastewater exiting manufacturing process units does not meet the spent solvent listing descriptions. Conversely, mixtures of wastewater and listed spent solvents are regulated unless the mixture meets the conditions set forth in the de minimis wastewater exemptions at §261.3(a)(2)(iv)(A) or (B). [RO 11193] Details of these exemptions are covered in Section 4.16.1. Other examples of process wastes that do not carry the F001–F005 listings follow.

Q *Organic liquid-liquid extractions use listed solvents that are nearly immiscible in water (e.g., methylene chloride) to remove an organic product from an aqueous stream. However, the stripped aqueous phase contains small amounts of the solvent that do dissolve into the water. When disposed, is the aqueous stream F002? The solvent is subsequently separated from the product and disposed. What is the regulatory status of the removed solvent?*

A The aqueous phase from organic liquid-liquid extractions is analogous to a process stream that has become minimally contaminated with solvent constituents (i.e., there was no intentional mixing of a water stream with a spent solvent stream). Therefore, it is not a regulated spent solvent and would be hazardous only if it is a K-listed manufacturing process waste or if it exhibits a characteristic. Conversely, the organic solvents separated from the product were used as extractants (which constitutes solvent use) and, if spent, are listed F001–F005 spent solvents. [RO 11384, 11437, 12658]

Q *Acetone, ethyl acetate, and xylene solvents are periodically used to clean out a reactor vessel. The spent solvents generated during cleanout are drummed and sent offsite for management as F003 wastes. However, trace amounts of solvent remain on the walls of the vessel. This coating is washed off with soap and water. How is this reactor vessel rinsewater regulated?*

A In this situation, EPA has determined that the resulting solvent-contaminated rinsewater is not covered by the F003 listing. It is generated from the washout of a reactor vessel containing solvent residues. Therefore, the waste is not a spent solvent, but a process wastewater contaminated with solvent constituents. "This waste is very different from a solvent stream that has been used and as a result of contamination can no longer be used as a solvent without further processing." [RO 11300; see also RO 11384, 11447]

Q *Metal parts are degreased using an F001-listed solvent. The degreased parts are then blasted and ground before finishing; the blasting and grinding grit contains small amounts of solvent residue. Is the grit also F001? What about rinsewater that contains carried-over solvent (dragout) from a metal degreasing operation; would it be hazardous due to the mixture rule?*

A Small amounts of solvent carried over on parts after cleaning/degreasing are not considered spent and, therefore, do not meet the spent solvent listing description. Therefore, neither the blasting grit nor the spent rinsewater would be F001. (Either of these two wastes could be hazardous via characteristic.) However, if more solvent than necessary is used during cleaning (i.e., it looks like the facility is trying to get rid of its spent solvent by excessive dragout on the metal parts), the excess solvent residues could be considered spent solvent wastes on a case-by-case basis. [RO 11440, 11638, 12896, 13007, 13149]

Q *During production of insulating and packaging foam, CFCs or methylene chloride are used as blowing agents to physically open the foam cells. Once they serve this function, the CFCs or methylene chloride are released and subsequently captured by a vapor-recovery system. Spent CFCs or methylene chloride so captured are shipped offsite for recycling or disposal. What is their regulatory status?*

A The CFCs or methylene chloride are not being used to solubilize or mobilize. Instead, they are being used to physically open the foam cell. Consequently, use of CFCs or methylene chloride as a blowing agent does not constitute use as a solvent, and the spent solvent listings would not apply. [RO 11494, 13322, 14323]

Q *Toluene is used as a carrier to transport a pharmaceutical product through a number of filtering stages. During the first stage of filtering, an iron impurity is filtered out. What is the regulatory status of the resulting iron-bearing filter cake when disposed? What is the status of the toluene that exits the last stage of filtering after the product has been removed?*

A Toluene in the stream is being used for its solvent properties (i.e., to solubilize and mobilize the product). However, at the first stage of filtering, the toluene is not spent. Therefore, the iron-bearing cake is neither a residue from the treatment of a spent solvent nor a mixture of a solid waste and a spent solvent. Instead, it is a process waste that is contaminated with a small amount of toluene—not F005. [RO 11311, 13060] Toluene that remains after the product has been separated from it would meet the definition of F005 spent solvent if disposed. [RO 11384]

3.2.4 Spent solvent mixtures

As noted in Section 3.2, the original spent solvent listings applied only when pure or technical grade solvents became spent; solvent mixtures were not subject to these listings when they became spent. EPA modified the F001–F005 spent solvent listings, effective January 1986, to close this regulatory loophole. [50 *FR* 53315]

It is easiest to explain how the spent solvent mixture rule works if we break it into two parts—one for F001, F002, F004, and F005 solvents, and the second for F003 solvents.

3.2.4.1 Mixtures containing F001, F002, F004, and F005 solvents

Solvent mixtures that contain F001, F002, F004, and/or F005 solvent constituents will carry one or more of the spent solvent waste codes if:

Case Study 3-1: "Before Use"

A generator purchased a product that contained 15% toluene and 85% water, then diluted the product with more water to prepare the material for use. The diluted product contained 8% toluene and 92% water. Is this material, which is used as a cleaning solvent, regulated as F005 when spent since it originally contained more than 10% solvent?

No. The F001–F005 listings include mixtures containing 10% or more total listed solvents (by volume) before use. Because EPA interprets "before use" to mean before use at the facility—not when purchased—the waste solvent would not meet the spent solvent listing description. [RO 11521] Although this guidance came from EPA headquarters, we would never recommend that a facility dilute its solvent as noted in this example for the purpose of getting below the regulatory threshold; we don't think such an activity could pass the "red-faced test." (If it were diluted for valid process reasons, that's a different story.)

---

1. The before-use solvent mixture contains 10% or more *total* F001, F002, F004, and/or F005 listed solvents (by volume). [RO 11527] The composition of the solvent mixture is evaluated before it is used (i.e., before use at the facility, not when purchased). [RO 11521] Case Study 3-1 illustrates this point.

2. The solvent mixture is used to solubilize or mobilize other constituents.

3. The solvent mixture is spent.

If these three criteria are satisfied, the spent solvent waste code for each F001, F002, F004, and/or F005 solvent chemical present is assigned to the spent mixture. A spent solvent mixture will be regulated unless the generator can prove that the before-use mixture contained less than 10% total F001, F002, F004, and F005 solvents. (The 10% threshold was based on data indicating that solvent mixtures used in commerce typically contain greater than 10% total solvents.) [50 *FR* 53316]

Examples of the solvent mixture rule for F001, F002, F004, and/or F005 solvents are given in Table 3-1.

3.2.4.2 Mixtures containing F003 solvents

The solvents included in F003 were all listed solely due to their ignitability. All of the other solvents included in F001, F002, F004, and F005 were listed in part or in whole for their toxicity. EPA thought this difference was significant, and so the agency made up a different solvent mixture rule for mixtures containing any of the F003

Table 3-1: Examples of F001, F002, F004, and/or F005 Spent Solvent Mixture Rule

Composition of solvent mixture before use (by volume)	Waste code(s) when solvent mixture becomes spent	Explanation
100% trichloroethylene	F001	The before-use solvent is pure (used for degreasing)
		The spent solvent mixture has a flash point >140°F
80% toluene 20% benzene	F005 D001	The solvent mixture contains only F005 listed solvents
		The spent solvent mixture has a flash point <140°F
60% 1,1,2-trichloroethane 40% *m*-cresol	F002 F004	The solvent mixture contains 10% or more total F001, F002, F004, and/or F005 solvent constituents, so the F-code for each such constituent is applied
		The spent solvent mixture has a flash point >140°F
40% chlorobenzene 30% isobutanol 30% petroleum distillates	F002 F005 D001	The solvent mixture contains 10% or more total F001, F002, F004, and/or F005 solvent constituents, so the F-code for each such constituent is applied (even though there is a nonlisted solvent included)
		The spent solvent mixture has a flash point <140°F
87% petroleum distillates 9% methylene chloride 4% methyl ethyl ketone	F001 F005 D001	The solvent mixture contains 10% or more total F001, F002, F004, and/or F005 solvent constituents, so the F-code for each such constituent is applied (note that the F-codes are used even though, individually, the listed solvents are each contained at <10%)
		The spent solvent mixture has a flash point <140°F
91% petroleum distillates 5% methylene chloride 4% methyl ethyl ketone	D001	F-codes do not apply because the solvent mixture does not contain 10% or more total F001, F002, F004, and/or F005 solvent constituents
		The spent solvent mixture has a flash point <140°F

Source: McCoy and Associates, Inc.

constituents. Again, the composition of the solvent mixture must be evaluated before use, and the solvent mixture must be used to solubilize or mobilize other constituents. Finally, the solvent mixture must be spent and, under the federal regulations, it must be ignitable at the point of generation. If those criteria are met, the F003 solvent mixture rule has three aspects:

1. If the before-use solvent mixture contains only F003 solvent constituents (e.g., a blend of 80% ethyl acetate and 20% xylene), the F003 code is applied to the spent mixture. In this situation, the mixture must contain 100% F003 solvents. If there are any non-F003 solvents (e.g., ethanol, mineral spirits) or other constituents (e.g., water, oil) present in the solvent mixture *before use*, the spent solvent mixture would not fall under the F003 listing. [RO 11249, 13041] However, the F003 code would apply if the solvent products contain small percentages of chemical impurities or contaminants that are inherent to the solvent manufacturing process. For example, "pure" xylene may contain 99.98% xylene and 0.02% benzene/toluene as manufacturing impurities. Even with the presence of these de minimis percentages of impurities, the F003 listing would apply to the spent xylene when it is discarded (after use for its solvent properties). EPA has not established specific percentages or other criteria for determining when manufacturing contamination is considered de minimis. Such decisions will be made on a case-by-case basis by the appropriate regulatory agency. [RO 13675]

2. If the before-use solvent mixture contains one or more F003 constituents *and* 10% or more total F001, F002, F004, and/or F005 listed solvents (by volume), the F003 code and the code for each F001, F002, F004, and/or F005 solvent present are assigned to the spent mixture. For example, a solvent mixture consisting of 15% xylene, 15% toluene, and 70% water (before use) would be F003 and F005. [RO 12863] The regulations do not specify a minimum 10% threshold for the F003-listed solvent(s) in the mixture in order for the F003 waste code to apply. Therefore, *any amount* of F003 may be present for the spent mixture to be considered F003 (as long as the mixture contains 10% or more of F001, F002, F004, and/or F005 solvents, and as long as the spent solvent is ignitable at the point of generation). [May 16, 2001; 66 *FR* 27266, RO 11521] See Case Study 3-2 for another example of this rule.

3. If neither of the two criteria above are satisfied, the spent solvent mixture will be hazardous only by characteristic (e.g., D001). For example, a solvent mixture containing 80% ethyl acetate and 20% petroleum solvents before use would not be F003 when spent, even though it contains significant quantities of the F003 constituent. [RO 12521]

Because we find it difficult to remember all of these rules, we have prepared the logic diagram in Figure 3-1 to make the spent solvent coding process easier.

Case Study 3-2: F003 Spent Solvent Mixtures

A paint thinner contains 80% xylene, 9% toluene, and 11% glycol ethers before use. Should it be classified as an F003 waste when spent?

No. Even though the solvent is used for its solvent properties, it would not be correctly classified as an F001–F005 spent solvent. Since the mixture contains an F003 constituent, it would not be a listed hazardous waste unless the total of all F001, F002, F004, and F005 constituents met the 10% threshold. Since the concentration of the F005 constituent (toluene) in the solvent blend before use is only 9%, the resulting waste stream would not be considered a listed spent solvent. However, the waste would be hazardous if it exhibits a characteristic (e.g., D001). [RO 11266]

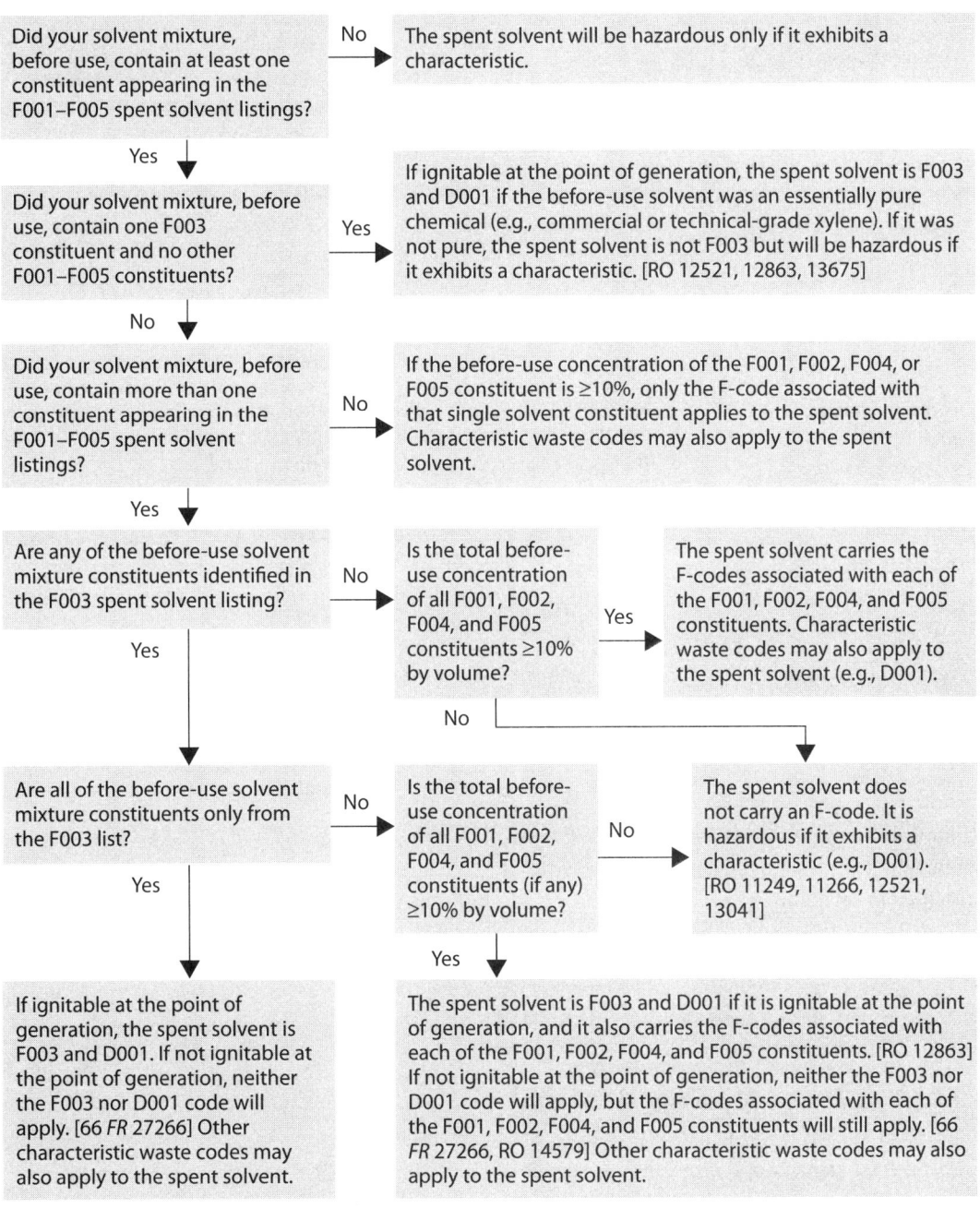

Figure 3-1: How Solvent Mixtures Are Regulated

Table 3-2: Examples of F003 Spent Solvent Mixture Rule

Composition of solvent mixture before use (by volume)	Waste code(s) when solvent mixture becomes spent	Explanation
80% acetone 20% xylene	F003 D001	The solvent mixture contains only F003 listed solvents The spent solvent mixture has a flash point <140°F
80% acetone 20% methyl ethyl ketone	F003 F005 D001	The solvent mixture contains an F003 solvent and 10% or more total F001, F002, F004, and/or F005 solvent constituents, so the F-code for each such constituent is applied The spent solvent mixture has a flash point <140°F
70% acetone 30% petroleum distillates	D001	F003 constituents mixed with only nonlisted solvents are not F003 The spent solvent mixture has a flash point <140°F
60% acetone 11% methyl ethyl ketone 29% petroleum distillates	F003 F005 D001	The solvent mixture contains an F003 solvent and 10% or more total F001, F002, F004, and/or F005 solvent constituents, so the F-code for each such constituent is applied (even though there is a nonlisted solvent included) The spent solvent mixture has a flash point <140°F
91% acetone 9% methyl ethyl ketone	D001	The solvent mixture contains an F003 solvent but the total F001, F002, F004, and/or F005 solvent constituents are <10%, so the F001–F005 codes do not apply The spent solvent mixture has a flash point <140°F
99.5% xylene 0.5% toluene (a manufacturing impurity)	F003 D001	The solvent mixture is an F003 waste because xylene is a technical-grade solvent (toluene is a manufacturing impurity) The spent solvent mixture has a flash point <140°F

Source: McCoy and Associates, Inc.

Examples illustrating how various F003 solvent-containing mixtures are regulated when spent are presented in Table 3-2.

3.2.4.3 Other oddities with F003 mixtures

Because F003 solvents were listed solely due to their ignitability, there are a couple of other unusual aspects to their management that are worth noting.

3.2.4.3.1 Mixtures and residues of F003 wastes

When F003 wastes are mixed with solid wastes, the special second element of the mixture rule applies. Consequently, a mixture of an F003 spent solvent with a solid waste (e.g., wastewater) is not regulated as a hazardous waste if the mixture is not ignitable and does not exhibit any other hazardous waste characteristic. [RO 12721] However, such mixing is considered treatment and requires a RCRA permit, unless the treatment is performed in a way that exempts it from such requirements (e.g., in a wastewater treatment unit regulated under the Clean Water Act). See Section 5.1.2.2 on the mixture rule for a complete discussion of this topic.

Similarly, due to the May 16, 2001 rule [66 FR 27266] described in Section 3.1.2, a residue from the treatment, storage, or disposal of an F003 waste is not an F003 waste if it does not exhibit the ignitability characteristic—see Section 5.2.2.2.

For example, a facility burns spent acetone solvent (F003 and D001), and an ash-like residue is produced. The ash is F003 waste under the derived-from rule only if it exhibits the ignitability characteristic. [§261.3(g)(2)(ii)] If the ash is not ignitable, it is neither F003 nor D001. It may carry another code if it exhibits a different characteristic (e.g., D008 for lead toxicity).

3.2.4.3.2 *F003-contaminated soil and the contained-in policy*

As mentioned above, the mixture rule applies when F003 wastes are mixed with a solid waste. The mixture rule does *not* apply, however, when F003 wastes contaminate environmental media, such as soil. In these situations, the contained-in policy applies instead (refer to Section 5.3). Once again, the May 16, 2001 rule [66 *FR* 27266] changed how F003 wastes are regulated under the contained-in policy:

> "Today's final rule does not directly affect the implementation of the contained-in policy. However, wastes that are contained in contaminated media are eligible for the 40 *CFR* 261.3(g) exemption for wastes listed solely for a characteristic. Therefore, under today's final rule, contaminated media that contain a waste listed solely for a characteristic would no longer need to be managed as hazardous waste when it no longer exhibits a characteristic." [66 *FR* 27286]

Hence, if F003-contaminated media is not ignitable at the point of generation, it is not F003.

3.2.5 Reclaimed solvents

Products that are reclaimed from spent solvents are generally exempt from further RCRA regulation. [§261.3(c)(2)(i)] This regulatory provision states that "materials that are reclaimed from solid wastes and that are used beneficially are not solid wastes and hence are not hazardous wastes…unless the reclaimed material is burned for energy recovery or used in a manner constituting disposal."

Q *A recycling company receives spent acetone from an offsite metal cleaning operation and regenerates it to produce a commercial gun cleaning solvent. The regenerated solvent is also sold as a fuel. What is the regulatory status of each product?*

A The reclaimed solvent that is sold for gun cleaning purposes meets the §261.3(c)(2)(i) criteria and is, therefore, exempt from regulation as hazardous waste. It is important to note, however, that even though the reclaimed material that will be reused as a solvent is exempt from hazardous waste regulation, any residues (e.g., still bottoms) generated from recovery of the reclaimed solvent are not. In this example, the residues would be F003 hazardous waste if ignitable and must be managed as such. [RO 12911] Section 261.3(c)(2)(i) specifically prohibits application of the product exemption for reclaimed materials that are burned for energy recovery. Therefore, if a portion of the reclaimed solvent was sold for use as a fuel, the fuel would be regulated as F003 hazardous waste. [RO 12850, 13208]

Q *Spent methanol generated during pharmaceutical manufacturing operations is reclaimed onsite to a minimum purity of 99.5%. The reclaimed methanol is then sent offsite to another facility where it is further reclaimed for final use. Does the shipment have to be managed as a hazardous waste?*

A Although the shipped material needs to be reclaimed further before use, EPA considers that the shipped methanol (with a purity of 99.5%) is more product-like than waste-like. As such, the agency determined that the initially reclaimed methanol is not regulated as a spent solvent and could be shipped offsite without a manifest. [RO 11109] Also note that based on the October 30, 2008 definition of solid waste rule [73 *FR* 64668], the spent methanol would not be a solid waste at its point of generation from the manufacturing operations if reclaimed in accordance with §261.2(a)(2)(ii).

Q *Waste ink from lithographic printers is mixed with spent solvents used to clean the equipment during printing operations. Since some of the solvents are included in the F001–F005 spent solvent listings, the waste ink mixture would be a regulated hazardous waste if disposed. However, the waste ink is instead manifested to an offsite facility where it is recycled to produce recovered ink and a solvent/water mixture. The recovered ink is immediately reused in* printing processes, and the solvent/water mixture is processed (i.e., surfactants are added) so that it can be sold to printers for use as a cleaner. Are these recovered materials hazardous waste?

A The recovered ink is clearly exempt from further RCRA regulation via §261.3(c)(2)(i). In order for the residual solvent/water mixture to be excluded from the definition of solid waste (and therefore not be hazardous waste), however, the facility will have to demonstrate that it is a legitimate product. This determination would be made on a case-by-case basis by the regulators. Factors that may be considered include: 1) how similar the recovered material is to the virgin product it is replacing (in terms of both its value and the presence of hazardous constituents not normally found in the virgin product), and 2) whether there are any product specifications that apply to the solvent/water "product." [RO 11765]

3.2.6 Miscellaneous spent solvent issues

Guidance discussing other issues associated with the management of F001–F005 spent solvents under RCRA is summarized below.

3.2.6.1 Spill cleanup residues

Spent solvents have been involved in incidences of environmental damage more often than any other waste type. Numerous facilities across the United States have soil and/or ground water contaminated with listed spent solvents. If such environmental media are contaminated with F001–F005 wastes, then they must be managed under the contained-in policy (see Section 5.3). Sometimes, however, facilities will find soil or ground water contaminated with a listed solvent chemical, but they are unsure as to the source of the chemical. Then what?

As discussed in Section 3.2.3, a generator must know how a solvent chemical was used before he/she can apply the F001–F005 listings to it. The solvent must have been used to solubilize or mobilize something, and, after it was used for that purpose, it must be too contaminated to use any more (i.e., it's spent). For example, soil at a site is found to be contaminated with trichloroethylene (TCE), but the facility owner/operator does not know the source of the contamination. If the TCE had been used as a degreasing solvent and was spent, the soil would be considered F001. If the contamination resulted from a spill of unused TCE, the soil would be U228. [RO 12906] Conversely, if the TCE contamination resulted from use of a product in which the solvents were used as reactants or ingredients, the F001–F005 codes would not apply. Similarly, if process wastes that became contaminated with small amounts of TCE during manufacturing operations were what contaminated the soil, it would not be listed.

So what does the site owner/operator do in such a situation? If the owner/operator does not know the source of the solvent chemical that is contaminating its soil and cannot determine the source after making a good-faith effort, the only way the soil can be hazardous is if it exhibits a characteristic. [RO 12171, 13586] Listed hazardous wastes are identified based on the source of the waste rather than the concentrations of hazardous constituents. Therefore, analytical testing alone, without information on a waste's source, will generally not produce information that will conclusively indicate whether a given waste is a listed hazardous waste. [RO 14291] See also Section 14.2.

Cleanup residues can also consist of contaminated absorbents. If a drum of listed spent solvent leaks, absorbent used to soak up the spill will be F001–F005 listed waste when sent for disposal, because it contains a listed hazardous waste. [RO 12906] (The mixture rule would not apply because the absorbent was a product, not a solid waste, when contact with hazardous waste occurred. Instead, when the spent solvent mixes with product absorbent, the resulting contaminated material, when discarded, would be F001–F005 via the contained-in policy. See the contained-in policy write-up in Section 5.3.) If F003 spent solvent leaks, the resulting contaminated absorbent is F003 only if it is ignitable per the May 16, 2001 rule. [66 *FR* 27266]

3.2.6.2 Used oil

As discussed in Section 12.1.4, used oil that contains more than 1,000-ppm total halogens is presumed to have been mixed with spent halogenated solvents. [§279.10(b)(1)(ii)] Accordingly, it is subject to regulation as F001 or F002 hazardous waste via the mixture rule. An entity in possession of used oil that contains more than 1,000-ppm total halogens can rebut the presumption of mixing with a listed spent solvent by showing through analysis or other means that the halogens didn't come from a listed waste source. However, "[a]ny person in possession of used oil containing more than 1,000-ppm total halogens must be able to provide documentation to support a rebuttal if the oil is not managed as hazardous waste." [RO 13282]

For example, a specific used oil contains more than 1,000-ppm total halogens, and the mixing-with-spent solvents presumption cannot be rebutted. If that used oil is mixed with coal (for dust suppression purposes), the used oil/coal mixture is hazardous waste fuel. [RO 13282]

3.2.6.3 Still bottoms

Each of the five spent solvent listing descriptions ends with the phrase "and still bottoms from the recovery of these spent solvents and spent solvent mixtures." Application of this phrase is usually straightforward: if a facility generates an F001–F005 spent solvent and sends it for either on- or offsite recovery via distillation or similar recycling technology, the still bottoms produced from recovering the solvent will retain the appropriate F001–F005 code.

However, this same result would have been produced by application of the derived-from rule. So why did EPA put this phrase in the listing descriptions? Although we have never seen any guidance where EPA has discussed this issue, one reason could have been to capture still bottoms from recycling such spent solvents in situations where the derived-from rule wouldn't apply. For example, if a spent solvent generated from a production process is reclaimed and then put back into the process in a closed-loop fashion, §261.4(a)(8) exempts the spent solvent from being a solid waste, so it could not be a hazardous waste. If the reclamation process consisted of a distillation column, what is going into that column is not a solid or hazardous waste, so the still bottoms could not be hazardous via the derived-from rule. Instead, those still bottoms are hazardous (if disposed) because they are solid waste that meets the last part of the F001–F005 listing descriptions. [July 14, 1986; 51 *FR* 25443, RO 11285, 12732, 13017, 13220] The same result will occur if a facility recycles spent solvent under EPA's October 30, 2008 final rule [73 *FR* 64668]; the spent solvent sent for onsite or offsite distillation is not a solid waste, but the still bottoms are F-listed because they meet the applicable listing description.

Also note that the May 16, 2001 rule [66 *FR* 27266, codified at §261.3(g)(2)(ii)] implies that still bottoms (treatment residues) from distilling F003 spent solvent are F003 wastes only if they are ignitable.

3.3 Solvent-contaminated rags/wipes

Solvent-contaminated rags generated from cleaning and degreasing activities may meet the F001–F005 listing criteria if they contain a listed spent solvent or spent solvent mixture. The rags are typically not solid wastes when they come into contact with the solvent—they are products being used for their intended purpose. Therefore, the mixture rule does not apply when trying to determine whether or not solvent-contaminated rags are hazardous waste. Instead, EPA's contained-in policy (discussed in Section 5.3) is more appropriate. Additionally, contaminated rags could exhibit a hazardous characteristic.

3.3.1 Background

Determining whether solvent-contaminated rags are hazardous is an area where EPA has really struggled. For the first 11 years of RCRA, the agency's position on whether such rags were F001–F005 wastes or not depended on how the listed cleaning solvent was applied to the rag. If the listed solvents were poured onto the equipment surface to be cleaned (and not onto the rag) and the rag was used to subsequently wipe up the spent solvent, the contaminated rag was classified as F001–F005 hazardous waste. The reasoning for this was that a solvent is considered spent when it has been used and is no longer fit for use without being regenerated, reclaimed, or otherwise reprocessed. Therefore, when solvents were applied to a surface or machinery (and used for their solvent properties), the solvents were spent before they were cleaned off with the rag. The resulting contaminated rag was covered by the F001–F005 listing. [RO 11249]

On the other hand, if the clean, listed solvents were poured onto the rag *first*, the resulting dirty rag *did not* meet the F001–F005 listing criteria. In this situation, the solvent was applied to the rag before it was spent, and the rag was, therefore, not covered by the spent solvent listings. [RO 11249]

3.3.2 Rags regulated on a case-by-case basis from 1991–2013

The above two paragraphs describe EPA's initial position on solvent-contaminated rags. As a practical matter, rags from both of these scenarios are identical in makeup and would pose similar hazards. In addition, generating and commercial hazardous waste management facilities were not able to verify how the rags became contaminated. Due to these considerations, EPA stated in January 1991 that the regulatory status of solvent-contaminated rags would henceforth be determined by regional EPA offices and authorized states on a case-by-case basis. [RO 11576]

EPA maintained that regulatory determinations or interpretations regarding solvent-contaminated industrial rags should be made by the appropriate regulatory agency since they are based on site-specific factors. These factors may include the degree of hazard posed by the contaminated rags, when a spent solvent is generated, and whether the mixture rule applies. [RO 11813, 11935, 14405] For example, one of EPA's concerns was that a facility might be improperly disposing spent solvents by mixing them with dirty rags (e.g., pouring a container of spent solvent into a dirty-rag drum) and then sending the mixture to a laundering facility or nonhazardous landfill.

As noted above, rags that pick up incidental amounts of solvent while being used for their intended purpose for years were regulated by regional EPA offices and authorized states on a case-by-case basis. Most regions and states developed specific policies for solvent-contaminated rags. If the rags will be laundered and then reused, many regions/states do not require hazardous waste management of the rags during storage at the generator's facility or while being shipped to the laundry. However, if the rags or wipers are contaminated with F001–F005 listed solvents and they will be disposed, many states require them to be managed and disposed as hazardous waste, no matter how the solvent got on the rag.

In regions/states that require hazardous waste management even for nondisposable solvent-contaminated rags, the RCRA regulations allow an exemption to the manifest requirements when small quantity generators (those that generate between 100 and 1,000 kg/mo of hazardous waste) ship hazardous waste offsite for reclamation. [RO 11178] The exemption is in §262.20(e) of the regulations. To take advantage of this exemption, the generation of *all* hazardous waste (not just the hazardous rags) must be between 100 and 1,000 kg/mo and the conditions in §262.20(e) must be met (see Section 6.6.2.1).

3.3.3 Exclusions for solvent-contaminated rags/wipes

In an attempt to provide national consistency for the management of solvent-contaminated rags and wipes, EPA issued a final rule on July 31, 2013. [78 *FR* 46448] The rule, which contains two conditional exclusions for solvent-contaminated wipes, applies to:

1. Wipes that contain one or more of the F-listed spent solvents or the corresponding P- or U-listed unused solvents (e.g., that were spilled and wiped up),

2. Wipes that exhibit a characteristic if that characteristic results from a solvent listed in Part 261 (e.g., ignitability due to acetone, toxicity due to methyl ethyl ketone), and

3. Wipes that exhibit the characteristic of ignitability due to the presence of one or more solvents that are not listed in Part 261.

Solvent-contaminated wipes that contain listed hazardous waste other than solvents, or exhibit a characteristic of corrosivity, reactivity, or toxicity due to contaminants other than solvents are not eligible for the new exclusions. For example, wipes that exhibit the toxicity characteristic because they're contaminated with RCRA metals won't be

Table 3-3: Regulatory Status of Solvent-Contaminated Wipes Under EPA's July 31, 2013 Rule[1]

Wipe contaminated with	Status if disposed	Status if laundered	Explanation
Spent solvent toluene	Not HW	Not SW	Meets paragraph (1)(i) of the definition of "solvent-contaminated wipe" in §260.10
Spent solvent methyl ethyl ketone failing TCLP for metals	HW	HW	Meets paragraph (2) of the definition of "solvent-contaminated wipe" in §260.10
Spent solvent trichloroethylene	F001/F002	Not SW	Meets paragraph (1)(i) of the definition of "solvent-contaminated wipe" in §260.10 but is not excluded from the definition of hazardous waste per §261.4(b)(18) when disposed
Spill of unused methyl ethyl ketone	Not HW	Not SW	Meets paragraph (1)(i) of the definition of "solvent-contaminated wipe" in §260.10
Spill of unused formaldehyde	U122	U122	Meets paragraph (2) of the definition of "solvent-contaminated wipe" in §260.10
Spent solvent acetone (wipe is dripping)	Not HW[2]	Not SW[2]	Meets paragraph (1)(i) of the definition of "solvent-contaminated wipe" in §260.10
Spent solvent acetone (spent wipe is not ignitable)	Not HW	Not HW	See Footnote 3
Spent solvent isopropyl alcohol (wipe is dripping)	Not HW[2]	Not SW[2]	Meets paragraph (1)(iii) of the definition of "solvent-contaminated wipe" in §260.10
Paint waste and wipe fails TCLP for metals	HW	HW	Meets paragraph (2) of the definition of "solvent-contaminated wipe" in §260.10
Spent chloroform from a lab	D022 if fails TCLP	D022 if fails TCLP	See Footnote 4

HW = hazardous waste; SW = solid waste; TCLP = toxicity characteristic leaching procedure.

[1] The status shown in this table assumes that all applicable conditions in §§261.4(a)(26) and 261.4(b)(18) are met.

[2] The wipes and the container they are in must not contain free liquids at the point of being transported for disposal or at the point of being sent for cleaning onsite or at the point of being transported offsite for cleaning.

[3] Wipes that are contaminated with an ICR-only listed waste but do not exhibit the ignitability, corrosivity, or reactivity characteristic are not hazardous wastes per §261.3(g)(1). Thus, these wipes would not need to be managed under the conditions of the wipes rule.

[4] Guidance available at http://www.epa.gov/osw/hazard/wastetypes/wasteid/solvents/sumry_chrt_wipes_fnl_rul_070913.pdf indicates that EPA did not intend to exclude wipes contaminated with solvent chemicals unless those chemicals are on the F001–F005 lists. However, the regulatory language is not entirely clear on this point.

Source: McCoy and Associates, Inc., unless otherwise noted; adapted from §§260.10, 261.4(a)(26), and 261.4(b)(18).

excluded. Table 3-3 contains additional examples that show when the exclusions for solvent-contaminated wipes can apply and when they won't. The next two sections summarize the conditions that must be met for the new exclusions to apply.

3.3.3.1 Exclusion from the definition of hazardous waste

To be excluded from the definition of hazardous waste, solvent-contaminated wipes that will be sent for disposal (to a landfill or combustion facility) must meet the following requirements:

- The wipes cannot be hazardous due to trichloroethylene.

- Containers used for accumulation and transportation must be 1) nonleaking and closed, 2) able to contain free liquids, and 3) labeled "Excluded Solvent-Contaminated Wipes." During accumulation, a container is considered closed when there is complete contact between the fitted lid and the rim, except when it is necessary to add or remove solvent-contaminated wipes. When the container is full, or when the solvent-contaminated wipes are no longer being accumulated, or when the container is being transported, the container must be sealed with all lids properly and securely affixed to the container and all openings tightly bound or closed sufficiently to prevent leaks and emissions.

- Generator accumulation time cannot exceed 180 days from the start date of accumulation for each container (i.e., the date the first solvent-contaminated wipe is placed in the container).

- At the point of being transported for disposal, the solvent-contaminated wipes must not contain free liquids as determined by the paint filter test (SW–846 Method 9095B), and there cannot be any free liquid in the container holding the wipes. Free liquids removed from solvent-contaminated wipes or from containers holding the wipes 1) are not excluded, 2) must be characterized, and 3) must be managed appropriately based on the hazardous waste characterization.

- Documentation per §261.4(b)(18)(v) must be maintained at the generator's site. See Section 3.3.3.3.

- The disposal facility must be one of the following:

    - A municipal solid waste landfill regulated under Part 258, including §258.40 (design criteria for new landfills);

    - A hazardous waste landfill regulated under Part 264 or 265;

    - A municipal waste combustor or other combustion facility regulated under Section 129 of the Clean Air Act; or

    - A hazardous waste combustor, boiler, or industrial furnace regulated under Part 264, 265, or Subpart H of Part 266.

3.3.3.2 Exclusion from the definition of solid waste

EPA excluded solvent-contaminated wipes that are sent for cleaning and reuse from the definition of solid waste because the agency believes they are more commodity-like than waste-like. To be excluded from the definition of solid waste, solvent-contaminated wipes that will be cleaned and reused must meet the following requirements:

- The exclusion is extended to wipes contaminated with any of the F- or U-listed solvents, including trichloroethylene.

- Containers used for accumulation and transportation must be 1) nonleaking and closed, 2) able to contain free liquids, and 3) labeled "Excluded Solvent-Contaminated Wipes."

- Generator accumulation time cannot exceed 180 days from the start date of accumulation for each container.

- At the point of being sent for cleaning, either onsite or offsite, the wipes and the container they are in must not contain free liquids.

- Documentation per §261.4(a)(26)(v) must be maintained at the generator's site. See Section 3.3.3.3.

- The wipes must be sent to a laundry or dry cleaner whose discharge, if any, is regulated under Sections 301 and 402 or Section 307 of the Clean Water Act.

3.3.3.3 Recordkeeping

In order for their solvent-contaminated wipes to be excluded from the definition of solid or hazardous waste, generators must maintain the following documentation at their site:

- The name and address of the laundry, dry cleaner, landfill, or combustor that is receiving the solvent-contaminated wipes;

- Documentation that the 180-day accumulation time limit is being met; and

- A description of the process the generator is using to ensure the solvent-contaminated wipes contain no free liquids at the point of being laundered onsite, or at the point of being transported for offsite laundering, dry cleaning, or disposal.

3.3.3.4 State implementation

The effective date of the rule is January 31, 2014. Because, the conditional exclusions discussed above are not HSWA regulations, they will not become effective in authorized states until those states adopt the rule. The rule includes requirements and conditions that are less stringent than those required under the base RCRA hazardous waste program. Thus, states, except as described below, are not required to adopt the conditional exclusions (although EPA encourages such adoption to reduce regulatory burden on businesses and maximize national consistency). Some conditions required in the final rule may be more stringent than certain existing state programs. As a result, authorized states whose programs include requirements that are less stringent than the final rule are required to modify their programs to maintain consistency with the federal program.

3.3.4 Ignitable rags

Sometimes, rags are contaminated with unlisted solvents (e.g., isopropyl alcohol). EPA has discussed how such rags containing no free liquids (from the paint filter test) may still be ignitable wastes. This may occur in two situations [RO 14285]:

1. When a number of solvent-contaminated rags are placed in a container, gravity (sometimes aided by compaction by plant personnel) causes solvent to be squeezed from the rags. As a result, free solvent liquid may form on the container bottom. Any such free liquid in the container should be tested for flash point; if the flash point is <140°F, the entire waste in the container should be considered ignitable.

2. When solvent-contaminated rags are placed in an environment where oxygen is present, they may meet the ignitable solids criteria in §261.21(a)(2).

In either situation, the exclusions for wipes discussed in Section 3.3.3 could still apply to the ignitable wipes if the conditions in §261.4(a)(26) or 261.4(b)(18) are met.

Also note that under the May 16, 2001 rule mentioned previously [66 *FR* 27266], if a rag is contaminated with F003 spent solvent but is not ignitable, it is not an F003 waste at the federal level (although it could exhibit a characteristic); state rag policies may be more stringent.

3.3.5 Rags contaminated with used oil

The used oil regulations [at §279.10(c)] state that rags contaminated with used oil that have been wrung out "to the extent possible such that no visible signs of free-flowing oil remain" are not used oil and are not subject to the Part 279 used oil regulations. Instead, wrung-out rags are solid waste (if they will be disposed), and a hazardous waste determination must be made for them just like any solid waste, considering any regional/state policy for rags. Used oil wrung from the rags, however, would continue to be subject to Part 279 standards.

3.4 F006, F007, F008, F009, and F019 wastes

Wastes that originate from electroplating operations are among the most complex from a RCRA regulatory perspective. The regulatory problems that are most common with these wastes involve determining which plating wastes are regulated, and deciding which hazardous waste codes apply.

Because the listing descriptions for the plating wastes may appear to overlap, a review of EPA guidance is critical to understanding the applicability of these waste codes.

3.4.1 The plating process

A diagram of a hypothetical electroplating process appears in Figure 3-2.

The metal parts (solid lines) are conveyed through a series of different process baths. The electroplating step is defined as the application of a surface coating. This process usually, but not always, involves electrodeposition to provide corrosion protection, erosion resistance, and/or friction-reducing characteristics or to improve appearance. [RO 11103] Chemical etching and milling are typically nonelectrodeposition processes that are also considered to be electroplating processes.

Wastes from each process bath (dashed lines) are sometimes collected in a process sewer. The process sewer flows to a wastewater treatment system that produces two residues: sludge and treated water. Besides the process flows identified above, electroplating facilities also produce wastes from floor washings, accidental spills, and general maintenance activities. The process sewer also typically captures these episodic wastes.

3.4.2 F006 wastes

The current listing description for F006 in §261.31 is: Wastewater treatment sludges from electroplating operations *except* from the following processes:

1. Sulfuric acid anodizing of aluminum;
2. Tin plating on carbon steel;
3. Zinc plating (segregated basis) on carbon steel;
4. Aluminum or zinc-aluminum plating on carbon steel;
5. Cleaning/stripping associated with tin, zinc, and aluminum plating on carbon steel; and
6. Chemical etching and milling of aluminum.

In addition to the six processes cited above that are specifically excluded in the regulatory description, F006 sludges will also not be produced by the following processes [December 2, 1986; 51 *FR* 43351]:

7. Chemical conversion coating (as we will discuss in Section 3.4.6, wastewater treatment sludges from chemical conversion coating of aluminum are considered to be F019 wastes);

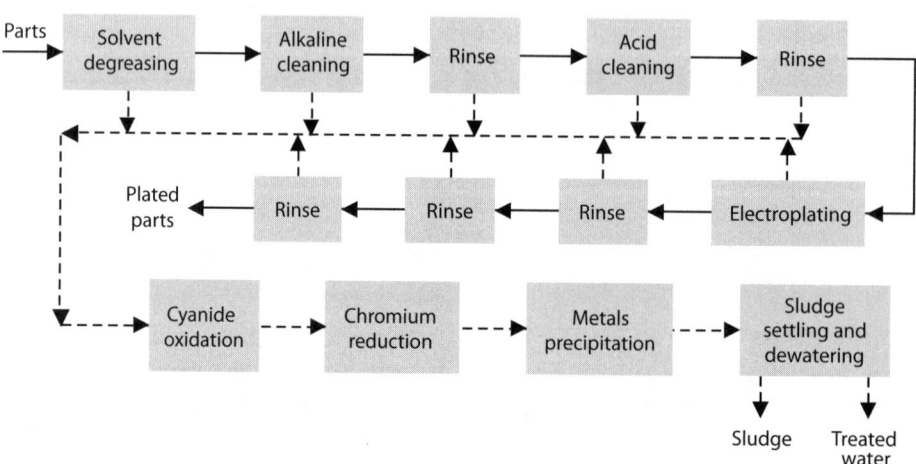

Figure 3-2: Hypothetical Electroplating Process

Source: McCoy and Associates, Inc.

8. Electroless plating; and

9. Printed circuit board manufacturing (unless the manufacturing process includes chemical etching or milling).

The nine exclusions given above describe processes that will *not* produce F006 wastewater treatment sludge. Perhaps what is more important is identifying the processes that *will* produce this listed waste. First, it is worth noting that F006 was originally listed because electroplating operations typically use cadmium, chromium, nickel, and complexed cyanides, all of which can partition into the wastewater treatment sludge. In order to be listed as F006, the sludge must originate from the following processes [51 *FR* 43351]:

- Common and precious metals electroplating [except tin, zinc (segregated basis), aluminum, and zinc-aluminum plating on carbon steel];

- Anodizing (except sulfuric acid anodizing of aluminum);

- Chemical etching and milling (except when performed on aluminum); and

- Cleaning and stripping associated with the above processes (except when associated with tin, zinc, and aluminum plating on carbon steel).

The following subsections describe important details of how various types of wastes associated with electroplating operations are regulated.

3.4.2.1 Baths and rinsewaters

In plating shops, the two most common wastes are baths and rinsewaters. The baths are essentially concentrated solutions used during the electroplating process. If the spent baths are kept segregated, they (and any residues produced from their treatment) will typically have their own hazardous waste code (e.g., F007, F008, or F009, each of which will be described later in this section). If the baths become mixed with electroplating wastewaters, however, sludges generated from treatment of the mixture would be F006. [RO 11315]

When parts pass through the various baths, they typically carry out residual droplets of the bath chemicals (called "dragout"), which are then removed by rinsing. Because these droplets are considered to be chemicals that are still in use (i.e., they are not spent chemicals), they are not solid wastes, do not have a waste code yet, and will not cause the rinsewater to be listed hazardous waste via the mixture rule. [RO 11198, 11269, 11339, 14314] In other words, rinsewaters are not themselves listed hazardous wastes (although they may exhibit a characteristic). Only when these rinsewaters are treated in a wastewater treatment system that produces a sludge does the F006 waste code apply. [RO 11269]

As mentioned previously, electroplating processes include common and precious metals electroplating, anodizing, chemical etching and milling, and cleaning and stripping associated with these processes. The following guidance relates to how EPA has interpreted the scope of the last two of these processes.

3.4.2.1.1 *Chemical etching and milling*

Chemical etching and milling processes utilize chemical solutions to dissolve metal layers from a part. [Etching and milling are essentially the same process, but etching only removes relatively small amounts of metal (1 to 5 mils).] According to EPA, this category of processes includes "chemical milling, chemical etching, bright dipping, electropolishing, and electrochemical machining." [RO 11214] "Bright dipping" of brass parts is an example of chemical etching; the process changes the brass surface appearance and/or improves coating adhesion. [RO 13387]

An example of chemical milling would be when a manufacturer produces gold beads by plating gold on a copper base metal. Nitric acid is then used to dissolve the copper (a chemical milling process). [RO 11315]

When parts are removed from the etching or milling process, they are typically rinsed to remove the etching/milling chemicals. If a sludge is produced when treating the rinsewater, it is F006.

Q: *Does the F006 listing apply to wastewater treatment sludges generated from chemical etching of magnesium?*

A: Yes. Magnesium etching is not excluded from the F006 listing description. While the December 2, 1986 *Federal Register* guidance [51 *FR* 43351] limited the F006 listing to electroplating processes on common and precious metals (except as noted above), there are no such limitations for the remaining processes (anodizing, chemical etching and milling, and cleaning and stripping). Therefore, there is no exemption for specific metal types from the F006 listing for any of the remaining processes, and wastewater treatment sludges generated from chemical etching of magnesium are subject to the F006 listing. [RO 14691]

Q: *Is brightening or polishing of copper parts using a hydrogen peroxide and sulfuric acid solution considered a process associated with the F006 listing?*

A: Yes. The copper bright dip solution containing hydrogen peroxide and sulfuric acid is used to descale/bright dip. Since descaling by an acidic solution removes metal from the surface, it is essentially a chemical etching process that meets the F006 listing description. [RO 14808]

3.4.2.1.2 Cleaning and stripping

Cleaning and stripping operations that are associated with electroplating processes can produce F006 sludge. The purpose of the cleaning/stripping steps is to remove any contaminants that might interfere with the electroplating process. Three phases of cleaning are typically considered to be integral steps in the electroplating process: precleaning (bulk removal of oil and dirt), electrocleaning, and pickling. Treatment of wastewaters from any of these steps, when conducted as part of an electroplating operation, would produce F006 sludge. For example, a pickling bath might be used to remove oxide scale from precious metals prior to electroplating. If the spent bath is neutralized, the resulting sludge is not F006 (because the bath is not wastewater). However, if the pickled part is subsequently rinsed, the sludge from treating the rinsewater would be F006. [RO 11315] In another example, a cleaning process associated with a bright dipping/chemical etching operation is also a process associated with the F006 listing description. [RO 14808]

Cleaning operations that are not associated with electroplating (e.g., routine cleaning by washing with soap and water) would not produce wastes carrying the F006 code.

Stripping steps in an electroplating operation are typically performed to remove metal from a base material. For example, if a part has been improperly plated, it might be stripped to dissolve the unwanted metal, followed by replating. Because the stripping solutions can contain significant metal concentrations and cyanides, sludges produced from the treatment of wastewaters from metal stripping operations are regulated under the F006 category. [RO 11340]

A twist occurs in cleaning and stripping operations when paint stripping is performed. For example, the U.S. Armed Services strip paint from aircraft parts in preparation for electroplating. EPA advised that paint stripping is not considered a specific part of the electroplating process. If solvents are used for paint removal, the spent solvents could be considered F001–F005 wastes, depending on their chemical composition. If the parts are then rinsed and the rinsewaters are kept segregated from subsequent metal cleaning and electroplating wastes, they would not be considered part of the F006 category. However, if rinsewaters from paint stripping are combined with wastewaters from the electroplating operations, they could produce F006 sludge. [RO 11340]

3.4.2.1.3 Confusion with K062—spent pickle-liquor wastes

Spent pickle liquor is essentially spent acid that has been used to remove metal oxide scale from a substrate. When these spent acids come from the iron and steel industry (SIC Codes 331 and 332), they are classified as K062 wastes. [RO 12972] If spent acids are produced as part of the electroplating process, and if these acids are dumped to a wastewater sewer (or if the metal is subsequently rinsed), the wastewater could produce an F006 sludge. [RO 11315] According to EPA:

"[T]he agency did not intend the K062 listing to include electroplating processes that generate spent pickle liquor. This would be duplicative since electroplating wastes are specifically covered under F006.... In considering petitions to delist electroplating waste, the agency has stated that the F006 listing includes acidic wastes (i.e., spent pickle liquor) from the electroplating process. Electroplating operations typically pretreat the metal using acidic baths prior to electroplating. The acidic wastes from this process are generally mixed with spent plating bath solutions and lime treated. Sludge generated from this process is considered F006. For example, an electroplater acid pickles metal parts as part of the electroplating process. The resultant wastewater (including spent pickle liquor and rinsewater) is neutralized with lime. Sludge generated from this process is F006. In another example, a galvanizer also pickles metal parts prior to galvanizing. Since galvanizing is not included under the electroplating category, spent pickle liquor from this process would be considered EPA hazardous waste K062. If the pickle liquor is lime treated prior to disposal, the sludge from this process is a hazardous waste by virtue of the [derived-from] rule. In cases where acidic wastes from the electroplating operation remain untreated or are segregated from other process waste and treated separately, the waste is then considered K062...." [RO 12268]

3.4.2.1.4 *Filtrate and supernatant*

In the final sludge settling and dewatering step shown in Figure 3-2, supernatant (liquid overlying solids) results from gravity separation. Similarly, filtrate (the liquid that passes through a filter) is produced by mechanical filtration. With regards to these wastes, EPA stated:

"[S]upernatant from F006 generation is not considered to be F006, but simply wastewater from treatment of electroplating wastewaters. Filtrate from F006 sludges could be hazardous under the derived-from rule, but if it is similar in terms of identity and concentration of constituents in the influent to the wastewater treatment process, it is not considered to be derived from F006. Rather, it is the original influent wastewater." [August 17, 1988; 53 *FR* 31153]

3.4.2.2 Sludge

The F006 listing applies to wastewater treatment "sludge." The term "sludge" is defined in §260.10 as "any solid, semisolid, or liquid waste generated from a municipal, commercial, or industrial wastewater treatment plant, water supply treatment plant, or air pollution control facility exclusive of the treated effluent from a wastewater treatment plant."

In most cases, F006 sludges are formed when heavy metals precipitate from electroplating wastewater as a result of adding sodium hydroxide, sulfides, lime, or some other precipitant to the wastewater. However, sometimes other types of materials are considered to be F006 sludge. For example, rinsewater from an electroplating process is pumped through a filter and ion-exchange column for removal of solids, metal ions, and cyanide. The treated water is then either recycled or disposed, and the used filter and ion-exchange resin are sent offsite for recycling. EPA considers the spent filter media and resin to be listed F006 sludges (even if they don't exhibit a characteristic), which must be manifested when sent offsite to a recycling facility. [RO 11244] Even if treated or reclaimed wastewater is reinserted into the process, contaminant removal is considered to be pollution control and spent resins are F006 sludge. [RO 11857]

3.4.2.2.1 *Sludge from sequential treatment of electroplating wastewaters*

Sometimes an issue arises concerning where F006 sludges will appear in a sequential wastewater treatment system. In Figure 3-3, electroplating wastewaters are treated, and F006 sludge is produced in a settling unit. Water (supernatant) from this unit is then mixed with sanitary wastes, and the combined mixture is treated in a surface impoundment (or it could alternatively be a tank). Sludge forms in this impoundment, and an issue arises as to whether this impoundment sludge is also regulated as F006.

Some facilities have tried to make the case that after some initial treatment (e.g., treatment to meet a Clean Water Act effluent limitation), the effluent is no longer "wastewater" and can, therefore, no longer generate F006 sludge. According to EPA, the F006 background documents provide no guidance as to when "wastewater" becomes a "treated effluent." [RO 11139] As a result, the F006 listing includes sludges derived from the treatment of electroplating wastewater, regardless of where the sludges are formed. Thus, if a sludge is formed in a wastewater treatment tank, filtration device, or surface impoundment, it is F006 sludge. The units would be subject to all applicable hazardous waste regulations; the tank and filtration device might be wastewater treatment units that are exempt from permitting, but the surface impoundment would be a hazardous waste unit requiring a permit. [RO 12267]

A question that arises occasionally is: Would the F006 code apply to sludge from a POTW if the POTW receives rinsewater from electroplating operations? Assume that the rinsewater was mixed with domestic sewage and arrived at the POTW by way of the sewer. EPA has implied that sludge from a POTW will not carry the F006 code under these circumstances but doesn't clearly explain why. [RO 11139, 14322]

3.4.2.2.2 Surprise! You've got F006

F006 sludges can be generated from nonhazardous wastes or wastes that are hazardous only for a characteristic (e.g., D002 corrosive wastewater). [RO 11269] For example, rinsewater from an electroplating process may not exhibit any characteristics and doesn't meet any listing descriptions. This nonhazardous waste could be shipped to an offsite facility without a manifest. However, if the offsite facility generates a sludge from processing the rinsewater, it has just generated F006 hazardous waste. [RO 11439] According to EPA:

> "The listing applies whether the sludge is generated at the electroplating facility or a commercial waste treatment facility. Thus, a commercial waste treatment facility must know the electroplating processes to identify the wastewater treatment sludge correctly as F006. This may require the treatment facility to obtain information from the waste generator regarding processes that produced the waste to be treated.... Once a treatment, storage, or disposal facility accepts the

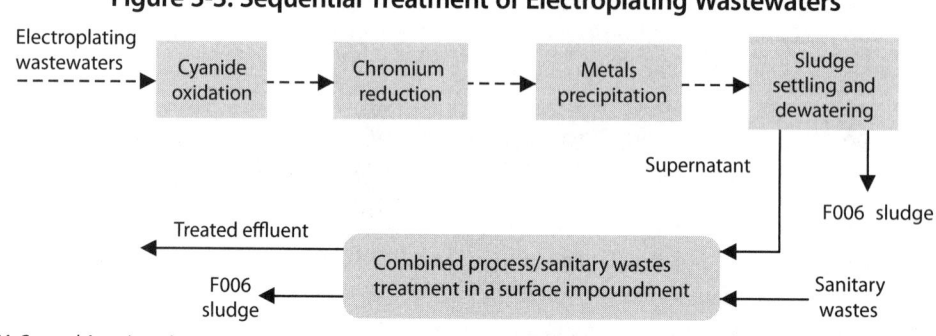

Figure 3-3: Sequential Treatment of Electroplating Wastewaters

Source: McCoy and Associates, Inc.

waste, it is their responsibility to accurately characterize any residual resulting from treatment." [RO 11375]

3.4.2.2.3 Court decision creates confusion

In a 1994 court decision (*U.S. v. Bethlehem Steel Corporation*, 38 F.3d 862), the U.S. Court of Appeals for the Seventh Circuit determined that when electroplating wastewater was mixed with other types of wastewater, any resulting sludge was not F006. EPA rejects this conclusion outside of the Seventh Circuit (the states of Illinois, Indiana, and Wisconsin), and in the Seventh Circuit continues to regulate sludges from mixed wastewaters as F006 via application of the mixture rule. (At the time of the court decision, the mixture rule was not in effect.) Therefore, according to EPA, the court decision has no practical effect. [RO 12849, 14391]

3.4.2.3 Examples of F006 wastes

EPA has determined that wastewater treatment sludges from the electroplating operations shown in Table 3-4 *are* F006 wastes.

3.4.2.4 Methods to avoid F006 generation

Two methods are available to help electroplaters minimize their production of F006 waste:

1. Physically separate the electroplating step (e.g., chemical etching) from the other process steps (e.g., cleaning and stripping). This can be accomplished by rinsing and drying components such that hazardous contaminants are not carried downstream. [RO 11851]

2. If a facility generates a waste that meets the listing description for F006, and if the sludge contains very low levels of contaminants, the facility can submit a delisting petition per §§260.20 and 260.22. Note that electroplating sludges are one of the most common delisted wastes in Part 261, Appendix IX.

3.4.2.5 Exemptions from the F006 listing

As noted in Section 3.4.2, nine different processes are excluded from the F006 listing. EPA has issued many interpretations concerning the scope of

Table 3-4: Electroplating Operations That Produce F006

Process, operation, or waste material	RO reference
Alkaline surface cleaning	11071
Zinc plating on carbon steel using cyanides	11152
Zinc plating on gray cast iron	14511
Iron plating on aluminum	14482
Precleaning activities prior to electroplating	11340
Pickling prior to electroplating	11315, 12268
Ion-exchange resins from rinsewater treatment	11244, 11269, 11857, 14017, 14331
Photoresist stripping	11851
Cleaning and stripping in the printing industry	12926
Bright dipping	11217, 12974, 13387
Silicon wafer etching	12183
Spent activated carbon from rinsewater treatment	14331
Chemical etching of copper	14808

Source: McCoy and Associates, Inc.

these exclusions, some of which are discussed on the next few pages.

3.4.2.5.1 Zinc plating

One of the processes that is not considered to involve electroplating is "zinc plating (segregated basis) on carbon steel." The phrase "zinc plating (segregated basis)" refers to zinc plating processes that do not use cyanide. For example, wastewater treatment sludges from zinc plating using baths formulated from zinc oxide and/or sodium hydroxide would be excluded from the listing. Similarly, hot-dip galvanizing on carbon steel is not considered to be an electroplating process and, when operated on a segregated basis, would not result in the production of F006. [RO 11068] However, wastewater treatment sludges from zinc plating processes utilizing zinc cyanide and/or sodium cyanide would be considered to be F006 sludges. Where both cyanide- and noncyanide-containing processes are used at the same facility, the noncyanide-containing sludges must be kept segregated from the cyanide-containing sludges in order to remain excluded from the F006 listing. [December 2, 1986; 51 FR 43351]

3.4.2.5.2 Chemical conversion coating

Chemical conversion coating includes chromating, phosphating, immersion plating, and coloring. While chemical conversion coating of aluminum can result in the production of F019 sludge (see Section 3.4.6), other types of conversion coating were specifically excluded from the F006 listing on December 2, 1986. [51 FR 43351] For example, when zinc plating on steel is followed by zinc phosphating, wastewater treatment sludges from the process are not considered to be F006. [RO 12972]

3.4.2.5.3 Printed circuit board manufacturing

Although a general exclusion from the F006 listing exists for wastes from printed circuit board manufacturing, two circumstances have been identified where the general exclusion does *not* apply.

1. When EPA decided that wastes from printed circuit board manufacturing were excluded from the definition of F006 [51 FR 43351], they added the following footnote:

   "Wastewater treatment sludges from printed circuit board manufacturing operations that include processes which are within the scope of the listing (e.g., chemical etching) are regulated as EPA Hazardous Waste No. F006."

2. Depending on how printed circuit boards are manufactured, photoresist "skins" may be an F006 waste. The principle to be applied is described as follows:

   "If there is any possibility of skins stripping solutions being mixed or commingled with other electroplating wastewater, or if the hazardous constituents generated by other electroplating processes can otherwise be 'carried forward' by the nature of the association of the two processes, the skins themselves could become contaminated with these hazardous constituents." [RO 11851]

For example, if circuit boards are sequentially etched and then stripped of photoresist material, contaminants from chemical etching (an F006-generating process) could enter the stripping bath and contaminate the photoresist solids. Under these circumstances, when the solids are filtered from the contaminated stripping solution, they would be F006. In EPA's language:

   "If the stripping operation is in line with or contiguous with an electroplating operation, then the stripper solution itself becomes an electroplating wastewater. 'In line with or contiguous with' in this case would mean the stripping operation is not physically separated from these [electroplating] operations and the printed circuit boards are not rinsed and dried prior to the photoresist stripping operation. The stripper solutions thus could be mixed or intermingled with electroplating wastewater." [RO 11851]

On the other hand, if the boards are etched, rinsed, dried, and then stripped, contaminants from the etching step would not be carried into the stripping step, and the photoresist solids would not be F006. [RO 11851]

EPA has determined that wastewater treatment sludges from the manufacturing operations shown in Table 3-5 are *not* F006 wastes.

Once you've figured out if you have F006 waste, the following subsections provide thoughts on its storage and recycling requirements.

3.4.2.6 F006 storage

On March 8, 2000 [65 FR 12378], EPA issued a rule allowing generators to accumulate F006 waste for up to 180 days (or 270 days if the waste will be shipped more than 200 miles to a metal recovery facility). Normally, large quantity generators would only be allowed to store the waste onsite for up to 90 days without obtaining a hazardous waste storage permit. In order for generators to utilize the

Table 3-5: Manufacturing Operations That Do Not Produce F006

Process, operation, or waste material	RO reference
Hot-dip galvanizing (segregated basis)	11068
Phosphating	11157, 12962, 12972
Steel pickling	12268, 12972
Cleaning not associated with electroplating	11157, 11245, 11273, 11340, 11458
Paint stripping	11340
Chemical conversion coating (including chromate conversion coating)	11273, 12764
Electroless plating	12962, 13102
Electrodeposition with a clear acrylic film	11940
Mechanical plating	12671
Metal hydroxide deposition into pores	11663
Electrostatic painting	11216
Zinc-cobalt alloy plating	11861
Air emissions filter on electroplating line	14108
Immersion coating of aluminum	12610
Immersion plating of bronze on steel wire	11371
Mechanical burnishing and polishing	11315
Contact cooling water	11315

Source: McCoy and Associates, Inc.

additional storage period, they must: 1) implement pollution prevention practices that reduce the amount of hazardous constituents in their F006 waste or otherwise released to the environment prior to recycling, 2) send the F006 waste to onsite or offsite metals recovery, 3) accumulate no more than 20,000 kg of F006 waste onsite at one time, and 4) keep certain documentation at their facility. These provisions are codified in §262.34(g).

3.4.2.7 F006 recycling

Because F006 waste disposal is relatively expensive, many facilities have looked at potential recycling methods for these wastes. Here are two examples:

1. Plating facilities sometimes want to send their F006 sludge to a copper smelter. When they sought EPA's concurrence that this type of recycling would be RCRA-exempt use as an ingredient under §261.2(e)(1)(i), the agency disagreed. The agency considers the copper smelter to be a reclamation device, and §261.2(e)(1)(i) does not allow reclamation. [50 FR 633, RO 11338, 11426, 11910, 14026]

2. Some F006 sludges are recycled by high-temperature metals recovery (HTMR) processes. These processes typically produce a slag residue, which was used as an anti-skid/deicing agent on roadways. This practice was specifically prohibited on August 24, 1994. [59 FR 43496] See §266.20(c). However, HTMR slags derived from F006 sludge may be used as road-base material or in other applications that involve land placement as long as the provisions of §266.20(b) are met; that is, the slag must meet the land disposal treatment standard for F006. [RO 14082]

EPA has frequently been asked about the regulatory status of other F006 recycling processes. The agency typically only gives general guidance on this issue and leaves specific determinations to state regulators. The primary concern for any F006 recycling scenario is that the operation may actually involve heavy metal treatment/disposal, which requires a RCRA permit. Issues that EPA typically evaluates when considering an F006 recycling scenario include:

- Whether a market actually exists for the product from the recycling operation [RO 11461];

- Whether the product contains high levels of hazardous contaminants when compared with similar (nonwaste-derived) products [RO 11461];

- Whether the recycling process is likely to release hazardous constituents that are different in type or concentration than those from the processing of analogous raw materials [RO 11426];

- Whether the F006 sludge is likely to be mismanaged prior to recycling [RO 11426];

- Whether the product is used in a manner constituting disposal (e.g., as road-base material); and

- Whether the recycling unit is a Bevill-exempt device (e.g., a primary smelter or cement kiln).

3.4.2.7.1 F006 exception from the derived-from rule

If F006 sludge is sent to an HTMR unit, any slag or other residue would normally be considered to be F006 waste via operation of the derived-from rule. However on August 18, 1992 [57 FR 37194], EPA codified an exception from the derived-from rule for nonwastewater residues resulting from HTMR processing of F006 (and K061 and K062). This exception, which is codified in §261.3(c)(2)(ii)(C), specifies that, as long as such residues meet certain concentration limits and are sent to a Subtitle D disposal unit, they are not hazardous wastes. In order to qualify for the exception, the residues must not exhibit any hazardous waste characteristics. Certain waste testing and recordkeeping requirements are also imposed by the exception. Section 5.2.3.4 contains additional details on the exception.

3.4.3 F007 wastes

The listing description in §261.31 for F007 wastes is: "Spent cyanide plating bath solutions from electroplating operations." Note that the plating bath solution is not an F007 waste until it is spent. Clearly, dumps of the bath are F007 wastes; however, EPA has given conflicting guidance on the status of bath dragout or carryover. In one interpretation, parts were transferred from a plating bath to a chlorination tank to stop the plating process. The agency stated that plating bath dragout is considered to be a spent bath solution; hence, any sludge that forms in the chlorination tank would be considered F007 waste. [RO 12269] In other interpretations, the agency concluded that dragout droplets are not spent materials and that rinsewater baths following the plating process are also not solid wastes. Therefore, these baths would not be F007, F008, or F009. [RO 11198, 11269, 11339, 14314] We believe that this second interpretation is the most defensible, and it certainly is the one EPA has confirmed most often.

F007 wastes can also be reactive due to their cyanide content. EPA gives this example: When anode bags are removed from a cyanide plating bath, they are considered to be a spent material. Washing the anode bags for reuse is considered to be reclamation; hence, according to Table 1 in §261.2(c), the bags are a solid waste. Because the bags are both reactive and have been soaked in spent plating bath, they carry both the D003 and F007 codes. [RO 13476] That's what EPA said, but, per §268.9(b), the D003 code would not have to be reported on the LDR notification. Also note that per EPA's October 30, 2008 final rule [73 FR 64668], spent materials sent for reclamation are not solid wastes if the provisions of that rule are met.

Finally, note that spent baths are only F007 wastes if they are "electroplating baths." As mentioned previously, electroless plating is not considered to be electroplating. [December 2, 1986; 51 FR 43351] EPA has stated that, even if electroless plating baths contain low concentrations of cyanide, they would not be considered F007 wastes when disposed. They could, however, be characteristic wastes. [RO 13311] Similarly, a cyanide-containing bath from an electrowinning process is not considered to be F007 waste. [RO 13304]

3.4.4 F008 wastes

F008 listed wastes are described in §261.31 as "Plating bath residues from the bottom of plating baths from electroplating operations where cyanides are used in the process." In one case, a filter was used to remove residue from a plating bath. When the filter media were removed for disposal, EPA stated that the proper waste codes would be D003 and F008. [Per §268.9(b), however, the D003 code will not have to be reported on the LDR notification.]

Although the filter residue could also contain spent plating bath (F007), the F008 listing is more specific and should be used instead of the F007 listing. [RO 13476]

3.4.5 F009 wastes

The listing description for F009 in §261.31 is "Spent stripping and cleaning bath solutions from electroplating operations where cyanides are used in the process." Where cleaning solutions do not contain cyanides (e.g., they are detergent solutions), the F009 listing is not applicable. [RO 13476]

An example of a stripping process that produces F009 waste involves a chemical etching process. In the process, a potassium ferricyanide solution at pH >12.0 is used to remove tungsten or molybdenum substrate material. Although the waste might meet the listing description for F007 (spent cyanide plating bath from electroplating), EPA decided that the correct waste code was F009. [RO 11251]

One confusing issue related to F009 wastes is the source of the cyanides in the baths. Two possibilities exist: 1) cyanide can be an ingredient in the stripping and cleaning solution, or 2) cyanide can be carried into a noncyanide-containing stripping and cleaning bath via dragout from a previous cyanide-containing step. Clearly, the F009 code was meant to apply in the first situation. Regarding the second situation, EPA stated:

> "A cleaning and stripping bath used prior to the cyanide plating bath would not contain cyanide contamination from carryover. Spent cleaning and stripping baths that follow cyanide plating baths at some point in the dip sequence would have levels of cyanide in them due to dragout. Therefore, it is EPA's intent to regulate only those spent cleaning and stripping baths from electroplating processes that are used at some point after the cyanide bath. However, if cleaning and stripping baths are commingled with other baths occurring during or after the cyanide baths or if cyanide-containing solutions or wastes are introduced or recycled in the process upstream of the

cyanide plating baths, then these cleaning or stripping baths would be F009." [RO 13301]

Finally, if any sludges form in the bottom of the cleaning tank, they are also considered to be F009. [RO 14276]

3.4.6 F019 wastes

When EPA developed the listing description for F006, it listed wastewater treatment sludges from chemical conversion coating of aluminum separately as F019. This distinction was made because the aluminum-based sludges were not expected to contain significant concentrations of cadmium and nickel (constituents typically found in other F006 wastes). The current listing description for F019 in §261.31 is "Wastewater treatment sludges from the chemical conversion coating of aluminum except from zirconium phosphating in aluminum can washing when such phosphating is an exclusive conversion coating process. Wastewater treatment sludges from the manufacturing of motor vehicles using a zinc phosphating process will not be subject to this listing...."

To EPA's thinking, F019 was, in effect, a subcategory of the F006 listing. [RO 11080] Later on [December 2, 1986; 51 *FR* 43351], EPA removed chemical conversion coating from the types of processes that would produce F006 wastes.

Because F019 was considered to be a subcategory of the F006 listing, many of the provisions applicable to F006 wastes also apply to F019 wastes. For example, supernatant and filtrate from processing F019 sludge are simply wastewaters and do not carry the F019 code. [RO 13323] Similarly, if wastewater from chemical conversion coating of aluminum is treated in sequential systems (e.g., two tanks in series followed by a surface impoundment), any sludge that appears in any of the units is F019. [RO 11507, 11961] Finally, wastewater treatment sludges generated from treatment of contaminated rinsewaters collected from stages prior to chromate conversion coating are not F019. [RO 11731]

3.4.6.1 Exemptions from the F019 listing

The exemption for zirconium phosphating was added in 1990. [February 14, 1990; 55 FR 5340] In the *Federal Register* notice, EPA stated:

"This final exclusion applies only to sludges from processes that exclusively use zirconium phosphating solutions that do not contain chromium or cyanides. Further, these processes are not associated with electroplating or conversion coating steps where hazardous constituents are used. For example, if a can maker employs a chromating step, separately or in conjunction with such zirconium phosphating, the wastewater treatment sludges would meet the F019 listing and would not be excluded under this rulemaking."

Other types of phosphating on aluminum (e.g., tin phosphating) are not covered by the zirconium phosphating exclusion and would result in the production of F019 wastes. [RO 11547, 14103]

EPA states that "[a]s to sludges from sulfuric acid anodizing, these wastes do not meet the F019 listing since anodizing is not considered to be a 'conversion coating' process. Anodizing is an electrical process wherein the part is made anodic, whereas conversion coating uses nonelectrical processes." [RO 11551]

EPA excluded wastewater treatment sludges generated from zinc phosphating in the automotive assembly process on June 4, 2008. [73 FR 31756] Until this rule was issued, the zinc phosphating (conversion coating) step in the manufacture of motor vehicles containing aluminum parts resulted in generation of F019 sludge from the wastewater treatment systems at auto plants. The new exclusion, which applies at the point of generation, is based on the following conditions:

1. The waste must be generated from motor vehicle manufacturing. This is defined to include the manufacture of automobiles and light-duty trucks/vehicles (NAICS codes 336111 and 336112, respectively). The exclusion to the F019 listing does not apply to other motor vehicle manufacturing industries (such as heavy-duty truck or motor home manufacturing).

2. Prior to offsite shipment, the waste must not be placed outside on the land.

3. The waste must be disposed in a landfill that is either 1) a state-permitted or authorized Subtitle D (municipal or industrial) unit that is equipped with at least a single clay liner; or 2) subject to, or otherwise meeting, the requirements in §258.40, 264.301, or 265.301.

4. Generators must maintain documentation onsite proving that they meet the above conditions excluding their wastewater treatment sludges from the F019 listing.

As a follow-up to the exclusion from F019 for wastewater treatment sludges generated from zinc phosphating in the automotive assembly process, EPA provided a May 2009 interpretation noting that sludges generated from zirconium oxide coating operations on automobile bodies containing aluminum are also excluded from the F019 listing. [RO 14806]

3.4.7 The big disclaimer

When EPA is asked to clarify how plating wastes are regulated under RCRA, they typically respond that they are providing hypothetical answers as to the general applicability of RCRA. The agency always goes on to state something like the following:

"For each individual facility, the appropriate region or RCRA-authorized state will have to make the final determination as to the applicability of RCRA regulations based on an analysis of the actual facilities and processes. Regardless of their RCRA authorization status, states may impose regulations more stringent or broader in scope than those in 40 CFR Parts 260–270 as a matter of state law." [RO 11385]

3.5 P- and U-listed hazardous wastes

People make more mistakes in identifying P- and U-wastes than any other type of listed hazardous waste. The most frequent mistake is that people decide they are managing P- or U-wastes when really they're not. We start our discussion of these wastes by reviewing their regulatory background.

3.5.1 Regulatory background

The P- and U-wastes are regulated under §261.33—Discarded commercial chemical products, off-specification species, container residues, and spill residues thereof. These commercial chemical products were first regulated on May 19, 1980; here is how EPA described them:

> "EPA intended to encompass those chemical products which possessed toxic or other hazardous properties and which, for various reasons, are sometimes thrown away in pure or undiluted form. The reasons for discarding these materials might be that the materials did not meet required specifications, that inventories were being reduced, or that the product line had changed. The regulation was intended to designate chemicals themselves as hazardous wastes, if discarded, not to list all wastes which might contain these chemical constituents. In drawing up these lists, the agency drew heavily upon previous work by EPA and other organizations identifying substances of particular concern [e.g., the Department of Transportation]." [45 *FR* 33115]

The so-called P-chemicals, which are identified in §261.33(e), possess "extremely hazardous properties" that make them lethal in very small quantities. [45 *FR* 33116] These chemicals meet the listing criteria of §261.11(a)(2). They are identified as acute hazardous wastes having a regulatory threshold of 1 kg/mo. If a facility generates a single P-waste (or multiple P-wastes) at a rate of >1 kg/mo, the facility is regulated as a large quantity generator.

The chemicals on the U-list [§261.33(f)] meet the listing criteria of §261.11(a)(3), which identifies various factors that indicate a waste is "toxic." These wastes are regulated at a regulatory threshold of 100 kg/mo. Thus, if a facility only generates one hazardous waste, and if this waste has a U-code, the facility would be regulated as a small quantity generator if the generation rate is between 100 and 1,000 kg/mo. It would be a large quantity generator if the plant generates 1,000 kg/mo or more of the waste.

3.5.2 Practical hints on using the P- and U-list

First, note that the chemicals on each list are listed twice: the first time in alphabetical order, the second time in numerical order by waste code. If, for some reason you have a waste code (e.g., P024), and you want to find out what chemical corresponds to this hazardous waste number, use the second list, which is formatted in numerical order by waste code. (It turns out that P024 is *p*-chloroaniline.)

Next, don't assume that EPA listed chemical names in a consistent manner. There are some inconsistencies in chemical nomenclature in the P- and U-lists. The risk here is that, if you can't find the chemical you are getting ready to discard on either list, you could conclude that your chemical is not regulated, when in fact you're looking in the wrong place or perhaps are using a different chemical name than EPA used. To avoid these problems, we like to use the "List of Lists." This publication not only consistently alphabetizes chemicals, but it also identifies chemicals by common synonyms. For example, 1,1,1-trichloroethane is also known as methyl chloroform, which is hazardous waste number U226. The "List of Lists" is accessible online at http://www.epa.gov/emergencies/tools.htm#lol.

3.5.2.1 CAS numbers, variants, and isomers

Although the P- and U-lists also cross reference chemical names to Chemical Abstracts Service (CAS) numbers, this information needs to be used with care. For example, you may be using a chemical name that is different from the one EPA uses. If you can look up the CAS number for your chemical in the CAS Ninth Collective Index or some other reference and compare it with the CAS number in the P- or U-list, the listed waste code will apply if the numbers are identical.

Note, however, that it is not safe to assume that only the CAS number given in the P- or U-listing is regulated. For example, anhydrous cyclophosphamide (CAS No. 50-18-0) is listed as U058. Because this compound is hygroscopic, it readily hydrates to cyclophosphamide monohydrate (CAS No. 6055-19-2). EPA says that because both compounds are known generically as "cyclophosphamide," they are both U058 when disposed. [RO 11687] Anhydrous chloral (U034) and chloral hydrate are also regulated in the same manner; that is, they are both U034 even though their CAS numbers are different. [RO 14175]

A related situation occurs with chemical isomers. For example, toluenediamine has many isomers, including toluene-2,4-diamine (CAS No. 95-80-7) and toluene-2,6-diamine (CAS No. 823-40-5). The U-list contains U221—toluenediamine (CAS No. 25376-45-8). At first glance, it might appear that neither toluene-2,4-diamine nor toluene-2,6-diamine are listed hazardous wastes because their CAS numbers don't correspond with the U-listing. However, the CAS number in the listing is for generic toluenediamine mixed isomers, which means that both isomers identified above are U221 when disposed. [RO 13760]

If EPA designates a specific isomer, then only that isomer is covered by the listing. For example, U140—isobutyl alcohol (CAS No. 78-83-1) covers only this isomer; *n*-butyl alcohol is not U140 (it's U031). [RO 13760]

On a related note, EPA attempted to make the P- and U-listings very specific. Some of the P- and U-listings specify a parent compound and "salts" or "esters" of that compound (e.g., P075—nicotine & salts). If a parent compound is listed, but the salt or ester of that compound is not, then only the parent compound is regulated. [RO 12155] For example, dinoseb is listed as P020; no salts are listed. Therefore, if a product consists of "sodium dinoseb," it is not P020. [RO 11489] Similarly, EPA has clarified that epinephrine salts are *not* included in the scope of the P042 hazardous waste listing. P042 applies only when epinephrine base (CAS No. 51-43-4) is discarded unused. [RO 14778] The agency has gone on to say the same reasoning applies to the disposal of unused phentermine. Only phentermine base (CAS No. 122-09-8) has the potential to meet the P046 listing description. [RO 14831] This guidance is significant because many of the pharmaceuticals in use in hospitals and other medical applications are in the form of salts or esters.

3.5.2.2 Hazard codes

Some of the chemicals on the P- and U-list are followed by one or more letters in parentheses; these are hazard codes. Per §261.33(e), all P-listed chemicals are acute hazardous wastes and have a hazard code of (H) [hazard codes are explained in §261.30(b)]. If a letter does not follow a chemical on the P-list, the chemical is an acute hazardous waste due to its acute toxicity. If a chemical on the P-list is followed by one or more hazard codes in parentheses, the chemical is acutely hazardous for that property or properties. For example, the first P-waste in the §261.33(e) alphabetical list, P023—acetaldehyde, chloro-, is not followed by any hazard code in parentheses and would, therefore, be an acute hazardous waste due to acute toxicity [with complete hazard code (H, T)]. On the other hand, P006—aluminum phosphide, is followed by "(R, T)," which means that this chemical is acutely hazardous due both to its reactivity and acute toxicity [and would have a complete hazard code of (H, R, T)].

Note that three of the P-wastes (P009, P081, and P112) have a hazard code of (R) following the chemical name. This means that they are acutely hazardous only due to their reactivity (e.g., P081 is nitroglycerine). "EPA has also defined this [P-listed] category of wastes to include wastes, such as explosives, which otherwise meet…the statutory definition of hazardous waste. This has been done in recognition that wastes may be acutely hazardous even if they are not toxic." [May 19, 1980; 45 *FR* 33106] These three P-listed wastes are handled in a special way if they are mixed with other solid wastes, as described in Section 5.1.2.2.

If a letter does not follow a chemical on the U-list, the chemical was placed on this list due to its toxicity and is assumed to have a (T) code. Conversely, if a letter in parentheses follows the chemical name, the chemical was added only for that reason [e.g., U001—acetaldehyde has an "(I)" following it, indicating that it was listed *solely* due to its ignitability—not toxicity and ignitability].

These hazard codes have three practical functions. First, all acute hazardous wastes [i.e., those with an (H) code, including all P-wastes] have a regulatory threshold of 1 kg/mo when determining generator status per §261.5(e)(1). (Readers should refer to the discussion in Section 6.1 on waste counting for further details.) Second, containers of acute hazardous wastes (including all P-wastes) must be triple rinsed before they are empty per §261.7(b)(3). (Readers should refer to the discussion in Section 9.5 regarding empty-container issues.) Third, P- and U-wastes followed only by an (I), (C), or (R) are eligible for the special provisions of §261.3(g). (Refer to Sections 3.1.2, 5.1.2.2, and 5.2.2.2 for details.)

3.5.2.3 Tip for identifying commercial chemical products

The most common mistake that we encounter with P- and U-wastes is that people tend to overuse these codes; that is, they erroneously conclude that they have a P- or U-waste. A typical scenario that leads to this outcome is as follows. A generator analyzes its waste and finds that it contains one or more of the chemicals on the P- or U-list. The generator then applies the waste codes associated with any such chemicals that it identifies and must then manage the waste according to all applicable RCRA rules. (Alternatively, not understanding how the RCRA regulations work, the generator starts looking for the chemicals in Appendix VIII of Part 261. Unfortunately, this appendix gives waste codes that *might* be associated with various chemicals.)

Here is a conceptual tool that we sometimes find useful in deciding whether a material is a listed P- or U-waste. Visualize the label on a container of the material in question. If the label uses one of the generic chemical names from the P- or U-list to describe the contents of the container, the material is a listed P- or U-waste when disposed. If the label uses a name or term that is not on the P- or U-list, the material is probably not P- or U-listed (unless it contains a P- or U-listed chemical as the sole active ingredient, as will be discussed later).

For example, consider an unused mercury thermometer that we intend to discard. Thermometers would typically be stored/shipped in a box, and the label on the box would say "Thermometer." When you look on the P- or U-list, the word "Thermometer" doesn't appear; hence, the thermometer is not a listed waste. (The thermometer is actually a manufactured article that contains mercury and would only be hazardous if it exhibits a characteristic—probably D009.) [RO 13310, 14012] Compare this against a vial of unused mercury that we intend to discard. The label on the vial would say "Mercury," which is listed as U151. The vial of mercury would be U151 hazardous waste.

Q *Creosote-containing railroad ties are to be disposed. Creosote appears on the U-list as U051; are the ties listed hazardous waste U051?*

A The label on a bundle of ties would read "Railroad ties." The label does not say "Creosote." Therefore, unless the ties exhibit a characteristic, they are not hazardous waste. [RO 11094, 11104, 11129, 11535, 12012] However, note the following twist. A facility bakes used railroad ties to recover creosote, and then uses the recovered creosote as a wood preservative without further processing. The reclaimed creosote is again a commercial chemical product (i.e., the label on a drum of this material would say "Creosote"). If spilled or discarded, this recovered material would be hazardous waste U051. [RO 13572]

Q *Unused paint that contains xylene is being discarded. Xylene appears on the U-list as U239; is the discarded paint U239?*

A No. The label on the container says "Paint" not "Xylene." If the paint exhibits the characteristic of ignitability, it would be D001. [RO 11180]

3.5.3 What is a "commercial chemical product"?

If you read the text of the regulations in §261.33(a) through (d), you will find references to a number of materials that are regulated as commercial chemical products (CCPs). These materials include:

1. Pure, unused chemicals having the generic name given in the P- or U-list;

2. Manufacturing chemical intermediates having these names;

3. Off-specification variants of 1 or 2 above;

4. Residues of these chemicals in containers that do not meet the definition of "RCRA-empty"; and

5. Residue and debris resulting from the cleanup of spills of these chemicals.

The only definition of these terms appears in the comment included in §261.33(d) that is part of the regulations:

> "*Comment:* The phrase 'commercial chemical product or manufacturing chemical intermediate having the generic name listed in…' refers to a chemical substance which is manufactured or formulated for commercial or manufacturing use which consists of the commercially pure grade of the chemical, any technical grades of the chemical that are produced or marketed, and all formulations in which the chemical is the sole active ingredient. It does not refer to a material, such as a manufacturing process waste, that contains any of the substances listed in [the P- or U-list]. Where a manufacturing process waste is deemed to be a hazardous waste because it contains a substance listed in [the P- or U-list], such waste will be listed [either as an F- or K-waste] or will be identified as a hazardous waste by the characteristics…."

Although this comment is helpful, it introduces several more terms that need interpretation:

6. Technical grades of the chemical;

7. Sole active ingredient; and

8. Manufacturing process wastes.

Each of these eight terms is discussed on the following pages.

3.5.3.1 Pure, unused chemicals

Pure, unused chemicals are CCPs. The general concept behind the regulation of pure, unused chemicals is that these are very concentrated (and, hence, very hazardous) materials. If improperly disposed, they would pose a significant threat to human health or the environment. Common situations that involve such chemicals include disposal of 1) excess or unsaleable inventory; 2) laboratory chemicals that are no longer needed, that are past an expiration date, or that are suspected of being contaminated or otherwise impure; and 3) feedstock or product from a process that is being shut down.

Unfortunately, EPA doesn't define the term "pure." As a result, a common concern in the regulated community is whether a waste that contains one or more of the P- or U-listed substances as a constituent must be managed as hazardous. EPA addressed this by noting:

> "The intent of the regulation was to encompass only those materials which were being thrown away in their pure form or as an off-specification species of the listed material…. [T]he regulatory language has been clarified to restrict the application of [the P- and U-listings] to chemical products, or their off-specification species, and not to wastes which contain these materials as constituents." [May 19, 1980; 45 *FR* 33115–6]

Where a chemical contains more than one constituent (i.e., it is impure), the agency captures such chemicals as CCPs through the use of other terms, such as "sole active ingredient" or "technical grade of a chemical," as will be discussed later. In most cases, if the main entry on the label on whatever you are discarding is the name of a chemical on the P- or U-list, it will be pure enough to be regulated.

In order to be regulated as this class of materials, the chemical *cannot have been used.* [RO 11202, 14686] "Only unused chemicals are considered commercial chemical products that could carry a 'P-listed' [or U-listed] waste code. Once a…chemical that is on the P-list [or U-list] has been used, it is not considered a commercial chemical product." [May 23, 2006; 71 *FR* 29725] This is an aspect to CCPs that people tend to frequently forget. In guidance, EPA has addressed several special cases or unusual applications of the "pure, unused" concept.

Q *A generator has two individual chemicals that it intends to discard: warfarin (concentration = 0.2%) and zinc phosphide (concentration = 8%). Each of these chemicals shows up on the P-list, but only for concentrations greater than what is being managed. Are the wastes nonhazardous?*

A No. Warfarin and zinc phosphide are unusual chemicals in that they appear on both the P- and U-lists. At the concentrations of the wastes, the generator would have U248—warfarin and U249—zinc phosphide.

Q *An incinerator is undergoing a trial burn, and the facility uses reagent-grade hexachloroethane to spike sand that will be test burned in the unit. This chemical shows up on the U-list as U131. Is the incinerator residue listed hazardous waste U131 by operation of the derived-from rule?*

A Yes. "When the sand was mixed with the chemicals, the sand becomes a solid waste and the chemical becomes a hazardous waste (U131) because the intent is to incinerate the mixture. Section 261.2(b)(2) clearly indicates these materials are solid wastes, unless excluded by §261.4(a) or by variance…. Accordingly, the mixture of U131 with sand is a hazardous waste by virtue of the 'mixture rule,' which provides that the mixture of a listed waste with a solid waste constitutes a hazardous waste. [See §261.3(a)(2)(iv).] As a result, the residue from the trial burn also would be a hazardous waste [via the 'derived-from' rule, §261.3(c)(2)(i)] because the residue is derived from a listed waste." [RO 11320; see also RO 13364, 14398]

Q *Glove boxes were used to form metal alloys, one component of which was beryllium powder. The P-list identifies beryllium powder as P015. Are wastes found in the glove boxes also considered to be P015 hazardous waste?*

A No. The P015 listing applies only to unused beryllium powder. The powder in this case has been used (when processed in the glove boxes) and, therefore, does not meet the P015 listing description. [RO 11706]

Q *Solid beryllium metal and alloys are machined, and resulting particles are captured by a vacuum exhaust system. The particles are subsequently found on the machining equipment, air filters, cyclones, and blowers. Would these be P015 wastes if disposed?*

A No. The P015 listing applies to beryllium dust that is a CCP. If machining operations create beryllium powder as a CCP, the P015 listing would apply. However, beryllium particles that are created in normal machining of beryllium-containing metals do not meet the P015 listing description. [RO 11706]

3.5.3.1.1 Chemicals used for their intended purpose

When a CCP is used for its intended purpose, it will not meet the listing description for any of the P- or U-wastes, since it has been used. [July 28, 1989; 54 *FR* 31336] Residual contamination that results from this use is only regulated if it meets the listing description for an F- or K-waste (an unlikely situation) or if it exhibits a characteristic. Numerous examples illustrate this principle.

Q *When a pesticide applicator sprays a U-listed pesticide from his/her crop duster, some of the chemical ends up on the side of the plane that is washed off upon landing. Is the washwater a U-listed hazardous waste when disposed?*

A No. "The agency does not believe that the pesticide residue left on the aircraft is a discarded commercial chemical product. The residue does not qualify as material discarded or intended to be discarded.... In listing commercial chemical products, EPA intended to cover those products which, for various reasons, are thrown away. The agency did not intend to cover these cases, as here, when the chemical is released into the environment as a result of use. Unless we take such a position, one could argue that the pesticide that is sprayed that does not fall directly on the crop (but falls on the ground next to the crop) would be disposal of an unused commercial chemical product; such an interpretation is a distortion of the commercial chemical product rule.... Rather, this rinsewater would be defined as hazardous only if it exhibits one or more of the characteristics." [RO 11096] See also RO 11160 that applies the same reasoning to washing the exterior of trucks and service vehicles.

When this interpretation was challenged by an EPA regional office, headquarters reaffirmed it and added the following twist: "Since the pesticide has been sprayed from the airplane, it technically has been used and, therefore, is not defined as a §261.33 commercial chemical product. (On the other hand, the pesticide residue that remains in the spray tanks after the spraying operation has not technically been used and, thus, would be defined as a commercial chemical product.)" [RO 11115]

In other examples where chlordane was used as a pesticide and resulted in soil contamination, EPA cited §261.2(c)(1)(ii), which states that P- or U-commercial chemical products are not solid wastes if they are applied to the land and that is their ordinary manner of use. The soil would be a hazardous waste only if it is excavated and exhibits a characteristic. [RO 11182, 12357] Similarly, spent ash from fumigation with aluminum phosphide (P006) is not a listed CCP because the chemical has been used (although the ash could exhibit a characteristic and be a regulated hazardous waste). [RO 11202]

Q: Methyl bromide (U029) is used in enclosed fumigation operations; vapors from the operation are captured in activated carbon that, when spent, is sent to a facility for regeneration. How is the spent carbon regulated?

A: The methyl bromide is being used for its intended purpose; hence, the vapors do not meet the listing description for U029. The spent carbon is a newly generated solid waste and must be tested for characteristics, but it does not meet the listing description for U029. [RO 14269]

Q: An industrial facility has problems with clogged sewer pipes. It uses a drain cleaner consisting of o-dichlorobenzene to clean the pipes by pouring the chemical into drains and toilets. The U-list identifies o-dichlorobenzene as U070 when discarded. Does this scenario involve hazardous waste management?

A: No. "[T]he normal use of these products requires that they be poured into toilets or drains. This action does not constitute disposal, as defined in...the hazardous waste regulations, nor is the product considered a solid waste because it has served its original intended use. Therefore, these products are not subject to regulation under RCRA until they are actually discarded." [RO 11011]

Q: Formaldehyde is added to ship ballast fluid to control biological growth. When the ballast fluid is disposed, is it listed as hazardous waste U122—formaldehyde?

A: No. The formaldehyde is being used for its intended purpose. The U122 code would apply only to pure, unused formaldehyde being discarded. [RO 12238]

Q: Are CFC refrigerants that are removed from a refrigeration system and sent for reclamation considered commercial chemical products and thus not solid wastes by the dash in Table 1 of §261.2(c)?

A: No. CFC refrigerants removed from a refrigeration system are classified as spent materials. Spent materials sent for reclamation are solid wastes. [July 28, 1989; 54 FR 31336] However, an exclusion in §261.4(b)(12) allows nonhazardous management of these materials if the CFCs will be reclaimed for reuse. Also, EPA's October 30, 2008 final rule [73 *FR* 64668] will keep the spent CFCs from being solid wastes if they are reclaimed under the conditions of that rule.

3.5.3.1.2 *Manufactured articles aren't P- or U-wastes*

One of the most common errors in applying the P- and U-codes is that generators analyze an object, find that it contains one or more of the listed chemicals, and then assign the associated hazardous waste code(s) to the object when they are getting ready to discard it. In general, EPA would term these objects "manufactured articles" and regulates them as follows:

"Manufactured articles that contain any of the chemicals listed [on the P- or U-list] are rarely, if ever, known by the generic name of the chemical(s) they contain and, therefore, are not covered by the [P- or U-] listings." [November 25, 1980; 45 *FR* 78541]

For example, wool blankets that were treated with DDT will be discarded. The blankets are manufactured articles containing DDT (which keeps moths from eating the blankets), and they would not be U061—DDT when disposed. [RO 11711] We mentioned earlier that thermometers, even if unused, are actually manufactured articles that contain mercury and would only be hazardous if they exhibit a characteristic. [RO 13310, 14012]

What about unused medicated patches? Are these unused items considered manufactured articles or commercial chemical products that would carry P- or U-listings when discarded? Recent guidance from EPA notes that nicotine patches are used to deliver the listed chemical in a certain dosage and may be generically referred to as the active chemical ingredient. Therefore, the use of these patches flows directly from the listed chemical and so they would be P075 when discarded in their unused form. [RO 14817]

3.5.3.2 Manufacturing chemical intermediates

When EPA regulated the CCPs, their intent was to capture those materials that are typically highly concentrated and that are sometimes disposed or spilled. The agency realized, however, that these chemicals might not always be sold as products. For example, in a chemical plant, a chemical might be produced in one process and then fed directly to another process as feedstock. In this case, the chemical is a "manufacturing chemical intermediate" rather than a "product." Nevertheless, the intermediate would pose the same threat to human health and the environment if discarded or spilled. Therefore, EPA decided to include these manufacturing chemical intermediates under the listing criteria for P- and U-wastes.

The agency has not provided any general guidance or definitions on how to identify a manufacturing chemical intermediate in the real world. Our advice is to rely on the knowledge of process engineers or operators who are familiar with the operation of a plant. Ask these people "What's in that pipe, tank, container, etc.?" If the answer is "Benzene" (or some other chemical on the P- or U-list), you are probably dealing with a manufacturing chemical intermediate or a CCP.

If their answer is more process related, like "Bottoms from the stripper tower," you will have to inquire about its composition. If the answer is something like "It's a mixture of benzene, toluene, and other stuff," you are not dealing with a regulated manufacturing chemical intermediate because it is not predominantly a single chemical. As always, if you're not sure whether the material is regulated, ask your regulator for a case-specific determination.

3.5.3.3 Off-specification commercial chemical products

The general concept of off-specification CCPs is that sometimes products contain contaminants that would prevent them from being sold or used as products. In guidance, EPA described an off-spec

CCP as "an unused material that would have been a CCP if it met specifications." [RO 14194]

Typically, this would occur at a manufacturing plant where something went awry in the production process, and the material is rejected by the quality assurance (QA) department as not meeting product specifications. Sometimes, a product might be rejected by a customer's QA department, and the product would be returned to the manufacturer. Because off-spec variants are probably as hazardous as on-spec products, EPA wanted to regulate them as P- or U-listed wastes when disposed or spilled. The following examples illustrate how off-spec products are regulated.

Q *A manufacturer stores product formaldehyde (CAS No. 50-00-0) in containers before use in a manufacturing process. While in storage, some of the formaldehyde polymerizes to form paraformaldehyde (CAS No. 30525-89-4), which is unusable in the manufacturing process. The manufacturer separates the paraformaldehyde from the usable formaldehyde and sends the paraformaldehyde for disposal. Unused formaldehyde is U122 when discarded, but unused paraformaldehyde is not on the P- or U-list. When discarded, is the unused paraformaldehyde a listed hazardous waste?*

A Yes. Paraformaldehyde is an off-specification form of formaldehyde and meets the U122 listing description. When a commercial chemical product listed in §261.33(e) or (f) undergoes a chemical change that renders the chemical off-specification, the applicable P- or U-listing for the original chemical applies, even in cases where the chemical composition has changed sufficiently to require assignment of a different CAS number. [RO 13658]

Q *A facility starts with product in bulk form and packages it in small containers on a packaging line. After the first product has been packaged (e.g., toluene), some residual product remains in the piping. When the next product is packaged (e.g., trichloroethylene), the first few containers are contaminated with toluene. This contaminated material is emptied into a holding tank for disposal. What is the regulatory status of this material?*

A The contaminated material is considered to be an off-specification CCP and would be hazardous waste U228—trichloroethylene. [RO 11471]

Q *If a customer receives an off-specification product that appears on the P- or U-list and returns it to the manufacturer for reprocessing, is it a hazardous waste?*

A No. As long as the chemical is reprocessed (reclaimed), according to Table 1 of §261.2(c), it is not a solid waste and therefore cannot be a hazardous waste. A more complete discussion of this issue is given in Section 3.5.5.

Q *A company recycles spent 1,1,1-trichloroethane (F002) by distillation. Sometimes the recovered solvent does not meet market specifications and is disposed. Is this waste F002 or U226?*

A The off-specification 1,1,1-trichloroethane would be considered an off-spec CCP and would be U226 when disposed. [RO 12276]

3.5.3.4 Container residues

The regulatory language dealing with container residues [§261.33(c)] is pretty much self-explanatory. If a container held a P- or U-chemical, the residue within the container or liner removed from the container retains the P- or U-listing unless the container or liner is "RCRA-empty" per §261.7. (Refer to Section 9.5 for a discussion of RCRA-empty containers.) Sometimes people overlook the *Comment* in §261.33(c), which is part of the promulgated regulations and reads as follows:

"*Comment:* Unless the residue is being beneficially used or reused, or legitimately recycled or reclaimed; or being accumulated, stored, transported, or treated prior to such use, reuse, recycling, or reclamation, EPA considers the residue to be intended for discard, and thus, a hazardous waste. An example of a legitimate reuse of the residue would be where the residue remains in the container and the container is used to hold the same commercial chemical product or manufacturing chemical intermediate it previously held. An example of the discard of the residue would be where the drum is sent to a drum reconditioner who reconditions the drum but discards the residue."

Two unique situations regarding residues in containers are discussed below.

3.5.3.4.1 *Partially used containers*

A bottle of P- or U-listed laboratory chemical is opened and some of the chemical is used. What is the regulatory status of the remaining chemical in the bottle when discarded? What about the bottle itself?

According to §261.33(c), the remaining chemical or residue in the bottle is regulated as P- or U-listed hazardous waste unless the container is rendered "RCRA empty" under §261.7(b). What's more, the bottle would be a hazardous waste container until it is rendered RCRA empty.

This regulatory language is bolstered by EPA guidance in RO 14656. In this guidance, the agency addressed unused solvent residues in spray cans that are being discarded. According to the agency, if the remaining unused solvent in the spray can is P- or U-listed, it would be a listed hazardous waste when discarded.

3.5.3.4.2 *Used vials and syringes*

Several U-listed chemicals are used as drugs in chemotherapy. These include chlorambucil (U035), cyclophosphamide (U058), daunomycin (U059), melphalan (U150), mitomycin C (U010), streptozotocin (U206), and uracil mustard (U237). The drugs are received in vials and dispensed with syringes; how would the used vials and syringes be regulated under RCRA?

If the vials and syringes are "empty" per §261.7(b)(1), they are not regulated under RCRA. Otherwise, they are. However, EPA became concerned about the safety issues associated with handling these drugs and provided the following guidance: "The agency is aware, however, that prudent practice dictates that materials contaminated with these chemicals (such as syringes, vials, gloves, gowns, aprons, etc.) not be handled after use. Therefore, to minimize exposure to these toxic chemicals, the agency recommends that the entire volume of waste be weighed and that there be no attempt to remove any residue from the vial before disposal." [RO 12946] The conditionally exempt small quantity generator exclusion (no more than 100 kg/mo) could apply in these circumstances. [RO 12549, 12946]

The agency came to a different conclusion when asked if a syringe containing epinephrine (P042) residues was a hazardous waste: "EPA considers such residues remaining in a dispensing instrument to have been used for their intended purpose. The epinephrine remaining in the syringe, therefore, is not a commercial chemical product and not P042 hazardous waste." [RO 13718] Because this is a more recent interpretation and leads to fewer regulatory issues, we prefer this approach rather than handling syringes as hazardous waste. Additionally, EPA clarified in RO 14788 that this epinephrine interpretation extends to other P- and U-listed pharmaceuticals administered by syringe. The agency went on to clarify (as it also had in RO 13718), that even though not listed, syringe residues exhibiting a characteristic must be managed in accordance with the hazardous waste regulations.

3.5.3.5 Cleanup residue and debris

Section 261.33(d) specifies that residues and contaminated soil, water, or debris resulting from cleaning up a P- or U-listed spill are also listed wastes. The regulations refer to spills "into or on any land or water"; however, EPA has stated in guidance that the listings apply to all spill residues, regardless of where the spill occurs. For example, debris from a spill that is contained entirely within a building would be subject to the listings. [RO 13335]

The term "spill debris" has not been defined by EPA, although there was a brief discussion of this issue in the May 19, 1980 *Federal Register.* [45 FR 33116] The concern expressed by someone in the regulated community was that wrecked rail cars or trucks that had been transporting P- or U-chemicals might be considered to be "spill debris." The agency stated:

> "EPA has chosen not to exclude such debris by definition. If contaminated, these items pose a substantial threat to human health and the environment and should be handled carefully. EPA presumes, however, that in virtually all cases, heavy equipment can be decontaminated and therefore will not become part of the contaminated debris."

The following examples taken from EPA guidance documents illustrate two issues associated with spill residues.

Q *A warehouse was used to store pesticides that are on the P-list. Wipe tests of the floor and structural I-beams in the building confirm the presence of these chemicals. If the residues resulted from spills of P-chemicals, can a state regulatory agency require cleanup of the warehouse under RCRA?*

A The spilled materials appear to have been abandoned by accumulating them in the warehouse rather than disposing them elsewhere. They may also have been abandoned by essentially being disposed within the building itself. This makes them solid wastes, and if they were listed P- or U-chemicals, the residues would also be hazardous wastes. If the spills occurred after November 19, 1980 (the effective date of the RCRA regulations), "treatment and containment of spills (except in immediate response to spills) must comply with applicable Part 265 requirements. Since it appears that this is not an immediate response situation, the facility would be subject to an enforcement action for treating, storing, or disposing of hazardous waste without interim status or a permit, and could be required to take appropriate action to clean up the residues." [RO 11161]

Q *If a manufacturer spills a listed CCP on soil and intends to reclaim the spill residue, is it listed hazardous waste?*

A If a spill residue can be returned to a process or otherwise put to beneficial use, it would not be a listed hazardous waste. However, the burden of proof is on the generator to show that legitimate recycling will take place. "In the absence of strong, objective indicators of recycling or intent to recycle a spill residue, 'the materials are solid wastes immediately upon being spilled because they have been abandoned' (November 22, 1989; 54 FR 48494) and must be managed in accordance with all applicable RCRA standards." [RO 13743]

3.5.3.6 What does "technical grades of the chemical" mean?

As stated earlier, CCPs include "technical grades of the chemical." Here is how EPA defines technical grades of a chemical:

> "The Office of Solid Waste uses a definition of technical grade which is in general usage by the chemical profession. Technical grade refers to all commercial grades of a chemical, which in some cases, may be marketed in various stages of purity. There are no exact criteria, such as percent purity, to define technical grade of a substance. The technical purity of a chemical substance will vary from compound to compound and may range from highly purified to very impure." [RO 11348]

> "Potentially, 'technical grade' or 'commercially pure grade' can refer to any and all grades of purity of a chemical that are marketed, or that are recognized in general usage by the chemical industry." [RO 11521]

3.5.3.7 What is a "sole active ingredient"?

If a chemical product is not pure (i.e., it is a mixture of several chemicals), it can still be a listed hazardous waste when discarded if the product has only one active ingredient and that ingredient is on the P- or U-list. EPA defines an active ingredient as a chemically active ingredient that performs the primary function of the product. [RO 11348,

11350, 14686, 14820] For example, a pesticide product might consist of a mixture of several chemicals; however, if only one of the chemicals kills bugs, it is the sole active ingredient. If that chemical is on the P- or U-list, the pesticide is that listed waste when disposed in its unused form. Fillers, solvents, surfactants, propellants, colorants, preservatives, etc. are considered to be inert, nonactive ingredients.

If a product to be discarded has more than one active ingredient, it would not be a P- or U-listed waste per the federal RCRA regs, even if both active ingredients are on the P- or U-list. There is one possible exception to this rule: EPA cites an example of a pesticide that contains chlordane (U036) and small amounts of heptachlor (P059). Both of these chemicals kill bugs; hence, under the guidance cited above, the pesticide should not be P- or U-listed when disposed. When EPA evaluated this product, however, it found that heptachlor is an impurity produced in the synthesis of chlordane that cannot be extracted economically (i.e., heptachlor is not mixed with chlordane to produce the product). Under these circumstances, EPA decided that the product had a sole active ingredient (chlordane) and would be U036 if disposed. [RO 11348]

Q: What is the regulatory status of unused nicotine patches when disposed?

A: "Nicotine in finished dosage forms, such as tablets or capsules, is regulated under RCRA, because it is a commercial chemical product formulation containing nicotine as the sole active ingredient…. We view transdermal patches as an analogous dosage form; therefore, unused patches are also listed P075 when discarded." Nicotine patches, gum, and lozenges are used to deliver the listed chemical in a certain dosage and may be generically referred to as the active chemical ingredient. Upon disposal, unused nicotine patches, gum, and lozenges will meet the P075 listing description. [RO 14817]

Q: A commercial product contains osmium tetroxide (P087) and pyridine (U196), both of which are active ingredients. Is this product a P- or U-listed waste when disposed?

A: No. The lists of CCPs apply only to formulations that are pure, technical grade, or contain a sole active ingredient; the osmium tetroxide and pyridine combination is none of these. [RO 11180]

Q: A product contains two active ingredients, only one of which is on the P- or U-list. Is the product a listed hazardous waste when disposed?

A: No. "It is not necessary for a chemical to be listed in §261.33(e) or (f) in order to meet the definition of an active ingredient…. If a formulation has more than one active ingredient, the formulation, when discarded, would not be within the scope of the listing in §261.33, regardless of whether only one or both active ingredients are listed." [RO 13530]

Q: A drug company manufactures tablets that contain a P-listed chemical as the sole active ingredient. What is the regulatory status of quality control (QC) residues from dissolving the tablets in distilled water or crushing the tablets to determine their physical strength?

A: The QC residues (dissolution solutions, dust, and fragments) pose the same threats as untested product and are therefore P-listed. [RO 14095]

Q: A U-listed chemical is used as a drug for cancer treatment. If diluted with water or saline solution and then discarded unused, is the chemical "spent"? Is it U-listed? If it is mixed with other pharmaceuticals for use but is then disposed unused, is it U-listed?

A: Section 261.1(c)(1) defines "spent material" as any material that has been used and, as a result of contamination, can no longer be used for its original purpose without processing. The unused, diluted drug does not meet these criteria and would be a U-listed waste when disposed. If the U-listed chemical is mixed with other pharmaceuticals before use, but is the sole active ingredient in the mixture, it would also be U-listed waste if disposed unused. If it isn't the sole active ingredient, the unused mixture would be hazardous only if it exhibits a characteristic. [RO 11343]

Q A laboratory either buys P- or U-listed CCPs to use directly as quality assurance (QA) standards, or they synthesize the standards by diluting them with water, acidic/basic solutions, organic solvents, or an inert media (such as soil) to the appropriate concentration levels. Each of the standards contains a single P- or U-chemical. If excess quantities of these standards are disposed, are they listed hazardous wastes?

A "Yes. Dilution of a commercial chemical product is not considered use of a commercial chemical product in this case. Thus, the excess QA standards intended for disposal would be listed hazardous wastes under §261.33." [RO 11437; see also RO 11523] This answer holds true even if the solvents used to dissolve the CCP are themselves listed in §261.33. [RO 11523]

Q A QA solution consisting of aldrin (P004) and dieldrin (P037) dissolved in methanol will be disposed. What waste code should be applied to the excess solution?

A Because the solution contains more than one active ingredient, neither the P004 nor the P037 waste codes would apply. When a solvent is used to formulate a compound or product, it does not meet the listing description for spent solvents; therefore, the F003 waste code is also not appropriate. Because the solution would probably exhibit the ignitability characteristic, D001 is the most likely waste code for the excess solution. [RO 11523]

Q Paint stripper contains methylene chloride, toluene, and other inert materials. Is the unused stripper a listed CCP when disposed?

A The answer in this case depends entirely on which ingredients are active when removing paint. If only the methylene chloride is active, it would be U080. If only the toluene is active, it would be U220. If both are active, the unused stripper would be hazardous only if it exhibits a characteristic. [RO 11787] See also RO 14686.

Q A pesticide manufacturer adds formaldehyde or paraformaldehyde to prevent deterioration of the product by bacteria or fungi. Does the presence of this chemical affect whether the pesticide is a P- or U-waste?

A "When formaldehyde or paraformaldehyde is added solely to preserve the activity of a pesticide formulation, it is not considered an active ingredient for purposes of the sole-active-ingredient requirement of §261.33." [RO 11405] Hence, if the pesticide has a sole active ingredient that appears on the P- or U-lists, it is a listed hazardous waste when disposed.

Q A chemical deodorant added to toilets contains low concentrations of formaldehyde to maintain sanitary conditions. When these chemicals are disposed unused, are they U122—formaldehyde? Is the answer the same if the chemicals are poured into toilets for sanitation purposes and ultimately end up in sewers and septic systems?

A Formaldehyde is not the sole active ingredient of the unused deodorant; hence, the deodorant would only be a hazardous waste if it exhibits a characteristic. [RO 12252] When poured into toilets for sanitation purposes, the deodorant is being used for its intended purpose and is also not a listed hazardous waste, even if formaldehyde was the sole active ingredient. [RO 12642]

Q A facility has an unused catalyst containing vanadium, the concentration of which is expressed as V_2O_5 (vanadium pentoxide—P120 when discarded). If the vanadium compound is the sole active ingredient, is the catalyst P120 when discarded?

A Unless the active ingredient is actually V_2O_5, the unused catalyst, when discarded, is not P120. Note, however, that the catalyst would have to be evaluated to determine if it exhibits any characteristics before concluding that it is not a hazardous waste. [RO 11016]

Q *A facility has some unused wood preservative product that it wants to discard. The label on the drum says that there are three active ingredients: tri-, tetra-, and pentachlorophenol. Is the plant safe in assuming that this isn't a listed hazardous waste, because it's a CCP with more than one active ingredient?*

A It turns out that EPA listed this specific chemical product as F027—Discarded unused formulations containing tri-, tetra-, or pentachlorophenol. No other chemical products show up on the F- or K-list.

3.5.3.8 Manufacturing process wastes

When asked why manufacturing process wastes containing P- and U-chemicals aren't considered to be P- or U-wastes, EPA provided the following rationale:

"Manufacturing process wastes…generally contain only low levels of these materials. Thus, expanding the hazardous waste identification regulations to encompass all manufacturing wastes containing the §261.33 compounds is likely to result in many false positives (i.e., wastes identified as hazardous which do not actually contain hazardous levels of the toxicants of concern) unless, and until, minimum concentration thresholds can be established for each compound. At this time, due to the lack of data, the agency is unable to set thresholds for all the compounds." [RO 11026, 12024]

Here is an example of how the P- and U-listings apply to manufacturing process wastes.

Q *A manufacturing facility has a release of "wet toluene" (i.e., toluene that contains 5% water) from a storage tank. How is this release regulated?*

A If the toluene is unused, it could be off-spec CCP, and the release would be U220. However, if the toluene picked up the water during use in a manufacturing process, it is a manufacturing process waste that would be considered a "spent material." Depending on how the toluene was used in the process, it could be F005 spent solvent, or it could be an unlisted manufacturing process waste that is hazardous only if it exhibits a characteristic. [RO 14194]

3.5.3.8.1 Manufacturing or product handling?

Sometimes, the issue of what constitutes a manufacturing process waste can become quite complex. In guidance, EPA has decided that "a point exists in the manufacturing process in which an operator creates either a commercial chemical product or manufacturing intermediates. When these chemicals meet a listing description under §261.33, any discard of these materials (including these materials captured on filters or mixed with other wastes) are considered hazardous wastes and must be handled accordingly…. Process wastes are generated prior to the creation of the product or intermediate and may be listed as F- or K-wastes under EPA's listing system." [RO 14095]

Let's start with manufacturing processes. Chemicals on the P- and U-list are often used as raw materials to make products; these products could be in the form of bulk materials, manufactured articles, etc. Wastes generated from storing or transferring the P- or U-chemicals *before they enter the first step of the manufacturing process* will carry the applicable P- or U-codes. Conversely, wastes generated during manufacturing operations where the P- or U-listed chemicals are *being used* are manufacturing process wastes—not P- or U-wastes. This concept is illustrated in the top diagram in Figure 3-4 and by the following example.

Q *Diethylhexyl phthalate (DEHP) is used to manufacture small capacitors. Rags, gloves, and other miscellaneous solid materials absorb DEHP (which is listed as U028) during the manufacturing operations. Are these materials listed hazardous wastes when disposed?*

A If the contamination results from using these materials during the manufacturing process (e.g., gloves become contaminated as a result of handling materials during manufacturing), they are *not* listed U-wastes. Note, however, that if the

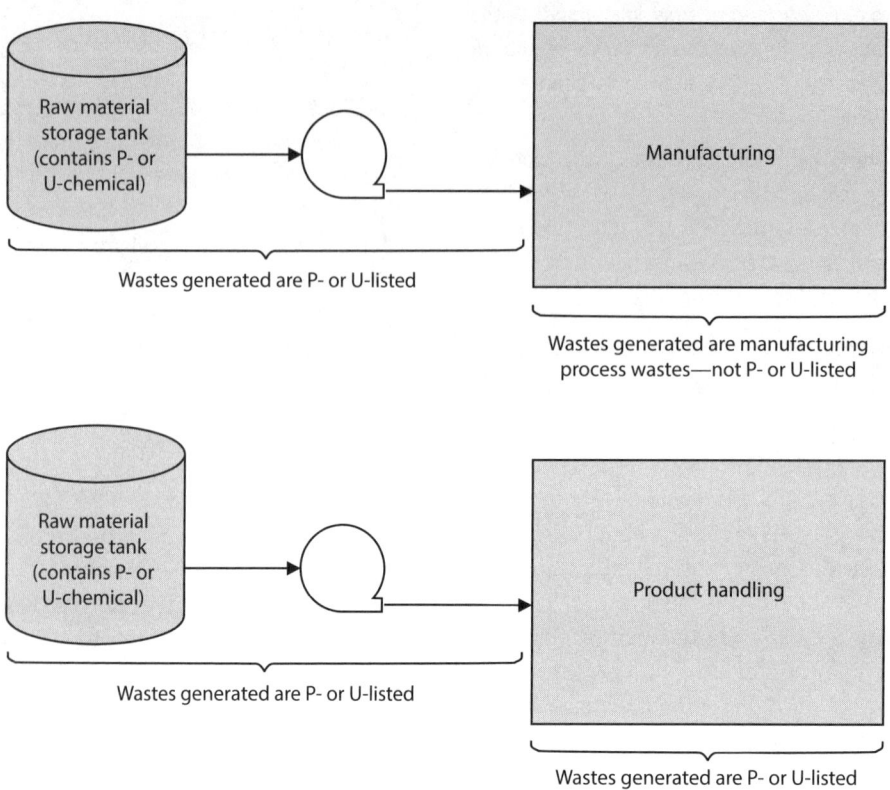

Figure 3-4: Commercial Chemical Products Used in Manufacturing Compared to Product Handling

Source: McCoy and Associates, Inc.

materials become contaminated when cleaning up spills or leaks of unused DEHP, such as from the DEHP raw material storage tank or transfer line, they would be U028 when disposed. [RO 12423]

What some facilities might consider to be "manufacturing," however, is simply "product handling" in EPA's eyes. Consider a drug company that purchases warfarin (P001) from a manufacturer and then packages the chemical into tablets of different strength. The tablets contain warfarin as a sole active ingredient at concentrations between 0.45% and 4.5%. It would seem logical that the drug company is producing the tablets in a manufacturing process, and that residues from the process would be considered manufacturing process wastes, not P001. The residues in question, all of which contain low concentrations of warfarin, include the following:

- Washdown water from cleaning machinery, containers, implements, and manufacturing rooms;
- Disposable gloves, gowns, and other personal protective equipment used by employees in the manufacturing area; and
- Airborne dust that is collected in air filters, which are periodically discarded.

EPA decided that such materials generated from manufacturing operations prior to the point where a CCP is produced would be manufacturing process wastes that are not covered by the §261.33 listing. In other words, the original warfarin manufacturer might generate these process wastes and not manage them as P- or U-wastes. However, the agency determined that the drug company is simply handling a CCP that has already been produced; that is, they are involved in product handling and do *not* have manufacturing process wastes. Therefore, all of the wastes noted above that are produced at the packaging facility would be listed hazardous wastes (see the bottom diagram of Figure 3-4). The washdown water would be P- or U-listed, but might be eligible for the de minimis loss exemption of §261.3(a)(2)(iv)(D). The personal protective equipment and air filters are listed debris because they "contain" warfarin. [RO 14095]

In a similar example, a company repackages toluene (U220) and methylene chloride (U080). Cartridge filters are used to remove dirt and other solids from these products during unloading, transfer, and packaging. When the filter cartridges become dirty or when there is a product change, they are flushed with water and discarded. The washed filters contain small amounts of the U-listed chemicals. As in the previous example, EPA decided that, because the filters contain U-chemicals that have already been produced (i.e., they already meet a listing description in §261.33) but are being discarded, they must be managed as U-listed hazardous waste. [RO 11325]

Once again, please note that in each of these examples, the materials being handled are already CCPs and are simply being repackaged.

3.5.3.8.2 *Unreacted reagent from reactors*

Personnel at chemical plants sometimes ask about the regulatory status of unreacted reagents that exit manufacturing process units (e.g., reactor vessels). If the reagents are on the P- or U-list of commercial chemical products, would unreacted chemicals exiting reactor vessels still carry the P- or U-code if disposed?

The agency first addressed this issue in preamble language as follows:

> "We also note that leftover, unreacted raw materials from a process are not spent materials, since they never have been used." [January 4, 1985; 50 *FR* 624]

This language would imply that, if they never have been used, they are still products and would carry the P- or U-codes if discarded.

Subsequent to that preamble discussion, EPA issued a letter noting that:

> "the term 'commercial chemical product' refers to a substance manufactured for commercial use which is commercially pure or a technical grade and formulations in which the chemical is the sole active ingredient. It does not refer to a material, such as a process waste, that contains any of the substances listed in §261.33(e) or [(f)]." [RO 11076]

EPA also issued three guidance letters to chemical companies [RO 11079, 11326, 12809], discussing the regulatory status of unreacted solvents exiting a manufacturing system, where the solvent was used as a reactant to manufacture a chemical product. In every case, the agency noted that the excess solvent exiting the reactor would not be a listed spent solvent, because it was used as a reactant—not just as a reaction (synthesis) media (refer back to Section 3.2.3.1.1). Additionally, EPA noted in one letter that the excess solvent would not carry the associated U-code, that the unreacted waste stream would not be listed in another letter, and in the final letter that the chemical company would be responsible for determining whether the unreacted waste stream exhibits a characteristic (implying that it would not be listed).

Based on the sum of the guidance available, we believe that once a chemical leaves its raw material storage area and crosses the manufacturing process boundary to any type of processing equipment, it has been used (see the top diagram in Figure 3-4). Molecules of unreacted chemical leaving a reactor have been used. Otherwise, all manufacturing process wastes would carry U-codes whenever a U-chemical was used as a feedstock, and this clearly is not the case in RCRA. See also RO 12004 and 14194.

The unreacted reagent stream could be hazardous by K-listing or if it exhibits a characteristic.

3.5.4 Equipment regulated under the contained-in policy

EPA's contained-in policy, which is discussed in Section 5.3, can apply to equipment contaminated with P- and U-listed chemicals. Typically, this policy is triggered when equipment is contaminated during use, not after it has been discarded. In the previous section dealing with product handling wastes, the spent filter cartridges and personal protective equipment are examples of materials that are regulated under the contained-in policy. In most cases, this type of equipment is considered to be "debris" and is subject to regulation under the land disposal restrictions (see Section 13.8).

An example of where the contained-in policy would apply to equipment involves unused carbon disulfide (P022) that is stored in a tank. Floating plastic balls are used to control vapor emissions from the tank. According to EPA, when the balls are disposed, they are hazardous waste "since carbon disulfide has been incidentally deposited on them." [RO 12778]

Simply put, equipment destined for disposal that has come in contact with P- or U-listed chemicals is regulated as P- or U-listed waste under the contained-in policy.

3.5.5 Generator issues

Because §261.33 deals with chemicals that are often sent to an end user but are occasionally returned to the manufacturer for various reasons, questions frequently arise as to if and when such returned material (sometimes called "returned goods") is regulated as hazardous waste. The following examples illustrate the concepts involved.

Q *If a P- or U-commercial chemical product becomes excess inventory or outlives its expiration date in the hands of a wholesaler, retailer, or an end user, is it a hazardous waste when returned to a supplier for resale or reprocessing?*

A No. The product is not being discarded by the wholesaler, retailer, or end user and is not a hazardous waste. [November 25, 1980; 45 FR 78540]

Q *If a container of P- or U-chemical breaks, and the supplier takes back the affected chemical, including recovered spilled chemicals, for repackaging or reprocessing, is the chemical a hazardous waste?*

A No. The chemicals are not being discarded and are not hazardous wastes. "If, however, some of the spilled chemicals are discarded or intended to be discarded because they cannot be returned (e.g., they are mixed with dirt or other materials), these spilled chemicals (and associated spill cleanup residues and debris) are hazardous wastes." [45 FR 78540]

Q *A manufacturer takes back P- or U-chemicals from an end user or branch operation and finds it necessary to discard some portion of the material because he/she is not able to reprocess, repackage, resell, or reuse it. Is the material a hazardous waste, and who is the generator?*

A "Where this occurs, that portion which is discarded becomes a hazardous waste when it is discarded or when a decision is made to discard the material. In this situation, the manufacturer or supplier is the generator of a hazardous waste because he/she is the 'person…whose act…produces hazardous waste' (see the definition of 'generator' in §260.10)." [45 FR 78541] See also RO 11012, 11606, and 12996.

Q *A supplier takes back P- or U-chemicals from a number of end users, but then sends them to a RCRA-permitted incinerator for disposal (i.e., none of the material can be reprocessed, repackaged, resold, or reused). Is the material a hazardous waste, and who is the generator?*

The material is a waste at the end users' facilities: "It is clear that end users who are accumulating [the material] before it is disposed are managing wastes." Since the material at the end users' facilities is a waste, they must determine if it is hazardous. If the material is included on the P- or U-lists or if it exhibits a characteristic, it is a hazardous waste. "Thus, the end users (who are the generators under these circumstances…) are responsible for managing their unused [material] consistent with the federal hazardous waste regulations." [RO 11492]

Sometimes facilities such as hospitals have excess or expired drugs that they can return to the manufacturer for credit through a reverse distribution system. Based on the EPA guidance summarized in this section, if there is a reasonable expectation that the returned product will be reprocessed, repackaged, resold, or reused by the manufacturer, it is not solid waste (see December 2, 2008; 73 *FR* 73525). However, we know of a state that has decided that outdated or overstocked pharmaceuticals that are sent back to a reverse distributor for credit are wastes, not products. If the drugs are P- or U-listed CCPs or exhibit a characteristic of hazardous waste, they must be managed as hazardous waste in that state. So, as always, keep in mind that states can have rules and policies that are more stringent than the federal rules and guidance presented in this book.

3.5.6 Recycling issues

Note that P- and U-chemicals are hazardous wastes when discarded or when they are intended to be discarded. Conversely, if a CCP (particularly an off-specification product) is reclaimed or speculatively accumulated, according to Table 1 in §261.2(c), the material is not a solid waste and cannot be a hazardous waste. (See Chapter 1 for further details on the applicability of Table 1.)

For example, a company that manufactures thermometers finds that some of its mercury is contaminated. The mercury has not been used yet and would be considered an off-spec CCP. (Mercury is listed as U151 when discarded.) If the mercury is sent to a reclamation facility to be purified, Table 1 of §261.2(c) indicates that CCPs being reclaimed are not solid wastes; hence, the mercury is not a hazardous waste. [RO 11378]

But keep in mind the following warning from EPA on the intent to recycle a CCP but never actually doing it: "[I]f unused CCPs were being stored for a long period of time without any foreseeable means of recovering the product, or if no foreseeable market existed for the recovered product, an overseeing regulatory agency might well conclude that they were abandoned, and thus subject to Subtitle C hazardous waste regulations. Determinations as to whether a CCP is abandoned are site-specific and are made by the regions and states implementing the RCRA program." [RO 14762]

3.5.6.1 Limitations on certain recycling practices

The introductory paragraph of §261.33 limits certain recycling practices for P- and U-wastes that might otherwise allow these chemicals to escape regulation. First, if these chemicals are mixed with used oil or other materials and then used for dust suppression, they are listed hazardous wastes. This prevents someone from claiming that the chemical is an ingredient in a product (dust suppressant), which might meet the recycling exemption in §261.2(e)(1)(i).

Second, if the chemicals are applied to the land, and that is not how they were originally intended to be used, they are hazardous wastes. For example, it would be acceptable to incorporate a P- or U-listed herbicide into a product and apply it to the land to kill weeds, because land application is an intended use for herbicides. On the other hand, if unused benzene (U019) is applied to the land (ostensibly to kill weeds), the benzene would be a hazardous waste. The practical effect of this limitation is that using benzene in this manner would be considered "use constituting disposal," which is an activity regulated by Part 266, Subpart C. The constraints imposed by Subpart C typically make land application of such chemicals impracticable.

Finally, if these chemicals are burned for energy recovery or used as a component of fuel, and if this is not their original intended use, they would be hazardous wastes. For example, burning unused methanol (U154) as fuel could be done outside of RCRA, because this is an intended use of methanol. (Some dragsters use methanol as fuel instead of gasoline.) However, blending creosote (U051) into fuel would make the fuel a hazardous waste. The practical effect of this limitation is that burning the creosote-containing fuel would be subject to the boiler and industrial furnace requirements of Part 266, Subpart H. Although many facilities burn hazardous wastes in compliance with Subpart H, the regulations are fairly complex and a RCRA permit is required.

Exclusions and Exemptions

Exclusions and exemptions that may keep you out of RCRA

At the beginning of Chapter 2, we stated that whenever people are trying to make a hazardous waste determination for a particular material, they should ask themselves four questions (see Section 2.1). We've discussed the first, third, and fourth questions in previous chapters, but we haven't addressed the second one: "Is the material exempt?" What we want to do in this chapter is to give you the final information you need to be able to answer all four questions completely.

In this chapter, we differentiate "exclusions" from "exemptions" as follows: Exclusions exclude a material from the definition of solid or hazardous waste, thereby making it beyond the reach of the hazardous waste management program. Conversely, a material that is a solid and hazardous waste may be partially or totally freed from the hazardous waste management rules because of an exemption available in the regulations.

There are numerous exclusions and exemptions in the RCRA hazardous waste regulations, most of which are included in §261.4. Table 4-1 provides a quick summary of the exclusions and exemptions available in the federal regulations, including where a detailed discussion of the exemption may be found in this book and the applicable regulatory citations. Note that some federal exclusions and exemptions may not be available in all states, as state hazardous waste programs may be more stringent than the federal RCRA program. You will note that we have broken the exclusions and exemptions in Table 4-1 into three groups:

1. Materials that are excluded from the definition of solid waste—These are materials that Congress and/or EPA have determined are outside the scope of RCRA; they are excluded from the definition of solid waste. Remember how this works: if something is not a solid waste, it cannot be a hazardous waste because hazardous wastes are a subset of solid wastes. The materials that are excluded from the definition of solid waste are identified in §261.4(a). Any material included in this section, even if it exhibits a characteristic or meets a listing description, is not subject to RCRA. We are not going to discuss all of those exclusions, but we discuss most of the major ones in Sections 4.1–4.9.

2. Solid wastes that are excluded from the definition of hazardous waste—Again, Congress and/or EPA have decided that certain materials, while solid wastes, are not subject to the expensive and complicated hazardous waste regulations. Solid wastes that are excluded from the definition of hazardous waste (even if they exhibit a characteristic, for example) are listed in §261.4(b). Most of the primary

Table 4-1: Exclusions and Exemptions Available in the Federal RCRA Regulations

Exclusion or exemption	Applicable regulations	Text reference
Materials that are not solid wastes		
Domestic sewage exclusion	§261.4(a)(1)	Section 4.1
NPDES discharge exclusion	§261.4(a)(2)	Section 4.2
Irrigation return flows	§261.4(a)(3)	Section 4.3
Source, special nuclear, and by-product material subject to the Atomic Energy Act	§261.4(a)(4)	Section 4.4
Materials subjected to in situ mining and left in place	§261.4(a)(5)	Section 4.5
Pulping liquors reclaimed in a recovery furnace and reused in the pulping process	§261.4(a)(6)	Section 4.6
Spent sulfuric acid used to produce virgin sulfuric acid	§261.4(a)(7)	Section 11.3.1.5
Hazardous secondary materials reclaimed and returned to the original production process for reuse	§261.4(a)(8)	Section 11.3.4
Spent wood preserving solutions reclaimed and reused for their original intended purpose	§261.4(a)(9)	Not covered
K060, K087, K141–K145, or K147–K148 and any wastes from the coke by-products processes that are hazardous only because they exhibit the toxicity characteristic if they are recycled to coke ovens, recycled to the tar recovery process, or mixed with coal tar prior to sale or refining	§261.4(a)(10)	Not covered
Nonwastewater splash condenser dross residue from the treatment of K061 in HTMR units	§261.4(a)(11)	Not covered
Oil-bearing hazardous secondary materials generated at a petroleum refinery and oil recovered from exploration and production, refining, marketing, and transportation of oil or oil products if inserted into the petroleum refinery process	§261.4(a)(12)	Section 4.7
Excluded scrap metal being recycled	§261.4(a)(13)	Section 4.21.2
Shredded circuit boards being recycled	§261.4(a)(14)	Section 4.21.5.2
Overhead gas condensate from kraft mill steam strippers	§261.4(a)(15)	Not covered
Comparable fuels	§261.4(a)(16)	Section 4.8
Secondary materials (i.e., spent materials) from the primary mineral processing industry that are recycled	§261.4(a)(17)	Sections 1.2.6.3.2, 4.15.2
Recovered oil from an associated organic chemical manufacturing facility, where the oil will be inserted into the petroleum refining process	§261.4(a)(18)	Section 4.7.2.2
Spent caustic solutions from petroleum refining that are used as feedstock to produce cresylic or naphthenic acid	§261.4(a)(19)	Section 4.9
Hazardous secondary materials used to make zinc fertilizers	§261.4(a)(20)	Sections 1.2.6.1.1, 11.4.1.1.1
Zinc fertilizers made from hazardous wastes or from excluded hazardous secondary materials	§261.4(a)(21)	Sections 1.2.6.1.1, 11.4.1.1.1
Used cathode ray tubes that are recycled	§261.4(a)(22)	Section 2.7.3
Hazardous secondary materials generated and reclaimed in the United States under the control of the generator	§§261.2(a)(2)(ii), 261.4(a)(23)	Section 11.2.3.1
Hazardous secondary materials transferred to a US third-party facility for reclamation	§261.4(a)(24)	Section 11.2.3.2
Hazardous secondary materials exported for reclamation	§261.4(a)(25)	Section 11.2.3.2.1
Solvent-contaminated wipes that are sent for cleaning and reuse	§261.4(a)(26)	Section 3.3.3.2

McCoy's RCRA Unraveled ©2015 McCoy and Associates, Inc.

Table 4-1: Exclusions and Exemptions Available in the Federal RCRA Regulations

Exclusion or exemption	Applicable regulations	Text reference
Hazardous secondary materials recycled by being used/reused as ingredients in an industrial process	§261.2(e)(1)(i)	Section 11.3.1
Hazardous secondary materials recycled by being used/reused as effective substitutes for commercial products	§261.2(e)(1)(ii)	Section 11.3.2
Hazardous secondary materials returned to the original process for use/reuse as a substitute for feedstock materials	§261.2(e)(1)(iii)	Section 11.3.3
Solid wastes that are not hazardous wastes		
Household wastes	§261.4(b)(1)	Section 4.10
Agricultural waste/manure returned to the soil as fertilizer	§261.4(b)(2)	Section 4.11
Mining overburden returned to the mine site	§261.4(b)(3)	Section 4.12
Fly ash, bottom ash, slag, and flue gas emission control waste generated during the combustion of fossil fuels	§261.4(b)(4)	Section 4.13
Drilling fluids, produced waters, and other wastes associated with the exploration, development, or production of crude oil, natural gas, or geothermal energy	§261.4(b)(5)	Section 4.14
Certain trivalent chromium wastes from the leather tanning industry and the titanium dioxide production process	§261.4(b)(6)	Section 2.6.4
Wastes from the extraction, beneficiation, and processing of ores and minerals	§261.4(b)(7)	Section 4.15
Cement kiln dust	§261.4(b)(8)	Not covered
Certain discarded arsenic-treated wood or wood products	§261.4(b)(9)	Not covered
Petroleum-contaminated media and debris from cleanup of underground storage tanks	§261.4(b)(10)	Section 2.6.4
Reinjected ground water from certain hydrocarbon recovery operations[1]	§261.4(b)(11)	Not covered
Used CFC refrigerants when reclaimed for further use	§261.4(b)(12)	Section 2.6.4
Non-terne-plated used oil filters that have been gravity hot-drained	§261.4(b)(13)	Sections 4.21.6, 12.1.5.1
Used oil re-refining distillation bottoms used as feedstock for asphalt	§261.4(b)(14)	Section 12.1.1.2
Leachate or gas condensate from certain solid waste landfills	§261.4(b)(15)	Not covered
Solvent-contaminated wipes sent for disposal	§261.4(b)(18)	Section 3.3.3.1
Carbon dioxide streams injected for geologic sequestration	§261.4(h)	Section 4.22
Hazardous wastes that are exempt from Subtitle C regulation		
Mixture of a de minimis amount of listed waste and wastewater that will discharge under a CWA permit	§261.3(a)(2)(iv)(A–G)	Section 4.16
Five specific residues from treatment of listed wastes	§261.3(c)(2)(ii)	Section 5.2.3.4
Hazardous debris treated using an extraction or destruction technology	§261.3(f)(1)	Section 13.8.2
Hazardous debris subject to a no-longer-contains determination by EPA/state	§261.3(f)(2)	Sections 5.3.2, 13.8.1
Radioactive mixed waste managed under the NRC regulations	§261.3(h)	Section 4.4.5
Wastes in active manufacturing process units	§261.4(c)	Section 4.17
Analytical samples	§261.4(d)	Section 4.18.1
Samples used in a treatability study	§261.4(e–f)	Section 4.18.2
Dredged material subject to a permit issued under the CWA or MPRSA	§261.4(g)	Section 4.2.4.1

Table 4-1: Exclusions and Exemptions Available in the Federal RCRA Regulations

Exclusion or exemption	Applicable regulations	Text reference
Waste containing precious metals being recycled	§261.6(a)(2)(iii)	Section 4.19
Spent lead-acid batteries being recycled	§261.6(a)(2)(iv)	Section 4.20
Industrial ethyl alcohol being recycled	§261.6(a)(3)(i)	Not covered
Scrap metal (other than excluded scrap metal) being recycled	§261.6(a)(3)(ii)	Section 4.21
Fuel derived from hazardous waste produced during petroleum refining, production, or transportation	§261.6(a)(3)(iii–iv)	Section 5.2.3.4
Used oil that is recycled	§261.6(a)(4)	Chapter 12
Residues in RCRA-empty containers	§261.7	Section 9.5
PCB-containing dielectric fluid or electrical equipment containing such fluid subject to TSCA regulation	§261.8	Section 2.6.4
Universal waste batteries, pesticides, mercury-containing equipment, and lamps	§261.9	Chapter 8
MGP waste is excluded from TCLP (D004–D043)	§261.24(a)	Section 2.6.2.6

CFC = chlorofluorocarbon; CWA = Clean Water Act; HTMR = high-temperature metals recovery; MGP = manufactured gas plant; MPRSA = Marine Protection, Research, and Sanctuaries Act of 1972; NPDES = national pollutant discharge elimination system; NRC = Nuclear Regulatory Commission; PCB = polychlorinated biphenyl; TCLP = toxicity characteristic leaching procedure; TSCA = Toxic Substances Control Act.

[1]This exclusion from the definition of hazardous waste expired January 25, 1993.

Source: McCoy and Associates, Inc.; adapted from §§261.2–261.4 and 261.6–261.9.

---

exclusions from the definition of hazardous waste are discussed in Sections 4.10–4.15.

3. Hazardous wastes that are exempt from Subtitle C regulation—There are a number of materials that meet the definition of a hazardous waste but are not regulated as such. They are solid and hazardous wastes, but they're exempt from regulation under RCRA Subtitle C. These exemptions are very useful to the regulated community. We have included a discussion of the most important of these exemptions in Sections 4.16–4.21.

Before we talk about the specific exclusions and exemptions, it's important to mention the documentation requirements. Based on RO 11678, all exclusions or exemptions from RCRA regulatory requirements that a facility claims should be documented: "For wastes that are conditionally exempt from RCRA, §261.2(f) requires that documentation be maintained to demonstrate that these wastes meet the terms or conditions of the exemption." See also January 2, 2008; 73 *FR* 67, July 28, 1994; 59 *FR* 38542, RO 11352, 11412, 11771, 11832, 11877, and 14677.

Also, §262.11 requires generators to make hazardous waste determinations for every solid waste they produce. This is the regulatory section on which the four questions are based. Question Number 2: "Is it exempt?" comes from §262.11(a). Such "determinations made in accordance with §262.11" have to be documented per §262.40(c) by both large and small quantity generators.

4.1 Domestic sewage exclusion

A mixture of hazardous waste and domestic sewage that passes through a sewer system to a publicly owned treatment works (POTW) ceases to be a solid or hazardous waste at the point the hazardous waste mixes with the sewage in the sewer line. Because a significant percentage of the hazardous waste produced in the country is in the form of wastewater, this exclusion is of great use and benefit to hazardous waste generators. A discussion of the domestic sewage exclusion is presented below.

4.1.1 Background

The domestic sewage exclusion is a carryover from the Solid Waste Disposal Act of 1965 (SWDA). In that statute, "solid waste" was defined so as to exclude from regulation solids in untreated domestic sewage. This approach avoided regulating mixtures of solid wastes and sewage under the SWDA, since Congress knew that such material was already subject to regulation under the Federal Water Pollution Control Act—the precursor to the Clean Water Act (CWA).

When Congress passed RCRA in 1976 to amend the SWDA, the statutory definition of solid waste was amended to exclude "solid or dissolved material in domestic sewage." [RCRA Section 1004(27)] Again, Congress was confident that industrial discharges to sewers would be adequately regulated under the CWA. To avoid duplicative regulation under both RCRA and the CWA, the domestic sewage exclusion applies to domestic sewage by itself or mixtures of domestic sewage and other wastes; such wastes are not solid wastes (so they cannot be hazardous wastes) under RCRA. The exclusion was codified in §261.4(a)(1) when the RCRA regulations were first promulgated on May 19, 1980. [45 FR 33120] The regulatory language of this exclusion has remained unchanged since that time.

Whether materials discharged to domestic sewers would otherwise be listed or characteristic hazardous wastes makes no difference as to qualifying for the domestic sewage exclusion. As long as such waste is mixed with domestic sewage that passes through a sewer to a POTW for treatment, the mixture is neither a solid nor hazardous waste. [RO 11181] Although the domestic sewage exclusion seems fairly straightforward, a number of questions arise when hazardous waste generators start to take advantage of it. Common questions are addressed below.

4.1.2 How do I qualify for the exclusion?

Sometimes people ask "Is it really legal to put hazardous waste into the sewer?" The answer is "Yes" under the federal regulations [EPA/233/B-00/002, May 2000, available from http://nepis.epa.gov/EPA/html/Pubs/pubtitleOther.html by downloading the report numbered 233B00002]. However, three basic requirements must be met: 1) what's flowing through the sewer line into which the hazardous waste is introduced must be domestic sewage, 2) the mixture of domestic sewage and industrial waste must be conveyed through a sewer system to a POTW for treatment, and 3) the discharge of the waste into the sewer line must be in compliance with all applicable CWA pretreatment regulations. Each of these three requirements is shown in Figure 4-1 and expounded upon in the remainder of this section.

Figure 4-1: The Domestic Sewage Exclusion

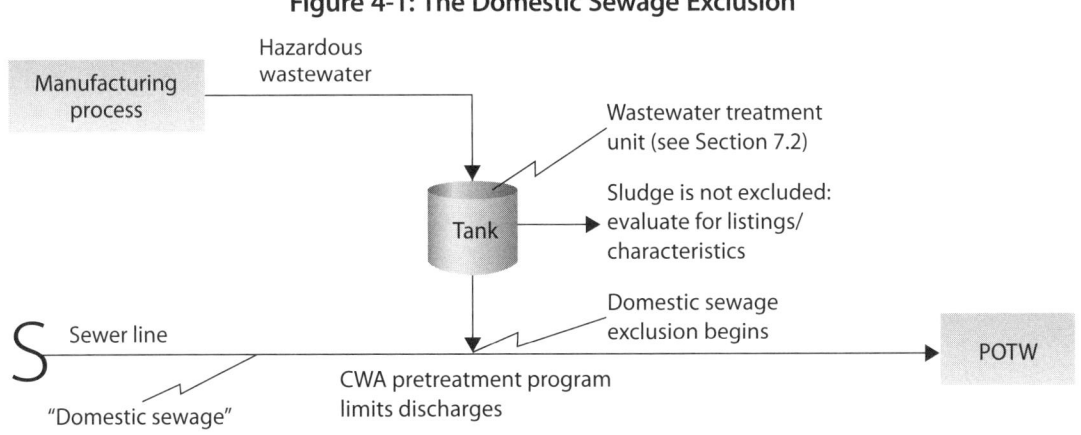

Source: McCoy and Associates, Inc.; adapted from §261.4(a)(1).

4.1.2.1 "Domestic sewage"

"Domestic sewage" is defined in the RCRA regs as "untreated sanitary wastes that pass through a sewer system." [§261.4(a)(1)(ii)] Note that this definition does not link domestic sewage to residences. By implication, discharging process wastes to an in-plant sewer line containing untreated sanitary wastes from restrooms in the plant might appear to qualify for the domestic sewage exclusion (assuming that the other requirements are met). As discussed in Section 4.1.4, mixing in-plant process and sanitary wastes probably will not trigger the domestic sewage exclusion.

4.1.2.2 "POTW"

A POTW is defined in §260.10 as "any device or system used in the treatment (including recycling and reclamation) of municipal sewage or industrial wastes of a liquid nature which is owned by a state or municipality. This definition includes sewers, pipes, or other conveyances only if they convey wastewater to a POTW providing treatment." "Municipality" is further defined in the CWA statute [Section 502(4)] to include a city, county, or special district.

A number of POTWs have recently been privatized (i.e., a private company now owns/operates the treatment facility for the city, county, special district, or state). The question comes up as to whether mixtures of hazardous waste and domestic sewage flowing to such privately owned facilities can also qualify for the exclusion. The answer is "No." EPA addressed this issue in 1980, stating that since "the treatment of sewage by privately owned treatment works is not similarly controlled through the agency's…pretreatment program, the exemption would not be available for mixed waste streams going to such treatment works." [45 *FR* 33097] This position was confirmed more recently:

> "The exception for municipal POTWs treating domestic sewage and other wastes does not extend to private or other nonmunicipal treatment works, because they are not subject to the same CWA requirements and, thus, need to be regulated under RCRA. Therefore, a mixture of sewage with other wastes en route to a non-municipal treatment works does not cease to be a solid waste. The waste's identity relevant to hazardous waste listings continues throughout the treatment works." [RO 14322]

A discussion of the domestic sewage exclusion as it applies to federally owned treatment works is included in Section 4.1.6.

4.1.2.3 CWA pretreatment program

POTWs are regulated under the CWA and also administer the CWA pretreatment program. Two different types of pretreatment standards are administered by the POTWs through permits or contracts issued to industrial users. First, general pretreatment standards include limits on the discharge of acids, flammable or ignitable materials, and oxygen-demanding pollutants. Second, categorical pretreatment standards apply to specific categories of industrial facilities determined to be the most significant sources of toxic pollutants discharged to POTWs. For example, a chemical plant would be subject to categorical pretreatment standards that limit what this type of facility can discharge to the sewer. These standards are usually concentration or mass limits. In addition to these two types of national pretreatment standards, the POTW can impose local limits to avoid 1) plant upsets, 2) generation of hazardous sludges, 3) health hazards to its employees, and 4) violation of its own (typically NPDES) discharge permit. [RO 11727]

A facility that wants to discharge hazardous waste to a domestic sewer line must comply with all applicable pretreatment standards. "The national pretreatment program, mandated by Section 307(b) of the CWA and implemented in 40 *CFR* Part 403, requires that industrial facilities pretreat pollutants discharged to POTWs to the extent that these pollutants interfere with, pass through, or are otherwise incompatible with the operations of POTWs. The exclusion avoids the redundancy of subjecting hazardous wastes mixed with domestic sewage to RCRA management requirements when these wastes are already subject to requirements under the CWA, including the pretreatment program." [52 *FR* 23478]

4.1.3 At what point does the exclusion apply?

The domestic sewage exclusion applies to mixtures of domestic sewage and other wastes. Since such mixtures are not solid wastes, they cannot be hazardous wastes. At what point exactly does the exclusion begin? In the preamble to the 1980 RCRA regulations, EPA noted that:

> "[t]he 'domestic sewage' exemption is only applicable to nondomestic wastes that *mix* with sanitary wastes in a sewer system leading to a POTW. An industrial waste stream that never mixes with sanitary wastes in the sewer prior to treatment or storage does not fall within the exemption, regardless of the public or private ownership of the treatment works. Defining the point at which 'mixture' occurs may seem to be a relatively straightforward task. Practical problems arise, however, in defining the point at which mixture of sanitary and other wastes occurs in a complex sewer system. Moreover it is particularly difficult to define this point for regulatory purposes in such a way that all parties understand when RCRA obligations begin and end.
>
> "EPA has, therefore, decided that a waste falls within the domestic sewage exemption when it first enters a sewer system that will mix it with sanitary wastes prior to storage or treatment by a POTW. EPA recognizes that this interpretation brings various wastes within the exemption before they are actually mixed with sanitary wastes. In light of the fact that the wastes will be mixed prior to treatment and that the mixture will be properly treated by the POTW, EPA believes that the need for administrative clarity in this otherwise complicated regulatory program warrants such an approach." [Emphasis in original.] [45 *FR* 33097] See Case Study 4-1.

Case Study 4-1: Domestic Sewage Exclusion Applicability

An industrial laundry accepts rags from a number of different customers, some of which are contaminated with listed spent solvents (F001–F005). After laundering, the washed rags are returned to the customers, and the washwater is discharged to a sewer flowing to a POTW. How does the domestic sewage exclusion apply in this situation?

The domestic sewage exclusion potentially applies to any hazardous spent solvents or other hazardous wastewaters that are discharged from the laundry to the sewer line, as long as the mixed domestic sewage/waste flows to a POTW. The discharges to the sewer line would be subject to general and categorical pretreatment standards. [RO 11777]

The guidance summarized above was issued before the exclusions for solvent-contaminated wipes were promulgated on July 31, 2013. See Section 3.3.3 for the details of that rule.

The agency expanded on this point in subsequent guidance, stating that the exclusion does not extend to wastes that do not pass through a sewer system prior to arriving at a POTW. Therefore, the exclusion does not apply to wastes that are transported to the POTW by way of truck, rail, or dedicated pipe. [51 *FR* 30166, RO 11204] For example, a hazardous waste shipment to a POTW via truck is subject to all appropriate Subtitle C accumulation and transportation requirements, including the use of a manifest. RO 11181] "A dedicated pipeline refers to a separate pipeline that is used to carry hazardous wastes directly to a POTW's property boundary without prior mixing with domestic sewage." [*Guidance Manual for the Identification of Hazardous Wastes Delivered to Publicly Owned Treatment Works by Truck, Rail, or Dedicated Pipe*, June 1987, available from http://nepis.epa.gov/EPA/html/Pubs/pubtitleOSWER.html by downloading the report numbered 5942.] Even if hazardous wastes, which were transported to the POTW from offsite, were mixed with the influent domestic sewage at the POTW before any treatment occurred, the exclusion would still not apply. EPA has interpreted that the exclusion is appropriate only to wastes that enter a sewer *leading* to a POTW. [RO 12963]

Any activities that occur prior to the actual introduction of the waste into the sewer system are not covered by the exclusion. Thus, hazardous wastewater or sludge treatment, storage, and/or disposal that occur upstream of the discharge point remain subject to RCRA regulation, unless another exemption applies (e.g., the wastewater treatment unit exemption discussed in Section 7.2). Additionally, dilution of wastewater in order to meet the CWA discharge limits is usually unacceptable. [EPA/233/B-00/002, cited previously] In Figure 4-1, the sludge generated in a wastewater treatment tank upstream of the discharge point is not excluded under the domestic sewage exclusion. If sent for disposal, it is a solid waste that must be evaluated for listings and characteristics.

The point where hazardous waste is discharged into the POTW's sewer line does not have to be on the facility's property to take advantage of the exclusion. That is, the facility's wastewater discharge pipe may intersect the POTW's sewer line outside of the facility's fence line. Sometimes, facilities would like to discharge their hazardous waste into an offsite manhole associated with the POTW's sewer line—is that legal? No. EPA noted on July 24, 1990 that POTWs may *not* designate discharge points outside of the POTW facility boundary for the introduction of hazardous wastes by trucked shipments to the sewer system. [55 *FR* 30099]

4.1.3.1 Notification for discharging hazardous waste to a POTW

The 1984 HSWA amendments require [in RCRA Section 3018(d)] that industrial dischargers taking advantage of the domestic sewage exclusion notify EPA or their state of any hazardous wastewaters discharged to POTWs and obtain an EPA ID number.

However, EPA has not changed its RCRA regulations [in either §261.4(a)(1) or 262.12] to implement this statutory provision. [EPA/000/R-85/002, September 1985, available at http://nepis.epa.gov/EPA/html/Pubs/pubtitleOther.html, by downloading the report numbered 000R85002] This notification requirement is in the CWA regulations at 40 *CFR* 403.12(p).

In addition, some permits or contracts issued by POTWs to industrial dischargers require that the discharger notify the POTW before any hazardous waste is discharged into the sewer line flowing to the POTW. Further, these permits or contracts may prohibit hazardous wastes from being discharged into the POTW's sewer line.

4.1.4 What if I mix hazardous waste into my facility's domestic sewage?

Sometimes an industrial facility will mix hazardous waste with untreated sanitary waste from the facility's own restrooms. Such mixing can occur well upstream of where the mixture is introduced into the POTW's sewer line. Where does the domestic sewage exclusion apply in this situation? Based on EPA's guidance, it appears that the exclusion does not apply until the point where the material becomes subject to the CWA pretreatment standards—that is, the point where the material enters the POTW's sewer line. "Although RCRA does not define 'sewer system,' it is not the agency's intent to include private sewers or wastewater treatment units upstream from the point where pretreatment standards (Section 307(b) of CWA) would apply to wastes going [through] a sewer to a publicly owned treatment works." [RO 12066]

Other agency guidance seems to corroborate the above statement. "The exclusion thus covers industrial wastes discharged to *POTW sewers* which contain domestic sewage…. The exclusion avoids the potential regulatory redundancy of subjecting hazardous wastes mixed with domestic sewage to RCRA management requirements *if these wastes are already subject to appropriate pretreatment requirements under the CWA."* [Emphasis added.] [51 *FR* 30166]

4.1.4.1 Does the exclusion apply to material removed from the sewer line?

If sludge (e.g., from sewer line low-point cleanout or from decommissioning) is removed from a sewer line downstream of the point where hazardous waste is introduced, it loses its excluded status under the domestic sewage exclusion. This occurs because such sludge has never reached the POTW (i.e., it does not "pass through a sewer system to a POTW"). Once removed, the sludge would be considered hazardous if the sludge exhibits a characteristic. If hazardous, the removed material would be subject to all Subtitle C standards. [RO 11490]

A similar situation occurs if the mixed hazardous waste/domestic sewage subsequently leaks from the sewer line downstream of the hazardous waste insertion point, but before it reaches the POTW. Such leakage would also not qualify for the domestic sewage exclusion. These leaks are solid wastes and, again, could be hazardous depending on operation of the mixture rule. [RO 14068]

Leakage from sewer lines could thus activate corrective action provisions. Two questions that govern corrective action requirements in this situation are: 1) does the facility have a RCRA permit that triggers corrective action (e.g., a RCRA operating permit or a post-closure permit), and 2) are the sewers part of the "facility"? ("Facility" has a special definition in §260.10 for purposes of implementing corrective action.) The answer to the second question is based on a number of site-specific factors, including: 1) whether the facility owner/operator or the POTW owns, operates, and/or maintains the line from which the leak occurred; and 2) whether the line transports wastes from a single industrial facility or from many, unrelated facilities.

Typically, sewer lines within the plant boundary are controlled by the facility owner/operator. If the owner/operator is subject to corrective action because of a RCRA permit, leaks from such lines would likely be subject to corrective action (e.g., addressed as a solid waste management unit). If the collection main is owned, operated, and/or maintained by the POTW, leakage would generally not be subject to corrective action (except as noted in the following section).

4.1.5 What about the sludge from the POTW?

A POTW treats domestic sewage and industrial wastewater using biological treatment. What is the regulatory status of the POTW's biological treatment sludge?

Two primary possibilities exist:

1. If all of the domestic sewage and industrial wastewater is received as a mixture in sewer lines, the domestic sewage exclusion is in effect, and no hazardous waste codes are associated with the influent to the POTW. The resulting sludge produced from treating the influent is considered a new point of generation and is a solid waste when disposed (i.e., the POTW's sludge is not covered by the domestic sewage exclusion). However, the only way the sludge could be hazardous is if it exhibits one of the characteristics. [51 *FR* 30166, RO 11181] In practice, POTWs set local pretreatment standards such that the sludge they produce will not exhibit any characteristics.

2. If the POTW accepts hazardous waste via truck, rail, or dedicated pipe, the domestic sewage exclusion does not apply, and any listed or characteristic hazardous waste codes would be carried with the waste shipment. Consequently, POTWs that manage hazardous wastes that have not passed through the sewer system would be subject to all applicable hazardous waste regulations. [RO 11204] POTWs are deemed to have a RCRA permit if they meet all notification, manifesting, and reporting requirements outlined under §270.60(c); thus, they can accept offsite hazardous waste shipments.

However, any listed hazardous waste codes associated with an offsite shipment would also have to be carried on the POTW's sludge via the derived-from rule. If the shipped waste exhibits a characteristic, the POTW's sludge would only be hazardous if it also exhibits a characteristic. [52 *FR* 23478, RO 12963] If a POTW receives hazardous waste by way of truck, rail, or dedicated pipe, one of the provisions in §270.60(c) is that the facility take corrective action for releases at their own solid waste management units. [51 *FR* 30169] Significant additional discussion of this option is contained in *Guidance Manual for the Identification of Hazardous Wastes Delivered to Publicly Owned Treatment Works by Truck, Rail, or Dedicated Pipe*, cited previously.

EPA implied that mixed radioactive/hazardous waste would be exempt from RCRA regulation under the domestic sewage exclusion if such waste was discharged to a domestic sewer system that leads to a POTW. [RO 12823] Presumably, however, the sludge produced at the POTW would still be radioactive, requiring compliance with Nuclear Regulatory Commission regulations. In most cases, it is doubtful that the POTW would allow a generator to discharge such mixed wastes to the sewer.

A question that arises occasionally has to do with the F006 listing. Would that waste code apply to sludge from a POTW if the POTW receives rinsewater from electroplating operations? Assume that the rinsewater was mixed with domestic sewage and arrived at the POTW by way of the sewer. EPA seems to imply that sludge from a POTW will not be F006 under these circumstances but doesn't clearly explain why. [RO 11139, 14322]

4.1.6 Can federally owned treatment works qualify for the exclusion?

Because federally owned treatment works (FOTWs) are not included in the definition of POTWs discussed previously, they were not eligible for the domestic sewage exclusion originally promulgated in 1980. However, when Congress passed the Federal Facilities Compliance Act (FFCA) in 1992, the act extended this exclusion to FOTWs if they comply with three requirements (which are taken from RCRA Section 3023—added by the FFCA). [RO 14169] First, the FOTW must be owned/operated by the federal government and must be treating wastewater, a majority of which is domestic sewage, before discharge under an NPDES permit. Second, the hazardous waste mixed into the domestic sewage must meet one of the following three conditions:

1. The hazardous waste is subject to a CWA Section 307(b) pretreatment standard and meets that standard before discharge into the FOTW sewer;

2. If the hazardous waste is not subject to a CWA pretreatment standard, it must meet land disposal restriction treatment standards and not be prohibited from land disposal before discharge into the FOTW sewer; or

3. The hazardous waste is generated by households or conditionally exempt small quantity generators (i.e., those that generate less than 100 kg/mo of nonacute hazardous waste) and is not acutely hazardous.

Finally, any hazardous waste meeting one of the three conditions noted above must only be sent to the FOTW via mixing into domestic sewage flowing to the FOTW (at which point in time the material ceases to be a solid waste).

4.1.7 Are there any LDR implications?

A facility was treating the domestic sewage it generated in a number of aboveground surface impoundments. Listed hazardous waste was also introduced into these impoundments, and EPA determined that the mixture was hazardous under the mixture rule. That is, the domestic sewage exclusion did not apply to the mixture because it was not passing through a sewer system to a POTW. [RO 12994] As such, the hazardous waste entering the surface impoundments was subject to the land disposal restrictions (LDR) program.

Conversely, if hazardous waste is not managed in a land disposal unit (e.g., a surface impoundment) from its point of generation to the point it is introduced into the sewer, land disposal is not occurring. [RO 12404] Thus, the waste would not be subject to LDR treatment standards. (Instead, it is subject to the CWA pretreatment standards.) However, a one-time notification must be placed in the facility's files per §268.7(a)(7). This requirement applies even if, prior to discharge of the waste to the sewer, the generator does not manage the waste in a manner that subjects it to substantive regulation (e.g., the generator does not accumulate the waste in 90-day tanks or containers per §262.34). [RO 13547]

The one-time notice should include a statement 1) that the hazardous waste was generated, 2) that the hazardous waste subsequently became excluded from regulation under §261.4(a)(1), and 3) identifying the disposition of the waste—that is, it was discharged to the sewer. Note that underlying hazardous constituents do not have to be identified or treated in characteristic wastes when these wastes are decharacterized and managed in wastewater treatment systems regulated under the CWA. [§268.1(c)(4), 61 *FR* 15661, RO 14216]

4.2 NPDES discharge exclusion

Wastewater discharged from an industrial facility to surface water of the United States ceases to be a solid or hazardous waste at the outfall point where the waste enters surface water. This exclusion works similarly to the domestic sewage exclusion discussed in the previous section and is extremely beneficial to facilities located near lakes, rivers, streams, or the ocean.

4.2.1 Background

The RCRA statutory definition of solid waste excludes "industrial discharges which are point sources subject to permits under Section 402 of [the Clean Water Act (CWA)]." [RCRA Section 1004(27)] Section 402 of the CWA outlines the National Pollutant Discharge Elimination System (NPDES) permitting program; such a permit is required whenever an industrial facility discharges wastewater into waters of the United States. This exclusion was created to avoid duplicative regulation of point-source wastewater discharges under both the CWA and RCRA. Without the provision, discharging wastewater into surface water would be disposal of solid (and potentially hazardous) waste and would potentially be regulated under both statutes. [May 19, 1980; 45 *FR* 33098]

The NPDES discharge exclusion is codified in §261.4(a)(2). Based on this regulatory language, point-source discharges that are subject to regulation under Section 402 of the CWA are excluded from the definition of "solid waste." As illustrated in Figure 4-2, this means that once wastewater from an NPDES-permitted discharge or outfall point enters waters of the United States, it is excluded from RCRA regulation. According to EPA, this is true even if the discharge could or should be regulated under a Section 402 permit—but is not. The agency believes that a point-source discharge without an NPDES permit would be a violation of the CWA, not RCRA. [RO 11125, 11408, 11895]

4.2.2 Where does the exclusion apply?

EPA notes in §261.4(a)(2) that the exclusion "applies only to the actual point-source discharge. It does not exclude industrial wastewaters while they are being collected, stored, or treated before discharge, nor does it exclude sludges that are generated by industrial wastewater treatment." In other guidance [RO 11139], EPA states:

"Since the Clean Water Act applies to discharges to navigable surface waters, point-source discharges cannot apply to some internal midway point in the wastewater treatment train on the grounds of a facility or another facility (unless it is a POTW) which treats, stores, or collects these wastewaters. Even if the wastewaters themselves were exempt from regulation while they were being treated, collected, or stored prior to

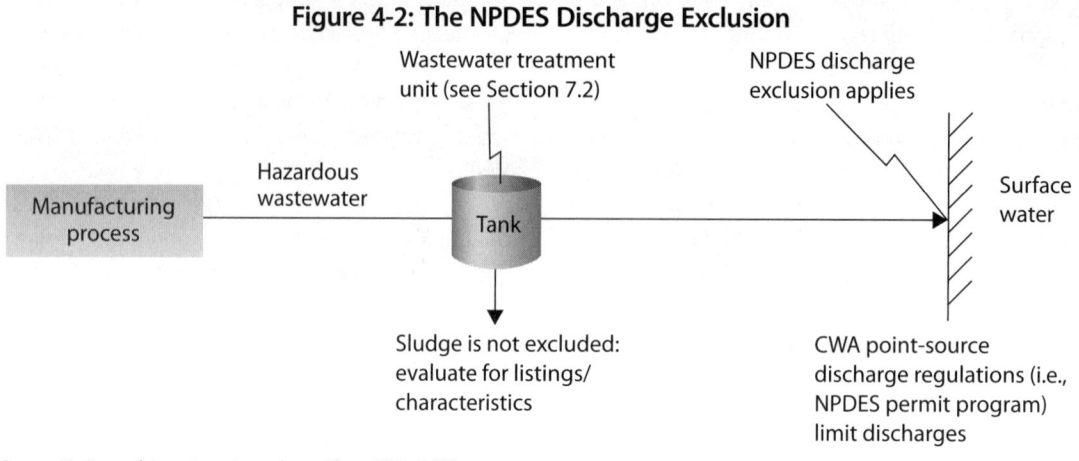

Figure 4-2: The NPDES Discharge Exclusion

Source: McCoy and Associates, Inc.; adapted from §261.4(a)(2).

discharge, the sludges are not exempt as the result of any exemption of the wastewater."

In other words, the exclusion does not apply to hazardous wastewater prior to its discharge point, because most of the potential environmental hazards (e.g., ground water contamination) are posed during upstream treatment and storage activities, which are not controlled under the CWA. [45 FR 33098, RO 11309] Thus, hazardous wastewater or sludge treatment, storage, and/or disposal that occur upstream of the discharge point remain subject to RCRA regulation, unless another exemption applies (e.g., the wastewater treatment unit exemption discussed in Section 7.2). In Figure 4-2, the sludge generated in a wastewater treatment tank upstream of the NPDES-permitted discharge point is not excluded under the NPDES discharge exclusion. If sent for disposal, it is a solid waste that must be evaluated for listings and characteristics.

A "point source" is actually a defined term in §122.2 of the CWA regulations: it is "any discernible, confined, and discrete conveyance, including but not limited to, any pipe, ditch, channel, tunnel, conduit...from which pollutants are or may be discharged." Typically, this would be a pipe discharging wastewater into a river, etc., and would be referred to as the "outfall point" by industry. Those CWA regulations also define "discharge" as the "addition of any pollutant or combination of pollutants to waters of the United States from any point source." Finally, "waters of the United States" are defined as "waters...susceptible to use in interstate or foreign commerce, including all waters which are subject to the ebb and flow of the tide; all interstate waters, including interstate wetlands; [and] all other waters such as intrastate lakes, rivers, streams.... Waste treatment systems, including treatment ponds or lagoons designed to meet the requirements of the CWA...are not waters of the United States."

An NPDES permit will occasionally specify intermediate, upgradient points at which certain conditions must be met in addition to the limits and conditions at the outfall. Facilities will sometimes argue that their wastewater ceases to be regulated under RCRA (i.e., it ceases to be a solid or hazardous waste) at the intermediate point. Based on the specific definitions of "point source" and "discharge" noted above, however, the exclusion does not take effect until the wastewater actually mixes with waters of the United States, as illustrated in Case Study 4-2 and the following example.

Q *A facility treats its wastewater and then discharges it to surface water under an NPDES permit. The treated wastewater is hazardous by listed*

> **Case Study 4-2:**
> **Wastewater Diverted Into a Basin for Fire Training**
>
> Hazardous wastewater discharged from a facility is subject to an NPDES permit and is therefore excluded from RCRA regulation at the discharge point. Periodically, the hazardous wastewater is diverted from the outfall to an onsite, nonhazardous waste surface impoundment for use in fire-fighting training exercises. Is the surface impoundment regulated under RCRA?
>
> Yes. The wastewater is not being discharged as that term is defined by §122.2. According to EPA, "[w]hile the diversion to the surface impoundment takes place after the water exits the pipe, the discharge must be mixed with waters of the United States in order to remain within the NPDES permit and thus be excluded from RCRA. Discharge to the surface impoundment would constitute illegal operation of a hazardous waste storage unit." [RO 13051]

not qualify for the RCRA exclusion (even if it is part of the [NPDES] permit). Therefore, the wastewater remains a solid and hazardous waste." [RO 14775]

Sometimes, it is confusing as to exactly where the discharge point is—especially if there is a surface impoundment as the last unit in an industrial wastewater treatment system. For example, it is at times difficult to distinguish where a wastewater treatment lagoon ends and waters of the United States begin. The agency notes that water bodies that are wholly within the facility's property boundary and are upgradient of the NPDES-permitted discharge point are clearly regulated as surface impoundments under RCRA. Conversely, a surface impoundment that was created by impounding a portion of a larger body of surface water may or may not be considered a hazardous waste unit; such determinations must be made by states or EPA regions on a case-by-case basis. Ponds that are located downgradient of an NPDES-permitted discharge point are, by definition, waters of the United States and are not subject to RCRA regulation. [RO 12826]

4.2.3 Discharges to ground water that has a direct connection to surface water

CWA Section 402 jurisdiction extends to point-source discharges to ground water if there is a direct hydrologic connection between the point source and nearby surface waters of the United States. However, discharges of leachate into ground water from leaking waste management units (e.g., a landfill) are *not* excluded from RCRA regulation under the §261.4(a)(2) NPDES discharge exclusion. (As mentioned previously, the exclusion applies only to traditional pipe outfall-type point-source discharges and not to discharges that occur upstream of that point.) Consequently, the RCRA requirements apply to discharges of leachate to ground water from waste management units, even when the ground water provides a direct hydrologic connection to a nearby surface water of the United States. [RO 11895]

---

waste code until it's excluded under §261.4(a)(2) at the point-source discharge. The facility would like to divert a portion of the treated wastewater for spray irrigation and maintenance of an onsite landfill cap. Would the diverted wastewater be exempt under the §261.4(a)(2) exclusion?

A No. EPA "determined that wastewater sprayed onto a landfill cap does not qualify for the industrial wastewater discharge exclusion under §261.4(a)(2). Although a portion of the effluent will continue to be discharged from [the NPDES]-permitted outfall to [the creek] (and thus permitted under Section 402), wastewater that is diverted to land application and is not discharged to waters of the United States is not a point-source discharge subject to regulation under the CWA and, therefore, does

4.2.4 Dredged sediments

In the 1980s and most of the 1990s, EPA's policy regarding the applicability of RCRA to contaminated dredged sediments was as follows. If the source of the contamination was a point-source discharge regulated under the NPDES program, the §261.4(a)(2) exclusion applied; therefore, hazardous wastes were not discharged into the surface water. Consequently, the resulting contaminated sediments were only regulated under RCRA if they 1) were dredged from the surface water, *and* 2) exhibited one or more hazardous waste characteristics. [RO 11125, 11455]

On the other hand, if evidence indicated that hazardous wastes had been dumped into surface water in a manner that did not trigger the NPDES regulations (e.g., by illegal dumping), the dumping constituted disposal under RCRA and was subject to the appropriate RCRA regulations. The resulting contaminated sediments were also regulated under RCRA as listed or characteristic hazardous waste under the contained-in policy. [RO 11125] See Section 5.3.1.2.

4.2.4.1 Dredged sediments now exempt

On November 30, 1998 [63 *FR* 65874], EPA excluded most dredged sediments from the definition of hazardous waste. Specifically, dredged material that is subject to the requirements of a permit issued under Section 404 of the CWA or Section 103 of the Marine Protection, Research, and Sanctuaries Act is not a hazardous waste. The permits must be issued by the U.S. Army Corps of Engineers or, in some cases, by the state. Note that this exclusion is for dredged material that will be disposed in the aquatic environment. It does not apply to dredged material destined for disposal in a landfill. See 63 *FR* 65921–2. As a result of the exclusion, which is codified in §261.4(g), contaminated sediments will rarely be subject to RCRA regulation.

4.3 Irrigation return flows exclusion

When agricultural land is irrigated, water not absorbed into the soil can return to water basins/reservoirs, either as surface water run-off or through ground water percolation. As noted in Figure 4-3, these return flows often pick up fertilizer or pesticide constituents, potentially rendering them hazardous by characteristic. [EPA/530/K-02/022I, October 2001, available at http://www.epa.gov/epawaste/inforesources/pubs/training/excl.pdf]

However, because this water may be reused for irrigation, it is excluded from the definition of solid waste per §261.4(a)(3). Although the regulatory language in that section specifies "irrigation return flows," *Federal Register* preamble language noted that this exclusion applies to solid or dissolved

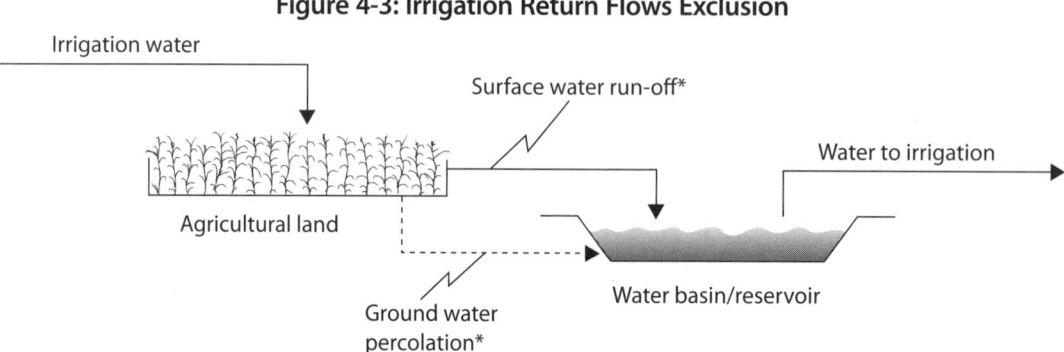

Figure 4-3: Irrigation Return Flows Exclusion

*May be contaminated with fertilizers/pesticides and/or exhibit a characteristic but is excluded from the definition of solid waste per §261.4(a)(3).
Source: McCoy and Associates, Inc.; adapted from §261.4(a)(3).

materials in these return flows. [May 19, 1980; 45 *FR* 33098]

4.4 Radioactive materials and mixed waste

Historically, uncertainty has surrounded the issue of whether or not RCRA is applicable to mixed wastes—wastes that are both hazardous and radioactive [i.e., they contain source, special nuclear, or by-product material regulated by the Atomic Energy Act of 1954 (AEA)].

This section describes radioactive and mixed wastes, as well as the exclusion for source, special nuclear, and by-product materials. The effects of a May 16, 2001 rule [66 *FR* 27218] on the storage, treatment, transportation, and disposal of mixed waste are incorporated throughout this section and are summarized at the end.

4.4.1 Two types of radioactive materials

There are two categories of radioactive materials:

1. Source, special nuclear, and by-product materials defined by the AEA; and

2. Naturally occurring and/or accelerator-produced radioactive material (NARM).

4.4.1.1 Source, special nuclear, and by-product materials

Source, special nuclear, and by-product materials are excluded from the RCRA definition of solid waste in §261.4(a)(4). These materials are defined as follows:

- *Source material*—Uranium, thorium, or any other material which is determined by the Atomic Energy Commission (AEC—the precursor agency to the NRC)…to be source material, or ores containing one or more of the foregoing materials, in such concentration as the AEC may by regulation determine from time to time. [AEA Section 11(z)]

- *Special nuclear material*—Plutonium, uranium enriched in the isotope 233 or 235, and any other material that the AEC…determines to be special nuclear material, or any material artificially enriched by any of the foregoing (but does not include source material). [AEA Section 11(aa)]

- *By-product material*—(1) Any radioactive material (except special nuclear material) yielded in or made radioactive by exposure to the radiation incident to the process of producing or utilizing special nuclear material; (2) the tailings or wastes produced by the extraction or concentration of uranium or thorium from any ore processed primarily for its source material content; (3) any discrete source of radium-226 or material made radioactive by use of a particle accelerator for commercial, medical, or research uses; and (4) any discrete source of naturally occurring radioactive material (other than source material) used for commercial, medical, or research activities that the AEC determines could pose a public health and safety or common defense and security threat similar to that of a discrete source of radium-226. [AEA Section 11(e)]

The specific regulations for radioactive material management developed under the AEA are administered by DOE at DOE facilities and by the NRC at all other facilities.

4.4.1.2 NARM

NARM is defined as radioactive materials that are naturally occurring or produced by an accelerator used in subatomic particle physics research. Technologically enhanced naturally occurring radioactive material (TENORM) is a subset of NARM and refers to materials whose radioactivity has been technologically enhanced by controllable practices, such as mineral extraction or processing activities. Such materials include exploration and production wastes from the oil and natural gas industry and phosphate slag piles from the phosphate mining industry. TENORM does *not* refer to the natural or background radioactivity of rocks and soils. [November 19, 1999; 64 *FR* 63466] (Before 1998, the term used for these materials was NORM; however, based on more current industry and regulatory practice, the term "TENORM" is considered more appropriate.)

TENORM is defined by the National Academy of Sciences as follows: "Technologically enhanced naturally occurring radioactive materials are any naturally occurring radioactive materials not subject to regulation under the Atomic Energy Act whose radionuclide concentrations or potential for human exposure have been increased above levels encountered in the natural state by human activities." Total amounts of TENORM wastes produced in the United States may be in excess of 1 billion tons annually. [*Evaluation of EPA's Guidelines for Technologically Enhanced Naturally Occurring Radioactive Materials*, EPA/402/R-00/001, June 2000, available at http://www.epa.gov/rpdweb00/docs/tenorm/402-r-00-001.pdf]

EPA has posted significant information about TENORM on its website at http://www.epa.gov/radiation/tenorm/about.html.

4.4.2 Mixed waste

"Mixed waste" is defined as "waste that contains both hazardous waste and source, special nuclear, or by-product material subject to the Atomic Energy Act of 1954." [RCRA Section 1004(41)] It is typically generated by DOE and certain military facilities, nuclear power plants, industrial facilities, research laboratories, and medical institutions. Mixed waste generated by commercial and non-DOE federal facilities is regulated by the NRC, while DOE regulates mixed waste generated at its own facilities. Such mixed waste is subject to both RCRA hazardous waste and AEA radioactive waste regulations. [RO 13004] Mixed waste may include any AEA-regulated radionuclide, regardless of whether that radionuclide is classified as high-level, transuranic, or low-level waste, as discussed further below. [RO 12935]

4.4.2.1 Commercially generated (non-DOE) mixed waste

Almost all commercially generated (non-DOE) mixed waste is made up of low-level mixed waste (LLMW) generated at nuclear power, industrial, research, and medical facilities. A large percentage of this non-DOE LLMW is liquid scintillation cocktails that contain small amounts of radioactivity in organic solvents. Other types of LLMW that may be generated at non-DOE facilities include spent solvents containing radionuclides; radioactive spent freon, acetone, or other solvents used to clean protective garments and equipment; filters used during radioactive solvent reclamation; still bottoms from the distillation of radioactive solvents; spent ion-exchange resins; adsorbents; residues from spill cleanups; lead shielding; lead-lined containers; welding rods; and batteries.

Spent fuel from nuclear power plants is categorized as high-level radioactive waste. Some of the elements of these wastes will remain radioactive for thousands of years. However, EPA noted that it does not believe that spent nuclear reactor fuels generally will be RCRA hazardous wastes. [RO 13641]

4.4.2.2 DOE mixed waste

Three main types of mixed wastes are generated or stored at DOE facilities:

- *High-level mixed wastes (HLMW)*—HLMW generated at DOE facilities includes waste from reprocessing spent nuclear fuel and other radioactive materials. Due to their high radioactivity, these wastes are very dangerous to handle. In addition, they contain highly corrosive components, organics, or heavy metals that make them RCRA hazardous. DOE currently stores its HLMW in large tanks at the Hanford Reservation in Washington, the Idaho National Laboratory in Idaho, the Savannah River Site in South Carolina, and the West Valley Demonstration Project in upstate New York. DOE plans to treat HLMW using vitrification, which would convert the waste into a solid, glass-like substance that will greatly limit the dispersion of hazardous and radioactive components into the environment.

- *Transuranic (TRU) mixed wastes*—TRU mixed wastes contain greater than 100 nanocuries per gram of radioactivity from elements with atomic numbers greater than 92 (the atomic number for uranium). These wastes typically pose greater radioactivity hazards than LLMW because they contain long-lived alpha radiation emitters. TRU mixed wastes are generated during nuclear weapons production, plutonium-bearing reactor fuel fabrication, and spent fuel processing. DOE is disposing its TRU mixed wastes at the Waste Isolation Pilot Plant (WIPP) near Carlsbad, New Mexico.

- *Low-level mixed wastes (LLMW)*—LLMW (radioactive mixed waste that is not classified as high-level or TRU waste) is generated or stored at 37 DOE sites in 22 states as a result of research, development, and nuclear weapons production. Typical LLMWs are 1) cleaning and degreasing solvents and scintillation liquids (which typically contain toluene or xylene); 2) oil mixtures used in operation and maintenance activities (including used oil from radiologically contaminated equipment); 3) heavy metal-contaminated wastes such as shielding, ion-exchange resins, corrosion inhibitors, and decontamination resins; and 4) aqueous corrosive liquids.

4.4.2.3 Hazardous NARM is not mixed waste

NARM is not regulated under the AEA. That fact, combined with Congress' definition of "mixed waste" in RCRA Section 1004(41), leads EPA to the conclusion that a NARM waste that is also RCRA hazardous does not meet the definition of a mixed waste. [June 1, 1990; 55 *FR* 22645] Such a waste would therefore only be regulated as a hazardous waste under RCRA and *not* a mixed waste subject to both RCRA and the AEA. [RO 14310] The radioactive portion of NARM, however, may be subject to state regulatory programs.

EPA's regulatory interpretation regarding NARM is in agreement with DOE policy contained in DOE Order 5820.2A. According to the order, DOE waste consisting of NARM that has been mixed with RCRA listed hazardous waste or which exhibits a characteristic must only be managed as RCRA hazardous waste; DOE radioactive requirements do not apply. See also RO 12309.

4.4.3 The RCRA exclusion for radioactive materials

The §261.4(a)(4) exclusion unmistakably excludes source, special nuclear, and by-product material regulated under the AEA from the definition of solid waste; thus, such materials are not subject to RCRA regulation. However, the applicability of the exclusion to wastes that contain these materials in addition to being RCRA hazardous wastes (i.e., mixed wastes) was far from clear in 1980. Initially, DOE determined that the §261.4(a)(4) exclusion also applied to mixed wastes and, in fact, that all wastes (radioactive as well as nonradioactive) generated in DOE defense-related activities authorized under the AEA were exempt from RCRA hazardous waste regulation. [OSWER Directive 9990.0, June 1983]

However, DOE's interpretation of the exclusion became untenable in the mid-1980s as a result of a federal court ruling (*LEAF v. Hodel*, 1984). In its ruling, the court concluded that the §261.4(a)(4) exclusion applies only to the radionuclides themselves within mixed wastes. If nonradioactive components of the waste make the material a RCRA hazardous waste, then the waste is regulated by both the AEA and RCRA. [EPA/530/K-02/022I, cited previously] Accordingly, the hazardous component of mixed wastes is subject to RCRA and the radioactive component is subject to the AEA. [July 3, 1986; 51 *FR* 24504, RO 13452] This position was reiterated when EPA and NRC jointly published *Guidance on the Definition and Identification of Commercial Mixed Low-Level Radioactive and Hazardous Waste*, October 4, 1989, available from http://www.epa.gov/radiation/mixed-waste/guidance-identification-llmw.html.

4.4.3.1 DOE acquiesces

Of the three types of radioactive materials (i.e., source, special nuclear, and by-product material), by-products are usually the materials that actually are wastes. Of particular concern at DOE facilities is the AEA Section 11(e)(1) portion of the "by-product" definition. These are the primary mixed wastes at these facilities. After EPA made its 1986 determination that only the radionuclides in mixed wastes are excluded under §261.4(a)(4), controversy erupted over the regulatory status of AEA Section 11(e)(1) by-product materials.

To clear up the confusion, DOE finalized an interpretation of AEA Section 11(e)(1) by-product material on May 1, 1987 [52 *FR* 15937] in 10 *CFR* 962.1–3, adopting EPA's position that only the actual radionuclides dispersed or suspended in DOE mixed wastes (not the entire waste stream) are considered by-product material and therefore are excluded from RCRA. These radionuclides in by-product material are regulated under the AEA; the nonradioactive components of DOE wastes are subject to RCRA regulation if they are listed RCRA hazardous wastes or exhibit a characteristic. In other words, any mixed waste that is a RCRA hazardous waste and a radioactive waste (except for NARM) is regulated under both RCRA and the AEA. DOE believes that this approach gives "RCRA and the AEA the greatest capacity to regulate effectively the special type of hazard that each statute was designed to control." [May 1, 1987; 52 *FR* 15940]

Based on the above interpretation, which is jointly shared by EPA, NRC, and DOE and is still current today, application of the exclusion for source, special nuclear, and by-product material is very limited.

4.4.3.2 AEA takes precedence over RCRA if conflicts occur

If the RCRA and AEA regulatory schemes conflict, RCRA Section 1006(a) specifies that RCRA must yield. In other words, the AEA requirement would take precedence, and the inconsistent RCRA requirement would be inapplicable. For example, a conflict may occur if compliance with a specific RCRA requirement would violate national security interests. In such situations, the AEA would take precedence and the RCRA requirement would be waived. [RO 12992]

4.4.3.3 Examples

Tritium (a radioactive isotope of hydrogen) is produced by activities regulated by NRC under the AEA and is considered a by-product material. The isotope is contained in toluene as a scintillation liquid. Is the scintillation cocktail a mixed waste when disposed?

Yes. Mixtures of tritium and toluene satisfy the definition of mixed waste because tritium is a low-level radioactive waste and the toluene is a hazardous spent solvent (F005) when the scintillation cocktail is discarded. [*Guidance on the Definition and Identification of Commercial Mixed Low-Level Radioactive and Hazardous Waste,* cited previously]

During operation of nuclear power plants, fission products, such as cesium-134 (Cs-134) and cesium-137 (Cs-137), build up in the cooling water systems. Potassium hexacyanocobalt(II)-ferrate(II), an insoluble granular chemical, is used as an ion-exchange media to remove Cs-134 and Cs-137 from the irradiated water before it can be released into the environment. Is unused potassium hexacyanocobalt(II)-ferrate(II) or the spent ion-exchange media subject to RCRA regulation?

The unused chemical is not currently listed as a P- or U-waste and does not appear to exhibit any hazardous characteristics. Therefore, discarded, unused potassium hexacyanocobalt(II)-ferrate(II) and any residues resulting from a spill of the chemical would not be RCRA hazardous wastes. However, the spent ion-exchange media may exhibit a hazardous characteristic (toxicity would be the most likely), depending on the composition of the water being treated. If the spent media exhibits such a characteristic, it would be a mixed waste and be subject to RCRA and the AEA. If the media does not exhibit a characteristic, it would be a radioactive waste subject only to AEA regulations. [RO 11965, 14016]

Chemicals are used in a production process at a DOE facility that are on the P- or U-lists of hazardous wastes or in the Part 261, Appendix VIII list of hazardous constituents. Should the facility assume that the low-level radioactive waste that is generated from this process is LLMW?

A Low-level radioactive waste that contains hazardous constituents is not necessarily LLMW. In order for the low-level radioactive waste to be a mixed waste, the waste must contain a known listed waste or exhibit a characteristic. The intent of the P- and U-listings was to encompass only those materials that are being thrown away in their pure, unused form or as an off-specification variant; these listings do not apply to wastes which contain these materials as constituents. [May 19, 1980; 45 *FR* 33115] EPA lists wastes as hazardous if they contain significant quantities of the Part 261, Appendix VIII hazardous constituents and the agency determines that these constituents are persistent, mobile, and pose a substantial or potential threat to human health and the environment. [§261.11, RO 11144, 12014, 12296, 13290] The presence of one or more of these hazardous constituents within the low-level waste does not by itself render the waste hazardous. If the waste does not exhibit any characteristics, and generator knowledge shows that it does not contain an F-, K-, P-, or U-waste, the waste is not hazardous. [*Guidance on the Definition and Identification of Commercial Mixed Low-Level Radioactive and Hazardous Waste,* cited previously]

4.4.4 Mixed waste management under RCRA

Unless a facility can take advantage of the Part 266, Subpart N exemptions from RCRA for mixed waste, any facility that generates, stores, treats, or disposes mixed waste is subject to the full set of RCRA regulations. As discussed below, this includes RCRA storage, manifesting, land disposal restrictions, and corrective action requirements. However, a facility that generates less than or equal to 100 kg/mo of nonacute hazardous waste or 1 kg/mo of acute hazardous waste, including LLMW, is categorized as a conditionally exempt small quantity generator. As a result, most of the RCRA regulations do not apply to the management of that waste, and the LLMW can be disposed as low-level radioactive waste if the materials meet the disposal site's waste acceptance criteria. [§261.5, August 7, 1995; 60 *FR* 40207] Case Study 4-3 evaluates the management of radioactive lead shielding under RCRA.

4.4.4.1 Storage/accumulation

Facilities that store mixed waste are subject to the same RCRA hazardous waste storage rules as for any other hazardous waste. As such, a RCRA storage permit is necessary if mixed wastes generated onsite are stored for more than 90 days (for LQG facilities), 180 days (for SQG facilities), or 270 days (for SQGs that ship their waste over 200 miles to a TSD facility). If these time limits are not exceeded, the mixed waste may be accumulated in 90- or 180-day accumulation units subject to the §262.34 requirements. In addition to accumulation, treatment of mixed wastes may occur in 90- or 180-day accumulation units to meet NRC disposal requirements and/or Department of Transportation (DOT) shipping rules (see Section 7.3). [RO 11598, 13297] Satellite accumulation units may be used for mixed waste accumulation if they comply with §262.34(c).

The daily and weekly inspection requirement for hazardous waste tanks and containers, respectively, remains in effect for accumulation of mixed waste. However, to minimize personnel exposure to radioactivity, such inspections may be made via remote monitoring devices, television monitors, etc. as opposed to walk-through inspections.

Draft guidance designed to help parties that store mixed waste comply with the RCRA and AEA requirements was issued in August 1995 by EPA and NRC (*Joint Guidance on the Storage of Mixed Low-Level Radioactive and Hazardous Waste*, August 7, 1995; 60 *FR* 40205, available from http://www.epa.gov/radiation/mixed-waste/guidance-storage.html).

Case Study 4-3: Is Lead Shielding Subject to RCRA?

Lead is often used as shielding material to contain radionuclides during use and subsequent disposal of radioactive materials. In almost all cases, such shielding will exhibit the toxicity characteristic for lead. Furthermore, due to its contact with radiation, the lead shielding may become radioactive. Consequently, lead shielding is often a mixed waste when disposed. What is the regulatory status of such shielding during use and subsequent disposal?

Shielding (e.g., lead bricks) used in an operating nuclear reactor is not a solid waste but a product being used for its intended purpose. When such shielding is to be discarded, however, it becomes a solid waste and a hazardous waste if it fails the TCLP for lead. If the radiation picked up by the lead has a short half-life (i.e., the time required for any radionuclide to lose half of its radioactivity), the shielding may be stored to allow radioactive decay to occur. Such storage can be conducted at a commercial (non-DOE) facility in accordance with the NRC regulations only (per the regulations in Part 266, Subpart N). Decay-in-storage of radioactive, hazardous shielding at a DOE facility would be subject to both RCRA and AEA storage regulations. After the radionuclides decay such that the material is no longer subject to the AEA regulations, the waste shielding would be managed only as a hazardous waste under RCRA after ensuring that all radioactive material labels are rendered unrecognizable. If decay-in-storage is not practicable, the shielding may still be managed under the NRC regulations (per Part 266, Subpart N) at commercial facilities, or as mixed waste subject to both RCRA and AEA standards at DOE facilities. Any storage of mixed waste for longer than 90 days (even during storage for decay) will require a RCRA permit at DOE facilities.

Lead containers or container liners used as shielding in low-level waste disposal operations are not regulated as hazardous waste when placed in a land disposal unit, even if they exhibit the toxicity characteristic for lead. Instead, they are products fulfilling their intended purpose. Assuming that the container contents do not meet the definition of hazardous waste, such containers of low-level waste could be disposed in a radioactive waste landfill without concern for the lead in the container or container liner. While the lead shielding is not a solid waste, however, EPA recommends that the outside be macroencapsulated (e.g., with a polymer coating) to prevent the shielding from leaching lead into the environment.

Conversely, if the radioactive contents are removed from the container or liner and now the shielding will be disposed or discarded, it would be considered a solid waste (and hazardous waste if it exhibits the toxicity characteristic). For example, shielding (that is part of the object being disposed) that is necessary during waste handling prior to disposal, but not after the object has been disposed, becomes a solid (and potentially hazardous) waste at disposal. Only shielding that is necessary for radiation protection *after* disposal (i.e., fulfilling its intended use) is exempt from RCRA regulation. [RO 12956, 13468, 13538]

4.4.4.1.1 *RCRA storage provisions may be inapplicable*

If certain conditions are met, storage and treatment of mixed waste at non-DOE facilities will not have to comply with the RCRA requirements noted above. Instead, such mixed waste storage/treatment will be subject only to NRC regulations. See Section 4.4.5 below.

4.4.4.2 Transportation

Mixed waste shipments must meet RCRA manifest requirements, unless shipping facilities are taking advantage of the Part 266, Subpart N conditional exemption from RCRA for mixed wastes. [§266.325] In that case, only NRC manifest and transportation regulations would apply. Of course, specific DOT provisions must be met for such radioactive mixed waste shipments, regardless of the applicability of RCRA manifesting requirements.

Q *A medical research facility plans to send scintillation vials containing radioactive D001 waste offsite for treatment at a hazardous waste management facility operating under interim status. What are the manifesting requirements? Does the receiving facility need to amend its RCRA Part A permit application in order to receive the mixed waste?*

A The mixed waste must be manifested in compliance with both RCRA and AEA requirements. (The conditional exemption from RCRA manifesting applies only if the waste already meets the LDR treatment standards.) The receiving facility does not need to modify its Part A permit application as long as the facility already manages D001 waste and the units or processes in which the D001 wastes are handled will not change. However, the facility must comply with any applicable NRC licensing requirements if it begins storing radioactive mixed waste. [RO 12710]

4.4.4.3 Land disposal restrictions

Because mixed waste is hazardous waste, it is subject to the land disposal restrictions (LDR) program regardless of its radioactivity. As such, mixed waste must meet the appropriate LDR treatment standards for all applicable waste codes prior to land disposal. (The LDR treatment standards are listed in §268.40.) This poses two problems: 1) it may be technically difficult to achieve the treatment standards due to the nature of the mixed waste, and 2) it may be hard to consistently achieve the treatment standards given the requirements imposed under the AEA. To minimize these concerns, EPA has established specific treatment standards for certain mixed wastes, such as D008 radioactive lead solids and D009 elemental mercury contaminated with radioactive materials. If no special standards are listed in the §268.40 table, the normal treatment standards for the particular waste code apply. [June 1, 1990; 55 *FR* 22626, RO 14079]

If the treatment technologies used to develop the LDR treatment standard for a specific hazardous waste code are inappropriate due to the radioactive hazard posed by the mixed waste (i.e., a different treatment technology is required), a site-specific variance from the treatment standard may be sought under §268.44. In such a case, site-specific alternative treatment standards would be established if a variance is granted. [55 *FR* 22626] See Section 13.13.4.

Q *Radioactive zirconium fines (that are pyrophoric) are generated at a DOE facility. What level of treatment would be required before land disposal of these wastes?*

A The material is both ignitable [under §261.21(a)(2)] and radioactive, making it a mixed waste. The LDR designation would be low-TOC D001 nonwastewater, with a treatment standard of deactivate and meet §268.48 standards for underlying hazardous constituents. Such wastes would have to be deactivated (e.g., stabilized) to remove the characteristic of ignitability and treated to reduce all underlying hazardous constituent concentrations to below universal treatment standards. Once this standard was achieved, the wastes could be legally land disposed, probably in an AEA landfill. [55 *FR* 22627]

For additional information on how the LDR treatment standards apply to mixed wastes, see Section 13.9.

4.4.4.3.1 *LDR storage prohibition*

RCRA Section 3004(j) prohibits the storage of any "prohibited" hazardous waste, including mixed waste, unless the waste is being stored to accumulate quantities necessary to facilitate proper waste recovery, treatment, or disposal. Section 268.50 implements this statutory requirement. A "prohibited" hazardous waste is a waste that will be land disposed but does not yet meet the LDR treatment standards. EPA has determined that when no viable treatment or disposal capacity for a hazardous waste exists, storage of the waste pending development of treatment/disposal capacity does *not* constitute storage to accumulate sufficient quantities to facilitate proper treatment or disposal. [August 29, 1991; 56 *FR* 42732] Therefore, the §268.50 storage prohibition applies to storage of mixed waste, even though treatment/disposal capacity for some of these wastes may not yet exist. The Federal Facilities Compliance Act of 1992 exempted certain DOE facilities storing mixed waste from §268.50; however, it did not exclude non-DOE and commercial facilities. [RO 14171] An example of the LDR storage prohibition is given in Case Study 4-4.

On August 29, 1991 [56 *FR* 42730], EPA issued an enforcement policy stating that facilities storing mixed waste in violation of §268.50 due to a lack of available treatment or disposal capacity would be considered a low enforcement priority (except where risk to public health or the environment was involved). The enforcement policy applied to facilities that generated less than 1,000 ft^3/yr of prohibited mixed waste. Although extended a number of times, EPA decided to let the policy expire on October 31, 2001. This date coincides (approximately) with the November 13, 2001 effective date of the conditional exemption from RCRA requirements for mixed waste managed at NRC-licensed commercial facilities. [May 16, 2001; 66 *FR* 27239]

4.4.4.4 Corrective action

Mixed waste is considered a solid waste for purposes of corrective action at solid waste management units (SWMUs). Therefore, units containing mixed wastes are SWMUs and subject to corrective action if there is a unit requiring a RCRA permit at the facility. [RO 12662, 12766, 12992]

4.4.5 2001 rule reduces dual regulation of LLMW

Regulated entities have long complained that the dual regulation of LLMW under both RCRA and NRC/DOE requirements is burdensome, duplicative, and expensive. Such facilities have also argued that the dual regulatory scheme provides

---

Case Study 4-4: Does the LDR Storage Prohibition Apply to Decay-in-Storage?

Mixed waste generated by hospitals and biomedical research facilities often contains radionuclides with relatively short half-lives. Under a decay-in-storage provision, the NRC generally allows medical facilities to store waste containing radionuclides with half-lives of less than 120 days if the radiation emitted from the unshielded surface of the waste is indistinguishable from background levels. The waste may then be disposed as nonradioactive waste if the facility removes or obliterates all radiation labels. [10 *CFR* 35.92]

A medical facility uses the NRC decay-in-storage provision to treat the radioactive portion of its mixed wastes, and the required decay period is usually 120 days. The facility has a RCRA storage permit, so it is allowed to store the mixed wastes for longer than 90 days. Is the facility violating the §268.50 LDR storage prohibition?

No. According to EPA and NRC, "decay-in-storage [is] a necessary and useful part of the best demonstrated available technology (BDAT) treatment process. 'Decay-in-storage' meets the definition of 'treatment' in 40 *CFR* 260.10, insofar as it is a method or technique designed to change the physical character or composition (amount of radioactivity) in the mixed wastes. Decay-in-storage subsequently makes the treatment of the hazardous constituents safer, and renders them safer for transport." Consequently, EPA does not apply the LDR storage prohibition to mixed waste held under an NRC-approved decay-in-storage program during the period of decay. After the waste has decayed, the RCRA §268.50 provisions apply to any additional storage of the (now) hazardous waste that occurs prior to completing the required BDAT treatment to meet LDR standards. [August 7, 1995; 60 *FR* 40209]

little or no additional protection of human health and the environment compared to what could be achieved under one program alone. Additionally, due to the limited treatment/disposal capacity for such wastes, many LLMW generators are forced to store their wastes onsite for long periods of time.

To address these concerns and meet the conditions of a consent decree concerning the hazardous waste identification rule, EPA promulgated a final rule on May 16, 2001 [66 *FR* 27218] that reduces the dual regulation of LLMW. The final rule includes two significant exemptions from the RCRA regulations: 1) a conditional exemption for LLMW during storage and treatment, and 2) an exemption for LLMW and hazardous NARM waste during transportation and disposal. These requirements are codified in 40 *CFR* Part 266, Subpart N.

4.4.5.1 The storage and treatment exemption at non-DOE facilities

The LLMW storage and treatment exemption excludes LLMW from the definition of hazardous waste *while it is in storage in tanks or containers at NRC-licensed facilities* if it meets certain conditions. Treatment of conditionally exempt LLMW may also occur within the tanks and containers if it is in compliance with NRC license requirements. However, once the LLMW is removed from storage for further management, it becomes subject to full RCRA regulation unless it qualifies for the transportation and disposal exemption (summarized in Section 4.4.5.2).

In order for LLMW to be eligible for the storage and treatment exemption from RCRA, it must be generated and managed under a single NRC or NRC agreement state license. Mixed waste generated at a DOE facility is not eligible for the exemption. Neither is LLMW generated at a commercial facility with an NRC license that is shipped to a second facility with a different license number. In addition, the exemption does not apply to NARM waste. To qualify for and maintain the exemption, the following conditions must be met [§§266.230, 266.235, 266.250]:

- LLMW must be stored in tanks or containers that are in compliance with the NRC or agreement state license requirements for the storage of low-level radioactive wastes (not including license requirements that pertain solely to recordkeeping). The tanks or containers must also be in compliance with RCRA chemical compatibility requirements in §264.177 or 265.177 (for containers) or §264.199 or 265.199 (for tanks).

- Treatment of exempt LLMW is limited to typical tank- or container-based waste treatment operations (e.g., stabilization, solidification, drying, neutralization, etc.) performed in accordance with the facility's NRC license. Such treatment can be conducted without a RCRA permit or time constraints. Treatment that would normally require a permit under RCRA (e.g., incineration, molten-salt oxidation, or supercritical water oxidation) is not allowed under this exemption.

- The owner/operator must certify that facility personnel who manage stored conditionally exempt LLMW have been properly trained to ensure safe management of the wastes, including training in chemical waste management and hazardous materials incident response that meets the §265.16(a)(3) RCRA personnel training standards.

- An inventory of the stored conditionally exempt LLMW must be conducted at least annually and an inspection must be performed at least quarterly. In addition, an emergency plan must be maintained and provided to all appropriate local authorities.

- The generator, treater, or handler of LLMW must notify EPA. The certified notification must include the facility's name, address, EPA ID number, NRC or NRC agreement state license number, the waste code(s) and storage unit(s) for which the exemption is sought, and a statement that the conditions of the final rule are met. Notifications must be submitted within 90 days of when a storage unit is first used to store conditionally exempt LLMW.

- Records pertaining to inventories and inspections, notifications, personnel training, and emergency plans must be maintained for three years after the waste is sent for disposal or in accordance with NRC regulations in 10 *CFR* Part 20, whichever is longer.

If the storage and treatment exemption from RCRA is lost, the LLMW becomes immediately subject to all RCRA Subtitle C requirements. Additionally, the waste reenters RCRA regulation if 1) the radioactivity has decayed to the point that the waste can be disposed as nonradioactive, or 2) the LLMW is removed from storage.

Q *Prior to storing LLMW under the storage and treatment exemption from RCRA, does the generator have to count the waste as hazardous? Does it have to be included on the biennial report?*

A Yes to both. "Prior to storage and/or treatment, all relevant regulations related to hazardous waste generators in Part 262 apply." [May 16, 2001; 66 *FR* 27225]

Q *A commercial facility is storing LLMW under the storage and treatment exemption from RCRA, but a quarterly inspection is missed, causing the exemption to be lost. Can the facility reclaim the exemption? What is the regulatory status of the waste while the exemption is lost?*

A A lost exemption can be reclaimed by once again meeting the exemption conditions and by notifying EPA that the exemption has been reclaimed. [§266.245] While the exemption is not in effect, the waste is subject to both RCRA and NRC requirements. However, EPA anticipates "that most generators will be able to correct a failure to meet the conditions within a 90-day period and reclaim the exemption, thus avoiding any practical effect of losing the storage and treatment exemption and becoming subject to RCRA Subtitle C regulations." [May 16, 2001; 66 *FR* 27238]

Q *A facility has a RCRA permit for a tank that has been storing the same LLMW since 1998. The facility begins taking advantage of the storage and treatment exemption from RCRA for that waste stream in first quarter 2002 but subsequently changes its operations, such that generation of the waste is discontinued. Does the facility have to close the tank under the RCRA-permitted closure plan?*

A No. Permitted storage units that have stored only LLMW before November 13, 2001, and which store conditionally exempt LLMW after that date, are not subject to RCRA closure requirements in Part 264. They would be subject only to NRC decommissioning requirements. [§266.260, May 16, 2001; 66 *FR* 27239] To effect this change, the facility should modify its RCRA permit via §270.42 at the time the LLMW becomes conditionally exempt.

4.4.5.2 The transportation and disposal exemption for all facilities

The transportation and disposal exemption excludes LLMW and NARM from the definition of hazardous waste when certain conditions are met. It exempts the following wastes from RCRA manifesting, transportation, and disposal requirements: 1) LLMW that meets the acceptance criteria of a low-level radioactive waste disposal facility (LLRWDF), and 2) hazardous NARM waste that meets the acceptance criteria of, and is allowed to be disposed in, a LLRWDF. Both DOE and commercial facilities can manage LLMW and hazardous NARM as solely radioactive wastes under this exemption.

The conditions that must be met in order for this second exemption to apply include [§§266.315, 266.330, 266.335, 266.340, 266.345, 266.350]:

- The waste must meet the applicable RCRA LDR treatment standards, including the alternative soil treatment standards.

- The waste must be packaged, labeled, manifested, and transported in compliance with NRC regulations. It is not necessary to package, label, or manifest the waste as RCRA hazardous waste when preparing the waste for transportation to the disposal facility.

- The waste has been placed on a transportation vehicle destined for the LLRWDF.

- The waste must be placed in containers *before disposal*. The containers must be one of the following: 1) a carbon steel drum, 2) an alternative container with containment performance equivalent to that of a carbon steel drum, or 3) a high-integrity container as defined by the NRC. [§266.340] Based on our conversations with EPA, it would be possible to ship a bulk load of LLMW/NARM (e.g., contaminated soil) in a tank or rail car and have the waste containerized at the LLRWDF.

- The waste must be disposed at a designated LLRWDF that is regulated and licensed by the NRC under 10 *CFR* Part 61 or by an NRC agreement state. In order to qualify for the exemption, wastes may not be sent for disposal at a DOE radioactive waste disposal facility.

- The facility originating the waste shipment must notify EPA of the exemption claim before the first shipment of an exempted waste to a LLRWDF. The written notification must include the facility's name, address, phone number, and EPA ID number. In addition to the EPA notification, *before shipment of each exempted waste*, the facility must notify the LLRWDF by certified delivery that they will be receiving the exempted waste. The wastes may only be shipped after a return receipt for the notification to the LLRWDF has been received.

- Copies of all notifications and return receipts must be kept for three years after the last exempted waste is sent for disposal. LDR program recordkeeping requirements specified in §§264.73, 265.73, and 268.7 must also be met. Finally, all other NRC documents related to tracking the transportation of the exempted waste must be maintained.

Since this transportation and disposal exemption may involve interstate transportation of conditionally exempt waste, the exemption must be adopted and authorized in both the state in which the generator is sited and the state in which the LLRWDF is located. If the waste travels through any transit states between the shipping and receiving state that have not become authorized for the exemption, the shipment must be in compliance with RCRA manifest provisions. EPA recommends that the generator note that the waste is subject to the Part 266, Subpart N transportation and disposal exemption in Block 14 of the uniform hazardous waste manifest.

Q: *A DOE facility treats the LLMW that it generates to meet LDR treatment standards. The treated waste is periodically analyzed per the facility's waste analysis plan (WAP). Will the facility remain subject to the WAP if it successfully claims the transportation and disposal exemption?*

A: Yes. Generators or owners/operators of permitted TSD facilities that plan to claim the transportation and disposal exemption remain subject to the waste analysis and WAP requirements of Part 268. [May 16, 2001; 66 *FR* 27258]

4.4.6 Mixed waste resources

A wealth of information on mixed wastes can be accessed via EPA's mixed waste website at http://www.epa.gov/radiation/mixed-waste/. Available information includes mixed waste guidance, mixed waste treatment options, pollution prevention information, the effect of the LDR program on mixed waste, mixed waste shipping guidance, and much more.

4.5 In situ mining exclusion

Oil shale, uranium, and other minerals may be mined using in situ techniques. A solvent is injected directly into mineral deposits, and the solvent solubilizes and mobilizes the mineral. The mineral/solvent mixture is then collected and removed using underground extraction wells. [May 19, 1980; 45 *FR* 33101, EPA/530/K-02/022I, cited previously]

EPA determined that soil, unrecovered minerals, and other underground materials that are not removed from the ground as part of the extraction process, although potentially contaminated with solvent, are excluded from the definition of solid waste. "Only when these materials are actually removed from the ground can it be reasonable to establish regulations governing the management of those materials." [45 FR 33101] The agency codified this exclusion at §261.4(a)(5). Nonrecovered solvents that remain underground are also excluded from the definition of solid waste. [EPA/530/K-02/022I]

4.6 Pulping liquors exclusion

Spent pulping liquor (usually called black liquor) at kraft paper mills is reclaimed (recovered) for reuse in pulp production. Because the spent alkaline liquor may have a pH >12.5, it could be considered a D002 waste prior to reclamation and reuse. However, EPA added an exclusion at §261.4(a)(6) to exclude these spent materials from the solid waste definition, unless they are speculatively accumulated. [January 4, 1985; 50 FR 641]

4.6.1 Kraft mill pulping process

The kraft pulping process converts raw materials (e.g., wood, plants) into fibers that can be formed into paper or paperboard. There are three main functions performed in the pulping process: pulp production (digestion), pulp processing, and weak black liquor processing to concentrate spent liquor for chemical recovery. [EPA/310/B-99/001, May 1999, available from http://nepis.epa.gov/EPA/html/Pubs/pubtitleOther.html, by downloading the report numbered 310B99001] The major equipment used in the kraft pulping process is shown in Figure 4-4 and further described below.

In the kraft pulping digestion step, wood chips are cooked or digested at elevated temperature and pressure with an alkaline pulping liquor that contains sodium sulfide (Na_2S) and sodium hydroxide (NaOH). During pulp processing, the raw pulp is cleaned of impurities prior to bleaching (if performed) or papermaking. The primary cleaning operations include deknotting, brown stock washing, and pulp screening and cleaning.

Brown stock washers recover spent cooking liquor (weak black liquor) for reuse in the pulping process. Weak black liquor, consisting of dissolved wood compounds and residual alkaline cooking chemicals, from the brown stock washers is stored in tanks or impoundments prior to concentration in multiple-effect evaporators. The liquor from the brown stock washers typically contains 13 to 17% solids and must be concentrated to 60 to 80% solids for efficient combustion in the recovery boiler.

Concentrated (strong) black liquor from the multiple-effect evaporators is burned in a recovery furnace to generate energy from combustion of the dissolved organic wood materials in the liquor, leaving a molten smelt consisting of sodium sulfide (Na_2S) and sodium carbonate (Na_2CO_3). The smelt is then dissolved in water to form green liquor. After clarification/filtration, the green liquor is causticized with lime, precipitating calcium carbonate and leaving an aqueous solution of sodium hydroxide and sodium sulfide (white liquor), which (after clarification/filtration) is reused in the digesters.

Recovery and reuse of spent black liquor can occur at a single paper mill or can involve two mills.

4.6.2 RCRA implications of pulping liquor recovery

Based on EPA's evaluation of the pulp and paper industry, the agency determined that the kraft process is not economically viable without recovering the black liquor. [January 4, 1985; 50 FR 641] Per §261.2(c)(3), however, the spent black liquor would normally be considered a solid waste because it is a spent material undergoing reclamation. Because the spent liquor may have a pH of >12.5, it could be considered a D002 waste prior to reclamation and reuse.

After review of the spent liquor management process, EPA noted that "[t]he whole operation is essentially an ongoing process, with chemicals being used, recovered, and returned in their original form to the same process in which they were generated, or to an analogous process at a different facility." [50 FR 641]

Figure 4-4: Typical Kraft Pulping Process

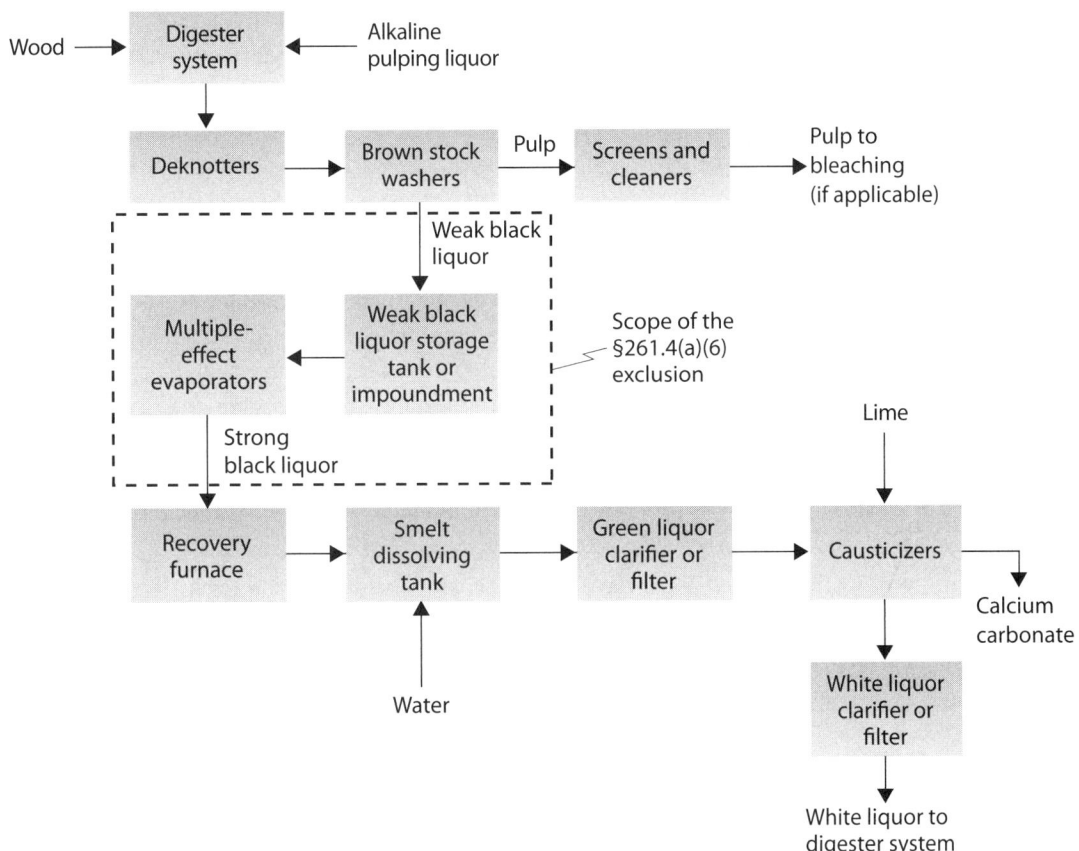

Source: Adapted from EPA/310/B-99/001.

EPA initially wanted to use the closed-loop recycling exclusion at §261.2(e)(1)(iii) to exclude spent black liquor, but that exclusion doesn't allow reclamation or land storage (surface impoundments). Then, the agency thought of using the closed-loop recycling with reclamation exclusion at §261.4(a)(8) to exclude spent black liquor, but that exclusion may not apply for two reasons: 1) recovery and reuse of spent black liquor can occur at two different mills, and 2) spent black liquor is sometimes stored in surface impoundments rather than tanks prior to recovery.

Since no existing exclusion was workable for this material and to promote waste minimization and recycling, EPA established an exclusion specific to spent black liquor being recovered. Codified at §261.4(a)(6), the exclusion says that pulping liquors that are reclaimed in a pulping liquor recovery furnace and then reused in the pulping process are not solid wastes. Implicit in the codification language is the following condition: "[b]lack liquor that is recycled in some other manner could be a waste…." [50 *FR* 642] The wording of the exclusion also subjects this material to the speculative accumulation requirements. When promulgating the exclusion, the agency cautioned that "black liquor that is disposed of and not recycled is a waste, and if hazardous, a hazardous waste. This includes black liquor that leaks, leaches, or overflows from an impoundment and is not recycled." [50 FR 642] Thus, spills of this material that are not sent through the recovery furnace and reused in the pulping process do not qualify for the exclusion.

4.7 Oil-bearing secondary materials and recovered oil exclusions

There are three exclusions from the definition of solid waste that encourage recycling of oily materials into a petroleum refinery. Some of these oily materials would otherwise be hazardous waste if disposed. As shown in Figure 4-5, these exclusions apply to various facets of the petroleum industry, as well as practices at co-located organic chemical manufacturing facilities. Each of these three exclusions is discussed below.

4.7.1 Oil-bearing secondary materials exclusion

The exclusion for hazardous oil-bearing secondary materials at §261.4(a)(12)(i) was promulgated on August 6, 1998 [63 FR 42110] and expanded slightly on January 2, 2008. [73 FR 57] This provision excludes from the definition of solid waste any oil-bearing materials *generated at a petroleum refinery*, including characteristically hazardous (e.g., D018) oil-bearing materials and materials otherwise meeting listing descriptions (e.g., F037–F038, K048–K052, and K169–K170). For the exclusion to apply, these materials must be recycled into the refining process, such as in distillation, fractionation,

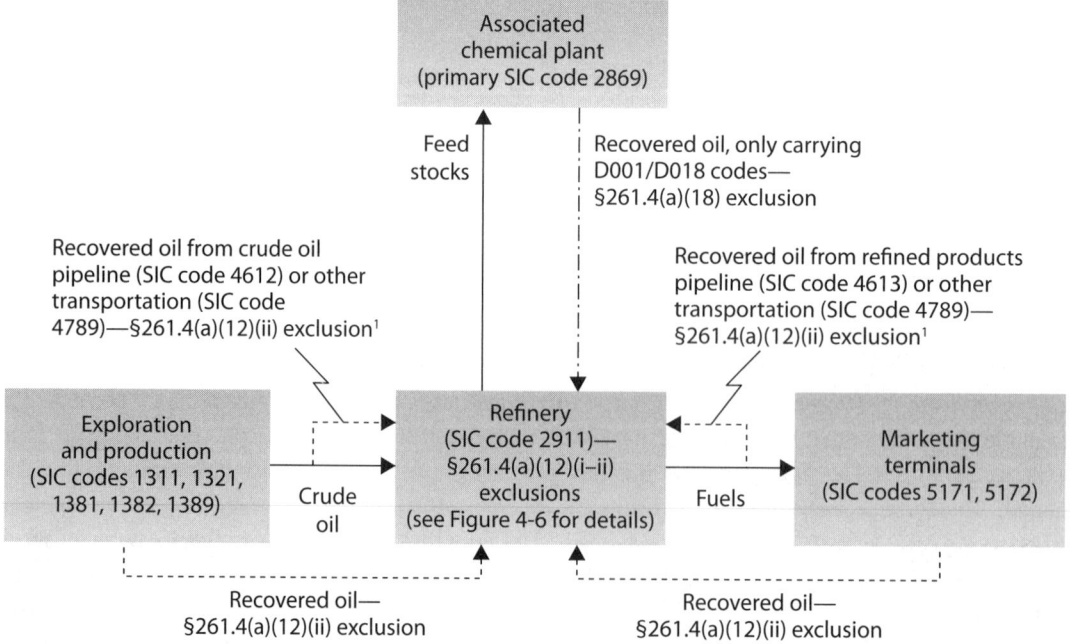

Figure 4-5: Oil-Bearing Secondary Materials and Recovered Oil Exclusions

[1] Oil recovered from natural gas transmission and distribution (SIC codes 4922 and 4923) is also included in the exclusion.

Source: McCoy and Associates, Inc.; adapted from §§261.4(a)(12), 261.4(a)(18).

catalytic cracking, gasification, or coking units. EPA noted that these materials are excluded because "they are viewed as in-process materials in a manufacturing process." [RO 14677]

On June 27, 2014, the U.S. Court of Appeals for the District of Columbia Circuit vacated the gasification exclusion for oil-bearing secondary materials in §261.4(a)(12)(i). (See *Sierra Club et al. vs. EPA*; Docket No. 08-1144, available from http://www.cadc.uscourts.gov/internet/opinions.nsf.) The court found that Congress, in Section 6924(q) of the RCRA statute, unequivocally required EPA to regulate hazardous-waste-derived fuels. The language implementing the gasification exclusion for oil-bearing secondary materials in §261.4(a)(12)(i) was added in the January 2, 2008 rule noted above. Although that language is still in the *CFR*, readers are advised to use extreme caution in using the gasification portion of the oil-bearing secondary materials exclusion as discussed in this section.

There is no minimum oil content for materials to be eligible for this exclusion, but there must be some recoverable amount of hydrocarbons for the recycling exclusion to be legitimate. Similarly, there is no minimum oil recovery efficiency requirement. [63 *FR* 42127, RO 14677]

Some oil-bearing hazardous secondary materials cannot be directly inserted into petroleum refining processes and may require processing beforehand. As shown in Figure 4-6, such processing could include thermal desorption, centrifugation, ball milling, or settling. Such processing does not affect the exclusion for the materials. Additionally, any processing of excluded oil-bearing hazardous secondary materials is exempt from RCRA hazardous waste management requirements because the input materials are neither solid nor hazardous wastes. However, the exclusion extends only to materials actually reinserted into the petroleum refinery process; any residues generated from the processing of oil-bearing hazardous secondary materials (that otherwise would have been listed hazardous wastes) *prior to* insertion into the petroleum refining process are designated as F037 waste if they cannot be inserted. [RO 14677]

If the hazardous materials are recycled into the coker (either as feedstock into the unit or as a quench stream during the quenching process), the resulting coke product cannot exhibit a hazardous characteristic for the exclusion to apply.

To be excluded, the oil-bearing materials must be either recycled into the refinery from which they were produced or shipped directly to another SIC code 2911 petroleum refinery. For example, if they are sent to an intermediate nonrefinery facility for processing, the exclusion is lost. [63 *FR* 42126–7] Such offsite transfers may occur between refineries within the same company, or between two different companies, without affecting the applicability of the exclusion. EPA also noted that a third-party contractor could own/operate the reclamation units/processes or even the wastewater treatment plant that recovers oil from oil-bearing hazardous secondary materials before it is inserted into the refinery process and that would not affect the exclusion. [RO 14444, 14677]

Other conditions that must be satisfied under the §261.4(a)(12)(i) exclusion include no placement of oil-bearing materials on the land (e.g., no storage in piles, land treatment units, or impoundments). EPA considers materials placed in such units to be discarded and thus subject to regulation as solid (and potentially hazardous) wastes. Speculative accumulation of the materials before they are recycled into the refining process will also void the exclusion. Additionally, EPA noted that "hazardous secondary materials that are released and not immediately recovered and used in a refining process would not be excluded, and thus would be subject to Subtitle C regulation as hazardous wastes that have been disposed." [63 *FR* 42126]

Fuels manufactured from these excluded materials are also excluded. Residues generated from the refining processes into which excluded materials

Figure 4-6: Oil-Bearing Secondary Materials Generated/Recycled at a Refinery

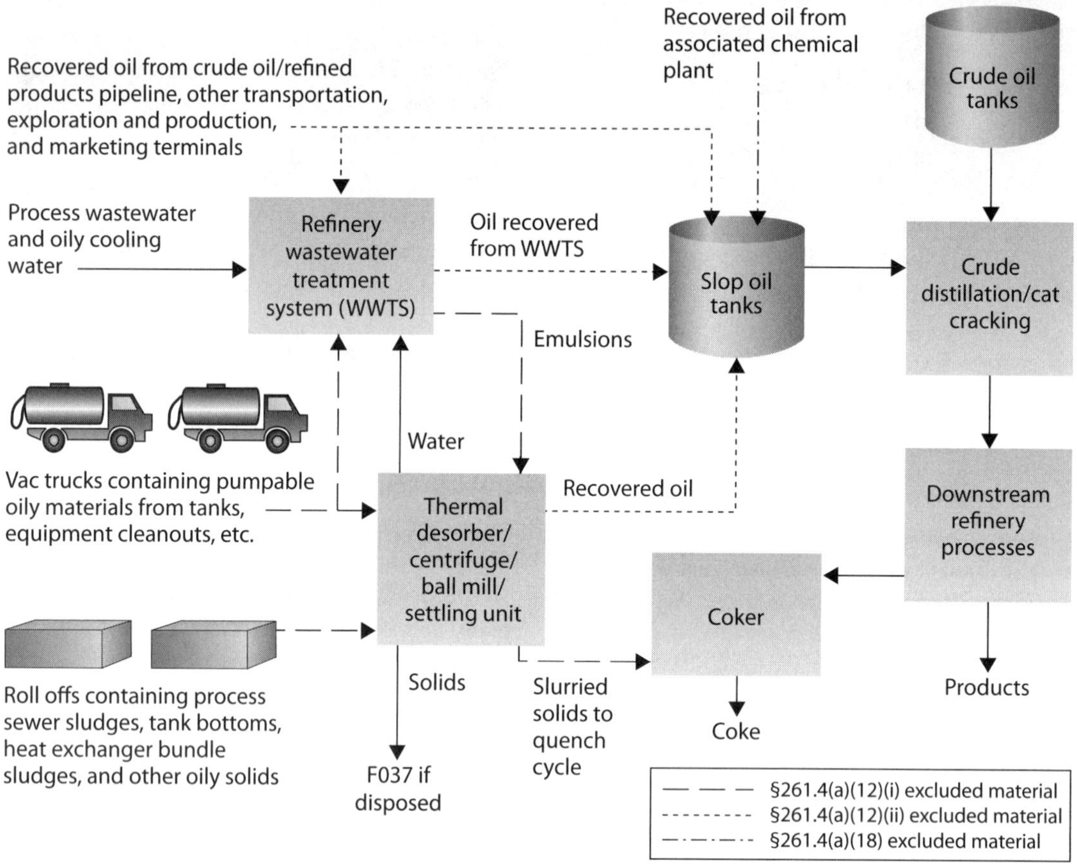

```
Inputs/sources:
- Recovered oil from crude oil/refined products pipeline, other transportation, exploration and production, and marketing terminals → Slop oil tanks
- Recovered oil from associated chemical plant → Slop oil tanks
- Process wastewater and oily cooling water → Refinery wastewater treatment system (WWTS)
- Vac trucks containing pumpable oily materials from tanks, equipment cleanouts, etc. → Thermal desorber/centrifuge/ball mill/settling unit
- Roll offs containing process sewer sludges, tank bottoms, heat exchanger bundle sludges, and other oily solids → Thermal desorber/centrifuge/ball mill/settling unit

Flows:
- Refinery WWTS → Oil recovered from WWTS → Slop oil tanks
- Refinery WWTS → Emulsions → Thermal desorber/centrifuge/ball mill/settling unit
- Thermal desorber/centrifuge/ball mill/settling unit → Water → Refinery WWTS
- Thermal desorber/centrifuge/ball mill/settling unit → Recovered oil → Slop oil tanks
- Thermal desorber/centrifuge/ball mill/settling unit → Solids → F037 if disposed
- Thermal desorber/centrifuge/ball mill/settling unit → Slurried solids to quench cycle → Coker
- Slop oil tanks → Crude oil tanks → Crude distillation/cat cracking → Downstream refinery processes → Coker → Coke
- Downstream refinery processes → Products

Legend:
— — — §261.4(a)(12)(i) excluded material
- - - - §261.4(a)(12)(ii) excluded material
—·—·— §261.4(a)(18) excluded material
```

Source: McCoy and Associates, Inc.; adapted from §§261.4(a)(12), 261.4(a)(18).

are recycled are classified as newly generated waste and would be hazardous only if they exhibit a characteristic or meet a listing description at the new point of generation. [73 *FR* 66]

4.7.2 Recovered oil exclusions

There are two exclusions from the definition of solid waste for recovered oil—one for the petroleum industry and one for associated organic chemical manufacturing plants—both of which promote recycling and reuse of hydrocarbons. Each of these exclusions is discussed below.

4.7.2.1 Recovered oil from petroleum operations

Oil recovered from secondary materials (including wastewater) generated from normal petroleum industry operations is excluded from the definition of solid waste if it is recycled into the refining process. [§261.4(a)(12)(ii)] This exclusion was initially promulgated on July 28, 1994 [59 *FR* 38536], but it was clarified and expanded on August 6, 1998. [63 *FR* 42110] The petroleum industry operations to which this exclusion applies include exploration and production, refining, transportation, and bulk storage/terminal activities. These operations, and the associated standard industrial classification (SIC) codes, are shown in Figure 4-5. Although not shown in this figure, oil recovered from natural gas transmission and distribution (SIC codes 4922 and 4923) is also included in the exclusion. Recovered oil from non-petroleum industry operations is not included in the exclusion.

The regulatory definition of "recovered oil" for this exclusion is "oil that has been reclaimed from secondary materials (including wastewater) generated from normal petroleum industry practices, including refining, exploration and production, bulk storage, and transportation incident thereto.... Recovered oil does not include oil-bearing hazardous wastes listed in subpart D of this part [261]; however, oil recovered from such wastes may be considered recovered oil." Agency guidance suggests that recovered oil is "comparable to normal feedstocks (i.e., oil) used in typical production processes." [63 FR 42128] Finally, recovered oil does not include used oil (which is addressed in Part 279 of the regs and Chapter 12 of this book), although de minimis quantities of used oil that are incidentally captured in a refinery wastewater treatment system would be eligible for the exclusion. [59 FR 38537]

Examples of recovered oil include secondary materials *consisting primarily of oil* such as oil-water separator skimmings from plant wastewaters, oil skimmed from ballast water tanks, slop oil and emulsions, oil recovered from oil and gas drilling operations, and oil from on- or offsite refinery process units (e.g., off-spec process streams). [59 FR 38537] EPA understands, however, that recovered oil may require processing (e.g., emulsion breaking) before it can be re-used in the refining process. The agency views these recovered oil processing steps as part of the petroleum refining process and not part of the waste disposal problem; thus, as with the oil-bearing secondary materials exclusion discussed in Section 4.7.1, such processing does not affect the exclusion for recovered oil. Additionally, any processing of recovered oil is exempt from RCRA hazardous waste management requirements because the input materials are neither solid nor hazardous wastes. Note, however, that residues from these processes that are not inserted into refining units are solid and potentially hazardous wastes. [59 FR 38539]

Because this material consists primarily of hydrocarbons, recovered oil from petroleum operations is often transferred directly into the refinery's slop oil system. However, sometimes recovered oil can have a significant water component to it, and so a refinery may alternatively transfer it to the wastewater treatment system (as noted in Figure 4-6), which includes significant oil-water separation steps.

Petrochemical processing units are sometimes located onsite at petroleum refineries. Wastewater from these units is typically discharged to the refinery's wastewater treatment system, although it rarely contains recoverable oil. EPA clarified that oil recovered from a wastewater treatment system at a refinery with onsite petrochemical processing units continues to qualify for the recovered oil exclusion. [RO 11904]

Similarly, an organic chemical manufacturing facility is often co-located with a refinery. The chemical plant may be owned by the same parent company as the refinery, and the two facilities often share a wastewater treatment system. However, the chemical manufacturing plant is often considered an offsite facility, because it has a separate EPA ID number and is considered a hazardous waste generator independent of the refinery. Even so, the agency noted in RO 11904 that, "in cases where petrochemical and petroleum refining operations are co-located and share a common wastewater treatment system, the integration between the two facilities is such that the petrochemical facility falls within scope of the [recovered oil] exclusion. In these situations, given the common wastewater treatment system and the predominance of petroleum refining wastewater, the petrochemical operation would be considered part of normal petroleum refining. The exclusion would therefore apply to oil recovered from a wastewater treatment system that a refinery shares with a co-located petrochemical facility."

Conditions that must be satisfied to claim the §261.4(a)(12)(ii) exclusion are similar to those discussed in Section 4.7.1 for the oil-bearing secondary materials exclusion:

- Recovered oil must be inserted into the refining process, such as in distillation, fractionation, catalytic cracking, gasification, or coking units.
- If recovered oil is recycled into the coker (either as feedstock into the unit or as a quench stream during the quenching process), the resulting coke product cannot exhibit a characteristic of hazardous waste.
- No placement of oil-bearing materials on the land is allowed (e.g., no storage in land treatment units or impoundments).
- No speculative accumulation of the materials can occur before they are recycled into the refining process.

Q: *Arctic-grade diesel is produced via distillation in crude oil "topping units." These units produce diesel for various uses in support of exploration and production operations on the Alaskan North Slope, including vehicle fuel, freeze protection in oil wells and pipelines, and pipeline pressure testing. After the diesel has been used in some of these exploration and production activities, it is introduced, along with crude oil feedstocks, either back into the topping units or into other petroleum refineries in North America via the Trans Alaska Pipeline. Can the used diesel be considered recovered oil that is excluded from the definition of solid waste per §261.4(a)(12)(ii)?*

A: Yes. EPA believes that used arctic-grade diesel that has been recovered from petroleum exploration and production activities and is inserted into the petroleum refining process meets the definition of "recovered oil." However, the exclusion does not apply to recovered oil that is managed on the land or speculatively accumulated before being recycled back into the petroleum refining process. Also, the exclusion does not apply to materials that are never recycled, such as unrecovered spills. [RO 14544]

4.7.2.2 Recovered oil from an associated organic chemical manufacturing facility

Oil recovered from an organic chemical manufacturing facility with primary SIC code 2869 that is physically co-located with a petroleum refinery *and* which receives hydrocarbon feedstocks from that refinery is excluded from the definition of solid waste if it is 1) returned and inserted into the petroleum refining process, and 2) hazardous only because it exhibits the characteristic of ignitability and/or toxicity for benzene. [§261.4(a)(18)] The exclusion applies to oil recovered from normal organic chemical manufacturing operations as well as to oil reclaimed from secondary materials (i.e., sludges, by-products, or spent materials including wastewater) that are produced in such processes. Examples of such recovered oil include hydrocarbon side streams that represent unreacted feedstock or other hydrocarbon by-products consisting almost entirely of oil. [63 *FR* 42130] Because this material is primarily hydrocarbons, it is often transferred from the chemical plant directly into the refinery's slop oil system (as noted in Figure 4-6).

The chemical plant must be physically co-located with the refinery, meaning that the two facilities are physically adjacent to one another or otherwise share a common boundary. Another means of demonstrating co-location is a high degree of integration between the two plants—evidenced by shared wastewater treatment systems, shared manufacturing units, shared emergency response equipment and procedures, transfer of materials between the two plants by dedicated pipelines, environmental permits covering both facilities, etc. There is no requirement for common ownership of the two plants. [63 *FR* 42130]

Although the organic chemical manufacturing facility must have a primary SIC code of 2869 (industrial organic chemicals), other operations at the plant may include SIC codes of 2821 (plastic materials and resins), 2822 (synthetic rubber), and 2865 (cyclic crudes and intermediates).

Finally, consistent with the other recovered oil exclusion discussed in Section 4.7.2.1, placing oil recovered from an associated organic chemical manufacturing facility on the land (e.g., in land treatment units or impoundments) or speculative accumulation of the recovered oil before it is recycled into the refining process will void the exclusion.

4.8 Comparable fuels exclusion

During the 1980s and 1990s, facilities generating materials with high-Btu content often wanted to burn them as fuel in onsite boilers, heaters, and furnaces. For example, a pharmaceutical company generating large amounts of spent methanol or ethanol solvents could burn these materials as a cheap, clean fuel source in onsite units. However, spent methanol or ethanol solvents meet the definition of spent materials and would, therefore, be solid wastes when burned for energy recovery per §261.2(c)(2). Additionally, these spent solvents would carry D001 and (in the case of methanol) F003 hazardous waste codes. Thus, the facility would need a RCRA permit to burn these solid and hazardous wastes in an onsite unit. Chemical manufacturing facilities generating high-Btu by-product streams were in the same situation.

Due to industry petitions and to encourage recovery of energy from high-Btu spent materials, sludges, by-products, and even unused commercial chemical products for which a facility no longer has a use, EPA promulgated an exclusion to the definition of solid waste for these materials on June 19, 1998. [63 *FR* 33782] The exclusion required a facility to show that its material is "comparable" to commercial fossil fuels and to ensure that such materials are burned in industrial furnaces, industrial or utility boilers, hazardous waste incinerators, or gas turbines. [§§261.4(a)(16), 261.38] As a result, this was often called the "comparable fuels" exclusion.

On June 27, 2014, the U.S. Court of Appeals for the District of Columbia Circuit vacated the comparable fuels exclusion in §§261.4(a)(16) and 261.38. (See *Natural Resources Defense Council et al. vs. EPA*; Docket No. 98-1379, available from http://www.cadc.uscourts.gov/internet/opinions.nsf.) The court found that Congress, in Section 6924(q) of the RCRA statute, unequivocally required EPA to regulate hazardous-waste-derived fuels. Although the language implementing the comparable fuels exclusion is still in §§261.4(a)(16) and 261.38 of the *CFR*, readers are advised to use extreme caution in using the comparable fuels exclusion.

4.9 Exclusion for spent caustic solutions from petroleum refining

Petroleum refining processes use considerable amounts of caustic, resulting in significant quantities of spent caustic. Depending on the process from which it is generated, the spent caustic may be contaminated with cresols, naphthenes, phenols, and/or sulfides. Some refineries have onsite caustic reclamation processes to recover caustic for continued use. Those that don't typically send large quantities of spent caustic offsite.

Two separate exclusions are available to keep these materials from having to be managed as solid and hazardous wastes. The first one is for spent cresylic, naphthenic, or phenolic caustic that is used to produce cresylic or naphthenic acids; the second one is for spent sulfidic caustic that is used as a substitute for virgin caustic in various manufacturing processes. Each of these exclusions is discussed below.

4.9.1 Spent cresylic, naphthenic, or phenolic caustic

When spent caustics from petroleum refinery liquid treating processes are used as feedstocks to produce cresylic or naphthenic acids, EPA considers the materials to be valuable commercial feedstocks used in the manufacture of commercial chemical products. [August 6, 1998; 63 *FR* 42115] Therefore, the agency added an exclusion to the definition of solid waste for these spent caustics in §261.4(a)(19).

As shown in Figure 4-7a, spent caustic contaminated with cresols, naphthenes, and/or phenols qualifies for this exclusion. Thus, storage of these

Figure 4-7: Spent Caustic From a Petroleum Refinery

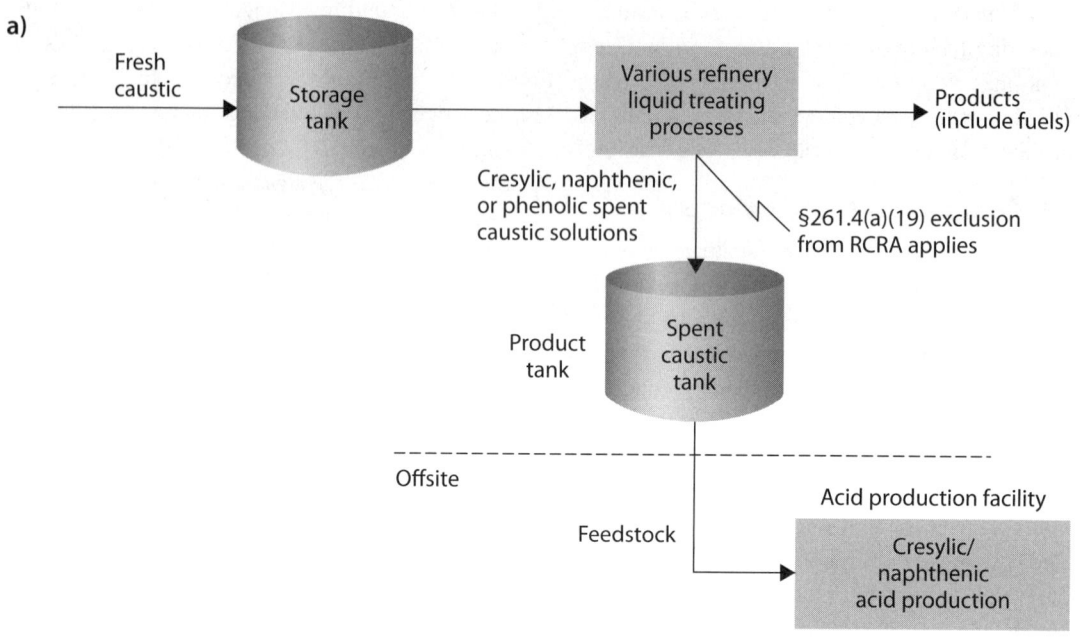

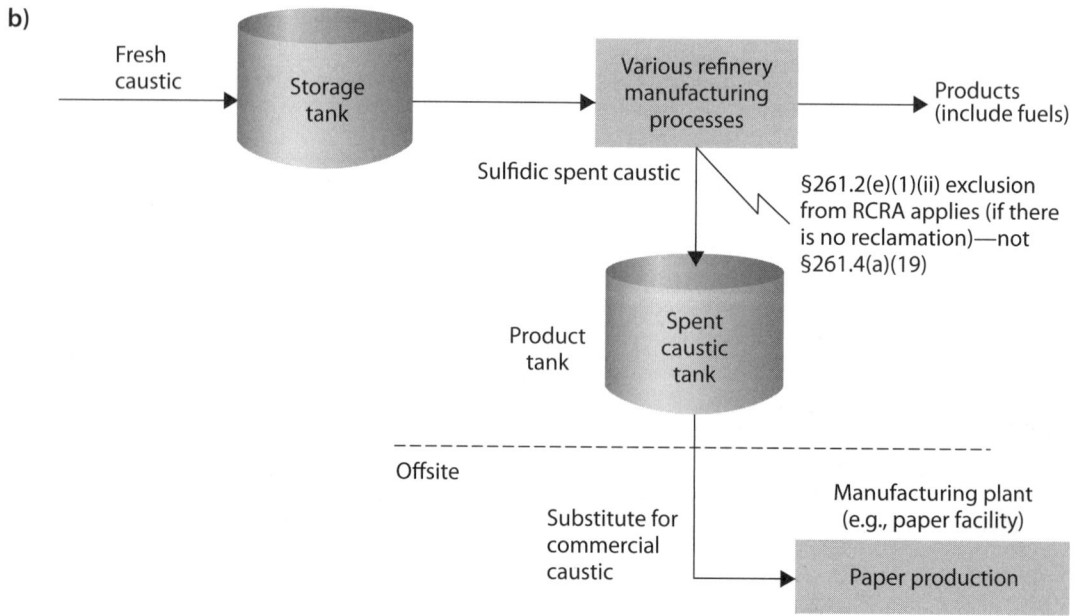

Source: McCoy and Associates, Inc.; adapted from §§261.2(e)(1), 261.4(a)(19).

spent caustics at the refinery before they are shipped offsite is in non-RCRA units.

Consistent with other refinery exclusions, placement of spent caustic solutions on the land (e.g., in surface impoundments) or speculative accumulation of the spent caustic before it is fed into the acid production process will void the exclusion.

4.9.2 Spent sulfidic caustic

Spent sulfidic caustic generated at a refinery is not suitable as a feedstock in the production of cresylic or naphthenic acids. Instead, refineries often contract with vendors who ship the spent sulfidic caustic (either directly or through one of the vendor's terminals) to a third-party facility (e.g., a paper facility) that uses the material as a substitute for commercial chemicals such as caustic soda or other sodium salt solutions. This option is shown in Figure 4-7b.

Because the spent caustic is not regenerated or treated before it is provided to these outlets (i.e., it goes from the refinery's tanks to the end user "as is"), the spent caustic is eligible for the §261.2(e)(1)(ii) recycling exemption from RCRA. See Section 11.3.2 for more detail and conditions applicable to this exemption.

4.10 Household hazardous waste exclusion

Under §261.4(b)(1), the 1.6 million tons of household hazardous waste generated annually in the United States is excluded from the RCRA Subtitle C regulations. The term household waste refers to any garbage, trash, and other waste from single and multiple residences and other residential units such as hotels and motels, bunkhouses, ranger stations, crew quarters, campgrounds, picnic grounds, and day-use recreation areas. In order for household hazardous waste to be excluded from regulation, it must meet two criteria: 1) the waste has to be generated by individuals on the premises of a temporary or permanent residence, and 2) the waste must be composed primarily of materials found in the waste generated by consumers in their homes. [November 13, 1984; 49 *FR* 44978]

For example, wastes generated from military base housing are excluded. [RO 11958, 13225] Wastes generated at hotels from hotel guests, room cleaning, or pesticide spraying in rooms are also excluded under the household exclusion; however, dry cleaning wastes from the hotel are not excluded because the spent solvents from such operations are not similar to wastes typically produced by consumers in their homes. [RO 12624, 13736] Household-type wastes generated at commercial facilities and office buildings are also not covered by the household exclusion [RO 11347], because they do not meet the first criterion above. Similarly, restaurants do not serve as temporary or permanent residences for individuals, and so wastes generated from restaurants will not qualify for the household waste exclusion. [RO 13744] EPA noted that wastes generated from retail stores and shopping centers are also not excluded. [49 *FR* 44978, RO 11958] Finally, hospitals, nursing homes, and day care centers are not considered generators of household hazardous wastes; although these facilities are residential in some ways, they generate amounts and types of wastes that are significantly different from wastes generated by consumers in their homes. [March 24, 1989; 54 *FR* 12339, RO 11958]

Although commercial facilities, office buildings, restaurants, retail stores, shopping centers, hospitals, nursing homes, and day care centers do not produce wastes that qualify for the household hazardous waste exclusion, hazardous wastes generated at these facilities may be exempt from most RCRA Subtitle C regulation as wastes generated by conditionally exempt small quantity generators (CESQGs). Hazardous wastes generated by CESQGs are subject only to the limited regulations in §261.5.

Where things get really complicated is when household products are taken to commercial facilities, office buildings, retail stores, etc. where they become wastes. Now what? In RO 13480, EPA noted that nickel-cadmium batteries removed from products/appliances by consumers in their homes are within the exclusion and are excluded from the hazardous waste regulations when disposed, while such batteries removed by service personnel from products/appliances that have been taken to service centers by consumers are not within the exclusion. The agency stated in RO 14084 "if the waste comes from a household, it would not be subject to the hazardous waste regulations even if it were later discarded on the premises of a business." The guidance in RO 11958 indicates that retailers may take advantage of the household waste exclusion for the management of products (that are known to be hazardous waste) returned from households. The conclusion that can be drawn from these three pieces of guidance is as follows: if the household material was a *product* when brought to the commercial facility, office building, retail store, etc. (such as the household appliance being repaired at a service center), then the household waste exclusion is not applicable to wastes (such as the batteries) that are generated from this material by the commercial facility; however, if the household material was a *waste* when brought to the commercial facility, office building, retail store, etc., the household hazardous waste exclusion has already kicked in and continues to apply to that material when disposed at the commercial facility.

Note that §261.4(b)(1) excludes household wastes from hazardous waste management requirements even when accumulated in quantities that would otherwise trigger regulation under RCRA Subtitle C. [RO 11958, 12547]

EPA has developed a website discussing household hazardous waste management options and collections facilities at http://www.epa.gov/osw/conserve/materials/hhw.htm.

4.10.1 Types of waste excluded

The exclusion applies to all household hazardous wastes, including electronics, appliances, medicinal drugs and ointments, waste oils, antifreeze, pesticides and other lawn and garden products, paint and paint thinner, batteries (including lead-acid batteries), fluorescent lamps, thermostats, spent filters from filtering water, aerosol cans, and cleaning fluids/solvents. [May 19, 1980; 45 *FR* 33099, June 12, 2002; 67 *FR* 40511, RO 11508, 11653, 11678, 11693, 11756, 11780, 11897, 12688, 13189, 13432, 13521] Contaminated media and debris generated from residential heating oil tanks [February 12, 1993; 58 *FR* 8505, RO 14701] and leaks from those tanks [RO 11563] are also excluded.

Unused flameless ration heaters (FRHs) disposed by individual soldiers in the field or at military installations are eligible for the household hazardous waste exclusion. This exclusion would also apply where FRHs are collected from a group of soldiers for disposal, as long the FRHs were initially issued to the soldiers for individual use in a permanent or temporary residential setting. [RO 14772, 14774]

Hazardous waste generated at a household by a person other than the homeowner (e.g., a contractor) is also excluded, provided that the waste is generated as part of daily living (e.g., routine residential maintenance). [RO 11897, 13358, 14673] Examples of such excluded waste include medical waste generated in private homes by a health care provider, even when the waste is removed from the home and transported to a physician's place of business. [54 *FR* 12339, RO 13277]

Conversely, natural gas regulators at residences are installed, replaced, and collected by utilities and gas suppliers—not by consumers in the home in the course of daily living. Therefore, EPA determined that natural gas regulators do not meet the two criteria discussed at the beginning of Section 4.10 and that, if they exhibit the toxicity characteristic for lead, they must be managed as hazardous waste. [RO 14115]

EPA originally determined that the household hazardous waste exclusion did not apply to wastes such as debris produced during construction, renovation, or demolition of houses or other residences, because these wastes are not similar to those generated by a consumer in the home in the course of daily living. [49 *FR* 44978, RO 11897] However, the agency more recently clarified that the exclusion *does* apply to lead-based paint waste generated as a result of renovation, remodeling, or abatement actions by residents of households, including their contractors. Excluded materials would include doors, window frames, painted woodwork, paint chips, dust, and sludges from these activities. [October 23, 2001; 66 *FR* 53537, RO 14459, 14673]

EPA does not consider RCRA to apply to contaminated human remains resulting from a chemical, biological, radiological, or nuclear event. Similarly, personal effects would likely be excluded from the hazardous waste regulations via the household hazardous waste exclusion. [RO 14811]

4.10.2 Exclusion may be lost if mixed with other wastes

If household hazardous waste is mixed with regulated hazardous wastes (i.e., hazardous waste generated by small or large quantity generators), the regulatory status of the mixture is subject to the hazardous waste mixture rule discussed in Section 5.1. In general, such mixtures will continue to be hazardous and subject to regulation under RCRA Subtitle C. [45 *FR* 33099, RO 12547]

Mixtures of household hazardous waste and CESQG hazardous waste are subject to the CESQG waste management standards of §261.5. [§261.5(h), RO 11377, 12547] However, §261.5(h) notes that, if the quantity of the mixture exceeds the §261.5 quantity limitations, then the mixture becomes subject to either small or large quantity generator standards if the mixture exhibits a characteristic. This put facilities collecting both household hazardous waste and CESQG waste in a quandary. EPA tried to ease this situation in 1992 by stating:

"[I]t is our interpretation that §261.5(h) applies to the CESQ generator and not to the subsequent managers of the CESQG waste described in §261.5(f)(3) and (g)(3). Programs and facilities receiving and mixing CESQG waste and [household hazardous waste] are subject to requirements imposed by states through the states' municipal or industrial waste permit, license, or registration programs, but are not subject to the full hazardous waste Subtitle C regulations, even if the mixed CESQG and household hazardous wastes were to exhibit a characteristic of a hazardous waste. The collection facility does not become the generator of the mixture merely by mixing CESQG waste with nonhazardous waste, and regardless of the quantity of the mixture of wastes, is not subject to the Part 262 generator regulations. By contrast, CESQ generators that mix hazardous and nonhazardous waste and whose resultant mixtures exceed the §261.5 quantity limitations and exhibit a characteristic, are no longer conditionally exempt and are subject to the applicable Part 262 hazardous waste generator regulations." [RO 11681]

4.10.3 Management/treatment of household hazardous wastes

Per the language of §261.4(b)(1), the collection, transportation, storage, treatment, disposal, and recovery or reuse of household hazardous wastes are not subject to Parts 262 through 270. "Because this exclusion attaches at the point of generation (i.e., the household) and continues to apply throughout the waste management cycle, household [wastes] collected in municipal recycling programs and subsequently managed in recycling programs continue to be excluded from the hazardous waste management regulations." [RO 11780] The household waste exclusion applies from collection through final disposition, including treatment and resultant residues. Thus, for example, landfill gas condensate derived from a fill that contains household waste exclusively is not a hazardous waste. [RO 12362]

However, household wastes are subject to federal, state, and local requirements concerning management of solid waste, including applicable RCRA Subtitle D requirements. [45 *FR* 33099, RO 11314, 11377] For example, household hazardous waste can be disposed in a municipal solid waste landfill or burned in a municipal solid waste combustor. [RO 14459] Although nonhazardous waste management is legal, EPA recommends that household hazardous waste be managed in the following order: 1) reused or recycled as much as possible, 2) treated in a hazardous waste treatment facility, or 3) disposed in a hazardous waste landfill. [RO 11377]

A recently published guidance document is available in which EPA recommends the incineration of pharmaceuticals collected from households during take-back events, mail-back events, or other collection programs. [RO 14833] The agency recommends that these household pharmaceuticals be combusted in a RCRA-permitted incinerator or cement kiln, or at the least, in a unit that meets EPA's large or small municipal waste combustor standards. Such units would not only be protective of the environment, but they could also meet the Drug Enforcement Administration's goal of preventing diversion of controlled substances.

4.10.3.1 Household hazardous wastes burned in resource recovery facilities

Household waste is often burned in resource recovery facilities to recover energy (these facilities are also often called waste-to-energy facilities). The household hazardous waste exclusion explicitly exempts such facilities from hazardous waste management regulations as long as such facilities receive and burn only 1) household waste, and 2) solid waste from commercial or industrial sources that does not contain hazardous waste. But what about the ash generated from these facilities?

EPA's initial assessment was that since "household waste is excluded in all phases of its management, residues remaining after treatment (e.g., incineration, thermal treatment) are not subject to regulation as hazardous waste." [45 FR 33099] This determination was challenged, however, and in a 1994 Supreme Court decision (*City of Chicago v. Environmental Defense Fund*, 511 U.S. 328), the court found that ash generated at resource recovery facilities burning household wastes and nonhazardous commercial wastes is *not* exempt from the hazardous waste requirements of RCRA Subtitle C if it exhibits a hazardous waste characteristic.

Because the court did not specifically determine the point at which regulation of the ash must begin,

EPA issued an interpretation stating that RCRA would first impose hazardous waste regulation at the point the ash leaves the "resource recovery facility," defined as the combustion building (including connected air pollution control equipment). Consequently, the point at which a hazardous waste determination must be made is the point at which ash exits the combustion building following the combustion and air pollution control processes. [February 3, 1995; 60 FR 6666, RO 11901]

4.10.4 Importing household hazardous wastes

Because imported waste is subject to the applicable domestic laws and regulations of the United States once it enters U.S. jurisdiction, imported household hazardous waste is excluded from the definition of hazardous waste in the same way as domestically generated household waste. U.S. importers may want to keep records of the foreign exporter and the origin of any imported household waste in case questions arise as to its regulatory status once it enters the United States. [RO 14308]

4.11 Exclusion for agricultural waste/manure returned to the soil as fertilizer

Solid wastes generated by crop or animal farming are excluded from hazardous waste regulation provided the wastes are returned to the ground as fertilizers or soil conditioners. Examples of such waste would be crop residues and manures. [May 19, 1980; 45 FR 33099, RO 12002] The agency codified this exclusion at §261.4(b)(2).

Congress did not intend to include silviculture waste (i.e., forestry waste such as foliage and branches) in this hazardous waste exclusion. As a result, generators of forestry waste need to determine whether their waste is hazardous. [45 FR 33099] Also, EPA considers food processing wastes that are applied to the land to be commercial or industrial waste. As such, they are not covered under the agricultural waste exclusion. [RO 12002]

This exclusion at §261.4(b)(2) is separate from the "farmer exemption" at §262.70. That regulatory paragraph allows a farmer to dispose hazardous waste pesticides from his own use on his farm. The farmer may mix, apply, and/or dispose unused pesticide solution on his property if done in accordance with the instructions on the pesticide label. All emptied pesticide containers must be triple rinsed, and the rinsate must also be disposed on his farm. [RO 11021, 12886]

4.12 Exclusion for mining overburden returned to the mine site

Reclamation of surface mines commonly involves returning waste overburden (i.e., earth and rocks), removed to gain access to ore deposits, to the mine. EPA excluded this waste at §261.4(b)(3) because mining overburden is not a discarded material within the scope of RCRA. This exclusion is limited to overburden, defined as "any material overlying an economic mineral deposit which is removed to gain access to that deposit and is then used for reclamation of a surface mine." [May 19, 1980; 45 *FR* 33100]

EPA does not intend that the definition of mining overburden given above be limited exclusively to the material directly above a mineral deposit. Some material is removed from the sides of a mining pit to gain safe access to the economically recoverable mineral, and such material is also excluded as overburden. [45 *FR* 33100]

4.13 Fossil fuel combustion wastes exclusion

Section 261.4(b)(4) of the federal RCRA regulations excludes fly ash waste, bottom ash waste, slag waste, and flue gas emission control waste from the definition of hazardous waste. These wastes are some of the so-called "Bevill wastes," temporarily excluded from hazardous waste regulation under the October 1980 Bevill amendment [Section 3001(b)(3)(A)] to the RCRA statute (named after the representative who introduced the amendment in Congress). The three classes of excluded Bevill wastes are:

1. Fly ash, bottom ash, slag, and flue gas emission control waste from the combustion of fossil fuels [§261.4(b)(4)];
2. Wastes from the extraction, beneficiation, and processing of ores and minerals [§261.4(b)(7)]; and
3. Cement kiln dust [§261.4(b)(8)].

The first of the above Bevill wastes is discussed in this Section 4.13, while the second is covered in Section 4.15. Note that the following discussion of fossil fuel combustion wastes is based on the regulations in effect and guidance available as of this writing. However, the regulatory status of coal combustion residues may change as a result of a significant rule proposed by EPA on June 21, 2010 [75 *FR* 35128] or legislation pending in Congress.

4.13.1 Four large-volume fossil fuel combustion wastes

Fly ash, bottom ash, slag, and flue gas emission control waste generated primarily from the combustion of coal or other fossil fuels are what EPA calls the four large-volume fossil fuel combustion (FFC) wastes. These Bevill wastes are specifically excluded from hazardous waste regulation per §261.4(b)(4). Figure 4-8 illustrates the scope of this Bevill exclusion for these four wastes at a coal-fired power plant.

Although these four wastes were excluded from the definition of hazardous waste in the original May 19, 1980 RCRA regulations [45 *FR* 33120], Congress required EPA to evaluate whether the temporary Bevill exclusion included in the 1980 regs should continue. After extensively evaluating these waste streams, the agency issued its Report to Congress in February 1988. [EPA/530/SW-88/002, February 1988, available from http://nepis.epa.gov/EPA/html/Pubs/pubtitleOSWER.html by downloading the report numbered 530SW88002]

Several years after issuing its Report to Congress, the agency published a final regulatory determination for these wastes on August 9, 1993. [58 *FR* 42466] In the August 1993 determination, EPA concluded that hazardous waste regulation is

Figure 4-8: Bevill Exclusion at a Coal-Fired Power Plant

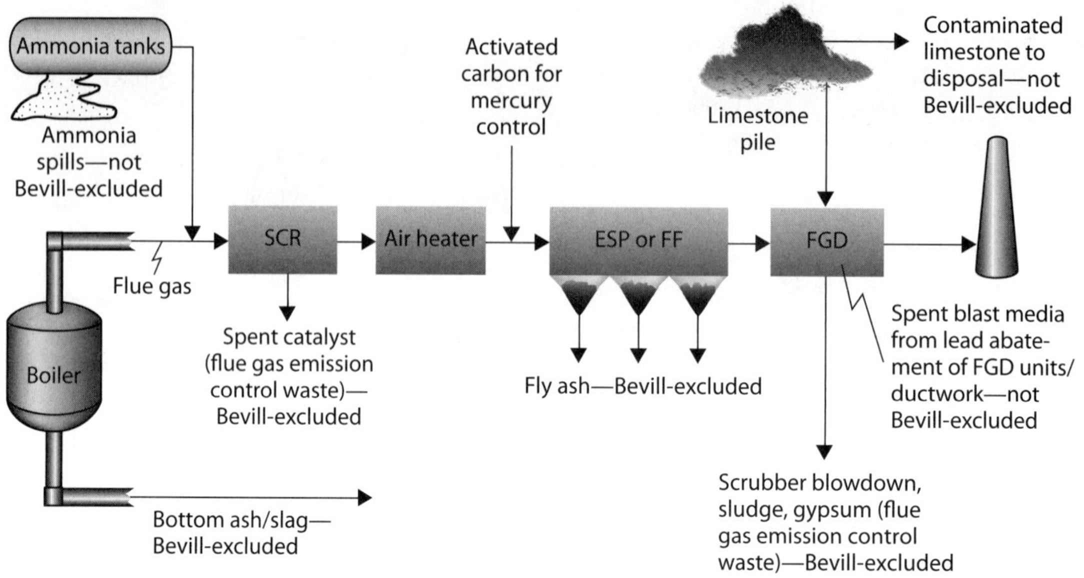

ESP = electrostatic precipitator; FF = fabric filter (baghouse); FGD = flue gas desulfurization; SCR = selective catalytic reduction.
Source: McCoy and Associates, Inc.; adapted from §261.4(b)(4).

inappropriate for these four large-volume FFC wastes. This determination was based on two qualifying criteria; these four wastes are excluded only if they are: 1) generated at coal-fired electric utilities or independent power producers, and 2) managed separately from other FFC wastes. These criteria are incorporated into Figure 4-9, which is a logic diagram for determining the regulatory status of FFC wastes

EPA noted that smelter slag is not the same thing as boiler slag (molten bottom ash) and so would not be covered under the §261.4(b)(4) exclusion. Instead, smelter slag might be addressed in the mining and mineral processing waste exclusion at §261.4(b)(7)—see Section 4.15. [RO 12259]

It is fairly clear from guidance [58 *FR* 42466, RO 11226, 12240] that wastes from a flue gas desulfurization (FGD) system are included in the term "flue gas emission control waste." What isn't so clear is whether spent catalyst and other wastes from a selective catalytic reduction (SCR) system used to control nitrogen oxide (NO_x) emissions is included in that term. In limited guidance, EPA has determined that spent SCR catalysts enjoy the §261.4(b)(4) Bevill exclusion. [*Performance of Selective Catalytic Reduction on Coal-Fired Steam Generating Units*, Final Report, June 25, 1997, available at http://www.epa.gov/airmarkets/progsregs/arp/docs/scrfinal.pdf]

Q *Water from quenching hot bottom ash from a utility boiler picks up the characteristic of corrosivity (pH >12.5). Is the resulting wastewater a hazardous waste subject to RCRA regulation?*

A No. The quench water becomes corrosive solely as a result of contact with the ash. Because the hazardous waste characteristic of the quench water is derived from an excluded Bevill waste, the resulting corrosive quench water retains the excluded status of that waste. In other words, whatever makes the water corrosive is already

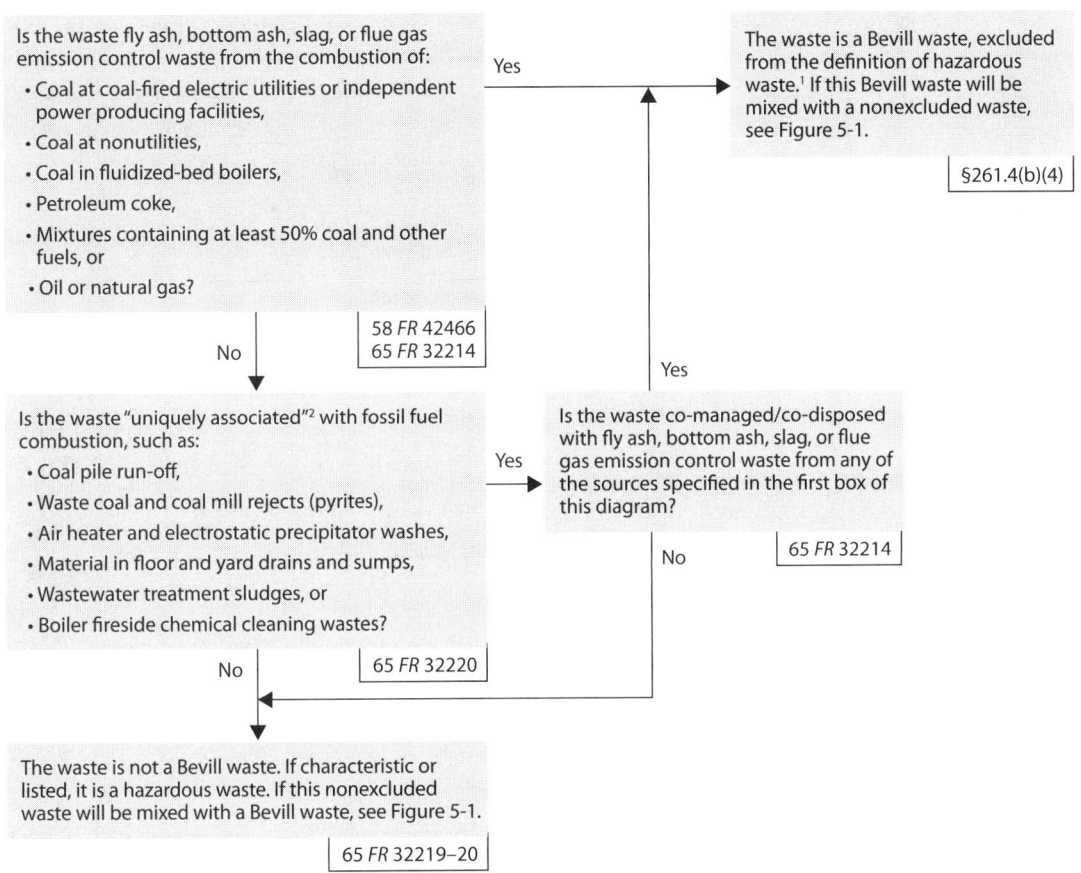

Figure 4-9: Regulatory Status of Fossil Fuel Combustion Wastes

[1] If disposed offsite, the company/facility needs to work closely with the nonhazardous waste (Subtitle D) landfill to ensure the waste (although excluded from the hazardous waste program) meets the landfill's acceptance criteria/permit conditions and will not cause any other problems that might result in future CERCLA or tort liability.

[2] EPA defined "uniquely associated" to mean a waste that "during the course of generation or normal handling at the facility, comes into contact with either fossil fuel (e.g., coal, oil) or fossil fuel combustion waste (e.g., coal ash or oil ash) and it takes on at least some of the characteristics of the fuel or combustion waste." [65 FR 32219]

Source: McCoy and Associates, Inc.

excluded, so the water is also excluded from regulation as a hazardous waste. [RO 11145, 11162, 12552]

4.13.2 Fossil fuel combustion wastes from other sources

The February 1988 Report to Congress and August 1993 regulatory determination reviewed in Section 4.13.1 did not address FFC wastes from sources other than coal-fired electric utilities or independent power producers. EPA addressed those other wastes in a second Report to Congress, issued in March 1999. [EPA/530/R-99/010, March 1999, available from http://nepis.epa.gov/EPA/html/Pubs/pubtitleOSWER.html by downloading the report numbered 530R99010]

In the regulatory determination that followed on May 22, 2000 [65 FR 32214], the agency determined that the following wastes are also Bevill wastes and, although not specifically excluded by the regulatory language in §261.4(b)(4), are excluded from the definition of hazardous waste:

- Wastes from the combustion of coal at non-utilities (e.g., paper mills, food processing facilities, chemical plants);
- Wastes from the combustion of coal in fluidized-bed boilers;
- Wastes from the combustion of petroleum coke;
- Wastes from the combustion of at least 50% coal with other fuels (e.g., petroleum coke, waste coal, tire-derived fuel) [RO 14452]; and
- Wastes from the combustion of oil and natural gas.

The regulatory determination for the wastes noted in the above five bullets is consistent with previous EPA guidance. [RO 11007, 12021, 12284, 12884]

4.13.3 Other, low-volume fossil fuel combustion wastes

In addition to the four large-volume FFC wastes discussed in Section 4.13.1 that are excluded from hazardous waste regulation by the specific wording in §261.4(b)(4), facilities that burn fossil fuels generate other wastes, in smaller quantities, from processes that are related to fuel combustion. Many of these low-volume FFC wastes are co-managed/co-disposed with the four large-volume FFC wastes, and so "their composition and character are 'masked' by the combustion [sic] and character of the combustion wastes; that is, they do not significantly alter the hazardous character, if any, of the combustion wastes." [RO 12021] Thus, EPA believes that Congress intended that these low-volume wastes should also be excluded as Bevill wastes under §261.4(b)(4) when they are mixed and co-managed with the four large-volume FFC wastes.

In a 1981 guidance letter [RO 12021], signed by Gary Dietrich of the Office of Solid Waste (hereinafter called the "Dietrich letter"), EPA noted that this extension of the Bevill exclusion was limited to low-volume wastes that are generated *in conjunction* with the combustion of fossil fuels and that are mixed and co-managed with the four large-volume FFC wastes. Until the agency could complete a more thorough evaluation of the hazardous properties of these other, low-volume wastes or their mixtures with the large-volume wastes, the Dietrich letter stated that the Bevill exclusion was extended to include, but not be limited to, the following wastes:

- Boiler cleaning solutions,
- Boiler blowdown,
- Demineralizer regenerant,
- Pyrites (coal mill rejects), and
- Cooling tower blowdown.

However, the Dietrich letter specifically said the following wastes are not excluded under §261.4(b)(4):

- Pesticide or herbicide wastes,
- Spent solvents,
- Waste oils,
- Other wastes that might be generated in construction or maintenance activities typically carried out at utility and industrial plants,
- Any hazardous waste listed in §261.31 or 261.32, and
- Any commercial chemicals listed in §261.33 that are discarded or intended to be discarded.

Years after the Dietrich letter was issued, EPA completed its evaluation of these other, low-volume FFC wastes and issued a March 1999 Report to Congress (cited previously) and subsequent regulatory determination for them on May 22, 2000. [65 FR 32214] In the regulatory determination, the agency distinguished between low-volume FFC wastes that were "uniquely associated" with fossil fuel combustion and those that weren't. EPA defined "uniquely associated" to mean a waste that "during the course of generation or normal handling at the facility, comes into contact with either fossil fuel (e.g., coal, oil) or fossil fuel combustion waste (e.g., coal ash or oil ash) and it takes on at least some of the characteristics of the fuel or combustion waste." [65 FR 32219] Table 4-2 summarizes the low-volume FFC wastes that EPA determined are uniquely associated with fossil fuel combustion vs. those that aren't.

A quick comparison of the low-volume wastes that were initially excluded by the Dietrich letter and the uniquely associated FFC wastes in Table 4-2 shows some significant differences. In its May 22, 2000 determination, EPA excluded from Bevill designation four wastes that were previously determined to qualify as Bevill wastes (when mixed and co-managed with the four large-volume FFC wastes): boiler waterside chemical cleaning wastes, boiler blowdown, water treatment and regeneration wastes, and cooling tower blowdown and sludges. Note, however, that EPA was quite tentative in this regulatory determination: "we are considering offering the following guidance concerning which low-volume wastes are uniquely associated with and which are not uniquely associated with fossil fuel combustion.... EPA solicits comments on this discussion of uniquely associated wastes in the context of fossil fuel combustion and will issue final guidance after reviewing and evaluating information we receive as a result of this request." [65 FR 32220] As far as we know, the agency has never finalized its uniquely associated wastes guidance for these low-volume FFC wastes. However, EPA's RCRA Online database notes the following for the low-volume wastes co-management portion of the Dietrich letter: "SUPERSEDED: See 65 FR 32213, 32219; 5/22/2000."

Based on the May 2000 regulatory determination, when low-volume uniquely associated wastes are mixed and co-managed with one of the four large-volume FFC wastes, they fall within the Bevill exclusion (see Figure 4-9). When managed separately, uniquely associated wastes are subject to regulation as hazardous waste if they exhibit a characteristic or are listed. [65 FR 32219, RO 12021, 13740]

Table 4-2: Low-Volume Wastes That Are "Uniquely Associated" With Fossil Fuel Combustion[1,2]

"Uniquely associated" FFC wastes	Not "uniquely associated" FFC wastes
■ Coal pile run-off	■ Boiler blowdown
■ Coal mill rejects (pyrites) and waste coal	■ Cooling tower blowdown and sludges
■ Air heater and electrostatic precipitator washes	■ Intake or makeup water treatment and regeneration wastes
■ Material in floor and yard drains and sumps	■ Boiler waterside chemical cleaning wastes
■ Wastewater treatment sludges	■ Laboratory wastes
■ Boiler fireside chemical cleaning wastes	■ General construction and demolition debris
	■ General maintenance wastes
	■ Spills or leakage of materials used in FFC processes (e.g., spills or leakage of unused boiler water treatment chemicals)

FFC = fossil fuel combustion.

[1] EPA was quite tentative in this regulatory determination: "we are considering offering the following guidance concerning which low-volume wastes are uniquely associated with and which are not uniquely associated with fossil fuel combustion.... EPA solicits comments on this discussion of uniquely associated wastes in the context of fossil fuel combustion and will issue final guidance after reviewing and evaluating information we receive as a result of this request." [65 FR 32220] As far as we know, the agency has never finalized its uniquely associated wastes guidance for these low-volume FFC wastes.

[2] EPA defined "uniquely associated" to mean a waste that "during the course of generation or normal handling at the facility, comes into contact with either fossil fuel (e.g., coal, oil) or fossil fuel combustion waste (e.g., coal ash or oil ash) and it takes on at least some of the characteristics of the fuel or combustion waste." [65 FR 32219]

Source: McCoy and Associates, Inc.; adapted from May 22, 2000; 65 FR 32219–20.

Low-volume wastes that are not uniquely associated with fossil fuel combustion get no breaks whatsoever. If one of these wastes is managed separately, it is also subject to regulation as hazardous waste if it exhibits a characteristic or is listed. If one of these nonexcluded wastes is mixed and co-managed with one of the Bevill-excluded large-volume FFC wastes, the Bevill exclusion may be lost, depending on application of the Bevill mixture rule (see Section 5.1.2.4 and Figure 5-1). [65 FR 32220]

Q *A cooling tower is cleaned out during a utility plant turnaround. The sludge removed contains 10% water, and TCLP analysis is as follows: chromium = 7.2 mg/L and lead = 1.2 mg/L. What is the regulatory status of this material when sent to a landfill?*

A The material is a solid waste, is not excluded, is not listed, but is characteristic for toxicity because of chromium (D007). Cooling tower cleanout sludge is not "uniquely associated" with fossil fuel combustion. Thus, if it is sent to an offsite landfill, it will be a hazardous waste. [May 22, 2000; 65 FR 32220, RO 13740]

4.13.3.1 Boiler chemical cleaning wastes

Utility boilers are generally cleaned every three years. In a typical scenario, the boiler is cleaned using an acid wash followed by one or more water rinses. The acid wash stream exiting the boiler is often toxic for chromium and lead and also is corrosive. However, the water rinses are usually nonhazardous. The acid wash and water rinses are collected in the same large tank or in interconnected tanks (e.g., interconnected frac tanks). Do the acid wash and water rinses have to be characterized for hazardous waste management individually or after commingling? EPA provided the answer to this question in *Federal Register* preamble guidance:

"The agency is today clarifying that, specific to power plant boiler cleanout (and potentially, to other sporadic cleaning activities involving multiple rinses), generation is at the completion of the entire cleanout process.... The agency views the cleanout of the boilers as one process and therefore does not consider the mixing of acid rinse and water rinse as impermissible dilution but as a single water rinsate resulting from the single cleanout process. This waste is subject to regulation if it exhibits a characteristic, and subject to LDR prohibitions if it exhibits a characteristic and is going to be land disposed. Today's clarification of the point of generation for boiler cleanout is limited to the situation in which the entire quantity of boiler cleanout rinses are contained in a single [tank] so that hazardous waste and LDR determinations can be made based upon the commingling of all the rinses together. If, for example, a temporary tank is brought onsite but does not have sufficient capacity to handle the estimated several hundred thousand gallons of rinsate at once, the waste will likely have to be managed in separate loads. In such instances, the generator will still be required to make hazardous waste and LDR determinations for each separate load." [May 12, 1997; 62 FR 26006–7]

The agency later clarified that this point of generation guidance applies to both permanent and temporary tank systems used to collect boiler cleanout solutions. [May 26, 1998; 63 FR 28623] Note that the last couple of sentences in the above agency quotation imply that, if the rinses are collected in separate permanent or temporary tanks that are not interconnected, the contents of each tank will have to be evaluated to determine if the collected material is hazardous.

Q *After a boiler is chemically cleaned, some unused cleaning fluids remain in the tank trucks. The trucks return to the service company's facility where they are flushed out with water. Would the resulting washwater qualify as a Bevill waste?*

A No. The washwater containing unused cleaning fluids from the tank trucks is not excluded. The unused chemicals are not uniquely associated with fossil fuel combustion or steam generation. Thus, they are not excluded, regardless of the intent of their preparation. [65 FR 32220, RO 12021]

4.13.4 Effect of co-burning wastes in boilers

The language in §261.4(b)(4) notes that the Bevill exclusion does not apply to FFC wastes if hazardous waste is burned or processed in the boiler, unless the following criteria in §266.112 are met:

1. The boiler must burn at least 50% coal on a total heat input or mass input basis; and

2. The FFC wastes are not significantly affected by the hazardous waste, demonstrated by passing one of the two tests given in §266.112(b).

If these two requirements are satisfied, fly ash, bottom ash, etc. generated from the boiler while hazardous waste is burned or processed continues to be excluded from the Subtitle C program and does not have to meet LDR treatment standards. [58 *FR* 42469, 63 *FR* 28574, RO 11881, 12021, 13401, 14344] The LDR paperwork requirements in this situation are discussed in Section 13.12.3.1.

EPA expects that there will be no impact on the Bevill status of FFC residues when used oil is co-fired with virgin fuel oil in utility boilers or furnaces. That assessment is based on the used oil comprising <1% of the total fuel mixture and meeting the used oil specification. Thus, the agency believes that the contamination levels in the residues will not be affected by the introduction of small quantities of used oil. [RO 11743]

4.13.5 Beneficial uses of FFC waste

Although significant quantities of FFC waste are disposed, these wastes are often beneficially used:

- Fly ash is used to make cement/concrete, bricks, and other building products [58 *FR* 42475, 65 *FR* 32229, RO 14247];

- Fly ash can also be used as a stabilizing medium for treatment of heavy metal-bearing wastes before they are landfilled [65 *FR* 32229, RO 13368];

- FFC wastes have been used for acid neutralization (e.g., neutralizing wastewater) [65 *FR* 32229–30];

- Flue gas desulfurization wastes are a source of synthetic gypsum used in manufacturing gypsum wall board [65 *FR* 32229, RO 14456];

- FFC wastes have been used as road bed material, snow and ice control material, structural fill, and blasting grit [65 *FR* 32229];

- FFC wastes have also been used as ingredients in insulation, paints, plastics, metals, and roofing materials [65 *FR* 32229];

- FFC wastes may be used in agricultural applications (e.g., as a substitute for lime) [58 *FR* 42475, 65 *FR* 32229]; and

- FFC wastes may also be used in the manufacture of absorbents and filter media. [65 *FR* 32229]

Generally speaking, EPA believes that these on- and offsite beneficial uses of FFC wastes can be conducted outside of the RCRA program. In the August 1993 regulatory determination, the agency noted that FFC waste utilization practices appear to be conducted in an environmentally safe manner. [58 *FR* 42468] In the May 2000 regulatory determination, EPA concluded that, except for placement of FFC wastes in mines, no RCRA regulation of the beneficial uses of these wastes is necessary. This determination included the use of FFC wastes as an agricultural soil amendment. [65 *FR* 32229–30, RO 14452, 14453]

EPA has developed a website discussing the beneficial reuse of these materials, available at http://www.epa.gov/waste/conserve/imr/ccps/benfuse.htm.

4.13.6 Importing fossil fuel combustion wastes

Once it enters U.S. jurisdiction, imported FFC waste is excluded from the definition of hazardous waste in the same way as domestically generated FFC waste. This is true regardless of whether the material will be beneficially reused as a cement admixture or disposed in an industrial landfill. [RO 14247] U.S. importers may want to keep records of the foreign exporter and the origin of any imported FFC waste in case questions arise as to its regulatory status once it enters the United States.

4.14 Oil and gas exploration and production exclusion

Due to their large volume and relatively low toxicity, EPA excludes drilling fluids, produced waters, and other wastes uniquely associated with the exploration, development, or production of crude oil, natural gas, and geothermal energy from the definition of hazardous waste. [§261.4(b)(5)] As such, even if these wastes exhibit a hazardous characteristic, they are not regulated as RCRA hazardous. Most states have adopted this federal exclusion, which (because of the relatively small size of the geothermal industry) is commonly referred to as the oil and gas exclusion or the E&P exclusion.

We want to emphasize at the outset, that even though certain crude oil, natural gas, and geothermal energy wastes are excluded from regulation as hazardous waste, they may still be regulated under state programs (e.g., oil and gas commission rules and regulations), the RCRA Subtitle D solid waste regulations, or other federal requirements (e.g., hazardous materials transportation regulations, reserve pit programs, National Pollutant Discharge Elimination System permits, and Safe Drinking Water Act provisions). Furthermore, the exclusion does not mean that these wastes do not present a threat to human health and the environment if improperly managed. [*Exemption of Oil and Gas Exploration and Production Wastes From Federal Hazardous Waste Regulations*, EPA/530/K-01/004, October 2002, available from http://www.epa.gov/epawaste/nonhaz/industrial/special/oil/index.htm]

This section provides background information on the history of the exclusion, a discussion of the types of wastes generated during oil, natural gas, and geothermal energy activities, and key information on how to determine which wastes are excluded under §261.4(b)(5) and which are not.

4.14.1 Background

In the RCRA statute, Congress conditionally excluded certain categories of solid waste from regulation as hazardous waste under RCRA. [Section 3001(b)(2)(A)] These wastes included drilling fluids, produced waters, and other wastes associated with the exploration, development, or production of crude oil, natural gas, and geothermal energy. This statutory requirement was implemented in the RCRA regulations on May 19, 1980. [45 *FR* 33120] Section 3001(b)(2)(B) required EPA to study these wastes and submit a report to Congress determining whether the conditional exclusion was warranted or whether such materials should be regulated as hazardous waste. In its December 1987 Report to Congress, EPA concluded that regulation of crude oil, natural gas, and geothermal energy wastes under the RCRA Subtitle C hazardous waste regulations was not justified. Instead, the agency recommended that environmental risks posed by these wastes be addressed through other regulatory strategies.

EPA's report was summarized in a subsequent regulatory determination for these wastes, published in the *Federal Register* on July 6, 1988. [53 *FR* 25446] EPA cited several reasons for its decision not to regulate these wastes as hazardous:

- Imposing hazardous waste regulations on all oil and gas wastes could subject billions of barrels of waste to RCRA regulation and would cause a severe impact on the oil and gas industry and production in the United States.

- Elimination of the exclusion would cause serious short-term capacity strains at commercial TSD facilities.

- Existing state and federal regulatory programs are "generally adequate" for controlling oil, gas, and geothermal wastes. Furthermore, EPA committed to help improve existing programs to close the gaps. Since 1988, the agency has taken several steps to improve non-RCRA regulatory programs applicable to oil and gas wastes. For example, effluent limitation guidelines have been promulgated under the Clean Water Act for offshore and coastal oil and gas operations. Revisions in the underground injection control program under the Safe Drinking Water Act have also been made to control migration of these wastes. Finally, EPA has provided funding to the Interstate Oil and Gas Compact Commission to develop guidelines for effective state regulatory programs for management of these wastes.

- If oil, gas, and geothermal wastes were subject to hazardous waste regulations, RCRA permitting delays would disrupt the search for new oil, gas, and geothermal energy deposits.

On March 22, 1993, EPA clarified the scope of the §261.4(b)(5) exclusion for waste streams generated by crude oil reclamation operations, crude oil pipelines, oil and gas service companies, and gas processing plants (and their associated field gathering lines). [58 *FR* 15284] These clarifications are included in our discussions below.

4.14.2 Exploration, development, and production processes

The primary processes associated with exploration, development, and production of oil, gas, and geothermal energy are described below.

4.14.2.1 Oil and gas processes

By far the largest quantity of wastes excluded under §261.4(b)(5) is produced during the exploration, development, or production of oil and gas. The processes that produce these wastes are described below. [*Profile of the Oil and Gas Extraction Industry*, EPA/310/R-99/006, October 2000, available at http://www.epa.gov/compliance/resources/publications/assistance/sectors/notebooks/oilgas.pdf]

4.14.2.1.1 *Exploration*

Exploration involves the search for geologic formations associated with oil or gas deposits using geophysical prospecting and/or exploratory drilling. The amount of wastes generated by exploration activities is relatively small when compared to development and production wastes.

4.14.2.1.2 *Development*

After an economically recoverable oil and/or gas field has been discovered, wells are constructed to retrieve the oil and/or gas (i.e., development drilling).

Drilling fluids are used during well construction to 1) cool and lubricate the drill bit; 2) remove drill cuttings from the drilling area and transport them to the surface; 3) counterbalance formation pressure to prevent formation fluids (i.e., oil, gas, and water) from entering the well prematurely; and 4) prevent the open, uncased wellbore from caving in. Typically, water- or oil-based drilling fluids (called drilling muds) are used that contain additives, such as corrosion inhibitors (e.g., aluminum bisulfate, iron oxide, zinc carbonate, and zinc chromate), dispersants (e.g., iron lignosulfonates), flocculants (e.g., acrylic polymers), surfactants (e.g., fatty acids and soaps), biocides (e.g., chlorophenols, formaldehydes, or organic amines), and fluid-loss reducers (e.g., organic polymers and starch).

During the drilling process, large quantities of drill cuttings coated with drilling fluids and oil are generated. Between 0.2 and 2.0 barrels (8.4 and 84 gallons) of total drilling waste are produced for each vertical foot drilled.

When drilling has been completed, several well-completion steps are needed before production may commence. Several tests are performed to verify oil or gas formations and pressures. Well-completion activities include those necessary to allow a well, once drilled, to produce oil or gas. These measures include installing and cementing casings, installing the production tubing and down-hole equipment, repairing damage that drilling may have caused to the formation, and stimulating the well. [If oil or gas flows are poor, stimulation fluids (acids and/or high-pressure liquids) may be pumped into the formation to open pores and fractures.] Perforation of the well casing at specific locations is conducted using explosive charges; these perforations allow the oil or gas to flow into the well.

4.14.2.1.3 *Production*

Production is the process of extracting the hydrocarbons from the subsurface and separating the mixture of liquid hydrocarbons, gas, water, and solids into saleable constituents. Oil and gas are frequently produced from the same reservoir. Primary production of these materials just relies on the natural reservoir pressure for recovery, sometimes supplemented with artificial lift equipment (pumps). Secondary or enhanced recovery of oil typically uses reinjection of produced water (water recovered with the oil) into the formation to repressurize the reservoir and reestablish fluid flow into the recovery wells. Steam, oil-miscible fluids, surfactants, and/or microbes are injected into the formation to extract oil and gas during tertiary recovery.

Most oil and gas production operations include tanks for the temporary storage of oil, natural gas liquids, and/or produced water. Tank bottoms containing oil and contaminants may be periodically removed from the tanks and disposed. Produced oil is typically processed at a refinery, while natural gas may be processed to remove impurities in the field or at a natural gas processing plant. After the hydrocarbons are extracted from the subsurface, the two primary production processes are discussed below:

1. Crude oil separation—When crude oil is brought to the surface, it may contain natural gas, water, sand/silt, and any additives used to enhance extraction. The primary by-product from the production process is produced water. Produced water may contain various contaminants, such as inorganic compounds (e.g., antimony, arsenic, barium, calcium, chloride, lead, magnesium, potassium, sodium, sulfur, and zinc), organic compounds (e.g., benzene, bromodichloromethane, naphthalene, phenanthrene, and toluene), and radionuclides. For wells nearing the end of their productive lives, water may make up 98% of the material brought to the surface. The American Petroleum Institute (API) estimates that over 15 billion barrels of produced water are generated each year from crude oil wells (almost 8 barrels of water for every barrel of oil produced).

   During crude oil separation, the gaseous components (e.g., natural gas and hydrogen sulfide) are separated using pressure chambers, if necessary. The remaining liquids and solids (a mixture of oil, water, and sand/silt) are separated using a process called "free-water knockout," in which free water (produced water) and the sand/silt (produced sand) are removed primarily by gravity separation. Finally, water/oil emulsions are broken by heating the fluid or by treating it with emulsion-breaking chemicals in "heater treaters." The resulting oil is about 98% pure, sufficient for storage and subsequent transportation to a refinery.

2. Natural gas processing—Gas wells usually generate much lower volumes of produced water than oil wells. Impurities are usually removed from the gas using two conditioning processes: dehydration (to remove water) and sweetening (to remove carbon dioxide and hydrogen sulfide). Dehydration can be performed using a liquid or solid desiccant or refrigeration (simply cooling the gas to a temperature below the condensation point of water). When a liquid desiccant is used, the gas is exposed to a glycol (such as triethylene glycol) that absorbs the water. The water can later be evaporated from the glycol by a process called heat regeneration, and the glycol can then be reused. (However, the glycol must be replaced periodically.) This process is most commonly used when processing occurs in the field. Solid desiccants (e.g., molecular sieves) are crystals with high surface areas and pore structures that attract and retain water molecules. The solids can be regenerated by heating them, but they too must be periodically replaced.

   Sweetening is commonly carried out using amine treatment or the use of iron sponge. Potential wastes include spent amine solution, iron sponge, and elemental sulfur.

4.14.2.1.4 Workover

From time to time, production wells require significant maintenance. These maintenance periods are called workovers and produce a number of wastes.

Salts (called scale) and paraffins that accumulate in the well during production are removed from production tubing or well casings using solvents. During such workovers, tubing, gathering lines, and valves are typically treated with corrosion inhibitors. Other procedures to maintain or enhance production (e.g., repairing or replacing down-hole equipment, stimulating the formation to restore or enhance production) are conducted at these times as well.

4.14.2.2 Geothermal energy processes

Geothermal energy sources include hot water and steam that are tapped from the earth. These streams are used as heating sources or to drive turbines for the generation of electricity. Drilling-related wastes and other wastes generated during exploration, development, or production of these energy streams are similar to those discussed above in the oil and gas industry.

In the United States, the geothermal energy industry is located mainly in California and Nevada; it is much smaller than oil and gas production. Therefore, the quantity of geothermal energy wastes subject to the §261.4(b)(5) exclusion is quite small in comparison. [July 6, 1988; 53 *FR* 25458]

4.14.3 Determining what wastes are excluded

Figuring out what wastes are excluded from hazardous waste regulation under §261.4(b)(5) isn't always straightforward. The first step is to review EPA's guidance on wastes "uniquely associated" with the exploration, development, or production of oil, gas, or geothermal energy.

4.14.3.1 "Uniquely associated" wastes

Drilling fluids and produced water represent the largest quantity of oil, gas, and geothermal energy wastes excluded from RCRA hazardous waste regulation. These wastes are estimated to constitute over 98% of the total industry waste stream. [*Associated Waste Report,* January 2000, available from http://www.epa.gov/epawaste/nonhaz/industrial/special/oil/execrep.htm] The remaining 2% consists of other wastes associated with the exploration, development, or production of oil, gas, or geothermal energy, which are also excluded from hazardous waste regulation under §261.4(b)(5).

The legislative history of the exemption implies that the term "other wastes associated with" includes wastes intrinsically derived from primary field operations associated with the exploration, development, or production of oil, gas, or geothermal energy. [EPA/530/K-01/004, RO 11610] EPA noted on March 22, 1993 that this concept includes wastes "associated with operations to locate or remove oil or gas from the ground or to remove impurities from such substances and [they] must be intrinsic to and *uniquely associated* with oil and gas exploration, development, or production operations." [Emphasis added.] [58 *FR* 15284] For example, methanol injected into natural gas production wells to keep them from freezing during the winter months produces a water/methanol waste stream when separated from the recovered natural gas; the water/methanol mixture is a "uniquely associated" waste and therefore is excluded under §261.4(b)(5). [RO 12258] Case Study 4-5 gives other examples of this concept.

4.14.3.1.1 *Primary field operations*

As mentioned above, uniquely associated wastes must be intrinsically derived from primary field operations associated with exploration, development, or production activities. Primary field operations include only those activities necessary to locate and recover oil, gas, and geothermal energy from the ground and remove impurities. Primary field operations consist of [EPA/530/K-01/004, RO 11610]:

- Exploration, development, and the primary, secondary, and tertiary production of oil or gas.

- Crude oil processing, such as water separation, demulsifying, degassing, and storage in tanks associated with a specific well or wells.

- Crude oil activities occurring at or near the wellhead and before the point where the oil is transferred from an individual field facility or

> **Case Study 4-5:**
> **Equipment Cleaning Wastes**
>
> I. Oil and gas drilling equipment (presumably contaminated with down-hole material) is steam cleaned at a separate location from the drilling site. The resulting decon stream exhibits a hazardous waste characteristic. Is the waste excluded from RCRA hazardous waste regulation under §261.4(b)(5)?
>
> "Since only water is used for steam cleaning, the drilling waste is still excluded from regulation. If another cleaning agent not uniquely associated with the exploration, development, or production of oil, gas, or geothermal energy was used, then the waste could be subject to regulation. For example, if methylene chloride was used to clean the equipment, the waste would be subject to regulation as F002." [RO 12197]
>
> II. In another situation, a solvent used to clean surface equipment or machinery is not excluded, because it is not uniquely associated with exploration, development, or production operations. However, if the same solvent were used in a production well (e.g., to remove scale or paraffin), recovered spent solvent would be excluded because it was generated via a down-hole procedure that is uniquely associated with production. [EPA/530/K-01/004]

- a centrally located facility to a carrier (i.e., pipeline or truck) for transport to a refinery.
- Natural gas activities occurring at or near the wellhead or at the gas plant and before the point where the gas is transferred from an individual field facility, a centrally located facility, or a gas plant to a carrier (i.e., pipeline or truck) for transport to market. Gas plants are considered to be part of production operations regardless of their location with respect to the wellhead, because natural gas often must be processed to remove water and other impurities prior to sale.

Primary field operations do not include transportation and manufacturing operations. Wastes generated during transportation and manufacturing are not excluded as discussed in more detail in Sections 4.14.4.1 and 4.14.4.2.

4.14.3.1.2 Rule of thumb for determining if wastes are "uniquely associated"

EPA has tried to clarify when oil, gas, and geothermal energy wastes are "uniquely associated" using a simple rule of thumb. According to the agency, a waste is considered "uniquely associated" with exploration, development, or production activities and, therefore, most likely excluded from hazardous waste regulation if:

- The waste came from down hole (i.e., it was brought to the surface during oil, gas, or geothermal energy exploration, development, or production operations); or

- The waste was otherwise generated by contact with the oil, gas, or geothermal energy production stream during the removal of produced water or other contaminants from the well or the product (e.g., waste demulsifiers, spent iron sponge). [March 22, 1993; 58 *FR* 15285]

4.14.3.1.3 Materials must be used in order to be "uniquely associated"

Only *used* materials are considered to be uniquely associated wastes when disposed. Discarded *unused* materials (such as unused chemicals) are not uniquely associated and therefore are not excluded from hazardous waste regulation when disposed, *regardless of their intended use.* [EPA/530/K-01/004]

Q *During oil field operations, excess unused drilling and workover fluids that exhibit the characteristic of corrosivity are disposed in a reserve pit. Are these unused materials subject to RCRA hazardous waste regulation?*

A Yes. According to EPA, unused chemical products (such as well-completion and workover fluids) are *not* excluded from hazardous waste regulation when they are disposed. However, if the unused materials are hazardous only due to their corrosivity, they can be neutralized by elementary neutralization or totally enclosed treatment in the same tanks used to hold the fluids prior to use without subjecting the tanks to hazardous waste regulation (see Section 7.5). The neutralized waste could then be disposed in the reserve pit, unless a state program prohibits such practice. [RO 11794, 12105]

4.14.3.2 Examples of wastes that are excluded

Based on the "uniquely associated" guidance discussed above, Table 4-3 lists examples of exploration, development, or production wastes that EPA has specifically said are excluded from regulation as RCRA hazardous waste.

4.14.3.3 Examples of wastes that are not excluded

EPA has determined that the oil and gas exclusion does *not* apply to transportation or manufacturing wastes (as discussed in more detail in Sections 4.14.4.1 and 4.14.4.2). To clarify this issue, the agency has provided examples of wastes that do not qualify for the §261.4(b)(5) exclusion (and therefore are hazardous wastes if they exhibit a characteristic or meet a listing description). Examples of these nonexcluded wastes are given in Table 4-4.

Using Tables 4-3 and 4-4 to determine if wastes are excluded is exemplified as follows.

Q. *Well stimulation, the process of acidizing or fracturing an oil or gas well, is sometimes performed to enhance production. Solutions of acid, chemicals, sand, and water may be pumped into a well to eliminate obstacles to the flow of oil and gas. Acidic solutions used in well stimulation are often partially prepared offsite at a service company by combining acid, chemical solutions, and water in tank trucks. During transfer of the acid to the tank trucks, small quantities of acid typically spill from the hose or overflow the tank truck. The spilled acid is rinsed with water on a concrete containment slab that drains into a wastewater holding tank. The service company believes that the wastewater should be classified as "well stimulation fluids" (an excluded waste under the oil and gas exclusion—see Table 4-3) because the sole purpose of preparing the solution is to stimulate or otherwise enhance oil and gas production. Furthermore, even though the wastes will not reach the wellhead, the company argues that the wastes are associated with production operations—not wastes related to transportation and manufacturing. Is this rationale correct?*

A. No. The acidic wastewater is not excluded because it is not intrinsically derived from the exploration, development, or production of oil or gas. It was never used in such operations, regardless of the intent in preparing the mixture. This type of waste fits under the category of "oil and gas service company wastes, such as…spilled chemicals, and waste acids" in Table 4-4, which are not excluded. [RO 13236]

Q. *After well stimulation is accomplished, some unused stimulation fluids remain in the tank trucks. The trucks return to the service company's facility where they are flushed out with water. Would the resulting wastewater qualify as "well stimulation fluids" and thereby be excluded?*

A. No. The wastewater containing unused stimulation fluids from the tank trucks is also not excluded. It is classified in the nonexcluded category of "unused fracturing fluids or acids" in Table 4-4. The unused fluids that remain in the trucks are not intrinsically derived from oil and gas exploration, development, or production operations, regardless of the intent of their preparation. [RO 13236]

Q. *Reserve and other types of pits are required for many E&P operations in order to efficiently extract oil and gas resources. Synthetic pit liners are used to keep E&P wastes from contaminating soil/water near pit locations. When removed, are spent synthetic pit liners excluded under the E&P exclusion of §261.4(b)(5)?*

Table 4-3: Examples of Exploration, Development, or Production Wastes Excluded From Hazardous Waste Regulation[1]

Drilling- and well-completion-related wastes[2]

- Drill cuttings
- Drilling fluids
- Drill cuttings and drilling fluids from offshore operations that are disposed onshore
- Used well-completion, treatment, stimulation, and workover fluids and wastes generated from these activities, including acidizing agents, biocides, corrosion inhibitors, damaged down-hole equipment, defoamers, detergents, fracturing media, gels, inert material from down-hole mechanical repair (including cement cuttings and slurries, pipe scale, and produced sand), packing fluids, paraffin, paraffin solvents and dispersants, sludges, solvents, surfactants, thinners, used filters, water- or oil-based muds, and weighting agents/viscosifiers

Oil and gas production wastes

- Accumulated materials, such as emulsions, hydrocarbons, sands, and solids from production separators, fluid treating vessels, and production impoundments
- Basic sediment (sand and other solids), emulsions, heavy hydrocarbons (asphaltic and paraffinic), water, and other tank bottoms from storage facilities that hold product prior to transport
- Boiler water
- Cooling tower blowdown and scrubber liquids associated with the cooling water system
- Gas-plant dehydration wastes, including backwash, glycol-based compounds, glycol filters and filter media, and spent molecular sieves
- Gas-plant sweetening wastes for sulfur removal, including amines, amine filters and filter media, backwash, caustic solutions, hydrogen sulfide scrubber liquid and sludge, precipitated amine sludge, slurries of sodium salts and sulfur, spent catalyst, spent iron sponge (water slurries/sludges and wood chips), and sulfinol
- Liquid hydrocarbons removed from the production stream, but not from oil refining
- Pigging wastes[3] from gathering lines, including debris, hydrocarbon solids, paraffins, rust, scale, and wax
- Pit sludges and contaminated bottoms from storage or disposal of excluded wastes
- Produced sand, including wet sludge containing oil and water
- Produced water, including produced water associated with natural gas dehydration and well workovers
- Produced water injected back into the formation for enhanced recovery[4]
- Spent caustic solutions and acids used as degreasers and solvents that came from down hole or were otherwise generated by contact with excluded wastes
- Spent filters (canisters, cartridges, socks), filter media (coal, diatomaceous earth, gravel, particulate, sand), drainage, and backwash (assuming the filter itself is not hazardous and the residue in it is from an excluded waste stream)
- Steam generator waste from formation steam-flood operations
- Untreatable emulsions, including brine, caustic solutions, dehydration chemicals, emulsion breakers (surfactants), hydrocarbons, scale, and silt
- Waste crude oil from primary field and production operations, including oil from well completion and workovers

Table 4-3: Examples of Exploration, Development, or Production Wastes
Excluded From Hazardous Waste Regulation[1]

Geothermal energy production wastes

- Flash tank solids (except those associated with electrical power generation) [RO 11226]
- Geothermal production fluids
- Hydrogen sulfide abatement wastes from geothermal energy production
- Most direct-use waste streams
- Piping scale [RO 11226]
- Precipitated solids from brine effluent [RO 11226]
- Reinjection well fluid wastes [RO 11226]
- Settling pond wastes [RO 11226]
- Waste streams produced from materials passing through the turbine in dry-steam power generation
- Waste streams resulting from a geothermal energy fluid or gas that passed through the turbine in flashed stream and binary power plants
- Waste streams resulting from the geothermal energy products passing through only the heat exchanger in binary operations or through the flash separator in the flash process

Miscellaneous wastes

- Constituents removed from produced water before it is injected or otherwise disposed
- Contaminated soil, including hydrocarbon-bearing soil, produced-water contaminated soil, soil contaminated with naturally occurring radioactive materials, sulfur-contaminated soil from sulfur-recovery units, and other soil contaminated from spills of excluded wastes
- Gases from the production stream, such as carbon dioxide, hydrogen sulfide, and volatilized hydrocarbons
- Light organics volatilized from excluded wastes in reserve pits, impoundments, or production equipment
- Materials ejected from a production well during blowdown
- Naturally occurring radioactive materials
- Oily rags and sorbent materials[5]
- Pipe scale, hydrocarbon solids, hydrates, and other deposits removed from piping and equipment prior to transportation
- Production line hydrotest/pressure fluids utilizing produced water
- Rig wash
- Wastes from subsurface gas storage and retrieval (except for nonexcluded wastes listed in Table 4-4)

[1]This list provides examples and is not comprehensive.
[2]These wastes are generated generically during the exploration and development of crude oil, natural gas, or geothermal energy.
[3]Pigging wastes are excluded from regulation as hazardous wastes if they are from gathering/production lines. They are not excluded if they are from transportation lines.
[4]Produced water used in this manner is considered to be beneficially recycled and is already regulated by the underground injection control program under the Safe Drinking Water Act. However, if produced water is stored in surface impoundments prior to injection, it may be subject to RCRA Subtitle D regulations. [July 6, 1988; 53 FR 25454]
[5]Oily rags and sorbent materials are excluded from hazardous waste regulation only if their generation is intrinsic to exploration, development, or production operations.

Source: McCoy and Associates, Inc.; adapted from 53 FR 25448, 25453–4 [July 6, 1988], EPA/530/K-01/004, and Associated Waste Report, unless otherwise noted.

Table 4-4: Examples of Wastes That Do Not Qualify for the Oil, Gas, and Geothermal Energy Exclusion[1]

- Boiler cleaning wastes
- Boiler refractory bricks
- Boiler ash, scrubber fluids, and sludges
- Caustic or acid cleaners that did not come from down hole and were not otherwise generated by contact with excluded wastes
- Drums, insulation, and miscellaneous solids, including packaging materials (such as sacks, plastic buckets, pallets, etc.)[2] [RO 14260]
- Gas-plant cooling tower cleaning wastes
- Incinerator ash
- Laboratory wastes
- Liquid and solid wastes generated by crude oil and tank bottom reclaimers[3]
- Off-specification, undetonated explosives expected to be used in crude oil, natural gas, and geothermal exploration activities [RO 12273]
- Oil and gas service company wastes, such as empty drums, drum rinsate, vacuum truck rinsate, sandblast media, painting wastes, spent solvents, spilled chemicals, and waste acids
- Painting wastes
- Pesticide wastes
- Radioactive tracer wastes
- Refinery wastes
- Sanitary wastes
- Unused, excess casing cement [RO 13236]
- Unused fracturing fluids or acids (both premixed fluids and unmixed raw ingredients) [RO 13236]
- Used equipment lubricating oils
- Used hydraulic fluids
- Vacuum truck and drum rinsate from trucks and drums transporting or containing nonexcluded waste
- Waste compressor oil, filters, and blowdown
- Waste in transportation-pipeline-related pits
- Waste (spent) solvents that did not come from down hole and were not otherwise generated by contact with excluded wastes

[1]These wastes are hazardous if they exhibit a characteristic or meet a listing description. This list provides examples and is not comprehensive.

[2]These materials most likely will be nonhazardous, unless they contain hazardous waste that is not uniquely associated with exploration, development, or production. [RO 14260]

[3]Nonexploration, nondevelopment, or nonproduction wastes generated from crude oil and tank bottom reclamation operations (e.g., spent solvents used to clean equipment) are not excluded from RCRA hazardous waste regulation. However, residues derived from excluded wastes (e.g., produced water separated from E&P exempt tank bottoms) are excluded. [March 22, 1993; 58 *FR* 15285] See Section 4.14.4.3.

Source: 53 *FR* 25454 and EPA/530/K-01/004, unless otherwise noted.

A No. "[S]ynthetic pit liners used in E&P operations are not covered by the RCRA exemption because they are not intrinsically derived from, or uniquely associated with operations associated with the exploration, development, or production of crude oil and natural gas." [RO 14815; see also RO 14816]

4.14.4 Additional guidance for determining if the exclusion applies

The agency has provided additional guidance in a number of different documents that may help determine the applicability of the §261.4(b)(5) exclusion to a particular waste stream. Below, we summarize EPA's interpretations regarding applicability of the §261.4(b)(5) exclusion to transportation and manufacturing activities, crude oil reclamation, service companies, and natural gas wastes. Based on this guidance, Figure 4-10 is a logic diagram for determining the general applicability of the oil, gas, and geothermal energy exclusion.

The regulatory status of mixtures that contain excluded oil and gas wastes and nonexcluded waste is discussed in Section 5.1.2.5 and Figure 5-2.

4.14.4.1 Transportation wastes

Wastes associated with transportation (e.g., wastes generated from a transportation pipeline) do not qualify for the §261.4(b)(5) exclusion. The question then becomes: when does transportation begin? This question is answered separately for crude oil and natural gas operations.

4.14.4.1.1 *Crude oil transportation*

In the oil field, crude oil and any produced water are directed from the wells to a series of tanks, called a tank battery, where the water and oil separate naturally due to gravity. (This process is sometimes enhanced by the use of heat.) Following water separation, the crude oil is metered prior to a change in custody or ownership and/or its transportation offsite.

As discussed previously, the oil and gas exclusion covers production-related activities that occur only as part of primary field operations at or near the wellhead. Only wastes generated before the end of primary field operations are excluded. Wastes generated as part of the process of transporting products away from primary field operations are *not* excluded. Examples of nonexcluded wastes resulting from transportation include pigging wastes from transportation pipeline pumping stations, contaminated water and snow resulting from spills from transportation pipelines or other forms of transport of the product, and soils contaminated from such spills. Conversely, storage of crude oil in stock tanks at production facilities is considered part of the production separation process, not transportation. Therefore, wastes from these tanks would be included in the exclusion, as noted in Table 4-3. [March 22, 1993; 58 *FR* 15284, 15286, EPA/530/K-01/004, RO 11610]

EPA generally defines the end of primary field operations and the beginning of transportation as 1) a custody transfer of the product from the producer (an individual field facility or a centrally located facility) to a carrier (pipeline or trucking company) for transport to a refinery; or 2) if no custody transfer takes place, the end of initial separation and dewatering of the crude oil (typically the point at which crude oil leaves the last vessel, including the stock tank, in the tank battery associated with the well or wells). Any waste generated after either of these two events is not included within the scope of the oil and gas exclusion. [March 22, 1993; 58 *FR* 15286, RO 11610]

4.14.4.1.2 *Natural gas transportation*

Some water may be separated from natural gas at the wellhead. However, gas from several wells is typically commingled and sent to a central gas plant where additional water and impurities are removed. Between the wellhead and the gas plant, ownership of the gas may change.

For natural gas, EPA defines primary field operations as production-related activities at or near the wellhead and at the gas plant (regardless of whether or not the gas plant is located near the wellhead), but

Figure 4-10: Regulatory Status of Oil and Gas E&P Wastes

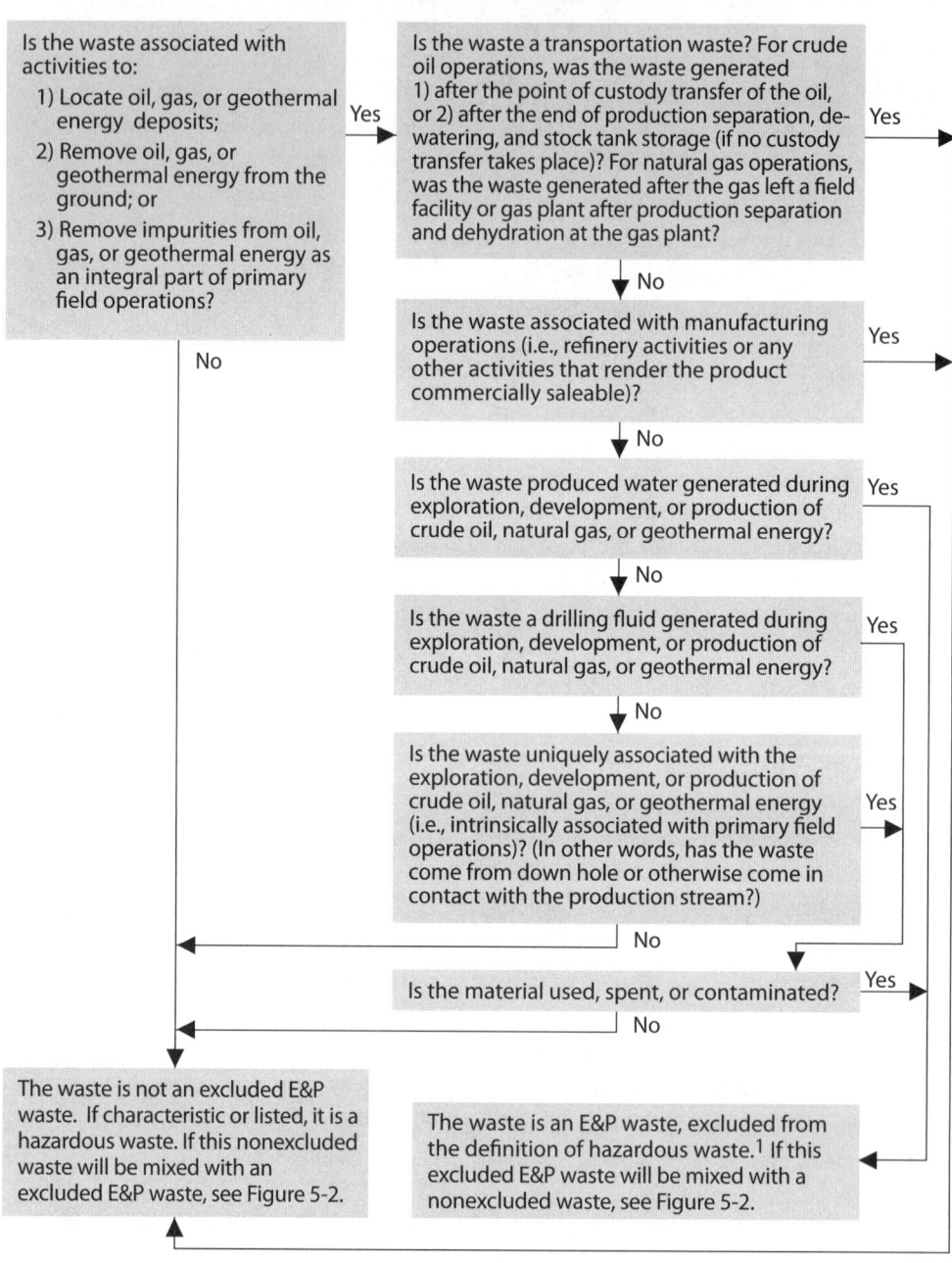

E&P waste = excluded oil and gas exploration and production waste.

[1] If disposed offsite, the company/facility needs to work closely with the nonhazardous waste (Subtitle D) landfill to ensure the waste (although excluded from the hazardous waste program) meets the landfill's acceptance criteria/permit conditions and will not cause any other problems that might result in future CERCLA or tort liability.

Source: McCoy and Associates, Inc.

prior to transport of the natural gas from the gas plant to market. According to EPA, "[b]ecause the movement of the natural gas between the wellhead and the gas plant is considered a necessary part of the production operation, uniquely associated wastes derived from the production stream along the gas plant feeder pipelines (e.g., produced water, gas condensate) are considered excluded wastes, even if a change of custody of the natural gas has occurred between the wellhead and the gas plant." [March 22, 1993; 58 FR 15286–7]

Thus, transportation begins at the point where the gas is transferred from an individual field facility, a centrally located facility, or a gas plant to a carrier for transport to market. If the gas requires purification and dehydration, the point that transportation begins is *after* product separation and dehydration at the gas plant. [EPA/530/K-01/004, RO 11610] See Case Study 4-6.

Gas-plant and feeder-pipeline wastes that are *not* excluded include [March 22, 1993; 58 FR 15287]:

- Pump lube oil,

- Waste mercury from meters and gauges,

- Soil contaminated by spills of wastes that are not uniquely associated with production operations (e.g., soil contaminated by mercury from gauges), and

- Wastes generated by nonproduction-related activities (i.e., manufacturing) that may occur at a gas plant (e.g., wastes produced during 1) operations that go beyond the removal of impurities; 2) the physical separation of the gas into its component fractions, such as cracking and reforming the various gas fractions; and 3) the addition of odorants or other substances).

Production of elemental sulfur from hydrogen sulfide gas at a gas plant is considered treatment of an excluded waste because hydrogen sulfide gas is a uniquely associated waste (see Table 4-3). Thus, any residual waste derived from the hydrogen sulfide remains excluded. [March 22, 1993; 58 FR 15287]

Case Study 4-6: Natural Gas Compressor Stations

In some U.S. locations, natural gas does not require sweetening or extensive dehydration. In these situations, the gas generally is not transferred to a gas plant but is sent from the wellhead to a main transmission line and, in some cases, directly to the customer. Compressor stations are located along the pipelines that run between the wellhead and the main transmission line or the customer to maintain pressure in the lines. Do wastes generated at these compressor stations qualify for the §261.4(b)(5) exclusion?

In the absence of gas plants, EPA's position is that compressor stations handling only local production gas should be treated the same as gas plants. ("Local production" refers to gas from a single nearby gas field or several nearby fields, as determined by the state oil and gas regulatory agency.) Consequently, wastes generated by these compressor stations are excluded. However, once gas from outside the local production area is commingled with gas from within the local area, wastes from pipeline facilities and compressor stations beyond that point would no longer be excluded, even if additional local production feeds into the system downstream from the point of commingling. Therefore, compressor stations located along main gas transmission lines (that are transporting commingled local production and outside sources of gas) are considered to be part of the transportation process, and any wastes generated by these compressor stations are not excluded. [EPA/530/K-01/004, RO 11794]

4.14.4.1.3 *Status of excluded wastes shipped offsite*

Note that excluded wastes do not lose their excluded status if they undergo custody transfer and are transported offsite for treatment, reclamation, or disposal. [March 22, 1993; 58 *FR* 15285, EPA/530/K-01/004]

4.14.4.2 Manufacturing wastes

In addition to transportation wastes, wastes associated with manufacturing operations are not uniquely associated with primary field operations and therefore are not excluded under §261.4(b)(5). [EPA/530/K-01/004, RO 11610] Manufacturing (for the oil, gas, and geothermal energy industry) is defined as "any activity occurring within a refinery or other manufacturing facility the purpose of which is to render the product commercially saleable." Consequently, any wastes associated with oil refining, petrochemical-related manufacturing, and electricity generation are not excluded. [RO 11610, 13293] Case Study 4-7 illustrates the status of a manufacturing waste under this exclusion.

4.14.4.3 Crude oil reclamation

The crude oil reclamation industry recovers marketable crude oil and other hydrocarbons from produced water, crude oil tank bottoms, and other oily wastes that are generated by the production of crude oil and natural gas. Thermal and/or physical processes, such as heat and gravity separation, are used to recover saleable crude oil. If it is difficult to separate crude oil from produced waters, demulsifiers are added. Wastes resulting from crude oil reclamation include produced water and tank bottom solids.

In the July 1988 regulatory determination, EPA included "liquid and solid wastes generated by crude oil and tank bottom reclaimers" on the list of *nonexcluded wastes* (see Table 4-4). However, in its March 22, 1993 clarification, EPA explained that this entry refers only to nonexploration, nondevelopment, or nonproduction wastes generated by reclaimers (i.e., wastes derived from processing nonexcluded oil field wastes or wastes that contain materials that are not uniquely associated with exploration, development, or production operations). Examples of nonexcluded wastes are waste solvents generated from solvent cleaning of tank trucks that are used to transport oil field tank bottoms. [58 *FR* 15285, RO 11595]

Generally, crude oil reclaimer wastes that are derived from processing excluded oil field wastes

---

> **Case Study 4-7: Scrubber Waste Generated at Geothermal Power Plants**
>
> At a typical geothermal power plant, geothermal steam is passed through turbines to produce electricity. As the steam drives the turbine, it condenses, and the condensate is generated as a waste stream. Scrubbers are used to remove sulfides from this condensate before it is discharged to the environment. Do the resulting sulfide-bearing scrubber wastes qualify for the §261.4(b)(5) oil and gas exclusion?
>
> No. According to EPA, "exploration, development, or production" refers to "locating energy deposits and extracting the oil, gas, or geothermal energy (steam) from those deposits." Later processing or manufacturing operations are not included in the exclusion. Furthermore, "only wastes 'intrinsically derived from primary field operations,' i.e., derived from the process of extracting the geothermal steam itself, are covered by the exclusion."
>
> In this case, "[t]he scrubber wastes are not covered by the exclusion because these wastes result not from the physical extraction of the geothermal energy, but from a separate manufacturing process downstream from the production operations. The generation of electricity is a separate process because it uses the steam as fuel to drive turbines and generate electricity." The oil, gas, and geothermal energy exclusion is limited to the production of geothermal energy and "does not extend to subsequent uses of that geothermal energy in power plants or other industrial operations." [OSWER Directive 9441.50-1A]

(e.g., produced water and basic sediment) are not subject to hazardous waste regulation because of the §261.4(b)(5) exclusion. Wastes derived from treating an excluded waste (including recovery of product from an excluded waste) usually remain excluded from RCRA requirements. For instance, waste residues from the onsite or offsite recovery of crude oil from tank bottoms obtained from crude oil storage facilities at primary field operations (i.e., operations at or near the wellhead) are excluded from hazardous waste regulations because the crude oil storage tank bottoms at primary field operations are excluded. As noted above, offsite transport of excluded wastes from a primary field site for treatment, reclamation, or disposal does not negate the exclusion. [58 *FR* 15285, RO 11595]

4.14.4.4 Service companies

Oil and gas service companies are hired by the principal operating company to supply materials for use at a drilling or production site or provide other onsite or offsite assistance. Offsite services performed by these companies may include product formulation, materials transportation, laboratory analysis, and waste handling/disposal.

In the 1988 regulatory determination, EPA stated that "oil and gas service company wastes, such as empty drums, drum rinsate, vacuum truck rinsate, sandblast media, painting wastes, spent solvents, spilled chemicals, and waste acids" are not covered by the oil and gas exclusion. However, as clarified in the March 22, 1993 notice, the agency intended for that item to include only wastes generated by service companies that are not uniquely associated with primary field operations. EPA did not intend to imply that service companies would never generate an excluded waste. Any waste generated by activities uniquely associated with the exploration, development, or production of oil, gas, or geothermal energy during primary field operations is excluded from hazardous waste regulation, regardless of whether it is generated by a service company or the principal operator.

For example, waste acid generated during a well workover would be excluded, if the waste came from down hole. Applicability of the exclusion wouldn't hinge on whether that waste was generated by a service company or the principal operator. When a service company's trucks leave the production site and return to its company's facility, any unused frac or stimulation fluids would not be excluded when they are washed out (as was discussed previously). However, rinsates from the company's trucks or drums are excluded if the wastes contained within them are excluded, provided that the rinsing fluid is not subject to RCRA (e.g., solvents). [58 *FR* 15286]

4.14.4.5 Natural gas wastes

Additional guidance on natural gas condensate, gas-plant cooling tower wastes, gas regulators, and underground gas storage fields is provided below.

4.14.4.5.1 *Natural gas condensate*

Condensate (drip gas) that is collected from pipelines associated with movement of natural gas at an exploration, development, or production site is excluded from classification as a hazardous waste. Examples of excluded condensate would be 1) drip gas from gas gathering lines on the production site, or 2) condensate from pipelines carrying natural gas from the gas field to the gas plant. However, if the drip gas is collected from lines that are used for the offsite transportation of natural gas, the drip gas is not excluded under §261.4(b)(5), because it is not intrinsically derived from primary field operations. For example, drip gas collected from lines used to transport natural gas from the production site to an offsite distribution center or condensate generated in a pipeline transporting natural gas from the gas plant to market would be subject to hazardous waste regulations if it exhibits a characteristic and will be discarded. Any condensate generated by transportation or manufacturing operations beyond the production process is not excluded. [RO 13253, 13617]

If the drip gas is not excluded under §261.4(b)(5), it would be considered a by-product under RCRA. [See §261.1(c)(3).] If it exhibits a characteristic (such drip gas will often be ignitable) and will be disposed, it would be subject to all hazardous waste generation, transportation, and disposal regulations. If the drip gas exhibits a characteristic and will be recycled, its regulatory status would be determined by Table 1 in §261.2(c). For example, pouring nonexcluded drip gas down a well as a solvent to remove paraffin buildups would be considered use constituting disposal. From Table 1, such a characteristic by-product must be managed as a solid waste and, because the solid waste exhibits a characteristic, a hazardous waste. [RO 11767]

Due to its Btu content, natural gas pipeline condensate is sometimes burned for energy recovery. Again looking at Table 1, such a characteristic by-product would normally be a solid and hazardous waste when managed in that manner. However, EPA considers the condensate to be an off-specification fuel—not a by-product—when burned for energy recovery. Per §261.2(c)(2)(ii), off-spec products are not solid or hazardous wastes when burned for energy recovery *if they are themselves fuels*. [RO 11831, 13049]

4.14.4.5.2 *Gas-plant cooling tower wastes*

As noted in Tables 4-3 and 4-4, EPA has determined that cooling tower blowdown is excluded under §261.4(b)(5), but gas-plant cooling tower cleaning wastes are not. Blowdown is made up only of water, scale, or other wastes generated by the actual operation of a cooling tower. On the other hand, cleaning wastes contain any solvents, scrubbing agents, or other cleaning materials introduced into the process to remove buildup or otherwise clean the equipment and are not included as part of the functional operation of the cooling tower. "Since these cleaning wastes can come from any cooling tower, they are not intrinsically derived from primary field operations for natural gas production. The determining factor for defining the exclusion is not the frequency with which the cooling tower is blown down, either with or without cleaning agents, but whether the resulting waste is solely derived from the normal operation of the tower for natural gas production or from any added cleaning materials." [RO 13293]

4.14.4.5.3 *Natural gas regulators*

Natural gas regulators are used to regulate the flow and pressure of natural gas supplied to customers. The regulators are installed, repaired, and replaced by the gas supplier or utility. As such, they do not qualify for the RCRA household hazardous waste exclusion in §261.4(b)(1) because they are 1) not generated by individuals on the premises of a household, and 2) not composed primarily of materials found in the waste generated by consumers in their homes. Consequently, natural gas regulators that fail the toxicity characteristic for lead must be managed as hazardous waste when discarded. [RO 14115]

4.14.4.5.4 *Underground natural gas storage fields*

Wastes uniquely associated with operations to retrieve natural gas from underground gas storage fields qualify for the oil and gas exclusion because these operations are almost identical to those involved with the production of natural gas for the first time. [March 22, 1993; 58 *FR* 15287] For example, an underground cavern is used to store natural gas for future use in the winter months. Compressor stations associated with the storage facility generate wastes that exhibit a characteristic. The natural gas must be retrieved from storage in a similar manner to when it was produced originally prior to storage, and the wastes generated in both cases are virtually the same. Accordingly, wastes from subsurface gas storage and retrieval are excluded from hazardous waste regulation, except for wastes that are not intrinsically associated with removal of the gas. [RO 11794, 13217]

4.15 Exclusion for mining and mineral processing wastes

Section 261.4(b)(7) of the RCRA regulations excludes wastes from the extraction, beneficiation, and processing of ores and minerals from the definition of hazardous waste. These wastes are some of the so-called "Bevill wastes," temporarily excluded from hazardous waste regulation under the October 1980 Bevill amendment [Section 3001(b)(3)(A)] to the RCRA statute (named after the representative who introduced the amendment in Congress). The three classes of excluded Bevill wastes are:

1. Fly ash, bottom ash, slag, and flue gas emission control waste from the combustion of fossil fuels [§261.4(b)(4)];

2. Wastes from the extraction, beneficiation, and processing of ores and minerals [§261.4(b)(7)]; and

3. Cement kiln dust [§261.4(b)(8)].

The first of the above Bevill wastes is discussed in Section 4.13, while the second is covered in this Section 4.15.

The exclusion for wastes generated from the extraction, beneficiation, and processing of ores and minerals in §261.4(b)(7) was first promulgated (on a temporary basis) on November 19, 1980. [45 *FR* 76618] In the preamble to that rule, the agency noted that the exclusion applied to wastes from "the exploration, mining, milling, smelting and refining of ores and minerals." [45 *FR* 76619]

Congress required EPA to evaluate whether this temporary Bevill exclusion, included in the 1980 regs, should continue. The agency's subsequent evaluations were separated into two efforts: one for extraction and beneficiation wastes, and the second for mineral processing wastes. These two assessments are discussed in the following two subsections.

4.15.1 Extraction and beneficiation wastes

After extensively evaluating wastes from extraction and beneficiation operations in the early 1980s, the agency issued its Report to Congress in December 1985, followed by a regulatory determination on July 3, 1986. [51 *FR* 24496] In the July 1986 determination, EPA concluded that hazardous waste regulation is not warranted for extraction and beneficiation wastes, and that exclusion continues today.

What exactly are extraction and beneficiation wastes? The first step in determining the applicability of the Bevill exclusion is determining that the wastes are generated by primary mineral production operations, defined as "those using at least 50 percent ores, minerals, or beneficiated ores or minerals as the feedstocks providing mineral value." [*Identification and Description of Mineral Processing Sectors and Waste Streams*, EPA/530/R-99/022, April 1998, available from http://nepis.epa.gov/EPA/html/Pubs/pubtitleOSWER.html by downloading the report numbered 530R99022] The Bevill exclusion does not apply to wastes from the *secondary* production of mineral commodities, such as wastes from scrap metal recycling or baghouse dust metal recovery. [RO 11395, 12664, 14436]

In the 1985 Report to Congress, EPA identified four large-volume extraction and beneficiation wastes: 1) mine waste, 2) mine water, 3) tailings, and 4) dump and heap leach waste. Each of these is discussed briefly in the following subsections.

4.15.1.1 Extraction wastes

The primary extraction waste is "mine waste":

"Mine waste is the soil or rock that mining operations generate during the process of gaining access to an ore or mineral body, and includes the overburden (consolidated or unconsolidated material overlying the mined area) from surface mines, underground mine development rock (rock removed while sinking shafts, accessing, or exploiting the ore body), and other waste rock, including the rock interbedded with the ore or mineral body." [EPA/530/SW-85/033, December 1985, available from http://nepis.epa.gov/EPA/html/Pubs/pubtitleOSWER.html by downloading the report numbered 530SW85033]

These mine wastes are typically stored onsite in large disposal piles, ranging from 5–600 acres in size, near or adjacent to the mine.

Another waste that could be considered an extraction waste is mine water—the water that infiltrates a mine and is subsequently removed to facilitate mining. Mine water is normally pumped out and stored in onsite ponds, including the tailings ponds.

Q: *Drainage from an active coal mine is collected in a large pond to be treated. The drainage is quite acidic due to the high concentration of iron sulfides in the coal. The coal mine owner has developed a sodium hydroxide (caustic) feeder system to neutralize the acid mine drainage prior to discharge. Occasionally, a caustic sludge layer (with a pH generally above 12.5) forms on the bottom of the pond. The facility wants to remove the sludge and dispose of it. Would such removal be considered generation of a RCRA hazardous waste?*

A: No. Pond sludge from the treatment of drainage from an active coal mine is excluded under the §261.4(b)(7) Bevill exclusion for mining and mineral processing wastes, even if it exhibits a characteristic. [RO 12356]

Figure 4-11 illustrates the scope of the Bevill exclusion for extraction wastes at a copper mining/processing facility.

4.15.1.2 Beneficiation wastes

EPA defined "beneficiation" in the 1985 Report to Congress as "the treatment of ore to concentrate its valuable constituents" but went on to describe it more fully as follows:

"After the ore is mined, the first step in beneficiation is generally grinding and crushing. The crushed ores are then concentrated to free the valuable mineral and metal particles (termed values) from the matrix of less valuable rock (called gangue). Beneficiation processes include physical/chemical separation techniques such as gravity concentration, magnetic separation, electrostatic separation, flotation, ion exchange, solvent extraction, electrowinning, precipitation, and amalgamation.... All [beneficiation] processes generate tailings, another type of waste." [EPA/530/SW-85/033]

Most tailings (the wastes remaining after physical/chemical beneficiation operations) consist of slurries containing 30%–50% solids, and they are usually disposed in onsite tailings ponds. These ponds, averaging 500 acres in size, are the primary method of treating wastewater in the metal ore (e.g., copper) mining segment.

Additional beneficiation operations noted in the 1985 Report to Congress include froth flotation and dump, heap, and in situ leaching. Dump and heap leach wastes are the low-grade ore from which metal values have been leached using a suitable leaching solution (e.g., cyanide for gold, water for uranium, sulfuric acid for copper). Thus, EPA noted in *Federal Register* preamble guidance that even if a heap leach pile becomes a disposal unit because wastes remain there permanently, those wastes are beneficiation wastes and are Bevill-excluded. [May 26, 1998; 63 *FR* 28589]

Beneficiation operations typically serve to separate the mineral values from waste material, concentrate the minerals, remove impurities, or prepare the ore/minerals for further refinement. Beneficiation activities generally do not change the mineral values themselves other than by reducing (e.g., crushing or grinding) or enlarging (e.g., pelletizing or briquetting) particle size to facilitate processing. A chemical change in the mineral value does not typically occur in beneficiation. [EPA/530/R-99/022]

Most beneficiation operations generate wastes that are earthen in nature, are often physically/chemically similar to the ore or mineral that entered the operation, and are generated in high volume. Once beneficiation is complete, based on EPA's regulatory definition supplemented by agency guidance, all operations that follow are considered mineral processing. [54 *FR* 36619]

CHAPTER 4 *Exclusions and Exemptions* 199

Figure 4-11a: Extraction, Beneficiation, and Mineral Processing Wastes From Copper Production

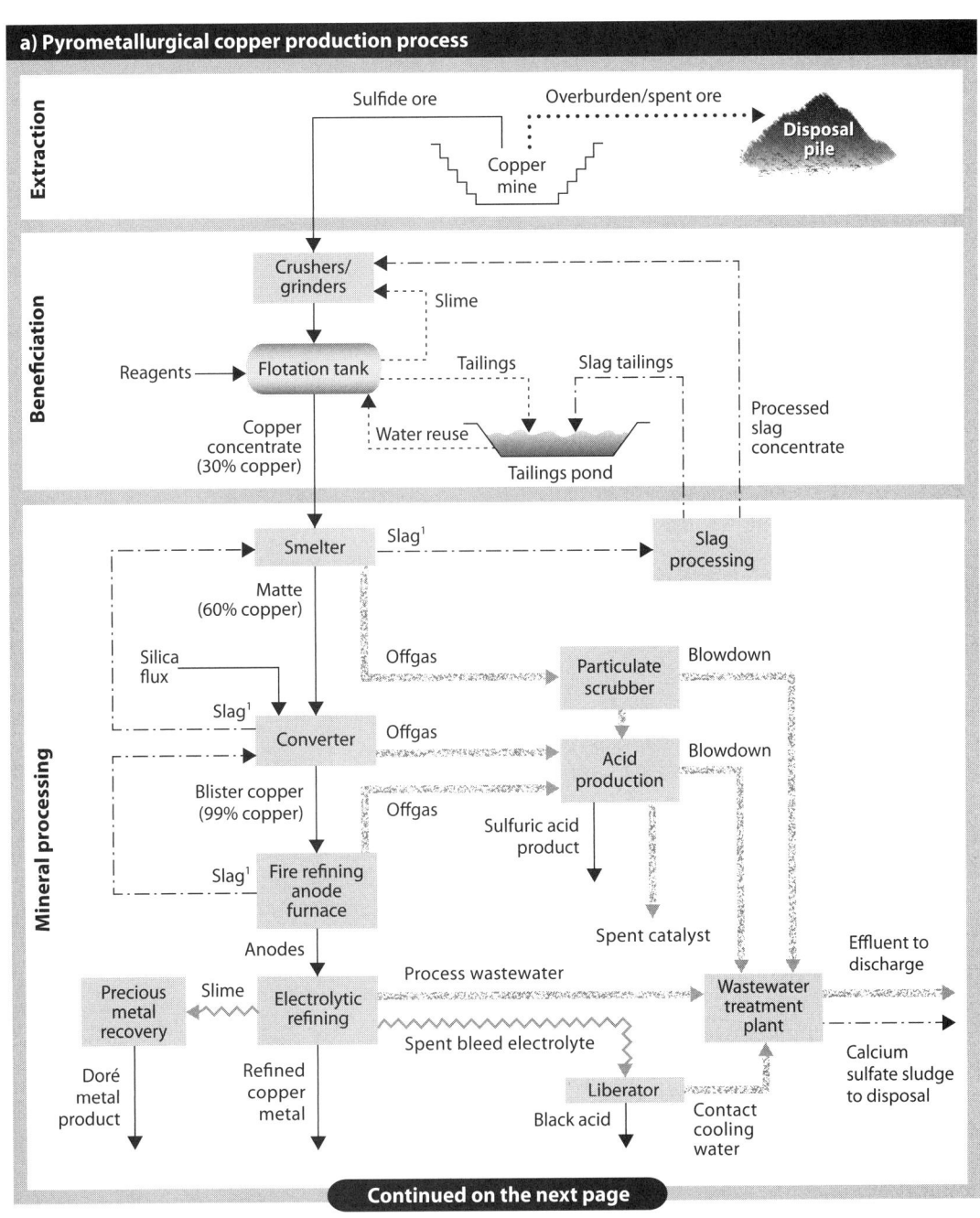

§4.15.1.2

©2015 McCoy and Associates, Inc. **McCoy's RCRA Unraveled**

Figure 4-11b: Extraction, Beneficiation, and Mineral Processing Wastes From Copper Production

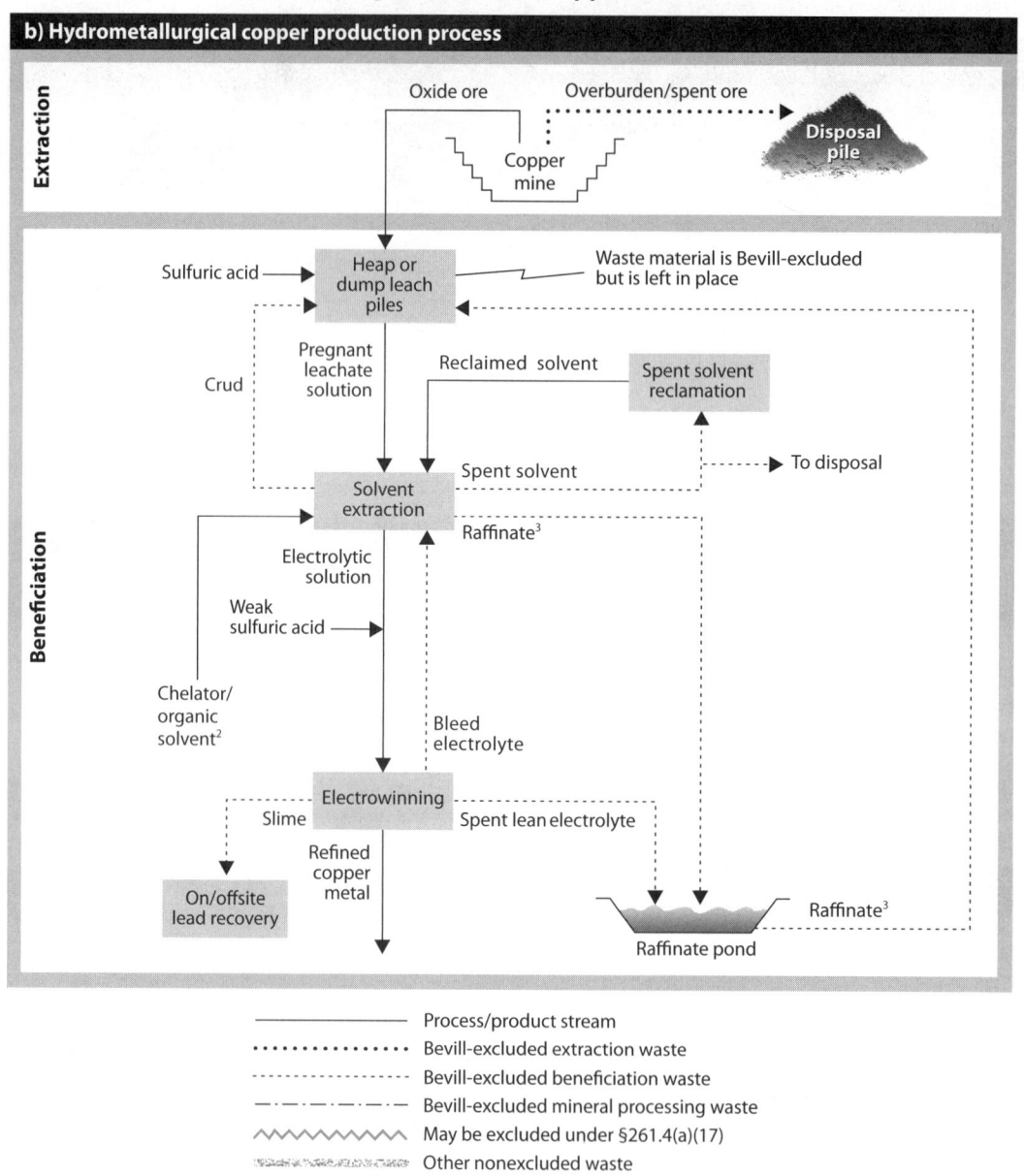

[1] EPA has elected to consider smelter slag, converter slag, and anode furnace slag as one waste stream, called "slag from primary copper processing." [54 FR 15334]
[2] Kerosene is often used as the organic solvent.
[3] Raffinate, also called barren leachate, is the acidic aqueous leachate solution remaining after copper is extracted from the pregnant leachate solution by the organic solvent. It is usually recycled to the leach piles because of its residual copper content and acidity.

Source: Adapted from EPA/530/SW-90/070C, EPA/530/R-99/022.

In the first of the mineral processing final rules [September 1, 1989; 54 FR 36592], EPA needed to distinguish between beneficiation and mineral processing. That distinction is important because all beneficiation wastes are excluded from the definition of hazardous waste, while only certain specific mineral processing wastes are excluded (as discussed in Section 4.15.2 below). Thus, the agency provided additional guidance on what exactly constitutes beneficiation. In that rule, EPA started with the beneficiation processes identified in the 1985 Report to Congress and added the following:

- Milling techniques, including crushing, grinding, washing, filtration, sorting, and sizing;
- Agglomeration techniques, including sintering, pelletizing, and briquetting;
- Tank and vat leaching;
- Heating steps, including calcining to remove water and/or carbon dioxide, fractional crystallization, drying, and roasting, autoclaving, and/or chlorination in preparation for leaching (except where the roasting, autoclaving, and/or chlorination/leaching sequence produces a final or intermediate product that does not undergo further beneficiation or processing); and
- Acid processes, including washing and dissolution (often accompanied by heat).

The resulting regulatory definition of beneficiation in §261.4(b)(7) that came out of that September 1989 rule is essentially the same one in the regs today. The line between beneficiation and mineral processing for 49 specific primary mineral production operations is discussed in some detail in EPA/530/R-99/022. Figure 4-11 also illustrates the scope of the Bevill exclusion for beneficiation wastes at a copper mining/processing facility.

Q *One of the preparatory steps for making cement involves the crushing of limestone. Would the washwater from rinsing off the limestone crushing equipment meet the Bevill exclusion?*

A The washwater is most likely excluded under §261.4(b)(7), because limestone is considered a mineral and its crushing is considered beneficiation. [RO 13188]

Q *Rainwater contacts an inactive gold heap leach pile, resulting in liquid draindown and seepage from the pile that could be toxic. Are the draindown and seepage regulated under RCRA Subtitle C?*

A No. Leach piles could fill with rainwater or ground water and in turn generate liquid wastes that could be toxic. However, EPA has concluded that such liquids are also Bevill excluded since their source was Bevill-excluded wastes. [RO 14499] EPA notes that these liquid wastes could be subject to CWA or SDWA regulations, the imminent hazard provisions in RCRA Section 7003, CERCLA, or state environmental regulations.

4.15.1.3 "Uniquely associated" wastes

As with the exclusion for fossil fuel combustion wastes discussed in Section 4.13 and the exclusion for oil and gas exploration and production wastes discussed in Section 4.14, EPA has provided guidance on which wastes are "uniquely associated" with extraction and beneficiation of ores and minerals:

- "This exclusion does not, however, apply to solid wastes, such as spent solvents, pesticide wastes, and discarded commercial chemical products, that are not uniquely associated with these mining and allied processing operations…. Therefore, should either industry generate any of these nonindigenous wastes and the waste is identified or listed as hazardous under Part 261 of the regulations, the waste is hazardous and must be managed in conformance with the Subtitle C regulations." [45 FR 76619]
- "Congress intended to put within the regulatory exclusion only wastes generated as a consequence of exploiting a natural resource, not wastes from other industrial activities, even if both occur at the same facility…." [54 FR 36616]
- "[T]he agency finds no compelling reason to provide exemptions for particular small volume wastes that may be associated with mineral processing operations, such as cleaning wastes. Many

other industrial operations also generate such wastes, and EPA does not believe that the fact that current management involving co-management justifies continued regulatory exclusion for wastes that are not uniquely associated with mineral processing (and therefore are not defined as mineral processing wastes) and would not, in any event meet the high volume criterion." [54 FR 36623]

- "The key consideration for establishing that a waste is uniquely associated is determining whether or not the waste originates primarily from, or, at the least, is significantly influenced by contact with ores, minerals, or beneficiated ores and minerals. Wastes that are essentially the same as analogous wastes generated by other industries or activities are not uniquely associated, and hence are not eligible for the mining waste exclusion. Even wastes that may come into contact with parts of the mineral feed stream, e.g., cleaning wastes, are not uniquely associated, because their fundamental character does not arise from such contact." [RO 13668]

- Any waste from ancillary operations (e.g., vehicle or machinery maintenance, cleaning, laboratory operations, painting, pesticide application, and plant trash incineration) is not uniquely associated [63 FR 28591], even if it is generated at a primary mineral production site. [EPA/530/R-99/022]

- In evaluating wastes from nonancillary operations, the extent to which the waste originates or derives from processes that serve to remove mineral values from the ground, concentrate or otherwise enhance their characteristics to remove impurities, and the extent to which the mineral recovery process imparts its chemical characteristics to the waste must be considered. The greater the extent to which the waste results from the mineral recovery process itself, and the more the process imparts to the waste its chemical characteristics, the more likely the waste is "uniquely associated." [63 FR 28591–2]

Based on these criteria, Table 4-5 gives examples from EPA guidance of extraction and beneficiation wastes that are uniquely associated with mining operations and therefore might qualify for the Bevill exclusion vs. those that aren't uniquely associated.

Q *The owner/operator of a taconite ore mining and processing facility uses a grinding and magnetic separation process to increase the concentration of the taconite ore. This beneficiation process uses Whitmore grease for mechanical lubrication. The grease, which exhibits the toxicity characteristic, is removed every 10 years and sent offsite for disposal. Does the spent Whitmore grease qualify for the §261.4(b)(7) exclusion?*

A No. According to EPA, "Whitmore grease is not limited to the mining industry but can be used on any industrial equipment where short-term grease applications are limited by difficult access and heavy use. Therefore, because Whitmore grease is not unique to mining operations, it is not excluded pursuant to §261.4(b)(7)." [RO 13247]

Table 4-5: Wastes That Are "Uniquely Associated" With Mining and Mineral Processing

"Uniquely associated" wastes[1]	Not "uniquely associated" wastes[2]
- Mining overburden, spent ore, and gangue; fugitive dust generated during screening and crushing - Mine water - Waste tailings, including waste slurries from milling, gravity concentration, flotation, and rock washing - Raffinate[3] from beryllium and copper production - Waste brine from borate, bromine, iodine, lithium, and soda ash production	- Wastes from vehicle and machinery maintenance operations and shops (e.g., used oil, used pinion gear grease, spent antifreeze, and shop floor drains) [RO 13247, 13615, 13661] - Cooling tower blowdown [RO 13661] - Noncontact cooling water - Pesticide wastes - Paint wastes

Table 4-5: Wastes That Are "Uniquely Associated" With Mining and Mineral Processing

"Uniquely associated" wastes[1]	Not "uniquely associated" wastes[2]
■ Process wastewater from washing in borax and iron production; from conveyance, flotation, mixing, and dissolution in copper production; from wet scrubbers in lithium, soda ash, and tungsten production; from sulfur deposit mining; and from thickeners and separators in tungsten production ■ Slimes from the steaming process in bromine production ■ Spent solvents used in boron, copper, and rubidium solvent extraction ■ Wastes from froth flotation in cesium production and spent flotation reagents in fluorspar, lead, manganese, and mercury production ■ Slime from copper ore flotation ■ Slime remaining after electrowinning in copper production ■ Crud generated during solvent extraction in copper production ■ Spent leaching solutions from gold heap leaching ■ Stripped carbon, carbon fines, and carbon water used to recover and concentrate values in gold production [RO 14489] ■ Solids from calciner offgas scrubbers in phosphorus production ■ Fines from roaster offgas scrubbers in lithium production ■ Sulfur compounds generated when hydrogen sulfide is removed from brine in iodine production ■ Sand and clay from phosphate beneficiation operations [RO 13540] ■ Spent ion-exchange solution and resins from rubidium and uranium production ■ Spent carbon and filter wastes from soda ash production ■ Filter muds from strontium production ■ Spent extraction/leaching solutions from uranium production ■ Filter cake from sulfur filtration	■ Cleaning wastes (e.g., spent solvents used for equipment maintenance and cleaning production vessels, acidic tank cleaning wastes) [RO 12314, 13661] ■ Laboratory wastes, including used crucibles and cupels, spent or contaminated reagents, used chemicals and liquid samples, and mineral sample residues from analytical testing [RO 13661, 14332, 14417, 14505] ■ Discarded commercial chemical products, including off-specification mineral-derived products and spills or leaks of unused beneficiation or processing reagents (e.g., sodium cyanide, and residues from cleaning up spills of unused solvents that were to be used in solvent extraction [beneficiation] operations) ■ Certain wastewater treatment sludges[4] ■ Spent batteries ■ Spent catalysts (e.g., vanadium pentoxide) used to produce sulfuric acid from sulfur-rich smelter emissions ■ Plant trash ■ Refractory bricks used at lime kilns [RO 13668] ■ Surface runoff [RO 13661] ■ Washdown water from facility cleaning operations [RO 13661] ■ Demineralized water plant discharge [RO 13661]

[1] In determining "uniquely associated," EPA requires consideration of "the extent to which the waste originates or derives from processes that serve to remove mineral values from the ground, concentrate or otherwise enhance their characteristics or remove impurities, and the extent to which the mineral recovery process imparts its chemical characteristics to the waste." [63 FR 28591]

[2] Any wastes from ancillary operations are not "uniquely associated" because they are not viewed as being "from" mining or mineral processing [63 FR 28591], even if they are generated at a primary mineral production site. [EPA/530/R-99/022]

[3] Raffinate, also called barren leachate, is the acidic, aqueous leachate solution remaining after beryllium or copper is extracted from the pregnant leachate solution by the organic solvent.

[4] Wastewater treatment sludges generated from mineral extraction and beneficiation operations plus any wastewater treatment sludges identified in §261.4(b)(7)(ii) are uniquely associated.

Source: McCoy and Associates, Inc.; adapted from 45 FR 76619, 63 FR 28591–3, EPA/530/R-99/022, unless otherwise noted.

Bevill-excluded extraction and beneficiation wastes sometimes are treated (e.g., acid mine drainage may be treated, generating a sludge). Residues generated from treatment of Bevill-excluded extraction and beneficiation wastes retain the Bevill exclusion. [EPA/530/R-99/022, RO 12356]

Figure 4-12 is a logic diagram that incorporates the above discussions, allowing the user to quickly determine the regulatory status of extraction, beneficiation, and mineral processing wastes.

4.15.2 Mineral processing wastes

Mineral processing involves changing the physical and/or chemical structure of ores and minerals. Heating operations, such as smelting (i.e., any metallurgical operation in which metal is separated by fusion from impurities) and fire refining (e.g., retorting) are mineral processing and not beneficiation. In these operations, the physical structure of the ore or mineral is destroyed, and neither the product nor the waste bear any close physical/chemical resemblance to the ore or mineral entering the operation. Similarly, acid or alkaline digestion are mineral processing that produce wastes that are not earthen, bear little resemblance to the feed materials, and are produced in relatively low volume. Any operations following the initial mineral processing step are considered processing, even though they might otherwise be included on the list of beneficiation activities in §261.4(b)(7)(i). [54 *FR* 36618–9]

Certain mineral processing wastes are also Bevill wastes that are specifically excluded from hazardous waste regulation in §261.4(b)(7). Although all extraction and beneficiation wastes are excluded from the definition of hazardous waste, only 20 specific mineral processing wastes enjoy this exclusion; these 20 wastes are identified in §261.4(b)(7)(ii)(A)–(T) and are listed in Figure 4-12. Thus, all other solid wastes from the processing of ores and minerals (that don't meet the definition of extraction or beneficiation wastes) that exhibit a hazardous waste characteristic or are listed in Part 261, Subpart D are subject to the Subtitle C hazardous waste regulations.

Why did EPA designate only 20 specific mineral processing wastes as Bevill-excluded? The full story is long and tortuous, but the abridged version is as follows. Based on Congressional intent, as interpreted by the judicial system, EPA determined that only those wastes from processing ores and minerals that meet "high volume, low hazard" criteria could remain excluded under the Bevill Amendment. In two final rules [September 1, 1989; 54 *FR* 36592 and January 23, 1990; 55 *FR* 2322], a July 1990 Report to Congress [EPA/530/SW-90/070, available from http://nepis.epa.gov/EPA/html/Pubs/pubtitleOSWER.html by downloading the report numbered 530SW90070], and a June 13, 1991 regulatory determination [56 *FR* 27300], EPA 1) defined the "high volume, low hazard" criteria, 2) determined that only these 20 specific mineral processing wastes meet those criteria, and 3) determined that hazardous waste regulation is inappropriate for these wastes.

Three of the Bevill-excluded mineral processing wastes are generated at copper mining/processing facilities, as shown in Figure 4-11.

Note that some of the mineral processing wastes identified in §261.4(b)(7)(ii)(A)–(T) that are excluded from hazardous waste management represent fairly broad categories. For example, "Process wastewater from primary magnesium processing by the anhydrous process" and "Process wastewater from phosphoric acid production" would seem to indicate that all wastewaters generated at those facilities are excluded mineral processing wastes—this is not necessarily true. In EPA's July 1990 Report to Congress, the agency expressly noted which waste streams are included in these descriptions and thus would qualify as Bevill-excluded. For example, only two wastewater streams generated at a magnesium processing facility that was evaluated by EPA (scrubber underflow process wastewater and scrubber liquor process wastewater) are included in the mineral processing regulatory exclusion. All other wastewater streams generated at that facility, except those that come from beneficiation processes, would not

Figure 4-12a: Regulatory Status of Mining and Mineral Processing Wastes

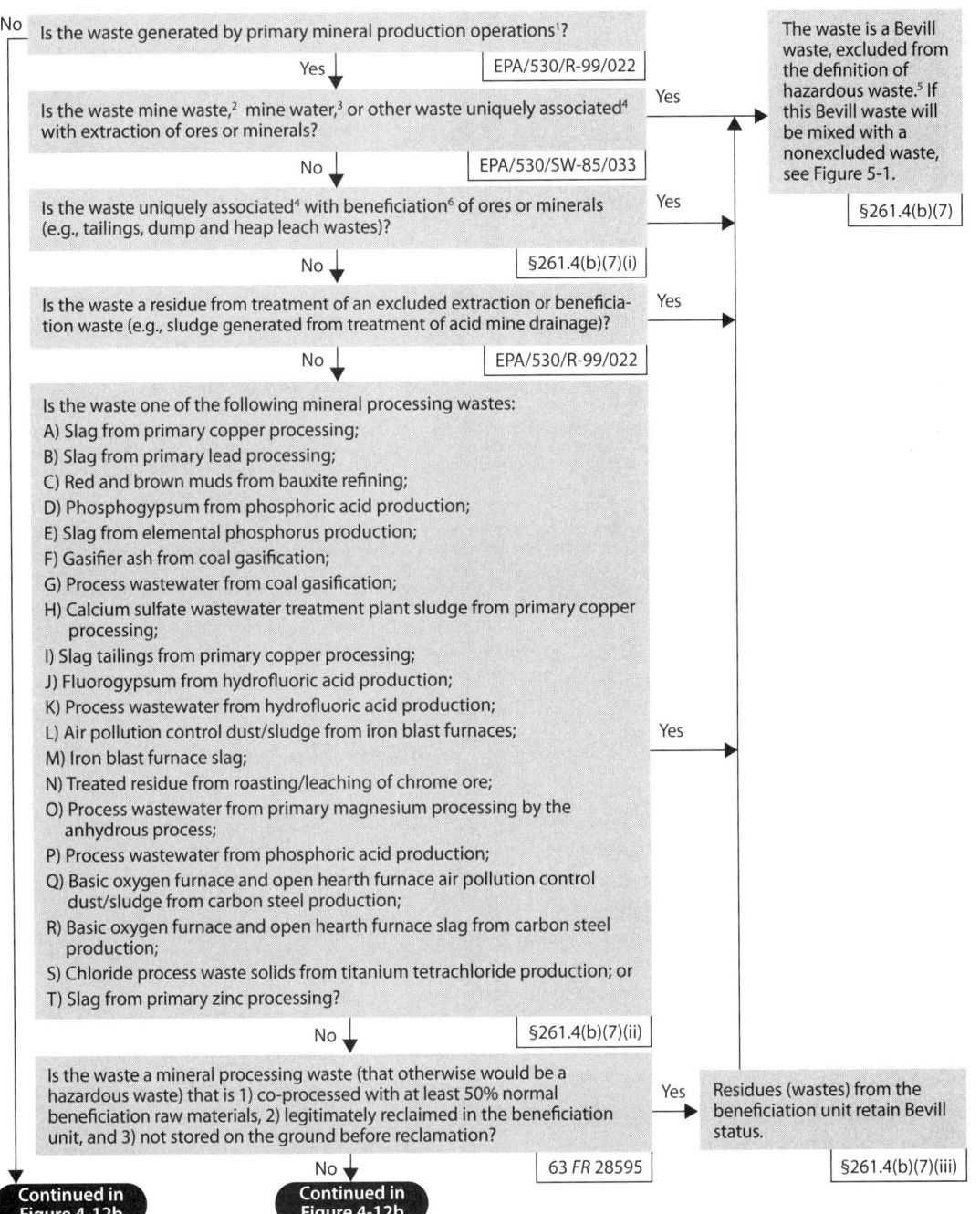

CHAPTER 4 Exclusions and Exemptions

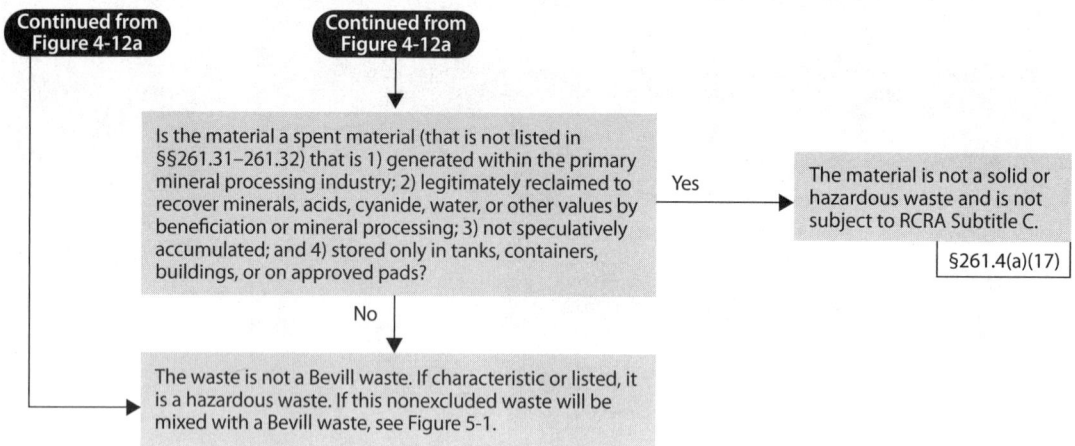

Figure 4-12b: Regulatory Status of Mining and Mineral Processing Wastes

[1] Primary mineral production operations are defined as "those using at least 50 percent ores, minerals, or beneficiated ores or minerals as the feedstocks providing mineral value." [EPA/530/R-99/022]

[2] "Mine waste" is "the soil or rock that mining operations generate during the process of gaining access to an ore or mineral body, and includes the overburden (consolidated or unconsolidated material overlying the mined area) from surface mines, underground mine development rock (rock removed while sinking shafts, accessing, or exploiting the ore body), and other waste rock, including the rock interbedded with the ore or mineral body." [EPA/530/SW-85/033]

[3] "Mine water" is "the water that infiltrates a mine and is subsequently removed to facilitate mining." [EPA/530/SW-85/033]

[4] In determining "uniquely associated," EPA requires consideration of "the extent to which the waste originates or derives from processes that serve to remove mineral values from the ground, concentrate or otherwise enhance their characteristics or remove impurities, and the extent to which the mineral recovery process imparts its chemical characteristics to the waste." [63 FR 28591]

[5] If disposed offsite, the company/facility needs to work closely with the nonhazardous waste (Subtitle D) landfill to ensure the waste (although excluded from the hazardous waste program) meets the landfill's acceptance criteria/permit conditions and will not cause any other problems that might result in future CERCLA or tort liability.

[6] The specific beneficiation activities that result in Bevill-excluded wastes are listed in §261.4(b)(7)(i).

Source: McCoy and Associates, Inc.; adapted from §§261.4(a)(17), 261.4(b)(7), 63 FR 28595, EPA/530/SW-85/033, EPA/530/R-99/022.

be excluded and, therefore, would be hazardous waste if they exhibit a characteristic or meet a listing description. If the facility mixes excluded and hazardous nonexcluded wastewater streams together before management in a surface impoundment, the Bevill mixture rule (see Section 5.1.2.4 and Figure 5-1) must be used to determine the regulatory status of the resulting combined wastewater influent to the impoundment. [RO 13661, 14505]

Q Is melting a beneficiation or mineral processing activity?

A Melting is a mineral processing operation. [RO 13362]

Q The tile production process uses various minerals as raw materials, such as talc, pyrophyllite, and ball clay. These materials are then subject to alloying (blending), fabrication (pressing), coating, and firing stages. Are the wastes generated from these tile production processes Bevill-excluded?

A No. The §261.4(b)(7) Bevill exclusion applies only to extraction, beneficiation, and primary processing of ores or minerals—not to subsequent shaping, alloying, or fabrication of materials derived from ores and minerals. [RO 12777]

Q Wastes generated from gold/mercury amalgam retorting were disposed on the ground in the 1950s and 1960s. If those wastes are dug up today, do they qualify for the Bevill exclusion?

A No. Retorting was specifically identified as a mineral processing operation in the September 1, 1989 mineral processing final rule. [54 FR

McCoy's RCRA Unraveled ©2015 McCoy and Associates, Inc.

36618] Because wastes generated from gold/mercury amalgam retorting are not one of the 20 specific mineral processing wastes excluded in §261.4(b)(7)(ii), these wastes would require hazardous waste management if they exhibit a characteristic or meet a listing description. However, wastes disposed prior to RCRA regulation are not subject to the hazardous waste program unless they are "actively managed." Active management includes physical disturbance of the wastes (e.g., excavation for disposal)—see Section 19.1.2.2. [RO 13602]

Q *Does process wastewater to be generated from a planned integrated coal gasification combined cycle project fall within the mineral processing waste exclusion of §261.4(b)(7)(ii)(G): "Process wastewater from coal gasification"?*

A No. All Bevill exclusions are limited to those wastes that were studied as part of the Bevill rulemaking processes in the late 1980s and early 1990s. Based on design data, some of the proposed process wastewaters that will be generated at the new facility may exhibit the toxicity characteristic for arsenic and selenium; the wastewaters from coal gasification EPA studied during the Bevill rulemakings did not exhibit those characteristics. Because it appears that process wastewaters that will be generated from the proposed coal gasification facility will differ significantly from those that were evaluated in the July 1990 Report to Congress (cited above), the wastewaters to be newly generated are not eligible for the Bevill exclusion. [RO 14809]

Q *A magnesium processing facility utilizes the §261.4(b)(7)(ii)(O) exclusion for process wastewater from primary magnesium processing by the anhydrous process. The facility claims all wastewater streams coming from the site are associated with primary magnesium processing and thus should enjoy the mineral processing exclusion. Is this appropriate?*

A No. Only wastes meeting the specific listing description in §261.4(b)(7)(ii)(O) enjoy the mineral processing exclusion. Wastes not specifically from that operation (e.g., wastes from hydrochloric acid manufacturing operations taking place at the site) and wastes that do not qualify as "wastewater" remain subject to Subtitle C management requirements. [*United States v. Magnesium Corporation of America et. al.,* U.S. Court of Appeals for the Tenth Circuit; Docket No. 08-4185, August 17, 2010]

4.15.2.1 Reclaiming mineral processing wastes

The regulations at §261.4(b)(7)(iii) allow secondary materials from mineral processing to be co-processed with normal raw materials in beneficiation operations that generate Bevill-excluded wastes, without changing the excluded status of the resulting waste. To claim this allowance, legitimate recovery of the mineral processing secondary material must be occurring, and normal beneficiation raw materials (primary ores and minerals) must account for at least 50% of the feedstock. Although not included in the language of the regulations, EPA noted in preamble guidance that this mineral processing secondary material cannot be stored on the ground (e.g., piles or impoundments) before it is co-processed. [63 *FR* 28595]

Finally, as noted in Section 1.2.6.3.2, an exclusion from the definition of solid waste exists in §261.4(a)(17) for spent materials from mineral processing operations that are reclaimed by beneficiation or mineral processing. This is not part of the Bevill exclusion—which only excludes solid wastes from the definition of hazardous waste—but is a separate exclusion. To qualify, the spent materials must be 1) legitimately reclaimed to recover minerals, acids, cyanide, water, or other values; 2) recycled without speculative accumulation; and 3) stored only in tanks, containers, buildings, or on approved pads.

4.15.3 Effect of co-processing hazardous wastes in furnaces

Beneficiation and especially processing of ores and minerals sometimes occur in industrial furnaces.

The language in §261.4(b)(7) notes that the Bevill exclusion is lost for the resulting beneficiation and mineral processing wastes if hazardous waste is also burned or processed in the industrial furnace, unless the criteria in §266.112 are met. Specifically:

1. At least 50% by weight of the input to the ore or mineral furnace must be from normal, nonhazardous raw materials; and

2. The beneficiation and mineral processing wastes from the furnace must not be significantly affected by the hazardous waste. This must be demonstrated by passing one of the two tests given in §266.112(b).

If these two requirements are satisfied, residues generated from the ore or mineral furnace while hazardous waste is burned or processed continue to be excluded from the Subtitle C program and do not have to meet LDR treatment standards. [RO 11881, 13462] The LDR paperwork requirements in this situation are discussed in Section 13.12.3.1.

4.16 De minimis wastewater exemptions

Under §261.3(a)(2)(iv)(A–G), mixtures of very small quantities of certain listed hazardous wastes and wastewater managed in a Clean Water Act (CWA) wastewater treatment system are exempt from regulation as hazardous wastes. These so-called "de minimis" wastewater exemptions prevent the mixture rule from applying to large-volume, nonhazardous wastewaters that 1) get mixed with small quantities of listed hazardous wastes that are not principal waste streams, and 2) are managed in a facility's onsite wastewater treatment system. Accordingly, if wastewater mixtures meet EPA's criteria (described below), they are not hazardous wastes and they will not produce wastewater treatment sludges that are listed hazardous waste. This is the primary reason the de minimis wastewater exemptions were promulgated: to protect wastewater treatment sludge from carrying listed waste codes.

EPA believes that these mixtures of wastewater and small amounts of listed wastes do not pose a substantial threat to human health and the environment because the relatively small amounts of listed wastes in these mixtures are at very low (often parts-per-billion) concentrations. Furthermore, the listed wastes are treated in the plant's onsite wastewater treatment system, further reducing their hazardousness. [November 17, 1981; 46 *FR* 56583, October 4, 2005; 70 *FR* 57771]

The §261.3(a)(2)(iv) de minimis wastewater exemptions are sometimes referred to as the "headworks exemptions" because they exempt eligible wastewater mixtures from the mixture rule *when they reach the headworks of the facility's wastewater treatment system.* [RO 11116, 11614] For instance, if a facility's wastewater treatment system leaks waste that meets a listing description before it reaches the headworks, the leaked material is classified as listed hazardous waste. [RO 14095] So where is the headworks of a specific wastewater treatment system? EPA says that it is where "final combination of raw or pretreated process wastewater streams typically takes place." [October 4, 2005; 70 *FR* 57775] "[H]eadworks can include a central catch basin for industrial wastewaters, a pump station outfall, equalization tank, or some other main wastewater collection area that exists in which transport of process wastewaters stops and chemical or biological treatment begins." [April 8, 2003; 68 *FR* 17242]

The de minimis wastewater exemptions cover six different types of listed wastes, but only if they are mixed with wastewater that is subject to regulation under Section 307(b) or 402 of the CWA. (Section 307(b) deals with discharges to POTWs; Section 402 deals with NPDES-permitted discharges.) Facilities that have eliminated their wastewater discharge as a result of NPDES or pretreatment program requirements (i.e., zero dischargers) may also take advantage of these exemptions. The agency has noted that facilities that have eliminated the discharge of wastewaters using permitted Class I injection wells can claim the de minimis exemptions: "EPA continues to believe that underground injection wells can meet the headworks' definition of zero discharge if the injection well is being used for the purposes of complying with a NPDES permit, other applicable effluent guideline, or pretreatment program requirements." [October 4, 2005; 70 FR 57777]

The six specific exemptions are discussed below.

4.16.1 Listed spent solvents

Two exemptions are provided for spent solvents that are discharged to a plant's wastewater system:

1. *Listed spent solvents that are carcinogens*—Four F001/F002/F005 spent solvents (benzene, carbon tetrachloride, tetrachloroethylene, and/or trichloroethylene) in wastewater mixtures at a concentration of no more than 1 ppm are not hazardous. [§261.3(a)(2)(iv)(A)]

2. *Listed spent solvents that are toxic*—Fourteen F001–F005 spent solvents (methylene chloride, 1,1,1-trichloroethane, chlorobenzene, *o*-dichlorobenzene, cresols, cresylic acid, nitrobenzene, toluene, methyl ethyl ketone, carbon disulfide, isobutanol, pyridine, chlorofluorocarbon solvents, and/or 2-ethoxyethanol) in wastewater mixtures at a concentration of no more than 25 ppm are not hazardous. [§261.3(a)(2)(iv)(B)]

These two regulatory sections exempt wastewater containing certain spent solvents listed in §261.31 if the appropriate concentration limit is not exceeded. What EPA has in mind here are spills or incidental losses from degreasing, maintenance, or manufacturing operations in which small amounts (not principal waste streams) of spent solvents are washed or otherwise released into a sump or drain and are subsequently managed in the onsite wastewater treatment system. [RO 11845] The agency believes that these small releases of organic solvents can be reasonably and efficiently managed by the chemical or biological wastewater treatment system associated with most affected facilities.

The exemption also applies to scrubber water generated during combustion of these spent solvents if it is discharged to the combustion facility's wastewater treatment system. Extending the de minimis exemption to such scrubber waters is a recent departure (see October 4, 2005; 70 FR 57777) from the agency's previous position (see RO 11116, 11845). EPA justified adding scrubber waters derived from such combustion, noting that "these scrubber waters would be comparable in expected constituents and concentration levels with the already exempted F-listed solvents." [70 FR 57777]

However, the de minimis exemption does not apply to spent solvent-contaminated sludges that are mixed or otherwise commingled with wastewater or to such sludges that generate a wastewater through dewatering. [RO 12283]

Two options are allowed to demonstrate compliance with the 1- or 25-ppm limits:

1. The concentration limit may be calculated from the maximum total weekly usage of these solvents (other than the amounts that can be demonstrated not to have been discharged to wastewater) divided by the average weekly flow of wastewater into the headworks of the facility's wastewater treatment system. Facilities must be able to prove by mass flow calculations that the applicable limits are not exceeded. The solvent-usage numbers can be determined via an audit of records already maintained at most facilities (e.g., invoices of solvent purchases, logs showing the quantities of solvents that were distributed to various locations throughout the plant, spent solvent quantities shipped offsite, and other operating records). New audits or calculations must be made whenever a change in the facility's operations could affect the amount of spent solvents in the wastewater. [November 17, 1981; 46 FR 56585]

2. The concentration limit may be determined by direct measurement of solvent concentrations, on an average weekly basis, at the headworks of the facility's wastewater treatment system. This option is conditioned on 1) the facility being subject to a CAA regulation (or an enforceable limit in a federal operating permit) that minimizes fugitive process or wastewater emissions, 2) the facility identifying the headworks of its wastewater treatment system, and 3) a sampling and analysis plan being developed and submitted to EPA or the state.

The agency noted that "it is not necessary for the receiving wastewater treatment unit itself to be subject to CAA regulations. However, EPA stresses that the process streams and wastewater streams that lead up to the headworks point must be subject to CAA regulations, or an enforceable limit federal operating permit, that minimizes fugitive emissions." [October 4, 2005; 70 *FR* 57774]

Wastewaters containing no more than 1 ppm benzene are exempt only if they are managed in an aerated biological wastewater treatment system that does not contain any unlined surface impoundments before secondary clarification.

Facilities may alternate between the two compliance options noted above or may use both methods and report the result of either method. [70 *FR* 57775]

Case Study 4-8 gives two examples of determining compliance with these spent solvent de minimis exemptions.

4.16.2 Refinery wastes

Mixtures of one or more of the following K-wastes and wastewater are not hazardous *if* the wastes are discharged to the refinery's oil recovery sewer upstream of primary oil/water/solids separation: K050 heat-exchanger bundle cleaning sludge, K169 crude oil storage tank sediment, K170 clarified slurry oil tank sediment, K171 spent hydrotreating

---

Case Study 4-8: De Minimis Wastewater Exemptions for Spent Solvents

I. A facility estimates that it uses a maximum of 700 kg of 1,1,1-trichloroethane and 150 kg of methylene chloride each week for solvent purposes. The weekly flow into the headworks of the plant's wastewater treatment plant is estimated to average 5 million gallons/day (132 million kg/week). Is the wastewater mixture exempt from RCRA regulation under the §261.3(a)(2)(iv)(B) de minimis exemption?

Using this information, the average 1,1,1-trichloroethane level in the wastewater is 5.3 mg/L (which is the same as 5.3 ppm) and the average methylene chloride level is 1.1 ppm, making the combined spent solvent concentration equal to 6.4 ppm. Since this combined concentration is less than the applicable 25-ppm limit for noncarcinogenic spent solvents, the mixture is eligible for the exemption. (This presumes that the wastewater is being discharged to a POTW or under an NPDES permit.)

II. At a certain facility, methylene chloride (which is eligible for the 25-ppm exemption) is used for paint stripping. The spent solvent is discharged to the plant sewer. The facility calculates the amount of solvent discharged by subtracting the amount of methylene chloride that evaporates during use from its weekly purchase records. Is this legitimate?

No. According to EPA, "[t]hat portion of solvents which is volatilized may not be excluded from the calculation of solvent usage." [November 17, 1981; 46 *FR* 56585, RO 11241] "This language was added to prevent facilities from qualifying for the exemption by volatilizing their solvents, and thus causing negative environmental impacts." [RO 11614]

catalyst, and K172 spent hydrorefining catalyst. [§261.3(a)(2)(iv)(C)]

This provision exempts wastewaters from refinery cleaning operations in certain situations. The exemption is not designed to allow the discharge of the entire waste stream (e.g., tank sediments or spent catalysts) into the wastewater collection and treatment system; rather, dilute wastewaters generated during tank or unit cleanout and dewatering operations are covered. [August 6, 1998; 63 *FR* 42120] The listed wastes excluded under this exemption are discussed below:

- *K050 heat-exchanger bundle cleaning sludge*—This hazardous waste is generated during periodic backflushing and/or hydroblasting of heat-exchanger bundles. The sludges from these routine maintenance operations (listed as hazardous due to the presence of hexavalent chromium) are usually discharged to the refinery's sewer system. EPA believes that mixtures of K050 sludges and nonhazardous wastewater do not pose a threat to human health and the environment because the hexavalent chromium from the sludge is almost completely reduced to the trivalent state by reducing agents, such as sulfides, in the raw wastewater. Furthermore, the chromium is present in very low concentrations.

- *K169 crude oil storage tank sediment and K170 clarified slurry oil tank sediment*—Wastewater containing these wastes is typically generated by dewatering (e.g., centrifuging) the sediment removed from tanks and by final rinsing of the tanks themselves. However, these residues are dilute and contain low levels of the listed wastes.

- *K171 spent hydrotreating catalyst and K172 spent hydrorefining catalyst*—Application of the headworks exemption to these catalysts allows refiners to continue to use water to cool and wash out spent catalysts from process units prior to further management. Wastewater recovered from "drilling out" the catalyst, steam stripping or washing, or pad drainage are all covered by the de minimis exemptions.

EPA noted that, although the wastewaters generated from the refinery cleaning operations noted above are excluded from hazardous waste management, any wastes discharged to the wastewater collection system that were not previously counted toward the total annual benzene quantity under the benzene NESHAP rule (40 *CFR* Part 61, Subpart FF) will have to be included in those calculations.

4.16.3 De minimis losses of commercial chemical products and F- and K-wastes

EPA recognizes that small amounts of products that are being produced by, or used as raw materials in, a manufacturing process are often unavoidably lost during normal material handling operations. Many of these materials are P- and U-listed hazardous wastes in §261.33 when discarded. Additionally, small amounts of F- and K-wastes are also inadvertently lost during normal material handling activities. All of these materials are typically disposed by draining or washing them into floor drains leading to the wastewater treatment system, a "reasonable and practical means of disposing of these lost materials." [November 17, 1981; 46 *FR* 56586]

Since these small quantities of listed wastes can be managed and treated in the facility's wastewater treatment system without posing a substantial hazard to human health and the environment, the agency allows mixtures of wastewater (which will be discharged under the CWA) and de minimis losses of §261.31 F-wastes, §261.32 K-wastes, and §261.33 P- and U-chemicals to be exempt from hazardous waste management. [§261.3(a)(2)(iv)(D)] The de minimis exemption for F- and K-wastes was added on October 4, 2005. [70 *FR* 57769] Examples of the de minimis exemption as it applies to commercial chemical products are included in Case Study 4-9.

Sometimes, a chemical is produced at a plant, not as an end product for sale, but as an intermediate in the manufacture of a different chemical. This wastewater exemption applies to de minimis losses of chemicals listed in §261.33(e) and (f), whether

> **Case Study 4-9: De Minimis Wastewater Exemptions for Commercial Products**
>
> I. Ground water at a manufacturing facility is contaminated by a leaking aboveground tank that contains a §261.33 commercial chemical product. When the contaminated ground water is pumped to the facility's wastewater treatment plant for treatment, does it qualify for the §261.3(a)(2)(iv)(D) exemption?
>
> No. According to EPA, "the fact that the flow has continued long enough to result in ground water contamination disqualifies it as a de minimis loss." [RO 12217]
>
> II. During the normal transfer of unused solvent into tanks at a manufacturing facility, the solvent is spilled onto a cement slab. The spill is collected in a sump and then discharged to the onsite wastewater treatment system, which is regulated under an NPDES permit. Does the spilled solvent/wastewater mixture qualify for the §261.3(a)(2)(iv)(D) de minimis loss exemption?
>
> Yes. "[A]lthough the material spilled is not a chemical intermediate used in a production process or a raw material used in a production process, it is a discarded commercial chemical product which has been spilled during normal material handling operations at a manufacturing site and is disposed of via drainage to the wastewater treatment process. The amount of material would not be counted against the 1- or 25-ppm exclusion level for spent solvents mixed with wastewater.... In this case, what is being discarded is not a spent solvent, but an unused commercial product and will meet the requirements of §261.3(a)(2)(iv)(D) de minimis losses." [RO 13097]

they are raw materials, end products, or chemical intermediates.

What constitutes a de minimis loss? First, when EPA promulgated the exemption in 1981, the *Federal Register* preamble contained the following discussion:

> "Data provided by [the Chemical Manufacturers Association] for several chemical manufacturing operations...show that the incremental amounts of §261.33 materials reaching the wastewater treatment system as a result of spills, leaks, maintenance, and laboratory activities usually constitute a small percentage (in all cases except one, less than one percent) of the total amount of such materials contained in the wastewater influent. For example, a plastics manufacturing plant using acrylonitrile [a commercial chemical product listed in §261.33(f)] discharges into its wastewater treatment system only 8 lb per day of discarded acrylonitrile resulting from equipment leakage and cleanup, relief-device discharges, and line rinsings, whereas the quantity of acrylonitrile introduced into its wastewaters from the manufacturing process per se amount to 800 lb per day." [November 17, 1981; 46 *FR* 56583]

Beyond this preamble language, the regs only provide a qualitative description. De minimis losses of F-, K-, P-, and U-wastes include "those from normal material handling operations (e.g., spills from the unloading or transfer of materials from bins or other containers, leaks from pipes, valves, or other devices used to transfer materials); minor leaks of process equipment, storage tanks, or containers; leaks from well maintained pump packings and seals; sample purgings; relief-device discharges; discharges from safety showers and rinsing and cleaning of personal safety equipment; and rinsate from empty containers or from containers that are rendered empty by that rinsing." [§261.3(a)(2)(iv)(D)]

Elsewhere, the agency said that large material losses would void the de minimis quantity exemption. [RO 14095] EPA added "inadvertent releases" to the §261.3(a)(2)(iv)(D) de minimis losses qualitative description on October 4, 2005. At that time, the agency clarified that de minimis losses "must be minor and must result from normal operating procedures at well-maintained facilities.... [A]ny large intentional losses of these wastes will not be considered as de minimis and, accordingly, will not be exempted under §261.3(a)(2)(iv)(D).... [L]osses, which result from mismanagement, neglectfulness, or carelessness during normal operating procedures are not (and have never been) included in the exemption." [70 FR 57778–9] Other than the above guidance, the dividing line between large and de minimis losses is probably going to be decided by state policy.

The de minimis losses exemption does *not* apply to wastewater mixtures created as a result of:

- Discarding off-specification §261.33 materials; or

- Discarding §261.33 materials during abnormal manufacturing operations (e.g., plant shutdowns or operational malfunctions resulting in substantial spills, leaks, or other releases). [46 FR 56586]

The agency believes that the above materials are not de minimis losses but can reasonably be segregated from wastewater and shipped offsite as hazardous wastes.

Note that EPA provided some disconcerting guidance about a de minimis loss being "rinsate from empty containers or from containers that are rendered empty by that rinsing." On January 31, 1991, the agency noted that only the "third rinse in triple rinsing" would meet the definition of rinsate from containers that are rendered empty by that rinsing. [56 FR 3869]

In addition to adding small losses of F- and K-wastes to the §261.3(a)(2)(iv)(D) de minimis exemption, the October 2005 final rule also expanded the types of facilities that are eligible to claim the exemption. Previously, this exemption was available only to manufacturing facilities. Now, the federal RCRA program allows nonmanufacturing plants, such as raw material storage terminals and hazardous waste TSD facilities, to claim the exemption. [70 FR 57772] (States must adopt the October 4, 2005 rule into their authorized programs for facilities to take advantage of this expansion.)

However, *before* a manufacturing facility can claim a de minimis exemption for any F- or K-waste and before a nonmanufacturing facility can claim a de minimis exemption for any F-, K-, P-, or U-waste, the facility must identify all expected constituents in these wastes that may be released at the CWA discharge. For de minimis releases of an F- or K-waste, the list of constituents includes: 1) all constituents in Appendix VII of Part 261 for which the waste was listed, and 2) all constituents of concern in the §268.40 table of LDR treatment standards associated with the waste code. For de minimis losses of a P- or U-waste, this list of constituents includes: 1) the specific chemical associated with the P- or U-code, and 2) all constituents of concern in the §268.40 table of LDR treatment standards associated with the waste code. These identified constituents must be submitted to the NPDES permitting authority (or pretreatment control authority if wastewater discharge is to a sewer line flowing to a POTW), and this notification must occur before the expanded portions of the de minimis exemption may be claimed. Once notified, the NPDES permit writer or pretreatment control authority will determine if permit limits or pretreatment standards need to be added to the CWA permit to account for the new de minimis losses into the facility's wastewater treatment system.

4.16.4 Laboratory wastes

Laboratory wastewaters containing listed wastes that were listed due to their toxicity are eligible for the de minimis exemption if: 1) the annualized average flow of laboratory wastewater does not exceed 1% of total wastewater volumetric flow into the headworks of the facility's wastewater treatment system, or 2) the combined annualized average concentration of toxic §§261.31, 261.32, and 261.33 wastes resulting from laboratory operations does not exceed 1 ppm at the headworks.

Toxic wastes used in laboratories that are demonstrated not to have been discharged to wastewater do not have to be included in this calculation. [§261.3(a)(2)(iv)(E)] See Case Study 4-10 for examples of the laboratory de minimis exemption.

The laboratory de minimis wastewater exemption applies to incidental losses of listed hazardous waste (e.g., small amounts of listed spent solvents, listed wastes brought in for analysis, §261.33 chemicals used during the analysis, etc. that are added essentially unavoidably to large volumes of water) from laboratory operations only. Examples include laboratory spills washed into a sink drain and residues from washing of glassware that are carried into the sewer. Deliberate bulk discharges of chemicals that are not part of laboratory operations are not covered. "The introduction of other listed wastes into the plant wastewater system [outside the conditions set forth in §261.3(a)(2)(iv)(E)] may void the exclusion for the facility." [RO 11727; see also RO 12257]

Both analytical and research and development laboratories may qualify for this exemption. However, EPA noted that a pilot plant is not considered to be a laboratory operation, and wastes generated from such a unit are not exempted under this section. [November 17, 1981; 46 *FR* 56587]

Case Study 4-10: De Minimis Wastewater Exemptions for Laboratory Wastes

I. A laboratory uses chloroform in quality control analyses. On occasion, unused chloroform as well as spent chloroform used as a solvent is disposed down the drain. Does the chloroform qualify for the laboratory de minimis wastewater exemption?

The laboratory wastewater exemption pertains only to listed hazardous wastes (designated as toxic) from laboratory operations. As such, the laboratory would qualify for the exemption if *unused* chloroform (U044 waste) is disposed in the laboratory sinks or drains, because U044 was listed due to toxicity. When the chloroform is used as a solvent prior to disposal down the drain, conversely, it would not meet the listing description for U044 (since it had been used). Chloroform is not included in the F001–F005 spent solvent listings; consequently, any spent chloroform would not be a listed hazardous waste, although it could be the characteristic waste D022. The §261.3(a)(2)(iv)(E) exemption does not apply to characteristic wastes disposed during laboratory operations. [RO 11727]

II. A laboratory generates wastes listed because they are toxic [hazard code of (T)], as well as acutely hazardous wastes [with a hazard code of (H)]. In addition, it generates wastes that are listed because they are both toxic and exhibit the hazardous waste characteristics of ignitability, corrosivity, or reactivity [hazard codes of (I, T), (C, T), or (R, T), respectively]. These wastes are discharged into the lab's sinks/drains that flow to the facility's CWA-regulated wastewater treatment system. Are the wastes that are listed because they are acutely hazardous or toxic and characteristic also exempt from the mixture rule under §261.3(a)(2)(iv)(E)?

Yes. "Wastes listed for being acutely hazardous [(H)] or both toxic and characteristic [(I, T), (C, T), or (R, T)] are also eligible for the wastewater treatment exemption from the mixture rule provided that the wastewater flow meets all the other conditions of §261.3(a)(2)(iv)(E).... This exemption does not apply, however, to wastewaters which were listed solely because they exhibit a characteristic [e.g., a hazard code of (I) only]. If wastes which were listed solely for exhibiting a characteristic were mixed with other solid wastes, such as wastewater, and ceased to exhibit any characteristic, they would, however, no longer be considered hazardous wastes via [§261.3(g)(2)(i)]." [RO 13784; see also RO 14633]

Facilities may prove compliance with the 1% wastewater flow limit by measuring (EPA's preferred approach) or conservatively calculating the annual average wastewater discharge from the laboratory and the annual average wastewater flow entering the wastewater treatment system.

EPA recognizes that, even though some laboratories may exceed the 1% limit, they still may not discharge enough listed hazardous wastes into their drains to warrant regulation under the mixture rule. As a result, the agency allows the 1-ppm compliance test to be used in lieu of the 1% wastewater flow limit. If a facility chooses this option, compliance may be demonstrated using an audit of laboratory chemical purchases, an estimate of the aggregate amounts of toxic §§261.31, 261.32, and 261.33 materials disposed, and estimates of wastewater flow into the headworks of the treatment or pretreatment facility. "Facilities must make the worst-case assumption that all listed hazardous wastes used in the laboratories will be discarded to wastewater, unless they can demonstrate through appropriate records that these materials were disposed of elsewhere." [November 17, 1981; 46 *FR* 56587]

4.16.5 Carbamate production wastes

De minimis losses of certain wastes (K156 and K157) from the production of carbamates and carbamoyl oximes are also exempt when they are mixed with wastewater and pass through the headworks of a facility's wastewater treatment system. [§261.3(a)(2)(iv)(F–G)] These exemptions operate similarly to those for losses of listed spent solvents discussed above; that is, the exemptions are predicated on meeting certain chemical concentration limits at the headworks of the wastewater treatment system.

4.16.6 Miscellaneous issues

The de minimis wastewater exemptions apply only to mixtures that are sent to a facility's *onsite* wastewater treatment system discharging under Section 402 or 307(b) of the CWA. Any mixtures of listed wastes and wastewater that are sent offsite must be managed as hazardous waste, including the use of a manifest. [RO 14181]

The exemptions in §261.3(a)(2)(iv) do not limit the manner in which the listed wastes are transported to the wastewater treatment plant. Therefore, the wastes can be transported via truck, for example, in addition to direct discharge to a sewer and still qualify for the exemption. [RO 13488, 14181]

The de minimis wastewater exemptions do not apply to facilities which discharge into *privately owned* treatment works. However, the privately owned treatment works may qualify for a de minimis exemption if its own discharge is subject to regulation under an NPDES permit or pretreatment program and any listed wastes that it generates meet the §261.3(a)(2)(iv) exemption criteria. [November 17, 1981; 46 *FR* 56584]

If a facility meets the criteria for one of the §261.3(a)(2)(iv) de minimis wastewater exemptions, does the exemption also apply to sludge produced by the wastewater treatment plant that treats the wastewater? Sludge generated from a wastewater that meets all of the criteria for a de minimis exemption would also be exempted from the hazardous waste listing. Since the wastewater is not listed at the headworks of the treatment facility (by virtue of the exemption from the mixture rule), the derived-from rule would not apply the listing to the resulting sludge. However, if the sludge exhibits a hazardous waste characteristic, it would be considered a hazardous waste for that reason. Additionally, if the influent wastewater meets a listing description not addressed by the de minimis exemptions, the sludge would be hazardous via the derived-from rule. [RO 13419, 13784]

As noted above, the de minimis wastewater exemptions apply to mixtures of listed hazardous wastes and wastewater at the headworks of the facility's wastewater treatment system. However, the listed wastes are still hazardous at their point of generation. By implication, all RCRA hazardous waste recordkeeping requirements apply to these wastes until they reach the headworks. For example, the wastes would be subject to the one-time LDR notification required in §268.7(a)(7). [April 8, 2003; 68 *FR* 17242, RO 11727]

4.17 Active manufacturing process unit exemption

Hazardous waste that is generated in an active manufacturing process unit or in a product or raw material storage tank, pipeline, or transport vehicle is exempt from RCRA regulation until one of two things occurs:

1. The hazardous waste is removed from the unit in which it was generated; or
2. The hazardous waste remains in the unit for more than 90 days after the unit is temporarily or permanently removed from service.

This exemption is extremely important to industry. It applies to residual material in raw material or product storage tanks, reactor vessels, separation units, pipelines, or other manufacturing process units or transport vehicles when such material exhibits a characteristic or meets a listing description. The material is not subject to the hazardous waste regulations until one of those two events takes place. Note that this exemption, which is codified in §261.4(c), applies only to manufacturing process units—not to waste storage or treatment units (e.g., surface impoundments).

This section reviews the background of this exemption and gives examples of the regulatory status of wastes in, and removed from, manufacturing process units, pipelines, and transport vehicles.

4.17.1 Background

After EPA promulgated the hazardous waste regulations on May 19, 1980, a significant regulatory problem surfaced. Based on the program as written, a solid waste became subject to Subtitle C regulation as soon as it met the definition of hazardous waste. Literally, this meant that active manufacturing process units, pipelines, and transport vehicles in which such residues were generated were actually hazardous waste storage facilities subject to RCRA storage permits and compliance with all Part 264 or 265 regulations for hazardous waste TSD facilities.

For example, sludge or sediment builds up in crude oil or petroleum product storage tanks. Such residues frequently exhibit a characteristic (e.g., toxicity for benzene) or meet a listing description (e.g., K169). Sludges and residues are also generated in tank trucks, rail cars, and ships or barges that carry products or raw materials. These materials are also sometimes hazardous. Finally, a large number of manufacturing processes generate hazardous residues in process units such as tanks, reactors, columns, etc. Although these hazardous residues may remain in such units and their associated pipelines for only minutes, hours, or days, the units were technically hazardous waste storage facilities subject to regulation.

When this was brought to EPA's attention, the agency responded as follows:

"Except for surface impoundments and nonoperating units, EPA did not intend to regulate product and raw material storage tanks, transport vehicles and vessels, or manufacturing process units in which hazardous wastes are generated…. Because of their design and operation, these units are capable of holding, and are typically operated to hold, the hazardous wastes which are generated in them, until the wastes are purposefully removed. Thus, these hazardous wastes are contained against release into the environment…and the risks they pose to human health or the environment are very low and are only incidental to the risks posed by the valuable product or raw material with which they are associated. Based on these conclusions, EPA believes it is not necessary…to require owners and operators of these units to obtain permits for these units or to comply with the requirements of §262.34 or Parts 264 or 265 with respect to these units." [October 30, 1980; 45 *FR* 72025]

To keep RCRA from applying to hazardous residues in active manufacturing process units, pipelines, and transport vehicles, the agency issued the manufacturing process unit exemption in §261.4(c) on October 30, 1980. [45 *FR* 72024]

Note that a very important application of this exemption involves shutting down or decommissioning process equipment. As long as the equipment is left in place (i.e., not disconnected or disassembled) it can be cleaned (e.g., washed or steamed out) within 90 days of becoming inactive without having to worry about material in the equipment being hazardous waste. Naturally, any cleaning residues that are removed from the equipment are subject to regulation if they exhibit a characteristic or meet a listing description.

Q: *A petroleum refinery operates a heat exchanger as part of the refining process. Sludge builds up in the heat exchanger and, when removed, is K050 hazardous waste. If the refining facility disconnects the heat exchanger and ships it offsite for cleaning within 90 days, would the exemption in §261.4(c) apply?*

A: No. EPA does not interpret the exemption as applying to manufacturing process units (that are stationary during operation) if these units are disassembled and shipped offsite for cleaning. There would be a loss of the unit's structural integrity if it were disassembled for offsite cleaning, with a potential for hazardous waste releases. [RO 13374]

4.17.2 What qualifies as a manufacturing process unit?

Section 261.4(c) specifies that the exemption applies to 1) a product or raw material storage tank, 2) a product or raw material transport vehicle or vessel (ship), 3) a product or raw material pipeline, 4) a manufacturing process unit, or 5) an associated nonwaste treatment manufacturing unit.

Although most people have a common, ordinary understanding of the term "manufacturing process unit," (MPU) there is no precise definition. The agency has provided some clarification, which is summarized in Case Study 4-11 and in the following guidance:

- Process units including distillation columns, flotation units, and discharge trays or screens are examples of MPUs. A cooling tower would be an associated nonwaste treatment manufacturing unit. [October 30, 1980; 45 *FR* 72025, RO 11935]

- MPUs would include tanks or tank-like units that are designed and operated to hold valuable products or raw materials in storage or transportation or during manufacturing. [October 30, 1980; 45 *FR* 72025, RO 11935]

- A cupola furnace at a foundry is directly connected via ductwork to an air pollution control device (i.e., a baghouse). The furnace is an MPU; the downstream baghouse, even though directly coupled to the production unit, is an air pollution control device—not an MPU. [RO 12824]

- Equipment used to process oil-bearing hazardous secondary materials for insertion back into the petroleum refining operations [these materials are excluded per §261.4(a)(12)(i)] are MPUs, since these streams are considered "in-process materials in a manufacturing process." [RO 14677]

4.17.2.1 What doesn't qualify?

EPA also gave some examples of things that don't qualify as MPUs. Because of their construction, surface impoundments are specifically not included in the §261.4(c) exemption because of the high potential that wastes will leach, leak, or otherwise escape from surface impoundments into the environment when compared to tanks or tank-like units. [October 30, 1980; 45 *FR* 72025]

4.17.2.2 What about pipelines?

Pipelines that are associated with active MPUs may contain hazardous residues. The §261.4(c) exemption works the same way for residues in these pipelines. As long as the pipeline is in manufacturing service, any hazardous wastes within it are not subject to regulation until the hazardous waste 1) is removed from the pipeline in which it was generated, or 2) remains in the pipeline for more

Case Study 4-11: Is It a Manufacturing Process Unit or RCRA Tank?

As shown in Figure 4-13, a facility uses a paint spray booth within its manufacturing process to apply coatings to products. Periodically, isopropyl alcohol (IPA) solvent is fed to the process to clean out the paint guns when paint colors are changed. The waste solvent, typically a D001 characteristic hazardous waste, is collected in funnels and piped to a small flow-equalization tank located within the process building. From the equalization tank, the solvent is pumped to an outdoor aboveground 90-day accumulation tank for eventual transfer to an offsite recycler. The accumulation tank and the outdoor spent solvent piping are protected by a secondary containment system. When does the IPA solvent become a hazardous waste? Is the equalization tank and the indoor solvent piping part of the manufacturing process and any hazardous waste contained within them exempt via §261.4(c)?

EPA's perspective is that the IPA solvent becomes a hazardous waste (i.e., its point of generation is) when the material leaves the manufacturing process and enters the drainage funnels. Thus, the equalization tank is *not* part of the manufacturing process. According to the agency, the "exemption at §261.4(c) applies where waste is generated and then contained for some period of time within process units (typically tank-like units), such as sludge that accumulates on the bottom of raw material product tanks. However, the [equalization tank] is not part of the production system, but serves solely to manage wastes." Therefore, the equalization tank is also a 90-day accumulation unit, subject to Part 265, Subpart J. [RO 14152]

This example suggests a good rule of thumb: if everything that goes into a tank comes out as waste, the unit is a waste tank—not a manufacturing process unit.

Figure 4-13: Identification of Manufacturing Process Units vs. RCRA Tanks

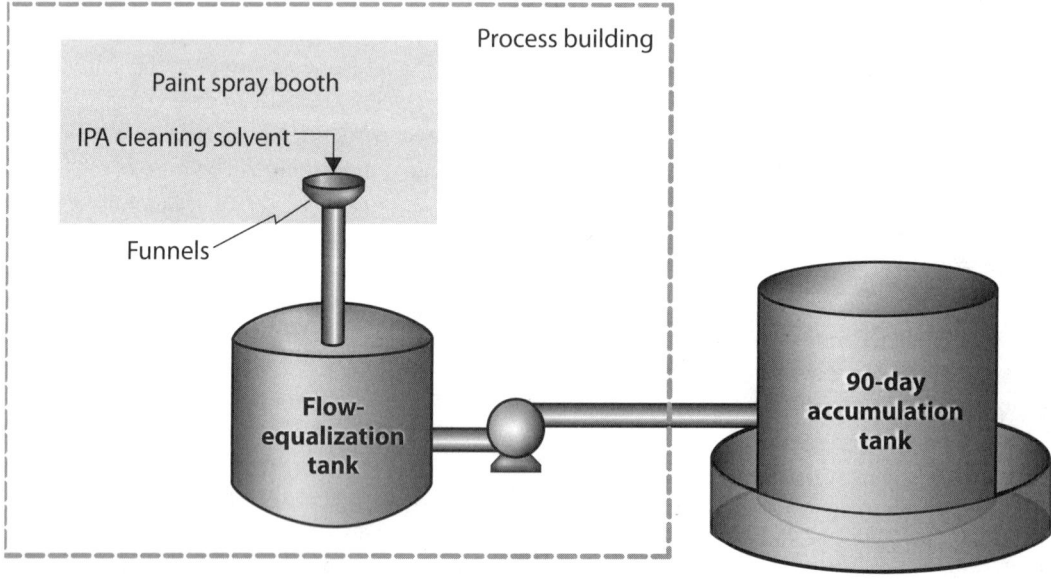

Source: McCoy and Associates, Inc.; adapted from RO 14152.

than 90 days after it is temporarily or permanently removed from manufacturing service.

This discussion raises the question of piping associated with an MPU that is sometimes used to transfer a liquid residue from the unit to a production operation, but at other times is used to transfer the liquid directly to a hazardous waste storage tank. A common situation is shown in Figure 4-14. The residue stream is pumped out of a process reactor (an MPU) to a distribution manifold. Based on residue stream chemistry and production demands, the residue may be recycled back to the production process or sent to an onsite hazardous waste storage tank. EPA has determined that the MPU exemption in §261.4(c) does *not* apply to the pumps and piping leading from the reactor to the distribution manifold, because this equipment is sometimes used to transport hazardous waste. Such ancillary equipment is subject to RCRA regulation, including Subpart BB requirements if it contains or contacts hazardous waste with at least 10% organics for at least 300 hours per calendar year, as well as Subpart J (requiring secondary containment). [RO 13790, 14469]

A related question concerns pipelines/hoses that are normally used in connection with manufacturing stream transfer but are also used as hazardous waste loading/unloading equipment. How do we keep hazardous waste codes off of process streams that are transferred after the equipment is used for hazardous waste purposes? Prior to returning such pipelines/hoses that were used for hazardous

Figure 4-14: Applicability of §261.4(c) Exemption to Pipelines Occasionally Used to Transfer Hazardous Waste

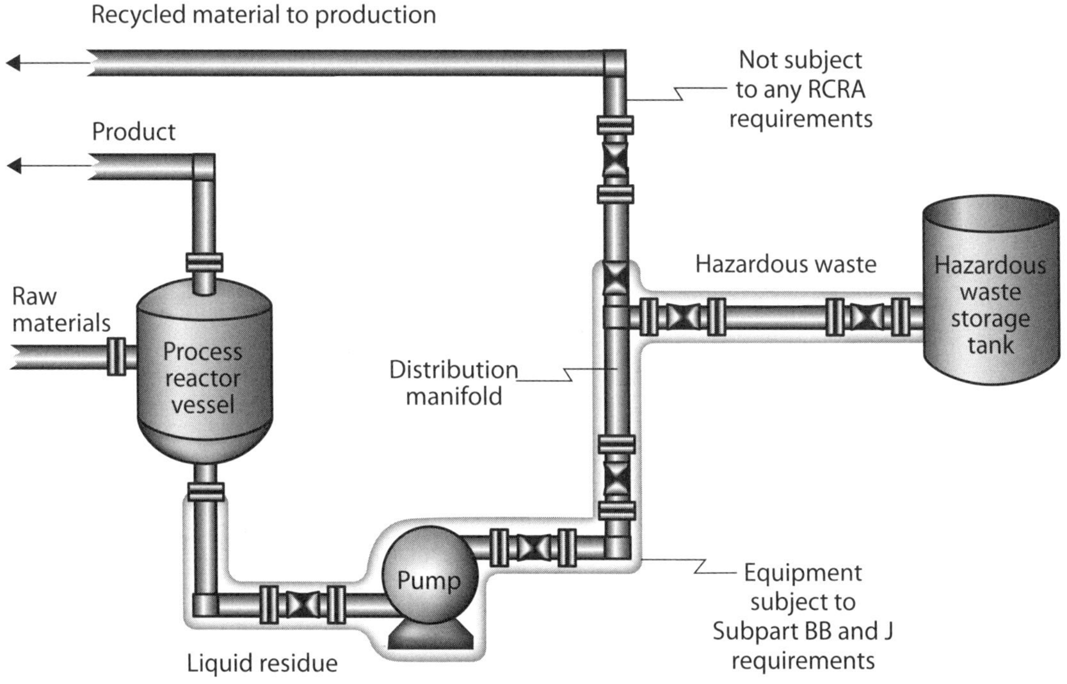

Source: McCoy and Associates, Inc.; adapted from RO 13790, 14469.

waste service to their normal use in manufacturing operations, "good practice would be to clean the hoses so that all hazardous waste residues are removed or decontaminated." [RO 13790]

4.17.2.3 And ships?

Merchant ships are considered to be product or raw material transport vessels. Thus, per §261.4(c), hazardous sludges and residues generated in tanks or holds on ships that carry products or raw materials are not subject to RCRA until: 1) the waste is transferred to a shore facility, or 2) the waste is stored on the ship for more than 90 days after the ship is taken out of service. [RO 11128, 11372, 11862] In RO 12727, EPA extended the §261.4(c) exemption to all hazardous wastes generated aboard ships (e.g., engine room wastes)—not just to wastes generated specifically from the storage of products or raw materials. See Section 6.5.6.6 for a discussion of the status of parties who remove such sludges and residues.

4.17.3 Wastes from manufacturing process units

As noted above, wastes in MPUs, even if characteristic or listed, are not subject to the hazardous waste regulations until:

1. The hazardous waste is removed from the manufacturing process unit, pipeline, or transport vehicle in which it was generated; or
2. The hazardous waste remains in the unit for more than 90 days after the unit is temporarily or permanently removed from service.

These situations are discussed separately in the next two subsections.

4.17.3.1 Wastes removed from the unit

Residues that are removed from an MPU, pipeline, or transport vehicle become subject to the hazardous waste regulations (including the land disposal restrictions program) if they are hazardous at the point of removal and will be disposed. [RO 12959]

"[I]t is only after the removal of hazardous wastes from these units that the wastes have the potential for releasing hazardous constituents into the environment and posing a substantial hazard to human health or the environment." [October 30, 1980; 45 *FR* 72025]

Cleanout residues that will be disposed are solid and potentially hazardous wastes. If residues will be reclaimed (such as bottoms from a raw material storage tank), they may be considered off-specification commercial chemical products destined for reclamation and, as such, would not be solid wastes via Table 1 in §261.2(c).

When characteristic residues are cleaned out of MPUs, pipelines, or transport vehicles for disposal, the residues will only be subject to hazardous waste regulation if they exhibit a characteristic at the exit of the unit. For example, if a large amount of water is used to flush out the residues and the washwater doesn't exhibit a characteristic when it is removed from the unit, the washwater is not subject to RCRA. Such activity is not considered hazardous waste treatment if conducted within 90 days of when the unit became inactive (as discussed in more detail below).

If the residues in the MPU, pipeline, or transport vehicle meet a listing description, they will continue to be listed when removed, regardless of whether water is used to wash them out. (However, ICR-only listed residues will be hazardous only if they exhibit a characteristic when they are removed from the MPU.)

If hazardous, residues cleaned out of MPUs can only be stored in satellite accumulation or 90/180-day accumulation units for short-term storage (see Sections 6.2 and 6.3) or interim status/permitted hazardous waste storage units. See Section 6.5 for a discussion of the status of parties who remove such residues.

Q *A tank is used to store spills and spill residues from MPUs. Is the spill tank subject to RCRA regulation?*

A If the material is or becomes a hazardous waste as a result of the spill, the spills and spill residues become subject to RCRA regulation at the point they exit the MPU. However, only the generator standards in §262.34 would apply to the spill tank and the waste contained therein, provided the waste is stored in the tank for less than 90 days. If the waste was stored for more than 90 days, the tank would become a hazardous waste storage tank subject to permitting and Part 264 or 265 TSD facility standards. [RO 12291]

4.17.3.2 Wastes remaining in the MPU for more than 90 days

The second situation in which residues in MPUs can become subject to the hazardous waste requirements is if they remain in an inactive unit for more than 90 days.

> "EPA believes that when operation ceases, the incentive to maintain the integrity of the unit to prevent leaks or other unintended release of products, raw materials, or manufacturing intermediates into the environment is substantially reduced. Consequently, the incentive to maintain the unit to prevent leaks or releases of hazardous wastes which may remain in the unit after cessation of operation would also be substantially reduced…. If hazardous wastes remain in these units more than 90 days after cessation of operation, EPA believes that these wastes should be fully regulated and that the units should be regulated as hazardous waste storage facilities." [October 30, 1980; 45 *FR* 72025]

For example, a gasoline storage tank is drained of the gasoline, but tank bottoms (that are characteristic for benzene) remain in the tank. On the 91st day after the storage tank has been removed from gasoline-storage service, the tank bottoms become subject to Subtitle C regulation.

On that 91st day, the gasoline storage tank is no longer an MPU but is a hazardous waste storage tank because it is storing hazardous waste that is subject to RCRA regulation. On that day, the storage tank must comply with all §262.34 accumulation tank requirements, which include compliance with Subpart J tank standards. That means the tank and ancillary equipment must have secondary containment and a release detection system, the tank must be marked with the words "Hazardous Waste," and it must be inspected every day. Additionally, a 90- or 180-day clock will begin running on that 91st day. [RO 11899, 11903, 12997]

Most MPUs (reactors, distillation columns, etc.) will meet the definition of a tank and so must be in compliance with Subpart J tank standards after the 90-day grace period runs out. If the MPU is not stationary, it may alternatively meet the definition of a container; in that case, it would be subject to Part 265, Subpart I standards on the 91st day.

Sometimes, large quantity generators think that the §261.4(c) exemption gives them six months after an MPU shuts down before they have to start worrying about RCRA for any hazardous residues remaining in the unit: 90 days before the waste becomes subject to RCRA and another 90 days associated with the §262.34 accumulation provisions. But, generators only get that second 90 days if the MPU is in full compliance with Subpart J standards (for MPUs that are tank-like) or Subpart I standards (for MPUs that are container-like) and the other §262.34 requirements. [RO 11903]

Q *An MPU is temporarily inactive and contains an acidic residue that would be a D002 hazardous waste if disposed. The operator wants to neutralize the residue to render it nonhazardous before removing it from the unit. Is this considered to be hazardous waste treatment requiring a permit?*

A The waste inside an MPU is not regulated until it is removed from the unit or until it remains in an inactive unit for more than 90 days. Neutralizing the waste would not be considered hazardous waste treatment as long as the waste is rendered nonhazardous before the unit has been inactive for 90 days.

Q *Residues exhibiting a characteristic are periodically cleaned out of a process tank. Is the* process tank subject to Subpart J standards, including secondary containment?*

A If the operator is "able to clean out [the] process tank within 90 days after production or product storage is stopped, that process tank would not be considered a waste accumulation tank and, therefore, would not be subject to secondary containment standards. The waste removed, however, is subject to the hazardous waste control system if it is determined to be a hazardous waste." [RO 13790]

Q *During a turnaround, a tank used for explosives manufacturing is cleaned out. The tank is rinsed with water, but an ignitable residue remains on the walls and floor of the unit. Plant personnel then use a torch to burn the residue off of the inner tank surfaces. Is this activity classified as hazardous waste treatment?*

A No, as long as the torching is conducted within 90 days of when the tank was removed from manufacturing service. If it is, the activity is considered tank cleanout operations—not hazardous waste treatment. If the torching isn't performed until after 90 days have passed, the activity would be hazardous waste treatment. Regardless of when the cleanout is accomplished, any residues that are removed from the tank are subject to regulation if they exhibit a characteristic or meet a listing description. [RO 13321]

Q *A tank-like MPU, which holds pure, unused methylene chloride, is located within a building that is to be demolished. How long does the site owner (who is a large quantity generator) have to ship the methylene chloride offsite?*

A If the unused methylene chloride will be disposed, it becomes U080 on the 91st day after the MPU becomes inactive (i.e., ceases to be used for manufacturing or product storage). Because the site is a large quantity generator, the MPU becomes a 90-day accumulation tank on that day, subject to §262.34 and Subpart J standards. Assuming that the tank meets those standards, the facility has a total of 180 days to ship the methylene chloride offsite before the facility would have to obtain a RCRA storage permit. [RO 12997]

Q *Oil-bearing hazardous secondary materials [excluded from RCRA regulation per §261.4(a)(12)(i)] are processed in manufacturing equipment before reinsertion into the petroleum refining process. (See Section 4.7.1.) What is the status of residues that remain in such equipment more than 90 days after it ceases to be operated for manufacturing?*

A If the equipment that is processing these oil-bearing hazardous secondary materials ceases to be operated, residues that remain in the equipment for more than 90 days become subject to RCRA regulation, and the unit becomes a RCRA-regulated waste management unit. [RO 14677]

4.17.3.2.1 Temporary or permanent removal from service

Many people believe that hazardous residues remaining in an MPU will become subject to RCRA regulation only after the unit has been permanently removed from service for 90 days. That is not true—hazardous waste requirements begin after the unit is *temporarily* or permanently removed from service:

> "For both temporary and permanent shutdowns, the agency will allow a reasonable time to remove any hazardous wastes that remain in the unit after operation ceases. Given the presumption that the unit has integrity before cessation of operation, the agency believes that a reasonable time is 90 days." [October 30, 1980; 45 FR 72025; see also EPA/530/K-02/022I, cited previously]

We frequently encounter people who operate processes on a cyclical basis that exceeds 90 days (e.g., 120 days on and 120 days off). Do they actually have to clean out all hazardous residues at the end of each cycle? All we can say is that there is considerable enforcement discretion associated with this provision. We have only seen this provision enforced under two circumstances: 1) a potential release to the environment (e.g., a leak) appears imminent, or 2) the regulators know that a unit will not be restarted (e.g., a plant is being closed, the owners are seeking bankruptcy protection, etc.).

4.17.4 MPUs as SWMUs

Can MPUs be considered solid waste management units (SWMUs) subject to corrective action? EPA has given conflicting answers to this question. When the agency proposed the corrective action program on July 27, 1990, it noted that a SWMU would include an "area at a facility at which solid wastes have been routinely and systematically released." [55 *FR* 30808] This definition would include a loading/unloading area at a facility where coupling and decoupling operations result in a small but routine amount of spillage that, over time, results in contaminated soil.

Conversely, EPA noted that:

"A one-time spill of hazardous wastes (such as from a vehicle traveling across the facility) would not be considered a solid waste management unit…. Similarly, leakage from a chemical product storage tank would generally not constitute a solid waste management unit; such 'passive' leakage would not constitute a routine and systematic release since it is not the result of a systematic human activity. Likewise, releases from production processes, and contamination resulting from such releases, will generally not be considered solid waste management units, unless the agency finds that the releases have been routine and systematic in nature." [July 27, 1990; 55 *FR* 30809]

EPA confirmed that this guidance was still valid in its May 1, 1996 discussion of the corrective action program. [61 *FR* 19443] In that preamble, however, the agency noted that one-time spills would have to be adequately cleaned up in order to not become SWMUs.

Q Would a leaking product tank holding unused plating solution be considered a SWMU subject to corrective action?

A No. "[I]t seems reasonably clear that the holding tank itself would not be considered a SWMU, since it appears that it was used exclusively to store product (i.e., plating solution), rather than solid or hazardous waste. However, the primary issue in this case is whether the area surrounding and underneath the holding tank, which was apparently contaminated from leakage from the tank, should be considered a SWMU." [RO 13441] Unfortunately, EPA never gave an answer to the last question it raised.

The last piece of guidance we have found on this subject is from August 1998. In it, EPA seems to have made a U-turn from its previous position:

"Manufacturing process units holding a hazardous waste are considered SWMUs for the purpose of corrective action. A SWMU is any discernible unit at which solid wastes have been placed at any time, irrespective of whether the unit was intended for the management of solid or hazardous waste…. Manufacturing process units often hold materials which can be classified as solid wastes and potentially hazardous wastes (e.g., precipitated residues). Even though these materials are exempt from hazardous waste regulation under Section 261.4(c), they are still considered solid wastes, thereby rendering the manufacturing process unit a SWMU. However, EPA may exercise differing statutory authority to require cleanup at the facility." [RO 14309]

We have two things to say about the above quotation: 1) It would be helpful if EPA issued additional guidance to clarify whether the August 1998 interpretation reflects the agency's current stance on MPUs as SWMUs; and 2) MPUs are potentially subject to the corrective action program as SWMUs, but only RCRA-permitted and interim status facilities are subject to corrective action.

4.18 Analytical sample/treatability study exemptions

Waste samples collected for analysis and/or treatability study purposes are conditionally exempt from the hazardous waste storage, manifesting, and treatment requirements. EPA's goal is to reduce the regulatory burden for people who manage these types of samples. However, once the excess sample or sample residues exit the analytical sample/treatability study loop, the exemption is lost, and, assuming the excess sample or sample residues will be disposed, they are solid wastes and possibly hazardous wastes.

Discussions of the analytical sample and treatability study exemptions are presented separately below.

4.18.1 Analytical sample exemption

To determine if a solid waste is hazardous, the generator can use testing or knowledge. If testing will be employed, the generator normally collects a sample of the solid waste and sends it to an analytical laboratory. Testing may include an evaluation of physical and chemical properties to help the generator determine whether the sample exhibits any hazardous characteristics, and it frequently includes an analysis for specific constituents. In addition to process residues and maintenance wastes, samples of ground water, surface water, or soil that are suspected of being contaminated can also be sent for testing. Residues from hazardous waste treatment units are also periodically collected and analyzed.

When the RCRA program became effective in 1980, if these samples were hazardous, they were subject to the same RCRA requirements as any other hazardous waste. Among other things, this meant that the samples had to be manifested for hazardous waste shipment, and they could only be shipped to a RCRA-permitted facility for testing. Storage of these samples at the testing facility would require a RCRA permit.

Shortly after the hazardous waste program was implemented, however, EPA came to the conclusion that applying the full set of Subtitle C regulatory requirements to these samples during storage, transportation, and treatment was unnecessary. The agency decided that there are incentives that encourage entities to handle analytical samples properly. The sample collector doesn't want to spend the money to collect another sample and may need the analytical results to comply with RCRA regulations. Also, EPA believes that the types of processing involved with these samples, which are generally less than one gallon in size, do not pose a significant hazard to human health or the environment that justifies the controls required by the full hazardous waste regulations. As such, the agency promulgated an exemption for such samples in §261.4(d) on September 25, 1981. [46 FR 47426]

4.18.1.1 Scope of the exemption

Section 261.4(d) exempts samples destined for characterization from the hazardous waste regulations in Parts 261–270 and from the requirement to obtain an EPA identification number, while the sample is:

1. Stored by the sample collector before transport to a laboratory for testing,
2. Transported from the sample collector to a laboratory for testing,
3. Stored in the laboratory before testing or between various analytical steps,
4. Analyzed at the laboratory,
5. Stored in the laboratory after testing (including storage for a specific purpose) but before its return to the sample collector, and
6. Transported from the laboratory back to the sample collector after testing (if applicable).

Despite the exemption from RCRA provisions, DOT requirements for packaging, marking, labeling, and placarding must be followed if the sample meets the definition of a hazardous material. Additionally, the sample must be packaged so that it does not leak, spill, or vaporize from its packaging. [§261.4(d)(2)(ii)(B)]

CHAPTER 4 *Exclusions and Exemptions*

Figure 4-15 illustrates how the analytical sample exemption works. As long as the sample is in the analytical loop from the sample collector to the laboratory and back, the sample is exempt from RCRA. As soon as the sample comes out of that loop, the exemption ends. Assuming that they will be disposed, the materials that exit the loop are considered to be solid wastes that must be evaluated to determine if they are hazardous (probably relying at least partly on the analytical results).

This §261.4(d) exemption does not apply to large-size samples that are used in treatability studies or other testing at pilot-scale or experimental facilities. Instead, the treatability study exemption will apply in those situations, as discussed in Section 4.18.2.

4.18.1.2 Is the sample collector the "generator"?

Section 261.4(d) specifically avoids the word "generator," but talks about the "sample collector." This was done intentionally so the exemption could apply to any person collecting a potentially regulated sample. Also, the sample collector may not be the person who is ultimately considered to be the generator of a waste when the sample is disposed. [RO 11053]

The term "generator" is defined in §260.10 as the "person, by site, whose act or process produces

Figure 4-15: Analytical Sample Activities Exempt From RCRA Requirements

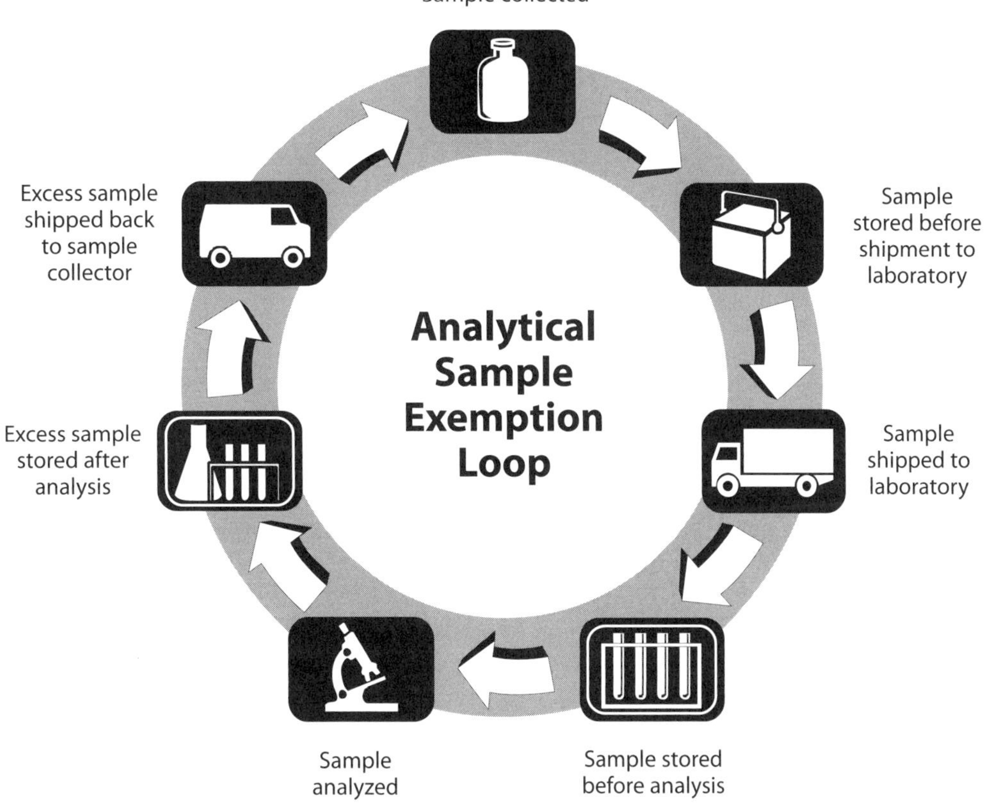

[Diagram: Analytical Sample Exemption Loop showing the following steps in a circular flow: Sample collected → Sample stored before shipment to laboratory → Sample shipped to laboratory → Sample stored before analysis → Sample analyzed → Excess sample stored after analysis → Excess sample shipped back to sample collector → (back to Sample collected)]

Source: McCoy and Associates, Inc.

hazardous waste...or whose act first causes a hazardous waste to become subject to regulation." Therefore, in many cases, waste samples covered by the analytical sample exemption have had no generator as long as the sample remains in the analytical loop of Figure 4-15.

However, if the laboratory disposes the sample, it exits the analytical loop and loses its exempt status. Hence, the sample is a solid waste and the laboratory must determine if it's hazardous. If it is hazardous, the laboratory is the generator, and the waste must be managed and disposed just like any other hazardous waste. [§261.4(d)(3), RO 12438]

If, conversely, the laboratory ships the sample back to the sample collector, the analytical sample exemption is over as soon as the sample collector receives the sample back. The analytical sample exemption does not apply to samples received and stored or disposed by the collector. At that point, they are solid wastes, and the sample collector must determine if they are hazardous. If they are, the sample collector is the generator, and the wastes are subject to all applicable RCRA requirements. [RO 12438, 12917] If the sample is sent from the laboratory back to a third party who collected the sample for analysis, that collector is the generator.

4.18.1.3 How should analytical samples be shipped?

As long as the shipment is in compliance with the documentation and packaging requirements of §261.4(d)(2), including DOT's hazardous materials regulations (which may include 49 *CFR* 172.101(c)(11) and/or 172.203(o)(3)), the analytical sample exemption applies and the mode of transportation used is at the discretion of the person shipping the samples. For example, samples of hazardous waste brought into the United States from Canada for analysis could be driven in personal or company vehicles. Because of the exemption, neither the hazardous waste import requirements in Part 262, Subpart F nor the manifest provisions in Subpart B would apply. [RO 11428]

Since samples are not wastes, the word "waste" should not be part of the DOT proper shipping description. A generic shipping description is often used (e.g., not otherwise specified or "n.o.s.") instead of technical names, and the word "sample" may be needed at the end of the DOT proper shipping name (e.g., Flammable liquid, n.o.s., sample) unless a specific proper shipping name is provided for a sample (e.g., Gas sample, non-pressurized, flammable). The highest packing group typically must be used in the shipping description, since the sample collector often does not know the exact properties of the sample. Based on this information, the required packaging can then be determined from the hazmat shipping table in 49 *CFR* 172.101. Sample collectors can use most major carriers: UPS, DHL, etc., but typically not the postal service.

4.18.1.4 What are the implications for laboratories?

Details of the exemption as it applies to the sample once it reaches the analytical laboratory are summarized below.

4.18.1.4.1 *Sample storage*

Per §261.4(d)(1)(iv), samples held for testing at the laboratory need not be managed as hazardous waste. Additionally, EPA noted that storage between various analytical steps was also exempt. [September 25, 1981; 46 *FR* 47427]

After analysis, storage of excess sample for a specific purpose qualifies for the exemption. The agency noted that "samples are sometimes saved for several years for additional and future analyses. Such analyses may be necessary to confirm original analytical results or to test for additional constituents or properties. Samples may also be stored by the laboratory for a specific purpose, such as when waiting until conclusion of a court case or an enforcement action." [46 *FR* 47427] Storing excess sample in case there is a problem with the material at the offsite facility or in case the analysis has to be rerun for any reason could also qualify under the exemption. [RO 11866] Conversely, if the laboratory stores samples on an indefinite or permanent basis for no specific purpose, such storage would not be exempt. [46 *FR* 47428]

4.18.1.4.2 *Sample analysis*

In the early days, some in the laboratory community were concerned that the process of analyzing a hazardous waste sample could be considered "treatment," as that term is defined in §260.10. EPA set the record straight by noting:

> "Testing is performed to identify the composition or characteristics of the sample, not for the purposes set forth in the definition of treatment in §260.10. Since laboratories are not conducting treatment of samples by the act of testing, they do not need a permit to conduct this testing." [46 *FR* 47428]

Nothing in §261.4(d) specifically says that all sample preparation/testing has to occur at a single laboratory. If sample preparation and splitting are conducted at a different facility than the laboratory that does the actual waste testing, this two-stage analytical process is allowable under the exemption. [RO 11362]

The addition of solvents in the F001–F005 listings to a sample by a laboratory following standard test procedures is allowable and does not affect the exemption. Hence, the laboratory is not regulated as a generator of spent solvent if the adulterated samples are returned to the sample collector. The laboratory may send the samples back without a manifest, as long as applicable DOT standards are met. [RO 12385] (Although this guidance document didn't discuss the alternative scenario of the laboratory directly disposing the solvent-bearing samples, this situation is examined in the next subsection.)

4.18.1.4.3 *Disposal of excess sample*

As mentioned previously, the analytical sample exemption ends if the laboratory disposes the excess sample instead of sending it back to the sample collector. If such solid waste is hazardous (as determined by the analytical results or the laboratory's knowledge), the laboratory is the generator and must manage the waste under all Part 262 generator requirements. (One of these requirements is to obtain an EPA identification number.) However, some laboratories may qualify for conditionally exempt small quantity generator status because of the relatively small amount of hazardous waste they generate. Facilities that generate no more than 100 kg/mo of nonacute hazardous waste and no more than 1 kg/mo of acute hazardous waste are conditionally exempt from the hazardous waste regulations (see Section 6.1.1). [§261.5(b), RO 13291]

If the laboratory does generate hazardous waste that will be disposed (including hazardous samples, spent solvents, and/or chemicals or calibration standards), the subsequent storage, treatment (e.g., neutralization), transportation, and disposal of the waste are subject to all Subtitle C regulations. [September 25, 1981; 46 *FR* 47428, RO 12438]

We think most laboratories that dispose excess sample/analytical residues simply check their waste containers for characteristics and send them for disposal as hazardous if they exhibit one or more of the characteristics. Often, labs do not know if samples should carry a listed waste code. If samples do carry a listed waste code, all excess sample/analytical residues should be disposed as listed waste. [RO 13375] For example, if the lab accepts a waste sample from one of the 13 industries identified in §261.32, it may be managing a K-listed waste. [EPA/233/B-00/001]

Sometimes, the laboratory knows that it is generating listed wastes. For example, laboratory solvents that are used during analysis are not covered by the analytical sample exemption if they are disposed. Depending on the chemicals involved, these materials could be F001–F005 listed spent solvents. Personnel protective equipment that contains a listed hazardous waste (e.g., protective gear contaminated with listed spent solvents) must also be managed as hazardous waste, unless it is decontaminated. [RO 12917]

Q *Samples of F004 spent solvent are sent to a laboratory to determine if they are ignitable. During analysis, disposable equipment (e.g., a pipet)* *becomes contaminated from the analysis of the spent solvent. What is the regulatory status of the pipet when it is disposed?*

A Any wastes generated from the analysis that contain the spent solvent are also identified as F004 under the contained-in policy (see Section 5.3). [RO 13375] In this specific case, however, an argument could be made that the pipet is an empty container, and any residues remaining in an empty container are exempt from RCRA.

Q *Crucibles and cupels are vessels used to assay precious metals (e.g., separate gold and silver from lead). During this procedure, both vessels absorb significant quantities of lead. When the vessels can no longer be used because of high lead contamination, they are sent to smelters for lead recovery. What is the regulatory status of the crucibles and cupels when shipped offsite to the smelter?*

A The crucibles and cupels meet the RCRA definition of spent materials. [§261.1(c)(1)] From Table 1 in §261.2(c), spent materials are solid wastes when reclaimed. If they fail the TCLP for lead, the analytical sample exemption does not apply, and the spent crucibles and cupels would be hazardous wastes when shipped to the smelter. [RO 14332] However, such spent materials may be excluded from the definition of solid waste when reclaimed under the conditions established in EPA's October 30, 2008 rule (see Section 11.2.3). [73 *FR* 64668]

4.18.1.5 Does the exemption include product samples?

The exemption applies to samples of solid waste, water, soil, or air—not to samples of products. Product samples, which sometimes are kept for many years, are regulated as commercial chemical products and are not eligible for the analytical sample exemption. [RO 11523] As products, these samples are not subject to RCRA regulations. If such a product sample is to be disposed, however, it would be a solid waste and a hazardous waste if it exhibits a characteristic or meets a listing description. To be listed, the product sample would have to be a pure, unused chemical on the P- or U-list, unused product containing a P- or U-chemical as the sole active ingredient, or F027.

Note that the only time we have seen people get into trouble under RCRA for storing product samples is when the labels have become illegible, essentially rendering the samples useless. Because these samples exhibited a characteristic, the state cited the facility for not storing these worthless (waste) samples in accordance with the RCRA storage standards.

4.18.2 Treatability study exemption

As noted previously, the analytical sample exemption does not apply to large samples of hazardous waste, water, or soil that are used in treatability studies or other testing at pilot-scale or experimental facilities. Therefore, samples used in these activities were subject to the full Subtitle C program, necessitating RCRA permitting for pilot-scale experimentation and research required to evaluate hazardous waste treatment options. In response to petitions from industry, EPA established the treatability study exemption in §261.4(e–f), and it works very much like the analytical sample exemption already discussed. [July 19, 1988; 53 *FR* 27290] Large-size samples used in pilot-scale treatability studies are exempt from hazardous waste storage, manifesting, and treatment regulations (including permitting), when certain conditions are met.

4.18.2.1 What is a treatability study?

EPA defines a "treatability study" in §260.10 as "a study in which a hazardous waste is subjected to a treatment process to determine:

1. Whether the waste is amenable to the treatment process,
2. What pretreatment (if any) is required,
3. The optimal process conditions needed to achieve the desired treatment,
4. The efficiency of a treatment process for a specific waste or wastes, or

5. The characteristics and volume of residuals from a particular treatment process."

Examples of the types of evaluations for hazardous wastes that are typically conducted under the exemption are physical/chemical/biological treatment, thermal treatment (incineration, pyrolysis, oxidation, or other combustion), solidification, sludge dewatering, volume reduction, toxicity reduction, and recycling feasibility. However, EPA considers the following types of analyses to also be exempt: 1) liner compatibility, corrosion, and other material compatibility studies (e.g., relating to leachate collection systems, geotextile materials, other land disposal unit requirements, pumps, and personal protective equipment); and 2) toxicological and health effects studies. [July 19, 1988; 53 FR 27293, RO 11366]

EPA notes that the exemption does not apply where the treatability study could result in a significant uncontrolled release of hazardous constituents to the environment. Therefore, it does not apply to open burning or to any type of treatment involving placement of hazardous waste on the land (e.g., in situ stabilization). [§261.4(f)(6)] The agency emphasizes that the exemption is provided for the evaluation of a treatment process and is *not* to be used as a means to commercially treat or dispose hazardous waste. [53 FR 27294]

4.18.2.2 Scope of the exemption

Treatability study samples or treatment residues are exempt from the hazardous waste program requirements while [§261.4(e–f), RO 11366, 13334]:

- Samples are collected, prepared, and stored by the generator or sample collector prior to transport;

- Samples are shipped to a laboratory or testing facility to conduct the treatability study;

- Samples are stored at the laboratory or testing facility before testing;

- Samples are treated by the laboratory or testing facility during the study;

- Excess sample and treatment residues generated from the treatability study are stored temporarily after testing; and

- Excess sample and treatment residues generated from the treatability study are shipped from the laboratory or testing facility back to the generator or sample collector.

The primary advantages offered by this exemption are that 1) the actual testing or treatment of the samples does not require a RCRA permit, 2) such testing does not require prior EPA approval, and 3) the transportation to and from the laboratory or testing facility does not have to be manifested. As with the analytical sample exemption, however, DOT packaging, marking, labeling, and placarding requirements must be followed. At a minimum, the sample must be packaged so that it does not leak, spill, or vaporize from its packaging. [§261.4(e)(2)(iii)]

A number of sample quantity and time limits are specified in the treatability study sample exemption; if these parameters aren't met, the affected sample and/or any treatment residues become subject to full RCRA regulation. Finally, once any unused sample and/or treatment residues are to be discarded, they are no longer exempt under §261.4(e–f). Such solid wastes must be evaluated to determine if they are hazardous and managed under RCRA Subtitle C if they are.

4.18.2.3 Treatability study sample quantity and time limits

EPA has placed limits on the amount of waste that can be stored and then evaluated in a treatability study and the time frames under which such storage and treatment must occur. These limits, which were significantly revised on February 18, 1994 [59 FR 8362], are summarized in Table 4-6. For situations in which the quantities or time frames given in the regulations are inadequate, EPA may grant variances on a case-by-case basis to raise the treatability sample amounts or increase the available time.

Table 4-6: Sample Quantity and Time Limits for Conducting Treatability Studies

Limits	Maximum quantity/time	Additional quantity/time available from case-by-case variance
Use in treatability study[1] (kg)		
■ Media[2] contaminated with nonacute hazardous waste	10,000	5,000
■ Media[2] contaminated with acute hazardous waste	2,500	2,500
■ Nonacute hazardous waste other than contaminated media	1,000	500
■ Acute hazardous waste	1	1
Shipment to a laboratory or testing facility (kg)		
■ Media[2] contaminated with nonacute hazardous waste	10,000	5,000
■ Media[2] contaminated with acute hazardous waste	2,500	2,500
■ Nonacute hazardous waste other than contaminated media	1,000	500
■ Acute hazardous waste	1	1
■ Single-shipment maximum	10,000	5,000
Per-day treatment rate[3] (kg/day)		
■ Media[2] contaminated with nonacute hazardous waste	10,000	NA
■ Media[2] contaminated with acute hazardous waste	2,500	NA
■ Nonacute hazardous waste other than contaminated media	250	NA
■ Acute hazardous waste	—	NA
Storage at the laboratory or testing facility[3] (kg)		
■ Media[2] contaminated with nonacute hazardous waste	10,000	5,000
■ Media[2] contaminated with acute hazardous waste	2,500	2,500
■ Nonacute hazardous waste other than contaminated media	1,000	500
■ Acute hazardous waste	1	1
■ Total waste storage inventory	10,000	5,000
■ Treatment residues that may be stored for future evaluation for up to 5 years after date of initial receipt[4]	500 per waste stream	NA
Time limits		
■ All excess sample/treatment residues must be returned to generator or sample collector	90 days after completion of treatability study	NA
■ All excess sample/treatment residues must be returned to generator or sample collector		
—Bioremediation treatability studies	2 years after initial sample shipment	2 years
—All other treatability studies	1 year after initial sample shipment	NA

NA = not applicable.
[1] Maximum sample quantity for each treatment process evaluated for each generated waste stream. "Waste stream" is not based on hazardous waste code alone; rather, the quantity limits apply for each medium (e.g., soil, water, or debris) or physical form in which the waste appears. [53 FR 27294]
[2] Media includes ground water, surface water, soil, sediment, and debris. [59 FR 8363]
[3] Quantities are based on hazardous waste "as received" in the shipment from the generator or sample collector and thus exclude any nonwaste material.
[4] Any quantity stored for future evaluation must be included in the total waste storage inventory.
Source: McCoy and Associates, Inc.; adapted from §261.4(e–f) and 59 FR 8365–6.

4.18.2.4 Sample import/export for treatability studies

A laboratory inside the United States that imports samples for a treatability study qualifies for the treatability study exemption as long as it manages the samples under all §261.4(e–f) requirements. The exemption extends to wastes that originate from outside of the United States. Therefore, the laboratory would not have to comply with the hazardous waste import requirements in Part 262, Subpart F. [RO 14460]

Similarly, generators or sample collectors who comply with the exemption requirements are exempt from the Part 262, Subpart E requirements if they export samples for treatability studies. [July 19, 1988; 53 *FR* 27293] There are two provisions within the exemption, however, that seem to make it impossible for a generator or sample collector to satisfy the exemption requirements for a treatability study sample bound for a foreign facility. First, §261.4(e)(2)(iv) requires the sample to be shipped to a permitted TSD facility or, alternatively, one that at least has an EPA identification number, as required in §261.4(f)(2). In addition, §261.4(e)(2)(v)(C)(*2*) requires the generator or sample collector to keep records of the facilities (and their corresponding EPA identification numbers) to which it sends samples for treatability studies. Laboratories or testing facilities outside the United States cannot be assigned an EPA identification number. However, EPA has noted that "persons who generate or collect samples for the purpose of conducting treatability studies outside the United States and who meet all of the requirements set forth in §261.4(e), except for §261.4(e)(2)(iv) and §261.4(e)(2)(v)(C)(*2*), meet the terms of the exemption." [RO 11667]

4.18.2.5 Laboratory/testing facility requirements

The exemption from RCRA storage requirements for laboratories or testing facilities conducting treatability studies does not apply to any intermediate storage that may occur between the generator or sample collector and the laboratory or testing facility. [RO 11366]

EPA has not limited the number of treatability studies that a laboratory or testing facility can perform per year. Typical hazardous waste treatment facility requirements, such as preparation of contingency plans and emergency procedures, are not necessary for laboratories or testing facilities conducting treatability studies. [July 19, 1988; 53 *FR* 27298]

As mentioned previously, a laboratory or testing facility conducting treatability studies must have an EPA identification number. [§261.4(f)(2)] However, large facilities (including federal facilities) with numerous laboratories or testing facilities are *not* considered by EPA to be a single laboratory or testing facility for purposes of the treatability study exemption. Therefore, each laboratory or testing facility within a larger facility could conduct treatability studies utilizing the sample quantities and time frames specified in Table 4-6, as long as it has its own, unique EPA identification number. [February 18, 1994; 59 *FR* 8364]

4.18.2.5.1 *Mobile treatment units*

Mobile treatment units (MTUs) conducting treatability studies qualify for the exemption. If more than one MTU is located at the same site, however, the group of MTUs is subject to the treatment rate, storage, and time limits and the notification, recordkeeping, and reporting requirements that are applicable to stationary laboratories or testing facilities. That is, a group of MTUs operating at one location will be treated as one MTU facility for purposes of §261.4(e–f). [July 19, 1988; 53 *FR* 27297, RO 11366]

4.18.2.6 Disposal of excess sample/analytical residues

The treatability study exemption ends if the laboratory or testing facility 1) does not return excess sample/treatment residues to the generator or sample collector within the time frames noted in Table 4-6, or 2) disposes the excess sample/treatment residues instead of sending them back to the generator or sample collector. In either situation, the laboratory or testing facility must determine if the excess sample/treatment residues are hazardous and, if they are, manage the waste under all Part 262 generator requirements. However, some laboratories or testing facilities may qualify for conditionally exempt small quantity generator status because of the relatively small amount of hazardous waste they generate. (If the laboratory or testing facility decides to directly dispose the excess sample/treatment residues instead of sending them back to the generator or sample collector, storage of the material before offsite shipment to the disposal facility remains exempt if such storage is within the time frames noted in Table 4-6.) [July 19, 1988; 53 FR 27292]

If the laboratory or testing facility returns excess sample/treatment residues, the treatability study exemption ends upon receipt by the generator or sample collector. At that point, excess sample/treatment residues that are still hazardous must be managed under Subtitle C. Ultimately then, all hazardous excess sample/treatment residues will be manifested to, treated, and/or disposed in a RCRA-permitted facility.

4.18.2.7 Reporting and recordkeeping requirements

Section 261.4(e–f) specifies reporting and recordkeeping requirements to document compliance with the quantity and time limits of the treatability study exemption. The generator or sample collector and the laboratory or testing facility conducting the treatability study must keep copies of all contracts and shipping documents for a minimum of three years after the completion of each study.

A generator or sample collector must also maintain the following records (and include this information in its biennial reports):

- The amount of waste (per waste stream and treatment process) shipped under the exemption;
- The name, address, and EPA identification number of the laboratory or testing facility;
- Shipment dates; and
- Whether or not any excess sample/treatment residues generated from the treatability study were returned.

The owner/operator of a laboratory or testing facility must:

- At least 45 days prior to conducting a treatability study, provide written notification to EPA or the authorized state of its intent to do so. (This is a one-time notification; a notice does *not* have to be sent before each treatability study. [RO 11695]) When treatability studies are no longer planned at its facility, the laboratory or testing facility must notify EPA or the authorized state of this fact, in writing.
- By March 15 of each year, submit a comprehensive report to EPA or the authorized state that documents the pertinent information for all studies conducted during the previous calendar year.
- Maintain records documenting compliance with the specified treatment and storage quantity and time limits for 3 years after the completion of each treatability study.

4.19 Precious metals exemption

An exemption is available in the RCRA regulations for hazardous wastes that are reclaimed to recover economically significant amounts of the following precious metals:

- Gold,
- Silver,
- Platinum,
- Palladium,
- Iridium,
- Osmium,
- Rhodium, and
- Ruthenium.

Examples of precious metals-bearing hazardous wastes include solutions and wastewater treatment sludges from electroplating and heat-treating operations; spent catalysts from the chemical and refining industries; used circuit boards and residues from processing these boards (e.g., shredded pieces, sweeps/ash, fluff, or baghouse dust); silver-bearing scrap, ion-exchange resins, and batteries; and silver-containing photographic films and fixer solutions. [RO 11158, 11197, 11210, 11689]

Note, however, that characteristically hazardous sludges and by-products (e.g., wastewater treatment sludges that are toxic due to silver) are not solid wastes when reclaimed (as long as the material is not otherwise a listed hazardous waste) per §261.2(c)(3). Because they are not solid wastes, they cannot be hazardous wastes, and the precious metals exemption is not necessary or applicable. Conversely, if the material is a solid and hazardous waste (e.g., a spent material or listed sludge containing regulated levels of silver that is sent for silver recovery), then the precious metals exemption is applicable. Used x-ray film and spent fixer solution are examples of spent materials to which the exemption may be applied. [RO 11156, 11541, 11744, 11814, 11912] Note that an October 30, 2008 rule [73 *FR* 64668] established recycling exemptions that allow spent materials, listed sludges, and listed by-products to be reclaimed and not be solid wastes if certain conditions are met (see Sections 11.2.3.1 and 11.2.3.2).

EPA is confident that the value of the precious metals in wastes eligible for this exemption is so high that they will not be mishandled. Instead, the agency's experience is that they are handled carefully from point of generation through metals recovery (e.g., these wastes are normally containerized as opposed to stored in piles or impoundments). As such, EPA does not believe that full compliance with the hazardous waste regulations is justified. [April 4, 1983; 48 *FR* 14494, January 4, 1985; 50 *FR* 648]

In §266.70(a), it says "[t]he regulations in this subpart apply to recyclable materials that are reclaimed to recover *economically significant* amounts of [precious metals]." [Emphasis added.] What is "economically significant" recovery? EPA has never given any quantitative guidance to answer this question. Instead, the agency has preferred to discuss this issue in terms of legitimate vs. sham recycling. See Section 4.19.2.2 below.

4.19.1 Extent of the exemption

The precious metals exemption is codified at §§261.6(a)(2)(iii) and 266.70. Based on those regulatory sections and EPA guidance, the RCRA hazardous waste management requirements from which a facility is exempt and those with which a facility must comply are listed in Table 4-7.

It is important to keep in mind that precious metals-bearing wastes (other than characteristic sludges and by-products and materials excluded from the definition of solid waste under the October 2008 DSW rule) sent for reclamation are solid and hazardous wastes if they exhibit a characteristic or meet a listing description. But they are hazardous wastes that receive a limited exemption from the RCRA Subtitle C regulations in Parts 262–267 and 270. [50 *FR* 648, RO 14092]

Because these materials are hazardous waste (if characteristic or listed), residues from their reclamation may also be hazardous, based on application of the derived-from rule (discussed in Section 5.2). Residues remaining after reclamation of the precious metals would no longer qualify for the precious metals exemption and so would be subject to full Subtitle C regulation if hazardous. [50 *FR* 649]

4.19.2 Recovery of precious metals

Facilities must send their precious metals-bearing wastes for reclamation to qualify for the exemption from hazardous waste management. Issues associated with such reclamation are addressed in EPA guidance, summarized in the following two subsections.

4.19.2.1 Precious metals recovery in industrial furnaces

Precious metals are often recovered by burning in thermal reduction furnaces. Because EPA believes that such furnaces are legitimately recovering

Table 4-7: Scope of the Precious Metals Exemption

Requirement	Exempt	Required	Reference
Reclaim the material to recover economically significant amounts of precious metals		X	§266.70(a)
Obtain a facility EPA ID number[1]		X	§266.70(b)(1)
Make hazardous waste determinations per §262.11 and assign hazardous waste codes	X[2]		
Use transporters and storage facilities with EPA ID numbers		X	§266.70(b)(1)
Use the uniform hazardous waste manifest		X	§266.70(b)(2)
Comply with LDR requirements, including LDR paperwork		X	§§261.6(a)(2), 268.1(b)[3]
Sign the waste minimization certification per §262.27		X	§266.70(b)(2)
Comply with Part 262, Subpart E when exporting precious metals-bearing hazardous waste		X	RO 11580, 12755
Comply with Part 262, Subpart F when importing precious metals-bearing hazardous waste		X	§266.70(b)(3)
Comply with Part 262, Subpart H and §265.12(a)(2) for precious metals-bearing hazardous wastes exported to or imported from OECD countries		X	§266.70(b)(3)
Keep records to document that the material is not being accumulated speculatively		X	§266.70(c)
Count the precious metals-bearing hazardous waste when making monthly generator-status determinations		X	50 FR 652, RO 14092
Include the precious metals-bearing hazardous waste on biennial reports		X[4]	
Use the derived-from rule to determine the regulatory status of residues produced from the recovery of precious metals		X	48 FR 14494
Accumulation time limit		X[5]	§266.70(d)
Accumulation quantity limit	X		
Use Part 265, Subpart I containers, Subpart J tanks, or Subpart DD containment buildings for storage	X		RO 14092
Use hazardous waste labels and mark containers with accumulation start dates	X		
Store in 90/180-day areas	X		

LDR = land disposal restrictions; OECD = Organization for Economic Cooperation and Development.

[1] An EPA ID number is obtained by completing the *Notification of Regulated Waste Activity*, EPA Form 8700–12, available at http://www.epa.gov/osw/inforesources/data/form8700/8700-12.pdf.

[2] These materials are not exempt from being hazardous wastes but are exempt from some hazardous waste management standards; waste codes will be needed for Block 13 of the manifest.

[3] According to §268.1(b), the LDR requirements of Part 268 apply unless specifically provided otherwise in Parts 261 or 268. The regulations governing precious metals recycling are in §261.6(a)(2)(iii) and state that Part 268 does apply. Therefore, the LDR requirements apply to recyclable precious metal wastes. [RO 11482, 13158] Confusion over this issue may exist because §266.70(b), which specifies requirements applicable for precious metal wastes, does not cite Part 268.

[4] In the biennial report instructions, EPA notes that a facility must report "each generated RCRA hazardous waste that is used to determine the site's generator status." Since a facility must count precious metals-bearing hazardous wastes towards its generator status [50 FR 652, RO 14092], they should be included on the biennial report. However, this conclusion is inferred and not supported by any EPA guidance.

[5] If the material is accumulated speculatively, the exemption is lost per §266.70(d). Records required to document that a facility is not accumulating speculatively are listed in §266.70(c).

Source: McCoy and Associates, Inc.; adapted from §§261.6(a)(2)(iii) and 266.70, unless otherwise noted.

precious metals from recyclable materials, the recovery falls under Part 266, Subpart F. Thus, the furnace is 1) exempt from RCRA permitting, and 2) not subject to Part 266, Subpart H boiler or industrial furnace requirements (except for the regulations regarding residue management at §266.112). However, a one-time written notice, sampling and analysis, and maintenance of records are required. [§266.100(g), August 27, 1991; 56 *FR* 42508, RO 11804, 13703, July 21, 2008 email from EPA to McCoy and Associates, Inc.]

4.19.2.2 Legitimate vs. sham recycling

Although EPA has never defined "economically significant" recovery of precious metals, the agency provided the following warning on operations that could be a sham:

> "We also note that sham recovery operations merely claiming to be engaged in precious metal reclamation are not exempt under this provision. Sham operations not only include those where no precious metals are present, but those where precious metals are present only in trace amounts, or in amounts too low to be economically recoverable. The regulations consequently state that the reclamation facility must be recovering economically significant amounts of precious metals from each waste for the waste to be conditionally exempt. For example, wastes from which small amounts of silver are recovered by a facility not ordinarily engaged in precious metal reclamation would not be exempt from regulation. Other factors indicating sham precious metal recycling are lack of strict accounting by either the generator or reclaimer of wastes to be reclaimed, storage (such as in open piles or impoundments) by either the generator or reclaimer not designed to protect wastes from release, payment to a reclaimer to accept wastes, or absence of efficient recovery equipment at the reclaimer's site." [50 *FR* 648–9]

The absence of air pollution control equipment to recover any precious metals contained in emissions from a furnace could be a potential indication of a sham operation. [56 *FR* 42509]

Finally, the agency noted "that generators and recovery facilities normally enter into written contracts before materials are transferred specifying compensation to the generator and when transfer is to occur, and that true precious metal recovery is characterized by net financial return to the generator (i.e., a price sufficient to cover all charges for transport, storage, and processing). Conversely, the absence of one or more of the features mentioned above, amounts of precious metals too low to be economically recoverable, or payment from the generator to a reclaimer to accept wastes could serve as potential indications of recycling operations that may not be eligible for the §266.70 exemption…." [RO 14267]

4.19.3 Importing and exporting precious metals

Because of the way EPA defines "primary exporter" in §262.51, the Part 262, Subpart E export requirements are applicable only to hazardous wastes that are subject to the Part 262, Subpart B manifesting requirements. Because precious metals-bearing hazardous wastes are subject to the manifest requirements per §266.70(b)(2), the export requirements of Part 262, Subpart E are applicable for precious metals-bearing wastes that are exported for reclamation. [RO 11580, 12755]

Hazardous waste, including precious metals-bearing hazardous waste, imported into the United States must comply with all applicable RCRA provisions as soon as it enters the country. Because the actual generator of the precious metals waste is a foreign entity, the shipment is not subject to RCRA regulations until it crosses the U.S. border. Therefore, when hazardous waste is imported into the United States, the U.S. importer is responsible for carrying out the RCRA generator requirements. As such, the U.S. importer of precious metals-bearing hazardous waste must have an EPA ID number, comply with appropriate DOT transportation requirements, prepare a hazardous waste manifest, use a transporter who has an EPA ID number, and comply with all other applicable requirements in Table 4-7. Regulations for hazardous waste importers, including instructions on completing manifests for imported hazardous waste, are found in Part 262, Subpart F.

4.19.3.1 Special requirements for OECD countries

For precious metals-bearing hazardous wastes exported to or imported from designated Organization for Economic Cooperation and Development (OECD) member countries for recovery, facilities must comply with Part 262, Subpart H and §265.12(a)(2) in lieu of the Part 262, Subparts E and F provisions (as discussed in the previous section). For precious metals-bearing hazardous wastes exported to or imported from non-OECD countries for recovery, facilities must comply with Part 262, Subparts E and F (discussed above).

4.20 Lead-acid battery exemption

Lead-acid batteries are the most widely used rechargeable batteries in the world. They are mainly used as starting, lighting, and ignition power batteries in automobiles, forklifts, and other vehicles. Such batteries are spent if they no longer perform effectively and cannot be recharged. [January 8, 2010; 75 FR 1239]

Used lead-acid batteries that can no longer be used for their intended purpose are considered spent materials. [RO 11101, 11822] Therefore, they are solid wastes whether they will be disposed [§261.2(b)(1)] or reclaimed. [Table 1 of §261.2(c)] (The exclusions for spent materials being reclaimed that were added by EPA in its October 30, 2008 rule [73 FR 64668] do not apply to spent lead-acid batteries.) Although not listed hazardous wastes, lead-acid batteries will typically exhibit the characteristics of corrosivity (D002) and toxicity for lead (D008). [RO 14147] Thus, they are hazardous wastes.

Assuming the spent lead-acid batteries are not from households and, therefore, are not covered under the household hazardous waste exclusion (see Section 4.10), these hazardous wastes can be managed in one of three ways under RCRA: 1) as hazardous waste under the full Subtitle C regulatory program, 2) as universal waste under the Part 273 management standards (good option), or 3) as hazardous wastes exempt from most RCRA management requirements under Part 266, Subpart G (even better option). Hazardous waste management requirements associated with Option 1 are discussed in Chapters 6, 7, 9, 10, and 13 of this book. The less-stringent universal waste provisions are discussed in Chapter 8. This section discusses the last (and possibly the best) option of managing these materials under Part 266, Subpart G.

4.20.1 The Part 266, Subpart G exemption

As noted just above, spent lead-acid batteries are hazardous waste at the point they are generated. [RO 11117] The hazardous waste provisions that apply to the management of spent lead-acid batteries under Part 266, Subpart G depend on the means of reclaiming the batteries.

Per the table in §266.80, spent lead-acid batteries reclaimed under Part 266, Subpart G are either regenerated (such as by recharging and/or replacing electrolyte) or otherwise reclaimed to recover metal values, ammonium sulfate, and plastic (such as by breaking, cracking, and/or smelting). Table 4-8 summarizes the RCRA Subtitle C regulatory requirements applicable to persons who manage spent lead-acid batteries that ultimately will be reclaimed using one of these two means. EPA adopted this dual regulatory approach in order to provide a balance between protecting human health and the environment and encouraging recycling of these batteries. [RO 12688, 12856]

4.20.1.1 Battery generators/collectors are exempt from most RCRA standards

As noted in Table 4-8, nearly everyone that manages these batteries (except for facilities that both store and reclaim them) are exempt from most RCRA requirements. Per §266.80(a), generators, transporters, and collectors of intact spent lead-acid batteries,

Table 4-8: RCRA Requirements for Managing Spent Lead-Acid Batteries

Person who:	Requirements if lead-acid batteries will be regenerated (e.g., recharging, replacing electrolyte)[1]	Requirements if lead-acid batteries will be reclaimed (e.g., breaking, cracking, smelting) to recover metal values
Generates, collects, transports, and/or stores intact batteries (but does not reclaim them)	■ Make hazardous waste determination per §262.11[2]	■ Make hazardous waste determination per §262.11[2] ■ Complete a one-time LDR notification per §268.7(a)(2)
Regenerates batteries	■ Make hazardous waste determination per §262.11[2]	NA
Stores and reclaims batteries at a single facility	NA	■ Obtain EPA ID number ■ Make hazardous waste determination per §262.11[2] ■ Comply with Subparts A–L of Part 264/265 (except §§264/265.13 and 264/265.71–2) ■ Obtain a RCRA storage permit and otherwise comply with Parts 124 and 270 ■ Manifest all hazardous residues that are shipped offsite ■ Comply with LDR for all residues
Reclaims batteries (but does not store them first)	NA	■ Make hazardous waste determination per §262.11[2] ■ Manifest all hazardous residues that are shipped offsite ■ Comply with LDR for all residues
Exports batteries for reclamation	■ Make hazardous waste determination per §262.11[2] ■ Ensure the following if the shipment is not to an OECD country[3] — The shipment complies with §§262.53, 262.56(a)(1–4, 6) and (b), and 262.57 — The batteries are exported only upon consent of the receiving country and shipped in conformance with the EPA Acknowledgement of Consent — A copy of the EPA Acknowledgement of Consent is provided to the transporter so it accompanies the shipment for export ■ Comply with Part 262, Subpart H if the shipment is to an OECD country[3]	■ Make hazardous waste determination per §262.11[2] ■ Ensure the following if the shipment is not to an OECD country[3] — The shipment complies with §§262.53, 262.56(a)(1–4, 6) and (b), and 262.57 — The batteries are exported only upon consent of the receiving country and shipped in conformance with the EPA Acknowledgement of Consent — A copy of the EPA Acknowledgement of Consent is provided to the transporter so it accompanies the shipment for export ■ Comply with Part 262, Subpart H if the shipment is to an OECD country[3]

Table 4-8: RCRA Requirements for Managing Spent Lead-Acid Batteries

Person who:	Requirements if lead-acid batteries will be regenerated (e.g., recharging, replacing electrolyte)[1]	Requirements if lead-acid batteries will be reclaimed (e.g., breaking, cracking, smelting) to recover metal values
Transports batteries for reclamation in a foreign country	■ Ensure the following if the shipment is not to an OECD country[3] — The shipment is rejected if it does not conform to the EPA Acknowledgement of Consent — A copy of the EPA Acknowledgement of Consent accompanies the shipment for export — The shipment is delivered to the facility designated by the exporter ■ Comply with applicable provisions in Part 262, Subpart H if the shipment is to an OECD country[3]	■ Ensure the following if the shipment is not to an OECD country[3] — The shipment is rejected if it does not conform to the EPA Acknowledgement of Consent — A copy of the EPA Acknowledgement of Consent accompanies the shipment for export — The shipment is delivered to the facility designated by the exporter ■ Comply with applicable provisions in Part 262, Subpart H if the shipment is to an OECD country[3]

LDR = land disposal restrictions; OECD = Organization for Economic Cooperation and Development.

[1]The battery casing is not broken or cracked to recover metal values. [RO 11934]

[2]Such spent lead-acid batteries will likely be D002 and D008. [RO 14147]

[3]OECD countries are given in §262.58(a)(1).

Source: McCoy and Associates, Inc.; adapted from §266.80.

---

and entities that store these batteries (but do not reclaim them), must make a hazardous waste determination. In addition, when these entities send batteries for reclamation (as opposed to regeneration), they must complete a one-time LDR notification. (Based on a February 25, 2008 email from EPA, the one-time LDR notice should include the information required in §268.7(a)(2), since these batteries do not meet the appropriate LDR treatment standard at the point they are shipped offsite.) But, the generating facility would not have to: 1) count these batteries when making a hazardous waste generator-status determination, 2) manifest them/use hazardous waste transporters (DOT requirements still apply), or 3) store them onsite per §262.34 or Part 265, Subpart I container standards (e.g., no 90-day clock, no labeling, no inspection requirements). [RO 13746, 14147]

Q Do the speculative accumulation provisions in §261.1(c)(8) apply to spent lead-acid batteries before they are sent offsite for reclamation under Part 266, Subpart G?

A No. The speculative accumulation provisions are not applicable to materials already defined as solid wastes, such as spent lead-acid batteries. [RO 11476]

Q Do spent lead-acid batteries have to be counted when making monthly generator-status determinations?

A No. Hazardous waste is counted for determining monthly generator status only if it is subject to substantive RCRA regulation (see Section 6.1.3). Entities who generate, transport, or collect intact spent lead-acid batteries destined for reclamation, and entities that store these batteries (but who do not reclaim them themselves), are not subject to substantive regulation. Therefore, spent lead-acid batteries destined for reclamation are not counted. [RO 13746]

Q *Are leaks and discharges from lead-acid batteries also covered under the Part 266, Subpart G exemption?*

A No. "Materials generated from a leak or discharge become newly generated wastes and, as such, are subject to a hazardous waste determination." [RO 14039]

4.20.1.2 Battery regeneration facilities are also exempt

Per §266.80(a), facilities that regenerate spent lead-acid batteries (such as by recharging and/or replacing electrolyte) under Part 266, Subpart G are also exempt from most RCRA requirements (see Table 4-8). EPA believes that the simple regeneration and subsequent resale of reconditioned batteries presents minimal environmental risk and is similar to the reclamation of commercial chemical products, which is not regulated under RCRA. Such regeneration can occur at any type of facility. [April 4, 1983; 48 *FR* 14496, RO 11167, 11934, 12449, 13709, 14039]

4.20.1.3 Battery reclamation facilities are not exempt

Although entities that generate, transport, and collect intact spent lead-acid batteries, and entities that store these batteries (but who do not reclaim them), are exempt from most hazardous waste management requirements, facilities that both store and reclaim them are not. [§266.80(b)] Facilities that both store and reclaim spent lead-acid batteries to recover metal values (such as by breaking, cracking, and/or smelting) under Part 266, Subpart G are subject to numerous TSD facility requirements, as noted in Table 4-8. [RO 11383, 12688, 12856, 14147] Thus, such facilities are subject to most Subtitle C provisions, including RCRA permitting, contingency planning, and hazardous waste storage, manifesting, and closure standards.

However, because battery reclamation is considered hazardous waste recycling, facilities that reclaim the batteries, but do not store them first, are not subject to substantive RCRA standards but are subject only to the same minimal provisions as generators of these materials. [RO 11383, 12856]

Q *The owner/operator of a facility that collects spent lead-acid batteries drains the acid and then manifests the acid offsite for reclamation. The battery shell, which still contains the lead plates, is sent to a facility that extracts the lead for smelting. Should these battery shells, when sent offsite, be managed as hazardous waste or as spent lead-acid batteries subject to Part 266, Subpart G?*

A EPA never really answered the question in this guidance. However, the agency did note that the act of draining the batteries is not considered part of the reclamation process. Therefore, the owner/operator would not be subject to requirements for facilities that both store and reclaim batteries in §266.80(b) (although the facility that smelts the batteries would be subject to these provisions). [RO 12836]

Q *A company operates a number of battery cracking and smelting facilities. A battery storage facility is owned by the same company, but the storage facility is not co-located with any of its battery reclamation facilities. Is the storage facility subject to the §266.80(b) requirements?*

A No. Only the minimal provisions in §266.80(a) apply to facilities at which spent lead-acid batteries are stored but not reclaimed, regardless of what battery management activities the owner of the storage area may conduct at other locations. [RO 11947, 14039]

4.20.2 Exporting spent lead-acid batteries

As of July 7, 2010, export shipments of spent lead-acid batteries (SLABs) for reclamation are prohibited in all states unless the exporter has 1) submitted a written notification to EPA, and 2) obtained the receiving country's consent. [§266.80(a), January 8, 2010; 75 *FR* 1236, RO 14825, EPA/530/F-10/007, April 2010, available at http://www.epa.gov/epawaste/hazard/international/slab-flyer.pdf] This revision harmonizes the notice and consent procedures for SLABs managed under Part 266, Subpart G with those required for SLABs managed under the universal waste regulations in Part 273.

Section 266.80(a) requires the SLABs exporter to follow select Part 262, Subpart E requirements. To summarize, the exporter must send a notification of intent to export to EPA's Office of Enforcement and Compliance Assurance 60 days prior to exporting the SLABs to a foreign facility. The notification must include information such as a description of the waste, estimates of the total quantity of waste that will be exported and the frequency at which it will be exported, and the name of the foreign destination facility. This notification may cover export activities over a 12-month period. EPA, in conjunction with the Department of State, will provide a complete notification to the receiving country and any transit countries. If the receiving country consents to the receipt of the SLABs, EPA will forward an Acknowledgement of Consent to the SLABs exporter. [§262.53] The SLABs exporter must ship the SLABs in conformance with the EPA Acknowledgement of Consent and provide a copy of that document to the transporter so it accompanies the shipment for export.

The SLABs exporter must file annual reports by March 1 of each year covering such exporting activities in accordance with §262.56(a)(1–4, 6) and (b). Finally, exporters of SLABs must keep copies of each notification of intent to export, Acknowledgement of Consent, confirmation of delivery from the consignee, and annual reports for at least three years. [§262.57]

In the case where the receiving country for the SLABs is one of the Organization for Economic Cooperation and Development (OECD) countries, the standards of Part 262, Subpart H rather than Subpart E apply. OECD countries are listed at §262.58(a)(1). Canada and Mexico are not on the OECD list, so Subpart E is applicable if shipping SLABs to either of these countries for reclamation.

4.21 Scrap metal exemption

EPA's regulatory scheme for scrap metal is a little tricky. First of all, the agency excludes a material called "excluded scrap metal" (as long as it will be recycled) from the definition of solid waste (and therefore from the hazardous waste regulations) under §261.4(a)(13). All other scrap metal sent for recycling/reclamation is a solid waste as noted in Table 1 in §261.2(c) and, therefore, is a hazardous waste if it exhibits a characteristic or has become contaminated with a listed waste. However, in a "hidden" section that some people aren't aware of [§261.6(a)(3)(ii)], EPA exempts all hazardous scrap metal that is sent for recycling/reclamation from RCRA regulation [if it is not already excluded under §261.4(a)(13)]. Finally, scrap metal that is to be disposed gets no breaks at all; it is a solid waste, and the generator must determine whether it is hazardous and manage it under full Subtitle C regulation if it is.

Details of these exclusions/exemptions/regulatory requirements are provided below, along with discussions of how EPA regulates specific types of scrap metal, such as printed circuit boards, used oil and fuel filters, punctured aerosol cans, and lead shot reclaimed from shooting ranges.

4.21.1 What is scrap metal?

We talked a little about scrap metal in Section 1.2.5.5. It is defined as "bits and pieces of metal parts (e.g., bars, turnings, rods, sheets, wire) or metal pieces that may be combined together with bolts or soldering (e.g., radiators, scrap automobiles, railroad box cars), which when worn or superfluous can be recycled." [§261.1(c)(6)] In other words, scrap metal is metal that has become worn or is no longer needed and is usually recycled to recover its metal content. Metal pieces that are generated from machining operations (i.e., turnings, stampings, etc.) are also considered scrap metal. "[T]o meet the definition of scrap metal, the material must have significant metal content, i.e., greater than 50% metal." [RO 13356]

Scrap metal does not include residues generated from smelting and metal refining operations (i.e., drosses, slags, and sludges). It also does not include liquid wastes containing metals (i.e., spent acids, spent caustics, or other liquid wastes with metals in solution), liquid metal wastes (i.e., liquid mercury), or metal-containing wastes with a significant liquid component, such as spent batteries. [January 4, 1985; 50 *FR* 624, RO 14183, 14195]

4.21.2 "Excluded scrap metal"

As mentioned above, "excluded scrap metal" being recycled is not a solid waste, so it can't be a hazardous waste. [§261.4(a)(13)] "Excluded scrap metal," which is defined in §261.1(c)(9–12), includes:

- *Processed scrap metal*—Scrap metal that has been baled, shredded, sheared, chopped, crushed, flattened, cut, melted, or separated by metal type (i.e., sorted), and fines, drosses, and related materials that have been agglomerated. This category applies to scrap metal that has undergone a manual or mechanical processing step, regardless of who does the processing (e.g., the generator, broker, scrap processor, or scrap recycler). [May 12, 1997; 62 *FR* 26011]

Drained, used oil filters may qualify as processed scrap metal if they are processed prior to recycling. Draining the oil by itself would not be considered processing, because processing requires physical alteration of the metal filter. However, hot-draining followed by crushing, dismantling followed by hot-draining, or shredding would constitute processing and enable the filters to qualify as processed scrap metal. [RO 14183, 14202]

Agglomerated drosses (i.e., solid chunks of metal in a physical state that does not allow them to be easily crushed, split, or crumbled) that are generated from operations such as sintering or melting can be classified as processed scrap metal if recycled. Thus, agglomerated drosses from primary and secondary metal processing operations may qualify for the exclusion. The exclusion would continue to apply even if the recycling involved agglomerated dross that is used in a manner constituting disposal. [RO 14195]

- *Unprocessed home scrap metal*—Scrap metal (such as turnings, cuttings, punchings, and borings) generated by steel mills, foundries, and metals refineries.

- *Unprocessed prompt scrap metal*—Scrap metal (such as turnings, cuttings, punchings, and borings) generated by metal-working/fabrication industries.

EPA decided to exclude these three types of scrap metal from solid and hazardous waste regulation because they are more product-like than waste-like, making RCRA jurisdiction over these materials unnecessary. Specifically, the agency believes that excluded scrap metal being recycled has: 1) established markets for its utilization, 2) an inherent positive economic value, 3) a stable physical form, and 4) no damage incidents attributable to these types of scrap metal. [May 12, 1997; 62 *FR* 26011]

4.21.3 Other recycled scrap metal is exempt

All material that meets the definition of "scrap metal" but not the definition of "excluded scrap metal" is what we call "other scrap metal." Other scrap metal sent for recycling/reclamation is a solid waste under §261.2(c)(3) and, therefore, is a hazardous waste if it exhibits a characteristic or has become contaminated with a listed waste. However, §261.6(a)(3)(ii) exempts hazardous scrap metal that is sent for recycling/reclamation from the RCRA regulations in Parts 262–270. [January 4, 1985; 50 *FR* 624, RO 11057, 11063, 11740, 11806, 11877]

One of the provisions from which generators of scrap metal are exempted under §261.6(a)(3)(ii) is the need to make a hazardous waste determination for this material (the requirement to make a hazardous waste determination is in §262.11). Therefore, it is unnecessary to make a hazardous waste determination (i.e., whether it exhibits a characteristic or is contaminated with listed hazardous waste) for material that meets the definition of scrap metal and will be recycled. [RO 11782, 11806, 11835, 14184]

Under the §261.6(a)(3)(ii) exemption, any hazardous scrap metal that is sent for recycling/reclamation is exempt from the hazardous waste management requirements, including storage prior to reclamation, manifesting, and the land disposal restrictions. [RO 11134, 11383, 11482, 11600, 14277] Furthermore, the exemption is viewed prospectively. In other words, as long as the generator intends to recycle the scrap metal at some point in the future, the scrap metal is exempt from the hazardous waste regulations *starting at the point of generation*. However, anyone who claims such an exemption must be able to document that legitimate recycling occurs. [RO 11877, 14277] Finally, scrap metal sent for recycling is also exempt from speculative accumulation provisions:

"The [speculative accumulation] provision does *not* apply to secondary materials that already are wastes when they are recycled, for example scrap metal...." [Emphasis in original.] [January 4, 1985; 50 *FR* 635]

Q During building demolition, building steel and certain metallic tanks, equipment, piping, and ductwork are generated. The building owner would like to scrap these materials, but they have been painted with numerous coats of lead-based paint, causing the components and equipment to potentially exhibit the toxicity characteristic for lead. Do these materials qualify for the §261.6(a)(3)(ii) exemption from hazardous waste management?

A Yes. As long as the building steel and metallic equipment will be recycled/reclaimed, they are not subject to any RCRA storage, transportation, treatment, or disposal regulations. Furthermore, as long as they meet the definition of scrap metal (e.g., there is no significant liquid component associated with them or the scrap metal isn't mixed with some other material), there is no need for the generator to make a hazardous waste determination (i.e., run the TCLP).

Other examples of hazardous scrap metal that qualify for the §261.6(a)(3)(ii) exemption when sent for recycling are listed in Table 4-9.

Table 4-9: Examples of Materials That Qualify for the §261.6(a)(3)(ii) Scrap Metal Exemption When Sent for Recycling/Reclamation

Material	RO reference
Automotive and electronic devices	11432
Brass particles	11769
Drained used oil and fuel filters	13498, 14183, 14184
Lead plates from batteries	11100, 11383
Lead shot	14070
Natural gas regulators	11860
Scrap metal containing precious metals	11117
Solder drippings generated during radiator repair operations	11740
Spent lead foil used in dental x-ray packaging	11742
Spent photoconductor drums (made of cadmium sulfide-coated aluminum) from photocopying machines	11710
Spent solder baths	11771, 11775, 13628
Steel aerosol cans that do not contain a significant amount of liquid (i.e., cans that have been punctured and drained)	11782, 11806
Steel battery cases from which the acid and lead components have been removed	11100
Used oxygen breathing apparatus canisters (i.e., canisters that have been punctured and rinsed)	11835
Used printed circuit boards (whole, not shredded)	11689, 14155
Used torpedo boilers	11134
Zinc bar, nickel plate, cadmium plate, and steel scrap removed from spent alkaline batteries	11184, 13376

Source: McCoy and Associates, Inc.

4.21.4 Scrap metal that is not exempt from RCRA

It is important to note that scrap metal that is not recycled (i.e., it is sent for disposal) is *not* excluded under §261.4(a)(13) or exempt under §261.6(a)(3)(ii). Instead, scrap metal that will be disposed is a solid waste that is subject to the full set of hazardous waste rules if it exhibits a characteristic or has become contaminated with a listed waste. [RO 11806, 11835, 13639, 14183]

4.21.4.1 Scrap metal that is mixed with something else

Hazardous scrap metal, even if sent for recycling, is not exempt from RCRA regulation if it is mixed with other materials (e.g., spent materials) that make the mixture hazardous by characteristic or listing. In this situation, the mixture doesn't meet the definition of scrap metal, and so the entire mixture would need to be managed as hazardous waste. For example, the following mixtures are not exempt scrap metal and are subject to regulation as hazardous waste:

- Used oil or fuel filters with a steel shell (scrap metal) that have not been punctured, crushed, or otherwise drained and still contain significant oil or fuel, where the mixture exhibits a characteristic. [RO 11566, 14183, 14184]
- Lead plates (scrap metal) mixed with lead oxide material, where both materials come from spent batteries, and where the mixture exhibits the toxicity characteristic for lead. [RO 11100, 11383]
- Lead shot (scrap metal) mixed with smokeless gunpowder, where the mixture exhibits the characteristics of ignitability and toxicity for lead. [RO 14070]
- Natural gas regulators (scrap metal) that contain liquid mercury, where the mixture exhibits the toxicity characteristic for mercury. [RO 11860]
- Steel aerosol cans (scrap metal) that have not been punctured and still contain significant liquids, where the mixture exhibits a characteristic and/or meets a listing description. [RO 11782, 11806]
- Zinc bar, nickel plate, cadmium plate, and steel scrap removed from spent alkaline batteries (scrap metal) mixed with a nonscrap metal material, where the mixture exhibits a characteristic. [RO 11184, 13376]

If the scrap metal in the above examples can be separated or segregated from the other materials in the mixture, then it could subsequently be managed as RCRA-exempt. [RO 11782, 11806, 14070, 14183, 14184]

The presence of an oily film on scrap metal (e.g., oil-coated steel turnings generated in a machine shop) would not be considered a mixture, and, therefore, the oily metal would qualify for the §261.6(a)(3)(ii) exemption if recycled. [RO 11184, 11783, 13639]

4.21.4.2 Scrap metal managed improperly loses the exemption

Even though hazardous scrap metal sent for recycling is exempt from RCRA regulation, persons managing such materials should be careful to make sure that any releases into the environment are minimized. According to EPA, if hazardous scrap metal management practices (e.g., the storage of scrap metal on the ground) cause hazardous constituents to leach into the soil or surface water or disperse into the air, such a release may be considered abandonment through disposal and trigger RCRA hazardous waste regulations. [RO 11771]

4.21.5 Printed circuit boards

Printed circuit boards are major components in personal computers and other electronic devices. As updated computers and other electronic equipment are constantly becoming available, large numbers of older machines are being scrapped. Such equipment is typically taken apart to salvage parts (such as circuit boards) for their monetary value via metal reclamation. In many situations, circuit boards are shredded prior to recycling, sometimes to protect proprietary design information. As discussed in the following sections, whole circuit boards and shredded circuit boards are regulated differently.

4.21.5.1 Whole circuit boards

According to EPA, whole, used circuit boards may be considered scrap metal and therefore qualify for the §261.6(a)(3)(ii) exemption when recycled. [May 12, 1997; 62 FR 26013, RO 11689, 14155] For a while, the agency's position was that whole, used circuit boards that contain mercury switches, mercury relays, nickel-cadmium batteries, or lithium batteries do not meet the definition of scrap metal because mercury (a liquid metal) and batteries (which may contain a significant liquid component) do not fall under the definition of "scrap metal." As such, the scrap metal exemption could not apply to these boards when recycled. [May 12, 1997; 62 FR 26013]

However, EPA later became aware that mercury switches and relays on circuit boards from typical applications contain minimal amounts of mercury (0.02–0.08 grams) and are encased in metal and coated in epoxy prior to attachment to the boards. Consequently, the agency changed its position, stating "[i]t is not the agency's current intent to regulate under RCRA circuit boards containing minimal quantities of mercury and batteries that are protectively packaged to minimize dispersion of metal constituents." However, once any switches or batteries are removed from the circuit boards during reclamation, the switches/batteries become a newly generated waste subject to hazardous waste determination. If they exhibit a characteristic, they must be managed as hazardous waste. [May 26, 1998; 63 FR 28630, RO 14155]

The above discussion is for whole, used printed circuit boards. Unused boards (e.g., ones that are found to be faulty after production at the manufacturing facility) would be considered off-specification commercial products. When such products are sent for reclamation, they are not solid wastes per Table 1 in §261.2(c). [RO 11689]

4.21.5.2 Shredded circuit boards

When circuit boards are shredded (a common practice prior to reclamation), small fines are generated that can be dispersed during subsequent handling. Accordingly, the shredded boards no longer meet the regulatory definition of "scrap metal" and therefore cannot qualify for the §261.6(a)(3)(ii) exemption. [May 12, 1997; 62 FR 26012, RO 11689]

However, EPA believes that shredding makes the circuit boards easier to handle, assay, and ship without generator confidentiality concerns. In addition, the shredded circuit boards are commodity-like due to their positive economic value. Smelters often pay $0.25/lb to $5/lb for shredded circuit boards, which may contain up to 18% copper and frequently contain precious metals (such as gold, silver, or platinum). As such, the agency concluded that shredded circuit boards qualify as secondary materials that should be excluded from RCRA regulation if certain conditions are met.

Therefore, on May 12, 1997 [62 FR 26012], EPA added an exclusion from the definition of solid waste specifically for shredded circuit boards being recycled if they 1) are stored in containers sufficient to prevent a release to the environment prior to recovery; and 2) are free of mercury switches, mercury relays, nickel-cadmium batteries, and lithium batteries. [§261.4(a)(14)] If mercury switches, mercury relays, nickel-cadmium batteries, and lithium batteries are removed from the whole circuit boards prior to shredding, the shredded boards remain within the scope of the exclusion. Conversely, shredded circuit boards that contain mercury switches, mercury relays, nickel-cadmium batteries, and lithium batteries are solid wastes when reclaimed and are subject to all applicable hazardous waste regulations if they exhibit a characteristic. [RO 14692]

4.21.6 Used oil and fuel filters

Depending on their material of construction and providing they have been hot drained, most used oil filters are excluded from hazardous waste regulation under §261.4(b)(13) (see Section 12.1.5 for a complete discussion of this topic). If they don't qualify for that exclusion, drained filters may alternatively be considered scrap metal [and therefore be eligible for the §261.6(a)(3)(ii) exemption] if they do not contain a significant amount of liquid and they will be recycled. Used oil filters do not contain a significant amount of liquid once the used oil has been drained or removed to the extent possible (e.g., such as by crushing) such that there are no visible signs of free-flowing oil. [§279.10(c), RO 11566, 13498, 14183, 14184]

The scrap metal exemption will apply to other types of drained filters as well. For example, metal fuel filters that are drained and no longer contain a significant liquid component meet the scrap metal definition and are exempt from hazardous waste regulation if recycled. [RO 14184]

4.21.7 Punctured aerosol cans

EPA's current regulatory interpretation is that a steel aerosol can that does not contain a "significant" amount of liquid would meet the definition of scrap metal in §261.1(c)(6). Consequently, if the can is recycled as scrap metal, it would be exempt from RCRA regulation under §261.6(a)(3)(ii). As such, a determination of reactivity or any other characteristic would not be necessary. [RO 11782, 11806]

According to the agency, aerosol cans do not contain "significant" liquids if they "have been punctured so that most of any liquid remaining in the can may flow from the can…and drain (e.g., with punctured end down)…." However, it is very important to note that this exemption applies only to the punctured can itself; any liquids and propellants removed from the aerosol cans are subject to regulation as hazardous wastes if they are listed or if they exhibit a characteristic. [RO 11782] (See Section 2.5 for more details.) Because the scrap metal exemption is viewed prospectively (as noted above), the puncturing step would not be considered hazardous waste treatment requiring a permit.

Similarly, EPA has determined that used oxygen breathing apparatus canisters (which typically exhibit the characteristic of ignitability) that are punctured and rinsed constitute scrap metal. Therefore, the rinsed canisters would be exempt from RCRA regulation under §261.6(a)(3)(ii) if they are recycled. Again, the process of emptying the canisters would be considered part of the scrap steel recycling process and not subject to RCRA permitting. However, if the canisters are disposed, they would be solid wastes, and a hazardous waste determination would have to be made. [RO 11835]

4.21.8 Lead shot reclaimed from shooting ranges

In guidance, EPA has noted that the discharge of ball and sport ammunition at shooting ranges does not constitute hazardous waste disposal. Although many of these bullets/shot are made of lead, the agency determined that the deposition of lead at shooting ranges was within the normal and expected use pattern of the manufactured product, and that the resultant contamination was not discard and so not subject to RCRA regulation. [RO 11368, 11700, 13525]

Based on limited guidance from EPA headquarters and Region II, it appears that lead shot and bullets reclaimed from shooting ranges are also not subject to RCRA regulation. Additionally, the reclamation step and subsequent redeposition of the soil or sand onto the range are also exempt.

Based on guidance in EPA Region II's document, *Best Management Practices for Lead at Outdoor Shooting Ranges*, EPA/902/B-01/001, June 2005, available at http://www2.epa.gov/sites/production/files/documents/epa_bmp.pdf, the agency is calling recovered lead shot/bullets at a shooting range "scrap metal" if they will be recycled or reused.

In addition to the Region II guidance, the agency has noted previously that lead shot/bullets, when recycled, are considered scrap metal and therefore are exempt from RCRA regulation. EPA does not consider such lead shot/bullets from weapons to be discarded, because the discharge is the normal and expected use of this manufactured product. As noted early on, RCRA does not apply to products that are used for their intended purpose. Therefore, lead shot/bullets would not be considered spent material but would be scrap metal when recycled. [RO 14070]

Finally, the process of recovering lead from a range and separating it from sand and soil would be addressed the same way it was in the military munitions rule. [February 12, 1997; 62 *FR* 6622] In that rule, EPA noted that range clearance activities (recovering lead shot/bullets) are an intrinsic part of range use and are, therefore, excluded from RCRA regulation. (See Section 14.5 for a discussion of the military munitions rule.) In guidance, EPA clarified that the military range clearance principles apply equally to nonmilitary ranges. [Letter from Elizabeth Cotsworth to John Cahill, April 29, 1997, found in Appendix D of *Best Management Practices for Lead at Outdoor Shooting Ranges*, referenced previously.]

4.22 Exclusion for geologic sequestration of CO_2

On January 3, 2014 [79 *FR* 350], EPA excluded waste carbon dioxide (CO_2) streams injected underground for geologic sequestration from the definition of hazardous waste. Generators and well owners/operators must meet the following conditions to claim the new exclusion: these CO_2 streams must be 1) captured from emission sources (e.g., power plants or other industrial sources), 2) transported via DOT-regulated (or state equivalently-regulated) pipelines or vehicles, and 3) injected into SDWA Class VI wells for purposes of geologic sequestration.

Generators of these excluded CO_2 streams and Class VI well owners/operators must separately certify that no hazardous wastes have been mixed, or otherwise co-injected, with the CO_2 stream. These certifications must be updated annually, kept onsite for three years, and be accessible on the facilities' publicly-available websites.

EPA noted in the preamble to the rule that the exclusion does not affect the regulatory status of CO_2 streams that are injected into wells other than SDWA Class VI wells for geologic sequestration. The agency specifically noted that "should CO_2 be used for its intended purpose as it is injected into UIC Class II wells for the purpose of [enhanced oil or gas recovery], it is EPA's expectation that such an injection process would not generally be a waste management activity." [79 *FR* 355]

The exclusion has been added as a new section, §261.4(h), effective March 4, 2014 in the federal RCRA regs. Because this exclusion makes the federal program less stringent, states are not required to adopt the new provision (although EPA encourages all states to do so). "EPA notes that in situations involving the interstate transportation of conditionally-excluded waste, the exclusion must be authorized in the state where the waste is generated, any states through which the waste passes, and the state where the UIC Class VI injection well is located, in order for that conditionally-excluded waste to be managed as excluded from Subtitle C from point of generation to injection in a UIC Class VI well. A state that has not adopted the conditional exclusion may impose state requirements, including the uniform hazardous waste manifest requirement (where applicable) if characteristically-hazardous CO_2 streams are being transported through that state." [79 *FR* 360]

Rules for Management

Regulatory status of hazardous waste mixtures, treatment residues, and materials that contain hazardous waste

Once you have determined that you have generated a solid and hazardous waste (based on the discussions in Chapters 1–4), subsequent management of the waste must comply with the RCRA regulations. In some cases, a generator may mix hazardous waste with other wastes or treat the generated waste before it is shipped. The rules for mixing hazardous waste with other materials are discussed in Section 5.1, while the regulatory status of residues that result from treating hazardous wastes is reviewed in Section 5.2.

Sometimes, hazardous waste contaminates environmental media (e.g., soil) or products (e.g., protective gloves). In such cases, EPA's contained-in policy can apply to capture these materials under RCRA. This policy is reviewed in Section 5.3.

5.1 The mixture rule

As noted in Section 2.1.1, there are four types of hazardous waste. We have previously discussed the first three ways that a waste may be hazardous: 1) if a generator declares it to be hazardous—see Section 2.1.1.1, 2) if it exhibits a characteristic—see Chapter 2, or 3) if it meets a listing description—see Chapter 3. The fourth way that a facility can have a hazardous waste is if "[i]t is a mixture of solid waste and one or more hazardous wastes listed in Subpart D [of Part 261]." [May 19, 1980; 45 *FR* 33119]

Hazardous wastes are frequently (although not always intentionally) mixed with other wastes. What is the regulatory status of the mixture? Is the mixture also hazardous? Many times it is. The RCRA regulations are quite prescriptive regarding the status of such mixtures—the rules are found in §261.3(a), (b), and (g). But before getting into the details of how the mixture rule works, it's important to consider first if such mixing is even legal. In other words, the mixture rule is used to determine the RCRA status of the mixture, but is it legal to mix the hazardous waste with some other material in the first place?

The concern with mixing is dilution. RCRA prohibits dilution of hazardous waste as a substitute for adequate treatment. When hazardous waste is mixed with something else, that other material tends to dilute the hazardous waste, and the dilution prohibition in the regulations may prohibit such mixing. The dilution prohibition is actually contained within the land disposal restrictions (LDR) regulations—see Section 13.11. Therefore, whenever you are thinking about mixing hazardous waste with something else, the first consideration should be "Is this activity permissible under the LDR program?" Section 5.1.3 discusses the LDR ramifications of mixing in more detail.

In addition to the LDR ramifications of mixing hazardous waste, the second concern is the issue of treatment. When a facility mixes hazardous waste with something else, there's a good chance that this activity will meet the definition of "hazardous waste treatment." In many cases, hazardous waste treatment requires a RCRA permit (although there are eight exemptions that excuse facilities from the permitting requirement). Section 5.1.4 and Chapter 7 deal with issues relating to hazardous waste treatment.

Therefore, before you mix hazardous waste with something else, make sure you evaluate the LDR ramifications and treatment-requiring-a-permit issue.

Let's start out with some background and then look at the details of how the hazardous waste mixture rule operates.

5.1.1 Background

When the RCRA regulations came out in 1980, the mixture rule was part of the new program, even though it had not been part of the December 1978 proposal. EPA explained this inconsistency by stating:

> "The waste mixtures provision is a clarification which has been added in response to inquiries about whether mixtures of hazardous and nonhazardous wastes would be subject to Subtitle C requirements. This is a very real issue in real-world waste management, since many hazardous wastes are mixed with nonhazardous wastes or other hazardous wastes during storage, treatment, or disposal.
>
> "Although it was not expressly stated in the proposed regulation, EPA intended waste mixtures containing listed hazardous wastes to be considered a hazardous waste and managed accordingly. Without such a rule, generators could evade Subtitle C requirements simply by commingling listed wastes with nonhazardous solid waste. Most of these waste mixtures would not be caught by the Subpart C characteristics because they would contain wastes which were listed for reasons other than that they exhibit the characteristics (e.g., they contain carcinogens, mutagens, or toxic organic materials). Obviously, this would leave a major loophole in the Subtitle C management system and create inconsistencies in how wastes must be managed under that system." [45 *FR* 33095]

Industry, however, was not as enthusiastic about the mixture rule as EPA; hence, they filed suit against the agency. One of industry's arguments was that EPA had failed to properly propose the rule before it was finalized. The case was finally decided by the U.S. Court of Appeals for the District of Columbia Circuit in December 1991 (*Shell Oil Company v. EPA*, 950 F.2d 741). That court agreed with industry that the 1978 proposal did not adequately provide notice of the rule and give industry sufficient opportunity to comment. Thus, the mixture rule was vacated (i.e., thrown out) by the court—not because the rule didn't make sense—but because of procedural reasons.

The court denied EPA's request to reconsider its decision but suggested that the agency could reinstate the rule on an emergency basis (under the "good cause" exemption of the Administrative Procedure Act), giving EPA time to propose and finalize replacement regulations. EPA acted on this advice and reinstated the mixture rule on an interim basis on March 3, 1992. [57 *FR* 7628] On May 16, 2001, EPA finalized a rule that retained the mixture rule essentially as it has been in the regs all along. [66 *FR* 27266] The rule also added a new section [§261.3(g)] to the regulations, addressing mixtures of ICR-only listed wastes (see Section 5.1.2.2).

5.1.2 Three elements to the mixture rule

There are three elements to the hazardous waste mixture rule:

1. If a characteristic hazardous waste is mixed with a solid waste, the resulting mixture is

hazardous unless it does not exhibit a characteristic. [§261.3(b)(3)] Although the regulatory language in that section is not very clear, EPA has been consistent in its guidance that "[m]ixtures containing only characteristic and nonhazardous wastes are hazardous only if the mixture exhibits the characteristic according to §261.3(b)(3)." [RO 11173] "Waste mixtures containing only wastes which meet the characteristics are treated just like any other solid waste; they will be considered hazardous only if they exhibit the characteristics." [45 FR 33095; see also RO 11140, 12030]

2. If a hazardous waste that was listed in §§261.31–261.33 solely because it exhibits the characteristic of ignitability, corrosivity, and/or reactivity (i.e., an ICR-only listed waste) is mixed with a solid waste, the resulting mixture is hazardous unless it does not exhibit a characteristic. [§261.3(g)(2)(i)] While there are 29 such listed wastes, as summarized in Table 5-1, one of the most commonly encountered ICR-only listed wastes is F003 spent solvent. In effect, this element of the mixture rule acts as a self-implementing delisting (i.e., a listed waste may become nonhazardous through mixing). [RO 11140, 11213, 12721] This second element was not part of the original 1980 RCRA regulations but was added November 17, 1981. [46 FR 56588]

3. If a listed hazardous waste (other than an ICR-only listed waste) is mixed with a solid waste, the resulting mixture is a listed hazardous waste. [§261.3(a)(2)(iv) and (b)(2)] This holds true even if the resulting mixture contains only very small amounts of listed waste and/or very low concentrations of hazardous constituents; that is, "[t]here is no de minimis amount below which a listed waste need not be identified." [RO 11327] The only way such a mixture ceases to be hazardous is if the generator petitions the agency for a delisting. [§261.3(d)(2)]

Table 5-1: Wastes Listed Solely Because They Exhibit a Characteristic

Waste code	Description
F003	Nonhalogenated spent solvents (I)
K044	Wastewater treatment sludges from explosives (R)
K045	Spent carbon from treating explosive wastewater (R)
K047	Pink/red water from TNT operations (R)
P009	Ammonium picrate (R)
P081	Nitroglycerine (R)
P112	Tetranitromethane (R)
U001	Acetaldehyde (I)
U002	Acetone (I)
U008	Acrylic acid (I)
U020	Benzenesulfonyl chloride (C, R)
U031	n-Butyl alcohol (I)
U055	Cumene (I)
U056	Cyclohexane (I)
U057	Cyclohexanone (I)
U092	Dimethylamine (I)
U096	Cumene hydroperoxide (R)
U110	Di-n-propylamine (I)
U112	Ethyl acetate (I)
U113	Ethyl acrylate (I)
U117	Ethyl ether (I)
U124	Furan (I)
U125	Furfural (I)
U154	Methanol (I)
U161	Methyl isobutyl ketone (I)
U186	1,3-Pentadiene (I)
U189	Sulfur phosphide (R)
U213	Tetrahydrofuran (I)
U239	Xylene (I)

(I) = ignitability; (C) = corrosivity; and (R) = reactivity.
Source: November 19, 1999; 64 FR 63390.

One very important aspect of the mixture rule is that it applies only to mixtures of hazardous wastes and solid wastes. Solid wastes are generally something we have decided to discard or dispose. The rule does not apply to mixtures of hazardous waste and products [RO 11175, 11198, 11219, 11320, 11387, 14334] or to mixtures of hazardous waste and soil, ground water, surface water, or other environmental media. [RO 11707]

Instead, the contained-in policy applies in those situations—see Section 5.3 of this chapter. Since hazardous wastes are by definition solid wastes, the mixture rule does apply when two or more hazardous wastes are mixed.

Based on operation of the three elements of the mixture rule, a mixture containing a hazardous waste may or may not be hazardous. If it is hazardous, all appropriate waste codes must be carried on the mixture. For example, if an arsenic characteristic waste (D004) is mixed with a solid waste and the mixture is still toxic for arsenic, the mixture will continue to carry the D004 code. If a non-ICR-only F-waste is mixed with a non-ICR-only U-waste, both codes must be carried on the resulting mixture. [RO 13024] Whether or not a characteristic code must be carried on a mixture that carries one or more listed waste codes will be determined by the LDR regulations—see Section 13.6.

Each of the three elements of the mixture rule is discussed in more detail below, with examples added for clarification.

5.1.2.1 Mixtures of characteristic wastes

Mixtures of characteristic hazardous wastes and solid wastes are hazardous only if they exhibit any of the characteristics. If such mixtures do not exhibit any characteristic, they are not hazardous. Keep in mind, however, that such decharacterized mixtures will likely still have to meet LDR treatment standards before they may be land disposed. This topic is discussed in more detail in Section 5.1.3.2.1. Also keep in mind that such mixing probably meets the "treatment" definition in §260.10, as noted in Section 5.1.4.

5.1.2.2 Mixtures of ICR-only listed wastes

Mixtures of ICR-only listed hazardous wastes and solid wastes are hazardous if the mixture exhibits any characteristic—not just the one exhibited by the ICR-only waste. [RO 11140, 11257] If the mixture exhibits no characteristic, it is not a hazardous waste. As with mixtures of characteristic wastes, however, it will still be subject to LDR treatment standards before land disposal (see Section 5.1.3.2.2 and RO 14836) and to the mixing-as-treatment considerations. A number of examples of this element of the mixture rule are given as follows.

Q *Spent solvent methanol (F003) is mixed with a nonhazardous waste, and the resulting mixture is no longer ignitable but contains silver at a concentration >5.0 mg/L TCLP. What is the regulatory status of the mixture?*

A It is D011 hazardous waste but not F003 as long as the mixture does not exhibit the ignitability characteristic. [RO 11140, 14749, August 28, 2008 email from EPA to McCoy and Associates, Inc.]

Q *An F003 waste is mixed with a K-, P-, or U-listed waste that is not an ICR-only material. What is the regulatory status of the mixture if it does not exhibit a characteristic?*

A It would carry the K-, P-, or U-code only—not the F003 code. However, the F003 LDR treatment standards must be met before the mixture can be land disposed, in accordance with §261.3(g)(3). [RO 14579]

Q *A drum of F003 spent solvent leaks; the leak is soaked up with an absorbent pad. The pad (which is not ignitable) is to be disposed. What is the regulatory status of the contaminated absorbent pad?*

A If the listed waste is mixed with "contaminated absorbents (a solid waste)" and the mixture of absorbent and F003 exhibits no characteristic, the contaminated absorbent is not an F-listed hazardous waste via the mixture rule in §261.3(g)(2)(i). [RO 11619] Note that EPA conditioned its application of the mixture rule in this situation on the absorbent already being contaminated and, therefore, designated as a solid waste. We would take a different approach. Absorbents and pads used to catch accidental spills are normally not wastes at all—they are products performing their intended function. Our spin on this scenario is that the second element of the mixture rule would not apply, because the product absorbent was not a solid waste when contact with hazardous waste occurred. Instead, we think that when the F003 spent solvent mixes with product absorbent, the resulting contaminated pad, when discarded, would be regulated under the contained-in policy (discussed in Section 5.3). (Under the contained-in policy, the pads are not F003 wastes if they are not ignitable.)

Q *Spent methanol solvent (F003) is discharged to a facility's wastewater treatment system. Is all of the mixed wastewater hazardous?*

A No. Unless the mixed wastewater stream at the headworks of the wastewater treatment system exhibits a characteristic, it would be nonhazardous. [RO 12126]

Q *Ground water contaminated with spilled commercial chemical product methanol carries the waste code U154 via the contained-in policy because the flash point of the contaminated ground water is <140°F. The contaminated ground water is pumped out of the ground and mixed with nonhazardous plant wastewaters, and the mixture is subsequently treated in the facility's wastewater treatment system. Treated water is discharged under an NPDES permit, but a sludge is produced in the treatment system that is land disposed. The sludge does not exhibit any hazardous waste characteristic. What is the regulatory status of the sludge?*

A The sludge would not be considered a hazardous waste. Since U154 (discarded commercial chemical product methanol) was listed solely because it exhibits the ignitability characteristic, it is an ICR-only listed waste. When the U154 ground water mixes with plant wastewater, if the resulting mixture is not ignitable (or otherwise characteristic), the mixture is not hazardous when it is introduced into the wastewater treatment system. Any sludge generated from the treatment of nonhazardous wastewater would not be regulated as U154. [RO 14207]

Q *Cumene hydroperoxide (U096 when discarded) is spilled onto soil. This chemical was added to the U-list solely because of its reactivity. The contaminated soil, when excavated, is not reactive. What is the regulatory status of the soil?*

A Since U096 is an ICR-only listed waste and the contaminated soil is not reactive, it is tempting to use the second element of the mixture rule to declare the soil nonhazardous. However, as noted above, the mixture rule cannot be used to determine the regulatory status of contaminated environmental media (including soil), because soil is not a solid waste. Instead, EPA requires use of its contained-in policy in this situation. [RO 14045] (The contained-in policy would say that the soil is not U096—see Section 5.3.)

5.1.2.3 Mixtures of other listed wastes

Mixtures of listed hazardous wastes (other than ICR-only listed wastes) and solid wastes are listed wastes. Such mixtures remain listed hazardous wastes unless and until they are delisted. [RO 11839] The first conclusion to draw from this element of the mixture rule is: Don't mix wastes together until you have evaluated them for any listed wastes codes; you could end up with more listed waste to manage than you started with. Listed waste mixture examples follow.

Q *A listed spent solvent (F001) is mixed with an ignitable waste (e.g., discarded ink, paint, or adhesive) and the mixture remains ignitable. What waste codes apply to the mixture?*

A Both F001 and D001. The F001 code applies because of the third element of the hazardous waste mixture rule, while the D001 code is carried in compliance with the first element. Even if the amount of spent solvent in the mixture is very small, no concentration levels are specified below which the mixture will no longer be listed. If the mixture does not exhibit the ignitability characteristic, the D001 code would not apply (but the mixture would remain F001). [RO 11220, 11327] Similar guidance for F005 is included in RO 11375.

Q *API separator sludge from a petroleum refinery is designated as K051. Conversely, sludges* generated from oil/water separation at petroleum bulk terminals are not typically within the scope of that listing. What is the regulatory status of oil/water separator sludge from a bulk terminal if the influent to this unit includes a small amount of slop-oil emulsion solids (K049) that were generated at a refinery?

A The sludge is K049 listed waste. The sludge from the terminal oil/water separator consists of a mixture of K049 and nonhazardous solid waste. [RO 11453]

Q Excess creosote is poured into a dumpster where it mixes with wood scraps and other solid waste. What is the regulatory status of the mixture? What about the disposal of unused creosote-treated railroad ties?

A Unused creosote is U051 when discarded. A mixture of U051 and solid waste would just be more U051. Discarded, creosote-treated railroad ties, conversely, would not be U051 via the mixture rule. In this scenario, the railroad ties are manufactured articles that contain creosote, but not as the sole active ingredient (see Sections 3.5.3.1.2 and 3.5.3.7). Thus, the unused railroad ties would be hazardous only if they exhibit a characteristic. Note that the disposal of creosote-treated ties may be subject to regulation under the Federal Insecticide, Fungicide, and Rodenticide Act (FIFRA). [RO 11094, 11104, 11129, 11535, 12012]

Q Metal parts are dipped into a tank containing cyanide-based electroplating solutions. The parts are then rinsed in a second tank containing only water. When disposed, the plating bath is F007. Are the small amounts of electroplating solution that are carried on the parts from the first to the second tank ("dragout") also F007? If so, is the rinsewater in the second tank F007 when disposed by virtue of the mixture rule?

A Trace amounts of electroplating bath solutions carried over to the rinse tank are not considered to be solid or hazardous waste; instead EPA considers such dragout to be "materials in use." Therefore, F007 is not being carried over into the rinse tank. Additionally, the rinsewater is not a solid waste when it becomes contaminated and, therefore, will not be a hazardous waste via the mixture rule; when disposed, it will only be hazardous if it exhibits a characteristic or is mixed with a hazardous waste. [RO 11269, 11339] EPA has rendered similar interpretations for carryover of solvents in parts-degreasing operations [RO 11283, 13007] and pickling liquor in steel finishing processes. [RO 14334]

5.1.2.4 Special mixture rule for Bevill wastes

Three classes of wastes are excluded from hazardous waste regulation under the so-called "Bevill amendment" (this amendment is named after the representative who introduced it in Congress):

- Fly ash, bottom ash, slag, and flue gas emission control waste from the combustion of fossil fuels [§261.4(b)(4)];

- Wastes from the extraction, beneficiation, and processing of ores and minerals [§261.4(b)(7)]; and

- Cement kiln dust [§261.4(b)(8)].

EPA established special rules that apply when hazardous waste is mixed with one of the excluded extraction, beneficiation, or processing wastes listed in the second item above. [September 1, 1989; 54 *FR* 36622 and May 26, 1998; 63 *FR* 28597] However, based on other guidance for fossil fuel combustion wastes [May 22, 2000; 65 *FR* 32220] and cement kiln dust [RO 13616], we believe the Bevill waste mixture rule would apply to any of the three types of Bevill wastes identified above.

The Bevill waste mixture rule works similarly to the standard mixture rule. In the following discussion, a "Bevill waste" refers to any of the three types of Bevill wastes identified above:

1. If a characteristic hazardous waste is mixed with a Bevill waste, the resulting mixture is hazardous only if it exhibits a characteristic not exhibited by the Bevill waste alone. [§261.3(a)(2)(i)]

2. If an ICR-only listed waste is mixed with a Bevill waste, the resulting mixture is hazardous only if it exhibits a characteristic not exhibited by the Bevill waste alone. [§261.3(g)(4)]

3. If a listed hazardous waste (other than an ICR-only listed waste) is mixed with a Bevill waste, the resulting mixture is a listed hazardous waste. [§261.3(a)(2)(iv) and 54 *FR* 36622]

These three points result in a special mixture rule for Bevill wastes, as shown in Figure 5-1.

Q *A waste that exhibits the toxicity characteristic for chromium (D007) is mixed with a Bevill waste that would exhibit the toxicity characteristic for lead. If the resulting mixture exhibits only the toxicity characteristic for chromium, would the mixture be hazardous? What if the mixture is toxic for both chromium and lead? If the mixture fails the TCLP only for lead, what is its regulatory status?*

A If the mixture exhibits only the characteristic for chromium, it is a D007 hazardous waste; if it is toxic for both chromium and lead, it is also a D007 waste (it would not be a D008 lead waste because lead originates only in the Bevill waste); if the mixture is toxic only for lead, it is excluded from the definition of hazardous waste. If, before mixing, both wastes only exhibit the toxicity characteristic for lead and the mixture continues to be toxic for lead, the mixture is hazardous (because it continues to exhibit the characteristic of the nonexcluded hazardous waste). [May 26, 1998; 63 *FR* 28597]

Q *F003 spent solvent is mixed with a Bevill waste that would exhibit the toxicity characteristic for cadmium. The resulting mixture is nonignitable but toxic for cadmium. Is it hazardous?*

A No. The only characteristic exhibited by the mixture is attributed to the Bevill waste alone. [63 *FR* 28597, RO 14644]

Q *F001 spent solvent is mixed with a Bevill waste that would exhibit the toxicity characteristic for arsenic. What is the regulatory status of the mixture?*

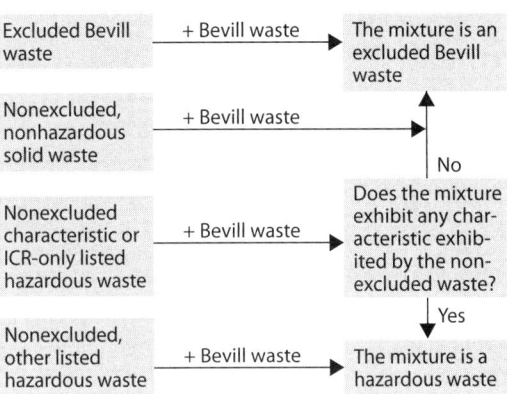

Figure 5-1: Regulatory Status of Mixtures Containing Bevill Wastes

Source: McCoy and Associates, Inc.; adapted from §261.3(a)(2)(i), (a)(2)(iv), and (g)(4) and 54 *FR* 36622, 63 *FR* 28597, and 65 *FR* 32220.

A F001. [63 *FR* 28597] The only way such a mixture ceases to be hazardous is if it's delisted using the procedures in §§260.20 and 260.22.

5.1.2.5 Special mixture rule for oil and gas wastes

Although not Bevill wastes, oil and gas exploration and production wastes are excluded from hazardous waste regulation per §261.4(b)(5)—see Section 4.14. EPA's guidance on the regulatory status of mixtures containing an excluded oil and gas waste and other solid wastes results in a special mixture rule for oil and gas wastes that is essentially identical to the mixture rule for Bevill wastes (see Figure 5-2).

Q *Nonexcluded waste caustic soda (D002) is mixed with an excluded oily waste in a pit. If the mixture has a pH >12.5, what is its regulatory status?*

A It is a D002 waste subject to hazardous waste regulation. [EPA/530/K-01/004]

Q *An excluded oil and gas waste that exhibits the toxicity characteristic for both benzene and lead is mixed with a nonexcluded, unlisted spent solvent that exhibits the characteristic for benzene*

Figure 5-2: Regulatory Status of Mixtures Containing Oil and Gas Wastes

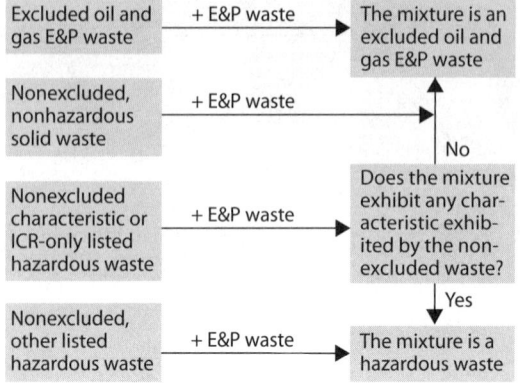

Note: E&P waste = excluded oil and gas exploration and production waste.

Source: McCoy and Associates, Inc.; adapted from EPA/530/K-01/004.

(D018). If the mixture exhibits only the toxicity characteristic for benzene, is it a hazardous waste? What if the mixture is toxic for both benzene and lead? If the mixture fails the TCLP for lead only, what is its regulatory status?

A From Figure 5-2, if the mixture exhibits a characteristic exhibited by the nonexcluded hazardous waste, it is a hazardous waste. Therefore, if the mixture exhibits only the characteristic for benzene or is characteristic for both benzene and lead, it is a hazardous waste. If the mixture exhibits only the characteristic for lead, it would not be a hazardous waste. [EPA/530/K-01/004]

5.1.3 LDR ramifications

Earlier in this section, we stated that it is important to think about the LDR ramifications of mixing hazardous waste with other stuff. Two primary LDR considerations are associated with mixing: 1) complying with the §268.3 dilution prohibition in the LDR regulations, and 2) complying with the LDR treatment standards even after mixing has taken place.

5.1.3.1 Dilution prohibition

The LDR program requires that hazardous wastes be treated to minimize threats to human health and the environment before they may be land disposed (e.g., in a landfill). Many of the resulting treatment standards that EPA developed are based on concentration limits; that is, a hazardous waste with concentration-based treatment standards must be treated such that all of the regulated hazardous constituents in the waste are at or below the concentration levels in the standard.

What is to prevent a generator from simply diluting the hazardous wastes it generates to meet these concentration-based standards? EPA doesn't think it's protective of human health and the environment if people simply mix their hazardous wastes with dirt or some other material to meet LDR standards before land disposal. Therefore, the agency added a dilution prohibition to the LDR regulations that prevents generators from diluting their wastes as a substitute for adequate treatment to comply with LDR standards. The dilution prohibition applies to the hazardous waste at its point of generation.

A complete discussion of the LDR dilution prohibition, which allows dilution under some circumstances, is presented in Section 13.11.

5.1.3.2 Complying with LDR treatment standards

As an additional means to keep generators from diluting or otherwise doing something inappropriate to their hazardous waste, EPA specifies that LDR regulations attach to the waste at its point of generation. Therefore, LDR treatment standards attach at the point of generation and must be satisfied before the hazardous waste *or a residue from treating the waste* can legally be land disposed. The implication is that even if a hazardous waste is rendered nonhazardous via mixing subsequent to its point of generation, the LDR standards must be satisfied before land disposal occurs.

5.1.3.2.1 Characteristic wastes

Going back to the first element of the mixture rule, if a characteristic waste is mixed with a solid waste so that the mixture is no longer characteristic (i.e., it is no longer hazardous), that mixture must still meet the LDR treatment standard for the original characteristic waste before it can be land disposed. "Formerly characteristic wastes are not excused from compliance with LDR treatment standards merely because they cease to exhibit a characteristic." [RO 14088] This is true even though the decharacterized waste will likely be disposed in a nonhazardous, Subtitle D unit.

Q: *An acid waste (D002) is mixed with nonhazardous wastewater such that the pH is >2.0. The nonhazardous wastewater subsequently flows to a surface impoundment that discharges under an NPDES permit. Do any LDR standards apply to this treated wastewater?*

A: Yes. LDR requirements attach to the acid waste at its point of generation. In this situation, three LDR requirements are invoked: 1) the waste must be decharacterized (it is)—see Section 13.4; 2) the mixing (dilution) must not violate the LDR dilution prohibition (it doesn't)—see Section 13.11; and 3) the appropriate LDR paperwork must be completed [the one-time notice of §268.7(a)(7)]—see Section 13.12.

5.1.3.2.2 ICR-only listed wastes

The second element of the mixture rule says that if an ICR-only listed waste is mixed with a solid waste so that the mixture is no longer characteristic, it is no longer a hazardous waste. However, §261.3(g)(3) states: "Wastes excluded under this section [i.e., ICR-only listed wastes] are subject to Part 268 of this chapter (as applicable), even if they no longer exhibit a characteristic at the point of land disposal."

In general, three Part 268 requirements appear to be "applicable" to ICR-only listed wastes that are rendered nonhazardous via mixing: 1) the waste must meet the treatment standard that applied to the ICR-only listed waste at its point of generation—see Section 13.5; 2) the mixing must not violate the LDR dilution prohibition—see Section 13.11; and 3) the appropriate LDR paperwork must be completed—see Section 13.12.

Q: *Cumene hydroperoxide was added to the U-list (U096) solely because of its reactivity. A bad batch of this chemical is mixed with a solid waste, and the resulting nonreactive mixture is to be disposed without further treatment in a Subtitle D landfill. Is this permissible?*

A: No. Even though the mixture is nonhazardous, the entire mixture must meet the nonwastewater LDR treatment standards for U096 (chemical oxidation, chemical reduction, or combustion) before it may be land disposed. The resulting treatment residues may be disposed in a nonhazardous waste landfill if they do not exhibit a characteristic. [RO 14045]

Q: *F003 spent solvent is mixed with a Bevill-exempt wastewater from the production of phosphoric acid. If the mixture is not ignitable, what is its regulatory status? Do LDR treatment standards apply to the mixture?*

A: If the mixture doesn't exhibit any characteristic attributable to the nonexempt F003 waste, then it is nonhazardous. However, the nonhazardous mixture remains subject to the LDR treatment standards for F003 (if the mixture will be land disposed), because these standards applied at the point of generation for that waste, which was before it was mixed with the exempt wastewater. "Thus, Bevill status of one waste cannot be used to immunize requirements applicable to prohibited wastes." (In this case, the prohibited waste was F003.) Additionally, the LDR recordkeeping requirements of §268.7(a) are applicable. Finally, mixing F003 with Bevill-exempt wastewater meets the definition of hazardous waste treatment (see Section 5.1.4 for ramifications); further, this mixing situation must be evaluated to see if it constitutes impermissible dilution. [RO 14644]

5.1.3.2.3 Other listed wastes

Finally, nothing unusual happens under the LDR program when non-ICR-only listed wastes are mixed with solid wastes. The same treatment standard applies to the mixture as to the original listed waste.

5.1.4 Mixing is treatment

When a facility mixes a hazardous waste with a solid waste for the purpose of rendering the hazardous waste nonhazardous (via either the first or second element of the mixture rule), that activity clearly meets the definition of "treatment" as that term is defined in §260.10. The general rule is: hazardous waste treatment requires a RCRA permit (although there are a number of ways a facility may treat a hazardous waste without a permit—see Chapter 7).

So, when you are thinking about mixing hazardous waste with something else, the second consideration (after any possible LDR implications) should be "Is this treatment that requires a RCRA permit?" The following examples illustrate this point.

Q: *When flue gas dust captured during scrap metal recycling is removed from a baghouse, it often exhibits the characteristic of toxicity for lead and/or chromium. The facility wants to mix it with nonhazardous foundry waste sands and dust and then dispose the mixture in a nonhazardous waste landfill. Is this permissible?*

A: The mixing would be considered treatment subject to RCRA permitting. Such treatment is essentially dilution as a substitute for adequate treatment and would probably be prohibited under the LDR regulations. However, if the regulatory agency with jurisdiction determined that there were effective reagents in the nonhazardous sands and dust resulting in legitimate treatment of the flue gas dust to achieve LDR treatment standards before landfilling, such treatment may be allowed. [RO 12561]

Q: *A crude oil exploration and production waste that is excluded from hazardous waste regulation under §261.4(b)(5) is mixed with a characteristic waste. If this mixing was performed for the purpose of rendering the hazardous waste noncharacteristic, is this treatment that requires a permit?*

A: Yes. Unless the facility conducts the mixing in a way that avoids this requirement, the mixing would be subject to RCRA permitting. [RO 14256]

5.1.5 Has mixing occurred?

The regulations say that a mixture must be managed as hazardous waste as soon as a listed waste is added to a solid waste. [§261.3(b)(2)] Sometimes, however, it is difficult to determine if mixing has occurred.

Take, for example, the situation where river water is used to transport listed spent carbon as a slurry. The spent carbon is separated from the water in a sump, after which the carbon is sent for regeneration and the water is sent for treatment prior to discharge or recycling.

The question posed to EPA was: "Is the separated water a listed hazardous waste, carrying the same codes as the spent carbon?" The agency's response was that, if the company could show that the hazardous constituents associated with the listed wastes do not desorb from the carbon into the water, the separated water would not be assumed to be hazardous. The demonstration should be made based on the relative flow rates and constituent concentrations for the spent carbon, influent water, and sump effluents. [RO 11146]

In other guidance, while addressing the regulatory status of waste pit liners that were used to contain excluded E&P waste, the agency clarified what it believes constitutes a mixture under RCRA. "EPA does not consider the placement of exempt E&P waste on a non-exempt pit liner to constitute the creation of a mixture. We believe mixtures constitute a commingling or blending of two or more substances. Pit liners are not blended or commingled with the pit contents into a single substance;

therefore, the mixture rule . . . is not applicable to synthetic pit liners contaminated with exempt E&P wastes." [RO 14815]

5.1.5.1 Petroleum refinery wastewaters in contact with listed sludges

A number of facilities generate sludges in their wastewater treatment operations that are listed wastes. For example, the petroleum refining industry generates several wastewater treatment sludges to which EPA has assigned listed waste codes (e.g., F037, F038, K051). Refineries have always been concerned about the possible application of the mixture rule to these sludges. What if one particle of listed sludge in the bottom of a tank or pond gets mixed with, and carried out by, the wastewater above the sludge? Would the (otherwise nonhazardous) wastewater then be hazardous via the mixture rule, causing all downstream units to be subject to RCRA regulation? EPA has consistently maintained (beginning in 1984) that the mixture rule is not generally applicable to such situations:

> "It is agency policy that no mixing occurs in a wastewater treatment unit that manages a non-hazardous [nonlisted] liquid waste even if that liquid generates a hazardous sludge that settles to the bottom of the unit, unless that sludge is in some way dredged up and physically mixed with the liquid. If the agency did not interpret the mixture rule in this manner, there would be no point in carefully limiting listings to include sludges but exclude wastewaters.

> "Under the policy explained above, for example, it is unlikely that any increased turbidity associated with the introduction of water from storm events would create the necessary scouring or physical mixing described above so as to convert nonhazardous wastewater to hazardous. Similarly, for example, the small amount of resuspension of primary sludge associated with the normal operation of a properly designed wastewater treatment system would not render the wastewater hazardous." [August 14, 1989; 54 *FR* 33387, RO 11626]

> "The mixture rule is relevant only in those cases where previously deposited sludge is scoured, resuspended, and then carried out of the unit with the wastewater." [RO 12348]

Although this guidance is specific to sludges generated in wastewater treatment systems at refineries, the same policy would presumably apply in other industries (e.g., electroplating, chemical conversion coating) where wastewater treatment *sludges* are listed, but the wastewater from which they are generated is not.

5.1.6 Miscellaneous issues

Guidance applying the mixture rule to other specific waste streams is presented below.

5.1.6.1 Precipitation run-off

Per §261.3(c)(2)(i), precipitation run-off from a facility that manages hazardous wastes is not presumed to be hazardous. However, if such run-off is known to have been mixed with a hazardous waste, the regulatory status of the mixture is determined by the standard mixture rule detailed above.

Precipitation run-off from the active portion of a hazardous waste landfill containing listed wastes that mixes with leachate from the landfill is a listed hazardous waste via the third element of the mixture rule. "Run-off from active portions of land-based hazardous waste treatment, storage, and disposal units will almost inevitably consist in part of leachate and may, therefore, be presumed initially to be a hazardous waste. However, the owner/operator can overcome this presumption by demonstrating that the collected liquid consists only of 'precipitation run-off.'" [RO 11035] (The essential distinction between leachate and run-off is that leachate results from liquids percolating through or draining from hazardous waste, while run-off is liquid that drains *over* land.) Run-off from closed (e.g., capped) units (except land treatment units) is unlikely to have mixed with leachate and is therefore not presumed to be hazardous. [RO 11035, 12332]

Even if the owner/operator of land-based hazardous waste treatment, storage, and disposal units demonstrates that run-off is not listed, it will be hazardous if it exhibits a characteristic. [RO 12332]

Note that in more recent guidance, EPA says that contaminated rainwater should be considered environmental media subject to the contained-in policy. See Section 5.3.1.3.

5.1.6.2 Rinsates from product tanks and containers

Residues generated from rinsing product tanks and containers are not solid or hazardous waste if they will be legitimately reused (e.g., for make-up water to prepare the next batch of the same product). Conversely, residues generated from rinsing product tanks and containers are solid and potentially hazardous waste if they are destined for disposal. Cleanout residues that are hazardous and are destined for disposal are not subject to RCRA regulation until they either 1) are removed from the unit, or 2) remain in an inactive unit for more than 90 days. [§261.4(c)]

If rinsates from product tanks and containers will be disposed, then the unused product in the bottom of these units (before rinsing) is a solid waste [i.e., it is being abandoned per §261.2(b)(1)]. Based on application of the mixture rule [RO 11004, 12299, 14827], the rinsate will be:

- P- or U-listed if the product is a P- or U-chemical (e.g., 2,4-D—U240);
- F027 if the product is an unused formulation containing tri-, tetra-, or pentachlorophenol or compounds derived from these chlorophenols;
- Hazardous (if it exhibits a characteristic) if the product is characteristically hazardous (e.g., hydrochloric acid—D002); or
- Nonhazardous (if it does not exhibit a characteristic) if the product is neither a P- or U-chemical nor an unused formulation containing chlorophenols or their derivatives.

If the mixture rule is problematic because the rinsewater is not considered a solid waste when it contacts the unused product, EPA has also used its "contained-in policy": "If the used rinsewater contains a pesticide listed in 40 *CFR* 261.33 that was not derived from an 'empty' container as defined in Section 261.7, the used rinsewater is a [listed] hazardous waste. If the pesticides do not meet a listing [description], the used rinsewater is a hazardous waste if it exhibits a characteristic…." [RO 11374; also see RO 11447, 11504]

5.1.6.2.1 *Rinsate from P-listed containers*

A commercial chemical product listed in §261.33(e) is a P-waste when discarded. To render a container that previously held a P-listed chemical RCRA-empty, §261.7(b)(3) requires that it be triple rinsed with a suitable solvent. What is the regulatory status of the rinses?

EPA used the mixture rule to answer that question. According to the agency, "[t]he rinsing solution is a 'solid waste' because it is being discarded. The rinsate is a mixture of solid waste and a hazardous waste listed in Subpart D [of Part 261]. Therefore, the rinsate is a hazardous waste." [RO 11004]

This provides that rinsates one, two, and three (none of which came from an empty container) are listed hazardous waste via the mixture rule. If a container were rinsed a fourth time, the rinsate would come from an empty container and would not be hazardous unless it exhibits a characteristic. [August 18, 1982; 47 *FR* 36095, RO 11004, 12299]

5.1.6.3 Sludges formed from mixtures of wastewaters

Industry has asked EPA about the regulatory status of sludges produced from mixtures of wastewaters. For instance, F006 is generated when sludges precipitate from wastewaters from electroplating operations. What if a sludge is formed from a mixture that consists partly of wastewater from electroplating operations and partly from wastewater from nonelectroplating activities?

"It has always been EPA's interpretation that sludges from wastewater mixtures of the type described above are covered by the listing description. When promulgating the wastewater treatment sludge listings, EPA contemplated that the listings applied to sludges that result from mixtures of precursor wastewaters. For example, the F006 listing covers 'wastewater treatment sludges from electroplating operations'; the listing is not modified in any way to suggest that it does not apply to sludges derived from combined wastewater streams…. Facilities with multiple operations routinely mix their wastewaters prior to treatment, and the agency intended the listings to cover sludges from these mixtures of wastewaters." [RO 14391; see also RO 12849]

5.1.6.4 Household and CESQG wastes

Hazardous wastes generated by households are exempt from RCRA regulations per §261.4(b)(1). Garden pesticides, unused paint or paint wastes, solvents, etc. are all exempt from Subtitle C regulation if they are generated by households. However, if exempt household hazardous waste (such as from a community hazardous waste collection program) is mixed with regulated hazardous waste, the regulatory status of the mixture will follow the standard hazardous waste mixture rule. Just substitute household hazardous waste for the solid waste in the three elements of the mixture rule. See Section 4.10.2 and RO 12547.

Conditionally exempt small quantity generators (CESQGs) also enjoy many exemptions from the hazardous waste regulations. Some of those exemptions include mixing CESQG hazardous waste with other materials and managing the mixtures outside of the hazardous waste management program—see §261.5(h, j).

However, "facilities receiving and mixing CESQG waste and [household hazardous waste] are subject to requirements imposed by states through the states' municipal or industrial waste permit, license, or registration programs, but are not subject to the full hazardous waste Subtitle C regulations, even if the mixed CESQG and household hazardous wastes were to exhibit a characteristic of a hazardous waste. The collection facility does not become the generator of the mixture merely by mixing CESQG waste with nonhazardous waste [e.g., household hazardous waste], and regardless of the quantity of the mixture of wastes, is not subject to Part 262 generator regulations. By contrast, [CESQGs] that mix hazardous and nonhazardous waste and whose resultant mixtures exceed the §261.5 quantity limitations and exhibit a characteristic, are no longer conditionally exempt and are subject to the applicable Part 262 hazardous waste generator regulations." [RO 11681] Thus, EPA's interpretation is that §261.5(h) applies to CESQGs but not to the subsequent facilities [described in §261.5(f)(3) and (g)(3)] that receive CESQG waste, such as state-approved facilities that collect both household and CESQG hazardous waste.

5.1.6.5 Used oil

When hazardous waste is mixed with used oil, the regulatory status of the mixture again will follow the standard hazardous waste mixture rule. Just substitute used oil for the solid waste. [One difference does exist, however, for mixtures of used oil and characteristic hazardous waste: If the mixtures are not hazardous, they are not subject to any LDR standards—see §279.10(b)(2)(ii).] A more complete discussion of used oil mixed with hazardous waste is contained in Section 12.1.3.

A provision in the used oil regulations says that used oil is presumed to have been mixed with listed spent solvent if it contains more than 1,000 ppm of total halogens (e.g., chlorinated compounds). Unless the generator can rebut the presumption by showing that the halogens did not come from mixing with such spent solvents, the used oil must be managed as listed hazardous waste. [§279.10(b)(1)(ii)] See Section 12.1.4 for more detail.

5.1.6.6 Mixtures of product and wastewater

In general, mixtures of product and water are considered to be off-spec product; off-spec product that will be reclaimed is not a solid waste per Table 1 in §261.2(c). Intentionally mixing product with wastewater to avoid regulation of the wastewater would not be considered legitimate recycling. [RO 11615]

For many years, some companies in the petroleum refining industry have claimed that refinery wastewater from which oil is removed is off-spec product, not waste. Interestingly, the DC Circuit Court of Appeals agreed with a similar argument. In that case, refiners claimed that oil-bearing wastewater was "in-process" material, not waste. Hence, the purpose of primary oil/water separation is to recover valuable product (oil), not treat wastewater. Because EPA did not adequately refute this argument, the court vacated EPA's designation of refinery wastewaters as solid wastes, and remanded the issue for further proceedings. [*American Petroleum Institute et al. v. EPA*, Docket Number 94-1683, U.S. Court of Appeals for the District of Columbia Circuit, decision dated June 27, 2000] The Spill Prevention, Control, and Countermeasures (SPCC) regs say the same thing: "The production, recovery, or recycling of oil is not wastewater treatment for purposes of this paragraph." [40 CFR 112.1(d)(6)]

5.1.6.7 Mixtures of excluded hazardous secondary materials and hazardous waste

Two recycling exclusions from the definition of solid waste were promulgated on October 30, 2008. [73 *FR* 64668] (See Sections 11.2.3.1 and 11.2.3.2.) The "hazardous secondary materials" to which this rulemaking applies are spent materials (that are hazardous either by listing or characteristic), listed sludges, and listed by-products that will be reclaimed. Examples are metal-bearing sludges and spent catalysts sent for metals recovery and organic-bearing liquids sent for recovery of solvents. If one of these two exclusions applies to a specific recycling situation, the hazardous secondary materials are not solid waste—they are manufacturing streams or otherwise valuable commodities that may be managed and recycled outside of the RCRA program. After these two exclusions were finalized, the regulated community asked EPA about the status of mixtures of the newly excluded materials

with regulated hazardous waste. The agency provided the following guidance.

Q *Can a hazardous secondary material be commingled with similar hazardous waste and still maintain the exclusion from the definition of solid waste?*

A No. Excluded hazardous secondary material cannot be commingled with regulated hazardous waste and still maintain the exclusion from the definition of solid waste. Excluded hazardous secondary material may be mixed with hazardous waste, but the resulting mixture would be hazardous waste. This follows the general principle that RCRA applicability cannot be avoided merely by mixing a hazardous waste with another material. [RO 14812, 14813, 14818]

Q *If hazardous waste is recovered by running it through a distillation column (still) and then that same still is to be used to reclaim excluded hazardous secondary materials, must the still be cleaned in between runs?*

A The same unit can be used to manage hazardous waste and excluded hazardous secondary materials, provided that the hazardous waste and associated residues are removed from the unit before processing the excluded hazardous secondary materials. The facility needs to ensure that the hazardous waste and associated residues are not commingled with the excluded hazardous secondary materials. Procedures for cleaning the still to ensure hazardous wastes are not commingled with excluded materials should be discussed with the state or regional implementing agency. [RO 14813]

5.2 The derived-from rule

Hazardous wastes often must be treated in order to meet the LDR treatment standards before they may be land disposed. What is the regulatory status of the treatment residues? What about residues remaining in the bottom of a hazardous waste storage tank or container? How about leachate that is generated from landfilling hazardous wastes?

The RCRA regulatory status of residues derived from treating, storing, or disposing hazardous waste can be determined from §261.3(c–d) and (g); EPA calls these regulations the "derived-from" rule. The basic concept of the derived-from rule is fairly straightforward, but there are some interesting twists as we'll discuss below.

5.2.1 Background

The background of this rule parallels our discussion earlier in this chapter for the mixture rule. When the RCRA regulations came out in 1980, the derived-from rule was included, even though it had not been part of the December 1978 proposal. EPA explained that the derived-from rule regulations "are new provisions which have been added both in response to comment and as a logical outgrowth of §261.3(b)." [45 FR 33096] (That section is the regulatory paragraph noting when a solid waste becomes a hazardous waste.)

The concern was whether residues from treating hazardous wastes should remain hazardous or, alternatively, should no longer be considered hazardous if they meet certain requirements. EPA addressed this issue as follows:

> "[T]reatment does not necessarily render a waste nonhazardous. It may only make it amenable for recovery, amenable for storage, or reduced in volume; or it may only eliminate one of several hazardous properties…. [The derived-from rule deals with] solid wastes generated by storage, disposal, and treatment—including leachate and treatment residues such as sludges and incinerator ash. Here, too, it is reasonable to assume that these wastes, which are derived from hazardous wastes, are themselves hazardous." [45 FR 33096]

> "We believe that without the…derived-from rule, some generators would alter their waste to the point it no longer meets the listing description without detoxifying, immobilizing, or otherwise actually treating the waste…. [W]ithout a derived-from rule, hazardous waste generators could potentially evade regulation by minimally processing or managing a hazardous waste and claiming that the resulting residue is no longer the listed waste, despite the continued hazards of the residue. It is therefore necessary for protection of human health and the environment to capture…derivatives of listed hazardous waste in the universe of regulated hazardous wastes. A hazardous waste regulatory system that allowed hazardous waste to leave the system as soon as it was modified to any degree by being mixed or marginally treated would be ineffective and unworkable. Such a system could act as a disincentive to adequately treat, store, and dispose of listed hazardous waste." [November 19, 1999; 64 FR 63389]

Industry, however, did not agree that the derived-from rule was "a logical outgrowth" of the original proposal. Hence, their court challenge of the mixture rule included a challenge of the derived-from rule. Again, one of industry's primary arguments was that EPA had failed to allow sufficient opportunity for public comment on the rule before it was finalized. As with the mixture rule, the U.S. Court of Appeals for the District of Columbia Circuit ruled in favor of industry in December 1991 (*Shell Oil Company v. EPA*, 950 F.2d 741), and the derived-from rule was vacated for procedural reasons.

EPA reinstated the derived-from rule on an interim basis on March 3, 1992. [57 FR 7628] The agency finalized an action on May 16, 2001 that retained the derived-from rule in the regulations, but with one significant modification as discussed in the next subsection. [66 FR 27266]

5.2.2 Three elements of the derived-from rule

Following the May 16, 2001 rulemaking [66 FR 27266], the derived-from rule parallels the mixture rule with three elements. (Prior to May 16, 2001, only the first and third elements given below were part of the federal regulations. The second element became effective at the federal level on August 14, 2001 but will not be effective in authorized states until they specifically adopt it. Furthermore, states are not required to adopt the new provision because it is less stringent than pre-existing rules.) The three elements are:

1. Residues from treating, storing, or disposing a characteristic hazardous waste are hazardous unless they do not exhibit any characteristic. [§261.3(c)(2)(i) and (d)(1)] EPA noted that "a [characteristic] hazardous waste remains a hazardous waste unless and until...it does not exhibit any of the characteristics." [45 FR 33096] See also June 1, 1990; 55 FR 22537, RO 11117, 12539, and 13730.

2. Residues from treating, storing, or disposing a waste listed in §§261.31–261.33 solely because it exhibits the characteristic of ignitability, corrosivity, and/or reactivity (i.e., an ICR-only listed waste) are listed hazardous wastes only if they continue to exhibit the characteristic for which they were listed. [§261.3(g)(2)(ii)] The 29 ICR-only listed wastes that are subject to this provision are given in Table 5-1, found in Section 5.1.2.

3. Residues from treating, storing, or disposing a listed hazardous waste (other than an ICR-only listed waste) are listed hazardous waste. [§261.3(c)(2)(i)] This holds true even if the resulting residue contains only very small amounts of listed waste and/or very low concentrations of hazardous constituents; that is, "[t]here is no de minimis amount below which a listed waste need not be identified." [RO 11327] The only way such a residue ceases to be hazardous is if the generator petitions the agency for a delisting. [§261.3(d)(2)] Note that this element of the derived-from rule applies "whether the waste is to be disposed or beneficially recycled." [RO 13730] The bottom line of this third element of the derived-from rule is: "listed into treatment, listed out of treatment" or (more colloquially) "listed in, listed out."

Based on operation of the three elements of the derived-from rule, a residue from treating a hazardous waste may or may not be hazardous. If it is hazardous, all appropriate waste codes must be carried on the residue. For example, if a characteristic waste is treated, the residues are hazardous only if they exhibit a characteristic (with one exception discussed in Section 5.2.2.1.1); therefore, the residues would carry the codes for whatever characteristics they exhibit. [RO 12539] If an ICR-only listed waste is treated, the residues carry the original listed waste code only if they exhibit the characteristic for which the waste was listed. If non-ICR-only listed wastes are treated, all residues that are generated carry the waste codes of the listed wastes from which they were derived. [RO 11500] If a listed residue also exhibits a characteristic, whether or not the characteristic code must also be carried depends on the LDR regulations—see Section 13.6. The three elements of the derived-from rule are discussed in more detail below, with clarifying examples added.

5.2.2.1 Residues from treating characteristic wastes

Residues derived from treating, storing, or disposing characteristic hazardous wastes are hazardous only if they exhibit any of the characteristics. Conversely, if such residues do not exhibit any characteristic (and do not carry any listed waste codes), they are no longer hazardous. [RO 12329] Keep in mind, however, that such decharacterized residues will likely still have to meet LDR treatment standards before they may be land disposed. This topic is discussed in more detail in Section 5.2.5.

Q *When an ignitable characteristic waste is incinerated, the resulting scrubber brine and sludge exhibit no characteristics. Can they be managed as nonhazardous?*

A *Yes, by application of the first element of the derived-from rule. [RO 12186]*

Q *Parts made of brass and other copper alloys are dipped into a chemical oxidizing bath* (containing hydrogen peroxide) prior to plating, lacquering, or other finishing step. Spent baths are considered to be D001 ignitable wastes due to the presence of hydrogen peroxide (an oxidizer). The spent baths are treated onsite to chemically decompose the hydrogen peroxide and do not exhibit any other characteristics. Are the treated, spent baths subject to further hazardous waste regulation?

A   If not characteristic, the treated spent baths are not hazardous wastes and may be shipped offsite under a bill of lading (i.e., no manifest would be required). [RO 11854]

5.2.2.1.1 Watch out for wastewater treatment sludges

Occasionally, a listed hazardous waste code will apply to residues from treating characteristic wastes, even though the original waste did not carry any listed codes. This occurs most often during the treatment of wastewaters, resulting in a wastewater treatment sludge that meets a listing description. For example, when rinsewaters from electroplating operations are treated to produce a wastewater treatment sludge, the F006 listing applies to that sludge. (The original rinsewaters are not listed hazardous wastes and may or may not exhibit a characteristic.)

The listing applies whether the sludge is generated at the electroplating facility or at an offsite commercial waste treatment facility. Thus, a commercial waste treatment facility must become knowledgeable about the processes that generate the wastewaters that they receive. [RO 11375]

5.2.2.2 Residues from treating ICR-only listed wastes

Residues derived from treating, storing, or disposing ICR-only listed wastes are hazardous only if they exhibit any characteristic. This provision was promulgated on May 16, 2001 [66 FR 27266], which added §261.3(g)(1–2) to the regs. As with residues from treating characteristic wastes, however, any nonhazardous residues will still have to meet LDR standards prior to land disposal. [§261.3(g)(3)] Two examples illustrate how this provision works.

Q   Spent carbon from treating wastewater that contains explosives (K045) is incinerated. K045 was listed solely because it is reactive. The resulting ash does not exhibit the reactivity (or any other) characteristic. What is the status of the ash?

A   The ash is not hazardous if it does not exhibit any characteristic. The ash is subject to two LDR requirements: 1) it must meet the treatment standard in §268.40 for K045 nonwastewater (the standard is DEACT [deactivate], which the ash will meet), and 2) the pertinent LDR paperwork must be prepared, which will depend on where the ash is disposed—see Section 13.12.

Q   Spent solvent acetone (F003) is distilled to recover the solvent. The still bottoms are not ignitable, but they contain >5.0 mg/L lead per the TCLP. Are the still bottoms hazardous?

A   Yes. The second element of the derived-from rule provides that the bottoms are not F003 because they do not exhibit the ignitability characteristic. Thus, the bottoms would carry only the D008 code. [August 28, 2008 email from EPA to McCoy and Associates, Inc.]

5.2.2.3 Residues from treating other listed wastes

Residues derived from treating, storing, or disposing listed hazardous wastes (other than ICR-only listed wastes) remain listed wastes. Unless the facility goes to the trouble of getting such residues delisted, "listed in, listed out" is in effect. Examples applying the third element of the derived-from rule are given below.

5.2.2.3.1 Residues from incineration/burning

Q   A dioxin-containing waste (F020) is incinerated, producing ash and scrubber water. What is the regulatory status of these residues?

A   They are F020. The residues are derived-from listed waste and carry the same waste code as the listed waste from which they were derived. [RO 11332] The incineration of dioxins and other acute hazardous wastes does not necessarily eliminate the hazardous constituents in the wastes. [RO 11761]

Q *While conducting an incinerator trial burn, sand is spiked with commercial chemical products that are burned as waste surrogates. If one of the surrogate chemicals used is hexachloroethane (which is U131 when disposed), what is the regulatory status of the ash generated from the test burn?*

A The ash is U131. Because the hexachloroethane is to be incinerated (i.e., disposed), it is considered to be a solid and hazardous waste. [§261.2(b)(2)] EPA also considers the sand to be a solid waste because it will be incinerated/disposed. The resulting mixture of spiked sand is U131 via the mixture rule. [§261.3(a)(2)(iv)] Residues derived from treating a listed hazardous waste are listed. [RO 11320, 13364, 14398]

Q *A warehouse full of acrylonitrile is accidentally destroyed by fire. Acrylonitrile is U009 when disposed. What is the regulatory status of the ash from the fire?*

A The ash is U009. Although the chemical was not a solid or hazardous waste prior to the fire, burning the product is viewed as discarding it (since burning is not its ordinary manner of use). [§261.2(b)(2)] Thus, the product was a solid waste listed in §261.33(f), and the ash produced as a residue from such a listed hazardous waste remains hazardous. [RO 12396]

Q *A service company picked up a shipment of waste from a facility that included 14 drums of rags contaminated with spent solvent methyl ethyl ketone (MEK). The service company inadvertently included the rags in a nonhazardous waste shipment to a second service company that shredded the 14 drums of rags along with 23 drums of nonhazardous waste. This combined material was then shipped to a waste-to-energy facility, mixed with other nonhazardous wastes, and burned. The resulting 1,410 yd³ of combustion ash was shipped to a regional nonhazardous waste landfill for use as daily cover. What is the regulatory status of the combustion ash?*

A The combustion ash is F005 listed hazardous waste. Rags contaminated with spent solvent MEK are F005. (These events occurred before the exclusions for solvent-contaminated wipes were promulgated on July 31, 2013. See Section 3.3.3 for details of that rule.) Since F005 was mixed with non-hazardous waste at the second service company, the combined material was F005 per the third element of the mixture rule (see Section 5.1.2.3). When that hazardous waste was combusted, the ash was F005 via the third element of the derived-from rule. Because of the mistake made by the first service company, 14 drums of listed rags turned into 1,410 yd³ of listed hazardous waste, a volume increase of 370 times. [September 11, 2003; 68 *FR* 53517]

5.2.2.3.2 Residues from reclamation

Q *Listed spent solvent (F001) generated from a metal cleaning operation is recycled to a solvent still. What is the regulatory status of the still bottoms?*

A F001. The bottoms are a derived-from waste and carry the same waste code as the spent solvent from which they were derived. [RO 11420, 12911] Note that spent solvents reclaimed in accordance with the provisions of EPA's October 30, 2008 DSW rule [73 *FR* 64668] would not be solid or hazardous waste. However, still bottoms generated from such reclamation would still be F001 because they meet the F001 listing description, which includes "still bottoms from the recovery of these spent solvents." [73 *FR* 64692, RO 11285, 12732, 13017, 13220]

Q *Rinsewater from electroplating operations is pumped into an ion-exchange column for metals removal. Spent ion-exchange resin from this water treatment step meets the definition of a "sludge" under RCRA and is considered F006 (even though the spent resin does not exhibit any hazardous waste characteristic). The listed resin is manifested to a regeneration facility for metals reclamation. What is the regulatory status of residues generated from the offsite regeneration process?*

A Any wastes generated during the processing of the F006 resin would also be F006, with the exception of the recovered metal that is sold as a product. [RO 11244] EPA "intends that residues derived from reclaiming listed by-products and sludges also be considered to be listed for purposes of this regulation." [January 4, 1985; 50 FR 619] This answer may change if the facility begins managing the spent resin as an excluded material that does not meet the definition of solid waste per §261.4(a)(24).

Q *Perchloroethylene (PCE—also known as tetrachloroethylene) is used extensively in the dry cleaning industry. At a typical dry cleaning facility, spent PCE is periodically distilled to recover usable solvent. Additionally, filter cartridges in PCE lines are steam stripped to recover the valuable material. Condensates produced from these distillation and stripping operations contain small quantities of PCE; are they nonhazardous?*

A No. Both of these aqueous streams are residues derived from treating F002 spent solvent and must be managed as F002 listed wastes. [RO 11224] This answer may change if the facility begins managing the spent PCE as an excluded material that is not a solid waste per §261.2(a)(2)(ii).

Q *Spent toluene that was used as a solvent is F005. If the spent solvent is filtered prior to distillation, what is the status of dirty filter cartridges that are removed from the lines?*

A They are also F005 by application of the third element of the derived-from rule. [RO 11325]

5.2.3 Exceptions to the derived-from rule

As noted in the above examples, the third element of the derived-from rule, "listed in, listed out," is applicable whenever non-ICR-only listed wastes are treated, stored, or disposed. However, EPA has decided not to apply the rule in a limited number of circumstances.

First, a facility can always ask EPA to "delist" a residue from treating, storing, or disposing a listed hazardous waste. Section 261.3(d)(2) gives facilities a procedure to remove their listed residues from hazardous waste management if they petition the agency per §§260.20 and 260.22 and demonstrate that the waste does not pose a hazard. [*Petitions to Delist Hazardous Waste*, EPA/530/R-93/007, March 1993, available at http://nepis.epa.gov/EPA/html/Pubs/pubtitleOSWER.html by downloading the report numbered 530R93007] The waste residues that have been delisted over the years are included in Appendix IX to Part 261. It is apparent from that appendix that a number of facilities have been able to take advantage of the delisting process. However, getting a waste delisted is very expensive and typically takes 2–5 years. EPA noted that about 13% of the delisting petitions they have received over the years have been approved. [*RCRA Hazardous Waste Delisting: The First 20 Years*, EPA/530/R-02/014, June 2002, available from http://www.epa.gov/epawaste/hazard/wastetypes/wasteid/delist/report.pdf]

5.2.3.1 Exempt in, exempt out

A corollary to the "listed in, listed out" provision of the derived-from rule is "exempt in, exempt out." That is, if a waste is exempt from hazardous waste management standards, residues generated from its treatment, storage, or disposal are also exempt from regulation. [March 22, 1993; 58 FR 15286] This concept must be applied carefully, but the following examples illustrate its usefulness.

Q *Gases produced in landfills are sometimes collected for their methane content. However, condensate from such collected gas can contain toxic constituents causing it to exhibit a characteristic. What is the regulatory status of the landfill-gas condensate?*

A It depends on the wastes disposed in the landfill. If only household wastes are contained in the landfill [household wastes being exempt from Subtitle C management per §261.4(b)(1)], the "household waste exclusion applies through its entire management cycle, from collection through final disposition including treatment and resultant residues." Therefore, landfill-gas condensate derived from a landfill containing only household waste is not hazardous. [RO 12362]

Q *Hazardous waste generated by a CESQG is exempt from almost all of the hazardous waste regulatory requirements. [§261.5(b)] If hazardous waste generated by a CESQG is incinerated, what is the regulatory status of the ash?*

A EPA has flip-flopped on this issue. Although §261.5(b) exempts CESQGs from most hazardous waste regulations, it does not exempt their waste from being classified as hazardous. The incinerator would not have to be RCRA-permitted to receive the CESQG's hazardous waste, but the incinerator could generate hazardous waste ash that would be subject to RCRA regulation. [RO 12892] However, this April 1987 guidance was contradicted in an August 17, 1988 *Federal Register* preamble [53 *FR* 31149], where EPA stated that the CESQG exemption applies cradle-to-grave just like the household waste exclusion discussed in the previous example.

Note that in one instance involving ash from burning household waste, the U.S. Supreme Court ruled that "exempt-in, exempt-out" does not apply. See Section 4.10.3.1.

Q *Fly ash, bottom ash, and other wastes generated from the combustion of coal or other fossil fuels are excluded from hazardous waste regulation under the "Bevill amendment." [§261.4(b)(4)] Water (from quenching the hot ash or simply from rainfall) that comes into contact with the excluded waste often picks up the characteristic of corrosivity (pH >12.5). Is the wastewater a hazardous waste?*

A No. "Since the hazardous waste characteristic of the [water] is derived from an exempt waste, the resulting corrosive water retains the exempt status of that waste (i.e., the water is also exempt from regulation as a hazardous waste)." [RO 11145; see also RO 11162, 12552]

Q *Per §261.4(b)(5), wastes uniquely associated with the exploration for and production of oil and natural gas are excluded from Subtitle C requirements. For example, bottoms from crude oil storage tanks located in the oil field are excluded from hazardous waste regulation. Are wastes that are generated from processes that recover crude oil from these tank bottoms also excluded?*

A Yes. Because the crude oil storage tank bottoms are excluded, wastes generated from their treatment are also excluded. [RO 11595]

Q *Acidic mine drainage is considered a waste from the extraction, beneficiation, and processing of ores and minerals; these wastes are excluded from hazardous waste regulation under §261.4(b)(7)—another one of the Bevill wastes. A caustic sludge (pH >12.5) is generated from treating the acidic mine drainage. Would it be hazardous?*

A No. Pollution control residues from the treatment of mining wastes are also excluded. [RO 12356]

Q *Are all wastes generated at facilities that treat or reclaim exempt wastes also exempt?*

A No. "The exemption applies only to those wastes derived from exempt wastes, not to additional wastes generated by the treatment or reclamation of exempt wastes. For example, if a treatment facility uses an acid in the treatment of an exempt waste, any waste derived from the exempt waste being treated is also exempt, but the spent acid is not." [EPA/530/K-01/004]

5.2.3.2 Exception for residues from treating petroleum refinery wastes

A number of listed hazardous wastes are produced at petroleum refineries (e.g., F037–F038, K048–K052, K169–K172). Common practice at a number of refineries is to treat these wastes to recover hydrocarbons and/or reduce their volume prior to disposal. Such treatment includes dewatering, filtration, and/or centrifugation. In addition to the recovered oil [which is exempt from regulation per §261.4(a)(12)(ii)], a water stream and solid residue are usually generated—see RO 14677. The solids remain listed hazardous waste (either retaining the waste code of the original listed waste or becoming F037—see Section 4.7.1). But what about the recovered water?

Refineries want to return that water back to their wastewater treatment systems. But, because the water is a residue from treating a listed hazardous waste, the derived-from rule would require it to be managed as hazardous waste. Once this listed wastewater mixes with other plant wastewaters, the combined water stream would be hazardous via the mixture rule, requiring all downstream residues to be managed as hazardous. EPA evaluated this situation and concluded that:

> "the derived-from rule is not uniformly applicable to the aqueous stream generated in a sludge dewatering process. Our interpretation is based on the presumption that properly conducted dewatering of a wastewater treatment residual will ensure that none of the listed waste is returned to the system, while simultaneously reducing the total amount of waste generated. It is our opinion that dewatering of the currently listed refinery wastes can be conducted in a manner that ensures the return of only the nonlisted wastewater which came into contact with, but was not mixed with, the listed waste."

This departure from the derived-from rule was published in 1985 guidance and is sometimes referred to as the "Skinner memo." [RO 11102] In that guidance, EPA established the criteria that a facility would have to meet to demonstrate that the returned water is not derived from the listed waste:

> "[I]f the refinery can show…that the return water stream is chemically equivalent to the nonlisted wastewater influent to the wastewater treatment device that originally generated the listed waste, then the return water stream is not 'derived from' the hazardous waste. It should be noted that this demonstration cannot be made if the influent to the waste treatment unit itself contained a listed hazardous waste. In this case, all waste derived from its treatment would be hazardous since the original wastewater was hazardous.

> "As an example, consider a refinery that generates an API separator sludge [K051]; suppose that the refinery pumps this listed hazardous waste to an impoundment for sludge dewatering, after which the sludge is sent to a land farm and the water supernatant is sent to the influent to the API separator. If the returned water stream is similar in composition of [Part 261] Appendix VIII hazardous constituents and total suspended solids (TSS) to the influent wastewater to the API separator, then only the nonlisted wastewater is being returned and the return wastewater is not a hazardous waste. On the other hand, if the level of some Appendix VIII constituents or the TSS is significantly higher than the level in the API separator influent, then hazardous waste is being returned to the wastewater treatment system and the mixture rule is triggered for the entire wastewater system.

> "What constitutes a significantly higher constituent level is obviously a case-by-case determination that is functionally dependent upon the amount of sampling data available."

This position was later formalized in preamble discussion. [November 2, 1990; 55 *FR* 46372] RO 12626 provides additional guidance on sampling requirements to take advantage of this exception from the derived-from rule.

Note that the 1985 Skinner memo predates the land disposal restrictions. Some of the waste management practices referred to in that memo would likely not be practical today. However, the memo still reflects EPA's policy on the application of the mixture and derived-from rules at petroleum refineries.

5.2.3.3 Exception for supernatant/filtrate from F006/F019

Wastewater treatment sludges generated from electroplating operations and from the chemical conversion coating of aluminum are F006 and

F019 listed wastes, respectively. These sludges typically form when heavy metals are precipitated from wastewaters generated in these operations. Questions have arisen with regard to the regulatory status of the effluent liquid (i.e., the supernatant) from the precipitation step. Is the supernatant hazardous via the mixture rule or derived-from rule?

EPA has consistently maintained that the mixture rule is not generally applicable to such situations:

> "It is agency policy that no mixing occurs in a wastewater treatment unit that manages a nonhazardous [nonlisted] liquid waste even if that liquid generates a hazardous sludge that settles to the bottom of the unit, unless that sludge is in some way dredged up and physically mixed with the liquid. If the agency did not interpret the mixture rule in this manner, there would be no point in carefully limiting listings to include sludges but exclude wastewaters." [August 14, 1989; 54 *FR* 33387, RO 11626]

Regarding application of the derived-from rule, EPA noted "as an interpretive matter, supernatant from F006 generation is not considered to be F006, but simply wastewater from treatment of electroplating wastewaters." [August 17, 1988; 53 *FR* 31153] EPA made a similar determination for supernatant from F019 generation, adding that it would only be hazardous if it exhibited a characteristic or if it becomes mixed with the listed sludge in some way:

> "There may be cases during wastewater treatment in which hazardous constituents that have settled out of wastewaters into a listed sludge become recombined and resuspended in the supernatant, resulting in a derived-from hazardous waste. This uncommon situation will generally occur due to improper design or malfunction of a wastewater treatment system." [RO 13323]

Both F006 and F019 sludges typically are dewatered before offsite disposal or metals recovery. Processing these sludges in a filter press generates filter cake, which continues to carry the listed codes due to the derived-from rule, and also produces a liquid effluent from the press called "filtrate." Wouldn't the filtrate be a solid waste generated from the treatment of a listed hazardous waste and therefore also be a listed hazardous waste? EPA decided the answer to that question was "No." Applying the same logic as developed for dewatering refinery sludges (discussed in Section 5.2.3.2), the agency noted:

> "Filtrate from F006 sludges could be hazardous under the derived-from rule, but if it is similar in terms of identity and concentration of constituents in the influent to the wastewater treatment process, it is not considered to be derived from F006. Rather, it is the original influent wastewater." [53 *FR* 31153]

The agency hinted that the above guidance for filtrate from F006 would also apply to F019. [RO 13323]

5.2.3.4 Regulatory exceptions

A number of specific exceptions to the derived-from rule are incorporated in the regulations. Section 261.3(c)(2)(ii) identifies five specific waste streams that are exempt from Subtitle C standards, even though they are residues from the management of listed hazardous wastes. (These five residues are exempt only if they do not exhibit any hazardous waste characteristic.) EPA guidance associated with each of the five streams is given below:

1. Lime-based waste pickle-liquor sludge generated from the iron and steel industry [June 5, 1984; 49 *FR* 23284]—Pickle liquor is an acidic solution used in the steel industry to remove oxide scale from steel surfaces. When spent, the solutions are typically corrosive due to low pH and may contain high levels of lead and hexavalent chromium. These spent solutions from the iron and steel industry are listed in §261.32 as K062.

    After generation, this listed waste generally is treated with lime, followed by flocculation and clarification. Sludge resulting from these wastewater treatment processes is dewatered and, most often, landfilled. Because the spent pickle liquor is a listed hazardous waste, the derived-from rule would apply to any treatment residues, including the sludge. After evaluating the sludge extensively in the early 1980s, however, EPA determined that the levels of lead and hexavalent chromium in this residue were relatively low. Thus, the agency exempted such lime-based waste pickle-liquor sludge from regulation as a hazardous waste (again, as long as it exhibits no characteristic). If the sludge exhibits a characteristic, it would have to be managed as K062. [RO 11526, 14097, 14100]

EPA has noted in guidance that the exemption from the derived-from rule applies even if the listed spent pickle liquor has been mixed with nonlisted process wastes generated from the iron and steel industry before it is neutralized with lime. [RO 11215] Conversely, if other listed wastes are treated along with K062, the lime-based sludge produced would not be exempt. [RO 11313] Further, if the spent pickle liquor is mixed with wastes from outside the iron and steel industry, the resulting sludge is not exempt. [RO 14100]

K062 is sometimes shipped from steel plants to offsite facilities for use or treatment. The agency indicated that lime-based sludges generated from such offsite use or treatment operations would also qualify for the exemption. [RO 11264, 11286]

Any effluent liquid (called the "supernatant") resulting from treating K062 waste does not qualify for the exemption. The supernatant from the flocculation and clarification units, and liquid from sludge dewatering would remain K062 via the derived-from rule. [RO 11299, 12239, 12250, 13065] Finally, if any reagents other than lime are used to treat K062 wastes, the resulting sludge would not qualify for the exemption. [RO 14189]

2. Residues generated from burning fuels derived from petroleum industry hazardous wastes [November 29, 1985; 50 FR 49190]—The petroleum refining, production, and transportation industries have always tried to recover hydrocarbon values from the oil-bearing hazardous wastes that they generate, some of which are listed hazardous wastes. A common practice is to produce fuels from such wastes that are then burned in on- or offsite industrial boilers and furnaces. EPA determined in the mid-1980s that these fuels were typically comparable to virgin fuels. Therefore, the agency provided a number of exemptions for these hazardous waste-derived fuels; they can be managed as products—not hazardous wastes (e.g., no hazardous waste storage or manifesting is required). [§261.6(a)(3)(iii–iv)]

Also, residues (e.g., fly ash, bottom ash) generated from burning these fuels would not be hazardous via the derived-from rule. "[W]astes from burning these fuels also would be no different than from burning virgin fuels, so the derived-from rule should not apply." [50 FR 49191]

3. Slag and other nonwastewater residues generated from recovering heavy metals from F006, K061, and K062 wastes [August 19, 1991; 56 FR 41170, August 18, 1992; 57 FR 37207]—Wastewater treatment sludges from electroplating operations (F006), emission control dust from the primary production of steel (K061), and spent pickle liquor from steel finishing operations (K062) all contain significant concentrations of heavy metals (e.g., cadmium, chromium, lead, nickel, zinc). Such wastes that contain economically recoverable levels of heavy metals (typically >1%) are often sent to high-temperature metals recovery (HTMR) units.

EPA wanted to encourage facilities to reclaim heavy metals from these waste streams (while also treating them to meet LDR treatment standards). The agency therefore blocked the derived-from rule from applying to non-wastewater residues produced when these three listed wastes are processed in HTMR units. (For purposes of this exception, "nonwastewater" takes on the LDR meaning: ≥1% by weight total organic carbon and/or ≥1% total suspended solids.) The agency placed a number of conditions on this exception: the slags or other residues must 1) be disposed in Subtitle D units, 2) meet health-based limits for 13 heavy metals (plus cyanides for F006), and 3) not exhibit any characteristic.

4. Biological treatment sludge produced from treating K156 or K157 [February 9, 1995; 60 *FR* 7826]—Organic wastes and wastewaters generated during the production of carbamate chemicals and carbamoyl oximes are listed manufacturing process wastes K156 and K157. When EPA listed these wastes in 1995, the agency decided that biological treatment sludges resulting from their treatment do not pose significant risks to human health or the environment. Only a small number of facilities produce carbamates and carbamoyl oximes, so this exception is not in widespread use.

5. Inert catalyst-support media that are separated from K171 or K172 spent catalyst [August 6, 1998; 63 *FR* 42184]—In 1998, EPA listed spent hydrotreating and hydrorefining catalysts from petroleum refining operations as K171 and K172, respectively. The inert ceramic or stainless steel material that is used to support the catalyst during operation does not, by itself, pose a threat to human health or the environment. These support media are often separated from the listed spent catalyst and managed independently; however, it appeared, absent this exception, that these materials would have to be managed as hazardous waste because they would be considered derived from listed waste. To keep this from happening, the agency promulgated this exception from the derived-from rule along with the hazardous listings.

5.2.3.5 Cement produced from hazardous waste fuels

A number of cement kilns burn hazardous waste-derived fuels. For example, some listed spent solvents (F003 and F005) can have significant heating value. The clinker produced from these kilns is used to make cement. A pure application of the derived-from rule would say that the clinker, as a residue from treating a listed hazardous waste, is a listed hazardous waste. When that clinker is incorporated into the cement, could cement be a hazardous waste?

EPA evaluated this scenario and stated that it had "no data indicating that there is a significant increase in risks posed by the use of cement product from clinker from kilns using hazardous waste fuel relative to cement produced from clinker from kilns using conventional fuels." [RO 13635] As such, the agency does not consider such cement to be derived from a hazardous waste based on the understanding that hazardous waste fuel residues do not end up in the cement product. [Note that cement kilns are considered to be "industrial furnaces." The boiler and industrial furnace (BIF) rules in Part 266 include a section (§266.112) that specifies how waste-derived residues are regulated.]

5.2.3.6 Exception for hazardous debris

One final exception to the derived-from rule is for hazardous debris. Decommissioned equipment, pieces of concrete slab, dirty filters, and personal protective equipment are examples of debris that may contain a listed hazardous waste or exhibit a characteristic. For debris that contains a listed hazardous waste, EPA has codified two exceptions to the derived-from rule in §261.3(f):

1. If the listed debris has been treated using one of the extraction or destruction technologies

specified in Table 1 in §268.45 and the performance standard in that table has been met, the treated debris is no longer a listed hazardous waste. This exception to the derived-from rule is predicated on the treated debris not exhibiting a characteristic.

2. If EPA or an authorized state determines that the debris is no longer contaminated with listed waste (at the point of generation or after treatment), then the debris is not subject to RCRA (again assuming it does not exhibit a characteristic).

5.2.3.7 No exception for residues from wastewater treatment units

Significant advantages exist for treating hazardous wastes in wastewater treatment units (WWTUs). Such units are not subject to substantive RCRA standards (e.g., no secondary containment is required). Plus, treatment of hazardous waste may occur in these units without a RCRA permit—see Section 7.2.

Because these units are exempt from RCRA standards, some people believe that the hazardous waste being managed in the units is also exempt from regulation. For example, if the hazardous waste being managed in a WWTU is listed, some people want to argue that the effluent from the unit or sludge generated in the unit is no longer hazardous. That is not true. Only the unit itself is exempt from RCRA regulation—not the waste managed within it. EPA has noted:

> "Only the wastewater treatment unit (i.e., the tank) is exempt; the exemption does not 'follow' or attach to the waste. Consequently, all applicable hazardous waste management standards apply to the waste prior to treatment in the WWTU, and to any residue generated by the treatment of that waste. In other words, solid waste resulting from the treatment of a listed hazardous waste in an exempt WWTU will remain a listed hazardous waste, and solid waste resulting from the treatment of a characteristic hazardous waste in an exempt unit will remain hazardous as long as the solid waste continues to exhibit a characteristic." [RO 13541]

5.2.3.8 Residues derived from recycling excluded materials

The derived-from rule gives us the regulatory status of residues resulting from the treatment (including recycling) of listed or characteristic hazardous wastes. But how does it work when the materials that are recycled are not solid or hazardous wastes? There are a number of recycling exemptions in the regs that keep materials from being regulated as solid wastes [e.g., §§261.2(a)(2)(ii), 261.2(c)(3), 261.2(e)(1), 261.4(a)(8), and 261.4(a)(23–25)]. Are residues that are derived from recycling these excluded materials hazardous?

EPA has given us limited guidance to answer this question. For example, if a spent solvent (in the F001–F005 listings) generated from a production process is reclaimed and then put back into the process in a closed-loop fashion, §261.4(a)(8) exempts the spent solvent from being a solid waste, so it couldn't be a hazardous waste. If the reclamation process consisted of a distillation column, what's going into that column is not a solid or hazardous waste, so the still bottoms could not be hazardous via the derived-from rule. Instead those still bottoms are hazardous (if disposed) because they are solid waste that meets the last part of the F001–F005 listing descriptions. [July 14, 1986; 51 FR 25443, RO 11285, 12732, 13017, 13220] Alternatively, residues that don't meet a listing description would be hazardous (if disposed) if they exhibit a characteristic.

More recently, the agency excluded spent materials, listed sludges, and listed by-products that are reclaimed under certain conditions from the definition of solid waste. [§§261.2(a)(2)(ii), 261.4(a)(23–25)] In the preamble to that rule, the agency stated the following: "EPA notes that the 'derived from' rule articulated in §261.3(c)(2) does not apply to residuals from the reclamation of hazardous secondary materials excluded under today's rule. These residuals are a new point of generation for the purposes of applying the hazardous waste determination requirements of §262.11. If the residuals exhibit a hazardous characteristic, or they themselves are a listed hazardous waste, they would be considered hazardous wastes (unless otherwise exempted) and would have to be managed accordingly. If they did not exhibit a hazardous characteristic, or were not themselves a listed hazardous waste, they would need to be managed in accordance with applicable state or federal requirements for nonhazardous wastes." [October 30, 2008; 73 FR 64692] This guidance is consistent with that previously given by the agency for residues generated from reclaiming materials in closed-loop recycling with reclamation per §261.4(a)(8).

5.2.4 Regulatory status of reclaimed materials

Occasionally, a residue produced from treating a hazardous waste can subsequently be used as a commercial product. This occurs, for example, when "clean" solvent is recovered from the distillation of spent solvent. If the spent solvent carried a listed waste code, does the derived-from rule (i.e., "listed in, listed out") make the recovered solvent a listed waste?

No. A parenthetical sentence in §261.3(c)(2)(i) provides relief from application of the derived-from rule for "materials that are reclaimed from solid wastes and that are used beneficially." Such products are not considered to be solid wastes, so they cannot be hazardous wastes. [RO 12911] However, this exception from the derived-from rule applies only if these reclaimed materials are not burned for energy recovery or applied to the land.

EPA provided the following clarification when it promulgated this provision:

> "[C]ommercial products reclaimed from hazardous wastes are products, not wastes, and so are not subject to the RCRA Subtitle C regulations.... Thus, regenerated solvents are not wastes. Similarly, reclaimed metals that are suitable for direct use, or that only have to be refined to be usable are products, not wastes.... We caution, though, as we did in the proposal, that this principle does not apply to reclaimed materials that are not ordinarily considered to be commercial products, such as wastewaters or stabilized wastes." [January 4, 1985; 50 FR 634]

Specific examples of materials reclaimed from hazardous wastes follow.

Q *Spent solvent methanol from pharmaceutical operations is reclaimed onsite to produce a usable methanol product. The reclaimed methanol is sent offsite to a chemical plant for use in various manufacturing processes. Does the parenthetical exemption from the derived-from rule apply if the reclaimed product will be sent offsite? Will it apply if the offsite facility must further reclaim the received methanol before using it?*

A Yes to both questions. A material that has been reclaimed from a hazardous waste and will be used beneficially ceases to be solid or hazardous waste; it is a product, and it doesn't matter if the product is used at the reclaiming facility or some other. Regarding the second question, materials that need to be reclaimed before they can be used are normally defined as wastes. [50 FR 633] However, in this situation, EPA decided that the methanol shipped offsite, which had a purity of 99.5%, is more product-like than waste-like; hence, the partially reclaimed methanol is a product and may be shipped without a manifest. [RO 11109]

Q *A solvent regeneration facility receives listed and/or characteristic spent solvents from a number of customers. The company reclaims two products from those solvents: 1) a commercial gun cleaning solvent, and 2) a waste-derived fuel. What is the regulatory status of the two products?*

A The commercial gun cleaning solvent is a product exempt from RCRA by §261.3(c)(2)(i). The waste-derived fuel, conversely, remains a solid waste under this regulatory section because it will be burned for energy recovery. If the fuel exhibits a characteristic or was formulated from a listed spent solvent, it will remain a hazardous waste subject to Subtitle C jurisdiction. [RO 13208]

Q: *Liquid mercury is recovered from switches and other dismantled equipment. The free-flowing metal (containing about 99% pure mercury) is transported to a refiner, who distills the mercury to at least 99.99% purity. Is the liquid mercury received at the refining facility a solid and hazardous waste? What if the switches and other equipment that contain the mercury are sent to the refiner?*

A: Based on the preamble language noted above, reclaimed metals that are suitable for direct use, or that only have to be refined to be usable, are products—not wastes. Therefore, the 99% pure mercury is not a waste and does not have to be manifested to the refiner. See also RO 11419. (Note, however, that the purity of the partially reclaimed material is key to this exemption, as noted in the next example.) If instead, mercury-containing equipment is shipped to the refiner, the regulatory status is different. Used switches and other equipment that are taken out of service and will be reclaimed are considered spent material. Spent material that exhibits a characteristic is regulated as a hazardous waste. [RO 11123, 11159, 11822] Note that EPA made mercury-containing equipment a universal waste on August 5, 2005. [70 *FR* 45508] Thus, at the federal level, mercury switches and other equipment can be managed as universal waste under Part 273 by handlers. LDR requirements do not apply to these materials until they are received at destination facilities. (See Section 13.13.3.)

Q: *Silver-recovery cartridges from photoprocessing operations are reclaimed for their silver content. The reclamation occurs in two steps at different facilities. Does the first reclamation facility have to manifest the cartridges to the second if the cartridges exhibit the toxicity characteristic for silver?*

A: Yes. If the cartridges are further processed to recover silver at the second facility, they are solid wastes and subject to the hazardous waste regulations (although the reduced requirements of Part 266, Subpart F for precious metals apply). [RO 11197] This is because the exception from the derived-from rule "does not apply to wastes that have been processed minimally, or to materials that have been partially reclaimed but must be reclaimed further before recovery is completed." [50 *FR* 634] As an aside, a state may require a precious metal recycler to obtain a RCRA permit, but there is no specific federal requirement for a RCRA permit at such a facility. [July 21, 2008 email from EPA to McCoy and Associates, Inc.; see also §266.100(g), August 27, 1991; 56 *FR* 42508, RO 11804]

Q: *F006, F019, and K061 are treated in an HTMR unit at a steel manufacturing facility to recover zinc and other metals. A partially reclaimed metal oxide is produced that is subsequently sold to an industrial smelter for further metal recovery. What is the regulatory status of the metal oxide?*

A: It is listed hazardous waste. "Normally, a partially reclaimed solid waste remains a solid waste as long as it must still be reclaimed before use, and the §261.3(c)(2)(i) 'derived-from' rule would make the partially reclaimed material a hazardous waste." [RO 11322] The line the agency has traditionally drawn between a partially and fully reclaimed material (when thermal metal recovery is involved) is that secondary materials remain wastes until smelting is completed. Thus, secondary materials destined for smelters remain hazardous wastes. After smelting, recovered metals that only need to be refined are products, not wastes. [August 19, 1991; 56 *FR* 41173, RO 11929, 11932 See also 56 *FR* 7144 (February 21, 1991) and RO 11385 and 11765.

Q *Spent catalysts from the petroleum industry are considered spent materials, and when these catalysts are sent for regeneration, they are solid wastes per Table 1 in §261.2(c). If these catalysts exhibit a characteristic or meet a listing description (e.g., K171 or K172), they must be managed as RCRA hazardous wastes until the regeneration is complete. (The exclusions for spent materials being reclaimed that were added by EPA in its October 30, 2008 rule [73 FR 64668] do not apply to K171 or K172.) If a company takes regenerated catalyst and further enhances it, would this be considered continued reclamation, such that the regenerated catalyst would have to be manifested from the first regeneration facility to the facility that provided the enhancement step?*

A EPA determined that the regenerated catalyst is already considered a viable commercial product per §261.2(e)(1)(ii); therefore, the enhancement step does not constitute further reclamation. So, the regenerated catalyst would not have to be manifested to the second facility. [RO 14586]

5.2.5 LDR ramifications

As noted in Section 5.1.3.2 discussing the mixture rule, EPA specifies that the appropriate LDR treatment standard attaches to a hazardous waste at its point of generation. Therefore, the treatment standard must be satisfied before the hazardous waste *or a residue from treating the waste* can legally be land disposed.

5.2.5.1 Characteristic wastes

Going back to the first element of the derived-from rule, if a characteristic waste is treated so that the residue is no longer characteristic (i.e., it is no longer hazardous), that residue may still need to meet the LDR treatment standard for the original characteristic waste code before it can be land disposed. According to EPA, "wastes that exhibit a characteristic at the point of generation may still be subject to the requirements of Part 268 [the LDR provisions], even if they no longer exhibit a characteristic at the point of land disposal." [§261.3(d)(1), January 31, 1991; 56 *FR* 3865] This is true even though the decharacterized waste may be disposed in a nonhazardous, Subtitle D landfill. For an in-depth discussion of how the LDR requirements apply to characteristic wastes, refer to Section 13.4.

5.2.5.2 ICR-only listed wastes

When EPA codified the second element of the derived-from rule for ICR-only listed wastes in §261.3(g), they also addressed how the LDR requirements would apply to noncharacteristic residues from treatment, storage, and disposal:

> "Wastes excluded under this section are subject to Part 268 of this chapter (as applicable), even if they no longer exhibit a characteristic at the point of land disposal." [§261.3(g)(3)]

Therefore, as with characteristic wastes, even though these derived-from residues are no longer hazardous, they may need to meet LDR treatment standards prior to disposal.

5.2.5.3 Other listed wastes

The third element of the derived-from rule says that residues from treating non-ICR-only listed waste remain listed. These residues must meet the applicable LDR treatment standards before being land disposed. The LDR treatment standards that apply are discussed in Section 13.5.

5.2.6 Miscellaneous issues

Guidance on applying the derived-from rule to other specific waste streams is presented below.

5.2.6.1 Landfill leachate

When hazardous wastes are land disposed (e.g., they are placed in a landfill), leachate will invariably result. "Leachate" is defined in §260.10 as "liquid that has percolated through or drained from hazardous waste." Since leachate is a solid waste generated from the disposal of a hazardous waste, it is clearly covered under the derived-from rule. For example, leachate from a sanitary or municipal landfill that has received listed wastes is considered a hazardous waste. [RO 12118] Leachate from a municipal landfill that has not received any listed hazardous wastes is subject to hazardous waste management only if it exhibits a characteristic. [RO 12149]

If only one listed hazardous waste is disposed in a hazardous waste landfill (i.e., a monofill), the leachate would carry only that code. But what if (more typically) numerous listed wastes are disposed in a Subtitle C landfill; wouldn't all of those listed waste codes apply to the leachate?

Until 1990, the answer was "Yes." In 1990, however, EPA decided that it would develop a special waste code for leachate derived from the land disposal of more than one type of listed hazardous waste. The "multisource leachate" waste code is F039. [June 1, 1990; 55 *FR* 22619] One exception exists for application of the F039 code: if the leachate results from disposing one or more of the dioxin wastes (F020–F022 and/or F026–F028) and no other listed hazardous wastes, then the leachate continues to carry each applicable dioxin waste code.

Ground water that becomes contaminated with leachate from the land disposal of multiple listed wastes must also be managed as F039 via EPA's contained-in policy (discussed later in this chapter). [RO 13438] Additionally, the agency has stated that rainwater percolating through soil contaminated with spills of several different listed hazardous wastes is also F039. ("Spillage and drippage are also forms of land disposal, albeit improper and illegal.") [RO 13509]

What about leachate from Subtitle D landfills that accept listed hazardous waste from CESQGs? EPA has addressed this subject as follows:

"EPA, however, does not read the derived-from rule as applying to small quantity generator hazardous wastes. Although the rules are not explicit on this point, the agency views this exemption, like other comparable provisions such as the household waste exclusion, as applying cradle-to-grave so that residues from managing the waste retain the exemption or exclusion. In this regard, the rules are explicit that the mixture rule does not apply to mixtures of small quantity generator wastes and solid wastes [see §261.5(h)]. EPA views the derived-from rule as similarly inapplicable." [August 17, 1988; 53 *FR* 31149]

5.2.6.2 Precipitation run-off

Per §261.3(c)(2)(i), precipitation run-off from a facility that manages hazardous wastes is not presumed to be derived from a hazardous waste. Run-off is liquid that quickly flows over and off of land surfaces, and the agency believes that it has little chance to contact hazardous waste or solubilize waste constituents. Hence, EPA specifically excluded precipitation run-off from the derived-from rule. [May 19, 1980; 45 *FR* 33096]

As discussed in Section 5.1.6.1, however, if such run-off becomes mixed with a hazardous waste (e.g., hazardous leachate), the regulatory status of the mixture is determined by the standard hazardous waste mixture rule. Also note that in more recent guidance, EPA says that contaminated rainwater should be considered environmental media subject to the contained-in policy. See Section 5.3.1.3.

5.3 Contained-in policy

EPA uses its contained-in policy to regulate materials under RCRA in situations where the material itself is not a solid waste, or at least it was not a solid waste when it became contaminated. The policy applies to contaminated environmental media (i.e., soil, ground water, surface water, and sediments) and to contaminated debris. The policy, as it applies to environmental media, has not been codified in the federal regulations but was upheld in 1989 by the U.S. Court of Appeals for the District of Columbia Circuit in *Chemical Waste Management, Inc. v. EPA* (869 F.2d 1526). The contained-in policy for debris was codified in 1992. In some interpretations, EPA has cited the contained-in policy as the basis for regulating materials that are neither environmental media nor debris. EPA's application of the contained-in policy to each of these types of materials is discussed below.

5.3.1 Contaminated environmental media

Soil, ground water, and other environmental media are not solid wastes; that is, they were never abandoned or discarded. Therefore, when such media become contaminated with hazardous waste, the hazardous waste mixture rule is not applicable. (The hazardous waste mixture rule applies only to mixtures of hazardous waste with solid waste.) However, if contaminated environmental media "contain" hazardous waste, EPA decided in the mid- to late-1980s that the media must be managed as if they are hazardous waste. [RO 11195, 11434, 11593] This determination became known as the "contained-in policy."

EPA generally considers contaminated environmental media to contain hazardous waste if: 1) the media exhibit a characteristic of hazardous waste, or 2) hazardous constituents from listed wastes are present in the media at concentrations that are above health-based levels. [RO 14283, 14291] Contaminated media that contain hazardous waste are subject to all applicable RCRA hazardous waste requirements until they no longer contain hazardous waste. [RO 11684]

The determination that any given volume of contaminated media does not contain or no longer contains hazardous waste is called a "contained-in determination." In the case of media that exhibit a characteristic of hazardous waste, the media are considered to contain hazardous waste as long as they exhibit a characteristic. Once the characteristic is eliminated (e.g., through treatment), the media are no longer considered to contain hazardous waste. "Since this determination can be made through relatively straightforward analytical testing, no formal 'contained-in' determination by EPA or an authorized state is required." [RO 14291]

For media contaminated with listed wastes, contained-in determinations must be made on a case-by-case basis by EPA or RCRA-authorized state regulatory agencies. [RO 11393] On May 26, 1998, EPA provided the following guidance:

"EPA has not, to date, issued definitive guidance to establish the concentrations at which contained-in determinations may be made.... [D]ecisions that media do not or no longer contain [listed] hazardous waste are typically made on a case-by-case basis considering the risks posed by the contaminated media. The agency has advised that contained-in determinations be made using conservative, health-based levels derived assuming direct exposure pathways." [63 FR 28622]

Additional guidance on contained-in determinations was given on April 29, 1996 [61 FR 18795]: "It has been the common practice of EPA and many states to specify conservative, risk-based levels calculated with standard conservative exposure assumptions (usually based on unrestricted access), or site-specific risk assessments." Although EPA does not believe it would be appropriate to perform a risk assessment at every site, particularly if the cleanup is of a relatively simple nature, the agency recommends that the following factors be considered in making contained-in decisions:

- Media properties;

- Waste constituent properties (including solubility, mobility, toxicity, and interactive effects of the constituents present that may affect these properties);

- Exposure potential (including potential for direct human contact and for exposure of sensitive environmental receptors, and the effect of any management controls which could lessen this potential);

- Surface and subsurface properties (including depth to ground water and properties of subsurface formations);

- Climatic conditions;

- Risk to human health and the environment; and

- Other site- or waste-specific properties or conditions that may affect whether residual constituent concentrations will pose a threat to human health and the environment.

Determinations that environmental media do not contain listed wastes must consider *all* Part 261, Appendix VIII constituents that are present in the listed waste, rather than just those constituents for which the waste was listed. [RO 11684]

Additional EPA guidance on the contained-in policy as it applies to contaminated environmental media is summarized below in question and answer format.

Q *Section 261.33(d) specifies that contaminated soil or water generated from the cleanup of releases of P- or U-listed chemicals is hazardous waste. Does that mean that a no-longer-contains determination is not applicable to such contaminated media?*

A No. The contained-in policy applies to P- and U-listed wastes in the same manner as for other listed wastes. Such soil or water would not be considered "contaminated" (i.e., a hazardous waste) if the jurisdictional agency determines that the media do not contain such a listed waste. [RO 13732]

Q *A tank containing virgin carbon tetrachloride leaks. (As a waste, carbon tetrachloride is listed as U211.) The soil around the tank is sampled and found to be contaminated with carbon tetrachloride but below state remedial levels. State policy and/or regulations do not require any remedial activity with respect to the in situ contaminated soil. If some of this contaminated soil is excavated incidentally to the removal of the tank, however, would the excavated soil have to be managed as hazardous waste or could it be returned to the excavation?*

A Under the contained-in policy, the authorized state or EPA has the discretion to determine contaminant-specific health-based levels such that, if the concentrations of the hazardous waste constituents are below those levels, the soil would no longer be considered to contain the waste. The health-based levels used in making no-longer-contains determinations are established on a site-specific basis, either in accordance with general state or federal guidelines or by means of a site-specific risk assessment. Given that the carbon tetrachloride levels in the contaminated soil are below the state's remedial requirements, it may be that the state would determine that the soil does not contain hazardous waste. If such is the case, and assuming that the state is authorized for the RCRA program, there would be no RCRA requirements applicable to the soil before or during excavations incidental to the removal of the tank. [RO 13568]

Q *Ground water contaminated with a non-ICR-only listed hazardous waste is pumped out of the ground, stored in a tank, used as a coolant in an onsite production process, and then treated to remove the hazardous waste constituents. Is the ground water storage tank a regulated hazardous waste tank, or does it qualify as an exempt manufacturing process unit?*

A The storage tank holds ground water contaminated with a listed hazardous waste, and therefore it is storing listed hazardous waste via the contained-in policy. As such, the tank is regulated as a hazardous waste storage tank. [RO 13373]

Q *Soil is contaminated with unused creosote. The facility owner wants to burn the creosote out of the soil. What is the regulatory status of the soil?*

A The soil is hazardous because it contains discarded commercial chemical product creosote (U051). A RCRA permit would therefore be required to burn the soil. [RO 14583]

Q *A tank contains spent cyanide plating bath solutions from electroplating operations (F007). After the tank is drained, it is dismantled and shipped offsite. What is the regulatory status of soil under the tank that is to be excavated and that is found to be contaminated with cyanides?*

A The excavated soil must be managed as F007 via the contained-in policy. [OSWER Directive 9938.4-03]

5.3.1.1 Contained-in policy for ICR-only listed wastes

When EPA promulgated the second element of the derived-from rule applicable to ICR-only listed wastes [May 16, 2001; 66 *FR* 27266], the agency was asked how this rulemaking would apply to the contained-in policy. EPA responded as follows:

"Today's final rule does not directly affect the implementation of the contained-in policy. However, wastes that are contained in contaminated media are eligible for the §261.3(g) exemption for wastes listed solely for a characteristic. Therefore, under today's final rule, contaminated media that contain a waste listed solely for a characteristic would no longer need to be managed as hazardous waste when it no longer exhibits a characteristic. However, consistent with the regulation of other decharacterized waste (and decharacterized contaminated media), it may remain subject to LDR requirements." [66 *FR* 27286]

Although not specifically stated by EPA, the application of the contained-in policy for ICR-only listed wastes appears to work similarly to the policy for characteristic wastes. That is, "since this determination can be made through relatively straightforward analytical testing, no formal 'contained-in' determination by EPA or an authorized state is required." [RO 14291]

Q Several hazardous wastes were listed solely because they are ignitable or reactive. For example, F003 spent solvents are listed solely because they are ignitable. The mixture rule in §261.3(g)(2)(i) says that if these wastes are mixed with solid wastes, the resulting mixture is only hazardous waste if it exhibits a characteristic. If F003 spent solvent is spilled on the ground, and the contaminated soil, when excavated, is not ignitable, how is the soil regulated?

A EPA's position is that the mixture rule does not apply in this situation because soil is not a solid waste. In this case, the soil is subject to regulation under the contained-in policy. As previously quoted for ICR-only listed wastes, if the soil is not ignitable at the point of generation, it is not a listed waste. However, it may remain subject to LDR requirements in §268.49. [66 *FR* 27286]

5.3.1.2 Contaminated sediments

Contaminated sediments will rarely be subject to the hazardous waste management standards of RCRA. On November 30, 1998 [63 *FR* 65874], EPA excluded most dredged sediments from the definition of hazardous waste. Specifically, dredged material that is subject to the requirements of a permit issued under Section 404 of the CWA or Section 103 of the Marine Protection, Research, and Sanctuaries Act is not a hazardous waste. The aforementioned permits would be issued by the U.S. Army Corps of Engineers or, in some cases, a state. [§261.4(g)] Note that this exclusion is for dredged material that will be disposed in the aquatic environment. It does not apply to dredged material destined for disposal in a landfill. See 63 *FR* 65921–2.

5.3.1.3 Contaminated rainwater

How is rainwater contaminated with hazardous waste regulated? In an October 14, 2004 memo [RO 14726], EPA noted that rainwater should be considered environmental media subject to the agency's contained-in policy. Thus, contaminated rainwater containing hazardous waste is subject to all applicable RCRA requirements until it no longer contains hazardous waste. The rainwater would no longer contain hazardous waste when it is no longer characteristic, or, in the case of rainwater contaminated with listed hazardous waste, when EPA or an authorized state makes a determination that the water no longer contains listed hazardous waste.

5.3.1.4 Applicability of the land disposal regulations

Contaminated environmental media that contain hazardous waste at the point of generation are subject to all applicable RCRA regulations, including LDR. In certain circumstances, the LDR regulations will continue to apply to contaminated media even after it has been determined not to contain hazardous waste. EPA's position is as follows:

> "If contaminated environmental media [that are determined to contain hazardous waste at their point of generation] are treated and then determined to no longer contain hazardous waste, the LDR treatment standards still must be complied with prior to land disposal. This means that the media would have to be treated to meet [the applicable treatment standards in §268.40 or 268.48] or a treatability variance would have to be obtained." [RO 11948] See also May 26, 1998; 63 FR 28618 and RO 14283.

As previously discussed, EPA recommends that no-longer-contains determinations be risk-based. Therefore, "the agency expects that, in most cases, a determination that soils do not (or no longer) contain hazardous waste will equate with minimize-threat levels and, therefore, encourages program implementors to combine contained-in determinations, as appropriate, with site-specific [treatability] variances." [63 FR 28622]

At many remediation sites, soil became contaminated with a listed waste before the waste in question was subject to LDR standards. In this situation, if EPA or an authorized state determines that the contaminated soil does not contain hazardous waste when it is first generated (excavated), the soil is *not* subject to any Subtitle C requirements, including the LDR program. [See §268.49(a).]

EPA has also indicated that it or an authorized state "could determine, at any time, that any given volume of environmental media did not contain (or no longer contains) any solid or hazardous waste (i.e., it's just media). These types of determinations might be made, for example, if concentrations of hazardous constituents fall below background levels, or are at nondetectable levels. Such a determination would terminate all RCRA Subtitle C requirements, including LDRs." [63 FR 28621]

The bottom line here seems to be that EPA has attempted to provide states with several mechanisms to administratively determine that a given volume of soil being managed as part of a remediation project is not subject to LDR regulations.

5.3.2 Contaminated debris

"Debris" is defined in §268.2(g) as "solid material exceeding a 60-mm particle size that is intended for disposal and that is: a manufactured object, or plant or animal matter, or natural geologic material...."

"Hazardous debris means debris that contains a hazardous waste listed in Subpart D of Part 261, or that exhibits a characteristic of hazardous waste identified in Subpart C of Part 261...." [§268.2(h)]

The contained-in policy, as it applies to hazardous debris, is codified in §261.3(f)(2); EPA or an authorized state can determine that debris is no longer contaminated with (or no longer contains) hazardous waste:

> "This involves a case-by-case determination by EPA, made upon request, that debris does not contain hazardous waste at significant levels, taking into consideration such factors as site hydrogeology and potential exposure pathways, but excluding management practices. Debris found not to contain hazardous waste (and not exhibiting a hazardous waste characteristic) would not be subject to further Subtitle C regulation, and so could be land disposed without further treatment." [August 18, 1992; 57 FR 37226]

EPA issued alternative land disposal treatment standards for hazardous debris on August 18, 1992. [57 FR 37194] The alternative standards are codified in §268.45 and allow hazardous debris to be treated using common extraction, destruction, or immobilization technologies (see Section 13.8). At the same time, the agency determined that debris treated to meet the standards in §268.45 using an extraction or destruction technology would be considered to no longer contain hazardous waste and, therefore, would no longer be subject to regulation under RCRA. This determination is predicated upon the treated debris not exhibiting any hazardous waste characteristic. This no-longer-contains determination is automatic—no agency action is needed. Note that this automatic no-longer-contains determination does not apply to debris treated using immobilization technologies identified in §268.45. [RO 14291]

The following examples illustrate how the contained-in policy applies to contaminated debris.

Q *During the closure of a chemical manufacturing facility, a contractor must dispose contaminated articles such as pallets, gloves, boots, coveralls, and aprons. If these articles came in contact with listed hazardous wastes while they were being used, how are they regulated?*

A These articles cannot be considered hazardous wastes via the mixture rule because they were not "solid wastes" when they became contaminated. However, if the articles are contaminated with listed waste, they are subject to regulation when disposed since they *contain* a hazardous waste. Therefore, the articles must be managed as if they themselves are hazardous wastes. If the articles are subsequently treated such that they no longer contain the hazardous waste (i.e., if the hazardous contamination can be removed), they would no longer be subject to regulation under Subtitle C of RCRA. [RO 11219, 11387]

Q *A generator sends a sample of F004 spent solvent to an analytical laboratory for testing to determine if it exhibits any hazardous waste characteristics. Equipment (e.g., a pipet) at the lab becomes contaminated with the spent solvent and is subsequently thrown away. Would the contaminated equipment be regulated as hazardous waste?*

A Yes. Any wastes generated from the analysis that *contain* the F004 spent solvent are also identified as F004 under the "contained-in" policy. However, it is possible that some of the wastes from the analysis (such as a pipet) may meet the definition of a container and therefore would not be subject to the hazardous waste regulations if they meet the definition of "empty" in §261.7. [RO 13375]

Q *Spent filter cartridges are generated during the packaging of commercial chemical product toluene. The filters remove dirt and other solids from the product. When the filter cartridges become dirty, they are flushed with water and replaced. The spent cartridges, still containing small amounts of toluene, are discarded. How are the spent filter cartridges regulated?*

A If the solvent contained in the filters is a discarded commercial chemical product listed in §261.33, then the filters are contaminated with, or *contain*, a hazardous waste and must be handled as hazardous waste (in this case U220) until they no longer contain the hazardous waste. [RO 11325]

Q *Refractory material and pieces of a scrubber system associated with a combustion facility that burned dioxin-containing waste (F020) are to be disposed. What is their regulatory status?*

A Materials that are contaminated with listed hazardous waste, and thereby *contain* hazardous waste (i.e., hazardous constituents from the waste), must be managed as hazardous waste for as long as they contain any of the listed waste. The generator must manage the materials as hazardous waste F020 or must be able to demonstrate that the materials have been decontaminated so that they no longer contain any listed waste. [RO 11332]

Q *A manufacturing facility purchases a chemical (warfarin sodium) that, if disposed, would be a listed commercial chemical product (i.e., it would be a P- or U-waste). The warfarin is blended with inert ingredients (lactose, starch, and water) and then dried and compressed into tablets that are marketed as "Warfarin." Employees at the plant use disposable gowns, gloves, and other personal protective equipment (PPE) to avoid any risk* of contaminating themselves or the product. When these items are disposed, they contain traces of warfarin. How is the PPE regulated?

A If the PPE comes in contact with small amounts of the actual formulated commercial chemical product or manufacturing chemical intermediate (not manufacturing process wastes), the discarded equipment is debris containing a listed hazardous waste. It, therefore, must be managed as a hazardous waste until it no longer contains the hazardous waste (e.g., it could be washed). This interpretation is based on the fact that the equipment would qualify as hazardous debris under §268.2(g) and (h). Under §261.3(f)(1), it would no longer be subject to regulation as hazardous waste if it is washed using one of the technologies described in §268.45, Table 1. Alternatively, EPA or a state could determine that the PPE is no longer contaminated with hazardous waste per §261.3(f)(2). [RO 14095]

5.3.3 Policy creep

The contained-in policy, as it applies to contaminated environmental media and debris, is well established. What is not clear is whether, or to what extent, the policy applies to solid wastes that are neither environmental media nor debris. EPA has provided the following guidance.

Q *Listed hazardous wastes are burned in a fluidized-bed incinerator. How is the spent fluidized-bed media from the incinerator regulated?*

A Spent fluidized-bed media contaminated with a listed hazardous waste must be managed as hazardous waste since it *contains* a listed waste. Treatment, storage, or disposal of the spent media must be handled as if the media itself were a hazardous waste. However, if the fluidized-bed media, as a result of the incineration process or as a result of other treatment, no longer contains a hazardous waste, it would no longer be subject to regulation under Subtitle C of RCRA. [RO 11205]

Q *An air stripping tower is being used to treat ground water contaminated with F001 spent solvent. How is plastic packing media removed from the air stripper regulated? Does the "derived-from" rule apply?*

A The plastic packing media, when removed from the air stripper for disposal, is considered a spent material that is subject to regulation as a hazardous waste because it *contains* a hazardous waste (i.e., F001). The derived-from rule [§261.3(c)(2)] is not directly applicable to the plastic packing media because the media "is considered to be an integral part of the treatment process, not a solid waste residue derived from the treatment of a hazardous waste. Therefore, when the media no longer contains the hazardous waste, it no longer is considered to be a hazardous waste and may be disposed in a Subtitle D landfill." [RO 11418]

Generator Issues

Counting, accumulating, and manifesting your waste; dealing with recordkeeping, employee training, and contractors

A "generator" is any person, by site, whose act or process produces hazardous waste or whose act first causes a hazardous waste to become subject to regulation. [§260.10] Generators must comply with many requirements under the RCRA regulations; some are easy—some are hard and expensive. This chapter looks at a number of issues that tend to confuse generators and/or result in enforcement actions.

We start by discussing which hazardous wastes must be counted when determining whether an entity is conditionally exempt, a small quantity generator, or a large quantity generator (Section 6.1). That section is followed by discussions of specific operating practices for generators who accumulate hazardous waste in satellite accumulation containers (Section 6.2) or 90/180-day accumulation areas (Section 6.3). Section 6.4 examines training requirements for facility personnel. The potential problems associated with hazardous waste generated by contractors on your site are reviewed in Section 6.5. Section 6.6 addresses manifesting issues, and the myriad generator recordkeeping requirements under RCRA are detailed in Section 6.7. Section 6.8 contains a discussion of the biennial report that is required of large quantity generators and Section 6.9 focuses on EPA ID numbers. Finally, Section 6.10 summarizes the alternative standards for managing unwanted materials in academic laboratories.

EPA has developed an excellent guidance document for hazardous waste generator requirements: *Hazardous Waste Generator Regulations—A User-Friendly Reference Document*, August 2012, available at http://www.epa.gov/osw/hazard/downloads/tool.pdf. It includes hyperlinks to *Federal Register* notices, letters and memoranda issued by EPA, and other relevant documents that provide clarification of the hazardous waste generator requirements.

6.1 Waste counting

Hazardous waste generators must calculate the quantity of hazardous waste they generate each month to establish their generator class, which then determines the RCRA regulations with which they must comply. (The regulations become increasingly stringent as the quantity of waste generated increases.) The rules outlining which hazardous wastes have to be included in these monthly calculations are found in §261.5(c) and (d). Please note that even though these rules are found in the regulations pertaining to conditionally exempt small quantity generators, they must be used by *all* generators to determine their RCRA generator status. [§262.10(b), March 24, 1986; 51 *FR* 10152]

6.1.1 Three classes of generators

Three classes of generators (each of which must comply with different sets of RCRA standards) are identified by EPA [RO 14703]:

1. Conditionally exempt small quantity generators (CESQGs)—These are entities who generate in a calendar month: a) no more than 100 kg (220 lb) of nonacute waste, b) no more than 100 kg of acute spill cleanup residue, and c) no more than 1 kg (2.2 lb) of other acute hazardous waste. (Acute hazardous wastes include the dioxin-containing F-wastes having an (H) code in §261.31 and the P-listed chemicals; nonacute hazardous wastes are all others.) CESQGs are exempt from the hazardous waste management regulations if they 1) identify and count all hazardous waste that they generate; 2) store no more than 1,000 kg (2,200 lb) of nonacute waste, no more than 100 kg of acute spill cleanup residue, and no more than 1 kg (2.2 lb) of other acute hazardous waste onsite at any time; and 3) ensure that the hazardous waste they produce is sent to an appropriate offsite treatment or disposal facility. The CESQG requirements are set forth in §261.5(a–b, e–j).

2. Small quantity generators (SQGs)—Entities who generate between 100 kg (220 lb) and 1,000 kg (2,200 lb) of nonacute waste, no more than 100 kg (220 lb) of acute spill cleanup residue, and no more than 1 kg (2.2 lb) of other acute hazardous waste per calendar month are SQGs. These generators must comply with many of the same requirements as the fully regulated large quantity generators, although some SQG requirements are less stringent than the rules for large quantity generators. In order to retain SQG status, these generators must not accumulate onsite more than 6,000 kg of nonacute hazardous waste, more than 1 kg of acute hazardous waste, or more than 100 kg of acute spill cleanup residue.

3. Large quantity generators (LQGs)—LQGs produce at least 1,000 kg (2,200 lb) of nonacute hazardous waste, more than 100 kg (220 lb) of acute spill cleanup residue, or more than 1 kg (2.2 lb) of other acute hazardous waste per calendar month. LQGs must comply with extensive management requirements set forth in Part 262.

As indicated above, the hazardous waste generator categories are based on quantities measured in kilograms or pounds. For facilities that generate liquid wastes, the generator must determine the density of the waste to evaluate his/her generator status. Roughly, 27 gallons (about half of a 55-gallon drum) of a waste with a density similar to water weighs about 220 lb; so, if all the hazardous waste you generate per month is less than half of a drum of nonacute liquid, you are likely a CESQG. It takes about 265 gallons of a waste (a little less than five drums) with a density similar to water to exceed 2,200 lb. For acute wastes, 1 kg is approximately equal to 1 quart.

Per §262.11, any person who generates a solid waste must make a hazardous waste determination. The solid waste generator must make such a determination before calculating what generator class he/she falls under. [November 19, 1980; 45 FR 76622] Chapter 18 will help in making these determinations.

6.1.2 Wastes must be counted monthly

At the federal level, generators are required to count all hazardous wastes that they generate on a calendar-month-by-calendar-month basis. As a result, depending on fluctuations in the amount of hazardous waste generated, a facility's generator status may change from one month to the next. If, for example, a facility generates less than 100 kg of nonacute hazardous waste during June, the facility would be considered a CESQG for that month, and the waste generated during that month would be subject to the CESQG requirements. However, if the waste generation rate increases to 500 kg in July, the facility's generator status would change; it would now be classified as an SQG, and the waste it generated in July would be subject to more-stringent requirements. [*Managing Your Hazardous Waste—A Guide for Small Businesses*, EPA/530/K-01/005, December 2001, available at http://www.epa.gov/epawaste/hazard/generation/sqg/sqghand.htm]

Generator status is based on the amount of waste *generated* per calendar month and the total amount accumulated onsite at any one time. The amount of waste manifested offsite at any one time does *not* affect generator status, provided the waste is shipped offsite prior to exceeding accumulation limits. [RO 14700] Typically, generators track monthly generation of each hazardous waste stream using spreadsheets or an environmental management system.

6.1.2.1 Episodic generators

What do you do if you are a so-called "episodic generator"—an entity that normally is a CESQG or SQG, but once or twice a year generates enough hazardous waste to push it into LQG status? For example, this could occur when cleaning out process units during a turnaround or getting rid of old chemicals from a laboratory. EPA addressed this issue in March of 1986:

> "[A]ny [hazardous] waste that is generated during a calendar month in which the 1,000-kg/mo cutoff is exceeded is subject to full regulation until it is removed from the generator's site. If such fully regulated waste is mixed or combined with waste [generated when the entity qualified as a CESQG or SQG], then all of the waste is subject to full regulation until the total mixture is removed from the generator's site."

> "If, on the other hand, the generator stores separately that waste generated during a month in which less than 1,000 kg (but more than 100 kg) of hazardous waste is generated, from waste generated during a month in which more than 1,000 kg is generated, the former is subject to [SQG standards], while the latter is subject to full regulation. Therefore, generators who expect to periodically exceed the 1,000-kg/mo cutoff for the reduced requirements…should be prepared to ship their waste offsite if they wish to avoid being subject to full regulation." [51 *FR* 10153–4; see also RO 12602]

Although this guidance is rather ambiguous, we interpret it to say that a generator should segregate its hazardous waste generated when it is an SQG from its hazardous waste generated when it is an LQG unless it wants to manage all of its waste under the LQG rules. Also, a facility must comply with the applicable SQG or LQG requirements for the waste generated in a given month, based on the facility's generation rate that month. [May 23, 2006; 71 *FR* 29721, RO 11688] For example, if an SQG generates enough hazardous waste to be an LQG one month out of the year, the waste generated that month could be stored onsite for only 90 days, and the facility would be subject to personnel training provisions in §265.16 and contingency plan requirements in Part 265, Subpart D. See RO 12265 for more on this topic.

At the federal level, EPA allows generators to switch from one generator status one month to another status the next month. However, the instructions to the Site ID form (EPA Form 8700-12 that is used to obtain an EPA ID number) note that facilities must use this form to submit a subsequent notification if they wish to change generator status. (Note that, if a CESQG switches to an SQG or LQG, that facility would need to obtain an EPA ID number using this form.) More discussion of this topic is contained in Section 6.9.1.3.1.

EPA has addressed reporting and recordkeeping requirements as follows: "The recordkeeping and reporting requirements of Part 262 apply, however, only to those periods in which the generator's hazardous waste is subject to full regulation under Part 262. Thus, for example, the annual report of a generator whose waste is subject to full regulation under Part 262 for three months in a year would cover the generator's activity only for those three months." [November 19, 1980; 45 *FR* 76621] On March 24, 1986 [51 *FR* 10160], the agency became more specific, noting that "episodic generators must comply with the biennial report requirements for those months in which they are 'large quantity generators'; that is, they must submit reports on their hazardous waste activities for those months in which their generator activities have changed and as long as the fully regulated waste remains onsite." However, the most current (2013) biennial report instructions note that, even if a facility is an LQG for one month out of the year, the biennial report must be completed "for all hazardous waste that was used to determine the site's generator status." [Instructions available at http://www.epa.gov/osw/inforesources/data/biennialreport/index.htm] Although that language is not completely clear, EPA seems pretty clear in RO 14842 when it notes that facilities need to "determine if, in any single calendar month, the site is an LQG, and if so, report all RCRA hazardous waste that is generated onsite…for the entire calendar year."

When asked if an SQG that becomes an LQG only once every five years must train his personnel annually in accordance with §265.16, EPA noted "the SQG need only comply with the §265.16(c) requirements when he is subject to all of §262.34 which, in this case, is once every five years." [RO 12245]

States may have more-stringent requirements than the federal provisions noted above. Many states require that a generator notify them within a certain number of days (typically 7 to 30) if they switch from one generator category to another. Some states require that, once a generator switches to a more-stringent status (e.g., from SQG to LQG), he/she remains at that level through the end of the calendar year; alternatively, a generator may have to remain an LQG for at least a year (12 months) before switching back to SQG status. Additionally, some states will encourage or even require generators who switch back and forth between categories to stay at the more-stringent level. This reduces compliance problems and simplifies state inspections. From a practical standpoint, a generator who is an SQG some of the year and an LQG the balance of the year may have to maintain compliance continuously with the LQG requirements, because the state will generally expect compliance with these provisions whenever LQG-regulated wastes are present onsite.

If a facility is on the border of a generator category, states often recommend that a simple written log be kept by each waste container that shows when and how much hazardous waste was generated per month. This will provide documentation to support the generator status level that the facility is claiming.

6.1.3 Wastes that must be counted

When making generator-status determinations, generators must count all hazardous wastes that they generate during each calendar month, except for hazardous wastes that are exempt from counting per §261.5(c) and (d). These counting exemptions are discussed in Sections 6.1.4 and 6.1.6, respectively.

Basically, hazardous waste must be counted if it is managed in units that are subject to "substantive RCRA regulation." Substantive regulation "includes regulations which are directly related to the storage, transportation, treatment, or disposal of hazardous wastes. Regulations which would not be considered 'substantive' for purposes of this provision would be requirements to notify and obtain an EPA identification number or to file a biennial report." [51 *FR* 10152]

In plain English, EPA is saying that you must count hazardous wastes that come out of your manufacturing process if they are managed in units that are subject to substantive RCRA standards. These units include tanks, satellite as well as 90- and 180-day accumulation containers, treatment units, land disposal units, etc. In general, generators must count all their hazardous wastes that are:

- Accumulated onsite in a RCRA-regulated unit for any period of time prior to treatment, recycling, or disposal. This includes hazardous wastes accumulating in satellite accumulation areas. [RO 11812, 13312, 14703]
- Packaged and transported offsite.
- Placed directly into an onsite RCRA-regulated treatment or disposal unit.

Note that the quantities of all hazardous wastes generated in a month must be aggregated. If an entity wants to dispose 0.5 kg of each of three different P-listed chemicals in the same month, that entity is an LQG for that month (because more than 1 kg of acute hazardous waste was generated).

In the following subsections, we address counting issues associated with various types of wastes and waste management activities.

6.1.3.1 Containerized waste

Hazardous wastes that are stored in containers must be counted, since containers are subject to substantive RCRA standards. Questions have arisen over whether or not the weight of the container itself should be included when a generator calculates the amount of waste he/she generates. According to EPA, the container itself is not considered a hazardous waste. Therefore, when making generator-status determinations (and for biennial reporting), the weight of the container does not have to be counted. However, it is customary to include the total weight (waste plus container) on manifests, since transporters often charge on the basis of total weight shipped. [RO 11803, 12151, 14827]

Weights that are listed on the manifest are often used by generators and inspectors to make estimates of generator status. If only the weight of the waste in a container is counted toward generator status, but the total weight is listed on the manifest, there could be some confusion about a generator's actual generator status. EPA recommends that when containers are manifested, the generator use Block 14 of the manifest (Special Handling Instructions and Additional Information) to indicate that although the total weight is included on the manifest, the weight of the containers was not included in determining its generator status. [RO 14827]

In RO 12946, EPA provides somewhat conflicting guidance. Several chemotherapy drugs are U-listed wastes; in order to minimize exposure, EPA recommends against rendering vials holding these drugs empty under §261.7. Instead, the agency recommends that the entire volume of waste, including the vials themselves, be weighed.

6.1.3.2 Spent filter cartridges and paint filters

EPA believes that spent filter cartridges (e.g., those used by dry cleaners to recover solvents such as tetrachloroethylene) are an integral component of the waste and therefore, are not "containers." Consequently, the entire cartridge is a hazardous waste (F002 in the above example), and the weight of the entire spent cartridge must be included when determining generator status. [OSWER Directive 9444.18(84)]

Similarly, EPA believes that the total weight of spent paint filters removed from spray booths (including the weight of the filter) must be included when calculating the quantity of hazardous waste generated. [RO 11024]

6.1.3.3 Batteries

The entire weight of a hazardous waste battery would be included in quantity determinations. [RO 12990]

6.1.3.4 Diluted P-listed chemicals

A material containing a chemical listed in §261.33(e) as its sole active ingredient is a hazardous waste when discarded, even though the chemical may have been diluted by the user. For instance, a laboratory may dilute a concentrated P-listed chemical to produce a calibration standard, or a facility may dilute a concentrated pesticide to make an application-strength solution. Even if the amount of chemical product originally used in the formulation was less than 1 kg, the weight of the disposed solution must be counted during generator-status determinations. If the acute hazardous waste threshold (1 kg/mo) is exceeded, the LQG requirements would apply. [RO 11021]

6.1.3.5 Wastes generated from cleaning out tanks and containers

Questions sometimes arise as to how to count tank or container cleanout/washout residues. If a RCRA-empty container is being cleaned, the residues are a new point of generation and, if they exhibit a characteristic, are subject to RCRA regulation and need to be counted. [RO 14708] For cleanout of tanks or nonempty containers, the following discussion applies.

6.1.3.5.1 Product tanks/manufacturing process units

Residues generated from rinsing product tanks and containers are not solid or hazardous waste if they will be legitimately reused (e.g., for make-up water to prepare the next batch of the same product). Additionally, if the residues are destined for reclamation rather than disposal, they may not be solid wastes via Table 1 in §261.2(c). This results from the dash at the intersection of "Commercial chemical products" and "Reclamation" in that table.

Conversely, when product storage tanks or manufacturing process units are being cleaned, the residues from cleaning the units will be solid and potentially hazardous wastes if they will be disposed. If the cleanout residues are solid and hazardous wastes, however, they are not subject to RCRA regulation until they either 1) are removed from the unit, or 2) remain in an inactive unit for more than 90 days. [§261.4(c)] (See Section 4.17.3.) Assuming the units are cleaned within the 90-day period, hazardous cleanout residues will have to be counted at the point they are removed from the unit.

Based on application of the mixture rule, the cleanout residues will be 1) P- or U-listed hazardous waste if the material being cleaned out of the unit is a P- or U-listed chemical, or 2) characteristic hazardous waste if the cleanout residues exhibit a characteristic. [RO 11004, 12299, 14827]

6.1.3.5.2 Hazardous waste tanks/containers

When cleaning out tanks or non-RCRA-empty containers that previously held characteristic or listed hazardous waste, care should be taken not to double count wastes. In other words, if the owner/operator already counted all the hazardous waste while it was stored in a unit, the cleanout residues shouldn't be counted again.

Sometimes though, the quantity of cleaning fluids added will have to be counted as hazardous. If the hazardous waste within the unit is hazardous only because it exhibits a characteristic, the quantity of any cleaning fluid used will have to be counted as hazardous only if the combined cleanout residues (unit residues plus any cleaning fluid) exhibit a characteristic. If the residues within the unit are listed, then any cleaning fluid used will usually become hazardous via the mixture rule and will have to be counted. From a practical standpoint, it may not be possible to separate the cleaning fluid quantity from the unit residues; in such a situation, counting all cleanout residues (unit residues plus any cleaning fluid) that exhibit a characteristic or meet a listing description would work.

6.1.3.6 Conditionally exempt small quantity generator wastes

An exemption exists at §261.5(h) for CESQGs that mix their hazardous wastes with nonhazardous wastes. If the resulting mixture does not exhibit a characteristic, only the original hazardous waste has to be counted, even though the quantity of the resultant mixture exceeds 100 kg/mo or 1,000 kg total accumulation. Thus, for listed residues cleaned out of a CESQG's hazardous waste tank, only the original listed residues within the unit are counted as hazardous, while any nonhazardous cleaning fluid added to them will not have to be counted (unless the resulting mixture exhibits a characteristic of hazardous waste, in which case the entire mixture must be counted). [OSWER Directive 9451.02(85)]

6.1.3.7 Imported waste

Hazardous waste imported from a foreign country must be counted in determining a facility's generator status if that facility is the U.S. importer. [*2013 Hazardous Waste Report Instructions and Form*, available at http://www.epa.gov/osw/inforesources/data/br13/br2013rpt.pdf]

6.1.4 Wastes exempt from counting

EPA has identified numerous hazardous wastes that are exempt from counting in §261.5(c). These wastes are listed in Table 6-1, along with appropriate regulatory citations and RO references.

In addition to the hazardous wastes specifically exempted from counting in the table, generators do not have to count materials that are not solid waste [included in §261.4(a)] or solid wastes that are not hazardous wastes [included in §261.4(b)]. For instance, counting is not required for hazardous wastes that are discharged directly to a publicly owned treatment works (POTW) without being stored or accumulated first. These wastes are exempt from RCRA regulation under §261.4(a)(1). [EPA/530/K-01/005, 51 *FR* 10152]

As mentioned in Section 6.1.3, §261.5(c)(2) and (3) exempt hazardous waste from counting if such waste is managed only in units that are not subject to "substantive RCRA regulation." Thus, you do *not* have to count hazardous wastes that come out of your manufacturing process and are managed (without prior storage or accumulation) in units that are not subject to substantive RCRA standards. These units include elementary neutralization units, wastewater treatment units, totally enclosed treatment facilities, and waste recycling units.

For example, listed wastes that are hazardous at their point of generation but are subsequently excluded from regulation under the §261.3(a)(2)(iv) de minimis wastewater exemptions may not have to be counted. Although these listed wastes are not specifically exempt from counting in §261.5(c), they are generally managed (without prior storage or accumulation) in wastewater treatment units within the facility's wastewater treatment system, and so such wastes are exempt from counting per §261.5(c)(2).

Note that some wastes that are exempt from counting for the purpose of determining generator status may still have to be included in biennial reports. If so, the instructions to the biennial report forms will identify these wastes—see Section 6.8.

6.1.5 Waste counting examples

Selected EPA guidance on when hazardous wastes have to be counted is summarized below.

6.1.5.1 Solvents in manufacturing process units

Section 261.5(c)(1) exempts hazardous waste in active tanks, vessels, pipelines, and other manufacturing process units from counting. Such wastes are not subject to RCRA management standards via §261.4(c). For example, solvents in dry cleaning machines equipped with closed-loop reclamation systems are only counted for purposes of determining regulatory status when they are physically removed from the dry cleaning process as wastes. Since the solvents in the dry cleaning machine are not considered to be solid or hazardous wastes while part of a closed-loop reclamation process, they are not subject to counting while in the machine. The still bottoms from the closed-loop reclamation do not have to be counted until they exit the unit and become subject to substantive regulation. [RO 12732]

However, when wastes such as spent filter cartridges are removed, they must be counted (unless they are reclaimed onsite without prior storage or accumulation). For example, if the filter cartridges were placed directly in an onsite distillation unit (which is exempt from substantive requirements) without intervening storage, the cartridges would not need to be counted according to §261.5(c)(3). The residues from the still would, however, need to be counted if they are hazardous. [RO 11359, 12201]

Certain parts washers consist of a cleaning apparatus attached to the top of a drum containing

Table 6-1: Hazardous Wastes That Do Not Have to Be Counted

Waste type[1]	Regulatory citation	RO reference
Hazardous waste in active tanks, vessels, pipelines, and other manufacturing process units[2] [§261.4(c)]	§261.5(c)(1)	11173, 12732
Analytical samples [§261.4(d)]	§261.5(c)(1)	13291
Treatability study samples [§261.4(e–f)]	§261.5(c)(1)	NA
Industrial ethyl alcohol that is reclaimed [§261.6(a)(3)(i)]	§261.5(c)(1)	NA
Scrap metal that is reclaimed [§261.6(a)(3)(ii)]	§261.5(c)(1)	NA
Fuels produced from hazardous wastes at oil refineries [§261.6(a)(3)(iii–iv)]	§261.5(c)(1)	NA
Hazardous waste remaining in RCRA-empty containers or in inner liners removed from RCRA-empty containers [§261.7(a)(1)]	§261.5(c)(1)	NA
PCB-containing dielectric fluid and electrical equipment containing such fluid regulated under TSCA that are hazardous only because they fail the toxicity characteristic for D018–D043 [§261.8]	§261.5(c)(1)	NA
Hazardous waste that is managed *without prior storage or accumulation* in onsite elementary neutralization units, wastewater treatment units, or totally enclosed treatment facilities as defined in §260.10 [§§264.1(g)(5–6), 265.1(c)(9–10), 270.1(c)(2)(iv–v)]	§261.5(c)(2)	11173, 13204
Hazardous waste that is recycled *without prior storage or accumulation* in onsite recycling processes subject to regulation under §261.6(c)(2)[3] [§261.6(c)(1)]	§261.5(c)(3)	11236, 11341, 11359, 11546, 11914, 12865, 14032, 14333
Used oil managed under §261.6(a)(4) and Part 279, even if it exhibits a characteristic as generated [§279.10]	§261.5(c)(4)	NA
Spent lead-acid batteries managed under Part 266, Subpart G [§266.80(a)]	§261.5(c)(5)	13746
Universal waste (i.e., batteries, pesticides, mercury-containing equipment, and lamps) managed under §261.9 and Part 273 [§273.1(b)]	§261.5(c)(6)	NA
Hazardous waste that is unused commercial chemical product that is generated solely as a result of a laboratory clean-out conducted at an eligible academic entity pursuant to §262.213	§261.5(c)(7)	51349

NA = none available; PCB = polychlorinated biphenyls; TSCA = Toxic Substances Control Act.

[1] The regulatory section that exempts the waste or unit from substantive RCRA standards is given in brackets.

[2] Wastes generated in these units are not subject to RCRA regulation until they either 1) exit the unit, or 2) remain in the unit more than 90 days after the unit ceases to be operated for manufacturing, storage, or transportation of product or raw materials.

[3] Section 261.6(c)(2) requires owners/operators of facilities that recycle materials without storing them before they are recycled to comply with 1) the RCRA Section 3010 notification requirements, 2) manifesting requirements set forth in §§265.71 and 265.72, and 3) §261.6(d) (which requires owners/operators of RCRA-permitted or interim status facilities that recycle hazardous waste in permit-exempt units to comply with Subparts AA and BB of Part 264 or 265).

Source: McCoy and Associates, Inc.

---

solvent. Solvent is drawn up into the washing apparatus for use and then is discharged back into the drum. When the solvent in the drum becomes too contaminated to clean effectively, plant personnel move the washer to a drum of clean solvent. When a number of drums of spent solvent have been produced, a contractor replaces them with fresh solvent and transports the spent drums for recycling. EPA has determined that the solvent becomes a solid (and potentially hazardous) waste when it has become too contaminated for further use. If hazardous, the plant owner/operator must count the spent solvent for generator-status-determination purposes when the washing apparatus is removed from the drum. [RO 12790]

6.1.5.2 Analytical samples

According to §261.5(c)(1), analytical samples exempt from hazardous waste regulation under §261.4(d) do not have to be counted in monthly generator-status calculations. The §261.4(d) exemption applies only while the analytical samples are: 1) stored prior to being sent to the laboratory, 2) transported to and from the laboratory, and 3) stored at the laboratory prior to and after testing. As stated in §261.4(d)(3), the exemption does *not* apply if the laboratory determines that the waste is hazardous and the three conditions listed above are no longer being met (e.g., excess sample is being discarded).

Consequently, EPA has determined that excess samples that have been determined to be hazardous must be included in generator-status calculations when they are discarded. [RO 13291] If the laboratory discards the samples, the laboratory would be the generator. If the samples were returned to the sample collector who subsequently discards them, the collector would be the generator.

6.1.5.3 Waste managed in elementary neutralization units

Corrosive waste is stored in a container prior to treatment in a separate elementary neutralization unit. Does the facility have to count the corrosive waste for purposes of determining generator status? What about corrosive waste that is pumped directly into an elementary neutralization unit but is stored there for two months before it is actually neutralized?

The §261.5(c)(2) exemption from counting applies only to hazardous waste that is managed *without prior storage or accumulation* in onsite elementary neutralization units, wastewater treatment units, or totally enclosed treatment facilities. [These units are exempt from substantive RCRA regulation via §§264.1(g)(5–6), 265.1(c)(9–10), and 270.1(c)(2)(iv–v).] Therefore, waste that is stored in a container (that *is* subject to substantive RCRA standards) prior to entering the elementary neutralization unit must be counted in the month that it is generated.

On the other hand, waste that exits a manufacturing process and enters the elementary neutralization unit without prior storage or accumulation (even if it is not treated in that unit for two months) qualifies for the §261.5(c)(2) exemption and is therefore not counted when determining generator status. [RO 11173, 13204]

6.1.5.4 Onsite solvent recycling

According to EPA, wastes that are continuously reclaimed onsite in a still or solvent cleaning machine without intervening storage and are reused onsite do not have to be counted when determining generator status. [§261.5(c)(3)] [Recycling units are exempt from substantive standards via §261.6(c)(1).] Take, for example, a facility that generates 1,500 kg of hazardous spent solvent per month and transfers the solvent directly to a still without intervening storage. The still produces 90 kg of hazardous still bottoms per month. In this case, because the spent solvent is not stored prior to recycling in the still, the 1,500 kg of spent solvent are not subject to counting. Only the still bottoms would need to be counted.

On the other hand, if the spent solvent is stored or accumulated in a unit subject to substantive regulation (e.g., in 55-gallon drums) prior to distillation, the 1,500 kg of spent solvent would have to be counted in the month in which it was generated. Following reclamation, the resulting still bottoms, however, would *not* need to be counted since the spent solvents (from which the bottoms were derived) were already counted once (see the double-counting discussion in Section 6.1.6). The regenerated solvent is not a solid or hazardous waste and therefore is not counted. [§261.3(c)(2)(i), 51 *FR* 10153, RO 11236, 11341]

EPA has clarified that generators do not have to install direct piping connections between the manufacturing process that generates the waste and the recycling unit in order to qualify for the §261.5(c)(3) counting exemption. As long as the waste is transferred immediately, it can be transferred in buckets or other containers and the "without prior storage or accumulation" provision will be satisfied. [RO 14333]

Q *An SQG generates a drum of spent solvent in October but doesn't recycle the solvent until November; another drum of spent solvent is produced in November. In what month(s) does the spent solvent have to be counted for the generator-status determination?*

A All waste counting is done on a calendar-month-by-calendar-month basis. Since the spent solvent is generated in both months and stored in units that are subject to substantive RCRA standards, it must be included in both October's and November's quantity determinations. The exemptions to avoid double counting (discussed in Section 6.1.6) apply only during a specific month (i.e., if the solvent is recycled a number of times in November producing a drum of spent solvent each time, it would have to be counted only once in November). [RO 12699, 12732]

6.1.5.5 Offsite solvent recycling

As mentioned above, the §261.5(c)(3) exemption from counting applies only to wastes that are reclaimed without prior storage *onsite*; it does not apply to recyclers that accept wastes from offsite sources. The requirements applicable to these recyclers are illustrated using the following example. A recycler regenerates F005 spent solvent received from offsite sources. The resulting still bottoms and part of the reclaimed solvent are burned onsite in an industrial furnace. The remaining reclaimed solvent is sold for use both as a fuel and as a solvent.

When the recycler makes its monthly hazardous waste quantity calculations, the still bottoms must be counted because they are hazardous wastes generated onsite and are burned for energy recovery (i.e., they are subject to substantive RCRA regulation). The reclaimed solvent that is burned in the onsite industrial furnace and the solvent that is sold for use as fuel must also be counted because they are hazardous waste via the derived-from rule [see the last sentence in parentheses in §261.3(c)(2)(i)] and subject to substantive regulations. The only material that does not have to be counted is the reclaimed solvent that will be sold for use as a solvent. The derived-from rule regulatory citation noted above exempts reclaimed materials that will be used beneficially from RCRA regulation as long as they are not burned for energy recovery or used in a manner constituting disposal. [RO 12850]

6.1.5.6 Wastes managed in totally enclosed treatment units

Totally enclosed treatment units are exempt from RCRA regulatory requirements. Only if the wastes exiting the unit are hazardous do they have to be counted towards the facility's generator status. [RO 12097]

6.1.5.7 Use wet weight for counting

Hazardous waste counting is based on the total weight of waste generated—not just the weight of hazardous constituents. For example, two facilities generate F006 wastewater treatment sludge from electroplating operations. One facility dewaters its sludge significantly in units meeting the wastewater treatment unit exemption and then stores the dewatered sludge in containers for offsite shipment; the other facility ships F006 sludge with a fairly high water content offsite in containers. The point where the hazardous sludge is counted at both facilities is in the first unit that is subject to substantive RCRA standards—the containers. Although both facilities are shipping roughly the same amount of metals and cyanides offsite, the first facility is an SQG, while the second is an LQG. There is nothing in the regs that specifically addresses water content in the generation of hazardous wastes; thus, the weight of the total volume of waste generated (hazardous constituents plus water) must be counted in the first unit that is subject to substantive RCRA standards. [RO 14619]

6.1.6 Don't double count your waste

EPA has taken steps to avoid double counting of hazardous wastes in the same month by including three exemptions in §261.5(d). These provisions specify that the following do *not* have to be counted:

1. Hazardous waste that is removed from onsite storage, if it was previously counted when it went into or was in storage;

2. Residues produced by onsite treatment (including reclamation) of hazardous waste, as long as the hazardous waste was counted before it was treated [RO 11308, 12865]; and

3. Spent materials that are generated, reclaimed, and subsequently reused onsite, as long as such spent materials have been counted once in a calendar month. [March 24, 1986; 51 *FR* 10152, RO 12699, 12732]

These exemptions apply only to the *onsite* treatment or storage of hazardous waste. If a permitted TSD facility or permit-exempt recycler receives waste from an offsite generator for treatment or storage, any hazardous waste generated by the treatment or storage process must be counted in the facility's or recycler's hazardous waste quantity determination. [November 19, 1980; 45 *FR* 76621] Specifically, EPA noted that:

"Owners and operators of hazardous waste management facilities may generate hazardous waste (e.g., residues created by treatment processes). With respect to the hazardous waste that these persons generate, they, like other generators, must comply with the applicable provisions of Part 262." [December 31, 1980; 45 *FR* 86969, RO 12865]

6.1.7 Acute hazardous waste counted separately

As stated in §261.5(e), if a generator produces more than 1 kg (2.2 lb) of acute hazardous waste or 100 kg (220 lb) of acute hazardous waste spill residue in a calendar month, such wastes are subject to full RCRA regulation under Parts 262 through 270.

Acute hazardous wastes are counted and may be managed separately from nonacute hazardous wastes. For example, consider a facility that generates more than 1 kg of acute hazardous waste and 900 kg of nonacute hazardous waste during a month. Since acute and nonacute hazardous wastes are counted separately, the generator can maintain SQG status for the nonacute hazardous waste, while managing the acute waste according to LQG standards. In other words, the generator could be an SQG and an LQG at the same time! The nonacute wastes could be managed in accordance with the less stringent SQG requirements of §262.34(d–f) (e.g., they could be accumulated onsite for up to 180 days before being shipped offsite). The generator would have to manage all of the acute waste as an LQG (e.g., the waste could be accumulated onsite for only 90 days before being shipped offsite). [RO 11288, 14031, 14700]

We question whether managing acute and nonacute wastes under different generator requirements would be worth the effort. Also the three pieces of guidance noted above conflict with a previous EPA discussion of the topic: "The agency believes that a generator who produces more than 1,000 kg of hazardous waste a month and is therefore subject to full regulation should handle his acutely hazardous wastes in the same manner as his other wastes." [November 19, 1980; 45 *FR* 76622]

6.2 Satellite accumulation

Four regulatory options are available to generators who want to accumulate or store hazardous wastes:

1. Both large and small quantity generators can accumulate small quantities of waste (up to 55 gallons of nonacute hazardous waste) in satellite accumulation areas. [RO 14618, 14703] No permit is required for this type of waste accumulation. The satellite accumulation provisions [§262.34(c)] are discussed in this section.

2. They can accumulate wastes for up to 90 days (large quantity generators) or up to 180 days (small quantity generators) without a permit. [Section 270.1(c)(2)(i) codifies this permitting exemption.] This type of accumulation is governed by §262.34 and is discussed in Section 6.3.

3. They can store wastes in interim status or permitted units. These units would typically be containers, tanks, containment buildings, waste piles, surface impoundments, etc. This is the least attractive option because of the cost and complexities involved with the RCRA permitting process.

4. They can accumulate or store hazardous wastes in units that are exempt from RCRA, such as elementary neutralization units or wastewater treatment units. These units are discussed in Chapter 7.

6.2.1 Satellite accumulation "areas"

When the RCRA regulations were first issued, only three types of storage or accumulation areas were allowed: 90-day areas, permitted storage areas, and exempt units. No special provisions were made for managing small quantities of wastes where they were initially generated and accumulated throughout a manufacturing facility. EPA termed these small accumulation sites "satellite accumulation areas" (SAAs) to distinguish them from centralized 90-day or permitted storage areas. (However, the term "satellite accumulation area" cannot be found in the RCRA regulations.) The agency recognized that satellite accumulation might occur at dozens of locations throughout a large plant (e.g., wherever sample purgings are collected), and requested comment on whether less stringent accumulation standards were appropriate for such locations on November 19, 1980. [45 FR 76625]

After four years of consideration, EPA issued a rule that allowed for reduced regulation of wastes accumulating in SAAs. [December 20, 1984; 49 FR 49568] The SAA regulations appear in §262.34(c). In order to quickly grasp the difference between satellite units and 90-day containers, refer to Table 6-2. There are many similarities, but some differences.

As we begin our discussion of SAAs, note that although the satellite accumulation regulations refer to "containers," the provisions actually relate more to "areas." The preamble establishing SAAs stated "Satellite areas are those places where wastes are generated in the industrial process or laboratory and where those wastes must initially accumulate prior to removal to a central area." [49 FR 49569]

A generator typically identifies an SAA by painting lines on the floor, hanging a sign or clipboard on the wall, roping or fencing off a small area, etc. Once designated, the SAA can include a number of containers, subject to the volume limits specified in the next paragraphs. [48 FR 119] The federal regulations don't require that SAAs be designated in any particular manner; however, this is a very good operating practice to make it clear to both the generator's personnel and to an agency inspector that satellite accumulation is occurring.

The regulations also impose no limits on the number of SAAs that can be designated at a generator's facility. [RO 12859]

The quantity of waste that can accumulate in a satellite accumulation area is limited to 55 gallons for nonacute wastes. Nonacute wastes are all characteristic wastes (D-coded), K-wastes, U-wastes, and F-wastes (except for the acutely hazardous dioxin wastes, coded F020–F023 and F026–F027). For the acutely hazardous P-wastes and the F-coded dioxin wastes just cited, the accumulation limit is one quart.

In actual practice, generators typically identify an SAA for individual points of waste generation. Consider the situation where two sample taps are located side-by-side; one tap is associated with Process A, while the other is associated with Process B. The generator can designate SAA-1 as being for sample purges taken from Process A and SAA-2 as being for sample purges from Process B. Assuming

Table 6-2: Federal Requirements for Satellite Accumulation Units vs. 90-Day Containers

Requirement	Satellite accumulation units	90-day containers
Must be in good condition	Yes	Yes
Must be compatible with the hazardous waste in the container	Yes	Yes
Must be closed at all times except when adding or removing waste	Yes	Yes
Inspection requirement	None	Weekly
Hazard marking requirement	"Hazardous waste" or other description	"Hazardous waste" only
Date marking requirement	On the date 55 gallons (or 1 quart for acute wastes) is exceeded	On the date waste first goes in the container
Maximum length of storage	Unlimited	90 days
Maximum waste volume in storage	55 gallons (or 1 quart for acute wastes)[1]	Unlimited
Personnel training required	No	Yes
Can treat hazardous waste in the unit	No	Yes
Special requirements for ignitable/reactive wastes	No	Yes
Special requirements for incompatible wastes	No	Yes
Must comply with Subpart CC air emission standards	No	Yes
Must comply with preparedness, prevention, contingency plan, and emergency procedures in Part 265, Subparts C and D	No	Yes

[1] If the maximum waste volume is exceeded, then the excess must be dated and moved within three days. [§262.34(c)(2)]

Source: McCoy and Associates, Inc.; adapted from §§262.34(a–c) and 265.171–178 and RO 14703, 14758.

that the wastes are not acutely hazardous, SAA-1 could contain up to 55 gallons of waste and SAA-2 could also contain up to 55 gallons. If only one SAA was identified, and wastes from both Processes A and B were placed in the area, the quantity limit would be 55 gallons total. [RO 11452, 14826]

SAA containers are not identified in the exceptions to the waste counting provisions of §261.5(c). These provisions are used to determine which wastes must be counted in determining whether a generator is conditionally exempt, a small quantity generator, or a large quantity generator. Therefore, a generator should include in his/her waste quantity calculations the amount of waste added to his/her SAAs on a monthly basis. [RO 11812, 13312, 14703] (Some facilities use detailed tracking sheets or clipboards so that operators can record the date and quantity added to each SAA.)

While containers are in an SAA, each must be marked with the words "Hazardous Waste" or some other words that identify the contents, such as "Spent Solvents." Although it is not required that a waste code be placed on a satellite accumulation container, this is certainly a good operating practice.

Q A generator places several containers within a cabinet. Does a hazardous waste label placed on the outside of the cabinet suffice, or must each individual container within the cabinet be labeled?

A Placing a label with the words "Hazardous Waste" on the outside of the cabinet will satisfy the SAA marking requirement if the cabinet meets the definition of a container. To be a satellite accumulation container, the cabinet must be portable, located at or near the point of generation and be under the control of the operator of the process generating the waste, in good condition, constructed of materials compatible with the waste, and closed at all times except when wastes are added or removed. If the cabinet does not meet all of the above requirements, each individual container within the cabinet would need to be labeled. [RO 14587]

 If a facility has very small containers (e.g., vials or tubes) of hazardous waste in an SAA that are too small to label with the words "Hazardous Waste" or other words that identify the contents, how should the labeling requirement be met?

A The small containers should be placed in properly labeled larger containers. [RO 14703]

6.2.2 At or near the point of generation

The SAA must be "at or near the point of waste generation" and "under the control of the operator of the process generating the waste." EPA allows the authorized regulatory agency to determine on a case-by-case basis what these terms mean. "Some Regions and states recommend utilizing an 'in sight of' approach to implement this requirement; however, site-specific conditions should generally determine compliance with what constitutes 'at or near any point of generation.'" [RO 14826] If a wall or closed door separates the operator from the SAA, it might not be under his/her control unless he/she uses a lock and key or some other means of preventing unauthorized access. In one case [RO 11317], EPA stated that for safety reasons an SAA could be located in a shed outside the building where the waste was generated. If the original process operator keeps the key to the shed, the SAA remains under his/her control.

Another implication of the term "under the control of the operator" deals with waste coding. At the point where a hazardous waste is generated, it is also assigned a waste code (e.g., F001 spent solvent). This code must follow the waste while it is in the satellite container (e.g., a red safety can), when it is transferred to a drum in a 90-day area, and when it is transferred from the drum to a tank truck for offsite transport. If someone other than the original process operator puts waste (e.g., F005 spent solvent) in the red safety can, how will the new waste code be tracked? (It might not be.)

In one interpretation, EPA indicated that the SAA provisions can apply to small quantities of like wastes that are generated throughout a facility. For example, maintenance personnel might generate small amounts of hazardous waste from many locations in a plant, and the SAA for these wastes could be located in the maintenance shop. [RO 11728]

In essence, the "at or near" and "under the control" terms are check-with-your-state issues. Many states have written policies defining these terms.

6.2.3 Physical limitations

Section 262.34(c)(1)(i) specifies that containers used in SAAs must 1) be in good condition (§265.171), 2) be compatible with the waste (§265.172), and 3) be closed except when adding or removing waste [§265.173(a)]. This third requirement has been the source of many notices of violation. When EPA initially discussed the closed-container provision, their intent was that waste containers be vapor tight and spill proof. [45 FR 33199] The agency provided more specific guidance on how to close satellite accumulation containers in RO 14826. That guidance is summarized in Section 9.2.1.

EPA interprets the SAA provisions to be available on a one-shot basis. Generators may not move hazardous waste between SAAs. Once a hazardous waste leaves an SAA, it must be destined for a 90/180-day accumulation area, a permitted area or unit, a permit-exempt unit, or offsite. However, a single SAA may have multiple points of generation. Movement or consolidation of hazardous waste within an SAA is permissible, as long as the hazardous waste remains "at or near the point of waste generation" and "under the control of the operator of the process generating the waste." EPA has also clarified that a generator may consolidate several partially full containers of the same hazardous waste into one container before transferring it to a 90/180-day accumulation area. [RO 14337, 14703, 14826]

The above guidance on multiple points of generation within a single SAA is particularly useful for laboratory settings. For example, in a laboratory, nonacute wastes from several analytical instruments drain into individual 1-gallon jugs under the instrument tables. When the jugs are full, lab techs empty them into a 55-gallon drum inside the lab. What is the regulatory status of the 1-gallon jugs and 55-gallon drum? Three interpretations can be made:

1. The 1-gallon jugs are integral parts of the analytical instruments (i.e., the instruments will not work without the jugs, or maybe the instrument manufacturer supplies the jugs). In this interpretation, the jugs are not waste units and the 55-gallon drum is the satellite accumulation unit. This interpretation is sometimes accepted by state/EPA inspectors, especially if there is a log associated with the 55-gallon drum that shows when and by whom each 1-gallon jug was emptied into the 55-gallon drum. This shows that the 55-gallon drum of waste (which the plant considers to be the satellite accumulation container) is "under the control of the operator of the process generating the waste." In 2004 guidance [RO 14703], however, EPA noted the following, which gives us limited confidence in this interpretation: "Even if the discharging unit [the instrument] is not regulated under RCRA, the attached *containers* that collect hazardous wastes from such equipment must be in compliance with the SAA regulations...." [Emphasis in original.]

2. The 1-gallon jugs are actually satellite accumulation containers (as opposed to an integral part of the instrument). The 55-gallon drum is a 90/180-day accumulation unit because, as noted above, wastes can't move from one satellite unit to another. This is a more conservative interpretation than the first one above.

3. Both the 1-gallon jugs and 55-gallon drum make up one satellite accumulation area. As long as the total volume of waste in the 1-gallon jugs and the 55-gallon drum into which these jugs are emptied doesn't exceed 55 gallons, the whole area could be considered a satellite area. This seems to be where EPA was headed when it stated: "It is possible for there to be multiple pieces of equipment within one SAA, and thus multiple points of generation within a single SAA, provided all the pieces of equipment are 'at or near' each other and 'under the control of the operator of the process generating the waste.' Under this scenario, the total amount of hazardous waste in the SAA would be limited to 55 gallons (or 1 quart of acute hazardous waste) and a generator would be allowed to consolidate like hazardous wastes from multiple discharging units." [RO 14703] The one concern with this approach is that the total volume of like hazardous waste (e.g., D001) in the several small jugs and 55-gallon drum could exceed 55 gallons. Physical container size or other controls must be implemented to ensure this doesn't occur.

6.2.4 Dating requirement

The provisions of §262.34(c)(2) dealing with excess accumulations of waste in an SAA are not very clear. Supposedly, any excess accumulation beyond the 55-gallon/1-quart limits must be removed within 3 days. The following example illustrates how this provision works in actual practice.

Consider a 55-gallon drum in an SAA associated with a sample tap. Every day, an operator purges 0.25 gallons of material from the sample tap and pours the material into the drum for disposal (i.e., the material is a hazardous waste). On June 1, the drum contains 54.75 gallons of waste, and when the operator adds the daily sample purge, it will be full. The SAA now contains the maximum allowable quantity of waste (55 gallons). On June 2, the operator brings in an empty drum and adds the daily sample purge to the new drum. On that day, the 55-gallon limit is exceeded for the satellite accumulation area, and the operator must comply with the provisions of §262.34(c)(2). He/she puts the current date (June 2) on the full drum because on this day he/she exceeds the 55-gallon limit. (This is called the "date of excess accumulation.")

Once the generator dates the drum, he/she now has three days (June 3, 4, and 5) to move the full drum from the SAA to one of the following destinations: 1) a 90/180-day accumulation area, 2) a permitted area or unit, 3) a permit-exempt unit (e.g., a recycling unit or a wastewater treatment unit), or 4) offsite.

Note that the above discussion is not exactly consistent with EPA's regulatory language. Section 262.34(c)(2) requires generators to date the container holding the excess accumulation of hazardous waste and move it within 3 days. [May 23, 2006; 71 *FR* 29733] In our example above, that would be the new drum holding just a little bit. Our experience indicates that generators date full drums, not the nearly empty ones, and move full drums within 3 days; state inspectors normally do not have any objections to that practice.

If the excess (i.e., whatever is over 55 gallons) is not moved within 3 days, it becomes subject to the 90/180-day accumulation container standards, depending on the generator's status. [RO 12503] If an inspector finds more than 55 gallons of waste in an SAA and the date of excess accumulation is missing or if more than three days has elapsed, an NOV will likely result (unless the generator is complying with the 90/180-day accumulation container standards). If the date of excess accumulation occurs on a Friday that will be followed by a 3-day weekend, tough luck. EPA expects you to plan ahead in meeting the 3-day deadline. Three days means 3 consecutive calendar days—not 3 working or business days. [May 23, 2006; 71 *FR* 29733, RO 14703]

Some facilities have standard operating procedures specifying that once the waste level reaches 2–4 inches from the top of the drum, the operator dates the container and moves it to a 90/180-day area within 24 hours. As such, the date on the container becomes the start date for the 90/180-day clock as well as (or in lieu of) the date of excess accumulation. This practice is not a regulatory requirement, however. In fact, if you move a satellite accumulation container to a 90-day area, you get a total of 93 days of accumulation time after the date of excess accumulation: 3 days in the SAA and 90 days from the date the waste enters the 90-day area. [RO 12503, 13410, 14703]

Q *If a 4-gallon container of nonacute hazardous waste is in an SAA, does the generator have to move the container from the SAA within 3 days of being filled?*

A No. There is no federal requirement that full containers of hazardous waste be moved within 3 days of being filled. Only when more than 55 gallons of nonacute hazardous waste is accumulated must waste be moved within 3 days. [RO 14703]

Note that sometimes an SAA can contain much more than 55 gallons of waste. For example, in a large chemical plant or refinery, a reactor might contain 50 cubic yards of catalyst. Hazardous spent catalyst is removed for disposal once every three years and is drained into two roll-off boxes. Because the SAA regulations don't specify container sizes, the roll-off boxes can be considered to be SAA containers. [RO 11442] The SAA regulations require that, as soon as more than 55 gallons of catalyst is drained to the first roll-off box, the current date be applied, and all of the catalyst needs to be moved within three days to one of the four destinations cited earlier. Use of the SAA provisions under these circumstances was specifically addressed in the proposed SAA rule. [48 *FR* 120] However, EPA recommends that generators contact their state agency for guidance/agreement on how such occurrences should be handled. [RO 14029]

The federal regulations do not specify how long wastes can remain in an SAA (if the 55-gallon/1-quart limits aren't exceeded). Conceivably, wastes might accumulate for several years before the quantity limit is reached. A regulatory concern can arise under these circumstances, however, because of the storage prohibition in the land disposal restrictions (LDR). As soon as a hazardous waste is generated, it is subject to LDR, including §268.50(c). This section states that if a waste is stored for more than one year, the burden of proof is on the owner/operator to show why this is necessary.

In 1990 guidance, however, EPA clarified that the accumulation time for the LDR storage prohibition starts when the waste is moved to a central accumulation area (i.e., a 90/180-day or permitted area). Therefore, EPA's perspective is that hazardous waste accumulation in satellite accumulation areas is *not* subject to the one-year storage prohibition. [OSWER Directive 9555.00-01] We are aware that some facilities (and some states) have a policy of dating SAA containers with the initial date of waste accumulation. The wastes are then moved within one year. A potential problem with this practice is that an inspector could confuse an accumulation start date with the date of excess accumulation.

6.2.5 Miscellaneous issues

Conditionally exempt small quantity generators do not have to comply with any of the SAA provisions. [§261.5(b)]

Personnel whose only hazardous waste activities involve placing wastes in SAA containers are not subject to the RCRA training requirements of §265.16 or 262.34(d)(5)(iii). [January 3, 1983; 48 *FR* 119, December 20, 1984; 49 *FR* 49570, RO 11373, 14703, 14758] In our opinion, such personnel should always be trained because their errors at the point of waste generation can create numerous RCRA problems. For example, personnel working in SAAs need to be familiar enough with the chemicals with which they are working to know 1) when they have generated a hazardous waste, and 2) that it must be managed in accordance with the RCRA regulations. Personnel who move hazardous waste from SAAs to 90/180-day areas must be RCRA-trained. [RO 14703]

The satellite accumulation container requirements in §262.34(c) reference three specific sections in Part 265, Subpart I, but none of the three is the weekly inspection requirement. EPA has confirmed in guidance that such satellite containers do not require weekly inspection. [RO 14418] "However, the SAA regulations do require that waste containers in an SAA must be under the control of the operator of the process generating the waste, in good condition (§265.171), compatible with its contents (§265.172), and closed except when adding or removing waste (§265.173), which should achieve the goal of inspections: containers that are free of leaks and deterioration." [RO 14703] Note that states may require weekly inspection as part of their programs.

A contingency plan is required in the regulations only for 90-day accumulation activities [see §262.34(a)(4), RO 14758]. The contingency plan is not required to address satellite accumulation areas [see January 3, 1983; 48 *FR* 119, December 20, 1984; 49 *FR* 49570, RO 11373, 14703, 14758]. However, it would be prudent (and some states require) that satellite areas be designated in the plan as well.

SAA containers are not subject to the emission control requirements of Part 264/265, Subpart CC. Note, however, that as soon as such a container is moved into a 90-day area (i.e., it is a 90-day container), it is subject to Subpart CC. [RO 13777, 14703]

Because the regulations establishing SAAs were amendments to the original RCRA regulations, and because they were less stringent than the pre-existing regulations, authorized states were not required to adopt the SAA provisions. Although we believe that nearly all states allow SAAs, we have not done a state-by-state survey to identify states without SAA provisions.

Although the SAA regulations predate most remediation regulations and policies, under many circumstances we see no reason why remediation wastes (such as ground water samples) could not be put into an SAA. Note that the SAA regulations did not envision operators generating one-time wastes (such as soil borings) and essentially abandoning the wastes in an "SAA." (We are not aware of any specific guidance on this issue, however.)

In general, treating hazardous waste requires a permit. As a matter of policy, EPA allows hazardous wastes to be treated in 90/180-day units without a permit. [51 FR 10168; see also Section 6.3.5] A similar policy does not exist for allowing treatment of waste in an SAA without a permit; "thus, the decision should be made by the appropriate implementing agency." [EPA/530/K-05/011, September 2005, available online at http://www.epa.gov/wastes/inforesources/pubs/training/gen05.pdf]

Finally, because many states have different regulations and policies concerning SAAs, do not rely solely on EPA's interpretations—check with your state regulators.

6.3 90- and 180-day accumulation

The regulatory provisions that allow generators to accumulate wastes without a permit are codified in §262.34. This section deals with three general topics. First, the standards applicable to LQGs are described in §262.34(a) and (b). Second, the satellite accumulation provisions are specified in §262.34(c). Finally, the standards applicable to SQGs are specified in §262.34(d), (e), and (f). Before discussing LQG and SQG requirements, a brief history of the regulations might be helpful.

6.3.1 Regulatory history

The provisions allowing LQGs to accumulate wastes for up to 90 days without a permit date back to the very beginning of the RCRA regulatory program. [May 19, 1980; 45 FR 33141] This so-called "90-day accumulation rule" originally provided manufacturers with a means to accumulate sufficient quantities of waste for onsite or offsite management without having to deal with burdensome permitting requirements. EPA justified this accumulation period as being necessary to ensure that the RCRA regulations did not interfere with industry's production processes.

On March 24, 1986 [51 FR 10146], EPA issued regulations applicable to SQGs. Because it takes a longer period of time for generators in this category to accumulate enough wastes for efficient treatment or disposal, they are allowed up to 180 days to accumulate wastes without a permit. If the waste is transported 200 miles or more for offsite treatment, storage, or disposal, the accumulation period extends to 270 days.

6.3.2 Summary of accumulation standards

Table 6-3 summarizes the hazardous waste accumulation requirements for large and small quantity generators. Note that many minor, but important, differences exist between the requirements applicable to these two groups.

6.3.3 Allowable facilities

The 90/180/270-day accumulation provisions apply to the original generator of the waste; hence, these units are typically located at generator sites.

A word of caution is in order regarding generator sites and the 90-day accumulation provisions. Let's assume that a consultant is collecting soil samples or doing cleanup work at a site that is not part of a treatment, storage, or disposal (TSD) facility or a RCRA generator's site. Can the consultant put hazardous soil borings or cleanup residues in drums or roll-off boxes and store the wastes on the site for up to 90 days without a permit?

The answer is "Yes," *if* all of the provisions of §262.34 are met. The problem arises with the need to comply with the Part 265, Subpart C requirements for preparedness and prevention, the Subpart D standards for a contingency plan and emergency procedures, and the §265.16 training provisions. Does the emergency equipment specified in

Table 6-3: Summary of Generator Waste Accumulation Provisions

Regulatory provision	Large quantity generators	Small quantity generators
Allowable accumulation period	90 days	180 days (270 days allowed if waste is shipped 200 miles or more)
30-day accumulation period extension availability	Case-by-case	Case-by-case
Allowable accumulation units	Tanks (per Part 265, Subpart J) Containers (per Part 265, Subpart I) Drip pads (per Part 265, Subpart W) Containment buildings (per Part 265, Subpart DD)	Tanks (per §265.201) Containers (per Part 265, Subpart I) Drip pads (per Part 265, Subpart W)[1] Containment buildings (per Part 265, Subpart DD)[1]
Maximum onsite waste accumulation	No limit	6,000 kg nonacute hazardous waste 1 kg acute hazardous waste 100 kg acute spill cleanup residue
Treatment allowed in accumulation units without a permit	Yes	Yes
Container location standards	At least 50 feet from property line per §265.176 for ignitables and reactives[2]	None
Air emission standards	Part 265, Subparts AA, BB, and CC for tanks and containers	None
Accumulation start date marking	Date appears on each tank[3] and container; recorded in facility logs for drip pads and containment buildings	Date appears on each tank[3] and container
"Hazardous waste" label	Applied to each tank and container	Applied to each tank and container
Preparedness and prevention	Per Part 265, Subpart C[4]	Per Part 265, Subpart C
Contingency plan and emergency procedures	Per Part 265, Subpart D[4]	Per §262.34(d)(5)
Closure requirements	Per §§265.111 and 265.114	Per §265.201(f)
Training requirements	Per §265.16[4]	Per §262.34(d)(5)(iii)
Waste analysis plans	Per §268.7(a)(5) for wastes being treated to meet LDR standards	Per §268.7(a)(5) for wastes being treated to meet LDR standards

[1]Small quantity generators can utilize containment buildings or drip pads only if they comply with the large quantity generator requirements in §262.34(a), including a limit for onsite accumulation of 90 days. [EPA/305/B-96/001, EPA/530/K-02/008I, EPA/530/K-05/008, RO 13696, 14662]
[2]Generators who physically cannot accumulate their ignitable or reactive waste 50 feet from their property line should work with their EPA region or state inspector to determine if the local fire department or fire marshal will grant a variance from the 50-foot buffer requirement. Any such variance should be in writing and maintained onsite. [RO 14840]
[3]The requirement to mark accumulation start dates on 90/180-day tanks comes from EPA guidance—not the regulations. [November 19, 1980; 45 FR 76624, March 24, 1986; 51 FR 10160, RO 11641, 14683]
[4]These requirements apply only to 90-day hazardous waste accumulation areas at large quantity generator sites; they do not apply to non-RCRA units (e.g., manufacturing units). [RO 14758]
Source: McCoy and Associates, Inc.; adapted from §262.34.

§265.32 exist at the drum or roll-off box accumulation area? Is a communication system per §265.34 available? Have arrangements with local emergency response authorities been made per §265.37? Is there a written contingency plan for the area (§265.51)? Is a copy of this plan maintained at the site, and has it been sent to local emergency response organizations (§265.53)? Is an emergency coordinator either at the site or on call at all times (§265.55)? Have all of the consultant's field personnel been trained in accordance with §265.16 requirements?

If a permitted or interim status TSD facility generates its own hazardous wastes (e.g., as treatment residues), they may also utilize the 90/180/270-day accumulation provisions. [December 31, 1980; 45 FR 86969, RO 12865] At permitted TSD facilities, these accumulation areas should appear on the facility map. [RO 12075] A sign or notice should be posted in a visible location to distinguish the 90/180/270-day area from permitted storage/treatment areas; construction of a 90/180/270-day accumulation area would not require modification of the facility's storage permit. [RO 12471] TSD facilities cannot utilize the 90/180/270-day accumulation provisions for wastes that they do not generate at their site. [December 31, 1980; 45 FR 86969, RO 11163, 11358, 14466]

Transporters receive wastes from other parties but typically do not generate their own wastes. (Bulking like wastes together to facilitate shipment may change the properties of a waste, but this activity is not necessarily considered to be waste generation. [RO 13272]) Section 263.12 specifies that transporters may store wastes in containers at a transfer facility for up to 10 days. However, transporters who mix wastes of different DOT shipping descriptions must remanifest the waste to accurately reflect the composition of the new waste. This would occur, for example, when transporters combine two RCRA ignitable wastes—one a DOT combustible material and the other a DOT flammable waste. Such mixing requires transporters to comply with Part 262 generator standards [per §263.10(c)(2)]. In this situation:

> "Although they may indicate on the manifest in Box [14] the name of the original generator(s) of the combined waste, they must represent themselves as the generator of the new waste. Although by creating or generating a new waste they have taken on some of the generator requirements, the transporter should continue to manifest the waste to the designated facility as indicated on the original manifest by the original generator." [RO 11567]

Note that mixing different hazardous wastes at transfer facilities does not generate a new waste and transporters are not considered generators. As such, transporters cannot make use of the 90/180/270-day accumulation provisions but must comply with the 10-day clock of §263.12. [RO 13272]

6.3.4 Allowable units

Generators practicing 90/180/270-day accumulation are allowed to use only certain types of units (see Table 6-3). In most cases, the units of choice are tanks and containers. The peculiarities associated with each type of unit are discussed in this section.

6.3.4.1 Tanks

When an LQG accumulates wastes in a tank, all of the Part 265, Subpart J tank standards apply, with the following exceptions [§262.34(a)(1)(ii)]:

1. The generator does not need to prepare closure plans and contingent post-closure plans for the tanks that do not have secondary containment [see §265.197(c)]; and

2. The generator does not need to conduct waste analyses and trial treatment tests per §265.200 when the waste properties change. [Note that if a waste is being treated in a 90-day tank for the purpose of meeting the land disposal treatment standards, a waste analysis plan is required per §268.7(a)(5).]

Probably the most important (and expensive) tank standards applicable to LQGs are the need to provide secondary containment and leak detection per §265.193. These standards imply that wastes can't simply be stored for up to 90 days in just any old tank. If the tank doesn't have impervious secondary containment, it can't be used for hazardous waste accumulation. Refer to Chapter 10 for a complete discussion of tank storage.

The tank standards applicable to 180/270-day accumulation at SQG sites are greatly simplified. These units are subject only to the standards in §265.201, which don't require that tanks be equipped with secondary containment and leak detection. An SQG would comply with the reduced tank standards in

§265.201, even if it is treating hazardous waste onsite in a 180/270-day accumulation tank. [RO 14684]

6.3.4.2 Containers

Both large and small quantity generators may use containers for waste accumulation. These containers are subject to the standards in Part 265, Subpart I. Note that per §262.34(d)(2), SQG containers are not subject to the Subpart I standards (in §265.176) specifying that ignitable or reactive wastes be stored at least 50 feet from the property line. SQG containers are also not subject to the air emission standards of Part 265, Subparts AA, BB, or CC.

Probably the most significant container standard is that containers must be closed except when adding or removing wastes. [§265.173(a)] When EPA initially discussed the closed-container provision, their intent was that waste containers be vapor tight and spill proof. [45 FR 33199] Refer to Chapter 9 for a complete discussion of container management issues.

6.3.4.3 Drip pads

Although LQGs are allowed to accumulate wastes for up to 90 days on drip pads, this is only of practical significance to companies in the wood preserving business. These companies utilize pressurized systems to force wood preservatives into wood products. When the products are removed from the pressurized systems, some of the preservatives "kick back" or drip from the products onto collection areas known as drip pads. The standards applicable to drip pads used for 90-day accumulation are found in Part 265, Subpart W.

SQGs may utilize drip pads only if they comply with the LQG requirements in §262.34(a) for the management of wastes on drip pads, including a limit for onsite accumulation of 90 days. However, SQGs using drip pads for 90-day accumulation may continue to accumulate hazardous waste in tanks and containers for up to 180 days per §262.34(d). [EPA/305/B-96/001, EPA/530/K-02/008I, RO 14662]

6.3.4.4 Containment buildings

Containment buildings are engineered structures that EPA intended to be used for storing/treating bulky solids such as contaminated debris. The standards for these buildings are specified in Part 265, Subpart DD. LQGs may use containment buildings for 90-day accumulation. The only way an SQG could use a containment building for waste accumulation would be by complying with the LQG requirements specified in §262.34(a) for the management of wastes in containment buildings, including a 90-day onsite accumulation limit. [EPA/530/K-05/008, RO 13696]

Some people mistakenly believe that "containment buildings" are container storage buildings that have secondary containment as specified in §264.175. Although such buildings might be able to meet the Subpart DD standards, this is typically not the case. For example, if a Subpart DD containment building is used to manage wastes containing free liquids, the building must have a primary barrier (e.g., a concrete floor), an underlying secondary barrier (e.g., an impermeable geomembrane), and a leak detection/collection system between the two barriers. [§265.1101(b)]

6.3.5 Allowable activities—accumulation and treatment

Although the 90/180/270-day accumulation provisions were originally developed to allow short-term accumulation of wastes, EPA has consistently maintained that wastes may also be treated without a permit in units that are in compliance with §262.34. (Note, however, that thermal treatment is not allowed, as is discussed below.)

The preamble to the SQG rule originally spelled out the treatment-in-accumulation-units exemption. [March 24, 1986; 51 FR 10168] The rationale for this treatment exemption was that the standards for containers (Part 264/265, Subpart I) and tanks (Part 264/265, Subpart J) don't distinguish between treatment and storage activities conducted in these units. Therefore, no justification existed for allowing storage, but not treatment, in accumulation units. Subsequent guidance extended the treatment exemption to LQGs and to activities in containment buildings. [RO 13553, 13782] A complete discussion of treatment in 90/180/270-day accumulation units is contained in Section 7.3.

6.3.5.1 Treatment in sequential units

EPA's guidance on the treatment of wastes in sequential units is confusing. EPA's original opinion, based on December 31, 1980 *Federal Register* preamble language [45 *FR* 86969], was that residues from treatment processes were new points of generation and were therefore eligible for a new 90-day clock. [RO 12865] For example, spent solvent stored in containers would be subject to the 90/180/270-day accumulation provisions. When the solvent is distilled onsite, the still bottoms are considered to be newly generated wastes, and are eligible for a new 90/180/270-day clock. [RO 11420, 12850, 12865, 13280] Note, however, that this guidance appears to apply to residues from either interim status/permitted units or RCRA-exempt units (e.g., recycling units or wastewater treatment units).

6.3.5.2 Thermal treatment

Thermal treatment is regulated by: 1) thermal treatment unit standards (Part 265, Subpart P), 2) incinerator standards (Part 264/265, Subpart O), 3) boiler and industrial furnace standards (Part 266, Subpart H), or 4) miscellaneous unit standards (Part 264, Subpart X). Thus, even if thermal treatment occurs in an accumulation unit, the permitting exemption of §262.34 does not apply. [RO 13553, 14662] For example, open burning/open detonation is cited as a type of thermal treatment that is not eligible for the §262.34 exemption. [RO 11310]

Clearly, combustion-type treatment may not be conducted in generator's accumulation units. The situation is less clear for noncombustion thermal processes such as evaporation (see Section 7.3.2.2).

6.3.6 The 90-day clock

Many issues arise with how the 90-day clock (180/270-day clock for SQGs) applies in different situations. The most common questions are answered below, and Case Study 6-1 illustrates numerous 90-day accumulation requirements under RCRA.

6.3.6.1 Unknown wastes

Q *How does the 90-day clock apply to an unknown waste placed in a drum that, after analysis, is found to be hazardous?*

A The 90-day clock starts when the waste is first generated and placed in the drum—not when the generator receives the waste analysis results. (EPA is concerned that if a waste isn't hazardous until the results of an analysis are obtained, generators could put off getting analytical information.) According to EPA, "If the date on which accumulation began was not marked on the drum [Section 262.34(a)(2)] or the drum was not marked 'Hazardous Waste' [Section 262.34(a)(3)], then the generator has not met the pre-conditions for the exemption from the permitting requirements…." [RO 11424; see also EPA/233/B-00/001]

In order to avoid confusion regarding the regulatory status of unknown wastes, a number of facilities use a label stating "Hazardous waste pending analysis." This wording makes it clear to the generator's personnel and to agency inspectors that the regulatory status of the waste is uncertain and may change. If analytical results indicate the waste isn't hazardous, the label may be removed. If the waste does prove to be hazardous, the 90-day labeling provision has been satisfied.

6.3.6.2 When does the clock start?

In general, the 90/180/270-day clock starts when the first drop of waste is placed into the accumulation unit. For wastes in satellite accumulation containers, the clock starts when 1) the containerized waste first enters the 90/180/270-day area, or 2) 3 days after the satellite unit is dated if it is not moved. [RO 13410, 14703]

Case Study 6-1: 90-Day Accumulation/Treatment Provisions

An LQG uses solvents in its manufacturing process that, when spent, are considered to be F001 hazardous waste. The generator accumulates the waste in tanks, and when a sufficient quantity has accumulated, brings in a mobile thin-film evaporation unit from a vendor. The spent solvent is pumped via pipes from the accumulation tanks to the evaporator; recovered solvent is pumped back to the generator's product tanks. Evaporator still bottoms are collected in drums and are retained by the generator.

Under these circumstances, if the initial spent solvent accumulation tanks are managed as 90-day tanks, no storage permit will be required. [§262.34(a)(1)(ii)] The tanks would have to meet Part 265, Subpart J standards, including secondary containment and leak detection. Because 90-day tanks are subject to Subpart CC air emission controls, the spent solvent accumulation tanks must comply with §265.1085 control standards.

The piping from the spent solvent tank to the evaporator is considered to be ancillary equipment in a 90-day tank system and must meet secondary containment standards. [§265.193(f)] If this pipe run contains spent solvent for more than 300 hours per year, equipment in the pipe run (e.g., pumps, valves, flanges, and sample taps) would have to comply with the fugitive emission standards of Part 265, Subpart BB because it is part of a 90-day tank system. [§265.1050(b)(3) and (e)]

The evaporator is considered to be a recycling unit and is not subject to permitting or other RCRA standards. [§261.6(c)(1)] Because the generator does not have interim status or a permit for any treatment, storage, or disposal units, the evaporator is not subject to Part 265, Subpart AA emission controls. [§265.1030(b)(2)]

EPA considers the original generator and the mobile equipment vendor to be cogenerators of the still bottoms. By mutual agreement, these two parties may determine who is responsible for the bottoms. [RO 13280] In this case, the generator assumes responsibility for the bottoms and retains them onsite. Still bottoms from the recycling unit are considered to be newly generated waste and, hence, are eligible for a new 90-day clock. [RO 12865] (However, the still bottoms do not have to be counted against the monthly generator-status determination if the spent solvent has already been counted once that month.) While a drum is connected to the evaporator for discharge of the bottoms, it is considered to be a satellite accumulation unit. [§262.34(c)] When it is disconnected and another drum begins filling, the full drum must be moved within 3 days to the 90-day accumulation area. On the day the drum enters the 90-day area, the 90-day clock starts. Within 90 days, the generator must take one of the five actions cited in Section 6.3.6.3.1 that will stop the 90-day clock for containers. [RO 12865]

Generators may transfer hazardous waste between containers in 90/180/270-day accumulation areas. However, the 90/180/270-day clock does *not* restart if the hazardous waste is transferred to another container. Additionally, the regulations do not prohibit the movement of hazardous waste from one 90/180/270-day area to another, as long as the waste remains onsite. However, the 90/180/270-day clock does *not* restart if the hazardous waste is moved to another 90/180/270-day area. [RO 14703]

For CESQGs, the 180-day clock starts when onsite accumulation equals or exceeds 1,000 kg. [§261.5(g)(2)]

6.3.6.3 When does the clock stop?

The action that stops the 90/180/270-day clock depends on the type of accumulation unit being used, as discussed in the following subsections.

6.3.6.3.1 Containers

For containers, the clock stops when:

- The container is moved from the 90/180/270-day accumulation area to an onsite interim status or permitted storage area [November 19, 1980; 45 *FR* 76625];

- The waste is transferred from a 90/180/270-day container to an interim status or permitted treatment or disposal unit [November 19, 1980; 45 *FR* 76625];

- The container is shipped offsite;

- The waste is transferred to a unit that is exempt from permitting (e.g., a recycling unit, a wastewater treatment unit, or an elementary neutralization unit); or

- The waste is rendered nonhazardous via treatment in the container.

As noted earlier, transferring wastes from a 90/180/270-day container to another container (assuming the second container is not in an interim status or permitted storage area) does not stop the clock. The clock continues to run on the second container until one of the five events cited above occurs.

6.3.6.3.2 Tanks

For tanks, the clock stops when the tank is "empty." According to EPA, "A tank will be considered 'empty' when its contents have been drained to the fullest extent possible. Since many tank designs do not allow for complete drainage due to flanges, screens, or syphons, it is not expected that 100% of the wastes will always be removed. As general guidance, a tank should be considered empty when the generator has left the tank's drainage system open until a steady, continuous flow has ceased." [January 11, 1982; 47 *FR* 1250]

Note that more recent EPA guidance allows flow-through tanks to be used for accumulation purposes by using a "turnover" or "mass balance" approach to determine whether a tank is emptied within 90 days:

"EPA is interpreting §262.34(a)(1)(ii) to allow for the turnover approach…. In the case of hazardous wastes flowing through tanks continuously, there is a means of demonstrating when a tank is 'emptied' within 90 days under §262.34(a)(1)(ii) that would not require completely emptying the tank, and that may be more suitable for tanks with continuous flow. More specifically, a mass balance approach (i.e., the 'turnover' approach, as you referred to it, in your letter) can be used for continuous flow tanks rather than the approach described above for batch process tanks. The key parameters in this mass balance approach are the volume of the tank (e.g., 6,000 gallons), the daily throughput of hazardous waste (e.g., 300 gallons per day) and the time period the hazardous waste 'resides' in the tank. In this example, the hazardous waste entering the tank would have a residence time of 20 days ((6,000 gallons/300 gallons per day) = 20 days) and meet the requirements of §262.34(a)(1)(ii) since the hazardous waste has been in the tank for less than 90 days." [RO 14763]

The guidance also discusses the kinds of records that a generator should maintain to demonstrate compliance with the 90-day time limit:

"Large quantity generators accumulating hazardous wastes through a continuous flow process must also demonstrate that the hazardous waste has not been stored for more than 90 days. This may be achieved by the use of inventory, or some form of accounting or monitoring data. For example, a generator could confirm that the volume of a tank has been emptied every 90 days by recording the results of monitoring equipment both entering and leaving a tank. This recordkeeping, in conjunction with the tank volume, would enable inspectors, as well as facility personnel to demonstrate compliance with §262.34(a)(1)(ii). Likewise, in marking the tank, a generator could mark both the tank volume and estimated daily throughput to allow inspectors to determine the number of days that hazardous waste resides in a tank to determine compliance with §262.34(a)(1)(ii)."

The 90/180/270-day clock would also stop in tanks if the waste is rendered nonhazardous.

As noted in Table 6-3, EPA wants accumulation start date marking for tanks to appear right on the tank. This is not a regulatory requirement [i.e., §262.34(a)(2) only requires start dates on *containers*]; instead, the agency has noted several times in guidance that this date must also be marked on tanks. [November 19, 1980; 45 *FR* 76624, March 24, 1986; 51 *FR* 10160, RO 11641, 14683] Despite this guidance, our experience is that many generators do not mark accumulation start dates on tanks, but track them in facility operating logs.

On March 18, 2010 [75 *FR* 12989], EPA published a direct final rule that, among other things, required generators to mark the accumulation start date on 90-day accumulation tanks. The agency received significant adverse public comment on this action and, as a result, withdrew that portion of the direct final rule on June 4, 2010. [75 *FR* 31716] In the withdrawal notice, the agency said that it had also proposed to require the start date on 90-day tanks on the same day that it had published the direct final rule [see 75 *FR* 13066] just in case there were adverse comments. Therefore, the current status of this issue is that EPA has proposed to require that the start date be marked on 90-day tanks, but the agency has not finalized this requirement. In the meantime, this issue is not fully resolved and we recommend that you ask your state what they require.

6.3.6.3.3 *Drip pads*

For drip pads, the clock stops when all wastes are removed from the pad and associated collection system. Owners/operators must maintain records describing the procedures that are used to ensure that all wastes are removed at least once every 90 days. The quantity, date, and time of each removal must also be documented. [December 6, 1990; 55 *FR* 50456]

6.3.6.3.4 *Containment buildings*

For containment buildings, the clock stops when each volume of waste is removed from the building. This requirement can be met by documenting that: 1) the containment building is emptied at least once every 90 days, or 2) procedures are in place to ensure that wastes are segregated by age and that no portion of the accumulated wastes remains in the containment building for more than 90 days. [August 18, 1992; 57 *FR* 37212]

6.3.6.4 Accumulation time extensions

Both large and small quantity generators can apply for a 30-day extension to the accumulation period where unforeseen, temporary, and uncontrollable circumstances prevent them from complying with the 90/180/270-day time limits. [§262.34(b) and (f)] Examples of such circumstances include refusals of waste shipments, transporter delays, and labor strikes [January 11, 1982; 47 *FR* 1249], as well as transporters and/or receiving facilities going out of business or otherwise closing. [RO 13013] Extensions are granted on a case-by-case basis by the authorized agency (EPA regional office or state RCRA agency).

After the 30-day extension expires, the facility must comply with interim status requirements or have a permit for continued storage.

If the generator is not granted an extension, he/she is the operator of a storage facility on the 91st or 181st day of storage, and the facility is subject to requirements in Parts 264, 265, and 270.

6.3.6.5 Generators get a new 90- or 180-day clock for returned shipments

To allow generators to receive shipments of non-RCRA-empty containers or rejected hazardous waste, EPA has amended the definition of "designated facility" in §260.10 to include generators receiving returned waste shipments. Generators receiving returned shipments have either 90 or 180 days (for large or small quantity generators, respectively) to send returned waste to an alternate facility. [§262.34(m)] This accumulation time limit is based on the generator's status when the rejected waste is received by the generator (not when the waste originally left the generator's site). If a generator was classified as an SQG when the waste was initially shipped offsite but is an LQG when the rejected waste is returned, the generator could accumulate the waste for no more than 90 days without a permit.

6.3.7 Preparedness and prevention

Large and small quantity generators are subject to the preparedness and prevention provisions of Part 265, Subpart C. Large quantity generators get there via reference from §262.34(a)(4)—small quantity generators from §262.34(d)(4). This subpart requires the following:

1. An accessible alarm/communication system capable of providing emergency instruction to facility personnel. Such instruction should enable rapid evacuation of the affected areas and initiate the emergency response.

2. Accessible telephones or two-way radios capable of summoning assistance from local authorities (e.g., police, fire departments, and emergency response teams).

3. Fire extinguishers, water hose stations or foam producing equipment, automatic sprinklers, and other fire control equipment, spill control equipment (e.g., pumps, absorbents, vacuum cleaners), and decontamination equipment (e.g., pumping or vacuum equipment).

4. Testing/maintenance of the above equipment/systems to ensure they are operational.

5. Adequate aisle space to allow emergency/spill response.

6. Operation/maintenance procedures that minimize the possibility of fire, explosion, or spills.

7. Coordination with local authorities on how emergencies will be addressed by both facility and offsite emergency personnel.

Because personnel safety may be compromised if these requirements are not met, preparedness and prevention issues should be a high priority at hazardous waste facilities. Additional details are provided in the following subsections.

Guidance on maintaining current and thorough preparedness and prevention information is given for permitted and interim status TSD facilities in RO 14832. Although this guidance does not directly address large and small quantity generators, we believe it may be useful to generators when developing and updating this information.

Emergency action plans, which are detailed in the OSHA regulations at 29 *CFR* 1910.38 and are triggered by OSHA requirements in 29 *CFR* Part 1910, may apply regardless of any applicable RCRA preparedness and prevention provisions and regardless of RCRA generator class.

6.3.7.1 Design and operation of facility

Section 265.31 requires, in general, that the design, construction, operation, and maintenance of the hazardous waste areas must "minimize" danger and contribute to preparedness and prevention.

According to RO 14036, state and EPA regional personnel have used the §265.31 regulatory language: "maintained and operated to minimize the possibility of fire, explosion…" to request generators to post "No Smoking" or "No Ignition Sources" signs near areas accumulating combustible hazardous wastes.

6.3.7.2 Internal communication or alarm system

Intercoms, internal telephones, or two-way radios can serve as internal communication devices. An alarm system should be a bell, siren, buzzer, or similar device audible from all hazardous waste areas; the alarms can be activated either manually or automatically. [OSWER Directive 9523.00-10, October 1983 (available from http://nepis.epa.gov/EPA/html/Pubs/pubtitleOther.html by downloading the report numbered SW968)]

6.3.7.3 External communication system

In an emergency situation, hazardous waste personnel may need some means of summoning assistance from local police and fire departments and state/local emergency response teams. A telephone or hand-held two-way radio are acceptable external communication devices. For example, a 90-day tank or container accumulation area is equipped with an intercom system or all personnel carry two-way radios to provide internal emergency communications. Any emergency in a 90-day area will be brought to the attention of a shift supervisor stationed at the facility operations control room. The supervisor evaluates the situation and places any necessary calls to police/fire departments and/or emergency response teams from one of the outside phones located in the control room.

6.3.7.4 Testing/maintenance of equipment

Regarding the testing/maintenance of emergency equipment [§265.33], there are no inspection or testing frequencies specified in the regs (but the regs say to inspect/test "as necessary"). It is up to the generator to determine inspection frequencies for this safety and emergency equipment and to document that such inspections/testing have been conducted. One method for documenting such inspections is to add inspections of specific safety and emergency equipment to daily tank inspection checklists and/or weekly container inspection checklists. Another way would be to add such emergency equipment inspection/maintenance to the general plant emergency equipment inspection/testing program.

6.3.7.5 Aisle space

Although no minimum aisle space is specified in the federal regulations (some state programs do specify a minimum), the facility must maintain adequate aisle spacing to allow unobstructed movement of personnel, fire protection equipment, and spill control equipment to any hazardous waste areas (e.g., container accumulation areas) at the facility. According to the *RCRA Inspection Manual*, a good rule of thumb is that the aisle space should be adequate to remove one drum or, alternatively, be at least 24 inches wide. Additionally, each individual container has to be able to be inspected.

[*RCRA Inspection Manual*, 1993 edition, available as document PB94-963605 from NTIS at (800) 553-6847]

6.3.7.6 Arrangements with local authorities

As required by §265.37, generators must familiarize local police, fire departments, emergency response teams, and hospitals with the hazardous wastes managed at the facility, the areas at the plant where these wastes will be managed, and the possible injuries, illnesses, or other hazards that could result from emergencies involving these wastes. Generators must attempt to ensure that they will get an immediate and appropriate emergency response from these authorities.

It is often hard to get hospitals and local authorities to read and comment on contingency plans or other emergency arrangement documents that the generator may prepare. However, it is the generator's responsibility to reach out to, communicate with, and update such responders, and all communication between generators and these local authorities should be documented. Even if these local authorities will not respond to a request to review an updated contingency/emergency plan, the generator should document that it sent the plan and tried to get feedback (e.g., with a follow-up phone call).

One good way of interfacing with these responders is to invite them out to the facility occasionally for a short walkthrough/tour and review of the chemicals used, types of wastes generated, and risks associated with your site (usually followed by you taking them out to lunch); these tours are then documented. Agreements between the facility and emergency responders are discussed further in Section 6.7.7.

In the unlikely event that a fire department, local hospital, or other agency refuses to enter into a shared agreement to respond to emergencies, §265.37(b) requires documentation of the refusal. If the generator's facility is not provided with fire protection coverage by a fire protection district or municipal fire department, the generator should prepare a fire protection and prevention plan of its own.

6.3.8 Contingency planning

Large quantity generators are required to prepare a formal contingency plan that outlines the procedures necessary to respond to fires, explosions, or releases of hazardous waste. This plan must meet the requirements of Part 265, Subpart D. It must demonstrate that facility-specific emergency procedures have been developed and will be implemented whenever a hazardous waste emergency situation occurs. The plan must be maintained at the facility and be available for inspection, and it also must be submitted to local police/fire departments and emergency response authorities.

The contingency plan should describe specific actions facility personnel must take in response to any unplanned sudden or nonsudden releases. The plan should be specific regarding what to do, who to notify, and in the case of off-plant assistance, what those groups will and will not do during the emergency. Accidental leaks and spills should be addressed immediately in accordance with the plan. EPA has not established a definition of what constitutes an immediate response to a spill situation. The time frames and extent of immediate response must be judged by persons responding to discharges on an individual basis. Extended responses which are not judged to be immediate in nature may result in: 1) a modification to the facility's contingency plan, or 2) an enforcement action for an inadequate contingency plan. [RO 12748]

According to OSWER Directive 9480.00.5, the most important aspect of an effective contingency plan is clear definition of responsibilities for execution. Plant management should be fully involved, and it is highly desirable to have a company officer be responsible for ensuring plan execution. The chain of command should be specified in advance, along with delegation of authority and backups where needed. A job description for each responsible party should be incorporated in the plan. [OSWER Directive 9480.00.5, September 1982, available from http://nepis.epa.gov/EPA/html/Pubs/pubtitleOther.html by downloading the report numbered OSWER9480005]

Per §265.52(b), the hazardous waste contingency plan may be a standalone document or part of an emergency plan prepared for the facility under another program (e.g., it may be hazardous waste management amendments to a Spill Prevention, Control, and Countermeasures [SPCC] plan previously developed for the facility).

Emergency action plans, which are detailed in the OSHA regulations at 29 *CFR* 1910.38 and are triggered by OSHA requirements in 29 *CFR* Part 1910, may apply regardless of any applicable RCRA contingency plan provisions and regardless of RCRA generator class.

On April 4, 2006 [71 *FR* 16862], EPA modified the regulations to indicate that facilities may consider developing one contingency plan for their entire facility, instead of having separate contingency plans to comply with RCRA, SPCC, OSHA, etc. Use of a single plan should eliminate confusion when facilities must decide which of its contingency plans is applicable to a particular emergency. If facilities decide to develop such a single plan, EPA recommends it be based on the *Integrated Contingency Plan (One Plan) Guidance*, available at http://www.epa.gov/osweroe1/guidance.htm. This guidance provides a mechanism for consolidating the multiple contingency plans that facilities have to prepare to comply with various government regulations.

A contingency plan is required in the regulations only for 90-day accumulation activities [see §262.34(a)(4), RO 14758]. For example, RO 12953 requires large, sudden failures of 90-day hazardous waste tank systems to be addressed; accidental spills in these areas should also be addressed in the plan. [RO 12748] The contingency plan is not required to address satellite accumulation areas [see January 3, 1983; 48 *FR* 119, December 20, 1984; 49 *FR* 49570,

RO 11373, 14703, 14758]. However, it would be prudent (and some states require) that satellite areas be designated in the plan as well.

The regs at §265.52(d) require the plan to list "names, addresses, and phone numbers (office and home)" of all persons qualified to act as emergency coordinator. In OSWER Directive 9523.00-10, cited previously, EPA shows a sample emergency coordinators' list with name, home address, and work and home phone numbers for each coordinator identified.

The regulations in §265.55 require that an emergency coordinator be available at all times. At a minimum, one additional employee must be designated and trained as emergency coordinator to provide around-the-clock and vacation coverage. [RO 13221] This means at least one employee who is either on the facility premises during peak operational periods or available to respond to an emergency by reaching the facility within a short period of time. [OSWER Directive 9523.00-10, cited previously]

If the contingency plan is activated due to fire, explosion, or releases of hazardous waste that could threaten human health or the environment outside the facility, the emergency coordinator must notify local emergency response authorities and call either the National Response Center or state or local emergency response line. [§265.56(d)]

Contingency plans should be reviewed regularly (some states require an annual review). Section 265.54 specifies five situations that require review of, and possible revisions to, the plan. Of those five, one of the most likely is changes to the list of emergency coordinators. Revised plans should be submitted to local emergency response authorities to keep them up-to-date.

Because §265.16(a)(3) requires large quantity generators to train facility personnel to respond effectively to emergencies, a good practice is to make the contingency plan provisions a part of both initial and annual training. Such training would include the location and use of equipment (e.g., fire extinguishers, spill response kits) used by personnel during an emergency.

Generators need to make sure that they will get an immediate and appropriate emergency response from local police, fire departments, emergency response teams, and hospitals. Thus, contingency plans should describe the arrangements agreed to with these authorities and the division of responsibilities between plant and offsite personnel. Written agreements between the facility and these local authorities are preferable; if such agreements are attained, they should be attached to the contingency plan.

Although not required by the federal regs, lists of the emergency coordinators included in the contingency plan (phone numbers, pager numbers, etc.) should be posted by telephones in the hazardous waste handling/storage areas. Also included on these lists should be phone numbers for the responding police department, fire department, and hospital. A sample emergency contacts telephone posting form is available from McCoy and Associates' website at http://www.mccoyseminars.com.

Small quantity generators are not required to develop a formal contingency plan as discussed above. Instead, they are required to comply with the emergency response provisions and procedures in §262.34(d)(5). As part of compliance with those regs, an emergency coordinator must be designated who will be at the facility or be available on call, and a list of emergency contacts with phone numbers must be posted next to telephones near the 180-day accumulation areas. This posting must also include locations of fire extinguishers and alarms and spill control material. A sample emergency contacts telephone posting form is available from McCoy and Associates' website at http://www.mccoyseminars.com.

6.3.9 Waste minimization program

Section 1003(b) of the RCRA law states the following: "[W]herever feasible, the generation of hazardous waste is to be reduced or eliminated as expeditiously as possible. Waste that is nevertheless generated should be treated, stored, or disposed of so as to minimize the present and future threat to human health and the environment." These statements form the statutory basis for waste minimization under RCRA. The RCRA regs implement this statutory language primarily in §262.27 and the generator certification statement in Block 15 of the manifest.

When a generator signs a hazardous waste manifest, he/she is certifying that the appropriate waste minimization statement in §262.27 is true. The regulatory wording in that section is slightly different for large quantity versus small quantity generators. A large quantity generator must certify that it has 1) a program in place to reduce the volume and toxicity of waste to the economically practicable degree; and 2) selected the practicable method of treatment, storage, or disposal currently available that minimizes the present and future threat to human health and the environment. A small quantity generator must certify that it has 1) made a good-faith effort to minimize waste generation, and 2) selected the best available and affordable waste management method. [§262.27, RO 12767] EPA has clarified that the certification applies to generators of hazardous waste both at manufacturing plants and in remediation settings. [RO 11855]

The regulations don't provide any details for generators who have to make these waste minimization certifications. Therefore, we find that some facilities don't give this regulatory requirement much thought or resources. Still, a state or EPA inspector may ask questions about your facility's waste minimization program, and you will need to be able to discuss how you are complying. In addition, biennial reports required of large quantity generators must include a description of the generator's waste minimization efforts. [§262.41(a)(6–7)]

The primary EPA guidance document on waste minimization is *Guidance to Hazardous Waste Generators on the Elements of a Waste Minimization Program*, which was published in the May 28, 1993 *Federal Register*. [58 *FR* 31114] A waste minimization Q&A document (EPA/530/F-93/036), also prepared by the agency in 1993, is available at http://nepis.epa.gov/EPA/html/Pubs/pubtitleOSWER.html by downloading the report numbered 530F93036.

EPA says that waste minimization includes source reduction and environmentally sound recycling. Source reduction means a reduction in the amount of a substance, pollutant, or contaminant entering any waste stream prior to recycling, treatment, or disposal, along with a reduction in the hazard to public health and the environment associated with the release of such substances. Environmentally sound recycling, the next preferable alternative for pollutants which can't be reduced at the source, involves recovering a usable product as opposed to conventional waste management activities like treatment or disposal. Waste minimization actions can encompass a variety of techniques—technology or process modifications, reformulation or redesign of products, replacing hazardous raw materials with less hazardous ones, and improvements in work practices (e.g., housekeeping, maintenance, inventory control). However, waste minimization does not include treatment of hazardous waste. [RO 12932]

6.3.9.1 What constitutes a waste minimization program?

The federal regulations do not require generators to have a written description of their waste minimization program, and they do not specify what the plan should contain (although some states do require a written plan). However, in the May 1993 guidance document cited previously, EPA recommends that a facility document its program and have that documentation signed by the corporate officer responsible for ensuring compliance. The agency suggests that a waste minimization program contain six general elements, as outlined in Table 6-4.

Generators that either recycle wastes onsite or send their wastes offsite to be recycled (e.g.,

Table 6-4: Elements of a Waste Minimization Program

Element / goal	Suggested techniques or methods to accomplish
Top management support	■ Make waste minimization part of organization policy. ■ Set explicit goals. ■ Commit to implementing recommendations. ■ Designate a waste minimization coordinator. ■ Publicize success stories. ■ Recognize individual and collective accomplishments. ■ Train employees on waste generating impacts of work practices.
Characterization of waste generation and waste management costs	■ Maintain a waste accounting system to track types and rates and/or amounts of wastes generated. ■ Determine the true costs of waste management and cleanup. This should include costs for 1) regulatory oversight; 2) paperwork and reporting; 3) loss of production potential and materials; 4) waste treatment, storage, and disposal; 5) employee exposure; and 6) insurance for potential future liability.
Periodic waste minimization assessments	■ Identify opportunities at all points in a process where materials can be prevented from becoming waste. Analyze waste minimization opportunities based on the true costs associated with waste management and cleanup. A variety of methods may be used to do this, including the following: —Identify sources of waste by tracking materials that end up as waste. —Perform mass balance calculations. —Establish a waste minimization team of independent experts outside the organization structure. —Establish an in-house waste minimization team.
Cost allocation	■ Allocate costs, "where practical and implementable," to the activities that generated the waste rather than simply to overhead.
Technology transfer	■ Encourage sharing of best practices. Seek or exchange technical information on waste minimization from 1) other parts of the organization/facility, 2) other companies/facilities, 3) trade associations/affiliates, 4) professional consultants, and 5) university or government technical assistance programs.
Program implementation and evaluation	■ Implement recommendations identified by the assessment process. ■ Periodically review program effectiveness. ■ Identify areas for improvement highlighted by these reviews.

Source: McCoy and Associates, Inc.; adapted from May 28, 1993; 58 *FR* 31114.

participation in a waste exchange, recycling of solvents) are exercising a form of waste minimization that may be used to satisfy the waste minimization certification requirements. [RO 12470, 12559] Conversely, burning hazardous waste for energy recovery or using hazardous waste in a manner constituting disposal (e.g., using hazardous waste to make fertilizer) do not qualify as waste minimization. Recycling activities such as these closely resemble conventional waste management practices and, therefore, do not qualify as waste minimization. [RO 13682] Similarly, diluting a characteristically hazardous waste into used oil to render the mixture nonhazardous does not reduce the volume of the waste and does not appear to reduce the amount of toxic constituents in the mixture; thus, it probably does not qualify as waste minimization. [RO 13570]

Note that agency guidance on the waste minimization program certification clearly states that determinations of what constitutes "economically practicable" are to be:

"made by the generator in light of his own particular circumstances. Thus, from an enforcement perspective, the agency will be concerned primarily with compliance with the certification signatory requirement…. The legislative history of HSWA makes clear that 'judgments made by the generator [for the purpose of the waste minimization certification] are not subject to external regulatory action….'" [RO 12629; see also May 28, 1993; 58 *FR* 31116]

EPA maintains a Waste Minimization website at http://www.epa.gov/epawaste/hazard/wastemin/, which provides technical and program information and resources.

Many states have their own pollution prevention requirements, which may include hazardous waste minimization. For example, the state of Tennessee requires that generators have formal hazardous waste reduction plans [Tennessee Statute 68-212-304], and failure to have such a plan is one of the top 10 generator violations in that state. If you are required by your state to have a pollution prevention program, that program could fulfill the §262.27 requirement by incorporating activities to reduce hazardous waste.

6.3.10 Training requirements

As specified in §262.34(a)(4), LQGs practicing 90-day accumulation must comply with the same training requirements (§265.16) as TSD facilities. The training requirements for SQGs are less prescriptive [§262.34(d)(5)(iii)]: "The generator must ensure that all employees are thoroughly familiar with proper waste handling and emergency procedures, relevant to their responsibilities during normal facility operations and emergencies." A discussion of these training requirements appears in Section 6.4.

6.3.11 Air emission standards

LQGs that accumulate hazardous wastes in 90-day tanks and containers are subject to the air emission standards of Part 265, Subparts AA, BB, and CC. [RO 13777, 14703] For example, if a 90-day tank were used as an air stripper, it would be subject to Subpart AA, which deals with emissions from process vents. Subpart BB, which controls fugitive emissions, would apply to equipment in pipe runs that are ancillary equipment to the tank. Subpart CC, which controls emissions from the tank itself, would apply only if Subpart AA does not. [§265.1080(b)(8)] Air emission standards for 90-day units are detailed in Chapter 15.

Although satellite accumulation containers are not subject to Subpart CC air emission standards for containers, when large quantity generators transfer wastes from satellite areas, the containers become subject to Subpart CC standards as soon as they enter the 90-day area. [RO 13777, 14703]

6.3.12 Corrective action

The corrective action provisions of RCRA apply only to interim status and permitted facilities. Therefore, corrective action does not apply to releases from 90/180/270-day accumulation units unless other units at the facility are subject to interim status provisions or require a permit. However, if a generator ever exceeds the applicable accumulation time or quantity limits without applying for a time extension, he/she could become subject to corrective action. A discussion of the corrective action program is contained in Chapter 17.

Also, note that generators who do not clean up spills from accumulation units can face enforcement actions for open dumping. If cleanup activities do not begin promptly, the spill is considered a land disposal unit subject to permitting and corrective action requirements. [RO 12698, 12748]

6.3.13 Closure standards

Closure requirements apply to LQGs for their 90-day units. [§262.34(a)(1)] General closure standards per §§265.111 and 265.114 essentially require that all units and surrounding soil be decontaminated and all hazardous waste/constituents be removed. Decontamination materials must be managed as hazardous waste (if they are listed or exhibit a characteristic). In addition, the need for further maintenance of the site and post-closure escape of hazardous waste or constituents to the environment must be minimized. The federal RCRA regs do not require generators to have a closure plan. [RO 14321]

Unit-specific closure standards for tanks, drip pads, and containment buildings provide additional closure details. When these units are closed, the owner/operator must remove all waste residues and decontaminate structures, equipment, secondary containment components, and contaminated soil. The level of decontamination of soils and equipment is not specified in the regulations, so best professional judgment/negotiation with the state are typically employed. If all of the contaminated soil cannot be decontaminated or removed at closure, the unit must be closed as a landfill (and the generator will have to comply with the post-closure requirements of Part 265, Subpart G and the financial assurance requirements of Part 265, Subpart H). [RO 13270] Presumably, such closure as a landfill would also make the facility subject to corrective action requirements.

SQGs must comply with minimal closure standards for 180/270-day tanks: "remove all hazardous waste from tanks, discharge control equipment, and discharge confinement structures." [§265.201(f)]

No unit-specific closure standards apply to 90/180/270-day containers. [RO 14321]

6.3.14 Financial assurance

Under normal conditions, LQGs don't need to meet the Part 265, Subpart H financial assurance requirements for closure or post-closure care because they are specifically exempt under §262.34(a)(1). This exemption applies regardless of the type of accumulation unit used. [RO 14321]

SQGs who use 180-day tanks are not subject to financial assurance requirements because these provisions are not specified in §265.201, which establishes SQG tank standards. No financial assurance requirements are associated with container storage.

Under abnormal conditions (i.e., where residual contamination at the accumulation unit cannot be cleaned up and the unit must be closed as a landfill), Subpart H requirements would presumably apply to both large and small quantity generators.

6.3.15 Converting permitted/interim status units to 90-day units

In some cases, it may be advantageous for a permitted or interim status storage/treatment facility to convert to 90/180/270-day status. This type of conversion lowers the regulatory profile of a facility, which means fewer agency inspections. Some of the costs associated with operating a permitted facility (such as waste analysis and reporting requirements) may also be reduced. Converting to 90/180/270-day status will not eliminate the need for a permitted/interim status facility to perform corrective action; however, the regulatory agency may assign a lower priority for corrective action at facilities that no longer are in the permitting program.

In order to convert to 90/180/270-day status, the facility must comply with the closure provisions for permitted units. However, since the permitted units will still be managing hazardous wastes, the permit closure requirements will not be triggered until the unit receives the last volume of hazardous waste as an accumulation unit. Financial assurance per Part 264/265, Subpart H will have to be maintained until closure of the unit is complete. Finally, "to indicate that hazardous waste management activities in the converted unit are no longer covered by the facility's permit, the facility must submit the appropriate permit modification." [RO 13774] The same procedures would apply to the conversion of an interim status unit. Interim status facilities should contact the regulatory agency to determine how to terminate interim status.

6.3.16 SQG standards

As Table 6-3 shows, many minor differences exist between the accumulation standards applicable to LQGs and SQGs. Many of these differences have been discussed above, but some require further explanation.

6.3.16.1 Accumulation units

Normally, the only units SQGs may use are tanks (meeting §265.201 standards) and containers (meeting Part 265, Subpart I requirements). The only way SQGs can utilize drip pads or containment buildings is if they comply with the LQG requirements in §262.34(a) for these units, including a limit for onsite accumulation of 90 days. [EPA/305/B-96/001, EPA/530/K-02/008I, EPA/530/K-05/008, RO 13696, 14662]

6.3.16.2 6,000-kg limit

While no limit is placed on the quantity of wastes that LQGs may have in 90-day accumulation, SQGs are limited to 6,000 kg. The 6,000-kg cap applies to all hazardous waste accumulated onsite. [RO 12602] The source of the 6,000-kg number is that SQGs cannot accumulate wastes for more than 6 months (180 days) and they also cannot generate more than 1,000 kg/mo of nonacute hazardous waste. The combination of accumulation time and rate produces the 6,000-kg cap. (The cap applies even if the wastes are stored for 270 days.)

If an SQG exceeds the accumulation limit and doesn't get a 30-day extension, it must have interim status or a permit for storage—it does not become an LQG. [§262.34(f), 51 *FR* 10161] (Exceeding the accumulation limit would also make the generator subject to corrective action.)

6.3.16.3 270-day limit

SQGs are allowed to accumulate wastes for up to 270 days if their wastes will be shipped more than 200 miles to an offsite facility. Several issues arise from this 270-day limit:

- SQGs may send wastes to a facility greater than 200 miles away (and accumulate for 270 days) even though another TSD facility is closer than 200 miles. [RO 13000]

- An SQG may accumulate some wastes for 180 days if they will be managed at a facility under 200 miles away, and other wastes for 270 days if they are shipped at least 200 miles. [51 *FR* 10161]

- An inspector could find it difficult to figure out if accumulation past 180 days is justified. Manifest copies can be used to check the location of the destination facility, which should be more than 200 miles away if the accumulation time exceeds 180 days. [51 *FR* 10161]

6.3.16.4 No contingency plan

Unlike LQGs, SQGs don't need a contingency plan. Instead, they must comply with the reduced emergency response requirements of §262.34(d)(5).

6.3.16.5 No buffer zone

Because many SQGs are located in urban areas with small lots, they are not subject to the LQG standard specifying that containers of ignitable or reactive wastes must be located at least 50 feet from the property line.

A quiz reviewing many 90/180/270-day RCRA requirements is given in Case Study 6-2.

6.3.17 Conditionally exempt small quantity generators

As noted in Section 6.1.1, CESQGs are entities who generate in a calendar month: 1) no more than 100 kg of nonacute waste, 2) no more than 100 kg of acute spill cleanup residue, and 3) no more than 1 kg of other acute hazardous waste. Because they generate such small quantities, CESQGs have very few RCRA obligations for the hazardous waste they produce. Some interesting rules apply to the management of CESQG waste, however, as fully discussed below.

6.3.17.1 Hazardous waste identification

CESQGs are required to make hazardous waste determinations for all solid waste they generate at their facility—just like all other generators [RO 11958, 14030]; however, they are not subject to the requirements for maintaining hazardous waste characterization data/other documentation in their files. Even so, we recommend that these conditionally exempt generators keep such documentation in

Case Study 6-2: 90/180/270-Day Accumulation Quiz

Answer the following questions:

1. Must containers used for 90/180/270-day accumulation meet DOT standards?

2. Must hazardous waste codes appear on 90/180/270-day containers?

3. A generator produces more than 1 kg/mo of acute wastes and between 100 and 1,000 kg/mo of nonacute wastes. Can he/she accumulate the acute wastes for up to 90 days and the nonacute wastes for up to 180/270 days without a permit?

4. A generator accumulates waste per §262.34 and then disposes the waste down a sewer that leads to a POTW. Does the generator need an EPA ID number?

5. Can an importer accumulate wastes for 90/180/270 days without a permit at or near the point of entry to the United States?

6. A person plans to recycle a characteristic sludge via a reclamation process. Per Table 1 in §261.2(c), the sludge is not a solid waste when reclaimed, but it would become a solid waste if it is accumulated speculatively (i.e., if less than 75% of it is reclaimed in a calendar year). If the sludge is not reclaimed by the end of the speculative accumulation period, could it be stored for 90/180/270 days in an accumulation unit?

7. Wastes can remain in inactive process units for up to 90 days before they become subject to hazardous waste regulation. [§261.4(c)] Once the 90 days expires and the waste becomes subject to RCRA regulation, can the wastes be held in the unit (e.g., a process tank) for an additional 90 days under the provisions of §262.34?

Case Study 6-2: Answers

1. No. Part 265, Subpart I does not specify the type of container that must be used. Of course, containers that are shipped offsite must meet DOT requirements; therefore, to avoid recontainerizing wastes, the use of DOT containers for accumulation is a good operating practice. [January 11, 1982; 47 *FR* 1249]

2. No. The words "Hazardous Waste" must appear on the containers. While putting hazardous waste codes on these containers may be a good operating practice, it is not a regulatory requirement. [January 11, 1982; 47 *FR* 1249]

3. Yes. According to EPA, "Acute hazardous wastes are counted and managed separately from [other] hazardous wastes [§261.5(e)]. In the example given, the generator would have 90 days to send the acute hazardous waste offsite but would have 180 days for the nonacute hazardous waste." [RO 11288]

4. Yes. Accumulation is a form of storage, and §262.12(a) states that a generator must not treat, store, dispose, transport, or offer for transport hazardous waste without having received an EPA ID number.

5. No. Accumulation under §262.34 applies only to hazardous wastes generated onsite; hence, the importer is not eligible for 90/180/270-day accumulation. [RO 13554]

6. Yes. The person becomes a generator when the sludge becomes a solid and hazardous waste. The generator can then utilize the 90/180/270-day accumulation provisions. [RO 14199]

7. Yes, but only if the unit meets the appropriate standards for 90-day accumulation units. For tanks, this would be Part 265, Subpart J standards, which include requirements for secondary containment and leak detection.

their files to prove that they legitimately can claim conditionally exempt status.

6.3.17.1.1 Hazardous waste mixture rule for CESQGs

An exemption exists at §261.5(h) for CESQGs that mix their hazardous wastes with nonhazardous wastes. If the resulting mixture does not exhibit a characteristic, the mixture remains conditionally exempt from RCRA under §261.5, even though the quantity of the resultant mixture exceeds 100 kg/mo or 1,000 kg total accumulation.

By contrast, CESQGs that mix hazardous and nonhazardous waste and whose resultant mixtures exceed the §261.5 quantity limitations and exhibit a characteristic are no longer conditionally exempt and are subject to the applicable Part 262 hazardous waste generator regulations. [RO 11681]

Additionally, for listed residues cleaned out of a CESQG's hazardous waste tank, only the original listed residues within the unit are counted as hazardous, while any nonhazardous cleaning fluid added to them will not have to be counted (unless the resulting mixture exhibits a characteristic of hazardous waste, in which case the entire mixture must be counted). [OSWER Directive 9451.02(85)]

6.3.17.1.2 Used oil mixture rule for CESQGs

The §279.10(b) used oil mixture rules do not apply to mixtures of used oil and hazardous waste generated by CESQGs regulated under §261.5. Such mixtures may be regulated as used oil under Part 279, even if the total halogen concentration of the mixture exceeds 1,000 ppm. [§§261.5(j), 279.10(b)(3)]

For example, a CESQG mixes a small amount of F005 listed spent solvent into its used oil. The resulting used oil contains 2,000-ppm total halogens. The presumption of mixing with listed hazardous waste can be rebutted by providing documentation to show that the mixture of CESQG waste and used oil is exempt under §§261.5(j) and 279.10(b)(3). [57 FR 41581, RO 14627]

While managing mixtures of used oil with CESQG hazardous waste as used oil is a regulatory possibility, prior to mixing, a CESQG should first check to see if the resulting mixture will still meet their used oil vendor's acceptance criteria.

6.3.17.1.3 Hazardous waste derived-from rule for CESQGs

EPA has addressed how the derived-from rule applies to CESQG hazardous waste as follows:

"EPA, however, does not read the derived-from rule as applying to [conditionally exempt] small quantity generator hazardous wastes. Although the rules are not explicit on this point, the agency views this exemption, like other comparable provisions such as the household waste exclusion, as applying cradle-to-grave so that residues from managing the waste retain the exemption or exclusion. In this regard, the rules are explicit that the mixture rule does not apply to mixtures of [conditionally exempt] small quantity generator wastes and solid wastes [see §261.5(h)]. EPA views the derived-from rule as similarly inapplicable." [August 17, 1988; 53 FR 31149]

Q: If hazardous waste generated by a CESQG is incinerated, what is the regulatory status of the ash?

A: EPA has flip-flopped on this issue. Although §261.5(b) exempts CESQGs from most hazardous waste regulations, it does not exempt their waste from being classified as hazardous. The incinerator would not have to be RCRA-permitted to receive the CESQG's hazardous waste, but the incinerator could generate hazardous waste ash that would be subject to RCRA regulation. [RO 12892] However, this April 1987 guidance was contradicted in the above August 17, 1988 Federal Register preamble language, where EPA stated that the CESQG exemption applies cradle-to-grave just like the household waste exclusion.

6.3.17.2 Onsite management

Onsite management of a CESQG's hazardous waste is exempt from almost all RCRA requirements. "Based on data presented in the agency studies and analysis of comments from the public, the agency concludes that these generators should be excluded from all of the Part 262 generator requirements except the hazardous waste determination requirement of §262.11." [EPA/530/R-80/018, April 1980, available at http://nepis.epa.gov/EPA/html/Pubs/pubtitleOSWER.html by downloading the report numbered 530R80018] Issues that CESQGs should consider as they manage their hazardous waste onsite are discussed below.

6.3.17.2.1 *Generation rate and accumulation quantity limits*

The monthly hazardous waste generation rate limits for CESQGs are as follows:

- Nonacute hazardous waste: 100 kg/mo—If a CESQG generates >100 kg/mo of nonacute hazardous waste, the facility loses CESQG status and becomes either a large or small quantity generator. [RO 11688]

- Acute hazardous waste: 1 kg/mo—If a CESQG generates >1 kg/mo of acute hazardous waste, the facility becomes a large quantity generator. [§261.5(e)(1)]

- Cleanup residues from spills of acute hazardous waste: 100 kg/mo—If a CESQG generates >100 kg/mo of acute cleanup residues, the facility becomes a large quantity generator. [§261.5(e)(2)]

CESQGs do not have accumulation time limits; however, they do have accumulation quantity limits:

- Nonacute hazardous waste: 1,000 kg—If a CESQG accumulates >1,000 kg of nonacute hazardous waste, the facility becomes a small quantity generator. As a small quantity generator, the beginning of the 180/270-day accumulation time limit is when onsite accumulation equals or exceeds 1,000 kg. [§261.5(g)(2)]

- Acute hazardous waste: 1 kg—If a CESQG accumulates >1 kg of acute hazardous waste, the facility becomes a large quantity generator. As a large quantity generator, the beginning of the 90-day accumulation time limit is when onsite accumulation exceeds 1 kg. [§261.5(f)(2)]

- Cleanup residues from spills of acute hazardous waste: 100 kg—If a CESQG accumulates >100 kg of cleanup residue from spills of acute hazardous waste, the facility becomes a large quantity generator. As a large quantity generator, the beginning of the 90-day accumulation time limit is when onsite accumulation exceeds 100 kg. [§261.5(f)(2)]

6.3.17.2.2 *Onsite treatment*

Per RO 11688 and 14662, CESQGs may treat hazardous waste onsite without a permit if the facility meets one of the conditions listed in §261.5(f)(3)(iii)–(vii) or 261.5(g)(3)(iii–vii). The only one of those regulatory provisions that is likely is if the state approves such treatment per §261.5(f)(3)(iii) or 261.5(g)(3)(iii). If a CESQG does not meet one of the §261.5(f)(3)(iii)–(vii) or 261.5(g)(3)(iii–vii) conditions, the facility must have a permit under Part 270, or operate under interim status per Parts 265 and 270, before treating hazardous waste onsite. However, a CESQG may treat hazardous waste in an onsite elementary neutralization unit without meeting the requirements in §261.5(f)(3) or 261.5(g)(3). [RO 13778]

6.3.17.3 Offsite treatment or disposal

Several issues arise when CESQGs ship their waste for offsite treatment or disposal, as discussed in the paragraphs below.

6.3.17.3.1 *EPA ID number exemption*

CESQGs are exempt from the requirement to obtain an EPA ID number when managing hazardous waste. [§261.5(b)] However, keep in mind that some states have programs that are more stringent than the federal regulations and require CESQGs to have EPA ID numbers or state ID numbers. Additionally, CESQGs are required to obtain an EPA ID number if they engage in one or more of the used oil or universal waste management activities identified in Table 6-8 or if they fail to satisfy the requirements of §261.5(f), (g), and (j).

6.3.17.3.2 Manifesting exemption

CESQGs are exempt from the manifest provisions at the federal level. [§261.5(b), RO 11589, 11772] However, these CESQGs can sometimes have a hard time finding a transporter who will accept such shipments of hazardous waste without a manifest. If a manifest is used, what information goes into Block 1—Generator ID Number? As noted above, CESQGs usually do not have EPA ID numbers. It is customary to put "CESQG" in Block 1 if a CESQG is not assigned an ID number.

If a CESQG uses manifests for offsite shipment of hazardous waste, it should keep copies of them to help prove its conditionally exempt status (i.e., that it is not generating more than 100 kg of nonacute hazardous waste/mo).

If a manifest is not used, a shipping paper may still be needed. If the generator's waste meets the definition of a hazardous substance or hazardous material, the waste must be accompanied by a shipping document that satisfies the DOT hazmat regulations.

6.3.17.3.3 Land disposal restrictions exemption

CESQGs are specifically exempt from the land disposal regulations of Part 268 in §§261.5(b) and 268.1(e)(1). Note, however, that these entities are only exempt if they send their wastes to permitted or interim status hazardous waste facilities, legitimate recycling facilities, or to other facilities permitted, licensed, or registered by the state to manage municipal or industrial solid wastes. [RO 12818] (In other words, if personnel at a CESQG dump their waste on the ground behind a building, the facility is not exempt from the LDR standards.)

6.3.17.3.4 Offsite transportation

The regulations state that a CESQG must ensure delivery of its hazardous waste to one of the facilities noted in §261.5(g)(3). The question frequently arises as to whether small amounts of hazardous waste can be transported from a CESQG site in a company vehicle without a manifest to a centralized consolidation location. For example, personnel conduct maintenance operations at pump or compressor stations (each of which is a CESQG) located every 20–30 miles along a pipeline. These personnel typically throw waste filters, rags, paint wastes, etc. into the back of their pickup and, at the end of the day, unload all of these wastes at a centralized consolidation location. These wastes are accumulated at that location, before they are shipped to an offsite TSD facility. EPA's guidance on this scenario (which is under headquarters' review) is summarized in RO 12894. A facility that serves as a central collection point to consolidate hazardous waste generated from numerous CESQGs would have to either:

- Qualify as one of the facilities specified in §261.5(g)(3) [realistically, the easiest approach would be for the centralized facility to receive state approval to manage the consolidated waste shipments per §261.5(g)(3)(iii)], or

- Qualify as a 10-day transfer facility per §263.12.

6.3.17.4 Generator issues

The following generator issues are relevant to CESQGs.

6.3.17.4.1 Container accumulation, labeling, and inspection

Federally, there are no container accumulation, labeling, or inspection requirements for wastes generated by CESQGs. For example, CESQGs do not have to comply with any of the satellite accumulation provisions in §262.34(c). State programs, OSHA regulations, or fire codes, however, may impose waste management requirements on these containers.

Although exempt from the RCRA hazardous waste management requirements, CESQGs should implement best management practices (BMPs) for accumulating hazardous waste. These BMPs include placing hazardous waste in containers that are closed, in good condition, compatible, and labeled with the words "Hazardous Waste."

6.3.17.4.2 Training

The training requirements of §§264/265.16 and 262.34(d)(5)(iii) do not apply to facility personnel at CESQG sites. [RO 14687] However, pertinent training would also be a BMP.

If the facility exceeds the CESQG generation or accumulation limits, the training requirements for large or small quantity generators will apply. Furthermore, if the facility is managing universal waste under the provisions of Part 273, the training requirements of §273.16 or 273.36 will apply.

6.3.17.4.3 *Recordkeeping and biennial reporting*

CESQGs are not subject to the requirements for maintaining hazardous waste characterization data/other documentation in their files. Even so, we recommend that these conditionally exempt generators keep such documentation, including manifests and shipping papers, in their files to prove that they legitimately can claim conditionally exempt status.

Federally, only large quantity generators are required to submit the biennial report. [§262.41] CESQGs are not subject to the federal biennial reporting requirements, but such generators should consult with their implementing agencies since states can have more stringent reporting requirements.

6.3.17.4.4 *Universal wastes*

CESQGs have the option of handling their hazardous batteries, lamps, etc. as hazardous waste under the reduced generator requirements codified in §261.5 or as universal waste under Part 273. [§273.8(a)(2)]

6.3.17.4.5 *Used oil*

The Part 279 used oil management standards apply to all facilities regardless of generator status. However, as noted above, CESQGs have the additional advantage of being able to mix their hazardous waste with their use oil and manage the mixture as used oil.

6.3.17.4.6 *Radioactive mixed waste*

At some CESQGs, low-level mixed waste (LLMW) is part of the 100-kg/month generation rate of nonacute hazardous waste. Based on the CESQG exemption to the RCRA requirements, the LLMW can be disposed as low-level radioactive waste if the materials meet the disposal site's waste acceptance criteria. [August 7, 1995; 60 *FR* 40207]

6.3.18 Signage requirements

The signage requirements under RCRA are dependent on facility status. Large quantity generators are not federally required to post any signage for their hazardous waste accumulation areas. While there is a "comment" within §265.176 which directs the reader to a "No Smoking" signage requirement in §265.17(a), EPA Region 5 has clarified that "comments" (and "notes") in the RCRA regulations are not legal requirements. The agency went on to say "[a] court may look at comments or notes to help interpret a provision, but the comments or notes are not themselves binding." [RO 14036] Could a state require a "No Smoking" sign? Yes, and here is the reasoning.

> "According to 40 *CFR* 265.31 of Subpart C…, with which a generator must comply pursuant to §262.34(a)(4)…, a facility must be maintained and operated to minimize the possibility of a fire, explosion, or any unplanned releases of hazardous waste or hazardous waste constituents which could threaten human health or the environment. Region 5 believes that 40 *CFR* 265.31…is broad enough to allow [a state] to request…that a generator post a sign near combustible [or reactive] waste advising that there is 'no smoking' or there are 'no ignition sources.'" [RO 14036]

Small quantity generators are not required to comply with §265.176; thus, the "comment" leading the reader to §265.17(a) seems to not be an issue. However, SQGs are also required to satisfy the requirements of Subpart C of Part 265; therefore, a state could use the previously quoted reasoning to require "No Smoking" signage. See Section 6.3.8 for emergency information posting requirements for SQGs.

CESQGs are not required to comply with Subparts C and I of Part 265. Thus, no RCRA signage is federally required.

Permitted TSD facilities and facilities operating under interim status have two signage requirements under RCRA. Sections 264/265.17(a) direct these facilities to post "No Smoking" signs wherever there is a hazard from ignitable or reactive waste. Furthermore, §§264/265.14(c) calls for "Danger-Unauthorized Personnel Keep Out" signs to be posted near each entrance to the hazardous waste management location and other locations in sufficient numbers to be seen from any approach to the active portion of the facility.

Q *If a TSD facility operates as a "tobacco-free" environment, is a "No Smoking" sign still required?*

A Yes. The fact that a facility operates as a "tobacco-free" environment may not fully address the human and environmental safety concerns stated in the regulation. Sections 264/265.17(a) clearly spell out that the purpose of the "No Smoking" signage provision is to prevent the exposure of ignitable or reactive wastes to open flames, smoking, cutting and welding, hot surfaces, frictional heat, sparks, spontaneous ignition, and radiant heat. [RO 14036]

Transporters of hazardous waste have no signage requirements while wastes are being held at a transfer facility providing wastes remain onsite for no greater than 10 days. If wastes are held at a transfer facility for greater than 10 days, the facility would be storing hazardous waste and would have to satisfy the requirements for permitted storage facilities, which include the signage requirements of §§264/265.17(a) and §§264/265.14(c).

Keep in mind we have only addressed the signage requirements under RCRA at the federal level. Additional requirements can be imposed by an authorized state, local regulations, other federal regulations (e.g., OSHA), or a site-specific facility permit.

6.4 Training

The RCRA regulations include personnel training requirements designed to reduce the potential for mistakes and accidents at hazardous waste facilities that might threaten human health or the environment. The purpose of these programs is to ensure that workers are adequately prepared to properly manage hazardous waste during normal operations and also respond to any emergencies. The following regulated parties must comply with RCRA training standards:

- Large quantity generators [§§262.34(a)(4) and 265.16];
- Small quantity generators [§262.34(d)(5)(iii)];
- Permitted treatment, storage, and disposal (TSD) facilities [§264.16];
- Interim status TSD facilities [§265.16]; and
- Universal waste handlers [§§273.16 and 273.36].

Note that conditionally exempt small quantity generators are not subject to any RCRA training requirements. [RO 14687]

The training requirements for large quantity generators, permitted TSD facilities, and interim status TSD facilities are identical (except that permitted facilities must include additional information in their Part B permit applications). Small quantity generators and small and large quantity handlers of universal waste are subject to less-stringent training standards designed to reduce their administrative burden (especially with regard to recordkeeping). Training requirements applicable to each of these groups are examined on the following pages.

6.4.1 Large quantity generators

Under §262.34(a)(4), large quantity generators who accumulate hazardous waste onsite for 90 days or less without a permit or interim status must comply with the same personnel training requirements as interim status TSD facilities that are set forth in §265.16. EPA justified making large quantity generators comply with the same training standards as TSD facility personnel by noting that "there is little difference between accumulation of hazardous waste for shipment offsite and storage so far as potential damage to human health and the environment is concerned. Therefore, the same standards…should apply." [May 19, 1980; 45 FR 33141] The §265.16 training requirements are summarized below.

6.4.1.1 Who must be trained?

First of all, §265.16(a)(1) requires that facility personnel must receive some type of training. EPA defines "facility personnel" in §260.10 as "all persons who work at, or oversee the operations of, a hazardous waste facility, and whose actions or failure to act may result in noncompliance with the requirements of Part 264 or 265…." If that definition seems a little unclear with regard to who needs to be trained, EPA was purposely vague: "Given the variability in waste types, management processes, and employee functions at hazardous waste facilities, the agency believes that it is neither necessary nor desirable to rigidly specify training courses in [the] regulations." [May 19, 1980; 45 FR 33182]

Instead, the type and degree of training required will depend on each employee's responsibilities associated with his/her job function/description. "[I]t would not make sense to require training in topics not germane to an employee's areas of responsibility since this would add considerable burden to some firms without corresponding environmental or health benefits…. [I]mplicit in the regulations is the requirement that the type and amount of training necessary for each employee stems from his specific responsibilities." [March 24, 1986; 51 FR 10165]

EPA noted in guidance that all personnel (supervisors as well as nonsupervisory personnel) who are actively engaged in the operation of the hazardous waste facility require training. [OSWER Directive 9523.00-10, cited previously] Certainly, workers that directly manage hazardous waste (including laboratory technicians) must receive training on the hazardous waste management procedures relevant to their individual responsibilities. [RO 12341] Those individuals that are labeling and moving drums, tracking 90-day clocks, performing inspections, preparing manifests, etc. should be specifically trained in those areas.

What about other plant workers (e.g., maintenance personnel) whose job functions preclude them from actually managing hazardous waste? EPA noted in preamble language that "[e]mployees who work in or adjacent to areas where hazardous wastes are generated, handled, or stored but do not handle hazardous wastes, must still be trained to be thoroughly familiar with basic emergency procedures." [March 24, 1986; 51 FR 10165]

It is important to note that personnel responsible for managing hazardous waste documentation should also receive training, even though they may not directly handle wastes. For example, clerical staff responsible for maintaining manifest and LDR records should be trained on the necessity of the paperwork and appropriate methods for maintaining the records.

At a minimum, all employees, regardless of their position, must be familiar with the facility's contingency plan and evacuation procedures so that they can respond effectively in an emergency. Clerical and office workers must receive this training as well as hazardous waste workers. [OSWER Directive 9523.00-10] If an incident occurs, most personnel will be responsible for vacating the premises according to the facility's evacuation plan. In the meantime, other facility personnel who have received more extensive emergency response training, will be responsible for containing any spills, informing the police and fire department, and/or operating firefighting or other emergency response equipment.

With regard to part-time or temporary employees, EPA noted that they also must receive appropriate training. [March 24, 1986; 51 FR 10165]

Finally, contractors working at large quantity generator sites fall under the §260.10 definition of "facility personnel" and consequently must receive RCRA training. [RO 14180] As such, contractors should be trained in certain areas (e.g., evacuation routes, emergency communications/

alarms, proper hazardous waste handling practices, etc.) before they begin work. One way to facilitate such training is for the facility to require contractors to take general hazardous waste and/or health and safety training as a condition of the contract. Although not required by the regulations, it is good practice for contractors to be supervised by facility personnel to make sure they properly manage any hazardous wastes that they generate or handle.

6.4.1.1.1 Training exception for satellite accumulation areas

Believe it or not, RCRA training is not required for personnel whose hazardous waste management activities are limited to working in or near satellite accumulation areas. EPA's position is that "since only one waste will normally be accumulated at each satellite area, and since only limited quantities are allowed to accumulate, contingency plans and training plans are not necessary." However, when waste generated in a satellite accumulation area is transported to a 90-day accumulation area regulated under §262.34(a) or to permitted or interim status storage, the RCRA training requirements will apply (including training for those who transfer the waste from satellite areas to central accumulation areas). [January 3, 1983; 48 *FR* 119, December 20, 1984; 49 *FR* 49570, RO 11373, 14703, 14758]

6.4.1.2 When must training be completed?

Personnel must successfully complete their initial training within 6 months after they are employed or assigned a new position at the facility. [§265.16(b)] New employees who will be performing hazardous waste management tasks must be supervised at all times until they have completed the initial training. This requirement is not extended to existing employees who transfer into the hazardous waste area; however, we recommend that transferees also not work unsupervised until their training is complete.

After the initial training, personnel must participate in an annual training review specific to the individual employee's hazardous waste duties. This refresher training is required to keep personnel up to date with changes in wastes managed at the facility, state and federal RCRA regulations, operating or paperwork procedures, the facility's contingency plan, etc. The regulations in §265.16(c) do not specify the amount of time associated with this annual review training. Case Study 6-3 provides EPA guidance on the definition of "annual" for this regulatory requirement.

6.4.1.3 What are the basic RCRA training requirements?

Section 265.16(a)(2) requires training in two basic areas: 1) hazardous waste management procedures, and 2) emergency response. As discussed previously, EPA does not specify the content of training programs; however, the following subsections will provide guidance that should prove helpful in setting up a training program. A good place to start is with guidance provided by the state of Colorado (see Table 6-5) on the training that should be provided for different job categories at large quantity generator facilities.A

6.4.1.3.1 Hazardous waste management procedures training

EPA requires facility personnel to successfully complete a classroom or on-the-job training program that teaches them to perform their duties in a way that ensures the facility's compliance with the RCRA regulations. This component of the training program must be designed to ensure that workers are able to safely handle hazardous wastes. Training must cover the hazardous waste management procedures relevant to the positions that the employees hold at the facility (e.g., accumulation standards, container and tank management, inspections, manifests, packaging, waste identification and determinations, etc).

These training sessions should cover topics such as [OSWER Directive 9523.00-10]:

Case Study 6-3: Definition of "Annual Review"

A large company has established an extensive RCRA training program for over 15,000 employees. Refresher training is conducted at any time within a 90-day period before each employee's training anniversary date (i.e., the date upon which that individual received the initial training). Accordingly, in any year, by the time the employee's training anniversary date occurs, he/she will have received refresher training. However, with this system, it is possible that an employee will go as long as 15 months between training sessions. For example, an employee with a June 1 training anniversary date could receive refresher training on March 1 in one year and not until June 1 the next year. Would this scheme comply with the annual review requirement for RCRA training?

EPA recognizes that it may not be feasible for companies with many employees to train each employee exactly one year after the last training. But, the agency "does expect companies to attempt to provide training so that personnel are trained every year." In this example, even though up to 15 months may elapse between training courses, an employee would, over the course of four years, receive four annual training reviews. Therefore, this training scheme would meet the RCRA annual training requirements, even though as much as 15 months passes between training sessions. However, the company should check with the relevant state(s) to ensure that they do not have more stringent requirements that would affect this situation. [RO 14286] For example, the state of Colorado allows only 13 months between training sessions.

- The chemical and physical characteristics of the wastes that the workers are assigned to manage (e.g., ignitability, reactivity, and/or incompatibility with other waste types);
- Knowledge of what to do if a spill or leak occurs;
- The types of protective equipment (e.g., respirators, self-contained breathing apparatus) or clothing to be worn during normal and emergency situations;
- Proper operation of trucks, forklifts, or any other machinery used in waste management;
- Basic first aid; and
- Who to inform if an emergency occurs.

During training, employees should be taught not only how, but why, certain operations are to be performed in a prescribed manner. Providing personnel with such explanations should reduce the use of short-cuts that may lead to problems. Furthermore, training should "be structured so that it parallels as realistically as possible the actual job in order that the 'real world' activities are approximated as much as possible." [OSWER Directive 9523.00-10]

6.4.1.3.2 Emergency response training

As mentioned in Section 6.4.1.3, all facility personnel must receive some type of emergency response training to ensure that they can effectively respond to emergencies. Section 265.16(a)(3) and associated guidance require that facility personnel be trained on the following topics, depending on the employee's job description:

- Contingency plan content and implementation;
- Communication and alarm systems;
- Standard operating procedures for using, inspecting, repairing, and replacing facility emergency and monitoring equipment;
- Key parameters for automatic-waste-feed cut-off systems;
- Use and limitations of personal protective equipment;
- Appropriate responses to fires, explosions, and ground water contamination incidents; and
- Shutdown of operations.

Table 6-5: Example Personnel Training Matrix for Large Quantity Generators

Training element	Generic job category[1]							
	Environ. manager	Environ. clerk	Production supervisor	Maint. supervisor	Haz. waste technician	Haz. waste generator	Gen. plant worker	Gen. office personnel
Hazardous waste management procedures training								
Accumulation (satellite/90 day)	✓	✓	✓	✓	✓	✓		
Container/tank management	✓			✓	✓	✓		
Hazardous waste regulations	✓							
Inspections	✓	✓		✓	✓	✓		
Making waste determinations	✓				✓			
Pretransportation (manifests/labels)	✓	✓			✓			
Properties of facility wastes	✓	✓		✓	✓	✓		
Reporting and recordkeeping	✓	✓			✓			
Waste minimization	✓	✓	✓	✓	✓	✓	✓	
Waste packaging	✓			✓	✓			
Emergency response training								
Communications, alarms, and evacuation routes	✓	✓	✓	✓	✓	✓	✓	✓
Contingency plan implementation/emergency response procedures	✓			✓	✓	✓	✓	
Emergency equipment use, inspection, and repair	✓			✓	✓			
Response to fire, explosion, and ground water contamination incidents	✓			✓	✓	✓	✓	
Site shutdown procedures	✓			✓	✓	✓	✓	

[1] Training requirements are representative of an average, hypothetical manufacturing facility. Job categories and associated training requirements will be specific to your facility and may not be identical to those presented in this table. Generic job categories are given as follows:
—*Environmental manager* is responsible for the overall hazardous waste management functions at the facility.
—*Environmental clerk* is responsible for hazardous waste paperwork and recordkeeping functions; has no direct contact with physical hazardous waste activities.
—*Production supervisor* is responsible for production activities; supervises staff that are hazardous waste generators.
—*Maintenance supervisor* is responsible for maintenance activities; supervises staff that are hazardous waste generators and has some environmental management responsibilities.
—*Hazardous waste technician* is responsible for some hazardous waste management functions; performs physical waste activities (e.g., moving and labeling containers, collecting samples, etc.).
—*Hazardous waste generator* generates hazardous waste as part of the production process; may utilize satellite containers.
—*General plant worker* generates no hazardous waste and has no direct contact with physical hazardous waste activities.
—*General office personnel* are responsible for general administrative and clerical activities but no hazardous waste functions.

Source: Adapted from *Personnel Training for Large Quantity Generators of Hazardous Waste*, Colorado Department of Public Health and Environment, March 1997, available from http://www.colorado.gov/pacific/sites/default/files/HM_hw-lqg-personnel-training.pdf.

EPA provided specific training requirements in §265.16(a)(3) for this aspect of the facility's training program because the agency believes that the ability to respond to emergencies is the most important skill that facility personnel must acquire to minimize the potential dangers associated with hazardous waste management.

Large quantity generators also have to satisfy the training requirements of the OSHA hazard communication program (29 *CFR* 1910.1200) and HAZWOPER program (29 *CFR* 1910.120). The hazard communication standard, which may be applicable to any facility, requires training on the physical and health hazards of chemicals in the workplace, protective measures (including work practices and use of personal protective equipment), labeling systems, and safety data sheets. On the other hand, HAZWOPER training is required only for facilities that have the potential for an emergency to occur due to an uncontrolled release of hazardous substances or hazardous raw materials. It requires training to make sure that employees can recognize and respond to an emergency.

In an April 4, 2006 final rule [71 *FR* 16862], EPA modified the regulations to give facilities the option of complying with either the RCRA or OSHA training requirements for emergency procedures. If a facility can meet all RCRA emergency response training requirements through an OSHA training course, the facility would be in compliance with RCRA. On the other hand, if a facility cannot meet all required RCRA emergency response training through such OSHA training, then it would be incumbent on the facility to address any gaps. [71 *FR* 16871] EPA recommends that generators work with their state agency or EPA regional office to ensure that the approach they take in developing an emergency response training program is in compliance with §265.16. Obviously, facilities not subject to OSHA training requirements would have to comply with the RCRA training standards.

In *RCRA-OSHA Training Requirements Overlap,* May 27, 1999 (available at http://www.epa.gov/epawaste/ inforesources/data/burdenreduction/), EPA noted that RCRA training requirements are more stringent than OSHA in the following areas:

- Reach of employees required to be trained,
- Scope of training on standard operating procedures,
- Extent of training on automatic-waste-feed cut-off systems,
- Response to ground water contamination incidents,
- Scope and length of time for training record retention, and
- Recordkeeping regarding who has been trained.

Therefore, although significant overlap exists, OSHA hazard communication or HAZWOPER training alone may not satisfy all of the RCRA training requirements. In addition, RCRA training must be tailored to the site-specific hazardous waste duties of facility personnel.

6.4.1.3.3 *Three training options available*

Initial and annual review training may be performed via 1) formal classroom programs offered offsite, 2) in-house classroom sessions, or 3) in-house, on-the-job learning. A combination of these three options may be used. Some facilities opt to send their supervisory personnel to formal, offsite training programs so that they can come back and train the remaining facility personnel in more focused, in-house training sessions. In addition to on-the-job (hands-on) learning, in-house training can consist of regularly scheduled meetings, such as monthly safety meetings, computer-based training, electronic memos, videos, or read-only training materials. The effectiveness of such training can be measured by performance or by written/oral tests.

EPA recognizes that supervised on-the-job training is a valid substitute for, or supplement to, formal classroom training. However, the agency stipulated in the preamble to the May 19, 1980 RCRA regulations that "the content, schedule, and techniques to be used in the on-the-job training program must be described in the training records maintained at the facility...." [45 FR 33182] Furthermore, on-the-job training should be conducted by a supervisor or other individual who is skilled in facility operations and the duties of the job.

6.4.1.4 Who may perform the training?

Per §265.16(a)(2), all RCRA training programs must be directed by a person trained in hazardous waste management procedures (qualified either by experience or education). The individual needs to be knowledgeable on hazardous waste regulatory requirements/implementation and the specific hazardous waste issues at the facility.

6.4.1.5 What are the recordkeeping requirements?

In enforcement actions, many RCRA training violations stem from failure to comply with recordkeeping requirements, rather than the content of the training program itself. Large quantity generators must keep the following training records at the facility [§265.16(d)]:

- The job title and name of each employee filling every position related to hazardous waste management. Job titles used to satisfy this requirement may or may not match those used for human resources purposes at the facility.

- A written job description (including the requisite skills, education level or other qualifications, and duties) for each position related to hazardous waste management; only duties directly related to hazardous waste management are typically included in these descriptions. This information allows EPA or the state to determine if each person is receiving a level of training that is commensurate with that person's duties and responsibilities.

- A written description of the type and amount of initial and annual review training that will be given to each person filling a position related to hazardous waste management. Outlines of training programs or training matrices (similar to the example matrix presented in Table 6-5) are often used to fulfill this requirement. The minimum number of hours for the training program should be included in this description. If a facility designs its own training program to be implemented in-house or on-the-job, a detailed written account of the material to be presented for each position, the techniques to be used, and a schedule to be followed by the instructors should be on file.

- Records (including the dates of training) documenting that the required initial and all subsequent annual review training sessions have been successfully completed by facility personnel. Employee participation documentation such as tests, class lists with signatures, individual training logs, or spreadsheets are typically employed.

These four recordkeeping requirements are depicted in Figure 6-1. Training records for current personnel must be kept until the facility closes. For former employees, training records must be maintained for at least 3 years from the date the employee last worked at the facility. [§265.16(e)]A

6.4.2 Small quantity generators

In order to reduce the administrative burden, EPA has developed a simpler set of training requirements for small quantity generators. The regulations say that small quantity generators must "ensure that all employees are thoroughly familiar with proper waste handling and emergency procedures, relevant to their responsibilities during normal facility operations and emergencies." [§262.34(d)(5)(iii)]

Very little guidance has been issued amplifying this provision. EPA noted in 1993 that small quantity generator personnel that handle hazardous wastes should have training in proper waste handling and emergency procedures appropriate to the types of waste handled, the management methods used, and the hazards presented by the waste. In addition, there needs to be at least one employee either on the premises or on call with the responsibility for coordinating all emergency response measures. [RO 11779]

Waste identification, manifest requirements, accumulation time limits, proper tank and container

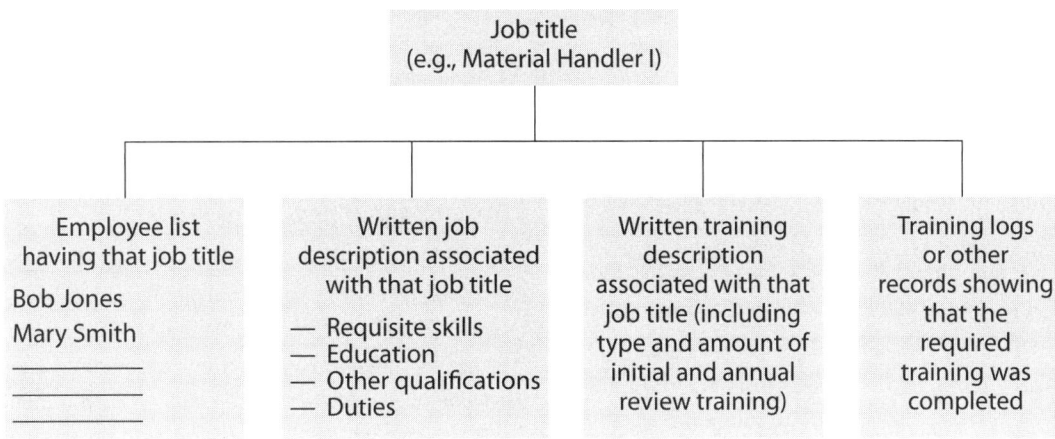

Figure 6-1: RCRA Training Recordkeeping Requirements for LQGs/TSD Facilities

Source: McCoy and Associates, Inc.; adapted from §§264/265.16(d).

management, and tank and container labeling should all be topics included in the small quantity generator's training program. Although not required by the regulations, a written training plan and records of employee names, training dates, and employee signatures would be helpful to demonstrate compliance with the training requirement. See Case Study 6-4 for EPA's thoughts on training requirements for facilities that bounce between large and small quantity generator status.

Regarding emergency procedures, employees must be trained to be familiar with 1) communication and alarm systems and procedures such as contacting emergency response personnel (e.g., the fire department), 2) how to extinguish a fire, and 3) how to contain and clean up hazardous waste spills. [RO 11429] However, no recordkeeping requirements apply to small quantity generator training, and there are no requirements for annual refresher training.

Does this mean that all small quantity generator employees, even clerical and office staff, need to be trained in hazardous waste management methods? EPA clarified this issue in the preamble to the March 24, 1986 small quantity generator rule.

"Employees who handle hazardous wastes as part of their normal job responsibilities or are likely to handle wastes in an emergency situation must be thoroughly familiar with proper waste handling and emergency procedures. Employees who work in or adjacent to areas where hazardous wastes are generated, handled, or stored, but do not handle hazardous wastes [e.g., office or clerical staff], must still be trained to be thoroughly familiar with basic emergency procedures." [51 FR 10165]

Case Study 6-4: Episodic Generators

A facility that is usually a small quantity generator produces enough hazardous waste to become a large quantity generator once every 5 years. Does that mean that the facility's personnel training must be updated annually?

No. "[T]he SQG need only comply with the §265.16(c) [LQG] requirements when he is subject to all of §262.34, which, in this case, is once every five years. Hence, personnel training would be updated every five years." [RO 12245]

6.4.3 Conditionally exempt small quantity generators

Conditionally exempt small quantity generators do not have to comply with any RCRA training requirements per §261.5. [RO 14687]

6.4.4 TSD facilities

Permitted and interim status TSD facilities must train their employees in accordance with the requirements of §§264.16 and 265.16, respectively. The requirements in these two sections are identical and are the same as discussed in Section 6.4.1 for large quantity generators.

In accordance with §270.14(b)(12), information on employee training must be submitted as part of a RCRA Part B permit application. An outline of both the initial and continuing training programs for facility personnel must be included. The application must also contain a brief description of how training will be designed to meet actual job tasks in accordance with requirements in §264.16(a)(3).

6.4.4.1 Training not required in post-closure permits

When EPA added specific items to be included in applications for post-closure permits on October 22, 1998, the agency purposefully left out any requirements for training programs. EPA noted that, in the case of a post-closure permit, the facility will not be operating, so personnel training is not required. [63 FR 56728]

6.4.5 Universal waste handlers

EPA has created less-stringent management requirements (including training provisions) for handlers of universal wastes. (Universal wastes include hazardous batteries, certain pesticides, mercury-containing equipment, and hazardous waste lamps and are regulated under the Part 273 universal waste management standards—see Chapter 8.) The level of training required for handlers' personnel depends on how much universal waste is stored onsite at any one time. Large quantity handlers (i.e., those that accumulate ≥5,000 kg total universal waste onsite at any time) are subject to more-stringent requirements than small quantity handlers (i.e., those that never have 5,000 kg or more total universal waste onsite at any one time). Although training requirements are specified for such handlers (as discussed in the next two subsections), EPA noted that "[t]raining that is required under other programs (such as OSHA or RCRA) will generally fulfill the Part 273 training requirements." [July 6, 1999; 64 FR 36475]

As noted just above, the potential for overlap exists between the universal waste training requirements and the OSHA programs. EPA stated "any training provided under other programs that would meet any or all of the Part 273 training requirements may be used to fulfill the RCRA requirements. As long as the substantive standards of the [RCRA] training provisions are met, the handler has fulfilled the training requirement. There is no requirement that training provided to meet the RCRA requirements be separate from other training given to employees." [May 11, 1995; 60 FR 25528]

Employees at TSD facilities that recycle, treat, and/or dispose universal wastes remain subject to the full §264.16 or 265.16 training requirements.

6.4.5.1 Large quantity handler training standards

Large quantity handlers of universal waste "must ensure that all employees are thoroughly familiar with proper waste handling and emergency procedures, relative to their responsibilities during normal facility operations and emergencies." [§273.36] (This language is identical to that applicable to SQGs of hazardous waste, as outlined in Section 6.4.2.) The preamble to the universal waste rule stated: "this does not require that any records be kept for training provided to employees, requires only that employees that have responsibilities for managing universal waste or for responding to emergencies be trained, and requires only that these employees be trained as is appropriate for their universal waste management responsibilities. Thus, employees who only minimally handle universal waste need only be trained to properly carry out that activity and to carry out their responsibilities, if any, in case of an emergency." [May 11, 1995; 60 *FR* 25528]

No annual review of the initial employee training is required for large quantity handlers, unless the facility is also a large quantity generator of hazardous waste. [RO 14598]

6.4.5.2 Small quantity handler training standards

The training requirements for small quantity handlers of universal waste are even less burdensome. These handlers are required only to *inform* all employees who handle or are responsible for managing universal waste of proper handling and emergency procedures for the types of universal waste handled at the facility. [§273.16] In the preamble to the May 11, 1995 universal waste rule, EPA stated that providing such information through oral communication is acceptable; however, the agency suggested that supplying employees with brochures and documents available from manufacturers, trade associations, etc. would probably be more effective. [60 *FR* 25528]

No annual review of the initial employee training is required, unless the facility is also a large quantity hazardous waste generator. [RO 14598]

6.5 Contractor relations

The issue of who is the generator of a hazardous waste when a facility owner/operator employs contractors can get quite sticky. EPA first addressed this issue on October 30, 1980. [45 *FR* 72026] In that discussion, the agency identified three parties who potentially could be considered the generator: 1) the owner/operator of a manufacturing process unit, product/raw material storage tank, transport vehicle, or vessel; 2) the owner of the product, raw material, or manufacturing material being stored that may generate a hazardous waste; and 3) the person who removes hazardous sludges, sediments, or other residues from a manufacturing process unit, tank, vehicle, or vessel.

EPA typically cites the definition of "generator" in §260.10 when determining the generator (or generators) when contractors are involved: "*Generator* means any person, by site, whose act or process produces hazardous waste identified or listed in Part 261 of this chapter or whose act first causes a hazardous waste to become subject to regulation." Based on this definition, more than one person can be considered the generator of a hazardous waste; in such situations, the agency typically refers to the two or more parties as cogenerators.

6.5.1 Cogenerator agreements

EPA's long-standing position has been that where two or more parties are considered to be cogenerators, they should mutually agree (by contract or other means) who will perform the duties of the generator on behalf of the other parties. EPA will then look first to the designated party to perform generator responsibilities. However, the agency reserves the right to enforce against any cogenerator in the event that the designated party fails to perform. [45 *FR* 72026, RO 11005]

Where cogenerators don't designate who has primary responsibility, EPA will initially look to the following parties [45 *FR* 72027]:

- The operator of a manufacturing process unit that generates hazardous waste;
- The operator of a product/raw material storage tank from which hazardous waste is removed;
- The operator of a central facility that removes hazardous waste from vehicles or vessels; and
- The operator of the vehicle or vessel where no central facility is involved.

In an example of how a cogenerator agreement might work, a chemical company wanted to retain control of specialty chemicals it sold to customers. The company would retain ownership of the chemicals, maintain a physical presence at the customer's site, and manage any hazardous wastes exiting the customer's process units in which these chemicals were used. EPA agreed that the chemical company would be considered a cogenerator and, under terms of a contract between the chemical company and the customer, would be the organization that EPA would look to first for compliance with RCRA generator standards. [RO 14027]

6.5.2 Manifesting and recordkeeping

Cogenerators can decide who will take responsibility for filling out manifests. For example, at a federally-owned facility, when a contractor that operates a portion of the site generates hazardous waste, the contractor's name, address, and EPA ID number can appear in Blocks 1 and 5 of the manifest. The facility owner's EPA ID number could be added to Block 14, Special Handling Instructions and Additional Information. If the contractor's name appears on the manifest, it does not mean that they have assumed liability for the owner. According to EPA, "the presence or absence of someone's ID number on the manifest is not the determining factor in assessing liability…. [B]oth parties may be liable for any violations or damages, depending on all the facts in question." [RO 11816]

If a contractor is being used by a facility to comply with recordkeeping, manifesting, and pretransport requirements, but the contractor is not a generator of the wastes, the facility's EPA ID number (not the contractor's) must be used on manifests and records. [RO 11372]

6.5.3 Waste counting for generator status

A contractor (such as a painter) should count the wastes that he/she generates at his/her place of work or at "individual work sites" toward his/her generator status. [RO 12661] Although not specifically stated, this could be read to imply that the facility where the contractor works need not count contractors' wastes when determining the facility's generator status. However, this issue needs to be worked out between the facility owner/operator and the contractor—one of the two parties would need to count the wastes and ensure that they are managed in compliance with all applicable regulations.

6.5.4 Contractor training

Contractors working at TSD facilities are held to the same standards as noncontract employees and must undergo the appropriate training per §§264/265.16. Contractors working at large quantity generator facilities that practice 90-day accumulation must also comply with the §265.16 training requirements. [RO 14180]

6.5.5 Accumulation requirements

In order for a contractor to accumulate wastes in 90/180/270-day units, it must be the generator (or cogenerator) of the waste. [RO 11372]

6.5.6 Examples of contractors/cogenerators

EPA guidance documents give several examples of how contractors are regulated under federal policy, as discussed in the following six sections.

6.5.6.1 Painters

Where lead-based paint is being removed by a contractor, the property owner and the contractor are considered to be cogenerators. The contractor is considered a generator because his/her act (paint removal) first causes a hazardous waste to become subject to regulation. [RO 11913, 14069]

6.5.6.2 Sample collectors

If a contractor collects a waste sample for analysis, the sample is exempt from regulation per §261.4(d). However, if the sample is returned to the contractor by the analytical laboratory, the contractor is the generator of a waste. [RO 12438]

6.5.6.3 Household contractors

When a homeowner hires a contractor to remove lead-based paint, the paint wastes are eligible for the household waste exclusion of §261.4(b)(1). Residential lead-based paint waste is defined as waste "generated as a result of activities such as abatement, rehabilitation, renovation and remodeling in homes and other residences. The term 'residential lead-based paint waste' includes, but is not limited to, lead-based paint debris, chips, dust, and sludges." [40 *CFR* 258.2] The exclusion is available to both contractors and do-it-yourselfers that generate and dispose lead-based paint wastes. [October 23, 2001; 66 *FR* 53537, RO 14459, 14673]

6.5.6.4 Mobile solvent recyclers

Conflicting guidance has been given where a contractor brings mobile solvent recycling equipment to a facility. In 1986 guidance, EPA stated: "Since the spent solvent is presumably already hazardous prior to [the contractor] coming onto the generator's site, and since the solvent is likely to have been accumulated prior to being recycled, the hazardous spent solvent would already have been subject to regulation under the accumulation provisions of Part 262 of the hazardous waste regulations. Thus none of the actions taken by [the contractor] would appear…to cause him to become subject to RCRA liability as a RCRA hazardous waste generator." [RO 12706]

In 1984 and 1989 guidance, however, EPA clarified that, if the mobile recycler pumps spent solvent out of the facility's storage tanks or containers and then pumps recovered solvent back into those units, the facility owner/operator and the contractor are cogenerators of the still bottoms from the recycling unit. If an agreement has not been reached that assigns generator responsibilities, EPA will look first to the recycling unit operator (the contractor) to fulfill the generator duties. [RO 12340, 13280]

Q *When a contractor that operates a mobile solvent recycling process visits a generator's site, who has to submit an 8700-12 notification to EPA or the state?*

A Typically, the generator's facility will already have submitted the notification. If this generator consents to performing the generator responsibilities associated with any still bottoms from the mobile unit, additional notification is not required. On the other hand, if the recycler will be the generator of record who will be responsible for manifesting, they would need to submit a notification form and obtain an EPA identification number for the site. [RO 13280]

6.5.6.5 Equipment maintenance contractors

If a maintenance contractor removes used oil from equipment at a facility, he/she is considered to be a used oil generator "because the contractor's act of servicing and removing used oil from equipment first causes the used oil to be subject to regulation. As a generator, such a contractor may self-transport up to 55 gallons of used oil without an EPA identification number pursuant to Section 279.24." [RO 14116]

6.5.6.6 Contractors dealing with shipboard wastes

The Federal Facilities Compliance Act of 1992 established basic rules for how hazardous wastes are managed on "public vessels." A "public vessel" is defined as "a vessel owned or bare-boat chartered and operated by the United States, or by a foreign nation, except when the vessel is engaged in commerce." Any hazardous waste generated on such a vessel is not subject to the storage, manifest, inspection, or recordkeeping requirements of RCRA until: 1) the waste is transferred to a shore facility; 2) the waste is stored on the vessel for more than 90 days after the vessel is taken out of service; or 3) the waste is transferred to another public vessel within the territorial waters of the United States, and the waste is stored on this second vessel for more than 90 days. [FFCA Section 106]

Given the requirements of the Federal Facilities Compliance Act, along with prior EPA guidance [RO 11571, 12727], it appears that if a contractor (or the operator of a port facility) moves the waste from ship to shore, he/she is a cogenerator along with the owner of the ship.

The situation is similar for merchant shipboard wastes. Merchant ships are considered to be product or raw material transport vessels, and wastes are exempt pursuant to §261.4(c) until removed from the vessel. According to EPA, "when any hazardous waste is removed from the vessel, the owner of the product or raw material, the operator of the vessel, and the person purposefully removing the hazardous waste from the vessel would all be considered 'generators.'" [RO 12727; see also RO 11128]

EPA will normally look to the operator of a port facility that removes hazardous wastes from ships to perform the generator duties, since it is the party best able to determine whether the waste is hazardous. Where wastes are not removed at a central facility, the agency would look to the operator of the ship to perform the generator duties. [RO 11128]

6.5.7 The states' role

Because the contractor/cogenerator guidance discussed above is federal policy, EPA emphasizes that interested parties should contact their state agency to determine if their state's position differs from EPA's.

6.6 Manifesting

When shipping hazardous waste offsite, generators (and all other facilities) must use a manifest. This document tracks hazardous waste from the time it leaves the originating facility until it reaches the offsite TSD facility that will treat, store, and/or dispose it. Provisions in the manifesting process enable the waste generator to verify that his/her waste has been properly delivered and that no waste has been lost or is unaccounted for in the course of transportation.

The manifest requirements for generators appear in §§262.20–262.27 and the requirements for TSD facilities appear in §§264.71, 264.72, 264.76, 265.71, 265.72, and 265.76.

6.6.1 Background

Before RCRA was enacted, U.S. Department of Transportation (DOT) shipping papers were required for tracking the movement of industrial and chemical waste. DOT did not require a specific form. Instead, the department simply required each transport vehicle to carry specific information, such as hazardous material name and hazard class.

When RCRA was signed into law, Congress required EPA to create a manifest system for tracking shipments of hazardous waste from a generator's site to the final management facility (i.e., "cradle-to-grave" tracking). In compliance with this mandate, EPA promulgated manifesting regulations designed to establish a paper trail for hazardous wastes on February 26, 1980. [45 *FR* 12724] The manifest shows who is in control of the hazardous waste at a given time and identifies where the waste is destined.

The manifest also identifies the hazards (in terms of DOT shipping description) and quantity of the waste—information that is necessary in the event of an emergency. The 1980 rule required only this information to accompany hazardous waste shipments, and EPA believed that existing DOT shipping papers could be modified to accommodate the new requirements. States were allowed to develop their own manifest forms, as long as they captured the information required by the 1980 rule. [45 *FR* 12729]

EPA's initial approach to manifests was short-lived. Soon after the 1980 rule was finalized, more than 20 states developed their own manifest forms, which required information in addition to the federal requirements. This led to widespread confusion over which manifest requirements were applicable. In many cases, generators had to prepare multiple manifests for interstate shipments in order to satisfy the requirements of all states through which the hazardous waste traveled. This made manifest compliance and enforcement very difficult. [May 22, 2001; 66 *FR* 28242]

In order to promote uniformity, create standardized manifesting procedures, and make enforcement easier, EPA and DOT finalized a rule on March 20, 1984 requiring the use of a Uniform Hazardous Waste Manifest. [49 *FR* 10490] The uniform manifest system was supposed to solve the aforementioned problems. On the same day [49 FR 10507], DOT added a requirement in 49 CFR 172.205 stating that all shipments of hazardous waste must be accompanied by a uniform manifest.

Although states were required to use the Uniform Hazardous Waste Manifest, the 1984 regulations allowed some variability; thus, if a state required additional information on the manifest to be included, a special manifest (available from the specific state) had to be used. That system resulted in 24 states requiring their own versions of the "uniform" hazardous waste manifest.

6.6.1.1 Uniform hazardous waste manifest

EPA established a truly uniform hazardous waste manifest in two 2005 rulemakings. [March 4, 2005; 70 FR 10776 and June 16, 2005; 70 FR 35034] The manifest is shown in Figure 6-2. Under these rules, the same manifest (i.e., EPA's national manifest) has been used for hazardous waste transportation in all states since September 5, 2006. Because the rules were also promulgated under DOT authority, the final manifest regs do not allow states to require generators to use a state form that is different from the federal manifest.

6.6.2 Exemptions from manifesting

Wastes that are excluded from the definition of solid or hazardous waste [see §261.4(a–b)] do not require manifesting when shipped offsite. Similarly, some hazardous wastes are exempt from RCRA requirements (including manifesting), such as analytical and treatability-study samples [§261.4(d–f)], lead-acid batteries [§261.6(a)(2)(iv)], and scrap metal and other recyclable materials in §261.6(a)(3). Finally, used oil being recycled (even if it is characteristically hazardous) and universal wastes, which are each subject to separate regulatory programs, are not subject to manifesting requirements. [§§261.6(a)(4) and 261.9, respectively] Other shipments of hazardous waste that are exempt from the RCRA manifesting requirements are outlined below.

6.6.2.1 Manifesting exemptions for CESQGs and SQGs

Conditionally exempt small quantity generators (CESQGs) are exempt from the manifest provisions at the federal level. [RO 11589, 11772] However, these CESQGs can sometimes have a hard time finding a transporter who will accept such shipments of hazardous waste without a manifest.

Small quantity generators are exempt from the manifesting requirements only for waste reclaimed under a contractual tolling agreement. (Under a tolling agreement, a generator typically arranges for an offsite facility to take the generator's waste, process or reclaim it, and then return the reclaimed or regenerated material to the generator. Examples would be where spent solvents or hazardous shop towels are sent offsite for recovery/cleaning.) To qualify for this exemption: 1) the vehicle used to transport the hazardous waste to the reclamation facility and regenerated product back to the generator must be owned and operated by the reclamation facility, and 2) the generator must maintain a copy of the written agreement for at least 3 years after it expires. [§262.20(e)]

Additionally, *all* of the nonacute hazardous waste generated by a small quantity generator in a calendar month must total between 100 kg and 1,000 kg—not just the hazardous waste subject to the tolling agreement. For instance, if a printer generates a total of 800 kg/mo of nonacute hazardous waste (some of which is hazardous spent solvent), he/she may send the spent solvent offsite to a solvent recycler without a manifest if all of the conditions of the exemption are met. [March 24, 1986; 51 FR 10156, RO 11178, 13171]

In addition, as noted in Section 6.6.9, small quantity generators who export hazardous wastes out of the United States under a tolling agreement are not subject to the Part 262, Subpart E export requirements. [RO 14015]

§6.6.2.1

Figure 6-2: Uniform Hazardous Waste Manifest

Please print or type. (Form designed for use on elite (12-pitch) typewriter.) Form Approved. OMB No. 2050-0039

UNIFORM HAZARDOUS WASTE MANIFEST | 1. Generator ID Number | 2. Page 1 of | 3. Emergency Response Phone | **4. Manifest Tracking Number**

GENERATOR

5. Generator's Name and Mailing Address — Generator's Site Address (if different than mailing address)

Generator's Phone:

6. Transporter 1 Company Name — U.S. EPA ID Number

7. Transporter 2 Company Name — U.S. EPA ID Number

8. Designated Facility Name and Site Address — U.S. EPA ID Number

Facility's Phone:

9a. HM	9b. U.S. DOT Description (including Proper Shipping Name, Hazard Class, ID Number, and Packing Group (if any))	10. Containers No.	10. Containers Type	11. Total Quantity	12. Unit Wt./Vol.	13. Waste Codes
	1.					
	2.					
	3.					
	4.					

14. Special Handling Instructions and Additional Information

15. **GENERATOR'S/OFFEROR'S CERTIFICATION:** I hereby declare that the contents of this consignment are fully and accurately described above by the proper shipping name, and are classified, packaged, marked and labeled/placarded, and are in all respects in proper condition for transport according to applicable international and national governmental regulations. If export shipment and I am the Primary Exporter, I certify that the contents of this consignment conform to the terms of the attached EPA Acknowledgment of Consent.
I certify that the waste minimization statement identified in 40 CFR 262.27(a) (if I am a large quantity generator) or (b) (if I am a small quantity generator) is true.

Generator's/Offeror's Printed/Typed Name — Signature — Month Day Year

INT'L

16. International Shipments ☐ Import to U.S. ☐ Export from U.S. Port of entry/exit:
Transporter signature (for exports only): Date leaving U.S.:

TRANSPORTER

17. Transporter Acknowledgment of Receipt of Materials
Transporter 1 Printed/Typed Name — Signature — Month Day Year
Transporter 2 Printed/Typed Name — Signature — Month Day Year

DESIGNATED FACILITY

18. Discrepancy
18a. Discrepancy Indication Space ☐ Quantity ☐ Type ☐ Residue ☐ Partial Rejection ☐ Full Rejection
Manifest Reference Number:
18b. Alternate Facility (or Generator) — U.S. EPA ID Number
Facility's Phone:
18c. Signature of Alternate Facility (or Generator) — Month Day Year

19. Hazardous Waste Report Management Method Codes (i.e., codes for hazardous waste treatment, disposal, and recycling systems)
1. 2. 3. 4.

20. Designated Facility Owner or Operator: Certification of receipt of hazardous materials covered by the manifest except as noted in Item 18a
Printed/Typed Name — Signature — Month Day Year

EPA Form 8700-22 (Rev. 3-05) Previous editions are obsolete. **DESIGNATED FACILITY TO DESTINATION STATE (IF REQUIRED)**

Source: EPA, 70 *FR* 35038, June 16, 2005.

6.6.2.2 Manifesting exemption for onsite shipments

When hazardous wastes are transported onsite (i.e., within the boundaries of a facility and not on public roads), no manifest is required. [RO 13315] "Onsite" is actually a defined term under RCRA; it is "the same or geographically contiguous property which may be divided by public or private right-of-way, provided the entrance and exit between the properties is at a cross-roads intersection, and access is by crossing as opposed to going along the right-of-way. Noncontiguous properties owned by the same person but connected by a right-of-way which he controls and to which the public does not have access, is also considered onsite property." [§260.10] So, transfer of hazardous waste within the onsite confines of a facility (e.g., via vacuum truck or forklift) is not a situation that requires a manifest. A manifest is required only for offsite transportation. [§262.20(a)]

The definition of "onsite" noted above raises the question of hazardous waste shipments within a large facility that is crisscrossed by public roads. The above definition suggests that such a shipment must move only *across* a public road (i.e., the gates or driveways to the facility must be directly across the road from each other such that the shipment crosses the public road at a right angle). If the gates or driveways were located such that the shipment of hazardous waste had to travel along the public road, even for a few feet, it would need a manifest based on the definition.

Owners/operators of large facilities (e.g., military installations, universities, industrial complexes) had a real problem with this provision in the RCRA regulations. Acknowledging this problem, EPA made a change to the manifesting regulations in its February 1997 military munitions rule to provide flexibility for consolidation of wastes without having to comply with hazardous waste manifesting and transporter requirements. [62 *FR* 6622] (Note that even though this provision was added in the military munitions rule, it is applicable to all generators.)

Under §262.20(f), which was added as part of the 1997 rulemaking, EPA now allows transport of hazardous waste without a manifest within or along the border of contiguous property controlled by the same person, even if the shipment travels along a public road. Contiguous properties that touch corners or are diagonally across from one another are considered by EPA to be contiguous properties separated by a right-of-way. Accordingly, hazardous waste transportation between such portions of the facility are included under the manifesting exemption. [February 12, 1997; 62 *FR* 6648]

Q *A company owns three contiguous properties along a public right-of-way: a manufacturing facility that qualifies as a large quantity generator, an industrial park, and a permitted TSD facility. (The industrial park is situated between the other two facilities.) Is a manifest required when hazardous waste is transported from the manufacturing facility to the TSD site?*

A If all three properties are under control of the same person, a manifest does not need to accompany hazardous wastes shipped along the border of contiguous property, even if such property is divided by a public or private right-of-way. [§262.20(f), RO 14343]

EPA considers different federal agencies and different services to be different "persons." For example, hazardous wastes could not be transported between adjacent Army and Air Force bases without a manifest. [February 12, 1997; 62 *FR* 6647]

In addition to being exempt from the manifesting requirements, use of a transporter with an EPA ID number is not required for such onsite movement of hazardous waste because of this manifest exemption. [62 *FR* 6645] It also eliminates the need to mark containers with the hazardous waste label as specified in §262.32(b).

Even though onsite shipments do not require a manifest or hazardous waste transporter, EPA believes that generators and TSD facilities taking advantage of the exemption must still be prepared for emergencies. As such, any spill or discharge that occurs on the public or private right-of-way is subject to the existing transporter response standards in §§263.30 and 263.31. In addition, generators that take advantage of the exemption must still comply with applicable contingency plan and emergency response requirements set forth in §262.34(a)(4) for large quantity generators and §262.34(d)(4–5) for small quantity generators. They must "consider how the emergency coordinator is to be kept informed of waste movement activities under the new circumstances involving shipments on public roads without a manifest, and how an emergency on a public road within, between, or on the perimeter of contiguous properties is to be managed so that it minimizes exposure to local areas surrounding the property." [February 12, 1997; 62 FR 6646]

6.6.2.2.1 DOT requirements may still apply

"If a material is not subject to EPA's RCRA manifest requirements, it is not considered a 'hazardous waste' by DOT. However, such material is still regulated as a [DOT] 'hazardous material' and is subject to [the DOT hazardous materials regulations] if it meets the defining criteria for one or more of the DOT hazard classes." [February 12, 1997; 62 FR 6646] Therefore, shipments of hazardous waste on public right-of-ways that do not have to be manifested under the §262.20(f) exemption may still be subject to DOT shipping paper, packaging, labeling, marking, and placarding requirements. One exception to that requirement is for shipments that consist solely of DOT Hazard Class 9 materials in amounts less than their reportable quantities. [RO 14151]

6.6.2.3 Manifesting exemption for immediate responses

As discussed in more detail in Section 7.5.4, transportation of hazardous wastes during an immediate response situation is exempt from manifesting requirements. [RO 11363, 11370, 12016, 12748, 13574] See Case Study 6-5.

Case Study 6-5: Transportation of Explosives That Pose a Safety Threat

The U.S. Bureau of Alcohol, Tobacco, and Firearms (BATF) discovers a bomb during a routine investigation. Since the bomb is a hazardous waste due to its reactivity, would a manifest be required when transporting the bomb to a remote site for destruction?

No. RCRA regulations exempt all activities taken in immediate response to a discharge of hazardous waste or an imminent and substantial threat of a discharge of hazardous waste from the RCRA permitting and substantive requirements. [§§264.1(g)(8), 265.1(c)(11), and 270.1(c)(3)] In this situation, any BATF actions taken to eliminate the imminent and substantial danger posed by the bomb would qualify for this exemption. Accordingly, transportation of the bomb to a remote site for destruction would not require a manifest. "However, the transportation is exempt only to the extent necessary to respond to the immediate threat. Hence, we expect the transportation would normally cover a relatively short distance and would occur in special transportation equipment such as bomb trailers." [RO 13574]

6.6.2.4 Manifesting exemption for noncharacteristic ICR-only listed wastes

Per a May 16, 2001 rule [66 FR 27268], ICR-only listed wastes (i.e., those hazardous wastes listed only due to their ignitability, corrosivity, and/or reactivity) are not hazardous at their point of generation if they do not exhibit any characteristic at that point. (See Section 3.1.2.) If they are nonhazardous, they are exempt from the hazardous waste regulations, including manifesting requirements. However, some states have not adopted the May 2001 rule, because its provisions are less stringent than the pre-existing RCRA program. This situation makes the interstate transportation of ICR-only listed wastes complicated, as noted in the following example.

Q *A facility is located in a state that is authorized for the May 16, 2001 rule. The facility generates an F003 spent solvent, which is listed solely for ignitability. However, the waste does not exhibit any hazardous waste characteristics and is therefore managed as a nonhazardous waste. The waste will be shipped for disposal to a state that has not adopted the ICR-only listed waste revisions. How will the shipment and disposal of this waste be regulated?*

A It is the generator's responsibility to know the requirements of the states in which the waste will be managed. Although the waste is nonhazardous in the generator's state, it will be considered hazardous in any state that has not adopted the ICR-only listed waste revisions, including the receiving state. Thus, while traveling through these states, the shipment must comply with all applicable RCRA requirements. Among other things, this means that the transporter must have an EPA ID number and a hazardous waste manifest must accompany the shipment. In addition, the disposal facility will be subject to the standards of the state where it is located. So, in this case, the disposal facility would have to manage the waste as hazardous. [RO 14716] In these situations, the originating facility should complete a manifest and send it to the transporter who will be carrying the waste through the first state that considers the waste to be hazardous. A good practice is for the originating facility to note in Block 14 of the manifest that the waste is not regulated as hazardous in the originating state but is in the receiving facility's state. The receiving facility would then sign the manifest and send it back to the originating facility.

6.6.2.5 Manifesting exemption for wastes shipped by pipeline

Transportation means "the movement of hazardous waste by air, rail, highway, or water." [§260.10] What about the movement of hazardous waste by pipeline? Is transportation by pipeline not allowed because it's not covered in the definition?

In April 1986, EPA clarified that the transport of waste by pipeline is allowed; however, the manifesting provisions of RCRA do not apply because only "transportation" requires a manifest:

"The fact that the definition of transportation does not include pipeline as a mode for the transportation of hazardous waste does not mean that hazardous wastes via pipeline [sic] is not acceptable under RCRA. Rather, because the definition of transportation does not include pipeline transport, the provisions of 40 *CFR* [Part] 263 do not apply...." [RO 11148]

The agency went on to say that while a manifest is not needed, other RCRA and CERCLA provisions, such as spill response and reporting requirements, can continue to apply.

6.6.3 Filling out the manifest

The Uniform Hazardous Waste Manifest contains information on the type and quantity of waste being transported, instructions for handling the waste, and signature lines for all parties involved in the transport process. The manifest appears in the Appendix to Part 262 (EPA Form 8700-22), as do instructions for completing it. Helpful hints on filling out certain blocks of this document are provided below.

6.6.3.1 Emergency response information

To aid responders in an emergency, an emergency response phone number must be identified clearly in Block 3 of the manifest for each waste shipment. This RCRA requirement stems from the DOT hazmat regulations at 49 *CFR* 172.604. This number must:

1. Be the number of the generator or an agency or organization who is capable of, and accepts responsibility for, providing detailed information about the shipment;

2. Reach a phone that is monitored at all times while the waste is in transportation, which includes transportation-related storage; and

3. Reach someone who is either knowledgeable of the hazardous waste being shipped and has comprehensive emergency response and spill cleanup/incident mitigation information for the material being shipped or has immediate access to a person who has knowledge and information about the shipment.

Most hazardous waste transport companies provide an emergency response number that can be used at the time of waste shipment. There are also a number of commercial emergency response companies that operate call centers in the event of a hazardous materials incident. If a commercial call center is used, however, their services must be paid for in advance (either a fee for a single shipment or a membership that can cover all shipments).

Q *If more than one emergency response telephone number is required because of the different wastes included on the manifest, how should these numbers be added?*

A The applicable telephone numbers should appear immediately following the shipping descriptions in each line of Block 9. [70 *FR* 10783]

On October 19, 2009 [74 *FR* 53413], the Department of Transportation published a final rule affecting shipping papers (e.g., bills of lading, manifests) in an effort to improve the effectiveness of arrangements between shippers and emergency response information services. This rule, which became effective October 1, 2010, amended the hazmat regulations primarily in two ways:

1. If a shipper has arranged with a service (e.g., CHEMTREC) to provide emergency response information, the shipper's name or contract number must be entered on the shipping paper with the emergency response information telephone number. The information must be located immediately before, after, above, or below the emergency response telephone number and be entered in a clearly visible manner that allows the information to be easily and quickly found. [49 *CFR* 172.604(b)] This information will enable the emergency response information service provider to identify the shipper on whose behalf it is accepting responsibility for providing emergency response information in the event of an incident and to obtain additional information about the hazardous material as needed.

Furthermore, subsequent entities in the transportation chain (e.g., freight forwarders) that prepare new shipping papers for the continued movement of a hazardous material are also responsible for satisfying the requirement mentioned above. [49 *CFR* 172.604(c)] If the emergency response information (i.e., phone number and corresponding name or contract number) from the original shipping paper will be used on any subsequent shipping papers, the new offeror must ensure that the name or the contract number is provided in association with the emergency response information telephone number. [74 *FR* 53418]

2. In those situations where the emergency response information telephone number service provider is located outside the United States, the international access code or the "+" (plus) sign, country code, and city code, as appropriate, must be included as part of the telephone number. [49 *CFR* 172.604(a)]

6.6.3.2 Generator's name and EPA ID number

Generators must enter their 12-digit EPA ID number in Block 1 of the manifest. If the generator does not have an EPA ID number, he/she should enter their state generator identification number. EPA ID numbers are discussed in Section 6.9.

The generator's name, mailing address, and telephone number are entered in Block 5 on the manifest. The mailing address to be entered is the one to which the designated facility should send the signed copy of the manifest when it receives the waste. The telephone number should be the normal business number for the generator, or the number where the generator may be reached to provide instructions in the event the designated facility rejects some or all of the shipment. There is also space in Block 5 to enter the address from which the shipment originates if this address is different than the mailing address.

If a facility hires a contractor to perform some work at the site that results in generation of a hazardous waste, whose name goes on the manifest as the generator? EPA has stated in guidance a number of times that both parties in this situation are considered cogenerators. Guidance on how to fill out the manifest when cogenerators are involved is discussed in Section 6.5.2.

6.6.3.3 Transporters' information

The hazardous waste transporter(s) involved with a shipment of hazardous waste must be identified on the manifest. The transporter's company name and EPA ID number must appear in Block 6 (and Block 7 if more than one transporter is used). A continuation sheet may be used to identify additional transporters as necessary. [RO 11242, 13781]

EPA defines "transporter" as "a person engaged in the offsite transportation of hazardous waste by air, rail, highway, or water." [§260.10] However, the term refers to the entire company involved in the transport of hazardous waste, not individuals within the company. So, for example, if a new driver from the same transportation company takes over from the first driver who actually picked up the load from the generator, that person does not also have to sign the manifest. [RO 14172] Similarly, the EPA ID number for a transporter is assigned to the transportation company as a whole, and all of the individual trucks in a given company use the same EPA ID number. [RO 14134] An interesting situation in which two transporters jointly move a hazardous waste shipment is discussed in Case Study 6-6.

If a transporter is unable to deliver a manifested shipment of hazardous waste to the designated facility, he/she must contact the generator for instructions and revise the manifest to reflect the approved changes to the prescribed chain of transport. [§263.21(b)] "Generators alone are responsible for identification of the complete chain of transportation and must, therefore, be apprised of and approve of all deviations from that plan." Transporters cannot make unapproved changes to the chain of transportation outlined on the manifest. [RO 13781]

6.6.3.3.1 *Transfer facilities*

Manifested shipments of hazardous wastes may be stored at a transfer facility for up to 10 days without requiring the facility to have a RCRA storage permit. A transfer facility includes "loading docks, parking areas, storage areas, and other similar areas where shipments of hazardous waste are held during the normal course of transportation." [§260.10] Use of more than one transfer facility during a single hazardous waste shipment could be considered incidental to normal transportation practices and, therefore, not a problem. However, EPA suggested three situations that ordinarily would be considered inconsistent with the normal course of transportation: 1) the waste is not transported from a site at all, but, rather, possession of the waste changes from one transporter to another while the waste remains at the site; 2) the waste is routed to the same geographic location more than once during the course of transportation; or 3) the waste is simply routed to numerous transporters for extended periods of time. [RO 11520, 11846]

The agency noted at the time of promulgation of the transfer facility requirements (December 31, 1980) that the transportation industry had indicated that shipments of hazardous waste normally take no longer than 15 days, including both on-the-road time and incidental temporary holding. While circumstances may occasionally justify periods significantly longer than 15 days, shipments lasting for long periods (e.g., 49 days in one case) strongly suggest that the intermediate transfer facilities were not holding the waste incidental to the normal course of transportation. However, determinations of what activities are consistent with the normal course of transportation must be made on a case-by-case basis. [RO 11520, 11846]

Case Study 6-6: Two Transporters Jointly Moving a Shipment

A shipment of hazardous waste is sent offsite to a commercial TSD facility via rail. However, the rail car in which the waste is transported is owned by one company and the train locomotive is owned by another. Which company should be listed as the transporter on the manifest?

Based on the regulatory definition of "transporter" in §260.10, when the owner/operator of the means of locomotion (e.g., locomotive, tractor) is different from the company that owns/operates the cargo-containing transport vehicle (e.g., rail car, truck trailer), EPA believes that both parties are potentially subject to the RCRA transporter requirements. The agency reasons that both parties are engaged in offsite transportation and qualify as transporters, because there could not be any movement of hazardous waste without the joint efforts of the companies providing the means of locomotion and the cargo-carrying unit(s).

In this situation, both companies could be identified in the separate transporter blocks when filling out the manifest. However, EPA believes that it would be better to identify on the manifest only the transporter company that is primarily responsible for movement of the waste. Typically, the primarily responsible company is the "employer of the individual who actually performs the positive acts required of transporters under the Part 263 regulations. These acts include signing and dating the manifest when received from the generator, returning a signed copy of the manifest to the generator, obtaining the date of delivery and handwritten signature of the representative of the facility or next transporter to which the waste is delivered, giving the remaining copies of the manifest to the facility or next transporter accepting the shipment, and retaining a copy of the manifest for the transporter company's files."

"EPA considers that the performance of these required acts connotes the custody, control, and responsibility for the movement of the waste, so that the employer of the individual performing these acts should generally be the company identified as the transporter on the manifest. If in particular instances, another company has agreed to assume responsibility for compliance with the transporter standards of Part 263, the other company may be identified as the transporter on the manifest, and the individual signing the manifest should do so in a manner that indicates that he is signing as agent for the other company. EPA will generally look first to this primarily responsible transporter for information about the movement of the waste, and for compliance with the Part 263 requirements." [RO 14134]

Under the federal requirements, when a transporter ships hazardous waste to and from a transfer facility that it owns/operates and the waste remains under the control of the transporter, the transfer facility does not need to be identified in the transporter blocks of the manifest (Block 6 or 7) or sign a transporter acknowledgment of receipt of materials (Block 17). In this situation, there has not been any change in custody of the waste. However, if the transfer facility accepts custody and control of a hazardous waste shipment from a separate transporter company that delivered the shipment to the transfer facility, the transfer facility must be identified in the next transporter block and sign a transporter's acknowledgment of receipt. [RO 11953]

Transporters routinely transfer shipments of hazardous waste from one vehicle to another at transfer facilities for redirecting or rerouting. For instance, a truck may arrive at a transfer facility carrying 50 drums of waste accompanied by a single manifest. The transporter then transfers half of the drums to another truck. Can the load covered under a single manifest be divided between two separately manifested shipments?

According to EPA, "[w]hen a shipment appearing on a single manifest must be divided by the transporter and split between outbound shipments, the transporter must obtain consent from the generator to amend the original manifest to show the correct number of drums and waste quantities and descriptions that will be placed on the first truck and complete a second manifest to indicate the number of remaining drums and corresponding information describing the wastes that will be placed on the second truck." The manifest document number of the second manifest should be noted on the first manifest in the discrepancy space (Block 18) so that the generator can reconcile both manifests with the original manifest retained by the generator. [RO 14279]

Hazardous wastes are sometimes transferred from one container into another or consolidated into larger containers at transfer facilities. For example, hazardous soil from remediation projects is often trucked from hazardous waste sites to rail sidings for rail delivery to TSD facilities. [December 6, 1995; 60 *FR* 62528] Or a rail car load of hazardous waste could be unloaded into trucks for final delivery. See Case Study 6-7.

Transporters who mix hazardous wastes at transfer facilities (e.g., consolidating drums of waste into a tank truck) may have to comply with the Part 262 generator standards. One such requirement is to ensure that the manifest accurately reflects the composition and volume of the waste mixture and reflects the type and number of containers. [§263.10(c)(2)] If hazardous wastes with different DOT shipping descriptions are mixed, a new manifest is required because the proper shipping name or hazard class on at least one of the original manifests is no longer accurate. For example, a new manifest would be necessary when two shipments of RCRA ignitable wastes (one a DOT combustible material and one a DOT flammable material) are mixed. The transporter must represent itself as the generator of the new waste mixture on the new manifest and indicate the names of the original generators in Block 14. The original manifests would then have to be attached to, and conveyed with, the new manifest. [RO 11567, 12458, 13272] EPA recommends that original manifest numbers also be referenced on any new manifests. [RO 14137]

If only one or two minor changes occur as a result of mixing (e.g., container changes), the original manifest could be marked to reflect the changes. This might occur, for instance, when drums containing wastes with the same DOT shipping description are consolidated into a tank truck at a transfer facility for shipment to a TSD facility. Combining different RCRA waste streams that are both classified by DOT as "hazardous waste solid, n.o.s." would also not require a new manifest. [RO 11567, 12087, 12458]

In any case, the waste may only be delivered to the designated facility indicated on the original manifest by the original generator. If the transporter consolidates wastes, requiring creation of a new

---

Case Study 6-7: Transfers From Rail Cars to Tank Trucks

A rail car containing 15,000 gallons of hazardous waste is loaded at a generator's facility. The car reaches a transfer facility, where the contents are transferred into three tank trucks (each carrying 5,000 gallons) for transport to the designated TSD facility. How should the manifest be filled out?

In order to reflect both the total quantity (15,000 gallons) and the individual tank truck quantities (5,000 gallons/truck), three line items of 5,000 gallons each can be noted on the manifest. Each tank truck driver would then receive a copy of the manifest with two of the entries lined out. This will give each truck driver a manifest that accurately reflects his/her load plus the total amount shipped. [RO 12141, 13050] Although not noted in the guidance, a notation in Block 14 regarding division of the shipment would be appropriate.

manifest, he/she cannot send them to a designated facility that was not indicated on the original manifest by the original generators. [RO 11567] EPA recommends that the original generators maintain control of the disposition of their hazardous wastes by requiring that transporters obtain generator consent before commingling occurs. [RO 14137]

Note that, even though transporters who mix different hazardous wastes at transfer facilities are subject to generator standards, mixing does not generate a new waste and the transporters are *not* considered generators. As such, the transporters cannot make use of the 90/180/270-day accumulation provisions associated with generating a waste but must comply with the 10-day clock associated with the transfer facility regulations in §263.12. [RO 13272]

6.6.3.4 Designated facilities' information

Under §262.20(b), generators must designate in Block 8 of the manifest a facility that is authorized to handle the waste described on the manifest (the designated facility). "Designated facility" is defined in §260.10 as a facility that is designated on the manifest and 1) has a RCRA permit or is operating under interim status in accordance with Part 270, 2) has a permit or is operating under interim status from an authorized state, or 3) is regulated under §261.6(c)(2) or Part 266, Subpart F (recycling activities at such facilities are exempt from RCRA permitting requirements). This means that a generator may list a recycling facility that is exempt from RCRA permitting as the designated facility on a manifest. [RO 13663]

In a rule published March 4, 2005 [70 *FR* 10776], EPA amended the definition of "designated facility" to include a generator site designated on the manifest to receive its waste as a return shipment from a facility that has rejected the waste in accordance with §264.72(f) or 265.72(f).

A generator cannot manifest its hazardous waste to another facility owned by the same company that happens to have extra 90-day tank or container storage capacity. Unless RCRA-permitted, the second facility does not meet the definition of "designated facility." [RO 14693]

The owner/operator of the designated facility, if permitted, must send written notice to the generator that it has the appropriate permits and will accept the hazardous waste that the generator is shipping. [§264.12(b)] A generator's hazardous waste shipment is often managed by more than one TSD facility; Case Study 6-8 explores the manifesting issues associated with this common situation.

A facility that consists of permit-exempt wastewater treatment units (WWTUs—as defined in §260.10) may qualify as a designated facility and receive manifested shipments of hazardous waste. EPA has determined that a facility "which has received a permit (or interim status)," as included in the definition of "designated facility," can refer to permit-exempt WWTUs. Therefore, such a facility can be considered a designated facility under the federal requirements. [RO 14206]

In some situations, a hazardous waste in one state is shipped to another state that does not regulate that waste as hazardous. This may occur with regard to newly listed wastes—states will begin regulating these wastes as hazardous under different schedules. EPA has stipulated that the manifest requirements and the definition of "designated facility" do not prohibit the shipment of hazardous waste from a state where the waste is hazardous to a state where the waste is not hazardous. In such situations, the receiving facility would not be required to have a RCRA permit or interim status as long as the state allows it to accept the waste. EPA recognizes that such a facility can be listed as the designated facility without violating state or federal law. The generator in the shipping state must require that the designated facility sign and return the manifest. [§262.23(e), RO 13297, 13500]

6.6.3.5 DOT shipping description

Blocks 9a and 9b are for generators to enter the DOT shipping description. This is usually the most challenging part of completing the hazardous waste

Case Study 6-8: Shipments Between TSD Facilities

A permitted TSD facility receives a shipment of hazardous waste from a generator. The TSD facility stores and partially treats the waste and then sends it offsite for final treatment and disposal. Must the owner/operator of the initial TSD facility fill out a new manifest or should the original manifest accompany the second shipment?

Owners/operators of TSD facilities that initiate shipments of hazardous wastes must comply with the generator requirements in Part 262 [per §§264/265.71(c) and 268.7(b)(5)]. Therefore, in this situation, the initial TSD facility must prepare a new manifest listing itself as the generator of the waste or treatment residue and comply with the exception reporting requirements in §262.42. After receiving the waste or treatment residue and signing the manifest, the second TSD facility must send a signed copy of the manifest back to the initial TSD facility. [RO 12287, 12539, 14408]

Note that the original generator in this process will have received a signed manifest from the initial TSD facility and, essentially, its RCRA obligations are fulfilled once the original manifest has been closed out. However, EPA notes that it would be a "prudent practice" for the original generator to request final, signed copies of the manifests used between the TSD facilities because "the generator retains potential liability under Superfund for future mismanagement of hazardous waste even after it has left his site and is out of his possession." [RO 11589]

---

manifest. For guidance on how to choose the proper DOT shipping name, generators must receive training on the DOT regulations in 49 *CFR* Parts 171–173.

Block 9a is commonly referred to as the hazardous materials column. It comes from the DOT shipping paper regulations found in 49 *CFR* 172.201(a)(1)(iii). Block 9a can be used when a DOT hazardous material is being shipped with a material not subject to the DOT hazardous materials regulations. The shipper would place an "X" in Block 9a preceding the proper shipping description of the DOT hazardous material found in Block 9b. For a material not subject to the DOT hazardous materials regulations, the corresponding box in Block 9a would be left blank. This is to aid emergency responders in the event of a hazardous material incident. The appearance of an "X" in Block 9a would inform an emergency responder that the material on that line is hazardous and requires special attention. If everything being shipped on the manifest is a hazardous material, completion of Block 9a is not necessary. Note that the "X" in the hazardous materials column may be replaced by "RQ" if the hazardous material is also a hazardous substance. Additional information and options can be found in 49 *CFR* 172.201(a)(1).

Will a RCRA hazardous waste be a DOT hazardous material? Yes. All wastes that are subject to the manifest requirements specified in Part 262 are DOT hazardous materials.

Block 9b is where the generator enters the shipping description for the hazardous waste. The generator is to select a proper shipping name from the Hazardous Materials Table (49 *CFR* 172.101) that most accurately describes the waste. Additionally, the DOT regulations require the word "waste" to precede the shipping name for a hazardous waste, unless the word "waste" is already in the shipping name. Four proper shipping names could be:

1. Waste Paint
2. Waste Cyclobutyl chloroformate
3. Hazardous waste, solid, n.o.s.
4. Waste Flammable liquids, toxic, n.o.s.

The initials "n.o.s." are used by DOT to represent "not otherwise specified."

Once the shipping name and corresponding hazard class(es), identification number, and packing group (if applicable) are selected from the Hazardous Materials Table, the generator enters the shipping information into Block 9b in the following sequence [49 *CFR* 172.202]:

1. Identification number,
2. Proper shipping name,
3. Hazard class or division number,
4. Subsidiary hazard class(es) or division number(s) in parentheses (if applicable), and
5. Packing group (if applicable).

Additional information such as technical name(s) and reportable quantities are to be included, if applicable. [49 *CFR* 172.203] The completed shipping descriptions would look like the following:

1. UN1263, Waste Paint, 3, II (RQ–D001)
2. UN2744, Waste Cyclobutyl chloroformate, 6.1 (8, 3), II
3. NA3077, Hazardous waste, solid, n.o.s., 9, III (D006, D008)
4. UN1992, Waste Flammable liquids, toxic, n.o.s., 3 (6.1), II (benzene, chloroform) (RQ–D018)

For a more thorough discussion of the DOT hazardous materials shipping paper requirements, consult 49 *CFR* Part 172, Subpart C (Shipping Papers).

When itemizing wastes on a manifest, different hazardous materials will need to be listed on separate lines. The following situations will typically require wastes to be placed on separate lines:

- Different proper shipping names;
- Different technical name/constituents;
- Different phases (liquid vs. solid);
- RQ is present in one container, but not the other;
- Different types of containers are used (e.g., DM vs. DF); and
- Different waste codes apply.

In summary, when multiple containers are being shipped, containers of the same type holding the same type of waste can be itemized together on the same line of the manifest. Any variations typically will warrant identification on separate lines of the manifest.

6.6.3.6 Container types

Block 10 of the manifest is for number and type of containers. The generator is to identify the number of containers that hold the waste identified in Block 9b. Container types are identified with two-digit capital letter abbreviations provided in the manifest instructions.

If a generator is shipping multiple containers of a single waste stream, the wastes must be packaged in the same type of container to be on one line of the manifest. For example, if a generator is shipping a plastic drum of spent solvent (container type is abbreviated DF) and a metal drum of the same type of spent solvent (abbreviated DM), the containers would need to be listed separately on the manifest (Block 9b1 and 9b2).

What if a generator has two containers of the same type but of different sizes, such as a metal 30-gallon drum of solvent and a metal 55-gallon drum of solvent? As long as the shipping descriptions and container types match, the two containers can be listed together on the same line on the manifest. Thus, "2 DM" would be found in Block 10 of the manifest.

If a generator is shipping 10 fiberboard drums shrink-wrapped on a pallet, how should this shipment be identified in Block 10 of the manifest? While there is no guidance from EPA, DOT requires the number and type of packages be identified on a shipping paper. In summary, a "package" is a receptacle that is performing the function of containing a hazardous material. The drums, not the pallet, are performing the function of containment. Thus, "10 DF" would be found in Block 10 of the manifest.

6.6.3.7 Weight designation

The generator enters the total quantity of waste in Block 11. Quantities on the manifest are to be reported as accurately as possible without using fractions or decimals. EPA understands that a scale is not always available, so estimated weights are acceptable and "EST" at the top of Block 11 is not needed. Container capacities cannot be used as the quantity shipped when it is known that a container is not filled to capacity. It is customary to use gross weight (waste weight plus container), since transporters usually charge on the basis of total weight shipped. [RO 12151]

6.6.3.8 Units of measurement

The generator must identify the units of measurement in Block 12. These are listed as single-digit capital letters provided in the manifest instructions. The generator must select a unit of measure that will accurately represent the waste in whole units in Block 11. Large units of measure (such as tons, cubic yards, and cubic meters) typically do not allow for precision and should be reserved for very large bulk containers, such as rail cars or barges. [70 *FR* 10791] If using the abbreviations "G" (for gallons) and "L" (for liters), the waste must be in liquid form. However, other abbreviations can also be used for liquids, such as "P" (for pounds).

6.6.3.9 Waste codes

Hazardous waste codes must be entered on the manifest in Block 13. For each waste listed in Block 9, up to six waste codes must be entered.

Recognizing that most waste streams carry six or fewer waste codes, EPA believes that six waste code fields in Block 13 would be adequate to describe typical wastes shipped on a manifest. The agency originally proposed a toxicological hierarchy (i.e., order) for entering waste codes, with the intent of the hierarchy to quickly alert manifest users (such as TSD facilities) and emergency responders to the relative hazards posed by each waste. However, EPA abandoned the proposed hierarchy because its use could miscommunicate hazards. Rather, the agency opted to let generators use their discretion as far as what order to list waste codes on the manifest. Also, TSD facilities or state regulatory agencies might have a preference.

If a waste carries more than six waste codes, the additional codes are not required on the manifest. [70 *FR* 10788] A generator is allowed to list additional waste codes in Block 14 (Special Handling Instructions and Additional Information), but their inclusion in Block 14 is not mandatory.

The instructions for Block 13 (in the Appendix to Part 262) require federal waste codes to be entered which are most representative of the properties of the waste. This general requirement allows generators to list waste codes they believe best describe the hazard(s) of their waste. EPA also requires state waste codes to be included on the manifest. However, if a state waste code is redundant with a federal waste code, the inclusion of the state code in Block 13 is not required. For example, some states have a state waste code that represents a corrosive liquid waste whose pH is less than or equal to 2 or greater than or equal to 12.5. If the generator's waste is an acid receiving D002 for corrosivity, the corresponding state waste code is redundant, warranting only D002 in Block 13. Some states have adopted additional P-codes for materials such as chemical warfare agents destined for disposal, or F-codes for the treatment residues from the destruction of such agents. Because there are no federal counterparts, these state waste codes would not be considered redundant and would be required in Block 13.

Although not clearly stated in the regulations, the generator and destination state waste codes are required in Block 13 (if not redundant with federal codes). [May 22, 2001; 66 *FR* 28258] Most state waste codes are three or four characters in length and easily fit into a field in Block 13. However, the 8-digit Texas state waste codes seemed to pose a problem. Texas generators have been authorized by EPA to use two fields for the applicable 8-digit state waste code.

In summary, if a state waste code has no corresponding federal waste code, or if the state waste code conveys additional information not expressed by the federal waste code, it must be included in Block 13. Because there are a wide variety of state waste codes, additional guidance from the generator and/or destination state might be required.

6.6.3.10 Special handling instructions and additional information

This block has many uses and may contain any additional information that has no home elsewhere on the manifest. The following information may be added to Block 14, Special Handling Instructions and Additional Information:

- Waste concentration, specific gravity, and pH data;
- Identification of an alternate designated facility;
- Instructions to return the shipment to the generator if any or all of it is rejected by the designated facility;
- Identification of waste product or handling codes or other numbers assigned by the designated facility;
- Identification of cogenerators (entities who share the generator responsibilities with the generator noted in Block 5);
- Indications that weight of container was not included in determining generator status [RO 14827];
- Indications by transporters that they have combined or divided loads at their transfer facilities and documentation of new or combined manifests;
- Identification that the shipment is universal waste covered under those regulations;
- Appropriate cross-references to the *Emergency Response Guidebook* or other emergency response information;
- Waste profile numbers;
- Bar codes or other container codes; and
- Specification of PCB waste descriptions and PCB out-of-service dates required under 40 *CFR* 761.207.

The use of Block 14 is limited primarily to waste handlers to record their site-specific or shipment-specific information. [70 *FR* 10781] Generators cannot be required to enter information in this space to meet state regulatory requirements.

6.6.3.11 Generator's/offeror's certification

After a manifest is filled out, the generator must sign it in Block 15, certifying that the manifest is accurate and that the shipment has been prepared according to appropriate EPA and DOT regulations. Unlike a generator's biennial reports, the generator's manifest certification does not need to be signed by an "authorized representative." An "authorized representative" is defined in §260.10 as "the person responsible for the overall operation of a facility or an operational unit (i.e., part of a facility), e.g., the plant manager, superintendent or person of equivalent responsibility." Instead, according to EPA, "the person signing the manifest certification should have first-hand knowledge of the information listed on the manifest." [RO 11199] EPA has noted in recent guidance that, in addition to RCRA training, persons signing manifests must have DOT training for hazardous materials employees specified in 49 *CFR* Part 172, Subpart H. Also, while "it is more typical for the company represented by the signor to be the subject of enforcement actions, in some situations (e.g., egregious or criminal violations), the signor could be held personally liable." [RO 14687]

The instructions for completing the certification in the appendix to Part 262 specifically allow a manifest to be signed by someone "on behalf of" the generator. Accordingly, the generator may authorize someone (including a contractor) to sign the certification. In such a situation, the generator and the contractor should establish the means to ensure that the contractor is properly authorized to sign on behalf of the generator. [RO 11372] However, it is the generator's responsibility to ensure that the information is correct. As such, the generator will always be ultimately responsible for the preparation of manifests. [RO 11030]

The certification also includes a reference to §262.27, which says that the generator has a waste minimization program in place. The regulatory wording is slightly different for large quantity versus small quantity generators. A large quantity generator must certify that it has

1) a program in place to reduce the volume and toxicity of waste to the economically practicable degree; and 2) selected the practicable method of treatment, storage, or disposal currently available that minimizes the present and future threat to human health and the environment. A small quantity generator must certify that it has 1) made a good-faith effort to minimize waste generation, and 2) selected the best available and affordable waste management method. [§262.27, RO 12767] EPA has clarified that the certification applies to generators of hazardous waste both at manufacturing plants and in remediation settings. [RO 11855] See Section 6.3.9 for a more complete discussion of waste minimization program requirements.

The wording of the waste minimization certification is in the first person (i.e., "I have a program in place to reduce the volume and toxicity of waste generated…"). Due to the way it is written, some employees may not feel comfortable signing the manifest since they, as individuals, do not personally have a waste minimization program in place. To solve this potential dilemma, EPA recommends that the employee sign the statement, then write the phrase "on behalf of [the company's name]" below the signature line (e.g., Harry Cooper on behalf of ABC Chemical Co.). This phrase can even be preprinted on manifest forms for convenience. [RO 11090] Alternatively, employees can fill in the company's name followed by the word "by" and then their signature (e.g., ACME Fireworks by John Smith). In this case, the company's name could also be preprinted on the manifests. [RO 11108]

Who is an "offeror" and when would they sign the manifest? "Offeror" is a term used in DOT's hazardous material regulations. An "offeror" is a person who performs, or is responsible for performing, any pretransportation functions required under the DOT hazardous material regulations, or a person who tenders or makes a hazardous material available to a carrier for transportation in commerce. "Pretransportation functions" include activities such as determining a material's hazard class, selecting a packaging, marking and labeling a package, filling a hazardous materials package, preparing a hazardous material shipping paper (including a hazardous waste manifest), providing emergency response information, and certifying that a hazardous material is in proper condition for transportation in conformance with the DOT hazardous material regulations.

It is not uncommon for TSD facilities to reject waste shipments either back to the generator or to an alternate facility that can better manage the waste. In many instances of waste rejection, a new manifest would need to be issued to accompany the shipment. The new manifest would require a generator signature in Block 15. However, the rejecting TSD facility was not the original generator of the waste. EPA believed the rejecting facility should not have to bear the liability of being the generator of record for that hazardous waste. To limit the liability of rejecting facilities, EPA broadened Block 15 to include offerors shipping hazardous materials. As an offeror, the rejecting facility would be responsible for the new movement of the waste, but would not be subject to the full range of generator requirements/responsibilities.

EPA was also aware that in some situations, the transporter making a waste pick-up helps prepare the waste for shipment. In these circumstances, the transporter may be more involved in the pretransportation functions (such as selecting packages, labeling containers, or completing the manifest) than the generator. In these situations, the transporter may sign the manifest. While a generator may certainly complete Block 15 in its capacity as the generator, another person, such as a transporter making a waste pick-up and helping with the pretransportation functions, may sign the certification statement on the manifest in their capacity as an offeror. [70 FR 10793]

Must the date of the generator's/offeror's signature match the date of the transporter's acceptance signature? EPA requires the generator/offeror to read and sign the certification at the initiation of each waste shipment, and the transporter is to sign the manifest and record the date of acceptance. [March 20, 1984; 49 *FR* 10498] While these instructions are somewhat vague, EPA has clarified in guidance that the generator's certification on a manifest does not necessarily have to be the date of shipment. [RO 12204] DOT has also noted that the use of different shipper/generator and initial carrier dates on the manifest is not prohibited. [Letter from Delmer F. Billings, Office of Hazardous Materials Standards to Daniel G. Fox, June 2, 2000]

6.6.3.12 Discrepancies

When the receiving facility accepts an offsite shipment of hazardous waste, the owner/operator must note in Block 18 of the manifest any significant discrepancies between the quantity or type of hazardous waste received versus that noted on the manifest.

Manifest discrepancies, as defined in §§264/265.72(a), that warrant completion of Block 18 are:

- Wrong type of waste—for example, manifested as ignitable, but found to be corrosive or manifested as solid, but found to be liquid.
- Wrong number of containers—for example, 25 drums manifested, but only 23 drums received.
- For bulk waste, which is defined as a container whose capacity is greater than 119 gallons, a variation greater than 10% in weight—for example, a tanker manifested as 25,000 pounds, but 29,000 pounds received.
- Rejected wastes, which may be a full or partial shipment of hazardous waste a TSD facility cannot accept.
- Container residues, which are residues that exceed the quantity limits for "empty" containers set forth in §261.7(b)—for example, a tanker heel that has congealed.

According to EPA, waste analyses do not have to be performed immediately after a waste is received in order to identify discrepancies. Only obvious discrepancies that can be immediately determined by counting or measuring the waste and comparing the manifest with waste labels need to be noted on the manifest. [OSWER Directive 9523.00-10]

After discovering a significant discrepancy, the TSD facility owner/operator must try to reconcile the discrepancy with the waste generator or transporter. If the discrepancy is not resolved within 15 days after receiving the waste, the TSD facility owner/operator must submit a letter to EPA describing the discrepancy and the attempts to reconcile it, along with a copy of the manifest.

Q *What should a transporter or TSD facility operator do if he/she knows that a generator has incorrectly identified a waste on a manifest?*

A It is the generator's responsibility to determine if his/her waste meets the listing descriptions for hazardous wastes or if it exhibits a hazardous characteristic. If the owner/operator of a transportation firm or waste treatment facility determines that a listed waste had been improperly labeled as a characteristic waste by a generator, then the generator should be advised to correct the error on the manifest or receipt of wastes from the generator should be refused. Given the requirements of the land disposal restrictions regulations, both the generator and the treatment facility could face penalties for improperly analyzing and managing a listed hazardous waste. [RO 11375]

For generators who receive a signed copy of a manifest from the TSD facility with a discrepancy noted in Block 18, we suggest that you document how that discrepancy was resolved between you and the TSD facility and attach that documentation to the signed copy of the manifest that you put in your files. Then, in two years, when a state inspector going through your old records asks you about it, the documentation will be there to refresh your memory.

Q *A transporter delivers a shipment of hazardous waste to the designated TSD facility. The manifest describes the shipment as 50 drums weighing 450 lb each. The owner of the TSD facility notes* *that all 50 drums are present, but that each drum weighs only 430 lb. Does this qualify as a significant discrepancy that must be reconciled?*

A This would not trigger the discrepancy requirements under the federal regulations. Because these containers have a capacity less than or equal to 119 gallons (commonly referred to as nonbulk), a significant discrepancy will exist only if the wrong number of drums arrived or if the wrong type of waste arrived. The manifest could be corrected to reflect the true shipment weight (which is a difference of 1,000 pounds), but it would not be federally required. It is important to note, however, that some states may require a discrepancy report to be filed in this situation. [RO 12191]

Q *A manufacturing facility sends a corrosive D002 hazardous waste to a permitted treatment facility. The manifest indicates the corrosive characteristic of the waste. However, by the time the waste gets to the treatment facility, the pH of the waste has changed so that it no longer exhibits the corrosivity characteristic. Should the treatment facility file a discrepancy report?*

A No. According to EPA, "[w]e believe that the manifest discrepancy regulation was intended to cover those situations where there is in fact a quantity of hazardous waste that is unaccounted for at the time of receipt…. [T]here is not really a deviation in the waste quantity…. Thus, the issue is really one of waste characterization, and not an issue of accountability for the waste quantities shipped and received…. The manifest was not intended to act as a certification that all shipped materials are indeed hazardous wastes. The regulations allow a generator to characterize its waste based on process knowledge, and it is understood that generators may at times characterize their wastes conservatively, rather than incur the costs of testing every batch or stream." In conclusion, EPA believes that the manifest discrepancy requirements should not apply in this situation. However, states may take a more stringent stance. [RO 11918]

While there may be situations in which the manifest must be corrected to accurately reflect the shipment, other than those situations identified above, completion of Block 18 is not required. For example, if it is discovered the facility identification number is incorrect in Block 1 on a manifest, it should be corrected, but it would not be considered a discrepancy under §§264/265.72(a). The receiving facility could simply line through the incorrect information, replacing it with accurate information.

If a TSD facility receives a shipment with a manifest that does not contain the waste minimization certification, this does not qualify as a significant manifest discrepancy. As such, the TSD facility would not need to attempt to reconcile the mistake or notify EPA. [RO 12451]

What if a discrepancy is found on a continuation sheet? Is the discrepancy identified in Block 35 of the manifest continuation sheet or Block 18 of the first manifest page? For discrepancies found on manifest continuation sheets, Block 18 of the first manifest page would be used. The discrepancy field on the manifest continuation sheet (Block 35) is additional space to more fully describe information on discrepancies identified in Block 18a.

6.6.3.12.1 Standardized procedures for rejected loads

The regulations contain discrepancy procedures for 1) hazardous waste residues that TSD facilities are unable to remove from containers when those non-RCRA-empty containers will be shipped offsite, and 2) hazardous waste shipments that are partially or fully rejected by TSD facilities. In both cases, the TSD facility must contact the generator for a decision about where to forward the waste.

The initial TSD facility can use the original manifest to forward a rejected *full* shipment to an alternate facility or back to the original generator, provided that the transporter attempting the delivery to the initial facility is still at the facility at the time of the rejection. In these limited circumstances, EPA considers that the rejected waste shipment is continuing in transportation. The initial TSD facility would note the discrepancy in Block 18a, enter the alternate facility or original generator information in Block 18b, and sign the original manifest in Block 20. The TSD facility must retain a copy for their records. If there aren't enough copies of the original manifest left to give to the transporter, the TSD facility must use photocopies. [70 FR 10806] Upon receipt, the alternate facility or original generator would sign and date Block 18c to indicate receipt of the shipment, thereby closing out the manifest.

The manifesting rules require the initial TSD facility to complete a *new* manifest if it: 1) rejects a partial load, 2) accepts a shipment of waste and unloads part of the shipment but can't get the container RCRA-empty, or 3) rejects a full load at a point in time after the transporter making the initial delivery has left the facility's premises. However, if the TSD facility rejects the waste back to the initial generator, the TSD facility's EPA ID number goes in Block 1 of the new manifest and the TSD facility's name/mailing address go in Block 5, since the facility originates the shipment of rejected waste; the original generator's information goes in Block 8. [§§264/265.72(f), RO 14770]

In the case of a TSD facility initiating a new manifest for rejected hazardous wastes, the TSD facility will sign the new manifest as the "offeror," not the generator or agent of the generator, of the waste shipment described in the second manifest. (An offeror is certifying only that the shipment has been properly described on the manifest and prepared for transportation in accordance with DOT regulations; an offeror is not certifying the waste minimization statement or that he/she is the generator of the waste in the shipment.) The TSD facility must ship the rejected load or non-RCRA-empty container to an alternate facility or back to the original generator within 60 days of the rejection or container residue identification. [§§264/265.72(d)(1)]

If a new manifest is required, the manifest tracking number of the original manifest will be entered in the special handling block (Block 14) of the new manifest, and the alternate facility or original generator will be designated in Block 8 of the new manifest and, upon receipt, will sign and date Block 20. The TSD facility must also amend the original manifest by writing the manifest tracking number of the new manifest in the Manifest Reference Number line in Block 18a of the old manifest. [§§264/265.72(e)(4)] If the TSD facility has already signed and returned a copy of the original manifest to the original transporter and generator, the TSD facility must amend its copy of the original manifest by completing Block 18a, including the addition of the manifest tracking number of the new manifest in the Manifest Reference Number line, and then resign, redate, and resend the amended original manifest to the transporter and generator. [§§264/265.72(g)]

6.6.3.12.2 Generators may receive manifested shipments

To allow generators to receive shipments of non-RCRA-empty containers or rejected hazardous waste, EPA has amended the definition of "designated facility" in §260.10 to include generators receiving returned waste shipments. Generators receiving returned shipments have either 90 or 180 days (for large or small quantity generators, respectively) to send returned waste to an alternate facility. [§262.34(m)] This accumulation time limit is based on the generator's status when the rejected waste is received by the generator (not when the waste originally left the generator's site). If a generator was classified as an SQG when the waste was initially shipped offsite but is an LQG when the rejected waste is returned, the generator could accumulate the waste for no more than 90 days without a permit.

When a generator receives a rejected shipment from a TSD facility, the generator must terminate the manifest. This is done by completing Block 20 of the manifest (or Block 18c if the original manifest is used by the TSD facility for the rejection). By completing this field, the generator is acknowledging receipt of the waste(s) listed on the manifest, except as noted in the discrepancy field (Block 18). The generator must provide the transporter with a copy of the signed manifest, send a copy back to the rejecting TSD facility within 30 days, and retain a copy of the manifest for a least three years. [§262.23(f)]

6.6.3.13 Management method codes

Block 19 is for identifying the Hazardous Waste Report Management Method Codes (formerly called Biennial Report System Type Codes). These codes (one assigned to each waste identified in Block 9) describe the type of hazardous waste management system used to treat, recover, dispose, or store the waste at the designated facility identified in Block 8. These codes do not describe the ultimate disposition of the waste. Rather, they describe the management method employed at the destination facility where the manifest is terminated. These codes are updated routinely and can be found in the instructions accompanying the current edition of the Hazardous Waste Report form (also known as the biennial report). The current Hazardous Waste Report instructions and forms can be found at EPA's website: http://www.epa.gov/epawaste/inforesources/data/biennialreport/.

6.6.3.14 TSD facility certification of receipt

By completing Block 20 of the manifest, the receiving facility acknowledges receipt of the waste(s) listed on the manifest, except as noted in the discrepancy field (Block 18). Because full and partial waste rejections are considered discrepancies on the manifest, receiving facilities are required to sign the certification field for all waste manifests brought into the facility for delivery. The signature serves to either acknowledge receipt of all the materials on the manifest, or to acknowledge that those materials identified in the discrepancy space were not received for management at that facility.

What if a transporter mistakenly pulls into the gates of a TSD facility that is not identified as the designated facility on the manifest? Would the TSD facility need to sign the manifest and reject all waste to the correct site? The TSD facility would not need to sign and reject all waste because it was not identified as the receiving facility on the manifest (in Block 8 or Block 18b). The driver would simply pull out of the gate and continue with the waste transport.

6.6.3.15 Continuation sheets

The instructions for completing the manifest in the appendix to Part 262 indicate that a continuation sheet must be used if more than two transporters are employed to transport a hazardous waste shipment or if more than four spaces are necessary for DOT shipping descriptions.

The manifest tracking number entered in Block 23 on the continuation sheets should be the same as the number entered in Block 4 on the first page of the manifest.

6.6.3.16 Making changes to the manifest

A manifest must properly reflect the waste shipment. Whether at time of shipment or at time of termination, erroneous information must be corrected. EPA spelled out in §§264/265.72 how a receiving facility will complete Block 18 for significant discrepancies, but the agency has not identified how to actually correct mistakes (such as incorrect container types or wrong phone numbers). How should a manifest correction be made? We have found no written guidance from EPA, but telephone conversations with agency personnel indicate that inaccurate information should be crossed out and replaced with accurate information. While some generators take a more conservative approach to resolving manifest mistakes, requiring the issuance of a new manifest, most in the regulated community agree that simply crossing out the mistake and supplying the correct information is acceptable. Must the generator/transporter/TSD facility initial next to the correction? Again, we have found no written guidance, but it is a common practice.

6.6.4 Exception reports

If a large quantity generator doesn't receive a signed copy of the manifest from the receiving facility within 35 days of shipping the waste, he/she must contact the transporter and/or the designated receiving facility to determine the status or location of the waste. If the generator still hasn't received a signed copy within 45 days, he/she must file an exception report with EPA or the authorized state. The exception report must include a legible copy of the appropriate manifest and a signed cover letter outlining the steps that the generator has taken to locate the hazardous waste and the results of those efforts. [§262.42(a)] It is also a good operating practice to keep information on how the problem was ultimately resolved with the original copy of the manifest. Per §262.42 and RO 12204, the date of shipment for exception report timing is based on the date of the initial transporter's signature, not on the date of the generator's certification signature.

Small quantity generators have an abbreviated exception reporting requirement under §262.42(b). If an SQG has not received a signed copy of the manifest from the receiving facility within 60 days after shipping the waste, the SQG must send a copy of the manifest to EPA or the state, with some indication that it has not received confirmation of delivery. [§262.42(b)]

Q: Are there time limits for submitting exception reports?

A: The exception reporting requirements of §262.42 do not specifically say how soon the exception report should be submitted to EPA. However, in *Federal Register* preamble guidance, EPA said the agency would like the exception report "at the earliest possible time." [February 26, 1980; 45 *FR* 12731]

6.6.4.1 Exception report for rejected shipment

On March 18, 2010 [75 *FR* 12989], EPA codified a new paragraph §262.42(c), requiring generators to comply with the manifest exception reporting requirements of §262.42(a) or (b) [for large quantity or small quantity generators, respectively] when hazardous wastes (or container residues in non-empty containers) are rejected from the designated TSD facility to an alternate facility using a new manifest. This amendment to the regulations was necessary because when a TSD facility rejects a shipment and initiates a new manifest, the TSD facility uses the original generator's name and EPA ID number on that new manifest and then signs the manifest as the offeror of the waste shipment. So when the alternate facility receives the shipment, they will send the signed copy of the manifest back to the original generator. But how will the generator track the 35/45-day or 60-day clock [for large quantity or small quantity generators, respectively] for this rejected shipment? The agency simultaneously amended §§264/265.72(e)(6), requiring the designated TSD facility to mail the generator a signed copy of the new manifest. Thus, the generator will know the date that the transporter picked up the rejected shipment.

6.6.5 Manifest retention

The RCRA regulations require facilities to keep copies of manifests for a certain amount of time. Under §262.40(a), a generator must keep the signed copy of each manifest received from the designated facility for three years after the date the waste was accepted by the initial transporter. The regulations do not specify that these copies must be kept onsite. Therefore, they may be retained at corporate headquarters or some other centralized location. However, the generator facility must be able to provide a state or EPA inspector with the appropriate records during their inspections. TSD facilities must retain copies of manifests *onsite* for at least three years from the date of delivery. [§§264/265.71(a)(2)(v), RO 12199]

EPA has determined that companies may implement automated, electronic manifest record systems under certain circumstances. In order to comply with the manifest retention requirements, such systems must: 1) store and print out high-quality image files that include handwritten signatures; 2) include design and operating controls that ensure record accuracy, integrity, and security; and

3) provide indexing and file retrieval features that allow reasonable access by inspectors. However, prior to the use of such systems, regulated parties must verify that the system will comply with the appropriate state agency's manifest retention regulations and the rules of evidence that govern the admissibility of computer-generated records in that state's courts and agencies. [RO 14105]

"[B]oth EPA and Department of Transportation (DOT) policies are consistent in this area, as the agencies regard an electronic image of a manifest as sufficient to constitute a record of a manifest that must be retained by RCRA waste handlers. While this policy addressed a fact pattern involving the retention of image files by designated facilities, we believe that image file storage…is also appropriate for generators and transporters as well, and, in fact, the [hazardous materials regulations] allow this result. Therefore, EPA concludes that a generator could satisfy its regulatory obligations with a faxed or scanned image file of a signed manifest delivered to the generator…if the image file is either printed and stored with the generator's other paper files or retained electronically...." [RO 14791]

6.6.6 Rail shipments

After the manifest requirements were finalized in February 1980, the rail industry complained that the new provisions would disrupt the normal operating practices of railroads with regard to intermodal shipments (i.e., shipments involving both rail and other types of transportation). In fact, several railroads indicated that they would refuse all shipments of hazardous waste. In order to accommodate rail transportation of hazardous waste, the agency modified the manifest requirements for intermodal transportation involving railroads. Under the scheme, promulgated on December 31, 1980 [45 *FR* 86970], a manifest must be filled out for rail shipments of hazardous waste, but the manifest does not need to accompany the shipment. Instead, each shipment must be accompanied by a shipping paper that contains certain information typically required on a manifest (excluding EPA ID numbers, generator certification, and signatures). [§263.20(f)] Although rail transporters are not required to carry manifests noting EPA ID numbers, they must have such numbers per §263.11. [RO 12192]

For rail shipments of hazardous waste within the United States that originate at the site of generation, the generator must send at least three copies of the hazardous waste manifest to 1) the next nonrail transporter (if any), 2) the designated facility if transported solely by rail, or 3) the last rail transporter to handle the waste in the United States if the waste will be exported by rail. [§262.23(d)] Therefore, if the waste is transported solely by rail, the generator must send the manifest copies to the designated facility. The manifest would need to include the name of the rail carrier in the transporter block, and the transporter's signature would be necessary. [RO 14791] Intermediate rail transporters do not need to sign the manifest or the shipping paper.

For other shipments involving rail transport (that do not originate at the site of generation), the initial rail transporter must perform the following when accepting hazardous waste from a nonrail transporter: 1) sign and date the manifest; 2) return a signed copy of the manifest to the nonrail transporter; 3) forward at least three copies of the manifest to the next nonrail transporter (if any), the designated facility (if the shipment is delivered to that facility by rail), or the last rail transporter designated to handle the waste in the United States; and 4) retain one copy of the manifest and shipping paper. [§263.20(f)(1)]

6.6.7 Universal wastes

The federal universal waste program in Part 273 allows four types of hazardous wastes (i.e., batteries, recalled pesticides, mercury-containing equipment, and lamps) to be managed under a simplified set of management standards in lieu of the full hazardous waste regulations. One of the primary advantages of this program is that shipments of universal wastes do *not* have to be accompanied by a manifest. However, some RCRA-authorized states have added wastes to their state universal waste programs that are not considered universal wastes in other states. This makes the interstate transportation of universal wastes complicated. Please see Section 8.5.2 for a discussion of what is required when shipping wastes from one state to another if the wastes are considered universal waste in some, but not all, of the states involved.

6.6.8 Imports

Hazardous waste imported into the United States must comply with all applicable RCRA provisions (including manifesting) as soon as it enters the country. Because the actual generator of the waste is a foreign entity, they are not subject to RCRA regulations, which can apply only to U.S. parties. Therefore, when hazardous wastes are imported into the United States, the U.S. importer is responsible for carrying out the RCRA generator requirements. [RO 11820] As such, the U.S. importer of the waste must have an EPA ID number; use appropriate DOT packaging, labeling, marking, and placarding; prepare a hazardous waste manifest; use a hazardous waste transporter; and comply with other requirements found in Part 262. Regulations for hazardous waste importers, including instructions on completing manifests for imported hazardous waste, are found in §262.60.

On the manifest for imported hazardous waste, Blocks 1 and 5 should contain the name and address of the foreign generator and the U.S. importer's name, address, and EPA ID number. The foreign generator's name and address can also be placed in Block 14. [RO 11820] The manifest may be signed by the importer or his/her agent (a party somehow legally affiliated with the EPA ID number used on the manifest). A foreign broker could sign the manifest only if the broker's company has an EPA ID number (requiring a U.S. address) or the broker is legally related to the importer (e.g., a subsidiary). [RO 13739]

If several importers are involved in a shipment, only one of them must perform the §262.60 importer requirements. In such situations, the importers should agree among themselves (e.g., through a contractual agreement) as to who will perform those duties. However, if that party fails to perform the required tasks, all of the parties could be subject to EPA enforcement actions. [RO 11085, 11820, 11953, 13725]

Under §263.10(c), transporters that bring hazardous waste into the United States must comply with the relevant generator requirements (i.e., the §262.60 importer provisions). In situations where the transporter is one of several parties who may be importers of a shipment, the transporter may arrange with the other parties (e.g., brokers or TSD facilities) to assume the importer responsibilities for the entire group. EPA requires only one of the parties to perform the importer duties, but all of the parties could be subject to enforcement for failure to comply. [RO 11085, 11953]

For hazardous waste being imported through U.S. territorial seas from a foreign country, EPA considers the U.S. coastal port as the "cradle" under the RCRA "cradle-to-grave" tracking system. Prior to bringing the hazardous waste ashore, the shipment would not need a hazardous waste manifest and the ocean carrier would not be required to have a transporter EPA ID number. [RO 14595]

Current regulations do not require importers to leave a copy of the manifest with U.S. Customs officials at the border. However, they are being collected voluntarily at some Customs checkpoints. [RO 11757]

Case Study 6-9 reviews the RCRA requirements for importers who consolidate shipments onto one manifest.

6.6.9 Exports

Exports of hazardous waste must comply with the provisions set forth in Part 262, Subpart E. These requirements include submitting a notification of intent to export to EPA, attaching an EPA Acknowledgement of Consent to the shipment's manifest once consent is received from the importing country, submitting an annual report documenting the shipment, and other requirements. [§262.52, RO 11723] The exporter must also submit a copy of the manifest to the Customs agent as the waste exits the United States; the manifest is then forwarded to EPA. [§262.54(i), RO 11863]

Because of the way EPA defines "primary exporter" in §262.51, the Subpart E export requirements are, in general, applicable only to hazardous wastes that are subject to the Part 262, Subpart B manifesting requirements. Consequently, if the waste to be exported is exempt from the manifesting requirements, it would also be exempt from Subpart E. For example, scrap metal sent for reclamation is exempt from the hazardous waste regulations (including the manifesting requirements) under §261.6(a)(3)(ii). Therefore, scrap metal is not subject to the Subpart E exporting requirements. In addition, small quantity generators who export hazardous waste under a reclamation (tolling) agreement qualify for the §262.20(e) exemption from manifesting requirements, and therefore, a manifest for the exported waste is not required. However, the exporter is advised to check with the receiving country to identify any applicable regulations before the waste crosses the border. [August 8, 1986; 51 *FR* 28669, RO 13719, 14015]

There are two exceptions to the general rule stated in the previous paragraph. First, a universal waste handler who exports universal waste to a foreign destination without first sending the waste to a U.S. consolidation point or destination facility must comply with the requirements applicable to primary exporters in Part 262, Subpart E, even

---

Case Study 6-9: Consolidating Several Shipments of Imported Hazardous Waste

Do the RCRA regulations allow hazardous waste importers to consolidate individual hazardous waste shipments before importing them into the United States? If so, how should an importer document consolidated shipments of imported hazardous waste on the manifest?

Federal RCRA regulations allow importers to physically consolidate hazardous waste shipments before they enter the United States, as long as the consolidated shipment is reflected accurately on the manifest. Thus, the importer will not have to complete separate manifests for each foreign hazardous waste source but could choose to document the individual hazardous waste sources on one consolidated manifest.

According to §262.60(b)(1), the import manifest must contain "the name and address of the foreign generator and the importer's name, address, and EPA identification number." Thus, the consolidated manifest must identify the individual foreign sources (i.e., generators) that contributed hazardous wastes to the consolidated shipment and their associated waste types and quantities.

For example, service companies or TSD facilities importing maquiladora hazardous waste into the United States from Mexico are not themselves "generators" as defined in §260.10. Instead, it is the individual maquiladora plants that are the generators, and, therefore, the name and address of each foreign generator of the maquiladora hazardous waste must be shown on the consolidated manifest along with the associated waste types and quantities. This detailed information could be shown in the waste description block of the manifest or by using a continuation sheet or other attachment. Another method would be to prepare a "cover" manifest that describes the contents of the entire shipment, and then attach the supporting manifests or shipping documents that indicate the contributions from each foreign generator. It is important to check with the appropriate state agencies to ensure that the importer's consolidated manifest arrangement is acceptable to them. [RO 14735]

though a manifest is not required. [April 12, 1996; 61 *FR* 16306, RO 14740] Sections 273.20 and 273.40 require specific Subpart E export regulations to be met when small or large quantity handlers, respectively, export universal waste.

The second exception is for spent lead-acid batteries (SLABs) that are exported for reclamation. Even though SLABs sent for reclamation are exempt from manifesting, they are subject to notification and consent requirements prior to export per §266.80, as amended on January 8, 2010. [75 *FR* 1236]

Canadian hazardous waste transportation regulations recognize the Uniform Hazardous Waste Manifest for shipments sent from the United States to Canada. As such, under the Canadian regulations, U.S. generators are required to prepare the U.S. manifest for hazardous waste shipments to Canada. [RO 12502] Note that additional requirements are contained in the 1986 bilateral agreement between the United States and Canada (and 1992 amendments) for the transboundary shipment of hazardous waste, available at http://www.epa.gov/epawaste/hazard/international/agree.htm. Also see RO 14442.

Q: *A shipment of hazardous nickel-cadmium batteries is sent from Mexico to Japan via the United States. Is a manifest required? If so, how would the generator and designated facility blocks need to be filled out?*

A: A manifest would be required as soon as the shipment enters the United States from Mexico and must accompany the shipment until it leaves for Japan. It would need to show the name and address of both the U.S. importer and the Mexican generator in Block 5. The EPA ID number of the U.S. importer would also be necessary in Block 1. The name and site address of the Japanese consignee would be listed in Block 8. [RO 11723]

Q: *A hazardous waste generator in the United States exports hazardous waste to another country via a broker who acts as an intermediary to arrange for the export. Must a manifest be filled out? Are any other forms necessary?*

A: All shipments of hazardous wastes, including those destined for other countries, must be manifested while in the United States. In addition, exporters of hazardous waste must attach an EPA Acknowledgement of Consent to the manifest (or shipping paper for exports by water) as required by §262.52(c). (For shipments by rail, the consent document must simply accompany the waste shipment; certain shipments by rail do not have to be accompanied by manifests if certain information accompanies the shipment.) An EPA Acknowledgement of Consent is a "cable sent to EPA from the U.S. Embassy in the receiving country that acknowledges the written consent of the receiving country to accept the hazardous waste and describes the terms and conditions of the receiving country's consent to the shipment." [§262.51] EPA then forwards the consent document to the "primary exporter," who is the person who is required to originate the manifest (e.g., the generator) and "any intermediary arranging for the export." [§262.51] Accordingly, a broker who arranges for an export may hold the EPA Acknowledgement of Consent. [RO 14421]

Q: *Would an ocean carrier be required to have an EPA ID number and carry a manifest for an export of hazardous waste from the United States to a foreign country through U.S. territorial waters?*

A: No. EPA would not require an ocean carrier in territorial seas to have an EPA ID number when it is importing or exporting hazardous waste cargo. This is true whether the carrier is a U.S.-flagged vessel or a foreign-flagged vessel. Additionally, a manifest would not be required at the time the ocean carrier leaves the port. [RO 14595]

Q: *A generator sends an F005 spent solvent to a U.S. recycling facility. Still bottoms generated from the recycling process (also F005 via the derived-from rule) are shipped from the recycler to a foreign TSD facility. Who is the primary exporter?*

A: The recycler will be the primary exporter. Section 262.51 designates the facility that manifests a hazardous waste shipment to a foreign TSD facility as the primary exporter. The initial generator would have designated the domestic recycler as the designated facility on his/her manifest and therefore would not meet the definition of "primary exporter." Although the initial generator is not the primary exporter, he/she may be subject to penalties under RCRA if he/she knowingly continues to send hazardous waste to a domestic facility that exports the waste without consent of the receiving country. [August 8, 1986; 51 *FR* 28671, RO 14593]

6.6.10 Electronic distribution of manifests

EPA issued its e-manifest rule on February 7, 2014. [79 *FR* 7518] The rule implements a portion of the Hazardous Waste Electronic Manifest Establishment Act (e-Manifest Act), which was signed into law on October 5, 2012. The act authorizes EPA to 1) develop an electronic hazardous waste manifest system; and 2) impose user fees as necessary to recover costs incurred in developing, operating, maintaining, and upgrading the system. EPA was given a one-year deadline to promulgate regulations to carry out the new law and a three-year deadline to have the system fully operational.

The primary purpose of the new rule is to clarify that e-manifests, completed and signed as specified in the rule, are legally valid for all RCRA purposes. When the e-manifest system is operational, e-manifests will be used for shipments of hazardous waste, state-only hazardous waste, and any other waste that a state requires be accompanied by a hazardous waste manifest. The benefits of an e-manifest include greater access by emergency responders to information about a waste shipment, higher quality and more timely waste shipment data, lower cost, and fewer burdens on the regulated community.

Under the e-Manifest Act, EPA is required to develop the e-manifest system by October 2015. Until the e-manifest system is ready, the paper manifest will continue to be used. Ultimately, operation of the e-manifest system will be paid for by fees collected from users. EPA plans to issue a separate rule to establish the fee structure in fiscal year 2015.

Waste handlers will be able to opt out of the electronic system and continue using paper manifests if they choose. If a paper manifest is used, the facility that receives the waste shipment and terminates the manifest will be required to send a copy to the operator of the e-manifest system so that the data from the manifest can be entered into the system. Thus, the electronic system will be a comprehensive repository for all information on hazardous waste shipments.

Newly codified §262.25 says that the electronic signature methods for the e-manifest must be: 1) legally valid and enforceable signatures under applicable EPA and other federal requirements pertaining to electronic signatures, and 2) designed so as to be cost-effective and practical. What constitutes a valid and enforceable signature is governed by EPA's Cross-Media Electronic Reporting (CROMERR) regulations, which are codified at 40 *CFR* Part 3. EPA believes that the first-generation system should support a PIN/password signature method and/or a digitized handwritten signature using a signature pad and stylus. Based on preamble language, EPA is recommending the PIN/password approach, with security questions, but the agency also expects to include the digitized handwritten signature method pending the outcome of studies to demonstrate its forensic reliability.

To satisfy DOT requirements, a generator originating an e-manifest must also provide the initial transporter with one printed copy of the manifest. After the initial transporter has signed the e-manifest to take custody of the waste shipment, if the e-manifest system goes down or if the e-manifest cannot be completed electronically for any reason, the transporter must make a copy of the printed manifest for each waste handler and two additional copies for the designated facility. These paper copies will then become the manifest for that shipment and must be signed and handled accordingly by each waste handler.

After the e-manifest rule becomes effective, the regulated community will no longer be able to assert that information entered on paper manifests, or on an e-manifest when the system is available, is confidential business information (CBI). EPA gave two reasons for this policy:

"First…as manifests are shared with several commercial entities while they are being processed and used, a business concerned with protecting its commercial information would find it exceedingly difficult to protect its individual manifest records from disclosure by all the other persons who come into contact with its manifests…. Second, we explained that much of the information that might be claimed by industry commenters to be CBI is already available to the public from a number of government and other legitimate sources, because a large number of states now require the submission of generator and/or TSDF copies of manifests to state data systems, and the data from these manifests are often made publicly available through state Web sites or reported and disclosed freely in federal and state information systems." [79 FR 7540]

The e-manifest rule became effective on August 6, 2014. However, the implementation and compliance date for these regulations will be delayed until the e-manifest system is ready for operation and the schedule of fees for manifest-related services has been issued. EPA will publish a further document subsequent to this rule's effective date to announce the user fee schedule—this document will also announce the date upon which compliance with this regulation will be required. Similar to HSWA regulations, the e-manifest rule will be effective in all states at the same time and will be implemented by EPA in RCRA-authorized states until the state adopts the rule.

6.7 Recordkeeping

Recordkeeping and reporting requirements are codified in §§262.40–262.43 for large quantity generators and §262.44 for small quantity generators (although other generator paperwork provisions are sprinkled throughout the RCRA regs). Conditionally exempt small quantity generators have almost no recordkeeping requirements.

A fairly comprehensive summary of the primary RCRA recordkeeping requirements for the three generator classes is given in Table 6-6. EPA guidance associated with the recordkeeping provisions given in that table is discussed in the following sections, but first we start off with an example of general applicability.

Q *A business that has been a large quantity generator of hazardous waste since 1980 is sold. The new owner assumes the generator responsibilities in June 2003, including keeping records of hazardous waste activities (e.g., signed manifests) for at least three years. Must the new owner keep the previous owner's records for a period of three years, or does the new owner begin the recordkeeping process on the date of sale?*

A The new owner must keep records of facility hazardous waste activities from the previous three years. Therefore, records of hazardous waste sent offsite from the facility in March 2002, prior to the change in ownership, should remain in the new owner's records until March 2005. [RO 14694]

6.7.1 Manifests/exception reports

Large and small quantity generators must keep copies of manifests (signed by the TSD facility) for at least three years from the date the waste was shipped offsite. The date of shipment is determined by the date of the initial transporter's signature—not the date the generator signs the generator's certification. [RO 12204] Nothing in §262.40(a) specifies that copies of manifests must be kept at the site of generation; instead, copies can be maintained at corporate headquarters or some other centralized location. However, RCRA Section 3007(a) requires generators to provide EPA or state personnel reasonable access to records regarding waste management activities during their inspections. [RO 12199]

Table 6-6: Primary Recordkeeping Requirements for the Three Generator Classes

Requirement[1]	Large quantity generator	Small quantity generator	Conditionally exempt small quantity generator
Manifests	Keep signed copies for 3 years from date the waste was shipped offsite [§262.40(a)]	Keep signed copies for 3 years from date the waste was shipped offsite [§§262.40(a); 262.44(a)]	Not required [§261.5(b)], but keep to prove CESQG status
Exception reports	Keep copies for 3 years [§§262.40(b); 262.42(a)]	A handwritten or typed note on or attached to the manifest is required [§§262.42(b); 262.44(b)]	Not required [§261.5(b)]
LDR forms	Keep copies for 3 years from date the waste was last sent to onsite or offsite treatment, storage, or disposal [§§268.1(b); 268.7(a)(6,8)]	Keep copies for 3 years from date the waste was last sent to onsite or offsite treatment, storage, or disposal [§§268.1(b); 268.7(a)(6,8)]	Not required [§§261.5(b); 268.1(e)(1)]
Waste analysis data	Keep test results, analyses, or other determinations in accordance with §262.11 for 3 years [§262.40(c)]	Keep test results, analyses, or other determinations in accordance with §262.11 for 3 years [§§262.40(c); 262.44(a)]	Not required, but should keep hazardous waste identification records to prove CESQG status
Waste analysis plans	Keep in onsite files for 3 years; required if treating waste in a 90-day accumulation unit for the purpose of meeting LDR standards [§§262.34(a)(4); 268.7(a)(5,8)]	Keep in onsite files for 3 years; required if treating waste in a 180-day accumulation unit for the purpose of meeting LDR standards [§§262.34(d)(4); 268.7(a)(5,8)]	Not required [§261.5(b)]
Inspection records	Keep 90-day accumulation unit inspections logs for 3 years from the date of inspection [§§262.34(a)(1)(ii); 265.195(c,g)][2]	Keep 180-day accumulation unit inspections logs for 3 years from the date of inspection[2]	Not required [§261.5(b)]
Training records	Keep training records on current personnel until closure of facility; on former employees for 3 years from departure date [§265.16(d,e)]	No specific training records are required by regulation [§262.34(d)(5)(iii)], but we recommend they be kept	Not required [§261.5(b)]
Emergency responder agreements	Arrangements/agreements with local/State emergency responders/hospitals [§§262.34(a)(4); 265.37]	Arrangements/agreements with local/State emergency responders/hospitals [§§262.34(d)(4); 265.37]	Not required [§261.5(b)]
Contingency plans	Maintained at facility [§262.34(a)(4); Part 265, Subpart D]	Not required, but comply with §262.34(d)(5)	Not required [§261.5(b)]
Biennial reports	Keep copies for 3 years [§§262.40(b); 262.41]	Not required [§262.44]	Not required [§261.5(b)]
Export records	Keep for 3 years [§§262.56–57]	Keep for 3 years [§§262.56–57]	Not required [§261.5(b)]
Additional reports	Per §262.43	Per §§262.43; 262.44(c)	Not required [§261.5(b)]

[1]Record retention times are extended automatically during any enforcement action. [§§262.40(d); 262.57(b); 268.7(a)(8)]

[2]Inspection logs for LQG 90-day containers and SQG 180-day tanks and containers are not required by regulation, but we recommend they be kept. The 3-year retention period for inspection records is not required by the regulations, but we recommend generators comply with §265.15(d).

Source: McCoy and Associates, Inc.

Conditionally exempt small quantity generators are not required to use hazardous waste manifests, much less keep copies of them. However, if a conditionally exempt generator uses manifests for offsite shipment of hazardous waste, it should keep copies of them to help prove to the state its conditionally exempt status (i.e., that it is not generating more than 100 kg of nonacute hazardous waste/mo).

EPA has determined that companies may use an electronic manifest record retention system, provided the system: 1) stores and prints high-quality image files that include handwritten signatures; 2) includes design and operating controls that ensure record accuracy, integrity, and security; and 3) provides indexing and file retrieval features that allow reasonable access by state or EPA inspectors. Prior to implementing such an electronic system, however, regulated parties must verify that it will comply with the appropriate state agency's manifest retention regulations and the rules of evidence that govern the admissibility of electronically generated records in that state's courts and agencies. [RO 14105]

"[B]oth EPA and Department of Transportation (DOT) policies are consistent in this area, as the agencies regard an electronic image of a manifest as sufficient to constitute a record of a manifest that must be retained by RCRA waste handlers. While this policy addressed a fact pattern involving the retention of image files by designated facilities, we believe that image file storage…is also appropriate for generators and transporters as well, and, in fact, the [hazardous materials regulations] allow this result. Therefore, EPA concludes that a generator could satisfy its regulatory obligations with a faxed or scanned image file of a signed manifest delivered to the generator…if the image file is either printed and stored with the generator's other paper files or retained electronically…." [RO 14791]

Large quantity generators are required to keep copies of exception reports (as discussed in Section 6.6.4) in their files for three years. There is no requirement for small quantity generators to submit or keep copies of formal exception reports. However, we recommend that small quantity generators document their compliance with §262.42(b) (as discussed in Section 6.6.4) and retain such documentation for three years.

6.7.2 LDR forms

A complete discussion of the land disposal restrictions (LDR) paperwork requirements is contained in Sections 13.12 and 18.4. Copies of this paperwork must be retained onsite by large and small quantity generators for at least three years from the date the waste was last sent to onsite or offsite treatment, storage, or disposal.

EPA noted in 1997 that "any records kept in connection with the LDR program may be stored electronically." [62 *FR* 26004] For example, scanned images of original paperwork that show handwritten signatures are acceptable.

6.7.3 Hazardous waste characterization records

Documentation requirements associated with making a hazardous waste determination for every solid waste produced at a facility are included in §262.40(c); small quantity generators get to this section via reference from §262.44(a). Section 262.40(c) requires records of test results, waste analyses, and other determinations developed for waste characterization purposes to be maintained in facility files for at least three years from the date the waste was last sent to onsite or offsite treatment, storage, or disposal. Test results and waste analyses would be records maintained if a generator is making analysis-based determinations. "Other determinations" is a reference to documentation reviewed or generated when making a knowledge-based assessment (e.g., SDSs). Both of these recordkeeping requirements are examined in the next two subsections.

Conditionally exempt small quantity generators are required to make hazardous waste determinations for all solid waste they generate at their facility—just like all other generators [RO 11958, 14030];

however, they are not subject to the requirements for maintaining hazardous waste characterization data/other documentation in their files. Even so, we recommend that these conditionally exempt generators keep such documentation in their files to prove to state personnel that they legitimately can claim conditionally exempt status.

6.7.3.1 Test results and waste analyses

If a facility samples a waste and sends the representative sample(s) for analysis, all sampling/analytical data should be retained in site files for the three-year period noted previously. These data include [EPA/530/R-02/003]:

- Sample location(s) and time(s);
- Number and size of samples;
- Sample type (e.g., single grab, composite);
- Sampling device (e.g., Coliwasa, weighted bottle, etc.);
- Sample collection and handling techniques (chain-of-custody procedures);
- Timing issues for sample handling and analysis;
- Analytical methods used and results obtained; and
- QA/QC protocols.

6.7.3.2 "Other determinations" (documentation of knowledge)

When a hazardous waste determination is made in part or in whole based on knowledge, it is critically important that generators document exactly what that knowledge is. In addition to test results and waste analyses, §262.40(c) requires large and small quantity generators to keep records establishing the basis for all knowledge-based hazardous waste determinations. [RO 13570]

If generators don't document their knowledge, they can't prove they made a hazardous waste determination for that solid waste. Relying on knowledge without adequate documentation ("But Joe told me it wasn't hazardous") may subject a generator to an enforcement action. Inadequate or insufficient documentation can be considered a violation of §262.11.

"EPA, in enforcement cases, looks for documentation that clearly demonstrates that the information relied upon is sufficient to identify the waste accurately and completely." [EPA/530/R-94/024]

One of the problems with documenting knowledge-based determinations is that EPA has never given detailed guidance on what constitutes adequate documentation. Any of the following information used for this purpose should be copied and put in a generator's files:

- SDSs or other manufacturing information;
- EPA's listing background documents for F- or K-wastes;
- Regulatory exemptions claimed [e.g., the alcohol exclusion in §261.21(a)(1)] and any guidance supporting such an exemption (e.g., RO documents);
- Material balances or other calculations for the source or process generating the waste, including raw materials or intermediate products fed to a process;
- Any knowledge used to eliminate constituents or characteristics that couldn't be in, or exhibited by, a waste stream;
- Test data not specified in the RCRA regs (e.g., pH paper results, total waste analyses in lieu of TCLP testing);
- Constituent-specific chemical test data for the waste from previous testing at the facility that are still applicable to the current waste;
- Previous test data from other locations using substantially similar processes and/or managing the same type of waste streams; and
- Other knowledge based on information in manifests, shipping papers, or waste certification notices.

We know of a facility that suffered an enforcement action for lack of documentation. The facility was unable to provide records to prove that a particular solid waste stream (that it was managing as nonhazardous) really was nonhazardous. Even though subsequent testing did provide that proof, the case for violating §262.11 remained valid.

Interestingly, EPA noted in a September 1996 guidance memo that "[i]n those cases where a waste generator declares their waste to be hazardous waste, the documentation is not needed, since the generator is subjecting its waste to hazardous waste management standards." [RO 14137]

6.7.4 Waste analysis plans

Under the federal RCRA regulations, large and small quantity generators are required to develop and follow a waste analysis plan only in one situation: if they are treating hazardous waste in a 90/180/270-day accumulation unit for the purpose of meeting an LDR treatment standard. These plans, which are discussed in some detail in Section 18.2.3.1, are required to be maintained in the facility's onsite files for at least three years from the date the waste was last sent to such onsite treatment.

6.7.5 Inspection records

As noted in Sections 9.6 and 10.5, respectively, generators must inspect 90/180/270-day hazardous waste accumulation containers weekly, and 90/180/270-day accumulation tanks must be inspected daily. These inspections are normally documented on checklists or in operating logs.

Documentation of large quantity generators' daily inspections of 90-day accumulation tanks is required per §265.195(g). Interestingly, the federal regs *don't* require documentation of weekly inspections conducted for large quantity generators' 90-day accumulation containers. Similarly, nothing in the federal regs requires small quantity generators to document inspections of 180/270-day accumulation tanks and containers. However, we recommend that these inspections be documented and recorded in the plant operating logs.

Although the federal regulations do not require generators to comply with the inspection records retention obligation in §265.15(d), we recommend that this provision be followed anyway—container and tank inspection records should be kept for three years from the date of inspection.

6.7.6 Training records

Records showing that large quantity generators have adequately trained facility personnel on an annual basis are required by §265.16(d) and must be maintained at the facility. These detailed requirements are summarized in Section 6.4.1.5. Section 265.16(e) specifies that training records for current personnel at large quantity generator facilities must be maintained until the facility closes. For former employees, training records must be kept for at least three years from the date the employees last worked at the facility.

There are no recordkeeping provisions in the small quantity generator training regulations at §262.34(d)(5)(iii). However, we recommend that small quantity generators keep records showing that the required training has been provided.

6.7.7 Emergency responder agreements

Large and small quantity generators conducting 90/180/270-day hazardous waste accumulation are required to comply with the preparedness and prevention requirements in Part 265, Subpart C. One of these requirements is coordination between the generator and local authorities (e.g., hospitals, fire departments) on how emergencies will be addressed by both facility and offsite emergency personnel. [§265.37]

It is often hard to get hospitals and local authorities to read and comment on contingency plans or other emergency arrangement documents that the generator may prepare. However, it is the generator's responsibility to reach out to, communicate with, and update such responders, and all communication between generators and these local authorities should be documented. Even if these local authorities will not respond to a request to review an updated contingency/emergency plan, the generator should document that it sent the plan and tried to get feedback (e.g., with a follow-up phone call).

One good way of interfacing with these responders is to invite them out to the facility occasionally for a short walkthrough and review of the chemicals used, types of wastes generated, and risks associated with the site (usually followed by taking them out to lunch); these tours are then documented.

In the unlikely event that a fire department, local hospital, or other agency refuses to enter into a shared arrangement to respond to emergencies, §265.37(b) requires documentation of the refusal. If the generator's facility is not provided with fire protection coverage by a fire protection district or municipal fire department, the generator should prepare a fire protection and prevention plan of its own.

6.7.8 Contingency plans

Large quantity generators are required to prepare a formal contingency plan that prepares for fires, explosions, or releases of hazardous waste. This plan must meet the requirements of Part 265, Subpart D. It must demonstrate that facility-specific emergency procedures have been developed and will be implemented whenever an emergency involving hazardous waste occurs. The plan must be maintained at the facility and be available for inspection, and it also must be submitted to local emergency response authorities.

Per §265.52(b), the hazardous waste contingency plan may be a standalone document or part of an emergency plan prepared for the facility under another program (e.g., it may be hazardous waste management amendments to a Spill Prevention, Control, and Countermeasures plan previously developed for the facility).

On April 4, 2006 [71 *FR* 16862], EPA modified the regulations to indicate that facilities may consider developing one contingency plan for their entire facility, instead of having separate contingency plans to comply with RCRA, CWA, etc. Use of a single plan should eliminate confusion when facilities must decide which of its contingency plans is applicable to a particular emergency. If facilities decide to develop such a single plan, EPA recommends it be based on the *Integrated Contingency Plan (One Plan) Guidance*, available at http://www.epa.gov/osweroe1/guidance.htm. This guidance provides a mechanism for consolidating the multiple contingency plans that facilities have to prepare to comply with various government regulations.

Contingency plans should be reviewed regularly (some states require an annual review). There are five situations identified in §265.54 that require review of, and possible revisions to, the plan. Of those five, one of the most likely is changes to the list of emergency coordinators. Revised plans should be submitted to local emergency response authorities to keep them up-to-date.

Generators need to make sure that they will get an immediate and appropriate emergency response from local police, fire departments, emergency response teams, and hospitals. Thus, contingency plans should describe the arrangements agreed to with these authorities and the division of responsibilities between plant and offsite personnel. Written agreements between the facility and these local authorities are preferable; if such agreements are attained, they should be attached to the contingency plan.

Small quantity generators are not required to develop a formal contingency plan as discussed above. Instead, they are required to comply with the emergency response provisions in §262.34(d)(5). As part of compliance with those regs, a list of emergency contacts with phone numbers must be posted next to telephones near the 180/270-day accumulation areas. This posting must also include locations of fire extinguishers and alarms and spill control material. A sample emergency contacts telephone posting form is available from McCoy and Associates' website at http://www.mccoyseminars.com.

6.7.9 Biennial reports

Large quantity generators must submit a biennial report to the EPA region and/or state by March 1 of each even-numbered year. [§262.41] The report covers the generator's hazardous waste generation and subsequent management activities for the previous odd-numbered calendar year. Each item specified in §262.41 must be included in the biennial report.

Large quantity generators are required to keep copies of biennial reports in their files for three years from the due date of the report. Nothing in §262.40(b) specifies that copies of biennial reports must be kept at the site of generation; instead, copies can be maintained at corporate headquarters or some other centralized location. Even so, RCRA Section 3007(a) requires that a generator must provide EPA or state personnel reasonable access to records regarding waste management activities during their inspections. [RO 12199]

Small quantity generators are not subject to biennial reporting under the federal regulations. [§262.44] Note that some states require an annual report, and small quantity generators may be pulled into this reporting requirement per more-stringent state regulations.

Section 6.8 contains additional details on biennial reporting.

6.7.10 Export reports

Both large and small quantity generators that export hazardous waste out of the United States are required to submit an annual report to the EPA region and/or state by March 1 of each year. [§262.56] The report covers the generator's hazardous waste export activities for the previous calendar year. Generators are required to keep copies of such annual reports in their files for three years from the due date of the report.

Additionally, copies of notifications of intent to export, EPA Acknowledgements of Consent, and confirmations of delivery should be maintained in the facility's files for three years from the date the hazardous waste was accepted by the initial transporter. [§262.57] These terms are discussed in Section 6.6.9.

Similar recordkeeping requirements are imposed on small and large quantity handlers of universal waste who export such waste. See §§273.20 and 273.40.

If generators of hazardous secondary materials export them to a foreign country for reclamation under the exclusion in §261.4(a)(25), the reporting requirements are similar to those noted above. In addition to the notification requirements of §260.42, such generators must file a report with EPA by March 1 of each year summarizing the types, quantities, frequency of shipments, and foreign destinations of all exempt materials exported for reclamation during the previous year. [§261.4(a)(25)(xi)] Additionally, these entities must keep copies of notifications of intent to export and Acknowledgements of Consent for three years. [§261.4(a)(25)(x)]

6.7.11 Miscellaneous recordkeeping

In addition to the primary recordkeeping requirements identified in Table 6-6, generators are required to maintain other miscellaneous records as noted in Table 6-7.

6.8 Biennial reports

Large quantity generators (LQGs) that ship hazardous waste to an offsite TSD facility located in the United States are required to submit a biennial report to their state and/or regional EPA office by March 1 of each even-numbered year, covering their hazardous waste management activities for the previous odd-numbered calendar year. For example, the biennial report submitted by March 1, 2014 covered each LQG's hazardous waste management activities during the 2013 calendar year. The information that LQGs must report biennially is given in §262.41.

Although EPA requires the biennial report to be submitted on EPA Form 8700-13A/B, available at http://www.epa.gov/osw/inforesources/data/biennialreport/index.htm, some states require use of a modified version of the EPA report and others (e.g., Arkansas, Indiana, Ohio) have their own instructions and forms for fulfilling the reporting requirements.

Per §262.41(b), any generator that also is a TSD facility that treats, stores, or disposes hazardous waste onsite must submit a biennial report covering those

Table 6-7: Miscellaneous Recordkeeping Requirements for the Three Generator Classes

Requirement	Large quantity generator	Small quantity generator	Conditionally exempt small quantity generator
Comparable fuels exclusion records	Per §261.38(b)(8–10)	Per §261.38(b)(8–10)	Per §261.38(b)(8–10)
Containment building records	Per §265.1101(b)(4), (c)(2–4), (d)(3)	NA[1]	NA
De minimis wastewater exemption records	Per §261.3(a)(2)(iv)	Per §261.3(a)(2)(iv)	Per §261.3(a)(2)(iv)
Drip pads records	Per §265.443(n)	NA[1]	NA
Extensions of accumulation period records	Per §262.34(b)	Per §262.34(f)	NA
Mixed waste exemption records	Per §§266.250, 266.350	Per §§266.250, 266.350	Per §§266.250, 266.350
Precious metal exemption records	Per §266.70(c)	Per §266.70(c)	NA
Recycling exemption records	Per §§260.42, 261.2(f), 261.4(a)(24)(v)(C–E)	Per §§260.42, 261.2(f), 261.4(a)(24)(v)(C–E)	Per §§260.42, 261.2(f), 261.4(a)(24)(v)(C–E)
Residues (e.g., slag) generated from recovery of F006, K061, and K062 exclusion records	Per §261.3(c)(2)(ii)(C)	Per §261.3(c)(2)(ii)(C)	Per §261.3(c)(2)(ii)(C)
Spent materials generated within primary mineral processing industry exclusion records	Per §261.4(a)(17)(v)	Per §261.4(a)(17)(v)	Per §261.4(a)(17)(v)
Spent wood preserving solutions exclusion records	Per §261.4(a)(9)(iii)(E)	Per §261.4(a)(9)(iii)(E)	Per §261.4(a)(9)(iii)(E)
Subpart AA air emission control records	Per §265.1035[2]	Not required [§262.34(d)(2–3)]	Not required [§261.5(b)]
Subpart BB air emission control records	Per §265.1064	Not required [§262.34(d)(2–3)]	Not required [§261.5(b)]
Subpart CC air emission control records	Per §§265.1087(c)(5), 265.1090	Not required [§262.34(d)(2–3)]	Not required [§261.5(b)]
Treatability study records	Per §261.4(e)(2)(v)	Per §261.4(e)(2)(v)	Per §261.4(e)(2)(v)
Universal waste tracking records	Per §273.39(c)[3]	Per §273.39(c)[3]	Per §273.39(c)[3]
Zinc fertilizers exemption	Per §261.4(a)(20)(ii)	Per §261.4(a)(20)(ii)	Per §261.4(a)(20)(ii)

NA = not applicable to this generator class.

[1] Small quantity generators can utilize containment buildings or drip pads only if they comply with the large quantity generator requirements in §262.34(a), including a limit for onsite accumulation of 90 days. [EPA/305/B-96/001, EPA/530/K-02/008I, EPA/530/K-05/008, RO 13696, 14662]

[2] Large quantity generators that are not subject to RCRA permitting will be subject to Subpart AA air emission controls only if they are conducting hazardous waste treatment (not recycling) in a 90-day unit using distillation, fractionation, thin-film evaporation, solvent extraction, or air or steam stripping. [§265.1030(b)]

[3] These recordkeeping requirements apply to large quantity *handlers* of universal waste; small quantity *handlers* of universal waste are not subject to waste tracking recordkeeping. Large quantity, small quantity, and conditionally exempt small quantity generators of *hazardous* waste can be large quantity *handlers* of universal waste.

Source: McCoy and Associates, Inc. and "RCRA Hazardous Waste Reporting Requirements," May 27, 1999, available from http://www.epa.gov/epawaste/inforesources/data/burdenreduction/.

waste management activities. The TSD facility biennial reporting requirements are in §§264.75, 265.75, 267.75, and 270.30(l)(9).

As noted in Section 6.1.2.1, episodic generators are required to complete and submit a biennial report if they are a LQG in any single calendar month of the odd-numbered calendar year. In terms of what wastes are to be reported, the instructions for the biennial report state that "all hazardous waste that was used to determine the site's generator status" must be reported. Although that language is not completely clear, EPA seems pretty clear in RO 14842 when it notes that facilities need to "determine if, in any single calendar month, the site is an LQG, and if so, report all RCRA hazardous waste that is generated onsite…for the entire calendar year."

6.8.1 Wastes to be included

A site required to file the biennial report must submit Form GM (Waste Generation and Management) if the site generated RCRA hazardous waste that was accumulated onsite; managed onsite in a treatment, storage, or disposal unit; and/or shipped offsite for management, consistent with the applicability requirements noted in the instructions. The instructions for Form GM are generally very clear as to which wastes should and should not be reported. Additional guidance on which wastes need to be identified on the biennial report is available in a document titled *Biennial Report: Reportable and Non-Reportable Wastes*. The document is available at http://www.epa.gov/osw/inforesources/data/biennialreport/index.htm.

Typical questions about which hazardous wastes must be included on Form GM of biennial reports are addressed in the following EPA guidance:

- As noted in Section 6.1.4, some hazardous wastes are not counted when determining a facility's generator status. For example, hazardous wastes that are generated and then managed in RCRA permit-exempt units (e.g., wastewater treatment units, elementary neutralization units, totally enclosed treatment facilities) without prior storage or accumulation are not included in generator-status calculations. EPA has decided that it does not want generators to report these hazardous wastes on the biennial report either. The latest biennial report form instructions that we have (from 2013) note that the following wastes are not to be reported on Form GM: "Wastes managed immediately upon generation only in onsite elementary neutralization units, wastewater treatment units, or totally enclosed treatment facilities as defined in 40 *CFR* 260.10. (40 *CFR* 261.5(c)(2)) *Any hazardous waste residues generated from these units, however, must be reported on the GM Form.*" [Emphasis in original.] See also RO 14487.

- When entering hazardous waste codes on the biennial report, one question that often crops up is "Do I list the same codes on the biennial report as I did on the manifest?" EPA has noted the following in response: "the federal biennial reporting requirement is not contingent upon which waste codes happen to appear on the hazardous waste manifest, but on which hazardous wastes are generated by the reporter during the reporting period." [RO 11741] The agency noted in RO 13455 that, for the purposes of biennial reporting, a generator is not required to determine if listed wastes also exhibit characteristics (although they do need to do this for purposes of completing LDR notifications—see Section 13.6). In other words, a LQG must put all applicable listed waste codes on biennial reports but not necessarily any characteristic codes associated with those listed wastes. For example, if unused chloroform was disposed, the U044 code is required on the biennial report, but not the D022 code.

- Radioactive mixed wastes generated at a site must be included in biennial reports. [RO 13535]

- Material shipped to a laboratory or testing facility as part of a treatability study must be included in a generator's biennial report [§261.4(e)(2)(vi)], even though these treatability samples are not included (counted) when making generator status determinations. [§261.5(c)(1)] Note that the instructions to Form GM state that such wastes should not be reported.

- EPA was asked, "[d]oes a large quantity generator who generates a hazardous waste in December of a non-reporting year but ships it in January of the reporting year need to include that waste in their hazardous waste report?" The agency's answer was: "No. The LGQ does not need to report the waste on their GM form, but the TSDF who received the waste does need to report it on their WR form." [http://www.epa.gov/osw/inforesources/data/br13/faq.pdf]

6.8.2 Weight designation

When entering waste quantities on the biennial report, the agency has clarified that the weight of the container does not have to be included (i.e., the container itself is not considered a hazardous waste—only the waste within it). However, it is customary to include the total weight (weight plus container) on manifests, since transporters often charge on the basis of total weight shipped. [RO 12151]

The biennial report instructions say that, when reporting quantities for lab packs, 1) include the weight of the containers if they are disposed (e.g., landfilled) or treated (e.g., incinerated) with the waste; but 2) exclude the weight of the containers if the waste is removed from the containers before treatment or disposal.

6.8.3 Waste minimization efforts must be included

Sections §262.41(a)(6) and (7) require biennial reports completed by LQGs to include waste minimization information. LQGs must describe their efforts undertaken to achieve waste minimization and the actual changes in the volume and toxicity achieved relative to other years. [RO 13747] This came out of a HSWA requirement promoting a national policy of reducing or eliminating the generation of hazardous waste as expeditiously as possible. [RO 12932]

6.8.4 Generator certification

Section 262.41(a)(8) requires the generator or its authorized representative to sign a certification associated with the biennial report. "Authorized representative" is defined in §260.10 as "the person responsible for the overall operation of a facility or an operational unit (i.e., part of a facility), e.g., the plant manager, superintendent or person of equivalent responsibility."

In EPA Form 8700-13A/B, the certification statement says:

"I certify under penalty of law that this document and all attachments were prepared under my direction or supervision in accordance with a system designed to assure that qualified personnel properly gather and evaluate the information submitted. Based on my inquiry of the person or persons who manage the system, or those persons directly responsible for gathering the information, the information submitted is, to the best of my knowledge and belief, true, accurate, and complete. I am aware that there are significant penalties for submitting false information, including the possibility of fine and imprisonment for knowing violations."

Thus, the language allows the generator or its authorized representative to certify to the truth, accuracy, or completeness of the report based on the signer's personal familiarity with the information and upon his or her personal inquiry of those responsible for obtaining the information. [RO 11199]

6.8.5 State-specific requirements

States may impose biennial reporting requirements above and beyond the federal provisions. In those states, additional information will generally be required on the federal form. Alternatively, some states (e.g., Arkansas, Indiana, Ohio) require use of their own instructions and forms for fulfilling the reporting requirements.

Some states may require an annual, as opposed to biennial, report to be submitted. Additionally, small quantity generators may be required to submit an annual or biennial report per more-stringent state regulations.

6.9 EPA ID numbers

Section 3010 of the RCRA statute requires hazardous waste generators, transporters, and TSD facilities to notify their state or EPA of the location and general description of hazardous waste management activities, and of the characteristic and/or listed hazardous wastes handled. This statutory requirement has been codified in §262.12 for generators, §263.11 for transporters, and §§264/265.11 for TSD facilities.

Thus, with the exception of conditionally exempt small quantity generators (CESQGs), a generator of hazardous waste must obtain an EPA identification (ID) number before it can manage (i.e., treat, store, dispose, transport, or offer for transportation) hazardous waste. [§262.12(a)] Additionally, generators (other than CESQGs) must use hazardous waste transporters and TSD facilities with EPA ID numbers. [§262.12(c)] The 12-character EPA ID number is used by states and EPA to track hazardous waste activities. EPA ID numbers must appear on manifests that are required to transport RCRA-regulated wastes offsite; see Section 6.6.3 for details.

6.9.1 Obtaining an EPA ID number

An EPA ID number is obtained by completing the *Notification of RCRA Subtitle C Activity*, EPA Form 8700-12, available at http://www.epa.gov/osw/inforesources/data/form8700/8700-12.pdf. [§262.12(b)] Table 6-8 summarizes the activities that trigger initial notification to obtain an EPA ID number and those activities that would trigger subsequent renotification. The completed Form 8700-12 (also known as the Site ID form) should be sent to the applicable state or EPA regional office, as noted in the contact list at http://www.epa.gov/osw/inforesources/data/form8700/BRcontacts2013.pdf.

This form (consisting of 14 detailed fields) gives states and EPA a thorough description of the site, lists activities taking place at the site, and describes the types of wastes handled at the site. After the completed notification form is received and processed, the generator is sent a written acknowledgement that will include the assigned EPA ID number.

Note that some states have notification requirements that differ from federal provisions. Those states may use the Form 8700-12 or they may use a similar state form that requires additional information not requested in the EPA form.

Table 6-9 provides a summary of EPA guidance on issuing EPA ID numbers for many different situations. Additional detail is given below for a few of the more complicated scenarios.

6.9.1.1 One EPA ID number for one site

The general approach used by state and EPA program staff is that for any given location, only one EPA ID number is issued. The EPA ID number assigned for that site does not change over time. Conversely, if a generator (or TSD facility) moves to another site, the new site is assigned a new EPA ID number. [RO 11028]

Since EPA ID numbers are location-specific, if a new owner/operator takes over the site hazardous waste activities, the ID number issued to the site would generally go to the next owner/operator of the site. [RO 11028] An exception to this approach can occur when a new owner/operator takes over a site with an existing EPA ID number—the state agency or EPA may assign a new EPA ID number to the new owner/operator to avoid association with the previous owner's/operator's environmental record. Additionally, a new owner/operator may be generating a waste stream that has no connection to the previous owner's/operator's RCRA Subtitle C activities; this may be another situation when a second EPA ID number is issued to the site. [*RCRA Subtitle C EPA Identification Number, Site Status, and Site Tracking Guidance*, March 21, 2005, available from http://www.epa.gov/wastes/hazard/tsd/permit/tsd-regs/general/win-inform.pdf]

Table 6-8: Activities That Trigger RCRA Subtitle C Notification or Renotification

RCRA Subtitle C activity	Initial notification	Renotification
Generate, store, transport, recycle, treat, or dispose hazardous waste	■	
Transport, process, or re-refine used oil	■	
Burn off-spec used oil for energy recovery[1]	■	
Market used oil fuel	■	
Accumulate ≥5,000 kg universal waste onsite (i.e., a large quantity universal waste handler)[2]	■	
Move the business to another location and conduct hazardous waste activities at the new location	■	
Opt into or withdraw from the Part 262, Subpart K alternative program for managing hazardous waste at an eligible academic laboratory	■ (opt into)	■ (withdraw from)
Begin or stop managing hazardous secondary materials outside of the RCRA program under §§261.2(a)(2)(ii), 261.4(a)(23), (24), or (25)[3]	■ (begin)	■ (stop)
Change the ownership of the site	■[4]	■
Change the contact or mailing address for the site		■
Add or replace an owner since the last notification was submitted		■
Change the type of regulated waste activity conducted at the site		■[5]

[1]Used oil generators are not required to notify EPA, even if they burn their own off-spec used oil in used oil-fired space heaters per §279.23.
[2]If the large quantity handler already has an EPA ID number, he/she does not have to notify. [§273.32(a)(2), RO 13783, 14088]
[3]Such facilities must submit an addendum to the Site ID Form.
[4]If a new owner/operator notifies for a location that already has an EPA ID number, it can still be an initial notification if the new owner/operator has not previously submitted a notification form. [*RCRA Subtitle C EPA Identification Number, Site Status, and Site Tracking Guidance*, cited in Table 6-9]
[5]Sites must use the Site ID form to submit a subsequent notification if they wish to change generator status (i.e., LQG, SQG, or CESQG).

Source: Adapted from *Notification of RCRA Subtitle C Activity*, EPA Form 8700-12, available at http://www.epa.gov/osw/inforesources/data/form8700/8700-12.pdf.

6.9.1.1.1 Multiple generators on a single contiguous property

Contiguous property is a property that is one continuous plot of land or several plots of adjoining land. Multiple operators can exist on a single contiguous property, such as numerous tenants at large manufacturing or industrial parks, airports, Department of Defense bases, or office buildings. In all such cases, where there is separate generation of waste streams, EPA recommends an individual EPA ID number be issued to each tenant. Although the location address will be the same for all operators, they should be differentiated by indication of area at the site, suite numbers, floors, or locations on a floor. The owner/operator of each company is responsible for the waste activities on the property as well as the legal owner of the property. [*RCRA Subtitle C EPA Identification Number, Site Status, and Site Tracking Guidance*, cited previously, RO 14031]

Not addressed by the above EPA guidance is the question of whether two or more tenants on the same contiguous property could have different generator status. For example, could a small Air National Guard operation that has its own EPA ID number (even though it is located on a large Air Force base) be a CESQG even though the Air Force base is a LQG? It would seem so, but the above EPA guidance isn't definitive. RO 12661 suggests that a painting contractor that generates >100 kg of hazardous waste at individual work sites would be an SQG or LQG and thereby subject to the applicable hazardous waste regulations. Conversely,

Table 6-9: Issuing EPA ID Numbers for Different Situations

Situation	EPA guidance on issuing EPA ID number
Site ownership changes—new owner is not involved in RCRA Subtitle C regulated activities	New owner is not required to notify and receive an EPA ID number
Site ownership changes—new owner is involved in RCRA Subtitle C regulated activities	New owner should notify the state or EPA; generally, the new owner is assigned the previous EPA ID number issued at that site [RO 11028], but sometimes a new EPA ID number is issued to the new owner (see Section 6.9.1.1)
Two or more adjacent sites merge into one site	The regulatory agency with jurisdiction typically determines which property's or properties' existing EPA ID number(s) would be inactivated
Two or more adjacent sites divided by public roads	EPA usually uses the definition of "on-site" in §260.10 to determine when one or more EPA ID numbers are warranted; properties that are "on-site" may be covered under a single EPA ID number (see Section 6.9.1.1.1)
Multiple generators on a single contiguous property (e.g., numerous tenants at large manufacturing or industrial parks, airports, Department of Defense bases, or office buildings)	Multiple operators can exist on a single contiguous property; in all such cases, where there is separate generation of waste streams, EPA recommends an individual EPA ID number be issued to each tenant (see Section 6.9.1.1.1)
Elongated contiguous sites (e.g., pipelines, sewer systems, and city transit stations without distinct addresses)	Elongated contiguous sites may typically cover large distances and may even span county, borough, or state lines; generally, all portions of an elongated contiguous site should be covered under a single EPA ID number; requests for such an EPA ID number should be evaluated based on the criteria of land ownership, allocation of the waste generation to entities, and context of the project (one-time versus ongoing waste generation); if the site is issued one EPA ID number, the site should document the decision and provide a map of the points of waste generation covered
Two or more cogenerators at the same site	Cogenerators should mutually agree (by contract or other means) who will perform the duties of generator on behalf of all the parties; EPA expects the designated party to perform generator responsibilities, including applying for a single site EPA ID number (see Sections 6.5.1 and 6.5.6.4)
Importers	The importer of the hazardous waste must have an EPA ID number assigned to a U.S. address [§262.60(b)(1)], with the first two characters of the EPA ID number the same as the state postal code of the importer (see Section 6.6.8); facilities may use their current EPA ID number for their import activities
Transporters	In general, one EPA ID number is issued to transport companies at the principal business location that covers all vehicles belonging to that company (see Section 6.6.3.3); EPA ID numbers are also issued to U.S. marine vessels when transportation of hazardous wastes is involved[1]
Hazardous waste generated on ships and vessels during normal transportation activities	For marine vessels, the point at which hazardous waste is typically considered generated is the point when the vessel reaches a port or dock located in U.S. waters and hazardous wastes are physically "off-loaded" and removed from the vessel to a shore facility—both U.S. and foreign-flagged vessels within the territorial waters of the United States may need an EPA ID number if ship-generated wastes are removed from the vessel to a shore facility (the issuing authority for the EPA ID number would typically determine whether the number should be permanent or temporary based upon the frequency of hazardous waste generation and other situation-specific facts)[2]; hazardous waste in vessels can also be considered generated for EPA ID number purposes when it is stored on vessels for more than 90 days after the vessels are put into reserve or are otherwise no longer in service (see Sections 6.5.6.6)

CHAPTER 6 *Generator Issues*

Table 6-9: Issuing EPA ID Numbers for Different Situations

Situation	EPA guidance on issuing EPA ID number
Hazardous waste generated on ships and vessels during maintenance activities	The identity of the generator depends on the agreement among the owners/operators of the 1) shipyard or port, 2) vessel, and 3) material in the vessel (if applicable); the person who performs the duties of the generator must have and use an EPA ID number for the site at which hazardous wastes are removed from the vessel[2] (see Section 6.9.1.1.2)
Superfund sites	A Superfund site would be issued a permanent EPA ID number in the same manner as any RCRA Subtitle C regulated generator or TSD facility; EPA ID numbers should be established at the beginning of Superfund remediation activities, since wastes generated by the cleanup can only be removed from the site (or stored for more than 90 days) using an active EPA ID number
Abandoned sites	The state or EPA region cleaning up the site (or their contractor) may be considered the generator of the waste and the contact person if responsibility/liability cannot be immediately assigned; typically, a generator EPA ID number is issued to this generator to, among other things, move hazardous wastes offsite
Offsite spills	The party causing the spill would be the generator who obtains an EPA ID number for the location of the spill; the name of the site should generally identify the activity and the responsible party (see Section 6.9.1.2)
Hazardous wastes that migrate offsite	Migrating wastes can contaminate property or bodies of water owned/operated by entities other than the owner/operator of the site whose hazardous wastes migrated; the EPA ID number of the site whose waste has migrated to offsite locations would generally be used for activities such as corrective actions resulting from the waste migration
Clandestine drug laboratories	As the party accepting responsibility for hazardous waste found at these sites, local law enforcement or federal law enforcement (i.e., the Drug Enforcement Agency) becomes the generator of the waste and is responsible for managing it in accordance with applicable RCRA Subtitle C regulations, including obtaining an EPA ID number[3] (see Section 6.9.1.2)

[1] As a general rule of thumb, an EPA ID number is not required or assigned to a transporter if a RCRA manifest is not required to accompany the hazardous waste. However, when a hazardous waste manifest is required, U.S. vessels that commercially transport hazardous wastes are subject to Subtitle C transporter requirements. Likewise, foreign-flagged vessels that engage in hazardous waste transportation activities within the territorial waters of the United States (when such transportation is not considered or otherwise associated with hazardous waste imports or exports) would also be subject to Subtitle C transporter requirements. U.S. or foreign-flagged vessels that are engaged in hazardous waste import or export activities generally do not require an EPA ID number while operating in the territorial waters of the United States. For these situations, since the port is generally considered to be either the "cradle" (for imports) or the "grave" (for exports), a manifest would not be used and, as such, an EPA ID number would not be issued or assigned to the transport vessel. [RO 11894, 14595]

[2] As an alternative to issuing an EPA ID number to a specific vessel, the state or EPA may consider allowing the dock or port operator to either assume the generator responsibilities or become a cogenerator with any U.S. or foreign-flagged vessel using their own EPA ID number. The state or EPA may also consider allowing a service contractor (other than the dock or port operator) hired to remove hazardous wastes from a vessel to assume the generator responsibilities or become a cogenerator with the U.S. or foreign-flagged vessel using their own EPA ID number.

[3] In cases of an imminent hazard, however, RCRA Subtitle C requirements can be waived (RCRA Section 7003).

Source: Adapted from *RCRA Subtitle C EPA Identification Number, Site Status, and Site Tracking Guidance*, March 21, 2005, available from http://www.epa.gov/wastes/hazard/tsd/permit/tsd-regs/general/win-inform.pdf, unless otherwise noted.

if the contractor generates ≤100 kg of hazardous waste, it would be a CESQG. Although not stated in that guidance, it seems reasonable that the painting contractor could be a CESQG while the facility as a whole is an LQG.

In RO 11028, EPA noted that "[c]are must also be taken to prevent subdivision of the waste at one location for the purpose of avoiding regulation, e.g., by slipping under the small generator limitation." A similar statement is in RO 11916. One might surmise from these statements that each generator gets their own independent generator status.

EPA usually uses the definition of "on-site" in §260.10 to determine when one or more EPA ID numbers are warranted for adjacent properties. Properties that are "on-site" may be covered under a single EPA ID number; for example, if access between adjacent sites is possible by going directly across (and not traveling along) a public road, the two properties are one site and need only one EPA ID number. [February 12, 1997; 62 FR 6647, document record detail to RO 11884, and RO 13314, 13351, 14031]

In another example, two autonomous divisions of a company occupy different portions of a contiguous piece of property. Since the company as a whole would meet the definition of "person" in §260.10, although autonomous, the divisions would not generally be considered separate generators if they operate on a geographically contiguous piece of property meeting the definition of "individual generation site" in §260.10. Although there is no specific prohibition in the regulations against a generator maintaining multiple EPA ID numbers for an individual generation site, EPA expects each individual generation site to have one EPA ID number. [RO 11916]

Q: A generator has multiple structures within the same contiguous property. Is each structure required to have an individual EPA ID number?

A: No. For the purposes of generator notification and obtaining EPA ID numbers, and assuming the structures are "on-site" as defined in §260.10, one ID number is sufficient for all structures at the site. EPA ID numbers are issued to facilities on a by-site basis. There is no regulatory definition for the term "by-site"; however, it refers to where a hazardous waste is generated.

For example, if a single company operates three laboratories on a single piece of property, all three laboratories may share one EPA ID number, and the waste from all three laboratories may be evaluated together. However, if the company operates three laboratories at three different locations that are not on contiguous property, each laboratory is viewed as a separate generator and is required to obtain an individual EPA ID number. Furthermore, when the laboratories or structures are owned by different people, the generator must obtain one ID number for each laboratory or structure even if the regulated activity is taking place on a contiguous piece of property. [http://www.epa.gov/osw/inforesources/data/br13/faq.pdf]

Q: Corporation A owns a large site. Corporation B, a wholly owned subsidiary of Corporation A, is a permitted TSD facility on the site with an EPA ID number associated with this site activity. Corporation C, another wholly owned subsidiary of Corporation A, is also located on this site and will be generating hazardous waste. Should Corporation C use the EPA ID number which is associated with the site (i.e., Corporation B's ID number), or is Corporation C required to obtain its own EPA ID number?

A: In this situation, Corporation B and Corporation C are two distinct entities (i.e., persons). They must each apply for a separate EPA ID number. Even though identification numbers are usually site-specific, where different people conduct different regulated activities on a site, a person conducting each regulated activity must obtain an EPA ID number. [RO 13129]

Q: A chemical company stores some of its products in containers at a warehouse owned by a separate company. Periodically, warehouse personnel will be the cause of a spill or damaged container resulting in generation of waste. For example, warehouse personnel driving a forklift cause a drum to fall off of a pallet or a fork to puncture a container, both of which cause a spill of product and generation of hazardous waste. "Generator" is defined in §260.10 as the "person, by site, whose act…produces hazardous waste…or whose act first causes a hazardous waste to become subject to regulation." Who is the generator of hazardous waste—the warehouse or the chemical company?

A The warehouse employee's act produces hazardous waste, so the warehouse is a generator. However, the chemical company would also be a generator, as EPA noted: "The owner of the product or raw material being stored or transported…also fits the definition of 'generator' of the hazardous waste…." [October 30, 1980; 45 *FR* 72026] Thus, the warehouse and chemical company would be cogenerators, and the two parties need to agree (in writing) who will take on the generator responsibilities (e.g., obtaining an EPA ID number, manifesting, 90/180/270-day accumulation, inspections, recordkeeping, etc.). The cleanest approach is probably to have the warehouse agree to be the generator of hazardous wastes that are generated at the warehouse as a result of storing the chemical company's products. Thus, the warehouse gets an EPA ID number from the state and represents itself as the generator whenever hazardous wastes are manifested offsite (i.e., the warehouse company's name/EPA ID number go on the manifest in the generator blocks—not the chemical company's name). The chemical company subsequently reimburses the warehouse for those hazardous waste disposal costs/activities. In this case, clearly the warehouse is a cogenerator of the hazardous waste and the two companies have decided that the warehouse will be responsible for complying with the RCRA generator responsibilities.

State hazardous waste programs may have specific provisions that are applicable in addition to the federal program, so requests for multiple EPA ID numbers for a single contiguous property would generally be evaluated on a case-by-case basis by the authorized state.

6.9.1.1.2 *Hazardous waste generated on ships and vessels during maintenance activities*

Periodic activities that generate large quantities of hazardous waste include major repair work done in shipyards, refurbishing portions of the ship while docked, and cleaning out sumps or other waste collection devices. The person who removes hazardous waste from a vessel will be jointly and severally liable as a cogenerator, along with the owner/operator of the vessel and the owner of the material in the vessel (if applicable). These three potential cogenerators should mutually agree (by contract or other means) who will perform the duties of generator on behalf of all the parties. [RO 11571, 12727] The person who performs the duties of the generator must have and use an EPA ID number for the site at which hazardous wastes are removed from the vessel.

The state or EPA may consider allowing a service contractor (other than the dock or port operator) hired to remove hazardous wastes from a vessel to assume the generator responsibilities or become a cogenerator using their own EPA ID number.

6.9.1.2 Obtaining a temporary, emergency, or other one-time-use EPA ID number

RCRA Subtitle C regulations do not distinguish between the generation of waste on an ongoing basis versus waste generated from a single event at a site. Therefore, all temporary and emergency small and large quantity generator sites should generally be assigned a unique EPA ID number.

A temporary generator site is where waste is not generated from ongoing industrial processes but rather through remediation or "one-time events," such as plant cleanout and closure or process equipment change. Unlike emergency generator sites, the state or EPA region usually receives an 8700-12 notification form from the temporary generator and issues an EPA ID number. However, the period of operation as a RCRA Subtitle C site is limited and typically short.

As noted earlier, the EPA ID number typically consists of 12 characters—although ID numbers assigned for state use can contain fewer characters—usually 3 letters followed by 9 numbers. Generally, the first two characters represent a state postal code or a two-letter code indicating a tribe or trust territory. The third character of the EPA ID number has changed over the years, but a "P" as the third character (e.g., LAP123456789) usually denotes a temporary EPA ID number issued by a state or EPA. [*RCRA Subtitle C EPA Identification Number, Site Status, and Site Tracking Guidance,* cited previously]

Q When a contractor that operates a mobile solvent recycling process visits a generator's site, who has to submit an 8700-12 notification to EPA or the state?

A Typically, the generator's facility will already have submitted the notification. If this generator consents to performing the generator responsibilities associated with any still bottoms from the mobile unit, additional notification is not required. On the other hand, if the recycler will be the generator of record who will be responsible for manifesting, they would need to submit a notification form and obtain an EPA identification number for the site. [RO 13280]

Emergency generator sites are those where the generation situation is unforeseen, uncontrollable, short-term, and not expected to exceed 30 days. Emergency generator sites need to be distinguished from typical RCRA Subtitle C regulated sites since the standards of quality and completeness applied to the Site ID form notification would probably be less stringent. Emergency generator sites typically do not have to submit a complete 8700-12 notification form. Rather, a brief telephone notification form is acceptable to take provisional details on the incident site and to ensure that an EPA ID number has not already been assigned. [*RCRA Subtitle C EPA Identification Number, Site Status, and Site Tracking Guidance,* cited previously] RO 12016 noted the need for such emergency EPA ID numbers to be issued for spill cleanup residues to be transported offsite. EPA regional offices may issue such ID numbers orally over the telephone.

EPA's policy regarding rapid issuance of EPA ID numbers for emergencies or unusual circumstances is extended to Drug Enforcement Agency agents to allow expedited shipment of waste from secured clandestine drug laboratories to an offsite TSD facility. The policy allows the law enforcement agency to obtain EPA ID numbers by telephone request. [RO 12855] Similar guidance is contained in RO 11363 for issuance of emergency EPA ID numbers by telephone to Bureau of Alcohol, Tobacco, and Firearms agents when they generate and need to transport explosives that present an immediate safety threat.

6.9.1.3 When renotification of hazardous waste activities is required

Even if a site has submitted an initial notification and received an EPA ID number, it may be required to submit a subsequent renotification. Renotification is required under the circumstances listed in Table 6-8. Of the conditions requiring renotification listed in Table 6-8, the last is the most misunderstood. Renotification is required if the type of regulated waste activity conducted at the site changes. But just how does this apply?

Item 10 on Form 8700-12 is the field used by the site to identify its regulated waste activities. These activities include, but are not limited to, hazardous waste generation, transportation, treatment, storage, recycling, or disposal; universal waste management; used oil processing; and underground waste injection. Quite simply, should any of these waste activities change (e.g., from No to Yes or Yes to No), subsequent renotification is required. For example, renotification would be required if a large quantity generator begins transporting hazardous waste or begins to recycle hazardous waste onsite.

What about renotification for the addition or deletion of hazardous waste codes managed at the site? Federally, the addition or deletion of waste codes managed onsite does not appear to trigger the renotification requirements. As noted previously, the regulated waste activities are found in Item 10, whereas the waste codes are included in Item 11. Thus, adding or dropping a hazardous waste code at the site does not, by itself, alter the regulated activity on the form:

"Persons who have provided proper notification of hazardous waste activity may later begin to handle additional hazardous wastes not included in the original notification…. EPA will not require these persons to file a new notification under Section 3010 with respect to those wastes." [February 26, 1980; 45 *FR* 12747]

However, we have found that this is a state-dependent situation. From phone conversations, we have discovered that many states do not require renotification when waste codes managed at a site change. Other states require renotification if any waste codes are added or deleted from a waste stream, while still other states require renotification only for the addition or deletion of acutely hazardous wastes (the P-codes and the dioxin F-codes).

Is there a time limit on when a site must renotify? There is no federal time limit that we could find regarding when a site must renotify, but it's best to check with the state for guidance. Many states have a preference of renotification within 15 to 30 days.

6.9.1.3.1 *Episodic generators*

The instructions for the Site ID form (see Table 6-8) require the owners/operators of a site to renotify if they wish to change generator status. However, EPA recommends leaving the generator status constant throughout the year. The status chosen (SQG, LQG, etc.) would presumably be the highest generator status that the facility would experience so the site is not under-reporting waste generation activities. [*RCRA Subtitle C EPA Identification Number, Site Status, and Site Tracking Guidance*, cited previously] Some states may have specific guidelines on when episodic generators should renotify.

6.9.2 Exemptions from having an EPA ID number

As discussed in the next two subsections, there are two situations where the federal RCRA program does not require the use of EPA ID numbers for entities managing hazardous waste.

6.9.2.1 Conditionally exempt small quantity generators

CESQGs are exempt from the requirement to obtain an EPA ID number when managing hazardous waste. [§261.5(b)] However, keep in mind that some states have programs that are more stringent than the federal regulations and require CESQGs to have EPA ID numbers or state ID numbers. Additionally, CESQGs are required to obtain an EPA ID number if they engage in one or more of the used oil or universal waste management activities identified in Table 6-8 or if they fail to satisfy the requirements of §261.5(f), (g), and (j).

6.9.2.2 Managing hazardous waste during an immediate response

Actions taken immediately to respond to spills, discharges, and other situations that pose an imminent hazard are termed an "immediate response." Such activities are not subject to RCRA permitting or substantive standards. [§§264.1(g)(8), 265.1(c)(11), 270.1(c)(3)] The immediate response exemption (discussed in Section 7.5.4) may also apply to the transportation of hazardous wastes in immediate response situations. However, the transportation is exempt only to the extent necessary to respond to the immediate threat. Hence, EPA expects the transportation would normally cover a relatively short distance. [RO 11363, 11370, 13574] In addition, officials may waive the EPA ID number requirements for generators and transporters engaged in immediate hazardous waste removal following a discharge incident. [§263.30(b), RO 12016, 12748]

6.10 Academic labs/Subpart K

Academic institutions often find it difficult to make hazardous waste determinations per §262.11 at the point wastes are generated in their laboratories. This is largely because most of the individuals in laboratories generating wastes are students, who are usually untrained to make hazardous waste determinations. To address this issue, EPA added a new, optional Subpart K to the hazardous waste generator standards in Part 262 that allows non-RCRA management of unwanted lab materials until a hazardous waste determination is made. These alternative requirements were published on December 1, 2008. [73 *FR* 72912] Minor technical corrections were made to Subpart K on December 20, 2010. [75 *FR* 79304]

To be eligible to operate under Subpart K, a facility must be a 1) college or university, 2) nonprofit research institute that is owned by or has a formal written affiliation with a college or university, or 3) teaching hospital that is owned by or affiliated with a college or university. However, "government facilities with laboratories that are operated by colleges and universities (such as many of the Department of Energy's laboratories) are not eligible to opt into Subpart K, because the government facility is not an eligible academic entity and the laboratories are not owned by an eligible academic entity." [73 *FR* 72924]

A laboratory is defined as "an area owned by an eligible academic entity where relatively small quantities of chemicals and other substances are used on a nonproduction basis for teaching or research (or diagnostic purposes at a teaching hospital) and are stored and used in containers that are easily manipulated by one person. Photo laboratories, art studios, and field laboratories are considered laboratories. Areas such as chemical stockrooms and preparatory laboratories that provide a support function to teaching or research laboratories (or diagnostic laboratories at teaching hospitals) are also considered laboratories." [§262.200]

Basically, Subpart K allows unused or used chemicals, solvents, and analytical residues that are no longer needed in a lab to be managed outside of the formal hazardous waste program until RCRA-trained personnel determine whether the materials are wastes and, if they are, whether they are hazardous wastes. These "unwanted materials" are defined as follows: "any chemical, mixtures of chemicals, products of experiments or other material from a laboratory that is no longer needed, wanted or usable in the laboratory and that is destined for hazardous waste determination by a trained professional." [§262.200] This definition allows some unwanted material to still be usable in another lab setting, thereby not meeting the definition of solid waste. Thus, unwanted materials are not subject to RCRA hazardous waste regulations but to performance-based standards while the unwanted materials remain in the lab. EPA establishes a framework for the performance-based standards in Subpart K, and the academic facility has to flesh out these standards by preparing a formal, enforceable laboratory management plan. [§262.214] For example, eligible facilities have up to six months to remove unwanted materials from labs; however, if the volume of these materials exceeds 55 gallons, then all unwanted materials have to be removed from the lab within 10 calendar days. [§262.208]

Finally, the new Subpart K regulations provide an incentive for eligible academic facilities to conduct periodic laboratory cleanouts of old, unneeded chemicals. Once a year, each laboratory will have 30 days to conduct a cleanout and move unwanted materials to an onsite central accumulation or permitted area. And, significantly, the facility will not have to count any listed or characteristic hazardous wastes (that are unused chemicals) generated during those 30 days towards the facility's generator status. [§262.213]

This final rule became effective in the federal regulations on December 31, 2008. EPA indicated that this rule is neither more nor less stringent than the normal RCRA standards; therefore, states are not required to adopt the rule. The changes will not be effective in RCRA-authorized states until those states adopt the rule. RO 51349 is a brochure entitled "Managing Laboratory Hazardous Waste, An Introduction to the Academic Laboratories Rule Subpart K."

Additional information on the Subpart K regulations is available at: http://www.epa.gov/waste/hazard/generation/labwaste/.

Hazardous Waste Treatment

Treatment issues and how to treat without getting a RCRA permit

Because many hazardous wastes are quite dangerous (e.g., they are toxic, flammable, carcinogenic, etc.), treatment of such materials can present serious risks to human health and the environment. To help ensure that hazardous wastes are treated safely and responsibly, EPA generally requires that facilities conducting such activities obtain a RCRA permit.

However, many facility owners/operators are reluctant to expose themselves to the process of obtaining and maintaining a RCRA permit. Getting a RCRA permit 1) is difficult and requires considerable resources; 2) exposes the facility to the RCRA corrective action program that requires owners/operators to investigate their entire site and implement extensive cleanup measures if contamination is found; and 3) raises the visibility of the facility on the state's "radar" in terms of regulatory oversight and inspection frequency (typically once a year). Thus, facility owners/operators frequently ask whether their handling of certain types of hazardous waste comprises "treatment" and, if so, whether such treatment requires a RCRA permit.

In this chapter, we examine (in Section 7.1) EPA's definition of what constitutes hazardous waste treatment. Numerous examples are given, illustrating the agency's interpretations for situations involving waste bulking and containerizing, mixing, fuel blending, recycling, and other waste management processes; in each case, we focus on *why* EPA does or does not consider a given activity to be treatment.

While most types of hazardous waste treatment require a RCRA permit, there are eight options for treating hazardous waste that are exempt from this requirement. That is, a hazardous waste generator could engage in one of these eight activities and remain just a generator. The remaining sections of this chapter, listed below, review these eight permit-exempt treatment methods, including the conditions for obtaining the exemption and examples of their occurrence in industry:

- Treatment in wastewater treatment units (Section 7.2),
- Treatment in 90/180/270-day accumulation units (Section 7.3),
- Recycling (Section 7.4),
- Elementary neutralization (Section 7.5.1),
- Treatment in totally enclosed treatment facilities (Section 7.5.2),
- Adding absorbents to wastes (Section 7.5.3),
- Immediate responses (Section 7.5.4), and
- Burning small quantities of wastes in onsite units (Section 7.5.5).

CHAPTER 7 Hazardous Waste Treatment

7.1 The definition of "treatment"

Before getting into the details of how facilities can treat hazardous waste without a permit, let's examine how "treatment" is defined. According to §260.10, "treatment" means:

> "[First part:] [A]ny method, technique, or process, including neutralization, designed to change the physical, chemical, or biological character or composition of any hazardous waste [Second part:] so as to neutralize such waste, or so as to recover energy or material resources from the waste, or so as to render such waste nonhazardous, or less hazardous; safer to transport, store, or dispose of; or amenable for recovery, amenable for storage, or reduced in volume."

This definition actually has two parts, which we have identified for clarity. [RO 12335, 13346] The first part of the definition focuses on whether a given activity changes the properties of a hazardous waste. Simply put, if an activity does not change a hazardous waste, it is not treatment.

The second part of the definition looks to the intent of the waste management activity. In other words, for a process to be treatment, it must not only change the hazardous waste (i.e., meet the first part of the definition), but it must do so for one of the reasons listed in the second part of the definition. Thus, for example, placing a granular hazardous waste in a large shipping container and moving that container from the generation point to a remote shipping dock for pickup by a transporter might compact the waste, thereby changing its density. This clearly meets the first part of the definition of treatment. Since the intent of this activity was simply to move the waste from one location to another, however, the second part of the definition is not met and the process is not treatment. In contrast, placing a hazardous waste in a vibrating container whose purpose is to compact that waste is treatment—in this case, the intent of the activity is to reduce the waste's volume.

Before moving to a more detailed discussion of the two parts of the treatment definition, it is worth mentioning that the definition of treatment in §260.10 has been upheld as reasonable in federal court (*Shell Oil v. EPA*, 950 F.2d 741, 753-56, U.S. Circuit Court for the District of Columbia Circuit, 1989). [RO 13651]

7.1.1 Does the activity change the hazardous waste?

In most cases, it is fairly straightforward to determine if an activity has changed the physical, chemical, or biological character or composition of a hazardous waste. To be considered treatment, however, a process must modify a material that is already considered a hazardous waste. This concept is discussed in the next subsection.

7.1.1.1 Point of generation issues

Over the years, EPA has had a difficult time defining the specific point when and where a hazardous waste is generated, called the point of generation (POG). Although a detailed discussion of this term is included in Section 14.1, the POG sometimes plays an important role in deciding whether hazardous waste treatment is occurring.

To illustrate this point, a company generated a corrosive (D002) waste stream from a reactor used in a polymer manufacturing process. The company decided to change its manufacturing operation and began adding reagents to the reactor so that the waste stream it produces is no longer corrosive (the reagents do not affect the finished product specifications). This change in manufacturing operations is not treatment of hazardous waste, because the waste is not regulated until it leaves the reactor vessel. "The hazardous waste regulations do not restrict the use of ingredients for the purpose of preventing waste from exhibiting a hazardous characteristic." [RO 11624] If, alternatively, the facility had added the reagents to the D002 waste after it exited the manufacturing process unit, that would have constituted hazardous waste treatment.

At a foundry, a baghouse is directly connected to a cupola furnace to minimize fly ash emissions. Because the ash exhibits the toxicity characteristic due to high levels of heavy metals, the facility wants to add treatment reagents in the ductwork between the cupola and baghouse. The company asked EPA if that would be treatment of a hazardous waste. The agency responded that the point of hazardous waste generation in this manufacturing setting is typically the bottom of the baghouse hoppers; thus, any processing that occurs prior to that point would not be hazardous waste treatment subject to RCRA requirements. [RO 11921, 12824, 14200]

In another example, consider a contractor who uses a mixture of iron-containing additives and other abrasives to remove lead-based paint (LBP) from structures. Because the iron in the additives interferes with the TCLP and "masks" the lead in the paint, the resulting LBP waste typically passes the TCLP, so it is nonhazardous. When asked if this process was hazardous waste treatment, the agency responded that the POG for the LBP waste was "once the paint has been removed from the surface of the structure." [RO 14069] Thus, although the process of adding the iron-containing additives clearly modified the resulting material's chemical properties, that material was not a hazardous waste at its point of generation, so treatment was not occurring. Note that EPA believes that the masking effect of the iron is temporary. Thus, if the LBP waste eventually fails the TCLP prior to disposal, it would be subject to RCRA regulations. Further, the generator could be held liable under CERCLA for environmental damages caused by any lead that eventually does leach out of the waste. [May 26, 1998; 63 *FR* 28568, RO 11624, 14069]

To summarize, EPA has stated the following regarding waste treatment and the POG:

"The [agency] will not exercise jurisdiction over in-plant production process modifications that result in less waste or less toxic waste. Modifications to the product manufacturing process are not within the purview of RCRA. A device that treats a waste stream in order to detoxify it or reduce its volume, however, is clearly within the authority of RCRA. In other words, production process changes are not regulated under RCRA; treatment of a waste stream is regulated under RCRA." [OSWER Directive 9560.12(85), available at http://nepis.epa.gov/EPA/html/Pubs/pubtitleOther.html by downloading the document numbered OSWERDIR95601285]

7.1.1.2 Container-related issues

Whether or not hazardous waste treatment is occurring sometimes has more to do with the container involved than the contents of that container. For example, the agency determined that a decanning process that simply aggregates waste pesticide from small containers into larger containers, with no intent to change the waste's characteristics or render it nonhazardous, was not treatment. [RO 12214]

In another situation, however, the agency noted that a drum-shredding unit processing containers filled with hazardous waste *was* a treatment unit; in this case, EPA's determination of treatment presumably was based on the fact that the collection/accumulation of wastes released during container shredding was actually changing the properties of those wastes. [RO 11466, 13202, 14241] In a separate example, water was added during the crushing of alcohol waste-bearing containers, ostensibly to reduce the risk of fires and explosions. Because the water also diluted the alcohol, rendering it nonhazardous, the physical character of the waste was changed and the agency found that treatment was occurring. [RO 12214]

Containers themselves are sometimes considered hazardous waste, as in the case of aerosol cans (which can explode when heated, potentially making them D003 reactive wastes). Although EPA is more likely to refer such situations to its regional offices or the states for case-by-case decisions, the agency maintains that puncturing and draining aerosol cans can be considered hazardous waste treatment. [RO 11414, 11466, 13225] Such processes modify the physical character of the waste

7.1.2 Why was the hazardous waste changed?

As noted in Section 7.1, the second part of the hazardous waste treatment definition addresses the more subjective issue of the intent of the waste handler. If managing the waste in a certain way modifies the properties of that waste (i.e., meets the first part of the definition) *and* the intent of the owner/operator is to achieve one of the following six results, then hazardous waste treatment is occurring:

1. Neutralize the waste;
2. Recover energy or materials from the waste;
3. Render the waste nonhazardous or less hazardous;
4. Render the waste safer to transport, store, or dispose;
5. Make the waste more amenable for storage or recovery; or
6. Reduce the volume of the waste.

In the sections that follow, we look at a variety of waste management activities and give EPA's opinion on whether such processes are or are not hazardous waste treatment. As detailed below, waste bulking and containerization are generally *not* considered hazardous waste treatment. In contrast, most instances of mixing, fuel blending, recycling, and other miscellaneous waste processing activities *are* considered treatment.

7.1.2.1 Waste bulking and containerization

Bulking or otherwise containerizing two or more hazardous wastes for transportation may change the physical and chemical properties of the waste. Certain physical properties, such as specific gravity, flash point, and viscosity may change, and the chemical compositions of the unmixed wastes may be different from that of the mixture. If so, the first part of the treatment definition is satisfied. But did the generator or transporter mix the wastes for any of the six reasons noted in Section 7.1.2? EPA has published the following guidance concerning the consolidation of multiple hazardous wastes into a single container (bulking) and related issues associated with waste containerization:

- Activities such as bulking, containerizing, consolidating, and deconsolidating are not considered treatment, as long as no blending (e.g., selective mixing to meet a fuel specification) is taking place. So, for example, the practice of repackaging wastes from larger to smaller containers is not treatment. If the intent of the deconsolidation is to simply make it more efficient and cost-effective to transport the shipment, that doesn't meet the second part of the treatment definition. [RO 13308, 13720]

- Placing different wastes, each of which is already considered a viable hazardous waste fuel, into the same tank truck is not considered treatment. [RO 11281]

- Bulking characteristic or listed hazardous waste shipments and/or consolidating compatible hazardous wastes to achieve efficient transportation may result in an incidental reduction of the hazards of the waste mixture. However, such activities are not treatment, because they are not designed to render the waste nonhazardous, less hazardous, or safer to transport, store, or dispose. In other words, as long as materials handled in this manner are still sent to a TSD facility, any reduction of the hazard due to combining the wastes is incidental (i.e., not the intent of the activity), and the process is not considered hazardous waste treatment. [RO 11497, 12458, 13764]

Even though they are generally not considered to be treating hazardous waste, transporters who combine materials with differing DOT shipping descriptions are subject to certain waste generator requirements (e.g., they must issue a new manifest when they combine certain wastes). This is spelled out in §263.10(c)(2); DOT shipping descriptions can be obtained in 49 *CFR* 172.101. [RO 11425, 11567, 12458, 13272]

7.1.2.2 Mixing

Hazardous wastes are often mixed with other materials in order to 1) render the waste nonhazardous; 2) effect a positive change to the waste stream's properties; or 3) facilitate storage, transport, recovery, or disposal of the waste. Clearly, any such action constitutes hazardous waste treatment based on the definition listed earlier since the properties of a hazardous waste are being modified and the goal or intent of the effort is to achieve one of the six results listed in Section 7.1.2.

Most of EPA's guidance on this issue revolves around the mixture rule. As explained in detail in Section 5.1, the mixture rule provides, in part, that a mixture of a characteristic hazardous waste or a hazardous waste listed solely because it exhibits the characteristics of ignitability, corrosivity, or reactivity (i.e., an ICR-only listed waste) with a solid waste is no longer hazardous if the mixture doesn't exhibit a characteristic. Such a combination of materials under the mixture rule is always considered hazardous waste treatment. [RO 11213, 11257, 11358, 12032, 12561, 14256]

In two related situations, EPA noted that 1) mixing two listed hazardous wastes is treatment if the process makes the combined wastes less hazardous or safer to transport [RO 13346], and 2) mixing a D001 ignitable hazardous waste with used oil is treatment if the intent is to make the D001 waste more amenable for recovery or less hazardous. [RO 13570]

As noted above, a facility's intent in mixing hazardous waste with other substances is at the heart of deciding whether treatment is occurring. Another example illustrating EPA's consideration of intent is presented in Case Study 7-1.

7.1.2.3 Fuel blending

Although not defined in the regulations, the term "fuel blending" is generally used to describe the

---

Case Study 7-1: Incidental Mixing vs. Purposeful Dilution

A facility uses a 50% alcohol/50% water mixture as a dip solution for certain small parts. Periodically, operators pour spent dip solution (which is D001 ignitable) into a plant sewer line that leads to an onsite wastewater treatment plant. About 12 gallons of spent solution are discarded each day, while the treatment plant processes nearly 2 million gallons of wastewater per day. By the time the alcohol waste reaches the treatment plant, it is (obviously) sufficiently diluted such that it is no longer hazardous. The lab asked EPA if this amounts to hazardous waste treatment in the sewer line.

EPA's response was that "pipes are designed and used to convey, not treat, wastes to the biological treatment plant that degrades the alcohol. Thus, the dilution is incidental to the transport of the waste to the wastewater treatment plant where treatment takes place. Therefore, in this case, the dilution is not treatment."

Elaborating on the issue, the agency stated that "dilution is not considered to be treatment when the characteristic waste is diluted while being conveyed to *acceptable* treatment…." [Emphasis added.] In other words, as long as the wastewater treatment plant is effective at treating the waste (and certainly biological treatment is effective at treating an alcohol), this practice is aggregation for centralized treatment and not impermissible dilution or treatment.

Finally, the lab asked EPA about the alternative practice of first diluting the alcohol/water waste in a sink prior to discharge to the sewer. EPA responded that "[i]f hazardous waste is diluted in the sink, it is hazardous waste treatment, since the dilution is intentional, rather than merely incidental to conveyance to the treatment plant. Intentional dilution of waste prior to discharge to decrease its incompatibility, ignitability, reactivity, etc. in the pipelines constitutes treatment." [RO 11173]

combination of hazardous wastes and other materials to create a material that is amenable to burning for energy recovery. Since processes that fit this description change the properties of a hazardous waste in order to meet a fuel specification and recover energy, both parts of the treatment definition are met. Hence, fuel blending is considered treatment by EPA. [RO 11497, 11881, 13577, 13764]

In a related matter, the agency has stated in guidance that it considers unit operations (beyond simple commingling) practiced during the preparation/processing of hazardous waste fuels to be treatment. This includes air stripping, centrifugation, and phase separation. Additionally, adding sodium hydroxide or other materials to hazardous wastes to form homogeneous mixtures/blends of fuels is treatment. [RO 13512, 13651]

Facilities conducting fuel blending have tried to argue that their hazardous waste management practices are a form of recycling (i.e., recovery of energy from a waste stream) and should be exempt from RCRA permitting or management standards under §261.6(c)(1). EPA's response is that "[t]here may be some recycling operations at a fuel blending facility that are exempt from permitting, even though the fuel blending process itself is not exempt." [RO 11881] See also April 13, 1987; 52 *FR* 11820, RO 11238, 11411, 11497, 13512.

7.1.2.4 Recycling

Similar to mixing and fuel blending, recycling is another method of managing hazardous waste that EPA considers to be treatment. However, it is a type of treatment that EPA wants to encourage. Therefore, recycling is generally not subject to RCRA permitting or management standards. [§261.6(c)(1)] See Section 7.4 for details.

Thus, for example, a firm that uses solvent to wash parts and periodically distills dirty solvent for reuse is changing the physical properties of the waste so as to recover material resources from the waste. Thus, the facility is conducting hazardous waste treatment in the form of recycling. [RO 12865]

Similarly, grinding a mixture of lead-based paint (LBP) chips and wood debris from LBP removal operations to prepare the waste for shipment to a lead reclaimer is a form of recycling and is considered hazardous waste treatment. [RO 11880] In both cases, however, the subject companies would qualify for the recycling exemption from RCRA requirements discussed in Section 7.4. Also, the regulatory status of the materials may change and the spent solvent or chips/wood debris might not be solid wastes if recycled in accordance with the provisions of EPA's October 30, 2008 rule. [73 *FR* 64668] See Section 11.2.3 for a discussion of this rule.

7.1.2.5 Miscellaneous processes

In addition to the above discussions, EPA considers many other processes to be hazardous waste treatment. Table 7-1 summarizes the agency's views on a number of miscellaneous activities.

7.2 Wastewater treatment units

Facilities sometimes find that more than one environmental program applies to a single waste stream. An example of this situation is a characteristic or listed wastewater, which is subject to RCRA regulation from its point of generation on, but whose management and discharge is subject to the CWA as well. This is a common occurrence as a significant percentage of the hazardous waste produced in the United States is aqueous waste or otherwise amenable to wastewater treatment. It would appear that equipment (e.g., tanks and other units) used to treat hazardous wastewater or hazardous wastewater treatment sludge would have to be permitted under both RCRA and the CWA and meet design and operating standards under both of these regulatory programs.

EPA recognized the above situation during the early stages of RCRA implementation and promulgated the so-called wastewater treatment unit (WWTU) exemption from RCRA requirements on November 17, 1980. [45 *FR* 76074] Designed specifically to avoid duplicative control standards under both RCRA and the CWA, the WWTU provision exempts

Table 7-1: Treatment Status of Miscellaneous Hazardous Waste Activities

Activity	Treatment? (Yes/No)	Comments/EPA references
Compaction within a drum	Yes	If the density of the hazardous waste in the drum is increased in order to reduce the volume of the waste, it is treatment. [RO 11609]
Shredding waste or debris	Yes	Shredding the waste or debris changes its physical character so as to reduce its volume or render it amenable for recovery. [RO 11466, 13202, 14241]
Evaporation	Yes	Evaporating water from a hazardous waste changes the waste's composition in order to reduce its volume. [RO 12923]
Volatilization	Yes	Volatilizing trichloroethylene from contaminated plastic packing media (an F001 listed waste) is treatment. [RO 11418]
Stabilization	Yes	Stabilizing F006 sludge is hazardous waste treatment. [RO 11422]
Stripping	Yes	Stripping benzene from a refinery's hazardous wastewater is treatment. [RO 13558]
Decanting	Yes	Decanting an aqueous phase from an organic hazardous waste is treatment. [RO 11885, 14834]
Washing spent electroplating bags	Yes	Cyanide-contaminated spent electroplating anode bags are hazardous wastes; washing cyanide plating solution off of the bags to allow their reuse is intended to recover material resources and is treatment. [RO 13476]
Water addition to dissolve packaging	No	Adding water to waste cyanide in dissolvable plastic bags will change the physical character of the waste to facilitate its disposal; this is not an intent listed in the second part of the definition of treatment (i.e., this activity does not make it safer to dispose the waste). [RO 12335]
Emulsifier addition to facilitate tank truck cleanout	No	Adding emulsifiers to tank truck heels to reduce flammable vapor levels will change the physical properties of the waste, but only for the purpose of removing them from a non-RCRA-empty container; this is not an intent listed in the second part of the definition of treatment. [RO 14125]

Source: McCoy and Associates, Inc.

equipment used to treat hazardous wastewater and wastewater treatment sludge from the 1) RCRA permitting requirements of Part 270, 2) general facility standards in Parts 264 and 265, and 3) specific tank design and operating standards in Subpart J of Parts 264 and 265. [RO 11066, 11379, 11408, 13526, 13727, 14122]

In our discussions below, we first summarize the specific applicability requirements of the three-part definition of a WWTU. A review of the scope of this exemption is then presented, illustrated by a number of WWTU examples.

7.2.1 The key definition

To qualify for the WWTU exemption, a piece of equipment must meet the three-part definition in §260.10, which reads as follows:

"*Wastewater treatment unit* means a device which:

"(1) Is part of a wastewater treatment facility that is subject to regulation under either Section 402 or 307(b) of the Clean Water Act; and

"(2) Receives and treats or stores an influent wastewater that is a hazardous waste as defined in §261.3 of this chapter, or that generates and accumulates a wastewater treatment sludge that is a hazardous waste as defined in §261.3 of this chapter, or treats or stores a wastewater treatment sludge which is a hazardous waste as defined in §261.3 of this chapter; and

"(3) Meets the definition of tank or tank system in §260.10 of this chapter."

In the sections below, we examine each of the three parts of this definition.

7.2.1.1 Part 1—The CWA must apply to the wastewater treatment system

In order to qualify for the WWTU exemption, treatment units must be subject to regulation under one of two CWA programs: 1) the National Pollutant Discharge Elimination System (NPDES) permitting program, which is outlined in Section 402 of the CWA; or 2) the national pretreatment program, mandated by Section 307(b) of the CWA for facilities that discharge to POTWs. (Note that EPA added a third option of being subject to an effluent guideline issued under Sections 301 and 402 of the CWA in RO 11408.) The agency's logic is as follows: "The underlying assumption used in justifying the wastewater treatment unit exemption was that tanks used to handle hazardous wastewaters at these facilities would be provided with EPA oversight under the Clean Water Act, thereby ensuring no significant decrease in environmental control afforded at these facilities." [RO 11519; see also RO 14122]

Although the NPDES program is based on source discharge permits and most POTWs issue permits to upstream facilities, EPA notes that WWTUs are not required to have CWA permits to qualify for the RCRA exemption:

"The agency also intends that the phrase 'subject to regulation under either Section 402 or 307(b) of the Clean Water Act' should be given a broad interpretation. This phrase includes all facilities that are subject to NPDES permits and encompasses facilities subject to either categorical pretreatment standards or general pretreatment standards. It is *not* necessary that the permits actually be issued or that pretreatment standards actually be in force. It is sufficient that the facility be subject to the requirements of the Clean Water Act." [Emphasis in original.] [RO 11020] See also RO 13526.

Q *Can zero-discharge facilities qualify for the WWTU exemption?*

A The WWTU exemption's applicability to zero-discharge facilities is based on whether a connection exists between CWA requirements and the facility's lack of a discharge. In other words, a facility that produces no effluent as a direct result of the CWA *is* eligible for the WWTU exemption. [RO 11036, 11374] This includes facilities that achieve zero discharge by onsite wastewater storage or disposal [RO 11139], or where applicable effluent guidelines or pretreatment standards specify zero discharge. [RO 14608] Stated another way, the WWTU exemption would be available to a facility that eliminated its discharge of treated effluent in order to meet an NPDES or pretreatment requirement, but the exemption would not be available to a facility that never had a discharge regulated under the CWA. [September 2, 1988; 53 *FR* 34080, RO 13526, 14608] In fact, EPA expects that some facility owners/operators may attempt to apply for a zero-discharge NPDES permit at an otherwise unregulated facility in order to qualify for the WWTU exemption from RCRA. [RO 13112]

Based on the above discussion, a hazardous wastewater treatment tank at a zero-discharge facility that does not meet the first part of the §260.10 definition *would* generally have to be RCRA-permitted (and would have to meet Part 264, Subpart J tank standards). Explaining this approach, EPA stated: "while it is true that a zero-discharge system does not require a NPDES permit, the absence of this permit (or an applicable effluent guideline or pretreatment standard specifying zero discharge) necessitates a RCRA Part B permit. Otherwise, a wastewater unit treating hazardous wastes could escape regulations developed to ensure protection of human health and the environment. Although this approach may, at first, be viewed as a disincentive to developing zero-discharge systems, a NPDES permit that specifies 'zero discharge' may be the most appropriate alternative to a RCRA Part B permit in industries without zero-discharge effluent guidelines, encouraging zero-discharge systems while being consistent with the agency's mandate to protect human health and the environment." [RO 11408]

7.2.1.2 Part 2—The unit must be managing hazardous wastewater or hazardous wastewater treatment sludge

The second part of the WWTU definition is pretty wordy, but the bottom line is that a WWTU must be receiving and treating or storing hazardous wastewater or hazardous wastewater treatment sludge. Guidance on how EPA defines those two materials is summarized below.

7.2.1.2.1 *Wastewater*

For the purposes of the WWTU definition, EPA has not promulgated a definition of "wastewater," although it notes that states can define it if they want. [RO 11582] The agency has, however, provided guidance on what wastewater is (and isn't). Wastewater is generally assumed to be wastes that "are substantially water with contaminants amounting to a few percent at most." [RO 11020, 14472]

The following examples apply the above logic in determining if a given stream is wastewater:

- A mixture of 50% alcohol/50% water is not wastewater, but a D002 corrosive waste containing 95% water and 5% total dissolved solids is. [RO 11173]

- Wastewater is not concentrated chemicals or nonaqueous wastes. [RO 11020, 11561, 11749] For example, free-phase tetrachloroethylene from a dry cleaner is not wastewater. [RO 11749]

- Petroleum tank bottoms are neither wastewater nor wastewater treatment sludge. [RO 11561]

- Spent solvents and D001 ignitable liquids are also not wastewater. [RO 12182]

- Hazardous landfill leachate may meet the definition of wastewater. [February 11, 1999; 64 *FR* 6807]

Note that the LDR definition of wastewater at §268.2(f) cannot be used to determine applicability of the WWTU exemption. [RO 11551]

7.2.1.2.2 *Wastewater treatment sludge*

Although not defined in the regulations, EPA has described the term "wastewater treatment sludge" in guidance, stating that "wastewater treatment sludge is any material that precipitates or otherwise is separated from wastewater during treatment." [RO 11551] For example, petroleum tank bottoms are not generated from the treatment of wastewater and so would not meet this definition. [RO 11561]

7.2.1.3 Part 3—Tanks and tank systems only

The third portion of the WWTU definition requires that a WWTU be either a tank or a tank system, both of which are defined in §260.10 as follows:

"*Tank* means a stationary device, designed to contain an accumulation of hazardous waste which is constructed primarily of non-earthen materials (e.g., wood, concrete, steel, plastic) which provide structural support."

"*Tank system* means a hazardous waste storage or treatment tank and its associated ancillary equipment and containment system."

Normally when we think of a tank, we think of an aboveground, cylindrical structure. But many other types of units may qualify for the WWTU exemption under the above definition of "tank."

Q *What is the RCRA status of an air stripping unit used to treat RCRA-hazardous contaminated ground water?*

A The air stripper may fit the definition of a tank in §260.10 and, assuming it meets the other two criteria in the WWTU definition, may qualify for the WWTU exemption. [RO 12783, 12880]

Q *What about benzene strippers operating in a refinery's wastewater treatment system?*

A Like various other air and steam strippers, benzene strippers may qualify as tanks and be eligible for the WWTU exemption. [RO 13558]

As described below, a variety of different types of equipment can be considered to be a tank or part of a tank system and, therefore, eligible for the WWTU exemption.

7.2.1.3.1 Evaporators, filter presses, dryers, etc.

Based on its belief that they fit the definition of tank or tank system (and assuming that the other two parts of the WWTU definition are met), EPA has determined that the following types of equipment qualify for the WWTU exemption: aeration tanks, blenders, clarifiers, dehydrators, dryers, evaporators, filters, grit chambers, presses, sludge digesters, thickeners, and many other types of processing equipment. [November 17, 1980; 45 *FR* 76078, RO 11020, 11118, 11379, 11561, 11749, 11752, 11840, 12527, 13003, 14472] Case Study 7-2 presents an example of determining whether a sludge dryer in a wastewater treatment system qualifies as a WWTU. Bear in mind, however, that EPA encourages facilities to check with their state RCRA authority for final decisions on the WWTU exemption for a given piece of equipment.

7.2.1.3.2 Sumps

Sumps that meet the definition of "tank" are WWTUs if they are managing hazardous wastewater or wastewater treatment sludge and are associated with a wastewater treatment system that is subject to regulation under the CWA. [RO 11134, 14028]

A sump meeting the definition of a tank that is used to collect and then convey hazardous wastewater to a separate WWTU could be considered ancillary equipment to the WWTU and would still be excluded from RCRA tank standards. [53 *FR* 34080, RO 11561]

If the sump does not meet the definition of a tank (i.e., its walls are not self-supporting), then it is a surface impoundment and, if managing hazardous waste, a hazardous waste surface impoundment subject to RCRA permitting and Part 264, Subpart K

Case Study 7-2: Sludge Dryers Are WWTUs

EPA was asked by a regulated facility whether adding a sludge dryer to process a hazardous sludge in an existing wastewater treatment system would jeopardize the system's current WWTU exemption. The agency responded that the addition of a sludge dryer to an existing treatment system does not subject the system to RCRA permitting. Regarding the sludge dryer itself, EPA evaluated it against the three criteria in the definition of WWTU.

First, the unit must be part of a wastewater treatment system that is subject to regulation under the CWA. EPA clarified that individual units (here a sludge dryer) that do not themselves have any discharge subject to regulation under Section 402 or 307(b) of the CWA, but that are part of the wastewater treatment system, qualify for the exemption if other tanks in the treatment train have discharges that are subject to these CWA provisions. [RO 12527]

Second, the sludge dryer must treat or store a wastewater treatment sludge that is a hazardous waste (i.e., the sludge is a listed waste, derived from treatment of a listed waste, or is hazardous because it exhibits a characteristic).

Third, the sludge dryer must meet the definition of a tank in §260.10; that is, it must be designed to contain hazardous waste and be constructed primarily of nonearthen materials that provide structural support. Because the dryer to be added at this facility was constructed of a steel shell, the agency agreed that it was a tank. Furthermore, the agency has clarified that the definition of tank—for the WWTU exemption—includes units such as "sludge digesters, thickeners, dryers, and other sludge processing tanks…in which hazardous wastewater treatment sludge is treated; and any…tanks used for the storage of such sludge." [November 17, 1980; 45 *FR* 76078]

Note also that the definition of incinerator (in §260.10) specifically excludes sludge dryers. (As detailed below, incinerators are not eligible for the WWTU exemption.)

standards. More on whether a sump meets the definition of a tank is contained in Section 10.2.

7.2.1.3.3 Sinks and sewers

Although the types of equipment listed in Section 7.2.1.3.1 that may qualify for the WWTU exemption are broad, several miscellaneous items merit additional clarification.

Q *A facility dips small parts in a 50/50 mixture of alcohol and water. Is the dilution of spent alcohol/water dip solution in a sink to render the D001 ignitable waste nonhazardous considered wastewater treatment, and, if so, is the process eligible for the WWTU exemption?*

A Since the dilution is intentional, rather than incidental to conveyance to the treatment plant, hazardous waste treatment is occurring. Although the sink is part of a CWA-permitted wastewater collection and treatment system and would probably meet the definition of a tank, the spent dip solution is not wastewater since it contains 50% alcohol. Thus, the sink does not qualify as a WWTU and the exemption would not apply. [RO 11173]

Q *If D002 corrosive hazardous waste from deionization units travels from a building to an onsite, industrial wastewater treatment plant via a sewer and the waste is neutralized in the sewer (so as to not exhibit the characteristic of corrosivity), could that sewer be a WWTU?*

A First, the wastewater treatment plant is subject to CWA requirements. Second, the corrosive hazardous waste is wastewater since it contains 95% water and only 5% total dissolved solids. Third, EPA considers that: "devices such as flumes, gutters, troughs, and pipes…are not commonly considered to be tanks, but…nevertheless meet the expansive definition of tank in §260.10." [November 17, 1980; 45 *FR* 76078] Thus, the sewer may qualify for the WWTU exemption (if state RCRA authorities concur). [RO 11173]

7.2.1.3.4 Ancillary equipment

On September 2, 1988 [53 *FR* 34080], EPA modified the WWTU exemption to ensure that it applies not just to tanks, but also to tank systems. The agency's intent was to extend the scope of the WWTU exemption to include ancillary equipment that is associated with an exempt hazardous waste storage or treatment tank. (Recall that the definition of "tank system" includes "associated ancillary equipment.")

The agency defines "ancillary equipment" in §260.10 as follows:

> "[A]ny device including, but not limited to, such devices as piping, fittings, flanges, valves, and pumps, that is used to distribute, meter, or control the flow of hazardous waste from its point of generation to a storage or treatment tank(s), between hazardous waste storage and treatment tanks to a point of disposal onsite, or to a point of shipment for disposal offsite."

Thus, the piping, fittings, flanges, valves, pumps, etc. associated with a WWTU also qualify for the exemption from RCRA permitting and Subpart J standards. [RO 13126]

Responding to questions from regulated facilities, EPA notes that the following items do *not* qualify as ancillary equipment:

- A container used to store wastewater prior to introduction to a tank used for treating wastewater is not considered ancillary equipment to that tank and cannot qualify for the WWTU exemption. Such containers would be 90- or 180-day containers subject to §262.34 and Part 265, Subpart I. [RO 11551]

- An unlined earthen ditch used to direct hazardous wastewater to a treatment tank (and any other device that is not designed to prevent leakage or discharge) cannot be considered ancillary equipment and cannot qualify for the WWTU exemption. EPA bases this conclusion on its opinion that any device that allows leakage or discharge of hazardous waste is actually disposing hazardous waste. Thus, tank systems (including associated ancillary equipment) must provide containment in order to qualify for the WWTU exemption. [RO 11631, 13653, 13669]

7.2.1.3.5 Incinerators, lagoons, and surface impoundments

While certain circumstances may sometimes render tanks, tank systems, and associated ancillary equipment ineligible for the WWTU exemption (see the section below for such examples), incinerators and lagoons/surface impoundments *never* qualify for the WWTU exemption, even if they are part of a wastewater treatment system subject to regulation under the CWA. [July 18, 1990; 55 *FR* 29230, RO 11020, 11134, 12605, 13147]

7.2.2 Scope of the WWTU exemption

This subsection addresses a number of questions that come up regarding the scope and reach of the WWTU exemption.

7.2.2.1 WWTU exemption applies to equipment, not to wastes

The regulatory exemption provided to a WWTU applies only to *the unit* (i.e., to the equipment that makes up the WWTU)—not to the waste within or discharging from the unit. Thus, if hazardous waste enters a WWTU, it remains a hazardous waste while in the unit (unless a characteristic waste is decharacterized), and the residues, including treated wastewater effluent and any sludge, may still be hazardous based on the derived-from rule. [RO 11066, 11551, 12527, 12605]

Because their WWTUs are exempt from RCRA standards, some owners/operators mistakenly believe that the hazardous waste managed in those units is also exempt from regulation. For example, if the hazardous waste being managed in a WWTU is listed, some people would argue that the effluent from the unit or sludge generated in the unit is no longer hazardous (as it would otherwise be under the derived-from rule). This is not true. Only the equipment itself is exempt from RCRA regulation—not the waste managed within it. EPA has noted:

"Only the wastewater treatment unit (i.e., the tank) is exempt; the exemption does not 'follow' or attach to the waste. Consequently, all applicable hazardous waste management standards apply to the waste prior to treatment in the WWTU, and to any residue generated by the treatment of that waste. In other words, solid waste resulting from the treatment of a listed hazardous waste in an exempt WWTU will remain a listed hazardous waste, and solid waste resulting from the treatment of a characteristic hazardous waste in an exempt unit will remain hazardous as long as the solid waste continues to exhibit a characteristic." [RO 13541]

Thus, sludge removed from a WWTU that exhibits a characteristic or meets a listing description is subject to all RCRA requirements if stored in a container and/or manifested offsite. [RO 14718] Any leaks from the WWTU could still be hazardous waste depending on the derived-from rule.

7.2.2.2 What are WWTUs exempt from?

Three specific provisions within the RCRA regulations implement the WWTU exemption:

- §264.1(g)(6) exempts WWTUs located at permitted TSD facilities from the requirements of Part 264, including the tank standards of Subpart J and the air emission standards of Subparts AA–CC. [December 6, 1994; 59 *FR* 62913]

- §265.1(c)(10) exempts WWTUs located at interim status TSD facilities and generator facilities from the requirements of Part 265, including the tank standards of Subpart J and the air emission standards of Subparts AA–CC.

- §270.1(c)(2)(v) exempts owners/operators from the requirement to obtain a RCRA permit for their WWTUs.

Additionally, EPA guidance notes that WWTUs are exempt from the generator accumulation standards in §262.34. Therefore, for example, no "Hazardous Waste" marking or 90- or 180-day accumulation start dates would be necessary for these units. [RO 13727]

As is evident from this list, the WWTU exemption provides owners/operators a significant advantage for treating hazardous wastes in qualifying units. As noted at the beginning of this section, however, the exemption applies only where the treatment unit is subject to certain provisions of the CWA.

7.2.2.3 WWTUs must be dedicated to onsite wastewater treatment systems

Having been asked repeatedly about this issue, EPA has stated that the WWTU exemption applies to "any tank system that manages hazardous wastewater and is *dedicated for use* with an *onsite* wastewater treatment facility." [Emphasis added.] [53 *FR* 34080] See also RO 11066, 11551, 13112, 13226.

For the purpose of the WWTU exemption, EPA allows a broad interpretation of the term "onsite." There need be no direct mechanical connection between tanks or sumps holding hazardous wastewater and other components of the wastewater treatment system. [RO 12354] The means of conveyance of the waste between storage and treatment units does not affect the applicability of the WWTU exemption. [RO 13112] "The applicability of the [WWTU] exemption does not depend on whether the wastewater is piped or trucked, or conveyed in any other manner to the wastewater treatment facility within the boundaries of the facility generating the wastewater." [53 *FR* 34080]

Q: *Hazardous wastewater is stored in a tank or sump that is emptied by a vacuum truck. The vac truck then drives across the facility and discharges the wastewater into the onsite wastewater treatment plant. Can the tank or sump be considered a WWTU?*

A: Yes. No requirement exists for WWTUs to be hard-piped together. Wastewater may be piped, trucked, or otherwise conveyed from one unit to the next and still be eligible for the exemption. [53 *FR* 34080, RO 12354]

Q: *Hazardous waste is treated in a RCRA-permitted tank subject to Part 264, Subpart J requirements. Can another tank that receives a residue from this treatment qualify for the WWTU exemption?*

A: Yes. "The [WWTU] exemption is not altered by the regulatory status of other storage tanks located at the same facility.…[I]f the storage of a waste results in the generation of a wastewater or wastewater treatment sludge, and a 'downstream' tank receives this wastewater or sludge, that downstream tank may be eligible for the exemption provided that it…meets the…criteria in §260.10." [RO 11561]

Based on the above guidance, EPA notes [in RO 11519] and we strongly recommend that any tanks or sumps that are considered WWTUs at a facility should be identified in the facility's CWA (NPDES or pretreatment program) permit—specifically in the process flow diagram associated with the permit.

7.2.2.4 Onsite handling with subsequent shipment for offsite treatment/disposal

As noted above, EPA clearly intends that the WWTU exemption apply only to units dedicated for use with an onsite wastewater treatment system that discharges under the CWA. "However, if a tank system, in addition to being used in conjunction with an onsite wastewater treatment facility, is used on a routine or occasional basis to store or treat a hazardous wastewater prior to shipment offsite for treatment, storage, or disposal, it is *not* covered by this exemption." [Emphasis added.] [53 *FR* 34080] Thus, facilities that generate hazardous wastewater, store it in tanks, and then ship it to offsite facilities for treatment or disposal cannot claim the WWTU exemption for those tanks. This limitation on the scope of the WWTU exemption holds 1) regardless of whether onsite treatment of the wastewater occurs at the generating facility prior to offsite shipment, and 2) even if the wastewater is shipped offsite to a POTW or another facility with an NPDES permit. [53 *FR* 34080, RO 13112, 13226, 13318]

As such, tanks that are used, even on a part-time basis, for the treatment and/or storage of hazardous wastewater prior to shipment offsite for treatment and/or disposal are *not* eligible for the WWTU exemption. Rather, these tanks would be RCRA units subject to Subpart J hazardous waste tank standards. [RO 11066, 11551, 13126, 13203, 14262] The agency explains its reasoning as follows:

> "[I]n order to satisfy the WWTU exemption, a tank must be dedicated solely for onsite wastewater treatment at all times and for no other purpose…. EPA did not intend the WWTU exemption to apply in situations involving 'dual use' of a tank (when a tank is concurrently used for wastewater treatment and for another purpose). Nor did EPA intend for the exemption to apply in situations…involving 'alternating use' of a tank. Since the purpose of this exemption is to avoid dual regulation under the [CWA] and [RCRA], EPA believes that a tank must be used only for wastewater treatment purposes at all times in connection with an onsite wastewater treatment facility in order to qualify for the exemption." [RO 14262]

EPA's concern behind this restriction is that tanks used to manage hazardous wastewater prior to offsite shipment are probably not subject to CWA regulatory requirements. Allowing the WWTU exemption to apply to such tanks would short-circuit the agency's belief that the CWA can adequately regulate the units in lieu of RCRA, since such tanks would be subject to neither program. [RO 11519]

This concept often brings up the question of maintenance problems associated with a facility's wastewater treatment system. If the onsite wastewater treatment plant is shut down due to malfunction or routine maintenance, hazardous wastewater may have to be shipped offsite. Would the hazardous wastewater storage tanks/sumps then lose their WWTU exemption? Although we've never seen any guidance on this issue, we suspect the exemption would not be lost for these units, subject to the enforcement discretion of the state agency.

7.2.2.4.1 *Offsite shipment of sludge for treatment/disposal*

If the offsite shipment of hazardous wastewater invalidates the WWTU exemption for a particular tank, what about removing sludge from a WWTU and shipping it offsite for disposal? EPA noted that the removal of wastewater treatment sludge or other residues from tanks for offsite disposal would not disqualify these tanks from the WWTU exemption "provided that this occurs as part of normal wastewater treatment activities. The removal and offsite disposal of treatment sludges and tank bottoms are not necessarily indications that the tanks in question are being used in a manner other than for onsite wastewater treatment; on the contrary, the generation of tank bottoms and filter cakes is a common process in wastewater treatment operations." [RO 13226; see also RO 14718]

However, we need to be careful about the regulatory status of units that are used to store hazardous wastewater treatment sludge prior to shipping the sludge offsite for treatment and/or disposal. If the sludge is mostly solid (e.g., filter cake), the last unit used to store it before offsite shipment will likely be a roll-off box or some other container. Such a unit cannot be an exempt WWTU because it doesn't meet the definition of a tank; thus, these containers will likely be 90- or 180-day accumulation containers if they are holding hazardous sludges.

Conversely, if the wastewater treatment sludge is a hazardous liquid (e.g., an oily waste removed from the wastewater), the last unit holding such material before offsite treatment/disposal (e.g., fuel blending) will more likely be a tank. Does this tank qualify as a WWTU? You might be able to make an argument that it does, based on the three-part WWTU definition. However, EPA's guidance notes that such tanks are not WWTUs:

> "While the storage of sludges may in many instances be part of ongoing wastewater treatment, the agency also wishes to emphasize that sludges may be generated and stored as part of activities

not related to wastewater treatment under the CWA. In cases where the sludges are to be recycled…sludges might be removed from treatment tanks and placed into tank-like units that mix or slurry and inject the sludges as part of the recycling process, in which case a judgment could be made that the mixing tank is part of the exempt recycling process (§261.6(c)), and is, therefore, not a WWTU…. Also, in instances where the sole purpose of the tank is to accumulate a sufficient volume of waste to facilitate the offsite transport of these sludges for further treatment and/or disposal, the [WWTU] exemption would not apply because that activity is separate and distinct from any wastewater treatment activities covered under the CWA. In this instance, a generator may choose (or be so directed by the implementing agency) to manage the tanks in accordance with the [90/180/270-day] generator accumulation provisions in §262.34, rather than as an exempt WWTU." [RO 14465]

7.2.2.5 Common wastewater treatment facilities

Given that onsite management of hazardous wastewater is key to the WWTU exemption, what about situations where multiple facilities use a common wastewater treatment facility or conveyance system? Figure 7-1 shows three such situations, which are discussed below.

In the first scenario, the same company operates a chemical plant and a refinery on adjacent sites. Both units generate hazardous process wastewaters that are collected and then piped to an NPDES-permitted wastewater treatment plant located at the refinery. (The treatment plant's permit limits are based on the waste loads from both facilities.)

Q *Is a tank at the chemical plant that stores hazardous process wastewater eligible for the WWTU exemption, along with the wastewater collection and treatment tanks at the refinery and the common treatment plant?*

A Yes. Per EPA, "[t]he fact that the NPDES permit is based on the waste loads of both the chemical plant and refinery is not necessarily the determining factor in deciding eligibility for the WWTU exemption…. In order to ensure that the reach of the NPDES permit is sufficient to adequately regulate the wastewater treatment tank at the chemical plant, the chemical plant and/or the tank itself needs to be specifically identified in the permit. This could be accomplished by stating expressly in the permit that it covers the chemical plant, or by making the operator of the chemical plant a co-permittee or a limited co-permittee on the permit with the operator of the refinery. This coverage would ensure adequate day-to-day control over the tank under the CWA to justify an exemption from RCRA requirements." [RO 11519]

Under the second scenario, Company A and Company B operate separate units within the same RCRA facility boundary. Hazardous process wastewater from each company is piped via a common sewer to an onsite wastewater treatment system operated by Company A. The treatment system has an NPDES permit for its outfall.

Q *Are the wastewater tanks operated by Company B WWTUs?*

A "The analysis for this scenario essentially is the same as for [the first scenario] above. To be eligible for the exemption, Company B must be a co-signatory to the NPDES permit and/or otherwise identified as a limited co-permittee on the permit issued to Company A, or the permit itself must expressly cover Company B (for example, the description of the facility covers the RCRA boundaries, and 'upstream' wastewater treatment processes and equipment are identified) so that CWA authorities can prescribe and enforce tank system requirements at Company B as well as at Company A." [RO 11519]

In the third scenario, Company A and Company B operate a joint venture at a Company A site that discharges hazardous process wastewater to a POTW.

CHAPTER 7 *Hazardous Waste Treatment*

Figure 7-1: WWTU Exemption Applicability at Common Wastewater Treatment Facilities

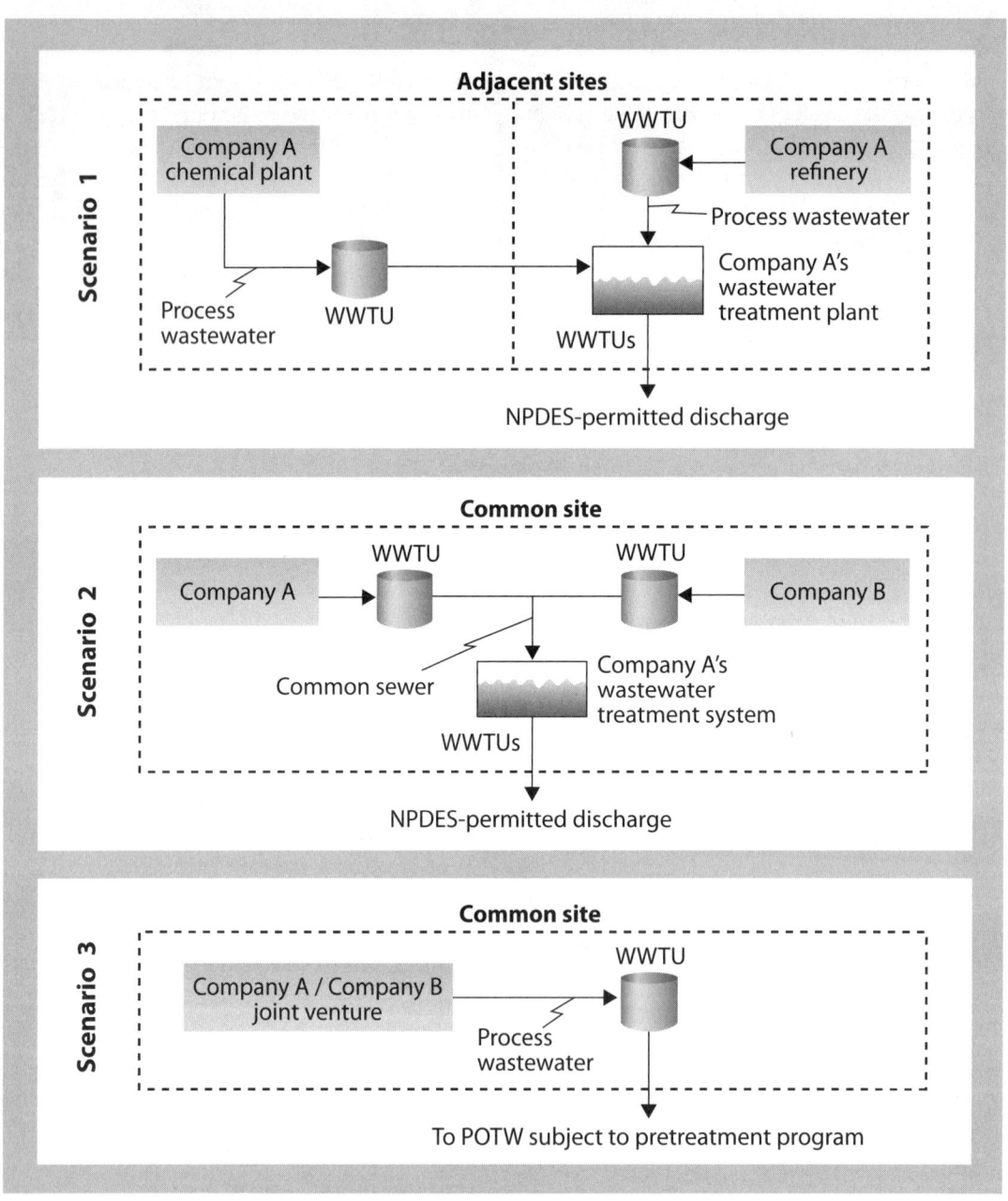

Source: McCoy and Associates, Inc.; adapted from RO 11519.

Q *Are tanks and tank systems within the joint venture facility eligible for the WWTU exemption?*

A "Both companies must comply with [CWA] Section 307(b) pretreatment requirements, since both are introducing pollutants directly into a POTW. Therefore, both companies are eligible for the WWTU exemption." [RO 11519]

7.2.2.6 Treatment of hazardous wastewater/sludge generated offsite

Since one of EPA's key goals in assessing eligibility for the WWTU exemption is ensuring that RCRA-exempt units are subject to CWA regulations, it follows that the agency would not be overly concerned about where the hazardous wastewater being managed comes from, as long as the unit is subject to CWA control. Thus, EPA has stated that, if a tank receives hazardous wastewater from *offsite* facilities for treatment and/or storage, it qualifies for the WWTU exemption if it meets all of the eligibility criteria. [53 *FR* 34080, RO 11020, 11561, 13112, 13318, 14206]

A facility may also receive hazardous wastewater treatment sludge from offsite and still be potentially eligible for the wastewater treatment unit exemption. As long as the sludge received is not a concentrated chemical or nonaqueous waste, the receiving facility may receive it. [RO 11561]

Three related issues are also addressed in guidance. First, such offsite hazardous wastewater or wastewater treatment sludge shipments must (of course) be manifested. [RO 11020, 11039, 11066, 14206] Second, an offsite facility with WWTUs that have qualified for the exemption may be listed as the "designated facility" authorized to receive shipments of hazardous waste on the manifest [per §262.20(b)]. In other words, such WWTUs meet the definition of "designated facility" promulgated at §260.10. [RO 11561, 14206] See Case Study 7-3. Third, the ownership of the offsite facility and any expense reimbursement for the facility accepting offsite material for treatment are irrelevant to determining the applicability of the WWTU exemption. [RO 11561]

7.2.2.7 Portable tanks

Only tanks and tank systems may qualify for the WWTU exemption discussed in this section. What about portable tanks such as frac tanks that are portable when empty but stationary (can't be moved) when full? Can they qualify as tanks under the WWTU exemption? The answer is "maybe."

In the preamble to a June 3, 1987 proposed rule [52 *FR* 20919], EPA stated the following:

"Several members of the regulated community have asked whether a mobile unit could ever qualify as a tank, because §260.10 defines tanks as 'stationary devices.' EPA confirms that a mobile tank would qualify as a tank under §260.10

Case Study 7-3: Receiving Offsite Shipments of Hazardous Wastewater Into a WWTU

A generator wants to send hazardous waste offsite to an industrial wastewater treatment facility. The wastewater will be received and treated in WWTUs operated by the industrial facility, but the facility has neither a RCRA permit nor interim status. Is this legal? Can the industrial facility's WWTUs be considered a "designated facility" as required on the hazardous waste manifest?

Yes to both questions. EPA believes that WWTUs are appropriate facilities to receive hazardous wastewater from offsite without a RCRA permit. Therefore, the agency interprets "designated facility" as defined at §260.10 to include RCRA permit-exempt WWTUs at industrial facilities. Note that the industrial facility would have to obtain an EPA ID number for generators to send hazardous wastewater to the facility's WWTUs per §262.12(c). Additionally, the facility would have to sign each manifest and indicate any discrepancies in hazardous waste shipments. [RO 14206]

and would be subject to the Subpart J tank standards of Part 264, as long as it was intended to be stationary during operation and it otherwise met the definition of a tank."

Although the above text addresses portable tanks in terms of TSD facility tank standards, the agency extended its June 1987 logic to apply to the WWTU exemption in several subsequent guidance letters. See RO 12220 (for portable filter presses), 12928 (for mobile sludge dewatering devices), and 13144 (for wheeled tanks). This guidance seems like pretty good evidence that the agency intends for portable tanks to qualify for the WWTU exemption under the right circumstances. However, RO 13144, which has been deleted from EPA's RCRA Online system, previously had been listed as "SUPERSEDED: NOT CURRENT EPA POLICY." (No such designation is noted for RO 12220 or 12928.) In addition, §260.10 gives the definition for a container as "any portable device in which a material is stored, transported, treated, disposed of, or otherwise handled." So, is a portable tank a tank, a container, or both?

The bottom line is that this is probably best left as a "check-with-your-state" issue. For example, we have seen letters written from different states to the manufacturer of a storage device that is portable when empty but stationary when full; some states say the device is a container—others that it is a tank. Bear in mind, however, that a portable tank set up with flexible hoses and quick-connect couplings to manage hazardous waste looks less like a tank than a portable tank with hard piping and permanent connections.

7.2.3 Miscellaneous WWTU exemption issues

EPA has dealt with a number of other issues involving the WWTU exemption, as noted below.

7.2.3.1 Eligibility decisions are unit-specific

Eligibility decisions for tanks and tank systems in terms of the WWTU exemption should be made on a unit-specific basis. For example, a given facility may have some tanks that handle hazardous wastewater prior to shipment to an offsite treatment plant and other tanks dedicated to an onsite wastewater treatment system. While the first set of tanks would not qualify for the WWTU exemption, this fact has no effect on the eligibility of the latter set of tanks. [RO 11551, 11561]

7.2.3.2 WWTUs that leak are not WWTUs

Besides contaminating the environment, WWTUs that leak could theoretically lose their designation as WWTUs and, therefore, their eligibility for exemption from Subpart J standards. [RO 11631, 13653, 13669] In short, such units might no longer be considered tanks (Part 3 of the WWTU definition), since a tank must be "designed to *contain* an accumulation of hazardous waste." [Emphasis added.] [§260.10] (See also November 17, 1980; 45 *FR* 76079 for details.)

Besides losing their WWTU exemption, leaking tanks could be subject to enforcement action based on disposal of hazardous waste without a permit. [RO 14470] We are aware of an enforcement case where a facility was using a leaking tank (sump) that it considered to be an exempt WWTU. Regulators, noting the leaks, determined that the unit was actually a surface impoundment that is not eligible for the WWTU exemption. A significant fine for operating an unpermitted RCRA surface impoundment was imposed.

7.2.3.3 WWTUs are SWMUs

Even though WWTUs are exempt from most RCRA requirements, they still meet the definition of a solid waste management unit (SWMU). [July 27, 1990; 55 *FR* 30808] Therefore, they may be subject to RCRA corrective action at facilities that have RCRA permits or operate under interim status.

7.2.3.4 LDR notices

Despite their exemption from Parts 264, 265, and 270 RCRA requirements, hazardous wastes managed in WWTUs are still subject to the LDR notification provision of §268.7(a)(7). Thus, when waste treated in a WWTU becomes nonhazardous, the owner/operator must place a one-time notice in the facility files describing 1) the generation of the subject hazardous waste, 2) the subsequent exclusion of the waste from the definition of hazardous or solid waste, and 3) the disposition of the waste. [RO 13547, 14216] More discussion of this topic is in Section 13.12.3.1.

7.3 Treatment in 90/180/270-day accumulation units

When EPA first promulgated the RCRA regulations in 1980, the agency recognized the potential impact that permitting would have on many generator facilities. To alleviate these concerns, EPA included an exemption from permitting requirements for accumulation of hazardous waste by large and small quantity generators (LQGs and SQGs, respectively). Specifically, §270.1(c)(2)(i) allows LQGs to accumulate hazardous waste for up to 90 days without having to get a RCRA permit. SQGs have, depending on circumstances discussed in Section 6.3.1, either 180 or 270 days. Although generator facilities must still meet certain management standards (as detailed below and in Section 6.3), the Part 270 permitting requirements and the bulk of the Part 264/265 TSD facility provisions do not apply to LQG or SQG facilities.

In addition to accumulation, EPA also allows generator facilities to perform certain types of hazardous waste *treatment* in 90/180/270-day accumulation units without the need for a RCRA permit. With one exception, the standards are the same for waste treatment or storage in 90/180/270-day accumulation units. That exception is the need (in some treatment situations) to establish and follow a waste analysis plan.

In the subsections that follow, we first discuss how EPA has developed the exemption for treatment in 90/180/270-day accumulation units and then examine the specific criteria under which facilities may qualify. As with other sections, we close with numerous examples described in EPA guidance. Remember that treatment residues may still have to be managed as hazardous waste (see Section 5.2), and residues destined for land disposal are subject to LDR treatment standards (see Chapter 13).

7.3.1 History of the accumulation unit treatment exemption

Unlike the other seven treatment-related permitting exemptions discussed in this chapter (which are clearly spelled out in regulatory text), the accumulation unit treatment exemption is not contained in the RCRA regulations [other than an obscure reference in §268.7(a)(5)]. Instead, EPA has explained in *Federal Register* preambles and other guidance documents how it believes the accumulation unit permitting exemption has applied all along. Here is a timeline of EPA's statements/actions on the matter:

- On January 12, 1981 [46 *FR* 2808], the agency stated in preamble language that "[t]he facility specific regulations promulgated today cover not only storage operations, but also many treatment facilities…. [A] determination that a facility is a 'treatment' facility is not relevant to a determination of whether it is, on one hand, a storage facility, or, on the other hand, a disposal facility." In other words (and as clarified in later guidance), just as treatment occurring at a facility does not affect whether it is a storage or disposal facility, treatment activities similarly do not affect the regulatory status of 90/180/270-day units at generator facilities under §262.34. [RO 11261, 14618]

- In the preamble to the March 24, 1986 final SQG rule [51 *FR* 10168], EPA clearly stated that generators in compliance with §262.34 can treat hazardous waste in accumulation tanks and containers without a permit. Although many took this verbiage to apply only to SQGs, the agency subsequently clarified [in RO 11163, 11641, 12811] that this policy applies to all generators: large and small.

- Guidance distributed by the agency in the second half of 1986 explained why treatment could be conducted in accumulation (i.e., storage) units:

"Since the agency has never developed standards specific to treatment in tanks and containers, the same technical standards applicable to such storage (i.e., Subpart I or J of Part 265) would also be applicable to treatment." [RO 11163]

"[EPA General Counsel] consistently dictates that treatment and storage which is regulated identically at permitted facilities also be regulated identically at generation sites…. Since the regulations do not impose additional standards for treatment when it occurs [at a permitted] storage facility, there is no basis for regulating treatment at an exempt storage facility." [RO 11207]

- EPA imposed a requirement for generators to develop and follow a waste analysis plan when treating hazardous wastes in 90/180/270-day accumulation tanks and containers for the purpose of meeting LDR treatment standards on June 1, 1990. [55 *FR* 22670] This requirement, which is codified in §268.7(a)(5), is the only citation in the federal RCRA regs that specifically notes that such treatment is allowed.

- A December 6, 1990 final rule [55 *FR* 50456] extended the 90-day accumulation unit permitting exemption to treatment on drip pads, an action of significance mainly to companies in the wood preserving industry.

- On August 18, 1992 [57 *FR* 37212], EPA added containment buildings to the list of units eligible for the 90-day accumulation/treatment permitting exemption. This action was taken principally to give generators a type of unit in which to treat hazardous debris without having to obtain a RCRA permit.

- Finally, more recent EPA guidance memos have reiterated the agency's positions as described above in response to a variety of questions and requests from the regulated community. [RO 11310, 11425, 11641, 11679, 13553, 13782, 14618, 14662]

7.3.2 Exemption qualifying criteria

While the bullet list above touches on some of the qualifying criteria for the 90/180/270-day accumulation unit permitting exemption for treatment, a more detailed discussion of this subject is presented in the following subsections.

7.3.2.1 Allowable units and associated management standards

Per §262.34, generators may accumulate (and, per EPA guidance, treat) hazardous waste in several types of 90/180/270-day accumulation units without having to obtain a RCRA permit. As discussed in Section 6.3, the facility must meet specific management standards when accumulating/treating waste in these units. LQGs may treat hazardous waste 1) in tanks, subject to the requirements of Part 265, Subpart J; 2) in containers, subject to Part 265, Subpart I; 3) on drip pads, subject to Part 265, Subpart W; and 4) in containment buildings, subject to Part 265, Subpart DD. In contrast, except as discussed in Section 7.3.2.1.1, SQGs may accumulate/treat hazardous waste only in 1) tanks, subject to §265.201; and 2) containers, subject to Part 265, Subpart I.

Note that besides the unit-specific requirements in the previous paragraph, §262.34 applies other nonunit-specific Part 265 provisions to treatment in permit-exempt accumulation units, especially at LQG sites. For example, air emission standards in Part 265, Subparts AA, BB, and CC apply to treatment in LQG 90-day tanks and containers. See Table 6-3 in Section 6.3.2 for details.

7.3.2.1.1 *Upward mobility*

While the above paragraphs emphasize the relationship between the class of the generator and which management standards apply for waste treatment in permit-exempt units, facilities should remember that there is some "upward" flexibility in these requirements. In other words, SQGs may also treat hazardous waste on drip pads or in containment buildings that are exempt from RCRA permitting, but only if they voluntarily comply with the §262.34(a) requirements for LQGs for these units (e.g., no more than 90-day accumulation, §265.16 training requirements, etc.). [RO 13696, 14662]

7.3.2.2 Prohibited treatment operations

The management standards EPA imposes on treatment in unpermitted 90/180/270-day accumulation units have the net effect of prohibiting certain types of treatment in these units. Examples of these situations follow.

Q *An ignitable waste is burned in a tank that meets Part 265, Subpart J standards or a container that meets Part 265, Subpart I standards. Can this practice be conducted in a 90/180/270-day unit and be exempt from a treatment permit?*

A No. Regardless of the timing of the action (i.e., whether it was done within 90/180/270 days of placement of waste in the tank or container), EPA considers this practice to be open burning or open detonation, which is a form of thermal treatment subject to the requirements of Part 264, Subpart X or Part 265, Subpart P. Therefore, this type of treatment is not eligible for the 90/180/270-day accumulation unit permitting exemption. [RO 11163, 11261, 11310, 11425, 11641, 13553, 14618, 14662]

Q *A solvent waste is stored in accumulation containers for less than 90 days. The solvents are allowed to evaporate from the containers. Is this practice eligible for the 90-day unit permitting exemption?*

A No. This practice would violate §265.173(a), which requires containers to be closed except when adding or removing waste, since a container would have to be left open to allow solvent to evaporate. Thus, the facility would lose its eligibility for the permitting exemption. [RO 11176]

Q *A facility employs a device that evaporates water from a hazardous waste. Would this treatment be allowed in unpermitted 90-day accumulation units?*

A According to EPA, no. The evaporator uses elevated temperatures to remove water, which constitutes thermal treatment. EPA initially determined that, because the evaporative unit meets the definition of a "tank," it could be a 90-day accumulation unit. [RO 12923] However, subsequent to issuing this guidance, the agency has marked this letter as "SUPERSEDED, no thermal treatment in generator units."

However, generators should check with their states regarding the use of evaporators as 90/180/270-day accumulation units to treat hazardous wastes without a permit. For example, under certain conditions, this practice is allowed as a matter of state policy in the three states noted below:

- Ohio—"Under certain circumstances, generators can evaporate the water from their hazardous waste, even without a hazardous waste permit.… Ohio EPA determined that the intent of the rules regulating thermal treatment are directed at thermal destruction or incineration, not the simple addition of heat to promote evaporation of water. Generators can use evaporators to reduce the volume of their hazardous waste under the generator treatment requirements. However, if there is a release of hazardous waste or hazardous waste constituent to the air or the environment, this would be considered disposal. A generator is not allowed to dispose of its hazardous waste in this manner." [http://epa.ohio.gov/portals/32/pdf/Summer2004Notifier.pdf]

- Indiana—"[T]he department has determined that the only types of units that may be used for evaporation purposes by generators without permits and without violation of generator management standards would be tanks or tank systems. The tanks must be designed, operated and maintained in accordance with the applicable requirements of 40 *CFR* Part 265, Subpart J and §262.34 if the status of the generator is SQG or LQG.… In conclusion, under limited circumstances the use of evaporators by generators for dewatering hazardous waste may be conducted without a permit only if applicable management standards, as discussed above, are complied with. The overriding requirement is that hazardous waste constituents are not released to the environment." [http://www.in.gov/idem/files/nrpd_waste-0045.pdf]

- Washington—"This document may be used by generators interested in treating their own waste by evaporation, onsite, in accumulation tanks or containers. Generators of hazardous waste who comply with these standards, and the standards in Technical Information Memorandum (TIM) #96-412, *Treatment by Generator*, will meet the requirements of the Dangerous Waste Regulations, Chapter 173-303 WAC." [https://fortress.wa.gov/ecy/publications/publications/96414.pdf]

7.3.2.3 Onsite generation versus offsite receipt

The provision that implements the 90/180/270-day accumulation unit permitting exemption applies to "*generators* who accumulate hazardous waste onsite for less than the time periods provided in §262.34." [Emphasis added.] [§270.1(c)(2)(i)] Since a generator is defined in §260.10 as the party "whose act or process produces hazardous waste," only the accumulation/treatment of hazardous waste generated onsite is eligible for the 90/180/270-day accumulation unit permitting exemption. Any party receiving hazardous waste from offsite would not be the generator of that waste and, thus, could never accumulate or treat such waste in a §262.34 permit-exempt unit. [December 31, 1980; 45 *FR* 86969, RO 11163, 11358, 11425, 14466]

7.3.2.4 Treatment in multiple vessels

Hazardous waste may be treated in multiple treatment units at a given facility where the entire process train is covered under the 90/180/270-day accumulation unit permitting exemption. However, the total time that the waste remains in the treatment train must not exceed the applicable §262.34 time limit (i.e., 90 days for LQGs and 180 or 270 days for SQGs). In other words, the 90-day clock (for LQGs) does not restart each time the waste enters a new accumulation/treatment unit in the treatment train. The 90/180/270-day clock stops only if the waste enters a permitted or exempt treatment unit (e.g., a wastewater treatment unit) or is rendered nonhazardous.

Also, every unit in the permit-exempt treatment train must meet the applicable Part 265, Subpart I or J management standards, which include provisions covering the addition and removal of waste from containers and tanks designed to prevent releases to the environment during such transfers. [RO 11791]

7.3.2.5 LDRs and WAPs

Hazardous waste generators conducting treatment in permit-exempt accumulation units in order to meet applicable LDR treatment standards (found in §268.40) must develop and follow a written waste analysis plan (WAP). Per §268.7(a)(5), the plan must describe the procedures the owner/operator will carry out to meet the treatment standards. Note that this is the only situation where generators are required by federal RCRA rules to prepare a WAP. See Section 18.2.3.1 for more detail on generator WAPs.

7.3.3 Check with your state

Since the 90/180/270-day accumulation unit permitting exemption for waste treatment is not codified in the regulations [other than the reference in §268.7(a)(5)], some states may simply not allow it. Also, states may choose to interpret federal guidance differently than EPA, possibly also affecting the terms of the exemption. Therefore, generators hoping to treat hazardous waste in a 90/180/270-day accumulation unit without obtaining a RCRA permit should be sure to check with their state RCRA authority to determine if such a practice is allowed. [RO 11546, 11791]

7.3.4 Examples

EPA is asked frequently to determine whether the 90/180/270-day accumulation unit permitting exemption applies to specific situations. In the following examples, the agency has concluded that the noted treatment processes *would* qualify for the permitting exemption, assuming the facility met the conditions of §262.34:

- Air or steam stripping wastewater or contaminated ground water [RO 12783, 13526, 13558];
- Blending hazardous contaminated soil and/or coal tar wastes from manufactured gas plant sites with coal or other fuels to produce a burnable, nonhazardous fuel [RO 11675, 11739, 14024, 14338, 14344];
- Blending hazardous waste fuels in tanks and/or containers [RO 11497, 11881];
- Chemical flocculation of hazardous aircraft engine washwater [RO 14104];
- Compacting hazardous waste within steel drums (i.e., containers) [RO 11609];
- Containment and treatment activities conducted subsequent to an immediate response to a release of hazardous waste or hazardous constituents [November 19, 1980; 45 *FR* 76629, RO 12748];
- Electrochemical oxidation of organic hazardous wastes [RO 14466];
- Ex situ (but onsite) treatment of hazardous soil during site remediation [RO 14112, 14291, 14471];
- Fixation/solidification of containerized hazardous liquids [RO 12617];
- Mixing an ignitable spent solvent with used oil to remove the solvent's hazardous characteristic [RO 13570];
- Neutralization of spent lead-acid batteries (that are both corrosive and toxic for lead) in accumulation tanks [RO 11763];
- Nonthermal treatment of hazardous debris in containment buildings [August 18, 1992; 57 *FR* 37242, RO 13553, 13696];
- Precipitation of heavy metals from solutions and oxidation/reduction reactions [RO 14618];
- Recycling hazardous CRT glass into lead and silica sand [RO 14839];
- Stabilization of 1) baghouse dust [RO 12883], 2) F006 hazardous wastewater treatment sludge in a ribbon blender [RO 11379, 11422], and 3) hazardous lead-based paint chips/dust [RO 11624]; and
- Treating mixed radioactive/hazardous waste to conform wastes to Nuclear Regulatory Commission disposal requirements and/or Department of Transportation shipping rules [RO 11598, 13297].

Another example of how waste might be treated in a 90/180/270-day accumulation unit is presented in Case Study 7-4.

One other example cited by EPA warrants further discussion. A facility was using a vacuum system to collect and transport characteristically hazardous abrasive waste from lead-based paint removal. Portland cement and proprietary materials added to the vacuum system duct stabilized the waste, rendering it nonhazardous. When asked if this form of treatment could qualify for the 90-day accumulation unit permitting exemption, EPA noted that the vacuum system didn't seem to meet either the tank or container definition. Therefore, the agency suggested that the treatment system appeared to be a miscellaneous Subpart X treatment unit (for which the exemption could *not* apply), pending a final determination from the state agency. [RO 14462]

7.4 Recycling is treatment

When EPA first issued the regulations that implemented the RCRA statute in 1980, most materials destined for recycling were not defined as solid waste and, therefore, were not regulated as hazardous waste. Unfortunately, some entities abused this recycling incentive in ways that damaged the environment. Recognizing that additional controls were needed, the agency revised the definition of solid waste in 1985 to capture many recycled materials under RCRA regulatory control. To continue to encourage recycling, however, an exemption exists in the regulations such that the units in

> ### Case Study 7-4: Treatment of Partially Full Cans of Unwanted Paint
>
> If a facility has partially full cans of unwanted paint, the best option is to find someone who can use the paint for its intended purpose. In that case, the paint is not a waste. If the paint will not be used for its intended purpose, the point of generation of a solid waste is when the person in control of the paint decides it can't be used anymore and will be discarded. When that happens, the generator must determine if the paint is hazardous. Per Section 3.2.3.1.2, unused paint will not carry any of the F001–F005 spent solvent codes [RO 11220, 11340, 11349, 12334, 12906, 13273], as long as listed spent solvent (e.g., spent toluene that was used to clean out paint guns) has not been mixed into the paint. Also, unused paint is not a U-listed waste [RO 11180]. So the paint must be evaluated to determine if it exhibits any of the hazardous waste characteristics, per Chapter 2. The likely characteristic codes are D001 for ignitability of certain oil-based paints, D004–D011 for heavy metals, and D035 for methyl ethyl ketone contained in some paints. If paint that was intended to be used is found to be dried out (i.e., there is no liquid when personnel get ready to use it), then the paint would not be ignitable at its point of generation but could still exhibit the toxicity characteristic.
>
> If the paint does exhibit a characteristic at the point of generation, it is subject to the hazardous waste regulations from that point forward. One option is to store the waste paint in either satellite or 90/180/270-day accumulation containers and then send it offsite to a RCRA-permitted TSD facility. (See a discussion of satellite or 90/180/270-day accumulation in Sections 6.2 or 6.3, respectively.) If the paint is shipped offsite, the shipment must be accompanied by a hazardous waste manifest. In addition, the generator must comply with the applicable LDR notification requirements.
>
> If instead of shipping the hazardous paint waste offsite to a TSD facility, the generator wants to treat the paint (e.g., solidify it), one way to do that without getting a RCRA permit would be to treat the waste in a 90/180/270-day accumulation container. For example, while the paint is in a 90/180/270-day accumulation area, the generator could add a solidifying/fixation reagent to the paint. Solidifying the paint would likely remove the D001 code but wouldn't necessarily remove other codes, such as D004–D011 or D035. While the hazardous paint is being treated, the generator would have to comply with Part 265, Subpart I (e.g., the container must be closed). Also, 90/180/270-day containers must be marked with the words "Hazardous Waste" and the accumulation start date. Finally, as noted in Section 7.3.2.5, generators who treat hazardous wastes in 90/180/270-day accumulation units in order to meet LDR treatment standards must prepare and follow a waste analysis plan per §268.7(a)(5).

which hazardous wastes are recycled do not need a RCRA permit and are not subject to any RCRA management standards (with the possible exception of air emission provisions).

But, make no mistake: recycling *is* treatment. [RO 11461, 11745, 11880, 12865, 13279] Recycling meets both parts of the treatment definition. This is actually the driving force behind the recycling exemption, since such operations would otherwise be subject to permitting under §270.1(c). Because it's a form of treatment that EPA wants to encourage, an exemption exists for recycling processes in §261.6(c–d).

In the subsections below, we examine the recycling exemption from RCRA permitting/standards in greater detail. First, we review the scope and applicability of the exemption. Then, we look at various treatment situations and note whether they qualify for the recycling exemption.

7.4.1 Scope and applicability

The discussions in this section apply to the regulations in §261.6(c–d) exempting hazardous waste recycling from RCRA permitting/management standards. (Other recycling exemptions are discussed in Chapter 11. For example, the effect of the recycling exclusions provided in EPA's October 30, 2008 DSW rule [73 *FR* 64668] are covered in Chapter 11 and are not discussed in this Section 7.4.)

The exemption from RCRA permitting/management standards for recycling processes in §261.6(c–d) is summarized in Figure 7-2. The figure represents a recycling facility that accepts hazardous spent solvents from a number of offsite generators. The solvents are recycled in a solvent still (distillation column), with the recovered solvent stored for resale back to industry. Still bottoms are sent to a RCRA incinerator; if the bottoms are hazardous via characteristic or listing, they are manifested to the incinerator. The regulatory status of the various pieces of equipment and processes noted in Figure 7-2 are discussed below.

7.4.1.1 Shipment of hazardous waste to a recycling facility

Facilities that generate hazardous waste sometimes ask whether they can legally ship their waste to offsite recycling facilities that are themselves exempt from RCRA permitting. The answer is yes. Although §262.20(b) notes that generators must designate on the manifest a receiving facility that is *permitted* to handle the waste, the definition of a designated facility in §260.10 includes permit-exempt recycling facilities. Therefore, a generator may list a recycling facility as the designated facility on a hazardous waste manifest. The agency notes, however, that the generator, transporter, and designated facility must comply fully with manifest and manifest discrepancy requirements for shipments of hazardous waste to recycling facilities. [RO 13663]

7.4.1.2 Storage before recycling

Facilities that conduct recycling typically store hazardous wastes prior to the recycling operation. If they do, such storage is fully regulated under RCRA, despite the exempt status of the recycling units. [§261.6(c)(1)] A facility's eligibility for the exemption from permitting/standards for the recycling process itself is independent of whether RCRA-regulated storage of hazardous wastes is being conducted before the wastes are recycled. [RO 12581] If storage is occurring, it is subject to, among other things, 1) permitting under Part 270; and 2) container or tank standards in Part 264/265, Subpart I or J, respectively. In Figure 7-2, for example, storage of F005 spent solvent prior to recovery in the still is subject to RCRA permitting and Subpart J tank standards.

A permit is not required for recyclers who do not store hazardous waste prior to processing. [RO 11131, 11404, 11411, 11765] Where a waste received at a recycling facility is fed immediately into the recycling process, storage is not occurring. Conversely, waste held in a container or tank for several days or weeks prior to recycling clearly is being stored. Uncertainty occurs when a facility holds waste for a short period of time after receipt but prior to recycling, as illustrated in the following examples.

Q *If hazardous waste is fed directly from tank trucks into the recycling unit, how long can the trucks remain onsite before they require permitting as storage containers?*

A According to EPA, "federal regulations do not specify an allowable 'holding time' prior to the waste being introduced to the recycling process; however, the appropriate EPA regional office or authorized state regulatory agency may specify such a holding time on a site-specific basis, defining a time at which storage begins…. [S]ome states and regions do allow up to 24 hours for the off-loading of a hazardous waste into the recycling process before the waste is considered to be stored, thus requiring a storage permit." [RO 11411] Additionally, when asked about a specific situation where it was understood that transport vehicles remain onsite for no more than 24 hours, EPA noted that it would not require the tank trucks to obtain a storage permit

Figure 7-2: Scope and Applicability of the Permit Exemption for Recycling

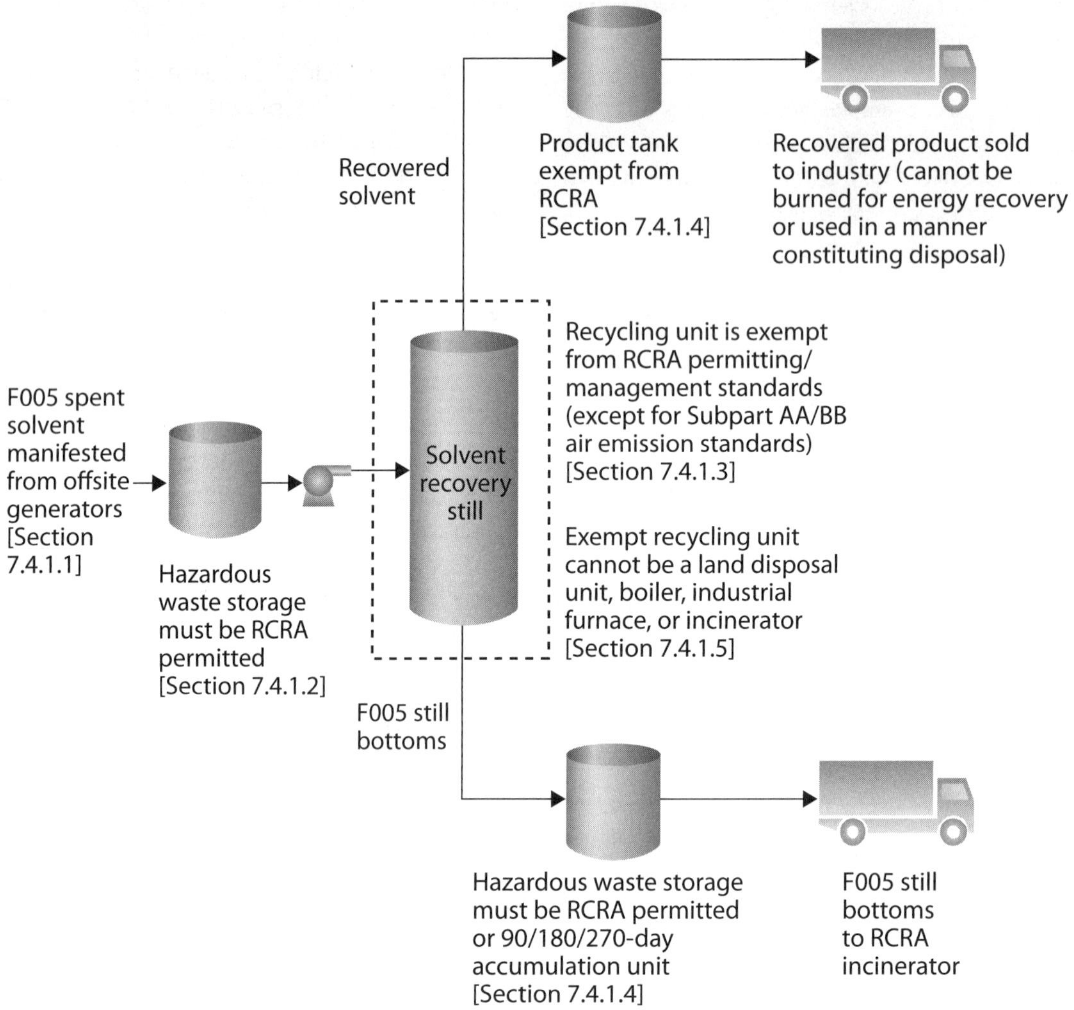

- Recovered solvent → Product tank exempt from RCRA [Section 7.4.1.4] → Recovered product sold to industry (cannot be burned for energy recovery or used in a manner constituting disposal)

- F005 spent solvent manifested from offsite generators [Section 7.4.1.1] → Hazardous waste storage must be RCRA permitted [Section 7.4.1.2] → Solvent recovery still

- Recycling unit is exempt from RCRA permitting/management standards (except for Subpart AA/BB air emission standards) [Section 7.4.1.3]

- Exempt recycling unit cannot be a land disposal unit, boiler, industrial furnace, or incinerator [Section 7.4.1.5]

- F005 still bottoms → Hazardous waste storage must be RCRA permitted or 90/180/270-day accumulation unit [Section 7.4.1.4] → F005 still bottoms to RCRA incinerator

Source: McCoy and Associates, Inc.

for off-loading the hazardous waste. [RO 11386, 13127] In other guidance, EPA stated "holding of drums for a few hours may not be storage," but this is a site-specific determination. [RO 11365] Bottom line: check with your state to determine if they set a time limit on waste transfers.

Bulk shipments and drums of spent solvent received at a recycling facility are pumped into a feed tank that is hard-piped to a distillation unit. When the distillation column is idle (i.e., at night), solvent remains in the feed tank. Does this practice constitute storage before recycling that is subject to a storage permit? What if any time there is spent solvent in the feed tank, a pump on the bottom of the tank operates and feeds solvent to the distillation unit such that the tank never contains solvent when the distillation unit is not in operation?

A In the first case, the feed tank is used to store spent solvent (albeit for short periods of time) and is considered a hazardous waste tank that needs a permit and is subject to Subpart J regulations. In the second case, the feed tank is not considered to be used for storage and is therefore not subject to permitting or regulation as a hazardous waste tank. Rather, it is used as a means of conveyance and is viewed as part of the exempt recycling unit. [RO 11365, 12895]

Q *Can hazardous waste be stored at a transfer facility at a recycling site for up to ten days and then be moved on the same site to the recycling process, where recycling begins immediately? Would this arrangement allow the facility to avoid getting a RCRA storage permit?*

A No to both questions. The transfer facility provisions of §263.12 apply to holding waste during the normal course of transportation. Arrival of the waste at the designated facility constitutes completion of the transportation phase, so the 10-day limit is not applicable at the recycling facility. A transporter who ships to a piece of property contiguous to a recycling facility has technically completed the transportation phase if no further "transportation" is to be conducted. Thus, a piece of property contiguous to a recycling facility must meet the definition of a designated facility, meaning that designated facilities cannot have transfer facilities on their property. If the recycling facility stores hazardous waste at the site for 10 days before recycling begins, a storage permit would likely be required. [RO 11365]

7.4.1.2.1 Storage at generator facilities

If the hazardous waste recycling is conducted at the same facility at which the waste was generated, the facility may be just a generator. Assuming that the recycling is *not* done in a manner that satisfies the §261.4(a)(8) closed-loop recycling exclusion, storage of such waste prior to recycling can be accomplished in 90/180/270-day accumulation units that are exempt from RCRA permitting but are subject to §262.34 requirements. [RO 11880] If the recycling at the generator facility does qualify for the §261.4(a)(8) closed-loop recycling exclusion, the material flowing around the closed loop is not a solid or hazardous waste, and so the tank-based storage would not be subject to RCRA at all (see Section 11.3.4).

7.4.1.3 Recycling units are exempt except for air emission standards

While storing hazardous waste prior to recycling is subject to RCRA permitting/management standards, the recycling unit itself is not. [RO 11383] "EPA does not presently regulate the actual process of recycling...only the storage, transport, and generation that precedes it." [August 20, 1985; 50 *FR* 33542] The parenthetical sentence in §261.6(c)(1) exempts the recycling process from permitting/management standards, with one exception. [RO 11238, 11814, 11880] Because EPA believes that processing recyclable materials may pose risks through direct emissions to the atmosphere, recycling processes are subject to organic air emission standards included in Part 264/265, Subpart AA (applicable to process vents) and Subpart BB (applicable to fugitive emissions from equipment leaks). [§261.6(c)(1) and (c)(2)(iii)] Therefore, the solvent recovery still and its associated ancillary equipment in Figure 7-2 will be subject to these standards.

Note, however, that recycling units at generator facilities (that do not require RCRA permits) are exempt from the Subpart AA and BB standards. [§§261.6(d), 265.1030(b)(2), 265.1050(b)(2)] Recycling units are subject to Subparts AA and BB only if there is a RCRA permit required for some other unit at the same facility. [RO 11881]

In addition to air emission standards, the recycler must obtain an EPA ID number, use the manifest, and report manifest discrepancies.

7.4.1.3.1 Mobile recycling units

Mobile equipment, such as truck-mounted solvent recovery units, can qualify under the §261.6(c) recycling exemption. Such units would still be subject to the RCRA air emission standards, however. [RO 12619, 13280]

7.4.1.3.2 Partial recycling by the generator is still exempt

Sometimes, hazardous waste is partially recycled by the generator before being sent offsite for further legitimate recycling. Per EPA guidance in RO 11880, processing activities at the generator's site are considered the first step in the recycling process and remain exempt under §261.6(c)(1).

7.4.1.4 Storage after recycling

Per the last, parenthetical sentence in §261.3(c)(2)(i), materials reclaimed from hazardous wastes that are used beneficially (e.g., reclaimed solvents) are not solid or hazardous wastes. Thus, the recovered solvent in Figure 7-2 is a product, not subject to any RCRA standards. (This exemption from the derived-from rule is based on the reclaimed material *not* being burned for energy recovery or applied to the land.) A tank holding such product is also exempt from RCRA.

Residues from recycling hazardous waste (such as the still bottoms in Figure 7-2) may continue to be hazardous, based on operation of the derived-from rule. Storage of hazardous residues from the recycling process will be subject to RCRA requirements: permitting under Part 270 (unless conducted in a 90/180/270-day accumulation unit), and container or tank standards in Part 264/265, Subpart I or J. (Residues from exempt recycling units are newly generated wastes and therefore may be stored in 90/180/270-day accumulation units—see Section 14.1 and RO 11420, 12865, 13280.)

7.4.1.5 Some processes can never be exempt recycling

The RCRA recycling exemption does not apply to processes analogous to land disposal or incineration. [January 4, 1985; 50 *FR* 643] Therefore, recycling in land disposal units (e.g., surface impoundments), burning for energy or material recovery, and incineration do *not* qualify for this exemption.

[RO 11131, 11873] A surface impoundment, waste pile, or other land disposal unit receiving hazardous waste must be permitted and comply with the unit-specific TSD facility standards in Parts 264 or 265. Burning hazardous waste in boilers or industrial furnaces is subject to RCRA permitting and the technical requirements in Part 266, Subpart H. Incineration of hazardous waste is subject to permitting and the RCRA management standards in Part 264/265, Subpart O.

7.4.2 Examples

EPA guidance addresses a multitude of examples involving potentially exempt hazardous waste recycling. Our coverage of some of the more interesting of these examples follows.

7.4.2.1 Qualifying processes

Listed below and in Case Study 7-5 are examples of recycling situations that qualify for the permitting exemption of §261.6(c–d). (Note that in stating its opinion that a given recycling process is permit-exempt, EPA always couches such pronouncements with the disclaimer that the facility should consult with its state or regional RCRA authority for binding determinations.)

- Crushing spent dry cleaning filters to remove solvents and the subsequent reclamation of the removed solvents. [RO 11670]

- Gasifying a mixture of listed hazardous sludges, oils, and other materials to recover syngas. EPA determined that the gasification unit would qualify for the recycling exemption since it was not an incinerator, boiler, or industrial furnace. [RO 11905, 11985]

- Grinding a mixture of lead-based paint (LBP) chips and wood debris from LBP removal operations to prepare the waste for shipment to a lead reclaimer. [RO 11880]

- Nonthermal reclamation of foundry sands with subsequent return of the material to the foundry process. [RO 12873, 13749]

- Processing baghouse dust to remove lead, with subsequent shipment of the dust to a fertilizer producer. [RO 13507]

- Recovering silver salt sludges from spent photo-processing fixer solutions. [RO 11814]

- Recovering usable solvent from a spent solvent via distillation, thin-film evaporation, etc. [RO 11200, 12422, 12581, 12865, 13280, 14089]

- Recovering zinc from K061 electric arc furnace dust via a process involving mixing with coal or coke. [RO 11353]

- Slurrying and subsequent removal of sulfur and chlorides from baghouse dust. [RO 13566]

7.4.2.2 Other examples

Q *Are facilities used to regenerate hazardous spent activated carbon eligible for the recycling permit exemption?*

A No. The definitions of "carbon regeneration unit" and "incinerator" in §260.10 make it clear that both direct-flame and nonflame carbon regeneration units are *not* incinerators but are thermal treatment devices. However, EPA has clarified [February 21, 1991; 56 *FR* 7200, RO 11955, 13491, 14242] that such carbon regeneration units are subject to either Part 264, Subpart X requirements or Part 265, Subpart P standards; thus, these processes would not be eligible for the §261.6(c–d) exemption from RCRA permitting.

Case Study 7-5: Metals Recovery in Industrial Furnaces

What about situations involving metals recovery in thermal treatment units? Generally, metals recovery is considered reclamation, which is a form of recycling [see the definitions in §261.1(c)(4) and (7)]. However, we noted previously that burning a hazardous waste in a thermal treatment unit such as a boiler or industrial furnace is not a permit-exempt recycling activity, but an undertaking that requires a permit and is subject to Part 266, Subpart H standards. Nevertheless, the following examples of thermal metals recovery indicate that such activities typically are permit-exempt under §261.6(c–d).

In one case, the agency was asked if burning precious-metal-bearing hazardous wastes such as paper, filters, and circuit boards in a thermal reduction furnace qualified for the recycling exemption. Because EPA believes that such furnaces are legitimately recovering precious metals from recyclable materials, the recovery would fall under Part 266, Subpart F, and the furnace would be exempt from RCRA permitting. Additionally, such furnaces would not be subject to Part 264/265, Subpart O incinerator standards or Part 266, Subpart H boiler or industrial furnace requirements. [§266.100(g), August 27, 1991; 56 *FR* 42508, RO 11804, 13703] A state may require a precious metal recycler to obtain a RCRA permit, but there is no specific federal requirement for a RCRA permit at such a facility. [July 21, 2008 email from EPA to McCoy and Associates, Inc.]

Roasting and/or retorting mercury-contaminated (D009) soils from natural gas pipeline meter leakage is also considered reclamation. However, the agency believes the equipment involved would fit the §260.10 definition of an industrial furnace as a smelting, melting, and refining furnace. Normally, hazardous waste burned in an industrial furnace would be subject to Part 266, Subpart H and would not be permit-exempt. However, a provision in Subpart H [§266.100(d)] specifically exempts industrial furnaces burning hazardous waste solely for metals recovery from permitting and most Subpart H requirements. Thus, this mercury reclamation process would escape RCRA permitting and most of the management standards. [RO 13642]

In a third situation, F006 electroplating sludge is commonly used as feedstock to metals recovery smelters. Although these units are considered industrial furnaces subject to Part 266, Subpart H, §266.100(d) is again applicable, and the process is permit-exempt via §261.6(c). [RO 11426, 12422]

Q *Is a RCRA permit required for recycling hazardous wastes by blending, mixing, or other physical processes in order to produce fuels, if no prior storage of the wastes occurs?*

A Yes. The agency notes that vessels used for storing/treating hazardous waste fuels that will be burned in boilers or industrial furnaces are subject to regulation/permitting as hazardous waste tanks. [§266.101(c)] This provision is applicable to storage/treatment by the burner as well as to such units at intermediate facilities between the generator and the burner. EPA states that there could be some recycling operations at a fuel blending facility that are exempt from permitting, even though the fuel blending itself is not exempt. [April 13, 1987; 52 *FR* 11820, RO 11238, 11411, 11497, 11881, 13512]

7.5 Other permit-exempt treatment options

In this section, we address five additional exemptions that facilities may use to avoid RCRA permitting when treating hazardous waste:

- Elementary neutralization units,
- Totally enclosed treatment facilities,
- Adding absorbents to wastes,
- Immediate responses, and
- Burning small quantities of wastes in onsite units.

7.5.1 Elementary neutralization units

Based on the type of hazards they pose to human health and the environment, EPA allows facilities to treat D002 corrosive hazardous wastes in elementary neutralization units (ENUs) without a RCRA permit. As described below, the ENU exemption is similar in implementation and scope to the WWTU exemption detailed in Section 7.2.

7.5.1.1 Implementation hinges on ENU definition

The ENU exemption is based on the term "elementary neutralization unit," which is defined in §260.10 as follows:

"[A] device which:

"(1) Is used for neutralizing wastes that are hazardous only because they exhibit the corrosivity characteristic defined in §261.22 of this chapter, or they are listed in Subpart D of Part 261 of the chapter only for this reason; and

"(2) Meets the definition of tank, tank system, container, transport vehicle, or vessel in §260.10 of this chapter."

Although the ENU exemption applies to the treatment of hazardous wastes that were listed solely because they exhibit the characteristic of corrosivity, there are currently no such listed wastes. A transport vehicle would be a vacuum truck, tank car, or rail car. When EPA uses the term "vessel" in this definition, it is referring to a ship, barge, or some other type of watercraft. We examine the two clauses in this definition in more detail below.

7.5.1.1.1 *Neutralization and ENU eligibility*

Although EPA does not provide a formal definition of the term "neutralize," the agency does offer the following guidance on applying this treatment process to hazardous wastes:

"[Y]ou cannot treat [neutralize] a waste that is both corrosive and otherwise hazardous (due to listing or by exhibiting a different hazardous characteristic) in an ENU since the influent must be corrosive only in order to meet the definition of an ENU." [RO 11551]

7.5.1.1.2 *Tanks, containers, etc.*

The ENU permitting exemption provides that tanks, containers, etc. that are used to neutralize D002 wastes are exempt from 1) the requirement to be RCRA-permitted, 2) general facility standards in Parts 264 and 265, 3) the specific RCRA standards in Part 264/265, Subparts J (for tanks) and I (for containers), and 4) the §262.34 accumulation unit standards. [RO 13727] The ENU eligibility criteria apply to (among other types of units) tanks and tank systems. Since the definitions of both of these terms are presented and discussed in detail in Section 7.2.1.3, we won't repeat that information here.

Readers should note the following points, however, with respect to ENUs.

First, the agency extends the ENU exemption "to include devices that are commonly considered to be tanks as well as devices such as flumes, gutters, troughs, and pipes which are not commonly considered to be tanks, but which nevertheless meet the expansive definition of tank in §260.10. It includes both units that are principally designed to neutralize corrosive wastes and units which achieve this and other waste treatment objectives such as precipitation of waste constituents." [November 17, 1980; 45 *FR* 76078; see also RO 11173] For example, the in-line neutralization of waste acid by addition of caustic in a sewer system would be considered treatment in an ENU. ENUs may consist of a series of tanks—not just a single unit (see Figure 7-3). [RO 11173]

Sumps meeting the definition of a tank (by passing the "parking lot" test—see Section 10.1.3.2) can qualify for the ENU exemption, along with their associated equipment. [RO 13170] Additionally, a sump meeting the definition of a tank that is used to convey corrosive wastewater to a separate ENU could be considered ancillary equipment to the ENU and would still be exempt from RCRA tank standards. [September 2, 1988; 53 *FR* 34080, RO 14470]

As with the WWTU exemption, the term "tank system" includes all ancillary equipment (see Section 7.2.1.3.4 for a discussion of this term). This would include piping, fittings, pumps, etc. associated with an eligible tank, but only where the ancillary

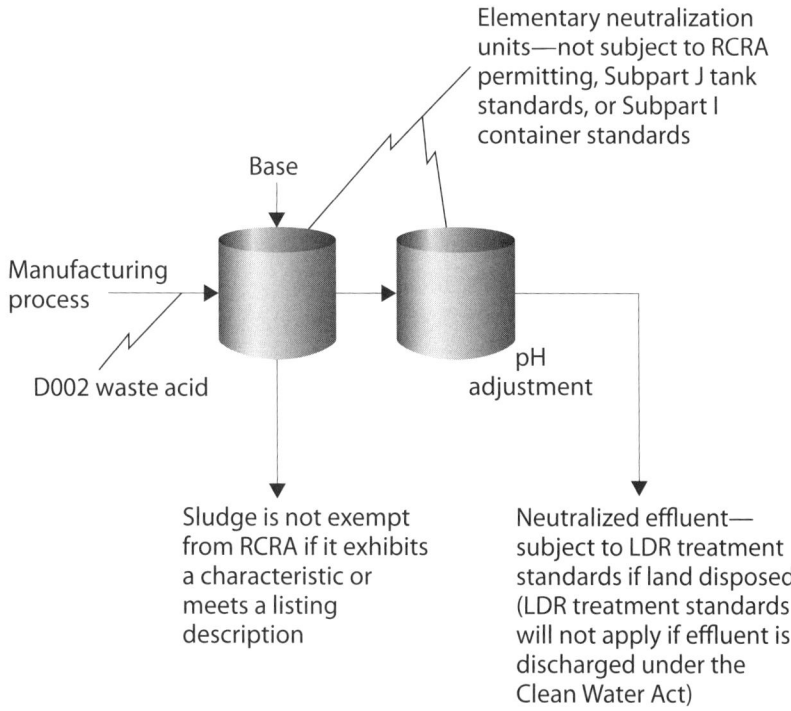

Figure 7-3: Applicability of Elementary Neutralization Unit Exemption

Source: McCoy and Associates, Inc.

equipment 1) is dedicated to exclusive use with an exempt tank, and 2) prevents leakage or discharge of the hazardous waste. [RO 11631, 13126, 13653] Equipment associated with an ENU that is not designed to contain wastes would not be eligible for the ENU exemption since it would be considered a type of disposal device and would be subject to permitting. Such discharges could also subject the facility to cleanup authorities. [RO 14470]

Finally, note that a container is never considered to be ancillary equipment to a tank. [RO 11551] This distinction is important because a container's eligibility for the ENU exemption depends on neutralization being conducted in the container, *not* on neutralization conducted in a tank the container might be associated with. For example, containers used to store corrosive wastewater at a generator's facility prior to neutralization in a tank are not ENUs but are subject to RCRA standards, including those in §262.34 and Part 265, Subpart I.

7.5.1.2 Scope of the ENU exemption

A number of issues associated with the ENU permitting exemption are addressed below.

Q *Are units eligible for the ENU permitting exemption subject to TSD facility standards?*

A No. Similar to the WWTU exemption, permit-exempt ENUs are also exempt from RCRA management standards. Specifically, §264.1(g)(6) excuses eligible ENUs from Part 264 requirements, including the tank and container standards of Subparts J and I, respectively, and the air emission standards of Subparts AA–CC. Similarly, §265.1(c)(10) exempts ENUs at interim status facilities from Part 265 standards. Finally, note that §270.1(c)(2)(v) is the provision that actually exempts ENUs from RCRA permitting. [RO 13717, 13727, 14745]

Q *Are ENUs subject to RCRA requirements at generator facilities?*

A No. ENUs are exempt from generator onsite time limits (i.e., no 90-day clock is running), daily or weekly inspections, marking requirements, and any other §262.34 provisions. [RO 13717, 13727, 14745]

Q *How is the onsite treatment and/or disposal of corrosive hazardous waste at conditionally exempt small quantity generators (CESQGs) regulated in terms of ENUs?*

A CESQGs actually get a bonus for treating D002 wastes in ENUs—they are not subject to the requirements of §261.5(f)(3) and (g)(3) that identify acceptable waste disposal options. Normally, CESQGs would have to meet these two provisions for their acute and nonacute hazardous waste, respectively, in order to escape other RCRA management standards. [RO 13778]

Q *Does the ENU exemption apply if the D002 waste remains in a tank for two months before it is neutralized?*

A Yes. The waste is not subject to the substantive RCRA standards as long as it remains in the ENU. [RO 13204]

Q *Does the ENU permitting exemption affect the identification and regulation of hazardous wastes generated and/or discharged from that unit?*

A No. These two concepts operate independently. In other words, residues generated during treatment of a corrosive waste in a permit-exempt ENU may still be designated as hazardous wastes; the "hazardousness" of any residues generated is independent of the permitting status of the unit in which the residues are formed. For example, sludge generated during the treatment of D002 electroplating wastewaters in an exempt ENU would be classified as F006 listed hazardous waste. [RO 11551] The exempt status of the unit does not attach to any discharged residues.

Addressing this issue from the opposite direction, eligibility for the ENU exemption is *not* affected by the generation of hazardous sludges or other wastes during the treatment process. Per EPA, "generation of a new, noncorrosive listed or characteristic hazardous waste during the neutralization process does not automatically bar the tank from the elementary neutralization unit definition." [RO 13717]

Q *Does neutralized waste have to be discharged under a CWA permit to claim the elementary neutralization unit exemption from permitting?*

A No. Wastes may be treated in an elementary neutralization unit and then (if treatment residues are nonhazardous) discharged to any nonhazardous waste management unit, such as a Class I nonhazardous waste injection well. Elementary neutralization units used in these applications are still exempt from RCRA permitting or management standards. [RO 14745]

Q *Do the LDR treatment standards apply to the treated residues after they have been neutralized?*

A Possibly. Wastes treated in an ENU would still be subject to LDR treatment standards for the D002 waste code if they will be land disposed after leaving the ENU. For example, D002 is neutralized followed by solidification of the neutralized effluent. Residues resulting from the solidification would have to meet universal treatment standards for any underlying hazardous constituents (identified at the point of generation of the original D002) before such material could be landfilled, even though the solidified material is no longer hazardous. [RO 13649] See Section 13.4.3 for more discussion of how LDR applies to ENU treatment residues.

Q *Can ENUs be considered solid waste management units (SWMUs)?*

A Yes. Even though they are exempt from permitting and most other RCRA standards, ENUs are still considered SWMUs and would be subject to investigation and possible remediation should a facility be subject to the corrective action provisions of Part 264, Subpart S. [RO 12371]

7.5.2 Totally enclosed treatment facilities

The three sections that provide the exemption from RCRA requirements for totally enclosed treatment facilities (TETFs) are §§264.1(g)(5), 265.1(c)(9), and 270.1(c)(2)(iv). As with the WWTU and ENU exemptions described above, qualified TETFs are exempt not only from RCRA permitting, but also from Part 264/265 standards. The definition of a TETF is as follows:

"Totally enclosed treatment facility means a facility for the treatment of hazardous waste which is directly connected to an industrial production process and which is constructed and operated in a manner which prevents the release of any hazardous waste or any constituent thereof into the environment during treatment. An example is a pipe in which waste acid is neutralized." [§260.10]

Although the exemption is defined to apply to a "facility," we think of it more in terms of a unit or a couple of units in series. EPA's rationale for providing this often-sought but seldom-granted exemption was discussed in the original May 19, 1980 RCRA rule:

"[T]o classify 'totally enclosed treatment systems,' such as pipes, as hazardous waste treatment facilities and to require them to meet Section 3004 standards and obtain a permit would not make a great deal of sense. These facilities by definition do not release wastes or waste constituents into the environment, and therefore stringent controls are not 'necessary to protect human health and the environment.' Such controls might also discourage the use of such facilities, which in many ways represent the optimum in good waste management practices. It may also be very difficult as a practical matter to permit or otherwise regulate these types of facilities—many are indoors, are part of complicated plumbing systems which do not fall within RCRA's jurisdiction, and do not have clearly defined starting and end points." [45 *FR* 33218]

7.5.2.1 Two characteristics of a TETF

Two specific items are key to EPA's interpretation as to whether various industrial situations qualify for the exemption:

1. The TETF must be directly connected to an industrial production process, and

2. There can be no release of any hazardous waste or any constituent thereof into the environment during treatment.

These two aspects are discussed below.

7.5.2.1.1 Directly connected

The first aspect of the definition of a TETF has to do with the treatment unit's direct connection to an industrial production process:

"The part of the definition requiring that totally enclosed treatment facilities be 'directly connected to an industrial production process' also generates some uncertainty. As long as the process is integrally connected via pipe to the production process, there is no potential for the waste to be lost. The term 'industrial production process' was meant to include only those processes which produce a product, an intermediate, a by-product, or a material which is used back in the production process. Thus, a totally enclosed treatment operation, integrally connected downstream from a wastewater treatment lagoon would not be eligible for the exemption because the process to which it is connected is not an 'industrial production process.' Neither would any totally enclosed treatment process at an offsite hazardous waste management facility qualify, unless it were integrally connected via pipeline to the generator's production process. Obviously, a waste transported by truck or rail is not integrally connected to the production process.

"Hazardous waste treatment is often conducted in a series of unit operations, each connected by pipe to the other. As long as one end of a treatment train is integrally connected to a production process, and each unit operation is integrally connected to the other, all qualify for the exemption if they meet the requirement of being 'totally enclosed.'" [RO 12097]

EPA guidance shows the application of this requirement to the TETF exemption:

- Ash from a hazardous waste incinerator is directed to a treatment device where the addition of a nonhazardous reagent binds otherwise leachable hazardous metals, reducing the toxicity of the ash. EPA notes that since the incinerator is itself a type of treatment device (and not an industrial production process), the ash stabilization system cannot qualify as a TETF. [RO 13022]

- In three similar situations, companies operate foundries in which dust emissions from a cupola are directed to a baghouse, which is itself connected to devices designed to render the dust nonhazardous. In each case, EPA determined that the treatment devices could not be called TETFs since they were connected to the baghouse (which is not a production device) instead of the cupola. Additionally, baghouses are open to the environment and do not remove 100% of the hazardous constituents. [RO 12561, 12824, 12883]

- A chemical agent/munitions destruction system was also determined not to qualify as a TETF because it was not connected to any industrial production process. (Like an incinerator, the system would also release emissions to the air, doubly disqualifying it.) [RO 12495]

- Similarly, a filter press proposed as part of a corrective action cleanup would not be directly connected to an industrial production process, rendering it ineligible for the TETF exemption. (EPA also found that the press could be opened to the environment, thus rendering it not totally enclosed.) [RO 12942]

7.5.2.1.2 No release of hazardous constituents

Regarding the second qualifying provision, EPA notes the following:

"The key characteristic of such a [TETF] is that it does not release any hazardous waste or constituent of hazardous waste into the environment during treatment. Thus, if a facility leaks, spills, or discharges waste or waste constituents, or emits wastes or waste constituents into the air during treatment, it is not a 'totally enclosed treatment facility.'" [45 FR 33218]

"The agency intends that a 'totally enclosed' treatment facility be one which is completely contained on all sides and poses little or no potential for escape of waste to the environment even during periods of process upset. The facility must be constructed so that no predictable potential for overflows, spills, gaseous emissions, etc. can result from malfunction of pumps, valves, etc. associated with the totally enclosed treatment or from a malfunction in the industrial process to which it is connected." [RO 12097]

The bottom line is that nothing can exit the TETF other than the treated waste. A pipe or tank that leaks is not totally enclosed and such leakage would disqualify it from the exemption. The following examples illustrate EPA's application of the above criteria. (As we noted above, the TETF exemption seems to be sought more than it is obtained; all of the examples below are cases where the TETF exemption was denied.)

- Corrosive wastewater is transferred via an open-channel sewer to a wastewater treatment system. The waste is neutralized as it flows through the sewer, a process that constitutes hazardous waste treatment. Since the sewer is located entirely within a building, EPA was asked if it qualified as a TETF. The agency responded that the sewer was not totally enclosed and that releases from the sewer could also escape from the building, disqualifying the sewer as a TETF. A TETF must be covered to eliminate gaseous emissions. [RO 11173]

- In a similar situation, the agency was asked if a wastewater collection sump and its associated equipment could qualify as a TETF. Again, EPA responded that the exemption would be unlikely to apply since the sump would allow waste emissions. [RO 13170]

- Two different types of emission control devices, an incinerator and a wet-air oxidation unit, also fail to qualify as TETFs since they both would allow airborne emissions. [RO 11263, 12558] Basically, no thermal treatment unit is going to qualify for this exemption, because gaseous emissions with low concentrations of hazardous constituents will inevitably be released to the environment: "continuous gaseous by-products emitted during treatment represent an open system that interacts significantly with the environment. In our opinion, extension of the exclusion to thermal treatment units would be inappropriate and unjustified." [RO 12558]

- Finally, a thermal evaporator would also fail to qualify as a TETF, because it releases the water vapor it removes from hazardous waste directly to the atmosphere. [RO 12923]

7.5.2.1.3 *What does qualify as a TETF?*

From a practical standpoint, a TETF is going to be limited to a pipe or a tank-like unit from which nothing leaks, spills, or is emitted. EPA noted that a TETF can have vents associated with it as long as such vents are trapped or provided with sorption columns to eliminate gaseous emissions. [RO 12097]

Are there any cases where EPA *has* found that a unit or facility qualified as a TETF? Yes. Here are several such examples:

- At a foundry, the owner/operator proposed to install a dust treatment system in the ductwork *between* the cupola furnace and its baghouse. Here the agency noted that the system's direct, closed connection to the industrial production process (i.e., the cupola furnace) would qualify the treatment system for the TETF exemption. [RO 14022]

- Provided it is totally enclosed and connected to an industrial production process, a *portable* hazardous waste treatment system could qualify as a TETF. [RO 12178]

- Production wastes that are treated in-pipe would also qualify for this exemption. [45 *FR* 33218]

- Although not discussed in guidance, we believe that hazardous waste treated via chemical oxidation or reduction in an enclosed system could qualify for this exemption. Adding reagents to a hazardous waste in this manner would physically consist of little more than a wide spot in the line, but it would likely be more sophisticated than just flanging two pipes together.

7.5.2.2 A final word on TETFs

Similar to the other exemptions we have already covered in this chapter, the exemption granted to equipment that qualifies as a TETF does not extend to the waste managed in the unit. Thus, for example, if the waste entering the TETF is a non-ICR-only listed hazardous waste, then any effluent from the facility is a listed hazardous waste via the derived-from rule and must be managed as such. [45 *FR* 33218, RO 12097]

7.5.3 Adding absorbents to wastes

An important provision of RCRA, codified in §§264.314 and 265.314 (for permitted and interim status TSD facilities, respectively), prohibits landfilling bulk and containerized hazardous wastes that either are liquid or contain free liquid. These regulatory sections result from a statutory ban [at RCRA Section 3004(c)(1)] on placing free liquids in landfills, which was enacted to minimize the migration (leaching) of wastes and waste constituents from landfills into ground water. The ban on liquids in landfills is discussed in Case Study 7-6.

In a nutshell, many wastes contain free liquid due to condensation, exposure to the elements, incomplete drainage of tanks and containers prior to waste addition, the type of production process from which they are generated, etc. Probably the simplest method to deal with these liquids (if they are

---

Case Study 7-6: Liquids-in-Landfills Ban

Since it is closely related to the absorbent addition permitting exemption, but not covered in detail elsewhere in this book, we thought it would be helpful to briefly review EPA's ban on the placement of liquid hazardous wastes in landfills. As noted above, this provision is codified in §§264.314 and 265.314 and prohibits the landfilling of bulk as well as containerized hazardous wastes that either are liquid or contain free liquid.

First, the placement of bulk or noncontainerized liquid hazardous waste or hazardous waste containing free liquids (whether or not absorbents have been added) in any landfill is prohibited. [§§264.314(b)/265.314(b)] EPA prohibits the use of absorbents to treat bulk liquid hazardous waste that will be landfilled. Instead, such liquid must be chemically, thermally, physically, or biologically treated without the use of absorbents; for example, chemical stabilization would be acceptable. [RO 12666, 13724] After treatment, a chemical stabilization test should be performed prior to placement in the landfill.

Similarly, spills of liquid hazardous waste that are cleaned up using absorbents cannot be placed directly in landfills; such absorbent/spill mixtures must be containerized first. However, this ban does not apply to soil that is contaminated with spills of liquid hazardous waste. If the contaminated soil passes the paint filter test (and meets LDR standards), it may be landfilled. [RO 12666]

For containerized hazardous wastes that contain free liquids, a likely method facilities would use for dealing with the ban is to add absorbents to their hazardous waste, thereby converting the liquid portion to solids that would pass the paint filter test. However, several conditions limit the applicability of this scenario. First, the hazardous waste/absorbent mixture must be placed in containers prior to landfilling per §§264.314(d)/265.314(c). Second, any material used to absorb liquid in containerized hazardous waste must be nonbiodegradable. [§§264.314(e)/265.314(f), RO 11798, 13724]

The bans on placing bulk or noncontainerized liquid nonhazardous wastes and on placing containers holding liquid nonhazardous wastes in municipal solid waste landfills are in §258.28(a) and (b), respectively.

containerized) is to add inert absorbent materials to absorb them. Once any liquids are absorbed, the hazardous waste will likely pass the paint filter test (Method 9095B in SW–846), which determines the presence of free liquids. If there are no free liquids present, a given waste can be landfilled. [RO 12443, 12452] However, EPA considers the deliberate mixing of absorbents with hazardous waste to be treatment and generally would require facilities conducting treatment to obtain a RCRA permit. [February 25, 1982; 47 *FR* 8304, RO 11619, 11694]

EPA recognized, however, that facilities may need to add absorbents to their containerized hazardous wastes to eliminate free liquids, and so early in the RCRA program (February 25, 1982), the agency exempted the addition of absorbents to hazardous waste in containers from RCRA permitting and management standards. [47 *FR* 8304] The permitting exemption is codified in §270.1(c)(2)(vii); the management standard exemptions are codified in §§264.1(g)(10) and 265.1(c)(13).

7.5.3.1 Implementation

The above-noted regulations that implement the absorbent addition exemption extend that provision to the following:

"Persons adding absorbent material to waste in a container (as defined in §260.10 of this chapter) and persons adding waste to absorbent material in a container, provided that these actions occur at the time waste is first placed in the container; and §§264.17(b), 264.171, and 264.172 of this chapter are complied with." [§270.1(c)(2)(vii)]

7.5.3.1.1 *Containers are key*

An important aspect of the absorbent addition exemption is that it applies only to containerized hazardous wastes, where a container is "any portable device in which a material is stored, transported, treated, disposed of, or otherwise handled." [§260.10] Thus, for example, standing liquid can be decanted off a load of bulk hazardous solids and placed into a container, where absorbents can be added without the need for a RCRA permit (or the application of Part 264/265 management standards). [RO 12443]

7.5.3.1.2 *Order does not matter, but timing does*

EPA does not care in what order hazardous waste and absorbents are added to containers. However, the agency is very concerned with *when* such additions are made. Section 270.1(c)(2)(vii) notes that the additions must occur at the time waste is first placed in the container. Thus, the absorbent addition exemption applies when hazardous waste is first added to containers, whether the absorbent is already in the container or is added at the same time as the waste.

EPA explains its reasons for the above requirements as follows:

"The agency does not believe that [mixing absorbents with hazardous wastes], when employed at the time hazardous wastes have been first placed in containers, poses a substantial hazard to human health or the environment…. [A]bsorbents commonly used are…[not] known to react in such a dangerous manner with any of the hazardous wastes identified in 40 *CFR* Part 261.

"However, the agency is not convinced that these treatment practices, when employed at a time after hazardous wastes have been emplaced in containers, pose no substantial hazard.

"[T]he potential hazards…do not derive from the actual addition of absorbents to wastes or wastes to absorbents but, instead, derive from the essential ancillary operation of opening containers." [47 *FR* 8305]

In other words, once wastes and absorbents are placed in a container, EPA believes it is safest for that container to remain closed. Therefore, the practice of adding absorbents at some time after the hazardous waste is placed in a container would constitute treatment requiring a permit (unless done in a 90/180/270-day unit) and would be subject to Part 264/265 management standards.

7.5.3.1.3 Other implementation issues

EPA applies several Part 264/265 provisions to the addition of absorbents, as follows:

- Per §§264.17(b)/265.17(b), owners/operators must ensure that the addition of absorbents to wastes in containers does not result in fires, explosions, violent reactions, or other such events (e.g., due to the mixing of incompatible materials). Addition of absorbent must take place in a container with solid structural integrity, and the waste, absorbent, and materials of construction of the container must be compatible.

- Per §§264.171/265.171, owners/operators must ensure that only containers that are in good condition (e.g., not rusting, deformed, or otherwise clearly damaged) are used for combining hazardous wastes and absorbents.

- Per §§264.172/265.172, owners/operators must ensure that the materials of construction of containers (or the materials containers are lined with, where applicable) are compatible with the hazardous waste/absorbents to be added to these containers.

7.5.3.2 Examples

Q *Is a permit required for a facility to repackage small containers of hazardous waste into large containers with absorbents (i.e., lab packs)?*

A No. The process of adding absorbents to lab packs prior to shipment offsite qualifies for the absorbent addition exemption. [RO 13117]

Q *If a facility transfers a hazardous waste from one container to another container and at the same time adds an absorbent material to the waste, is the act of placing the waste in the second container covered by the absorbent addition exemption?*

A Yes. The absorbent addition exemption is intended to reduce the amount of free liquids in containerized wastes by allowing anyone to add absorbent material to the waste at the time the waste is first placed in the container. Thus, a hazardous waste can be transferred from one container to another container and absorbent material added at the time of transfer without the need for a RCRA permit. [RO 12116]

Q *Must a generator add absorbent to its waste the first time the waste is containerized, or may he/she treat or store the waste in other units before performing absorption in a specified container?*

A Storing or treating hazardous waste in tanks or other containers before absorbent is added would not cause a generator to lose this exemption. Such preceding storage and/or treatment, however, is not included in the exemption and would be subject to permitting or §262.34 requirements. [RO 13391]

Q *Can cement be added as an absorbent to a container of D001 waste and qualify for the permit exemption?*

A We have heard anecdotally that EPA considers use of cement as an absorbent to actually be solidification or stabilization, which is a form of waste treatment. Hence, the absorbent addition exemption would not apply. If the container met 90/180/270-day standards, however, the treatment could be exempt from permitting as discussed in Section 7.3.

Here's what EPA has to say about simply adding absorbent to a waste as a substitute for treating or recycling it more appropriately:

> "Based on the recent amendments to the Resource Conservation and Recovery Act (RCRA), we believe the Congress intended that liquid wastes that can be safely incinerated or otherwise treated or that can be reclaimed and reused, especially organic liquids, should be so treated or reclaimed. Further, we believe the language of Section 3004(c)(1) of RCRA prohibiting the landfilling of liquids that are solely treated by the use of absorbents is intended to encourage such treatment or reclamation. Therefore, generators should be discouraged from simply adding absorbent materials to such wastes." [RO 12443]

Per §265.1081, waste stabilization includes mixing hazardous waste with binders or other materials and curing the resulting hazardous waste and binder mixture. Waste stabilization doesn't include addition of absorbent materials to the surface of a waste to absorb free liquid if no mixing, agitation, or curing occurs. Stabilization is generally not occurring if absorbent is added to the surface of the waste (if no mixing occurs) at the end of a work day or at the completion of a waste transfer. [December 8, 1997; 62 FR 64651–2]

7.5.4 Immediate responses

Certain actions need to be taken to respond to spills, discharges, and other situations that pose an imminent hazard. For example, absorbents may be used to contain spilled hazardous material or a corrosive waste spill may be neutralized. Similarly, hazardous waste spills may be diked or otherwise contained. Some of these actions constitute storage and/or treatment of hazardous waste and, thus, would normally require a RCRA permit. In order to encourage timely and effective responses to such incidents and to avoid placing persons conducting such responses in the uncomfortable position of violating RCRA regulations, EPA created the immediate response exemption. [November 19, 1980; 45 FR 76626, RO 12748]

Although EPA does not specifically define the term "immediate response," the agency lists in the regulations the following four situations where this exemption applies:

- A discharge of a hazardous waste;

- An imminent and substantial threat of a discharge of hazardous waste;

- A discharge of a material which, when discharged, becomes a hazardous waste; and

- An immediate threat to human health, public safety, property, or the environment from the known or suspected presence of military munitions, other explosive material, or an explosive device. [§270.1(c)(3)]

Similar to several of the other exemptions discussed in this chapter, the immediate response exemption excuses facilities from permitting [via §270.1(c)(3)] and RCRA substantive management standards [see §§264.1(g)(8) and 265.1(c)(11)]. Going one step further, the exemption also applies to the transportation of wastes in immediate response situations (i.e., generators and transporters are exempt from obtaining EPA ID numbers and from following manifesting rules for hazardous waste shipments during immediate responses). [RO 11363, 11370, 12016, 12748, 13574] However, certain Part 264/265 standards continue to apply to facilities during immediate response actions: Subpart C—Preparedness and Prevention, and Subpart D—Contingency Plan and Emergency Procedures. [RO 12748]

7.5.4.1 Scope of the exemption

The scope of the immediate response exemption is limited to activities that are taken "immediately." Once immediate response actions are complete, the exemption ends. In other words, containment and treatment activities conducted *after* the initial response period *would* be subject to RCRA permitting and substantive management standards. Cognizant that the distinction between immediate and ongoing responses is subjective, EPA notes that "the time frames and extent of immediate response must be judged by persons responding to discharges on an individual basis." [RO 12748] (EPA also has authority under §270.61 to issue emergency RCRA permits for ongoing response activities.) [RO 13296]

As an example, a tank car of liquid hazardous waste ruptures at a generator's site, and the waste spills on the ground. The facility immediately builds an emergency dike to contain the spilled waste and then pumps the liquid into drums. The drums of hazardous waste remain onsite for several weeks before being shipped offsite to an incinerator. EPA's evaluation of this scenario indicates that the construction and operation of the dike are not subject to RCRA standards or permitting, since it was an immediate response. However, once the liquid was contained, the immediate response was over, and the subsequent storage in drums is subject to 90/180/270-day accumulation provisions specified in §262.34. If the spilled waste and/or contaminated soil in the dike are treated as part of the immediate cleanup, such treatment would not be subject to RCRA regulation. However, if the treatment occurred beyond the immediate response (a judgment call between the facility and the state), an emergency permit would be required unless the generator treated the waste in a way that is exempt from permitting (e.g., in a 90-day unit). [November 19, 1980; 45 *FR* 76629]

One final note: EPA understands that facilities often build structures (including tanks) designed to provide emergency secondary containment or similar functionality. Addressing the application of the immediate response exemption to such equipment, the agency states the following:

> "To qualify for the exemption, a unit must be intended exclusively for immediate responses to discharges of hazardous wastes, such as burst pipes, ruptured containers or tanks, breached dikes, and the like. Structures used for responding to discharge events which occur periodically or repeatedly, or in which containment or treatment extends beyond the immediate response period, do not qualify for the exemption." [RO 12298]

7.5.4.1.1 Sumps

EPA discussed the application of the immediate response exemption to sumps in the September 2, 1988 *Federal Register*. [53 *FR* 34084] There are three types of sumps from a RCRA perspective, and each one is subject to different regulatory requirements:

1. *Temporary sumps*—Hazardous waste is stored in such units only in response to a leak, spill, or other temporary, unplanned occurrence. The immediate response exemption to RCRA permitting and Subpart J tank standards applies to these sumps. "A sump that may be used to collect hazardous waste in the event of a spill, whether accidental or intentional, and that is not designed to serve as a secondary containment structure for a tank storing hazardous waste, is generally exempt from regulatory and permitting requirements so long as it is used to contain hazardous waste only as an immediate response to such a spill." [53 *FR* 34085]

2. *Secondary containment sumps*—These units are used to collect spills from a primary containment vessel storing hazardous waste. They are not immediate response units and are subject to Subpart J tank standards (with the exception of the requirement for secondary containment since that's what they are). Secondary containment sumps are specifically designed to serve as the collection device for spills of hazardous waste from an adjacent primary containment vessel. The immediate response exemption "is inapplicable to units constructed for the purpose of containing hazardous waste from routine and systematic discharges of hazardous waste or designed to serve as the secondary containment for a tank system treating or storing hazardous waste." [53 *FR* 34085]

3. *Primary containment sumps*—Sumps that provide primary containment for the storage of hazardous waste are primary containment sumps. Such sumps are subject to all Subpart J tank standards, including the need for secondary containment. For example, sumps used to store hazardous waste generated from periodic boiler cleanouts are primary containment vessels. However, as noted in Sections 7.2.1.3.2 and 7.5.1.1.2, such primary containment sumps may meet the definition of a wastewater treatment unit or elementary neutralization unit and, thus, be exempt from RCRA permitting and tank standards under one of those exemptions—not the immediate response exemption. Leachate collection sumps associated with landfills are also exempt from Subpart J tank standards. [RO 14011]

7.5.4.2 Examples

Besides explosives, which are addressed in Case Study 7-7, EPA has identified in guidance three

Case Study 7-7: Explosives, Explosive Devices, and Other Shock-Sensitive Materials

Much of EPA's guidance on immediate responses and the associated RCRA exemption addresses explosives and related materials. Generally these actions involve law enforcement (bomb squad) activities or removal of obsolete, reactive laboratory chemicals. Situations where the immediate response exemption would apply include: 1) responses to bombings and the discovery of explosive devices, including recovery of undetonated explosives; 2) serving a search warrant on an improvised bomb factory or illegal explosives manufacturing facility; 3) recovering explosives stored illegally in a residential neighborhood; 4) recovering deteriorated explosives deemed unsafe for transportation; and 5) removing shock-sensitive or otherwise explosive chemicals such as picric acid and/or ethyl ether (usually from a laboratory). In general, any activities undertaken to eliminate the imminent and substantial danger from these materials would qualify for the exemption. [RO 11363, 11370, 13237, 13574]

Another issue involving explosives and shock-sensitive materials is whether the destruction of such materials (usually via detonation or open burning) is covered under the immediate response exemption. In general, the agency defers to law enforcement officials' judgment as to whether an emergency exists and immediate action is required (in which case the exemption would apply). In such cases, the destruction of these explosives, etc. is exempt from RCRA permitting and substantive requirements. Where no "imminent threat" exists, however, the detonation or open burning would be subject to RCRA requirements. [RO 11363, 11370, 13603]

Finally, EPA offers the following four situations where detonation/open burning would most likely qualify for the immediate response exemption:

- Unusable explosive wastes from land clearing operations that, for safety reasons, cannot be stored or transported offsite;

- Leaking or damaged explosives in a temporary or permanent magazine operated by an explosives distributor that are not transportable offsite;

- Unusable and nontransportable explosives at an onsite magazine for a project with changing locations (e.g., a road or pipeline construction operation); and

- Undetonated and unusable explosives from a mining operation that remain after the initial firing (wherein most of the explosives did detonate). [RO 13237]

---

specific instances where the immediate response exemption can be applied:

1. Actions taken to remove media contaminated by incidental, infrequent drippage of hazardous preservatives from treated wood at a wood preserving plant, where the wood has already been moved from drip pads designed to contain normal drippage. [RO 11612, 14608]

2. Excavating soil contaminated by a hazardous waste spill and subsequent storage in containers prior to shipment of the soil to an offsite facility. [RO 13296]

3. Treating ground water at a facility (this situation is tricky in that the difference between an immediate and ongoing response might be unclear). [RO 12880]

7.5.5 Burning small quantities of hazardous waste in onsite units

Section 266.102(a)(1) of the boiler and industrial furnace (BIF) regs in Part 266, Subpart H requires facilities that burn hazardous waste in a BIF to get a RCRA permit or operate under interim status for that activity. However, there is a little-known exemption to that permitting requirement for generators who burn small quantities of hazardous waste for energy recovery in onsite burners per §266.108. EPA calls this permit exemption the "small quantity onsite burner exemption."

The statutory citation for this exemption is RCRA Section 3004(q)(2)(B). Based on that authority, EPA promulgated the exemption at §266.108 in the 1991 BIF rule. [February 21, 1991; 56 *FR* 7208] That section notes that burning small quantities of hazardous waste in onsite units is exempt from all of the BIF requirements, in addition to being a permit-exempt activity. Thus, facilities qualifying for the exemption will not have to comply with the Part 266, Subpart H design and air emission standards for units burning hazardous waste. However, CAA requirements (e.g., potential modifications to the BIF air permit) may still apply. To claim the exemption, four primary conditions must be met, as reviewed in the following subsections.

In *Little Known But Allowable Ways to Deal With Hazardous Waste*, EPA/233/B-00/002, May 2000, EPA notes that facilities that want to burn small quantities of hazardous waste without a permit should consult with both their state RCRA regulators and their air pollution control authority. The above-cited report is available online at http://nepis.epa.gov/EPA/html/Pubs/pubtitleOther.html by downloading the report numbered 233B00002.

7.5.5.1 Waste must be burned at the site of generation

The first condition is that the hazardous waste must be burned at the same facility at which it is generated. Therefore, the exemption does not apply to hazardous waste burned at a facility that receives the waste from offsite, even if the offsite facility is under the same ownership and operational control as the facility at which the waste will be burned. [56 *FR* 7190]

There is one exception to the above requirement of burning the waste at the same facility at which it was generated: hazardous waste generated by conditionally exempt small quantity generators (CESQGs) can be burned in a BIF at an offsite facility under the §266.108 exemption, provided that the quantity of CESQG waste is included in the maximum hazardous waste burning rate discussed below. [§266.108(c)]

7.5.5.2 Quantity/specification limitations on hazardous waste burned

When Congress added the small quantity burner exemption at RCRA Section 3004(q)(2)(B), it directed EPA to issue implementing regulations such that the onsite burner would be "operated at a destruction and removal efficiency sufficient such that protection of human health and environment is assured." To achieve this mandate, the agency's regulations at §266.108(a) specify four requirements:

1. The maximum hazardous waste burning rate (from 0 to 1,900 gallons/month) is limited based on the terrain-adjusted effective height of the stack associated with the onsite BIF. [The terrain-adjusted effective stack height is calculated per §266.106(b)(3).] The higher the effective stack height, the more hazardous waste can be burned per month, due to dispersion of potential airborne contaminants.

2. The maximum hazardous waste burning rate *at any time* is limited to 1% of the total fuel input to the BIF. This is calculated based on the heating value (Btu) input or mass input, whichever results in the *lower* mass firing rate.

3. The heating value of the hazardous waste must be at least 5,000 Btu/lb as generated.

4. The hazardous waste cannot include any of the dioxin listed wastes (F020, F021, F022, F023, F026, or F027).

If a facility burns hazardous waste in more than one onsite BIF, the equation given in §266.108(c) is used to determine the maximum firing rate into each BIF. An example of using the equation is given at 56 *FR* 7192.

7.5.5.3 Accumulation before burning

Facilities accumulating/storing unmixed hazardous waste fuel are responsible for complying with all applicable RCRA standards. This includes §262.34 requirements for large and small quantity generators accumulating the hazardous waste in 90/180-day units before mixing it with other fuels (typically virgin fuel oil). After such mixing, however, accumulation/storage units would be exempt from the RCRA standards. Thus, at facilities that are eligible for the small quantity burner exemption, tanks that are used to store the mixed fuel would be exempt from RCRA regulation. [§266.101(c)(2), 56 *FR* 7192]

7.5.5.4 Notification and recordkeeping requirements

Another condition that a facility must satisfy to claim the small quantity burner exemption is to send a one-time written notification to EPA. Per §266.108(d), the following must be included:

- The facility's status as a small quantity burner of hazardous waste,

- An indication that the facility is in compliance with the requirements of §266.108, and

- The maximum allowable hazardous waste quantity that may be burned per month as determined by §266.108(a)(1).

The recordkeeping requirements include at a minimum [§266.108(e), 56 *FR* 7192]:

- Quantities of hazardous waste and other fuel burned in each unit per month,

- Quantities of hazardous waste and other fuels burned at any time to demonstrate conformance with the 1% hazardous waste firing rate limit, and

- The heating value of the hazardous waste.

These records must be maintained for at least three years.

Universal Wastes

Simplified procedures for hazardous batteries, bulbs, and mercury-containing equipment

Some hazardous wastes are so common and seemingly innocuous that the dangers associated with them are often virtually ignored. Take spent batteries and light bulbs for example—we threw them in the dumpster for years before we figured out that they were hazardous. Now that we know, we still sometimes throw them away illegally either because we're lazy or because we have an attitude of "if it wasn't a problem before, why is it a problem now?" Some commercial establishments and even industrial facilities are still throwing their hazardous dead batteries and burned-out light bulbs in the dumpster because they are unaware that they are subject to RCRA Subtitle C regulation.

Attempting to address this situation, EPA established a program governing the generation, handling, and transportation of these hazardous wastes on May 11, 1995. [60 *FR* 25492] Because they are common to all industry types and facilities, the agency calls them "universal" wastes. Since the program is different from the traditional hazardous waste regulations, the agency promulgated the universal waste requirements in a separate Part 273 of Title 40 of the *CFR*. The program allows less-stringent procedures for people who generate, store, and transport these wastes. However, as detailed in the sections that follow, the federal universal waste program applies to only four types of hazardous wastes: batteries, pesticides, mercury-containing equipment, and lamps.

EPA identified two primary goals that it hopes to achieve under the universal waste program:

- Increase recycling—By reducing the scope and complexity of the waste management requirements for generators, consolidators, and transporters of universal wastes, the agency believes that participation in universal waste collection programs will increase significantly. In turn, the availability of large quantities of universal wastes from generators may make recycling more economically feasible as opposed to disposal of these wastes.

- Reduce illegal disposal of universal waste in municipal waste landfills and combustors—EPA believes the universal waste management system is simple and cheap enough that handlers of such wastes will manage them under the universal waste program instead of illegally disposing them into dumpsters. [60 *FR* 25501–2, RO 11960]

In this chapter, we discuss the simplified management system EPA has implemented for universal wastes. First, we examine what materials qualify as universal wastes. Then, we take a "big-picture" look at the structure of the universal waste program, followed by more detailed reviews of the requirements applicable to the four types of regulated entities (small and large quantity handlers—Section 8.3, and transporters and destination facilities—Section 8.4). The last two sections of this chapter address state implementation issues peculiar to universal wastes (Section 8.5) and the Mercury-Containing and Rechargeable Battery Management Act (Section 8.6).

8.1 What are universal wastes?

Under the federal universal waste management standards, EPA defines "universal waste" as follows [§273.9]:

"[A]ny of the following hazardous wastes that are subject to the universal waste requirements of this Part 273:

(1) Batteries as described in §273.2;

(2) Pesticides as described in §273.3;

(3) Mercury-containing equipment as described in §273.4; and

(4) Lamps as described in §273.5."

EPA replaced mercury thermostats with mercury-containing equipment as the third type of universal waste on August 5, 2005. [70 FR 45508] The category of mercury-containing equipment includes mercury thermostats, thermometers, barometers, switches, etc.

As noted in the above definition, universal wastes *are* hazardous wastes. For example, certain batteries contain lead or cadmium that make them exhibit the toxicity characteristic. Mercury-containing equipment will typically exhibit the toxicity characteristic for mercury. Spent lamps typically will be hazardous due to toxicity for mercury, lead, or other heavy metals. Thus, absent the universal waste program, these materials would generally be subject to full regulation under RCRA Subtitle C.

Although some states have added other hazardous wastes to their state universal waste programs, only the four types of hazardous wastes noted above qualify as universal wastes under the federal regulations. Each type of universal waste is defined and discussed in Table 8-1.

Table 8-1: Types of Universal Wastes in the Federal Program

Universal waste	Definition/guidance
Battery	A device consisting of one or more electrically connected electrochemical cells that is designed to receive, store, and deliver electrical energy. An electrochemical cell is a system consisting of an anode, cathode, and an electrolyte, plus such connections (electrical and mechanical) as may be needed to allow the cell to deliver or receive electrical energy. The term battery also includes an intact, unbroken battery from which the electrolyte has been removed. [§273.9]
	A used battery becomes a waste when taken out of service for discard or reclamation; an unused battery becomes a waste when the handler decides to discard it. [§273.2(c)]
	All sizes and types of waste batteries are included in this definition. Both "wet" and "dry" batteries, which are distinguished based upon whether a liquid or nonliquid electrolyte is used, are included. Because all waste batteries can be managed as universal waste, the requirement to determine whether individual batteries exhibit a characteristic is eliminated. [60 FR 25504]
	Spent lead-acid batteries that are managed under Part 266, Subpart G are *not* universal wastes. Such batteries that are not managed under Part 266, Subpart G, however, may be managed under the universal waste program. [60 FR 25505, RO 13772, 14088, 14124]
	Waste batteries that are broken or damaged may still be managed as universal waste if they are properly containerized. [§§273.13(a)(1)/273.33(a)(1), RO 14146] However, a "handler of universal waste may only manage broken or damaged hazardous waste batteries as universal wastes if the breakage or damage does not constitute a breach in the cell casing…. Therefore, universal waste batteries are intended to be intact (i.e., where the casing of each individual battery cell is not breached)." [RO 14634]

Table 8-1: Types of Universal Wastes in the Federal Program

Universal waste	Definition/guidance
Pesticide	Any substance or mixture of substances intended for preventing, destroying, repelling, or mitigating any pest, or intended for use as a plant regulator, defoliant, or desiccant. [§273.9] A recalled pesticide becomes a waste when the pesticide holder agrees to participate in the recall and the person conducting the recall decides to discard the pesticide. [§273.3(c)] Only pesticides that have been recalled or come from stocks of unused products gathered as part of a waste pesticide collection program are classified as universal wastes. Recalled or unused pesticides managed by farmers in compliance with §262.70 are *not* universal wastes. [§273.3] Pesticides spilled onto soil, once recontainerized, may continue to be managed as universal waste if the residues meet the definition of universal waste pesticide in §273.9; if not, a hazardous waste determination must be made per §273.17(b) or 273.37(b). [RO 14560]
Mercury-containing equipment (MCE)	A device or part of a device (including thermostats, but excluding batteries and lamps) that contains elemental mercury integral to its function. [§273.9] MCE includes equipment that may contain mercury as open-ended tubes (e.g., barometers and manometers), as well as ancillary parts that may contain mercury (e.g., valves). These devices vary in size and function, but, for the most part, the mercury is 1) a relatively small amount of the complete piece of equipment, 2) encapsulated in some way in an ampule or other housing, and 3) used for delicate measuring of temperature or pressure or for completing an electrical circuit. MCE includes mercury thermometers, manometers, barometers, flow meters, light switches, regulators, pressure-relief gauges, water-treatment gauges, and gas-safety relays. [August 5, 2005; 70 *FR* 45512] Equipment and devices from which mercury-containing components have been removed are not MCE and may be managed as nonhazardous waste. [§273.4(b)(3), 70 *FR* 45516] Used MCE becomes a waste when taken out of service for discard or reclamation; unused MCE becomes a waste when the handler decides to discard it. [§273.4(c)]
Lamp	The bulb or tube portion of an electric lighting device. A lamp is specifically designed to produce radiant energy, most often in the ultraviolet, visible, and infrared regions of the electromagnetic spectrum. Examples of common universal waste electric lamps include, but are not limited to, fluorescent, high-intensity discharge, neon, mercury-vapor, high-pressure-sodium, and metal-halide lamps. [§273.9] A used lamp becomes a waste when taken out of service for discard or reclamation; an unused lamp becomes a waste when the handler decides to discard it. [§273.5(c)] All waste lamps that exhibit a characteristic are universal wastes, including incandescent lamps (even though they are not specifically included in the definition). [July 6, 1999; 64 *FR* 36477] Facilities conducting TCLP testing of fluorescent tubes to be discarded must use a "representative selection of lamps randomly chosen" to determine if they are hazardous and, therefore, subject to universal waste regulations. (Chapter 9 of SW–846 provides guidance on how to develop a sampling plan and gather a representative waste sample.) [RO 11907] It is not completely clear from regulatory language [§§273.13(d)/273.33(d) and 273.17(b)/273.37(b)] or preamble language [64 *FR* 36474, 36479] whether incidentally broken lamps (e.g., broken fluorescent light tubes) can still be managed as universal waste. However, in Topic #: 23002-14524 on EPA's RCRA FAQ Database (http://waste.supportportal.com/), the agency clearly states that broken lamps may be managed as universal waste (although state regulations can be more stringent). In light of this guidance, it is our opinion that they can—just as damaged batteries are still universal waste. EPA mentions that "[h]andlers also must contain any *universal waste lamps* that show evidence of breakage, leakage, or damage that could cause the release of mercury or other hazardous constituents to the environment." [Emphasis added.] [64 *FR* 36479]

Source: McCoy and Associates, Inc.

8.1.1 State universal wastes

Although there are currently only four types of universal waste in the federal program, states that have adopted the universal waste regulations are allowed to add additional wastes to their state universal waste program without first requiring those wastes to be added at the federal level. [§273.80, RO 14468] A number of states have taken advantage of this option. For example, Colorado has added hazardous aerosol cans as universal wastes in its state program. California considers CRTs and other electronic devices to be universal waste within its state borders. The state of Texas has added hazardous paint and paint-related wastes to its universal waste program, as long as certain container and tank standards are observed.

EPA has a great website dedicated to universal wastes. As part of that site, the agency has a map showing what states have added universal wastes to their state programs that are different than those in the federal program. See http://www.epa.gov/epawaste/hazard/wastetypes/universal/statespf.htm.

8.1.2 Other issues

Three other items are worth mentioning in terms of the definition of universal waste. First, some people want to argue that if they recycle used batteries, mercury-containing equipment, and/or burned-out light tubes, those materials never become wastes. However, EPA notes that used nickel-cadmium batteries, mercury lamps, etc. that can no longer be used for the purpose for which they were produced clearly meet the definition of spent material. Per Table 1 in §261.2(c), spent materials sent for reclamation *are* solid wastes and are hazardous wastes as well if they exhibit a hazardous characteristic. If hazardous, such batteries, mercury equipment, and lamps must be managed as universal wastes under Part 273 or as hazardous wastes under Parts 260–270. [RO 11789, 14468]

Secondly, the universal waste program applies to the four types of wastes described above regardless of whether those wastes will ultimately be recycled or disposed. [May 11, 1995; 60 *FR* 25501–2, EPA/530/K-05/019, September 2005, available from http://www.epa.gov/wastes/inforesources/pubs/training/uwast05.pdf]

Finally, not many people manage any significant quantity of hazardous recalled pesticides and/or mercury-containing equipment. From a practical standpoint, then, the federal universal waste program really boils down to a simplified approach for managing hazardous batteries and burned-out lamps.

8.2 Universal waste program structure

The regs governing the management of universal wastes are codified in Part 273. Before examining the details of what these rules require of specific facilities, however, we take a look at the overall regulatory structure of the universal waste program.

First of all, managing hazardous batteries, burned-out lamps, etc. under the less-stringent universal waste program is a choice. If desired, generators may alternatively manage such materials under the full hazardous waste regulatory program spelled out in Parts 260–270. [§273.1(b), RO 14088] Why would anyone not take advantage of the simplified requirements associated with the universal waste program? There are a couple of reasons:

- Some facilities have mentioned to us that, because the amount of universal waste they generate is fairly small, they choose to continue managing their hazardous batteries and lamps under their formal hazardous waste management system. They don't want to deal with the expense and potential confusion that a separate waste management structure might cause.

- In some states, crushing hazardous fluorescent light tubes is not allowed under the universal waste program. In order to continue crushing these lamps, facilities in those states will have to manage them as hazardous wastes. [July 6, 1999; 64 *FR* 36477–8]

If facilities choose to manage the spent batteries and lamps that they generate under the universal waste program, Figure 8-1 is a "big-picture" view of the management and regulation of these wastes. The top half of the figure shows the flow of universal wastes from their point of generation to their final destination, while corresponding boxes below show the associated universal waste scheme and applicable Part 273 regulation(s). Each of the stopping points on the flowlines is described below.

The regulation of universal wastes starts where such wastes are generated—typically the point at which the generator of the battery, thermometer, lamp, etc. decides to discard that item or send it for reclamation. The generator of the universal waste is called a *handler* under the Part 273 structure. Each separate facility (e.g., generating facility or collecting facility) is considered a separate handler. [RO 14081] Universal waste handlers come in two varieties, small and large. Management standards for these entities are codified in Subparts B and C, respectively, of Part 273 and are reviewed in Section 8.3.

Once generated, universal wastes are usually shipped directly to the reclamation or disposal facility, but they are sometimes transported to an intermediate collection or consolidation facility; these intermediate facilities are also called handlers and are subject to the small or large quantity universal waste handler provisions of Subpart B or C. [RO 14081] Based on the dual use of the term, handlers of universal waste may generate such wastes, receive shipments of universal waste from offsite, accumulate universal wastes for prescribed periods of time, and ship wastes to another handler, a destination facility, or a foreign destination. Note that handlers cannot treat, recycle, or dispose universal wastes.

The entity that moves universal wastes between handlers and from a handler to a destination facility or foreign destination is called a *transporter*. Transporter requirements are codified in Subpart D of Part 273 (these are discussed in Section 8.4.1). Universal wastes may be transported through multiple intermediate handlers between the generator and destination facility or foreign destination.

Upon leaving the original or intermediate handler, universal wastes are transported to a facility where the wastes are treated, recycled, and/or disposed. This treatment, recycling, or disposal location is

Figure 8-1: Management and Regulation of Universal Wastes

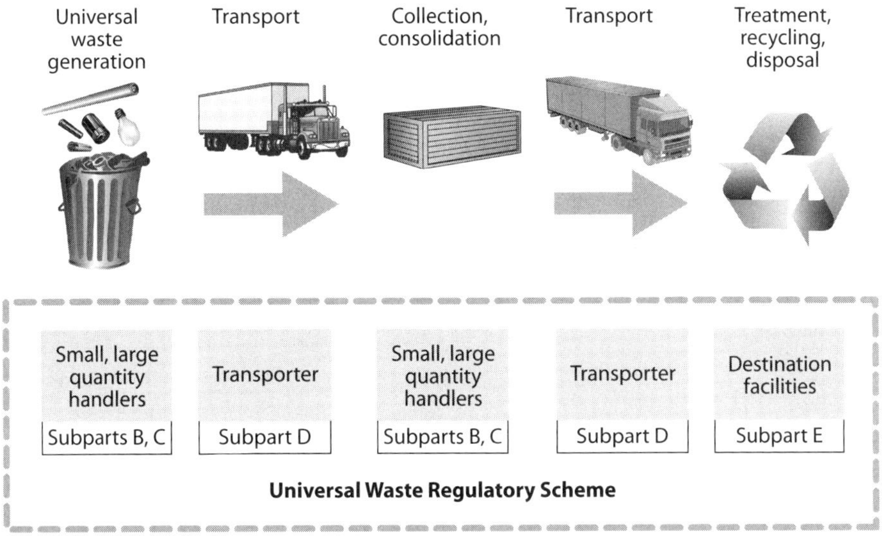

Source: Adapted from 60 *FR* 25497.

called a *destination facility* and is subject to Part 273, Subpart E. (See Section 8.4.2 for details.) As mentioned in Section 8.1.2, the same universal waste regulations apply whether these wastes will ultimately be recycled or disposed at the destination facility. [May 11, 1995; 60 *FR* 25501–2]

As noted in §273.60(a) in Subpart E, destination facilities get no breaks at all. They remain subject to the hazardous waste TSD facility requirements in Parts 264–270. Thus, the universal waste program provides a simplified approach for generators (handlers) to store and transport universal wastes, but the destination facilities who actually perform the treatment, recycling, or disposal continue to be regulated under the full Subtitle C program, ultimately resulting in a high standard of care for these hazardous wastes.

8.3 Universal waste handler requirements

A small quantity handler of universal waste is a handler that always has less than 5,000 kg (11,000 lb) of total universal waste (batteries, pesticides, mercury-containing equipment, and lamps, calculated collectively) onsite at any time. Any handler that has more than this amount in storage is a large quantity handler. All universal waste handlers start out on January 1 each year as a small quantity handler (assuming they have <5,000 kg of total universal waste onsite on that date). As long as they don't have 5,000 kg or more of these wastes onsite at any time throughout the year, they remain small quantity handlers the entire year. If they have 5,000 kg or more onsite at any one time, they become large quantity handlers and remain in that status for the remainder of the calendar year. (They will start over as small quantity handlers if they have <5,000 kg of total universal waste onsite on the succeeding January 1st.) [RO 14107]

Q A facility that normally has about 1,000 kg of universal waste onsite receives an unusual shipment of 10,000 kg of such material in August. Less than a month later, the 10,000 kg of universal waste is sent to a destination facility, and the facility resumes its practice of storing only about 1,000 kg onsite. What is the designation of that facility with regard to universal waste?

A Until the date the 10,000-kg shipment was received, the facility is a small quantity handler. From the date the shipment was received until December 31 of that year, the facility would be a large quantity handler. Beginning January 1st of the following year, the facility reverts to being a small quantity handler. [RO 14107]

EPA clarified that, for handlers managing universal waste mercury-containing equipment, if the mercury has not been removed from the equipment, then the weight of the entire device is counted toward the 5,000-kg limit. If, however, the mercury-containing component has been removed, only the component being managed as mercury-containing equipment is counted. [70 *FR* 45516]

Sometimes, an owner/operator will hire a contractor to conduct activities that generate universal waste (e.g., relamping a building) instead of using their own personnel. Contractors who remove universal waste from service at an owner's/operator's site are considered handlers and cogenerators of the waste (with the owner/operator as the other cogenerator). As cogenerators, both have liability as universal waste handlers, and the two parties should mutually agree as to which will perform the universal waste handler duties specified in Part 273. [July 6, 1999; 64 *FR* 36474, RO 14719]

As mentioned in Section 8.2, handling spent hazardous batteries and lamps as universal waste is an alternative to full hazardous waste management of these materials. As such, there are separate accounting systems for these two programs. Universal wastes managed under Part 273 do not have to be counted in a facility's monthly hazardous waste generator status determination. [§261.5(c)(6)] Thus, you can be a large quantity handler of universal waste while a small quantity generator of hazardous waste (and vice versa). That is, your status under one of the programs is independent of the other.

Conditionally exempt small quantity generators of hazardous waste have the option of managing their hazardous waste batteries and lamps as 1) hazardous waste under the reduced requirements of §261.5 (in which case such wastes would be hazardous wastes and *would* have to be counted when determining if the facility is generating less than 100 kg of hazardous waste/month), or 2) universal waste under the Part 273 program (in which case such wastes would not be subject to hazardous waste management requirements and would *not* have to be counted against the 100-kg monthly hazardous waste generation limit). [§§261.5(c)(6), 273.8(a)(2)] Therefore, if a generator manages spent hazardous waste batteries and lamps as universal wastes under Part 273 and does not generate any other hazardous waste, he/she is not subject to any other RCRA Subtitle C regulation. [64 *FR* 36475]

8.3.1 Specific handler requirements

Table 8-2 summarizes the requirements applicable to the two classes of universal waste handlers. The table notes that the requirements for both types of handlers are identical with the exception of

Table 8-2: Requirements for Small and Large Quantity Handlers of Universal Wastes

Requirement	Small quantity handlers	Large quantity handlers
Notification of universal waste activity and receipt of EPA ID number	Not required	Required[1]
Universal waste accumulation time limit	One year from the date the waste was generated or received from another handler[2]	One year from the date the waste was generated or received from another handler[2]
Releases from universal wastes	Must prevent releases of universal waste components to the environment; must immediately contain any such releases	Must prevent releases of universal waste components to the environment; must immediately contain any such releases
Waste/container labeling/marking (e.g., "Universal Waste—Batteries," "Waste Lamps")	Required	Required
Training[3]	Employees must be informed of proper waste handling and emergency response procedures	Employees must be thoroughly familiar with proper waste handling and emergency response procedures
Use of hazardous waste manifests/LDR notification forms for offsite shipments of universal waste	Not required—bills of lading only [60 *FR* 25501, RO 14088], but DOT packaging, labeling, marking, and placarding rules can still apply[4]	Not required—bills of lading only [60 *FR* 25501, RO 14088], but DOT packaging, labeling, marking, and placarding rules can still apply[4]
Waste tracking/recordkeeping	Not required[5]	Must keep a record of each shipment received at, or sent from, the facility (e.g., log, copies of bills of lading)[6]

LDR = land disposal restrictions.
[1]If the large quantity handler already has an EPA ID number, he/she does not have to renotify. [§273.32(a)(2), RO 13783, 14088]
[2]The one-year time limit may be exceeded where necessary to facilitate proper recovery, treatment, or disposal, but the handler bears the burden of proving this need. Adherence to the time limit must be demonstrated by 1) marking the earliest date of generation or receipt on individual universal waste items (or containers of such wastes), 2) maintaining an inventory system that identifies the accumulation start date, 3) segregating universal wastes to specific areas whose time-limit start date is known, or 4) using any other method that identifies the date of waste generation/receipt. The one-year accumulation time limit applies to each successive universal waste handler; if universal waste is transported from one handler to another, the second handler gets another year from the date it received the material to send it on, and so on. [RO 14088, 14179]
[3]See Section 6.4.5 for a more detailed discussion of training requirements.
[4]DOT requirements for shipping batteries are discussed in Section 8.4.1.1.
[5]A small quantity handler does not need to keep records of universal waste shipments. However, the one-year accumulation time limit must be tracked.
[6]The record, which must be retained for at least three years, must list the name and address of the source or destination facility, the quantity and type of universal wastes in the shipment, and the date the shipment was received or shipped.

Source: McCoy and Associates, Inc.; adapted from §§273.12–273.39, unless otherwise noted.

notification, training, and waste tracking/recordkeeping requirements. A useful document for reviewing a handler's compliance with the universal waste regulatory requirements is *Protocol for Conducting Environmental Compliance Audits of Used Oil and Universal Waste Generators Under RCRA*, EPA/300/B-00/002, March 2000, available from http://infohouse.p2ric.org/ref/14/13639.pdf.

8.3.1.1 Waste management

Waste-specific management requirements for all handlers are summarized below:

- Universal waste batteries must be stored in a way that prevents releases to the environment. This requirement doesn't necessitate the use of a Part 265, Subpart I container for battery accumulation; instead, you could just use a box or, per EPA guidance, just shrink-wrap a bunch of batteries on a pallet. [RO 14146] Any universal waste battery that shows evidence of leakage, spillage, or damage must be placed in a closed, structurally sound container. DOT requirements for shipping batteries are discussed in Section 8.4.1.1.

  As long as the universal waste battery case is not breached, handlers may conduct the following activities without becoming destination facilities: discharging or regenerating batteries, removing batteries from consumer devices, disassembling multibattery packs, and removing electrolyte from batteries (in which case the battery case may be opened while the electrolyte is being removed). Any battery electrolyte removed that is a characteristic hazardous waste is subject to full RCRA regulation. See Case Study 8-1.

  If the batteries are damaged, EPA's guidance is as follows: "A handler of universal waste may only manage broken or damaged hazardous waste batteries as universal wastes if the breakage or damage does not constitute a breach in the cell casing…. Therefore, universal waste batteries are intended to be intact (i.e., where the casing of each individual battery cell is not breached)." [RO 14634] Where batteries are damaged but the casing is not breached, §§273.13(a) and 273.33(a) allow them to continue to be managed as universal waste if they're appropriately containerized.

  Spent lead-acid batteries can alternatively be managed under the minimal requirements of Part 266, Subpart G. Generally speaking, generators, collectors, and transporters of *intact* spent lead-acid batteries are exempt from most RCRA standards under Part 266, Subpart G. [RO 14147] Therefore, we think the most prudent manner of managing *broken* lead-acid batteries is to manage them as D002/D008 hazardous waste under the full hazardous waste management program.

- Universal waste pesticides must be kept in closed, structurally sound containers, tanks, or transport vehicles that prevent releases to the environment. [Tanks must also meet Part 265, Subpart J, except for §§265.197(c), 265.200, and 265.201.]

- Storage of universal waste mercury-containing equipment is similar to batteries. Any mercury-containing equipment with noncontained elemental mercury or that shows evidence of leakage, spillage, or damage must be placed in a closed, structurally sound container that will prevent mercury emissions into the environment. The agency has included additional handling requirements for removal and management of ampules, open tubes, and ancillary parts containing mercury. Generally, removal of these mercury-containing devices must be done 1) in a manner that prevents breakage, 2) over or in a containment device (e.g., tray or pan), and 3) in a well-ventilated area that meets OSHA exposure levels for mercury. [§§273.13(c)(2), 273.33(c)(2)] For equipment that does not contain mercury in an ampule, the mercury housing must be sealed air-tight. [§§273.13(c)(3), 273.33(c)(3)]

- All unbroken universal waste lamps must be stored in closed, structurally sound containers that are adequate to prevent breakage. Two- or three-ply cardboard boxes may be used for this purpose. (By far the majority of people we run into use the boxes that new lamps come in for storage of

Case Study 8-1: Decharacterized Universal Waste Disposed in Municipal Landfills

Lithium-sulfur dioxide (Li-SO$_2$) batteries used extensively by the military are considered hazardous waste at their point of generation (time of removal from service) because of their potential reactivity. Military personnel often discharge these batteries to remove the electric charge; if conducted in accordance with the universal waste regs, such discharge is not considered hazardous waste treatment. [§§273.13(a)(2)(iii), 273.33(a)(2)(iii)] After discharging the batteries and rendering them nonhazardous, can personnel dispose them in municipal landfills?

As noted in Section 8.1.2, EPA allows batteries to be managed under the universal waste program even if they are headed for disposal rather than recycling (although some state programs require recycling for certain universal wastes). However, because these batteries are hazardous at their point of generation, personnel managing them are required to comply with land disposal restrictions (LDR) requirements. That is, even if discharged Li-SO$_2$ batteries are no longer hazardous, they may not be disposed in solid waste landfills until LDR treatment standards (which attached at the point of generation) are met.

The LDR requirements include identification of any underlying hazardous constituents (UHCs) present in characteristic wastes at levels above the universal treatment standards (UTS). [§268.9] Characteristic wastes must be treated so that UTS levels are met for all UHCs before land disposal is allowed, even if the waste has been decharacterized. EPA notes that the SDS for Li-SO$_2$ batteries lists acetonitrile as a component, present at 5–6%. The UTS for acetonitrile of 38 mg/kg, as well as UTS for any other applicable UHCs, must be met before land disposal can occur. Once these LDR standards are achieved, decharacterized universal waste batteries can be disposed in municipal solid waste landfills. [RO 14756]

Note that this guidance seems to conflict with previous agency guidance in RO 14088, which says that "[t]he destination facility is the first entity that handles a [universal waste] that is responsible for compliance with any of the LDR requirements, including recordkeeping." This older guidance is applicable to universal waste sent from a handler to a RCRA-permitted TSD facility or recycling facility. However, if the destination facility is a nonhazardous waste landfill that is normally not required to comply with the LDR program, the handler must take on the responsibility of ensuring that decharacterized universal wastes or nonhazardous residues from treating universal wastes comply with LDR treatment standards before they enter the landfill.

---

burned-out lamps; however, other handlers are using shipping containers specially designed for spent lamps, often supplied by the destination facility to which the lamps are shipped.)

Broken or leaking lamps must be placed in a closed, nonleaking container capable of preventing the release of mercury and/or other hazardous constituents (e.g., a waxed-fiberboard drum). [64 FR 36479] Regarding management of accidentally broken lamps under the universal waste program, the universal waste regulations and preamble are somewhat vague. The agency clearly states that broken lamps may be managed as universal waste; however, state regulations may not allow that practice. [Topic # 23002-14524 on EPA's RCRA FAQ Database, http://waste.supportportal.com] Some states (including California) allow leaking, broken, or otherwise damaged lamps to be managed as universal (as opposed to hazardous) wastes, if repackaged according to the regulations. This allowance is often conditioned on whether the transporter and destination facility agree to accept such broken lamps under the universal waste program. For example, some receiving facilities will accept up to 10% broken lamps in a shipment of universal waste. However, some states take the position that, if (on a site-specific basis) the universal waste lamps are being broken intentionally, the lamps may be subject to all hazardous waste regulations and/or appropriate enforcement action may be taken. So, check to see if your state or receiving facility has a specific policy regarding incidentally broken lamps.

EPA has funded the development of a training module for the management of fluorescent and mercury-containing lamps, which could be used as partial fulfillment of the universal waste training requirement. Titled *Training Module (1-hour version) for Generators and Handlers of Fluorescent and Mercury-Containing Lamps (and Ballasts)*, it is available online at http://www.almr.org/1hourtrainingmodule.pdf.

8.3.1.2 No treatment allowed

When EPA promulgated the universal waste program, the agency determined that handlers (who are not required to comply with the full set of RCRA standards for the universal wastes that they manage) should *not* treat universal wastes.

Except for certain battery management activities and mercury ampule removal, which normally would be considered treatment but is specifically exempted from such designation by the regulations, handlers may not treat universal wastes except when responding to releases. [RO 14124] This raises the ugly question of crushing burned-out fluorescent light tubes.

When people found out in the early 1990s that fluorescent light tubes often failed the TCLP for mercury and were, therefore, subject to RCRA Subtitle C regulation, many began crushing the bulbs. Crushing facilitated significant volume reduction, minimizing hazardous waste transportation costs. However, facilities and EPA soon began worrying about the potential mercury vapor exposure to personnel who were operating the crushing equipment, and so a number of facilities quit crushing lamps because of personnel safety issues.

When EPA added hazardous lamps to the list of universal wastes in July 1999, the agency made it clear that the prohibition against treatment for universal waste handlers includes a prohibition on crushing lamps. [64 *FR* 36477–8] EPA believes that crushing lamps in containers meeting the minimal universal waste provisions would not control emissions of mercury. Therefore, handlers are not allowed to crush lamps under the federal universal waste regulations. If a facility wants to crush lamps, it will have to continue to manage them under the full hazardous waste standards in Parts 260–270 (with the exception noted below).

Some states allow crushing of lamps as part of their universal waste programs (including, for example, Colorado). EPA addressed this issue as follows: "EPA will consider authorization of state programs that include provisions for controlling treatment or crushing of universal waste lamps, where the state program application includes a demonstration of equivalency to the federal prohibition…in controlling emissions of hazardous constituents." [64 *FR* 36478]

8.3.1.3 Accumulation location

The Part 273 universal waste regulations do not limit the location or number of locations at which a handler may accumulate universal wastes. [60 *FR* 25527] Thus, spent batteries or burned-out bulbs may be accumulated at the point of generation, or they may be accumulated in a maintenance shop or any other convenient location. Satellite accumulation provisions do not apply to universal wastes; thus, universal wastes may be accumulated in as many areas as a facility wishes, provided the waste management requirements specified above and the one-year time limit are met for accumulated materials. [RO 13783]

8.3.1.4 Exporting universal waste

If a handler wants to export universal waste, he/she must comply with the export standards for universal waste at §273.20 or 273.40 (for small or large quantity handlers, respectively). These sections reference the applicable hazardous waste export requirements for generators in Part 262, Subpart E. The export standards in Part 262, Subpart E normally apply only to a person designated as a "primary exporter," defined as "any person who is required to originate the manifest for a shipment of hazardous waste…." [§262.51] However, shipments of universal waste are not required to be shipped under a manifest, whether they are sent to domestic or foreign destination facilities. [April 12, 1996; 61 FR 16306] Even though a manifest is not required for universal waste shipments, including exports, §273.20 or 273.40 requires specific Part 262, Subpart E export regulations to be met for universal waste exports. [RO 14740]

The Part 262, Subpart E standards include notification, annual reporting, and recordkeeping requirements. To summarize, the exporter (universal waste handler) must send a notification of intent to export to EPA's Office of Enforcement and Compliance Assurance 60 days prior to exporting the universal waste to a foreign facility. The notification must include information such as a description of the waste, estimates of the quantity and frequency the waste will be shipped, and the name of the foreign destination facility. This notification may cover export activities over a 12-month period. [§262.53]

EPA, in conjunction with the State Department, will provide a complete notification to the receiving country and any transit countries. If the receiving country consents to the receipt of the universal waste, EPA will forward an Acknowledgement of Consent to the handler. The handler must file annual reports by March 1 of each year covering such activities. Finally, exporters of universal waste must keep copies of each notification of intent to export, Acknowledgement of Consent, confirmation of delivery from the consignee, and annual reports for three years. [§262.57]

In the case where the destination country for the universal waste is one of the Organization for Economic Cooperation and Development (OECD) countries, the standards of Part 262, Subpart H rather than Subpart E apply. The exception to the application of Subpart H is if the waste is being shipped for disposal rather than recovery; in that case, Part 262, Subpart E requirements would still apply. [§262.58(b)] OECD countries are listed at §262.58(a)(1). Canada and Mexico are not on the OECD list, so Subpart E is applicable if shipping universal wastes to either of these countries.

8.4 Universal waste transporter and destination facility requirements

The function of transporters and destination facilities in the universal waste program is described in the next two sections, respectively.

8.4.1 Transporters

A universal waste transporter is a person/entity engaged in the offsite transportation of universal waste by air, rail, highway, or water. [§273.9] If a handler transports any amount of universal waste, he/she is a transporter subject to Part 273, Subpart D; there is no de minimis cut-off amount like there is for used oil generators who can transport up to a certain amount and still remain just a generator. [RO 14088] Companies that are not hazardous waste transporters can be universal waste transporters. This is true because no manifests are required for universal waste shipments—just bills of lading.

Like universal waste handlers, transporters are prohibited from treating, recycling, or disposing universal wastes. They also must immediately contain any spills or releases of materials and must manage any cleanup residue that fits the definition of hazardous waste under full RCRA requirements. Finally, they may only transport universal wastes to a universal waste handler, a destination facility, or a foreign destination.

Transporters may only hold universal wastes at a transfer facility for ten days or less. If they exceed this time limit, they become universal waste handlers subject to the small or large quantity handler requirements described earlier. [§273.53(b), RO 14186]

8.4.1.1 DOT requirements

Universal waste transporters are fully subject to DOT requirements, spelled out in 49 CFR Parts 171–180. Note, however, that even though they are hazardous wastes under RCRA, universal wastes are *not* hazardous wastes under DOT rules (because shipments of these wastes don't require a manifest). Thus, the word "waste" should not be used in front of or in a shipping description. [§273.52(b)] Instead, transporters of universal waste shipments must comply with the DOT requirements that would be applicable if the waste were being transported as a product. For example, in an October 1, 1998 DOT letter (Ref. No. 98-0250), DOT noted that "Mercury contained in manufactured articles" may be used to describe shipments of universal waste fluorescent light bulbs. Table 8-3 lists DOT shipping descriptions for batteries that are commonly managed as universal waste.

When transporting universal waste batteries:

"[T]he transporter must comply with the appropriate DOT requirements, which are based on whether the particular battery type is a DOT hazardous material, and if so, which DOT hazardous material requirements apply to the specific battery type." [May 11, 1995; 60 *FR* 25501]

On April 3, 2009, DOT wrote a public letter (available at http://www.phmsa.dot.gov/staticfiles/PHMSA/DownloadableFiles/Files/2009_Battery_Safety_Compliance_Advisory.pdf) to battery recyclers and other battery handlers to emphasize the need to properly package spent batteries before shipping them offsite. In that letter, DOT noted an ongoing

Table 8-3: DOT Shipping Descriptions for Batteries Commonly Managed as Universal Waste

Battery type	DOT shipping description
Alkaline, common AA, C, and D cells	Non-DOT Regulated Solid, or Battery, dry[1]
Lead-acid	Batteries, wet, filled with acid, 8, UN2794, III
Lead-acid (non-spillable)	Batteries, wet, non-spillable, 8, UN2800, III
Lithium-ion button, commonly used in watches and calculators	Lithium battery, 9, UN3090, II[2]
Lithium-manganese dioxide, commonly used in phones and digital cameras	Lithium battery, 9, UN3090, II[2]
Lithium (in equipment)	Lithium batteries, contained in equipment, 9, UN3091, II[2]
Mercuric oxide	Non-DOT Regulated Solid, or Battery, dry
Nickel-cadmium (dry), commonly used in appliances, toys, and cordless phones	Non-DOT Regulated Solid, or Battery, dry[3]
Nickel-cadmium (wet)	Batteries, wet, filled with alkali, 8, UN2795, III
Nickel-metal hydride	Non-DOT Regulated Solid, or Battery, dry[4]
Silver oxide, commonly used in watches	Non-DOT Regulated Solid, or Battery, dry

[1]See July 7, 1999; 64 *FR* 36744.
[2]Some lithium batteries (and equipment containing lithium batteries) may be exempt from most hazmat regulations, providing the conditions of Special Provision 188, 189, or 190 in the DOT regulations at 49 *CFR* 172.102 are met.
[3]See DOT Interpretation 08-0145, available at http://phmsa.dot.gov/hazmat/regs/interps.
[4]See DOT Interpretation 08-0019, available at http://phmsa.dot.gov/hazmat/regs/interps.

Source: McCoy and Associates, Inc.; adapted from DOT Interpretation 08-0202, available at http://phmsa.dot.gov/hazmat/regs/interps, unless otherwise noted.

trend of serious safety problems and noncompliance related to the classification, packaging, marking, labeling, documentation, and transportation of spent batteries. Common violations and safety problems noted include:

- Large numbers of used batteries, of many different types, are collected in large containers that do not adequately prevent damage to the batteries or prevent their release during transportation.

- Outer packages are not marked and labeled as required to indicate that they contain batteries, and the shipments are not described properly on the accompanying shipping documents.

- No action is being taken to prevent a short circuit, such as separating the batteries by placing each one in a separate plastic "baggie" or taping the terminals of the battery.

The April 3, 2009 letter contains a summary of the DOT requirements that apply to ground shipments of batteries for recycling or disposal.

On January 14, 2009 [74 FR 2200], DOT issued a final rule that 1) clarifies the requirement that batteries and battery-powered devices be offered for transportation and transported in a manner that prevents short circuiting, the potential of a dangerous evolution of heat, and/or damage to terminals; and 2) includes several examples of packaging methods that can be used to meet this requirement. This may be achieved by packing each battery in fully enclosed inner packagings made of nonconductive material or separating the batteries from each other and other conductive material in the same package.

Q *What requirements apply to lithium batteries when they are shipped offsite to a battery reclaimer?*

A A hazardous waste manifest is not needed for shipments of universal waste. However, the DOT hazmat regulations will continue to apply if the universal waste meets the definition of a hazardous material or a hazardous substance. Lithium batteries meet the definition of a hazardous material (Class 9); thus, full compliance with the DOT regulations will be required (e.g., shipping papers, markings, packaging requirements, etc.).

For ground shipments of lithium batteries (including lithium-ion batteries), batteries have to be separated or packaged so as to prevent short circuits and must be packed in a strong outer packaging or contained in equipment. [Special Provisions 188, 189, and 190 of 49 CFR 172.102 and 49 CFR 173.185(d)] For example, the batteries could be individually packaged in plastic baggies or have their terminals taped to prevent short circuits.

8.4.2 Destination facilities

According to §273.9, a universal waste destination facility is any entity that treats, recycles, or disposes a particular category of universal wastes. Because these activities pose significant risks to human health and the environment, destination facilities are subject to the same management standards as other TSD facilities managing hazardous waste. First and foremost, they must have a RCRA permit [unless they are a §261.6(c) permit-exempt recycling facility]. Additionally, all equipment must be designed and operated in accordance with the Part 264 or 265 management standards, and the substantive and administrative LDR program requirements are applicable, including recordkeeping. [64 FR 36481] The destination facility is typically the first facility handling universal waste that is responsible for compliance with any LDR provisions. [RO 14088] However, as discussed in Case Study 8-1, if the destination facility is a solid (nonhazardous) waste landfill, the handler will have to ensure that any decharacterized universal wastes or nonhazardous residues from treating universal waste meet LDR treatment standards before they go into that landfill. [RO 14756]

Destination facilities are subject to requirements in Part 273, Subpart E. In addition to the above major requirements, they are required to maintain a record of all universal wastes they receive via a log or by keeping copies of invoices, bills of lading, or other shipping documents.

When containers of universal waste are shipped from a handler, they are labeled as universal wastes—not hazardous wastes. When these wastes are received at destination facilities, though, they are managed under the facility's hazardous waste permit. EPA has noted that such containers of universal waste received at destination facilities do not have to be relabeled as hazardous waste. [RO 14088]

8.5 State authorization issues

The universal waste program was not promulgated to implement the HSWA amendments to RCRA. Therefore, the universal waste program took effect on May 11, 1995 only in those states that were not authorized for the base RCRA program. In authorized states, the universal waste rules weren't effective until states revised their hazardous waste program to incorporate the regulations. [60 *FR* 25536, RO 11952] Similarly, when EPA added hazardous lamps to the program (effective January 6, 2000 at the federal level), that rule was not effective in authorized states until they adopted it.

For handlers and transporters, the universal waste program is *less stringent* than pre-existing RCRA hazardous waste rules. Therefore, states are not obligated to add this program to their existing regulatory structure (although EPA encourages states to do so). Thus, the universal waste program has taken effect only in those authorized states that have chosen to adopt and administer either the federal universal waste program or their own (no-less-stringent) version of the program.

8.5.1 Current state status

The map in Figure 8-2 shows the current adoption status of the base (May 1995) universal waste program in the United States. All states have 1) become authorized to administer this program, 2) adopted the universal waste program, or 3) EPA administering the program (Alaska and Iowa).

8.5.2 Interstate transport

States may add other materials to their own universal waste program (e.g., paint and paint-related wastes are universal wastes in Texas, as was noted in Section 8.1.1). These situations present logistical problems for the interstate transportation of these wastes.

When shipping to another handler or a destination facility, a universal waste handler must know whether the states that his/her waste is being shipped to or through regulate the specific wastes being shipped as universal waste. In those states that do regulate the specific wastes in the shipment as universal waste, compliance with only the Part 273 standards will be required. In those states that do not regulate the wastes as universal waste, compliance with the full hazardous waste provisions is required, including the use of manifests and hazardous waste transporters. [June 12, 2002; 67 *FR* 40520] There have been instances where a truck has broken down in a state that has not adopted the waste as universal waste, and the handler/transporter was fined for having unmanifested hazardous waste onboard.

In some cases, a waste regulated as universal waste in one state may be sent to a state where it is still subject to the full set of hazardous waste regulations. For the part of the trip through the originating state and any other states where the waste is regulated as universal waste, neither a hazardous waste transporter nor manifest would be required. However, when the waste travels through the receiving state and any other states that still consider the waste to be hazardous, a manifest is required and transport must be accomplished by a hazardous waste transporter. In these situations, the originating facility should complete a manifest and send it to the transporter who will be carrying the waste through the first state that considers the waste to be hazardous. EPA recommends that the originating facility note in Block 14 of the manifest that the waste is covered under the universal waste regulations in the originating state, but not in the receiving facility's state. The receiving facility would then sign the manifest and send it back to the originating facility. [May 11, 1995; 60 *FR* 25537] Although EPA's guidance is silent on the matter, an LDR form

Figure 8-2: Authorization Status of the State Universal Waste Programs

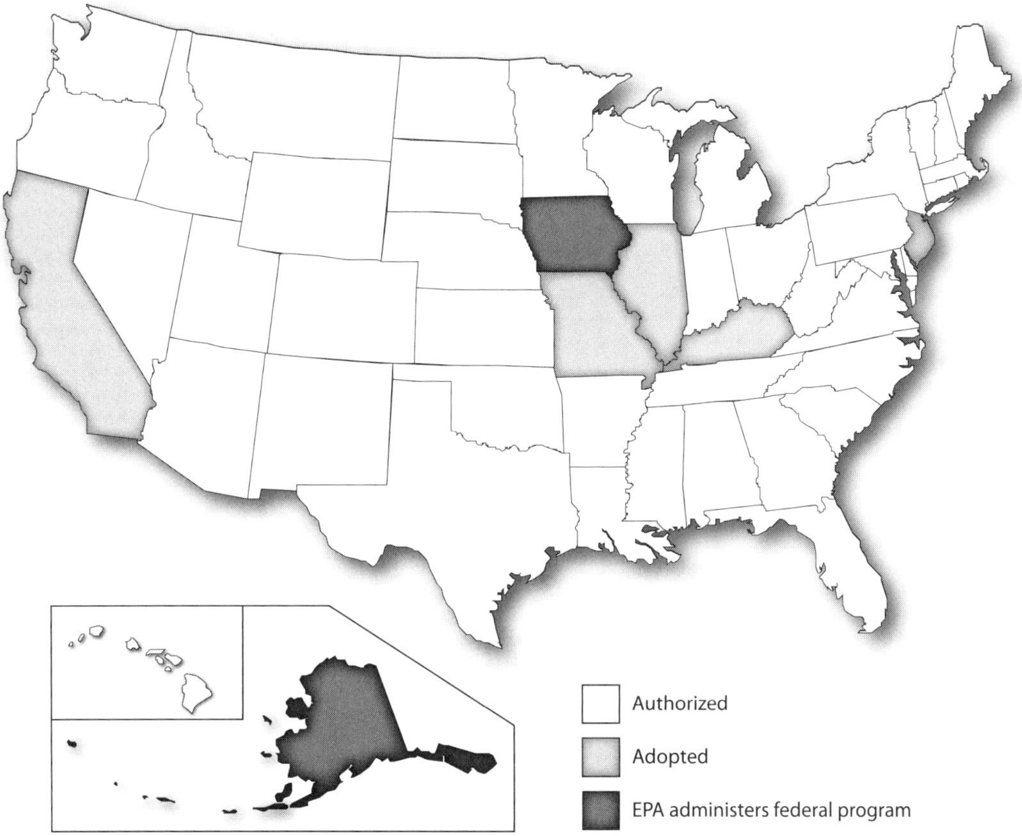

- Authorized
- Adopted
- EPA administers federal program

NOTE: There is no significant difference between Authorized and Adopted to a regulated entity in a given state. An authorized universal waste program has gone through EPA's formal authorization/approval steps; an adopted program has the same regulatory effect, but has not been formally approved by EPA.

Source: Adapted from http://www.epa.gov/epawaste/laws-regs/state/stats/authall.pdf.

---

would also be required by the receiving facility; thus, it would be necessary for the originating facility to also forward an LDR form to the receiving facility (e.g., sending it to the transporter who will be carrying the waste through the first state that considers the waste to be hazardous).

Hazardous waste that is not regulated as universal waste in the originating state may sometimes be sent to a state where it qualifies as universal waste. In this case, the waste must be carried by a hazardous waste transporter in the originating state and any other states where it is not a universal waste. The originating facility would need to complete a manifest and give copies to the transporter who picks up the waste. However, transportation in the receiving state and any other states where the waste is universal waste would not require a manifest or need to be carried out by a hazardous waste transporter. Even so, the originating facility is responsible for ensuring that the manifest is forwarded to the receiving facility and for obtaining a signed copy back from the receiving facility. Again, EPA recommends that the originating facility note in Block 14 of the manifest that the waste is covered under universal waste regulations in the receiving state, but not in the originating state. [May 11, 1995; 60 *FR* 25537, RO 14088] Although the receiving facility may not require an LDR form from the generator in this situation, the generator's state RCRA program will require that the generator complete the appropriate LDR form per the state equivalent of §268.7(a). Additionally, even though EPA's guidance suggests that LDR forms are not required for shipments of universal waste, some receiving facilities may require them anyway if they will be treating or recycling the waste.

8.6 Mercury-Containing and Rechargeable Battery Management Act

In May of 1996, the Mercury-Containing and Rechargeable Battery Management Act became law. This law was designed to 1) standardize efforts already underway in over a dozen states to promote the recycling or proper disposal of lead, mercury, nickel-cadmium, and other types of rechargeable batteries; and 2) limit the mercury content of consumer batteries. In a nutshell, the act:

- Requires all 50 states to implement collection, storage, and transportation provisions that are identical to the universal waste program requirements promulgated on May 11, 1995 for 1) rechargeable nickel-cadmium batteries, 2) lead-acid batteries not covered by Part 266, Subpart G, 3) rechargeable alkaline batteries, 4) certain mercury-containing batteries banned from domestic sale, and 5) used consumer products containing rechargeable batteries that are not easily removable [RO 14020, 14088, 14290];

- Limits the mercury content of alkaline-manganese button-type batteries;

- Prohibits the sale of nonbutton-type alkaline-manganese batteries and all zinc-carbon batteries that contain intentionally added mercury;

- Prohibits the sale of button-type mercuric-oxide batteries and other types of mercuric-oxide batteries, except where nonbutton battery manufacturers or importers have identified an approved recycling collection site and informed battery purchasers of this site;

- Mandates that rechargeable nickel-cadmium and certain small sealed lead-acid batteries in consumer devices be "easily removable"; and

- Establishes national, uniform labeling requirements for rechargeable and other regulated batteries and associated packaging.

On September 22, 1998 [63 *FR* 50569], EPA approved a new label for rechargeable batteries that helps consumers identify recyclable batteries and locate the nearest battery collection center.

EPA has noted that the §§273.13(a)(2)/273.33(a)(2) provision that allows handlers to remove electrolyte from certain batteries is covered under the auspices of the battery act. This means that states may not alter this provision when implementing their own version of the universal waste program. [RO 14124]

States that have battery management standards are required by the law to have programs identical to the federal universal waste program for these materials. Hazardous waste batteries that meet the definitions in the law must be managed as universal waste. [EPA/530/K-05/019] Although the law requires all states to have a national, uniform set of collection, storage, and transportation regulations (identical to the universal waste program) for the noted types of batteries, the law does not address the subsequent recycling or disposal of those batteries. Thus, states may adopt battery recycling or disposal standards that are more stringent than existing federal requirements under RCRA.

A good summary of this law is available in *Implementation of the Mercury-Containing and Rechargeable Battery Management Act*, EPA/530/K-97/009, November 1997, available from http://www.epa.gov/epawaste/hazard/recycling/battery.pdf.

Container Issues

Everything you need to know about hazardous waste containers

We begin this chapter with three short quotes from EPA's May 19, 1980 rule that first implemented RCRA controls on hazardous waste containers:

The good—"Drums and other containers provide an inexpensive means for generators of hazardous wastes to accumulate and store the wastes, in a form which will be easy and relatively inexpensive to carry away."

The bad—"All too frequently, generators and others storing hazardous waste drums have simply put them somewhere out of sight, without any further concern about what would eventually happen to the wastes…. The drums eventually weather and corrode, releasing their contents."

The ugly—"Dumps of decaying drums have seriously contaminated surface water and ground water; have emitted fumes which have killed vegetation and nauseated and sickened nearby residents, facility operators, and enforcement officials; and have burned or exploded, injuring and killing facility personnel and sending clouds of toxic smoke and fumes over adjacent heavily populated areas, disrupting the activities and threatening the health of thousands of people." [45 *FR* 33199]

Obviously, EPA viewed container mismanagement as a serious issue requiring substantial action back in 1980. However, the agency believed that a simple, performance-based set of regulations it terms "good management practices" could eliminate most of the problems. It was with this philosophy in mind that EPA crafted its requirements for hazardous waste containers.

In this chapter, we review the rules EPA has promulgated to regulate the treatment, storage, and/or disposal of hazardous waste in containers. Because containers are portable, generators often use the same unit for accumulation, transportation, and ultimate disposal. [RO 12500] Note that the same standards apply regardless of whether the containers are used for treatment or storage. [*Introduction to Containers*, EPA/530/K-05/010, September 2005, available from http://www.epa.gov/wastes/inforesources/pubs/training/cont05.pdf]

These regulations are codified in Part 264, Subpart I—applicable to permitted TSD facilities, and Part 265, Subpart I—applicable to interim status facilities and generators' 90/180/270-day waste accumulation containers. (Generators operating 90/180/270-day accumulation containers get to Part 265, Subpart I by reference from §262.34.) The Part 264 and 265 standards for containers are very similar.

Our coverage addresses the following issues:

- What is a container? (Section 9.1),
- Containers must be closed (Section 9.2),
- Container labeling requirements (Section 9.3),
- Secondary containment requirements (Section 9.4),
- Empty containers (Section 9.5),
- Container inspection requirements (Section 9.6),
- Lab packs (Section 9.7), and
- Containers holding incompatible wastes (Section 9.8).

Hazardous wastes may be accumulated in 90/180/270-day containers, as we previously discussed in Section 6.3.4.2, and may be treated in these containers without a permit, as discussed in Section 7.3. Additionally, absorbent may be added to hazardous waste in containers without a permit as was discussed in Section 7.5.3.

9.1 What is a container?

EPA's regulation of containers begins with a definition of the term; §260.10 defines a "container" to be:

> "[A]ny portable device in which a material is stored, transported, treated, disposed of, or otherwise handled."

EPA has made this definition intentionally broad to encompass all different types of portable devices that may be used to manage hazardous waste. Examples of containers range from rail cars, tanker trucks, and roll-on/roll-off transport boxes to steel, plastic, and fiberboard drums of assorted sizes. Containers also include small buckets, cans, and laboratory test tubes.

The primary difference between a container and a tank is that a tank is stationary (usually bolted down), while a container is portable. See Section 7.2.2.7 for guidance on the difference between containers and portable tanks.

Inclusion of the phrase "or otherwise handled" in the definition given above would tend to make nearly any portable device in which hazardous waste is placed a hazardous waste container. There are some gray areas, however, as the next section illustrates.

9.1.1 Regulation of container-like equipment

Under EPA's "contained-in" policy (which is discussed in detail in Section 5.3), equipment to be disposed that contains or has otherwise been in contact with hazardous waste must be managed as hazardous waste under the RCRA Subtitle C regulations. [RO 11219, 11325, 11387] However, if this hazardous equipment can be considered a container, it will not be subject to regulation if it is shown to be RCRA-empty. (See Section 9.5 for more on "RCRA-empty.")

Based on the definition of "container" given in the previous section, many types of container-like equipment are regulated as containers: examples include mercury-containing items such as pumps, thermometers, manometers, thermostats, batteries, and ampules. [RO 14685] For instance, a pump containing elemental mercury is taken out of service. The pump is used to store and transport the mercury prior to its eventual treatment and disposal. EPA determined that the pump would meet the definition of a container if it is portable and, thus, would be subject to Part 264/265, Subpart I provisions. However, if the mercury were removed from the pump, rendering it empty per §261.7, the pump would be exempt from regulation. [RO 11647, 13788]

Similarly, disposable laboratory equipment (e.g., a pipet) used to analyze listed hazardous waste is normally subject to RCRA regulation under the contained-in policy when it is discarded. However, the agency indicated a pipet may meet the definition of a container and would therefore be exempt from hazardous waste regulation if it is RCRA-empty. [RO 13375]

McCoy's RCRA Unraveled ©2015 McCoy and Associates, Inc.

Another common example of container-like equipment is the shell of an intact battery whose contents, when discarded, would be considered hazardous waste (e.g., a lead-acid automobile battery). EPA generally considers such battery shells to fit the definition of a container and regulates them as such. [March 22, 1982; 47 *FR* 12318, June 1, 1990; 55 *FR* 22637, RO 13339, 13638, 14675]

In RO 14587, EPA indicated that a cabinet could be a container (e.g., a satellite accumulation container) as long as it meets all of the container standards (which are discussed in the rest of this chapter).

In contrast to the above examples, EPA does *not* consider automotive oil filters to be containers, because they are designed to filter particulates from oil—not store it. As a result, a drained filter could not be an empty container. However, as discussed in Sections 4.21.6 and 12.1.5, drained or crushed filters would be considered scrap metal and would be exempt from most RCRA regulations if recycled. [§261.6(a)(3)(ii), RO 11566, 13498, 14183, 14184]

9.1.1.1 Containers versus debris—LDR implications

When it added the debris rule as part of the land disposal restrictions (LDR) program on August 18, 1992, EPA noted in the preamble that it had received several comments/questions concerning the relationship between containers and debris (e.g., could empty containers be debris?). The agency stated "intact containers are never considered to be debris, and thus would never be subject to [LDR] treatment standards for debris." EPA noted that the term "intact container" means a container that could "still function as a container," and was "unbroken and still retains at least 75% of its original holding capacity." [57 *FR* 37225, RO 13638, 14675] The agency reaffirmed this position in 2003, noting that mercury-containing items such as pumps, thermometers, manometers, thermostats, dental amalgam collection devices, and ampules are containers and thus do not fall under the debris definition. Instead, these items, when disposed, would be subject to nondebris LDR treatment standards for mercury wastes. [RO 14685] (Note that some of these wastes would be universal wastes under the federal regulations.)

Subsequent agency guidance has since expressed this logic in its reverse form, noting that containers that are ruptured, broken, or deteriorated (or have holes in them that allow material to flow in or out) can no longer function as containers and (if they contain or are contaminated with hazardous wastes) are hazardous debris subject to LDR treatment standards. [RO 14215, 14241, 14675]

9.1.2 Container-related issues

In this section, we address situations where the RCRA hazardous waste container requirements interact with other regulations.

9.1.2.1 DOT standards

First, most containers used for hazardous waste storage are ultimately slated for transportation to a TSD facility. EPA has therefore concluded that containers meeting Department of Transportation (DOT) standards [49 *CFR* Part 171–180] are also acceptable from an environmental protection prospective, as long as the Subpart I standards are met. [RO 12500, 12543]

9.1.2.2 Counting containerized waste

Hazardous waste accumulated in containers must be counted. As noted in Section 6.1.3.1, questions have arisen over whether the weight of the container itself should be included when a generator calculates the amount of waste he/she generates monthly. According to EPA, a container's weight should *not* be included in calculations of waste generation quantities or in generator's biennial reports. However, since transporters usually charge based on the total weight they are carrying, hazardous waste manifests customarily show total (gross) weights (i.e., wastes plus containers). [RO 11803, 12151, 14827]

Weights that are listed on the manifest are often used by generators and inspectors to make estimations of generator status. If only the weight of the waste in a container is counted toward generator status, but the total weight is listed on the manifest, there could be some confusion about a generator's actual generator status. EPA recommends that when containers are manifested, the generator use Box 14 of the manifest (Special Handling Instructions and Additional Information) to indicate that although the total weight is included on the manifest, the weight of the containers was not included in determining its generator status. [RO 14827]

In RO 12946, EPA provides somewhat conflicting guidance. Several chemotherapy drugs are U-listed wastes; in order to minimize exposure, EPA recommends against rendering vials holding these drugs empty under §261.7. Instead, the agency recommends that the entire volume of waste, including the vials themselves, be weighed.

9.1.2.3 Containers are not ancillary equipment to tanks

As noted in Chapter 7, certain tank systems, including their ancillary equipment, are exempt from RCRA permitting and Part 264/265 management standards if they qualify as wastewater treatment or elementary neutralization units. Some facilities were interested in whether containers used to store hazardous waste prior to its treatment in the exempt tanks could be considered part of the tank's ancillary equipment and, therefore, enjoy the exemption from permitting/management standards. EPA determined that containers do *not* meet the definition of ancillary equipment and can never qualify for the wastewater treatment unit exemption. If such containers are used for storage—but not neutralization—of hazardous waste, they are also not exempt from regulation as elementary neutralization units. [RO 11551]

9.1.2.4 Containers and the liquids-in-landfills ban

As detailed in Section 7.5.3, §§264/265.314 prohibit landfilling containerized hazardous wastes that either are liquid or contain free liquid. A likely method facilities would use for dealing with the ban is to mix their containerized liquid hazardous waste with absorbents, thereby converting those materials to solids that would pass the paint filter test. This test is required to show the absence of free liquids. [§§264.314(b)/265.314(c)] However, several conditions limit the applicability of this scenario. First, the hazardous waste/absorbent mixture must be placed in containers prior to landfilling per §§264.314(c)/265.314(b). Second, any absorbents used to treat liquid in containerized hazardous waste must be nonbiodegradable. [§§264.314(d)/265.314(e), RO 11798, 13724]

EPA included an exception to the liquids-in-landfills ban for small containers (such as ampules [§§264.314(c)(2)/265.314(b)(2)]) and where a "container is designed to hold free liquids for use other than storage, such as a battery or capacitor." [§§264.314(c)(3)/265.314(b)(3)] Explaining its logic, the agency noted that the difficulty of opening and emptying batteries and capacitors appears to outweigh the small benefit gained from eliminating the liquid content of such containers. In a similar vein and subject to certain limitations, EPA also exempts lab packs from the liquids-in-landfills ban in §§264.314(c)(4)/265.314(b)(4). [RO 12832]

Finally, the agency noted that an out-of-service pump that contains free liquid may still be landfilled because, even though it meets the definition of a container, it is designed to hold free liquids for use other than storage (just like a battery or capacitor). Thus, the facility would be exempt from the requirement to remove or absorb free liquids. [RO 13788] (Note that we find this logic somewhat tenuous and would caution facilities to check with their state before trying to use this exemption.)

9.1.3 Container management 101

While the remaining sections of this chapter deal with the most significant management requirements for hazardous waste containers, a brief description of all container regulations in Part 264/265, Subpart I is listed in Table 9-1.

9.2 Containers must be closed

One of the primary causes of notices of violations for noncompliance with the RCRA regulations is that satellite and 90/180/270-day hazardous waste containers are not closed. The requirement

Table 9-1: Design and Management Standards for Hazardous Waste Containers

Part 264/265, Subpart I reference	Description
§§264.171/265.171	Containers must be in good condition. Where rusting, cracking, leaking, etc. are observed, wastes must be transferred to a different container or otherwise managed to prevent releases.
§§264.172/265.172	Materials of construction of containers or liners must be compatible with the waste placed in them, so as not to react with the waste or otherwise affect the ability of the containers to hold the waste.
§§264.173/265.173	Containers must be closed at all times, except when adding or removing wastes—see Section 9.2. Additionally, the container must be managed in a way that prevents rupturing or leaking.
§§264.174/265.174	Containers must be inspected weekly for leaks or other deterioration—see Section 9.6.
§264.175[1]	Containers in permitted storage areas must be provided with secondary containment—see Section 9.4.[2]
§§264.176/265.176[3]	Containers holding D001 ignitable or D003 reactive waste must be located at least 15 meters (50 feet) from the facility's property line.[4]
§§264.177/265.177	Incompatible wastes must not be placed in the same container (this includes a restriction on placing wastes in unwashed, RCRA-empty containers that have incompatible residues in them). [45 FR 33200] Containers holding wastes that are incompatible with other wastes or materials stored nearby must be segregated by dikes, walls, berms, or other devices. "Incompatible waste" is defined in §260.10—see Section 9.8.
§264.178[1]	At closure of a permitted container storage area, all hazardous wastes and residues must be removed, and containers, liners, and contaminated soil must be decontaminated or removed.
§§264.179/265.178[3]	Management of hazardous waste in containers must conform with the air emission standards of Part 264/265, Subparts AA, BB, and CC.

[1]There is no comparable section in Part 265.
[2]Secondary containment is not required if the container is 1) used to store a hazardous waste that contains no free liquids; *and* 2) the storage area is sloped or otherwise designed and operated to drain and remove run-on/precipitation, or containers are elevated or are otherwise protected from contact with accumulated liquid.
[3]Small quantity generators are not subject to this requirement. [§262.34(d)(2)]
[4]This is a National Fire Protection Association code requirement to protect adjacent properties from the effects of an explosion or fire. Some or all of the required 50-foot buffer zone may include leased land, but the lease agreement should specify that the lease will remain in force even if the land is sold to another owner. [RO 12076] For generators who physically cannot meet this 50-foot buffer requirement, EPA recommends that they work with their regional or state inspector to determine if the local fire department or fire marshal will grant a variance. Any such variance should be in writing and maintained onsite. [RO 14840]

Source: McCoy and Associates, Inc.

---

that containers of hazardous waste be closed is stated clearly in §§264/265.173(a):

"A container holding hazardous waste must always be closed during storage, except when it is necessary to add or remove waste."

While no definition of "closed" is provided in the regulations, EPA explained that the purpose of this requirement is "to minimize emissions of volatile wastes, to help protect ignitable or reactive wastes from sources of ignition or reaction, to help prevent spills, and to reduce the potential for mixing of incompatible wastes and direct contact of facility personnel with waste." [May 19, 1980; 45 FR 33199] Based on this preamble language, we have generally interpreted the agency's intent as requiring containers to be vapor tight and spill proof.

EPA issued considerable guidance on this issue on December 3, 2009, which it updated on November 3, 2011 and reissued as RO 14826. Developed to assist federal and state regulators and the regulated community with the closed-container provisions of §§264/265.173(a), this guidance is relevant to both small and large quantity generators accumulating waste in satellite accumulation and 90/180/270-day accumulation areas. Similarly, it is applicable to containers being stored at RCRA permitted and interim status facilities. Sections 9.2.1–9.2.4 summarize the guidance in RO 14826.

9.2.1 Satellite accumulation containers

Satellite accumulation containers holding free liquids or liquid hazardous wastes are closed "when all openings or lids are properly and securely affixed to the container, except when wastes are actually being added to or removed from the container." The objective of ensuring that the lid is securely fixed to the container or securely covers the container is to prevent vapor emissions and prevent a spill if the container is tipped over. According to EPA, if hazardous waste will frequently be added to the container, it may not be practical to secure snap rings, securely cap the bungholes, etc. during working hours. However, the container must still be covered tightly to prevent spills and air emissions.

In those situations where funnels are used, EPA recommends that they be screwed tightly into the bunghole and be equipped with manually or spring-closed lids; such lids may be fitted with a gasket or lid locking mechanism to seal the funnel lid firmly closed. Funnels with a one-way valve that allows waste to enter the container but prevents waste or emissions from exiting may also be used.

EPA noted that liquid hazardous wastes can also be accumulated in open-head drums or containers (e.g., where the entire container lid is removable and typically secured with a ring and bolts or a snap ring). These containers meet the definition of "closed" if the rings are clamped or bolted to the container. However, the agency indicated that a container could be considered closed in some situations if the lid covers the container top securely even though the rings are not clamped or bolted.

For containers that continuously receive liquid hazardous wastes (e.g., laboratory instruments draining to small bottles or jugs), even though the containers may require a vent system, they must still be closed to minimize releases. In these situations, EPA recommends secondary containment, such as a pan, and/or securing the containers to prevent overturning. When the process or laboratory instrument is not in operation, the containers should be closed.

Satellite containers holding solid and semisolid hazardous wastes are closed when there is complete contact between the lid and the rim all around the top of the container. This includes containers with covers opened by a foot pedal. Per EPA, having a tight seal prevents VOC emissions. Seals on containers can erode over time, so the agency recommends inspecting any such seals and replacing them if necessary. EPA noted that containers continuously or intermittently receiving solid or semisolid wastes often remain open while connected to the process unit (e.g., a baghouse or filter press). Such containers should be capable of catching and retaining all material during transfer to avoid spills or releases. Roll-off containers should be closed with either manufacturer-supplied lids that provide a good seal around the rim or covered with tarps with no visible holes or gaps/open spaces.

9.2.2 90/180/270-day containers

For wastes in central accumulation areas, EPA takes a less flexible approach. The agency noted that containers in central accumulation areas are closed if bungholes are capped and lids are secured with rings that are tightly clamped or bolted. So again, these containers must be vapor tight and spill proof. Where appropriate, pressure-relief valves should be used to maintain the container's internal pressure to avoid explosions.

Q: *Some equipment (e.g., a baghouse or filter press) generates waste nearly continuously. Where a large container is being used to collect hazardous waste which is continually exiting from the equipment, it might be necessary to leave the container open to collect waste until the process stops. Is it acceptable to leave the container open the entire time it is connected to or positioned to collect waste from the baghouse or filter press?*

A No. Section 265.173(a) requires that a hazardous waste container must always be closed during storage, except when it is necessary to add or remove waste. Where a container is being used to collect hazardous waste which is continually exiting from the baghouse or filter press, then it might be "necessary" to leave the container open to collect waste until the process is stopped. In the more common circumstance, however, where the deposition of hazardous waste into containers is a batch process, a container of hazardous waste must be kept closed during times when the process is not depositing hazardous waste into the container. [RO 14826]

9.2.3 Preventing spills

EPA recommends that to prevent spills, containers should be located in areas with little or no vehicular traffic (e.g., forklifts). In those situations where container lids are not securely affixed to the container, the agency recommends the practice of securing containers with a chain or strap to a wall or building support column to prevent accidental overturning of the container. Furthermore, the agency supports the use of secondary containment systems, although their use is not mandatory under the federal regulations except for permitted container storage areas. Secondary containment requirements for hazardous waste containers are discussed in Section 9.4.

9.2.4 Miscellaneous container closure guidance

Other types of containers used for managing hazardous waste in satellite or central accumulation areas include: bags, boxes, roll-off boxes, and plastic or steel totes. The agency considers these containers closed when they are "sealed to the extent necessary to keep the hazardous waste and associated air emissions inside the container."

Many containers are subject to the Subpart CC air emission standards. In RO 14826, EPA indicated that for roll-off containers that are not "in light material service," the use of a tarp with no visible holes, gaps, or open spaces is an example of a suitable Subpart CC Level 1 control device. However, if using a tarp under Subpart CC, it would also have to be suitable for weather conditions, including exposure to wind, moisture, and sunlight. If a roll-off container is subject to Level 2 controls, the use of a tarp would not be an acceptable closure device. See Section 15.3.5 for more information on complying with the Subpart CC air emission control requirements for containers.

EPA noted that this guidance does not supersede or replace any existing state closed-container guidance. Different states (and even different state inspectors) may have their own, more specific definition of "closed." State inspectors often use the following rule of thumb: if the contents would spill if the container was overturned, then the container is considered open. (Relief from this rule of thumb is sometimes available if the container holds physically solid material.)

9.3 Container labeling requirements

The container standards in Part 264/265, Subpart I have no specific labeling requirement. However, §262.34(a)(2–3) and (d)(4) require generators who accumulate hazardous waste for 90/180/270 days to clearly mark each container with:

- The words "Hazardous Waste," and
- The date upon which each 90/180/270-day accumulation period begins.

This labeling keeps personnel at the site aware of the hazardous waste contents and assists state or EPA inspectors in determining compliance. While there is no federal requirement to further label hazardous waste containers being managed onsite, adding hazardous waste codes and/or other descriptions of the container contents is considered good container management practice

[January 11, 1982; 47 *FR* 1249] and may be a state program requirement.

Containers used for offsite shipment of hazardous waste must be marked and labeled according to DOT specifications for the appropriate hazard class. [§§262.31 and 262.32(a)] Additionally, generators who ship hazardous waste in containers of 119 gallons or less must label each container with the following [§262.32(b)]:

- HAZARDOUS WASTE—Federal Law Prohibits Improper Disposal. If found, contact the nearest police or public safety authority or the U.S. Environmental Protection Agency.
- Generator's name and address.
- Generator's EPA ID number.
- Manifest tracking number.

DOT regulations [49 *CFR* Part 172] also require that markings and labels be durable. Typed labels, other than vinyl labels, are considered durable. If vinyl labels are used, the typing can smear. Special marking pens for vinyl labels are recommended to prevent smearing. [RO 11031]

If your facility uses any commercially available labels, be sure that the wording is verbatim as required in the regulations. Even when proper labels are used, misspellings and improper codes may result in citations and/or fines. However, the development and use of computer software and printing has significantly reduced label preparation errors.

Inspectors can be particular about labeling requirements. Some insist, as an example, that containers be oriented during accumulation so that the inspector can see the label without moving the container.

Q *A facility palletizes and shrink wraps expired consumer goods in their original packaging. The expired consumer goods are hazardous waste. Does the facility need to label every container with a hazardous waste label, or can it label each pallet with a single label? Would a single label placed on the shrink-wrapped pallet be sufficient to meet the label requirements for accumulation of hazardous waste?*

A "Pursuant to the RCRA regulations, each container used for onsite hazardous waste accumulation must be labeled or marked clearly with the words 'Hazardous Waste' and the date on which accumulation began. [§262.34(a)(2–3)] Therefore, the generator must label each individual container on the pallet. A generator may also label the pallet, if he so chooses, but the pallet label does not in and of itself satisfy the requirements of §262.34(a)(3)." [Topic #: 23002-13620 on EPA's RCRA Wastes – Frequent Questions Database (http://waste.supportportal.com/)]

9.3.1 Labeling requirements for satellite accumulation containers

While containers are in a satellite accumulation area (SAA), each must be labeled with the words "Hazardous Waste" or other words that identify the contents, such as "Spent Solvents." [§262.34(c)(1)(ii)] Although it is not federally required that a waste code be placed on an SAA container, this is certainly a good operating practice.

Q *A generator places several containers within a cabinet. Does a hazardous waste label placed on the outside of the cabinet suffice, or must each individual container within the cabinet be labeled?*

A Placing a label with the words "Hazardous Waste" on the outside of the cabinet will satisfy the SAA marking requirement if the cabinet meets the definition of a container. To be a satellite accumulation container, the cabinet must be portable, located at or near the point of generation and be under the control of the operator of the process generating the waste, in good condition, constructed of materials compatible with the waste, and closed at all times except when wastes are added or removed. If the cabinet does not meet all of the above requirements, each individual container within the cabinet would need to be labeled. [RO 14587]

Q *If a facility has very small containers (e.g., vials or tubes) of hazardous waste in an SAA that are too small to label with the words "Hazardous*

Waste" or other words that identify the contents, how should the labeling requirement be met?

A The small containers of hazardous waste should be placed in properly labeled larger containers. [RO 14703]

On the day the 55-gallon limit is exceeded in an SAA, the operator must comply with the provisions of §262.34(c)(2). He/she usually puts the current date on the label of the full drum because on this day he/she exceeds the 55-gallon limit. (This is called the "date of excess accumulation.")

9.4 Secondary containment requirements

EPA imposes secondary containment standards on permitted facilities that handle hazardous waste containers. The agency's logic in including this provision is explained as follows:

> "Containers are relatively thin-walled, can be punctured by fork-lift trucks, and are prone to break open when dropped or knocked over. They tend to corrode or otherwise deteriorate relatively rapidly both from the inside as a result of reaction with the waste, and from the outside as a result of exposure to the environment. The agency believes, therefore, that it is prudent to require a secondary containment system under container storage areas." [January 12, 1981; 46 *FR* 2829]

Only RCRA-permitted facilities are required to provide secondary containment for hazardous waste containers. Thus, although many generators and interim status TSD facilities do provide secondary containment for these units as a matter of good housekeeping (i.e., to catch drips and spills), it is not required per *federal* regulations. Such facilities should check with their state RCRA authority to determine any state program requirements for container secondary containment. [RO 11909]

9.4.1 Technical details

A typical containment system consists of a poured concrete pad with curbing to prevent a release of hazardous waste into the environment; the pad is often sloped to a sump for removal of any leakage or other liquid drainage. Secondary containment requirements for permitted facilities are spelled out in §264.175 and are summarized below.

- Containment systems must have a base that is sufficiently impervious so as to contain leaks, spills, container failures, and run-on/precipitation until the collected material is detected and removed. Typically, such bases are made of concrete or asphalt, but other materials may be used. The regulations do not require that a liner or coating be used in conjunction with a concrete containment system. [RO 11909]

- The capacity of the secondary containment must hold the larger of 1) 10% of the volume of all containers protected by the system, or 2) the volume of the largest single container protected by the system. Note that containers not holding free liquids need not be considered in this determination (under either approach), unless the solids carry the following waste codes: F020, F021, F022, F023, F026, or F027.

- Run-on/precipitation must be prevented from entering the containment system, unless enough capacity (in addition to that required above) has been built into the system to hold any run-on/precipitation that might enter.

- The containment must be sloped or otherwise designed and operated to drain and remove liquids resulting from leaks, spills, container failures, and run-on/precipitation, unless the containers are elevated or otherwise protected from contact with accumulated liquids. The agency expects that by one hour after a leak or storm, no standing liquid will remain on the base in contact with the containers. [46 *FR* 2830]

- Spilled or leaked waste and any collected run-on/precipitation must be removed from the sump or collection area in a timely manner to prevent the collection system from overflowing.

Guidance on container secondary containment systems is summarized in the following questions and answers.

Q *How should the required capacity be computed for secondary containment areas that protect both hazardous waste tanks and containers?*

A Per §264.193, containment systems for tanks must be sized to hold 100% of the volume of the largest hazardous waste tank in the system, plus 24 hours of precipitation from a 25-year storm. The secondary containment system needs to be sized to meet this tank requirement and the above-noted container system capacity requirement, considered *separately*. If the secondary containment volume calculated for the tank equals or exceeds that calculated for the containers, both requirements have been satisfied. In other words, such systems do *not* have to be sized to hold the summation of the tank and container volume requirements. [RO 13190]

Q *Does the common practice of placing drums on pallets qualify as a means of elevating containers above accumulated liquids, thereby obviating the need for a sloped storage pad?*

A Yes. [46 *FR* 2830]

Q *Do any aspects of a secondary containment system design for containers need to be certified by a professional engineer?*

A No. [RO 12094]

9.4.1.1 Secondary containment exemption

EPA provides an exemption to container secondary containment for permitted TSD facilities. Container storage areas that hold only hazardous waste with no free liquids (as demonstrated by the paint filter test) do not need secondary containment if 1) the storage area is sloped or otherwise designed and operated to drain and remove liquids resulting from run-on/precipitation; or 2) containers are elevated or otherwise protected from contact with accumulated liquids. [§264.175(c)]

Note that this exemption *does not apply* to areas storing containers holding the following types of listed dioxin wastes, regardless of whether the containers hold free liquid or not: F020, F021, F022, F023, F026, and F027.

9.5 RCRA-empty containers

On November 25, 1980, EPA decided that "except where the hazardous waste is an acutely hazardous material listed in §261.33(e), the small amount of hazardous waste residue that remains in individual empty, unrinsed containers does not pose a substantial hazard to human health or the environment." [45 *FR* 78525] Thus, such residues in "RCRA-empty" containers are not subject to hazardous waste management requirements. In that *Federal Register*, the agency promulgated rules for what constitutes a RCRA-empty container in §261.7.

EPA issued these regulations to let facilities know how to empty their containers so they would no longer be subject to RCRA regulation, even if some residue remains in them. Therefore, these regs allow owners/operators to reuse containers meeting the provisions in §261.7, since the container is no longer considered to hold hazardous waste. [EPA/530/K-05/010, cited previously] However, the Department of Transportation has certain restrictions on the reuse of containers for the shipment of hazardous materials. Under the DOT regulations, a previously used container may be reused *once* for the shipment of waste material without reconditioning if the conditions in 49 *CFR* 173.12(c) are met. A container that previously held a hazardous material (including a hazardous waste) cannot be reused for shipment of a nonwaste hazardous material unless it is reused/reconditioned per 49 *CFR* 173.28.

It is important to note that the §261.7 regulations exempt only the *residue* remaining in an empty container; the regs do not apply to the container itself. Consequently, if a RCRA-empty container that is being discarded itself exhibits a characteristic (e.g., it is made out of lead or is an unpunctured aerosol can that exhibits the reactivity characteristic), the container would have to be managed as a hazardous waste, even though the residue would be deemed nonhazardous. [RO 11219, 13027, 13435]

Of course, hazardous waste residue in any container that is not RCRA-empty remains subject to full Subtitle C regulation, unless some exemption applies. If a container holding hazardous waste is not RCRA-empty and is being sent to a drum reconditioner for cleaning and reuse, for example, the container must be manifested and may only be shipped to a facility with a RCRA permit (or interim status) or to a recycling facility. [45 *FR* 78527, RO 11089]

The empty-container requirements apply to three different types of containers: 1) containers or inner liners that held nonacute hazardous waste, 2) containers or inner liners that held acute hazardous waste, and 3) containers that held hazardous compressed gas. The RCRA-empty criteria for each of these three types of containers are examined below.

9.5.1 Containers that held nonacute hazardous waste

Containers or inner liners that have held any nonacute hazardous waste (other than a compressed gas) qualify as RCRA-empty if they meet two requirements: 1) use of a common container emptying practice, and 2) either a residue depth or weight limit of residue remaining in the container. (Nonacute hazardous wastes are all D-, F- (other than dioxin-containing F-wastes having an (H) code in the right-hand column of §261.31), K-, and U-wastes.)

If the depth limit is used, all wastes must have been removed that can be removed using "commonly employed practices" (e.g., pouring, pumping, and aspirating) *and* no more than one inch of residue may remain on the bottom of the container or inner liner. Alternatively, containers may be deemed empty if all wastes have been removed that can be removed using commonly employed practices *and* no more than 3% by weight of the total capacity of the container remains in the container or inner liner. This 3% limit is for containers with capacities of less than or equal to 119 gallons. For containers greater than 119 gallons in size, no more than 0.3% by weight of the total capacity of the container may remain in the container or inner liner. [§261.7(b)(1)]

9.5.1.1 Commonly employed practices

The first requirement that must be met for a container of nonacute hazardous waste to be considered empty is that residues must have been removed using practices commonly employed for emptying that type of container. The regs at §261.7(b)(1)(i) call out pouring, pumping, and aspirating as common practices; EPA/530/K-05/010 adds "draining" to the list. "Commonly employed" refers to the normal practice of industry, not to what a given person does. Thus, containers that have not been subjected to all commonly employed methods of emptying are still subject to regulation. [RO 11048]

EPA has provided scattered guidance on commonly employed practices for different types of containers, including the following:

- *Tank cars*—EPA expects bottom valves to be used (when present) if they provide maximum removal of hazardous waste. [RO 11048]
- *55-gallon drums*—According to EPA, "[i]f pouring from an inverted drum removes more residue than a hand pump does, then pouring is obligatory." [RO 11048] See Case Study 9-1.
- *Bags containing solid hazardous wastes*—Pouring out the contents from a bag, then shaking and tapping the outside of the bag is a common emptying practice. [RO 12307]

9.5.1.2 The one-inch requirement

As EPA noted in the preamble to the August 18, 1982 amendments to the empty-container provisions, one inch of waste "may remain in an empty container only if it *cannot be removed* by normal means…. On the other hand, if extraordinary means are necessary to remove the waste to lower

> **Case Study 9-1: Use of Commonly Employed Practices**
>
> A generator arranged to send 55-gallon drums containing less than one inch of waste to a TSD facility. However, when the drums reached the facility, plant personnel discovered that the drums could be emptied further by inverting them and pouring out the residue. In this situation, did the drums actually meet EPA's definition of "RCRA-empty" when they arrived at the TSD facility?
>
> No. The containers were not empty according to RCRA. EPA stipulates that one inch of waste is a secondary constraint and may remain in an empty container only if all wastes cannot be removed by normal means. A container must be emptied by pouring, pumping, and aspirating, etc.; only then, if it holds less than one inch, is it considered to be empty per §261.7. [RO 11559, 12342]

the contents of the container down to a depth of one inch, then they must be employed." [Emphasis in original.] [47 FR 36093]

Q: *Two inches of nonacute pesticide residue is rinsed out of a 55-gallon drum, and the rinsewater is disposed. What is the regulatory status of the rinsewater?*

A: Because the drum was not RCRA-empty (i.e., it retained more than one inch of residue), the residue was still subject to regulation. If the rinsewater contains a pesticide listed as a P- or U-waste when discarded, the rinsewater is a P- or U-listed hazardous waste. If the pesticide is not P- or U-listed, the rinsewater is only hazardous if it exhibits a characteristic. [RO 11374]

Q: *A generator sends a sample of a spent solvent (F004) waste to a laboratory for analysis. During analysis, laboratory equipment, such as a pipet, is used. After the analysis is performed, would the empty pipet (which will be discarded) be considered hazardous waste?*

A: At the lab, any wastes generated from the analysis would still be considered F004 wastes under the "contained-in" policy. However, some of the laboratory equipment used in the analysis, such as a pipet, may meet the definition of a container and therefore would not be regulated as hazardous waste if it met the §261.7 empty criteria. [RO 13375]

Q: *If a spent filter cartridge has been drained, is it considered to be an empty container?*

A: No. EPA does not consider spent filter cartridges to be containers, so the empty-container provisions would not apply to spent cartridges contaminated with hazardous waste. [OSWER Directive 9444.18(84), October 1984, available from http://nepis.epa.gov/EPA/html/Pubs/pubtitleOther.html]

9.5.1.3 The weight-limit alternatives

In some situations, it may be very hard to measure the depth of residue in the bottom of containers due to their shape (rounded or conical bottoms) or the position of the container opening. Nevertheless, EPA has stipulated that the residue depth must be measured from the deepest point of the bottom of the container. [August 18, 1982; 47 FR 36093] Due to these difficulties, EPA allows the generator to choose a weight limit as an alternative to the one-inch criterion.

Two different weight criteria are specified, depending on the size of the container. Containers with capacities of less than or equal to 119 gallons can contain no more than 3% by weight of their total capacity; no more than 0.3% by weight of the total capacity of the container can remain for containers greater than 119 gallons in size. The 3% limit for containers with capacities of less than or equal to 119 gallons is based on the fact that 3% by weight of the contents of a standard 55-gallon drum (one of the most common containers for hazardous waste) is equal to 1 inch of residue on the bottom of that drum.

Typically, when hazardous wastes are transported in containers larger than 119 gallons in size, they are transported in tank-like containers of at least 5,000-gallon capacities (e.g., portable tanks, cargo tanks, and tank cars). Consequently, EPA based the 0.3% by weight limit on the fact that 0.3% by weight of a 10,000-gallon tank (30 gallons) is roughly equivalent to 1 inch of residue. [August 18, 1982; 47 *FR* 36093]

9.5.1.3.1 *Empty aerosol cans may still be hazardous*

Rendering aerosol cans empty so they will exit RCRA hazardous waste management is problematic. As noted in Section 2.5.1, even an empty aerosol can may still be hazardous waste. This is the federal interpretation; some authorized states have taken a different approach to managing these wastes. For example, Colorado has added aerosol cans as a universal waste. Additionally, some states (e.g., California) define an empty aerosol can to be nonhazardous. Consequently, it is important to contact the appropriate state agency to find out if they have different regulations or policies that apply to these wastes.

Q *A furniture manufacturer generates spent aerosol paint and solvent cans from painting and cleaning activities. The cans are emptied using common industry practices and contain less than 3% by weight of the total can capacity. However, they may still be capable of detonation if subjected to a strong initiating source or if heated under confinement. Do the cans meet EPA's definition of empty and, as such, can they be disposed as nonhazardous waste?*

A Even though the aerosol cans are RCRA-empty (i.e., commonly employed emptying practices have been used and they contain less than 3% by weight hazardous wastes), they still may demonstrate the RCRA hazardous characteristic of reactivity per §261.23(a)(6). Accordingly, they may be hazardous wastes, and if they are, they must be managed as such. [RO 13027]

9.5.2 Containers that held acute hazardous waste

Containers or inner liners that held acute hazardous waste listed in §261.31 (the dioxin F-wastes) or 261.33(e) (the P-wastes) are RCRA-empty if:

1. They have been triple rinsed using a solvent capable of removing the hazardous waste,

2. They have been cleaned by another method that has been proven to achieve removal equivalent to triple rinsing, or

3. The inner liner that prevented contact of the hazardous waste with the container has been removed. [§261.7(b)(3)]

These requirements are more stringent than those for containers holding nonacute hazardous waste because acute hazardous wastes pose a greater hazard to human health and the environment.

Q *Epinephrine (P042 when discarded) is a drug that is injected using syringes. Since some of the drug typically remains in the syringe after injection, should residual drug in such syringes be managed as P042 until triple rinsed?*

A No. EPA considers drug residues remaining in used syringes to have been used for their intended purpose and are not discarded commercial chemical products. Therefore, the residues in this example would not be P042 but would be hazardous waste only if they exhibit a characteristic. [RO 13718] Additionally, EPA clarified in RO 14788 that this epinephrine interpretation extends to other P- and U-listed pharmaceuticals administered by syringe. The agency went on to clarify (as it also had in RO 13718), that even though not listed, syringe residues exhibiting a characteristic must be managed in accordance with the hazardous waste regulations.

9.5.2.1 Triple rinsing

The typical method of rendering empty a container that held an acute hazardous waste is to triple rinse it with a suitable solvent. Water is often used if the chemical is water soluble. Otherwise, another appropriate solvent must be employed.

The rinsate from containers being triple rinsed is a hazardous waste if it 1) exhibits a hazardous waste characteristic, or 2) comes out of a nonempty container. This last provision means that rinsates one, two, and three (none of which came from an empty container) are P-listed hazardous waste via the mixture rule. If a container were rinsed a fourth time, the rinsate would come from an empty container and would not be hazardous unless it exhibits a characteristic. [August 18, 1982; 47 *FR* 36095, RO 11004, 12299, 14827]

The regulations don't specify the volume of solvent that should be used in each of the three rinses. When the regs were proposed, however, EPA said that "a volume of diluent at least equal to ten percent of the container's capacity" should be used for each rinse. [December 18, 1978; 43 *FR* 58955] Based on this proposed language (which was never finalized), an industrial "rule of thumb" is that a rinse volume of about 10% of the container capacity should be used. It turns out that this "rule of thumb" is actually codified in the TSCA regs for decontaminating a PCB container. According to 40 *CFR* 761.79(c)(1), "[e]ach rinse shall use a volume of the flushing solvent equal to approximately 10 percent of the PCB Container capacity."

Note that triple rinsing is *not* considered treatment in the federal regs, and, consequently, a permit is not required. [November 25, 1980; 45 *FR* 78528]

Q *A drum recycler burns the hazardous waste residue left in drums that have been triple rinsed (i.e., the drums are RCRA-empty). Does this practice constitute hazardous waste incineration subject to RCRA regulation?*

A No. Since the residue is not regulated as a hazardous waste under §261.7(a)(1), burning it would not be considered hazardous waste treatment, and the activity would not be subject to RCRA regulation or require a permit. [RO 12535]

Q *Do the triple rinsing requirements apply to containers that hold residues (e.g., ash) from incinerating acute hazardous wastes?*

A Yes. The residue from incinerating an acute hazardous waste is itself an acute hazardous waste via the derived-from rule (see Section 5.2.2.3). The triple rinsing standards apply to these types of containers in order to render them empty under RCRA. However, the agency points out that §261.7(b)(3)(ii) allows the use of alternative cleaning methods in place of triple rinsing. [RO 11761]

9.5.2.2 Alternative cleaning methods

No formal EPA approval is required for alternative cleaning methods under §261.7(b)(3)(ii); that is, the alternative is self-implementing. However, EPA suggests that anyone using an alternative method be sure to document the equivalency of the method and its use and keep such records as part of the facility's operating log. Discussing use of the method with the appropriate agency responsible for administering hazardous waste regulations might also be helpful. [RO 11761, 11803, 14827]

Q *How should a generator empty paper bags that contain acute hazardous waste that is physically solid?*

A EPA has determined that repeated beating of the outside of the inverted paper bags after emptying them can be an alternative to triple rinsing, since paper bags cannot be triple rinsed. [RO 12407, 14827]

9.5.2.3 Regulatory status of warfarin residues in containers

Warfarin appears on both the P- and U-lists: warfarin (and its salts) at a concentration of >0.3% is listed as P001 in §261.33(e), while warfarin (and its salts) at a concentration of ≤0.3% is listed as U248 in §261.33(f). Some warfarin-based pharmaceuticals (brand names Coumadin and Jantoven) will meet the P001 listing description while others will meet the U248 listing if disposed. EPA clarified in RO 14827 that the regulatory status of the residue in an empty bottle that previously held a warfarin-based pharmaceutical is to be based on the warfarin concentration in the residue—not the warfarin concentration of the pills previously held in the container. Thus, if analysis indicates that the residue in a container that previously held a P001 pharmaceutical actually contains warfarin at concentrations ≤0.3%, the residues would be U248.

9.5.3 Containers that held compressed gas

As set forth in §261.7(b)(2), a container that held a hazardous compressed gas is empty when the pressure in the container approaches atmospheric pressure. Products on the P- and U-lists (i.e., commercial chemical products that are hazardous wastes when discarded) that are gases at standard temperature include P031—cyanogen, P033—cyanogen chloride, P063—hydrogen cyanide, P096—phosphine, U001—acetaldehyde, U043—vinyl chloride, U045—chloromethane, U122—formaldehyde, and U135—hydrogen sulfide. Even if the gaseous material in the container is an acute P-waste, the container is empty when the internal pressure approaches atmospheric pressure. [RO 12138]

Q: *Discarded vinyl chloride gas (U043) is transported to a TSD facility in a rail car. When it arrives at the TSD facility, the gas is removed from the rail car, and the rail car is opened to the atmosphere, rendering it empty in accordance with §261.7(b)(2). The empty rail car is then sealed and shipped back to the generator. During shipment, the empty rail car is heated by the sun, and the remaining gaseous residue inside the rail car subsequently develops pressure. Is the pressurized residue subject to regulation as hazardous waste?*

A: No. "It is not EPA's intent to regulate containers in such situations where an incidental rise in pressure occurs resulting from ambient environmental conditions. In this specific rail-car scenario, the 'empty' status of the rail car does not change due to the heating from the sun." [RO 14120]

9.5.3.1 Cylinders returned to gas suppliers and removal of residues

Due to unique ownership issues surrounding compressed gas containers or cylinders, EPA has provided guidance on who is responsible for managing compressed gas remaining in cylinders. Compressed gas cylinders, which are typically owned by the gas supplier, are usually returned to the supplier when they are empty or when the customer no longer needs them. The purpose of this shipment is for refilling and/or to return the supplier's property—not to discard the remaining contents of the cylinder. Therefore, the customer does not have any input on the final disposition of the residue in the cylinder, which occurs at the supplier's facility at the supplier's discretion.

Consequently, EPA has determined that returning a compressed gas cylinder to the supplier does not constitute generation of waste under RCRA. Neither the returned cylinder nor the residue it contains is a "solid waste," even if the cylinder is not empty (i.e., it is still pressurized). As such, the shipment of cylinders from the customer back to the supplier does not have to be manifested. However, DOT requirements will apply, and the cylinders may have to be transported as DOT hazardous materials. [August 18, 1982; 47 *FR* 36094, RO 14759, 14760, 14762]

If the gas supplier decides to discard the contents of returned cylinders, any liquid or physically solid waste removed from the cylinders is regulated if it is hazardous waste. If the supplier sends the cylinders offsite for treatment, storage, or disposal, they must be manifested if they are not empty and contain hazardous waste.

"However…the handling of gaseous residues removed from the cylinders and neutralization or scrubbing of gases prior to release are not subject to RCRA regulation [because they do not meet the definition of 'solid waste']. Any liquid or physically solid wastes derived from the treatment of hazardous compressed gas is still subject to RCRA regulations, if it is derived from listed waste or if the residue is hazardous under Part 261, Subpart C (characteristics)." [RO 12350]

Although the above quotation from EPA guidance might lead a facility to believe it is acceptable to vent unwanted gas from compressed gas cylinders, we have significant reservations about that practice. First, RO 11835 suggests that, if all of the materials generated by the venting (including the cylinders) will be discarded, then the practice might be considered treatment requiring a RCRA permit. Second, such venting may violate state air regulations or the conditions in a facility's air permit. These concerns argue against this practice and we strongly recommend not venting gases from compressed gas cylinders to the atmosphere.

9.5.4 Beneficial reuse of residues

Sometimes, residues in containers are not intended for disposal but rather for reuse. For example, if a container that has held a P- or U-listed commercial chemical product is to be reused to contain the same material that it previously held (and the initial residue is not discarded), that constitutes a beneficial use or reuse of that residue. In such situations, EPA notes that the residues never became wastes, and the accumulation, transportation, and treatment prior to use or reuse is exempt from hazardous waste management. [§261.33(c)]

Another beneficial reuse situation is when a material is mixed into a container holding the residue of a P- or U-listed product, where the mixing constitutes a beneficial use or reuse of that residue. [§261.33(c), November 25, 1980; 45 *FR* 78527] An example of this situation might be where a drum contains a U-listed solvent residue, and unused paint is mixed with the residue in the drum in order to make it "sprayable."

Residues generated from rinsing product containers are not solid or hazardous waste if they will be legitimately reused, such as for make-up water to prepare another batch of the same product. This occurs frequently in the agriculture industry. Finally, product container rinsewaters are not solid or hazardous waste if they are reclaimed to recover product. This results from the dash at the intersection of "Commercial chemical products" and "Reclamation" in Table 1 in §261.2(c).

9.5.5 Regulatory status of residues removed from empty containers

Under §261.7(a)(1), any hazardous waste remaining in either an empty container or an inner liner removed from an empty container is not regulated as hazardous waste (i.e., the RCRA hazardous waste regulations, including the land disposal restrictions, do not apply). [RO 12793] Since residues remaining in RCRA-empty containers are not regulated, it would seem reasonable that any management practices associated with residues removed from empty containers would not constitute hazardous waste management or treatment.

This was EPA's original (1980) position regarding residues or rinses removed from empty containers. However, even then, the agency was concerned that drum reconditioners and other facilities that clean out large numbers of RCRA-empty containers might accumulate, treat, and/or dispose significant quantities of these unregulated residues, which could pose a hazard to human health and the environment. Therefore, in the November 25, 1980 *Federal Register*, EPA sought input on a number of approaches for regulating residues *removed* from RCRA-empty containers. [45 *FR* 78526] After reviewing comments submitted, the agency decided not to amend the §261.7 provisions and not to regulate such removed residues. [August 18, 1982; 47 *FR* 36096]

Throughout the 1980s and early 1990s, EPA reiterated in guidance that, when residues are removed from RCRA-empty containers, the resulting residues, rinsates, or washwaters are also exempt from RCRA under §261.7(a)(1). Accordingly, generators would not have had to determine whether they exhibit a hazardous waste characteristic. However, residues, rinsates, or washwaters that result from cleaning out containers that do not meet the §261.7 definition of empty (or result from activities that render the container RCRA-empty) are *not* exempt but are fully subject to RCRA Subtitle C if they either exhibit a characteristic or are a mixture containing a listed hazardous waste. [RO 11374, 11447, 11504]

In 2004, however, EPA reverted back to its earlier concerns about residues removed from RCRA-empty containers. In RO 14708, the agency noted that "when residue is removed from an 'empty' container, the residue is subject to full regulation under Subtitle C if the removal or subsequent management of the residue generates a new hazardous waste that exhibits any of the characteristics identified in Part 261, Subpart C...." (See also October 4, 2005; 70 *FR* 57779.)

This April 2004 letter makes EPA headquarters consistent with the conservative positions previously taken by some of the regions and states. For example, EPA Region I has for some time required characterization of residues/rinsates from RCRA-empty containers, as discussed in Case Study 9-2.

RO 14708 adds that "if the rinsing agent includes a solvent (or other chemical) that would be a listed hazardous waste when discarded, then the rinsate from an 'empty' container would be considered a listed hazardous waste." This would be due to the nature of the rinsing agent (e.g., toluene used to clean out a RCRA-empty drum would be F005 listed spent solvent), not due to the residues remaining in the empty container. (See also October 4, 2005; 70 *FR* 57779.)

9.5.5.1 Discharging rinsates into wastewater treatment systems

EPA believes that small amounts of chemicals that, when discarded, are hazardous wastes listed in §261.33 and small amounts of F- and K-wastes are often "unavoidably lost" during normal material

---

Case Study 9-2: EPA Region I Position on Tank Truck Washwater

A 5,000-gallon tank truck was emptied using commonly used procedures associated with off-loading trucks but still contained 10 gallons of hazardous waste. As such, it was considered empty because it held less than 0.3% by weight of the total tank capacity. The regulated party, however, wanted to know if the truck could be sent to a commercial truck wash facility (whose discharge is regulated by the Clean Water Act) to completely wash out the remaining residue.

In its response, Region I indicated that the truck is not prohibited under RCRA from being washed out at a commercial truck wash facility. (The region noted, however, that DOT generally recommends steam cleaning as the minimal requirement in order for a tanker to qualify as "empty" under the DOT regulations.) The region then stated "if the rinsate exhibits the characteristic of a hazardous waste, the entire volume is subject to the applicable provisions of RCRA." Even if the tank originally contained listed hazardous waste, the region clarified that the rinsate would only be hazardous based on characteristics.

If a trucking company employee performs the cleaning, the trucking company becomes the generator of any rinsate that exhibits a characteristic. If a commercial truck wash employee performs the cleaning, that facility becomes the hazardous waste generator.

Furthermore, the region stated "if the allowable concentration limits for hazardous waste/constituents found in the 'regulated' discharge [under the Clean Water Act] can be achieved in a manner not constituting improper dilution, the discharge would not be regulated under RCRA." (Tank cleaning/rinsing methods that are not beneficial and do not contribute to the cleaning process would be considered improper dilution.) [Letter from Gerald Levy, EPA Region I to Daniel Gillingham, Franklin Environmental Services, Inc., 1991]

handling operations; such activities include emptying and rinsing containers. Facilities frequently dispose such chemicals and F- and K-wastes rinsed out of containers by draining or washing them into the facilities' wastewater treatment systems. According to the agency, "[t]his typically is a reasonable and practical means of disposing of these lost materials. Segregating and separately managing them often would be exceedingly expensive and may not be necessary because the small quantities can be assimilated and treated in the wastewater treatment system." [November 17, 1981; 46 FR 56586] Due to the small quantities of wastes involved, the agency does not believe that this practice poses a substantial threat to human health or the environment.

Based on this reasoning, EPA exempts mixtures of wastewater and such de minimis losses of P- and U-listed chemical rinsates and F- and K-waste container rinsates from RCRA regulation under §261.3(a)(2)(iv)(D) if they are discharged to wastewater treatment systems regulated under the Clean Water Act. (See Section 4.16.3.) Included in this exemption is rinsate from empty containers or from containers that are rendered empty by that rinsing. This exemption lets facilities discharge rinses from containers holding §261.31 F-wastes, §261.32 K-wastes, and §261.33 chemicals into their wastewater treatment system without worrying about hazardous waste management. The wording of the exemption indicates that, as long as the rinsate flows into the wastewater treatment system, it doesn't matter if the container was RCRA-empty or not before it was rinsed (although the container must be RCRA-empty after rinsing).

Note that EPA provided some disconcerting guidance about these de minimis losses from container rinsing. On January 31, 1991, the agency noted that only the "third rinse in triple rinsing" would meet the definition of rinsate from containers that are rendered empty by that rinsing. [56 FR 3869]

9.5.5.2 Miscellaneous empty-container examples

Three additional examples illustrating the complex considerations of managing and emptying a hazardous waste container are given in Case Study 9-3.

9.5.6 RCRA and DOT definitions of "empty" are different

Even though an empty container may be exempt from RCRA requirements, DOT shipping requirements may still apply because the RCRA and DOT definitions of an "empty" container are not the same. Under the 49 CFR 173.29 DOT standards, any packaging (e.g., container) that contains residues of hazardous materials must be handled and transported as if it were full unless it meets *all* of the following criteria:

- The packaging 1) is unused, 2) is cleaned of residue and purged of vapors so that no potential hazard remains, 3) is refilled with a nonhazardous material so that any remaining hazardous material residue no longer poses a hazard, *or* 4) contains only the residue of an ORM-D material (other regulated material, such as consumer commodities) or a Division 2.2 nonflammable gas (other than anhydrous ammonia) with no subsidiary hazard at a pressure of <29.0 psig at 68°F.

- The material that remains after cleaning does not meet the definition of a RCRA hazardous waste, DOT hazardous substance, or marine pollutant.

- All hazardous material shipping name and identification number markings, hazard warning labels or placards, and any other markings indicating that the material is hazardous must be removed, obliterated, or securely covered.

9.5.6.1 Manifesting nonempty shipments back to a generator

Due to the sticky nature or other properties of some hazardous waste, it is often very difficult, if not impossible, for a TSD facility to render a generator's tank truck or rail car RCRA-empty. Therefore, when the TSD facility sends the container back to the generator to pick up more of the same

Case Study 9-3: Container Management Examples

I. Used oxygen breathing apparatus (OBA) canisters and/or component parts are likely to exhibit at least one characteristic of a hazardous waste (e.g., D001 ignitability). How can these canisters (which meet the definition of a container) be managed under RCRA?

If the canisters are disposed, they will be a hazardous waste not only due to the properties of the contents, but they also may be D003 reactive wastes due to the properties of the canister itself (as is the case with aerosol cans). If the canisters are punctured and rinsed out, the puncturing process may require a treatment permit. Any rinsate will have to be evaluated to determine if it exhibits a characteristic, and if so, managed as hazardous waste.

Another option is to empty the canisters and recycle them under the scrap metal exemption. In this case, the process of emptying the canisters (e.g., by puncturing and rinsing) is part of the scrap metal recycling process and would not be subject to permitting as a treatment process. A determination of ignitability or any other characteristic would not be relevant for the canisters if recycled as scrap metal. Any liquids or contained gases removed from the canisters would have to be evaluated to determine if they exhibit a characteristic prior to disposal. [RO 11835]

II. Tank trucks containing "heels" or residues of hazardous waste sometimes arrive at truck cleaning facilities with high levels of flammable vapors. (The tank trucks are not RCRA-empty.) If emulsifiers are added to the tank truck during cleaning to reduce the danger of the vapors, would this be considered to be treatment requiring a permit?

"EPA would not consider the use of emulsifiers in this way to constitute hazardous waste treatment. This use of emulsifiers is not regulated treatment because it is simply part of the overall rinsing process, which EPA has previously stated does not amount to treatment when the intent is to remove the waste (see 45 *FR* 78528, November 25, 1980). Failing to add such an emulsifier agent may jeopardize the safety of the persons removing the residues, making it a necessary step in certain cleaning operations." [RO 14125]

III. If a facility has partially full cans of unwanted paint, the best option is to find someone who can use the paint for its intended purpose. In that case, the paint is not a waste. If the paint will not be used for its intended purpose, the point of generation of a solid waste is when the person in control of the paint decides it can't be used anymore and will be discarded. When that happens, the generator must determine if the paint is hazardous, typically by exhibiting characteristics.

To render cans that hold characteristically hazardous paint RCRA-empty per §261.7(b)(1), all liquid paint that can be removed needs to be removed by pouring it out of the cans. Leaving ¾ of an inch of liquid paint in the bottom of a paint can with the lid off so it can dry out is not an allowable practice, because the container does not meet the definition of RCRA-empty since all of the liquid paint that could be removed hasn't been. Non-RCRA-empty containers of hazardous waste must be closed at all times except when adding or removing waste. If all of the liquid has been poured out (and there is ≤1 inch of nonpourable residue remaining), the paint cans are RCRA-empty and can be managed as nonhazardous (e.g., crushed and recycled as scrap metal or disposed in a nonhazardous waste landfill). However, in RO 14708, EPA stated that any residues that drain out of or otherwise exit a RCRA-empty container are considered a new waste and the generator must determine whether they exhibit a hazardous waste characteristic. If they do, the residues are subject to the hazardous waste regulations from that point forward.

waste (i.e., the container is dedicated to this service), the shipment consists of a nonempty container of hazardous waste that should be manifested.

In a March 4, 2005 final rule that changed the manifest system, EPA modified the definition of "designated facility" in §260.10 to include generators receiving non-RCRA-empty containers returned under a new manifest from a TSD facility. [70 *FR* 10776] Generators receiving such returned shipments must sign in Block 20 of the manifest, and they then have either 90 or 180 days (depending on their generator status) to send returned waste to a designated facility under yet another manifest. [§262.34(m)]

9.5.7 Waste counting issues

Per §261.5(c)(1), residues in RCRA-empty containers are not counted when making monthly generator status determinations.

9.5.8 CERCLA issues

Containers that are rendered RCRA-empty can be disposed in a nonhazardous waste landfill. However, keep in mind that liquid residues (e.g., <1 inch of liquid remaining in a drum that previously held a nonacute hazardous waste) that may escape into and contaminate a nonhazardous waste landfill could potentially expose the generator to future cleanup liability, such as under CERCLA. Rendering containers as empty as possible before disposal is a good environmental and liability minimization practice. Other options include sending empty drums to a drum reconditioner or metal recycler.

9.5.9 EPCRA reporting

Material contained in RCRA-empty containers that is sent offsite for disposal must be reported as an offsite transfer for purposes of disposal on Form R of the toxic release inventory (TRI). This situation occurs because material remaining in empty containers is still considered a toxic chemical under the Emergency Planning and Community Right-to-Know Act (EPCRA) Section 313. The status of a toxic chemical as nonhazardous waste under RCRA has no impact on the applicability of EPCRA regulations. Therefore, the quantity of such material should be reported in Section 8.1 of Form R, unless the facility is exempt from reporting. [EPCRA Hotline question and answer dated February 1996]

9.6 Container inspection requirements

Sections 264/265.174 require owners/operators of TSD facilities, as well as generators, to conduct weekly inspections of hazardous waste container storage areas:

> "At least weekly, the owner or operator must inspect areas where containers are stored...[and] must look for leaking containers and for deterioration of containers and the containment system caused by corrosion or other factors."

Per §§264/265.15(d), permitted and interim status facility owners/operators are required to document container area inspections. Generators conducting 90/180/270-day accumulation operations are not subject to those documentation regulations and are not required by the federal RCRA regs to document their weekly container area inspections. However, without documentation, generators cannot prove to state inspectors that they have performed the weekly inspections. Therefore, we strongly recommend that generators document their weekly inspections of hazardous waste container accumulation areas. McCoy and Associates modified an EPA Region IX container area checklist (covering both satellite and 90/180/270-day accumulation areas), and this checklist is available for generators' use at http://www.mccoyseminars.com/forms.cfm.

Weekly inspections should document the following [§§264/265.15(d)]:

- The date and time of the inspection,
- The name of the inspector,
- A notation of any observations made (e.g., leaks, rusting), and
- The date and nature of any repairs or other remedial actions initiated.

The regulatory language requires the *name*—not the initials—of the inspector to be documented.

Inspections should be recorded in the plant operating log and should be kept for a minimum of three years from the date of inspection.

Inspection records are virtually certain to be scrutinized in a regulatory inspection. Facility personnel can become bored with making the inspections and/or filling out inspection logs and sometimes forget to make entries in the log. Facilities need to give special attention to ensuring that inspections are conducted and documented.

Examples of items that should be considered for inclusion in a typical container inspection checklist or record include [OSWER Directive 9523.00-10, October 1983, available from http://nepis.epa.gov/EPA/html/Pubs/pubtitleOther.html by downloading the report numbered SW968]:

- Appropriate container labeling;
- Tightness of lids, nuts, rings, drain valves, etc.;
- Leaks in containers, flanges, plug seals;
- High liquid level/overflow;
- Structurally intact with no puncturing of container walls or inner liners;
- Evidence of corrosion, excessive rusting, or damage;
- Any wheels properly braked/chocked;
- Overfilling controls operative;
- Containers located so as to minimize potential for external puncture;
- Storage area pad free of chips, cracks, etc.; and
- Any sump pump operative.

Do the regs require container area inspections once a calendar week or every seven days? This will typically be a determination based on state policy. However, OSWER Directive 9523.00-10 notes for permitted facilities that "there must not be more than 6 calendar days between inspection dates."

9.6.1 Satellite containers do not require weekly inspection

The satellite accumulation container requirements in §262.34(c) reference three specific sections in Part 265, Subpart I, but none of the three is the weekly inspection requirement. EPA has confirmed in guidance that satellite containers do not require weekly inspection. [RO 14418] "However, the SAA regulations do require that waste containers in an SAA must be under the control of the operator of the process generating the waste, in good condition (§265.171), compatible with its contents (§265.172), and closed except when adding or removing waste (§265.173), which should achieve the goal of inspections: containers that are free of leaks and deterioration." [RO 14703] Note that states may require weekly inspection of satellite accumulation containers.

9.7 Lab packs

Lab packs are a particularly interesting type of containerized hazardous waste. Because lab pack drums may contain dozens of individual wastes of varying waste codes, they present unique challenges for EPA and the regulated community.

The federal regulations only define "lab packs" as "small containers of hazardous waste in overpacked drums." [§§264/265.316] In the preamble to the November 18, 1992 liquids-in-landfills ban, EPA offered the following description:

> "Lab packs are small containers of liquids (typically of one gallon or less), most commonly used for laboratory wastes, that are placed in a drum and surrounded by sufficient sorbent material to sorb the liquids should the containers fail." [57 FR 54457]

With this description in mind, we turn to a number of issues that this container-within-a-container concept spawns.

Does the liquids-in-landfills ban apply to lab packs?

No. EPA is satisfied that free liquids in small laboratory containers, when properly overpacked and surrounded by absorbent material, will not be released into landfills. Per the terms of §§264/265.316(b), however, the absorbent must be nonbiodegradable. In addition, reactive wastes, other than cyanide- and sulfur-bearing compounds, must be rendered nonreactive prior to their placement in a lab pack. [§§264/265.316(e)]

Q *Is a permit required for a facility to repackage small laboratory containers (that hold characteristic and/or listed hazardous waste) into large containers with absorbents (i.e., lab packs)?*

A No. The process of adding absorbents to lab packs prior to shipment offsite qualifies for the absorbent addition exemption from permitting in §270.1(c)(2)(vii)—described in Section 7.5.3. This is true even if a service company does the lab packing. [RO 13117]

Q *When lab packs are shipped to TSD facilities, does a generator have to list all of the hazardous waste codes for the contained wastes on manifests? How about on biennial reports?*

A EPA believes that six waste code fields in Block 13 would be adequate to describe typical wastes shipped on a manifest. If a waste carries more than six waste codes, the additional codes are not required on the manifest. [70 *FR* 10788] A generator is allowed to list additional waste codes in Block 14 of the manifest, but their inclusion in Block 14 is not mandatory. See Section 6.6.3.9 for additional guidance.

So that the agency knows what types of waste a given facility generates, all applicable waste codes (including the contents of lab packs) must be listed on generators' biennial reports. [RO 11196]

Q *To take advantage of the §268.42(c) LDR provision for incinerating lab packs, many generators and service companies use fiber drums for lab packs. Must such drums be overpacked into steel drums per §§264/265.316(b)?*

A No. Sections 264/265.316(f) allow the packaging of lab containers and nonbiodegradable sorbent in fiber drums, when the resulting lab pack will be disposed via incineration. In effect, the laboratory container is the inside container and the fiber drum is the overpack. However, the fiber drum must meet DOT specifications listed at 49 *CFR* 173.12(b), and sufficient nonbiodegradable absorbent must be placed in the drum to absorb any liquids from failed inner containers. [RO 13397, 13435, 13522] Additional discussion on how lab packs are managed under the LDR program is contained in Section 13.13.1.

Q *When lab packs are destined for incineration, can fiber or wooden boxes be substituted for fiber drums under the terms of §§264/265.316(f), without the need of a steel overpack?*

A No. Although DOT rules at 49 *CFR* 173.12(b) list several types of drums that would qualify for the overpack exemption, such containers *must be drums*. Hazardous laboratory wastes can be packed into boxes, crates, etc., along with absorbent, but such containers would have to be overpacked into drums prior to incineration. [RO 13429]

9.8 Segregation of incompatible wastes in containers

Three regulatory requirements address incompatibility issues for hazardous wastes managed in containers:

1. Incompatible wastes must not be placed in the same container. This §§264/265.177(a) requirement notes that incompatible wastes may not be stored together in the same container unless such storage results in no deleterious reaction as described in §§264/265.17(b).

2. Hazardous waste must not be placed in an unwashed container that previously held an incompatible waste or material. Guidance for this §§264/265.177(b) requirement notes that there are many situations in which it is acceptable to place waste in unwashed containers. Therefore, it would be unnecessary and burdensome to require that all containers be washed between each use. EPA expects that owners and operators will institute some means of identifying the previous contents of empty containers, such as labels, records, segregated storage, or tests if empty containers are not routinely

washed. Further, it is incumbent upon owners and operators to train their employees sufficiently to recognize incompatible wastes if they are going to place hazardous wastes in unwashed containers. [EPA/530/R-80/037, available at http://nepis.epa.gov/EPA/html/Pubs/pubtitleOSWER.html by downloading the report numbered 530R80037]

3. A container holding a hazardous waste that is incompatible with any other waste or materials stored nearby in other containers must be separated from the other waste or materials by means of a dike, berm, wall, or other device. Since leakage of containers could cause incompatible waste to commingle in storage areas, §§264/265.177(c) require that containers holding incompatible wastes be separated from one another by physical barriers (e.g., dikes, berms, walls, or other structures). EPA has noted in guidance, however, that if separated sufficiently, leaking wastes will not commingle. Containers holding incompatible wastes may be separated from one another by a sufficient distance to prevent commingling in the event of leakage. This relatively inexpensive precautionary measure will help prevent one source of potentially dangerous reactions, which can cause fires, explosions, and gas emissions as a result of mixing of incompatible wastes from leaking containers. [EPA/530/R-80/037]

Tank Issues

Providing the details for storing and treating hazardous waste in tanks

Because large volumes of hazardous waste are stored and treated in tanks, mismanagement can result in serious incidents (large spills, fires, vapor releases, etc.). Another potential concern is that the design and/or location of many tanks (e.g., underground) makes the detection of leaks difficult.

Based on the above risks, EPA has put in place stringent design and management standards for hazardous waste tanks. In general, these standards ensure that such tanks are 1) properly designed, installed, and operated to prevent hazardous waste releases; 2) equipped with secondary containment systems that will prevent leaks from reaching soil, ground water, and/or surface water; 3) inspected so that system problems and possible leaks are detected promptly; 4) subject to response procedures to contain releases from tanks and prevent or mitigate similar future releases; and 5) emptied and decontaminated upon closure.

RCRA tank standards are codified at Part 264, Subpart J, which is applicable to RCRA-permitted facility tanks, and Part 265, Subpart J, applicable to interim status facilities and 90/180/270-day accumulation tanks. (Generators operating 90/180/270-day accumulation tanks get to Part 265, Subpart J by reference from §262.34.) Unlike many other parts of RCRA where permitted and interim status facility standards differ markedly in stringency, the Parts 264 and 265 tank standards are very similar. Explaining this similarity, EPA notes its belief that there is no significant difference with respect to the risks posed by 90-day accumulation tanks, permitted tanks, etc. [July 14, 1986; 51 *FR* 25444] Thus, although somewhat less-stringent standards are required for tanks used by small quantity generators, all other hazardous waste tanks face roughly the same level of regulation.

In the sections that follow, we first discuss how EPA defines the term "tank," and the associated term "tank system" (Section 10.1). This section will also look at various tank-like equipment with an eye on whether EPA regulates such equipment as a tank, describe the few exemptions the agency has provided from the tank requirements, and review some of the basic management requirements for hazardous waste tanks. Focusing on an important, specific instance of tank-like equipment, Section 10.2 discusses whether sumps are tanks subject to tank standards. Section 10.3 details secondary containment and release detection/response requirements, while Section 10.4 addresses the nebulous issue of RCRA-empty tanks. Finally, Section 10.5 reviews inspection requirements. We previously discussed 90/180/270-day accumulation in tanks in Section 6.3.4.1 and the possibility of treating hazardous wastes in 90/180/270-day tanks without a permit in Section 7.3.

10.1 What is a hazardous waste tank?

According to Subpart J of Parts 264 and 265, RCRA tank standards apply to "owners and operators of facilities that use tank systems for storing or treating hazardous waste...." [§§264/265.190] In turn, the agency defines the terms "tank" and "tank system" in §260.10 as follows:

> "*Tank* means a stationary device, designed to contain an accumulation of hazardous waste which is constructed primarily of non-earthen materials (e.g., wood, concrete, steel, plastic) which provide structural support."

> "*Tank system* means a hazardous waste storage or treatment tank and its associated ancillary equipment and containment system."

The primary difference between a hazardous waste tank and container is that a tank is stationary (usually bolted down), while a container is portable. See Section 7.2.2.7 for guidance on the difference between containers and portable tanks.

The definition of tank requires that a tank must "contain an accumulation of hazardous waste" and "provide structural support." As such, questions arise as to whether an in-ground basin or sump would be a tank or not. EPA uses the "parking lot" test to distinguish between in-ground tanks and surface impoundments. We discuss this in more detail in Section 10.1.3.2. The second definition above emphasizes that equipment ancillary to a hazardous waste tank as well as its secondary containment structure are subject to Subpart J standards—not just the tank itself. [September 2, 1988; 53 *FR* 34080]

People have been asking us lately if lift stations meet the definition of tanks. We have searched for but have been unable to find any guidance from EPA on this question. We sent the agency a query on this topic, and EPA's email response was "There is no available federal clarification from the Office of Solid Waste regarding whether lift stations meet the definition of a tank." So, talk with your state.

In an effort to answer the question of what is a hazardous waste tank, we have broken the universe of stationary units that manage wastes into three groups: 1) units that meet the "tank" definition and are hazardous waste tanks subject to Subpart J tank standards, 2) units that meet the "tank" definition but are exempt from Subpart J, and 3) units that don't meet the "tank" definition and so are exempt from Subpart J. These three types of equipment are discussed in the next three subsections.

10.1.1 Hazardous waste tanks and tank systems

The following examples and Case Study 10-1 illustrate EPA's designation of hazardous waste tanks that are subject to Subpart J tank standards.

Q *Facilities often seek an exemption under RCRA to avoid compliance with tank regulations. For example (and as noted in Section 7.2), tanks used to store and/or treat hazardous wastewater may qualify for the wastewater treatment unit (WWTU) exemption and be exempt from Subpart J standards. If such storage and/or treatment occur prior to shipment of the wastewater to an offsite treatment facility, would the WWTU exemption apply?*

A No. If a tank storing/treating hazardous wastewater is not dedicated for use with an onsite wastewater treatment system that discharges under the CWA, the unit would be a hazardous waste tank subject to Subpart J tank standards in Part 265 if the facility is a hazardous waste generator only or an interim status TSD facility, Part 264 otherwise. [53 *FR* 34080, RO 11066, 11519, 11551, 13112, 13126, 13203, 13226, 13318, 14262]

Q *Are tanks used to blend hazardous waste fuel regarded as recycling units exempt from RCRA permitting and substantive standards, or are they hazardous waste tanks subject to Subpart J standards?*

A Such tanks are subject to permitting requirements unless they are 90/180/270-day accumulation units. Either way, they are hazardous

Case Study 10-1: Storage Prior to Recycling—Two Scenarios

According to §261.6(c)(1), recycling hazardous waste is exempt from most RCRA requirements. Sometimes, however, recyclers store wastes for some period prior to actually processing the material. Two scenarios involving a recycler who reclaims solvents via distillation illustrate when holding wastes is storage (subject to regulation under Subpart J tank standards) and when it isn't.

In the first scenario, bulk shipments and drums of spent solvent received at the recycling facility are pumped into a feed tank that is hard-piped to the distillation unit. When the distillation column is idle (i.e., at night), solvents may remain in the feed tank. In this case, the feed tank is used to store spent solvent (albeit for short periods of time) and is considered a hazardous waste tank subject to Subpart J regulations. Our discussions in Section 7.4.1.2 note that eligibility for the recycling exemption is independent of whether the recycler conducts RCRA-regulated storage of wastes prior to processing.

The second scenario is similar to the first, except that any time there is spent solvent in the feed tank, a pump on the bottom of the tank operates and feeds solvent to the distillation unit; the tank never contains solvent when the distillation unit is not in operation. In the second case, the feed tank is not used for storage and is therefore not subject to regulation as a hazardous waste tank. Rather, it is used as a means of conveyance and is viewed as part of the exempt recycling unit. [RO 11365, 12895]

---

waste tanks subject to Subpart J requirements. [April 13, 1987; 52 *FR* 11820, RO 11238, 11411, 11497, 11881, 13512]

Q: *Are tanks used for the treatment of D002/D008 battery waste considered exempt elementary neutralization units or hazardous waste tanks subject to Subpart J provisions?*

A: Because of the D008 code, such tanks cannot be elementary neutralization units. Instead, they are hazardous waste tanks that will have to either be 90/180/270-day accumulation units or subject to permitting. In either event, they will be Subpart J units. [RO 11763] (This answer could change if the tanks meet the criteria for the WWTU exemption.)

Q: *Would air strippers used to treat contaminated ground water be regulated under RCRA?*

A: Possibly. If the ground water contains hazardous waste, it is a hazardous waste via the contained-in policy (see Section 5.3.1). Since such air strippers likely fit the §260.10 definition of "tank," the strippers would be subject to Subpart J regulation unless the WWTU exemption is applicable. [RO 12783]

Q: *When a large quantity generator mixes D001 ignitable mineral spirits and used oil in the same tank, which regulations take precedence: Part 265, Subpart J standards for 90-day accumulation tanks, or Part 279 used oil tank standards?*

A: Neither takes precedence—they both apply. Regardless of whether the resulting mixture is hazardous waste or used oil, the tank would have to comply with Subpart J because it must be a 90-day accumulation unit to allow hazardous waste treatment without a permit, and the tank would also have to comply with the Part 279 used oil tank requirements. Most of the tank design and management requirements would come from Part 265, Subpart J; the only additional requirement imposed by Part 279 relates to labeling the tank with the words "Used Oil." [RO 11708]

Q: *Hazardous waste is stored and/or treated in tanks that are located within a RCRA containment building (the containment building is subject to Part 264, Subpart DD standards). Are the tanks* considered ancillary equipment to the containment building?

A No. The tanks are distinct units subject to separate regulation; their location within a containment building does not change their RCRA status. Thus, they remain subject to Subpart J. Note, however, that depending on its design, the containment building may meet the Subpart J secondary containment requirements for the tanks. [July 14, 1986; 51 *FR* 25452, April 4, 2006; 71 *FR* 16880, RO 13633, 14119, 14463, 14632]

Q Is a hazardous waste transporter who stores manifested hazardous waste in tanks for less than 10 days eligible for the transfer facility exemption codified in §263.12?

A No. The 10-day storage exemption for transporters applies only to waste held in containers. Transporters who hold manifested hazardous wastes in tanks at some offsite location must have a RCRA permit or be operating under interim status. [RO 13786]

Q Are only tanks holding hazardous waste liquids subject to Subpart J?

A No. Tank systems storing or treating liquid hazardous waste, nonliquid waste (e.g., solid hazardous waste, residue, dried sludge), and even gaseous hazardous waste must comply with Subpart J requirements (unless they qualify for a variance). [EPA/530/SW-86/044, December 1986, available from http://nepis.epa.gov/EPA/html/Pubs/pubtitleOSWER.html as report number 530SW86044]

10.1.2 Exempt/unregulated tanks and tank systems

This section deals with equipment that meets the definition of tank or tank system, but escapes regulation under Subpart J for some reason. In some cases, this is because the equipment in question is regulated under other EPA programs. In other cases, the equipment qualifies for an exemption from environmental regulations in general.

10.1.2.1 Permit-exempt treatment can exempt tanks as well

Chapter 7 of this book describes a number of important exemptions from permitting and other RCRA regulatory requirements for facilities conducting hazardous waste treatment. While that chapter touched on how several of these exemptions affect the regulation of tanks under Subpart J, a more complete review of this relationship is presented below. Table 10-1 lists the eight exemptions covered in Chapter 7 and whether the scope of each exemption extends to Subpart J tank standards. Thus, for example, the WWTU exemption exempts such tanks from Subpart J regulations, while the 90-day accumulation unit exemption for large quantity generators does not.

A word of warning: while most of the exemptions discussed below exempt tanks from Subpart J requirements, the reach of these exemptions should be checked carefully. Only those activities conducted in tanks that are part of the permit-exempt treatment operation receive the Subpart J exemption. For example, the storage of hazardous waste in tanks prior to recycling of that material *would* be subject to Subpart J requirements, even though most recycling activities conducted in tanks would not. [RO 11131, 11404, 11411, 11765] On the other end of the spectrum, tanks used to store or treat hazardous sludges generated by a RCRA-exempt WWTU may also be exempt from Subpart J requirements. [RO 11118, 12527, 13003, 14465]

10.1.2.2 Waste exclusions/exemptions can apply to tanks

Besides tank standard exemptions based on permit-exempt hazardous waste treatment, facilities may also escape Subpart J requirements based on waste-based exclusions and exemptions that were discussed in Chapter 4. The list below summarizes some of these exclusions and discusses whether and how they extend to involved tanks.

- The domestic sewage exclusion applies to hazardous waste discharged to a domestic sewer line leading to a publicly owned treatment

Table 10-1: Treatment Exemptions Applicability to Tanks

Exemption	Eligible tanks exempt from Subpart J?	Comments
Wastewater treatment units (Section 7.2)	Yes per §§264.1(g)(6)/ 265.1(c)(10)	"[I]f a wastewater treatment…unit is not subject to the RCRA Subtitle C hazardous waste management standards, the ancillary equipment connected to the exempted unit is likewise not subject to the Subtitle C standards." [September 2, 1988; 53 *FR* 34080]
Treatment in 90/180/270-day accumulation units (Section 7.3)		
—Large quantity generators (90-day units)	No	
—Small quantity generators (180/270-day units)	Partially	Small quantity generators are exempt from all Subpart J requirements except for §265.201.
Recycling (Section 7.4)	Mostly per §261.6(c)(1)	Most tanks used for hazardous waste recycling are exempt from Subpart J standards. A notable exception includes tanks used to manage hazardous wastes that are subsequently burned in boilers and industrial furnaces. [April 13, 1987; 52 *FR* 11820, RO 11238, 11411, 11497, 11881, 13512]
Elementary neutralization units (Section 7.5.1)	Yes per §§264.1(g)(6)/ 265.1(c)(10)	"[I]f an…elementary neutralization unit is not subject to the RCRA Subtitle C hazardous waste management standards, the ancillary equipment connected to the exempted unit is likewise not subject to the Subtitle C standards." [September 2, 1988; 53 *FR* 34080]
Totally enclosed treatment facilities (Section 7.5.2)	Yes per §§264.1(g)(5)/ 265.1(c)(9)	
Adding adsorbents to wastes (Section 7.5.3)	NA	The adsorbent addition exemption does not apply to waste handling in tanks.
Immediate responses (Section 7.5.4)	Yes per §§264.1(g)(8)/ 265.1(c)(11)	Temporary tanks and sumps are exempt from Subpart J. See additional discussion in Sections 10.1.2.4 and 10.2.1.
Small quantity burner (Section 7.5.5)	No	The exemption applies to the onsite boiler or industrial furnace (BIF) burning hazardous waste—it does not apply to any tanks holding hazardous waste upstream of the BIF.

Source: McCoy and Associates, Inc.

works (POTW—Section 4.1). Where the hazardous waste is handled in tanks prior to discharge to the sewer line, those tanks are probably not subject to Subpart J requirements. However, such exemption from Subpart J is based on the tanks meeting the WWTU exemption; the domestic sewage exclusion does not apply until the hazardous waste enters the sewer line.

- The National Pollutant Discharge Elimination System (NPDES) discharge exclusion (Section 4.2) applies only to actual point-source discharge locations. Similar to the domestic sewage exclusion, however, any tank-based storage, treatment, and/or other activities that occur prior to the discharge point would likely be exempt from RCRA tank requirements due to the WWTU exemption.

- Radioactive mixed wastes are wastes that are both hazardous and radioactive. Because of the limited scope of the exclusion for radioactive materials (Section 4.4), tank-based storage or treatment of mixed waste is normally subject to Subpart J tank standards. Tanks storing low-level mixed waste (LLMW) under the Part 266, Subpart N storage and treatment exemption promulgated on May 16, 2001, however, do not have to comply with Subpart J. [66 FR 27218]

- The oil and natural gas exploration and production exemption (Section 4.14) applies to "drilling fluids, produced waters, and other wastes associated with the exploration, development, or production of crude oil, natural gas, or geothermal energy." [§261.4(b)(5)] Although such materials are classified as solid waste, they are excluded from the definition of hazardous waste. Thus, tanks holding these materials are not subject to Subpart J standards.

- The de minimis wastewater exemptions (Section 4.16) prevent the mixture rule from applying listed hazardous waste codes to a facility's wastewater stream that discharges under the CWA. Any tanks in the CWA system would be exempt from Subpart J tank standards due to the WWTU exemption.

- The active manufacturing process unit exemption (Section 4.17) applies to hazardous wastes generated in a 1) product or raw material storage tank, 2) product or raw material transport vehicle or vessel, 3) product or raw material pipeline, and 4) manufacturing process unit or an associated nonwaste-treatment-manufacturing unit. Under this exclusion, hazardous waste remaining in a manufacturing process unit is not subject to RCRA regulation for up to 90 days after that unit ceases operation; thus, if the owner/operator is able to clean out the process unit within 90 days after production or product storage stops, such a tank would not become subject to Subpart J. [RO 13790] After the 90-day clock runs out, however, the inactive unit does become subject to Subpart J as a 90/180/270-day accumulation unit if it meets the definition of "tank." Case Study 10-2 discusses manufacturing vs. RCRA tanks.

Residues generated from rinsing product tanks are not solid or hazardous waste if they will be legitimately reused, such as for make-up water to prepare another batch of the same product. Conversely, residues generated from rinsing product tanks and containers are solid and potentially hazardous waste if they are destined for disposal. Cleanout residues that are hazardous and are destined for disposal are not subject to RCRA regulation until they either 1) are removed from the unit, or 2) remain in an inactive unit for more than 90 days. [§261.4(c)] If rinsates from product tanks and containers will be disposed, then the unused product in the bottom of these units (before rinsing) is a solid waste [i.e., it is being abandoned per §261.2(b)(1)]. Based on application of the mixture rule, tank cleanout residues sent for disposal will be 1) P- or U-listed hazardous waste if the material being cleaned out of the unit is a P- or U-listed chemical, or 2) hazardous if they exhibit a characteristic. [RO 11004, 12299, 14827]

- The analytical/treatability sample exemptions of §261.4(d–f) (Section 4.18) apply to samples undergoing analysis, waste characterization, or treatability studies. Generally, these exemptions extend to Subpart J, meaning that such samples are not subject to RCRA tank standards while they are in the analytical/treatability study loop. Tank-based storage of hazardous treated residues or excess sample material that exits the loop, however, must comply with RCRA tank standards.

10.1.2.3 Subtitle I USTs

Subpart J of Parts 264 and 265 regulates all types of tanks that store/treat hazardous waste, including underground storage tanks (USTs) containing hazardous waste. Subtitle I of the RCRA statute gives EPA the authority to regulate USTs that contain CERCLA hazardous substances (other than hazardous wastes) and petroleum products. The agency promulgated the Subtitle I UST regulations in Part 280.

The Part 280 regulations apply to all USTs holding hazardous substances and petroleum products,

> ### Case Study 10-2: Is It a Manufacturing Unit or RCRA Tank?
>
> As shown in Figure 10-1, a facility uses a paint spray booth within its manufacturing process to apply coatings to products. Periodically, isopropyl alcohol (IPA) solvent is fed to the process to clean out the paint guns when paint colors are changed. The waste solvent, typically a D001 characteristic hazardous waste, is collected in funnels and piped to a small equalization tank located within the process building. From the equalization tank, the solvent is pumped to an outdoor aboveground accumulation tank for eventual transfer to an offsite recycler. The accumulation tank and the outdoor spent solvent piping are protected by a secondary containment system. When does the IPA solvent become a hazardous waste? Is the equalization tank and the indoor solvent piping part of the manufacturing process? What is the regulatory status of the outdoor solvent piping and the accumulation tank?
>
> According to EPA, the IPA solvent becomes a hazardous waste (i.e., its point of generation is) once the material leaves the manufacturing process and enters the collection funnels. Thus, according to EPA, the equalization tank "is not part of the production system, but serves solely to manage wastes." [RO 14152] Therefore, both the equalization and accumulation tanks are hazardous waste storage tanks. Because the facility has no RCRA permit, each tank is a 90-day accumulation unit subject to Part 265, Subpart J.
>
> This example suggests a good rule of thumb: if everything that goes into a tank comes out as waste, the unit is a waste tank—not a manufacturing process unit.
>
> Besides the tanks, the indoor and the outdoor piping and any associated pumps, valves, flanges, etc. are considered ancillary equipment that are also subject to Subpart J regulation as part of a hazardous waste tank system. Piping upstream of the equalization tank and between the two tanks is subject to RCRA provisions (including Subpart BB air emissions standards) even though it is contained within the process building. Both tanks and all piping from the point of generation through the accumulation tank need secondary containment. [RO 14152, 14604, 14632]
>
> This regulatory interpretation was challenged in court. On April 2, 2004, the U.S. Court of Appeals for the DC Circuit disagreed with the petitioner's challenge and dismissed the petition.

but they do not apply to USTs holding hazardous waste (which are subject instead to Part 264/265, Subpart J). While a complete review of Part 280 requirements is beyond the scope of this book, these provisions are comparable in many respects to those found in Part 264/265, Subpart J. For example, Part 280 includes design and management standards for USTs, leak detection, release notification and response, and closure plans.

10.1.2.4 Temporary tanks

As detailed in Section 7.5.4 and Table 10-1, equipment used to respond to spills, discharges, and other situations that pose an imminent hazard is exempt from both RCRA permitting and Part 264/265 design and management standards. This exemption extends to tanks and tank systems used for immediate responses; such tanks are often called temporary tanks or sumps.

Temporary tanks or sumps are fully exempt from Subpart J requirements. To qualify for the exemption, the unit must be intended exclusively for immediate responses to unexpected discharges of hazardous wastes, such as burst pipes, ruptured containers or tanks, breached dikes, etc. Tanks used for responding to discharge events that occur periodically or repeatedly, or in which containment or treatment

Figure 10-1: Regulation of Spent Solvent Management

a) Facility's perspective

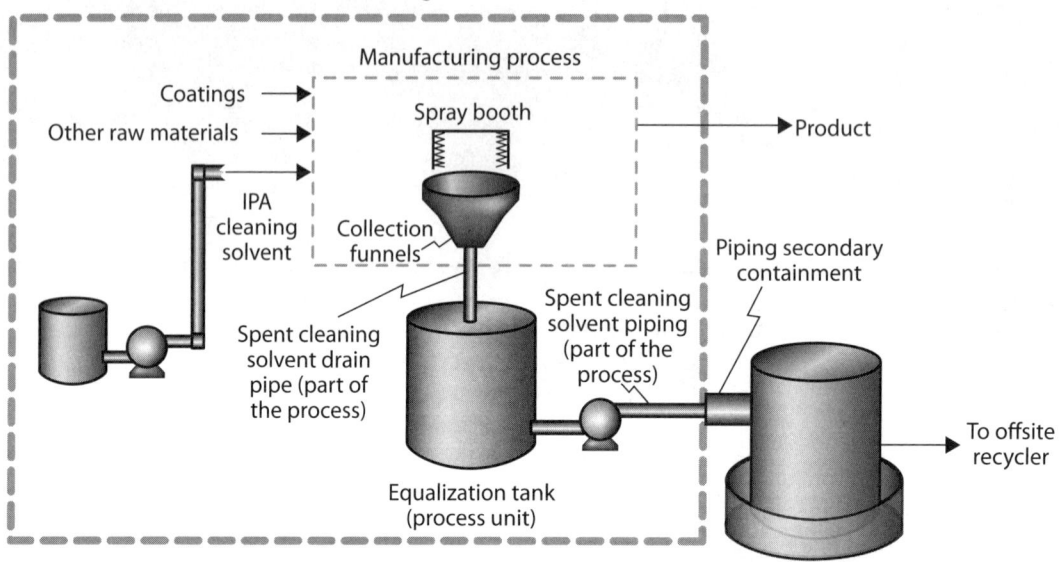

b) EPA's perspective

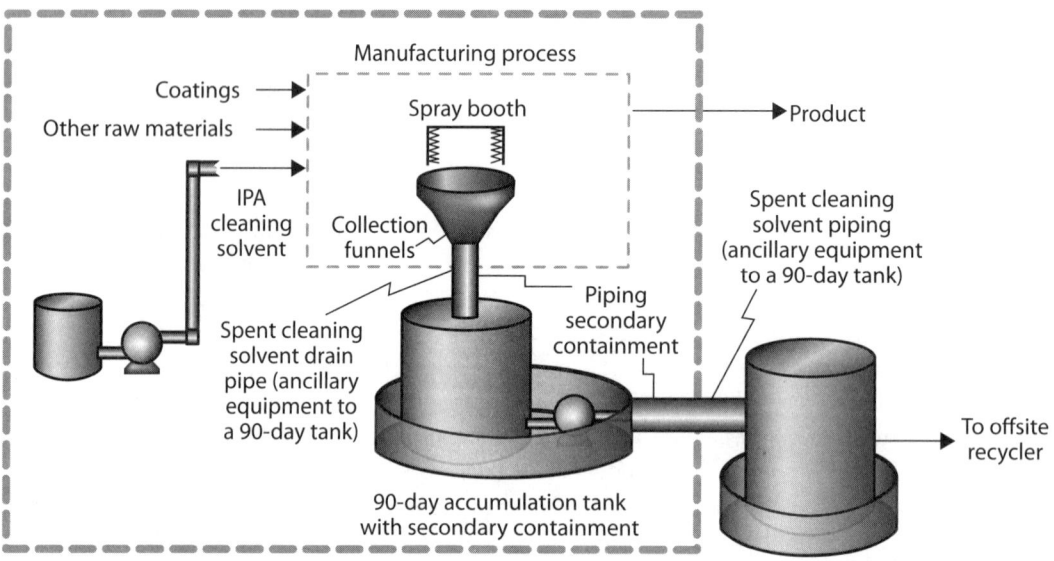

Source: McCoy and Associates, Inc.; adapted from RO 14152, 14604, 14632.

extends beyond the immediate response period, do not qualify for the exemption. [RO 12298]

"Generally speaking, any tank system into which hazardous waste is routinely and systematically introduced, regardless of frequency or duration of storage, is not considered either a temporary tank or part of the secondary containment system and therefore must be provided with secondary containment." [EPA/530/SW-87/012, August 1987, available from http://nepis.epa.gov/EPA/html/Pubs/pubtitleOSWER.html as report number 530SW87012] Examples of units that do *not* qualify for the temporary tank exemption and, therefore, are subject to Subpart J are listed below [EPA/530/SW-87/012, RO 12291, 12298, 12913]:

- A stand-by tank designed to catch material in the event of an overfill of the primary tank due to operator error or level control malfunction is clearly not functioning in an unexpected or emergency situation.
- A tank used to contain spilled residue from a truck loading/unloading area, where spills are routine and predictable, is not exempt.
- A tank installed in parallel with another tank that is brought on-line when the first tank is shut down for normal production cycling, mechanical integrity inspections, or most other reasons is also not a temporary tank.
- Finally, tanks used to collect routine maintenance cleanouts of sludge and other residual material are also functioning in situations that are periodic and expected and, therefore, are not eligible for the immediate response exemption.

10.1.2.5 Closed-loop recycling tanks

The closed-loop recycling exemption applies to tanks that are part of manufacturing production processes—not distinct waste management operations. As spelled out in §261.4(a)(8), this provision applies to tanks that are used to store secondary material so it can be reclaimed and then recycled back to the process or processes from which it was originally generated. See Section 11.3.4 for additional detail.

Technically speaking, the closed-loop recycling exemption is not really an exemption at all. Tanks that meet the above conditions are actually part of the manufacturing processes that generate the secondary material eventually recycled. Such tanks are not handling solid or hazardous wastes, and, thus, aren't hazardous waste tanks subject to RCRA. [July 14, 1986; 51 *FR* 25441]

10.1.3 Equipment that isn't a tank or tank system

In this section, we describe equipment that does *not* fit the definition of tank or tank system and, therefore, is not subject to the Subpart J standards.

For example, a drum shredding unit that processes containers filled with hazardous waste does not fit the definition of tank or tank system and, therefore, would not be subject to regulation under Subpart J. EPA determined that the unit was not designed to contain an accumulation of hazardous waste. The unit, however, would face regulation as a miscellaneous treatment unit under Part 264, Subpart X. [RO 13202]

At another facility, a vacuum conveyance system is used to collect spent abrasive wastes (that are hazardous due to heavy metal content) from a surface coating removal system. While in the conveyance system, additives are introduced to stabilize the toxic metals in the abrasive wastes prior to discharge of the waste to a containment unit. EPA was asked if the vacuum system could qualify as a 90-day accumulation tank. Again, the agency responded that the vacuum system seemed to be functioning as a Subpart X miscellaneous hazardous waste treatment device; it was not a tank or tank system. [RO 14462]

In the subsections below, we address four other types of equipment that EPA has found, for various reasons, generally do not fall under the category of tanks or tank systems and so would not be subject to Subpart J.

10.1.3.1 Waste piles

An interesting area of potential regulatory overlap occurs when one considers a hazardous waste pile in comparison to a hazardous waste tank. First, EPA defines a waste pile in §260.10 as:

> "[A]ny non-containerized accumulation of solid, nonflowing hazardous waste that is used for treatment or storage and that is not a containment building."

Clearly, a freestanding mass of hazardous soil on a concrete pad would be a hazardous waste pile. But what if that soil were placed inside a four-walled metal structure; is the soil still a waste pile, or is the structure a hazardous waste tank?

EPA addressed this question by focusing on the concept of containerization. As noted above, a pile consists of "non-containerized" waste, while a tank is a device designed to "contain" hazardous waste. In guidance on the subject, the agency stated the following:

> "Is containerization a function of the structure or is it a function of the waste itself? If the waste is contained within the unit by virtue of the fact that it is a cohesive solid, the unit is a waste pile. If the unit would contain any waste, including a free-flowing liquid, it is a tank." [RO 12899]

For example, hazardous waste contained within a four-sided structure with openings in it would be a waste pile. Since the waste would escape from the unit if it were a liquid, only the characteristics of the waste allow it to be contained.

10.1.3.2 Surface impoundments

A surface impoundment is defined in §260.10 as follows:

> "[A] facility or part of a facility which is a natural topographic depression, man-made excavation, or diked area formed primarily of earthen materials (although it may be lined with man-made materials), which is designed to hold an accumulation of liquid wastes or wastes containing free liquids, and which is not an injection well. Examples of surface impoundments are holding, storage, settling, and aeration pits, ponds, and lagoons."

An aboveground tank is not likely to be confused with a surface impoundment, but what about a large in-ground basin constructed with concrete walls. Unlike waste piles, where their distinction from tanks depends on the properties (i.e., the cohesion) of the waste in question, EPA distinguishes between in-ground tanks and surface impoundments based on the properties of the structure. In this case, the agency has developed what is commonly referred to as the "parking lot" test to differentiate between the two. The key concepts are that a tank must "provide structural support" and "contain an accumulation of hazardous waste," both of which are included in the definition of "tank."

The parking lot test essentially requires that, in order for an in-ground basin or sump to be considered a tank, it must have freestanding walls. The unit must be evaluated (by engineering calculations) as if it were freestanding and filled to design capacity with the material it is intended to hold. In other words, if the unit were placed fully loaded on a flat, open parking lot, would the walls continue to provide structural support? If the walls or shell of the unit provide sufficient structural support to maintain the structural integrity of the unit under these conditions, then the unit can be considered a tank. Conversely, if the unit needs the supporting earthen materials to maintain its structural integrity, it is a surface impoundment.

EPA's use of this test has been discussed in guidance documents [RO 12104, 12224, 13669] and has been upheld by a federal court of appeals (*Beazer East, Inc. v. EPA*, May 12, 1992, No. 91-1692, U.S. Court of Appeals for the Third Circuit).

Therefore, facility personnel can expect EPA inspectors to carefully record and photograph any visual evidence of cracking or other integrity concerns for an in-ground unit handling hazardous wastewater that the facility is managing as a tank (and not a surface impoundment). Additionally, facility personnel should expect the following questions:

- What is the age of the unit?

- Have any repairs been made to the unit?
- Has an integrity assessment been conducted for the unit?
- Does the unit leak?
- Have any studies been conducted of the sewer system and associated sumps, including the unit in question?
- Is there any ground water monitoring in place around the unit?
- What construction diagrams/drawings are available for the unit?
- Are any calculations available showing that the unit passes the parking lot test?

10.1.3.3 Vaults

EPA has taken the position that vaults and vault-like structures may be designed to resemble tanks but, as a practical matter, are intended for disposal rather than temporary storage and/or treatment. Thus, such units are not regulated under Subpart J as tanks or tank systems; they must be permitted as land disposal units (e.g., landfills). [OSWER Directive 9481.00-9]

10.1.3.4 Drip pads

The idea that a drip pad cannot qualify as a tank would be part of our common understanding—except for two potentially complicating issues. First, EPA notes that drip pads are not part of tank systems, because they are not ancillary equipment to tanks. [December 6, 1990; 55 FR 50480] However, any tanks and sumps that are used in conjunction with drip pads *are* subject to Subpart J requirements. [§§264/265.190(c)] For example, drip pad *sumps* that contain hazardous wastewater generated at wood preserving facilities would normally be subject to Part 264/265, Subpart J tank standards [except where the sump is part of the facility's wastewater treatment system and, thus, exempt from RCRA standards under the WWTU exemption (see Section 7.2)]. [RO 14028]

10.1.4 Tank systems include ancillary equipment

As we noted in Section 10.1, a tank system includes the tank itself, its secondary containment structure, and equipment ancillary to the tank. Ancillary equipment is defined in §260.10 as:

"[A]ny device including, but not limited to, such devices as piping, fittings, flanges, valves, and pumps, that is used to distribute, meter, or control the flow of hazardous waste from its point of generation to a storage or treatment tank(s), between hazardous waste storage and treatment tanks to a point of disposal onsite, or to a point of shipment for disposal offsite."

This definition lists various types of equipment that qualify as ancillary equipment. However, the definition also makes clear EPA's desire that the scope of this term and that of a tank system starts at the point a given material becomes a hazardous waste (i.e., its point of generation). Ancillary equipment then extends to the point that another regulated activity kicks in (i.e., when the waste enters another storage or treatment tank, an onsite disposal point, or a shipping point to offsite disposal). Although point-of-generation determinations are site-specific, EPA normally considers the point of generation to be the point of exit from the process tank or area, and this point will also be the introductory point for the hazardous waste into a hazardous waste tank system. [July 14, 1986; 51 FR 25441, RO 13790, 14469] See Section 14.1.

The following examples give EPA's position as to what types of equipment are and are not ancillary equipment:

- Piping, valves, pumps, and other material transfer equipment normally used for process activities are ancillary equipment subject to Subpart J regulation if they are 1) occasionally used to transfer hazardous waste residue from process tanks to hazardous waste storage/treatment tanks during washout/cleanout operations, or 2) used to transfer a material that is sometimes reused or recycled and sometimes disposed as a hazardous waste. The

§261.4(c) active manufacturing process unit exemption would not exempt such equipment. If such ancillary equipment contains or contacts hazardous waste with at least 10% organics for at least 300 hours per calendar year, it is also subject to Subpart BB organic air emission standards. [RO 13790, 14469]

- Piping associated with a loading/unloading station is used to unload raw materials from transport vehicles into raw material storage tanks. However, the same piping system is used to unload hazardous waste from a hazardous waste tank to transport vehicles. Therefore, the piping system is ancillary equipment to that hazardous waste tank and is subject to RCRA standards. [Memo from Bruce Weddle, EPA Headquarters to Stanley Siegel, EPA Region II, March 17, 1988]

- Building floor drains and outdoor trenches used to transfer hazardous wastewater to an in-ground storage tank are ancillary equipment to that tank. [RO 12829]

- An unlined earthen ditch used to direct hazardous wastewater to a treatment tank is not ancillary equipment because it is not designed to prevent leakage or discharge of the waste; instead, EPA considers such leakage or discharge to be disposal of hazardous waste. "[I]n order for a device to be defined as 'ancillary equipment,' it must be designed to prevent leakage or discharge." [RO 13653; see also RO 11631, 13669]

- Ancillary equipment associated with an exempt unit is not subject to Subpart J standards. For example, piping systems and equipment associated with wastewater treatment or elementary neutralization units would not be subject to RCRA standards. "However, if this equipment is used, even intermittently, for storage/treatment of a hazardous waste or wastewater prior to shipment offsite as a hazardous waste, the exemption is not applicable. The...hazardous waste tank system standards...would then apply." [RO 13126] Additionally, piping associated with surface impoundments would not have to comply with Subpart J

requirements. "It was indeed EPA's intent not to regulate ancillary equipment itself." [RO 13126]

- Finally, containers are *not* ancillary equipment to tanks. [RO 11551]

10.1.5 Hazardous waste tank design and management standards

While secondary containment and inspection requirements for hazardous waste tanks are addressed separately in Sections 10.3 and 10.5, respectively, we summarize in this section the basic Subpart J design and management requirements for these tanks. Table 10-2 lists the various requirements included in Part 264/265, Subpart J, along with the applicability of those provisions to existing and new hazardous waste tanks for four types of facilities: 1) permitted TSD facilities, 2) interim status TSD facilities, 3) large quantity generators, and 4) small quantity generators.

A useful document for reviewing compliance with hazardous waste tank regulatory requirements is *Protocol for Conducting Environmental Compliance Audits of Storage Tanks Under RCRA*, EPA/300/B-00/006, March 2000, available from http://www.epa.gov/compliance/resources/policies/incentives/auditing/apcol-rcratanks.pdf.

10.1.5.1 New vs. existing hazardous waste tanks

Although most of Table 10-2 is self-explanatory, the issue of new versus existing tanks requires further explanation. For new tank systems (differentiated from existing tank systems as described in this subsection), applicable requirements must be in place before the systems begin hazardous waste service.

The cut-off date for new versus existing hazardous waste tanks is July 14, 1986. With a couple of exceptions, any tank system whose installation commenced on or before this date is existing; hazardous waste tanks installed after that date are new. EPA expanded upon this issue in preamble language:

"[T]he term 'new tank system' means not only newly-manufactured tank systems that will be put into service for the first time but also those

Table 10-2: Design and Management Standards for Hazardous Waste Tanks

Requirement (*CFR* citation)	Permitted TSD facilities		Interim status TSD facilities		Large quantity generators		Small quantity generators[1]	
	Existing	New	Existing	New	Existing	New	Existing	New
Integrity assessments for existing tank systems via PE-certified check of design standards, leak test, or other integrity examinations (§§264/265.191)[2]	Y	—	Y	—	Y	—	—	—
Design and installation standards for new tank systems via PE-certified check of design standards, external corrosion protection, equipment inspections, tightness check, etc. (§§264/265.192)	—	Y	—	Y	—	Y	—	—
Secondary containment systems with release detection for tank and ancillary equipment (§§264/265.193)	Y[3]	Y[3]	Y[3]	Y[3]	Y[3]	Y[3]	—[4]	—[4]
General operating procedures to avoid spills, overflows, ruptures, leaks, corrosion, and other tank/ancillary equipment failures (§§264/265.194)	Y	Y	Y	Y	Y	Y	—	—
Periodic inspection programs for tanks, overfill controls, corrosion protection systems, etc. (§§264/265.195)	Y	Y	Y	Y	Y	Y	—	—
Spill response programs, including containment procedures, notifications, and prescribed repair/preventive measures (§§264/265.196)	Y	Y	Y	Y	Y	Y	—	—
Closure and post-closure care for tank systems (§§264/265.197)	Y	Y	Y	Y	Y[5]	Y[5]	—	—
Equipment setback requirements and special management procedures for ignitable and reactive wastes (§§264/265.198)	Y	Y	Y	Y	Y	Y	—	—
Waste segregation procedures for incompatible wastes (§§264/265.199)	Y	Y	Y	Y	Y	Y	—	—
Subpart AA/BB/CC organic air emission standards (§§264.200/265.202)	Y	Y	Y	Y	Y	Y	—	—
Waste analysis and trial storage/treatment tests for tank-based hazardous waste activities (§265.200)	—[6]	—[6]	Y[6]	Y[6]	—	—	—	—
Special small quantity generator tank requirements, including waste management standards, inspections, tank closure provisions, and incompatible waste segregation procedures (§265.201)	—	—	—	—	—	—	Y	Y

Y = Yes, required.

[1]Conditionally exempt small quantity generators (facilities generating no more than 100 kg/month of hazardous waste) become subject to the small quantity generator standards listed in this table if they accumulate more than 1,000 kg of hazardous waste onsite.

[2]Per §§264/265.191(a) and EPA/530/SW-88/004, an annual integrity assessment does not apply to existing tanks that are fitted with secondary containment meeting Subpart J standards but does apply to tanks without such secondary containment. Also see RO 13038.

[3]Secondary containment is not required if the tank is 1) used to store or treat a hazardous waste that contains no free liquids, *and* 2) located inside a building with an impermeable floor. Requirements for when secondary containment must be installed are discussed in §§264/265.193 and Section 10.3.

[4]Small quantity generator tanks do not require secondary containment systems except under very limited circumstances. See §265.201(b)(3) for details.

[5]Closure and post-closure care plans, as well as financial assurance, are not required for 90-day accumulation tanks. [EPA/530/SW-88/004]

[6]Periodic waste analysis is required at this type of facility per the nontank-specific provisions of §§264/265.13.

Source: McCoy and Associates, Inc.; adapted from Part 264/265, Subpart J.

other tank systems that even if in existence and in use prior to [July 14, 1986] are then reinstalled and used as replacement tank systems for existing hazardous waste tank systems. Likewise, an existing tank system that is not being used for the storage or treatment of hazardous waste, but is then put into service or converted to use as a hazardous waste storage or treatment tank system subsequent to [July 14, 1986] is considered to be a new tank system." [July 14, 1986; 51 FR 25446]

Based on the above guidance, if an aboveground or underground tank used to store hazardous waste is moved to another location at the same facility, the tank would be classified as a new tank after being moved and reinstalled. [RO 12920] Similarly, if a new tank were installed in an existing sump to facilitate compliance with the secondary containment requirements, the installation would be considered a new tank system, subject to the Subpart J standards for new systems. [RO 13174]

Finally, what is the status of a tank system that is being used to store a material that was not a hazardous waste until a recent EPA action added it to the definition of a hazardous waste? Such a tank system that was in existence prior to the time that its contained material became a hazardous waste is considered an *existing* tank system (even if installed after July 14, 1986). Conversely, an existing tank that begins to store a hazardous waste after the time that the waste was added to the definition of a hazardous waste (but that was not used to store the material before it became hazardous) is a new tank system. [RO 13137]

10.1.5.2 Tank integrity assessments

The previous section presented EPA's guidance for determining whether a hazardous waste tank is existing or new. If it is existing, a tank system integrity assessment per §§264/265.191 is required if the tank has not been provided with secondary containment meeting the requirements of §§264/265.193. [EPA/530/SW-88/004 (cited in Section 10.2.2), RO 13038] If an existing tank system includes a compliant secondary containment system, there is no federal requirement for an integrity assessment.

If the tank is new, conversely, §§264/265.192(a) require an integrity assessment of the tank and its components upon installation. This assessment must be certified by a qualified professional engineer (PE): "The requirement for [PE] certification, in fact, extends to all new tank system components." [RO 13244] Significant guidance on conducting tank integrity assessments is available in EPA/530/SW-86/044, cited previously.

Additionally, recertification is required if, after initial operation, modifications are completed that affect the structural integrity of the tank system. The following paragraphs from RO 13192 sum up EPA's intent:

"The standards in §§264/265.192 require that the correct installation of new tank systems or components be certified by an independent registered professional engineer or independent qualified installation inspector. The agency's intent in promulgating this provision was that such a certification provides EPA with a means of knowing that hazardous waste tank systems were initially installed in a correct manner. EPA was concerned that many tank systems were being improperly installed thereby resulting in failure of the tank, piping, etc. The failures were of particular concern because in the absence of secondary containment many of these releases could go undetected indefinitely.

"Proper installation on new tank systems and components is an ongoing concern to the agency. However, it was not EPA's intent that every minor or routine replacement of a tank system component need recertification each time it is replaced. Replacement of valves, pumps, or even small sections of piping were not envisioned as needing recertification since they do not affect the structural integrity of the tank system. Rather, the agency intends this requirement to apply to components affecting the system's structural integrity, e.g., the more major, non-routine and complex retrofit/replacement tasks. For example, the installation of new tanks including reinstallation of existing tanks, the installation of new secondary containment systems, and the replacement of extensive piping are relatively complex tasks that are critical to structural integrity and require oversight to ensure proper installation. This oversight is supplied by the independent registered professional engineer or independent qualified installation inspector.

"It is not feasible for the agency to lay out a detailed menu of the items that do or do not need certification of installation. Facility owners and operators should contact the appropriate EPA regional or state authorities to determine which new tank system components need certification of proper installation." [RO 13192]

Although the guidance quoted above from RO 13192 notes that an "independent" registered professional engineer provide tank integrity assessments, EPA revised this requirement in an April 4, 2006 rule: "we have decided to delete the independent qualification for certification made by a professional engineer. EPA continues to believe that this proposed modification retains the most important requirements: That the engineer is *qualified* to perform the task and is a professional engineer (i.e., licensed to practice engineering under the title Professional Engineer.) We believe that a professional engineer, regardless of whether he/she is independent is able to give fair and technical review because of the programs established by the state licensing boards." [Emphasis in original.] [64 *FR* 16869]

10.2 Are sumps tanks?

Yes—if they pass the parking lot test described in Section 10.1.3.2. A look at EPA's definition of the term "sump" notes that:

"*Sump* means any pit or reservoir that meets the definition of tank and those troughs/trenches connected to it that serve to collect hazardous waste for transport to hazardous waste storage, treatment, or disposal facilities; except that as used in the landfill, surface impoundment, and waste pile rules, 'sump' means any lined pit or reservoir that serves to collect liquids drained from a leachate collection and removal system or leak detection system for subsequent removal from the system." [§260.10]

There are several things going on in this definition, but the primary issue is that sumps that pass the parking lot test are regulated exactly like tanks under RCRA Subtitle C (with a couple of exceptions as noted below). Comprehensive guidance on determining the integrity of concrete sumps is given in EPA/530/R-93/005, available from http://nepis.epa.gov/EPA/html/Pubs/pubtitleOSWER.html by downloading the report numbered 530R93005. Based on this definition and agency guidance, we have concluded that there are three types of sumps from a RCRA perspective (and each one is subject to different regulatory requirements):

1. Temporary sumps,
2. Secondary containment sumps, and
3. Primary containment sumps.

10.2.1 Temporary sumps

Temporary sumps are regulated just as temporary tanks discussed in Section 10.1.2.4. Hazardous waste is stored in temporary sumps only in response to a leak, spill, or other temporary, unplanned occurrence. The immediate response exemption (see Section 7.5.4) applies to these "dry" sumps and, therefore, RCRA permitting and Subpart J tank standards do not. "A sump that may be used to collect hazardous waste in the event of a spill, whether accidental or intentional, and that is not designed to serve as a secondary containment structure for a tank storing hazardous waste, is generally exempt from regulatory and permitting requirements so long as it is used to contain hazardous waste only as an immediate response to such a spill." [September 2, 1988; 53 *FR* 34085]

A temporary sump is *not* a unit designed to serve as secondary containment to a primary hazardous waste containment vessel (e.g., a storage tank) or its associated ancillary equipment. Such a unit is a secondary containment sump, as described below.

10.2.2 Secondary containment sumps

Sometimes, sumps aren't the primary vessel used to manage hazardous waste but are utilized as a secondary containment structure. These units are specifically designed to contain spills from a primary containment vessel (or its associated ancillary equipment). Such secondary containment sumps are not immediate response units but, because they meet the definition of a tank, are subject to Subpart J tank standards (e.g., they must be impermeable, they must be inspected, etc.). However, they are not themselves required to have secondary containment (since that's what they are). EPA included the phrase "[t]ank systems, *including sumps*" in the provision that precludes duplicative secondary containment for these secondary containment sumps. [§§264/265.190(b); see also EPA/530/SW-88/004, January 1988, available from http://nepis.epa.gov/EPA/html/Pubs/pubtitleOSWER.html by downloading the report numbered 530SW88004]

The immediate response exemption "is inapplicable to units...designed to serve as the secondary containment for a tank system treating or storing hazardous waste." [53 *FR* 34085]

10.2.3 Primary containment sumps

Sumps that provide primary containment for storing hazardous waste are primary containment sumps. Just like primary containment tanks, such sumps are subject to all Subpart J tank standards, including the need for secondary containment. When EPA overhauled the hazardous waste tank regs on July 14, 1986, it stated in the preamble that "sumps may present the same potential for leaks and releases as hazardous waste storage and treatment tanks.... Thus, EPA concludes that sumps generally should be subject to the same standards as tanks." [51 *FR* 25441]

The first part of the "sump" definition, then, applies tank standards to these units that collect and transport routine and systematic discharges of hazardous waste if they meet the definition of a tank. [53 *FR* 34085] For example, sumps used to store hazardous waste generated from periodic reactor cleanouts are primary containment vessels and would be subject to all Subpart J tank standards. However, as noted in Sections 7.2.1.3.2 and 7.5.1.1.2, such primary containment sumps may meet the definition of a WWTU or elementary neutralization unit and, thus, be exempt from RCRA permitting and tank standards under one of those exemptions.

The agency understands that sumps sometimes function differently from tanks and need to be regulated accordingly. For example, sumps that operate as part of a leachate collection and removal system for a landfill, surface impoundment, or waste pile are an integral part of the land disposal unit's liner system, and imposing Subpart J tank standards (e.g., secondary containment) would be impractical and unnecessary. [January 29, 1992; 57 *FR* 3471, RO 14011] As such, even though these units provide primary containment for hazardous leachate, they are exempt from these standards.

Sumps holding hazardous waste that don't pass the parking lot test (see Section 10.1.3.2) are surface impoundments and would be regulated under Part 264/265, Subpart K. Imposing tank standards on these structures would be illogical. [RO 11134]

10.2.4 Examples

The examples below show how EPA applies the definition of sump in various situations.

 Can sumps be used as 90-day units for hazardous waste accumulation?

 Yes. Assuming they pass the parking lot test and qualify as tanks, sumps may be used to accumulate hazardous waste for 90 days or less without the need for a RCRA permit. However, such sumps would be subject to Subpart J standards, including the need for secondary containment. [RO 12442]

A sump is used to collect intentional discharges of hazardous waste from a centrifuge. Is it a

temporary, secondary containment, or primary containment sump?

A It is a primary containment sump, since it collects routine and systematic discharges of hazardous waste. [July 14, 1986; 51 *FR* 25441]

Q *How are trenches/floor drains that transport hazardous wastewater from a periodic cleaning operation to a sump regulated?*

A If the sump meets the definition of a tank (i.e., it passes the parking lot test), the sump and associated trenches/floor drains comprise an integrated tank system. Since the sump is receiving hazardous waste on a routine and systematic basis, it is a primary containment sump and does not qualify for the §§264/265.190(b) exemption for sumps or tanks that are part of a secondary containment system. Therefore, both the sump itself and the ancillary trenches/floor drains would be subject to Subpart J tank standards, including the need for secondary containment. [RO 12829, 13653]

Q *Is hazardous waste treatment allowed in sumps? How is this regulated?*

A EPA makes no distinction between hazardous waste treatment in sumps versus tanks. If hazardous waste treatment in the sump is routine and systematic, the sump would be considered a primary containment sump subject to all Subpart J standards. Where treatment in sumps is found to be occurring (using the two-part definition of treatment described in Section 7.1), the agency allows either the WWTU or elementary neutralization unit exemptions to apply if the sump meets the definition of a tank and other exemption conditions are met. Where it fails the parking lot test, the sump would face more stringent regulation as a surface impoundment under Part 264/265, Subpart K. [RO 11134, 13170, 14028, 14470]

Q *Would a tank that receives hazardous leachate from a collection sump associated with a hazardous waste landfill also be exempt from Subpart J tank standards?*

A No. Only the actual leachate collection sump is excluded from hazardous waste tank standards. Any subsequent tank that manages the hazardous leachate will be subject to Subpart J standards, unless it qualifies for an exemption (e.g., the WWTU exemption). [RO 14011]

Q *Can a manhole (sump) in a hazardous wastewater distribution system be classified as a tank?*

A Yes. Note, however, that since the manhole/sump is part of a secondary containment system, it is a secondary containment sump. Per §§264/265.190(b), it would have to meet all Subpart J requirements, except that it would not have to be protected by another secondary containment system. [RO 12788]

Q *What is the regulatory status of a sump associated with a secondary containment system surrounding a hazardous waste-derived fuel truck unloading area at a cement facility?*

A The sump will contain hazardous waste only in the event of a spill during the offloading of hazardous waste-derived fuel into the cement kiln. It cannot be a primary containment sump, because it doesn't receive routine or systematic discharges of hazardous waste. In addition, it is not serving as a secondary containment structure for spills from a primary containment vessel storing hazardous waste; the trucks containing the hazardous waste-derived fuel are not storage vessels when located onsite for short periods during the transfer of hazardous waste into the kiln. Since the unit is only used to contain hazardous waste as an immediate response to a spill, it is a temporary sump and qualifies for the immediate response exemption. [RO 13127]

10.3 Secondary containment and release detection requirements

Sections 264/265.193 require permitted and interim status TSD facilities and large quantity generators to install secondary containment and release detection for their hazardous waste tanks. While this requirement is a cornerstone of EPA's goal of preventing hazardous waste releases to the environment, it is also one of the most expensive provisions that facilities must implement as part of their RCRA compliance efforts for hazardous waste tanks.

10.3.1 Applicability

As noted earlier in Table 10-2, secondary containment/release detection must be provided for all existing and new tanks and their associated ancillary equipment at the following locations: 1) permitted TSD facilities, 2) interim status TSD facilities, and 3) large quantity generators. Notably absent from this list is 180/270-day accumulation tanks operated by small quantity generators. Small quantity generators are not required to provide secondary containment for their tanks except under very limited circumstances. [§265.201(b)(3)] Also, secondary containment is not required for a hazardous waste tank at any type of facility if the tank system is 1) used to store or treat a hazardous waste that contains no free liquids, and 2) located inside a building with an impermeable floor. [§§264/265.190(a)]

10.3.1.1 Avoiding an endless loop

One common way for a facility to provide secondary containment for a Subpart J regulated tank is to install a concrete pad and walls under and around the tank. However, such a concrete structure meets the definition of "tank" in §260.10, and so itself is subject to Subpart J standards, including the requirement for secondary containment. Obviously, providing additional containment for a secondary containment structure was not EPA's intent, and so, in §§264/265.190(b), the agency notes that any tank system (including a sump) that serves as part of a secondary containment need not be provided with another secondary containment system. See also EPA/530/SW-88/004, cited earlier.

10.3.2 Technical details

According to §§264/265.193(b), secondary containment systems must be:

"(1) Designed, installed, and operated to prevent any migration of wastes or accumulated liquid out of the system to the soil, ground water, or surface water at any time during the use of the tank system; and

"(2) Capable of detecting and collecting releases and accumulated liquids until the collected material is removed."

Toward these goals, secondary containment systems must meet the following general technical requirements. They must be:

- Constructed of, or lined with, materials that 1) are compatible with the hazardous waste(s) that may enter the system; and 2) have sufficient strength and thickness to prevent failure due to the weight of the waste(s), physical contact with the waste, external hydrogeological forces, climatic conditions, and/or the stresses of day-to-day operations;

- Placed on a foundation or base capable of supporting the containment system; resisting pressure gradients above and below the system; and preventing failures due to settlement, compression, or uplift; and

- Sloped or otherwise operated to drain and remove liquids resulting from leaks, spills, or precipitation. EPA noted that liners should have a minimum slope of 1/4 inch per linear foot to a dry well or sump to allow liquids to drain for detection and removal. [EPA/530/SW-86/044, cited previously]

EPA believes that most secondary containment systems for tanks will consist of one of three primary system designs: external liners (see Figure 10-2), vaults (see Figure 10-3), or double-walled tanks. Specific design requirements for these three systems are listed in Table 10-3.

10.3.2.1 More on required containment system capacity

As noted in Table 10-3, external liner systems and vaults must be sized for the largest hazardous waste tank they contain plus 24 hours of rainfall from a 25-year storm (e.g., 10 inches in Houston, 4.5 inches in Chicago). In contrast, a specific capacity standard

Figure 10-2: Aboveground Tank With External Liner Secondary Containment

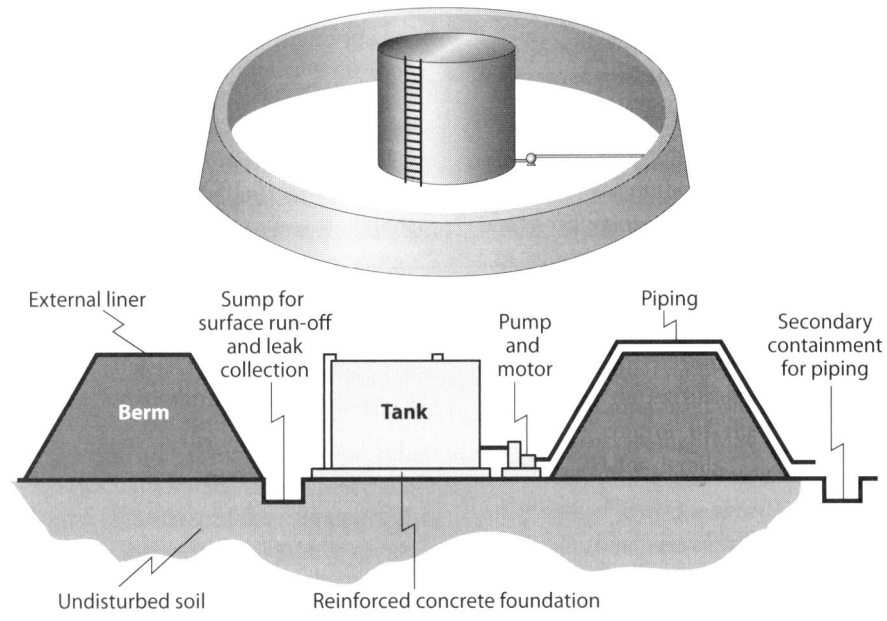

Source: Adapted from EPA/530/K-05/018.

Figure 10-3: Secondary Containment Vault for Underground Tanks

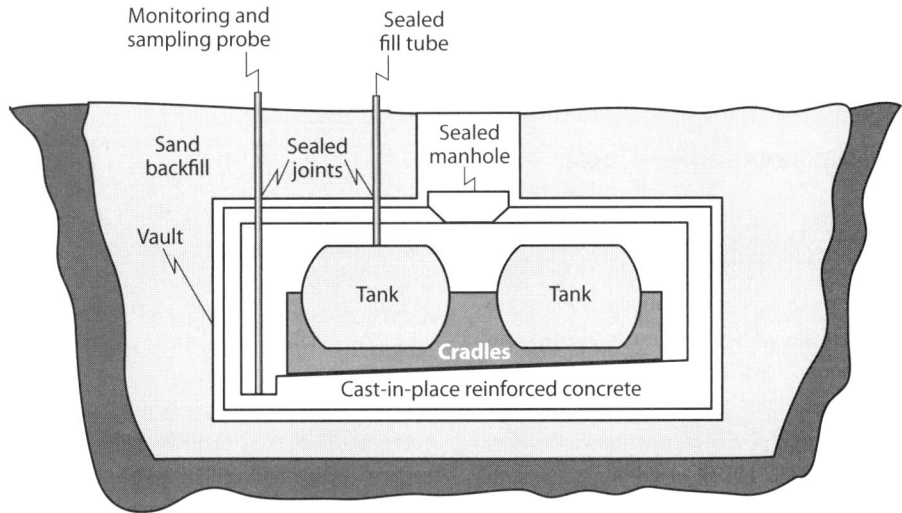

Source: Adapted from EPA/530/K-05/018.

Table 10-3: Specific Design Requirements for Tank Secondary Containment Systems

Secondary containment system type	Requirements
External liners	Sized to contain 100% of the capacity of the largest hazardous waste tank within its boundary *plus* 24 hours of rainfall from a 25-year storm, unless precipitation is precluded from entering the containment system.
	Free of any cracks or gaps.[1]
	Provided with an impermeable liner or interior coating compatible with the stored waste that will prevent waste migration into the concrete and/or the environment.[2]
	Constructed with chemical-resistant water stops at all joints. [53 *FR* 34084, RO 14395]
	Designed and installed to completely surround the tank and be capable of preventing lateral and vertical migration of waste.[3]
Vaults[4]	Sized to contain 100% of the capacity of the largest hazardous waste tank within its boundary *plus* 24 hours of rainfall from a 25-year storm, unless precipitation is precluded from entering the containment system.
	Provided with an impermeable liner or interior coating compatible with the stored waste that will prevent waste migration into the concrete and/or the environment.
	Constructed with chemical-resistant water stops at all joints and an external moisture barrier to prevent migration of moisture into the vault (if the vault is subjected to hydraulic pressure).
	Provided with a means of avoiding the formation and ignition of flammable/explosive vapors within the vault, where wastes stored/treated in the tank are either D001 ignitable or D003 reactive.
Double-walled tanks	Designed as an inner hazardous waste tank completely enveloped by an outer shell so that any release from the inner tank is fully contained by the outer shell.
	If constructed of metal, both the primary tank interior and the external surface of the outer shell must be protected from corrosion.

[1]Techniques that may be used to minimize external liner cracks and gaps are discussed in Section 7.5 of EPA/530/SW-86/044, cited previously.
[2]"[G]iven the relative permeability of concrete, the agency believes that most secondary containment concrete structures, vaults or otherwise, will require an impermeable coating or lining that will prevent migration of waste into the concrete." [53 *FR* 34084; see also RO 14395]
[3]Containment design need not account for a total, catastrophic failure of the hazardous waste tank. [RO 13244]
[4]A tank contained within a building may be considered to be within a vault. For this arrangement to meet regulatory requirements, the building, aboveground or inground (e.g., with a basement structure), must meet all requirements of §§264/265.193(e)(2) concerning design capacity, run-on and infiltration, water stops, interior lining, moisture barrier, and ignitable/explosive vapors. [EPA/530/SW-86/044]

Source: McCoy and Associates, Inc.; adapted from §§264/265.193(e), unless otherwise noted.

---

is not given for double-walled tanks. EPA believes the inherent design of such systems makes such a requirement unnecessary. [RO 13162]

Some facilities have severe space constraints, such that building an external liner of sufficient capacity around their hazardous waste tank would be difficult. One such facility asked EPA if they could alternatively use a pump system to remove accumulated liquids in combination with a smaller-volume containment structure to satisfy the capacity requirement. The agency was skeptical in its reply, noting that mechanical breakdowns, power failures, and other problems greatly reduce the reliability of operational controls, such as pumps, versus passive systems such as concrete walls. However, EPA suggested that such systems might work in certain situations, if the facility can demonstrate to the state or EPA region that the operational system does not increase the risk of a release of hazardous waste into the environment when compared to using a passive secondary containment barrier. [RO 13341]

The agency also notes that secondary containment systems that serve multiple tanks need be sized only for the largest contained tank that is in hazardous waste service. As is implied, this means that there are no restrictions against placing hazardous waste and other types of tanks (e.g., petroleum storage tanks) in the same containment area. Where this is the case, however, the entire area must meet applicable Subpart J requirements. [RO 12965, 13336]

10.3.2.2 Containment structures must be impermeable

EPA believes that concrete external liner and vault structures used as secondary containment systems will require an impermeable coating or lining (e.g., an epoxy coating, high-density polyethylene, stainless steel plate) to prevent migration of waste into the concrete. Although this requirement is specified only for vaults in the regulations, the agency didn't intend to limit such a coating or lining requirement to concrete vaults. The coating or lining must entirely and uniformly cover the surface of the concrete structure that could come into contact with released waste. [EPA/530/SW-88/004, 53 FR 34084, RO 13152, 14395]

The coating or lining must be compatible with the stored hazardous waste. [RO 12953] A concrete secondary containment structure to which an impermeable chemical-resistant coating or lining is applied must be certified by a qualified professional engineer. [RO 13250]

Any wear, cracks, etc. in the concrete or coating or lining must be repaired promptly. [RO 13152] We have encountered situations where maintenance personnel scuff or scrape off the epoxy coating from a secondary containment structure, resulting in nonimpermeable containment and notice of violation.

 Can a building serve as secondary containment for a tank system located inside it?

Yes. A building could function as secondary containment for a tank, providing the structure meets the requirements of §§264/265.193(b–c). [July 14, 1986; 51 FR 25452, RO 13633, 14119, 14463, 14632] "In cases involving buildings serving as secondary containment, authorized states necessarily have the ultimate authority to make the determination that secondary containment requirements are met (taking into account all relevant site-specific considerations)." [April 4, 2006; 71 FR 16880]

10.3.2.3 USTs

Although usually a more expensive proposition than for their aboveground counterparts, underground storage tanks (USTs) that store hazardous waste must also be provided with secondary containment. [RO 12774] Normally, this would take the form of either a vault or a double-walled tank system.

10.3.2.4 Ancillary equipment

Because ancillary equipment is considered part of a tank system, it is subject to Subpart J requirements including secondary containment that meets the general design conditions listed above. Pumps, valves, flanges, fittings, etc. that are associated with a hazardous waste tank require secondary containment. Such secondary containment usually consists of jacketing, double-walled piping, the process building itself [51 FR 25452, RO 14632], or lined trenches underneath the equipment that flow to a regulated unit. However, the following ancillary equipment does not require secondary containment:

- Aboveground piping that is visually inspected for leaks on a daily basis is exempt from secondary containment. In addition, pipe runs that have all welded fittings and that are inspected daily do not require secondary containment. [§§264/265.193(f)(1–2), RO 12953, 12998] Threaded fittings on pipe runs outside of the tank's secondary containment structure require their own secondary containment, because these fittings are where EPA thinks most leakage occurs. A flange is considered "welded" if the pipe-to-flange joint is welded (as opposed to threaded)—the flange faces at the gasket do not have to be seal welded. Therefore, weld-neck, lap-joint, socket-weld, and slip-on flanges are all considered welded flanges. [September 2, 1988;

53 *FR* 34081, RO 12953, 12973, 13221] For plastic piping, solvent-cemented and heat-fusion connections would be considered welded and would not require secondary containment if inspected daily; mechanical joints, however, would need secondary containment. [53 *FR* 34082, RO 12953, 13221] Secondary containment is required for soldered and brazed joints, compression fittings, and flared fittings for hazardous waste tubing.

- Sealless or magnetic-coupling pumps and sealless valves that are visually inspected for leaks on a daily basis do not require secondary containment. [§§264/265.193(f)(3)] Packing-type pump seals require secondary containment. [RO 12973] EPA notes that valves that use a metal bellows to seal the valve stem, and essentially any other design that achieves containment within the valve body, would qualify as sealless. [RO 12720, 13047] This exemption from secondary containment applies only to the seals; the connections of this equipment to the pipeline must be provided with secondary containment unless they are welded. [RO 12720]

- Secondary containment is not required for pressurized aboveground piping systems with automatic shut-off devices (e.g., excess flow check valves, flow-metering shutdown devices, shut-off devices actuated on a loss of pressure) that are visually inspected for leaks on a daily basis. [§§264/265.193(f)(4)] This provision applies even if welded fittings, sealless or magnetic-coupling pumps, and/or sealless valves are not used. [53 *FR* 34084, RO 13005] EPA has not assigned a specific numerical value to distinguish pressurized from nonpressurized piping. The agency generally considers only that piping located on the discharge side of pumps to be "pressurized." Although technically it is under pressure due to static head pressure, piping connected to the bottom of tanks is not considered pressurized. [RO 13244]

- Ancillary equipment associated with non-Subpart J regulated units, such as surface impoundments or exempt tanks, is not subject to secondary containment requirements. [RO 13126]

The agency notes in guidance that secondary containment requirements *do* apply to underground hazardous waste piping, stating its reasoning as follows:

"EPA believes that secondary containment with leak detection is an absolute necessity. Not only is underground piping more prone to failure due to corrosion and other soil-related stresses, than is aboveground piping, but it is also impossible to visually observe any impending or actual failure of the piping system, particularly for less than catastrophic releases." [RO 12973]

Secondary containment systems for ancillary equipment must be capable of "collecting releases and accumulated liquids until the collected material is removed." [§§264/265.193(b)(2)] Where secondary containment systems serve tank system piping in transport vessel loading stations, required secondary containment capacity would be based on the largest projected spill that could occur prior to cleanup responses, plus incidental precipitation. Such a system might consist of a concrete pad with 6-in. curbing. [Memo from Bruce Weddle, EPA Headquarters to Stanley Siegel, EPA Region II, March 17, 1988]

Q *A facility wants to use a trench that normally has nonhazardous wastewater flowing through it as secondary containment for a hazardous waste pipeline. The capacity of the trench will hold both the nonhazardous wastewater flow and the contents of the pipe should it burst. Can the trench be used for this purpose?*

A No. Secondary containment must normally be dry in order to detect a release from the hazardous waste line. [RO 13221]

10.3.3 Release detection

Besides the general design requirements noted above, secondary containment systems must also be provided with release detection such that a failure of the primary containment system will be detected within 24 hours of its occurrence, or at the earliest practical time. [§§264/265.193(c)(3)] Once detected, the facility must ensure that spilled or leaked waste (and precipitation) is removed from the secondary containment system within 24 hours, or "in as timely a manner as is possible to prevent harm to human health and the environment." [§§264/265.193(c)(4)]

Release detection systems include wire grids, observation wells, and U-tubes. The types of leak sensors used in these systems include thermal conductivity sensors, electrical resistivity sensors, and vapor detectors. [EPA/530/K-05/018, EPA/530/SW-86/044] Significant technical guidance on these systems is provided in EPA/530/SW-86/044.

We have seen a number of tank systems with built-in sumps in the bottom of the secondary containment structures (see Figure 10-2). These sumps are provided with liquid level instrumentation that triggers an alarm to appropriate facility personnel if liquid is detected and/or they are inspected every day for signs of liquid. A drain line is often installed in the sump, allowing accumulated liquids to be withdrawn from the containment area. Should a valve be installed in that drain line? The agency noted in EPA/530/SW-86/044 that containment systems should be:

> "equipped with manual release valves, siphons, or pumps to permit removal of collected liquids. Valves should be chained and locked in a closed position when not in use.... If the collected material is hazardous...it must be managed in accordance with all applicable requirements of Parts 262 through 265 of RCRA...."

Except for double-walled tanks, daily visual inspections may be conducted as an alternative means of satisfying the release detection requirements for secondary containment systems, subject to the restrictions discussed below. Of course, the tank and ancillary equipment must be aboveground and accessible for such inspections to be an acceptable alternative. Where, for example, a tank is resting directly on a concrete pad, the owner/operator must demonstrate (to his/her RCRA authority) that visual inspections would reliably indicate leakage from the bottom of the tank within 24 hours of its occurrence. To make such a case, the concrete pad would have to be impermeable and free of cracks (e.g., lined or coated) and sloped or otherwise designed to facilitate the flow of released waste from under the tanks (e.g., equipped with ridges or grooves). Another possibility is raising the tank above the pad onto structural support steel; again, a daily visual inspection would fulfill the release detection requirement. [April 4, 2006; 71 *FR* 16879, RO 12701, 12921, 13173, 13195]

Also, note that where daily visual inspections are being used as a means of leak detection, the inspections must be conducted every day that the subject tank contains hazardous waste, even if the manufacturing system that generates waste and sends it to the tank is not in operation every day. [RO 13063] (This inspection requirement, which is meeting the release detection provisions, is independent of the general tank inspection requirements in §§264/265.195—see Section 10.5.)

Secondary containment system requirements for double-walled tanks specify built-in, continuous leak detection. This is specified because it is a standard design feature of these types of tanks. [RO 12701]

Release detection is also required for ancillary equipment. Leak-detection sensors along the lengths of piping enable an owner/operator to detect even small leaks within a piping system. Sensors used in these applications are the same as those described above.

Finally, note that release detection systems monitor for leaks from primary containment devices (i.e., the tank and its ancillary equipment) into the secondary containment structure. Leak detection is not required outside of the secondary containment system in order to detect failure of that equipment. [September 2, 1988; 53 *FR* 34084]

10.3.3.1 Responses to releases

In order to minimize effects on human health and the environment, EPA has spelled out a number of mandatory steps owners/operators must take when leaks or spills occur from tank systems, or when tanks are found to be unfit for use. These steps are as follows [§§264/265.196]:

- The flow of hazardous waste into the tank or secondary containment system must be stopped immediately and the system must be inspected to determine the cause of the release.

- If the release was from a tank system, enough waste material must be removed from the tank system within 24 hours after release detection, or as soon as practical, to prevent further releases and facilitate inspection and repairs.

- Where released material was captured in a secondary containment system, the waste must be removed from the containment system within 24 hours, or in as timely a manner as possible.

- An immediate visual inspection of the leak or spill must be made so that steps may be taken to prevent or minimize migration of material to soil and/or surface water, and so that visibly contaminated soil and/or surface water may be removed and disposed.

- Tanks, ancillary equipment, and secondary containment systems must be repaired prior to being placed back into service. Where such repairs are major, they must be certified by a qualified professional engineer. If needed repairs cannot be made, the tank system must be closed per the terms of §§264/265.197.

- If the source of the leak or spill is a component of a tank system currently without secondary containment, that component must be provided with secondary containment before being placed back into service (unless the component is aboveground and can be inspected daily for leaks).

10.3.3.1.1 Reporting

If any leak or spill enters the environment, the facility owner/operator must verbally notify the state or EPA within 24 hours of its detection. If the National Response Center has been notified pursuant to CERCLA requirements in 40 *CFR* Part 302, that call satisfies this reporting requirement. A written report summarizing the circumstances of the release and its subsequent investigation (e.g., cause, extent of contamination, corrective action, etc.) must subsequently be sent to the state or EPA within 30 days of detecting the release. Note that reporting is not required where no more than one pound of waste was released and the released material was immediately contained and cleaned up.

Releases completely contained within secondary containment systems are not reportable since they never entered the environment. [EPA/530/SW-87/012, cited previously]

10.3.4 Implementation timing for secondary containment

Sections 264/265.193(a) list the following deadlines by which secondary containment must be installed on new and existing hazardous waste tank systems ("new" versus "existing" tanks are discussed in Section 10.1.5.1):

1. Prior to placing new tank systems into hazardous waste service; and

2. For existing tank systems that store or treat material that becomes a hazardous waste (i.e., as a result of a newly identified or listed hazardous waste), within 2 years of the new hazardous waste identification or listing, or when the tank system has reached an age of 15 years, whichever is later.

In the July 14, 1986 final rule that implemented the secondary containment requirement for hazardous waste tanks [51 *FR* 25422], EPA noted that 15 years is the approximate median time to failure for underground steel tanks the agency studied in developing the rule. Thus, this is the trigger age for existing tanks to install secondary containment.

In evaluating equipment in terms of the 15-year trigger, the agency notes that it is the age of the tank itself, not the length of time it has been in hazardous waste service that is pertinent. Thus, a 14-year old tank in hazardous waste service for just three years would require secondary containment within one year. [RO 13191] For ancillary equipment in pipe runs, the age of the piping, valves, fittings, etc. should be used against the trigger date, rather than the age of the hazardous waste tank with which the piping is associated. This is especially important where piping was completely replaced after the installation date of its associated tank. [RO 12702]

Q: *A tank system installed in July of 1978 was used to store a previously nonhazardous material that was listed as a hazardous waste on May 15, 2001. When is secondary containment required?*

A: The unit is considered an existing hazardous waste tank. Although the tank system was already more than 15 years old as of May 15, 2001, the owner/operator has two years from that date to install secondary containment. The deadline, therefore, was May 15, 2003.

10.3.4.1 Interim measures for existing tanks without secondary containment

Tank systems subject to secondary containment requirements, but which have not yet been so equipped, are subject to several interim requirements:

- Aboveground and enterable USTs must be leak tested annually or be subject to an agency-approved integrity assessment prepared by a professional engineer. If the latter assessment is conducted, it must be adequate to detect cracks, leaks, and corrosion or other conditions that might lead to cracks and leaks. The frequency of the integrity assessment must be based on the age of the system, material of construction of the tank and its ancillary equipment, characteristics of the wastes stored or treated in the system, type of corrosion protection in use, and rate of corrosion or erosion noted in previous inspections.

- Nonenterable USTs must be leak tested annually per the terms of §§264/265.191(b)(5).

- Ancillary equipment must be leak tested or have their integrity checked annually via an agency-approved method.

Q: *Owners/operators of certain hazardous waste storage tanks that do not have secondary containment must conduct leak tests "at least annually" until secondary containment is provided. [§§264/265.193(i)(1)] What is the definition of "annually"?*

A: EPA interprets this term to mean once every 12 months, as opposed to once every calendar year. [RO 14168] For example, if the owner/operator conducted a leak test on August 1 of one year, the test would have to be conducted by August 1 of the second year. If the owner/operator tested the tank on July 1 of the second year, the next test would need to be performed on or before July 1 of the third year.

10.3.5 Variances may be requested

Owners/operators of tank systems may request a variance from EPA or the state from the secondary containment requirements. Such a variance must propose an alternative design and/or operating scheme that, in conjunction with site-specific characteristics, will provide similar spill and leak containment and control. Without going into too much detail [see §§264/265.193(g) for specifics], the facility must petition the agency 1) at least 24 months prior to the date secondary containment is required for existing systems, or 2) at least 30 days before entering into a contract to install a new tank system. The agency will weigh the petition and an associated demonstration of the alternative approach, along with site-specific factors before issuing a decision whether to grant the variance. [RO 12754] The factors evaluated when a variance is proposed using a process building for secondary containment are summarized in RO 14632. Significant guidance on obtaining either a technology-based or risk-based variance from the secondary containment requirements is contained in EPA/530/SW-87/002A or B, February 1987, available from http://nepis.epa.gov/

EPA/html/Pubs/pubtitleOSWER.html by downloading the reports 530SW87002A or 530SW87002B, respectively.

10.4 RCRA-empty tanks

In Section 9.5, we discussed the regulatory definition of a RCRA-empty container. If a container of hazardous waste is emptied using the provisions in §261.7, any residues remaining in the container are exempt from further RCRA regulation; that is, the waste code is eliminated on any residues remaining in the container. Thus, some other waste can be put into that container without the previous waste code attaching to the new contents.

What about a hazardous waste tank? Can we do the same thing? Doubtful. EPA has never promulgated standards on the matter—there is no mention of a RCRA-empty tank anywhere in the federal hazardous waste regulations. "[T]here is no 'empty' rule for anything but containers...." [RO 11387] The implication is that we don't know how to get the waste code off residues left in hazardous waste tanks. Except for some hints, which we describe below, there is also a lack of guidance on the subject.

10.4.1 Emptying a 90-day accumulation tank

Hazardous waste may be accumulated in large quantity generators' tanks for up to 90 days without a permit. Each time a tank is emptied, the 90-day clock resets. But what does it mean to empty the tank? The only guidance available on the subject of RCRA-empty tanks concerns emptying 90-day accumulation tanks for purposes of restarting the 90-day clock. Specifically, the preamble to a January 11, 1982 final rule states the following:

"Questions have been raised concerning the applicability of the 90-day accumulation provision to accumulation in tanks.... A tank will be considered 'empty' when its contents have been drained to the fullest extent possible. Since many tank designs do not allow for complete drainage due to flanges, screens, or siphons, it is not expected that 100% of the wastes will always be removed. As general guidance, a tank should be considered empty when the generator has left the tank's drainage system open until a steady, continuous flow has ceased." [47 FR 1250; see also RO 14763]

Unfortunately, the above guidance says nothing about the residues. Are they exempt from further RCRA regulation, which is the case for residues left in empty containers, or do they still carry the waste code? We don't know. Check with your state.

Note that a recent EPA guidance memo allows flow-through tanks to be used for accumulation purposes by using a "turnover" or "mass balance" approach to determine whether a tank is emptied within 90 days:

"EPA is interpreting §262.34(a)(1)(ii) to allow for the turnover approach.... In the case of hazardous wastes flowing through tanks continuously, there is a means of demonstrating when a tank is 'emptied' within 90 days under §262.34(a)(1)(ii) that would not require completely emptying the tank, and that may be more suitable for tanks with continuous flow. More specifically, a mass balance approach (i.e., the 'turnover' approach...) can be used for continuous flow tanks rather than the approach described above for batch process tanks. The key parameters in this mass balance approach are the volume of the tank (e.g., 6,000 gallons), the daily throughput of hazardous waste (e.g., 300 gallons per day) and the time period the hazardous waste 'resides' in the tank. In this example, the hazardous waste entering the tank would have a residence time of 20 days ((6,000 gallons/ 300 gallons per day) = 20 days) and meet the requirements of §262.34(a)(1)(ii) since the hazardous waste has been in the tank for less than 90 days." [RO 14763]

10.4.2 What about heels?

In August 1982 guidance, EPA stated "there may be cases where a tank is never 'completely empty.' We recognize this problem but believe a deviation from 'completely empty' is a satisfactory compromise in a real world of day-to-day operations."

[RO 12062] Unfortunately, this says nothing about the heel of material left in a tank that cannot be completely emptied. Does that heel still carry the waste code of the stored hazardous waste? Probably. Would solid waste added to the emptied tank become a hazardous waste under the mixture rule? If the heel was a non-ICR-only listed hazardous waste, probably; otherwise, the resulting mixture would probably not be hazardous if it did not exhibit a characteristic.

10.4.3 Check with your state

In lieu of any other guidance on this topic, we suggest you determine if your state has its own policy on rendering a hazardous waste tank RCRA-empty.

10.5 Tank inspection requirements

Subpart J, specifically §§264/265.195 and 265.201, imposes a number of inspection requirements on hazardous waste tanks at permitted and interim status TSD facilities, as well as tanks at large and small quantity generators. These requirements are summarized in Table 10-4.

As noted in Section 10.3.3 and subject to certain restrictions, EPA will allow daily inspections of tank and secondary containment systems as a way for owners/operators to meet the release detection system requirements of §§264/265.193. Recall that the release detection system standard is that the system be able to detect a release within 24 hours of its occurrence. Therefore, if such tank inspections also fulfill the release detection system requirement, they must be conducted every day that the system contains hazardous waste, regardless of whether the facility is conducting manufacturing operations. [RO 13063]

Addressing another area of possible confusion, it is clear that the bottom of tanks resting on concrete pads are technically aboveground. However, EPA recognizes that they cannot be visually inspected. Thus, the agency encourages facilities to carefully look for leakage around the base of such tanks, which could indicate releases from the tank bottom. [RO 12921]

Finally, although EPA would probably frown on the use of a "web cam" as a means to meet the tank system inspection requirement, the agency in some cases would allow video monitoring for such inspections. However, such a video system must "provide a level of performance comparable to actual close-up visual inspection of the entire system and the capability of effectively detecting leaks within 24 hours." [RO 12868; see also EPA/530/SW-87/012, cited previously]

Owners/operators of permitted and interim status TSD facilities and large quantity generators must document that the above-noted inspections were conducted. [§§264.195(h)/265.195(g)] Nothing in the regulations requires small quantity generators to record their obligatory inspections, but we would recommend that such entities do that. Although the federal regulations do not require generators to comply with the inspection records retention obligation in §265.15(d), we recommend that this provision be followed anyway. Log entries for inspections should include the following:

- The date and time of the inspection,
- The name of the inspector,
- A notation of any observations made (e.g., leaks, rusting), and
- The date and nature of any repairs or other remedial actions initiated.

The regulatory language requires the *name*—not the initials—of the inspector to be documented. Inspections should be recorded in the plant operating log and should be kept for a minimum of three years from the date of inspection.

Inspection records are virtually certain to be scrutinized in a regulatory inspection. Facility personnel can become bored with making the same inspection day after day and sometimes want to skip the inspection itself or filling out inspection logs. Facilities need to give special attention to ensuring that inspections are conducted and documented each operating day.

Table 10-4: Inspection Requirements for Hazardous Waste Tanks

Required inspection	Inspection frequency	
	Permitted and interim status TSD facilities and large quantity generators	Small quantity generators
Inspect aboveground portions of tank system for corrosion and signs of waste releases[1]	Each operating day[2,3]	Weekly
Inspect construction materials and areas surrounding externally accessible portions of tank system for erosion and signs of waste releases, including cracks, wear, etc., in concrete or coatings/linings	Each operating day[2,3]	Weekly
Inspect overfill/spill control equipment to ensure system is in good working order	Each operating day[2,3,4]	Each operating day[3,5]
Inspect data from release detection and other monitoring equipment to ensure operation in accordance with system design	Each operating day[3]	Each operating day[3,5]
Inspect ancillary equipment that is not required to have secondary containment (e.g., aboveground piping with welded connections)	Each operating day[3,6]	Not required
Inspect cathodic protection systems for proper operation	Within six months of installation and annually thereafter	Not required
Inspect sources of impressed current for proper operation	Every other month	Not required
Inspect level of the waste in the tank to ensure adequate freeboard	Not required	Each operating day[3,5]

Tank system = The tank, its secondary containment system, and its associated ancillary equipment.

[1]Particular attention should be given to bottom-to-shell welds/connections, flanges, rivet holes, welded seams, valves, nozzles, pumps, bypass piping, welded brackets, and secondary containment systems. Guidance for inspection of specific equipment and components is given in EPA/530/SW-86/044. Recommendations for inspection of metal, fiberglass-reinforced plastic, and concrete tanks are included in this guidance document.

[2]Facilities may inspect this equipment weekly if 1) leak detection equipment is in place, or 2) established work practices are in place that ensure any leaks or spills will be promptly identified.

[3]EPA guidance for the term "each operating day" is contradictory. RO 13063 (dated October 1987) clearly states that "each operating day" means each day during which manufacturing operations are being conducted; that is, tank inspections need only be conducted on days when the subject facility is in operation. In contrast, the much more recent *Introduction to Tanks* [EPA/530/K-05/018, September 2005, available from http://www.epa.gov/epawaste/inforesources/pubs/training/tanks05.pdf] notes that the agency has clarified that "each operating day" means "every day the tank is in operation (i.e., storing or treating hazardous waste) and not necessarily just on days the facility is open for business." This definition would require tank inspections on days the facility may not actually be operating (e.g., on weekends). Absent an unequivocal (e.g., written) statement from EPA or your state RCRA authority, we believe it would be prudent to follow the more recent and conservative EPA report approach and inspect tank systems every day they contain hazardous waste. Daily internal inspections are not required. [RO 12921]

[4]Except that permitted TSD facilities may have a different inspection frequency, as required in their permits.

[5]Facilities may inspect this equipment weekly if the tank has full secondary containment *and* either 1) leak detection equipment is in place, or 2) established work practices are in place that ensure any leaks or spills will be promptly identified.

[6]In situations where ancillary equipment is located inside a building that has been determined to provide secondary containment, that equipment may be inspected weekly if either 1) leak detection equipment is in place, or 2) established work practices are in place that ensure any leaks or spills will be promptly identified. [April 4, 2006; 71 *FR* 16880]

Source: McCoy and Associates, Inc.; adapted from §§264/265.193(f), 264/265.195, and 265.201(c), unless otherwise noted.

Examples of items that should be considered for inclusion in a typical tank inspection checklist or record include [OSWER Directive 9523.00-10]:

- Appropriate tank and piping labeling/marking;
- Leaks from tanks, pumps, and inlet/outlet nozzles or flanges;
- Leaks from or damage to ancillary equipment;
- Seal integrity on manholes, gauge hatches, weather shields, and level gauges;
- High liquid level/overflow;
- Structurally intact with no damage to tank walls;

- Evidence of corrosion, excessive rusting, or damage;
- Overfilling controls operative;
- Secondary containment free of chips, cracks, etc.; and
- Any sump pump operative.

10.5.1 Weekly tank inspections allowed under certain conditions

Under an April 4, 2006 final rule [71 *FR* 16862], large quantity generators and TSD facilities are allowed to reduce the inspection frequency of some components of their hazardous waste tank systems (as noted in Table 10-4) from daily to weekly when either of two conditions are met: 1) tank owners/operators employ leak detection equipment per §§264/265.193(c)(3); or 2) in the absence of leak detection equipment, tank owners/operators employ established work practices that ensure that when any leaks or spills occur, they will be promptly identified and remediated per §§264/265.193(c)(3) and (4).

EPA gave the following examples of established work practices that could be used to warrant weekly rather than daily tank inspections of certain tank components [71 *FR* 16879]:

- Use of an environmental management system that includes plans and practices to ensure that any releases are promptly identified, contained, and cleaned up.
- If the tank system is in an area frequented by employees, where releases will be immediately obvious, all employees are trained to watch for releases and report them to appropriate facility personnel.
- An employee is assigned to check the tank secondary containment on a daily basis without conducting a full tank system "inspection."

The tank regulations in §265.201 do not require small quantity generators to have secondary containment for their 180/270-day hazardous waste tanks. However, under the April 4, 2006 rule [71 *FR* 16862], small quantity generators are allowed to reduce the inspection frequency of some components of their hazardous waste tanks (as noted in Table 10-4) from daily to weekly if they provide full secondary containment for the tank or tank system *and* either use leak detection equipment or implement established work practices to ensure leaks are promptly identified.

If a facility reduces its hazardous waste tank inspection schedule to weekly for certain components, it must document the alternate inspection schedule in its operating record. If the reduced inspection frequency is based on established work practices to ensure leaks are promptly identified, these practices must be included in the documentation.

Recycling

11

If we recycle this stuff, is it solid waste?

In Section 1.2 of this book, we evaluated whether a secondary material that is recycled in certain ways is a solid waste. This decision is key since the agency's statutory authority under RCRA Subtitle C is limited to materials defined as solid wastes (i.e., the RCRA hazardous waste regulations cannot be applied to materials that are not solid wastes). As noted in that section, our initial consideration of recycling was limited to §261.2(c), which defines whether materials recycled via use constituting disposal, burning for energy recovery, reclamation, and/or speculative accumulation are solid wastes. In this chapter, we broaden our coverage of this issue to include the remaining recycling provisions of Part 261—these deal with the use/reuse exclusions, reclamation exclusions, legitimacy, documentation requirements, and other issues.

We begin our second, more comprehensive look at recycling with three definitions:

1. "A material is 'recycled' if it is used, reused, or reclaimed.

2. "A material is 'used or reused' if it is either:

    "(i) Employed as an ingredient (including use as an intermediate) in an industrial process to make a product (for example, distillation bottoms from one process used as feedstock in another process). However, a material will not satisfy this condition if distinct components of the material are recovered as separate end products (as when metals are recovered from metal-containing secondary materials); or

    "(ii) Employed in a particular function or application as an effective substitute for a commercial product (for example, spent pickle liquor used as phosphorous precipitant and sludge conditioner in wastewater treatment).

3. "A material is 'reclaimed' if it is processed to recover a usable product, or if it is regenerated. Examples are recovery of lead values from spent batteries and regeneration of spent solvents...." [§261.1(c)(4), (5), and (7)]

Deciding whether reclaiming and/or recycling a residue back into a production process is simply a manufacturing operation outside the scope of RCRA or, conversely, waste management subject to Subtitle C requirements is not obvious. The RCRA recycling provisions are complicated and confusing. For example, is reuse of a process residue excluded from RCRA if it must be neutralized or filtered before it can be recycled? What if the residue has a lower active ingredient content than the virgin feedstock it is meant to replace? Can a residue be legitimately reclaimed and reused in production operations without regard to RCRA? These and other issues create problems when deciding whether materials being recycled are legitimately excluded from RCRA regulation.

In studying the Part 261 regulations and guidance on the subject, we have come up with a five-step procedure for determining whether materials being recycled are subject to regulation under RCRA. The remainder of this chapter discusses this procedure. In Section 11.1, we introduce the five individual steps and present a logic diagram that integrates these steps into a coherent scheme. Then, in Sections 11.2–11.6, we take a detailed look at each step and examine many examples documented in EPA guidance.

11.1 Overview of the five-step recycling determination process

This whole chapter (and Section 1.2 earlier) discusses recycling of a *hazardous secondary material*. We start our examination of recycling by discussing what is meant by this term.

11.1.1 What is a hazardous secondary material?

When EPA redefined "solid waste" on January 4, 1985, the agency used the term "secondary material" in many of its discussions. That term was defined in preamble language as a material that is potentially a solid and hazardous waste when recycled. [50 *FR* 616] More recently, EPA added a definition of "hazardous secondary material" in §260.10 as a "secondary material (e.g., spent material, byproduct, or sludge) that, when discarded, would be identified as hazardous waste...." [October 30, 2008; 73 *FR* 64757] Determining the RCRA status of hazardous secondary material that is recycled is the subject of this chapter and our five-step process.

11.1.2 What are the five steps?

The five steps of our recycling determination process involve asking the following questions:

- Step 1—Is the material determined to be a solid waste using Table 1 of §261.2(c)? This step includes an evaluation of whether the material is excluded from the definition of solid waste under one of the two reclamation exclusions listed in §§261.2(a)(2)(ii) and 261.4(a)(23–25). [Section 11.2]

- Step 2—Is the secondary material excluded from the definition of solid waste under one of the four use/reuse recycling exclusions listed in §§261.2(e)(1)(i–iii) and 261.4(a)(8)? [Section 11.3]

- Step 3—If the material is excluded under one of the §261.2(e)(1)(i–iii) use/reuse recycling exclusions, does the process meet the additional qualifying criteria listed in §261.2(e)(2)(i–iv)? [Section 11.4.1] If the material is excluded under one of the §§261.2(a)(2)(ii)/261.4(a)(23–25) reclamation exclusions, does the reclamation meet the additional qualifying criteria listed in these sections and 73 *FR* 64669–70? [Section 11.4.2]

- Step 4—Is the recycling legitimate? This step includes an evaluation of the legitimate recycling criteria codified in §260.43 and contained in previous EPA guidance, as well as a possible discussion with the facility's state or regional RCRA authority. [Section 11.5]

- Step 5—For all materials determined not to be solid wastes, can the facility meet the prescribed documentation requirements of §261.2(f) in the event of an inspection, evaluation, or enforcement action by the state or EPA? This step includes documentation that the recycling is legitimate. [Section 11.6]

11.1.3 Putting it all together

Figure 11-1 organizes the five-step recycling determination process into a logic diagram. For clarity, we have indicated which boxes correspond to which steps.

With the logic diagram at the ready, we turn now to the first step of our recycling determination process.

11.2 Step 1—Check the table with the asterisks

The first step in determining the status of a hazardous secondary material that will be recycled is to

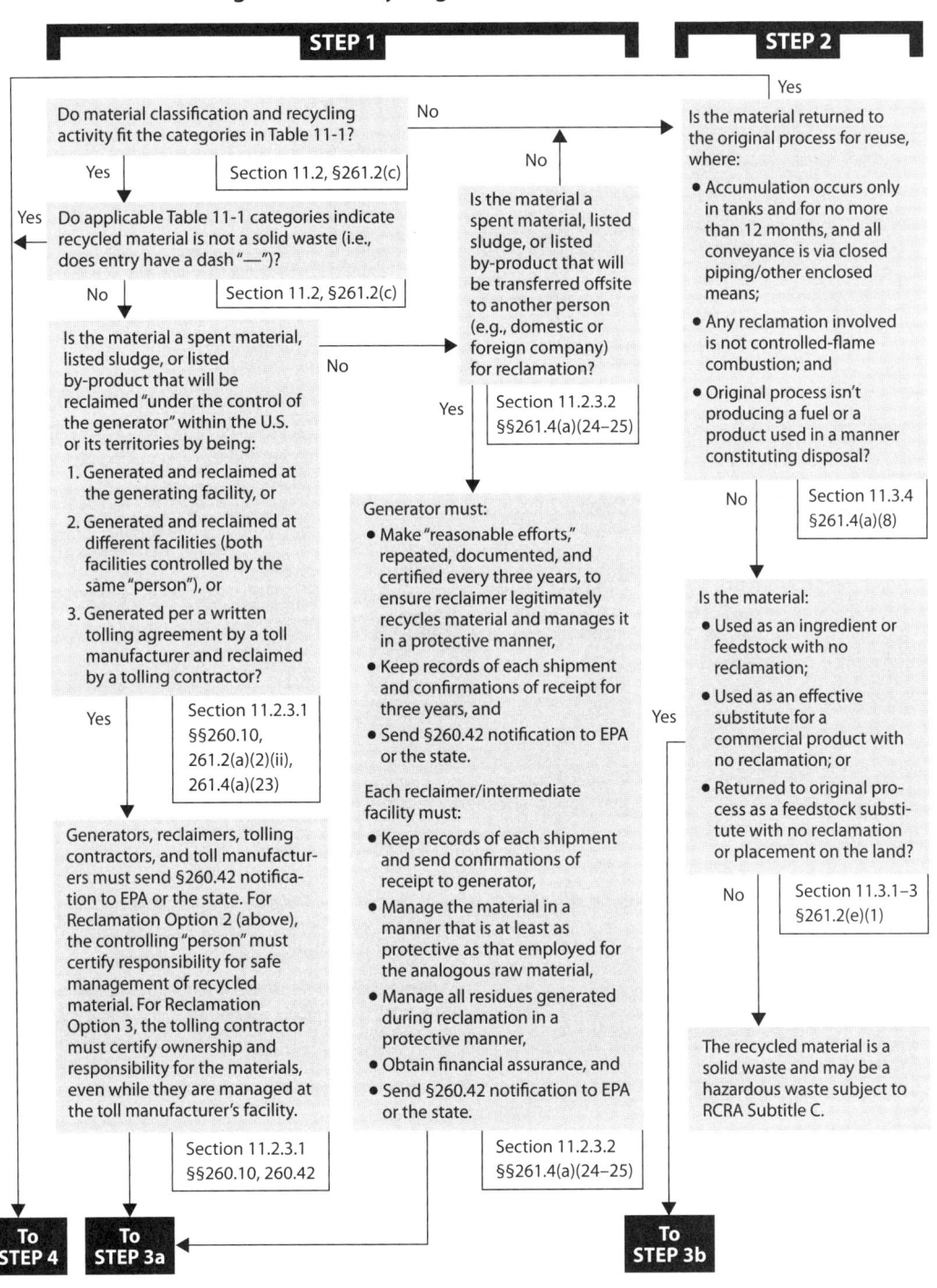

Figure 11-1: Recycling Determination Process

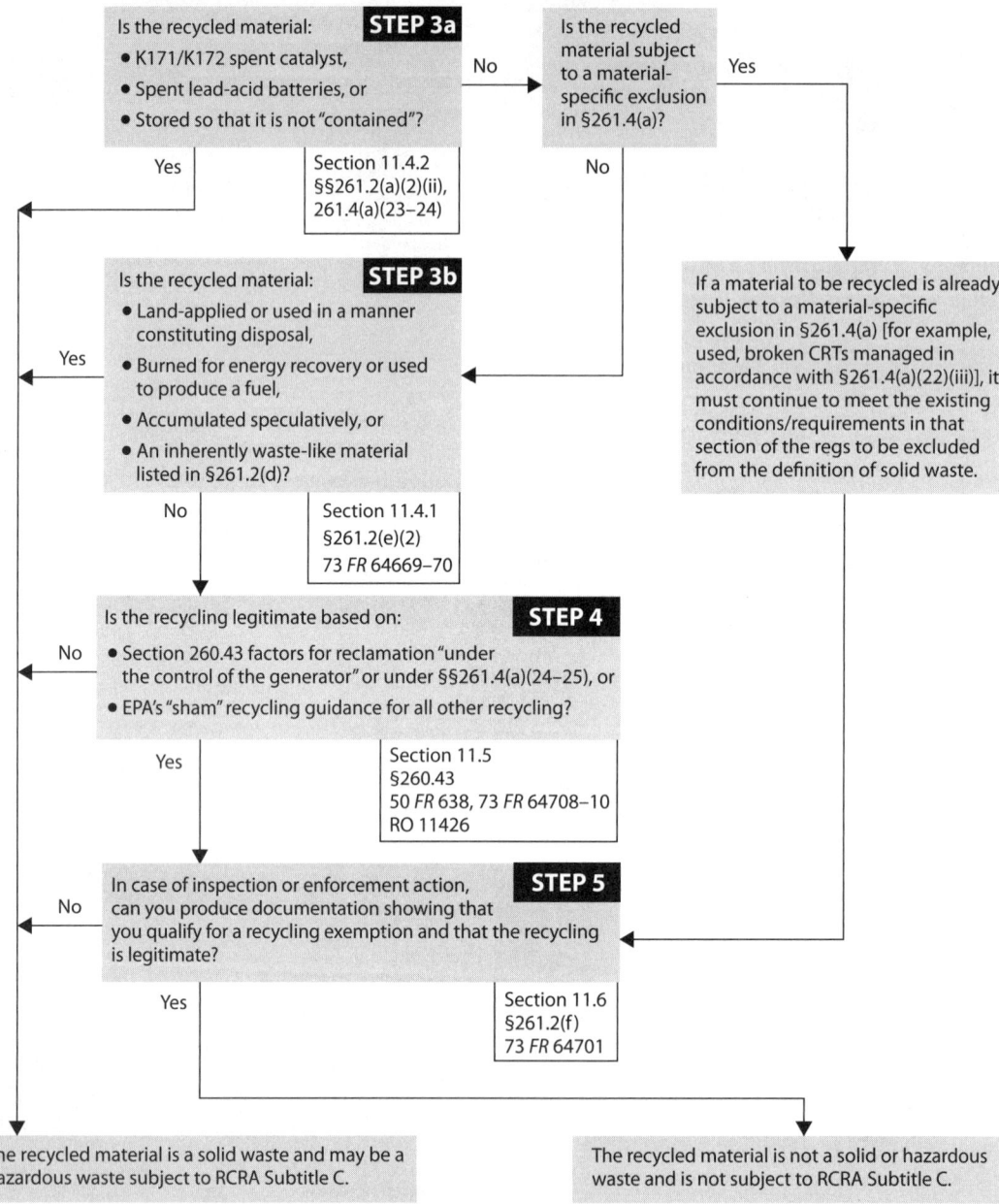

Figure 11-1 (continued): Recycling Determination Process

Source: McCoy and Associates, Inc.

check Table 1 in §261.2(c), which we have reproduced here as Table 11-1. This table is always the place to start when you're trying to figure out if a material is a solid waste when recycled. In this step, you determine if there are categories in Table 11-1 that fit the specific recycling situation at your facility. In effect, this step involves two separate questions: 1) Can the material in question be classified into one of the material categories in the seven rows of the table *and* can the recycling process of interest be categorized as one of the four activities listed in the table column headings? and 2) If the answer to the first question is "Yes," is there a "—" at the intersection of the appropriate row and column? If the answer to either of these questions is "No," a frequent occurrence, you must look to other possible exemptions to potentially escape RCRA regulation.

If a material classification and recycling activity in Table 11-1 do apply, you check the intersection of the proper row and column. If there is an "*" at the intersection, the secondary material *is* a solid waste (and potentially a hazardous waste) when recycled in that manner; since the table with the asterisks didn't help you get out of RCRA, you move on to check for other possible recycling exemptions. If there is a "—" at the intersection of the appropriate row and column, then the hazardous secondary material is *not* a solid waste if you can show the recycling is legitimate and can meet prescribed documentation requirements (as required by Steps 4 and 5).

11.2.1 Classifying hazardous secondary materials

In the section above, we make reference to seven categories (rows) of materials and four categories (columns) of recycling processes. Section 1.2 of this publication covered each of these 11 items in significant detail and presented numerous examples that EPA has provided in guidance documents. We briefly review these materials and activities below while referring readers wanting more detail back to Section 1.2.

The regulations that govern whether recycled hazardous secondary materials are solid wastes are based on classifying those materials into one of the following categories:

Table 11-1: Determining If Recycled Materials Are Solid Wastes

Secondary material	Use constituting disposal [§261.2(c)(1)]	Energy recovery/fuel [§261.2(c)(2)]	Reclamation [§261.2(c)(3)] [except as provided in §§261.2(a)(2)(ii), 261.4(a)(17), 261.4(a)(23), 261.4(a)(24), or 261.4(a)(25)]	Speculative accumulation [§261.2(c)(4)]
Spent materials	(*)	(*)	(*)	(*)
Sludges (listed in 40 *CFR* 261.31 or 261.32)	(*)	(*)	(*)	(*)
Sludges exhibiting a characteristic of hazardous waste	(*)	(*)	—	(*)
By-products (listed in 40 *CFR* 261.31 or 261.32)	(*)	(*)	(*)	(*)
By-products exhibiting a characteristic of hazardous waste	(*)	(*)	—	(*)
Commercial chemical products listed in 40 *CFR* 261.33	(*)	(*)	—	—
Scrap metal that is not excluded under §261.4(a)(13)	(*)	(*)	(*)	(*)

Note: The terms "spent materials," "sludges," "by-products," and "scrap metal" are defined in §261.1.
Source: §261.2(c).

- A *spent material* is "any material that has been used and as a result of contamination can no longer serve the purpose for which it was produced without processing." [§261.1(c)(1)] Examples include spent solvents, depleted catalysts, clogged filters, and burned-out light bulbs.

- Simply put, a *sludge* is any residue from a water or air pollution control device (a more formal definition is codified in §260.10). This category includes materials not normally thought of as sludge, such as baghouse dust, oil from an oil-water separator, and spent resin from an ion-exchange column. Note that in terms of Table 11-1, EPA actually splits sludges into two categories: 1) sludges that are listed in §261.31 or 261.32, and 2) sludges that exhibit a characteristic of hazardous waste.

- A *by-product* is "a material that is not one of the primary products of a production process and is not solely or separately produced by the production process. Examples are process residues such as slags or distillation-column bottoms." [§261.1(c)(3)] "By-product" does not include co-products (e.g., gasoline, kerosene) that are intentionally produced and have a ready market without additional processing. As with sludges, EPA splits by-products into separate listed and characteristic categories in Table 11-1.

- *Commercial chemical products* are 1) unused, essentially pure chemicals listed in §261.33 (i.e., the P- and U-lists); and 2) for purposes of determining whether a recycled material is a solid waste, all types of unused products such as circuit boards, batteries, gasoline, etc., whether or not they are actually chemicals and regardless of whether they are listed in §261.33. [April 11, 1985; 50 *FR* 14219, RO 11713, 11726, 13356, 13490]

- *Scrap metal* is "bits and pieces of metal parts (e.g., bars, turnings, rods, sheets, wire) or metal pieces that may be combined together with bolts or soldering (e.g., radiators, scrap automobiles, railroad box cars), which when worn or superfluous can be recycled." [§261.1(c)(6)]

11.2.2 Categorizing recycling activities

Usually it isn't too hard to classify the hazardous secondary material we want to recycle into one of the seven material categories noted above. What is difficult at times, however, is trying to figure out which (if any) of the four recycling activities in Table 11-1 matches the recycling scenario at our facility. Four types of recycling activities are included in this table:

- *Use constituting disposal*—Materials are placed directly on the land (e.g., road oiling) or are used to produce products that are applied to the land (e.g., fertilizer or asphalt).

- *Burning for energy recovery*—Materials are burned directly in boilers or industrial furnaces (BIFs) or are blended into fuels burned in BIFs. Note that materials burned in incinerators and other waste destruction units are considered to be discarded, *not* burned for energy recovery.

- *Reclamation*—Materials are either 1) processed to recover something of value, or 2) regenerated. Examples include recovery of lead from dead batteries and regeneration of spent catalyst.

- *Speculative accumulation*—Materials that are accumulated prior to recycling are speculatively accumulated if a facility does not recycle at least 75% of them in a calendar year. EPA introduced this concept to prevent facilities from avoiding RCRA regulation by merely claiming that they were going to recycle the wastes they were collecting. The definition of this term, codified in §261.1(c)(8), allows persons to accumulate recyclable materials for limited periods of time and requires them to demonstrate that 1) a material has a feasible means of being recycled, and 2) at least 75% of the accumulated material is being recycled each calendar year. Failure to meet these provisions automatically makes the accumulated material solid waste potentially subject to full RCRA regulation.

11.2.3 Exceptions to the asterisks

Note that there are numerous exceptions to the asterisks in Table 11-1. First, commercial chemical products are normally solid wastes when applied to the land or used to produce a product that is applied to the land (use constituting disposal)—that is, there is an asterisk at the intersection of "Commercial chemical products" and "Use constituting disposal." However, per §261.2(c)(1)(ii), commercial chemical products that are applied to the land are not solid wastes if that is their ordinary manner of use. An example would be an unused P- or U-listed herbicide that is recycled into a product that is applied to the land to kill weeds. Since that is an intended use for herbicides, such recycling would be outside of RCRA.

Second, commercial chemical products that are burned for energy recovery or used to produce fuels are solid waste per the asterisk at the intersection of "Commercial chemical products" and "Energy recovery/fuel." Per §261.2(c)(2)(ii), however, commercial chemical products are not solid wastes when burned for energy recovery or used to produce a fuel if they are themselves fuels. Thus, unused commercial chemicals that are normal components of fuel (e.g., benzene, toluene, and xylene) are not solid wastes when burned for energy recovery. [RO 12505] Similarly, off-spec fuels (e.g., gasoline, jet fuel, kerosene, diesel) are not solid wastes when burned for energy recovery or used to produce fuels. [RO 11138, 11449, 11713, 11848, 12825, 14503]

Third, there are exceptions in five regulatory sections noted in the column heading under "Reclamation" in Table 11-1. The five sections are:

1. §261.2(a)(2)(ii)—Reclamation "under the control of the generator" within the United States or its territories, where the materials are managed in non-land-based units (discussed in Section 11.2.3.1);

2. §261.4(a)(17)—Reclamation of spent materials generated in the primary mineral processing industry (discussed in Section 1.2.6.3.2);

3. §261.4(a)(23)—Reclamation "under the control of the generator" within the United States or its territories, where the materials are managed in land-based units (discussed in Section 11.2.3.1);

4. §261.4(a)(24)—Materials are transferred to another company for reclamation within the United States or its territories and certain conditions are met (discussed in Section 11.2.3.2); and

5. §261.4(a)(25)—Materials are exported for reclamation in a foreign country and certain conditions are met (discussed in Section 11.2.3.2.1).

Items 1 and 3–5 above implement two self-implementing exclusions from the definition of solid waste that EPA finalized on October 30, 2008. [73 FR 64668] If one of these two exclusions applies to a specific recycling situation, and you meet the criteria in Steps 3–5 of our five-step process, the hazardous secondary materials being recycled are not solid wastes—they are manufacturing streams or otherwise valuable commodities. Note that the two exclusions became effective on December 29, 2008 at the federal level. However, the exclusions, which are explained in Sections 11.2.3.1 and 11.2.3.2 of this book, are not effective in RCRA-authorized states unless and until the state adopts them. Furthermore, EPA proposed significant changes to these exclusions on July 22, 2011. [76 FR 44094]

11.2.3.1 Materials reclaimed "under the control of the generator"

The first reclamation exclusion is for materials that are generated and reclaimed "under the control of the generator." The hazardous secondary materials to which this exclusion applies are spent materials (that are hazardous by either listing or characteristic), listed sludges, and listed by-products that will be reclaimed. Examples are metal-bearing sludges and spent catalysts from which metals will be reclaimed and organic-bearing liquids that will be processed to recover solvents. Such materials are not solid or hazardous wastes if reclaimed under the control of the generator.

The "under the control of the generator" exclusion actually consists of three options:

1. The material is generated and reclaimed at the generating facility (located in the United States or its territories); or

2. The material is generated and reclaimed at different facilities, both of which are located in the United States or its territories and both of which are controlled by the same "person"; or

3. The material is generated pursuant to a written contract between a tolling contractor and a toll manufacturer (both of which are located in the United States or its territories) and is reclaimed by the tolling contractor.

Because the hazardous secondary materials remain under the control of the generator from the point of generation through completion of reclamation, EPA is fairly confident that there will be minimal mismanagement or discard resulting in a threat to human health or the environment. "This is because the hazardous secondary material is being treated as a valuable commodity rather than as a waste.... EPA continues to believe that when a generator legitimately recycles hazardous secondary material under its control, the generator has not abandoned the material and has every opportunity and incentive to maintain oversight of, and responsibility for, the hazardous secondary material that is reclaimed." [73 FR 64676] Therefore, the agency has placed very few restrictions on these recycling alternatives, as discussed in the next three subsections.

11.2.3.1.1 *Materials generated and reclaimed at the generating facility*

The first way hazardous secondary materials can be reclaimed under the control of the generator (and thus be excluded from RCRA control) is for the materials to be generated and reclaimed at the generating facility. This is illustrated in Figure 11-2a. For purposes of this exclusion, "generating facility" means all contiguous property owned, leased, or otherwise controlled by the hazardous secondary material generator. [§260.10]

Under this scenario, hazardous secondary materials are generated in a process. Although Figure 11-2a shows the materials being generated in a production process, the regulations do not limit the type of process from which the materials are generated. Thus, it could presumably be a production, pollution control, maintenance, or laboratory operation. The materials, if disposed, would be hazardous waste. Instead of disposal, however, the facility wants to reclaim something of value from these materials.

Before being reclaimed, these materials can be stored in either non-land-based units (e.g., tanks, containers, and containment buildings) [see §261.2(a)(2)(ii)] or land-based units (e.g., surface impoundments and piles) [see §261.4(a)(23)]. In either case, the materials must be "contained" in such units (basically, this means that the material is not released into the environment—see Section 11.4.2). The materials are then sent to onsite reclamation, producing a reclaimed material. This material can be reused at the generator's facility or sent to another facility for use. However, there are limitations on how this reclaimed material can be used (see Section 11.4.1). There undoubtedly will be some residues generated in the onsite reclamation process. Assuming they will be discarded, these residues represent the first point of generation of a solid waste in the entire recycling process. If they are hazardous, the residues must be managed as such under RCRA Subtitle C.

A facility that collects hazardous secondary materials from other persons (for example, when mercury-containing equipment is collected through a special collection program) is not the hazardous secondary material generator of those materials. However, if a generator contracts with a different company to reclaim hazardous secondary materials at the generator's facility, either temporarily or permanently, the materials would be considered under the control of the generator. For example, spent pickle liquor is sometimes reclaimed at a steel plant by a company that is different from the company operating the plant. This activity would qualify for the exclusion. [73 FR 64725]

Figure 11-2: Materials Reclaimed "Under the Control of the Generator"

a) Materials are generated and reclaimed at the generating facility

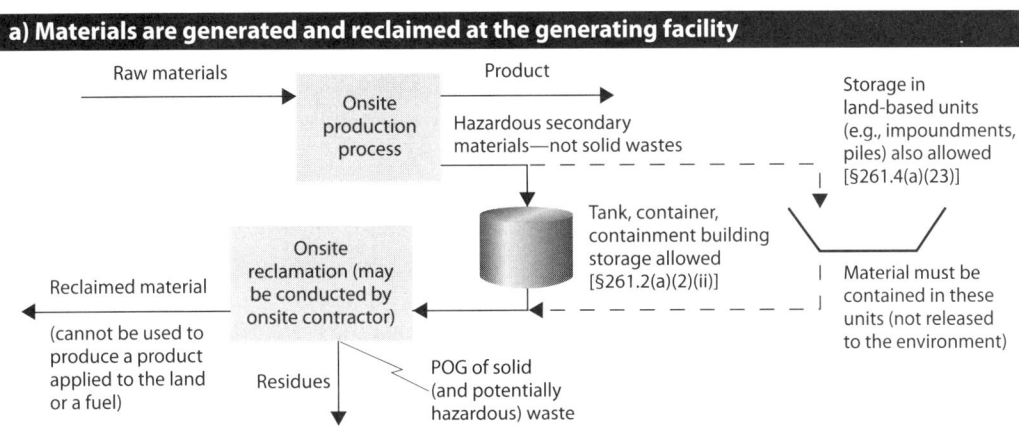

b) Materials are reclaimed at a different facility controlled by the same "person"

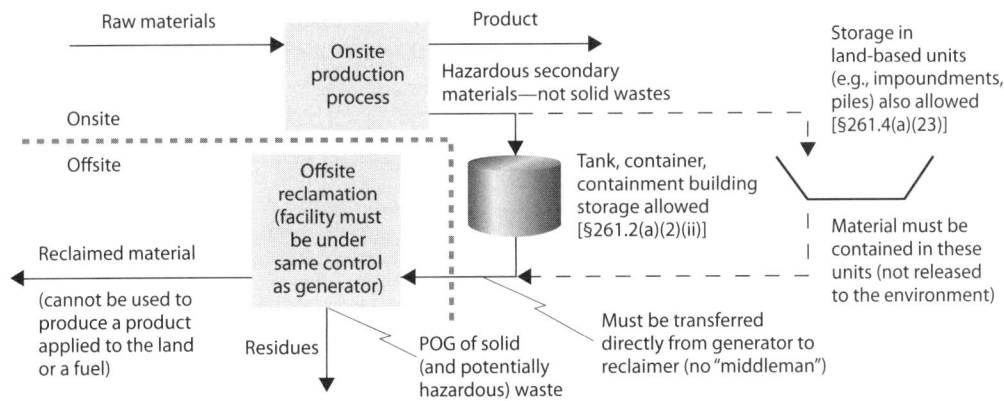

c) Materials are reclaimed at a different facility under a written tolling agreement

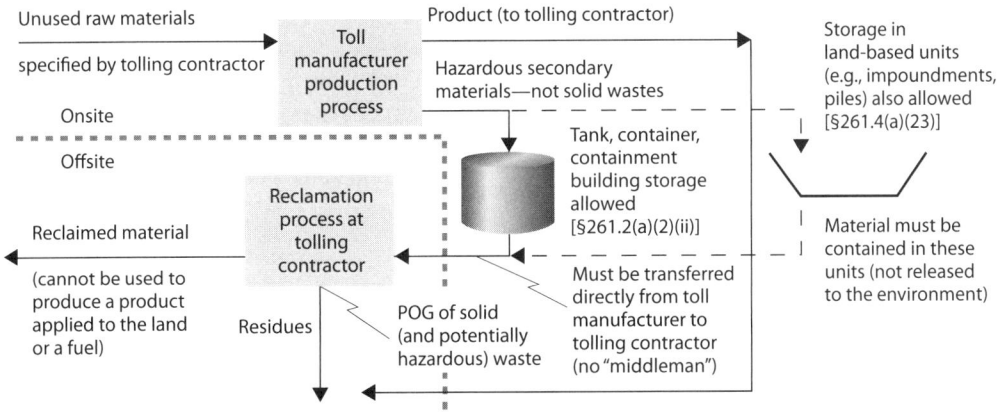

Source: McCoy and Associates, Inc.; adapted from §§261.2(a)(2)(ii), 261.4(a)(23).

Generators sometimes contract with a second company to collect hazardous secondary materials at the generating facility for subsequent reclamation at the second company's facility. In that situation, the hazardous secondary materials would no longer be considered under the control of the generator and would instead have to be managed under the transfer-based exclusion discussed in Section 11.2.3.2. [73 FR 64680]

11.2.3.1.2 Materials generated and reclaimed at different facilities

Reclamation of a hazardous secondary material may also be under the control of the generator if the materials are generated and reclaimed at different facilities. However, this option requires that the reclaiming facility be controlled by the generator, or that the same "person" as defined in §260.10 must control both the generating facility and the reclaiming facility (see Figure 11-2b). For purposes of this option, "control" means the power to direct the policies of the facility, whether by the ownership of stock, voting rights, or otherwise, except that contractors who operate facilities on behalf of a different person shall not be deemed to "control" such facilities. Thus, where a contractor operates two facilities, each of which is owned by a different company, hazardous secondary materials generated at the first facility and reclaimed at the second facility are not considered under the control of the generator; such materials could only be excluded under the transfer-based exclusion (see Section 11.2.3.2). However, in those situations where the generating facility and the reclaiming facility are both owned by the same organization (e.g., a government agency or university), the two facilities would be under common control because the organization has the power to direct the policies of both the generating facility and the reclaiming facility. Under this scenario, both facilities would therefore be eligible for the same-company exclusion, even if operated by different contractors. [73 FR 64727]

In this scenario, secondary materials from a production, pollution control, maintenance, or laboratory operation are generated at a facility that has no onsite reclamation capacity. However, an offsite, sister facility (or other offsite facility under the same control as the generating facility) has a recycling process that can reclaim the hazardous secondary material. Before being reclaimed, these materials can be stored in either non-land-based units (e.g., tanks, containers, and containment buildings) or land-based units (e.g., surface impoundments and piles) at either facility. In either case, the materials must be "contained" in such units (basically, this means that the material is not released into the environment—see Section 11.4.2).

The secondary materials, which are not solid waste, are then reclaimed at the second facility, producing a reclaimed material. This material can be used at the generator's or reclaimer's facility or sent to another facility for use. However, there are limitations on how this reclaimed material can be used (see Section 11.4.1). Again, there undoubtedly will be some residues generated in the reclamation process. Assuming they will be discarded, these residues are solid waste. If they are hazardous, they must be managed as such under RCRA Subtitle C.

As noted in Figure 11-2b, "under the control of the generator" means (in this scenario) that hazardous secondary materials may not be sent to an intermediate facility (i.e., a facility, other than the generator or reclaimer, that stores the materials for more than 10 days). If these materials are sent to an intermediate facility before they are sent to the reclaiming facility, they would not meet the definition of hazardous secondary materials reclaimed under the control of the generator, and they would be subject to the conditions of the transfer-based exclusion, discussed in Section 11.2.3.2 below.

The controlling "person" must certify responsibility for safe management of the recycled material, using the language given in §260.10. This certification should be made by an official familiar with the corporate structure of both the generating and the reclaiming facilities and should be retained at the generating facility. [73 FR 64680]

11.2.3.1.3 Materials generated and reclaimed per a tolling agreement

The third option for reclamation under the control of the generator is for a tolling contractor to arrange for the production of a product or intermediate made from specified unused raw materials through a written contract with a toll manufacturer. As shown in Figure 11-2c, the toll manufacturer produces the product or intermediate and sends it, along with the hazardous secondary materials generated during its manufacture, to the tolling contractor. At this point, the secondary materials are not solid waste. Before being reclaimed, the hazardous secondary materials can be stored in either non-land-based units (e.g., tanks, containers, and containment buildings) or land-based units (e.g., surface impoundments and piles) at either facility. In either case, the materials must be "contained" in such units (basically, this means that the material is not released into the environment—see Section 11.4.2). The hazardous secondary materials must be transferred directly from the toll manufacturer to the tolling contractor and may not be sent to an intermediate facility.

The tolling contractor then reclaims the secondary materials at its site, producing a reclaimed material. This material can be reused at the tolling contractor's facility or sent to another facility for use. However, there are limitations on how this reclaimed material can be used (see Section 11.4.1). Again, there undoubtedly will be some residues generated in the reclamation process. Assuming they will be discarded, these residues are solid waste. If they are hazardous, they must be managed as such under RCRA Subtitle C.

The tolling contractor must certify that it has a written contract with the toll manufacturer to manufacture the product or intermediate and that the tolling contractor will reclaim the hazardous secondary materials generated during its manufacture. The tolling contractor must also certify that it retains ownership of, and liability for, the hazardous secondary materials that are generated during the course of the manufacture, including any releases of hazardous secondary materials that occur during the manufacturing process at the toll manufacturer's facility. This certification should be made by an official familiar with the terms of the written contract and should be retained at the tolling contractor's site. [73 *FR* 64680]

11.2.3.1.4 Notification requirement for hazardous secondary materials

Generators, reclaimers, and any tolling contractors/toll manufacturers that manage hazardous secondary materials that have previously been subject to regulation as hazardous waste, but which are excluded as materials reclaimed "under the control of the generator," must send a notification to EPA or the state. [§260.42] The notification, which must be sent prior to managing the material under one of the new exclusions and by March 1st of each even-numbered year thereafter, must identify: 1) the name, address, EPA ID number, and NAICS code of the facility; 2) the name and phone number of a contact person; 3) the exclusion [i.e., §261.2(a)(2)(ii) or 261.4(a)(23)] under which the material will be managed and whether the material will be managed in any land-based units; and 4) the materials that will be excluded, when the materials will begin to be managed in accordance with the exclusion(s), and the quantities managed annually. The notification must be submitted using EPA Form 8700-12, and the certification included on that form must be signed by an authorized representative of the facility.

"The specific information included in [the] notification requirement will enable regulatory agencies to monitor compliance adequately and to ensure hazardous secondary materials are managed according to the exclusion and not discarded. For example, in the notification, EPA requires facilities to include the quantity of hazardous secondary materials that will be managed according to the exclusion and whether certain types of hazardous secondary materials will be managed in land-based units. This information can be used to assist RCRA inspectors in determining which facilities may warrant greater oversight and provides a basis for setting enforcement priorities." [73 FR 64682]

"We note that the requirement to provide this notification is not a condition of the exclusion. Thus, failure to comply with the requirement constitutes a violation of RCRA, but does not affect the excluded status of the hazardous secondary materials." [73 FR 64682]

11.2.3.2 The transfer-based exclusion

The second exclusion promulgated in the October 30, 2008 rule is for materials that are transferred from the generator to another "person" for reclamation. EPA calls this the "transfer-based exclusion," and it is illustrated in Figure 11-3. As with the "under the control of the generator" exclusion, this transfer-based exclusion applies to spent materials (that are hazardous by either listing or characteristic), listed sludges, and listed by-products that will be reclaimed. These hazardous secondary materials are not solid wastes if they are managed as described below.

The generator may send the hazardous secondary materials to an intermediate facility, which may hold the materials for more than 10 days, before they are sent on to the reclaimer. The reclamation facility then reclaims the materials at its site, producing a reclaimed material. This material can be reused at the generator's or reclaimer's facility or sent to another facility for use. However, there are limitations on how this reclaimed material can be used (see Section 11.4.1). There undoubtedly will be some residues generated in the reclamation process. Assuming they will be discarded, these residues are solid waste and, if they are hazardous, they must be managed as such under RCRA Subtitle C.

Because the generator has relinquished control of the material, EPA believes that the following conditions are needed under this exclusion to ensure the hazardous secondary material will not be discarded [see §261.4(a)(24)]:

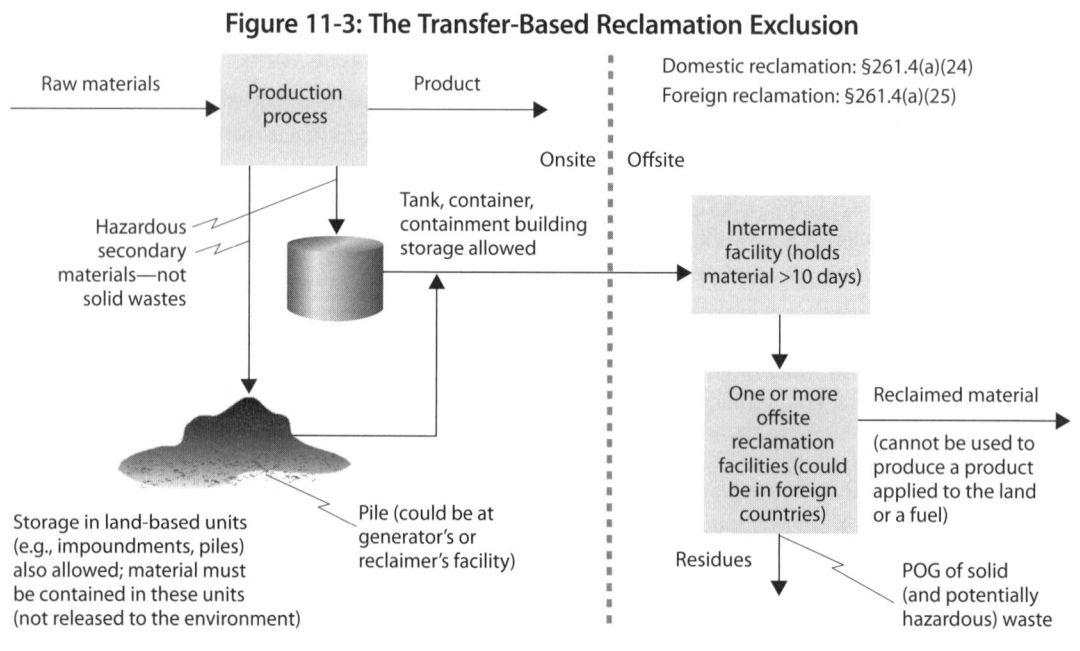

Figure 11-3: The Transfer-Based Reclamation Exclusion

Source: McCoy and Associates, Inc.; adapted from §§261.4(a)(24–25).

1. The material must not be handled by any person or facility other than the generator, the transporter, an intermediate facility, and one or more reclaimers. An intermediate facility is allowed to store the material for more than 10 days, but the generator/intermediate facility must ensure that the material is sent to the generator-designated reclamation facility.

   The transfer-based exclusion is available for hazardous secondary materials that are recycled in one or more reclamation processes, including when they occur at more than one reclamation facility. However, "[w]here recycling of a hazardous secondary material involves more than one reclamation step at more than one facility, generators should be well informed as to how the materials will be reclaimed, and by whom, throughout the recycling process." [73 FR 64684]

2. The generator must make "reasonable efforts," repeated, documented, and certified every three years, to ensure that each reclaimer will legitimately recycle the material and that any intermediate facility and reclaimer will manage the material in a manner protective of human health and the environment. This condition effectively requires that generators perform a type of environmental "due diligence" on the reclaimer and any intermediate facility to ensure that those facilities intend to properly manage the hazardous secondary materials as commodities and legitimately recycle rather than discard them. A generator may use any evidence available in making "reasonable efforts," including information gathered by the generator (e.g., during an audit), provided by the reclaimer and any intermediate facility, and/or provided by a third party. For example, the hazardous secondary material generator might hire an independent auditor to review the operations of a reclaimer, obtain audit reports from a consortium of generators (e.g., CHWMEG, Inc.), or rely on an assessment of a recycler or intermediate facility made by a parent corporation or trade association that is used by several generating facilities.

"Reasonable efforts" are only required for reclaimers and any intermediate facilities where management of the material is not addressed under a RCRA permit or interim status standards. If a reclamation or intermediate facility has a RCRA permit or complies with the interim status standards for an onsite operation unrelated to the management of the hazardous secondary materials sent by the generator, then the generator is required to make a "reasonable efforts" inquiry of the facility. [73 FR 64686]

"EPA intends that if a hazardous secondary material generator has met the reasonable efforts condition prior to transferring hazardous secondary materials to the reclamation or intermediate facility, then the reclaimer or intermediate facility, not the generator, would be liable under RCRA if the materials were discarded (i.e., not properly and legitimately recycled). However, if the generator does not meet the reasonable efforts condition, then the generator is ineligible for the transfer-based exclusion and would be potentially liable in the event its hazardous secondary materials were discarded by a reclamation or intermediate facility." [73 FR 64687]

Generators must maintain documentation showing that they satisfied the "reasonable efforts" requirement before they ship materials to an offsite reclaimer and any intermediate facility. [§261.4(a)(24)(v)(C)] These records would include audit reports and other relevant information. The generator is also required to certify that these reasonable efforts were made; EPA believes that requiring a certification creates a necessary level of oversight from an authorized representative, who can be any appointed company representative, and who must affirm that the condition is met and that hazardous secondary materials will not be discarded. Such documentation and certification may be maintained at the generator's company headquarters or other offsite facility. [73 FR 64690]

Additional guidance is available in "Revisions to the Definition of Solid Waste Final Rule: The Reasonable Efforts Condition." [RO 51393]

3. The generator, any intermediate facility, and each reclaimer must keep records of each shipment of hazardous secondary material for at least three years. [§261.4(a)(24)(v)(D) for the generator and §261.4(a)(24)(vi)(A) for any intermediate facility and each reclaimer] Additionally, generators must receive confirmations of receipt from each intermediate facility and reclaimer and maintain them for at least three years. [§261.4(a)(24)(v)(E)] This requirement may be satisfied by routine business records (e.g., financial records, bills of lading, copies of DOT shipping papers, or electronic confirmations of receipt). No specific template or format is required for these records; thus, reclaimers and any intermediate facilities may make these confirmations of receipt available to the generator via an extranet. [RO 14812]

4. Before shipment offsite, the generator may store hazardous secondary materials in either non-land-based units (e.g., tanks, containers, and containment buildings) or land-based units (e.g., surface impoundments and piles). In either case, the materials must be "contained" in such units (basically, this means that the material is not released into the environment—see Section 11.4.2).

The reclaimer and any intermediate facility must manage the material in a manner that is at least as protective as that employed for the analogous raw material. In other words, these materials must be managed as valuable commodities. EPA discussed this condition as follows:

"An 'analogous raw material' is a material for which a hazardous secondary material substitutes and which serves the same function and has similar physical and chemical properties as the hazardous secondary material. A raw material that has significantly different physical or chemical properties would not be considered analogous even if it serves the same function. For example, a metal-bearing ore might serve the same function as a metal-bearing air pollution control dust, but because the physical properties of the dust would make it more susceptible to wind dispersal, the two would not be considered analogous. Similarly, hazardous secondary materials with high levels of toxic volatile chemicals would not be considered analogous to a raw material that does not have these volatile chemicals or that has only minimal levels of volatile chemicals." [73 FR 64691]

5. Any residues generated from the reclamation process must be managed in a manner that is protective of human health and the environment. Obviously, any hazardous wastes generated must be managed in accordance with RCRA Subtitle C.

EPA noted that the derived-from rule (see Section 5.2) will not apply to residues generated during reclamation of these excluded hazardous secondary materials. Instead, they will be considered a new point of generation for determining any applicable characteristic or listed waste codes. [73 FR 64692] Informal discussions with EPA headquarters on this point resulted in the following example: Materials that previously met the listing description for F006 (wastewater treatment sludges from electroplating operations) will now be sent as commodities (not solid wastes) under a bill of lading to offsite reclamation facilities. Residues from the reclamation process, which previously had to be managed as F006 due to the derived-from rule, will now be hazardous only if they exhibit a characteristic or meet the original F006 listing description (which they likely will not).

6. The reclaimer and any intermediate facility must obtain financial assurance (similar to that

required for permitted TSD facilities) per Subpart H of Part 261. EPA is requiring such financial assurance to ensure these offsite facilities will 1) not abandon the secondary material, 2) properly decontaminate equipment used to manage these materials, 3) clean up any releases or other contamination caused by managing these materials, and 4) remove any stored materials upon closure of the reclamation operations.

The following examples from EPA guidance illustrate several of the above conditions.

Q: *If a ten-day transfer facility is used as an intermediate facility, can excluded hazardous secondary materials be bulked into stationary tanks?*

A: No. Hazardous secondary materials may not be managed in stationary tanks at transfer facilities. Under §261.4(a)(24)(ii), hazardous secondary materials may not be stored for more than 10 days at a transfer facility and must be packaged according to applicable DOT regulations at 49 *CFR* Parts 173, 178, and 179 while in transport. A transfer facility is defined in §260.10 to mean "…areas where shipments of hazardous waste or hazardous secondary materials are held during the normal course of transportation." Tanks are not portable and thus are not part of the "normal course of transportation." In addition, the DOT regulations referred to above do not apply to stationary tanks. Therefore, storage of hazardous secondary materials in tanks would not be allowed unless the facility in question had a RCRA permit or interim status, or meets the conditions for an intermediate facility in §261.4(a)(24)(vi). [RO 14812]

Q: *A RCRA permit for the reclamation facility and any intermediate facility satisfies the "reasonable efforts" requirements. However, recycling units located at permitted facilities are often not part of the facility RCRA permit because of the permitting exemption found in §261.6(c)(1). As long as the storage and associated handling equipment for the hazardous secondary material is RCRA-permitted and the reclamation equipment is used as intended for the hazardous secondary materials, will this meet the reasonable efforts requirements of the generator?*

A: Yes. Reasonable efforts are not required if the intermediate facility or reclaimer's RCRA permit or applicable interim status standards extend to the management of the hazardous secondary materials in question. Recycling units may or may not be addressed in a facility's permit or applicable interim status standards due to the permitting exemption found in §261.6(c)(1). [RO 14812]

Q: *Can a hazardous secondary material be commingled with similar hazardous waste and still maintain the exclusion from the definition of solid waste?*

A: No. Excluded hazardous secondary material cannot be commingled with regulated hazardous waste and still maintain the exclusion from the definition of solid waste. If an excluded hazardous secondary material is mixed with hazardous waste, the resulting mixture would likely be a hazardous waste. [RO 14812, 14813]

Q: *If hazardous waste is reclaimed in a distillation column (still) and then that same still is to be used to reclaim a hazardous secondary material, must the still be cleaned first?*

A: To maintain the exclusion for a hazardous secondary material being reclaimed, any hazardous waste and associated residues must be removed from the still before processing the hazardous secondary material. The agency recommends that each facility discuss with their state (or EPA region) the procedures they will need to implement to clean the still in order to ensure that hazardous secondary materials are not commingled with hazardous wastes. [RO 14813]

In the preamble to the October 30, 2008 rule promulgating these exclusions for hazardous secondary materials being reclaimed, EPA noted the following in terms of enforcement:

"If a hazardous secondary material generator fails to meet any of the above-described conditions that are applicable to the generator, then the hazardous secondary materials would be considered discarded by the generator and would be subject to the RCRA Subtitle C requirements from the point at which such material was generated. In addition, if a reclaimer or an intermediate facility failed to meet any of the above-described conditions, then the hazardous secondary materials would be considered discarded by the reclaimer or intermediate facility and would be subject to the RCRA Subtitle C requirements from the point at which the reclaimer or intermediate facility failed to meet a condition or restriction, thereby discarding the material.

"It should be noted that the failure of the reclaimer or intermediate facility to meet the conditions of the exclusion does not mean that the hazardous secondary material was considered waste when handled by the generator, as long as the generator can adequately demonstrate that it has met its obligations, including the obligation…to make reasonable efforts to ensure that the hazardous secondary material will be reclaimed legitimately and properly managed.… In such situations, and where the generator's decision to ship to that reclaimer or intermediate facility is based on an objectively reasonable belief that the hazardous secondary materials would be reclaimed legitimately and otherwise managed in a manner consistent with this regulation, the generator would not have violated the terms of the exclusion." [73 *FR* 64699–700]

11.2.3.2.1 *Materials exported for reclamation*

EPA has also excluded hazardous secondary materials that are exported for reclamation from the definition of solid waste. To use this exclusion, the following conditions must be met [see §261.4(a)(25)]:

- The generator must notify EPA of its intent to export the material to a foreign reclamation facility and obtain, through EPA, written consent from the receiving country (Acknowledgment of Consent). The Acknowledgment of Consent must accompany each shipment. The notification to EPA must be issued at least 60 days before the first shipment is scheduled.

- The material cannot be handled by any person or facility other than the generator, the transporter, an intermediate facility, and one or more reclaimers. The generator must ensure that the material is sent to the generator-designated reclamation facility.

- The generator must make "reasonable efforts," repeated and documented every three years, to ensure that each reclaimer will legitimately recycle the material and that each reclaimer and any intermediate facilities will manage the material in a manner protective of human health and the environment.

- The generator must keep records of each shipment of hazardous secondary material and all Acknowledgments of Consent for at least three years and must file with EPA an annual report summarizing all such exports for reclamation.

11.2.3.2.2 *Notification required*

Generators, reclaimers, and any intermediate facilities that manage hazardous secondary materials that have previously been subject to regulation as hazardous waste, but which are excluded from the definition of solid waste under the transfer-based exclusion in §261.4(a)(24–25), must send a notification to EPA or the state. The notification, which must be sent prior to managing the material under the exclusion and by March 1st of each even-numbered year thereafter, must identify: 1) the name, address, EPA ID number, and NAICS code of the facility; 2) the name and phone number of a contact person; 3) the exclusion [i.e., §261.4(a)(24) or (25)] under which the material will be managed and whether the material will be managed in any land-based units; 4) the materials that will be excluded, when the materials will begin to be managed in accordance with the exclusion(s), and the quantities managed annually; and 5) documentation that reclaimers and any intermediate facilities have the required financial assurance. The notification must be submitted using EPA Form 8700-12, and the certification included on that form must be signed by an authorized representative of the facility.

"The requirement to provide this notification is not a condition of the exclusion. Thus, failure to comply with the requirement constitutes a violation of RCRA, but does not affect the excluded status of the hazardous secondary materials." [73 FR 64685]

11.2.3.2.3 State authorization for DSW rule

The October 30, 2008 definition of solid waste (DSW) rule described in Sections 11.2.3.1 and 11.2.3.2 became effective as part of the federal regulations on December 29, 2008. However, the exclusions are not effective in RCRA-authorized states until the state adopts them. Furthermore, states are not required to adopt the provisions of the DSW rule because the rule made RCRA less stringent or reduced its scope.

As of August 4, 2013, the exclusions in the DSW rule were applicable in only six states: Alaska, Idaho, Illinois, Iowa, New Jersey, and Pennsylvania. Puerto Rico has passed a resolution allowing application of the rule. Updated information on which states have adopted the rule can be found at http://www.epa.gov/wastes/hazard/dsw/statespf.htm.

Q: *Can an excluded hazardous secondary material be transported through a state that has not adopted the DSW rule without a hazardous waste manifest?*

A: RCRA Section 3009 allows states to impose standards more stringent than those in the federal program. Thus, a state that has not adopted the DSW rule may impose state requirements, including the hazardous waste manifest requirement, on hazardous secondary materials while the materials are being transported through that state. Transporters should contact the state in question to ascertain its policy about shipments that would be excluded in other states. [RO 14812]

11.2.4 Examples of Step 1 recycling scenarios

Using Step 1 (checking the table with the asterisks) to determine whether a secondary material is subject to RCRA or not is exemplified as follows.

Q: *A facility generates a by-product that exhibits a characteristic. Since the facility cannot afford the cost of shipping the material directly to a reclaimer, it instead ships the material to a nearby TSD facility that, in turn, sends the material on to the reclaimer. Is the by-product a solid and hazardous waste at the generator facility? How about after it is shipped to the TSD facility?*

A: Recycling activities are viewed prospectively; that is, the status of the material is determined based on how it *is going* to be recycled. Thus, when it is known that the by-product is eventually headed for legitimate reclamation, we look at Table 11-1. A dash is found at the intersection of the row for "By-products exhibiting a characteristic" and the column for "Reclamation," indicating that the material is not a solid waste when recycled in that manner. Thus, the material does not have to be managed as a solid or hazardous waste at the generator or TSD facility.

"The term 'when' as it is used in §261.2(c) for recycling activities (e.g., when reclaimed) is not meant to refer only to the moment in time when that activity occurs, in order to determine the regulatory status of a material…. [I]f the generator intends to have his/her characteristic by-product reclaimed at some point in the future, he/she would not be deemed to be managing a solid or hazardous waste, according to [Table 11-1]." [RO 11877]

If, however, the eventual disposition of the by-product were not known (or not certain enough that the generator or the TSD facility could adequately document the planned reclamation of the material), then storage at the generating facility and shipment to the TSD facility would be subject to hazardous waste regulations. Additionally, "when secondary materials are excluded or exempt based on a claim of recycling, the material will no longer be excluded or exempt if it is accumulated speculatively prior to recycling." [RO 11877] If at least 75% of the by-product is not recycled within the calendar year, the material is considered to be speculatively accumulated and would be a solid and hazardous waste per Table 11-1. [RO 11747]

Q *An unused pesticide will be recycled by burning it for energy recovery. What is the regulatory status of the pesticide?*

A An unused product, such as a pesticide, is considered a "Commercial chemical product." Although the wording in Table 11-1 seems to cover only those chemicals in §261.33 (i.e., the P- and U-listed chemicals), EPA has noted that "we meant also to include commercial chemical products that are not specifically listed in §261.33 as well (for example, a commercial chemical product that exhibits the ignitability characteristic identified in §261.21)." [RO 11713] Thus, whether or not the unused pesticide in question is on the P- or U-list in §261.33, it is regarded as a "Commercial chemical product" in Table 11-1 for purposes of determining whether it is a solid waste.

Because there is an asterisk at the intersection of "Commercial chemical products" and "Burning for energy recovery," the pesticide is a solid waste when put into a fuel blending program. If the pesticide is included in the P- or U-list of chemicals or if it exhibits a characteristic, it would have to be managed as a hazardous waste. Per §261.2(c)(2)(i)(B), the resulting fuel will also be a solid and hazardous waste. Note, however, that §261.2(c)(2)(ii) exempts from RCRA commercial products that are burned for energy recovery if they are themselves fuels. For example, off-spec gasoline or jet fuel would not be a solid waste when reclaimed and burned for energy recovery. [RO 11713]

Q *A characteristically hazardous sludge is generated in an air pollution control device. The sludge is sent to another facility to recover its copper content; lead that is also recovered from the sludge is processed into a low-grade solder. What is the status of the sludge when shipped to the offsite recovery facility?*

A Recovering copper and lead from a sludge would meet the definition of "Reclamation." Checking Table 11-1, there is a dash at the intersection of a "Characteristic sludge" (third row) and "Reclamation" (third column). This indicates that the material is not a solid waste (so it cannot be a hazardous waste) when recycled in this fashion. The sludge will be subject to RCRA only if it is speculatively accumulated. [RO 11113]

Q *Wastewater from electroplating nickel and chrome bumpers is treated in tanks. Periodically, a metal hydroxide sludge is removed from the tanks and sent to an offsite copper smelter for metal recovery. How is the sludge regulated?*

A Sludge from treating wastewater from electroplating operations meets the F006 listing description. Metal smelters are reclamation units. There is an asterisk at the intersection of "Listed sludge" and "Reclamation" in Table 11-1, indicating that the material is a solid waste when sent for reclamation. [RO 11910] As noted in Section 11.2.3, however, there are three sets of exceptions to the asterisks; the third set consists of the exceptions noted in the column heading under "Reclamation" in that table. Since the metal sludge is not generated in the primary mineral processing industry, the applicable exceptions are discussed in Sections 11.2.3.1 and 11.2.3.2. Reading through those sections results in selection of the exclusion for materials reclaimed under the transfer-based exclusion. Note that to qualify for the exclusion, the metal recovery from the sludge must meet the requirements in §266.100(d)(1)–(3) and the residues from the reclamation process must meet the requirements in §266.112. [§261.1(c)(4)] If the recycling process meets these requirements as well as all other conditions discussed in Section 11.2.3.2, the sludge *may* not be a solid waste, contingent upon passing Steps 3–5 (discussed later in this chapter).

Q *Must gasoline recovered from ground water that will be put into a fuel blending program be managed as a D001 hazardous waste?*

A No. The recovered gasoline is considered an off-spec commercial chemical product. (Fuel is not "used" until it is burned.) Section 261.2(c)(2)(ii) exempts from RCRA commercial products that are burned for energy recovery if they are themselves fuels. "Recovered free product that is normally used as a fuel is not regulated as a RCRA Subtitle C waste if it is used as an ingredient to make a fuel (e.g., by blending into fuel)." [RO 11713; see also July 28, 1994; 59 FR 38539] Be careful, however, that what you are managing outside of RCRA is truly an off-spec product. If the ground water contains just a few ppm of hydrocarbons, that looks more like a solid waste to EPA: "The management of petroleum-contaminated ground water in separation and treatment units is clearly solid waste (and potentially hazardous waste) management, essentially wastewater treatment." [59 FR 38540]

11.2.5 Why "—" results may be exempt

In the first example described under Section 11.2.4, the dash at the Table 11-1 intersection indicates that the subject hazardous secondary material recycled in the manner described is not a solid waste. This is *not* the end of the process for these materials, however. As shown in Figure 11-1, when you exit the second Step 1 box via the "Yes" answer, you still must complete Steps 4 and 5 before you may conclude that your hazardous secondary material is not a solid waste. Details of these steps are discussed in Sections 11.5 and 11.6 below.

Note that sometimes, more than one of the recycling activities listed in Table 11-1 may occur in sequence. For example, what if a characteristic sludge undergoes reclamation followed by use constituting disposal. There is a dash (indicating that the material is not solid waste) associated with reclamation, but an asterisk (indicating that the material is a solid waste) when such a material is applied to the land. So, is it a solid waste or not? EPA notes that when a secondary material will ultimately be incorporated into a product that will be placed on the land, the entire recycling activity is defined as "use constituting disposal." [RO 11113, 13723, 14348] Thus, a characteristic sludge or by-product that would not be a solid waste when reclaimed (per Table 11-1) is a solid waste if the reclaimed material is subsequently burned for energy recovery, applied to the land, or speculatively accumulated. [RO 14195]

11.2.5.1 Managing exempt materials

When a hazardous secondary material is recycled in a way that results in a dash in Table 11-1, the material is not a solid waste (subject to the limitations discussed just above). Such materials are not subject to RCRA hazardous waste regulations in terms of storage, transportation, etc. A characteristic by-product destined for reclamation, for example, could be placed/stored in piles on the ground prior to reclamation and not be subject to Subtitle C requirements, including any LDR provisions dealing with land placement (as long as it is not speculatively accumulated). [RO 11087, 14099, 14268]

11.2.6 Don't quit if you hit an "*"

As noted previously, checking Table 11-1 in Step 1 of the recycling determination process may seem to result in the recycled material being designated as a solid waste. Nevertheless, don't quit the process if you hit an asterisk. As noted in Figure 11-1, if you hit an asterisk under the "Reclamation" column, evaluate whether you qualify for one of the two reclamation exclusions that EPA added to the regs on October 30, 2008. If one of those exclusions applies, as discussed in Sections 11.2.3.1 and 11.2.3.2, you move to Step 3, which is discussed in Section 11.4. If neither of those apply, or you have hit any other asterisk in Table 11-1, you continue with Step 2 as discussed in Section 11.3 below. Even though the material might seem to be a solid waste per the table with the asterisks, it may still qualify for one of the use/reuse recycling exclusions described in Step 2 of our process. If it does, it may yet be excluded from the definition of solid waste.

11.3 Step 2—The four use/reuse recycling exclusions

Sometimes, Step 1 (i.e., checking the table with the asterisks) won't give you an answer because the recycling activity that you are involved in is not one of the four listed across the top of the table. Other times, Table 11-1 gives you an answer, but not the one you wanted (i.e., the table notes that the material is a solid waste when recycled in that manner). If either of these outcomes results from Step 1, you may continue with Step 2. The example below illustrates this situation.

Q *A company plans to use an ignitable waste resin from a production process to manufacture a separate product, and it wants to know if use of the resin is subject to management under the RCRA regulations. The waste resin will not be reclaimed prior to its use.*

A The waste resin in question most likely qualifies as a by-product that exhibits a characteristic. Thus, the company would use the fifth row in Table 11-1. However, the recycling activity that the facility is proposing is not 1) land application or otherwise use in a manner constituting disposal, 2) use as a fuel (burning for energy recovery), 3) reclamation, or 4) speculative accumulation. Since the company wants to use the resin without reclamation, no recycling activity in Table 11-1 applies. In this case, Step 1 doesn't help, so the facility would have to move to Step 2 to decide if the use is regulated under RCRA.

In Step 2 we examine four use/reuse recycling exclusions in §§261.2(e)(1)(i–iii) and 261.4(a)(8). Each of these four exclusions is addressed in the subsections that follow.

If one of these four exclusions applies to a specific recycling situation, and you pass the tests in Steps 3–5, the hazardous secondary materials are not solid waste—they are just manufacturing streams in ordinary production operations. [January 4, 1985; 50 *FR* 619]

11.3.1 Use/reuse as an ingredient or feedstock

The first use/reuse recycling exclusion applies to secondary material recycled for use/reuse as an ingredient or feedstock in an industrial process. Referring back to our logic diagram, Figure 11-1, this exclusion is found in the first bulleted item in the lower of the two Step 2 boxes. Where a facility believes that this exclusion applies to a given situation, a further check of the Step 3–5 conditions must be made *before* the secondary material in question can be declared exempt from RCRA (i.e., not a solid waste). Failure to meet the conditions in any of these steps will disqualify the secondary material from a solid waste exclusion.

11.3.1.1 Regulatory nuts and bolts

The text of the first use/reuse recycling exclusion, as codified in §261.2(e)(1)(i), states that recycled secondary materials are not solid waste when:

> "Used or reused as ingredients in an industrial process to make a product, provided the materials are not being reclaimed."

As shown in Figure 11-4, secondary material recycled directly (with no reclamation) from its point of generation (POG) and used as an ingredient (along with nonwaste raw materials) in an industrial process that produces a product is excluded from the definition of solid waste and, therefore, cannot be a RCRA hazardous waste. "When secondary materials are *directly* used as an ingredient or a feedstock, we are convinced that the recycled materials are usually functioning as raw materials and therefore should not ordinarily be regulated under Subtitle C." [Emphasis in original.] [January 4, 1985; 50 *FR* 619]

We'll discuss POG issues and several other limitations of the exclusion later, but first, we examine the concept of reclamation—if reclamation is occurring, you can't take advantage of this use/reuse recycling exclusion. You can conduct reclamation, but the secondary material remains a solid (and potentially hazardous) waste, unless one of the recycling exclusions involving reclamation applies (see Sections 11.2.3.1 and 11.2.3.2).

11.3.1.2 Reclamation vs. incidental processing

We touched on the concept of reclamation in Section 11.2.2 above. In the preamble to the January 4,

Figure 11-4: Exclusion for Use as an Ingredient or Feedstock

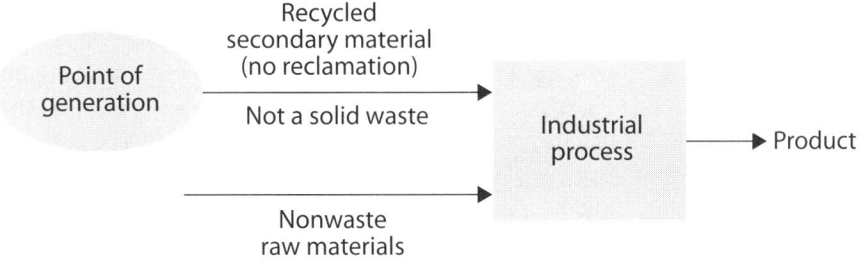

Source: McCoy and Associates, Inc.; adapted from §261.2(e)(1)(i).

1985 rule, EPA offered the following guidance on the concept of "reclamation":

"[R]eclamation involves regeneration or material recovery. Wastes are regenerated when they are processed to remove contaminants in a way that restores them to their usable original condition.

"We also [draw] a distinction…between situations where material values in a spent material, by-product, or sludge are recovered as an end product of a process (as in metal recovery from secondary materials) as opposed to situations where these secondary materials are used as ingredients to make new products without distinct components of the materials being recovered as end products. The former situation is reclamation; the latter is a type of direct use that usually is not considered to constitute waste management." [50 *FR* 633]

If we recover something of value from the secondary material before we use it as an ingredient, we don't get the benefit of this particular recycling exclusion. Similarly, if we process a material to remove contaminants and restore it to an original condition, such reclamation voids the exclusion (although another exclusion might apply). Applying the above logic, the agency offers the following examples of reclamation:

- Recovering solvent or other organics from spent solvents or other organic chemicals *is* reclamation. [50 *FR* 633]

- Smelting F006 electroplating sludge to recover copper, nickel, or other metals *is* reclamation. [50 *FR* 633, RO 11338, 11426, 11910, 14026]

- Re-refining off-specification jet fuel from fueling operation overflows and tank/line draining to produce a salable jet fuel *is* reclamation. [RO 11449] (Note, however, that the off-specification fuel is considered to be a "Commercial chemical product" in Table 11-1; commercial chemical products being reclaimed are not solid wastes.)

- Dewatering a chrome-coated zinc and charcoal solution to facilitate the return of the chrome-coated particles to a molten zinc bath used in a galvanizing process *is* reclamation. Along the same line, dewatering wastewater treatment sludges and any treatment of wastewater prior to recycling are also reclamation. [50 *FR* 639, RO 11415, 11910]

Conversely, "incidental processing" of a material does not void this first use/reuse recycling exclusion. EPA distinguished incidental processing from reclamation as "processing steps that do not themselves regenerate or recover material values and are not necessary to material recovery." [50 *FR* 639] More recent guidance notes that incidental processing changes the material's physical form with no or only minor changes in the mass or chemical composition of the material. [RO 14748] The following processes are incidental processing and are *not* considered reclamation:

- Changing the particle size of a material, such as crushing, grinding, sizing, and repackaging spent refractory bricks for reuse as an ingredient in making new refractory bricks. [RO 11201, 14748]
- Shredding and grinding of waste leather trimmings to reduce their particle size to market the material as an absorbent for spilled liquid. [RO 14025]
- Processing that changes the physical properties of secondary materials but that does not recover material values or regenerate the materials (e.g., impurities or contaminants are not removed). Such processing includes viscosity adjustment, minor physical or gravity separation, wetting of dry wastes to avoid wind dispersion, thermal agglomeration or sintering, melting of base materials, and briquetting of dry wastes to facilitate resmelting (although the smelting itself is reclamation). [50 FR 639, RO 11271, 14748]
- Triple distilling 99% pure mercury to produce a higher purity. [RO 11159]
- Screening or filtering to protect the integrity of downstream pumps or equipment from foreign material (e.g., screws, nuts, bolts). [RO 14748]
- Ensuring purity by separating minor amounts of foreign material (e.g., grit, ash, or water). [RO 14748]
- Final processing (e.g., clarifying or settling) of a material that closely resembles a finished product to remove minor impurities. [RO 14748]

"These incidental processing activities may take place at any step during the use/reuse process. In addition, a process may involve more than one such processing step and still not be considered reclamation when the cumulative effect of all processing activities results in only the kinds of processing changes that would be considered 'incidental.'... [A]nother indicator to consider in determining whether an activity may be incidental processing, as opposed to reclamation, is to evaluate whether an analogous process using raw materials includes the same or similar 'incidental'...activities at the same point in the process. For example, some processes that use virgin materials filter out small amounts of ash, other inert process residues, or water as a polishing step.... Thus, it may be appropriate to view similar steps to be incidental processing when making analogous products from secondary materials." [RO 14748]

EPA offers a warning to facilities looking to use past guidance and others' examples to decide the status of their secondary material handling. "The determination of whether the hazardous secondary materials processed in your recycling process are more appropriately defined as 'reclaimed' or 'used as an ingredient' is a case-specific determination, more appropriately made by the state regulatory agency or the EPA regional office. Also, the state regulatory program may have regulations that differ from the federal program, so you should contact them for a more definitive determination." [RO 11886]

11.3.1.2.1 The reclamation process

Unless one of the reclamation exclusions discussed in Sections 11.2.3.1 and 11.2.3.2 applies, hazardous secondary materials (spent materials, listed sludges, and listed by-products) being reclaimed remain hazardous until the reclamation process is complete. [50 FR 633] Thus, they would be subject to the hazardous waste generator standards in Part 262 (including use of the manifest), the hazardous waste transporter standards in Part 263, and the land disposal restrictions requirements in Part 268. [§261.6(b)] The recycling (reclamation) facilities would be subject to Parts 264 or 265, 266, 268, and 270 for storage of the hazardous wastes prior to recycling, but the recycling (reclamation) process itself is exempt from RCRA standards (with the possible exception of Subparts AA–BB). [§261.6(c)(1); see also Section 7.4.1]

Materials reclaimed from secondary materials are considered products—not wastes [§261.3(c)(2)(i)]; however, this principle does not apply to wastes that have been partially reclaimed but must be reclaimed further. [50 FR 633, RO 11109, 13125] When thermal metal recovery is involved, EPA notes that secondary materials destined for smelters remain hazardous waste. After smelting, recovered metals that only need to be refined are products, not wastes. For example, a lead- and copper-bearing secondary material that is 92–99% lead and only needs to be refined prior to use would meet the definition of a fully reclaimed material. [January 4, 1985; 50 *FR* 634, RO 11123, 11159, 11644, 11929, 11932]

Additionally, EPA noted that reclaimed materials that are not ordinarily considered commercial products, such as wastewater or stabilized waste, generally remain solid wastes. [50 *FR* 634] However, the agency suggested that reclaimed wastewater that is reused (e.g., in a cleaning or rinsing process) may lose its status as a solid waste, provided it really is reclaimed and that it is an effective substitute for what would otherwise be used in the process. Such a determination would be made by state or EPA regional personnel. [RO 11374, 11546]

Although the occurrence of reclamation will invalidate this first use/reuse recycling exclusion (i.e., the secondary material will be a solid and potentially hazardous waste), the reclamation process itself is not subject to RCRA regulation. Per §261.6(c–d), hazardous waste reclamation is exempt from the RCRA management standards (except for Subparts AA and BB air emission standards) and permitting requirements. See Section 7.4.1.3 for a complete discussion of this topic.

Q: *A facility reclaims spent xylene solvent to produce marketable xylene solvent. The spent solvent is stored before it is reclaimed, and the reclaimed material is stored in a tank prior to sale to offsite facilities. During one reclamation campaign, a batch of product was produced that was not suitable for solvent use and was instead sent offsite as a fuel. What is the regulatory status of the 1) spent solvent storage units, 2) reclamation unit, 3) reclaimed material when reused as a solvent versus used as a fuel, and 4) tank used to hold the reclaimed material?*

A: A spent solvent is a spent material; per Table 11-1, the spent solvent is a solid waste prior to reclamation. Because it meets a listing description, the spent solvent is F003 prior to reclamation, and the spent solvent storage units would be hazardous waste tanks or containers. Note that this material would not be a solid waste if one of the reclamation exclusions discussed in Sections 11.2.3.1 or 11.2.3.2 applies. If the spent solvent is a solid and hazardous waste, the reclamation process itself is exempt from RCRA permitting or substantive standards. [§261.6(c)(1)]

The RCRA regulations [at §261.3(c)(2)(i)] note that materials that are reclaimed from solid wastes and that are used beneficially are not considered solid or hazardous wastes. However, this exception from the derived-from rule applies only if the reclaimed materials are not burned for energy recovery or applied to the land. Thus, when material is reclaimed from F003 and subsequently reused as a solvent, it is not F003 but a product, and the holding tank is a product storage tank (i.e., a manufacturing unit).

Conversely, reclaimed solvent that is burned for energy recovery remains F003 per the derived-from rule and must be manifested offsite to the user. Regarding the status of the holding tank, EPA noted that "[b]ecause your operation normally produces reclaimed solvent, the [material] actually became a hazardous waste at the time you determined that it was not suitable for solvent use (and that it therefore had to be marketed as fuel).... EPA does regulate the storage of hazardous waste fuel as well as fuel blending tanks. In your case, however, it appears that the tank was really a product (solvent)…tank, and so not subject to regulation. This determination is based on your assurance that the fuel production was an isolated incident, and that your original intent in placing reclaimed xylene in the tank was to produce solvent, not fuel." [RO 11238]

Q: *A facility generates wood chips and other materials from lead-based paint abatement projects that exhibit the characteristic for lead (D008). These materials can be sent to a hazardous waste recycler if they meet certain size specifications. Can the generator process (e.g., cut, chop, shred, grind) the materials at its facility without a RCRA permit before manifesting them to the offsite recycler?*

A Sometimes, hazardous waste is partially recycled by the generator before being sent offsite for further legitimate recycling. The processing activities at the generator's site are considered the first step in the recycling process and remain exempt from RCRA permitting and management standards under §261.6(c)(1). However, "any storage of [lead-based paint] hazardous waste before or after processing is subject to RCRA Subtitle C regulation (e.g., §262.34 for generator accumulation…)." [RO 11880]

11.3.1.2.2 Some materials can be reclaimed without being solid wastes

The question of whether reclamation is occurring is a go/no-go issue in terms of qualifying for this first use/reuse recycling exclusion (and, as we'll see later, the second and third exclusions as well). Thus, we've tried to provide a clear understanding of reclamation in the discussions above.

It's important to remember, however, that there are cases where a hazardous secondary material can be reclaimed without triggering RCRA regulations. Specifically, EPA allows facilities to reclaim characteristic sludges, characteristic by-products, and commercial chemical products outside the reach of RCRA by excluding these materials from the definition of solid waste in §261.2(c)(3) (i.e., there are dashes for these materials in the "Reclamation" column in Table 11-1).

For example, zinc oxide dust captured in air pollution control equipment that exhibits the toxicity characteristic for lead is considered a characteristic sludge by EPA. Per Table 11-1, this stream would not be solid waste when reclaimed. The same logic would apply to the characteristic by-product we discussed earlier: chrome-coated zinc and charcoal particles that are reclaimed and then recycled to the molten zinc galvanizing bath. [RO 11412, 11415] Referring back to Figure 11-1, these materials would escape the definition of solid waste via Step 1 of our determination process (but they would still have to pass through Steps 4–5).

Additionally, either of the two reclamation exclusions codified on October 30, 2008 can apply to hazardous secondary material being reclaimed (see Sections 11.2.3.1 and 11.2.3.2). Finally, we'll see in the fourth use/reuse recycling exclusion another way that a hazardous secondary material can be reclaimed and still not be a solid waste (see Section 11.3.4 below).

11.3.1.3 POG considerations

Although not specifically stated in the regulations, EPA has noted repeatedly in guidance that the determination of whether a recycled secondary material is a solid waste or, conversely, excluded from RCRA applies at its POG. [RO 11412, 11644, 11747, 11774, 11877] (For more on the POG, see Section 14.1.) Per the agency, "the solid waste determination for a recycled material is made at the point of generation of the waste, and takes into account the entire waste recycling process, not just the first step in a waste recycling train." [RO 11644]

Thus, if a hazardous secondary material qualifies for this first use/reuse recycling exclusion, its entire downstream storage/transportation/processing/use is not subject to RCRA Subtitle C control. For example, secondary material that qualifies for this exclusion may be temporarily stored in piles on the land or in surface impoundments and still be exempt from the definition of solid waste. [RO 11087, 14099, 14268]

In addition, recycled secondary material need not remain at a single location in order to qualify for the first use/reuse recycling exclusion. The material can be shipped from the generating site to an offsite facility where it is used as an ingredient or feedstock. Again, as long as it is not speculatively accumulated, the material is excluded from the definition of solid waste while it is in storage at either the generating or offsite facility. [RO 11271]

11.3.1.4 Examples

Examples of the first use/reuse recycling exclusion follow. (When reading these examples, don't forget that even if a material qualifies for the recycling exclusion, Steps 3–5 must subsequently be evaluated per Figure 11-1.)

Q: As shown in Figure 11-5, a spent copper chloride etchant stream is used as an ingredient to produce an animal micronutrient product. This industrial process also generates a hazardous residue. The residue is reclaimed prior to regenerating fresh alkaline etchant. Does the first recycling exclusion apply to the spent etchant, even though a residue stream from the process undergoes subsequent reclamation?

A: If the entire operation is considered to be one industrial process, the reclamation would disqualify the facility from claiming the exclusion; if the operation consists of two separate, sequential processes, the reclamation in the subsequent process would not nullify the exclusion for the initial production step. For this specific case, EPA allowed separate determinations for the two processing steps. The first step (micronutrient production) is not considered to be reclamation and, thus, qualifies for the first use/reuse recycling exclusion. In contrast, the agency believes the second step (alkaline etchant production) is reclamation, making the handling of the intermediate residue downstream of its POG subject to RCRA [RO 14102], unless one of the reclamation exclusions discussed in Sections 11.2.3.1 and 11.2.3.2 applies.

Q: K061 electric arc furnace (EAF) dust is used as an ingredient to produce glass frit that is then used as an abrasive blasting agent or as an ingredient in roofing granules, glass ceramic, and ceramic glaze. How is the dust regulated when recycled in this manner?

A: This scenario does not involve any of the recycling activities in Table 11-1. In Step 2, using the K061 waste as an ingredient to produce those products would qualify for the first use/reuse recycling exclusion as long as Steps 3–5 are completed successfully. Per §261.2(e)(1)(i), the EAF dust would not be a solid waste from its POG. [RO 11714]

Q: K061 EAF dust is mixed with a sodium silicate binder and pressed into briquettes. The briquettes are then reused as an ingredient in steel production. Is this exempt recycling?

A: Because EPA has noted that simply briquetting a secondary material is not considered reclamation but is incidental processing (see Section 11.3.1.2), the table with the asterisks provides no information as to the status of the EAF dust used in this way. Looking at the first recycling exclusion [§261.2(e)(1)(i)] in Step 2, the dust will be used as an ingredient in an industrial process to make a product and thus will not be subject to RCRA as long as Steps 3–5 are completed successfully. [RO 11271]

Q: K061 EAF dust is recycled by roasting, leaching, and electrowinning to convert the zinc

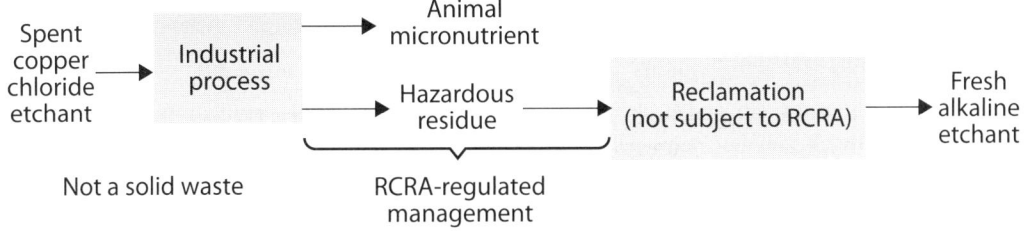

Figure 11-5: Recycling That Includes Both Reclamation and Use as an Ingredient

Source: McCoy and Associates, Inc.; adapted from RO 14102.

salt into metallic zinc for use as an ingredient in an industrial process to make a product. Is the EAF dust a solid waste?

A A preliminary look at Table 11-1 indicates that the material is a solid waste per Step 1, since there is an asterisk at the intersection of "Sludges (listed in 40 CFR 261.31 or 261.32)" and "Reclamation." Such processing would not qualify for the §261.2(e)(1)(i) use/reuse recycling exclusion because the dust is being reclaimed to produce metallic zinc. This reclamation would void the exclusion for use as an ingredient. [RO 11932] However, one of the recycling exclusions for materials being reclaimed might be used to exclude the EAF dust from the definition of solid waste (see Sections 11.2.3.1 and 11.2.3.2).

Q Copper chloride and copper ammonium chloride are by-products from circuit board manufacturing. If these by-products are used directly as ingredients in the production of copper sulfate and copper hydroxide, did they ever become solid waste?

A No. The first use/reuse recycling exclusion applies. [RO 11334]

Q Baghouse dust and other sludges from air pollution control equipment associated with brass production contain zinc oxide and also exhibit a characteristic. The materials are fed to a process where virgin hydrochloric or sulfuric acid is added, thereby producing solutions of zinc chloride or zinc sulfate. The zinc chloride or sulfate solution is then purified. No reclamation of the secondary materials occurs before they're fed into the production process. Are the baghouse dust and other sludges solid wastes?

A No. Step 1 is check the table with the asterisks. If the characteristic sludge is not used to produce a product that is placed on the land and is not accumulated speculatively, Table 11-1 does not address this specific recycling situation. Moving to Step 2, this process would qualify for the first recycling exclusion (use as an ingredient) since the sludges are used (along with nonwaste ingredients) to produce zinc chloride or sulfate products. Thus, the characteristic sludges are not solid wastes when recycled in this way, even though the zinc chloride/sulfate solutions are subsequently purified in the process. [RO 11056] (Steps 3–5 would also have to be completed successfully, however.) [RO 11933]

The fact that the *products* need to be reclaimed (purified) does not void the recycling exclusion for use as an ingredient; if the secondary materials were reclaimed prior to use as ingredients, however, the recycling exclusion would be voided. [RO 14736] If the secondary materials did have to be reclaimed prior to use as an ingredient in the production process, they still would not be solid wastes—but not because of the exclusion for use as an ingredient. In this alternative situation, Step 1 would exclude them because, by Table 11-1, characteristic sludges sent for reclamation are not solid wastes (as long as the reclaimed sludge is not used to produce a product applied to the land and the sludge is not speculatively accumulated). [RO 11933]

Q When off-spec jet fuel is recycled as a feedstock to produce new jet fuel, is the exclusion for use as an ingredient applicable?

A Off-spec jet fuel is a commercial chemical product. Looking at Table 11-1, commercial products that will be burned for energy recovery are solid wastes (i.e., there is an asterisk at the intersection). However, §261.2(c)(2)(ii) excludes from RCRA commercial products that are burned for energy recovery if they are themselves fuels (as noted in Section 11.2.3). "Thus, this off-spec jet fuel, if used to produce jet fuel, is not a solid waste (i.e., an off-spec fuel is being reclaimed to be used as a fuel—its intended purpose)." Therefore, you don't have to proceed to Step 2. [RO 11449]

Q K088 spent aluminum potliner is recycled into a vitrification system to produce various glass products. Can this scenario qualify for the §261.2(e)(1)(i) recycling exclusion?

A Yes. Again, the table with the asterisks gives no guidance on the status of this spent material when recycled in a vitrification system. Moving to Step 2, reuse of the entire secondary material stream as an ingredient in the production of glass products would qualify under §261.2(e)(1)(i). [RO 14153]

Q *When distillation bottoms from the manufacture of carbon tetrachloride (K016) are used directly as feedstock in the production of tetrachloroethylene, does the first use/reuse recycling exclusion apply?*

A Yes. The table with the asterisks gives us no information when this listed by-product is recycled in this way. Moving to our Step 2 evaluation, the distillation bottoms are not solid waste when used as an ingredient to produce tetrachloroethylene, as long as Steps 3–5 are completed successfully. [January 4, 1985; 50 FR 619]

Q *Is the use of K085 chlorobenzene distillation bottoms as an ingredient to a hydrodechlorination process that makes other chlorinated compounds and muriatic acid subject to RCRA standards?*

A No. K085 is a listed by-product. However, EPA does not view the hydrodechlorination process as controlled-flame combustion or as reclamation. Since no use constituting disposal or speculative accumulation is involved, Table 11-1 does not address this recycling situation. Evaluating the first recycling exclusion in Step 2, the secondary material is used as an ingredient to make products. Therefore, the processing is eligible for the §261.2(e)(1)(i) recycling exclusion, and the material is not K085 or even a solid waste when recycled in this way. [RO 11342]

Q *Acrylic plastic abrasive is used to blast paint and coatings from aircraft. Waste produced from this process includes acrylic plastic dust mixed with paint/coating chips that often contain cadmium and chromium. When the waste is sent overseas for reuse as an ingredient in the manufacture of acrylic plastic sheets, the plastic/chips mixture is heated in a molten lead bath, wherein the metal contaminants partition to the lead and the acrylic polymer cracks into methyl methacrylate monomer. Does the reuse of the material qualify for the first use/reuse recycling exclusion?*

A No. Heating the dust as described above clearly is recovering something of value from the secondary material and is considered to be reclamation. The plastic/chips mixture is a spent material, and a spent material undergoing reclamation is a solid waste per Table 11-1, unless one of the recycling exclusions that allows reclamation is applicable (see Sections 11.2.3.1 and 11.2.3.2). If Step 1 doesn't exclude the material, moving to Step 2, reclamation is not allowed under the first use/reuse recycling exclusion. As such, the secondary material is a hazardous waste when exported. [RO 11937]

Q *Can a facility recycle waste colored glaze that exhibits the toxicity characteristic for cadmium, chromium, and/or lead as an ingredient in new pottery without meeting RCRA standards?*

A Yes. Reuse of the colored glaze in total, including any leachable metals, indicates that the material is not being reclaimed. Although Table 11-1 provides no information about recycling the spent material in this way, the waste glaze would qualify for the exclusion for use as an ingredient. [RO 13183]

Q *A facility receives K062 spent pickle liquor from offsite, treats the material with ferric oxide and lime to decrease its acidity, then recovers/reclaims the resulting material to produce saleable ferric chloride/ferrous chloride. Does the K062 qualify for the §261.2(e)(1)(i) exclusion?*

A No. The process of mixing K062 with ferric oxide produces ferrous chloride. Although it could be argued that the K062 is used as an ingredient to produce a product, EPA believes that the primary purpose of the mixing is to reduce the acidity of the spent pickle liquor (if the acidity is below a certain level, this step would likely not be conducted). Thus, the agency considers that any ferrous chloride produced is incidental—not the primary purpose of the activity—and that the mixing step amounts to reclamation. A spent material that is reclaimed is a solid waste per Table 11-1, unless one of the reclamation exclusions discussed in Sections 11.2.3.1 and 11.2.3.2 applies. In our Step 2 evaluation, reclamation voids the first use/reuse recycling exclusion. Based on the terms of §261.6, storing the waste prior to the reclamation would be subject to Part 264/265, Subpart J tank standards, but the reclamation process would not be subject to RCRA permitting or management standards. [RO 11093]

Q: *A facility reclaims K047 "red water," producing a sodium sulfite product that is used as an ingredient to make a product. Can the plant take advantage of the §261.2(e)(1)(i) recycling exclusion?*

A: No. When a listed by-product is reclaimed, it is a solid waste per Table 11-1. The §261.2(e)(1)(i) exclusion doesn't apply because reclamation voids the exclusion. The sodium sulfite stream, however, is a nonwaste product and, thus, no longer subject to RCRA. [RO 11253] Again, the facility may still qualify for a recycling exclusion from RCRA if one of the reclamation exclusions discussed in Section 11.2.3.1 or 11.2.3.2 applies.

11.3.1.5 Special provision for recycling spent sulfuric acid

Sulfuric acid is one of the highest-volume chemicals produced and used in the world. Thus, it is not surprising that EPA has provided a specific recycling exclusion for this type of material in §261.4(a)(7). This provision says that spent sulfuric acid used to produce virgin sulfuric acid is not a solid waste, unless it is accumulated speculatively. Because of the specific nature of this exclusion, Steps 3–5 aren't applicable (see RO 14348).

The reason EPA promulgated this exclusion is that certain parties believed that the reuse of spent sulfuric acid could be considered reclamation, which would result in a solid waste per Table 11-1 and would void the first use/reuse exclusion. Rather than add several *million* tons annually of this spent acid to the hazardous waste realm, EPA decided "the spent sulfuric acid recycling process more closely resembles a manufacturing operation than a reclamation process." [January 4, 1985; 50 *FR* 642] In that 1985 rule, the agency added the §261.4(a)(7) provision. The rulemaking record for this exclusion notes that the general concentration range of spent sulfuric acid is 5–100%. [RO 14713]

The §261.4(a)(7) exclusion is specific to spent sulfuric acid being burned in an industrial furnace to produce sulfur; the sulfur is then used to produce sulfuric acid. [RO 14570] Other recycling schemes for spent sulfuric acid are not excluded under §261.4(a)(7) and, therefore, must be judged on their own merits. [RO 11351, 11352, 11468, 12551, 14570] Similarly, high-sulfur residues burned in a sulfuric acid regeneration furnace are not excluded if they are not spent sulfuric acid. [RO 11856, 14086] If the spent sulfuric acid is stored in a land-based unit, such as a surface impoundment, that portion which leaches into the ground, if not recovered, has been disposed and would not be excluded. If such spent sulfuric acid exhibits a characteristic, the surface impoundment would be a hazardous waste management unit. [RO 11351]

Q: *At a manufacturing plant, hydrocarbon feedstock is reacted with sulfur trioxide in the presence of concentrated sulfuric acid to produce sulfonic acid. In addition to the product, a waste solution is generated that contains 4–15% sulfuric acid, as well as sulfonation residues. Would this waste stream be covered under the §261.4(a)(7) exclusion if it is sent to a facility that uses it as feedstock in a sulfuric acid production furnace?*

A: Yes. When the waste stream is removed from the manufacturing process, the contained sulfuric acid can no longer serve its intended purpose and is, therefore, spent sulfuric acid. Since the sulfuric acid concentration in the waste is within the 5–100% range, it qualifies for the exclusion when sent to a sulfuric acid production furnace. [RO 14713]

11.3.2 Use as an effective substitute for a commercial product

The second use/reuse recycling exclusion applies to hazardous secondary materials that are used or reused as substitutes for commercial products. The actual exclusion states that materials are not solid wastes when:

"Used or reused as effective substitutes for commercial products." [§261.2(e)(1)(ii)]

A hazardous secondary material that is used "as is" (with no reclamation) as an effective substitute for a product that a facility would otherwise have to buy is excluded from the definition of solid waste and, therefore, cannot be a RCRA hazardous waste. "When secondary materials are *directly* used as substitutes for commercial products, we also believe these materials are functioning as raw materials and therefore are outside of RCRA's jurisdiction and, thus, are not wastes." [Emphasis in original.] [January 4, 1985; 50 *FR* 619]

Referring back to our logic diagram, Figure 11-1, this provision is listed as the second bulleted item in the lower of the two Step 2 boxes. Following the logic lines in the diagram, where this exclusion applies to a given situation (i.e., the answer to the question is "Yes"), a further check of the Step 3–5 conditions is required to ensure that the recycled material in question is excluded (i.e., not a solid waste). Failure to meet the conditions in any of these steps would disqualify the hazardous secondary material stream from a solid waste exclusion. Again, Steps 3–5 are described below in Sections 11.4–11.6, respectively.

11.3.2.1 No reclamation please

While the provision implementing the first use/reuse recycling exclusion [in §261.2(e)(1)(i)] explicitly prohibits reclamation, the regulatory language for the second exclusion does not. Facilities may not take consolation in this fact, however, as EPA has stated repeatedly in guidance that reclamation is *not* allowed. [RO 11374, 11395, 11468, 11868, 13539] This position goes all the way back to the preamble to the January 4, 1985 final rule implementing the use/reuse recycling exclusions; in that document, the agency stated several times that secondary materials must be *directly* used or reused as substitutes for commercial products to qualify under §261.2(e)(1)(ii). [See 50 *FR* 619 and 637.] What does and does not constitute reclamation was discussed previously in Section 11.3.1.2.

For example, a company recycles K062 spent pickle liquor. As shown in Figure 11-6, the facility treats the waste to extract saleable iron sulfate that is subsequently used as a substitute for the commercial product. However, the recovery step is considered reclamation. Checking Table 11-1 indicates that a spent material sent for reclamation is a solid waste unless one of the reclamation exclusions discussed in Sections 11.2.3.1 and 11.2.3.2 applies. If Step 1 doesn't get the material out of RCRA, Step 2 is

Figure 11-6: Reclamation Voids the Use as an Effective Substitute Recycling Exclusion

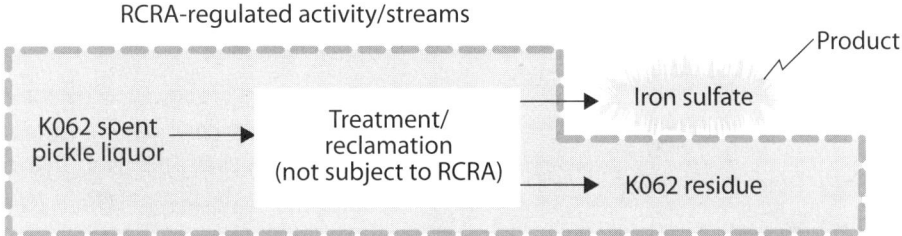

Source: McCoy and Associates, Inc.; adapted from RO 11468.

initiated. Again, the reclamation process invalidates the second use/reuse exclusion for the K062. The iron sulfate reclaimed in the process is considered a product when used in place of a virgin commercial product, while the remaining K062 liquor is still subject to RCRA. [Per §261.3(c)(2)(i), the reused iron sulfate is not K062 as long as it is not used in a manner constituting disposal or burned for energy recovery.] [RO 11468]

11.3.2.2 Other limitations/considerations

This second use/reuse recycling exclusion is very similar to the first one. As such, the same limitations, conditions, and POG considerations discussed in Section 11.3.1.3 apply to this exclusion as well.

11.3.2.3 Examples

The list below gives straightforward examples from EPA guidance of materials that are eligible for the second use/reuse recycling exclusion in §261.2(e)(1)(ii).

- K062 spent pickle liquor is used directly as a substitute for ferric chloride in a wastewater conditioning operation. [January 4, 1985; 50 FR 637, RO 11081]
- Spent acids are used directly in the treatment of wastewater or irrigation water as a substitute for virgin acids. [RO 11154, 11185, 13023, 14330]
- Spent deionization acid is reused "as is" in place of virgin acid. [RO 11154]
- Hydrofluorosilic acid (an air pollution control sludge) is used without reclamation as a substitute for a commercial fluoridating agent to fluorinate drinking water. [50 FR 637]
- Used refrigerant is collected during equipment servicing and then reused without filtration or other processing in place of virgin refrigerant. [RO 11565]
- Spent solvent from cleaning circuit boards is still pure enough to be used "as is" in a metal degreasing operation. [50 FR 624]
- Used chemicals that still have very high purity, such as isopropyl alcohol, sulfuric acid, hydrochloric acid, hydrofluoric acid, and phosphoric acid, are collected and resold for further use in a manner consistent with their intended purpose without reclamation. [RO 11868]
- Hydrochloric acid by-product is used without reclamation as a substitute for commercial hydrochloric acid as pickling liquor. [50 FR 619, RO 13671]

Agency guidance also examines situations that are not so clear cut as to whether they qualify for the second use/reuse recycling exclusion. These are detailed in Case Study 11-1 and the examples below.

Q: Can off-spec fuels from aircraft maintenance, spill cleanups, etc. be reused under the second use/reuse exclusion in place of virgin fuels?

A: As discussed in Section 11.3.1.4, off-spec fuels are considered commercial chemical products in Table 11-1. Although there is an asterisk at the intersection of "Commercial products" and "Burning for energy recovery," §261.2(c)(2)(ii) notes that commercial products that are themselves fuels are not solid wastes when burned for energy recovery. See Section 11.2.3 for additional discussion. As such, you don't have to proceed to Step 2. The off-spec fuels are not wastes when burned so long as the fuels to be reused do not contain nonfuel contaminants that are listed or characteristic hazardous wastes. In that case, the off-spec fuel would be regulated as a hazardous waste, even when burned for energy recovery. [RO 11938]

Q: A facility blends virgin sodium hydroxide (NaOH) with a spent 5% NaOH solution from a metal cleaning operation in order to increase the NaOH concentration to 10%. Can the upgraded solution then be reused at another facility under the substitution exclusion?

A: Yes. In Step 1, a review of the table with the asterisks does not provide an answer as to the status of the spent material. In Step 2, EPA does not consider the blending to be a form of reclamation, so the application qualifies for the §261.2(e)(1)(ii) exclusion (as long as Steps 3–5 are completed successfully). [RO 12918]

Case Study 11-1: Is Spent Solvent Reuse for Drum Washing Legitimate?

A company collects used parts washing solvent from customers. In the past, the company would take the spent solvent back to its facilities and recover fresh solvent for reuse at its customers' plants. Because a spent material was being reclaimed, however, the used parts washing solvent was a solid waste per Table 11-1. Because it was usually a listed solvent or exhibited the ignitability characteristic, the spent solvent was hazardous as well and had to be manifested to the company's facilities.

The company recently changed its practice for some of the collected used parts washing solvent. It now reuses that portion for in-house drum washing, and wants to take advantage of the §261.2(e)(1)(ii) exclusion. After review by EPA, the agency concluded that these collected solvents are not secondary materials since their reuse in drum washing fits their original intended use as solvents:

> "[W]hen a used solvent is employed for another solvent use, this continued use indicates that the solvent remains a product. The used solvent in this case is a material continuing to be used as a solvent, the purpose for which it is intended, rather than a spent material being reused. Consequently, the used solvent to be employed for drum washing would not be considered a solid waste and would not be subject to the [RCRA] Subtitle C hazardous waste regulations when generated, transported, or used." [RO 14281]

Thus, the agency concluded that consideration of this practice under the second use/reuse recycling exclusion was moot. Instead, the collected solvents used for drum washing would be exempt from RCRA under the agency's "continued-use policy." In order to prevent abuse of this policy, however, EPA attached the following conditions. The used solvents must be:

- Effective in the drum washing operation, especially if used in lieu of solvents that would otherwise have to be purchased;

- Used only for washing drums that actually need to be washed; and

- Used in quantities that would normally be used to wash drums (no excessive use of the solvent would be allowed as a means to simply dispose that material outside the purview of RCRA).

After the solvents have been used for drum washing, any spent solvent would be a solid waste subject to a hazardous waste determination.

The agency is suggesting that, because the solvent is continuing to be used for its intended purpose, it is a product—not a secondary material. Therefore, you don't have to evaluate the use/reuse recycling criteria in §261.2(e)(1). Our problem with this analysis is that, if you don't have to go to §261.2(e)(1), then you aren't subject to the qualifying criteria of §261.2(e)(2) or the documentation requirements of §261.2(f). That means that the used parts washing solvent could be speculatively accumulated [since §261.2(e)(2)(iii) doesn't apply]. Also, the agency added some conditions onto its continued-used policy—in this case to make sure the solvent isn't sham recycled. EPA has no authority under RCRA to add such restrictions to products being used for their intended purpose. Thus, we think this so-called "continued-use policy" is essentially the same as the §261.2(e)(1)(ii) exclusion, and we recommend that the §261.2(e)(2) and 261.2(f) requirements be satisfied.

Note that some states we've talked to aren't excited about this policy. In one case, when inspectors followed the continued-use solvents back to the company's recycling facility, the driver and facility personnel weren't able to keep track of which drums were which, and the facility never used any of the solvent for drum washing. At that point, the generator has just shipped potentially hazardous waste without making a waste characterization, counting it, manifesting it, etc.

11.3.3 Closed-loop recycling with no reclamation

The third use/reuse recycling exclusion involves closed-loop recycling of secondary material. According to §261.2(e)(1)(iii), recycled secondary materials are not solid wastes when:

"Returned to the original process from which they are generated, without first being reclaimed or land disposed. The material must be returned as a substitute for feedstock materials. In cases where the original process to which the material is returned is a secondary process, the materials must be managed such that there is no placement on the land. In cases where the materials are generated and reclaimed within the primary mineral processing industry, the conditions of the exclusion found at §261.4(a)(17) apply rather than this paragraph."

When hazardous secondary materials are returned to the original production process from which they were generated without first being reclaimed, EPA considers this to be manufacturing operations: "we likewise believe this recycling activity does not constitute waste management." [January 4, 1985; 50 FR 620] In Figure 11-1, this provision is listed in the third bulleted item in the lower of the two Step 2 boxes. As with the first and second use/reuse recycling exclusions, satisfying the Step 3–5 conditions is required to ensure that the recycled material in question is RCRA-excluded. Failure to meet the conditions of the exclusion *or* Steps 3, 4, and 5 would make the material a solid waste subject to regulation under RCRA (if the stream is hazardous).

To take advantage of this third exclusion, several conditions must be met, as discussed below.

11.3.3.1 Condition 1—Reuse must be in the original process

The first condition to the closed-loop recycling exclusion is that the recycled hazardous secondary material must be reused in the original process from which that material was generated (see Figure 11-7). Note that this condition does not say that the secondary material must be recycled to the exact same process unit from which it was generated. Manufacturing plants commonly use multistep production processes, and EPA requires only that the unreclaimed secondary material be:

"returned to the same *part* of the process from which it was generated. The material need not be returned to the same unit operation from which it was generated. It is sufficient if it is returned to any of the unit operations associated with production of a particular product, if it originally was generated from one of those unit operations. For example, an emission control dust from a primary zinc smelting furnace could be returned to any part of the process associated with

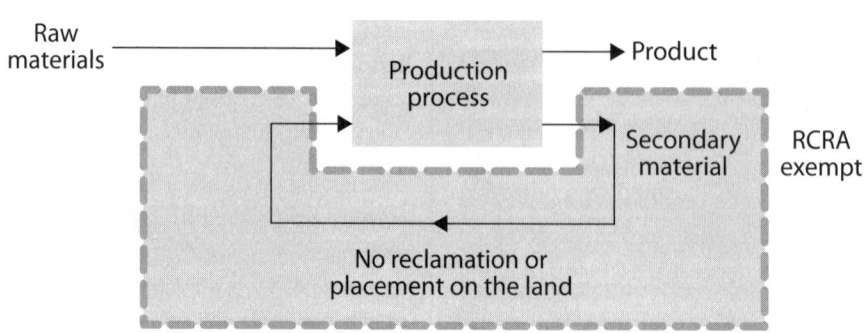

Figure 11-7: Closed-Loop Recycling With No Reclamation

Source: McCoy and Associates, Inc.; adapted from §261.2(e)(1)(iii).

zinc production, such as the smelting furnace in the pyrolytic plant, or the dross furnace. A spent electrolyte from the primary copper production process could be returned to any part of the process involved in copper production—including the roaster, converter, or tank house. An emission control dust from steel production could be returned to the sintering plant for processing before charging to the blast furnace.

"However, in the first example, if the emission control dust from the zinc smelting furnace was sent to by-product cadmium recovery operations, it would not be considered to be returned to the same [part] of the process from which it was generated. This is because the cadmium production processes produce a different product from zinc production operations. For the same reason, if the spent electrolyte in the second example were sent to by-product recovery operations for recovery of nickel sulfate, they would not be considered to be returned to the original process. Note that this principle holds even if the by-product recovery operation is located at the same plant site." [Emphasis in original.] [January 4, 1985; 50 *FR* 640]

In other guidance, EPA noted that some facilities use multiple, identical reactors in banks with common feed and discharge piping. Segregating secondary materials for closed-loop recycling to the individual generating reactor would not be feasible. Thus, recycling to the same type of reactor is allowed. [RO 11361]

Although the condition of reuse in the original process would seem to eliminate the option of transporting the hazardous secondary material to another facility for recycling, EPA noted in guidance that "nothing in §261.2(e)(1)(iii) limits the closed-loop exclusion to onsite recycling." The last sentence in the 1985 quotation cited just above also hints at offsite recycling. For example, K061 EAF dust from one steel plant could be recycled without reclamation into offsite steel production processes. Obviously, maintaining a closed-loop system is somewhat more difficult if two separate facilities are involved, but whether hazardous secondary material is recycled on- or offsite is not a limiting factor for this exclusion [consistent with the other two exclusions in §261.2(e)(1)]. [RO 11271]

11.3.3.2 Condition 2—No reclamation can be occurring

As shown in Figure 11-7, the closed-loop recycling exclusion is conditioned on there being no reclamation of the hazardous secondary material from its point of generation to the point it is reinserted back into the production process. Again, we refer readers to our earlier discussion in Section 11.3.1.2 of what constitutes reclamation.

11.3.3.3 Condition 3—Reuse must be as a feedstock

Secondary material recycled under the closed-loop exclusion must be returned to its generating production process for *use as a feedstock*. The recycled material is being reused in a production process as a substitute for raw materials that the facility would otherwise have to buy. For example, recycling spent degreasing solvent within a parts cleaning process would not qualify for the third use/reuse recycling exclusion because the returned solvent is not involved as a feedstock in a production operation—it merely cleans equipment. [50 *FR* 639]

11.3.3.4 Other eligibility issues

Other eligibility issues associated with the closed-loop recycling exclusion are discussed in the subsections below.

11.3.3.4.1 *No placement on the land*

According to the language of the closed-loop recycling provision, recycled hazardous secondary materials are not solid wastes when "returned to the original process from which they are generated, *without first being…land disposed*." [Emphasis added.] Thus, facilities wanting to take advantage of this exclusion cannot store the material in a waste pile or surface impoundment prior to recycling it. Storage in a unit designed to contain the material or otherwise prevent its release to the environment, such as in a containment building or tank, is permissible. [September 19, 1994; 59 FR 48015]

11.3.3.4.2 Primary and secondary processes on equal footing

If you've worked with RCRA a long time, you may recall that the closed-loop recycling exclusion was originally promulgated with another limiting condition. That condition, which was removed in a September 19, 1994 final rule [59 FR 47982], required that the secondary material be recycled to a *primary* production process. [A primary process is one that uses raw materials as the majority of its feedstock; in contrast, a secondary production process uses mostly spent or waste materials (e.g., scrap metal) as its feedstock.] EPA's goal with the original provision was to ensure that the secondary material recycled within a process should be no less of a waste than the material originally introduced to the process.

EPA removed the primary process restriction based on its belief that proper management of hazardous secondary materials within all processes (primary and secondary) is more important to protecting the environment than preventing their recycling to secondary processes. Therefore, recovery operations from secondary production processes are now on the same regulatory footing as such operations from primary processes. An important beneficiary of this change is the secondary lead smelting industry—see RO 13566.

11.3.3.5 Examples

The following examples illustrate applications of the closed-loop recycling exclusion:

- A smelting facility generates a dry emission control dust that it collects, stores (but not on the land), and resmelts in the original smelting furnace. [January 4, 1985; 50 FR 640; RO 13566]

- A maker of alkylate products from alkylation of paraffins and olefins directly recycles used sulfuric acid to the alkylation reactors in order to fully utilize that acid. [RO 11361]

- A scrap metal recycling foundry recycles baghouse dust to the process cupola furnace without reclamation to recover metals in the dust. [RO 12561]

11.3.4 Closed-loop recycling with reclamation

None of the first three use/reuse recycling exclusions discussed in Sections 11.3.1–11.3.3 allow reclamation of the secondary material. The fourth use/reuse recycling exclusion allows certain materials to be excluded from the definition of solid waste if they are:

> "Secondary materials that are reclaimed and returned to the original process or processes in which they were generated where they are reused in the production process...." [§261.4(a)(8)]

The basic premise of this recycling exclusion is that reclaiming and then recycling a process residue are "best viewed as part of the production process, not as a distinct waste management operation." [July 14, 1986; 51 FR 25442] This fourth exclusion is included as the upper of the two Step 2 boxes in the Figure 11-1 logic diagram. If you believe that this exclusion applies to your situation, a further check of the Steps 4 and 5 conditions must be made *before* your hazardous secondary material can be considered excluded from RCRA (i.e., not a solid waste). Failure to meet the conditions in any of these two remaining steps will disqualify the secondary material from the exclusion.

As is evident from the regulatory language, reclamation of the hazardous secondary material is allowed in the fourth use/reuse recycling exclusion. The material is run through a reclamation process, and it is cleaned up in some way or something of value is recovered from it. However, "the act of reclamation must be directly related to the act of production." [51 FR 25442] A simplified representation of this fact and the other qualifying conditions for this exclusion are shown in Figure 11-8. Four primary conditions must be satisfied for this exclusion to apply, as codified in §261.4(a)(8) and discussed in the subsections that follow.

Figure 11-8: Closed-Loop Recycling With Reclamation

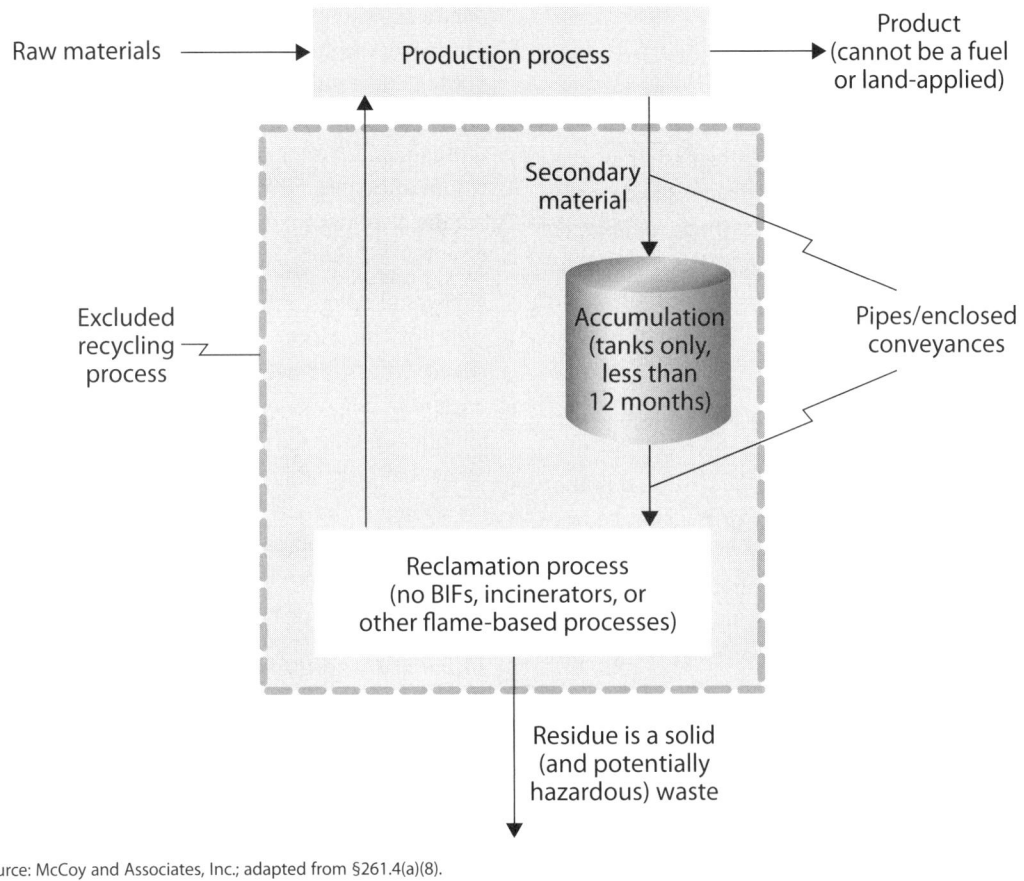

Source: McCoy and Associates, Inc.; adapted from §261.4(a)(8).

11.3.4.1 Condition 1—Reuse must be in original process

The recycled hazardous secondary material must be returned to the original production process from which it was generated. In the preamble to the July 14, 1986 final rule in which this exclusion was promulgated, EPA noted:

"[T]he material that is returned after having been reclaimed can be reused as a feedstock, as a purifying agent to remove contaminants from feedstock, and can also be reused for other purposes, including as a reaction medium to dissolve or suspend chemicals, or as a reactant to facilitate chemical reactions. To be considered as being 'returned to the original process,' the reclaimed material need not be returned to the same unit operation from which it was generated, but only to the same *part* of the process. In addition, if the same material is reused in a number of production operations at an integrated plant, and the secondary material is reclaimed in a common reclamation operation, the reclaimed material can be returned to any process which originally used the material....

"By production process, the agency intends to include those activities that tie directly into the manufacturing operation or those activities that are the primary operation at an establishment; it does not include ancillary or secondary activities that are carried out as part of the total activities at the facility....

"EPA believes that solvents returned for use as cleaning agents in dry cleaning operations will be considered to be reused in the production process...since they are used as the basic raw material in the process.... On the other hand, materials used to clean equipment (for example, solvents returned and reused as degreasers) are not normally considered to be reused in a production process." [Emphasis in original.] [51 *FR* 25442]

11.3.4.2 Condition 2—Storage and conveyance limitations

According to §261.4(a)(8)(i) and (iii), this use/reuse exclusion applies when 1) any storage of recycled materials is in tanks and for no more than 12 months prior to reclamation; and 2) the entire recycling train through completion of reclamation is connected via hard piping or other comparable closed conveyances.

While piping and other conveyances must be closed, EPA does not prohibit the use of open-top tanks in eligible closed-loop recycling operations. Normally, however, the agency would expect facilities to use closed tanks to 1) reduce contamination due to rain or dust entering the tank, and 2) preclude safety problems such as the threat of explosion due to ignition of flammable vapors. The agency further noted that in order to qualify as legitimate recycling, the hazardous secondary material must be managed as a valuable commodity (i.e., in a manner designed to avoid loss or release). Thus, the use of open-top tanks for volatile material might cause the exclusion "to be inapplicable because [evaporated] secondary materials are not being reclaimed and returned to the process." [51 *FR* 25443; see also RO 13591] Obviously, the accumulation of recycled secondary materials under this fourth exclusion is not allowed in nontank units such as containers, piles, or surface impoundments.

In terms of the 12-month accumulation limit, you would not need to physically empty a given tank once a year. Rather, you would only need to show via recordkeeping or other means that at least 100% of the tank's content was turned over within a 12-month period. This concept allows flow-through tank systems and other continuous processes to qualify for the recycling exclusion without the need to shut down and empty tanks once a year. [51 *FR* 25443]

Piping and other conveyances used in processes eligible for this fourth recycling exclusion must be closed, which the agency defines as "hard connections from point of generation to point of return to the original process." [51 *FR* 25443] Thus, conveyances such as troughs, gutters, and trenches would not qualify under this condition. In addition, trucks, even if loaded/unloaded via hard-piped connections and kept closed during transport, would not qualify as a closed conveyance. [RO 11468]

11.3.4.3 Condition 3—No flame-based reclamation

Although reclamation is allowed under this fourth use/reuse recycling exclusion, such reclamation may not involve any type of controlled-flame combustion. This would preclude the use of BIFs and incinerators in the recycling process. [§261.4(a)(8)(ii)]

If it meets the limitations just noted, the reclamation process itself is not subject to RCRA regulation. Instead, it would be a manufacturing operation.

11.3.4.4 Condition 4—No production of fuels or products applied to the land

If the production process to which a reclaimed material is returned produces a fuel as the product, the exclusion is voided. Also, to preserve its authority to regulate hazardous secondary materials and products that contain these wastes when they are used in a manner constituting disposal, the recycling exclusion is not available where the hazardous secondary material is recycled to a process that produces a product that is applied to the land (e.g., fertilizer, cement). [§261.4(a)(8)(iv), RO 11732]

11.3.4.5 Additional details

Several additional issues regarding the scope of the closed-loop recycling with reclamation exclusion are worthy of mention:

- Secondary materials that meet the eligibility criteria described above are excluded from the definition of solid waste. Such materials, therefore, cannot be hazardous waste, and equipment in which they are managed (i.e., conveyed, accumulated, and/or reclaimed) is exempt from Subtitle C regulation. Thus, for example, tanks used to store recycled material per the terms of the exclusion are not subject to Part 264/265, Subpart J standards. [EPA/530/SW-87/012, cited in Section 10.1.2.4]

- The recycling with reclamation exclusion does not extend beyond the dashed lines shown in Figure 11-8. In other words, material that exits the closed loop loses the solid waste exclusion. For example, a residue generated by the reclamation process would be a solid waste (assuming no other exclusion applies to it) and subject to RCRA regulation if it meets the definition of hazardous waste. A common example of this situation would be still bottoms exiting from a closed-loop solvent reclamation process that are hazardous by characteristic or listing. [July 14, 1986; 51 *FR* 25443, RO 11285, 12732, 13017, 13220]

- Similar to residues, releases from excluded closed-loop recycling processes that are hazardous are also subject to full regulation under RCRA. For example, releases at RCRA-permitted facilities would be subject to possible corrective action under Part 264, Subpart S if routine and systematic. Such releases would also be subject to possible notification requirements under CERCLA Part 302 regulations. [EPA/530/SW-87/012]

11.3.4.6 Examples

Q: *An automotive paint manufacturing process generates spent acetone solvent that is recovered via distillation in a closed-loop reclamation system and then returned to the original production process. Under certain circumstances, some of the recovered acetone cannot be recycled to the paint manufacturing area (for process chemistry and mass balance reasons); during those times, the excess acetone is sold as commercial-grade solvent. Is this recycling scheme eligible for the fourth use/reuse recycling exclusion?*

A: No. In RO 14089, EPA said that failure to recycle *all* of the reclaimed solvent to the original production process violates the intent of §261.4(a)(8)—the loop is not closed. Therefore, the spent acetone is a solid and hazardous waste. However, the spent acetone solvent would probably be excluded from the definition of solid waste under the exclusion for materials reclaimed "under the control of the generator." See Section 11.2.3.1 for details.

Q: *Water is used to slurry listed emission control dust and convey it to a dust fixation system that consists of a filter press. Hazardous wastewater generated in the fixation system (filtrate from the filter press) is returned to the baghouse for reuse. Is this reclamation process eligible for the closed-loop recycling with reclamation exclusion?*

A: No. The fourth use/reuse recycling exclusion applies to secondary material that is reclaimed and then recycled to a manufacturing process. Since the baghouse is a treatment unit, not a manufacturing process, the filter press operation is ineligible for the exclusion. [RO 12942]

11.4 Step 3—Additional qualifying criteria

If one of the two reclamation exclusions codified in §§261.2(a)(2)(ii) and 261.4(a)(23–25) from Step 1 of the recycling determination process, or one of the first three use/reuse recycling exclusions codified in §261.2(e)(1)(i–iii) from Step 2 appears to apply to your recycling situation, you cannot stop at that point. You must move on to Step 3 and review the qualifying criteria in §§261.2(a)(2)(ii), 261.2(e)(2), and 261.4(a)(23–25), as applicable, before your tentative conclusion that the material is not a solid waste can be finalized. Note that, as shown in

Figure 11-1, you don't have to go to Step 3 if you believe you qualify for the fourth use/reuse recycling exclusion, because the qualifying criteria are built into the §261.4(a)(8) exclusion language. The basic intent of §§261.2(a)(2)(ii), 261.2(e)(2), and 261.4(a)(23–25) is to limit the applicability of the reclamation exclusions and the first three use/reuse recycling exclusions. Specifically, several types of recycled hazardous secondary materials are not excluded from the definition of solid waste, even though the recycling involves reclamation, use, reuse, or return to the original process.

As noted in Figure 11-1, there are actually two sets of qualifying criteria: 1) a general set that applies to the two reclamation exclusions (discussed in Sections 11.2.3.1 and 11.2.3.2) and the first three use/reuse recycling exclusions (discussed in Sections 11.3.1–3), and 2) a specific set that applies only to the two reclamation exclusions. These two sets are discussed in more detail in the next two subsections.

11.4.1 General qualifying criteria

There are four general qualifying criteria that must be met to claim either of the two reclamation exclusions and any of the first three use/reuse recycling exclusions. They are codified in §261.2(e)(2) for the first three use/reuse recycling exclusions, and they are referenced in preamble language at 73 *FR* 64669–70 as also applying to the two reclamation exclusions. These four general qualifying criteria are shown in the box labeled Step 3b in Figure 11-1, and they identify four types of hazardous secondary materials that will always be solid (and potentially hazardous) waste:

1. *Materials used in a manner constituting disposal or used to produce products that are applied to the land [§261.2(e)(2)(i)]*—"The agency is thus asserting jurisdiction over all hazardous secondary materials, and over products that contain these wastes, when they are applied to the land. Thus, fertilizers, asphalt, and building foundation materials that use hazardous wastes as ingredients and are then applied to the land are subject to RCRA jurisdiction. Secondary materials applied directly to the land likewise are within the agency's Subtitle C regulations…." [50 *FR* 628]

According to EPA, "if a secondary material is ultimately used in formulating a product to be placed on the land, then it is a solid waste from the point of generation, through transportation, and including any storage prior to being used in formulating a product." [RO 11395] Another example of a secondary material that is a solid waste because it is used in a manner constituting disposal is hydrochloric acid (a by-product generated during production of a primary product) that is sold for use as a fracturing agent in oil and gas wells. During fracturing, the acid is injected through a wellbore into the earth where it reacts with limestone formations. This activity is similar to deep-well injection and clearly is a form of land disposal; thus, it is subject to RCRA regulation as use constituting disposal. [RO 13671]

2. *Materials burned for energy recovery, used to produce a fuel, or contained in fuels [§261.2(e)(2)(ii)]*—"[R]esidual materials such as tank bottoms…are by-products and are considered to be wastes when used as fuels or when incorporated into fuels." [50 *FR* 630] For example, light hydrocarbon streams that are by-products of a production process are solid wastes if they are used as fuel additives and/or burned as supplemental fuel in onsite boilers. [RO 11793]

3. *Materials that are accumulated speculatively [§261.2(e)(2)(iii)]*—"[H]azardous secondary materials accumulating before recycling are wastes unless the person accumulating is able to show on request that he is indeed recycling sufficient volumes of the materials on an annual basis." [50 *FR* 634] According to EPA, persons who accumulate speculatively are those who fail to recycle at least 75% of a recyclable material during the calendar year or fail to

demonstrate that a feasible means of recycling exists. Hazardous secondary materials that are otherwise excluded from the definition of solid waste become regulated as solid waste (and potentially hazardous waste) if they are accumulated speculatively. [RO 13755]

4. *Materials that are inherently waste-like listed in §261.2(d)(1–2) [§261.2(e)(2)(iv)]*—"We continue to think that certain residual materials are inherently waste-like, either because: a) they are typically disposed of or incinerated on an industry-wide basis, or b) they contain toxic constituents in concentrations not ordinarily found in the raw materials or products for which they substitute, which toxic constituents are not used, reused, or reclaimed during the recycling process. In addition, recycling of the materials must have the potential to pose a substantial hazard to human health and the environment." [50 *FR* 637] The two types of materials that the agency has designated as inherently waste-like are 1) dioxin wastes that will be recycled [hazardous waste codes F020, F021 (unless used as an ingredient to make a product at the site of generation), F022, F023, F026, and F028], and 2) hazardous secondary materials fed to a halogen-acid furnace (except for certain brominated materials). EPA recently provided guidance as to what types of units do not constitute halogen-acid furnaces. [RO 14576]

The first two general qualifying criteria of §261.2(e)(2) are examined in the examples that follow.

11.4.1.1 Use constituting disposal

Use constituting disposal occurs when materials are applied directly to the land or are used as ingredients to produce products that are applied to the land. When hazardous secondary materials are used in a manner that constitutes disposal, this will usually void any recycling exclusion from RCRA. The following subsections review the regulatory status of hazardous secondary materials that are recycled in this manner.

11.4.1.1.1 *Use as an ingredient to produce fertilizer*

Using hazardous secondary materials as feedstocks in fertilizer production is generally not considered to be reclamation (although there are exceptions [RO 11113, 11932]). Instead, the agency's position is that fertilizers are used in a manner constituting disposal (i.e., they are land-applied). As such, secondary materials that will be recycled into fertilizers must be managed as solid waste from their point of generation (unless they are being used to make zinc fertilizers). [RO 11112, 11124, 11275, 11932, 14148] If such a secondary material is also hazardous, its storage and transportation would be subject to RCRA waste management standards (e.g., tank and container standards, air emission standards, manifesting, etc.). However, the fertilizer manufacturing unit itself would be exempt from regulation as a recycling unit. [RO 11113, 14148] Case Study 11-2 reviews these concepts.

As of July 24, 2002, secondary materials used to produce zinc fertilizers are not solid (or hazardous) waste if the provisions in §261.4(a)(20) are met. [67 *FR* 48393, RO 14658] Thus, zinc-rich dusts from brass foundries and fabricators and from other industries will no longer have to be managed as hazardous waste from their point of generation if they will be incorporated into such fertilizers.

Sometimes, hazardous secondary materials are reclaimed, and the resulting reclaimed products are then used to produce fertilizer. For example (as shown in Figure 11-9), a facility processes a characteristic copper oxide baghouse dust (a sludge under RCRA) to produce copper sulfate, which is then used as an ingredient to produce fertilizer. Since the fertilizer is land-applied (use constituting disposal), the copper oxide baghouse dust is a solid and hazardous waste. "The agency has consistently interpreted this provision to apply without regard to whether the…sludge as a whole (or some portion of it that is recovered) is used as a product that is placed on the ground or used to

Case Study 11-2: Secondary Material Used as a Fertilizer Ingredient

I. Used sulfuric acid from a chlorine dehydration process is too dilute to reuse in that manner. Consequently, it no longer can serve the purpose for which it was purchased without processing (i.e., it is a spent material). Is the spent acid considered a solid waste if it is used without being reclaimed to make nonzinc fertilizer?

In Step 1, we check Table 11-1, which indicates that a spent material being used in a manner constituting disposal is a solid waste. Moving to Step 2, the sulfuric acid is used as an ingredient in an industrial process to make a product, so the first use/reuse recycling exclusion in §261.2(e)(1)(i) would appear to apply. Before we stop the recycling determination process, however, we must review the qualifying criteria in §261.2(e)(2). When we do this, we find that materials that are used to produce products that are applied to the land are solid wastes when so recycled. Therefore, if the spent acid is used to produce nonzinc fertilizer, it would be considered a solid waste and is a hazardous waste if it exhibits a characteristic. However, if the spent acid is reused directly as an ingredient in a product that is not placed on the land or burned [and the other conditions in §261.2(e)(2) are met], the material would not be a solid waste. [RO 11361]

II. A brass mill produces brass dross skimmings that exhibit a characteristic and are sent to a recycling facility to separate metals from the oxides. If the oxides (which still exhibit a characteristic) are then sent to a nonzinc fertilizer maker, do they have to be manifested? How about the original skimmings?

The brass dross skimmings are a characteristic by-product. Checking Table 11-1 in Step 1 indicates that characteristic by-products that are reclaimed are not solid wastes. However, as we noted previously, this exclusion is nullified and overridden and the entire process is subject to RCRA, including management of the brass dross skimmings, since the oxides are destined for use constituting disposal as a nonzinc fertilizer. Although this process appears to meet the recycling exclusion under §261.2(e)(1)(i) in Step 2, reviewing the criteria in Step 3 disqualifies the material from the exclusion since the product produced will be used in a manner constituting disposal. Therefore, RCRA regs apply to the skimmings from their POG at the outlet of the manufacturing process, meaning they would have to be stored as hazardous waste and manifested for offsite shipment. Similarly, the recycling facility would have to store the skimmings and resulting oxide material as hazardous waste, and the shipment of oxides to the fertilizer company would have to be manifested. Finally, the fertilizer company would have to manage the oxides as hazardous waste until they enter the fertilizer manufacturing unit. [RO 11083]

Figure 11-9: Use of Reclaimed Materials in Fertilizer Production

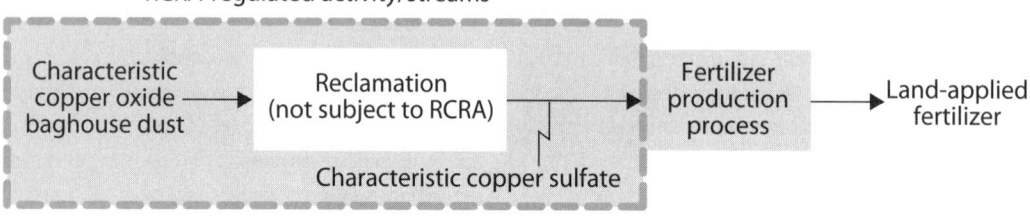

Source: McCoy and Associates, Inc.; adapted from RO 11113.

produce a product that is placed on the ground." [RO 14158] We must evaluate the entire recycling process, not just the first reclamation step in the process; if a characteristic sludge or some portion of it will be recycled in a manner constituting disposal, the sludge is a solid and hazardous waste. [RO 11113, 13723, 14195, 14348] However, the reclamation process (converting the copper oxide into copper sulfate) and the fertilizer production unit would be exempt from regulation as recycling units. [RO 11113, 11302, 14158]

In a situation where the copper oxide dust was converted to copper sulfate and a portion of the resulting copper sulfate is sent for fertilizer production and a portion is sent for metals recovery, the copper oxide sludge would still be a solid and hazardous waste because some of the sludge was used in a manner constituting disposal. If none of the copper sulfate is used to produce fertilizer or otherwise applied to the land, the copper oxide dust would not be a solid waste, because reclamation of a characteristic sludge is not subject to RCRA per Table 11-1. [RO 11275]

One final note concerning fertilizer production from hazardous secondary materials and the scope/breadth of RCRA coverage. Not only is the hazardous secondary material that is recycled into the fertilizer a solid and hazardous waste, but [per §261.2(c)(1)(i)(B)] the fertilizer itself remains a solid waste (and hazardous waste if it exhibits a characteristic). However, EPA has provided a special exemption for products applied to the land at §266.20(b). Thus, fertilizer made from hazardous secondary materials would not be subject to RCRA if 1) it is produced for the general public's use, 2) the recycled materials undergo a chemical reaction so as to be inseparable from the fertilizer by physical means, and 3) the fertilizer meets land disposal restrictions (LDR) treatment standards (e.g., if the fertilizer is produced from a secondary material that is characteristic for lead, the fertilizer would have to meet the LDR treatment standard for D008, including all underlying hazardous constituents). If the fertilizer does not meet the conditions of §266.20(b), use of the product would be subject to §266.23 (i.e., full Subtitle C regulation), unless it is a zinc fertilizer meeting §261.4(a)(21) requirements. [RO 11644, 11683, 14148, 14158, 14437, 14566]

11.4.1.1.2 *Use as an ingredient to produce animal feed supplements*

As noted above, fertilizers are used in a manner constituting disposal (i.e., they are land-applied), but the agency's position is that animal feed supplements are not. Thus, when hazardous secondary materials are used (without prior reclamation) as ingredients in feed supplement production, they are not addressed in Table 11-1, and they are not solid wastes because of the first use/reuse recycling exclusion. [RO 11275, 11932, 14148]

Looking back to Figure 11-4, when a process qualifies for the first use/reuse exclusion (such as production of animal feed supplements), the recycled material downstream of its POG is not considered a solid waste. All associated storage and transportation will be exempt from RCRA (e.g., no manifesting is required).

11.4.1.1.3 *Use as an ingredient to produce cement/aggregate*

Cement/aggregate is a product normally applied to the land and, therefore, used in a manner constituting disposal. Checking Table 11-1 per Step 1, there is an asterisk for every entry under "Use constituting disposal." Therefore, no matter what type of secondary material is being recycled into the cement, that material would be a solid waste.

Q *A facility wants to produce Portland cement out of K061 EAF dust—a listed sludge—captured in an emission control system. Will the K061 have to be managed as hazardous waste?*

A Probably. Using the material as an ingredient in an industrial process to produce a product qualifies for the first use/reuse recycling exclusion. However, using a hazardous secondary material to produce a product that is applied to the land voids the exclusion [per §261.2(e)(2)(i)].

Therefore, the EAF dust would have to be managed according to RCRA standards as a hazardous waste from its point of generation to the cement production unit. (Note that the presumption that cement is applied to the land is rebuttable; if the facility can demonstrate that the cement product will not be applied to the land, the recycling would be exempt.) [RO 11426, 11491]

The same result would be attained for the following materials recycled as ingredients in cement:

- Spent sandblast grit (a spent material that commonly exhibits the toxicity characteristic for lead due to blasting lead-based paint off of structures and equipment). [RO 11433]
- K048/K052 filter cake (a listed by-product). [RO 11573]
- Aluminum dross (a characteristic by-product). [RO 11395]

11.4.1.1.4 Use as an ingredient to produce pesticides

Hazardous secondary materials recycled into pesticides should be managed in the same way as such materials that are used as an ingredient in fertilizer. That is, the hazardous secondary materials would be solid (and potentially hazardous) wastes because they are being used to make a product that is applied to the land. In RO 14437, EPA discussed the regulatory status of fertilizers and pesticides containing hazardous waste on the same footing.

11.4.1.1.5 Use as an ingredient to produce absorbent material

EPA has noted that the use of absorbent products for cleaning up spilled liquids on the ground would be considered use constituting disposal. Therefore, hazardous secondary materials used as ingredients to produce absorbents that may be used on the ground would be solid wastes per §261.2(e)(2)(i). [RO 14025] If such hazardous secondary materials exhibit a characteristic (e.g., of toxicity for chromium), they would have to be managed as hazardous wastes from their point of generation to the absorbent production unit.

11.4.1.1.6 Spent material reuse as wastewater conditioner

Facilities often take advantage of the §261.2(e)(1)(ii) exclusion by reusing hazardous secondary materials as wastewater conditioners. For example, a facility generates spent ferric chloride from an etching process and sends it to another plant for use as a substitute for commercially available conditioning agents. EPA has addressed the issue of whether wastewater conditioning is considered use constituting disposal and has decided it is not: "the secondary material is chemically combined as part of a conditioning process and is subsumed as an ingredient in the conditioned water." [January 4, 1985; 50 *FR* 628, RO 11081]

11.4.1.2 Burning for energy recovery/use to produce a fuel

People often want to burn residues from manufacturing operations in onsite boilers or industrial furnaces (BIFs). However, burning secondary materials for energy recovery will generally void any recycling exclusion from RCRA per §261.2(e)(2)(ii). The following examples review the regulatory status of secondary materials that are recycled in this way.

Q: Before performing repairs on an airplane, the fuel contained in the fuel cells (JP-4) is drained. A mixture of JP-5 fuel and lubricating oil is then injected into the fuel cells to remove any remaining JP-4. This JP-5/lubricating oil mixture is used until the flash point of the mixture is approximately 120°F. At this point, the material is pumped to a tanker truck and sent to a nearby refinery where the fuel mixture is placed in the refinery process to produce petroleum products. Are there any RCRA implications to this practice?

A: Yes. The JP-5/lubricating oil mixture is used like a solvent to remove the remaining JP-4 from the fuel cells; as such, the contaminated JP-5/lubricating oil is a spent material. Spent materials are solid wastes when they are used to produce a fuel. Therefore, if this solid waste exhibits a hazardous characteristic, it is subject to EPA's authority under the hazardous waste rules. [RO 11105]

Q *A process combusts a characteristic secondary material with a high-Btu and sulfur content in order to recover the sulfur for reuse as an ingredient in the production of sulfuric acid. Is the hazardous secondary material eligible for the first use/reuse recycling exclusion?*

A Checking Table 11-1, any hazardous secondary material that is burned for energy recovery is a solid waste. Moving to Step 2, even though sulfur is used as an ingredient to produce sulfuric acid, EPA considers the process to be burning for energy recovery rather than use as an ingredient to produce a product if the heating value of the hazardous secondary material is greater than 5,000 Btu/lb. [§266.103(a)(5)(ii)(B)] Therefore, such burning voids the recycling exclusion per §261.2(e)(2)(ii). [RO 11856]

Q *A petrochemical facility uses a catalytic extraction process to produce synthesis gas (syngas) from a variety of hazardous and nonhazardous secondary materials (including F024, K019, and K020 listed wastes). Are these materials subject to RCRA regulation before they are processed?*

A It depends. These listed by-products are not recycled in a way noted in Table 11-1. When the secondary materials are used as feedstocks to produce a product (the syngas), EPA noted that they would qualify for the §261.2(e)(1)(i) recycling exclusion. However, if the syngas is burned for energy recovery, a likely option, the secondary material feed streams would be classified as solid/hazardous wastes by §261.2(e)(2)(ii) and would be fully subject to RCRA. [RO 11985] Conversely, if the syngas is used to make a product that is not burned as a fuel (e.g., used to make methanol that is subsequently used in the production of photographic film or similar products), then the hazardous secondary materials used as feedstock to the unit would not be solid/hazardous wastes. [RO 14643]

Q *Coal tar decanter sludge (K087 if disposed) is recycled to produce coke. Can the facility take advantage of the §261.2(e)(1)(iii) exclusion for the recycled sludge?*

A No. The manufacture of coke from the sludge is producing a fuel, which is then burned as a reducing agent during the production of iron. Per §261.2(e)(2)(ii), the coal tar decanter sludge is not excluded when recycled and is K087 and subject to hazardous waste regulation. [RO 11290]

11.4.2 Specific qualifying criteria

There are four specific qualifying criteria that must be met to claim either of the two reclamation exclusions discussed in Sections 11.2.3.1 and 11.2.3.2. These criteria are codified in §§261.2(a)(2)(ii) and 261.4(a)(23–25) and are shown in Step 3a in Figure 11-1. They state that 1) three types of hazardous secondary materials are not eligible for these two reclamation exclusions, and 2) all hazardous secondary materials must be "contained" in management units before reclamation. Each of these criteria is discussed below.

1. *K171/K172 spent catalyst from petroleum refineries*—The two reclamation exclusions are not applicable to K171/K172 spent catalyst from petroleum refineries. EPA is planning to address whether these materials are solid wastes when reclaimed in a future rulemaking, where the agency will address the potential pyrophoric properties of these materials, particularly during transportation and storage prior to reclamation. [73 FR 64714]

2. *Spent lead-acid batteries*—EPA wants spent lead-acid batteries to continue to be managed under either Part 266, Subpart G or as universal waste under Part 273. That is because these are actually hazardous waste regulations, although regulations with less-stringent management standards, that are appropriate due to the unique nature of these batteries. Therefore, spent lead-acid batteries are not eligible for either of the two reclamation exclusions. [73 FR 64714]

3. *Materials already subject to an exclusion in §261.4(a)*—The two reclamation exclusions do not apply to materials that are otherwise subject to material-specific management conditions

under §261.4(a) when reclaimed; such materials must continue to meet the existing conditions/requirements in that section of the regs to be excluded from the definition of solid waste. For example, the exclusion for used, broken cathode ray tubes (CRTs) at §261.4(a)(22)(iii), that requires these CRTs to be managed under conditions that are more specific than those specified in the reclamation exclusions, is not changed or superseded. [73 FR 64713]

4. *Materials must be "contained"*—For hazardous secondary materials to qualify for either of the two reclamation exclusions, the materials, when managed in either non-land-based units (e.g., tanks, containers, and containment buildings) or land-based units (e.g., piles and surface impoundments), must be "contained" in such units before reclamation. Although the regs don't give any specific definition for this term, EPA noted in preamble language that "[g]enerally, such material is 'contained' if it is placed in a unit that controls the movement of the hazardous secondary material out of the unit and into the environment." [73 FR 64681] This requirement is consistent with normal manufacturing processes, which are designed to use valuable materials rather than allow them to be released to the environment.

The agency gave the following example of a material that was not contained, thereby voiding any reclamation exclusion: "an acidic hazardous secondary material undergoing reclamation could be stored in a tank that experienced a failure. A facility might fail to monitor the structural integrity of the tank, as most product tanks are monitored, or the tank might not be constructed to contain acidic hazardous secondary materials, causing a significant release of such materials into the environment that is not immediately recovered. The unit itself would consequently be considered a hazardous waste management unit because the hazardous secondary materials were not being managed as a valuable raw material, intermediate, or product, as evidenced by the failure to monitor it for structural integrity, resulting in the release. Thus, the unit and any remaining waste would be subject to Subtitle C controls because the hazardous secondary materials in the unit have been discarded. In addition, any of the released materials that were not immediately recovered would also be considered discarded and, if hazardous, subject to appropriate federal or state regulations and applicable authorities." [73 FR 64681]

Additional guidance is available in "Revisions to the Definition of Solid Waste Final Rule: The Contained Standard." [RO 51392]

11.5 Step 4—Recycling must be legitimate

If you believe that you can take advantage of one of the recycling exclusions available in the RCRA regs, as identified in Step 1 or 2 of our process, and the qualifying criteria in Step 3 are satisfied, you still can't stop but must move on to Step 4. This step requires you to determine if your recycling activity is legitimate. "EPA retains its long-standing policy that all recycling of hazardous secondary materials must be legitimate." [73 FR 64707] Hazardous secondary materials are not solid wastes when managed in accordance with one of the recycling exclusions because "recycling of such materials often closely resembles normal industrial manufacturing rather than waste management. However, since there can be a significant economic incentive to manage hazardous secondary materials outside the RCRA regulatory system, there is a potential for some handlers to claim that they are recycling, when, in fact, they are conducting waste treatment and/or disposal in the guise of recycling. To guard against this, EPA has long articulated the need to distinguish between 'legitimate' (i.e., true) recycling and 'sham' (i.e., fake) recycling...." [73 FR 64700]

EPA's October 30, 2008 rule codified four factors in §260.43 for determining whether hazardous secondary materials are being legitimately recycled under the two reclamation exclusions. Previously, such legitimacy determinations were made on a case-by-case basis by regulators using EPA's "sham" recycling guidance (see January 4, 1985; 50 *FR* 638 and RO 11426). Although the newly codified legitimacy factors appear to be different in some respects from EPA's previously published recycling guidance, the agency noted in the October 30, 2008 rule preamble that the four factors being codified "are equivalent to the existing legitimacy policy that applies to all recycling." [73 *FR* 64710] Thus, current exclusions and other prior solid waste determinations or variances, including determinations made in letters of interpretation and inspection reports, remain in effect. EPA does not expect regulatory agencies to revisit past legitimacy determinations based on the factors codified in the October 30, 2008 rule.

Despite what EPA said about the codified factors being equivalent to existing legitimacy policy, the agency currently requires the four codified factors to be evaluated only when determining the legitimacy of one of the reclamation exclusions in §§261.2(a)(2)(ii) and 261.4(a)(23–25). The legitimacy of a recycling activity that is not subject to RCRA because of one of the dashes in Table 11-1 or due to one of the four use/reuse exclusions in §261.2(e)(1) or 261.4(a)(8) needs to be evaluated under EPA's pre-existing legitimacy policy/guidance cited above. "[T]he agency has decided not to codify the legitimacy factors for existing exclusions and, thus, states and other implementing agencies will continue to apply the existing legitimacy policy to all recycling as they have in the past in order to ensure that recycling is real and not a sham." [73 *FR* 64708] Thus, we discuss both the new factors and pre-existing policy/guidance in the next two subsections.

11.5.1 Recycling legitimacy factors codified

Section 260.43 contains four factors for determining if hazardous secondary materials are being legitimately recycled under the two reclamation exclusions promulgated in 2008. The first two are mandatory; that is, they are required to be addressed to demonstrate that recycling is legitimate. The second two factors are not mandatory but should be "considered" when making a legitimacy determination. The two sets of factors are discussed below. Additionally, EPA's thoughts on the role of economics in determining recycling legitimacy are given in Section 11.5.1.3.

11.5.1.1 Mandatory recycling legitimacy factors

Two factors must be satisfied to demonstrate that recycling of hazardous secondary materials under the two reclamation exclusions is legitimate. The two mandatory legitimacy factors are:

1. The secondary material must provide a useful contribution to the recycling process or to a product or intermediate of the recycling process. A "useful contribution" means: 1) contributing valuable ingredients to a product or intermediate, 2) replacing a catalyst or carrier in the recycling process, 3) providing a valuable constituent that is recovered or regenerated, or 4) being used as an effective substitute for a commercial product.

   This factor is intended to prevent the practice of adding hazardous secondary materials to a manufacturing operation simply as a means of disposing them. EPA noted the following situations that would be considered legitimate under this factor [73 *FR* 64702]:

   - A recycling operation involving precious metals might not recover all of the components of the hazardous secondary material, but it would recover precious metals with sufficient value to consider the recycling process legitimate.

   - Not every constituent or component of the hazardous secondary material has to make a contribution to the recycling activity to be considered legitimate. Additionally, a recycling activity does not have to involve the hazardous component of the hazardous secondary materials if the contribution of the nonhazardous component (e.g., zinc) justifies the legitimacy of the recycling activity.

- In a situation where more than one hazardous secondary material is used in a single recycling process and the hazardous secondary materials are mixed or blended as a part of the process, the recycler would have to be able to satisfy the useful contribution factor for each hazardous secondary material to show legitimacy. This requirement prevents situations where a worthless hazardous secondary material is mixed with valuable and useful hazardous secondary materials in an attempt to disguise and dispose it.

- A recycling scenario with a low recovery rate could still be legitimate. For example, if the concentration of a valuable metal in a hazardous secondary material is low (e.g., 2%–4%), but the recycling process was able to recover a large percentage of the target metal, the recycling may be legitimate. Conversely, if there is a constituent in the hazardous secondary material that could add value to the recycling process, but, due to process design, most of it is not being recovered but is being disposed in the residues, this would be a possible indicator of sham recycling.

2. The recycling process must produce a valuable product or intermediate that is 1) sold to a third party, or 2) used by the recycler or the generator as an effective substitute for a commercial product or as an ingredient or intermediate in an industrial process.

This factor is intended to prevent the practice of running a hazardous secondary material through an industrial process to make something just for the purpose of avoiding the costs of hazardous waste management, rather than for the purpose of using the product or intermediate from the recycling activity. EPA provided the following examples and discussion of how this factor should be applied [73 *FR* 64703]:

- A product of the recycling process may be sold at a loss in some circumstances, but the recycler would have to be prepared to show how the product is clearly valuable to the purchaser.

- Some recycling processes may consist of multiple steps that may occur at separate facilities, such as when a metal-bearing hazardous secondary material is processed to reclaim a precious metal and is then put through another process to reclaim a different mineral. When each step in the process yields a valuable product or intermediate that is salable or usable in that form, the recycling activity would be legitimate.

- If the product or intermediate from the recycling process replaces a virgin product that would otherwise have to be purchased, or the product or intermediate of the recycling process meets specific product specifications or specific industry standards, the product or the intermediate is valuable.

- If a recycling process produces building materials that are sold to customers, it would be considered legitimate. However, an example of sham recycling would be the situation where a recycler reclaims a hazardous secondary material and then uses the reclaimed material to make blocks or building materials for which it has no market and then "uses" those building materials to build a warehouse in which it stores the remainder of the building materials that it is unable to sell.

11.5.1.2 Nonmandatory recycling legitimacy factors

Two factors are not mandatory but should be "considered" when making a legitimacy determination: "the agency knows that there will be some situations in which a legitimate recycling process does not conform to one or both of these two factors, yet the reclamation activity would still be considered legitimate.… In the event that the process does not conform to one of the two factors…the facility should be able to show that it considered that factor and why the recycling activity overall remains legitimate." [73 FR 64701] The two nonmandatory legitimacy factors are:

1. The secondary material should be managed as a valuable commodity by the generator and recycler. Where there is an analogous raw material, the hazardous secondary material should be managed in a manner consistent with the management of the raw material. Where there is no analogous raw material, the secondary material should be contained. (See the discussion of "contained" in Section 11.4.2.)

EPA believes a recycler will value hazardous secondary materials that provide an important contribution to its process or product and, therefore, will manage those materials in a manner consistent with how it manages a valuable feedstock. The agency gave the following example [73 FR 64703–4]:

- A manufacturer uses a dry raw material managed in Super Sacks®. If a hazardous secondary material that is a similar dry material will be used as a legitimate substitute for the virgin material, it should also be managed in Super Sacks® or in a manner that would provide equivalent protection. If, on the other hand, the hazardous secondary material was managed in an outdoor pile without appropriate controls in place to address releases to the environment, such management may indicate that the material is not being handled as a valuable commodity. But what if the situation is changed so that the hazardous secondary material is managed in Super Sacks®, whereas the analogous raw material is shipped and stored in drums? A strict reading of this factor may indicate that the hazardous secondary material is not being managed in a manner consistent with the raw material, even though the differences in management are not likely to cause a release.

2. The product of the recycling process should not contain "toxics along for the ride." That is, the product should not 1) contain significant concentrations of hazardous constituents (in Part 261, Appendix VIII) that are not found in analogous products, 2) contain significantly higher concentrations of hazardous constituents than those found in analogous products, or 3) exhibit hazardous characteristics not exhibited by analogous products.

Any of the three situations described in this factor could indicate that sham recycling is occurring because, in lieu of proper hazardous waste disposal, the recycler may simply be incorporating (discarding) hazardous constituents into a final product when they are not needed to make that product effective for its purpose. Evaluating what constitutes "significant," may "involve taking into consideration several variables, such as the type of product, how it is used and by whom, whether or not the elevated levels of hazardous constituents compromise the efficacy of the product, the availability of the hazardous constituents to the environment, and others." [73 FR 64706] The agency gave the following examples of how this factor might be used in a legitimacy determination [73 FR 64701–6]:

- "[R]euse of lead-contaminated foundry sands may or may not be legitimate, depending on the use. The use and reuse of foundry sands for mold making in a facility's sand loop under normal industry practices has been found to be legitimate because the sand is part of an industrial process where there is little chance of the hazardous constituents being released into the environment or causing damage to human health and the environment when it is kept inside, because there is lead throughout the foundry's process, and because there is a clear value to reusing the sand. However, in the case of lead-contaminated foundry sand used as children's play sand, the same high levels of lead would disqualify this use from being considered legitimate recycling."

- If a recycling process produces paint, the levels of hazardous constituents in the paint will be compared with the levels of the same constituents found in similar paint made from virgin raw materials. "[T]he central question is whether or not (and in what amount) hazardous constituents pass through the recycling process and become incorporated into the products of recycling." [73 FR 64710]

- A recycler is also allowed to compare the hazardous constituents in the hazardous secondary material feedstock with those in an analogous raw material feedstock. If the hazardous secondary material feedstock does not contain significantly higher concentrations of hazardous constituents than the raw material feedstock, then the end product of the recycling process would not contain excess hazardous constituents "along for the ride" either. There may be cases in which it is easier to compare feedstocks than it is to compare products.

- Recycling may be legitimate when the product has concentrations of toxics that could be considered "significantly higher" than the analogous product made from virgin raw materials, but the product made using recycled materials meets industry specifications for the product that include specifications for the hazardous constituents of concern.

- If zinc galvanizing metal made from reclaimed hazardous secondary materials contains 500 ppm of lead, while the same zinc product made from raw materials typically contains 475 ppm, this difference in concentration would likely not be considered "significant." If, on the other hand, the lead level in the zinc product made from reclaimed

hazardous secondary materials was 1,000 ppm, it may indicate that the product was being used to dispose lead and that the activity is sham recycling, unless other factors would demonstrate otherwise. In another example, if a virgin solvent contains no detectable amounts of barium, while spent solvent that has been reclaimed contains a minimal amount of barium (e.g., 1 ppm), this difference might not be considered significant. If, however, the barium in the reclaimed solvent was at 50 ppm, it may indicate discard of the barium and sham recycling.

11.5.1.3 Economics to be considered when determining recycling legitimacy

Although the October 30, 2008 rule did not codify specific regulatory language regarding economics as part of a legitimacy determination:

"EPA believes that consideration of the economics of a recycling activity can be used to inform and help determine whether the recycling operation is legitimate. Positive economic factors would be a strong indication of legitimate recycling, whereas negative economic factors would be an indication that further evaluation of the recycling operation may be warranted in assessing the legitimacy factors.

"Considering the economics of a recycling activity can also inform whether the hazardous secondary material inputs provide a useful contribution and whether the product of recycling is of value. Economic information that may be useful could include 1) the amount paid or revenue generated by the recycler for recycling hazardous secondary materials; 2) the revenue generated from the sale of recycled products; 3) the future cost of processing existing inventories of hazardous secondary materials; and 4) other costs and revenues associated with the recycling operation. The economics of the recycling transaction may be more of an issue when hazardous secondary materials are sent to a third-party recycler, but even when the hazardous secondary materials are recycled under the control of the generator, the generator must still show that the hazardous secondary materials are, at a minimum, providing a useful contribution and producing a valuable product." [73 *FR* 64706]

The agency provided more detail on using the four types of economic information noted in the quote above when making legitimacy determinations in the October 30, 2008 rule preamble at 73 *FR* 64706–7.

Two extensive compilations of EPA guidance on what constitutes "legitimate recycling" are available in RO 51394 and 51395.

11.5.2 Previous legitimacy guidance

The legitimacy of a recycling activity that is not subject to RCRA because of one of the dashes in Table 11-1 or due to one of the four use/reuse exclusions in §261.2(e)(1) or 261.4(a)(8) needs to be evaluated under EPA's long-standing legitimacy policy (i.e., sham recycling guidance). The following are the sham recycling criteria against which you should evaluate your recycling process:

- *Waste materials that are being "recycled" are ineffective or only marginally effective for the claimed use.* In this case, the activity is not recycling but surrogate disposal. For instance, this would be the case if heavy metal-bearing sludges were used as an ingredient for concrete, even though the sludges do not significantly contribute to the concrete's properties. [January 4, 1985; 50 *FR* 638, RO 11426] In order for a process to be legitimate recycling, the waste materials must actually play a part in the manufacturing process, instead of being simply treated or disposed by incorporating them into a product. According to EPA, "the fact that a material can be inserted into a production process without detriment to the quality of the end product does not mean that the waste is actually being used as an ingredient." [RO 11491] The recycled waste must contribute to the effectiveness of the product; for example, wastes used in fertilizer must contain nutrients or micronutrients, and wastes used in cement must have pozzolanic properties. [EPA/530/SW-86/015, March 1986, available from http://nepis.epa.gov/EPA/html/Pubs/pubtitleOSWER.html by downloading the report numbered 530SW86015] In addition, it is not enough to say that use of the waste materials does not have an adverse impact on the environment. [RO 11573, 14001]

- *Waste materials are used in excess of the amount necessary to operate a process.* An example would be when chlorine-containing wastes are used (in an amount that significantly exceeds process requirements) as an ingredient in a process that requires chlorine. [50 *FR* 638, RO 11426] Burning wastes in an industrial furnace would be considered disposal rather than recycling if the quantities of waste burned provide materials in excess of what can feasibly be recovered and used. [EPA/530/SW-86/015] If two or three times more secondary material is being used than is required to make a good product, it looks like waste disposal.

- *The hazardous secondary material is not an effective substitute for the virgin material that it is replacing.* It is not enough to say that because a hazardous waste can be used as an ingredient and still result in a marketable product, such usage is legitimate recycling. Rather, a demonstration must be made that the hazardous waste itself is an effective substitute for a nonwaste material. [RO 11461, 11491, 11573] If the secondary material is as effective as the virgin material it is replacing, the activity is much more likely to be considered legitimate recycling. For instance, spent pickle liquor is recognized as an effective substitute for virgin phosphorous precipitants in wastewater treatment operations. [50 *FR* 638]

- *Situations in which waste materials are apparently reused, but the generator or recycler is unable to document how, where, and in what volumes the materials are being recycled.* Documentation is normally kept when using virgin materials and products; the absence of records when a hazardous secondary material is recycled is considered to be evidence of sham recycling. [50 FR 638, RO 11426]

- *Hazardous secondary materials are not handled in a manner consistent with their use as raw materials or commercial product substitutes.* If secondary materials are stored or otherwise managed in a way that does not guard against significant economic loss (e.g., they are stored in open piles or leaking surface impoundments), EPA will suspect that the material is not valued as a raw material or feedstock, and, therefore, the use or reuse is not legitimate recycling. [50 FR 638, RO 11426] The manner in which a material is managed and stored is definitely indicative of its value and the legitimacy of the recycling.

- *Activities where wastes are being used as feedstocks or substitutes for commercial chemical products, but the wastes contain high levels of contaminants that are not found in the original (nonwaste) raw material or commercial product.* An evaluation must be made of whether hazardous constituents found in the secondary material but not the analogous raw material/product are necessary for the materials' end-use functionality or just "along for the ride." What is a "high" level of contaminants? That has to be determined on a case-by-case basis by the state. [RO 11426, 11461, 11491, 11573, 14330]

Such a situation may occur, for instance, if industrial wastewater is used as a substitute for fresh water at a cement manufacturing plant. In this case, the wastewater would not be an effective substitute due to the presence of significant concentrations of hazardous constituents, which are not a desired ingredient in the final product. "Note that this determination is not based on the qualities of the final product (cement) but on the qualities of the water sources. This approach determines whether the actual secondary material is an 'effective substitute.'... The issue is whether the constituents in the substitute water source are a desired ingredient of the final product or are being, in some fashion, treated." [RO 14001; see also RO 11585, 11618] Particular focus should be given to the fate of hazardous constituents in the waste that are incorporated into the product (compared to the fate of constituents in raw materials that would otherwise be used). [RO 11491]

11.5.2.1 Questions to ask to determine if sham recycling is occurring

The task of determining if an activity should be classified as legitimate or sham recycling is not an easy one. To provide some guidance to the regulated community, states, and EPA regions, the agency has put together a list of questions that can be used to distinguish between legitimate and sham recycling. This list, which is shown in Table 11-2, reflects EPA's basic consideration of whether or not the secondary material is "commodity-like."

11.5.2.2 Some activities are under close scrutiny

One of the areas that EPA and the states are targeting for close scrutiny is the use of corrosive wastes as neutralizing agents. The agency has stated that it "will not accept a claim that a corrosive secondary material is being used as a substitute for virgin acid or caustic unless indicia of legitimate recycling are present. These include that the secondary acid or caustic meet relevant commercial specifications, that they be as effective as the virgin material for which they substitute, that they be used under controlled conditions, and that in a two-party transaction there be consideration (usually monetary) for use of the material. In addition, the more contaminated the acid or caustic is in relation to virgin material, the less likely the agency is to view its application as legitimate recycling." [January 4, 1985; 50 FR 638; see also RO 11154]

EPA and the states are also on the lookout for sham precious metal recovery operations. The agency will consider precious metal reclamation activities to be sham recycling if precious metals are present only in trace amounts or in amounts too low to be economically recoverable. [50 FR 648, EPA/530/SW-86/015 (cited previously)]

Table 11-2: Questions That May Be Used to Distinguish Between Legitimate and Sham Recycling

Similarity of the secondary material to the analogous raw material/product it replaces

- Does the secondary material contain Part 261, Appendix VIII constituents that are not found in the analogous raw material/product (or at higher levels)?
- Does the secondary material exhibit hazardous characteristics that the analogous raw material/product does not?
- Does the secondary material contain levels of recoverable material similar to the analogous raw material/product?
- Is much more of the secondary material used as compared with the analogous raw material/product it replaces? Is only a nominal amount of it used?
- Is the secondary material as effective as the raw material/product it replaces?

Degree of processing required to produce a finished product

- Can the secondary material be fed directly into the process (i.e., direct use) or is reclamation (or pretreatment) required?
- How much value does final reclamation add?

Value of the secondary material

- Is the secondary material listed in industry newsletters, trade journals, etc.?
- Does the secondary material have economic value comparable to the raw material that it replaces?

Market for the end product

- Is there a contract in place to purchase a product produced from the secondary material?
- If the type of recycling is reclamation, is the product used by the reclaimer or the generator? Is there a batch tolling agreement?
- Is the reclaimed product a recognized commodity? Are there industry-recognized quality specifications for the product? Is a legitimate safety data sheet (SDS) available?

Handling of the secondary material compared to the raw material/product it replaces

- Is the secondary material stored on the land?
- Is the secondary material stored in a similar manner as the analogous raw material (i.e., to prevent loss)?
- Are adequate records regarding the recycling transactions kept?
- Do the companies involved have a history of mismanagement of hazardous wastes?

Other relevant factors

- What are the economics of the recycling process? Does most of the revenue come from charging generators for managing their wastes or from the sale of the product?
- Are the toxic constituents actually necessary (or of sufficient use) to the product or are they just "along for the ride"?

Source: Adapted from RO 11426.

11.5.2.3 Economics—not the key factor in determining legitimate recycling

As noted in Table 11-2, the relative economics (who pays who) are evaluated when determining whether a recycling scenario is legitimate; however, the economics themselves are not determinative. The fact that secondary materials sent for reclamation are solid wastes under RCRA does not change "just because a reclaimer has purchased or finds value in the components." [*United States v. ILCO*, 996 F.2d 1131 (1993)] For example, a mixture of petroleum product and water is sent offsite to a recycler for product recovery. The recycler charges the generator a fee that includes the cost of treatment/disposal of the hydrocarbon-contaminated water that remains after the product is recovered. If the cost of that treatment/disposal is higher than the value of the recovered product, does that mean that the product recovery operation is sham recycling?

In general, "the relative profitability of the processing of a specific material is not the determining factor [when distinguishing between legitimate and sham recycling], although it certainly is one consideration. Rather, the key focus is whether the material being processed has recoverable levels of the constituent intended for recovery and the extent to which other hazardous constituents are being treated, however incidental to recycling that treatment may be. These factors indicate the legitimacy of the claimed recycling process and whether the intent is actually to treat or otherwise dispose of nonrecoverable hazardous constituents." [RO 11615]

Section 11.5.1.3 contains additional EPA guidance on considering economics when making a recycling legitimacy determination.

11.5.3 Legitimacy evaluation may involve regulator input

Evaluating the legitimacy of a recycling operation may involve input from your state or EPA regional regulators. Although this is not a regulatory requirement, we recommend that you talk with your regulators before recycling a hazardous secondary material under one of the exclusions discussed in this chapter for the following reasons:

1. Every time someone from the regulated community writes EPA a letter asking if their particular recycling scheme is excluded from RCRA, the agency responds with general comments based on the facts it was given. However, EPA typically ends each such letter with the following qualifier: "[I]t is the responsibility of EPA regional offices or RCRA-authorized states, using specific criteria related to a particular site, to determine whether or not a particular process is a legitimate recycling operation or whether it is a form of waste treatment." [RO 11985] EPA wants the states or regions to review each recycling situation to make sure there are no significant threats to human health or the environment.

2. By conferring with the state or region up front, you can avoid implementing a recycling process that you believe is excluded from RCRA regulation—but is actually not. Making mistakes in this regard could lead to significant enforcement penalties.

3. One of the issues on which the state or region will focus is whether the recycling activity is legitimate or a sham. Whether you are using the codified legitimacy factors discussed in Section 11.5.1 or the pre-existing legitimacy guidance discussed in Section 11.5.2, regulatory personnel will help you with the legitimacy determination.

11.6 Step 5—Documentation requirements

To recap, once you've found a recycling exclusion that you think you can use in Steps 1 and/or 2, you read through the qualifying criteria in Step 3 to make sure you're not doing something that will keep you from using the exclusion. You then show that the recycling is legitimate per Step 4. At that point, you still need to move to Step 5 and read the documentation requirements in §261.2(f). Anyone who claims that a certain material is not a solid waste when recycled must keep appropriate documentation proving that claim (e.g., contracts showing that a second party uses the material as an ingredient in a production process). In other words, parties making the claim have the burden of proof to show, through appropriate documentation, that they meet the conditions of the exclusion. [October 30, 2008; 73 *FR* 64700, January 2, 2008; 73 *FR* 67]

The regulations do not specify exactly what records must be kept. However, EPA stated that "persons must keep whatever records or other means of substantiating their claims that they are not managing a solid waste because of the way the material is to be recycled. They also must show that they are not overaccumulating their secondary materials." [January 4, 1985; 50 *FR* 642–3] In order to claim an exclusion from the definition of solid waste, the generator must be able to demonstrate that the material is being used in a manner that qualifies for the exclusion. Information stating that it is merely capable of such use or that it has been used for such purposes in the past would not be sufficient. [RO 11185] "Absence of documentation not only would make it difficult or impossible for a respondent to carry its burden of proof, but also would itself be evidence that the claimed recycling is a sham." [50 *FR* 643]

Regarding documentation of the legitimacy of a recycling activity, the agency noted: "Although there is no specific recordkeeping requirement that goes with the ability to demonstrate legitimacy, EPA would expect that in the event of an inspection or an enforcement action by an implementing agency, the recycler would be able to show how it made the overall legitimacy determination per §261.2(f)." [73 *FR* 64701]

Our position on this issue is: *No documentation means no exclusion.*

Other information facilities should keep to support their claim of a recycling exclusion includes:

1. The process generating the secondary material,
2. The chemical and physical composition of the secondary material,
3. The process in which the secondary material will be used,
4. Any applications in which the secondary material will be used,
5. Storage and accountability procedures for the secondary material,
6. Any product that is derived from the secondary material and its use, and
7. All applicable state and federal regulations used in the solid waste determination.

To illustrate the documentation requirements in Step 5, consider the situation in which a generator claims that wastewater treatment sludge (which exhibits a hazardous characteristic) is not a solid waste because it is reclaimed (in Table 11-1, there is a dash at the intersection of "Sludge exhibiting a characteristic" and "Reclamation"). To meet the burden-of-proof requirement, the generator must have documentation proving that the sludge is not a solid waste (e.g., documents that show that the sludge contains recoverable levels of metals and is processed by an appropriate metals reclaimer). [RO 11546] The use of batch tolling agreements (e.g., in which recovered materials that are reclaimed from secondary materials are returned to the original generator for reuse) would probably help a generator satisfy the burden of proof that the material is not a solid waste. [RO 11271]

In a situation where used chemicals will continue to be used for their intended purpose at a separate facility, proper documentation of the claim that those chemicals are not solid wastes includes records of the used chemicals onsite storage, shipments, and of the second facility's use (to demonstrate that the chemicals were not burned or used in a manner constituting disposal). [RO 11868]

The §261.2(f) provisions also require owners/operators of facilities claiming to recycle materials to show that they have the necessary equipment to do so.

Case Study 11-3 illustrates all five steps of the RCRA recycling determination process.

Case Study 11-3: Putting the Five Steps Together

A facility wants to use petroleum-contaminated soils from a cleanup site as an ingredient in the production of asphalt. Does this qualify as legitimate recycling that is exempt from RCRA?

Probably not. Asphalt is a product that will be applied to the land (i.e., used in a manner constituting disposal). When we check Table 11-1 in Step 1, no matter how we classify petroleum-contaminated soil, it will be a solid waste when recycled in this manner (because there are asterisks for every type of secondary material in the column under "Use constituting disposal"). It also does not qualify for either of the two reclamation exclusions promulgated in October 2008. Moving to Step 2, the contaminated soil is used as an ingredient to make a product; therefore, the first use/reuse recycling exclusion in §261.2(e)(1)(i) appears to be applicable. However, in reading over the qualifying criteria in Step 3, we note that the contaminated soil would remain a solid waste since the asphalt product will be applied to the land. [§261.2(e)(2)(i)] The only exception to this is that commercial chemical products are not solid wastes if they are applied to the land and that is their ordinary manner of use. [§261.2(c)(1)(ii)] Thus, if the soil is contaminated with an unused petroleum product that is normally used in asphalt production, the petroleum-contaminated soil would not be a solid waste.

When determining if this activity is legitimate recycling in Step 4, the facility (possibly with the state's input) will consider several factors. First, the contaminated soil will be compared with the analogous raw materials normally used in asphalt batching. If the contaminated soils contain hazardous constituents that are not found in the analogous raw materials or contain hazardous constituents in significantly higher concentrations than in the raw materials, the batching process would be considered sham recycling (unless the hazardous constituents can be demonstrated to be useful in the production of the product or in the product itself). Second, the facility/state would look at whether the contaminated soils are legitimately replacing a raw material or ingredient normally used in the process. For instance, if the contaminated soils are being used in excess of the amount of raw materials that would otherwise be used, sham recycling would be suspected.

Step 5: If the soil is contaminated with petroleum products normally used in the production of asphalt, the owner/operator needs to keep documentation of this fact and any other documentation of legitimacy. In addition, they need to keep contracts showing that the asphalt producer uses the soil as an ingredient in his/her production process and that the asphalt is used commercially. [RO 11616]

Used Oil

The used oil regs are simple if you are just a generator— they get more complicated if you burn or process the oil

From the largest industrial facility to the smallest manufacturing plant, just about every site generates used oil. This chapter deals with the generation, management, and burning of used oil. The bottom line is: if you generate used oil and have a recycler come in and haul your used oil offsite, your requirements are minimal. If you do other things with the used oil that you generate, your requirements may increase significantly.

In the mid- to late-1980s, EPA was trying to determine whether it should regulate used oil destined for disposal as a listed hazardous waste. The agency swayed back and forth on this issue a couple of times but finally decided not to list it on May 20, 1992. [57 *FR* 21524] (Used oil must be managed as hazardous waste in the state of California and maybe some other states, but it isn't listed at the federal level.) A major reason EPA didn't list used oil as hazardous was the presence of numerous existing federal programs pertaining to mismanagement (disposal) of this oil. For example, the RCRA underground storage tank regulations, the Spill Prevention, Control, and Countermeasures (SPCC) program under the CWA, and storm water provisions all address used oil management in some way. Plus, used oil destined for disposal that exhibits a characteristic must be managed under the full Subtitle C program.

Instead of listing it, EPA issued management standards for used oil that will be recycled, providing additional safeguards against mismanagement. These regulations, which are codified in 40 *CFR* Part 279, apply to generators, collection centers and aggregation points, transporters, processors and re-refiners, burners, and marketers of used oil. A quick guide to the Part 279 standards for each of these entities is available in *Part 279 Requirements—Used Oil Management Standards*, EPA/530/H-98/001, available from http://www.epa.gov/epawaste/conserve/materials/usedoil/poster.pdf. The most stringent requirements apply to used oil processors and re-refiners, because they handle the largest quantities of used oil. [September 10, 1992; 57 *FR* 41593, RO 11736]

12.1 Applicability of the used oil management standards

The used oil management standards are structured such that used oil that will be recycled is subject to management under the Part 279 standards, which are less stringent than the hazardous waste requirements in Parts 260–270. Such used oil can be managed under the Part 279 requirements, even if it exhibits a characteristic at its point of generation (before it is mixed with any other material). [§§261.6(a)(4), 279.10(a)]

Conversely, used oil that will be disposed is a solid waste and must be characterized for its hazardousness just like any other solid waste. The management of used oil that is destined for disposal and that exhibits a characteristic or has been mixed with listed hazardous waste is subject to the more-stringent hazardous waste regulations. [§279.81(a), September 10, 1992; 57 *FR* 41578, RO 11811]

Determining the applicability of the used oil standards can be confusing at times. Therefore, we have put together some logic diagrams that should provide some assistance. Figures 12-1 and 12-2 will help you determine if the Part 279 requirements apply in your specific situation. The issues that appear in the boxes on these two figures are discussed in further detail throughout Section 12.1. Once you know if the Part 279 standards apply from these figures, other logic diagrams presented later in this chapter will help you determine your compliance requirements.

12.1.1 Definition of used oil

In §279.1, EPA defines "used oil" as "any oil that has been refined from crude oil, or any synthetic oil, that has been used and as a result of such use is contaminated by physical or chemical impurities." This definition "covers the majority of oils used as lubricants, coolants (noncontact heat-transfer fluids), emulsions, or for similar uses [that] are likely to get contaminated through use." [September 10, 1992; 57 *FR* 41574]

In order to meet the definition of "used oil," a substance must meet each of the three criteria listed below [*Managing Used Oil—Advice for Small Businesses*, EPA/530/F-96/004, November 1996, available from http://www.epa.gov/epawaste/conserve/materials/usedoil/usedoil.pdf, RO 14090]:

- *Origin*—The substance must be derived from crude or synthetic oil;

- *Use*—The material must have been used as a lubricant, coolant, noncontact heat-transfer fluid, hydraulic fluid, buoyant, or other similar purpose (to be determined by authorized states or EPA regions); and

- *Contamination*—The oil must be contaminated with physical impurities (e.g., water, metal shavings, sawdust, or dirt) and/or chemical impurities (e.g., lead, solvents, halogens, or other hazardous constituents) as a result of use.

12.1.1.1 Examples of used oil

Based on the above definition and criteria, examples of substances that, when used, are used oil are listed in Table 12-1.

12.1.1.1.1 *Synthetic oil is used oil*

EPA has determined that synthetic oils are included in the definition of used oil because these oils 1) are generally used for the same purposes as petroleum-derived oils, 2) are usually mixed and managed in the same manner after use, and 3) present the same level of hazard as petroleum-based oils. The agency's intent is to include all synthetic oils that function similarly to petroleum-based lubricants, oils, and surface agents under the definition of used oil. Synthetic oils that may be considered used oil include oils that are not petroleum-based (i.e., those produced from coal or oil shale), oils that are petroleum-based but are water soluble (e.g., concentrates of metal-working oils/fluids), oils that are polymer-type, and nonpolymer-based oils. [September 10, 1992; 57 *FR* 41574, RO 11724]

12.1.1.1.2 *Consolidated used oil is just used oil*

In general, "EPA does not consider the consolidation of different sources of used oil to be a mixture of used oil. EPA regulates the consolidated used oil as used oil under the Part 279 used oil management standards." [RO 11786] The implication of this guidance is that used oil from anywhere in a facility can be consolidated into one or more used oil storage tanks without regulatory concern. This would not apply, however, to metal-working oils/fluids and CFC-contaminated used oil that are not subject to the rebuttable presumption (see Section 12.1.4.2).

Figure 12-1: Applicability of the Used Oil Management Standards

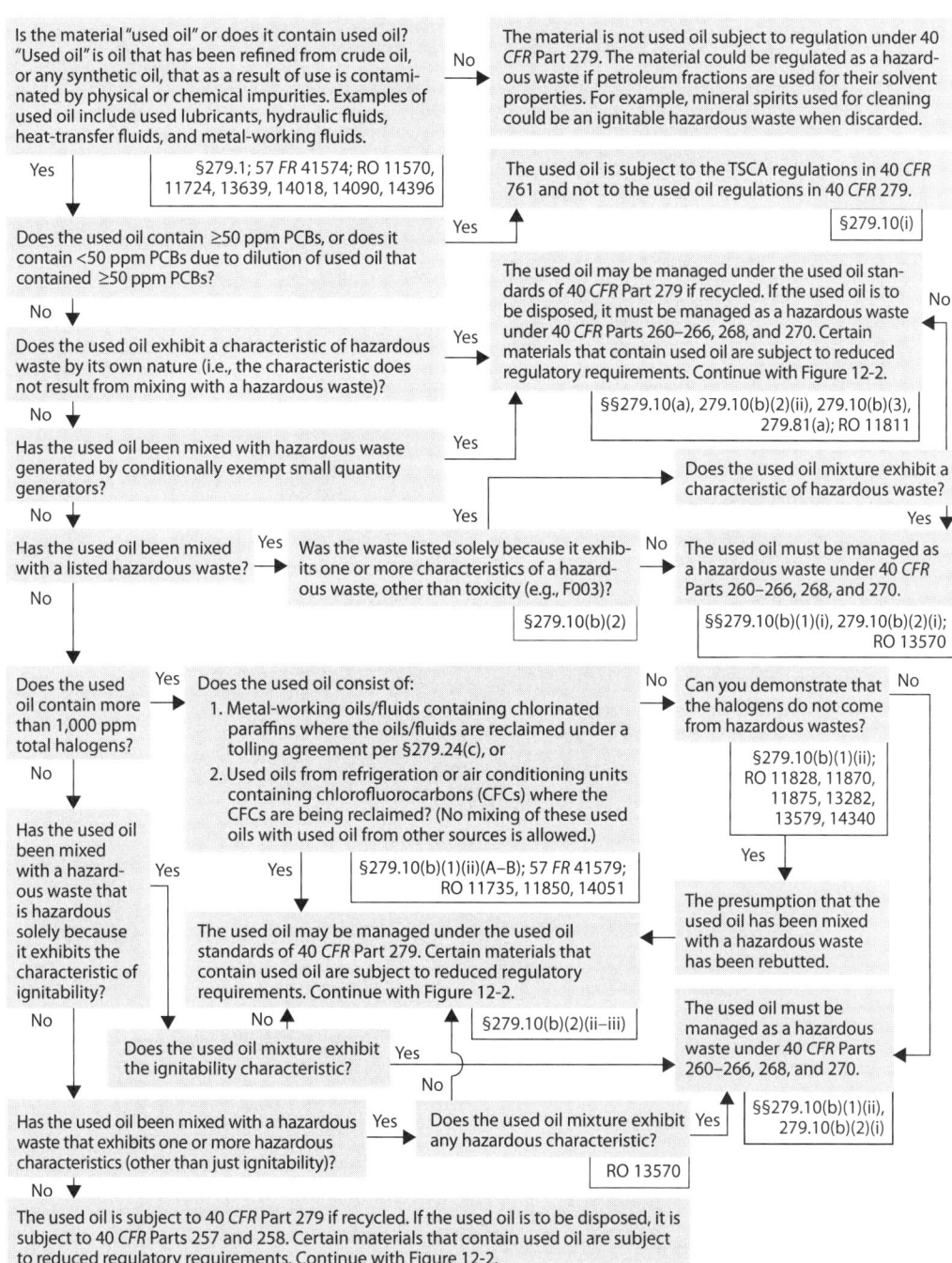

Source: McCoy and Associates, Inc.; adapted from 40 CFR Part 279, Subpart B.

Figure 12-2: Used-Oil-Containing Material Eligible for Reduced Regulation

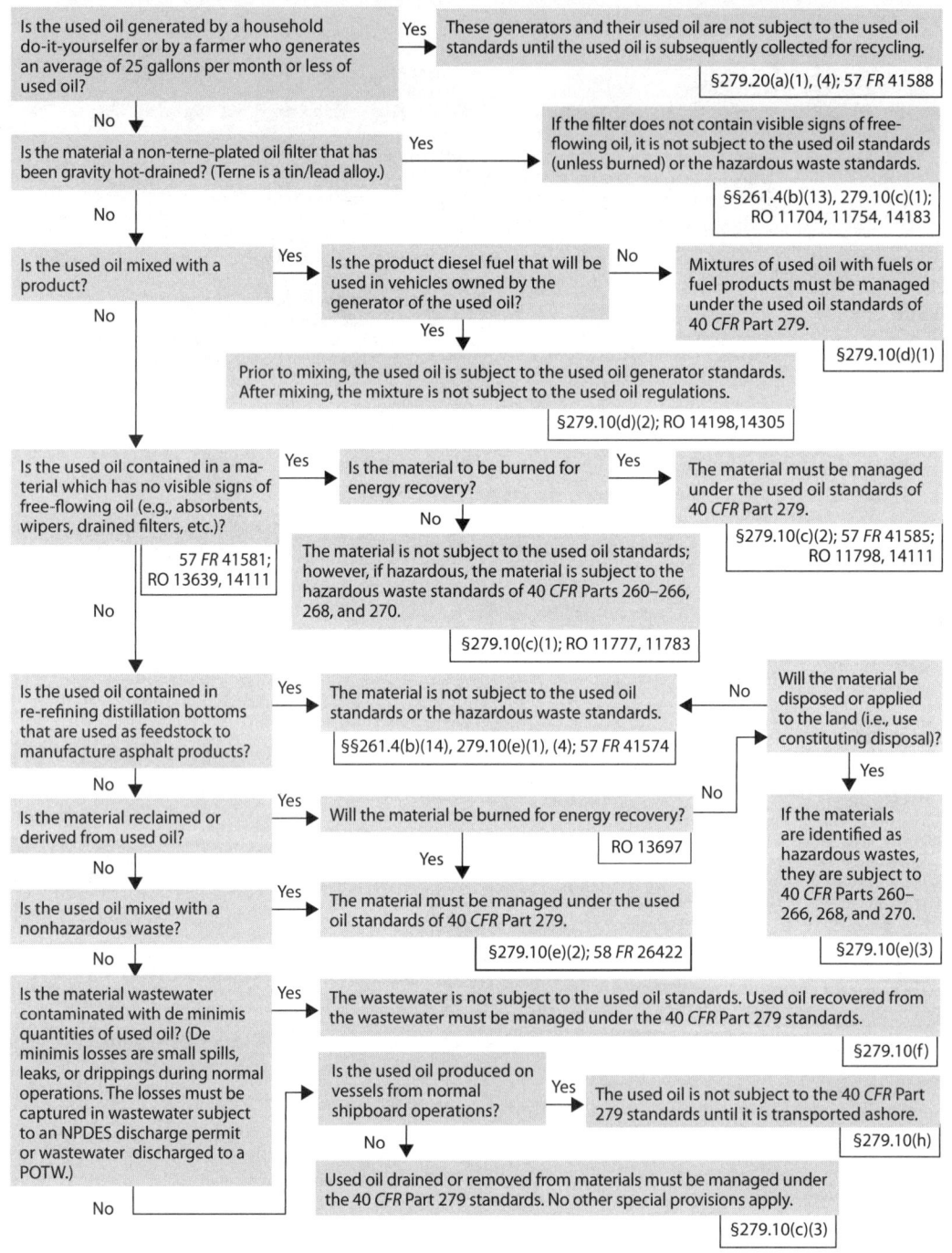

Source: McCoy and Associates, Inc.; adapted from 40 *CFR* Part 279, Subpart B.

Table 12-1: Examples of Materials That, When Used, Qualify as Used Oil[1]

- Compressor oils
- Coolants
- Copper- and aluminum-wire drawing solution
- Electrical insulating oil
- Emulsions used as lubricants
- Engine oil (typically includes gasoline and diesel engine crankcase oils and piston-engine oils for automobiles, trucks, boats, airplanes, locomotives, and heavy equipment)
- Heating media
- Industrial hydraulic fluid
- Industrial process oils
- Laminating oils
- Lubricant sprayed onto the bull gears of cement kilns
- Metal-working fluids and cutting oils
- Mineral oil
- Oils used as buoyants
- Refrigeration oil
- Synthetic oil (typically derived from coal, shale, or polymer-based starting material)
- Transmission fluid
- Used oil residues or sludges resulting from the storage, processing, or re-refining of used oils (when recycled by burning for energy recovery)

[1]This list is not comprehensive; other materials may also qualify as used oil when used.

Source: McCoy and Associates, Inc.; adapted from EPA/530/F-96/004, November 1996, May 3, 1993; 58 FR 26422, RO 11570, 13639, 13697, 14090.

Table 12-2: Examples of Materials That, When Used, Do Not Qualify as Used Oil[1]

- Animal and vegetable oil (even when used as a lubricant)
- Antifreeze
- Kerosene
- Petroleum distillates used as solvents to solubilize or mobilize (e.g., mineral spirits, petroleum naphthas) and solvents manufactured from synthetic materials
- Used oil re-refining distillation bottoms that are used as feedstock to manufacture asphalt products[2]
- Waste oil resulting from cleanout of fuel storage tank bottoms, spills of virgin fuel oil, or other oil wastes that have not been used

[1]This list is not comprehensive; other materials may also not qualify as used oil when used.

[2]These wastes are exempt from hazardous waste regulation per §261.4(b)(14).

Source: McCoy and Associates, Inc.; adapted from §279.10(e)(4), EPA/530/F-96/004, November 1996, September 10, 1992; 57 FR 41574–5, RO 14018, 14090, 14396, 14550.

12.1.1.1.3 *Is spent grease used oil?*

We have seen no guidance at the federal level since the Part 279 management standards were issued (in 1992) on the question of whether spent grease could be managed as used oil. We think this is probably a question best answered by authorized states or EPA regions. Speaking of which, EPA Region VIII weighed in on this question in 1993, concluding that used grease meets the three criteria listed in Section 12.1.1 (i.e., grease is derived from crude or synthetic oil, used as a lubricant, and contaminated during use). Thus, that region would allow generators of spent grease to manage it as used oil, subject to the Part 279 management standards. [RO 14597]

12.1.1.2 Materials that are not used oil

The materials listed in Table 12-2 do *not* meet the definition and/or three defining criteria for used oil. Therefore, when these materials become solid wastes, they must be managed as hazardous wastes if they exhibit a characteristic or are listed.

Q *A petroleum-based solvent is used to clean oily machined parts. When spent, the solvent contains significant quantities of oily residues. Can this spent solvent be managed as used oil?*

A No. The oily residues are present because of the solvent's use in cleaning. EPA does not consider spent petroleum-based solvent to be used oil and does not allow it to be regulated as such. When the solvent is spent, it is a spent material and would be a solid waste when burned for energy recovery, sent for disposal, or reclaimed. A hazardous waste determination would have to be made for this solid waste using testing or knowledge. [RO 14396]

Note that "used oil" and "waste oil" are not interchangeable terms. Although not defined in the federal hazardous waste regulations, waste oil does not usually meet the definition of used oil. As noted in Table 12-2 for instance, tank bottoms from a fuel oil storage tank or residues from a spill of virgin oil may be contaminated with impurities but still not meet the used oil definition, because these materials have never been used. [EPA/530/K-02/025I, October 2001, available from http://www.epa.gov/wastes/inforesources/pubs/training/uoil.pdf]

12.1.2 All used oil is recyclable until a decision is made to dispose

EPA has based the Part 279 used oil management standards on a presumption that all used oil is recyclable and therefore should be regulated under one set of standards—the Part 279 standards. This presumption applies regardless of whether the used oil exhibits a hazardous characteristic (at its point of generation) and regardless of whether the used oil will ultimately be recycled or disposed. [§279.10(a), RO 14739] "In other words, the generator (or any other person who handles the oil prior to the person who decides to dispose of the oil) need not decide whether the used oil eventually will be recycled or disposed and thus need not tailor its management of the oil based upon that decision (and, if destined for disposal, whether the used oil is hazardous). Rather, the Part 279 standards apply to all used oils until a person disposes of the used oil, or sends it for disposal." [September 10, 1992; 57 FR 41578]

According to EPA, "[t]he recycling presumption allows a used oil handler or any other person who handles the oil prior to the person who decides to dispose of the oil, to presume that his/her used oil will be recycled regardless of its final disposition." [RO 14054] As such, these persons (generators, transporters, processors, burners, and marketers) would only be subject to Part 279, unless one of them decides to dispose the used oil. See Case Study 12-1.

Interestingly, even if a generator knows it will send its used oil for disposal, EPA notes that used oil (including used oil that exhibits a characteristic) should be managed under the Part 279 management standards until it is actually disposed or sent offsite for disposal. "Thus, a generator ultimately sending used oil offsite for disposal would be required to label, store, and otherwise manage it in accordance with Part 279, Subpart C, until it is shipped offsite." [RO 14739]

"If used oil is recycled, however, no characteristic determination is required, but all parties handling

---

Case Study 12-1: Generator Requirements When Processor Disposes Used Oil Instead of Recycling It as Originally Planned

A generator sends used oil to a processor to be recycled, but the processor disposes the used oil instead. In this situation, is the generator required to determine if the used oil is hazardous and revise its records since it was not recycled as originally planned?

No. The generator does not have to perform a hazardous waste determination for used oil originally sent to the processor for recycling. The used oil is presumed to be recyclable (and therefore subject to Part 279) until the decision is made to dispose it. At that time, the person who is in control of the used oil (the processor in this case) is responsible for performing a hazardous waste determination and complying with the hazardous waste regulations if necessary. Anyone who handles the used oil prior to that time can presume that the used oil will be recycled, regardless of its final disposition. [RO 14054, 14739]

the used oil must comply with the Part 279 management standards." [57 *FR* 41579] Hence, the generator of used oil does not need to test it for characteristics unless he/she is going to dispose it. [EPA/530/H-98/001] (If a used oil generator is sending its used oil to a processor or fuel blender, there is also no need for the generator to determine if the used oil meets the used oil specification in §279.11; a generator would only need to make such a determination if it was burning it onsite as on-spec used oil or was sending the oil directly to a burner—see Section 12.3.)

This recycling presumption applies only for "as-generated" used oil (i.e., used oil that has been drained from equipment or service vehicles and has not been mixed with any other material). Its status may change if the used oil is mixed with hazardous waste or other materials (see Sections 12.1.3–12.1.6).

Since used oil is not a listed hazardous waste at the federal level, used oil to be disposed would be hazardous only if it 1) exhibits a hazardous waste characteristic (by its own nature or by mixing characteristically hazardous waste into it), 2) has been mixed with ICR-only listed waste and continues to exhibit a characteristic, 3) has been mixed with non-ICR-only listed hazardous waste (other than listed waste generated by conditionally exempt small quantity generators), or 4) contains greater than 1,000-ppm total halogens and the presumption that it has been mixed with hazardous waste cannot be rebutted. Used oil disposal is subject to §279.81, which requires hazardous used oil to be managed in accordance with the hazardous waste regulations.

12.1.3 Mixing used oil with hazardous waste

The regulations governing mixtures of used oil and hazardous waste are codified in §279.10(b). The standards differ for used oil mixed with characteristic waste versus used oil mixed with listed waste, as discussed below.

Remember that used oil exhibiting a hazardous characteristic by its own nature (before it is mixed with any other material) may be handled under the less-stringent Part 279 standards and is exempt from hazardous waste regulation if recycled. Consequently, generators should think long and hard about mixing used oil with any waste that could be hazardous; doing so could jeopardize the applicability of Part 279 and may require compliance with the full set of hazardous waste regulations. Additionally, such mixing is generally considered hazardous waste treatment, and a permit may be required for this activity—see Chapter 7.

Note that the §279.10(b) used oil mixture rules do not apply to mixtures of used oil and hazardous waste generated by conditionally exempt small quantity generators regulated under §261.5. Such mixtures may be regulated as used oil under Part 279, even if the total halogen concentration exceeds 1,000 ppm. [§§261.5(j), 279.10(b)(3), RO 14627]

In general, mixtures of used oil and hazardous waste follow the three elements of the hazardous waste mixture rule (discussed in Section 5.1), with used oil as the solid waste.

12.1.3.1 Used oil mixed with characteristic hazardous waste

The regulations contain two scenarios for mixtures of used oil and characteristic wastes:

1. *Used oil is mixed with a waste that is hazardous solely because it exhibits the characteristic of ignitability*—In this situation, if the mixture does not exhibit the characteristic of ignitability, it can be managed as used oil under Part 279, regardless of any other hazardous characteristics that the mixture may exhibit. [§279.10(b)(2)(iii)] For example, used oil that inherently exhibits the toxicity characteristic for lead is mixed with an ignitable hazardous waste (D001), and the resulting mixture is no longer ignitable. Under §279.10(b)(2)(iii), this mixture may be managed as used oil under Part 279, even though it may still exhibit the toxicity characteristic for lead. This provision represents an expansion of

the hazardous waste mixture rule. [EPA/530/K-02/025I, RO 11776, 13590]

2. *Used oil is mixed with a waste that is hazardous because it exhibits one or more hazardous characteristics (other than just ignitability)*—When this occurs, if the mixture exhibits *any* hazardous characteristic, it is regulated as hazardous waste. [§279.10(b)(2)(i), RO 13570] On the other hand, if the mixture does not exhibit any hazardous characteristic, it may be managed under the Part 279 used oil standards. [§279.10(b)(2)(ii)] If the same used oil exhibiting the toxicity characteristic for lead is mixed with a D003 reactive waste instead of an ignitable-only waste, the resulting mixture would need to be void of both the toxicity and reactivity characteristics (and all other characteristics) in order to be managed as used oil. Otherwise, it would have to be managed as a hazardous waste. [EPA/530/K-02/025I, RO 11776, 13590]

12.1.3.1.1 *EPA trying to close a loophole*

As noted above, the used oil regulations in Part 279 allow characteristic hazardous waste to be mixed with used oil. If the resulting mixture doesn't exhibit a characteristic, the mixture is subject to regulation as used oil—not hazardous waste. Used oil that is recycled (including burning for energy recovery) is not subject to any hazardous waste regulation in Parts 260–266, 268, and 270. [§§261.6(a)(4), 279.10] This "loophole" allows characteristic hazardous wastes to be mixed with used oil and relieved of Subtitle C regulations, specifically including the land disposal restrictions (LDR). [September 10, 1992; 57 *FR* 41605] EPA attempted to close this loophole in 1995 [October 30, 1995; 60 *FR* 55202], but the agency had to back off a few months later due to a ruling issued by the U.S. Court of Appeals for the District of Columbia Circuit. [June 28, 1996; 61 *FR* 33691]

Even though the loophole still exists, mixing used oil with characteristic hazardous waste is considered treatment (as noted earlier) and, therefore, would have to be performed in a 90- or 180-day accumulation unit or a permitted unit (see Case Study 12-2); in addition, the used oil generator storage tank standards in Part 279, Subpart C would apply. [RO 11679, 11708, 13570]

12.1.3.2 Used oil mixed with listed hazardous waste

If used oil is mixed with a listed hazardous waste (except ICR-only listed waste), the mixture is subject to regulation as hazardous waste (under Parts 260–266, 268, and 270). It is not regulated as used oil under Part 279. [§279.10(b)(1)] For example, if listed hazardous waste (e.g., waste brake cleaner or other chlorinated spent solvent) is blended into used oil, the mixture would be a hazardous waste subject to all Subtitle C regulations. If the mixture is

---

Case Study 12-2: Mixing Hazardous Waste and Used Oil Is Treatment

A facility mixes used oil and spent mineral spirits that exhibit the characteristic of ignitability only. Is the tank where mixing occurs regulated as a hazardous waste treatment tank or a used oil tank?

When the spent mineral spirits and used oil are placed in the same accumulation tank, the tank is regulated as both a hazardous waste tank under §262.34 and a used oil tank under §279.22. "Regardless of whether the resultant mixture is used oil or hazardous waste, both sets of standards apply as the used oil and hazardous waste are being mixed in the same tank. However, the only additional requirement that is added in Part 279 is that the tank must be labeled with the words 'used oil.' This mixing [is] considered treatment, since the purpose of the mixing is to make the waste more amenable for recovery (i.e., energy recovery), and/or to make the waste less hazardous (i.e., to remove the solvent's ignitable characteristic)." [RO 11708; see also RO 11679, 13570]

burned for energy recovery, the boiler and industrial furnace (BIF) rules in Part 266, Subpart H will apply. [RO 13570]

There are 29 listed wastes that were listed solely because they exhibit a characteristic of ignitability, corrosivity, and/or reactivity (ICR-only listed wastes—see Table 5-1). If one of those 29 listed wastes is mixed with used oil, the rules are the same as for mixtures of used oil and characteristic wastes discussed in Section 12.1.3.1. [§279.10(b)(2)]

12.1.4 Rebuttable presumption

During the development of the Part 279 standards, EPA sampled and analyzed used oil from all around the United States. The agency found that the used oil often contained high concentrations of halogens—especially chlorinated constituents. After researching this issue, EPA determined that the source of these halogenated compounds was often spent chlorinated solvents (F001 or F002) that were being mixed into used oil. From the used oil mixture provisions noted in the previous section, mixing listed spent solvents into used oil causes the mixture to be a listed hazardous waste.

During inspections, it is often difficult for regulatory personnel to determine if a hazardous waste has been mixed into used oil. Therefore, the agency established a rebuttable presumption in the Part 279 standards that states: if used oil contains more than 1,000-ppm total halogens, it is presumed to have been mixed with listed spent solvents, and so it is presumed to be F001 or F002 listed hazardous waste under the mixture rule. [§§261.3(a)(2)(v), 279.10(b)(1)(ii), November 29, 1985; 50 *FR* 49176] Such used oil must be managed as hazardous waste, unless the handler can rebut the presumption as discussed in Section 12.1.4.1. [RO 12608, 13282] "When the total halogen concentration is greater than 1,000 ppm, the burden of proof (rebutting the rebuttable presumption) is on the used oil handler." [*Guidance and Summary of Information Regarding the RCRA Used Oil Rebuttable Presumption*, EPA/905/R-03/005, March 2005, available online at http://www.epa.gov/epawaste/conserve/materials/usedoil/oil-rebut.pdf]

Because of this rebuttable presumption, when the used oil transporter comes onto a facility's site to haul away used oil, the transporter will usually check for total halogen content (typically using a colorimetric test kit). [§279.44(a) and (b)] If the total halogen level exceeds 1,000 ppm, "[t]he transporter may accept such shipments of used oil as a hazardous waste transporter…." For situations when the transporter doesn't test for total halogens before accepting the used oil and subsequently determines it has greater than 1,000-ppm total halogens, "if the original generator of the hazardous waste cannot be identified, the transporter may have to assume hazardous waste generator responsibilities and comply with both the generator standards of 40 *CFR* Part 262 as well as the hazardous waste transporter requirements of 40 *CFR* Part 263." [September 10, 1992; 57 *FR* 41592]

The rebuttable presumption applies to all used oil regardless of whether it will be recycled or disposed, and must be applied to the used oil before it is blended with other materials. Used oil containing more than 1,000-ppm total halogens may not be blended to lower the halogen level below 1,000 ppm in order to meet the rebuttable presumption and call the used oil nonhazardous. [September 10, 1992; 57 *FR* 41579, EPA/530/K-02/025I]

Q *My used oil contains greater than 1,000-ppm total halogens, but I am planning on disposing rather than recycling it. Since I will be managing my used oil as a solid waste rather than used oil under the Part 279 standards, do I have to worry about the rebuttable presumption?*

A Yes. Even when disposing used oil as a solid waste, the rebuttable presumption will still apply. The rebuttable presumption is found at §261.3(a)(2)(v) in addition to the Part 279 standards. Thus, for used oil managed as a solid waste instead of used oil under Part 279, the generator will have to ask three more questions (see Section 2.1). When you ask "Is it listed?," you are going to have to presume that used oil containing greater than 1,000-ppm total halogens is listed unless you determine, based on testing or knowledge, that the halogens in the used oil could not have come from mixing with F001/F002 hazardous waste.

12.1.4.1 How to make a rebuttal

The agency understands that some used oils may exceed the 1,000-ppm total halogen limit without having been mixed with hazardous waste. [RO 12319] In these cases, the generator or other entity with possession can rebut the §279.10(b)(1)(ii) presumption by demonstrating through analysis or other documentation that the used oil does not contain hazardous waste. If a rebuttal is successful, the used oil will be subject to the used oil regulations instead of the hazardous waste standards.

Rebuttals can, but do not have to, be based on actual and documented testing of the used oil. For example, SW–846 analytical methods could be used to prove that the used oil does not contain "significant concentrations" of individual halogenated hazardous constituents listed in Appendix VIII of Part 261 that are used at the facility and those which could reasonably be expected to enter the used oil waste stream. [RO 12567] EPA recommended that SW–846 Method 8010 be used when rebutting the presumption based on testing. [September 10, 1992; 57 *FR* 41579] That method was subsequently deleted from SW–846 in December 1996. However, a table listing 9 common SW–846 methods that may be used to test used oil for halogens can be downloaded from http://www.epa.gov/epawaste/hazard/testmethods/pdfs/uoil.pdf.

The regulations do not define "significant concentrations." However, regarding hazardous halogenated solvents, "EPA has stated that a level of 100 ppm of individual solvent compounds is generally considered a significant concentration. Thus, one may try to rebut the presumption by showing that less than 100 ppm of any individual hazardous halogenated constituent listed as a hazardous spent solvent in 40 *CFR* 261.31 is present." However, other site-specific factors will also have to be considered. [November 29, 1985; 50 *FR* 49176, RO 12567, 13579]

Thus, if none of the halogenated solvents in the F001 and F002 spent solvent listings (e.g., trichloroethylene, 1,1,1-trichloroethane, methylene chloride) are contained in the used oil at concentrations of ≥100 ppm, you can successfully rebut the presumption that listed spent solvent was mixed into the used oil, and your material is just used oil subject to Part 279. If, conversely, analysis indicates that one or more of the halogenated solvents in the F001 or F002 spent solvent listings are present in the oil at ≥100 ppm, the presumption is not rebutted, and the material must be managed as a hazardous waste.

EPA has noted in guidance that "significant concentrations" of non-solvent listed wastes (e.g., chlorinated pesticides) could be revealed by concentrations of <100 ppm in used oil. "However, mixing of used oil with other hazardous halogenated waste could be indicated by concentrations of Appendix VIII halogenated compounds at levels lower than 100 ppm, especially if the hazardous halogenated waste is not generated at the same site as the used oil or would not be expected to be formed during use of the oil. The example we used in the preamble (also at 50 *FR* 49176) was mixing of chlorinated pesticides with used oil." [RO 12567; see also EPA/905/R-03/005, cited earlier]

The presumption could also be rebutted by demonstrating that all halogens are inorganic (e.g., salts). [RO 12319, 12738] For example, it's not uncommon for ships and coastal facilities to generate used oil containing salt water.

Another way of rebutting the presumption would be to document the source of the halogens (i.e., show that the halogens are not attributable to regulated, listed hazardous wastes). Four examples of this approach are as follows:

1. Used oil that contains less than 1,000-ppm total halogens is placed in RCRA-empty drums. However, residues remaining in the drums cause the total halogen level of the used oil to exceed

1,000 ppm. In this case, if the containers meet the definition of RCRA-empty (see Section 9.5), information attributing the source of the halogens to residues remaining in the drums would be sufficient to rebut the presumption. The halogens would not be attributable to mixing with listed spent solvents because the drums do not, by definition, contain RCRA-regulated hazardous waste. [RO 11870]

2. A company collects do-it-yourselfer oil waste from households, and the used oil contains more than 1,000-ppm total halogens. In order to rebut the presumption of mixing with listed spent solvents, the firm could provide documentation that the source of the used oil is exclusively household waste (which is excluded from the definition of hazardous waste) and prove that the chain of custody has been maintained to prevent mixing with hazardous waste after collection. [RO 11828, 11875]

3. A conditionally exempt small quantity generator (CESQG) mixes a small amount of F005 listed spent solvent into its used oil. The resulting used oil contains 2,000-ppm total halogens. The presumption of mixing can be rebutted by providing convincing documentation to show that the mixture is an excluded CESQG used oil mixture covered under §§261.5(j) and 279.10(b)(3). [57 FR 41581, RO 14627]

4. Used oil that contains less than 1,000-ppm total halogens is mixed with dielectric fluid containing PCBs. Such PCB-containing fluids are exempt from RCRA regulation provided they are 1) regulated under TSCA, and 2) hazardous only because they exhibit the toxicity characteristic from D018 to D043. [§261.8] If the total halogens in the mixture exceed 1,000 ppm solely because the used oil was mixed with such a RCRA-exempt PCB waste, the used oil mixture will not be regulated as hazardous waste. Instead, the mixture will be subject to regulation under both Part 279 (if the PCB concentration in the mixture is <50 ppm) and TSCA (see Section 12.1.7). [EPA/905/R-03/005, cited earlier]

To rebut the presumption using knowledge, the agency says sources must provide convincing documentation to: 1) show the exclusive source of the used oil, and 2) demonstrate that the chain of custody has been maintained so as to preclude mixing. [RO 11875] Appropriate documentation could include product (e.g., lubricant, hydraulic fluid) composition before use from the manufacturer, a sufficiently detailed description of the process generating the used oil in order to eliminate listed hazardous wastes by knowledge, etc. For example, safety data sheets (SDSs) have been used in rare instances to show that a halogenated constituent is present in the virgin oil formulation and not introduced by mixture of the used oil with a listed hazardous waste. [EPA/905/R-03/005, cited previously]

Even if a rebuttal is successful, if the total halogen level in the used oil exceeds 4,000 ppm, the used oil will not meet the used oil specification limit for total halogens (see Case Study 12-3 and Section 12.3.1). The material is still used oil subject to Part 279, but it is off-spec used oil. Consequently, if the used oil is destined to be burned for energy recovery, it will have to either be 1) processed to lower the total halogen level prior to burning, or 2) burned as off-specification used oil fuel (in which case the used oil fuel burner must comply with Part 279, Subpart G).

12.1.4.2 Exemptions from the rebuttable presumption

EPA recognizes that some oils can become contaminated with more than 1,000-ppm total halogens simply through normal use (without mixing them with hazardous waste). Accordingly, the agency has exempted two specific types of used oil from the rebuttable presumption:

1. Metal-working oils/fluids containing chlorinated paraffins are not subject to the presumption if they are processed to reclaim the oils/fluids via a tolling arrangement as described

> **Case Study 12-3: Presumption May Be Rebutted Regardless of Halogen Level**
>
> Used oil that is burned for energy recovery is considered to be off-specification if it contains more than 4,000-ppm total halogens. [§279.11] If an automotive service station generates used oil that has a total halogen level of 5,000 ppm, can the owner still rebut the presumption that the used oil was mixed with listed spent solvents?
>
> Yes. The rebuttable presumption operates independently of the used oil specification level for total halogens. "There is no halogen level over which it is impossible to rebut the presumption of mixing.... Regardless of the specification of the used oil...a handler always has the option of seeking to rebut the presumption of mixing with a listed hazardous waste." [RO 14340]

in §279.24(c). [§279.10(b)(1)(ii)(A)] EPA believes that tolling agreements "restrict the handling of the oils/fluids and provide for a mutual interest in preventing any potential contamination of the oils/fluids to assure that the oils/fluids can be recycled." Generators and/or handlers of metal-working oils who recycle them in any other manner (e.g., process the used oil into fuel or lubricant) or dispose them remain subject to the rebuttable presumption. Thus, they must document that the oils are not mixed with chlorinated hazardous wastes if the total halogen levels are greater than 1,000 ppm. [September 10, 1992; 57 FR 41579]

2. Used oils removed from refrigeration units that are contaminated with chlorofluorocarbons (CFCs) are not subject to the rebuttable presumption if the CFCs are destined for reclamation. [§279.10(b)(1)(ii)(B)] This exemption extends to used oil contaminated with chemicals (such as hydrochlorofluorocarbons or HCFCs and hydrofluorocarbons or HFCs) that are used in a manner similar to CFCs in refrigeration units. [RO 14400] The used oil is exempt from the rebuttable presumption at the point of draining from the compressors as long as the CFCs (or HCFCs or HFCs) are eventually reclaimed (i.e., removed from the used oil). CFC-contaminated used oils that have been mixed with used oil from other sources remain subject to the rebuttable presumption, as do oils from which the CFCs are not reclaimed. [RO 11735, 11850, 14051] To avoid any enforcement issues, we recommend that a generator seeking the CFC-reclamation exemption from the rebuttable presumption keep the records noted in Section 12.1.4.1 as well as documentation showing that the CFCs were reclaimed from the oil.

We are not aware of any facilities that reclaim CFCs from used oil. Because of this lack of CFC/oil reclamation capacity, this rebuttable presumption exemption is rarely exercised. Instead, the generator is left with two options: 1) follow the Part 279 regulations, which would require the generator to rebut the presumption that the halogens came from mixing F001/F002 into the used oil; or 2) manage the used oil as F001/F002 hazardous waste if it contains greater than 1,000-ppm total halogens.

Generators of used oil that qualify for one of these two exemptions do not have to prove that the oil has not been mixed with a hazardous waste even though it may contain more than 1,000-ppm total halogens. Such oil remains subject to the Part 279 standards.

12.1.5 Used oil filters

Depending on their material of construction and whether or not they have been drained, most used oil filters are excluded from the hazardous waste regulations. [§261.4(b)(13)] As discussed below, if they don't qualify for that exclusion, used oil filters may be subject to the Part 279 used oil management standards, RCRA Subtitle D nonhazardous waste provisions, or, under a worst-case scenario, applicable hazardous waste regulations.

12.1.5.1 Exclusion for non-terne-plated filters

Under §261.4(b)(13), EPA excludes non-terne-plated used oil filters from automobiles, trucks, off-road vehicles, diesel-powered locomotives, and heavy equipment from the RCRA hazardous waste regulations, as long as they are not mixed with listed hazardous wastes. [RO 11704] In order to qualify for the exclusion, used oil filters must be gravity hot-drained by 1) puncturing the filter anti-drain back valve or the filter dome and hot-draining, 2) hot-draining and crushing, 3) dismantling and hot-draining, or 4) any other equivalent hot-draining method that will remove the used oil. Prior to draining, the storage and handling of used oil filters must comply with the used oil management standards in Part 279. [September 10, 1992; 57 *FR* 41581, RO 11704, 14183]

"Hot-drained" means that the oil filter is drained near engine operating temperature and above room temperature (i.e., 60°F). EPA recommends a minimum 12-hour hot-drain time for punctured or pierced used oil filters. The agency notes that if an oil filter is picked up by hand or lifted by machinery and used oil simply drips or runs from the filter, the filter is *not* considered to have been drained. [May 20, 1992; 57 *FR* 21531] Used oil removed from filters is subject to the Part 279 used oil management standards if it is recycled. [September 10, 1992; 57 *FR* 41581] Furthermore, any residual oil that leaks out of processed used oil filters (e.g., after the filters have been drained and crushed) remains subject to Part 279. [RO 14202]

Used oil filters that qualify for the §261.4(b)(13) exclusion (i.e., they've been hot-drained) may be disposed in a nonhazardous waste landfill without performing a hazardous waste determination. [May 20, 1992; 57 *FR* 21531, RO 11754, 14183]

12.1.5.2 Scrap metal exemption also available

An environmentally preferable option to the §261.4(b)(13) exclusion is managing drained oil filters as scrap metal. Drained filters may be considered scrap metal [and therefore eligible for the §261.6(a)(3)(ii) exemption] if they do not contain a significant amount of liquid and they will be recycled. Used oil filters do not contain a significant amount of liquid once the used oil has been drained or removed to the extent possible such that there are no visible signs of free-flowing oil. [§279.10(c), RO 11566, 13498, 14183, 14184]

Metal fuel filters that have been drained for a sufficient period of time so that they do not have a significant liquid component also meet the definition of scrap metal. Commingling spent metal fuel filters with used oil filters appears to reduce safety concerns. [RO 14184] See Section 4.21.6 for additional discussion.

12.1.5.3 Nonexcluded oil filters

Terne-plated used oil filters are not included in the §261.4(b)(13) exclusion because the terne (tin/lead alloy) plating makes the filter exhibit the toxicity characteristic for lead. [May 20, 1992; 57 *FR* 21531] Terne may be used on filters found in heavy trucks or equipment. It may be difficult to differentiate between terne- and non-terne-plated filters. In those situations, generators may have to contact the filter's manufacturer to obtain this information.

Other types of filters, such as fuel filters, transmission oil filters, hydraulic fluid filters, and specialty filters are also not included in the scope of the §261.4(b)(13) exclusion, because EPA lacked quantitative data on these types of filters. [May 20, 1992; 57 *FR* 21532, RO 11808]

Used oil filters that do not qualify for the used oil filter exclusion in §261.4(b)(13) or scrap metal exemption in §261.6(a)(3)(ii) are subject to the Part 279 used oil standards if they will be recycled. For example, if the filters (drained or undrained) are to be burned for energy recovery, they are regulated as used oil under Part 279. [§279.10(c)(2), RO 11808, 13639] If the filters cannot be recycled, they must be disposed. If these solid wastes are not hazardous, they may be disposed in accordance with Parts 257 and 258. [§279.81(b), RO 11808] If filters to be disposed exhibit a hazardous characteristic and are not excluded under §261.4(b)(13), they must be managed as hazardous waste.

12.1.6 Other mixtures of used oil and solid wastes/other nonhazardous materials

Used oil is often mixed with, or contaminates, many types of material besides hazardous waste and filters. The regulatory status of such mixtures and contaminated materials is detailed in §279.10(c–f) and described below.

12.1.6.1 Used oil mixed with fuels

With one exception, mixtures of used oil and fuels or fuel products are regulated as used oil under Part 279. [§279.10(d)(1)] This provision assumes that the mixture will be recycled (i.e., burned for energy recovery).

The one exception is for mixtures of used oil and diesel fuel. Under §279.10(d)(2), mixtures of used oil generated onsite and diesel fuel that are used as fuel in the generator's own vehicles are not subject to the Part 279 regulations after the mixing occurs. However, all other applicable regulations (e.g., Clean Air Act standards) pertaining to the management or burning of such mixtures apply. Prior to mixing, the standards for used oil generators (Part 279, Subpart C) apply to the used oil. [RO 14198, 14305] EPA has decided that this exception at §279.10(d)(2) also applies to mixtures of used oil and JP-8. [RO 14305]

12.1.6.2 Used oil-contaminated materials

Materials (e.g., absorbents, rags and wipers, scrap metal, etc.) that are contaminated with used oil fall into two regulatory categories [§279.10(c), September 10, 1992; 57 FR 41581, May 3, 1993; 58 FR 26423, RO 11704, 13639, 14111]:

1. Materials that contain visible signs of free-flowing oil—As long as they are destined for recycling, such materials are regulated as used oil under Part 279 until the used oil is removed from the material (i.e., the material no longer contains visible signs of free-flowing oil).

2. Materials that do not contain visible signs of free-flowing oil—These materials are not used oil and are not subject to Part 279.

EPA recommends that used oil be separated from used oil-contaminated materials before the mixture is managed. Prior to separation, the mixtures (e.g., dripping rags) must be stored and handled in compliance with Part 279.

After separating used oil from these materials so that they show no visible signs of free-flowing oil, the materials are no longer subject to Part 279. As such, if the materials will be disposed, they are solid wastes, and the generator must determine whether or not they are hazardous. Because used oil is not listed (at the federal level), the only way these used oil-contaminated solid wastes could be hazardous is if they exhibit a characteristic or if they were mixed with listed hazardous waste. If so, they must be managed as hazardous waste. If these materials do not exhibit a characteristic (and were not mixed with listed hazardous waste), they are nonhazardous solid waste. (A hazardous waste determination is not required for drained non-terne-plated used oil filters, as discussed in Section 12.1.5.1.) [57 FR 41585, RO 11777, 11783]

However, even if a used oil-contaminated material to be disposed does not exhibit a characteristic (e.g., a wrung-out oily rag) and, therefore, is not considered a hazardous waste under the federal regulations, some states may regulate them as "special wastes" (e.g., certain petroleum-contaminated, nonhazardous wastes) when disposed. [RO 11798]

Q *Steel turnings generated during machine shop operations are coated with cutting oil. As noted in Section 12.1.1, used cutting oil is considered used oil. What is the regulatory status of the used oil-coated turnings?*

A The steel turnings coated with used oil would be regulated as used oil if they were visibly dripping with used oil. Once the used oil has been removed so that they no longer contain visible signs of free-flowing oil, they may be managed as scrap metal if recycled. [RO 11184, 11783, 13639]

Used oil that is separated from such used oil-contaminated materials is subject to the Part 279 regulations unless it is destined for disposal (in which case it is a solid waste and must be evaluated to determine if it is a hazardous waste). For instance, used oil recovered from sorbents, industrial wipers, and scrap metal is subject to Part 279 when recycled. [57 *FR* 41581]

12.1.6.2.1 *Contaminated materials are used oil when burned for energy recovery*

Materials that contain or are contaminated with used oil are regulated as used oil if they are burned for energy recovery. [§279.10(c)(2), 57 *FR* 41585] This applies regardless of whether the materials contain visible signs of free-flowing oil. More detail on this issue is provided in Case Study 12-4 and Section 12.3.4.

12.1.6.3 Residues from managing used oil

The regulatory status of materials that are reclaimed or derived from used oil depends on how they are managed. If the materials are used beneficially (e.g., re-refined into lubricants) and are not 1) burned for energy recovery, 2) disposed, or 3) used in a manner constituting disposal, they are products—not used oil (and therefore are not regulated under Part 279). [§279.10(e)] In addition, such materials are not solid wastes and, therefore, cannot be hazardous wastes. On the other hand, any materials produced from used oil that are burned for energy recovery are regulated as used oil. Finally, any materials derived from used oil (except used oil re-refining

---

Case Study 12-4: Used Oil Storage Tank Bottoms Burned for Energy Recovery

An automotive repair shop stores used oil in an onsite storage tank. The used oil is not mixed with any other waste streams. Over time, the solids and heavy fractions have settled out of the used oil, causing a thick tar-like layer to accumulate at the bottom of the used oil storage tank. The tank bottoms typically exhibit the toxicity characteristic for benzene, cadmium, chromium, and lead. If the owner sends these tank bottoms offsite to be burned for energy recovery, must they be handled as used oil or hazardous waste?

According to §279.10(e)(2), residues or sludges resulting from the storage, processing, or re-refining of used oil are considered used oil when they are recycled by burning them for energy recovery. Therefore, when the repair shop's tank bottoms are sent offsite for energy recovery, they qualify as used oil and may be handled according to the Part 279 regulations. The fact that the tank bottoms exhibit several hazardous characteristics does not alter their status as used oil. In this case, the tank bottoms would have to be handled as hazardous waste only if they fail the rebuttable presumption [§279.10(b)(1)(ii)] or they have been mixed with hazardous waste [§279.10(b)(1–2)]. However, if the tank bottoms are not burned for energy recovery and are disposed instead, they must be managed as characteristic hazardous waste. [May 3, 1993; 58 *FR* 26422, RO 13697]

distillation bottoms that are used as feedstock to manufacture asphalt products) that are disposed or used in a manner constituting disposal are solid wastes and, therefore, are subject to the hazardous waste regulations if they exhibit a characteristic or are listed.

12.1.6.4 Used oil-contaminated wastewater

Unless contaminated with de minimis amounts, wastewater that contains used oil meets the §279.1 definition of "used oil" and therefore is subject to the Part 279 standards. The used oil regulations also apply to [58 FR 26422, RO 11818]:

- Used oil recovered during wastewater treatment to meet a CWA-permitted discharge;
- Oil recovered from wastewater generated from a used oil recovery process; and
- Residues or sludges resulting from CWA treatment of wastewater containing used oil.

However, wastewater contaminated with de minimis levels of used oil is *not* regulated under Part 279 if it is subject to Section 402 (the NPDES program) or 307(b) (discharges to POTWs) of the CWA. This includes wastewater at facilities that have eliminated their wastewater discharge as a direct result of the CWA. EPA defines "de minimis" quantities of used oil as "small spills, leaks, or drippings from pumps, machinery, pipes, and other similar equipment during normal operations, or small amounts of oil lost to the wastewater treatment system during washing or draining operations." The exception does not apply if the used oil enters the wastewater system as a result of abnormal manufacturing operations that cause substantial leaks, spills, or other releases. [§279.10(f)] For example, pouring collected used oil into any part of a wastewater treatment system is not covered under the de minimis exemption. [RO 11858]

12.1.6.4.1 *Used oil-contaminated wastewater at petroleum refineries*

Per §279.10(g)(5), used oil that incidentally enters and is captured by a petroleum refinery's wastewater treatment system that is subsequently recovered along with other oil in the refinery's recovered oil system (for insertion into the refining process) is not regulated as used oil under Part 279. Such oil recovered from this system is more properly characterized as crude feedstock than used oil. Such incidental losses of used oil would include drippage and minor spillage. [March 4, 1994; 59 FR 10555, July 28, 1994; 59 FR 38543]

However, this exemption is void if used oil is deliberately poured into a refinery's wastewater treatment system; if this occurs, the recovered oil from the wastewater treatment system would not be exempt from Part 279 standards under §279.10(g)(5). For example, used oil collected from equipment or vehicle maintenance activities that is intentionally introduced into a refinery's wastewater treatment system would not be exempt from the Part 279 standards once recovered. Similarly, used oil that is generated offsite and brought into the refinery may not be added to the refinery's wastewater treatment system and be exempt from Part 279 under §279.10(g)(5) once recovered. [March 4, 1994; 59 FR 10555] Section 12.1.8 contains more information about managing used oil in the petroleum industry.

12.1.6.5 Used oil mixed with nonhazardous waste

People sometimes ask about the regulatory status of mixtures of used oil and nonhazardous solid waste. EPA has clarified in EPA/530/K-02/025I that mixtures of used oil with nonhazardous solid waste are regulated as used oil under Part 279. However, we caution entities about mixing solid waste into used oil, as your used oil recycler may not want the mixture.

12.1.7 PCBs in used oil

Polychlorinated biphenyls (PCBs) can sneak into used oil in a couple of different ways. Since PCBs are still prevalent in electrical transformers, capacitors, and other older equipment, PCB-containing oil shows up frequently at utility power plants and other industrial facilities. Additionally, some facilities use recycled lubricating oil; PCBs can sometimes be unknowingly contained in such lubricants.

How do I know if PCBs are in my used oil? Neither RCRA nor TSCA requires generators to test for PCBs—instead, knowledge is widely used. For example, if used oil is drained from service vehicles or rotating equipment wherein new lubricating oil was recently introduced, knowledge would indicate that there is no chance of PCBs being in the used oil. On the other hand, if the used oil is drained from any electrical equipment or old equipment in which the source of lubricating oil, hydraulic fluid, heat-transfer fluid, mineral or refrigeration oil, etc. is unknown, then the presence of PCBs should be suspected and analyzed for. The only conclusive means of demonstrating used oil does not contain quantifiable levels of PCBs is direct testing. Direct testing of used oil for PCBs may be a cost-effective insurance policy.

Q *If I want to test my used oil to determine the PCB concentration, what test method should I use?*

A The only defensible analytical method for demonstrating the concentration of PCBs for purposes of disposal or burning for energy recovery is EPA Method 8082 in SW–846. There are no field test kits that EPA has found acceptable for this purpose.

Materials and wastes that contain PCBs are subject to standards promulgated under the Toxic Substances Control Act (TSCA); these requirements are codified in 40 *CFR* Part 761. A question often raised is whether PCB-containing used oil is regulated under the TSCA regulations, RCRA Part 279 standards, or both.

EPA clarified the status of PCB-contaminated used oil in a July 30, 2003 final rule. [68 *FR* 44659] From a regulatory perspective, the agency has divided the universe of PCB-containing used oil into three categories, as shown in Figure 12-3. The RCRA Part 279 used oil standards will apply to only two of the three categories, while the TSCA PCB regs in Part 761 potentially apply to *all* used oil.

First, used oil shown to contain <2 ppm PCBs (via testing or through "other information") is subject to the Part 279 used oil standards and also to the TSCA recordkeeping requirements in §761.20(e)(2,4) if burned for energy recovery. ("Other information" consists of personal knowledge of the source and

Figure 12-3: Regulation of PCB-Containing Used Oil

Requirements

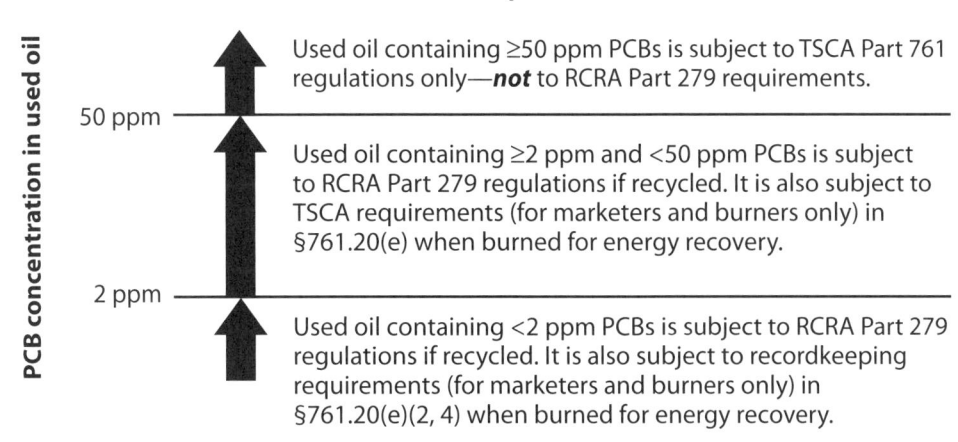

Source: McCoy and Associates, Inc.; adapted from 68 *FR* 44660, RO 14606.

composition of the used oil, or a certification that the used oil contains <2 ppm PCBs from the person generating the used oil. [§761.20(e)(2)(iii)]) However, these TSCA recordkeeping requirements apply only to marketers and burners of the used oil—not to entities that are generators only (although a marketer may request the generator to certify that the used oil contains <2 ppm PCBs).

Used oil that contains greater than or equal to 2 ppm but less than 50 ppm of PCBs is also subject to the RCRA Part 279 standards and, for marketers and burners of used oil, the marketing, testing, burning, and recordkeeping requirements codified in the TSCA regs in §761.20(e), when such used oil is burned for energy recovery. [§279.10(i), 68 FR 44660, RO 14117] As an aside, EPA has noted in guidance that, when two environmental programs both apply—in this case RCRA and TSCA—the more stringent requirements govern. [§761.1(e), July 8, 1987; 52 FR 25770, August 18, 1992; 57 FR 37237, RO 11470]

Used oil that contains less than 50 ppm PCBs that is recycled in ways other than burning for energy recovery is regulated under Part 279 but is excluded from TSCA Part 761 requirements *unless* 1) it was diluted to below the 50-ppm threshold; or 2) the PCBs were not legally manufactured, processed, distributed in commerce, or used under TSCA. [§§761.3 "Excluded PCB Products," 761.20(a)(1), 761.20(c), 68 FR 44660] However, §761.20(d) prohibits the use of used oil that contains any detectable level of PCBs as a sealant, coating, or dust-control agent (e.g., road oiling).

Finally, used oil containing 50 ppm PCBs or more must be managed only under TSCA; therefore, Part 279 doesn't apply to such high-PCB used oil. [§761.60] EPA believes that the current requirements in TSCA for PCB-contaminated wastes adequately control the management and disposal of such used oil. [68 FR 44660]

Used oil that contains PCBs may not be diluted to avoid a particular regulatory requirement. [§761.1(b)] When PCB concentrations in used oil are diluted, the used oil is subject to the requirements applicable at the original PCB concentration. So, for example, used oil that contains >50 ppm PCBs at its point of generation that has been diluted so that its PCB concentration is <50 ppm is still subject only to regulation under TSCA. [68 FR 44660]

However, used oils of unknown PCB concentration can be mixed with other such used oils in a common container and subsequently tested to determine the PCB concentration. (However, no chemicals or other nonused oils may be added to the container to take advantage of this provision.) [§§761.20(e)(2), 761.60(g)(2), 68 FR 44660]

12.1.8 Used oil managed in the petroleum industry

Used oil produced in the petroleum industry is often recycled by mixing it with crude oil or other petroleum feedstocks and re-refining it into fuel and lubrication products. Before used oil is inserted back into the refining process, however, the Part 279 regulations may apply to its management. Section 279.10(g) is the portion of the used oil regulations that applies to used oil introduced into a crude oil pipeline or petroleum refinery.

12.1.8.1 Used oil inserted into crude oil pipelines

Used oil that is generated at oil and gas exploration and production (E&P) operations is often mixed with crude oil or natural gas liquids in production separators, stock tanks, or other tank-based units that are connected via pipeline to a petroleum refining facility—see Figure 12-4. Once the used oil mixes with crude oil in one of these units, it becomes exempt from Part 279. [§279.10(g)(1)] Thus, these production separators, stock tanks, etc. that introduce the used oil/crude oil mixture to the pipeline are also exempt from Part 279. Tank bottoms removed from stock tanks containing mixtures of used oil and crude oil/natural gas liquids [that are exempt under §279.10(g)(1)] are also exempt from all Part 279 standards per a separate provision codified at §279.10(g)(6).

Figure 12-4: Used Oil Inserted Into Crude Oil Pipelines

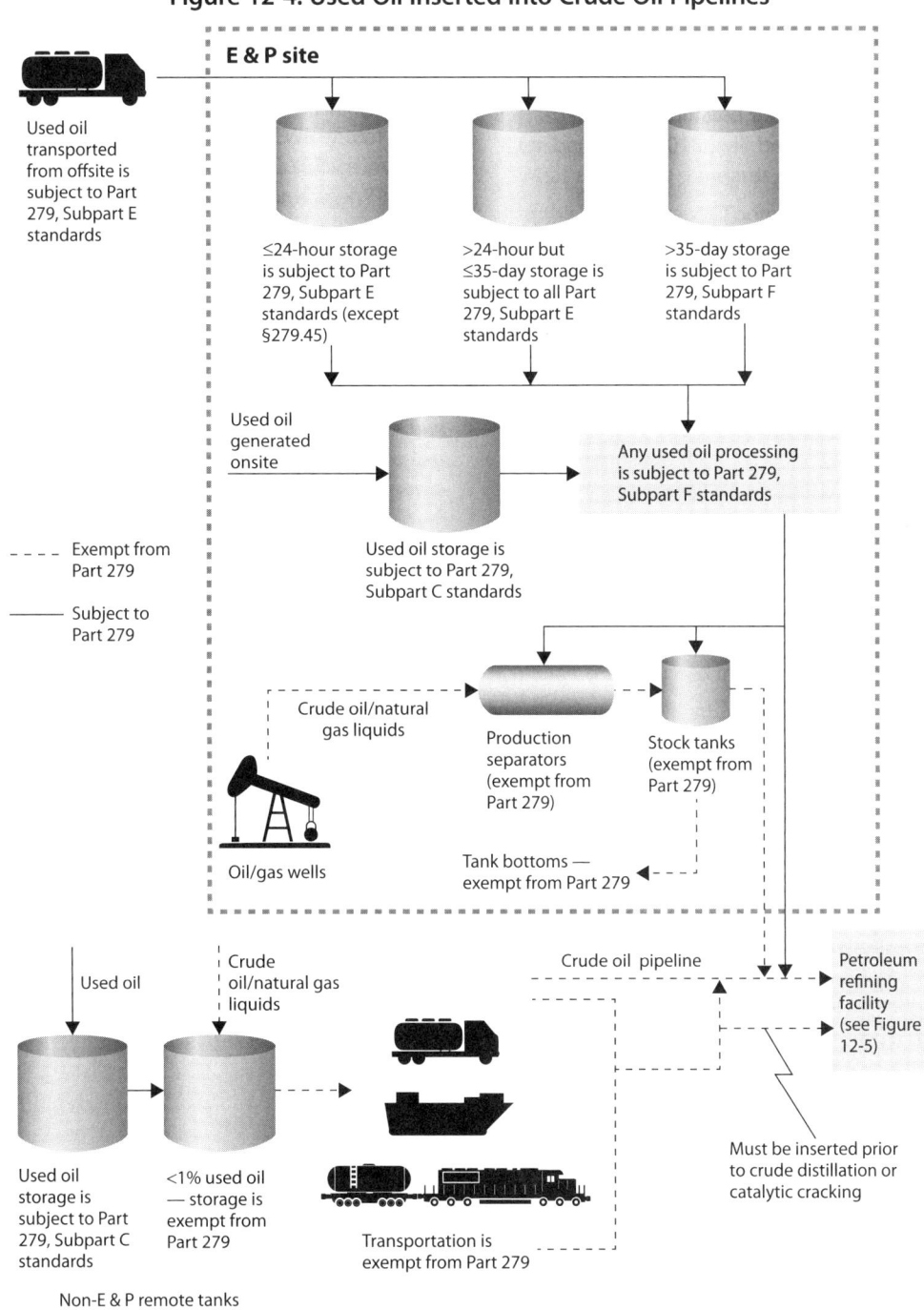

E&P = exploration and production.

Source: McCoy and Associates, Inc.; adapted from §§279.10(g)(1–2),(6) and 279.45, 59 FR 10552–3, RO 11844.

Used oil can also be directly inserted into a crude oil pipeline, at both E&P sites and non-E&P sites. For example, used oil generated at crude oil pipeline pumping stations is often placed in a sump and pumped directly into the pipeline without premixing with crude oil.

Any tanks or containers that store used oil generated onsite before it is mixed with crude oil in production separators, stock tanks, etc. or pumped directly into the pipeline do not enjoy the exemption and are subject to the Part 279, Subpart C generator standards. Also, if processing of the used oil is performed prior to mixing with crude oil, such processing is subject to the Part 279, Subpart F standards for used oil processors.

Although EPA notes at 59 *FR* 10552 [March 4, 1994] that the used oil is typically <1% of the crude oil, the regs at §279.10(g)(1) do not limit the ratio of used oil to crude oil. Thus, the exemption applies to any amount of used oil that mixes with crude oil in production separators, stock tanks, etc. that are connected via pipeline to a petroleum refining facility.

Used oil that is transported from offsite for insertion into a crude oil pipeline, either directly or via mixing with crude oil in production separators, stock tanks, etc., is subject to the Part 279, Subpart E transporter standards. Once at the pipeline, storage of such offsite used oil is regulated as follows:

- Stored at the pipeline for no more than 24 hours prior to mixing with crude oil—subject to all Part 279, Subpart E transporter standards except for §279.45 that applies to transfer facilities.

- Stored at the pipeline for more than 24 hours and no more than 35 days prior to mixing with crude oil—subject to all Part 279, Subpart E transporter/transfer facility standards.

- Stored at the pipeline for more than 35 days prior to mixing with crude oil—subject to all Part 279, Subpart F processor standards. [§279.45, RO 11844]

12.1.8.1.1 Storage/transportation of used oil/crude oil mixtures before insertion into crude oil pipelines

EPA also provides an exemption for mixtures of used oil with crude oil/natural gas liquids during remote storage and transportation to the crude oil pipeline. Codified at §279.10(g)(2), this provision exempts these mixtures from Part 279 storage and transportation standards as long as they contain <1% used oil. This exemption, which is shown in the lower portion of Figure 12-4, applies whether the mixture (containing <1% used oil) is ultimately inserted into a crude oil pipeline or transported directly to a petroleum refinery. One additional caveat applies: the mixture must be inserted into the refining process prior to crude distillation or catalytic cracking. (This is discussed more in Section 12.1.8.2.)

For example, used oil generated during off-shore drilling activities (e.g., from compressors and other heavy equipment) is routinely mixed with crude oil in units (e.g., production separators, stock tanks, etc.) located at the E&P site, and then the mixture is transported to a pipeline or petroleum refining facility. Conveyance of the mixture (containing <1% used oil) may involve multiple modes of transportation (i.e., from the off-shore platform to land by vessel or pipeline and then to the crude oil pipeline or petroleum refining facility by land-based transport), but all modes are exempt from Part 279 standards per §279.10(g)(2). [59 *FR* 10553]

In essence, because of the high ratio of crude oil to used oil, EPA considers these mixtures to be equivalent to crude oil for regulatory purposes. [59 *FR* 10553] As with the §279.10(g)(1) exemption, however, used oil that is generated at E&P or other remote sites continues to be subject to the Part 279, Subpart C generator standards prior to being mixed with crude oil/natural gas liquids.

12.1.8.2 Used oil inserted into petroleum refineries

EPA has defined petroleum refineries in §279.1:

"*Petroleum refining facility* means an establishment primarily engaged in producing gasoline, kerosene, distillate fuel oils, residual fuel oils, and lubricants, through fractionation, straight distillation of crude oil, redistillation of unfinished petroleum derivatives, cracking or other processes (i.e., facilities classified as SIC 2911)."

This definition allows EPA to distinguish a petroleum refinery from a used oil re-refiner, which the agency defines as "a facility that processes used oil to produce lube base stocks and greases, industrial fuels, asphalt extenders, diesel like fuels, and other products." In order for a facility to be considered a petroleum refining facility, the material fed to the front end of the refining process must be composed primarily of crude oil. [59 *FR* 10552]

There are several exemptions in §279.10(g) that apply to used oil inserted into refinery units. Each of these is discussed in the following subsections.

12.1.8.2.1 *Used oil inserted prior to crude distillation or catalytic cracking*

As mentioned in Section 12.1.8.1.1, mixtures of used oil and crude oil/natural gas liquids (containing <1% used oil) transported directly to a petroleum refinery are not subject to Part 279 standards if they are inserted into the refining process prior to crude distillation or catalytic cracking. [§279.10(g)(2)] This exemption extends to any storage of these mixtures at the petroleum refining facility prior to insertion (see Figure 12-5).

EPA clarified at 59 *FR* 10555 that this same §279.10(g)(2) exemption from Part 279 applies to the storage of mixtures of used oil and slop (recovered) oil (containing <1% used oil) prior to insertion upstream of crude distillation or catalytic cracking. Therefore, slop oil units at refineries are not subject to Part 279 standards as long as the used oil is <1% of the mixture.

Used oil is sometimes inserted directly into the refining process prior to crude distillation or catalytic cracking without mixing it beforehand with either crude or slop (recovered) oil. The regulations at §279.10(g)(3) exempt the used oil at the point it enters the refining process. To take advantage of the §279.10(g)(3) exemption, two requirements must be met: 1) the used oil must be inserted into the refining process prior to crude distillation or catalytic cracking, and 2) the volume of used oil must be less than 1% of the crude oil feed to any given process unit. The §279.10(g)(3) exemption makes no distinction between on-spec or off-spec used oil as differentiated by §279.11. However, "…because we specifically stated that it does not make any sense to limit the amount of on-specification used oil inserted after the crude distillation or catalytic cracking unit, which provides additional contaminant reduction, we did not intend to limit the amount that can be inserted before." [RO 14829] Thus, the <1% volume restriction in §279.10(g)(3) applies only to off-spec used oil; there is no quantity restriction for on-spec used oil.

Petroleum refining facilities that receive used oil from offsite for direct insertion into the petroleum refining process are subject to the used oil processor requirements from the point at which they receive the used oil up until the point at which the used oil is inserted into the petroleum refining process. [59 *FR* 10554] These requirements, which include notification, contingency planning, waste analysis plans, tracking, and recordkeeping (§§279.51–279.53, 279.55–279.57) apply to such refineries even if the used oil is stored for less than 24 hours before mixing or insertion into the refining process. If the used oil is stored before mixing, §279.54(a–f) and (h) also apply. [RO 11844] One of the primary requirements noted above is preparation of a waste analysis plan to ensure that the used oil has not been mixed with hazardous waste; additionally, such facilities must maintain an operating record to document compliance with the plan. [59 *FR* 10554]

12.1.8.2.2 *Used oil inserted after crude distillation or catalytic cracking*

Section 279.10(g)(4) exempts on-spec used oil when used as a feedstock after crude distillation or catalytic cracking. "The basis for this determination is that on-specification used oil is equivalent to

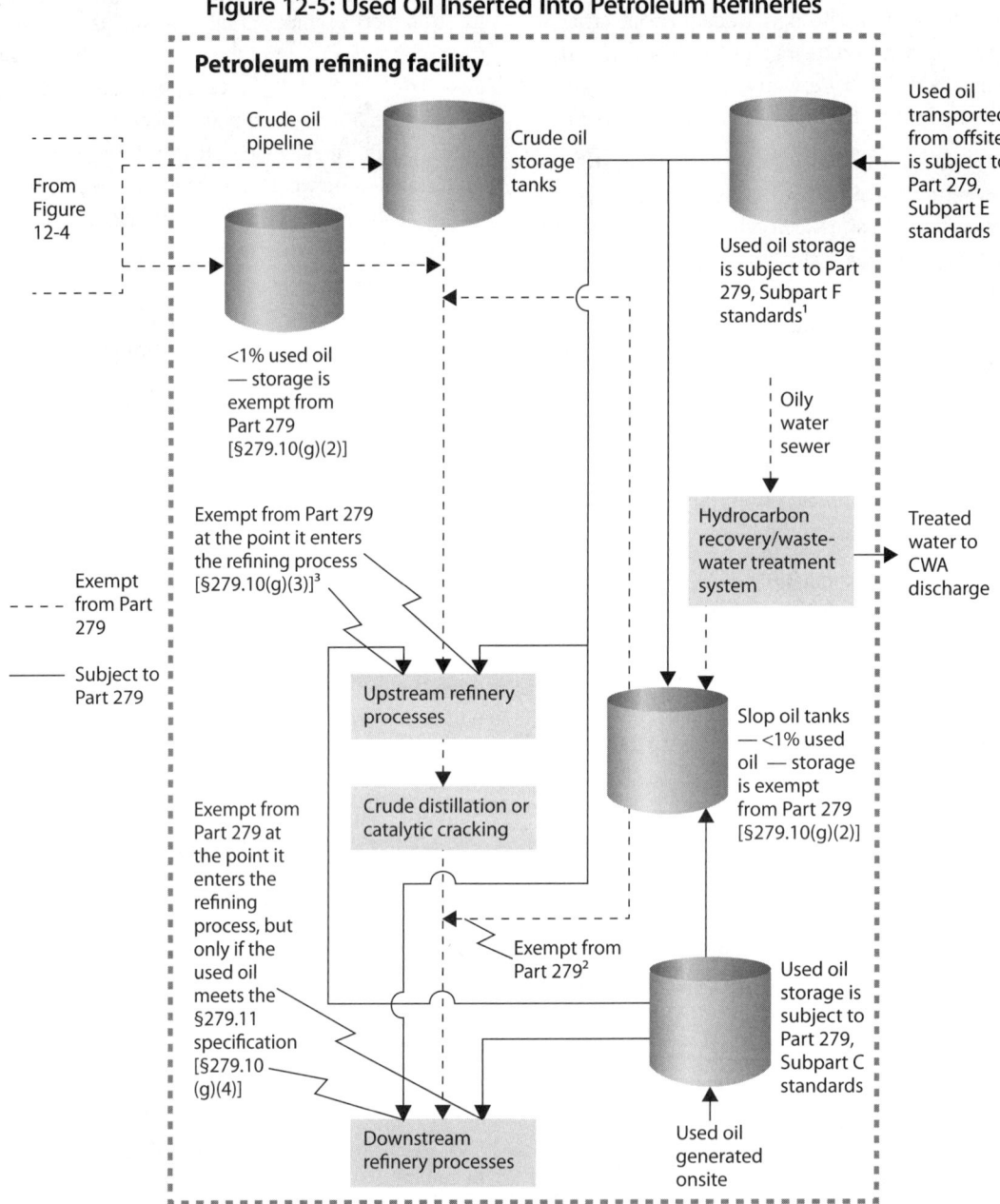

Figure 12-5: Used Oil Inserted Into Petroleum Refineries

1 However, used oil that meets the §279.11 specification is exempt from Part 279 standards per RO 14829.

2 Mixutres of used oil and slop (recovered) oil that contain <1% used oil and are inserted into the refining process after crude distillation or catalytic cracking are exempt from Part 279 only if the used oil meets the §279.11 specification prior to mixing with slop (recovered) oil. [§279.10(g)(4), 59 FR 10555]

3 Although §279.10(g)(3) requires that the used oil be inserted into the refining process at a volume of <1% of the crude oil feed at any given time, RO 14829 eliminates this quantity restriction if the used oil meets the §279.11 specification.

Source: McCoy and Associates, Inc.; adapted from §§279.10(g)(2–5), 59 FR 10553–5, RO 14829.

virgin fuel oil for regulatory purposes." [RO 14829] Therefore, EPA's interpretation is that on-spec used oil is exempt from Part 279 when used as a feedstock in petroleum refinery operations—regardless of whether it is inserted before or after crude distillation or catalytic cracking.

However, if off-spec used oil is inserted as a feedstock after crude distillation or catalytic cracking (e.g., into a coker or asphalt tower), the facility would be subject to the used oil processing requirements in Part 279, Subpart F. [59 FR 10554]

The same used oil processor requirements noted in Section 12.1.8.2.1 for used oil received from offsite and then inserted prior to crude distillation or catalytic cracking would apply to used oil received from offsite and then inserted after crude distillation or catalytic cracking.

12.1.8.2.3 *Used oil captured in hydrocarbon recovery/wastewater treatment system*

The final provision regulating used oil at refineries is §279.10(g)(5), which exempts from Part 279 used oil that is incidentally captured in the hydrocarbon recovery/wastewater treatment system and inserted into the petroleum refining processes. To qualify for the exemption, the used oil must be generated from routine refinery process operations (e.g., drips, leaks, and spills from compressors, valves, and pumps). Oil (that contains these small amounts of used oil) that is recovered from a refining facility's hydrocarbon recovery or wastewater treatment system is typically pumped to the slop oil system and then used as feedstock to produce more petroleum products.

The §279.10(g)(5) exemption does not apply to used oil that is intentionally put into the hydrocarbon recovery/wastewater treatment system (e.g., pouring collected used oil into any part of the hydrocarbon recovery system, storm or process sewer system, or into wastewater treatment units). [59 FR 10555]

The examples cited in the de minimis wastewater exemption [§279.10(f)] provide guidance on what types of releases to a refinery's hydrocarbon recovery or wastewater treatment system would be considered "routine" or "incidental":

"small spills, leaks, or drippings from pumps, machinery, pipes, and other similar equipment during normal operations or small amounts of oil lost to the wastewater treatment system during washing or draining operations."

12.2 Used oil management standards for generators

According to §279.1, a "used oil generator" is:

"[A]ny person, by site, whose act or process produces used oil or whose act first causes used oil to become subject to regulation."

The first part of this definition is fairly straightforward. The second part, "...whose act first causes used oil to become subject to regulation," is a little vague; we'll clear that up in our discussions below. Examples of used oil generators include vehicle maintenance shops, automobile service stations, quick-lube shops, government motor pools, grocery stores, metal-working industries, and boat marinas. [EPA/530/F-96/004] Additional guidance on who qualifies as a used oil generator was given in preamble language:

"[Used oil] generators include all persons and businesses who produce used oil through commercial or industrial operations and vehicle services, including government agencies, and/or persons and businesses who collect used oil from households and 'do-it-yourself' oil changers." [September 10, 1992; 57 FR 41584]

As both of these quotations imply, certain entities that generate used oil are exempt from regulation under Part 279. These entities, along with the point where the used oil they generate becomes subject to regulation, are as follows [§279.20(a)]:

- Household do-it-yourselfers (DIYs) are exempt from Part 279. The used oil these households generate (from changing their own oil in their personal vehicles) becomes subject to regulation when it is accepted/aggregated at a used oil collection center (e.g., operated by a service station or quick-lube shop). Under this scenario, the collection center is considered the generator of the used oil. [May 3, 1993; 58 FR 26423, RO 11828]

- Vessels at sea or in port that produce used oil are not subject to Part 279, until the used oil they generate is transported ashore. Once ashore, the used oil is subject to Part 279 regulation. In this case, the vessel owner/operator and the person who removes or accepts the used oil from the vessel are cogenerators of the used oil. Under this arrangement, both parties are responsible for managing the oil under the applicable provisions of the Part 279, Subpart C generator standards. "The cogenerators may decide among them which party will fulfill the requirements of this subpart." [§279.20(a)(2)]

- Generators who mix used oil generated onsite with diesel fuel for use in their own vehicles are not subject to Part 279 once the mixing has occurred. In this case, the used oil *is* subject to regulation prior to mixing.

- Farmers who generate an average of 25 gallons per month or less of used oil in a calendar year from vehicles or machinery used on the farm are not subject to the requirements of this part. As long as this threshold is not exceeded, the farmer remains exempt from Part 279. Similar to household DIYs, used oil generated by these farmers becomes subject to Part 279 regulation when collected at a used oil collection center.

Note that there is no such thing as a small quantity used oil generator. Thus, no need exists for a generator to measure the quantities of used oil collected and stored each month. [57 FR 41585]

12.2.1 Used oil generator standards

Figures 12-6a and 12-6b show the issues that generators of used oil must address in terms of management requirements. As referenced in these two logic diagrams, most of the management standards applicable to used oil generators are codified within Subpart C to Part 279. Additional requirements are applied to generators via references to other Part 279 subparts, other RCRA regulations, and CWA rules.

A useful document for reviewing a generator's compliance with the used oil regulatory requirements is *Protocol for Conducting Environmental Compliance Audits of Used Oil and Universal Waste Generators Under RCRA*, EPA/300/B-00/002, March 2000, available at http://infohouse.p2ric.org/ref/14/13639.pdf.

12.2.1.1 Used oil storage

The used oil management standards for generators address storage in three primary categories of units: 1) containers and aboveground tanks, 2) underground storage tanks (USTs), and 3) other units.

12.2.1.1.1 *Containers and aboveground tanks*

The wording in §279.22(a) could be misinterpreted, requiring containers and tanks used to store used oil to comply with Part 264/265, Subparts I and J, respectively. This was not EPA's intent; used oil storage containers and tanks need not comply with those requirements unless the used oil has been mixed with hazardous waste. [RO 14118] Instead, such storage units must meet the following.

According to §279.22(b–c), containers and aboveground tanks used to store used oil must 1) be in good condition with no severe rust, structural defects, or deterioration; 2) have no visible leaks; and 3) be labeled clearly with the words "Used Oil." Our experience indicates that the number one source of noncompliance with the used oil regulations is facilities forgetting to label their containers with the words "Used Oil."

We get asked questions about small-capacity containers of used oil such as buckets and drip pans. Our experience indicates that buckets used to store used oil should meet the above requirements including labeling with "Used Oil." Drip pans, which are often very shallow and are used to catch drips of oil from equipment, are usually not considered used

Figure 12-6a: Regulatory Requirements for Used Oil Generators

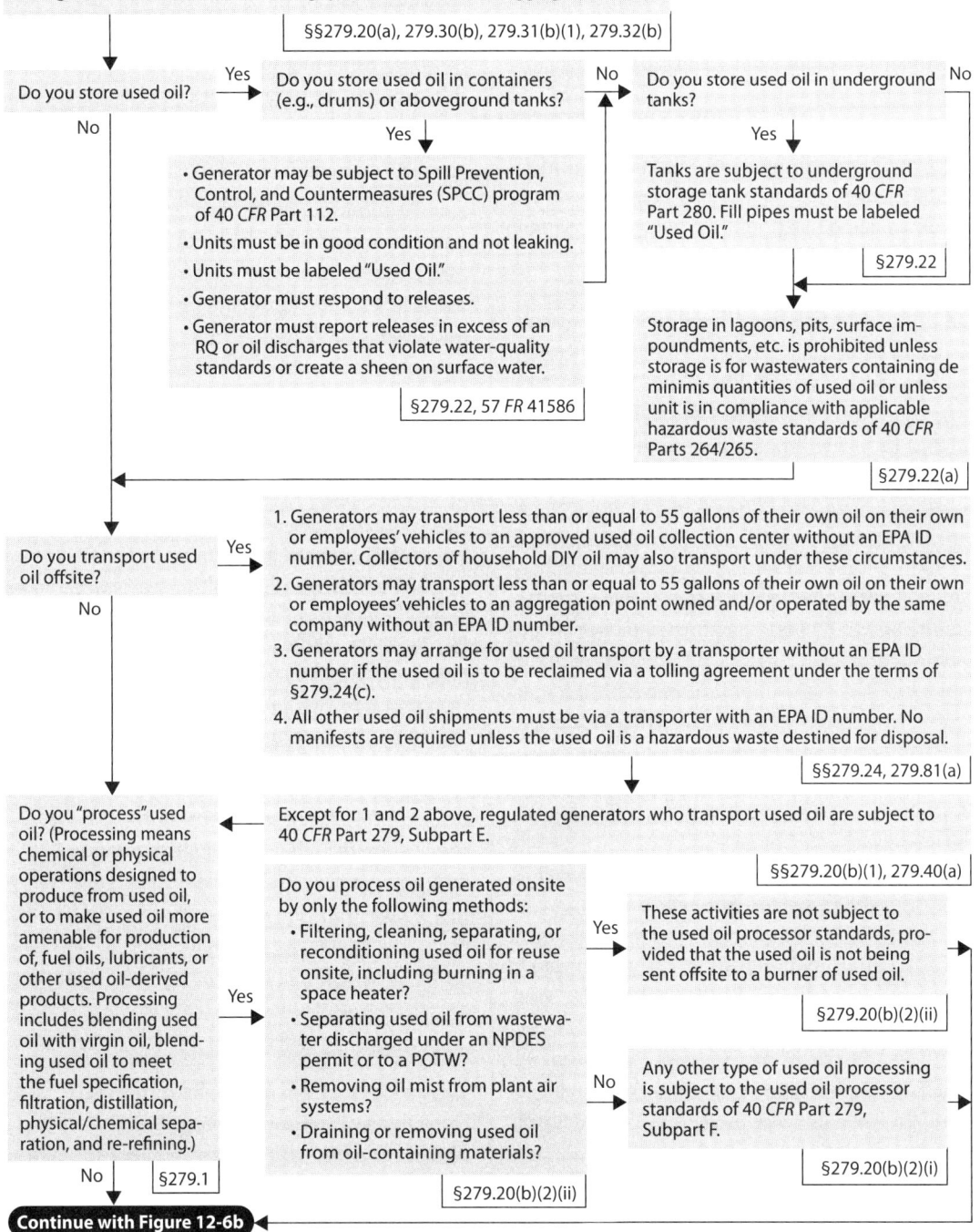

Figure 12-6b: Regulatory Requirements for Used Oil Generators (Continued)

Continued from Figure 12-6a

- Do you burn used oil for energy recovery?
  - **Yes** → Do you burn the used oil in space heaters where:
    - You generate the used oil or it comes from household DIYs,
    - The heater's capacity is less than or equal to 0.5 million Btu/hr, and
    - Combustion gases are vented to the ambient air?
    - **§279.23**
    - **Yes** → Burning is not subject to oil burners' standards of 40 CFR Part 279, Subpart G. [§§279.11, 279.12(c)(2)(iii), 279.23]
    - **No** → Does the used oil meet all of the following specifications:
      - Arsenic . . . . 5 ppm max
      - Cadmium . . . . 2 ppm max
      - Chromium . . . . 10 ppm max
      - Lead . . . . 100 ppm max
      - Flash point . . . . 100°F min
      - Total halogens . . 4,000 ppm max
      - **§279.11**
      - **Yes** → Generators who first claim used oil meets specifications must keep analyses or other information for 3 years. Such generators must have an EPA ID number. [§§279.72, 279.73]
      - **No** → Off-specification used oil may only be burned in boilers or industrial furnaces, utility boilers, or RCRA hazardous waste incinerators. Standards of 40 CFR Part 279, Subpart G apply. See Figure 12-7. [§§279.12(c), 279.20(b)(3)]
  - **No** → Do you direct shipments of used oil to a used oil burner?
    - **Yes** → You must comply with the used oil fuel marketer standards of 40 CFR Part 279, Subpart H. See Figure 12-8. [§279.20(b)(4)]
    - **No** → Generators who dispose used oil or who use it as a dust suppressant must comply with the standards of 40 CFR Part 279, Subpart I. [§279.20(b)(5)]

Source: McCoy and Associates, Inc.; adapted from 40 CFR Part 279, Subparts C and D.

---

oil storage containers subject to the above, but that call is somewhat dependent on how long the used oil remains in the pan and is up to the enforcement discretion of the state inspector.

Should a release to the environment from a container or an aboveground storage tank occur, the owner/operator must 1) stop the release; 2) contain the released used oil; 3) clean up any released used oil and other materials; and 4) if necessary, repair or replace any leaking storage container or tank prior to placing it back into service. A notice of violation will likely result if a facility fails to promptly clean up even a small leak (e.g., from a used oil storage tank) or if oil-soaked absorbent material used to contain prior releases is still on the floor. According to EPA guidance, these requirements would not apply to releases within contained areas, such as concrete floors or impervious containment areas. However, facilities would normally clean up such spills as a matter of good housekeeping practices; in any event, they would be obligated to clean up contained spills or leaks before they reach the environment. [September 10, 1992; 57 FR 41586, RO 14339]

Used oil generators are also required to report a release of hazardous substances to the environment under other environmental programs. Releases of used oil that is contaminated with

CERCLA hazardous substances (e.g., lead) must be reported to the National Response Center (NRC) if the reportable quantity (RQ) of the contaminating substance is released within a 24-hr period. [§302.6(a)] Additionally, the Clean Water Act requires facilities to report certain oil spills. Discharges of oil to navigable waters or adjoining shorelines must be reported immediately to the NRC if they: 1) violate applicable water-quality standards, 2) cause a film or "sheen" on the surface of the water or adjoining shorelines, or 3) cause a sludge or emulsion to be deposited beneath the surface of the water or upon adjoining shorelines. [40 CFR 110.3, 110.6] Finally, a discharge must be reported to the EPA regional administrator and appropriate state agency when there is a discharge of:

- More than 1,000 gallons of oil in a single discharge to navigable waters or adjoining shorelines, or

- More than 42 gallons of oil in each of two discharges to navigable waters or adjoining shorelines occurring within any 12-month period. [40 CFR 112.4]

When determining the applicability of this SPCC reporting requirement, the gallon amount specified (either 1,000 or 42) refers to the amount of oil that actually reaches navigable waters or adjoining shorelines, not the total amount of oil spilled. [*Spill Prevention, Control, and Countermeasure (SPCC) Regulation, 40 CFR Part 112—A Facility Owner/Operator's Guide to Oil Pollution Prevention*, EPA/540/K-09/001, June 2010, available at http://www.epa.gov/oem/docs/oil/spcc/spccbluebroch.pdf]

Besides the common-sense requirements noted above, owners/operators using containers and aboveground tanks for oil storage may also be required to prepare and follow an SPCC plan for these units. See Section 12.2.1.1.4 for more details.

Finally, it is interesting to note what requirements are *not* imposed on container and aboveground tank storage of used oil by Part 279. First, because EPA views used oil as a marketable commodity and wants to encourage its recycle and reuse, there is no accumulation time limit and no limit on the quantity of used oil that may be stored. [EPA/530/K-02/025I, RO 14739] Also, there are no secondary containment, inspection, or closure requirements (although one or more of these may be imposed under SPCC plans). Finally, used oil generators are not required to obtain EPA ID numbers.

12.2.1.1.2 *USTs*

EPA believes that storage of used oil in USTs poses similar risks to the underground storage of gasoline and other petroleum products. Thus, the agency requires underground used oil storage tanks to meet the requirements of 40 *CFR* Part 280, which apply to USTs managing CERCLA hazardous substances and petroleum products. Additionally, fill pipes associated with USTs storing used oil must be labeled with the words "Used Oil." [§279.22(c)(2)]

While a complete review of Part 280 requirements is beyond the scope of this text, these provisions are comparable in many respects to Part 264/265, Subpart J requirements for hazardous waste storage tanks (which are covered in Chapter 10). These include design and management standards, corrosion protection, leak detection, spill response and notification, and closure plans. Thus, used oil storage in USTs may be more involved compared to the requirements for storage in containers and aboveground tanks.

There is an exemption from Part 280 requirements, however, for certain USTs storing used oil. That exemption is for the underground storage of used oil that is being utilized as heating oil to fire heating equipment, boilers, or furnaces. Based on §280.12, a tank used for storing heating oil for consumptive use on the premises where stored does not meet the "UST" definition and so is not subject to the Part 280 requirements. "Heating oil" is also defined in §280.12 and includes "other fuels when used as substitutes for one of these fuel oils." Used oil that is burned in heating equipment, boilers, or furnaces would meet this "heating oil" definition.

Thus, USTs used to store used oil that is consumed onsite as a substitute for fuel oil (e.g., burned in an onsite space heater) are excluded from Part 280 requirements; conversely, USTs that store used oil awaiting recycling pickup are not heating oil tanks and are not excluded. [September 23, 1988; 53 FR 37117]

12.2.1.1.3 Other units

Regarding the storage of used oil in units other than tanks and containers, EPA noted:

"Storage of used oil in lagoons, pits, or surface impoundments is prohibited, unless the generator is storing only wastewaters containing de minimis quantities of used oil, or unless the unit is in full compliance with 40 CFR Part 264/265, Subpart K." [57 FR 41586]

In other words, the agency will allow the storage of used oil in alternate units only if they meet stringent Subtitle C RCRA requirements for hazardous waste management units.

12.2.1.1.4 SPCC plans

As part of the CWA regulations, SPCC requirements, which are spelled out in 40 CFR Part 112, apply to facilities that manage oil (including petroleum, fuel oil, and used oil) and are designed to prevent discharges to navigable waters. All facilities handling oil must prepare a plan if they are located in an area where a release to a navigable waterway could be expected and they:

- Have an underground oil storage capacity of more than 42,000 gallons; or
- Have a container and aboveground tank oil storage capacity of more than 1,320 gallons. When determining aboveground storage capacity, only containers of oil with a capacity of ≥55 gallons are counted.

SPCC plan requirements are quite extensive, so we won't try to cover them in detail here. To summarize, they include tank and piping design, construction, and inspection requirements; secondary containment requirements; personnel training provisions; unit security and vehicle control requirements; storm water diversion and control requirements; contingency planning; and emergency response and notification procedures. Besides the regulations themselves, a good reference document for more detail on SPCC plans is Appendix D to *Environmental Regulations and Technology—Managing Used Motor Oil* [EPA/625/R-94/010, December 1994, available from http://nepis.epa.gov/EPA/html/Pubs/pubtitleORD.html by downloading the report numbered 625R94010].

12.2.1.2 Offsite shipments

Part 279 rules governing the offsite transportation of used oil are very simple: "generators must ensure that their used oil is transported only by transporters who have obtained EPA identification numbers." [§279.24] When the transporter comes to the generator's facility to pick up the used oil, no manifest is required. EPA believes that the information maintained by used oil transporters will be sufficient to track used oil shipments without the need for a manifest. [57 FR 41587]

The Part 279 regs provide three alternatives to using a used oil transporter with an EPA ID number. First, used oil generators may self-transport small amounts of used oil offsite to used oil collection centers, without first obtaining an EPA ID number, provided:

- No more than 55 gallons of used oil is transported at any one time;
- The used oil is transported in a vehicle owned by the generator or one of its employees; and
- The used oil is transported to a used oil collection center that is registered, licensed, permitted, or otherwise recognized/allowed to manage used oil by a state, county, or municipal government.

This option allows generators of small quantities of used oil and generators who have several, separate generation points (each producing small quantities of used oil) to recycle their used oil without significant cost. A typical application is given in Case Study 12-5. Note that used oil collection centers must use used oil transporters with EPA ID numbers when shipping the collected used oil offsite for recycling.

A second option allows generators to self-transport small amounts of used oil without an EPA ID number to an aggregation point. Again, no more than 55 gallons of used oil may be transported at any one time and a generator- or generator employee-owned vehicle must be used. Most importantly, the generating facility and the aggregation point must have the same owner. Again, the aggregation point would have to use a transporter with an EPA ID number to ship collected used oil offsite.

Under the third option, a generator may arrange for used oil transportation offsite by a transporter without an EPA ID number if the used oil is to be reclaimed and eventually returned to the generator for reuse as a lubricant, cutting oil, or coolant. The reclaiming arrangement must be spelled out in a contractual tolling agreement that indicates 1) the type of used oil involved and the frequency of the offsite shipments, 2) that the reclaimed oil will be returned to the generator, and 3) that the vehicles used to transport the used oil to the processor and the reclaimed oil back to the generator are owned by the processor.

Note that the above discussion is for offsite shipments of used oil. Onsite movement of used oil is not subject to either §279.24 or to the Part 279, Subpart E used oil transporter and transfer facility standards. [RO 11762]

12.2.1.3 Other generator operations

Used oil generators may manage used oil in ways other than simple storage and offsite shipment to a processor. Such operations include used oil processing, burning for energy recovery, and disposal, each of which is addressed later in this chapter. However, processing, burning, and/or disposal of

---

Case Study 12-5: Servicing a Customer's Fleet of Vehicles

A service company employee drives to a customer's site, where the employee services the customer's fleet of vehicles. Part of this service involves changing the oil in the vehicles. The fleet's used oil is collected in an empty 55-gallon drum—never more than 45 gallons of used oil are collected in any one day. The company employee leaves the customer's site with 45 gallons or less of used oil in a drum in the service company's vehicle. Because the customer's fleet is often serviced at night, the used oil drum may be stored in the service company's vehicle until the next business day. At that time, the employee delivers the used oil drum to either the customer's used oil aggregation point or a third-party, government-registered, used oil collection center. Can the service company transport the used oil in this manner without having an EPA ID number and without complying with the used oil transporter standards in Part 279, Subpart E?

The service company and the customer are co-generators of the used oil, and, in this scenario, the service company has elected to fulfill the used oil generator requirements. Because the service company is handling shipments of used oil totaling 55 gallons or less from the generation site to a used oil aggregation point or collection center, the activity would not be regulated under the Subpart E standards for used oil transporters and transfer facilities. Rather the transportation would be regulated under the used oil generator standards in Part 279, Subpart C. See the requirements at §§279.24, 279.31, and 279.32.

The fact that the used oil is sometimes stored in the service company's vehicle until the next business day does not preclude the company from being regulated as a used oil generator. However, the drum must be labeled with the words "Used Oil." [RO 11923]

used oil will subject the generator to more-stringent standards. In addition, used oil generators may ship their oil directly to used oil burners. In this case, the generator would also be subject to requirements applicable to used oil fuel marketers, which are contained in Part 279, Subpart H and are discussed in Section 12.3.5.

A facility sometimes hires a contractor to conduct activities that generate used oil instead of using its own personnel. In such a situation, both entities are cogenerators of the used oil. An example of used oil activities conducted by contractor personnel is given in Case Study 12-6.

12.2.1.4 No recordkeeping required

No specific tracking or recordkeeping are required in the used oil generator standards. [EPA/530/H-98/001] When the agency issued the Part 279 standards, it noted:

> "EPA has determined that information maintained by used oil transporters will provide sufficient records of used oil transport activities without burdening used oil generators with additional tracking requirements. Information collected when accepting used oil shipments, such as quantities and type of used oil collected, the name and location of used oil generators, and analytical data for the rebuttable presumption, would be maintained by the used oil collectors/transporters as part of the recordkeeping requirements…. Using this information maintained by used oil transporters, the agency can track a used oil generator, if needed. Therefore, the agency has eliminated the…tracking requirements for used oil generators." [September 10, 1992; 57 *FR* 41587]

The agency noted in that same preamble, however, that "EPA believes that used oil generators maintain used oil collection and shipment records as standard business information." [57 *FR* 41587]

Case Study 12-6: Used Oil Activities Conducted by Contractors

A facility hires a contractor to come onsite and service equipment that contains oil. If the contractor removes oil from the equipment during servicing, is the contractor considered the used oil generator, since the contractor's act first causes the used oil to become subject to regulation?

According to EPA, the contractor and the facility owner/operator are cogenerators of the used oil and both are responsible for managing the oil under the applicable provisions of Part 279. As with the situation where a separate entity offloads used oil from ships, the cogenerators must decide between themselves who will fulfill the used oil generator requirements. For example, the contractor could self-transport up to 55 gallons of the removed used oil to a collection center or aggregation point without an EPA ID number. [RO 14116]

12.2.2 Generators: don't process the oil you generate

Just as facilities that generate and accumulate hazardous waste can avoid many of the more-stringent RCRA requirements by not *treating* their waste, used oil generators should avoid *processing* their oil lest they face the tough provisions for processors and re-refiners in Part 279, Subpart F. EPA defines used oil processing in §279.1 as:

> "[C]hemical or physical operations designed to produce from used oil, or to make used oil more amenable for production of, fuel oils, lubricants, or other used oil-derived product. Processing includes, but is not limited to: blending used oil with virgin petroleum products, blending used oils to meet the fuel specification, filtration, simple distillation, chemical or physical separation, and re-refining."

If a generator does something to its used oil to make it more amenable for someone else to take and burn for energy recovery, that's "processing" subject to the Subpart F requirements. For example, a generator that blends off-spec used oil with fuel oil to meet the specification is a processor. [RO 14110]

There are a few specific things that a used oil generator can do to its used oil and remain just a generator. The following four activities are *not* considered processing [§279.20(b)(2)(ii)], provided the used oil is generated onsite and, after the activity, is not sent directly to an offsite used oil burner:

1. Filtering, cleaning, separating, or otherwise reconditioning used oil before reusing it *onsite*, including burning it in a space heater pursuant to §279.23—For example, onsite maintenance and reconditioning activities designed to extend the life of used oil are not considered processing. [March 4, 1994; 59 *FR* 10555] A typical example is filtering contaminants from used metal-working fluids and then recycling the fluids back into machining, grinding, and/or boring operations; such reconditioning is not considered processing. [59 *FR* 10556, RO 11783, 14055] Additionally, the removal of CFCs and/or HCFCs from drained compressor oil (so that an owner/operator can take advantage of the exemption from the rebuttable presumption) is not processing. [RO 11850, 14051] Finally, filtering off-specification used oil for subsequent burning for energy recovery in an *onsite* industrial furnace is also not processing. (Had the filtered used oil been shipped offsite for burning, the generator would have been subject to Subpart F processor requirements.) [RO 13666]

2. Separating used oil from wastewater generated onsite to make the wastewater acceptable for discharge under an NPDES permit or POTW pretreatment standards—The separation of used oil from wastewater can also be accomplished to recover the used oil for reuse. [59 *FR* 10556, RO 11783, 11818] For example, recovery of metal-working fluids from a facility's wastewater treatment system for onsite recycling is not processing. This activity is incidental or ancillary to normal manufacturing operations (i.e., used oil processing is not the facility's primary purpose). [RO 11792] Note that entities conducting oil-water separation on wastewater that is received from offsite would be considered used oil processors. [59 *FR* 10557, RO 11818]

3. Using oil-mist collectors to remove small droplets of used oil from plant air systems.

4. Draining or otherwise removing used oil from used oil-contaminated materials—Removing or separating used oil from materials, such as draining used oil from non-terne-plated filters or separating used oil from sorbent materials is not processing. [May 3, 1993; 58 *FR* 26421, RO 11874] In another situation, onsite dewatering of used oil-based coolant so that the coolant can then be sent to an offsite re-refiner or fuel blender (but *not* directly to an offsite burner) is not processing. [RO 13757] Wringing out a rag or absorbent pad that contains used oil is not processing.

As noted above, these four activities are not considered processing as long as any used oil produced is not sent directly to an offsite used oil burner. Additionally, such produced used oil can be burned onsite without triggering processor requirements. EPA noted that it:

"[I]s allowing onsite but not offsite burning of used oil generated from designated onsite activities because the agency believes that this approach best enables EPA to strike a reasonable balance between encouraging beneficial onsite reuse and recycling activities that should pose very limited risks, on one hand, and ensuring that activities undertaken primarily to make used oil more amenable for burning (i.e., used oil processing) are adequately controlled under the more-stringent used oil processing standards." [59 *FR* 10556; see also RO 11874, 13666]

Mixing used oil generated onsite with diesel fuel for use in a generator's own vehicles as fuel (discussed in Section 12.1.6.1) is also excluded from the processor standards. [EPA/530/K-02/025I]

What additional requirements does a processor of used oil face versus a simple generator? Processors must have an EPA ID number and implement preparedness and prevention plans to prevent and respond to fires, explosions, spills, and other incidents. They must provide secondary containment for containers and tanks used to store used oil, and they must meet closure requirements for tanks and containers upon cessation of operations. An analysis plan must be prepared and followed to determine compliance with the used oil fuel specification and to ensure their compliance with the rebuttable presumption. Finally, used oil processors must track their receipts and shipments of used oil, maintain a written operating record for their facility, and report annually to EPA on their activities.

12.2.3 Used oil disposal

Subpart I to Part 279 details requirements for generators (and others) who dispose used oil. First of all, EPA will allow the use of used oil as a dust suppressant only in those states that successfully petition the agency under the terms of §279.82(b). As of September 2013, no such petitions have been approved.

Aside from the dust suppressant issue, used oil that will be disposed rather than recycled is a solid waste. Therefore, before a facility may dispose used oil, it must first determine whether that material is a RCRA hazardous waste. Used oil is a hazardous waste if it 1) exhibits a hazardous waste characteristic (by its own nature or by mixing characteristically hazardous waste into it), 2) has been mixed with ICR-only listed wastes and continues to exhibit a characteristic, 3) has been mixed with non-ICR-only listed hazardous waste (other than listed waste generated by conditionally exempt small quantity generators), or 4) contains greater than 1,000-ppm total halogens and the presumption that it has been mixed with hazardous waste cannot be rebutted.

Hazardous used oil that will be disposed (i.e., it will not be recycled) must be managed in accordance with the hazardous waste management requirements of Parts 260–266, 268, and 270. Such used oil will typically be burned in a RCRA-permitted hazardous waste incinerator.

Used oils that are not RCRA hazardous and cannot be recycled must be disposed per the terms of 40 CFR Parts 257 and 258.

12.3 Burning used oil

Used oil is burned for energy recovery at numerous facilities throughout the United States. Available information indicates that about 82% of recycled used oil is burned as a fuel, with the balance being re-refined into recycled lube oil or other products. The units burning the greatest amounts of used oil as fuel are asphalt plants, industrial boilers, utility boilers, steel mills, and cement kilns. [*Used Oil Re-refining Study*, July 2006, available at http://www.fossil.energy.gov/epact/used_oil_report.pdf] This section discusses the regulatory aspects of such combustion. Figure 12-7 shows the Part 279 issues that must be considered when burning used oil.

12.3.1 The used oil specification

In the 1970s and 1980s, numerous facilities were burning used oil for energy recovery, because it was a cheap, high-Btu fuel. However, EPA became concerned about the potential human health effects associated with air emissions from such used oil burning. Some used oil contains fairly high levels of carcinogenic or toxic constituents that could be emitted in stack gas. Schools, hospitals, and apartment buildings that burn contaminated used oil typically use low-efficiency boilers, have little or no air pollution controls associated with them, and are located in residential areas.

In writing the Part 279 provisions dealing with used oil burning, EPA attempted to balance its desire to encourage recycling by energy recovery and its concern over the potential impact of air emissions from used oil combustion. The primary means that the agency formulated to balance these conflicting issues is the used oil fuel specification. EPA established the

Figure 12-7: Regulatory Requirements for Used Oil Burners

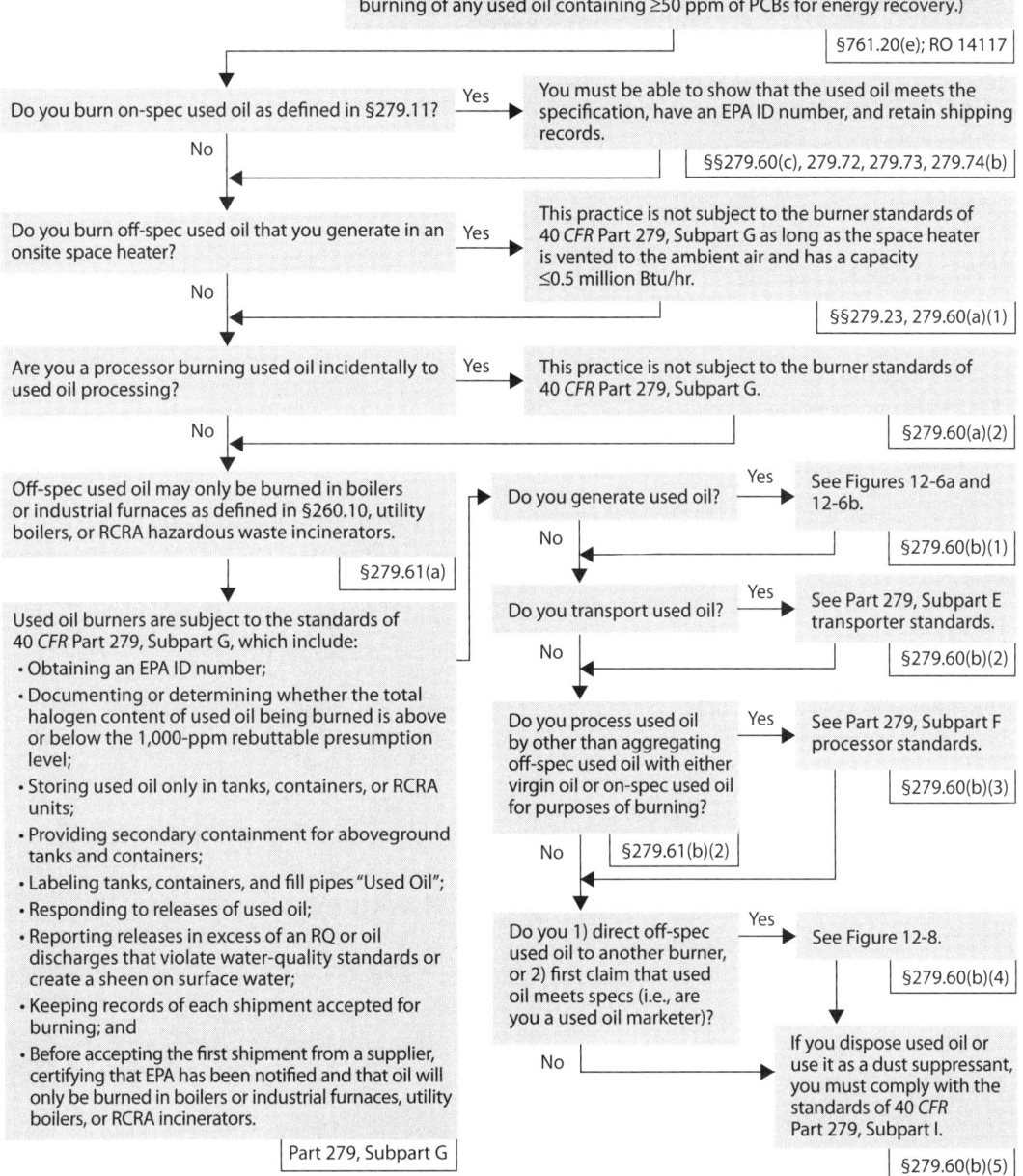

Source: McCoy and Associates, Inc.; adapted from 40 CFR Part 279, Subpart G.

specification for used oil that will be burned for energy recovery on November 29, 1985. [50 *FR* 49164] As shown in Table 12-3, the specification addresses the heavy metal and total halogen content, as well as the flash point, of used oil destined for energy recovery.

Arsenic, cadmium, and chromium are known carcinogens, and increased ambient concentrations would cause an increased risk of cancer to exposed individuals. Specification levels for these three metals were set based on metal concentrations found in dirty virgin fuel oils. The specification level for lead, conversely, was set based on meeting the national ambient air quality standard (NAAQS) for lead in densely populated areas. [November 29, 1985; 50 *FR* 49184–6]

The total halogen specification will minimize hydrogen chloride emissions that can increase ambient levels of hydrochloric acid and contribute to acid rain. [50 *FR* 49181] Note that the total halogen limit of 4,000 ppm operates independently of the 1,000-ppm rebuttable presumption level for determining if listed spent solvents have been mixed into used oil. In other words, used oil for which the presumption of mixing with hazardous waste has been rebutted may be burned as on-specification used oil if the total halogen content is no greater than 4,000 ppm, or as off-spec used oil if the content is greater than 4,000 ppm. [RO 14340]

Although used oil typically has a flash point of greater than 200°F, a minimum flash point of 100°F is included in the specification because that is the American Society for Testing and Materials (ASTM) specification level for virgin fuel oils. [50 *FR* 49187]

The presence of PCBs in used oil is not one of the used oil fuel specifications. Thus, used oil can be on-spec, even if it contains between 2 and 50 ppm PCBs. In a footnote to the specification contained in §279.11, however, EPA has noted that used oil containing PCBs that will be burned for energy recovery is subject to §761.20(e). That section of the TSCA regs notes that PCB-contaminated used oil (containing between 2 and 50 ppm PCBs) may

Table 12-3: The Used Oil Fuel Specification

Constituent/property	Allowable level
Arsenic	5 ppm maximum[1]
Cadmium	2 ppm maximum[1]
Chromium	10 ppm maximum[1]
Lead	100 ppm maximum[1]
Total halogens	4,000 ppm maximum
Flash point	100°F minimum

[1] Total analysis, not TCLP. [RO 14584]
Source: Adapted from §279.11, Table 1.

only be combusted in certain types of units (TSCA or RCRA incinerators, TSCA high-efficiency boilers, industrial furnaces or boilers as defined in §260.10, or utility boilers); certain notification requirements apply as well. "Although the RCRA regulations do not identify the presence of PCBs in used oil as relevant to the determination of whether the used oil is on- or off-specification, the presence of PCBs in used oil is relevant for determining the applicability of the TSCA regulations for the burning of used oil." [RO 14606] Note that burning used oil that contains ≥50 ppm PCBs for energy recovery is prohibited. [§761.20(a), RO 14117]

The existence of the used oil specification does *not* require generators to analyze their used oil to determine if it meets spec before the oil is transported offsite to a processor or fuel blender—that's the receiving facility's job. [EPA/530/H-98/001] A generator would only have to determine compliance with the used oil specification if it was burning used oil onsite or was sending used oil directly to a burner; in either of these situations, the generator would also be a used oil marketer, as discussed in Section 12.3.5. We find that some generators (especially at utility power plants) do analyze their used oil on a routine basis; however, such practice is typically required because of TSCA concerns or as a business arrangement with the used oil collector.

In another footnote to the used oil specification, EPA notes that the specification is not applicable for used oil that is mixed or contaminated with hazardous waste. [RO 14110, 14606] With the exception of noncharacteristic mixtures of characteristic waste and used oil (and noncharacteristic mixtures of ICR-only listed waste and used oil), mixtures of hazardous waste and used oil are regulated under the BIF regs in Part 266, Subpart H when burned for energy recovery.

12.3.2 Burning on-specification used oil

Once it's established that used oil meets the fuel specification, the scope of regulations that apply to the oil diminishes dramatically. Provided the entity who asserts that the used oil is on-spec meets the following three qualifying conditions, neither Part 279 nor any other RCRA requirement applies to that material [§279.11, RO 14606]:

- Documents the finding that the used oil meets the specification via testing or other methods and retains records of these documents for three years [§279.72],
- Has an EPA ID number [§279.73], and
- Maintains records of any shipments of the on-spec oil to used oil burners [§279.74(b)].

Once these conditions are met, the on-spec used oil is not subject to RCRA regulation and may be managed the same as any conventional fuel, such as virgin fuel oil. [November 29, 1985; 50 *FR* 49189, RO 14755] On-spec used oil may be burned in any type of combustion unit, including those in schools, hospitals, apartment complexes, etc. In fact, the agency will even allow on-spec used oil to be used in the manufacture of products such as ammonium nitrate fuel oil blasting agents. [RO 11807]

Q: *Bilge water collected from ships contains small amounts of used oil. Once the used oil is separated from the water (using filtration, centrifugation, and demulsification), the recovered used oil is burned for energy recovery. However, because the bilge water meets the used oil spec and the generator complies with §§279.72, 279.73, and 279.74(b), does the processing of the bilge water escape Part 279 standards per §279.11?*

A: No. What escapes Part 279 management [if it meets the used oil spec and the generator complies with §§279.72, 279.73, and 279.74(b)] is "used oil that is to be burned for energy recovery." According to RO 12738, it is the as-burned used oil that must meet the specification to be unregulated. In this example, it is not the bilge water but the recovered used oil that will be burned for energy recovery. "On-specification used oil that is re-refined…rather than burned for energy recovery, is subject to all applicable requirements of Part 279." [RO 14110] Thus, bilge water processing would be subject to Part 279, Subpart F standards.

Although its management is outside of RCRA, on-spec used oil is still subject to other applicable regulations. For example, on-spec used oil stored in USTs will be subject to Part 280 requirements (unless the "heating oil" exemption discussed in Section 12.2.1.1.2 applies), and used oil transport must comply with DOT rules. Also, facilities handling on-spec used oil may be required to prepare and implement SPCC plans. In addition, when an action is taken at the generating facility that may affect the chemical or physical properties of the on-spec used oil, that oil must be reevaluated in terms of the specification. [RO 14110] On-spec used oil that will be processed/re-refined or disposed, rather than burned for energy recovery, remains subject to Part 279 standards. [RO 14110]

Also note that on-spec used oil that contains PCBs at a concentration between 2 and 50 ppm and that will be burned for energy recovery is subject to certain marketer and burner requirements as discussed in Sections 12.1.7 and 12.3.1. These requirements, which come from TSCA—not RCRA, reference some of the standards in Part 279 for marketers and burners of *off*-spec used oil. "Therefore, by operation of the TSCA rules, used oil that is on-specification under the RCRA rules may nevertheless be subject to certain requirements specified in the RCRA rules for off-specification used oil…. The fact that the TSCA rules incorporate by reference these RCRA standards does not mean that PCB-containing [on-spec] used oil is regulated under RCRA authority or that such used oil is off-specification as defined by Part 279." [68 FR 44661; see also RO 14117] Specifically, burners of on-spec used oil containing between 2 and 50 ppm PCBs must comply with a number of the Part 279, Subpart G requirements, including restrictions on burning, notification, certification, tracking, and recordkeeping, in addition to requirements specified in §761.20(e)(3–4).

Used oil generators that burn their own on-spec used oil for energy recovery in onsite boilers or industrial furnaces (i.e., without complying with the off-spec burner requirements in Part 279, Subpart G) are, by definition, used oil fuel marketers. In order to avoid the Subpart G burner standards, they must determine that their used oil meets the used oil fuel specification. That makes them a marketer per §279.70(a)(2). [RO 14280]

12.3.2.1 For CAA purposes, on-spec used oil is not solid waste when combusted

In response to court decisions, EPA had to rework two Clean Air Act (CAA) rules: 1) the commercial and industrial solid waste incineration (CISWI) rule, and 2) the industrial boilers maximum achievable control technology (MACT) rule. The court concluded that the agency erred by excluding some units that combust solid waste for the purpose of energy recovery from the CISWI rule, instead regulating these units under the boiler MACT rule.

As EPA revised these two air rules for repromulgation in accordance with court mandates, the agency was obligated to articulate which nonhazardous secondary materials constitute "solid waste" under RCRA, since the CAA requires that these rules be based (in part) on that definition. Any unit combusting any solid waste must be regulated as a CISWI unit, regardless of whether the material is being burned for energy recovery or destruction. Conversely, if a nonhazardous secondary material is not a solid waste under RCRA, then any unit combusting that material will be subject to other CAA standards, such as the boiler MACT rule.

In a March 21, 2011 final rule [76 FR 15456], EPA determined that on-spec used oil meets the definition of a traditional fuel and is, therefore, not solid waste when used in a combustion unit. (Of course, it would be a solid waste if discarded.) [§241.2] Thus, burning on-spec used oil will not subject a combustion unit to CISWI standards, but other CAA controls, such as the boiler MACT rules, may apply.

12.3.3 Burning off-specification used oil

Facilities that combust used oil that does *not* meet the §279.11 fuel specification face a much different outlook than their counterparts that burn on-spec oil. Burning off-spec used oil subjects a facility to the burner standards in Part 279, Subpart G. Those standards limit the types of units in which off-spec used oil may be burned and impose additional requirements on such facilities. These requirements are reviewed below.

12.3.3.1 Allowable units

Off-spec used oil may be burned only in a boiler or industrial furnace as defined in §260.10, utility boiler, or RCRA hazardous waste incinerator. In the case of a boiler or industrial furnace, such units do not have to be BIFs subject to regulation under Part 266, Subpart H; they simply must meet the definition of boiler or industrial furnace in §260.10. [57 FR 41599] Hazardous waste incinerators, of course, must be in compliance with Part 264/265, Subpart O. There is one exception to the general rule stated above, which we address in the next subsection.

The four types of units noted above that are allowed to burn off-spec used oil address EPA's air emission concerns. These units typically use high-efficiency boilers, have significant air pollution controls associated with them, and often are not located in residential areas.

The agency has noted in guidance that facilities do not have to analyze their used oil to see if it meets the fuel specification if it will be combusted for energy recovery or incineration in one of the units noted in the paragraph above. They may simply voluntarily assume it is off-spec and choose to meet the Part 279, Subpart G requirements for burning off-spec used oil. [RO 11811] This would be similar to a generator of solid waste declaring its waste to be hazardous (in lieu of analyzing it) and managing it as such.

12.3.3.2 The space heater exception

A used oil generator may burn off-spec used oil *onsite* in space heaters (without complying with the Part 279, Subpart G burner standards) provided [§279.23]:

- Only used oil that the facility generates or receives from household DIYs is burned in such heaters,
- The space heater is rated at not more than 0.5 million Btu/hr, and
- Combustion gases from the space heater are vented to ambient air.

Generators taking advantage of this exception do not have to determine (analyze) whether the used oil meets the fuel specification. Also, a conditionally exempt small quantity generator (CESQG) could burn a mixture of its used oil and hazardous waste that is exempt per §§261.5(j) and 279.10(b)(3) in such onsite space heaters whether or not the mixture meets the used oil specification.

Note that an onsite space heater that exceeds the Btu capacity limit given above is considered a nonindustrial boiler. Thus, the owner/operator is prohibited from burning off-spec used oil in such a unit. [RO 14280] See Case Study 12-7 for an example of the applicability of this exception.

12.3.3.3 Off-spec used oil burner requirements

Per Part 279, Subpart G, facilities that burn off-spec used oil must meet a number of conditions. They must:

Case Study 12-7: Burning Used Oil in a County Maintenance Facility Space Heater

A county highway maintenance garage wants to burn off-spec used oil generated at three offsite sources, along with its own oil, in an onsite space heater. The three offsite sources are 1) other, non-related businesses, 2) other county maintenance facilities, and 3) county-run DIY collection centers. (The onsite space heater itself meets the Btu-limit and combustion-gas-venting provisions noted in §279.23.) Is this allowed?

The space heater exception allows generators to burn only their own used oil or that received from DIYs. [§279.23(a)] The first source of offsite used oil, from nonrelated businesses, would not qualify for the space heater exception. Here, the county would be accepting used oil generated by other entities; this is not allowed under §279.23(a).

The second source of used oil would qualify for the space heater exception. In this case, the maintenance garage would be considered a used oil aggregation point, because the garage aggregates used oil from other generation sites that are owned by the same entity (the county). Thus, all of the used oil from those sources is considered to be generated by the same entity. Note, however, that the used oil from sites other than the garage would have to be transported per the terms of §279.24(b) (i.e., in quantities of no more than 55 gallons in county- or county employee-owned vehicles), unless the county chose to meet used oil transporter requirements of Part 279, Subpart E.

The third source of offsite used oil would also qualify for the space heater exception under the same terms as the second source [i.e., the garage is an aggregation point and shipments must follow §279.24(b)]. Note, however, that only used oil from *county*-run DIY collection centers may be burned in the space heater. Used oil from a state- or privately-run DIY collection center would not qualify. [RO 11944]

- Obtain an EPA ID number;

- Document or determine whether the halogen content of the used oil they burn is above or below 1,000 ppm and, if above, rebut the presumption that the used oil has been mixed with a hazardous waste;

- Store used oil only in tanks, containers, or RCRA units;

- Provide secondary containment for containers and aboveground tanks used to store used oil;

- Label tanks, containers, and UST fill pipes "Used Oil";

- Respond to releases of used oil;

- Report releases of used oil contaminated with hazardous substances in excess of CERCLA reportable quantities and releases that violate water-quality standards or create a sheen on surface water;

- Keep records of each shipment of used oil accepted for burning; and

- Before accepting the first shipment from a supplier, certify that EPA has been notified and that oil will only be burned in §260.10 boilers and industrial furnaces, utility boilers, or hazardous waste incinerators.

12.3.3.4 For CAA purposes, off-spec used oil is solid waste when combusted

As noted in Section 12.3.2.1, EPA issued a rule on March 21, 2011 [76 FR 15456] to delineate which nonhazardous secondary materials are solid waste under RCRA when combusted. Any unit combusting any solid waste must be regulated under the CAA as a CISWI unit, regardless of whether the material is being burned for energy recovery or destruction. EPA determined in the March 2011 rule that off-spec used oil is a solid waste, because it contains higher levels of contaminants than traditional fuels. [76 FR 15502] Thus, units burning off-spec used oil (other than small space heaters per Section 12.3.3.2) will be subject to CISWI standards.

Although off-spec used oil is a solid waste when combusted as noted above, it can be processed (under the Part 279, Subpart F provisions for processors and re-refiners) to meet the §279.11 used oil fuel specification. Once it is on-spec, it will no longer be solid waste if combusted. [§241.3(b)(4)]

12.3.4 Burning used oil-contaminated and derived materials

Materials that contain or are contaminated with used oil are regulated as used oil if they are burned for energy recovery. [§279.10(c)(2), 57 FR 41585] This applies regardless of whether the materials contain visible signs of free-flowing oil. For instance, hydraulic fluid filters and other used oil filters are regulated under Part 279 if they are burned for energy recovery, regardless of the degree of oil removal. [RO 11808] Similarly, sorbents containing used oil (without any free-flowing oil visible) are subject to the Part 279 used oil regulations if they are to be burned for energy recovery. [RO 11798] However, EPA has provided a caveat to the above:

"[S]ome sorbents have a high [Btu] value and once contaminated with used oil are managed by burning for energy recovery and, therefore, are regulated under Part 279. Contaminated materials (after draining) which provide little or no energy when burned, such as soil or clay-based sorbents, are not subject to Part 279. Whether a material is 'burned for energy recovery' depends on the type of materials being burned and the combustion equipment being used. For purposes of the EPA regulations governing boilers and industrial furnaces, burning for energy recovery is limited to materials that have a heating value of at least 5,000 Btu/pound…. EPA believes it is reasonable and consistent with the regulations to apply the same interpretation under Part 279. Of course, an authorized state may interpret what constitutes 'burning for energy recovery' more stringently than EPA and that interpretation could be controlling…." [RO 14111]

12.3.5 Used oil fuel marketer requirements

Besides regulating used oil burned for energy recovery, Part 279 also addresses used oil fuel "marketers." Any entity that conducts one of the following two activities is subject to Subpart H requirements as a used oil fuel marketer:

1. Directs a shipment of off-spec used oil from their facility to a used oil burner; or

2. First claims that used oil to be burned for energy recovery meets the used oil fuel specification.

Subpart H does not apply to used oil generators and transporters who send shipments of off-spec used oil to processors, even if such processors incidentally burn used oil. The marketer regs also don't apply to persons who direct shipments of on-spec used oil, but are not the first person to claim the oil meets the specification.

Note that used oil generators that burn their own on-spec used oil for energy recovery in onsite boilers or industrial furnaces (i.e., without complying with the Part 279, Subpart G requirements) are, by definition, used oil fuel marketers. In order to avoid the Subpart G burner standards, they must determine that their used oil meets the used oil fuel specification. That puts them into the marketer category via the second activity above. [RO 14280]

Used oil fuel marketer requirements focus on analytical results, tracking, and recordkeeping. Specifically, marketers must:

- Obtain an EPA ID number;

- Ensure that off-spec used oil is shipped only to used oil burners who 1) have an EPA ID number, and 2) plan to combust that oil in a §260.10 boiler or industrial furnace, utility boiler, or hazardous waste incinerator;

- Obtain a certification of compliance from the burner prior to the first shipment of off-spec used oil to that party; and

- Maintain records of 1) on-spec used oil analyses, and 2) shipments of on-spec and off-spec used oil. Marketers must keep records of shipments only to the initial facility to which it delivers the oil. [§279.74(b), 68 *FR* 44662] Marketers are not required to maintain records of subsequent transfers of this used oil to other entities.

For example, a service station that generates used oil to be burned for energy recovery and claims that it meets the fuel specification is a used oil fuel marketer. The facility must obtain an EPA ID number and test the oil to show that it is on-spec. These requirements must be met prior to the used oil being shipped offsite as on-spec—the oil cannot be shipped under the assumption that it is or will be blended into on-spec used oil. [RO 14110] Finally, the service station must maintain records of used oil analyses and on-spec oil shipments. Conversely, if the service station simply gives the used oil it collects to a used oil recycler, the station would be a used oil generator—but not a used oil fuel marketer—and the marketer requirements would not apply.

Although the second bullet above seems to imply that marketers can ship used oil only to used oil burners, they in fact can ship used oil to burners, processors, or other marketers. [RO 14755]

Section 761.20(e)(2) requires marketers (and burners) to presume that used oil to be burned for energy recovery contains ≥2 ppm PCBs and is, therefore, subject to the TSCA requirements in §761.20(e). The presumption can be overcome if a marketer determines through testing or "other information" that the used oil contains <2 ppm PCBs. [RO 14606] ("Other information" consists of personal knowledge of the source and composition of the used oil, or a certification that the used oil contains <2 ppm PCBs from the person generating the used oil. [§761.20(e)(2)(iii)]) If the used oil contains between 2 and 50 ppm PCBs, marketers must comply with a number of Part 279, Subpart H requirements, including notification, certification, tracking, and recordkeeping, in addition to the requirements specified in §761.20(e)(1) and (4). [68 *FR* 44661, RO 14117]

CHAPTER 12 *Used Oil*

In guidance, EPA noted that the frequency of testing used oil to ensure it meets the fuel specification depends on a number of site-specific considerations. For example, if some action, mixing, or storage conditions affect the physical or chemical composition of the used oil, a marketer must reevaluate whether it meets the specification. [RO 14110, 14626] Entities making a claim that used oil meets the fuel specification should provide documentation of testing and sampling methods used as well as the frequency of sampling/testing in the facility's records. [September 10, 1992; 57 *FR* 41597]

The applicability requirements, management standards, and other Part 279 provisions that apply to used oil fuel marketers are shown in Figure 12-8.

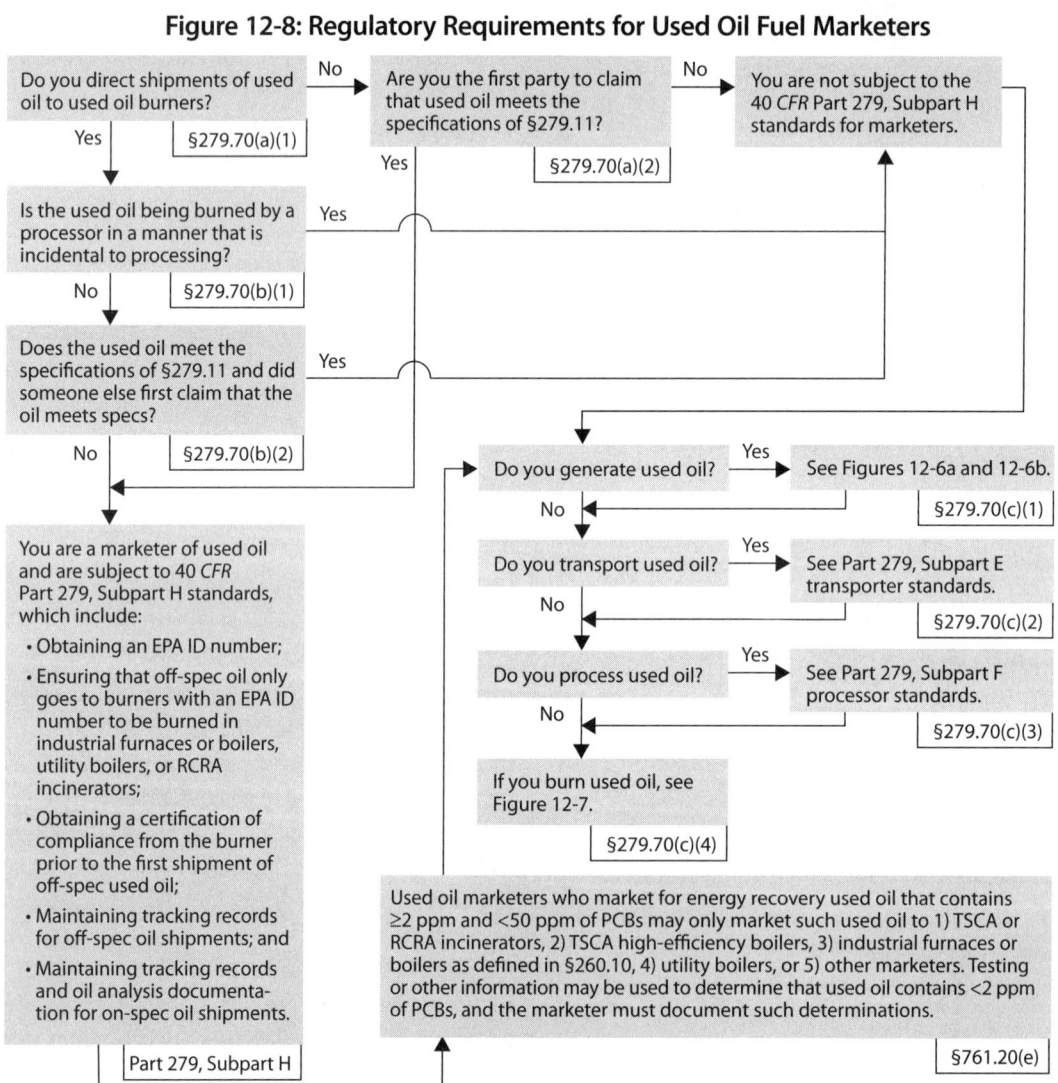

Figure 12-8: Regulatory Requirements for Used Oil Fuel Marketers

Source: McCoy and Associates, Inc.; adapted from 40 *CFR* Part 279, Subpart H.

Land Disposal Restrictions

Simplifying the most complicated RCRA regulations

Between 1985 and 1998, most of the new RCRA regulations imposed by EPA were associated with the land disposal restrictions (LDR) program. This program was mandated by the Hazardous and Solid Waste Amendments of 1984 (HSWA), and it reflects Congress' concern with the disposal of untreated hazardous wastes. After Congress became familiar with the environmental problems (and costs) associated with cleaning up abandoned waste sites under the Comprehensive Environmental Response, Compensation, and Liability Act (CERCLA or Superfund), it decided to establish a program under RCRA that would prevent such sites from being created in the future.

In general, the LDR program requires that any hazardous waste destined for land disposal be treated to reduce the toxicity and/or mobility of its hazardous constituents. Treatment standards were established for every hazardous waste; wastes meeting the treatment standards may be land disposed.

Almost everyone who handles hazardous waste is subject to some aspect of the LDR program. In many cases, all that is required for compliance is that the proper paperwork be prepared and submitted to the appropriate party or filed in onsite records. However, considerable knowledge is required to correctly fill out the paperwork.

In this chapter we will examine the requirements of the LDR program for all affected parties. We begin with an overview of how the program works.

13.1 Overview of the LDR program

The regulations implementing the LDR program are found in two places: for wastes being underground injected, the regulations are codified in 40 *CFR* Part 148; for wastes being disposed in other ways, the pertinent regulations are in 40 *CFR* Part 268. These regulations are *not* user friendly. Without some other source of information (such as this chapter), it is nearly impossible to read the regulations and determine how to comply. A good place to start is with a discussion of whether you are subject to this program.

13.1.1 Does the LDR program apply to you?

The regulations [§268.1(b)] state that the LDR program applies to "persons who generate or transport hazardous waste and owners and operators of hazardous waste treatment, storage, and disposal facilities." We prefer to say that the LDR program applies to everyone who handles hazardous waste except conditionally exempt small quantity generators (CESQGs) and pure transporters.

CESQGs are specifically exempt from the land disposal regulations of Part 268 in §§261.5(b) and 268.1(e)(1). Note, however, that these entities are only exempt if they send their wastes to permitted or interim status hazardous waste facilities, legitimate recycling facilities, or to other facilities permitted, licensed, or registered by the state to manage municipal or industrial solid wastes. [RO 12818] (In other words, if personnel at a CESQG dump their waste on the ground behind a building, the facility is not exempt from the LDR standards.)

Transporters who simply pick up hazardous waste at one location and deliver it to another are not subject to the LDR program. However, transporters who manage hazardous wastes in a way that results in a change in waste codes (e.g., by blending hazardous wastes) or who generate hazardous wastes (e.g., by cleaning up a hazardous waste spill) have to comply with the LDR requirements applicable to generators.

13.1.2 Summary of requirements under the LDR program

The activities required of various hazardous waste handlers are summarized in the following paragraphs.

Generators must:

1. Determine if they are managing a hazardous waste and assign the proper hazardous waste code(s);
2. Determine if the waste will be managed in a manner that triggers compliance with the LDR treatment standards;
3. Classify the waste according to LDR parameters (i.e., by subcategory, if any, and treatability group);*
4. Identify the appropriate treatment standards and determine if their waste meets those standards;*
5. Comply with the LDR dilution prohibition;
6. Prepare the proper paperwork for wastes shipped offsite; and
7. For wastes stored onsite for more than 90 days (necessitating a RCRA permit), comply with the LDR storage prohibition. For wastes treated/disposed onsite, comply with LDR requirements associated with treatment or disposal facilities.

*Steps 3 and 4 will not be necessary if generators choose not to determine if their hazardous waste requires treatment prior to land disposal. If generators choose this approach, they must manifest the waste to a RCRA-permitted hazardous waste treatment facility that will have the responsibility for determining if treatment is required. [§268.7(a)(1), April 4, 2006; 71 *FR* 16872]

Treatment facilities (including recyclers) must:

1. Concur with the waste classification provided by the generator;
2. Treat the waste to comply with the applicable LDR treatment standards or ship the partially treated waste to another treatment facility for additional processing;
3. Comply with the LDR dilution prohibition;
4. Prepare the proper paperwork for wastes shipped offsite for further treatment or disposal; and
5. For wastes stored onsite, comply with the LDR storage prohibition. For wastes disposed onsite, prepare paperwork confirming that the waste has been treated to the appropriate treatment standard and comply with the requirements applicable to disposal facilities.

Disposal facilities must:

1. Concur with the waste classification provided by the generator or treatment facility;
2. Assure that the waste meets the applicable treatment standards prior to land disposal, and
3. Maintain records associated with their LDR activities.

The following examples deal with the applicability of the land disposal restrictions and the responsibilities of the regulated entities.

Q *A state regulates hazardous wastes generated by households and CESQGs. Are these wastes subject to federal LDR requirements?*

A No. The state regulations are "broader in scope" than the federal regulations and are not subject to EPA oversight and enforcement. Therefore, the federal LDR standards do not apply, unless they are adopted and enforceable under state law. [RO 11481]

Q *A generator operates an onsite solvent recycling unit that is not subject to permitting. Is this facility subject to the LDR requirements?*

A Yes. LDR requirements are associated with wastes and how they are managed; LDR standards are independent of the RCRA permit program. [RO 13280]

Q *Do the LDR requirements apply to recyclable materials from which precious metals are reclaimed?*

A Yes. According to §268.1(b), the LDR requirements of Part 268 apply unless "specifically provided otherwise" in Parts 261 or 268. The regulations governing precious metals recycling are identified in §261.6(a)(2)(iii) and state that Part 268 does apply. Therefore, the LDR requirements apply to recyclable precious metal wastes. [RO 11482, 13158] A practical implication of this interpretation is that shipments of precious metal-containing wastes would have to be accompanied by the appropriate LDR paperwork (i.e., the appropriate notification/certification—see Section 13.12.3.5). Confusion over this issue may exist because §266.70(b), which specifies applicable requirements for precious metal wastes, does not cite Part 268.

Q *If the RCRA permit at a TSD facility does not include any LDR requirements, is the facility able to use the permit-as-a-shield provisions of §270.4(a) to avoid complying with the LDR requirements?*

A No. The LDR requirements are self-implementing by statute. In fact, §270.4(a)(1)(ii) specifically states that LDR requirements apply to permitted facilities.

Q *What liability does a TSD facility have if it disposes a waste that doesn't meet LDR standards due to misclassification of the waste by the original generator?*

A According to EPA, "[a] rule of strict liability applies under RCRA, so that a disposal facility can be liable for improper disposal of untreated waste even if it does so in the good-faith belief that the treatment standard does not apply.... [G]enerators and disposers may enter into indemnification agreements to allocate liability between them in the event that prohibited wastes are land disposed.... [W]hile good-faith efforts to comply are not a defense to liability, they may be considered in the assessment of penalties under EPA's…civil penalty policy." [RO 13630] See Chapter 16 for more details on enforcement issues.

13.1.3 What triggers the land disposal requirements?

The land disposal restrictions are triggered if a generator's waste or a residue from treating the waste will ultimately be disposed in a land disposal unit. The regulations [§268.2(c)] identify specific land disposal units within the overall context of land disposal: "*Land disposal* means placement in or on the land, except in a corrective action management unit or staging pile, and includes, but is not limited to, placement in a landfill, surface impoundment, waste pile, injection well, land treatment facility, salt dome formation, salt bed formation, underground mine or cave, or placement in a concrete vault, or bunker intended for disposal purposes."

In many cases, the fact that land disposal is occurring is obvious. For example, a generator sends his/her waste to a commercial TSD facility that will treat the waste to meet appropriate treatment standards and then dispose the waste (or treatment residue) in a landfill. Similarly, it is obvious that land disposal occurs when a generator sends its hazardous wastewater for disposal in an underground injection well.

In other cases, the fact that land disposal is occurring is not obvious. For example, when a generator sends spent solvent to a recycler, the recycler typically is not operating a land disposal unit. However, when the recycler recovers usable solvent from the original waste, he/she is left with a treatment residue (still bottoms). Typically, the still bottoms contain high levels of organics that must be destroyed before the bottoms can be disposed. A RCRA incineration facility would typically burn the still bottoms and send the incinerator residue (ash) to yet another facility for disposal. This last facility is typically a landfill, which is a land disposal unit.

Hence, even though the generator and recycler do not operate any land disposal units, a residue from treating the generator's waste will ultimately be land disposed. The way the LDR program works is that everyone from the generator to the final disposal facility must be kept informed (via LDR paperwork) that they are managing a waste that is subject to treatment standards. The treatment facility is responsible for meeting the treatment standards, and the disposal facility may only dispose waste that it knows meets the standards. Thus, the management of this one waste pulls the generator, recycler, incinerator, and landfill into the LDR program.

The general rule is that treatment of hazardous wastes must take place in non-land-based units before a prohibited waste is land disposed. Thus, hazardous wastes are typically treated in units such as incinerators, tanks, containers, and containment buildings before they are land disposed. The one exception to this is that hazardous waste may be treated in a surface impoundment or series of surface impoundments if the conditions in §268.4 are met. Note that storage areas in or on a landfill are defined in the statute and the Part 268 rules as land disposal. See RCRA Section 3004(k) and 40 *CFR* 268.2(c). The statute draws no distinction in the duration of disposal. Temporary placement in a land disposal unit is "land disposal" just as much as is permanent disposal. Containers located in or on a landfill are also considered land disposal. [RO 14843]

Q: *Is open burning or open detonation, which often takes place in earthen pits, considered to be land disposal?*

A: No. The open burning and open detonation of waste explosives is considered a treatment process rather than a waste disposal operation. [RO 11363, 12793, 13184, 13265]

Q: *In a related question, wastewater that is not hazardous flows into an earthen pit from which water evaporates. The remaining residue is D003 reactive. The residue is then destroyed by open burning. Do LDR requirements apply to the pits?*

A: EPA considers these earthen pits to be surface impoundments used for treating hazardous wastes. LDR requirements apply to wastes being managed in surface impoundments. [RO 13265]

Q: *A facility wants to treat a hazardous waste to meet LDR treatment standards by land farming the waste. Is this allowable?*

A: Generally no. Because a land farm is considered to be a land treatment unit, which is a land disposal unit, wastes must meet the LDR treatment standards before they are placed in a land farm. The exception to this rule is if the land farm has a no-migration exemption. As far as we know, only one facility has such an exemption (a refinery in Billings, Montana.) [RO 14289]

Although the storage prohibition of §268.50, which will be discussed in Section 13.10, can be read to apply to all restricted wastes, including those not destined for land disposal, that does not appear to be EPA's intent. For example, on September 6, 1989, the agency noted: "The general principle is that the dilution and storage prohibitions apply only if the waste is disposed by a prohibited method of land disposal." [54 *FR* 36968]

13.1.3.1 Avoiding LDR requirements

With one or two exceptions [a paperwork requirement in §268.7(a)(7) and possibly the storage prohibition in §268.50], no LDR requirements attach to wastes that are not land disposed. In other words, it pays to know how to manage hazardous wastes without triggering the LDR requirements.

First, and most importantly, hazardous wastewater that is handled in pipes and tanks prior to discharge under an NPDES permit or prior to discharge to a sewer line leading to a POTW is not subject to the LDR standards. [Note, however, that the one-time notice of §268.7(a)(7) describing these practices must be kept in the facility's files.] To avoid LDR requirements, these wastewaters must not be conveyed in earthen ditches or stored/treated in surface impoundments, which are land disposal units.

Second, if you are managing remediation (cleanup) wastes, you can avoid the LDR requirements by managing the wastes in a corrective action management unit (CAMU) as described in §§264.551 and 264.552, by disposing CAMU-eligible wastes in offsite hazardous waste landfills (§264.555), or by leaving the waste in the area of contamination (AOC). [RO 11954, 11970, 13442, 14112] See Case Study 13-1. You may also manage remediation waste in a staging pile (§264.554) without being subject to LDR requirements. However, since staging piles may only be used for temporary storage, wastes removed from these units would become subject to LDR standards unless moved to a CAMU.

Finally, wastes can be moved within a RCRA unit (e.g., a landfill) without triggering the LDR requirements. [RO 11950]

Q: *Investigation-derived waste is temporarily stored in drums within an AOC. When the waste is removed from the drums and redeposited in the AOC, are any LDR requirements triggered?*

A: If waste is placed in a drum within the AOC and the drum is not in a distinct "container storage area," the waste may be removed from the drum and redeposited in the AOC without triggering any LDR requirements. On the other hand, if drums are managed in a separate storage or treatment area either on land or on a pad within the AOC, redepositing wastes would trigger LDR requirements. This hair-splitting distinction occurs because a drum is not in itself a "hazardous waste management unit" as defined in §260.10. Multiple containers in an area or on a pad could constitute a "container storage area," which is a RCRA storage unit. Moving wastes from a RCRA unit to an AOC is what triggers the LDR requirements. [RO 11597]

Q: *Ground water contaminated with a listed waste is treated in an air stripper. The treated water is then used for spray irrigation. Is this considered to be land disposal?*

Case Study 13-1: When Does "Placement" Occur for Remediation Wastes?

The LDR standards are triggered if a waste is "placed" in a land disposal unit. EPA believes that placement occurs when remediation wastes are:

- Consolidated from different AOCs into a single AOC;

- Moved outside of an AOC (e.g., for treatment or storage) and returned to the same or a different AOC; or

- Excavated from an AOC, placed in a separate unit, such as an incinerator or tank that is within the AOC, and redeposited into the same AOC.

Placement does not occur when wastes are:

- Treated in situ,

- Capped in place,

- Consolidated within an AOC, or

- Processed within an AOC (but not in a separate unit, such as a tank) to improve its structural stability (e.g., for capping or to support heavy machinery). [RO 11954]

Q Yes. Spray irrigation is a form of land treatment and would be considered land disposal. [RO 12783] Therefore, if hazardous waste is spray irrigated, it would first have to meet the LDR treatment standards or the land treatment unit would have to have a no-migration exemption per §268.6. [RO 12957]

Q In some cases, hazardous ground water at CERCLA sites and RCRA corrective action sites can be reinjected into or above the uppermost aquifer. [See RCRA Section 3020(b).] Do the LDR requirements apply to this type of injection well?

A EPA's position on this issue is as follows:

"[C]ontaminated ground water reinjected during the course of RCRA or CERCLA cleanups in a manner consistent with the RCRA Section 3020(b) exemption is not subject to RCRA land disposal restrictions." [RO 14512]

"Section 3020(b) specifies that such prohibition [against reinjection into an uppermost aquifer] does not apply to contaminated ground water which is reinjected into the aquifer from which it was withdrawn if three criteria are met: 1) it is part of corrective action required under RCRA or CERCLA intended to clean up such contamination; 2) the contaminated ground water is treated to substantially reduce hazardous constituents prior to reinjection; and 3) the proposed corrective action will be sufficient to protect human health and the environment upon completion…. Thus, if the implementing agency at a particular site finds that the treatment of ground water as part of the response action has 'substantially reduced' the hazardous constituents and the response action is 'sufficient to protect human health and the environment,' then the ground water may be reinjected even if it does not otherwise meet [LDR] requirements." [RO 13463]

Q Is in situ treatment considered to be "placement" that would trigger LDR standards?

A No. "Placement does not occur when waste is consolidated within an AOC (area of contamination), when it is treated in situ, or when it is left in place." [March 8, 1990; 55 *FR* 8759; see also RO 13442, 13643]

Q Hazardous soil is treated to a level below soil remediation standards used by a state. The state standards are less stringent than LDR standards. May the soil be land disposed without complying with the LDR standards?

A No. If the excavated soil is a hazardous waste via the contained-in policy (see Section 5.3), it must meet LDR treatment standards prior to land disposal. As an alternative, a treatability variance per §268.44(h)(3) or (4) is available under certain conditions. [RO 14112]

Q Contaminated soil from excavation or construction activities is temporarily moved within the AOC and is subsequently redeposited in the excavated area. Are any LDR requirements triggered?

A These activities do not constitute treatment, storage, or disposal of a hazardous waste under RCRA. Moving wastes within an AOC does not constitute land disposal; hence, LDR requirements do not apply. [RO 11671]

13.1.4 What is the objective of the LDR program?

The overall objective of the LDR program is that hazardous wastes must be treated to protect human health and the environment prior to land disposal. HSWA required EPA to evaluate every hazardous waste and determine how it should be treated prior to disposal. A three-tiered hierarchy was used to select a treatment approach for each waste. First, in a few limited cases, the agency chose recycling, rather than treatment, as a method for handling hazardous wastes. The problem with most recycling technologies is that a residue is typically produced (such as still bottoms) that requires disposal. Second, technologies were selected that remove or destroy toxicants. For example, incineration was selected as a treatment technology for many concentrated organic wastes. Finally, as a last option, solidification/stabilization or some other form of immobilization was selected to tie up toxicants in a form that cannot readily enter the environment.

When evaluating each waste under this hierarchy, EPA typically relied on its experience to identify the technology that was best suited to the waste. They termed their selected technology "Best Demonstrated Available Technology" — BDAT. At one time, people found it useful or interesting to know what BDAT was for their wastes. Today, however, that information is of little practical significance. In general, EPA chose incineration as BDAT for concentrated organic wastes, biological treatment for organic-containing wastewaters, chemical precipitation for inorganic-containing wastewaters, and stabilization for inorganic-containing sludges and solids.

13.1.5 How were treatment standards established?

Because Congress gave EPA very little time to develop treatment standards for all of the various hazardous wastes, the agency chose a quick-and-dirty approach to setting treatment standards:

1. It selected BDAT as described above.

2. It identified the "hazardous constituents" that were of concern in each waste. These hazardous constituents are typically toxic chemicals; the number of hazardous constituents in a waste ranges from one for each P- and U-waste to a dozen or so for F- and K-wastes. F039, multisource leachate, is the record holder at 211 hazardous constituents.

3. EPA got a representative sample of the waste in question and ran that waste through the BDAT process to see how well it performed—how well it reduced the toxicity and mobility of the hazardous constituents. The agency accomplished this by looking at the residues that came out of the BDAT process to determine how much of each hazardous constituent remained. The measured concentrations in the residues became the LDR treatment standards. However, any technology (except dilution) can be used to meet the concentration levels—not just the BDAT technology. (This type of treatment standard became known as a "concentration-based standard.")

4. If a waste contained hazardous constituents that were "unanalyzable" (i.e., they could not be analyzed using conventional analytical methods), the agency simply picked a treatment technology that they knew would deal with the hazardous constituents. For such a waste, EPA specified the chosen technology as the treatment standard, and the agency required that it be used to treat the waste. (This type of treatment standard became known as a "specified method standard.")

5. The agency then determined if sufficient treatment capacity for each type of waste was available on a nationwide basis. If sufficient capacity existed for a specific waste, the treatment standards for that waste took effect immediately, and the waste could not be land disposed unless it met the standards. Where inadequate capacity existed, the effective date of the treatment standard could be postponed for up to four years. All such capacity extensions expired several years ago.

6. Once a treatment standard takes effect for a specific waste, that waste may no longer be land disposed unless it meets the treatment standard.

Several general observations will help put this treatment-standard-setting process into perspective. Treatment standards that are established as described above are referred to as "technology-based standards" (i.e., they are based on the performance of a particular type of technology). As a result, the LDR standards are not "risk-based standards" (i.e., they have no connection to risk assessments, acceptable exposure levels, etc.).

Also, when the agency ran their treatment tests, it used processes that were simply "well designed and well operated." The implication of this terminology is that EPA did not run its tests on processes that were the best in the country. Hence, the standards are not "technology forcing" (i.e., they are not set at such stringent levels that waste treaters are forced to upgrade their processes to the highest performance level available).

Because EPA was extremely pressed for time, it didn't actually test every hazardous waste in the manner described above. It would be more accurate to say that the agency tested every hazardous constituent using the selected technology and then transferred the treatment standard for each constituent to every hazardous waste in which that constituent was of concern. For example, EPA may have burned one benzene-containing waste in an incinerator, and then transferred the resulting treatment standard for benzene to every other benzene-containing waste for which incineration was BDAT.

Case Study 13-2 exemplifies the process EPA went through to set hazardous waste treatment standards.

13.1.5.1 Treatability groups established

During the process of setting treatment standards, EPA decided that every hazardous waste should have two treatment standards—one for wastewater forms of the waste and one for nonwastewaters. The term "wastewater" is defined in §268.2(f) as "wastes that contain less than 1% by weight total organic carbon (TOC) and less than 1% by weight total suspended solids (TSS)." Conversely, "nonwastewaters" are defined in §268.2(d) as all waste forms that are not wastewaters. These two waste forms are referred to as "treatability groups" in the LDR program.

Because of the way "wastewater" and "nonwastewater" are defined, every hazardous waste, no matter where it exists, is either wastewater or nonwastewater. For example, at the point of generation, a K-waste might be a wastewater. If the generator wants to dispose this waste, it would be subject to the wastewater treatment standard. If, however, the generator treats the waste to produce an aqueous (wastewater) residue and a sludge (nonwastewater) residue, each of these residues would be subject to the treatment standard for the appropriate treatability group before it could be land disposed.

The important point to remember is that the LDR treatment standard depends on the form of the waste being handled (i.e., wastewater or nonwastewater). Because the form of the waste can change during storage or treatment, it's critical that the treatability group be correctly assessed to avoid meeting the wrong treatment standard.

When a waste is in one treatability group at its point of generation but changes to the other during treatment, the applicable LDR treatment standards are based on the form of the waste being placed on the land—not the one at the point of generation. [August 17, 1988; 53 *FR* 31209, June 1, 1990; 55 *FR* 22537, May 24, 1993; 58 *FR* 29872, May 11, 1999; 64 *FR* 25411, June 19, 2000; 65 *FR* 37944] For example, a D005 barium-containing waste is a wastewater at its point of generation. However, the waste is stabilized to render it noncharacteristic and meet the LDR treatment standard, resulting in nonwastewater treatment residues. Even though the waste was a wastewater at its point of generation, the applicable LDR treatment standard that must be met before the nonwastewater residues may be land disposed is the nonwastewater standard for D005 of 21 mg/L via the TCLP—not the wastewater treatment standard of 1.2 mg/L total analysis. As long as a waste is treated in an appropriate and effective manner (and not just diluted), such a change in treatability group is not a form of impermissible dilution intended to avoid treatment.

13.2 Requirements apply at the point of generation

LDR requirements attach to a hazardous waste at its point of generation (see Section 14.1.5). The practical implication of this arrangement is that after a waste is generated, it must meet treatment standards prior to disposal; the LDR dilution and storage prohibitions are in effect from the point of generation, as are any other requirements imposed by the Part 148/268 regulations. Even if a hazardous waste is rendered nonhazardous subsequent to the point of generation, the treatment standards attached at the

Case Study 13-2: How LDR Treatment Standards Were Developed

Hazardous waste K021—aqueous spent antimony catalyst waste from fluoromethane production contains three hazardous constituents: carbon tetrachloride, chloroform, and antimony. This waste can exist in two forms (treatability groups): wastewater and nonwastewater. For the wastewater form, EPA would have chosen biological treatment as BDAT to remove the carbon tetrachloride and chloroform, followed by chemical precipitation to remove the antimony (refer to Figure 13-1).

In principle, EPA would have treated a representative sample of this K021 wastewater using these two technologies and then analyzed the treated wastewater for the hazardous constituents. The observed concentrations of carbon tetrachloride, chloroform, and antimony in the treated wastewater were 0.057, 0.046, and 1.9 mg/L, respectively. These concentrations became the wastewater treatment standard. Therefore, before a K021 wastewater can be placed in a surface impoundment (a land disposal unit), these concentration limits must be met. Any technology (except impermissible dilution) may be used to meet the treatment standard.

In Figure 13-1, sludge from the biological treatment and chemical precipitation units meets the definition of K021 nonwastewater due to high TOC or TSS concentrations. For this form of the waste, EPA would have chosen a different BDAT technology: incineration to destroy the organics, followed by stabilization of the ash to immobilize the antimony. After running the waste through these processes, EPA determined that the stabilized incinerator ash contained 6.0 mg/kg of carbon tetrachloride and 6.0 mg/kg of chloroform. The leachable concentration of antimony in the stabilized waste via the TCLP was 1.15 mg/L. These concentrations became the treatment standard for nonwastewater forms of the waste. If K021 nonwastewater is to be sent to a landfill (a land disposal unit), it would have to meet this standard prior to disposal. Any technology (except impermissible dilution) can be used to meet the standard.

Note that whenever a hazardous waste is treated, the derived-from rule applies (see Section 5.2). In this example, the K021 waste was a wastewater as originally generated. When it is treated, a sludge could form during biological treatment or chemical precipitation; this sludge would also be K021 and would meet the definition of a "nonwastewater." If the sludge were to be sent to a landfill, it would have to meet the nonwastewater treatment standard. The sludge might meet the treatment standard at the point where it is removed from the wastewater treatment system. If so, it could be land disposed without further treatment. If the K021 sludge didn't meet the nonwastewater treatment standard, it would have to be treated to meet that standard prior to land disposal.

Also note that even after K021 wastes meet the LDR treatment standards, they are still hazardous wastes. Therefore, if the wastes are placed in a land disposal unit, the unit would be a RCRA-regulated Subtitle C hazardous waste unit.

In another example, EPA determined that hazardous waste K116 (organic condensate from the solvent recovery column in the production of toluene diisocyanate via phosgenation of toluenediamine) contains an unanalyzable constituent (phosgene). Because a concentration-based treatment standard cannot be established where it is impossible to analyze for a constituent, EPA established a treatment standard that is a specified method. Wastewater forms of K116 that will be land disposed must first be treated by activated-carbon adsorption or combusted in a RCRA incinerator or a RCRA boiler or industrial furnace (BIF). Nonwastewater forms of K116 must also be combusted in a RCRA incinerator or a BIF. Once the wastes are treated in these units, they automatically meet the treatment standards (i.e., no analytical work is required) and the treatment residues may be disposed in a Subtitle C land disposal unit. [§268.7(a)]

Figure 13-1: How LDR Treatment Standards Were Developed

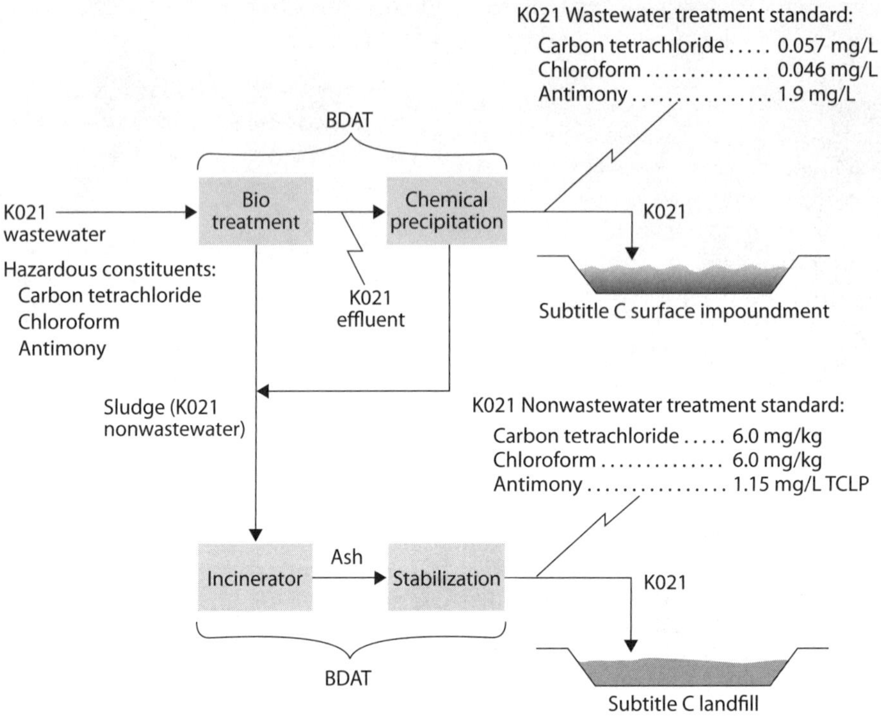

K021 Wastewater treatment standard:
- Carbon tetrachloride ..... 0.057 mg/L
- Chloroform .... 0.046 mg/L
- Antimony .... 1.9 mg/L

K021 wastewater → Bio treatment → Chemical precipitation (BDAT) → K021 → Subtitle C surface impoundment

Hazardous constituents:
- Carbon tetrachloride
- Chloroform
- Antimony

K021 effluent

Sludge (K021 nonwastewater) → Incinerator → Ash → Stabilization (BDAT) → K021 → Subtitle C landfill

K021 Nonwastewater treatment standard:
- Carbon tetrachloride ..... 6.0 mg/kg
- Chloroform .... 6.0 mg/kg
- Antimony .... 1.15 mg/L TCLP

Source: McCoy and Associates, Inc.

point of generation and must be satisfied prior to land disposal. Case Study 13-3 summarizes the point of generation for LDR purposes.

A review of all of the activities that occur at the point of generation may be helpful:

1. The generator of any solid waste is required to determine if that waste is hazardous. Only wastes that are RCRA hazardous at the point of generation are subject to the LDR program. EPA has clarified that generators can determine if their wastes must be treated to meet LDR treatment standards (before they can be land disposed) at the same time as they make hazardous waste determinations. [§268.7(a)(1), April 4, 2006; 71 *FR* 16872]

For example, a generator could make a single request to a laboratory to perform analytical testing to determine whether a sample of its waste is hazardous and if it meets LDR treatment standards for primary waste codes and any underlying hazardous constituents.

2. The generator assigns the appropriate waste code(s) to the waste. EPA has clarified that generators can determine each waste code applicable to a hazardous waste [in order to determine the applicable treatment standard(s)] at the same time as they make hazardous waste determinations. [§268.9(a), April 4, 2006; 71 *FR* 16872] Because the waste coding process has an LDR component, we discuss this topic in detail in Section 13.6.

> **Case Study 13-3: How Is "Point of Generation" Defined Under the LDR Program?**
>
> Surprisingly, the definition of "point of generation" is different for LDR purposes than it is for waste coding purposes (see Section 14.1). For LDR purposes, the point of generation of a hazardous waste is when/where:
>
> - A waste is removed from a manufacturing process unit;
> - A remediation waste is actively managed or removed from an area of contamination (AOC);
> - For listed wastes that are listed due to toxicity, a treatment residue is in a different treatability group than the original waste;
> - For characteristic wastes and wastes listed solely due to ignitability, corrosivity, or reactivity, if the wastes are managed in Clean Water Act (CWA) or CWA-equivalent systems or Class I Safe Drinking Water Act (SDWA) systems and a treatment residue is in a different treatability group than the original waste; or
> - For any characteristic waste, a treatment residue exhibits a new characteristic not exhibited by the original waste.

3. The generator determines if the waste falls into any "subcategories" established under the LDR program. Only a few wastes have subcategories. For example, there are four subcategories for D008 lead-containing wastes: 1) radioactive high-level wastes generated during the reprocessing of fuel rods, 2) radioactive lead solids, 3) lead-acid batteries, and 4) all other wastes that exhibit the toxicity characteristic for lead.

4. The generator classifies the waste according to its treatability group (i.e., wastewater or nonwastewater).

5. From a table of treatment standards (to be discussed later), the generator determines if a specified treatment method (e.g., combustion) applies to the waste. If a specified method is identified, the method attaches to the waste at the point of generation and continues in force until the waste is treated by that method.

    A practical implication of this provision is that mixing different hazardous wastes may have unanticipated consequences. For example, if a waste having only concentration-based treatment standards is mixed with a waste having "combustion" as the specified method, the entire mixture must now be combusted prior to land disposal.

6. The LDR storage prohibition attaches to the waste as discussed in Section 13.10.

7. The dilution prohibition attaches to the waste as discussed in Section 13.11.

8. For some characteristic wastes, underlying hazardous constituents must be identified at the point of generation. This topic will be discussed in the next section.

Point of generation is discussed in detail in Section 14.1 of this book. A few examples will hopefully clarify point of generation for LDR purposes.

Q *How does the change-in-treatability-group principle apply to residues from treating listed wastes?*

A The change-in-treatability-group principle does apply to residues from treating most listed wastes. However, such residues are typically also listed under the derived-from rule and thus remain subject to treatment standards for the new treatability group.

For example, if F001 spent solvent wastewater is combusted, the resulting ash (a newly generated waste) is also F001 under the derived-from rule. The ash must meet the treatment standard appropriate to its treatability group (F001 nonwastewater). [RO 14448]

A complication to the foregoing discussion occurs for wastes that are listed solely because they exhibit the characteristic of ignitability, corrosivity, or reactivity. According to a May 16, 2001 rule [66 *FR* 27266], residues from treating these wastes are not listed if they do not exhibit a characteristic. In the case of F003 spent solvent wastewaters that are combusted (F003 was listed for ignitability only), the ash would not be F003 under the May 16, 2001 rule if it doesn't exhibit any characteristic. However, the preamble to this rule stated: "Wastes that are characteristic at the point of generation and then are subsequently decharacterized are still subject to LDR requirements." [66 *FR* 27269] This implies that the F003-derived ash must also meet the treatment standard appropriate to its treatability group (F003 nonwastewater).

EPA clarified in the same rule (in the same paragraph) that when a waste has been listed solely because it exhibits a characteristic of ignitability, corrosivity, and/or reactivity, and that waste does not exhibit any characteristic at its point of generation, then that waste is not subject to the LDR requirements (because it wasn't hazardous at its point of generation).

The only way to make sense of the foregoing guidance is to conclude that the change-in-treatability-group principle does not apply to ICR-only listed wastes when they are not managed in CWA or CWA-equivalent systems or Class I underground injection wells.

Q *At the point of generation, environmental media (e.g., soil) is determined to be a hazardous waste under the contained-in policy (see Section 5.3). If the media is treated and then determined to no longer contain hazardous waste (i.e., a no-longer-contains determination has been made by EPA or the state), is it still subject to the LDR requirements?*

A Yes. LDR requirements attach at the point of generation. Therefore, even if a no-longer-contains determination is obtained subsequent to that point, the LDR requirements still apply. [RO 11948]

Q *An electroplating facility generates wastewater treatment sludge that meets the F006 listing description. Subsequent to its generation, the facility submits a delisting petition to EPA and is successful in getting the sludge delisted per §260.22. Must the generator comply with the LDR standards before land disposing the delisted waste?*

A Yes. The facility must comply with LDR treatment standards before land disposing the delisted waste, because the standards attached at the point of generation of the hazardous waste. Although a delisted waste may be managed as nonhazardous, a hazardous waste that is generated and subsequently delisted is subject to Part 268 requirements before disposal. Conversely, if the delisting is granted prior to generation, no LDR requirements would apply because the waste is not hazardous at its point of generation. [RO 14699; see also December 1, 2011; 76 *FR* 74714]

Q *A facility operator plans to excavate and dispose soil that he/she knows is contaminated with listed waste due to facility operations that occurred in the 1970s (prior to promulgation of RCRA regulations). Because the concentrations of contaminants are very low, a state issues a no-longer-contains determination for the soil when it is first generated. Do any LDR requirements apply?*

A No. If a no-longer-contains determination is obtained when the soil is first generated, LDRs do not apply because the soil is not hazardous at its point of generation. [§268.49(a)]

Q *A waste is generated in Canada that would be hazardous for lead (D008) in the United States. This waste is decharacterized in Canada and then exported to the United States for disposal. Do any LDR requirements apply?*

A No. If the waste is not hazardous at the point it enters the United States, it is not subject to LDR requirements. [RO 14496]

13.3 Underlying hazardous constituents

When EPA established treatment standards for characteristic wastes in 1990, a new issue arose that has greatly complicated the LDR program: the regulation of "underlying hazardous constituents" (UHCs). In Figure 13-2, we illustrate the origin of this issue. At a chemical plant, a corrosive (D002) wastewater is produced. Because the waste is generated in a chemical manufacturing process, many chemicals are present at low concentrations in the wastewater. None of these chemicals (which the agency later termed "UHCs") is present at a concentration that exceeds a toxicity characteristic level. For example, benzene might be present, but below the characteristic level of 0.5 mg/L TCLP.

The plant treats this wastewater in its wastewater treatment system, which is regulated under the CWA. In actuality, what happens in this system is that the D002 wastewater is mixed with large volumes of other process wastewater in an equalization tank. As a result of this mixing, the corrosivity characteristic is simply diluted away before the combined water enters a bio pond (surface impoundment) that treats the organics in the waste. As long as the wastewater isn't hazardous when it enters the surface impoundment, 1) land disposal of a hazardous waste isn't occurring, and 2) the surface impoundment is not a Subtitle C (hazardous waste) unit. Finally, treated water from the impoundment is discharged under a CWA NPDES permit.

When EPA evaluated this system in the context of the LDR program, it decided that diluting away a characteristic like corrosivity (or ignitability or reactivity) was okay. Additionally, limits in the NPDES permit could be relied upon to protect the environment from toxic constituents. Therefore, EPA initially established a treatment standard of deactivation (DEACT) for D002 wastes. This standard specifically allowed for dilution as described above.

When this standard was challenged in court, it was overturned because the court found that DEACT did not meet the statutory requirement that a treatment standard must "minimize threats to human health and the environment." Specifically, constituents that enter the impoundment may (as shown in Figure 13-2) be vaporized to the air, may leak into ground water because the impoundment is not required to be lined, and may concentrate in sludge that may ultimately be removed and land disposed. The court remanded the DEACT standard to the agency, with instructions that these constituents must be considered.

Figure 13-2: Hypothetical Wastewater Treatment System at a Chemical Plant

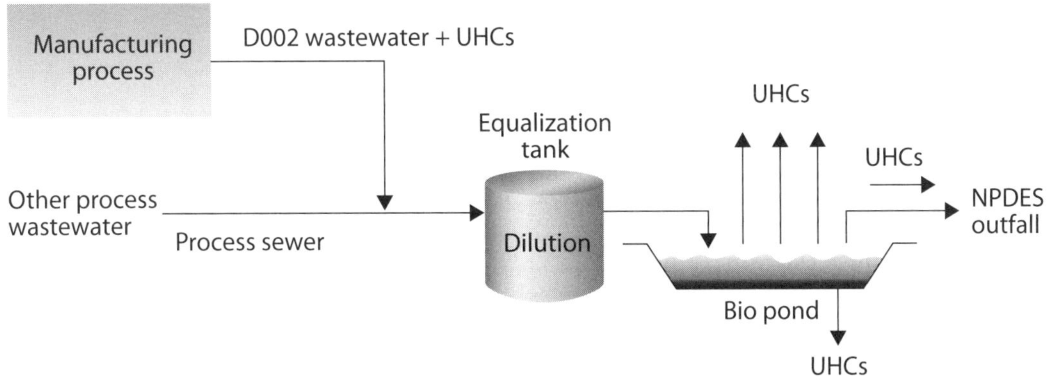

UHCs = underlying hazardous constituents.
Source: McCoy and Associates, Inc.

Ultimately, as will be discussed in the next section, the agency established treatment standards for these constituents in certain characteristic wastes. They also established the following definition in §268.2:

> "*Underlying hazardous constituent* means any constituent listed in §268.48, Table UTS—Universal Treatment Standards, except fluoride, selenium, sulfides, vanadium, and zinc, which can reasonably be expected to be present at the point of generation of the hazardous waste at a concentration above the constituent-specific UTS treatment standards."

This definition cites a new table that we have not yet discussed—the universal treatment standards (UTS) table in §268.48. The origin of this table is simple: it identifies every chemical that has a concentration-based LDR treatment standard. Notice that the table starts with a list of organic constituents and ends with a list of inorganic constituents. The term "universal" refers to the fact that where treatment standards are given for a specific chemical (e.g., benzene) those standards will apply to every waste in which that chemical is a constituent of concern (e.g., for benzene: D018, F005, K048, U019, etc.).

When a treatment standard refers to "and meet §268.48 standards," this is EPA's way of requiring that we identify and treat UHCs in our waste. Via testing or knowledge, we consider the chemicals in the UTS table. If we believe that one (or more) of these chemicals is present in our waste at the point of generation at a concentration above that given in the UTS table, it (or they) is a UHC in our waste.

In the above definition, the language "reasonably expected to be present" often causes concerns. What does this mean? When EPA first discussed this concept, the agency noted that generators "may base this determination on their knowledge of the raw materials they use, the process they operate, and the potential reaction products of the process, or upon the results of a one-time analysis for the entire list of constituents at §268.48." [September 19, 1994; 59 *FR* 48015]

In one guidance document [RO 13748], EPA stated that where no institutional knowledge exists concerning waste contaminants (e.g., for soil contaminated by a previous party), analysis should be conducted for the entire list of §268.48 constituents. Although this would certainly be a bulletproof approach to identifying UHCs, it doesn't reflect what we observe as actual practice. The problem with this approach is the extremely high analytical cost associated with analyzing a waste for all §268.48 constituents (anecdotally reported to us as being in the range of $3,000 to $5,000 per sample).

In other guidance [RO 14325], EPA states that the generator "may use testing or knowledge to ascertain if [UHCs] are present at levels above UTS. Where this determination is based on testing, any of the constituents not shown to be below UTS and not otherwise known to be below UTS should be listed as a 'constituent of concern' [on LDR paperwork]."

In our experience and in discussions with regulators, we find that most generators rely on analytical results that have been collected for other purposes (e.g., for waste identification) and on process knowledge to identify UHCs in their wastes. The regulators don't tend to second-guess the generators' conclusions, unless they (the regulators) have specific knowledge that the generator has overlooked. For example, if everyone in a specific industry identifies benzene as a UHC in a particular waste and you don't, expect the regulators to question your reasoning. We are not aware of any instance where regulators have analyzed a waste for all 200+ constituents in the UTS table and then cited a generator for not having identified an obscure constituent. (Such an enforcement approach has two problems: 1) it costs the agency $3,000 to $5,000 to test a single sample of the waste, and 2) the definition of UHC specifically says "can reasonably be expected to be present" not "must be tested for.")

Q *I have a solid waste which is not hazardous at the point of generation, but it does contain phenol above the §268.48 UTS concentration. Do I have to worry about UHCs in my waste?*

A No. Underlying hazardous constituents only apply to your waste if it is hazardous at the point of generation. [RO 14216]

Q *If a waste is toxic due to lead (thus carrying the D008 code), should lead also be identified as a UHC on the LDR paperwork?*

A If one simply follows the plain language of the UHC definition in the regs, then the answer would be "Yes." The constituent causing the waste to exhibit a characteristic can be a UHC if its concentration is above the UTS level in §268.48. However, two documents clarify EPA's intent. First, in preamble language, the agency described UHCs as "hazardous constituents other than those for which the waste exhibits the characteristic." [September 14, 1993; 58 FR 48115] The second document describes UHCs as constituents that "are not what causes the waste to exhibit a characteristic, but they can pose hazards nonetheless." [*RCRA Orientation Manual*, 2011 edition, available at http://www.epa.gov/osw/inforesources/pubs/orientat/] From reviewing the two documents identified above, a contaminant causing a waste to exhibit a toxicity characteristic would not need to be identified as a UHC.

Q *The definition of UHC (and Footnote 5 in the §268.48 UTS table) specifies that fluoride, sulfide, vanadium, and zinc are not "underlying hazardous constituents." Why not?*

A According to EPA, these chemicals are not UHCs because they are not "hazardous constituents" based on the definition in §268.2(b) (i.e., they do not appear in Part 261, Appendix VIII). [RO 14514] Nevertheless, the agency has set treatment standards for fluoride, sulfide, vanadium, and zinc in certain listed wastes, which is why these chemicals are included in the UTS table. For example, the treatment standard for K088 has a fluoride standard; D003 has a reactive sulfide standard; K171–K172 have vanadium standards; and K061 has a zinc standard.

13.3.1 Wastes for which UHCs have to be identified

Per §268.9(a), EPA requires generators of characteristic wastes (except for high-TOC D001 wastes treated by CMBST, RORGS, or POLYM) to determine the UHCs in their waste. This requirement is eliminated in one situation: "generators with decharacterized wastewaters that are being managed in CWA or CWA-equivalent systems or injected into Class I nonhazardous injection wells do not have to identify underlying hazardous constituents." [April 8, 1996; 61 FR 15661]

The implication is that UHCs must be identified by the generator in all other situations where characteristic wastes are generated. That is, UHCs will have to be identified for all characteristic wastes that are not managed in CWA/CWA-equivalent systems or Class I injection wells unless the wastes are high-TOC D001 that will be treated by CMBST, RORGS, or POLYM. This is true even if "and meet §268.48 standards" is not part of the treatment standard in §268.40. Whether or not these identified UHCs must be put on LDR paperwork depends on whether the TSD facility will monitor and treat for all of them [Item 3 in the §268.7(a)—Generator Paperwork Requirements Table].

Section 268.40(e) indicates that characteristic wastes that are not managed in CWA/CWA-equivalent systems or Class I injection wells must be treated to meet UTS for UHCs, regardless of whether treatment of UHCs is specified for that waste in the §268.40 table of treatment standards. However, EPA clarified in preamble language that "[t]he universal treatment standards apply to all…characteristic wastes for which treatment standards have been promulgated, with two exceptions…. The second exception is for those wastes for which the treatment standard is a specified method of treatment. Most of these wastes must continue to be treated using those required technologies." [September 19, 1994; 59 FR 47988] For example, D008 radioactive lead solid nonwastewaters have to be treated using macroencapsulation (which has the five-letter code of MACRO in the §268.40 table of treatment standards), but UHCs in these wastes do not need to be treated to below UTS levels prior to land disposal. EPA also clarified that the concentration-based treatment standards for D003 reactive cyanides in §268.40 do not require treatment of UHCs. [April 8, 1996; 61 FR 15568]

Based on application of §§268.7(a), 268.9(a), and 268.40(e), we're left with a situation where generators have to identify UHCs and include them on LDR notifications even for wastes that don't require their treatment (e.g., for D008 radioactive lead solid nonwastewaters that will be treated by macroencapsulation). This doesn't seem to make sense, but we haven't seen any guidance providing relief from this requirement.

13.4 Treatment standards for characteristic wastes

The key to understanding the treatment standards for characteristic wastes (or any waste for that matter) is knowing how to interpret the treatment standards table in §268.40. All hazardous wastes are subject to the treatment standards given in this table; however, alternative treatment standards are available for the following three classes of wastes:

1. Hazardous soil may be treated by the alternative standards of §268.49 (refer to Section 13.7),

2. Hazardous debris may be treated by the alternative standards of §268.45 (refer to Section 13.8), and

3. Lab packs may be treated by the alternative standards of §268.42(c) (refer to Section 13.13.1).

Before examining the treatment standards, some additional terminology is helpful in understanding the LDR regulations. When the agency uses the term "restricted" wastes, it is referring to hazardous wastes that are subject to the LDR program; all hazardous wastes currently fit into this category. When it uses the term "prohibited" wastes, it is referring to hazardous wastes having treatment standards that are in effect. The treatment standards for all hazardous wastes are currently in effect. Therefore, no distinction currently exists between "restricted" and "prohibited" wastes; every hazardous waste is both restricted and prohibited.

When EPA identifies new hazardous wastes via new listings, the wastes are typically subject to LDR treatment standards; hence, they are "restricted." However, the effective date of the treatment standards for some forms of the waste (e.g., radioactive variants) may be delayed. Under these circumstances, the radioactive variants would be "restricted" but not "prohibited."

13.4.1 Understanding the table of treatment standards

Figure 13-3 shows the start of the table of treatment standards as given in §268.40. Points of interest in this table are described below.

❶ This column contains the waste codes. You must know the waste code(s) for your waste to use this table. Section 13.6 explains what you need to know about waste coding for these purposes.

❷ This column either specifies the characteristic or presents the listing description associated with the waste code. It also identifies any subcategories of the waste. Note that two subcategories are identified for D001 wastes: high TOC (the second entry) and all other ignitables (the first entry). Be very careful to select the right subcategory for your wastes when using this column. As you can see from the two right-hand columns, the treatment standards for the two D001 subcategories are different. Complying with the wrong treatment standard can involve significant enforcement penalties.

❸ If a concentration-based treatment standard is established for the waste, this column will identify the constituents subject to treatment and will also give their Chemical Abstract Service (CAS) number. For example, D001 wastes are regulated due to their flash point, not because they contain any specific

Figure 13-3: Key to Understanding the §268.40 Table of Treatment Standards

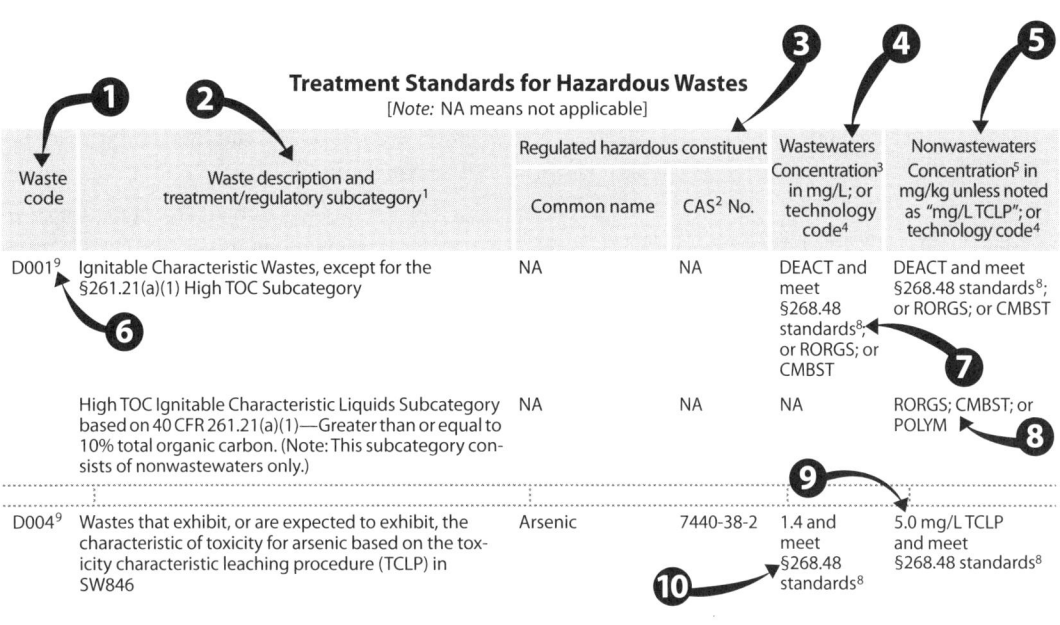

Source: McCoy and Associates, Inc.; adapted from §268.40.

chemical; therefore, no chemicals are identified in column ❸. On the other hand, D004 wastes are hazardous because they contain arsenic, which is identified along with its CAS number in column ❸.

❹ If the waste being handled is a wastewater, the applicable treatment standard will appear in this column. (Refer to the discussion on treatability groups in Section 13.1.5.1 for a definition of "wastewater" and "nonwastewater.") Footnote 3 at the head of this column essentially says that when the agency tests wastewaters to determine if they meet concentration-based treatment standards, they will collect and test a composite sample (rather than a grab sample) of the wastewater. Also note that any concentrations given in this column are based on an analysis of the total concentration of a constituent in the waste (not an analysis of the leachable concentration of the constituent). If the treatment standard is a specified method (as evidenced by a five-letter code such as CMBST), Footnote 4 directs you to definitions of these specified methods in Table 1 of §268.42.

❺ If the waste being handled is a nonwastewater, the applicable treatment standard will appear in this column. Footnote 5 at the head of this column essentially says that when the agency tests nonwastewaters to determine if they meet concentration-based treatment standards, they will collect and test a grab sample (rather than a composite sample) of the nonwastewater. Where concentration values are given in this column (refer to the standard for D004), they may be based on two different analytical methods. If the standard says "mg/L TCLP," the concentration is measured in a leachate from the waste obtained using the TCLP. If the standard does not say "mg/L TCLP," the treatment standard refers to the total concentration of the constituent in the waste, not the leachable concentration. Concentration standards for organics will almost always be total concentration (mg/kg), but concentrations for metals will always be mg/L TCLP. Again, if the treatment standard is a specified method (as evidenced by a five-letter code such as CMBST), Footnote 4 directs you to definitions of these specified methods in Table 1 of §268.42.

❻ Footnote 9, which appears on every characteristic waste code except radioactive high-level waste variants, has major implications for wastes that are to be disposed by underground injection. In essence, this footnote says that if a waste is decharacterized, it may be disposed in a Class I underground injection well regulated under the SDWA without having to comply with the LDR treatment standards. The basis for this footnote is §268.1(c)(3), which reads:

"(3) Wastes that are hazardous only because they exhibit a hazardous characteristic, and which are otherwise prohibited under this part, or part 148 of this chapter, are not prohibited if the wastes:

"(i) Are disposed into a nonhazardous or hazardous injection well as defined under 40 *CFR* 146.6(a) [we believe the correct citation is §146.5(a)]; and

"(ii) Do not exhibit any characteristic of hazardous waste identified in 40 *CFR* Part 261, Subpart C at the point of injection."

The statutory basis for Footnote 9 (as well as Footnote 8) is the Land Disposal Program Flexibility Act of 1996 (Public Law 104-119).

Although not specifically stated in the regulations, a characteristic may be removed by dilution through aggregation of different waste streams prior to underground injection. [April 8, 1996; 61 *FR* 15661, RO 14547]

❼ Footnote 8, which appears with the treatment standards for most characteristic wastes, is perhaps the most important footnote in the table of treatment standards. It is based on §268.1(c)(4), which reads:

"(4) Wastes that are hazardous only because they exhibit a hazardous characteristic, and which are otherwise prohibited under this part, are not prohibited if the wastes meet any of the following criteria, unless the wastes are subject to a specified method of treatment other than DEACT in §268.40, or are D003 reactive cyanide:

"(i) The wastes are managed in a treatment system which subsequently discharges to waters of the U.S. pursuant to a permit issued under Section 402 of the Clean Water Act [i.e., NPDES-permitted discharges]; or

"(ii) The wastes are treated for purposes of the pretreatment requirements of Section 307 of the Clean Water Act [i.e., discharges to POTWs]; or

"(iii) The wastes are managed in a zero discharge system engaged in Clean Water Act-equivalent treatment as defined in §268.37(a); and

"(iv) The wastes no longer exhibit a prohibited characteristic at the point of land disposal (i.e., placement in a surface impoundment)."

In other words, when most characteristic hazardous wastes are treated in one of the systems identified in §268.1(c)(4) cited above, they only have to be decharacterized before they enter a land disposal unit (i.e., a surface impoundment). Under these conditions, the treatment standards of §268.40 do not apply. [RO 14547] Note that this "exemption" does not apply to any of the following wastes:

- D001—High-TOC ignitables,
- Any of the characteristic radioactive high-level waste variants,
- D003—Reactive cyanides,
- D006—Cadmium-containing batteries,
- D006—Radioactive cadmium-containing batteries,
- D008—Lead-acid batteries,
- D008—Radioactive lead solids,
- D009—High-mercury organics,
- D009—High-mercury inorganics,
- D009—Elemental mercury contaminated with radioactive materials,
- D009—Hydraulic oil contaminated with mercury radioactive materials,
- D009—Radioactive mercury-containing batteries,

- D011—Radioactive silver-containing batteries, or
- D012–D017—Pesticide wastewaters.

Zero-discharge systems described in §268.1(c)(4)(iii) above are termed "CWA-equivalent systems," and are defined in §268.37(a) as follows:

> "CWA-equivalent treatment means biological treatment for organics, alkaline chlorination or ferrous sulfate precipitation for cyanide, precipitation/sedimentation for metals, reduction of hexavalent chromium, or other treatment technology that can be demonstrated to perform equally or greater than these technologies."

❽ This is an example of a treatment standard that is a specified method (RORGS, CMBST, or POLYM). Definitions for each of these methods can be found in §268.42, Table 1. One of these methods must be used before the waste may be land disposed.

❾ This is an example of a concentration-based treatment standard (5.0 mg/L TCLP). A waste may be treated by any method (except impermissible dilution) to achieve this standard.

❿ The language "and meet §268.48 standards" in this table, is EPA's shorthand method of saying that the waste must be treated for any UHCs that might be present in the waste. (Refer to Section 13.3 for a discussion of this topic.)

Figure 13-4 summarizes how the §268.40 table of treatment standards should be used and interpreted. The information that follows provides an explanation of how the LDR treatment standards apply to specific characteristic wastes. Useful examples are also presented.

13.4.2 D001 ignitable wastes

In the table of treatment standards, D001 wastes are divided into two subcategories: high-TOC ignitable liquids (the second entry in the table), and all other ignitable wastes. The high-TOC subcategory applies to ignitable liquid wastes containing greater than or equal to 10% TOC; hence, all wastes in this subcategory are nonwastewaters. When these wastes or residues from treating these wastes are destined for land disposal, they must be: 1) treated to recover the organics (RORGS), 2) combusted (CMBST), or 3) polymerized (POLYM).

Table 1 of §268.42 requires that high-TOC ignitables managed per RORGS be recycled in a recovery unit of some kind, such as in a distillation or liquid-liquid extraction column. If CMBST is chosen, the ignitable waste would have to be burned in a RCRA-permitted or interim status incinerator or BIF. If and only if the waste is a monomer that originates in the plastics industry, POLYM can be used to form high-molecular weight solids that may be disposed. Regardless of whether RORGS, CMBST, or POLYM is chosen, UHCs would not have to be identified or treated. [§268.9(a)]

All D001 wastes other than high-TOC ignitable liquids have a treatment standard of deactivate (DEACT) and meet §268.48 standards, or RORGS, or CMBST. The RORGS and CMBST treatment standards would be the same as discussed above for high-TOC D001 wastes. The DEACT and meet §268.48 standards option requires that the D001 waste be treated without dilution to render it nonignitable, and all UHCs must additionally be treated to meet UTS standards in §268.48.

Footnote 9 applies to both subcategories of D001, while Footnote 8 applies only to the low-TOC ignitables. Many of the nuances associated with these D001 treatment standards are illustrated in Case Study 13-4 and the following examples.

Q *A facility generates waste paint that is toxic for cadmium (i.e., D006) in addition to being a high-TOC D001 waste. Will UHCs need to be identified on the LDR paperwork when the waste is shipped offsite for incineration?*

A The wording of §268.9(a) seems to exempt the waste from UHC determination, because the waste is a high-TOC ignitable liquid treated by CMBST. EPA notes that if the waste in question is characterized only as high-TOC D001 (and does not exhibit any other characteristic) and is treated by

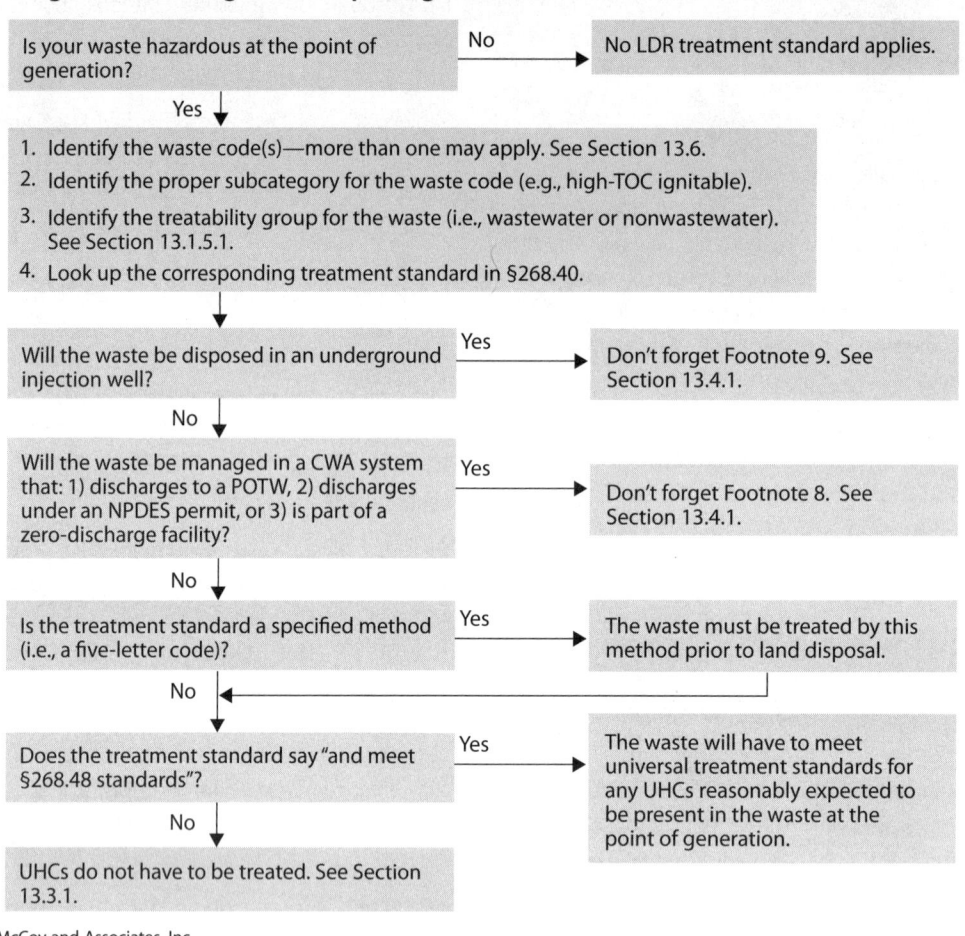

Figure 13-4: Using and Interpreting the §268.40 Table of Treatment Standards

Source: McCoy and Associates, Inc.

CMBST, RORGS, or POLYM, then UHCs do not need to be identified or treated. However, if the high-TOC D001 waste also carries other characteristic codes (e.g., D006), UHCs would have to be identified and treated. [March 31, 2009 email from EPA to McCoy and Associates, Inc.]

Q *A D001 high-TOC waste contains both solids and liquids. The treatment standard for this waste is RORGS, CMBST, or POLYM. If the solids are removed by filtration or decanting and do not exhibit any characteristics, are they still subject to LDR requirements?*

A EPA originally answered this question in 1990, stating that the solids are not subject to LDR standards. The agency considered the act of removing solids from an ignitable liquid to be an acceptable form of pretreatment that can be considered part of a "recovery of organics" treatment method. As long as the liquid phase is treated by one of the specified methods, the nonhazardous solids pass from Subtitle C to Subtitle D regulation. No LDR treatment standards or paperwork attach to the solids. LDR notification/certification requirements continue to apply to the liquid phase. [RO 13404]

Today we are a little uneasy with this 1990 answer because it predates the principle established in 1993 (58 *FR* 29872) that wastes that exhibit a characteristic at the point of generation remain subject to the land disposal restrictions even if they are subsequently decharacterized. (As discussed previously, this principle does not apply to most characteristic wastes when they are managed in CWA treatment systems or injected into Class I SDWA wells.)

Case Study 13-4: Treatment Standards for D001 Wastes

I. D001 high-TOC wastes are introduced into a tank-based wastewater treatment system (no ponds) where the resulting effluent is either discharged under an NPDES permit or to a POTW—see Figure 13-5. What LDR standards apply?

As long as the wastewater treatment system does not contain a land-based unit (i.e., a surface impoundment), the LDR requirements do not apply. In other words, the treatment standard of RORGS, CMBST, or POLYM does not attach to the waste because it is not being land disposed. Similarly, because no land disposal occurs in an entirely tank-based system, the LDR dilution prohibition does not apply. Hence, there are no LDR regulations that specifically prohibit the management of D001 high-TOC wastes in such a system. [RO 14214]

With regard to any sludge generated in the tanks, EPA issued some creative guidance. Even though the D001 high-TOC waste is a nonwastewater, it is converted to a wastewater when it mixes with other wastewaters in the treatment system. Any sludge that is generated is a nonwastewater; hence, EPA's "change-in-treatability-group principle" applies. This principle states that for characteristic wastes managed in Clean Water Act systems, each change of treatability group in the treatment train marks a new point of generation for determining if a characteristic waste is prohibited from land disposal. The sludge is, therefore, a newly generated waste, and if it does not exhibit any characteristics, it is not subject to any LDR standard. [May 11, 1999; 64 *FR* 25408, RO 14718]

Finally, the generator of the waste is subject to one LDR paperwork requirement: the one-time notice of §268.7(a)(7), which is discussed in Section 13.12.1.1 of this chapter.

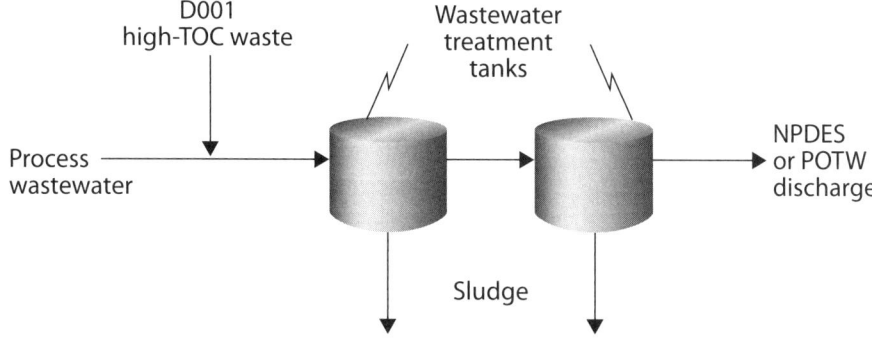

Figure 13-5: Treatment of D001 High-TOC Wastes in a CWA System Without a Land Disposal Unit

Source: McCoy and Associates, Inc.

Case Study 13-4 (Continued): Treatment Standards for D001 Wastes

II. As shown in Figure 13-6, the same D001 high-TOC waste as in the previous example is introduced into a wastewater treatment system that includes a surface impoundment. Because the surface impoundment is a land disposal unit, the D001 waste must meet treatment standards prior to entering the impoundment. Because the waste has a treatment standard that is a specified method (RORGS, CMBST, or POLYM), and because this treatment is not being provided prior to land disposal, this would be illegal disposal of the high-TOC waste.

Note that because the treatment standard for D001 high-TOC wastes does not contain a reference to "§268.48 treatment standards," UHCs are not of concern.

III. Figure 13-6 also shows D001 low-TOC waste being discharged to the same wastewater treatment system. The treatment standard for this subcategory is "DEACT and meet §268.48 treatment standards[8], or RORGS, or CMBST." The DEACT (deactivation) treatment standard appears for a number of D001, D002, and D003 wastes, and essentially means "treat the waste to remove the characteristic," in this case, ignitability. Appendix VI to Part 268 recommends technologies that can be used to deactivate various wastes; however, use of these technologies is not mandatory. Also note that dilution in a CWA system is an acceptable method of deactivation [RO 14547]—see §268.3(b) and Section 13.11 for a discussion of dilution.

Therefore, in this example, as long as the wastewater entering the surface impoundment is not ignitable, we have deactivated the waste. It is very important to note, however, that Footnote 8 appears on the treatment standard "DEACT and meet §268.48 standards." Because this is a CWA system (it involves an NPDES discharge), Footnote 8 specifies that the waste only has to be rendered nonhazardous and then no treatment standards apply. Because our waste has been rendered nonhazardous (i.e., deactivated via dilution or treatment in tanks), the requirement to treat underlying hazardous constituents does not apply. [April 8, 1996; 61 FR 15661] Therefore, the operator of this CWA system does not have to monitor the inlet to the impoundment to assure that UHCs are below §268.48 levels.

Figure 13-6: Treatment of D001 Wastes in a CWA System With a Land Disposal Unit

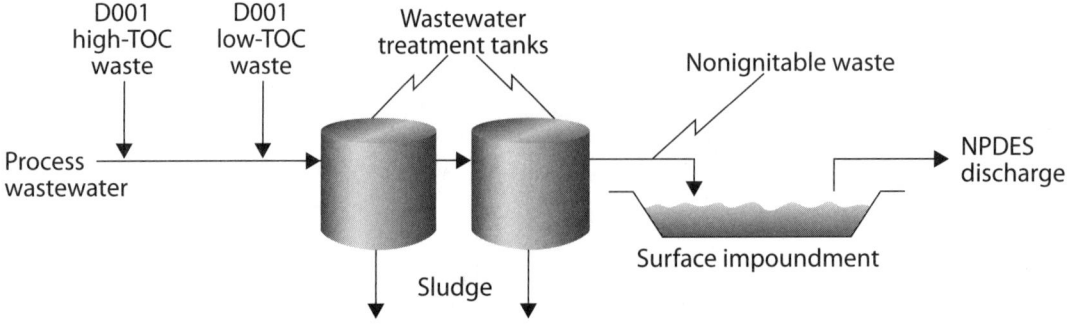

Source: McCoy and Associates, Inc.

Case Study 13-4 (Continued): Treatment Standards for D001 Wastes

IV. Figure 13-7 shows D001 wastes being mixed (diluted) in a tank such that they are no longer ignitable prior to being injected in a Class I injection well. Note that Footnote 9 appears on the D001 waste code in the §268.40 table of treatment standards. This footnote says that as long as the waste is rendered nonhazardous prior to land disposal, no treatment standards apply. Hence, this is legal disposal of both high- and low-TOC D001 wastes. Refer to Section 13.11 and RO 14547 for a discussion of dilution in this type of system.

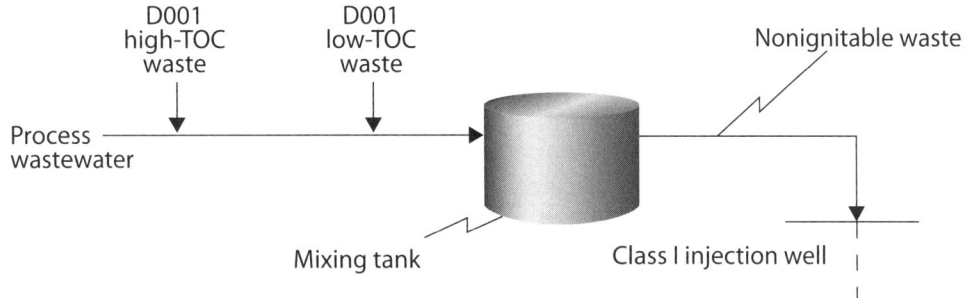

Figure 13-7: Disposal of D001 Wastes in a Class I Injection Well

Source: McCoy and Associates, Inc.

The only way we can rationalize EPA's 1990 answer as still being valid is if a case can be made that the filtration/decantation step is actually a well designed, well operated part of a "recovery of organics" treatment method. For a more recent interpretation (October 1994) on a similar issue, refer to the following question.

Q: *A waste is received at a TSD facility that is bilayered and carries the waste codes D001—high-TOC subcategory and D002. Can the waste be phase-separated such that the upper layer (D001) is treated by combustion and the lower layer (D002) is sent to wastewater treatment?*

A: According to EPA, the waste can be treated as separate hazardous wastes: "The phase separation is considered an appropriate pretreatment step for these wastes. Therefore, the D001 high-TOC portion can be treated to meet its LDR treatment standards, and the other phase can be sent to wastewater treatment, provided it is not an ignitable waste containing greater than 10% TOC." [RO 13704] Note that EPA later determined that if the wastewater treatment system contains no land-based units, it can also be used to treat D001 high-TOC wastes. [RO 14214]

Q: *When I read the definition of CMBST in Table 1 of §268.42, it says that hazardous waste incinerators and BIFs meet the definition. However, the definition then identifies "other units operated in accordance with applicable technical operating requirements." What does this mean?*

A: EPA believes that certain thermal units other than incinerators and BIFs are suitable for treating organic-containing hazardous wastes. They therefore included general language in the CMBST definition to make this position clear. For example, when a company inquired about whether a gasification and vitrification unit met the definition of CMBST, EPA stated: "[this technology is] achieving two measures that describe parameters which are characteristic of good combustion conditions for the destruction of organic compounds in this type of unit: 1) temperatures in the process chamber consistently are in the range of 2,000°F; and 2) temperatures in the air pollution dry particulate control devices are maintained outside the temperature window optimum for surface-catalyzed dioxin/furan formation (450°F to 750°F). Furthermore, the measured destruction and removal efficiencies...exceed 99.9999%." [RO 14535]

A determination as to whether a thermal treatment technology will meet the definition of CMBST will have to be made on a case-by-case basis. One way to initiate such a determination is to submit an application for approval of an alternative treatment method as described in §268.42(b). See Section 13.13.4.3.

13.4.3 D002 corrosive wastes

The treatment standard for D002 wastewaters and nonwastewaters is "DEACT and meet §268.48 treatment standards[8]." Both Footnotes 8 and 9 apply to these wastes in the §268.40 table of treatment standards; hence, these wastes only have to be rendered nonhazardous when treated in a CWA/CWA-equivalent system with a land disposal unit or when disposed in a Class I injection well. Much of the discussion in the previous section dealing with D001 wastes also applies to D002 wastes.

Note that corrosive radioactive high-level wastes generated during the processing of fuel rods have a specified method of vitrification (HLVIT).

The following examples and the ones in Case Study 13-5 illustrate the management of D002 wastes under the LDR program.

Q *If D002 corrosive waste is treated to remove the characteristic, can the treated waste be used as a dust suppressant?*

A Yes. But only if the treated waste meets LDR treatment standards per §266.20(b). Not only must the waste no longer exhibit the corrosivity characteristic, but it must also meet treatment standards for UHCs. Note that the prohibition on the use of hazardous wastes in dust suppressants codified in §266.23(b) does not apply in this case because the decharacterized waste is no longer hazardous. [RO 14315]

Q *Sulfuric acid produced in an air emission control system at a primary lead smelter is used as an ingredient in the manufacture of fertilizer and pesticides. The sulfuric acid may contain heavy metals, including lead, mercury, and selenium. Do any LDR requirements apply to this activity?*

A The sulfuric acid is considered a "sludge" because it is a residue from a pollution control device. According to Table 1 in §261.2(c), characteristic sludges that are used in a manner constituting disposal (fertilizer and pesticides are typically applied to the land) are solid wastes. If the sulfuric acid exhibits the corrosivity characteristic and/or is characteristic due to heavy metal content, it is a hazardous waste, and when used in a manner constituting disposal, must comply with the requirements of §266.20(b). This section specifies that the products (fertilizer and pesticides) would have to meet LDR treatment standards. Thus the treatment standard for any characteristic exhibited by the sulfuric acid would have to be met, along with the treatment standard(s) for any UHCs. [RO 14437, 14566]

13.4.4 D003 reactive wastes

Reactive wastes are divided into a number of subcategories. Only reactive cyanides have a concentration-based treatment standard; all other reactive wastes have a treatment standard of DEACT, some with a requirement to treat UHCs, some without. Note that Footnote 9 allows all D003 wastes to simply be deactivated prior to injection in a Class I well, while Footnote 8 allows explosives, water reactives, and other reactives to simply be decharacterized in CWA/CWA-equivalent systems (i.e., UHCs do not have to be treated

> **Case Study 13-5: Treatment Standards for D002 Wastes**
>
> I. As shown in Figure 13-8, corrosive D002 wastewater is used to neutralize other nonhazardous wastewaters in a neutralization tank (i.e., a wastewater treatment unit), followed by treatment in a surface impoundment. Treated wastewater is discharged to a sewer leading to a POTW; nonhazardous sludge generated in the neutralization tank is disposed in a sanitary landfill. What LDR requirements apply?
>
> Because the D002 waste will be decharacterized in a tank that is part of a CWA system, per Footnote 8, no LDR treatment standards apply to the D002 wastewater. It is also unnecessary to identify underlying hazardous constituents in the D002 waste under these circumstances. [April 8, 1996; 61 *FR* 15661]
>
> Under the change-in-treatability-group principle, the sludge is a newly generated waste. Because it is nonhazardous when removed from the wastewater treatment unit, no LDR requirements attach to this waste. (It is also not necessary to identify UHCs in this waste.)
>
> A one-time notice as specified in §268.7(a)(7) must be placed in the generator's files. The notice must state that a D002 waste was generated, that the waste was subsequently excluded from regulation under the domestic sewage exclusion of §261.4(a)(1), and that the waste was discharged to a sewer leading to a POTW. [RO 14216]
>
> II. Spent acid from an industrial operation contains low levels of heavy metals (cadmium, lead, and zinc). The concentrations of these metals in the acid (which is a wastewater) are: cadmium – 0.10 mg/L, lead – 2.0 mg/L, and zinc – 10.0 mg/L. The facility wants to treat the acid with lime to remove the corrosivity characteristic and then dispose the waste in a nonhazardous waste landfill. What LDR standards apply?
>
> Because a treatment residue from the D002 waste will be land disposed, the treatment standard "DEACT and meet §268.48 standards" applies. Neither Footnotes 8 nor 9 have any bearing in this case, because we aren't involved with treatment in a CWA system or disposal in an injection well. Because UHCs are subject to treatment, the heavy metal concentrations in the original waste are compared against the wastewater values in §268.48, with the following results:
>
> 1) cadmium is below the UTS concentration of 0.69 mg/L and is not a UHC,
>
> 2) lead exceeds the UTS concentration of 0.69 mg/L and is therefore a UHC,
>
> 3) zinc is never a UHC (see Footnote 5 in the UTS table of §268.48).
>
> Before the treated waste can be sent to the Subtitle D landfill, it must be noncorrosive and must meet the treatment standard of 0.75 mg/L TCLP for lead. This is a situation where the treatment residue is in a different treatability group than the original waste. The LDR standard that must be met is for the treatability group of the waste that is land disposed (in this example, nonwastewater). Paperwork as described in §268.9(d) must be kept in the facility's files.

under these circumstances). [April 8, 1996; 61 *FR* 15661]

13.4.5 TC wastes (heavy metals, pesticides, and organics)

For the most part, the treatment standards for the D004–D043 toxicity characteristic (TC) wastes are very similar to those discussed previously for D001–D003:

- Footnote 9 allows all D004–D043 wastes to simply be deactivated (without treatment for UHCs) prior to injection in a Class I well [April 8, 1996; 61 *FR* 15661];

Figure 13-8: Treatment of D002 Wastes in a CWA System

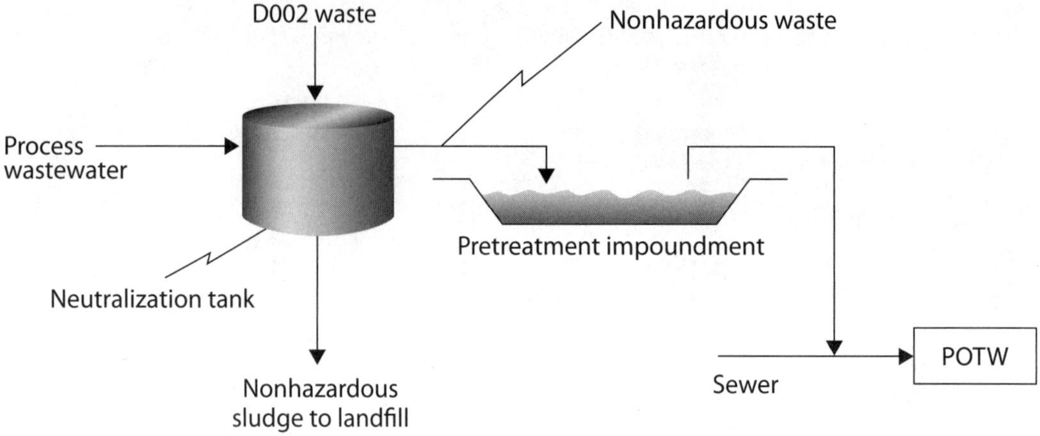

Flow: D002 waste and Process wastewater enter Neutralization tank → Nonhazardous sludge to landfill. Nonhazardous waste goes to Pretreatment impoundment → Sewer → POTW.

Source: McCoy and Associates, Inc.

- Footnote 8 allows wastes with concentration-based treatment standards to simply be decharacterized in CWA/CWA-equivalent systems (i.e., UHCs do not have to be treated under these circumstances); and
- Certain special wastes have specified treatment methods, including D004–D011 radioactive high-level wastes, cadmium-containing batteries, lead-acid batteries, radioactive lead solids, high-mercury and radioactive-mercury wastes, and pesticide wastewaters.

Examples of the applicability of LDR treatment standards to TC wastes are given in Case Study 13-6 and as follows.

Q: Zinc-carbon batteries are found to contain leachable levels of cadmium greater than 1.0 mg/L via the TCLP. What LDR treatment standard applies to the batteries?

A: Assuming that the batteries are D006 wastes when disposed, they would be subject to the D006 nonwastewater treatment standard. Even though a D006 cadmium-containing batteries subcategory appears in the §268.40 table of treatment standards, this subcategory does not apply to zinc-carbon batteries. EPA's BDAT document for D006 identifies three types of batteries subject to the cadmium battery subcategory: cadmium-nickel, cadmium-mercury, and cadmium-silver. [RO 13608] Note that the batteries could be recycled for their zinc content and might qualify for the scrap metal exemption—see Sections 1.2.5.5 and 4.21.

Q: A mercury waste as generated fits into the D009 high-mercury organic subcategory (i.e., the mercury concentration is ≥260 mg/kg total mercury and the waste contains organics). The treatment standard for this waste is IMERC (incineration) or RMERC (retorting to recovery mercury). After the waste is treated for another constituent, the waste fits into the low-mercury subcategory. The treatment standard for low-mercury wastes is concentration based (0.025 mg/L TCLP and meet §268.48 treatment standards). Can the partially treated waste now be treated to meet the low-mercury treatment standard?

A: No. The treatment standard for high-mercury wastes attaches at the point of generation and must be met at the point of disposal. [RO 14513]

Case Study 13-6: Treatment Standards for TC Wastes

I. Wastewater that is contaminated with low concentrations of gasoline exhibits the toxicity characteristic for benzene (D018). As shown in Figure 13-9, the wastewater is treated in a CWA system consisting of biological treatment in a surface impoundment. What treatment standards apply?

The treatment standard for D018 wastewater is "0.14 and meet §268.48 standards[8]." Because of the presence of Footnote 8, the wastewater in this CWA system must simply be rendered nonhazardous (dilution from other plant wastewater is acceptable) prior to entering the surface impoundment (i.e., the benzene concentration must be reduced to less than 0.5 mg/L TCLP). (This is the concentration that makes a waste hazardous for benzene—see §261.24, Table 1.) Under these circumstances, no need exists to identify or treat UHCs. Also note that under these conditions, the surface impoundment is not a Subtitle C unit because it is not receiving hazardous waste (assuming that hazardous sludges are not being formed in the impoundment).

II. Gasoline-contaminated soil that is hazardous for benzene (D018) will be sent to a landfill. What treatment standards apply?

In this case, because Footnotes 8 and 9 have no bearing, the treatment standard is "10 and meet §268.48 standards." Two items are worth noting: 1) UHCs must be identified and treated prior to land disposal; and 2) the benzene treatment standard (10 mg/kg) is a total concentration, not a TCLP leachable concentration. Potential UHCs would be toluene, ethyl benzene, xylene, and lead if the gasoline was leaded. If any of these chemicals are present in the soil at concentrations above those given for nonwastewaters in §268.48, they would be UHCs subject to treatment. Note that the contaminated soil could also be managed under the alternative treatment standards for soil in §268.49, which are discussed in Section 13.7.

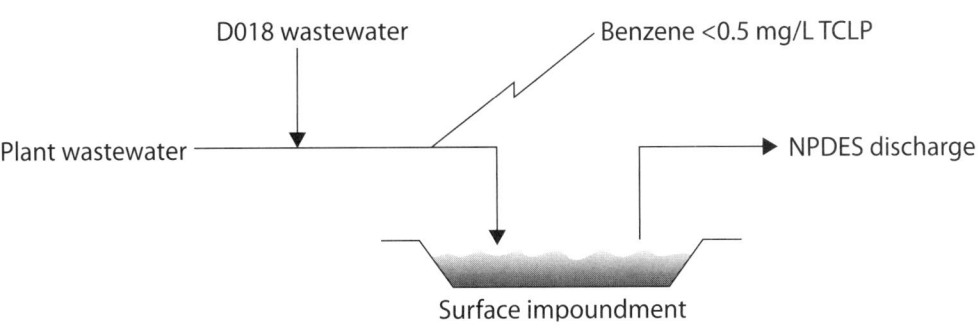

Figure 13-9: Treatment of D018 Wastewater in a CWA System

Source: McCoy and Associates, Inc.

Q: The §268.40 treatment standard for low-mercury (<260 mg/kg total mercury) D009 nonwastewaters is 0.025 mg/L via the TCLP (or 0.20 mg/L TCLP for residues from retorting), plus treatment of UHCs to meet UTS levels. Can these standards be met using stabilization of the waste?

A: Yes. When concentration-based standards are specified in §268.40, any appropriate treatment technology (excluding simple dilution) can be used to achieve the standards. For D009 low-mercury nonwastewaters, stabilization is an appropriate treatment technology. [RO 14685]

Case Study 13-6 (Continued): Treatment Standards for TC Wastes

III. As shown in Figure 13-10, a D008 wastewater that also contains low levels of cadmium is treated in a CWA treatment system having a surface impoundment. A metal-precipitating reagent, containing small amounts of bis(2-ethylhexyl) phthalate (DEHP), is used to remove the heavy metals. Sludge produced in the system is hazardous for lead (D008) and cadmium (D006). The wastewater is no longer hazardous, but, along with the sludge, contains DEHP at concentrations greater than the UTS levels of 0.28 mg/L for wastewaters and 28 mg/kg for nonwastewaters. What treatment standards apply to the wastewater leaving the tank, the wastewater leaving the surface impoundment, and the sludge from the tank?

The treatment standard for the original wastewater (D008) is "0.69 and meet §268.48 standards[8]." Because this is a CWA system, Footnote 8 says that the waste only has to be rendered nonhazardous before it is land disposed. Therefore, the lead concentration must be less than 5.0 mg/L TCLP in the wastewater before it enters the impoundment. DEHP is not a UHC in the wastewater leaving the tank because it was not present in the D008 waste at the original point of generation. Water leaving the surface impoundment is only subject to the permit levels specified in the NPDES permit. No LDR treatment standards apply at this point.

Not only is the sludge hazardous for lead, it also exhibits a new characteristic (D006). Because of this new characteristic, and also because of the change-in-treatability-group principle, this is considered to be a newly generated waste that contains the UHC DEHP. As a result, the treatment standard for the sludge is: lead—0.75 mg/L TCLP, cadmium—0.11 mg/L TCLP, and DEHP—28 mg/kg. [RO 50866]

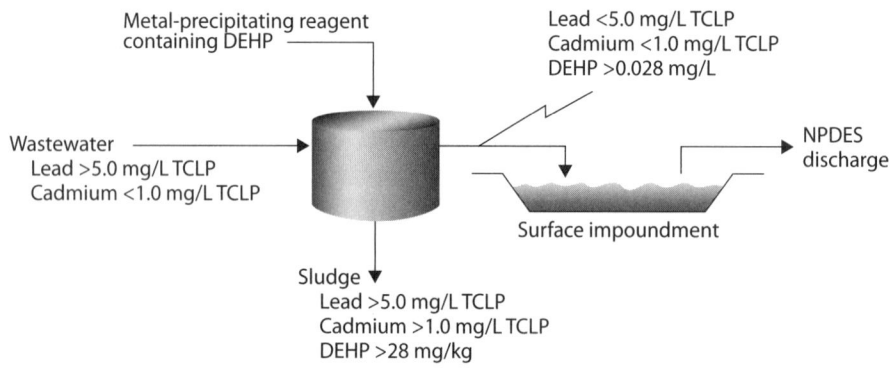

Figure 13-10: Treatment of a Lead-Containing Wastewater

Source: McCoy and Associates, Inc.

Q The treatment standard for an aqueous D009 high-mercury inorganic subcategory nonwastewater is retorting. Can this waste be treated to precipitate mercury salts prior to retorting?

A Precipitation is an acceptable pretreatment step to retorting. "Any residue that exceeds 260 mg/kg of mercury from the treatment of the aqueous phase must also be retorted. This

Case Study 13-6 (Continued): Treatment Standards for TC Wastes

IV. Waste paint is high-TOC ignitable (D001) and TC hazardous for lead (D008). Can this waste be burned as hazardous waste fuel?

All hazardous components in a waste stream must be treated to meet the applicable LDR standards before they are land disposed. The subcategory for the D001 code would be high-TOC ignitable; the associated treatment standard is recovery of organics (RORGS), combustion (CMBST), or polymerization (POLYM, which is only applicable to plastics wastes). Table 1 in §268.42 defines combustion to include burning the wastes in RCRA boilers and industrial furnaces regulated under Part 266, Subpart H. Burning the paint waste in one of these units as fuel would satisfy the CMBST standard; however, this would not be sufficient because the waste must also be treated to meet the D008 standard. Hence, the combustion residue (ash) would have to be treated to meet the lead standard. [RO 11815]

V. An aqueous waste that is hazardous for chromium (D007) is treated in a tank-based wastewater treatment system that discharges treated wastewater under an NPDES permit. Sludge from the system is hazardous for chromium and is sent offsite to a metals recovery firm for chromium reclamation. What LDR standards apply to the sludge?

Don't forget the RCRA fundamentals when addressing this question. As discussed in Section 1.2 of this book, characteristic sludges that are reclaimed are not solid wastes (refer to Table 1 in §261.2). If the sludge isn't a solid waste, it cannot be a hazardous waste and no LDR standards apply. [RO 13030]

---

pretreatment step cannot be used to avoid the D009 retorting treatment standard." [RO 13613]

See Case Study 13-7 for the LDR implications of managing cans of unwanted paint.

13.5 Treatment standards for listed wastes

Treatment standards for listed wastes are much easier to interpret and apply than the standards for characteristic wastes. Footnotes 8 and 9 that are critically important for characteristic wastes don't appear on any listed wastes. Also, none of the treatment standards for listed wastes refer to §268.48 standards; hence, UHCs don't have to be identified/treated in listed wastes unless the waste must also carry a characteristic code (see Section 13.6). [RO 14628] Finally, very few listed wastes have any subcategories. (F001–F005 spent solvents have four subcategories. Other listed wastes with multiple subcategories are F025—condensed light ends, spent filters and filter aids, and spent desiccant wastes from the production of certain chlorinated aliphatic hydrocarbons; K069—emission control dust from secondary lead smelting; K071—brine purification muds from the mercury cell process in chlorine production; K106—wastewater treatment sludge from the mercury cell process in chlorine production; P047—4,6-dinitro-o-cresol; P065—mercury fulminate; P092—phenyl mercuric acetate; U151—mercury; and U240—2,4-D.)

Most of the time, using the table of treatment standards in §268.40 for listed wastes simply involves looking up the proper waste code. Some listed wastes have rather complicated specified methods as treatment standards. For example, the treatment standard for K123 wastewater is "CMBST; or CHOXD fb (BIODG or CARBN)." Translated, this means 1) the waste may be combusted, 2) the waste may be chemically oxidized followed by (fb) biodegradation, or 3) the waste may be chemically oxidized followed by activated carbon adsorption.

Case Study 13-7: LDR Applicability to Partially Full Cans of Unwanted Paint

For cans of unwanted paint, the point of generation of a solid waste is when the person in control of the paint decides it can't be used for its intended purpose and will be discarded. When that happens, the generator must determine if the paint is hazardous. Unused paint will not generally carry any listed waste codes. [RO 11180, 11220, 11340, 11349, 12334, 12906, 13273] But, the paint must be evaluated to determine if it exhibits any of the hazardous waste characteristics. The likely characteristic codes are D001 for ignitability of certain oil-based paints, D004–D011 for heavy metals, and D035 for methyl ethyl ketone. If paint that was intended to be used is found to be dried out (i.e., there is no liquid when personnel get ready to use it), then the paint would not be ignitable at its point of generation but could still exhibit the toxicity characteristic.

If the paint does exhibit a characteristic at the point of generation, it is subject to the hazardous waste regulations from that point forward. One option is to store the waste paint in either satellite or 90/180/270-day accumulation containers and then send it offsite to a RCRA-permitted TSD facility. Such shipment must be accompanied by a hazardous waste manifest and the applicable LDR notification.

If instead of shipping the hazardous paint waste offsite to a TSD facility, the generator wants to treat the paint (e.g., solidify it), one way to do that without getting a RCRA permit would be to treat the waste in a 90/180/270-day accumulation container (see Section 7.3). For example, while the paint is in a 90/180/270-day accumulation area, the generator could add some kind of solidifying/fixation reagent to the paint. Solidifying the paint would likely remove the D001 code but wouldn't necessarily remove other codes, such as D004–D011 or D035. While the hazardous paint is being treated, the generator would have to comply with §262.34 and Part 265, Subpart I (e.g., the container must be closed). Finally, generators who treat hazardous wastes in 90/180/270-day accumulation containers in order to meet LDR treatment standards must prepare and follow a waste analysis plan per §268.7(a)(5).

Even if the paint is rendered nonhazardous (e.g., any D001, D004–D011, and D035 codes have been eliminated via treatment in a 90/180/270-day accumulation container), LDR standards still apply. This is because the LDR standards attach at the point of generation of the waste, before treatment. For example, if the paint is a high-TOC D001 waste (≥10% TOC) at its point of generation, the treatment standard is combustion (CMBST) or recovery of organics (RORGS). That means the waste paint (even if it has been solidified) must be treated by one of those methods before it can be placed in a landfill. If the paint was a low-TOC D001 waste (<10% TOC) at its point of generation, the treatment standard is "deactivation (DEACT) and meet §268.48 standards." Solidifying the paint will achieve deactivation of the ignitability characteristic, but any underlying hazardous constituents (UHCs—identified at the point of generation) would still have to meet universal treatment standards prior to land disposal. Additionally, the treatment standards for any other applicable waste codes (e.g., D004–D011 and D035) would have to be met before land disposal.

In one case (F024), the waste has *both* a specified method (CMBST) and concentration-based treatment standards. EPA chose this unusual treatment standard because the waste occasionally contains dioxins. Because the permitted treatment capacity for dioxin-containing wastes is extremely limited, the agency did not want to identify dioxin as a constituent of concern. Instead, they specified combustion (which destroys the dioxin) and then established concentration-based standards for the other constituents.

Because questions and complications seem to arise more often for listed spent solvent wastes than for

Case Study 13-8: Treatment Standards for Spent Solvent Wastes

I. 1,1,1-Trichloroethane (TCA) is used for degreasing. When spent (F001), it will be sent to a solvent recycler. Do any LDR requirements apply to this waste?

Figure 13-11 illustrates the "cradle-to-grave" path that such a spent solvent is likely to take. Recall that LDR standards are triggered if a hazardous waste or a residue from treating such a waste is placed in a land disposal unit. In this case, incinerator ash is a residue from treating the original spent solvent, and the ash is disposed in a landfill. Therefore, each of the entities shown in Figure 13-11 will have to comply with certain LDR requirements:

- The generator will have to fill out an LDR notification that will accompany the first shipment of the waste to the solvent recycler.

- The recycler will have to fill out an LDR notification as the generator of the still bottoms. This notification will accompany the first shipment of the waste to the incineration facility.

- The incineration facility will burn the F001 still bottoms, producing an ash that will be land disposed. From §268.40, the treatment standard for 1,1,1-TCA nonwastewaters in F001 is 6.0 mg/kg. The incineration facility will fill out a notification/certification that will accompany the first shipment of the waste to the landfill. The certification will assure the landfill that the ash meets treatment standards and can be land disposed.

- The landfill will keep the notification/certification received with the waste and will test the waste to assure that it meets the treatment standards.

We will provide a more thorough discussion on the paperwork requirements in Section 13.12.

II. Toluene is used to clean paint spray guns. The paint contains xylene and methyl ethyl ketone (MEK). Xylene is identified in the F003 listing; MEK and toluene are F005 constituents. Which spent solvent treatment standards apply to the waste?

The treatment standards for F001–F005 only apply to spent solvents. Toluene is the only spent solvent component in the waste. Xylene and MEK were ingredients in the paint and are not spent solvents. Therefore, only the treatment standard for toluene (10 mg/kg) applies. [RO 11815]

III. Xylene is used for cleaning purposes and, when spent, is classified as F003 spent solvent waste. The spent solvent also contains traces of methanol, which was an ingredient in fuel that was solubilized by the xylene during cleaning. (Methanol is also an F003 constituent when used for its solvent properties.) Does the spent solvent have to meet LDR treatment standards for both xylene and methanol?

No. Because the methanol was not used for its solvent properties, only the treatment standard for xylene applies to this F003 waste. [RO 13359]

Figure 13-11: Cradle-to-Grave Path for Recycling a Spent Solvent

Source: McCoy and Associates, Inc.

Case Study 13-8 (Continued): Treatment Standards for Spent Solvent Wastes

IV. The listing description in §261.31 for F001 spent solvents identifies "chlorinated fluorocarbons" as regulated constituents. The treatment standard for F001 in §268.40 identifies only two chlorinated fluorocarbons: 1,1,2-trichloro-1,2,2-trifluoroethane and trichloromonofluoromethane. If a chlorinated fluorocarbon other than one of the two identified in §268.40 is used for degreasing, what treatment standard applies?

Only the treatment standards for the two chlorinated fluorocarbons identified as regulated hazardous constituents in §268.40 for F001–F005 wastes apply. [RO 11877] Presumably, if a spent solvent contains only chlorinated fluorocarbons that are different than the two cited above, a generator could certify that the spent solvent meets the treatment standard at the point of generation.

V. The second subcategory of F001–F005 listed spent solvents is for F003 and/or F005 spent solvent wastes that contain only carbon disulfide, cyclohexanone, and/or methanol. If one or more of these three solvent constituents is contained in a nonwastewater spent solvent with one or more of the 27 other solvents included in the first subcategory of F001–F005 wastes, do the carbon disulfide, cyclohexanone, and/or methanol components have to be treated to meet treatment standards?

Probably not. The treatment standards for the nonwastewater form of these three solvents is NA when they are in a spent solvent mixture with other F001–F005 constituents; therefore, they do not have to be monitored or treated. However, if the spent solvent mixture also carries a characteristic code whose treatment standards require treatment of UHCs, then the carbon disulfide, cyclohexanone, and/or methanol must be treated to meet UTS levels in §268.48, because they will be UHCs. [RO 14569]

VI. A facility generates spent acetone solvent, which is correctly coded and managed as F003. At its point of generation, the waste is ignitable (D001). Subsequent to the point of generation, the F003 is mixed with water, resulting in a non-characteristic mixture. Does the generator have to send the mixture to a hazardous waste TSD facility for disposal?

No. "Since the waste does not exhibit a characteristic, it is no longer hazardous and does not have to be disposed of in a Subtitle C facility.... However, [since] the waste was hazardous at the point of generation, the waste must meet any LDR treatment requirements, which are listed in §268.40, prior to land disposal.... Thus, you would have to meet the treatment standard for the F003 constituents, as well as for any underlying hazardous constituents (UHCs) that were reasonably expected to be present in the D001 waste at the point of generation." [RO 14749] Note that UHCs would normally be required to be identified only if the waste was not high-TOC D001, per §268.9(a).

---

other types of listed wastes, we give a number of F001–F005 examples in Case Study 13-8 and then conclude this section with miscellaneous questions and answers dealing with listed wastes.

Q *Under what circumstances are listed wastes subject to the universal treatment standards in §268.48 for underlying hazardous constituents?*

A For listed wastes, the treatment standards in §268.40 address all of the hazardous constituents that are of concern; therefore, listed wastes treated to these standards do not need to be treated for additional underlying hazardous constituents in §268.48. If, however, a listed waste also carries a characteristic code (see Section 13.6), the characteristic code could bring with it a requirement to treat underlying hazardous constituents. [RO 13779]

 Hazardous waste U096 (alpha, alpha-dimethylbenzylhydroperoxide) was listed

solely because it exhibits the characteristic of reactivity. Under the mixture rule, if this waste is mixed with a solid waste and the mixture is no longer reactive, the mixture is no longer hazardous. Under the old mixture rule at §261.3(a)(2)(iii), and the new mixture rule promulgated May 16, 2001 at §261.3(g)(2)(i), such mixtures are still subject to the land disposal restrictions. What does this mean?

A The mixture, even if nonhazardous, must be treated to meet LDR treatment standards in §268.40 which attached to the U096 waste at its point of generation. [RO 14045] For U096 wastewaters, the treatment standards are the specified methods of 1) chemical oxidation, 2) chemical reduction, 3) activated carbon adsorption, 4) biodegradation, or 5) combustion. For U096 nonwastewaters, the specified methods are 1) chemical oxidation, 2) chemical reduction, or 3) combustion. Also, since the original waste was hazardous at the point of generation, LDR paperwork requirements will apply (see Section 13.12).

Q *According to §268.40, K069—emission control dust/sludge from secondary lead smelting is divided into two subcategories: low lead and high lead. Does EPA establish a lead concentration level that distinguishes between these two subcategories?*

A No. The distinction between the K069 subcategories is made based on the process generating the dust/sludge, not the lead concentration. The low-lead subcategory includes only dust/sludge generated as calcium sulfate in secondary wet scrubbers using lime neutralization. High-lead dust/sludge is not generated as calcium sulfate in the secondary wet scrubbers. [RO 14495]

Q *During corrective action at a hazardous waste landfill, 400 gallons of leachate-contaminated ground water is recovered for offsite disposal in an injection well. Numerous listed wastes were disposed in the landfill. What is the regulatory status of the recovered ground water?*

A Under the contained-in policy, the ground water meets the definition of multisource leachate, F039. [RO 13438] The treatment standards for this waste code in §268.40 are by far the most difficult to meet because over 200 chemicals are identified as constituents subject to treatment.

13.6 Waste coding

Prior to issuance of the land disposal regulations, the waste identification process was relatively simple: a waste was either listed or characteristic, but not both. Once LDR treatment standards were finalized for characteristic wastes, a question arose about how listed wastes that also happened to exhibit a characteristic were regulated. EPA decided that it didn't make sense to allow land disposal of a waste that met the appropriate listed waste treatment standards, but that didn't meet any characteristic treatment standards that might apply. They therefore changed the generator waste identification rules in §262.11.

After a generator concludes that he/she is managing a solid waste that is not otherwise exempt from regulation [§262.11(a)], he/she then determines if the waste is listed in Part 261, Subpart D. [§262.11(b)] If a listed code is applicable, it is always applied to the waste. If the waste *isn't* listed, the generator then determines if the waste exhibits any characteristics. [§262.11(c)]

A potential compliance problem occurs, however, if a generator determines that he/she *is* managing a listed waste. Section 262.11(c) states that "for purposes of compliance with 40 *CFR* Part 268" (the LDR standards), the generator must determine if the listed waste also exhibits a characteristic. At this point, a casual reader of the regulations could reasonably conclude that a listed waste also carries any characteristic codes that are exhibited by the waste. However, buried in the LDR regulations [§268.9(b)] is a rule that specifies when characteristic codes are carried on listed wastes.

©2015 McCoy and Associates, Inc. **McCoy's RCRA Unraveled**

The rule in §268.9(b) requires the generator to look up the LDR treatment standard for the listed waste in question and then do one of two things:

1. If the treatment standard for the listed waste addresses the constituent causing the waste to exhibit the characteristic, only the listed waste code (and treatment standard) applies; or
2. If the treatment standard for the listed waste does not address the constituent causing the waste to exhibit the characteristic, both the listed and characteristic waste codes (and treatment standards) apply.

Before we look at some examples to see how these waste coding rules work, we want to point out a disadvantage to adding a characteristic waste code to a waste that already carries a listed code. Per our discussions in Sections 13.3 and 13.4, assigning characteristic waste codes often requires the generator to determine if the waste contains UHCs and, if so, add them to LDR paperwork. Receiving TSD facilities then have to treat the waste to meet the listed and characteristic waste treatment standards and also ensure that all identified UHCs are treated to meet UTS levels. Thus, arbitrarily overcoding listed wastes with unnecessary characteristic codes can increase your analytical and treatment costs.

Q: *F037—petroleum refinery primary oil/water/solids separation sludge is also characteristically hazardous for benzene (D018). Which waste code or codes are used on the LDR paperwork?*

A: Only F037. The treatment standard in §268.40 for F037 wastes includes a standard for benzene. Therefore, because the listed waste treatment standard specifically addresses the D018 characteristic, only the listed code (F037) applies (see Rule #1 above). [RO 14545]

Q: *A spent solvent mixture carrying the listed codes F003 and F005 is also ignitable. Should the waste also carry the D001 code?*

A: Yes. F003 and F005 wastes that are also ignitable should also be identified as D001, because the treatment standards for F003/F005 do not specifically address ignitability (see Rule #2 above). [RO 11877] "[T]he generator would need to list both the F003 and D001 waste codes.... The 'in lieu of' principle [in §268.9(b)] is not applicable in this situation. As you point out, the 'in lieu of' principle applies where the treatment standard for the listed waste includes a treatment standard for the constituent that causes the waste to exhibit the characteristic. Part 261, Appendix VII, Basis for Listing Hazardous Waste, includes no hazardous constituents for F003. Instead, the term N.A. appears, which is defined as: 'Waste is hazardous because it fails the test for the characteristic of ignitability, corrosivity, or reactivity.' Therefore, because there is no constituent specified as the basis for the F003 listing, it is not possible to ascertain that the LDR treatment standard for the F003 waste would address the constituent(s) that caused the D001 waste to display the characteristic of ignitability." [RO 14749]

Q: *Listed waste K061—electric arc furnace dust is also characteristically hazardous for lead. How is the waste coded for purposes of the LDR paperwork and generator's biennial report?*

A: Because the treatment standard for K061 includes a standard for lead, §268.9(b) requires only that the K061 code be used on LDR paperwork and the generator's biennial report (see Rule #1 above). The D008 code for lead is not required. [RO 11545]

Q: *An F001 spent solvent is also characteristic for trichloroethylene (D040). For D040 wastes, UHCs are of concern and are not addressed by the treatment standard for F001 spent solvents. How should this waste be coded?*

A: Because the treatment standard for F001 includes a standard for trichloroethylene, only the F001 waste code applies. No requirement is imposed to treat UHCs associated with the D040 code. [RO 13721]

How do these rules for determining which waste codes to put on LDR notifications and certifications affect other generator paperwork requirements, such as the generator notification requirement in §262.12 and the biennial report? Here is what EPA said: "When a generator determines that a solid waste meets a Part 261, Subpart D hazardous waste listing, he/she is not required to determine whether the listed waste exhibits any characteristics for purposes of filling out Part 262 paperwork such as generator notification forms (8700-12) and biennial reports. (However, the generator may elect to determine whether the waste exhibits a characteristic for his/her own information or for other reasons.) On the other hand, the paperwork of Part 268 must reflect the amended language of Section 262.11(c) which states that for the purposes of compliance with Part 268 a generator must determine if a listed waste is also characteristically hazardous." [RO 13455]

We suggest that whatever codes appear on LDR forms also appear on manifests in Block 13. When agency inspectors look at LDR paperwork, they often check to see if information on manifests matches information on the associated LDR forms. Where discrepancies are identified, notices of violation (NOVs) may be issued or a discussion will be initiated as to why the documents don't match.

13.7 Contaminated soil standards

When EPA initially developed the LDR treatment standards, they were concerned primarily with process wastes that were being generated on an ongoing basis by industry. The agency was not initially concerned with soil that might be contaminated with hazardous waste. Over time, it became obvious that hazardous soil was frequently produced as the result of spills of product/waste and as a result of remediation activities. If this hazardous soil was to be land disposed, it had to meet the treatment standards in §268.40 that had been developed for process wastes. The end result of this regulatory approach was that contaminated soil was frequently being incinerated (an expensive proposition) because many of the §268.40 treatment standards were based on that technology. The high costs of treating contaminated soil in this manner adversely affected the willingness of many parties to undertake site remediation activities. Realizing that soil is a completely different matrix than process wastes, the agency eventually developed alternative treatment standards for contaminated soil. These standards are codified in §268.49.

Contaminated soil will be subject to the LDR program only if 1) it is hazardous (i.e., it exhibits a characteristic or contains a listed waste), and 2) it is generated (i.e., excavated). If hazardous soil is excavated and will be land disposed somewhere, three options are available for treating the soil under the LDR program:

1. The soil can be treated to meet the nonwastewater treatment standard for the hazardous waste contaminating the soil. For example, if product benzene (U019) is spilled on soil, the contained-in policy (see Section 5.3) applies the U019 code to the soil. The associated nonwastewater treatment standard for this waste code is 10 mg/kg benzene. [Treating soil to meet a §268.40 standard is generally not the best option because, as mentioned previously, meeting a standard of 10 mg/kg benzene may require the use of a very efficient (but very expensive) treatment technology, such as incineration.]

2. The generator/TSD facility may request a treatability variance for the contaminated soil per the provisions of §268.44. While this might have been a viable option when the only other alternative was to treat soil to the §268.40 standards, it is generally not an attractive option today. Any time a party requests special consideration from EPA or the state, the process is likely to be expensive and time consuming, and significant risk exists that the agency will not be willing to issue a variance. See also Section 13.13.4.4.

3. By far the best option is to use the alternative soil treatment standards of §268.49. EPA refers to the alternative treatment standards as "90% removal capped at 10 times UTS." As illustrated in Figure 13-12, the initial concentration

of "constituents subject to treatment" is determined. Treatment must achieve a 90% reduction in total concentration (refer to constituent A), unless that level of treatment would reduce the concentration to below 10 times the UTS level for the constituent. In no case does treatment have to reduce the concentration to below this 10-times-UTS floor (refer to constituent B). Therefore, the treatment provided under this option allows residual contaminant concentrations in soil to be at least 10 times greater than for Option 1 above. Case Study 13-9 exemplifies this option.

The required 90% removal efficiency under the alternative treatment standards allows the use of less expensive technologies than would be required for Option 1, which relies on incineration-like technologies, typically performing at 99.99% or greater efficiency.

"Constituents subject to treatment" for these alternative soil treatment standards are defined as any constituents in §268.48 that are reasonably expected to be present in the soil at concentrations greater than 10 times UTS levels. [§268.49(d)] As with UHCs discussed earlier, fluoride, selenium, sulfides, vanadium, and zinc are not constituents subject to treatment.

Generators of hazardous soil are not required to monitor the soil for the entire list of constituents in §268.48 to determine the "constituents subject to treatment." Instead, EPA allows generators of hazardous soil to reasonably apply knowledge of the contaminants that are likely to be present in the soil and use that knowledge to select appropriate constituents, or classes of constituents, for monitoring. [May 26, 1998; 63 *FR* 28609]

Note that the alternative soil treatment standards provide different data collection options for remediation projects:

- If the 90% treatment standard is selected and analytical testing is used to confirm compliance,

Figure 13-12: Alternative Treatment Standards for Soil

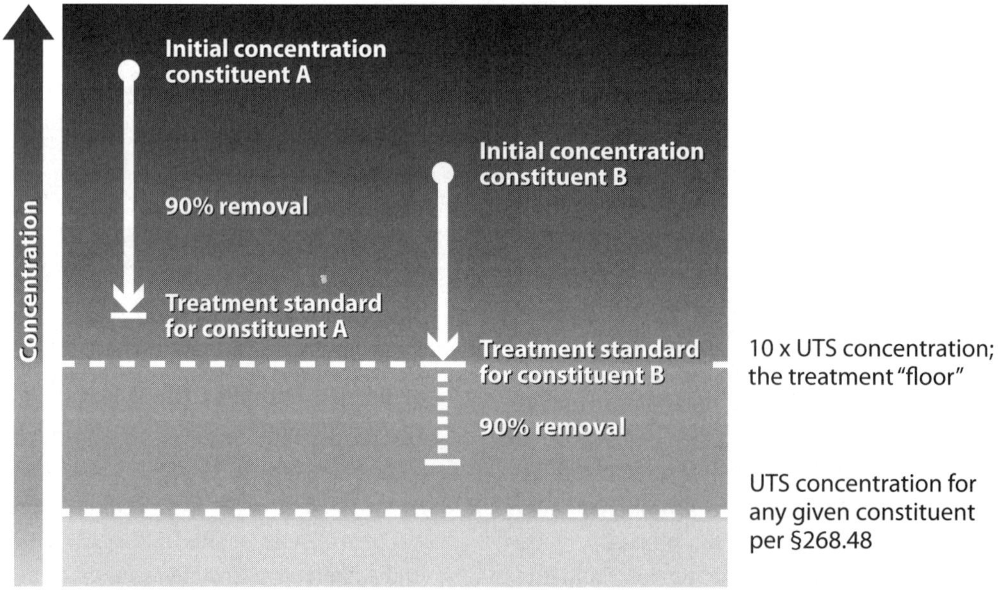

Source: McCoy and Associates, Inc.

Case Study 13-9: Alternative Treatment Standards for Soil

Contaminated soil from a surface spill of gasoline along a roadside exhibits the toxicity characteristic for benzene (D018). Analysis indicates that the untreated soil, when excavated, contains 2,000 mg/kg benzene, 600 mg/kg xylene, and 4.0 mg/L lead via the TCLP.

What is the treatment standard for each constituent when using the alternative treatment standards of §268.49?

Table 13-1 simplifies the process of identifying the appropriate treatment standard for each constituent. The table gives the 90% removal and 10-times-UTS concentrations for each constituent. The right-hand column gives the most favorable treatment standard of 90% removal capped at 10 times UTS. Note that no treatment of the soil for lead is needed. Therefore, a low-tech treatment technology (such as thermal desorption) could be used to treat the organic constituents.

Table 13-1: Alternative Treatment Standards for Gasoline-Contaminated Soil

Constituent	Initial concentration	UTS concentration per §268.48	10 X UTS	90% removal	Treatment standard
Benzene	2,000 mg/kg	10 mg/kg	100 mg/kg	200 mg/kg	200 mg/kg
Xylene	600 mg/kg	30 mg/kg	300 mg/kg	60 mg/kg	300 mg/kg
Lead	4.0 mg/L TCLP	0.75 mg/L TCLP	7.5 mg/L TCLP	0.4 mg/L TCLP	7.5 mg/L TCLP

Source: McCoy and Associates, Inc.

---

two sets of samples are required—one at the point of generation and one after treatment.

- If the 90% treatment standard is selected and process data are used to show that the process always operates at greater than 90% efficiency, no routine sample analyses would be required. Instead, compliance could be confirmed by monitoring process variables, controls, and operating conditions.

- If the 10-times-UTS option is chosen, only one set of samples is required—after waste treatment.

A very helpful reference on how the alternative treatment standards for soil are applied is *Guidance on Demonstrating Compliance With the Land Disposal Restrictions (LDR) Alternative Soil Treatment Standards*, EPA/530/R-02/003, July 2002, available from http://www.epa.gov/osw/hazard/tsd/ldr/soil_f4.pdf.

 If hazardous soil is to be treated and then used as an ingredient to make asphalt, can the alternative soil treatment standards of §268.49 be used?

A No. "A hazardous-waste-contaminated soil that is going to be used in products which are subsequently used in a manner constituting disposal must meet the treatment standards developed for as-generated industrial waste at §268.40." [EPA/530/R-02/003]

At a number of areas around our country, natural background concentrations of certain constituents in §268.48 (e.g., arsenic) are high. As a result, such constituents may be identified as constituents subject to treatment in hazardous soil (e.g., resulting from a recent nonarsenic spill), and the alternative soil treatment standards may require treatment of these constituents to levels below natural background concentrations. EPA doesn't think it would have the authority to require that. Therefore, in situations where treated soil will continue to be managed onsite (e.g., as backfill) or in an area with similar natural background concentrations, such constituents will not have to be treated below background levels. Conversely, if the soil will be sent offsite for land disposal, full compliance with the alternative soil treatment standards is required since "the agency believes that natural background concentrations onsite will not automatically correspond to natural background concentrations at a remote land disposal facility." [63 FR 28609] (For the purpose of this provision, EPA considers natural background concentrations to be those in soil which has *not* been influenced by human activities or releases.) The agency will require that individuals who want to cap LDR treatment at natural background levels apply for and receive a treatment variance per §268.44(h)(4).

13.7.1 Soil contaminated with characteristic wastes

Soil contaminated with characteristic waste is subject to the land disposal restrictions if it exhibits a characteristic when it is excavated. "It is conceivable that soil could be contaminated with a characteristic hazardous waste [e.g., due to a spill] and yet not display a hazardous characteristic at the point of generation due to dilution in the soil matrix or to breakdown or alteration of the constituents in the soil environment. If this soil does not exhibit a characteristic when it is generated (excavated), then LDR requirements do not apply." [RO 14547]

If contaminated soil exhibits a characteristic of ignitability, corrosivity, or reactivity, these characteristics must be eliminated prior to disposal per §268.49(c)(2). In addition, if any constituents in §268.48 are reasonably expected to be present in the soil at concentrations greater than 10 times UTS, they must also be treated to meet the 90% removal capped at 10-times-UTS standards.

Examples of LDR standards for characteristic soils are given in Case Study 13-10.

13.7.2 Soil contaminated with listed wastes

EPA included in the regulations [§268.49(a)] a rather complex table to use in determining if soil contaminated with listed wastes is subject to LDR standards. This chart can be simplified to a single statement: Soil contaminated with listed waste must comply with LDR treatment standards unless the soil was contaminated before LDR standards applied to the listed waste and a no-longer-contains determination has been obtained from the agency at the point of generation (excavation) of the soil. Table 2 in Appendix VII of Part 268 provides these effective dates.

Once it is determined that soil is subject to treatment standards, the next step is to determine which contaminants are subject to treatment. Per §268.49(d), constituents subject to treatment are those constituents in the UTS table (§268.48) that are reasonably expected to be present in the soil at concentrations greater than 10-times-UTS levels.

When using the alternative treatment standards for soil containing a listed waste, the constituents subject to treatment may include constituents that do not require treatment under the §268.40 treatment standard for the applicable waste code(s). The treatment standards in §268.40 for listed wastes do not require identification or treatment of UHCs. However, under the alternative treatment standards for soil, generators must identify and treat all constituents subject to treatment that are reasonably expected to be present in listed and characteristic soils. [RO 14628]

Q *When a hazardous waste is treated to meet the standards of §268.40, underlying hazardous constituents are only of concern in certain characteristic wastes; UHCs are never of concern in listed wastes (unless the listed wastes must also carry characteristic codes—see Section 13.6). Does this same approach also apply to contaminated soil?*

A No. If soil is treated to meet the alternative treatment standards of §268.49, "all UHCs present at levels greater than 10 x UTS must be treated regardless of whether the soil contains a listed waste or exhibits a characteristic when the soil is generated." [EPA/530/R-02/003, cited earlier]

Case Study 13-10: Alternative Treatment Standards for Characteristic Soil

I. Contaminated soil is hazardous for lead (D008), but has no UHCs at the point of generation. The original lead concentration is 40 mg/L TCLP. The soil is treated by washing with a metal chelating agent to meet the alternative soil treatment standards of §268.49; the lead concentration must be reduced by 90% capped at 10 times the UTS lead standard in §268.48. Because 90% reduction would lower the lead concentration to 4.0 mg/L TCLP, the treatment floor of 10 times UTS (i.e., 7.5 mg/L TCLP) is the LDR standard for this soil. At this concentration level, the treated soil is still a hazardous waste because the lead concentration is greater than 5.0 mg/L TCLP. Therefore, even though the soil meets treatment standards, the treated soil must be disposed in a Subtitle C hazardous waste disposal facility. In order for the soil to be disposed in a Subtitle D unit, it would have to be treated to reduce the lead concentration to below 5.0 mg/L TCLP. [RO 14409]

The spent chelating agent contains lead at concentrations significantly higher than 5.0 mg/L TCLP and must also be managed as a hazardous waste. This treatment residue is subject to the treatment standard for D008 wastes in §268.40 before it can be land disposed. Because the soil as originally generated contained no UHCs, no UHCs in the chelating agent are subject to treatment as long as this residue does not exhibit any new characteristics.

II. Soil from a remediation site is hazardous due to the presence of chromium (100 mg/L TCLP), and it also contains 800 mg/kg PCBs. What treatment standards apply to the soil?

The contaminated soil is D007; the UTS level for D007 nonwastewater is 0.60 mg/L TCLP. Per the alternative treatment standards of §268.49, the soil must be treated to remove 90% of the chromium or to reduce the leachable chromium concentration by 90%, but in no case does it have to be treated to below 10 times UTS. Because 90% reduction would result in a chromium concentration of 10 mg/L TCLP, this is the treatment standard for chromium rather than 10 times UTS (6.0 mg/L TCLP). At a residual concentration of 10 mg/L TCLP chromium, the treated soil will still be hazardous and must be disposed in a Subtitle C landfill. Although PCBs could also potentially be a constituent subject to treatment (the 10-times-UTS level for total PCBs in §268.48 is 100 mg/kg), EPA has temporarily deferred this standard for soils that are hazardous only due to D004–D011 constituents. (Refer to §268.32 and Footnote 8 at the end of the UTS table.)

III. Contaminated soil is hazardous for lead, and the remediation firm is considering two treatment technologies: solidification/stabilization and soil washing. Are the treatment standards identical for both technologies?

No. When solidification/stabilization is used to treat metals, compliance with the 90%-reduction treatment standard is based on TCLP analyses. When a metal removal technology is used, compliance with the 90%-removal treatment standard is based on total constituent (not leachable) analyses. [§268.49(c)(1)(B)]

IV. Contaminated soil contains >260 mg/kg mercury and no organic contaminants. The treatment standard in §268.40 for the D009 high-mercury, inorganic subcategory is RMERC. Can the generator/treater use the less-stringent §268.49 treatment standards, or does the mercury-contaminated soil have to be retorted?

Mercury has a concentration-based standard in §268.48, so it is considered an analyzable constituent. Therefore, the mercury-contaminated soil can be treated to the numerical treatment standards specified in §268.49. [May 11, 1999; 64 *FR* 25410] One way to accomplish this would be to retort the contaminated soil.

If soil is contaminated by a waste that consists only of nonanalyzable constituents, the soil must be treated by the specified method for the nonwastewater form of the waste as given in §268.40. A constituent is nonanalyzable when 1) the appropriate §268.40 listing specifies a treatment technology, and 2) there is no concentration-based limit in the §268.48 UTS table. [May 11, 1999; 64 *FR* 25410–1] For example, if product formaldehyde (U122) is spilled on soil, formaldehyde is a nonanalyzable constituent. (We can tell this because the treatment standard for U122 in §268.40 is a specified method—CMBST, *and* formaldehyde does not have a concentration-based limit in §268.48.) Therefore, our U122-contaminated soil must be combusted; it cannot be treated by the alternative treatment standards of §268.49.

If soil is contaminated with both analyzable and nonanalyzable constituents, the alternative standards of §268.49 may be used, and only analyzable constituents present at greater than 10 times UTS would be subject to treatment. [See §268.49(c)(3).] EPA expects that treatment of the analyzables will also provide adequate treatment of the nonanalyzable constituents. Where the analyzable and nonanalyzable constituents are not both organics, however, EPA believes that treating the analyzable constituents may not serve as a surrogate for treatment of the nonanalyzables. Such cases would have to be addressed with regulators on a site-specific basis. [May 11, 1999; 64 *FR* 25410]

Q *Acrylamide-contaminated soil, U007, is excavated and containerized for offsite treatment and disposal. The U007 treatment standard in §268.40 has a specified method of treatment—CMBST. Can the generator use the alternative soil treatment standards of §268.49 for this contaminated soil?*

A Yes. Because acrylamide has a concentration-based standard in §268.48, it is considered to be an analyzable constituent. Soil that is contaminated only with analyzable constituents must be treated to the numerical treatment levels; because the soil has numerical treatment levels, the alternative treatment standards of §268.49 are applicable. [May 11, 1999; 64 *FR* 25410]

13.7.3 Dealing with residues from soil treatment

After contaminated soil is treated, two treatment residues remain: 1) treated soil (which may still contain relatively high contaminant concentrations), and 2) nonsoil treatment residues such as washwater. The regulatory status of these residues is determined using the derived-from rule. For example, if the soil was originally contaminated with a listed spent solvent (e.g., F005), the derived-from rule (see Section 5.2) states that both the treated soil and nonsoil residues continue to be F005 listed wastes. Hence, the treated soil would have to be disposed in a Subtitle C landfill (unless a no-longer-contains determination is obtained from the state). The nonsoil residue must be treated to meet the §268.40 standards prior to land disposal.

Case Study 13-10 also illustrates how residues from soil contaminated with a characteristic waste would be managed.

13.8 Contaminated debris standards

As with contaminated soil, debris is very different in physical form from the process wastes that EPA evaluated when establishing LDR treatment standards. The definition of debris in §268.2(g) is:

"*Debris* means solid material exceeding a 60 mm particle size that is intended for disposal and that is: a manufactured object; or plant or animal matter; or natural geologic material. However, the following materials are not debris: any material for which a specific treatment standard is provided in Subpart D, Part 268, namely lead-acid batteries, cadmium batteries, and radioactive lead solids; process residuals such as smelter slag and residues from the treatment of waste, wastewater, sludges, or air emission residues; and intact containers of hazardous waste that are not ruptured and that retain at least 75% of their original volume. A mixture of debris that has not been treated to the standards provided by §268.45 and other material is subject to regulation as debris if the mixture is comprised primarily of debris, by volume, based on visual inspection."

Several aspects of the debris definition warrant discussion:

1. Debris is material that is bigger in one dimension than 2.5 inches (60 mm).

2. Debris is material that is *intended for disposal*. When manufacturing operations are shut down, a very common material that is encountered is metallic equipment and components that might be contaminated with hazardous constituents. If this material is recycled instead of disposed, it is not "debris" subject to the LDR debris standards. In other words, recycling rather than disposing material may eliminate the need to comply with the debris standards.

3. Examples of manufactured objects that are "debris" are gloves and other personal protective equipment, solder paste wipes, pipes, pumps, valves, etc. that will be disposed (perhaps due to radioactive contamination that makes them nonrecyclable). [RO 14660]

4. An example of plant matter that would be "debris" is a tree stump.

5. An example of geologic material that would be "debris" is a rock.

6. The language about intact containers addresses drums that might be dug up as part of a remediation project. If the drums contain hazardous waste and are intact, the contents of the drums must be treated to meet §268.40 standards for whatever waste codes are associated with the contained wastes. In other words, intact drums of hazardous waste are not eligible for the more lenient alternative treatment standards for debris. On the other hand, if a ruptured drum is excavated, it may be managed under the debris standards.

7. The language about material being primarily debris based on visual inspection implies that it is not necessary to run a sieve analysis to determine if a waste is mostly debris, as opposed to something else (e.g., soil). A visual assessment is all that is required.

A few examples illustrate issues associated with the definition of debris.

 Can mercury batteries be considered to be "debris"?

If the batteries are deteriorated such that material can flow into and out of the batteries, they may be considered "debris." If the batteries are intact, they may more closely meet the definition of an "intact container," which is excluded from the definition of debris in §268.2(g). The distinction between "debris" and "intact containers" is significant. If the batteries are "debris," they may be treated by the alternative treatment standards of Table 1 in §268.45; if the batteries are "intact containers," they must be treated to meet the appropriate D009 standard of §268.40. [RO 13638, 14215, 14675, 14685]

What is the status under Part 268 of other mercury-containing items sent for disposal, such as thermometers, manometers, switches, pumps, jars of elemental mercury, dental amalgam collection devices, and ampules?

Again, these items are considered containers and, if intact, meet the definition of an "intact container." Therefore, these mercury-containing items do not fall under the debris definition, and they are thus subject to the nondebris treatment standards in §268.40 if the items to be disposed are U151 or D009. [RO 14685] Note that EPA added mercury-containing equipment as universal waste on August 5, 2005. [70 *FR* 45508] Thus, per the federal RCRA regs, mercury thermometers, manometers, switches, ampules, etc. can be managed as universal waste by handlers. LDR requirements do not apply to these materials until they are received at destination facilities. (See Section 13.13.3.)

Q *Can empty drums headed for disposal be managed as debris under §268.45?*

A No. "[I]ntact containers are never considered to be debris, and thus would never be subject to treatment standards for debris. Intact containers are either empty or nonempty. If empty they are not subject to regulation, as provided by §261.7(a)(1). If nonempty, the hazardous waste within the container is subject to the land disposal prohibitions (as well as the rest of Subtitle C regulations)." [August 18, 1992; 57 FR 37225]

Q *Are intact tanks considered to meet the definition of "debris"?*

A No. Intact tanks do not meet the definition of debris. If a tank is not intact, it may meet the definition of debris if it is ruptured and will not retain at least 75% of its original volume. [August 18, 1992; 57 FR 37225, RO 14402]

Q *Is debris (e.g., tanks and piping) from removing old petroleum-contaminated underground storage tanks subject to any LDR standards?*

A Section 261.4(b)(10) exempts such debris from regulation as a hazardous waste as long as the debris is only hazardous for D018–D043 constituents. If the debris were hazardous for some other constituent (e.g., lead), it would not qualify for this exemption and would be subject to LDR standards. [Note, however, that in most cases, the metallic components of underground storage tanks are recycled and are exempt from the LDR standards under the scrap metal exemption in §261.6(a)(3)(ii).]

Q *A ruptured drum retains a considerable quantity of hazardous waste. Can it still be treated by the debris standards?*

A EPA intended for the debris standards to apply in cases where the debris and the waste are inseparable. Therefore, wastes in a ruptured drum can be left in the drum and the entire matrix treated as debris only if the wastes are not readily separable from the drum. [RO 14241]

Q *The definition of "debris" in §268.2(g) concludes with the following sentence: "A mixture of debris that has not been treated to the standards provided by §268.45 and other material is subject to regulation as debris if the mixture is comprised primarily of debris, by volume, based on visual inspection." How does EPA define the term "primarily"?*

A EPA has not specifically defined the term "primarily." However, in guidance, the agency has provided the following two examples: 1) if a mixture is comprised of three components (debris, soil, and sludge), the mixture would be classified as debris if the volume of debris is greater than soil and greater than the volume of sludge; 2) if a mixture is comprised of two components (e.g., debris and soil), the mixture would be classified as debris if the debris component exceeds 50% by volume of the mixture based on visual inspection. [RO 13705] EPA noted in RO 14685 that, if intact containers are mixed with true debris [i.e., with material that meets the §268.2(g) definition of debris] and the mixture is hazardous, the intact containers would have to be removed and managed separately as nondebris hazardous waste. The agency also recommends that mercury-contaminated piping or broken gauges be removed from debris and managed under the nondebris treatment standards for D009 hazardous waste.

13.8.1 Three options for managing hazardous debris

Debris can be hazardous if it 1) exhibits a characteristic (e.g., toxicity), or 2) contains a listed hazardous waste (see Section 5.3.2). [§268.2(h)]

Generators or TSD facilities have three options for managing hazardous debris:

1. They can manage the debris according to the treatment standards of §268.40 for whatever waste codes apply to the debris. This is generally a very poor option because not only are the §268.40 standards relatively stringent, but most of them are concentration-based. This means that debris treated to these

standards would have to be sampled and analyzed. Sampling and analysis of nonhomogeneous materials can be very difficult. Also, if this option is chosen, listed debris that is treated remains a listed waste under the derived-from rule.

2. If the debris contains very low levels of contaminants, a no-longer-contains determination can be requested from the regulators. This is codified in §261.3(f)(2) of the regulations.

3. The debris can be managed under the alternative treatment standards of §268.45, which were specifically crafted for dealing with this difficult class of wastes. This option is almost always the best approach for dealing with contaminated debris and is discussed in the next subsection.

13.8.2 Interpreting the alternative treatment standards table for debris

If the alternative treatment standards for debris are chosen, numerous treatment options are provided in Table 1 of §268.45. For our discussion, consider the intact concrete slab illustrated in Figure 13-13. The slab has been contaminated by spills of F001 spent solvent and will be demolished prior to disposal. Since the slab itself will be hazardous waste under the contained-in policy once it is demolished, it must meet LDR standards prior to disposal. Because the alternative debris standards of §268.45 are almost always the best option, we consult Table 1 in that section of the regulations for treatment options.

Table 1 is divided into three main categories: A) extraction technologies, B) destruction technologies, and C) immobilization technologies. Our first step

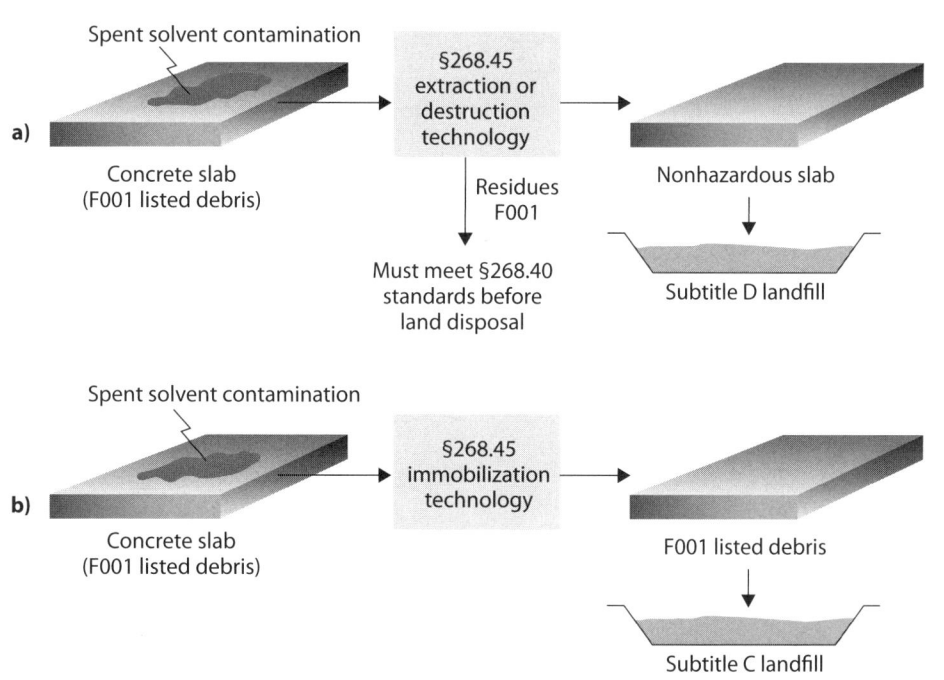

Figure 13-13: Contaminated Concrete Slab to Be Treated Via the Alternative Debris Standards

Source: McCoy and Associates, Inc.

is to think of a really easy and inexpensive way to clean a concrete slab, and then check to see if that method is given in Table 1. Water washing seems like an easy treatment method, and it appears as technology A.2.a: "Water washing and spraying: application of water sprays or water baths of sufficient temperature, pressure, residence time, agitation, surfactants, acids, bases, and detergents to remove hazardous contaminants from debris surfaces and surface pores or to remove contaminated debris surface layers."

Before we can use this technology, however, we must check the right-hand column of Table 1 to see if water washing can be used where F001–F005 spent solvents are involved. This column has a restriction for concrete: "Contaminant must be soluble to at least 5% by weight in water solution or 5% by weight in emulsion…." In order to comply with this restriction, we will have to add a surfactant or some other type of cleaning agent to the water that is capable of solubilizing the solvent constituents.

Finally, we must check the performance standards of the center column to determine how well water washing must work. This column also has a standard for concrete: "Debris must be no more than 1.2 cm (0.5 inch) in one dimension…." Clearly, we don't want to have to comply with this standard because it would require crushing the concrete slab to 0.5-inch particles prior to washing.

With these concepts in mind, we browse through the other technologies of Table 1 and decide that "Scarification, Grinding, and Planing" can be done relatively easily while meeting both the contaminant restrictions (none) and the performance standard (removal of at least 0.6 cm of the surface layer, and treatment to a clean debris surface).

When the slab is treated using this extraction technology, if we meet the treatment standard in the center column, the slab is rendered nonhazardous (refer to Figure 13-13). We can then break up the slab and haul it to a Subtitle D landfill for disposal.

[§268.45(c)] The other residue from treatment (scarification dust) carries the F001 code and must meet the appropriate treatment standard in §268.40 before land disposal.

Note that if we started with a type of debris for which we choose an immobilization technology in Table 1, the treated debris would still be a hazardous waste and could only be disposed in a Subtitle C unit. [§268.45(c)]

EPA notes that the treatment technologies listed in Table 1 that are most applicable to mercury-containing debris are macroencapsulation and microencapsulation, although retorting can also be effective. The technology options in Table 1 do not distinguish between debris containing high and low levels of mercury, as the §268.40 standards do for nondebris mercury wastes. [RO 14685]

Use of the alternative standards for treating toxicity characteristic debris and listed debris are illustrated in Case Studies 13-11 and 13-12, respectively.

Case Study 13-11: Alternative Treatment Standards for TC Debris

I. If debris is contaminated with D004–D043 characteristic wastes, do treatment standards for UHCs associated with these waste codes also have to be met?

No. According to §268.45(b)(1), the contaminants subject to treatment for TC-contaminated debris are only those contaminants for which the debris exhibits the characteristic. The debris is not otherwise subject to the universal treatment standards of §268.48. [RO 14220]

II. Can debris that exhibits a characteristic and that is treated using an immobilization technology in Table 1 of §268.45 be disposed in a Subtitle D landfill?

Only if the treated debris no longer exhibits a characteristic. [RO 13681, 14446]

> **Case Study 13-12: Alternative Treatment Standards for Listed Debris**
>
> I. If debris is contaminated with a listed waste for which a treatment method is specified, can the debris be treated by one of the alternative debris treatment standards of §268.45?
>
> Yes. However, consider the following example. Debris is contaminated with P040, for which the nonwastewater treatment standard in §268.40 is combustion (CMBST). If the debris is treated by water washing, §268.45(d)(1)(ii) requires that any treatment residue (i.e., washwater) be treated according to the waste-specific treatment standard (i.e., for wastewater, CARBN or CMBST). [RO 11815, 14220]
>
> II. Can debris that is contaminated with a listed waste be treated by an immobilization technology and then disposed in a Subtitle D facility?
>
> No. This is prohibited by §268.45(c).

13.8.3 Constituents subject to treatment

If debris exhibits the toxicity characteristic (i.e., it carries waste codes D004–D043), the constituents subject to treatment are those that create the characteristic. [§268.45(b)(1)]

If debris is contaminated with listed wastes, the contaminants subject to treatment are those identified in the §268.40 table of treatment standards for the associated listed waste code. For example, if concrete is contaminated with F006 waste, the contaminants subject to treatment are cadmium, chromium, cyanides, lead, nickel, and silver.

13.8.4 Is debris treatment subject to permitting?

One of the most difficult aspects of dealing with contaminated debris under the LDR program is dealing with the "treatment requires a permit" issue. When hazardous debris is treated, including treatment by the alternative methods of §268.45, a permit is required, unless one of the permitting exemptions applies. (Refer to Chapter 7 for a discussion of the various permitting exemptions.)

Q *An intact building will be decontaminated before it is demolished. The building or demolition residue of the building would meet the definition of "debris." If contaminants are removed from a building prior to demolition, is this considered to be hazardous waste treatment requiring a permit?*

A No. EPA considers the actual removal of the contaminants to be the point of waste generation. An intact, standing building continues to perform the essential function of a building and is not considered to be "discarded" until it is actually destroyed. [RO 11841] "[M]aterials that might at some later time become debris, such as equipment or building structures, but that are still in use are not subject to the treatment standards. Such in-use material is not a solid waste because it has not been discarded or intended for discard, as these terms are used…." [August 18, 1992; 57 *FR* 37222]

Although we haven't seen any guidance on this topic, our experience is that regulators will usually allow building components, that have been treated (before demolition) to meet the §268.45 alternative debris treatment standards, to be considered nonhazardous once the building components are demolished and become "debris." This would be a self-implementing no-longer-contains determination.

A September 2004 report issued by EPA addresses construction and demolition (C&D) debris wastes. The publication, entitled *RCRA in Focus: Construction, Demolition, and Renovation*, EPA/530/K-04/005, is available from http://www.epa.gov/osw/inforesources/pubs/infocus/rif-cd.pdf. The publication is intended to provide a basic understanding of the regulatory requirements for hazardous C&D waste. Helpful features include information on managing typical hazardous C&D wastes and a hazardous waste requirements checklist for C&D projects.

13.8.4.1 Containment buildings

EPA's strategy for dealing with contaminated debris involves "containment buildings" as defined in Parts 264/265, Subpart DD. These are essentially enclosed buildings where debris can be treated. The buildings are designed and constructed to prevent releases of fugitive dust emissions and, when used to manage liquids, typically have an impermeable floor, a collection system to remove liquids from this primary barrier, and an underlying secondary containment system with leak detection/liquid collection systems.

In our experience, these units are somewhat uncommon; however, we do find them at facilities that are managing large quantities of hazardous debris (e.g., government facilities and large manufacturing sites that frequently shut down and decommission equipment/processes).

Containment buildings can be put into operation at the three types of RCRA facilities (generator sites, interim status facilities, and permitted facilities) using different approaches, as discussed below. [RO 13609]

13.8.4.1.1 Generator sites

Under the 90-day accumulation provisions in §262.34(a)(1)(iv), generators may accumulate/treat hazardous debris in containment buildings that are in compliance with the Part 265, Subpart DD standards. A certification by a qualified professional engineer that the building meets design standards must be placed in the facilities records prior to operating the unit.

13.8.4.1.2 Interim status facilities

Interim status facilities may add new or additional treatment or storage capacity (such as containment buildings) using the provisions of §270.72(a)(2–3). The facility must submit a revised Part A permit application and a justification explaining the need for the change, which must be approved by the authorized agency before starting construction.

13.8.4.1.3 Permitted facilities

If a permitted facility is merely converting an enclosed hazardous waste pile into a containment building, a Class 2 permit modification is required. [§270.42, Appendix I, Item I-6] Construction to implement Class 2 changes can generally start within 60 days after submitting the modification request. [§270.42(b)(8)] If the agency does not respond within 90 days (or 120 days if it requests an extension), the facility is automatically authorized to operate the containment building for 180 days if it complies with Part 265, Subpart DD standards. If the agency still hasn't responded after this automatic authorization expires, the facility is authorized to operate for the life of the permit. [§270.42(b)(6)(iii–v)]

If a permitted facility already has a containment building and wants to increase its capacity by up to 25%, it may also use the Class 2 permit modification process to do so. [§270.42, Appendix I, Item M-1-b]

Class 3 permit modifications are required at facilities with no existing containment building capacity. [§270.42, Appendix I, Item M-1-a] Class 3 permit modifications require formal agency approval and no deadline is imposed on the agency for taking action. Temporary authorizations for 180 days (with a 180-day extension) are also provided under §270.42(e).

13.8.5 Miscellaneous debris issues

The following questions and answers explore some details of the LDR debris program that have caused concern.

Q *Table 1 in §268.45 identifies microencapsulation as an immobilization technology. The performance standard specifies: "Leachability of the hazardous contaminants must be reduced." Is a specific test required for demonstrating reduced leachability?*

A No. The regulations do not require that a particular method be used. One way to demonstrate reduced leachability would be to "determine the potential leachability of the toxicants before and after treatment by the TCLP test. If the leachability of the toxicants has been reduced, you have met the performance standards." [RO 13575] Information on specific factors that should be considered when evaluating microencapsulation for treatment and disposal of mercury-contaminated hazardous debris is contained in RO 14685. Note that free liquids (including mercury) are prohibited under RCRA from land disposal in microencapsulated debris. [August 18, 1992; 57 FR 37235, RO 14685, 14711]

Q: *Is it permissible for either a TSD facility or a generator to shred hazardous debris prior to macroencapsulation?*

A: There is no prohibition against shredding debris prior to macroencapsulation. However, EPA guidance states that material with a particle size less than 60 mm is amendable to conventional treatment methods rather than the alternative methods for debris. [August 18, 1992; 57 FR 37235] Furthermore, §268.45, Table 1, Footnote 5 also applies to macroencapsulation and states that if the particle size is reduced so that the material no longer meets the 60-mm minimum size limit for debris, then the most stringent §268.40 treatment standard applies. [RO 14241]

Q: *Personal protective equipment (PPE) has come in contact with a listed commercial chemical product or manufacturing chemical intermediate (i.e., a P- or U-chemical). Does EPA recognize an exemption for discarded PPE contaminated with de minimis losses of these chemicals?*

A: No. The PPE would typically be considered debris contaminated with a listed chemical under the contained-in policy. It must be managed as hazardous waste until it no longer contains the listed waste. This could be accomplished by treating the PPE using one of the alternative technologies described in Table 1 of §268.45 (such as washing). [RO 14095]

Q: *Refractory brick in a hazardous waste incinerator has come in contact with numerous listed wastes and, when disposed, carries the associated listed waste codes via the contained-in policy. How should the brick be handled to comply with the LDR standards?*

A: First, check the facility's closure plan to determine if it specifies how the brick must be handled. If the closure plan is silent, you have four options: 1) treat the brick to meet the §268.40 standards for each waste code, 2) use one of the alternative debris treatment standards in §268.45, 3) obtain a no-longer-contains determination from EPA or the state, or 4) request a treatment variance per §268.44. [RO 14515]

Q: *Macroencapsulation (MACRO) shows up in two places: 1) as a specified method in §268.40 for D008 radioactive lead solids, and 2) as an alternative treatment standard (i.e., one of the immobilization technologies) for hazardous debris in Table 1 of §268.45. When comparing the definition of MACRO in Table 1 of §268.42 with that in Table 1 of §268.45, one significant difference is apparent: the use of tanks or containers is not allowed for D008 radioactive lead solids. Can tanks or containers be used to macroencapsulate hazardous debris?*

A: Yes. A commercial waste management firm asked EPA's approval of a process described as follows: "A jacket of inert inorganic material is placed around the hazardous debris as the encapsulating agent in a high-density polyethylene vault. The lid of the vault is secured and the unit is disposed in a Subtitle C (hazardous waste) landfill."

EPA's response was: "It is EPA's determination that your treatment process meets the definition of macroencapsulation for hazardous debris, subject to an evaluation that the tank or container is structurally sound and resistant to degradation, in order to substantially reduce exposure to potential leaching media. As you allude to in your letter, merely placing hazardous debris in a tank or container, except under special circumstances where the container is made of noncorroding materials (e.g., stainless steel), would not fulfill the macroencapsulation treatment standard…. [EPA] is clarifying that for the treatment of hazardous debris, the definition of macroencapsulation in §268.45 should be used, and for the treatment of D008 radioactive lead solids, the definition in §268.42 should be used." [RO 13762; see also RO 13655]

Methods for ensuring that the encapsulating material completely encapsulates the debris for compliance with the §268.45, Table 1 performance standard are specific to the technology used. For example, "leak-tightness or pressure testing of high-density polyethylene pipes or containers has been approved for testing of treated debris. Visual inspection may be appropriate for verifying that sprayed-on or applied coatings have complete integrity, without cracks, voids, or protruding waste to ensure that the hazardous debris is completely encapsulated." [RO 14685] This memo also provides information on specific factors that should be considered when evaluating macroencapsulation for treatment and disposal of mercury-contaminated debris.

13.9 Mixed waste standards

Mixed waste is defined as a mixture that contains both hazardous waste and radioactive materials (i.e., source, special nuclear, or by-product material) subject to the Atomic Energy Act (AEA). The hazardous waste components are subject to the same LDR treatment standards as any other hazardous waste. [§268.42(d)] The real problem with mixed wastes is that limited treatment and disposal capacity is available for dealing with these wastes.

13.9.1 Mixed wastes at DOE facilities

Because of a lack of treatment capacity, large quantities of mixed wastes are in long-term storage at DOE facilities awaiting the development of additional capacity. This extended storage of mixed wastes was made possible by the Federal Facilities Compliance Act (FFCA) of 1992 (Public Law 102-386). This law allowed DOE to continue storing mixed waste as long as 1) the waste is managed in accordance with other applicable requirements; and 2) an existing permit, agreement, or administrative/judicial order does not apply to the waste. DOE facilities were required to develop site treatment plans for treating mixed wastes so they can be disposed in accordance with the LDR program. EPA subsequently approved these plans and, in essence, DOE facilities have sovereign immunity from violations of the LDR storage prohibition as long as they are in compliance with their plan. [RO 14416]

Waste treatment subcategories have been developed for seven types of mixed wastes:

1. D002 and D004–D011 radioactive high-level wastes generated during the reprocessing of fuel rods,
2. D006 radioactive cadmium-containing batteries,
3. D008 radioactive lead solids,
4. D009 elemental mercury contaminated with radioactive materials,
5. D009 radioactive hydraulic oil contaminated with mercury,
6. D009 radioactive mercury-containing batteries, and
7. D011 radioactive silver-containing batteries.

All other radioactive mixed wastes are subject to the treatment standards that apply to the hazardous portion of the waste.

13.9.2 Mixed wastes at commercial sites

Mixed wastes that are generated at commercial sites (power plants, hospitals, laboratories, etc.) are also subject to the LDR program. However, the long-term storage allowed at DOE sites by the FFCA does not apply to these facilities. Instead, EPA issued a "Policy on Enforcement of RCRA Section 3004(j) Storage Prohibition at Facilities Generating Mixed Radioactive/Hazardous Wastes." [August 29, 1991; 56 *FR* 42730] This policy, which was last extended to October 31, 2001 [November 6, 1998; 63 *FR* 59989], says that enforcement of the storage provision for radioactive mixed wastes will be a low priority for commercial generators who produce <1,000 cubic feet per year of mixed waste as long as the mixed waste is managed in an environmentally responsible manner.

Much of the pressure on commercial generators of mixed waste was relieved by a May 16, 2001 rule. [66 *FR* 27218] This rule allows NRC-licensed facilities to store mixed waste in tanks and containers in compliance with NRC license conditions. Wastes stored under these conditions will be conditionally exempt from RCRA. When these wastes are ultimately disposed, however, they will have to meet LDR treatment standards. Section 4.4.5 discusses the details of this exemption from RCRA at NRC-licensed facilities.

Q: *A drained lead-acid battery is radioactively contaminated. Would the battery be subject to the lead-acid battery treatment standard (lead recovery), or the radioactive lead solids treatment standard (macroencapsulation)?*

A: According to EPA, "the appropriate treatment standard is macroencapsulation. This treatment standard applies not only to lead shielding, but to other elemental forms of lead. Thus, there is latitude in the treatment standard to permit its application to radioactive lead-acid batteries. We also believe that macroencapsulation is appropriate because it would require less worker handling than lead recovery, and we want to minimize worker exposure to radioactivity. Furthermore, lead recovery of these batteries would radioactively contaminate the entire mass of lead that was recovered, making it unusable." [RO 14554]

Q: *Mercury-contaminated soils (>260 mg/kg mercury) are also contaminated with low-level radioactive material. The LDR treatment standard for this waste is RMERC (recovery of mercury); however, any mercury recovered from this waste will be unusable because it will be radioactive. Therefore, this recovered mercury will have to be disposed and will fit into a different D009 subcategory—elemental mercury contaminated with radioactive materials. The treatment standard for this waste is amalgamation (AMLGM). Are any other options available to the generator of this waste?*

A: Yes. The generator may submit an application to EPA for approval of an alternative treatment method as described in §268.42(b). In this specific case, the generator followed these procedures and was granted an alternative treatment standard of 0.2 mg/L TCLP for mercury. [RO 14270]

Q: *Submarine reactor compartments consist of radioactive components, surrounded by lead shielding, all of which is encased in a thick, sealed steel jacket. Do the compartments, as generated, meet the D008 radioactive lead solids treatment standard of macroencapsulation (MACRO), and is any testing for leachable lead required?*

A: The definition of MACRO in §268.42 specifically allows for a "jacket of inert inorganic materials," such as a steel jacket. Due to size and structure, the jacket does not meet the definition of a drum or container; therefore, it meets the standard of MACRO. Furthermore, this is a technology-based standard and does not require a TCLP analysis for lead. Such an analysis would require crushing or grinding of components of the reactor compartments, and could pose a high radiation exposure risk. Avoiding these risks was the whole purpose of establishing the MACRO standard. [RO 13393]

Q: *What is the treatment standard for D008 radioactive lead solids that happen to be tanks or containers?*

A: Although the MACRO standard cannot be met by placing radioactive lead solids in a tank or container, if the solids are themselves tanks or containers, they may be treated with an application of surface coatings or jacketing to reduce exposure to leaching media. [RO 14091]

Q: *Plastic-coated, lead lined gloves will be disposed as D008 radioactive lead solids. Do they comply with the MACRO standard as generated?*

A: Yes. Provided that none of the lead is exposed (i.e., the entire surface of the lead is coated) and provided that the coating provides a substantial reduction in surface exposure to potential leaching media, the LDR standard is met at the point of generation. [RO 13437]

Q *A commercial facility that generates mixed waste is relying on EPA's low-enforcement-priority policy to continue storing the waste. If a legitimate recycling facility is now able to recycle the waste, can the commercial facility continue to store the mixed waste?*

A No. The enforcement policy does not extend to generators who do not avail themselves of legitimate recycling (or treatment) opportunities. [RO 14101, 14171]

13.10 The LDR storage prohibition

Because treating wastes to meet LDR standards can be relatively expensive, EPA was concerned that generators would be tempted to put their wastes into indefinite storage so as to avoid these costs. In order to forestall this activity, the agency promulgated a storage prohibition in §268.50. A generator may accumulate wastes onsite in tanks, containers, or containment buildings "solely for the purpose of the accumulation of such quantities of hazardous waste as necessary to facilitate proper recovery, treatment, or disposal." [§268.50(a)(1)]

A similar storage prohibition applies to TSD facilities in §268.50(a)(2). A one-year time limit is imposed on such storage. (Note that such waste storage at TSD facilities is subject to permitting.) During this one-year storage period, the appropriate regulatory agency has the burden of proof to show that storage is not "necessary" as described above. After the one-year storage period expires, the burden of proof shifts to the TSD facility. Hence, storage of wastes subject to LDR standards for more than one year is allowed only for good cause.

The federal regulations do not specify how long wastes can remain in a satellite accumulation area (if the 55-gallon/1-quart limits aren't exceeded—see Section 6.2.4). Conceivably, wastes might accumulate for several years before the quantity limit is reached. A regulatory concern can arise under these circumstances, however, because of the above-mentioned storage prohibition. As soon as a hazardous waste is generated, it is subject to LDR, including §268.50, which states that if a waste is stored for more than one year, the burden of proof is on the owner/operator to show why this is necessary. In 1990 guidance, however, EPA clarified that the accumulation time for the LDR storage prohibition starts when the waste is moved to a central accumulation area (i.e., a 90/180-day or permitted area). Therefore, EPA's perspective is that hazardous waste accumulation in satellite accumulation areas is not subject to the one-year storage prohibition. [OSWER Directive 9555.00-01]

The storage prohibition doesn't apply to certain wastes:

1. Wastes subject to a case-by-case extension, no-migration exemption, or national capacity variance [§268.50(d)];
2. Wastes that meet LDR treatment standards [§268.50(e)]; and
3. Remediation wastes stored in a staging pile [§268.50(g)].

Q *If a facility is storing wastes subject to the LDR standards for more than one year, is it required to notify EPA or the state of this storage?*

A For storage more than one year, the burden is on the facility owner or operator to demonstrate that such storage time is necessary. The owner/operator does not have to notify the agency of storage for more than one year. The burden of proof only applies in the event of an EPA inspection or for enforcement purposes. [RO 12845, 12851]

Q *A characteristic hazardous waste has been treated to remove the characteristic and is being stored at a TSD facility. Even though the waste is no longer hazardous, UHCs have not been treated to meet treatment standards. Is the waste still subject to the LDR storage prohibition?*

A The waste remains subject to the storage prohibition even though it is no longer hazardous. Because the LDR treatment standards apply at the point of generation, the storage prohibition applies until the waste is treated to fully meet the standards. One year after the waste was originally generated, the facility storing the waste will have the burden of proving that the waste is still being stored to facilitate proper recovery, treatment, or disposal. [RO 14048]

Q *Lead-acid batteries that are to be recycled are stored on an earthen pad. Is this allowable under the LDR storage prohibition, or would the batteries have to meet treatment standards (lead recovery) before being placed on the pad?*

A Wastes may be stored in tanks or containers without meeting treatment standards. The shell surrounding an intact lead-acid battery is considered to be a container. Therefore, as long as the batteries are being stored for the purpose of facilitating proper recovery, treatment, or disposal (as specified in §268.50), the storage is allowable. The batteries do not have to meet treatment standards before being stored. [RO 13339]

13.10.1 Mercury Export Ban Act

The Mercury Export Ban Act (MEBA or the Act, Public Law No. 110-414), which became law on October 14, 2008, is intended to reduce the availability of elemental mercury in domestic and international markets. With a few exceptions, the export of elemental mercury from the United States has been prohibited effective January 1, 2013. To avert unsafe accumulation of waste elemental mercury, the Act requires the Department of Energy (DOE) to designate a facility to be used for long-term storage of waste elemental mercury generated in the United States.

This ban can directly affect hazardous waste TSD facilities that manage waste elemental mercury. Here's how.

The LDR treatment standard for the high-mercury subcategories of D009 and U151 is recovery of mercury via roasting or retorting. The concept when this treatment standard was set in the early 1990s was that the mercury would be recovered and reused rather than land disposed.

However, domestic uses for mercury are declining. Therefore, since the effective date of the MEBA, elemental mercury recovered by mercury recovery facilities will likely have to be stored as hazardous waste. In general, storing hazardous waste for longer than one year is prohibited under the LDR storage prohibition in §268.50. However, MEBA specifically excludes elemental mercury being stored in the yet-to-be-identified DOE repository from the one-year storage limit.

The Act also allows elemental mercury to be stored for longer than one year at any RCRA-permitted TSD facility under the following conditions:

1. DOE is unable to accept the mercury at its yet-to-be-designated repository for reasons beyond the control of the owner or operator of the TSD facility,

2. The owner or operator of the TSD facility certifies in writing that it will ship the mercury to DOE's repository when it is able to accept the mercury, and

3. The owner or operator of the TSD facility certifies that it will not sell or otherwise place the mercury in commerce.

13.11 The LDR dilution prohibition

EPA has always been uneasy about allowing hazardous wastes to be diluted prior to disposal. This is a valid concern, because one of the easiest ways to meet a concentration-based treatment standard would be to simply dilute the waste. Therefore, the LDR program includes a dilution prohibition in §268.3 stating that dilution as a substitute for adequate treatment is prohibited.

Note that this dilution prohibition applies only where an element of land disposal is involved. For example, if listed wastewater is treated in tanks and discharged under an NPDES permit, no land disposal is involved, no LDR treatment standard applies, and the LDR dilution prohibition is not triggered. If a hazardous sludge is generated in the wastewater treatment process, the dilution prohibition would apply from its point of generation to its point of disposal.

In some circumstances, dilution occurs in conjunction with adequate treatment, as discussed in the next two sections.

13.11.1 Aggregation for centralized treatment

For some large-scale operations, it is common to mix different wastes together prior to treatment. One such situation is illustrated in Figure 13-14, where different process wastes are consolidated and delivered to a RCRA-permitted incinerator. As each of the process wastes is consolidated, it is diluted; in some cases, this dilution might be sufficient to meet LDR treatment standards even though no treatment has occurred.

EPA terms this type of system "aggregation for centralized treatment." The only way to avoid the inherent dilution in this system would be to treat each waste stream individually—this would clearly be uneconomical in large plants. Instead, the agency allows this inherent dilution as long as the type of treatment ultimately provided will remove or destroy the contaminants being diluted. Referring to Figure 13-14, incineration will destroy benzene, trichloroethylene, and methylene chloride from the combined waste; hence, dilution of these waste streams is permissible. On the other hand, incineration will not destroy P011—arsenic pentoxide. If this waste were consolidated with the rest, the agency would consider impermissible dilution to have occurred.

Q *During a site cleanup, contaminated soil from various onsite locations is consolidated. Could this be considered impermissible dilution?*

A "If mixing occurs through the normal consolidation of contaminated soil from various portions of a site that typically occurs during the course of remedial activities or in the course of normal earthmoving and grading activities, then the agency does not consider this to be intentional mixing of soil with nonhazardous soil for the purposes of evading LDR treatment standards. Therefore, it is not viewed as a form of impermissible dilution…."

"Some situations may require soil mixing, as part of a pretreatment process, to facilitate and ensure proper operation of the final treatment technology to meet the LDR treatment standards. If the mixing or other pretreatment is necessary to facilitate proper treatment in meeting LDR standards, then dilution is permissible. For example, addition of less-contaminated soil may be needed to adjust the contaminated soil Btu value, water content, or other properties to facilitate treatment. These adjustments would be for meeting the energy or other technical requirements of the treatment unit to ensure its proper operation. The agency views this type of pretreatment step as allowable, provided the added reagents or other materials produce chemical or

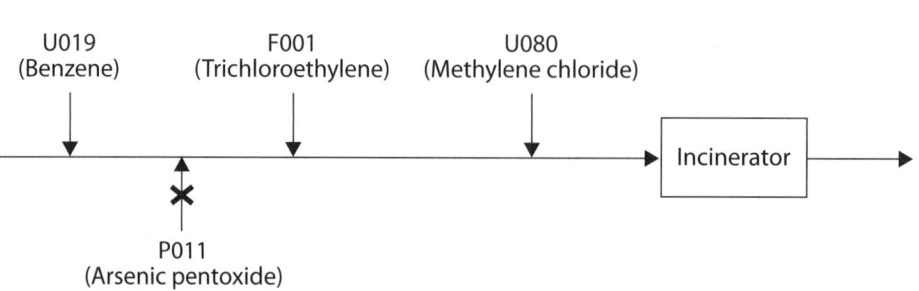

Figure 13-14: Aggregation for Centralized Treatment

Source: McCoy and Associates, Inc.

physical changes and do not 1) merely dilute the hazardous constituents into a larger volume of waste so as to lower the constituent concentration, or 2) release excessive amounts of hazardous constituents to the air." [EPA/530/R-02/003 (cited earlier), RO 14338]

Q: *Fifteen thousand gallons per day of F039—multisource leachate is produced at a manufacturing facility. The leachate contains organic and heavy metal constituents above LDR treatment standards. The leachate is mixed with 4.5 million gallons per day of process wastewater that is not subject to LDR standards. The combined wastewater undergoes primary treatment via mixing, equalization, neutralization, and clarification in tanks prior to biological treatment in surface impoundments. (The neutralization/clarification consists of simple settling, not chemical precipitation.) When the combined water enters the biological impoundments, it meets the LDR treatment standards for organics and metals. Does this system involve impermissible dilution under the LDR rules?*

A: Yes. EPA decided that the primary treatment was simple aggregation that produces a homogeneous mixture for biological treatment of organics. Adequate treatment of metals would require some type of treatment like chemical precipitation; biological treatment only provides incidental removal of metals. Therefore, the LDR treatment standards for metals in F039 are being achieved by impermissible dilution. [RO 13414]

13.11.2 Dilution as a consequence of treatment

Figure 13-15 illustrates another type of permissible dilution. In this case, large volumes of treatment reagent (stabilization agents) are added to a waste. Any dilution that is inherent in a process of this sort is permissible as long as the reagents are effective in treating the waste. In this example, the stabilization agents would have to reduce the leachability of the waste. If the reagent is simply sand, no effective treatment would be occurring, and the agency would consider the process to be impermissible dilution.

The regulations specifically identify a case of impermissible dilution of this sort: "It is a form of impermissible dilution, and therefore prohibited, to add iron filings or other metallic forms of iron to lead-containing hazardous wastes in order to achieve any land disposal restriction treatment standard for lead." [§268.3(d)]

Q: *The treatment standard for F006, wastewater treatment sludges from electroplating operations, identifies both cyanides and heavy metals as hazardous constituents. Would stabilization of this waste be considered to be impermissible dilution?*

A: EPA considers stabilization of cyanide to be impermissible dilution because, while the leachability of the cyanide may be reduced, the compound is not destroyed. The agency believes that

Figure 13-15: Dilution as a Consequence of Treatment

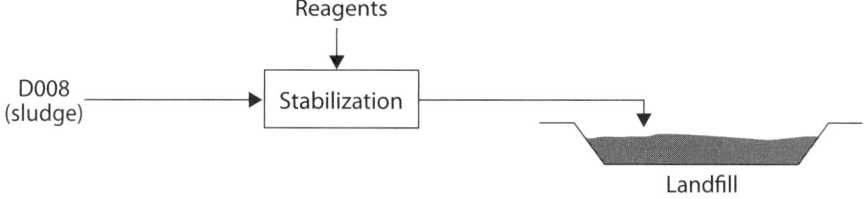

Source: McCoy and Associates, Inc.

Congress intended that cyanides be destroyed as a precondition to land disposal. [RO 11545, 13444]

Q *When organic-containing hazardous wastes are solidified, would EPA consider this to be a form of impermissible dilution?*

A Maybe. Stabilization is a well-recognized process for treating metal-containing wastes. Stabilizing organic wastes, however, may be considered to be impermissible dilution when:

1. During stabilization, hazardous organic constituents are released to the air in an uncontrolled manner at levels that pose a risk to human health and the environment.

2. Organics in the waste prevent the stabilization reagents from bonding with metals in the waste.

3. Stabilization reagents such as lime or cement kiln dust increase the pH, solubility, and leachability of arsenates, chromates, and other metals.

4. Analytical tests after stabilization do not show adequate binding of organic constituents. LDR treatment standards for organic constituents in nonwastewaters are based on a totals (not leachable) analysis. For untreated wastes, an aggressive extractant (such as a solvent) is used to extract and quantify organics. Stabilization reagents (such as clays, silica, alumina, zeolites, and/or activated carbon) must be chosen that will bond with organics such that they are not extracted after stabilization. As long as an appropriate analytical method with proper QA/QC is chosen for "before" and "after" testing, EPA will not consider impermissible dilution to be the cause of concentration reductions because of the known adsorptive capability of the reagents mentioned earlier.

5. Site-specific leachate properties at the disposal site break down the adsorptive capacity of the reagents over time.

This information is extracted from "Draft Interpretive Memorandum on Stabilization of Organic-Bearing Hazardous Wastes to Comply with RCRA Land Disposal Restrictions," September 2001, available at http://www.understandrcra.com/rccd/StabilizationOfOrganics.pdf.

13.11.3 McCoy's dilution diagram

EPA's rules and policies regarding when dilution is permissible are very fragmented. Therefore, we have prepared the logic diagram in Figure 13-16 to simplify the issue. The diagram is self-explanatory and allows the user to determine when dilution is allowable and when it is prohibited. Pertinent references to the regulations or *Federal Register* preambles are included. Dilution examples using this diagram are given in Case Study 13-13.

Here are some additional examples that involve dilution issues.

Q *Waste paint is both ignitable and TC hazardous for lead (D001, D008). When blended with other wastes into a hazardous waste fuel, would this be considered illegal dilution of the lead component?*

A Impermissible dilution, as it relates to combustion of wastes, is addressed in §268.3(c). For this specific example, §268.3(c)(4) applies: the waste (D008) is cogenerated with wastes for which combustion is a required method of treatment (D001). Under these circumstances, blending with other wastes to produce a hazardous waste fuel is not impermissible dilution. [RO 11545, 11815]

Q *A generator produces an explosive, heavy metal bearing waste (D003, D004–D011). The explosive component of the waste is organic. The generator would like to incinerate this waste. Does the dilution prohibition in §268.3(c) forbid this practice?*

A No. EPA believes that incineration of this material is allowed under §268.3(c)(5): "The waste is subject to federal and/or state requirements necessitating reduction of organics (including biological agents)." [RO 14501]

13.12 LDR paperwork requirements

Paperwork and recordkeeping are extremely important components of the LDR program. In many

Figure 13-16: Determining When Dilution Is Permissible

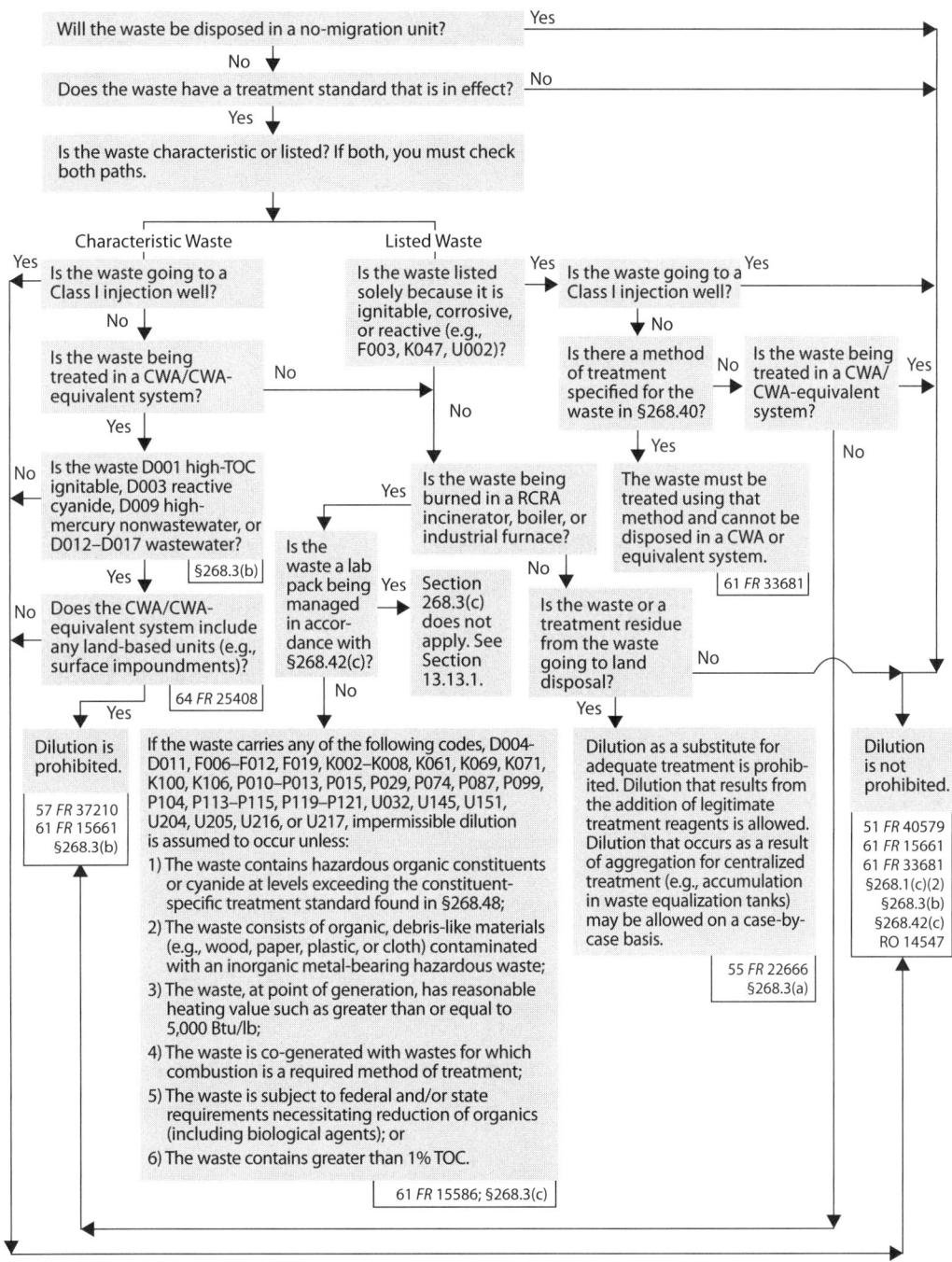

Source: McCoy and Associates, Inc.

Case Study 13-13: Dilution Prohibition

Using the dilution diagram (Figure 13-16), determine whether dilution is allowed under any of the following circumstances:

a) A facility generates F005 spent solvent. The generator wants to dilute the solvent with water until it meets LDR treatment standards and then send it to an underground injection well that does not have a no-migration exemption.

b) F003 spent solvent (less than 10% TOC) is directed to a process sewer where it is diluted with other plant wastewaters, rendering the commingled mixture nonignitable. The noncharacteristic wastewater flows to a lagoon that discharges to surface water through an NPDES-permitted outfall.

c) D001 high-TOC waste is added to a plant wastewater system that includes a surface impoundment. Water from the impoundment is discharged under an NPDES permit.

d) Spent acid (D002) is mixed with other wastewaters before being discharged to a surface impoundment that is part of a CWA system.

e) Lead-contaminated soil (D008) is mixed with clean soil to render it nonhazardous prior to disposal in a municipal landfill.

Answers: Dilution is allowed for b) and d); it is prohibited for a), c), and e).

---

cases, the only way that regulators can determine if a facility is in compliance with LDR requirements is by examining the various forms and data required by the regulations. In other words, if a facility's LDR paperwork is in order, the likelihood that they will encounter LDR enforcement problems is greatly reduced.

The basic intent of LDR paperwork is illustrated in Figure 13-17. In this case, a generator's waste is sent offsite to a treatment facility that will have the responsibility of treating the waste to meet LDR standards. The treatment facility then sends the treated waste (which meets LDR standards) to a disposal facility.

Whenever a generator sends a hazardous waste offsite, the initial shipment of the waste must be accompanied by an LDR notification. The purpose of the notification is to inform the receiving facility that the waste is subject to LDR requirements. In some cases, this notification also informs the receiving facility of the specific chemicals (e.g., UHCs) that are subject to treatment. The paperwork requirements applicable to generators are found in §268.7(a).

Using information contained in the generator's notification (e.g., waste codes and identified UHCs), the treatment facility treats the waste to meet the appropriate treatment standard and then sends the treated waste to an offsite disposal facility. The initial shipment of this waste is accompanied by a notification/certification. The notification informs the disposal facility that the waste is subject to

Figure 13-17: Basic Intent of the LDR Paperwork Requirements

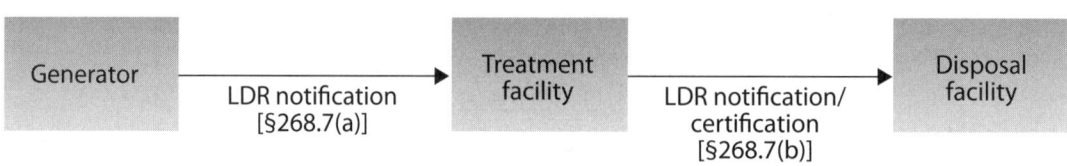

Source: McCoy and Associates, Inc.

LDR treatment standards and the certification stipulates that the standards have been met. This certification (along with confirmatory testing by the disposal facility) is the paperwork that allows the waste to be land disposed. [RO 13181] The paperwork requirements applicable to treatment facilities are found in §268.7(b).

An April 4, 2006 final rule [71 FR 16862] allows generators to choose not to determine if their hazardous waste requires treatment prior to land disposal. [§268.7(a)] If the generator chooses this approach, the waste must be manifested to a RCRA-permitted hazardous waste treatment facility who will have the responsibility for determining if treatment is required. In this case, the LDR notification sent with the waste will include only 1) the waste code(s), 2) the manifest tracking number of the first shipment, and 3) the following statement: "This hazardous waste may or may not be subject to the LDR treatment standards. The treatment facility must make the determination."

In the case where a generator has not made the determination of whether the hazardous waste requires treatment to meet LDR treatment standards, the treatment facility must make this determination. If the treatment facility determines that the waste, as received, meets LDR standards, no treatment is required; conversely, if the waste needs treatment to meet LDR standards, the treatment facility will provide it. Standard paperwork will then be used to accompany the treated waste to the disposal facility.

The disposal facility generally doesn't have to prepare any LDR paperwork. Instead, they must be sure that the paperwork provided by others is in order before they dispose the waste. [§268.7(c)] The disposal facility must also test the waste in accordance with its waste analysis plan to confirm that the waste meets the treatment standards.

The notifications/certifications described above are not standardized forms. Instead, EPA simply specifies the information that must be included in the paperwork. In most cases, the source of the forms used to complete the paperwork is the offsite facility that will receive the waste. These facilities typically require forms to be filled out that are part of their waste acceptance procedures; the information required by the LDR program is often included on these forms. In a few situations (to be described later) no outside source of forms for paperwork will be available, and the documents will have to be prepared by the facility managing the waste.

13.12.1 Figuring out what forms to use

Unfortunately, the most difficult aspect of complying with the LDR program involves figuring out what the exact paperwork requirements are for a given waste. The difficulty derives from the fact that §268.7, which specifies paperwork requirements, is very poorly written and difficult to follow. Additionally, not all paperwork requirements appear in this section; §268.9(d) also specifies important paperwork associated with characteristic wastes.

To make the paperwork identification process easier, we developed a series of logic diagrams that may be used by generators and treatment/recycling facilities. Figure 13-18 and Figure 13-19 apply to generators; Figure 13-20 applies to treatment and recycling facilities. For convenience in using these figures, we reproduced the *Generator Paperwork Requirements Table* from §268.7(a)(4) as Table 13-2 and the *Treatment Facility Paperwork Requirements Table* from §268.7(b)(3) as Table 13-3.

13.12.1.1 The §268.7(a)(7) one-time notice

Perhaps the most often overlooked LDR form is specified in §268.7(a)(7). This form is overlooked because EPA provided no trigger mechanism elsewhere in the regulations that leads to this requirement. Additionally, this form is required even if no land disposal is occurring.

Another reason for low compliance with this paperwork provision is that no one sends the generator a form to complete when the §268.7(a)(7) requirement is triggered. Instead, the generator could complete Figure 13-21 to fulfill this requirement.

CHAPTER 13 Land Disposal Restrictions

Figure 13-18: Generators' LDR Paperwork Requirements for Wastes Shipped Offsite

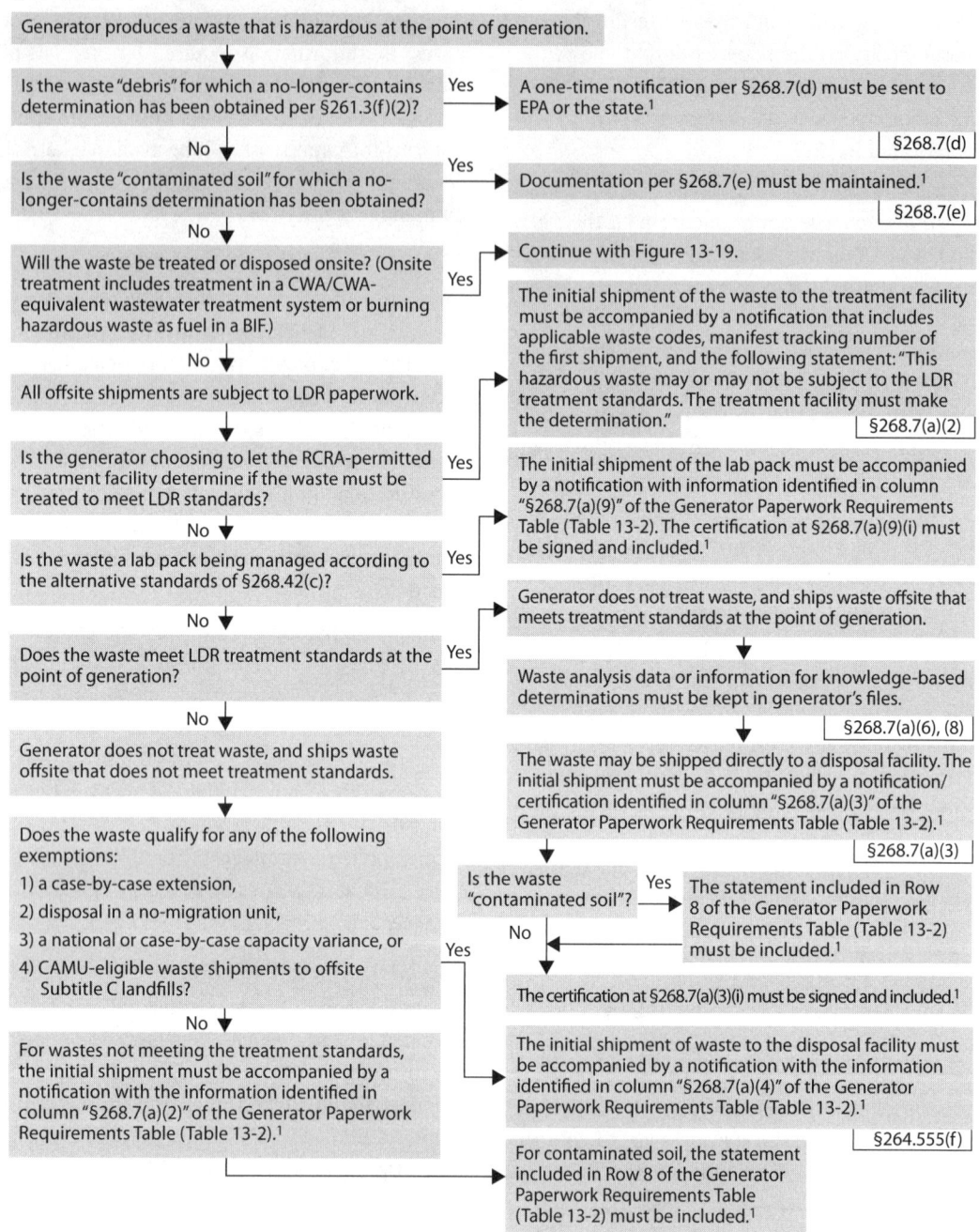

[1] Copies of LDR paperwork should be retained for at least 3 years after the waste was last disposed onsite or shipped offsite. SQGs shipping waste offsite under a tolling agreement per §262.20(e) must retain paperwork for 3 years after expiration of agreement.

Source: McCoy and Associates, Inc.

CHAPTER 13 Land Disposal Restrictions

Figure 13-19: LDR Paperwork Requirements for Wastes Treated or Disposed Onsite

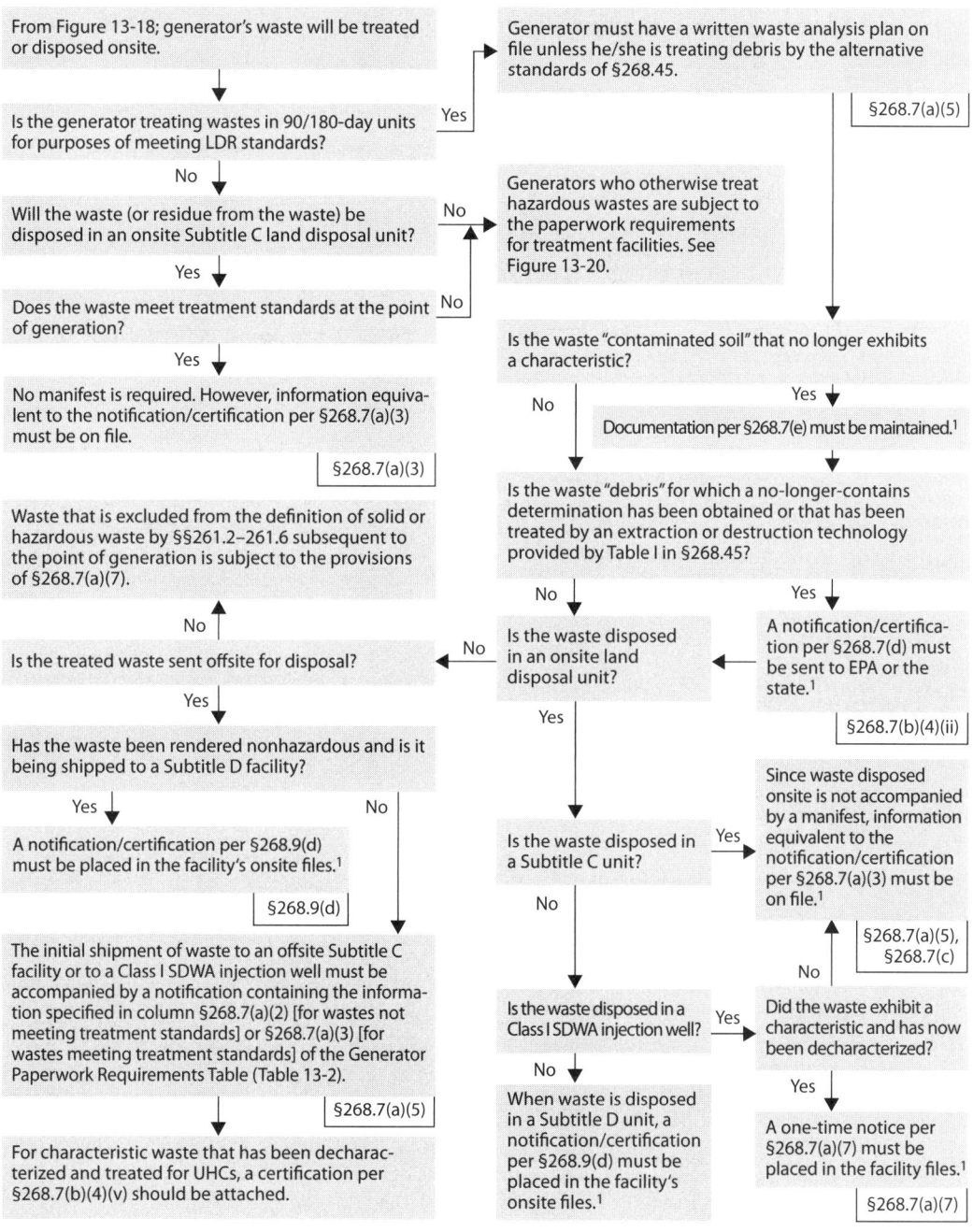

[1] Copies of LDR paperwork should be retained for at least 3 years after the waste was last disposed onsite or shipped offsite. SQGs shipping waste offsite under a tolling agreement per §262.20(e) must retain paperwork for 3 years after expiration of agreement.

Source: McCoy and Associates, Inc.

Figure 13-20: Treatment Facilities' LDR Paperwork Requirements

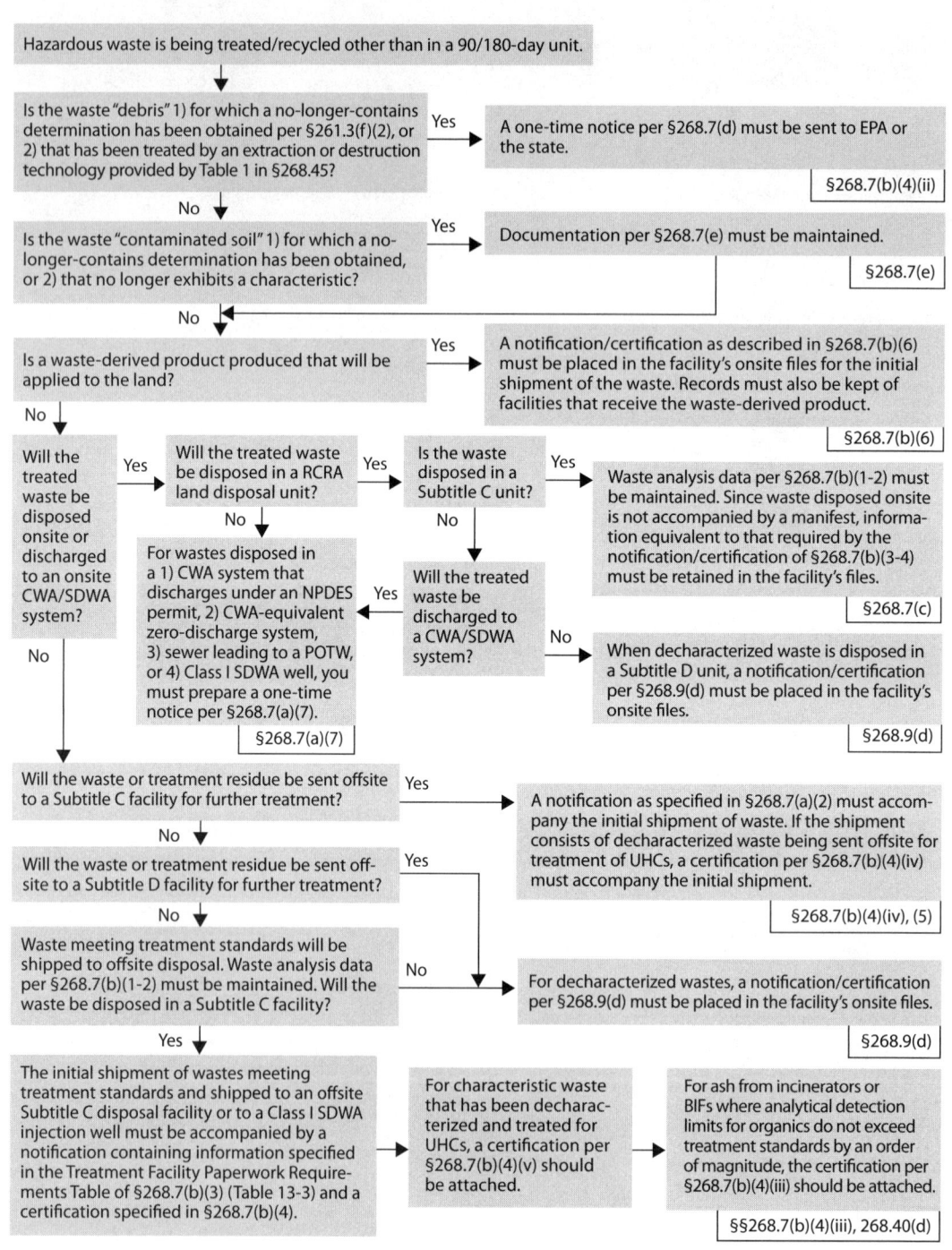

Source: McCoy and Associates, Inc.

Table 13-2: Generator LDR Paperwork Requirements Table

Required information	§268.7(a)(2)	§268.7(a)(3)	§268.7(a)(4)	§268.7(a)(9)
1. EPA hazardous waste numbers and manifest number of first shipment	✔	✔	✔	✔
2. Statement: This waste is not prohibited from land disposal			✔	
3. The waste is subject to the LDRs. The constituents of concern for F001–F005 and F039, and underlying hazardous constituents in characteristic wastes, unless the waste will be treated and monitored for all constituents. If all constituents will be treated and monitored, there is no need to put them all on the LDR notice	✔	✔		
4. The notice must include the applicable wastewater/nonwastewater category (see §§268.2(d) and (f)) and subdivisions made within a waste code based on waste-specific criteria (such as D003 reactive cyanide)	✔	✔		
5. Waste analysis data (when available)	✔	✔	✔	
6. Date the waste is subject to the prohibition			✔	
7. For hazardous debris, when treating with the alternative treatment technologies provided by §268.45: the contaminants subject to treatment, as described in §268.45(b); and an indication that these contaminants are being treated to comply with §268.45	✔		✔	
8. For contaminated soil subject to LDRs as provided in §268.49(a), the constituents subject to treatment as described in §268.49(d), and the following statement: This contaminated soil [does/does not] contain listed hazardous waste and [does/does not] exhibit a characteristic of hazardous waste and [is subject to/complies with] the soil treatment standards as provided by §268.49(c) or the universal treatment standards	✔	✔		
9. A certification is needed (see applicable section for exact wording)			✔	✔

Source: Adapted from §268.7(a)(4).

Table 13-3: Treatment Facility LDR Paperwork Requirements Table

Required information	§268.7(b)
1. EPA hazardous waste numbers and manifest number of first shipment	✔
2. The waste is subject to the LDRs. The constituents of concern for F001–F005 and F039, and underlying hazardous constituents in characteristic wastes, unless the waste will be treated and monitored for all constituents. If all constituents will be treated and monitored, there is no need to put them all on the LDR notice	✔
3. The notice must include the applicable wastewater/nonwastewater category (see §§268.2(d) and (f)) and subdivisions made within a waste code based on waste-specific criteria (such as D003 reactive cyanide)	✔
4. Waste analysis data (when available)	✔
5. For contaminated soil subject to LDRs as provided in §268.49(a), the constituents subject to treatment as described in §268.49(d), and the following statement: This contaminated soil [does/does not] exhibit a characteristic of hazardous waste and [is subject to/complies with] the soil treatment standards as provided by §268.49(c)	✔
6. A certification is needed (see applicable section for exact wording)	✔

Source: Adapted from §268.7(b)(3).

Figure 13-21: Section 268.7(a)(7) One-Time Notice to File

1. Generator information

Name   XYZ Corporation

Address   555 Industrial Parkway, Houston, TX 77002

EPA ID No.   TXD123456789

Section 268.7(a)(7) states: "If a generator determines that he is managing a prohibited waste that is excluded from the definition of hazardous or solid waste or is exempted from Subtitle C regulation under 40 CFR 261.2 through 261.6 subsequent to the point of generation (including deactivated characteristic hazardous wastes managed in wastewater treatment systems subject to the Clean Water Act (CWA) as specified at 40 CFR 261.4(a)(2) or that are CWA-equivalent, or are managed in an underground injection well regulated by the SDWA), he must place a one-time notice describing such generation, subsequent exclusion from the definition of hazardous or solid waste or exemption from RCRA Subtitle C regulation, and the disposition of the waste, in the facility's on-site files."

2. Waste description at point of generation

Waste stream(s)   Waste acid from boiler cleaning operations

Waste codes   D002

Description of waste generation   Waste acid is generated from cleaning plant boilers with hydrochloric acid.

3. Waste disposition

Subsequent exclusion or exemption   Waste acid is rendered nonhazardous in an elementary neutralization unit by the derived-from rule in §261.3(c–d). The neutralized wastewater is excluded from the definition of solid waste by the domestic sewage exclusion in §261.4(a)(1).

Current disposition   The neutralized wastewater was discharged to the sewer leading to the POTW.

Joe Smith	Joe Smith, Environmental Engineer	10/30/2014
Generator's signature	Printed/typed name & title	Date

Source: McCoy and Associates, Inc.

We prepared this form for facilities to use, and a blank form is available in electronic format on our website: http://www.mccoyseminars.com/forms.cfm. Once completed, the one-time notice is simply placed in the facility's files. The §268.7(a)(7) form isn't sent to any other entity.

In general, this form applies to any waste that was hazardous at the point of generation, but somehow dropped out of RCRA. The reason that the waste exits RCRA would be found somewhere in the regulations between §§261.2 and 261.6. The most common situations include:

1. Wastes are discharged to POTWs and are excluded from regulation under the domestic sewage exclusion of §261.4(a)(1),
2. Wastes are discharged under an NPDES permit per §261.4(a)(2),
3. Wastes are managed in a CWA-equivalent zero-discharge facility,
4. Wastes are managed in a SDWA Class I injection well,
5. De minimis losses of listed wastes lose their listed codes at the headworks of a CWA-permitted wastewater treatment system per §261.3(a)(2)(iv),
6. Hazardous waste is burned as fuel in a Bevill-exempt unit per §261.6(a)(2)(ii), and
7. Wastes are rendered nonhazardous via the mixture or derived-from rules in §261.3 and do not end up being land disposed (e.g., D003 aerosol cans that are punctured and managed as scrap metal).

The best way for a generator to determine if they have wastes subject to §268.7(a)(7) is to make a list of all hazardous wastes generated onsite. For any hazardous wastes shipped offsite for treatment/recycling/disposal, an LDR form will have been prepared and a copy should be in the generator's files. When the list of hazardous wastes is matched up with the LDR forms, any waste without a form is probably subject to the §268.7(a)(7) requirements.

13.12.2 Forms that are not available from someone else

As mentioned earlier, many of the LDR forms are obtained by generators from offsite recycling or other TSD facilities to which they will send their hazardous waste. However, in a number of circumstances, generators must prepare notifications or certifications, but forms will not be available from any offsite source. A good example is the §268.7(a)(7) one-time notice described in the preceding section. Another example is treatment of a characteristic waste by a generator, with nonhazardous treatment residues being sent to a Subtitle D landfill; the LDR paperwork in §268.9(d) is required in this situation. We have prepared a "master form" in Figure 13-22 that generators may use to satisfy these unique requirements. This form is also available in electronic format on our website: http://www.mccoyseminars.com/forms.cfm.

13.12.3 Common paperwork examples

The following questions and answers explore EPA's interpretation of LDR paperwork requirements.

13.12.3.1 The §268.7(a)(7) one-time notice

Q *A manufacturer generates a listed waste that is piped directly to a wastewater treatment unit that discharges to a POTW. What LDR notification requirements apply, if any?*

A The one-time notice at §268.7(a)(7) applies. The waste was hazardous at the point of generation and was subsequently excluded from the definition of hazardous waste at the point of discharge to the sewer leading to the POTW. This exclusion appears at §261.4(a)(1) of the regulations. [RO 13547]

Q *A D002 waste acid is generated from boiler cleaning operations. The waste acid is neutralized without a permit in an elementary neutralization unit, and the neutralized wastewater is discharged to a POTW. Does a §268.9(d) one-time LDR notice have to be completed?*

A No. Only the §268.7(a)(7) one-time notice needs to be completed and put in the

Figure 13-22: Master Form for LDR Notifications/Certifications That Are Not Available From Offsite Facilities

1. Generator information

Name _____

Address _____

_____

EPA ID No. _____

2. Receiving facility information (if applicable)

Name _____

Address _____

Manifest No. _____

EPA ID No. _____

3. Waste description at point of generation

Line item	Waste description	Hazardous waste code(s)	LDR subcategory	WW/ NWW	Underlying hazardous constituents [§268.2(i)][1]
1					
2					
3					
4					

4. Waste disposition

Line item	Subtitle C exclusion subsequent to point of generation (if applicable)	Current disposition of waste	§268.45, Table 1 technology used to treat debris (if applicable)	Date shipped (if applicable)
1				
2				
3				
4				

5. Was the waste hazardous at the point of generation but subsequently became excluded from the definition of hazardous waste or exempt from Subtitle C regulation (including characteristic wastes managed in wastewater treatment systems discharging under the CWA)? ❏ Yes ❏ No (If yes, this constitutes the §268.7(a)(7) one-time notification.)[2]

6. Was the waste characteristic at the point of generation, treated onsite to remove all characteristics, and treatment residues then shipped to a Subtitle D land disposal facility? ❏ Yes ❏ No (If yes, complete Certification 1, 2, 3, or 4.)[2]

7. Was the waste "debris" that was hazardous at the point of generation but subsequently became excluded from the definition of hazardous waste under §261.3(f)(1) by treating it using an extraction or destruction technology in §268.45, Table 1? ❏ Yes ❏ No (If yes, complete Certification 5.)[3]

8. Was the waste "debris" that was hazardous at the point of generation but subsequently became excluded from the definition of hazardous waste under §261.3(f)(2) by receiving a "no-longer-contains" determination from EPA or the authorized state? ❏ Yes ❏ No (If yes, this constitutes the §268.7(d)(1) one-time notification.)[4]

9. Was the waste "soil" that was hazardous at the point of generation but subsequently became excluded from the definition of hazardous waste via a "no-longer-contains" determination from EPA or the authorized state or by the generator determining that the soil no longer exhibits a characteristic? ❏ Yes ❏ No (If yes, this notice and all supporting information and documentation must be maintained in the facility files for at least three years per §268.7(e).)

10. Is the waste residue from treating K061, K062, and/or F006 wastes in high-temperature metals recovery (HTMR) units that 1) meets the generic exclusion levels in §261.3(c)(2)(ii)(C), 2) does not exhibit any characteristics, and 3) is shipped to a Subtitle D land disposal facility? ❏ Yes ❏ No (If yes, complete Certification 6.)[5]

11. Certifications:	
1. ☐ **Waste that has been treated to remove characteristics and that did not contain underlying hazardous constituents at the point of generation.** I certify under penalty of law that I have personally examined and am familiar with the treatment technology and operation of the treatment process used to support this certification. Based on my inquiry of those individuals immediately responsible for obtaining this information, I believe that the treatment process has been operated and maintained properly so as to comply with the treatment standards specified in 40 CFR 268.40 without impermissible dilution of the prohibited waste. I am aware there are significant penalties for submitting a false certification, including the possibility of fine and imprisonment.	Applies to line items: Reference: §§268.7(b)(4) and 268.9(d)
2. ☐ **Waste that has been treated to remove characteristics and to meet universal treatment standards for underlying hazardous constituents.** I certify under penalty of law that the waste has been treated in accordance with the requirements of 40 CFR 268.40 to remove the hazardous characteristic and that underlying hazardous constituents, as defined in §268.2(i) have been treated on-site to meet the §268.48 Universal Treatment Standards. I am aware that there are significant penalties for submitting a false certification, including the possibility of fine and imprisonment.	Applies to line items: Reference: §§268.7(b)(4)(v) and 268.9(d)
3. ☐ **Waste that has been treated to remove characteristics but does not meet universal treatment standards for underlying hazardous constituents.** I certify under penalty of law that the waste has been treated in accordance with the requirements of 40 CFR 268.40 or 268.49 to remove the hazardous characteristic. This decharacterized waste contains underlying hazardous constituents that require further treatment to meet treatment standards. I am aware that there are significant penalties for submitting a false certification, including the possibility of fine and imprisonment.	Applies to line items: Reference: §§268.7(b)(4)(iv) and 268.9(d)
4. ☐ **Soil that has been treated to meet the alternative treatment standards.** I certify under penalty of law that I have personally examined and am familiar with the treatment technology and operation of the treatment process used to support this certification and believe that it has been maintained and operated properly so as to comply with treatment standards specified in 40 CFR 268.49 without impermissible dilution of the prohibited wastes. I am aware there are significant penalties for submitting a false certification, including the possibility of fine and imprisonment.	Applies to line items: Reference: §268.7(b)(4)
5. ☐ **Debris that has been treated to meet the alternative treatment standards.** I certify under penalty of law that the debris has been treated in accordance with the requirements of 40 CFR 268.45. I am aware there are significant penalties for making a false certification, including the possibility of fine and imprisonment.	Applies to line items: Reference: §268.7(d)(3)(iii)
6. ☐ **HTMR residue from treating K061, K062, and/or F006 wastes.** I certify under penalty of law that the generic exclusion levels for all constituents have been met without impermissible dilution and that no characteristic of hazardous waste is exhibited. I am aware that there are significant penalties for submitting a false certification, including the possibility of fine and imprisonment.	Applies to line items: Reference: §261.3(c)(2)(ii)(C)
Generator's signature	Printed/typed name & title	Date

[1] Use an attachment if necessary. If all underlying hazardous constituents will be treated and monitored, there is no requirement to list any on this notification.

[2] This one-time notification is placed in the facility's onsite files only. For compliance with the 268.9(d) one-time notification and certification, if the waste does not meet universal treatment standards for underlying hazardous constituents (i.e., Certification 3 above), the generator must somehow communicate the need for UHC treatment to the Subtitle D facility; the notification and certification must be updated if the process or operation generating the waste changes and/or if the Subtitle D receiving facility changes.

[3] This one-time notification must be sent to EPA or the authorized state and placed in the facility's files. It is not sent with the shipment to the Subtitle D receiving facility. The notification must be updated if 1) a different type of debris is treated, 2) a different §268.45, Table 1 technology is used to treat the debris, and/or 3) the Subtitle D receiving facility changes. The certification (Certification 4 above) must be placed in the facility's files for each shipment of treated debris.

[4] This one-time notification must be sent to EPA or the authorized state and placed in the facility's files. It is not sent with the shipment to the Subtitle D receiving facility. The notification must be updated if the Subtitle D receiving facility changes.

[5] This one-time notification and certification must be sent to EPA or the authorized state and placed in the facility's files. It is not sent with the shipment to the Subtitle D receiving facility. The notification and certification must be updated if the process or operation generating the waste changes and/or if the Subtitle D receiving facility changes.

Source: McCoy and Associates, Inc.

facility's file. An example of the completed §268.7(a)(7) one-time notice for this situation is given in Figure 13-21. The §268.9(d) notice is completed only when characteristic wastes are treated to render them nonhazardous and the nonhazardous residues are subsequently shipped to a Subtitle D nonhazardous waste facility. [June 1, 1990; 55 FR 22531, 22662, 22688]

Q *If hazardous wastes are to be burned as fuel in a Bevill device (e.g., a cement or light-weight aggregate kiln), what LDR requirements (including paperwork requirements) apply?*

A Here is EPA's somewhat complex answer to this question:

"If the wastes are burned for energy recovery in a Bevill device that processes normal Bevill raw materials as well, and the Bevill device can show that its residues were not significantly affected by its hazardous-waste-burning activities (the 'significantly affected' test is found in §266.112), then the residues can retain Bevill-exempt status and not have to meet LDR treatment standards. Further, if the Bevill device produces a product that is used in a manner constituting disposal [e.g., cement or light-weight aggregate], and the hazardous waste is burned for energy recovery rather than for destruction or as an ingredient, then the product is not required to meet LDR treatment standards.

"In these situations where neither residues nor products are subject to LDR treatment standards, the original generator's wastes would not be considered prohibited from land disposal. According to §268.7(a)(7), if such a generator can assure that the conditions discussed above are all true regarding the disposition of its otherwise prohibited waste, then the generator is only required to prepare a one-time notice for its facility records documenting this disposition and not to comply with other tracking/notification requirements. If a generator is not in a position to know that this is the case, then the full notification/certification requirements under §268.7(a) would apply." [RO 11881]

Q *A laboratory discharges wastes/chemicals down sinks under the de minimis lab waste exclusion cited in §261.3(a)(2)(iv)(E). Do any LDR paperwork requirements apply?*

A Since these wastes are hazardous at the point of generation and subsequently become excluded from regulation under the above-cited exemption, the LDR form specified in §268.7(a)(7) should be prepared and placed in the lab's files. [April 8, 2003; 68 FR 17242, RO 11727]

13.12.3.2 Wastes sent to fuel blenders

Q *If a generator sends hazardous waste to a fuel blender, is an LDR notification required?*

A Yes. Whenever a generator ships a hazardous waste to another entity for eventual land disposal, an LDR form is required. In this case, even though the fuel blender will send the waste for combustion, a combustion residue (i.e., ash) will be produced that is typically land disposed. [RO 11881]

Q *Does a fuel blender, who sends hazardous waste-derived fuels to boilers or industrial furnaces, have to prepare any LDR paperwork?*

A Yes. The fuel blender is considered to be a treatment facility and must send the appropriate notification/certification to the fuel burner (e.g., a BIF). The fuel burner is also considered to be a treatment facility. In cases where the fuel blender has not treated the waste to meet treatment standards, he/she will prepare the notification that is required of generators at §268.7(a)(2). If the blender only blends D001 high-TOC wastes, combustion is the specified method, and no requirement is imposed to identify underlying hazardous constituents [see §268.9(a)]. For other types of characteristic wastes, even those that have been decharacterized by the blender, UHCs may have to be identified so that ash from the BIFs is properly tested prior to land disposal. [RO 11881]

Q: A characteristic waste is blended for fuel substitution and the resulting fuel no longer exhibits a characteristic. Are any LDR paperwork requirements triggered?

A: Whenever a characteristic hazardous waste loses its characteristic and is sent to a nonhazardous (Subtitle D) facility, the notification/certification specified in §268.9(d) must be completed and placed into the facility's files. [RO 11545]

13.12.3.3 Wastes sent to storage facilities

Q: Does a generator have to send an LDR notification to a facility that will simply store the waste prior to sending it to another facility?

A: Yes. An LDR notification is required when wastes are shipped to an offsite storage facility. [RO 13070]

13.12.3.4 Wastes sent to TSD facilities

Q: A generator decides, based on process knowledge, that his/her waste does not meet the LDR treatment standards and sends the appropriate notification [per §268.7(a)(2)] to the treatment facility. The treatment facility analyzes the incoming waste according to its waste analysis plan and finds that the waste does meet treatment standards. What paperwork requirements apply?

A: The treatment facility should document its data showing that the waste meets the treatment standard. The treatment facility must also complete a notification per §268.7(b)(3) and a certification that the waste met the applicable treatment standards as generated [see §§268.7(b)(5) and 268.7(a)(3)]. The treatment facility would then send the waste with the notification and certification and the analytical data to the disposal facility. The treatment facility would retain copies of the analytical data and the notification/certification in its files. [RO 13522]

Q: Are TSD facilities required to keep LDR notifications and certifications that they receive from generators? If so, how long are they required to keep them?

A: Per §§264/265.73(b), TSD facilities must maintain copies of LDR notifications and certifications in their operating record. The operating record must be kept at the facility until closure of the facility. [Topic # 23002-14239 on EPA's RCRA FAQ Database, http://waste.supportportal.com/ics/support/KBAnswer.asp?questionID=14239]

13.12.3.5 Wastes sent to recycling facilities

Although recycling facilities may be exempt from RCRA regulation, the hazardous wastes they receive are subject to the Part 268 regulations, including LDR notification forms. [RO 13181]

Q: A recycling firm is located adjacent to a TSD facility; the facilities are sister companies but have separate EPA ID numbers. Still bottoms and washwaters from the recycling facility are piped either intermittently or continuously to the TSD facility. What LDR paperwork requirements apply?

A: Because the two facilities have different EPA ID numbers, they are considered separate facilities. The recycling facility is subject to the generator paperwork requirements of §268.7(a); the TSD facility is subject to the requirements of §268.7(b). Questions on how frequently the required paperwork should be sent from the recycling facility to the TSD facility (i.e., what constitutes a "shipment") should be directed to the RCRA-authorized agency. [RO 13522]

Q: Do shipments of precious metal-containing wastes have to be accompanied by any LDR paperwork?

A: Yes. The requirements for recyclable materials from which precious metals are reclaimed in §261.6(a)(2)(iii) subject the generator to regulations in Part 266, Subpart F. Further, §261.6(a)(2) states that generators are subject to Part 268 regulations. Therefore, the LDR requirements, including paperwork, apply to shipments of recyclable precious metal wastes. [RO 11482, 13158]

13.12.3.6 Wastes sent to Subtitle D facilities

Q: If a generator treats its lead waste (D008) to meet treatment standards (and renders the waste nonhazardous), should he/she send the notification/certification in §268.7(a)(3) to the Subtitle D municipal solid waste landfill?

A No. LDR notifications and certifications should not accompany shipments from generators to Subtitle D facilities. Instead, the notification/certification required in §268.9(d) should be placed in the facility's onsite files only. [April 4, 2006; 71 *FR* 16889] This form is for characteristic wastes that were hazardous at the point of generation, were subsequently decharacterized, and residues are being sent to a Subtitle D facility. [RO 14546]

Q *Another generator treats a characteristic waste onsite, decharacterizing it. The generator sends the decharacterized residues to a Subtitle D facility for further treatment of UHCs and subsequent land disposal. What LDR paperwork is required of the generator and the Subtitle D facility?*

A As in the previous example, the generator must place the one-time notification/certification of §268.9(d) in the facility's onsite files. If Figure 13-22 is used for this purpose, the generator would have to check Certification Box 3 in Item 11 of this form. However, the generator is not required to send any LDR paperwork to the Subtitle D facility.

The Subtitle D facility is not subject to any LDR notification/certification requirements, even though they must treat the UHCs in the residues to meet universal treatment standards. Such facilities are not required to verify compliance with treatment standards. [September 14, 1993; 58 *FR* 48135, September 19, 1994; 59 *FR* 48016, RO 14585]

13.12.3.7 Decharacterized wastes sent to Subtitle C facilities

Q *If a characteristic waste is rendered nonhazardous, but decharacterized residues are sent to a Subtitle C (hazardous waste) facility, should the §268.9(d) notification/certification be completed and placed in the facility's onsite files only?*

A No. The notification/certification should be sent to the Subtitle C facility. [RO 11545] The paperwork specified in §268.7(b)(3–4) for treatment facilities sending wastes to disposal facilities seems to better fit the conditions in this example, rather than the paperwork specified in §268.9(d).

13.12.3.8 Wastes treated in 90/180/270-day units

Q *If a generator treats a hazardous waste onsite in a 90/180/270-day accumulation unit (i.e., a tank, container, or containment building regulated under §262.34) with the intent of meeting an LDR treatment standard, is any special paperwork required?*

A Yes. The generator must have a waste analysis plan per §268.7(a)(5) describing the procedures used to comply with the treatment standards. A copy of this plan must be kept in the generator's onsite records.

13.12.3.9 Wastes used in a manner constituting disposal

Q *A waste that contains heavy metals is used to produce fertilizer. Are any LDR paperwork requirements imposed on this operation?*

A Yes. Such wastes are subject to Part 266, Subpart C—Recyclable Materials Used in a Manner Constituting Disposal. Products (such as fertilizer) that are produced for the general public's use must meet the applicable LDR treatment standards. [§266.20(b)] In addition, the LDR paperwork requirements of §268.7(b)(6) are applicable. In essence, the fertilizer manufacturer must place a notification/certification in its onsite files for the initial shipment stating that the product meets treatment standards. The manufacturer must also keep records of the name and location of each entity receiving the fertilizer. [April 4, 2006; 71 *FR* 16889, RO 11645]

13.12.3.10 Wastes sent to Canadian facilities

Q *Do generators exporting hazardous wastes to Canadian TSD facilities need to send LDR paperwork with the shipment?*

A Yes. If a generator is shipping a hazardous waste that is subject to the LDR regulations to Canada, although it "is not legally necessary for the Canadian disposal facility…the agency still requires the notification and/or certification for each shipment of restricted waste. Unforeseen circumstances may arise during the transportation of the restricted waste, and it might need to be handled by a domestic hazardous waste treatment, storage, or disposal facility. The notification and/or certification documentation will allow waste handling in accordance with the land disposal regulations should this situation arise." [RO 13052]

13.12.3.11 Miscellaneous paperwork issues

Q *What paperwork is required for contaminated soil shipped offsite?*

A A generator of contaminated soil that does not meet applicable treatment standards at the point of generation is subject to the notification requirements of §268.7(a)(2). The notification must include all of the elements in the column headed "§268.7(a)(2)" of the Generator Paperwork Requirements Table presented in Table 13-2.

For soil that meets applicable treatment standards at the point of generation, the generator must prepare a notification as cited in §268.7(a)(3)(ii). The information included in the notification is specified in the column headed "§268.7(a)(3)" on the Generator Paperwork Requirements Table in Table 13-2. Note that Row 8 of this table includes the following statement that must be included:

"This contaminated soil [does/does not] contain listed hazardous waste and [does/does not] exhibit a characteristic of hazardous waste and [is subject to/complies with] the soil treatment standards as provided by §268.49(c) or the universal treatment standards."

In addition, the generator certification statement in §268.7(a)(3)(i), which states that the soil meets the applicable treatment standards, must be signed by an authorized representative and included with the LDR paperwork. [RO 14516]

Q *Item 3 on the Generator Paperwork Requirements Table (our Table 13-2) states that "constituents of concern for F001–F005 and F039, and underlying hazardous constituents in characteristic wastes" must be identified unless the waste is treated and monitored for all the constituents. How do these provisions apply to spent solvent constituents carbon disulfide, cyclohexanone, and methanol?*

A First, note that in §268.40, the F001–F005 wastes are actually broken up into four subcategories, one of which is for "F003 and/or F005 solvent wastes that contain any combination of one or more of the following three solvents as the only listed F001–5 solvents: carbon disulfide, cyclohexanone, and/or methanol." If a waste fits into this subcategory, each of the three chemicals is considered to be a constituent of concern and, if present, should appear on the LDR notification. If these three constituents are present along with other solvents, they are not constituents of concern and should not be included on the LDR notification. [RO 14325]

13.13 Miscellaneous issues

13.13.1 Lab packs

The federal regulations define "lab packs" as "small containers of hazardous waste in overpacked drums." [§§264/265.316] The packing requirements for lab packs are also found in those sections. When EPA was establishing the LDR program in the late 1980s, the agency addressed the applicability of treatment standards to hazardous wastes in lab packs as follows: "[L]ab packs are typically used by industry to dispose of small quantities of commercial chemical products (U- and P-wastes) and residues from analytical samples. These lab packs may contain hundreds of restricted wastes, and the applicable treatment standards must be achieved for each waste code contained in the lab pack. The commenters stated that these requirements pose an administrative burden that is incommensurate with the amount of waste being land disposed. In the Second

Third final rule (54 *FR* 26594), the agency restated its position that all restricted wastes placed in lab packs and land disposed must comply with the land disposal restrictions." [June 1, 1990; 55 *FR* 22629] See also November 7, 1986; 51 *FR* 40585.

In an effort to alleviate the burden of treating each small volume of waste in a lab pack according to its corresponding §268.40 LDR treatment standard, EPA promulgated an alternative treatment standard for lab packs on June 1, 1990. [55 *FR* 22700] The alternative treatment standard for lab packs is found in §268.42(c) and requires that the lab pack be incinerated in a hazardous waste incinerator. If the lab pack contains D004, D005, D006, D007, D008, D010, or D011 wastes, the incinerator residues must be treated to meet the treatment standards specified for those wastes in §268.40. In other words, none of the treatment standards for any organic wastes in the lab pack are applicable after incineration.

In order to take advantage of the alternative treatment standard, the lab pack must not contain any wastes listed in Part 268, Appendix IV. Metal-bearing wastes, other than the Appendix IV waste codes, can be lab packed and sent for incineration per §268.42(c). Some in the regulated community asked EPA if incinerating the allowed metal-bearing wastes in lab packs would violate the agency's dilution prohibition at §268.3(c), which considers the combustion of certain metal-bearing wastes to be impermissible dilution. EPA responded that the "dilution prohibition does not supersede the streamlined treatment standards promulgated in [§268.42(c)]. Therefore, metal-bearing inorganic wastes may be included in a lab pack unless it is prohibited under the list of wastes in Part 268, Appendix IV." [April 8, 1996; 61 *FR* 15587]

The paperwork requirements for lab packs are found in §268.7(a)(9). Where lab packs contain characteristic wastes (codes D001–D043), UHCs need not be determined. [§268.7(a)(9)(iii)] Note that §268.7(a)(9)(iii) refers to D001–D043, but it actually should refer to D001–D008 and D010–D043 because lab packs cannot contain D009 and be incinerated per §268.42(c).

The lab pack regulations don't include any provisions for consolidating hazardous waste codes on LDR notification forms. In other words, LDR paperwork must identify all of the waste codes associated with all of the individual containers in a lab pack. EPA suggests that generators of lab packs contact their state agency to determine if any other approaches are allowed. [RO 11196] McCoy and Associates has developed an LDR lab pack notification/certification form, which is available for your use from our website at http://www.mccoyseminars.com/forms.cfm.

The lab pack standards reference certain landfill requirements in §§264/265.316(f) that deal with how lab packs are to be disposed. This paragraph contains the following confusing statement: "Persons who incinerate lab packs according to the requirements of 40 *CFR* 268.42(c)(1) may use fiber drums in place of metal outer containers. Such fiber drums must meet the DOT requirements in 49 *CFR* 173.12 and be overpacked according to the requirements of paragraph (b) of this section." Paragraph 264/265.316(b) states that lab wastes must be overpacked in an "open head DOT-specification metal shipping container." It therefore appears that fiber drums containing lab packed wastes must be overpacked in metal drums. However, EPA has stated that fiber drums used for lab packs do *not* have to be overpacked in metal drums. [RO 13397, 13435, 13522]

13.13.2 Used oil

The general rule is that once a hazardous waste is generated, the LDR standards attach to that waste and must be met before any residue from the waste is land disposed. One exception to this rule occurs with mixtures of characteristic hazardous waste and used oil as specified in §279.10(b)(2). Under these provisions, a characteristic hazardous waste may be mixed with used oil, and, if the resulting mixture does not exhibit a characteristic, the mixture is regulated as used oil (and is not subject to Part 268 LDR standards).

Although EPA at one time attempted to eliminate this anomaly, their activities were successfully challenged in court and the LDR loophole still exists. Bear in mind that mixing used oil with characteristic hazardous waste is considered treatment. See Section 12.1.3.1.1.

13.13.3 Universal wastes

Universal waste handlers and universal waste transporters are normally exempt from the LDR notification/certification requirements of §268.7. [§268.1(f)] Destination facilities, which treat, recycle, or dispose universal wastes, are the first entities handling universal waste that are responsible for complying with any LDR standards, including paperwork. [RO 14088] The destination facilities are subject to all applicable LDR requirements. [§273.60] Therefore, they must determine the appropriate hazardous waste codes for the universal wastes and manage them according to the corresponding LDR standards.

However, as discussed in Case Study 13-14, if the destination facility is a solid (nonhazardous) waste landfill, the handler will have to ensure that any decharacterized universal wastes or nonhazardous residues from treating universal waste meet LDR treatment standards before they go into that landfill. [RO 14756]

13.13.4 Variances, extensions, and exemptions

The LDR regulations allow for several types of variances whereby affected parties can address problems or circumstances that are unique to their operations. In our experience, the LDR variance provisions are rarely used. This occurs for two reasons: 1) whenever an affected party seeks special treatment from the regulators, the approval process tends to be complex, time consuming, and expensive; and 2) some types of variances were important during early years of the LDR program, but are no longer of significance. A brief synopsis of the available LDR variances follows.

13.13.4.1 Case-by-case extensions to the effective date

In the early years of the LDR program, EPA was establishing treatment standards for every hazardous waste. At the same time, they determined when the treatment standards would take effect; after the effective date, the subject hazardous waste was barred from land disposal unless the treatment standards were met. In some cases, generators of hazardous waste might not have been able to find anyone capable of treating their waste to meet the standards. Under these circumstances, they could petition the agency for an extension of the effective date. The agency could provide up to two additional years for the generator to find (or build) treatment capacity. The terms of this provision are codified at §268.5.

This extension is of no practical significance today because all treatment standards have been in effect for more than two years. If EPA decides to identify new hazardous wastes in the future, the case-by-case extension could again be of use.

13.13.4.2 No-migration exemptions

Some land disposal units might be designed/constructed such that any hazardous waste placed in the unit could never migrate into the environment. If the waste can never migrate, it does not matter whether the waste has been treated to meet an LDR standard. Therefore, entities can dispose untreated hazardous wastes into units that have been granted a no-migration exemption from EPA (see §268.6).

Underground injection wells have been the primary beneficiaries of the no-migration exemption. Using computer modeling, many of these facilities demonstrated that wastes disposed in the injection zone will not migrate to an underground source of drinking water for more than 10,000 years. We believe many hazardous waste injection wells have such an exemption.

> ### Case Study 13-14: Decharacterized Universal Waste Disposed in Municipal Landfills
>
> Lithium-sulfur dioxide (Li-SO$_2$) batteries used extensively by the military are considered hazardous waste at their point of generation (time of removal from service) because of their potential reactivity. Military personnel often discharge these batteries to remove the electric charge. If conducted in accordance with the universal waste regs, such discharge is not considered hazardous waste treatment. [§§273.13(a)(2)(iii), 273.33(a)(2)(iii)] After discharging the batteries and rendering them nonhazardous, can personnel dispose them in municipal landfills with no LDR implications?
>
> As noted in Section 8.1.2, EPA allows batteries to be managed under the universal waste program even if they are headed for disposal rather than recycling (although some state programs require recycling for certain universal wastes). However, because these batteries are hazardous at their point of generation, personnel managing them are required to comply with LDR requirements. That is, even if discharged Li-SO$_2$ batteries are no longer hazardous, they may not be disposed in solid waste landfills until LDR treatment standards (which attached at the point of generation) are met.
>
> The LDR requirements include identification of any underlying hazardous constituents (UHCs) present in characteristic wastes at levels above the universal treatment standards (UTS). [§268.9] Characteristic wastes must be treated so that UTS levels are met for all UHCs before land disposal is allowed, even if the waste has been decharacterized. EPA notes that the SDS for Li-SO$_2$ batteries lists acetonitrile as a component, present at 5–6%. The UTS for acetonitrile of 38 mg/kg, as well as UTS for any other applicable UHCs, must be met before land disposal can occur. Once these LDR standards are achieved, decharacterized universal waste batteries can be disposed in municipal solid waste landfills. [RO 14756]
>
> Note that this guidance seems to conflict with previous agency guidance in RO 14088, which says that "[t]he destination facility is the first entity that handles a [universal waste] that is responsible for compliance with any of the LDR requirements, including recordkeeping." This older guidance is applicable to universal waste sent from a handler to a RCRA-permitted TSD facility or recycling facility. However, if the destination facility is a nonhazardous waste landfill that is normally not required to comply with the LDR program, the handler must take on the responsibility of ensuring that decharacterized universal wastes or nonhazardous residues from treating universal wastes comply with LDR treatment standards before they enter the landfill.

Many facilities (primarily refineries) tried to obtain no-migration exemptions for their land treatment units. With one or two exceptions, these efforts were unsuccessful.

Today, the only facilities likely to pursue a no-migration exemption would be new injection wells or perhaps underground mines.

13.13.4.3 Alternative treatment methods/determinations of equivalent technology

Sometimes, the treatment standard for a waste is not appropriate for a specific facility's waste. For example, a DOE facility might generate a radioactive mercury waste (D009) for which the treatment standard is RMERC (retorting followed by mercury recovery). However, when this mercury waste is retorted, the recovered mercury cannot be reused or recycled because of radioactive contamination. Hence, the recovered mercury will have to be treated and disposed, and the cost associated with retorting will be wasted. DOE could (and did—see RO 14270) apply for a Determination of Equivalent Technology (DET) that would allow for a more practical concentration-based treatment standard.

The provisions for determinations of equivalent technology are codified in §268.42(b). In addition to being granted for unusual wastes (such as radioactive mercury), DETs have also been granted to process developers who want their technology to be an acceptable method of waste treatment.

On May 14, 2008, EPA published a final rule granting a Utah-based TSD facility a site-specific treatment variance for certain P- and U-listed hazardous mixed wastes. [73 *FR* 27761] The usual treatment standard for the RCRA-regulated portion of the waste is expressed as a required method of treatment: combustion (CMBST). The variance granted the petitioner the ability to perform vacuum thermal desorption (VTD) in lieu of combustion prior to placing the waste in a land disposal unit. With a removal efficiency of at least 99.99%, the applicant demonstrated that it was technically inappropriate to require the mixed waste to be treated by the specified method (combustion) once vacuum thermal desorption was performed.

Guidance on obtaining a DET is available in *Variance Assistance Document: Land Disposal Restrictions Treatability Variances & Determinations of Equivalent Treatment*, http://www.epa.gov/epawaste/hazard/tsd/ldr/guidance2.pdf.

13.13.4.4 Treatability variances

Originally, the regulated community was very concerned that EPA would establish treatment standards that could not be met at every operating facility. To address this concern, §268.44 was codified to allow facilities to obtain a treatability variance if for some reason their specific waste could not be treated to meet treatment standards. A handful of facilities [see §268.44(o)] have obtained site-specific treatability variances for their process wastes. In most cases, however, the LDR treatment standards have not proved difficult to meet and treatability variances are rarely needed for as-generated process wastes.

Much more commonly, treatability variances were aimed at facilities conducting remediation projects. Contaminated soil at these facilities originally had to meet the treatment standards for process wastes—a difficult and expensive undertaking. EPA indicated that it was their intent to issue treatability variances for contaminated soil until the alternative treatment standards for soil were promulgated. When these soil standards were issued in 1998, much of the need for treatability variances was eliminated.

While issuance of treatability variances was never very common, it is even less so today. Anyone needing additional information on the procedures involved in obtaining a treatability variance is referred to the following for guidance:

- *Variance Assistance Document: Land Disposal Restrictions Treatability Variances & Determinations of Equivalent Treatment*, available online at http://www.epa.gov/epawaste/hazard/tsd/ldr/guidance2.pdf;

- *Federal Register* preamble discussions at August 17, 1988; 53 *FR* 31199, April 29, 1996; 61 *FR* 18828, December 5, 1997; 62 *FR* 64504, and May 26, 1998; 63 *FR* 28606; and

- RO 14078.

Q: *An electroplating facility generates wastewater treatment sludge that meets the F006 listing description. Subsequent to its generation, the facility submits a delisting petition to EPA and is successful in getting the sludge delisted per §260.22. Because the LDR treatment standards attach at the point of generation, however, the delisted waste must meet LDR standards prior to land disposal. Can the generator seek a treatability variance from meeting these standards?*

A: Yes. Although the generator of the wastewater treatment sludge remains subject to Part 268 requirements, it may petition the state/EPA to receive a variance from the F006 treatment standards per §268.44. In the petition, the generator would need to demonstrate that the waste cannot be treated to the specified standard levels in §268.40 because, for example, its physical or chemical properties differ significantly from the waste used to establish the LDR treatment standard. [RO 14699; see also December 1, 2011; 76 FR 74714]

Point of Generation

Discussing the difficult issue of when a material becomes a hazardous waste

This chapter looks at several point of generation issues that sometimes come up when dealing with the hazardous waste regulations. RCRA requires generators of solid wastes to determine if they are hazardous. In some cases, generators are also required to ascertain if their wastes contain underlying hazardous constituents. These determinations are to be made at the "point of generation" of the waste. Where is the point of generation of a hazardous waste? How is this term defined? Section 14.1 takes a look at these questions.

Determining if unknown wastes are hazardous can be a tricky proposition. Section 14.2 contains EPA's guidance on point of generation issues associated with unknowns.

Spills of hazardous waste and products happen all the time. How should these be handled? Are they hazardous? What is their point of generation? This hot topic is included in Section 14.3. The regulatory status and point of generation of gases and vapors from manufacturing and waste management operations and residues generated from controlling gaseous emissions are discussed in Section 14.4. Finally, the point of generation for waste military munitions is defined in Section 14.5.

14.1 Point of generation issues

The point of generation (POG) of a hazardous waste is a fundamental (and important) concept under RCRA. Surprisingly, EPA has had a very difficult time clearly and consistently explaining when and where a hazardous waste is generated, thus subjecting it to RCRA Subtitle C regulation. Unfortunately, as the regulations have evolved over the last 30 plus years, the concept that any given waste has a single POG has been lost. Today a waste can have different points of generation, depending on the RCRA regulatory issue under consideration. In this section, we will describe the POG from the following perspectives: 1) the fundamentals of when/where a waste is generated, 2) waste coding, 3) waste counting, 4) treatment in 90-day units, 5) land disposal restrictions (LDR) requirements, and 6) spills of wastes/products.

14.1.1 The fundamentals: when and where

The POG of a hazardous waste involves two issues that are linked together:

1. When is a hazardous waste generated?
2. Where, physically, does the hazardous waste become subject to RCRA regulation?

14.1.1.1 When is a hazardous waste generated?

The original RCRA rulemaking [May 19, 1980; 45 *FR* 33095] addressed the POG of a hazardous waste from the standpoint of *when* generation occurs. EPA gave the following guidance:

- For characteristic wastes, the POG is whenever the waste exhibits a characteristic and becomes subject to the hazardous waste management standards of Parts 262 through 265. An example was given of a company storing acid waste onsite prior to offsite transport for disposal. The company would have to make a hazardous waste determination when the waste 1) is poured into a container, 2) enters a neutralization tank via pipes, or 3) enters a tank truck for offsite shipment. For waste piped to a surface impoundment, a determination must be made when the waste enters the impoundment and when any sludge is formed in the unit.

- For listed hazardous wastes, the POG is when the waste first meets the listing description. (This presumes that the waste already meets the definition of a solid waste—see Chapter 1 for a discussion of solid waste identification.) For listed manufacturing process residues, emission control dusts, or wastewater treatment sludges, the POG is when they are created. (EPA stated that this point in time is generally well defined.) For spent solvents and P- and U-wastes, the POG is "when their intended use has ceased, and they begin to be accumulated for disposal, reuse, or reclamation." For example, the POG for solvents used in parts washers is when the parts washer apparatus is removed from a drum of used solvent. [RO 12790]

- For mixtures of listed waste and nonhazardous solid waste, the POG is when the listed waste is added to the solid waste. Prior to mixing, the listed waste would have to be managed as hazardous waste.

These concepts for determining when a hazardous waste is generated were codified in §261.3(b).

Sometimes a waste is generated that undergoes a physical or chemical change after the original point of generation, causing what was originally a nonhazardous waste to become hazardous. For example, some printing shops generate a wash water stream that contains solvent. At the point of generation, the wash is well-mixed, homogeneous, not listed, and not characteristic. After some time, the waste separates into two phases: organic solvent and water. When this separation occurs, the solvent phase exhibits the characteristic of ignitability, requiring a D001 code. EPA was asked "when should the hazardous waste determination be made in this situation?" The agency replied:

"[A] generator's responsibility to make a hazardous waste determination may continue beyond the determination made at the initial point of generation. In the case of a nonhazardous waste that may, at some point in the future, exhibit a hazardous waste characteristic or meet a hazardous waste listing description, there is an ongoing responsibility to monitor and reassess if changes occur that may cause the waste to become hazardous." [RO 14834]

14.1.1.2 Where is the POG of a hazardous waste?

The physical POG of a waste is site-specific. As a result, EPA initially hesitated to provide guidance on where Subtitle C regulations first attach to a waste. On July 14, 1986 [51 *FR* 25441], the agency noted that "the point at which the material will initially be considered to be a hazardous waste is the point at which the material leaves a process tank or area." Later that year, EPA confirmed that:

"We consider the point of exit from the process [unit] to be the introductory point for the hazardous waste into a hazardous waste tank system. Therefore, any process transfer equipment, even if normally used for production purposes, that is also used to transfer hazardous waste…to a hazardous waste storage/

treatment tank, would be considered part of a hazardous waste tank system and thus subject to the standards for such." [RO 13790]

In response to numerous inquiries that arose under the LDR program, EPA addressed POG on June 11, 1987 [52 *FR* 22356]:

"Where the wastes are not generated in an enclosed system and there is no normally occurring aggregation of waste streams, wastes would be prohibited at the point where the waste is initially generated. For example, wastes not aggregated for treatment or for management to facilitate treatment would be prohibited at the point where the waste leaves the manufacturing or other process that generates the waste."

Where wastes are generated in enclosed systems, EPA stated that sample taps can typically be installed for waste characterization purposes, or, in extreme cases, wastes can be sampled when they leave closed systems. [July 8, 1987; 52 *FR* 25766]

EPA expounded further on when and where wastes are generated in *Federal Register* preamble language [October 30, 1980; 45 *FR* 72025]:

- A waste is generated upon removal from a manufacturing process or waste treatment unit.

- In the case of wastes generated in manufacturing process units, the waste becomes subject to regulation when it is removed from the unit, or 90 days after the unit is taken out of service if the waste remains in an inactive unit [see §261.4(c)].

In general, the point of generation is that point where a material first meets the definition of a "solid waste." At that point and time, generators are required to determine if the waste meets a listing description or exhibits a characteristic prior to commingling (mixing) with other waste streams. [March 29, 1990; 55 *FR* 11830] For example, the POG for wastewater generated in a pulp and paper mill bleach plant is at the outlet of the plant, prior to mixing with other wastewater streams. [RO 11631]

Sometimes, manufacturing residues from a reactor are sent to a distribution manifold where, depending on product and process chemistry considerations, they can be 1) recycled back into the process, 2) sent for onsite recovery, or 3) managed as hazardous waste. EPA noted in guidance that the "liquid removed from the reactor may be reused or recycled, but it may also be sent directly to hazardous waste storage tanks…. [T]he manufacturing process unit exemption in Section 261.4(c) does not apply to the pipes and pumps leading from the reactor to the distribution manifold, and those pieces of ancillary equipment are subject to RCRA regulation, including Subpart BB." [RO 14469]

14.1.2 POG for waste coding

Perhaps the most important concept associated with the POG of a hazardous waste is that at this point the appropriate characteristic or listed codes attach. In the beginning, EPA was primarily concerned with when/where manufacturing process wastes become regulated. However, they did note in the May 19, 1980 rule [45 *FR* 33095] that treatment residues from listed wastes, such as landfill leachate, wastewater treatment sludge, and incinerator ash remain a listed waste. This was simply a restatement of the derived-from rule for listed wastes. They hadn't yet thought of whether the POG for these residues is different from the POG of the original manufacturing process waste.

On December 31, 1980 [45 *FR* 86969], EPA recognized that treatment residues have a new POG:

"Owners and operators of hazardous waste management facilities may generate hazardous waste (e.g., residues created by treatment processes). With respect to the hazardous waste that these persons generate, they, like other generators, must comply with the applicable provisions of Part 262 [dealing with onsite accumulation and offsite transport]."

Note that this statement is not specific to onsite or offsite treatment. Clearly, if an offsite TSD facility treats or recycles a generator's hazardous waste, the TSD facility's treatment residues are newly generated wastes, and the TSD facility is the generator. [RO 12287, 12539] Although not specifically stated by EPA, the same situation applies to a generator who treats his/her waste onsite; residues from the treatment process are newly generated waste. This outcome is mandated by the derived-from rule (see Section 5.2).

For instance, if a characteristic waste is treated and a residue is produced that does not exhibit a characteristic, the residue is not a hazardous waste. What is the POG of this nonhazardous waste? Where it comes out of the treatment process. Similarly, if a D001 ignitable waste that contains low levels of lead is incinerated, the lead will be concentrated in the ash. The derived-from rule specifies that, if the treatment residue exhibits any characteristic (e.g., it is toxic for lead), it remains a hazardous waste. What is the POG of the D008 lead waste? Where the ash comes out of the incinerator.

The agency noted in the December 31, 1980 preamble, however, that if an offsite facility simply stores a generator's waste and then removes it from storage for shipment to another facility, "the storage facility does not 'generate' a waste simply by removing it from storage." [45 *FR* 86969] The POG of the stored waste remains the same as established by the original generator.

In summary, by the end of 1980, EPA had recognized that the POG of a waste is where it leaves a manufacturing process or where it leaves a treatment process. Over the years, EPA issued guidance that further defined the POG for process wastes, treatment residues, and discarded chemicals. The examples that follow and Case Study 14-1 illustrate the agency's thinking.

Q: *After spray painting operations at automobile assembly plants, pure solvent is pumped through the spray painting guns for cleaning when changing paint color. What is the POG of the spent solvent?*

A: EPA's contention is that the POG of the spent solvent is at the emergence from the spray paint gun. The spent solvent is a solid waste (and a hazardous waste determination must be made) at that point. [RO 14152, 14604, 14632]

Q: *A facility reclaims spent xylene solvent (F003) to produce marketable xylene solvent. The reclaimed material is stored in a holding tank prior to sale to offsite facilities. Per §261.3(c)(2)(i), material reclaimed from F003 and subsequently reused as a solvent is not F003 but a product, and the holding tank is a product storage tank (i.e., a manufacturing unit). During one reclamation campaign, however, a batch of product was produced that was not suitable for solvent use and was instead sent offsite as a hazardous waste fuel. What is the POG of the fuel?*

A: Reclaimed solvent that is burned for energy recovery remains F003 per the derived-from rule and must be manifested offsite to the user. Regarding the status of the holding tank, EPA noted that "[b]ecause your operation normally produces reclaimed solvent, the [material] actually became a hazardous waste at the time you determined that it was not suitable for solvent use (and that it therefore had to be marketed as fuel).... EPA does regulate the storage of hazardous waste fuel as well as fuel blending tanks. In your case, however, it appears that the tank was really a product (solvent)...tank, and so not subject to regulation. This determination is based on your assurance that the fuel production was an isolated incident, and that your original intent in placing reclaimed xylene in the tank was to produce solvent, not fuel." [RO 11238] Thus, the POG appears to be when the fuel exited the holding tank.

Q: *Spent solvents are reclaimed in a distillation unit, which produces still bottoms. What is the POG of the bottoms?*

A: The recycling unit is exempt from regulation [§261.6(c)(1)]; however, the still bottoms are considered to be a newly generated waste. The time of generation of the waste is when it is removed from the distillation unit.

Case Study 14-1: Point of Generation for Off-Shore Oil Platform Wastes

Do the RCRA hazardous waste regulations apply to wastes generated on off-shore oil platforms, or do the regulations kick in once the wastes are transferred to a shore facility? We found the following information in guidance from FAQ ID 23002-15246 and 23002-17240 on EPA's FAQ Database (http://waste.supportportal.com/).

In states authorized to administer RCRA, state regulations apply to platforms located within the boundary of state waters. If states are not authorized to administer RCRA, federal regulations apply to platforms located in state waters. Platforms located on the Continental Shelf beyond the bounds of state waters are subject to regulation under the Outer Continental Shelf Lands Act, which generally provides for the application of federal law to such platforms.

The Bureau of Ocean Energy Management within the U.S. Department of Interior notes that international law, as reflected in the United Nations Convention on the Law of the Sea, recognizes that the "seaward limit [of the Outer Continental Shelf] is defined as the farthest of 200 nautical miles seaward of the baseline…." See the Minerals Management Service: Outer Continental Shelf website for further explanation and additional details. [http://www.boem.gov/Oil-and-Gas-Energy-Program/Leasing/Outer-Continental-Shelf/Index.aspx]

Thus, the RCRA hazardous waste regulations apply to wastes generated on off-shore platforms located within the jurisdiction of either the adjacent state or the federal government. Off-shore drilling platforms are considered separate generation points subject to Part 262 generator standards. [RO 11329] That means that each platform would independently determine if it was a large quantity generator, small quantity generator, or conditionally exempt small quantity generator under RCRA. While some wastes produced on an oil drilling platform may be excluded under §261.4(b)(5), not all wastes are. See Section 4.14.

When it comes to transferring hazardous wastes from the off-shore platform to shore, EPA would generally require a water transporter to carry a manifest on its vessel for nonbulk (i.e., containerized) shipments of hazardous waste per §263.20. However, if the transporter delivers a bulk shipment directly to a designated facility, the transporter may comply with the applicable manifest provisions for water (bulk) shipments at §263.20(e). Water (bulk) shipments are defined as "the bulk transportation of hazardous waste which is loaded or carried on board a vessel without containers or labels." [§260.10]

Hazardous waste transported to a port from an off-shore oil rig located beyond the boundary of state or federal waters would not be subject to the manifest regulations before the shipment was off-loaded at port. In this situation, EPA considers the port as the "cradle" under the RCRA "cradle-to-grave" manifest tracking system and would not require a manifest at the time the shipment entered U.S. territorial seas (although EPA would require a manifest at the port). However, the transportation of hazardous waste is regulated jointly under the RCRA Subtitle C regulations and DOT hazardous materials transportation regulations; therefore, DOT may require a shipping paper containing the same information as the manifest (excluding EPA ID number, generator certification, and signatures) to accompany the shipment on the vessel.

[RO 11420, 12850, 12865, 13280] This example establishes the principle that when hazardous residues are produced by a unit that is exempt from RCRA permitting (e.g., recycling units, wastewater treatment units, elementary neutralization units, or totally enclosed treatment units), the residues are newly generated waste.

Q *What responsibility, if any, does a generator of hazardous waste have for residues produced from the recycling of his/her waste? For example, corrosive hazardous wastes produced by numerous generators are transported to the same recycler. The recycler subsequently generates by-products that are also hazardous wastes. Are the original generators liable under RCRA if hazardous wastes generated by the recycler are mismanaged?*

A No. Hazardous wastes generated during the recycling operation are newly generated wastes and the recycler is viewed as the generator. Thus, the recycler is responsible for compliance with Part 262 and for the hazardous waste by-products produced from his/her recycling activities. [RO 12287] (Note that the original generators may retain liability for the recycler's actions under CERCLA.)

Q *A recycling facility treats spent photographic fixer solution (D011) in a tank to precipitate a silver-containing sludge that is then thermally refined. How is the sludge regulated?*

A EPA considers the sludge to be a newly generated waste exhibiting a characteristic. Characteristic sludges that are reclaimed are not solid wastes per §261.2(c), Table 1; hence, the sludge is not a hazardous waste. [RO 11814] Because the tank is part of the silver recycling process, this example follows the principle that treatment residues from exempt recycling units are newly generated wastes.

Q *If a hazardous sludge forms in an impoundment, is the POG when the sludge forms, or when the sludge is removed from the impoundment for disposal?*

A The agency has always maintained that sludges are generated at the moment of their deposition at the bottom of the unit (point of generation). Note that "deposition" is defined as a condition where there has been at least a temporary cessation of lateral particle movement. [November 2, 1990; 55 *FR* 46380, RO 11102, 11588]

Q *What is the POG for a P- or U-listed commercial chemical?*

A "[T]he commercial chemical becomes a hazardous waste instantly when the act of discarding takes place." [RO 12012] "EPA believes that an unused product becomes 'discarded' when an intent to discard the material is demonstrated." [February 12, 1997; 62 *FR* 6626]

Q *A D002 acidic waste and a D002 basic waste from two different manufacturing processes are individually piped to a collecting pipe. The two wastes neutralize each other in the collecting pipe and the result is a nonhazardous waste. The entire system is enclosed. Is there a POG of a hazardous waste in this case?*

A Each of the corrosive wastes has a POG that is upstream of the collecting pipe. [RO 13395]

Q *A utility boiler is cleaned using an acid wash followed by two water rinses. The acid wash stream exiting the boiler is toxic for chromium and lead and is also corrosive. Rinsates from the two water rinses are nonhazardous. All three streams are collected in the same large tank. Do the three rinses have to be characterized for hazardous waste management individually or after commingling?*

A "The agency is today clarifying that, specific to power plant boiler cleanout (and potentially, to other sporadic cleaning activities involving multiple rinses), generation is at the completion of the entire cleanout process.... The agency views the cleanout of the boilers as one process and therefore does not consider the mixing of acid rinse and water rinse as impermissible dilution but as a single water rinsate resulting from the single cleanout process." [62 *FR* 26006; May 12, 1997] A similar interpretation was rendered by EPA for the regeneration of a demineralizer, which generates several rinses (some of which have a pH <2 or >12.5) that are all routed to a tank where they neutralize each other. In its March 28, 2002 interpretation, the agency noted that it considers a single water rinsate to be produced from the single regeneration process. [Letter from James Berlow, EPA headquarters to Douglas Green, Utility Solid Waste Activities Group]

Q *An automobile service center drains antifreeze from car radiators and commingles the antifreeze in a tank prior to shipment to a recycling facility. The antifreeze from any given radiator is hazardous for lead about 40% of the time, but the commingled antifreeze in the tank is nonhazardous. The recycling facility distills the antifreeze and sends the still bottoms (which are hazardous for lead) to an incinerator (provided they have sufficient Btu value). How are POG issues resolved in this scenario?*

A EPA determined that the POG for the antifreeze is at each individual radiator. However, for policy (i.e., encouraging recycling) and practicality reasons, the agency allowed the service center to assume that 40% of the antifreeze would be hazardous and would count toward the center's monthly generator status. The service center could also assume that the accumulated total volume of spent antifreeze was nonhazardous and could ship it to the recycler without a manifest. The recycler would have to characterize the still bottoms, and, if hazardous, they would be a newly generated hazardous waste that would have to be manifested to a hazardous waste combustion facility. [RO 14003]

Q *A D001 high-TOC ignitable waste is filtered to remove solids prior to incinerating the liquid. The solids are nonhazardous; how are they regulated?*

A EPA states that the solids resulting from pretreatment would be nonhazardous wastes (i.e., they are newly generated). "This would be the case for any aqueous, liquid, or solid material, which, as a result of pretreatment, no longer exhibits a characteristic." [RO 13404]

Q *An emission control system connected to an electric arc furnace captures dust in a baghouse. Dust from the baghouse is piped to a silo, from which the dust is loaded into trucks or rail cars. Is the POG of the waste where the dust a) leaves the furnace, b) is removed from the baghouse, or c) is removed from the silo?*

A Although the POG would normally be where the dust is removed from the baghouse [RO 11921, 12824], EPA considers the piping from the baghouse to the silo and the silo itself all to be part of the "dust handling system." Hence the POG is where the dust leaves the silo. [RO 14200]

Q *What is the POG for waste generated by paint removal operations?*

A Paint waste is considered to be generated once the paint has been removed from the surface of the structure. [RO 14069]

Q *What is the POG for materials seized by Drug Enforcement Administration (DEA) or state/local law enforcement agency personnel that shut down a clandestine drug laboratory?*

A Materials seized at a clandestine drug laboratory become waste when DEA/law enforcement personnel decide what to keep as evidence and what to dispose. Seized items not required as evidence or for analysis must be managed as solid and potentially hazardous waste. [*Guidelines for Law Enforcement for the Cleanup of Clandestine Drug Laboratories*, 2005 edition, available at http://www.justice.gov/dea/resources/img/redbook.pdf]

In summary, the following are considered to be points of generation of a hazardous waste:

- When the waste exits a nonwaste unit or piece of equipment;
- When the waste exits a manufacturing process unit;
- When a waste-containing material is spent and a decision is made to discard or recycle it;
- When a decision is made to discard a P- or U-listed chemical;
- When a treatment residue exits a treatment unit, such as an incinerator;
- When a hazardous sludge is deposited in a waste management unit, such as an impoundment; and

- When a residue exits a unit that is exempt from RCRA permitting requirements (e.g., a recycling unit, wastewater treatment unit, elementary neutralization unit, or totally enclosed treatment unit).

Waste codes attach at these points. Once a waste code attaches, several RCRA requirements are triggered:

1. If the waste is hazardous, any unit receiving the waste must be either a RCRA-regulated unit or specifically exempt from the regulations (e.g., a wastewater treatment unit). Consider, for example, a wastewater that exhibits a characteristic at the point where it leaves a manufacturing process (e.g., it carries a D-code). The wastewater is treated in a tank, where it is rendered nonhazardous. The tank is exempt from RCRA because it is a wastewater treatment unit (see Section 7.2). When the wastewater leaves the tank, it is a newly generated waste, and it carries no hazardous waste code. If the water flows into a surface impoundment, the impoundment is a Subtitle D (nonhazardous waste) unit.

    Note that in this example, the treated wastewater leaving the tank is an "intermediate-step treatment residue." For waste coding purposes, it is critical that we characterize these residues, because the regulatory status (i.e., Subtitle C or D) of the unit receiving these residues is based on this characterization.

2. When hazardous wastes are shipped offsite, manifests are required. Generators are required to put waste codes on the manifest. Consider a TSD facility that incinerates a characteristic organic waste, thereby producing an ash that is characteristic for heavy metals. The TSD facility is the generator of the ash and applies the appropriate waste codes before manifesting it to a hazardous waste landfill.

3. The waste code or codes determine what LDR treatment standards apply. Note, however, that the POG of a waste for LDR purposes is not necessarily the same as described above. More about this in Section 14.1.5.

The POG for waste coding purposes appears relatively straightforward. However, POG issues became progressively more complicated as the RCRA rules evolved.

14.1.3 POG for waste counting

When determining generator status (i.e., large quantity, small quantity, or conditionally exempt), the waste counting rules of §261.5(c–d) apply. These rules take a different approach to POG than for waste coding. On November 19, 1980 [45 FR 76621], EPA decided that, for counting purposes, a generator does not have to count hazardous wastes (i.e., residues) that are produced from onsite treatment of hazardous wastes.

This issue was addressed further on March 24, 1986 [56 FR 10153], when EPA revised §261.5(d)(2) to state that a generator need not count treatment residues generated from onsite treatment as long as the original hazardous waste was counted once. A generator also does not have to count hazardous wastes that come out of a manufacturing process and that are managed immediately in units that are not subject to substantive RCRA standards (e.g., a recycling unit, wastewater treatment unit, elementary neutralization unit, or totally enclosed treatment unit). Section 6.1 discusses waste counting requirements in detail.

In summary, the POG of a waste for counting purposes is very different than for coding purposes.

14.1.4 POG for residues from 90-day units

We have just seen that, when a residue exits a treatment unit, it is a newly generated waste. EPA recognized a problem with this approach when they considered residues from treating hazardous wastes in 90-day units. Consider the following examples.

Q *If a large quantity generator treats a waste in a 90-day container, can the treatment residue be stored for an additional 90 days without a permit?*

A No. "[R]esidue removed from a 90-day treatment unit remains subject to the 90-day accumulation time limit of the original waste placed in the unit, i.e., a new 90-day limit does not begin for the residue when it is removed from a 90-day unit."

Q *Hazardous wastes are treated in multiple 90-day tanks in series. Are the treatment residues from each treatment tank newly generated wastes?*

A "If the waste remains hazardous as it moves through the treatment process, the 90-day period applies to the entire waste stream. In other words, as a waste moves from tank to tank, the 90-day time period will not begin again each time the waste enters a new treatment tank. The 90-day clock stops only if the tank in question is a permitted or exempt treatment unit."

Accordingly, residue removed from a 90-day unit in series is not newly generated waste unless the residue is from the last unit in the series.

(The EPA guidance that is the source for the above two questions and answers is no longer available on RCRA Online.)

14.1.5 POG issues associated with the LDR program

When EPA began issuing regulations under the LDR program in late 1986, the POG of a waste became a critical issue. This resulted from the agency's decision that, in general, the LDR requirements attach to a waste at its POG. From this time onward, POG issues became inextricably linked to LDR issues.

EPA's initial pronouncement on LDR/POG issues came on November 7, 1986. [51 *FR* 40620] EPA stated: "The agency is requiring that applicable Part 268, Subpart D treatment standards for a restricted waste be determined at the point of generation." This was reaffirmed on July 8, 1987 [52 *FR* 25766] and more recently on May 23, 2006. [71 *FR* 29721]

On June 11, 1987, EPA addressed the status of treatment or disposal residues in the context of the LDR program [52 *FR* 22357]:

"[W]here a waste generated before a land disposal prohibition effective date is later removed from storage or disposal, it then becomes subject to the land disposal prohibitions…. By the same logic, residues generated from such wastes, such as leachate or contaminated ground water containing F001–F005 [spent] solvent wastes disposed prior to November 8, 1986, would be viewed as newly generated wastes."

14.1.5.1 The change-in-treatability-group principle

For our purposes, the most important aspect of the June 11, 1987 rule is that here EPA first introduced what is currently referred to as the "change-in-treatability-group principle." [52 *FR* 22357] Under the LDR program, all hazardous wastes fall into one of two treatability groups: wastewater or nonwastewater. Wastewaters contain less than 1% by weight total organic carbon (TOC) and less than 1% by weight total suspended solids (TSS); nonwastewaters are everything else. EPA decided that whenever a waste undergoes a change in its treatability group (typically as a result of treatment), the waste in the new treatability group is newly generated and, hence, has a new POG. An example would be where a wastewater is incinerated and an ash (nonwastewater) is produced. The ash is a newly generated waste.

Up until 1990, the LDR program dealt only with listed wastes, and the derived-from rule specifies that any treatment residues from listed wastes remain listed. The POG of these listed residues could be different than the POG of the original wastes depending on whether a change in treatability group occurred. On June 1, 1990, EPA explained how the change-in-treatability-group principle would apply to both listed and characteristic wastes. [55 *FR* 22661–2] In each of its examples, the POG is tied to the concept that the waste or treatment residue is to be land disposed:

1. If a listed wastewater is treated and produces a nonwastewater residue, the nonwastewater is in a different treatability group and is a newly generated hazardous waste.

2. If a listed nonwastewater is treated and produces a nonwastewater residue, no change in treatability group has occurred; hence, the treatment residue is not a newly generated waste.

3. A characteristic wastewater is treated and produces a nonhazardous sludge (nonwastewater) and nonhazardous treated wastewater. The sludge is in a new treatability group, is a newly generated waste, and is nonhazardous at its POG. The treated wastewater is in the same treatability group as the original waste, and is therefore not a newly generated waste.

4. Characteristic electroplating wastewater is treated and produces a noncharacteristic sludge. The sludge is a newly generated waste because it is in a different treatability group and, because it meets the listing description for F006, is a listed hazardous waste.

The change-in-treatability-group principle had to be revised in 1993 when EPA began regulating underlying hazardous constituents in characteristic wastes. To see why, consider the following example: D002 wastewater is burned in an incinerator. Any ash produced will be nonwastewater. If the ash is considered a newly generated waste and is nonhazardous, it would not be subject to LDR standards. But EPA wanted to apply LDR standards to the ash (i.e., the ash must meet universal treatment standards for underlying hazardous constituents). Therefore, on May 24, 1993, the agency decreed that changes in treatability group are not a new point of generation for characteristic wastes if they are not managed in CWA or CWA-equivalent treatment systems or Class I SDWA systems. [58 FR 29871] If characteristic wastes are managed in CWA or CWA-equivalent treatment systems or Class I SDWA systems, each change in treatability group is still considered a new point of generation. These same principles also apply to wastes listed solely because they are ignitable, corrosive, or reactive. [June 28 1996; 61 FR 33681, May 16, 2001; 66 FR 27269]

Note that the change-in-treatability-group principle was a significant departure from EPA's earlier position on the POG of a waste. As mentioned earlier, for coding purposes, residues from treatment processes are newly generated wastes. In the late 1980s and early 1990s, conversely, the agency appeared to be saying that treatment residues are only newly generated wastes if they are in a different treatability group. Thus, if an incinerator burns a high-TOC ignitable waste (nonwastewater) and produces ash (nonwastewater) that is characteristic for heavy metals, the ash is not a newly generated waste. This conflict with earlier POG guidance was addressed in a May 11, 1999 *Federal Register* notice, where EPA stated that whenever a residue from treating a characteristic waste exhibits a new characteristic, this represents a new POG for LDR purposes. This new POG occurs without regard to there being a change in treatability group. [64 FR 25411–2]

In summary, for LDR purposes, the POG of a hazardous waste is when/where:

1. A waste is removed from a manufacturing process unit;

2. A remediation waste is actively managed or removed from an AOC;

3. For listed wastes that are listed due to toxicity, a treatment residue is in a different treatability group than the original waste;

4. For characteristic wastes and wastes listed solely due to ignitability, corrosivity, and/or reactivity, if the wastes are managed in CWA or CWA-equivalent systems or Class I SDWA systems and a treatment residue is in a different treatability group than the original waste; or

5. For any characteristic waste, a treatment residue exhibits a new characteristic not exhibited by the original waste.

14.1.5.2 Examples of POG under the LDR program

Q *A D001 high-TOC ignitable waste is treated in wastewater treatment tanks that generate a sludge. The original waste is a nonwastewater, and the sludge is a nonwastewater. Since the sludge is not in a different treatability group, it doesn't appear to* be a newly generated waste. Does that mean that the sludge must be treated by the same treatment standards as for D001 high-TOC ignitables before it can be land disposed?

A "It is EPA's view that the sludge in this situation should be viewed as a new treatability group. Put another way, the change-of-treatability-group principle applies to situations where liquid wastes which are technically nonwastewaters are inadvertently placed in wastewater treatment systems in small quantities for legitimate wastewater treatment, thereupon becoming wastewaters [as defined in §268.2(f) of the rules], and subsequently generating a sludge…. Consequently, because the sludge generated from the tank-based wastewater treatment system is a different treatability group from the wastewater from which it is generated, it would be considered to be a newly generated waste that should be evaluated at its point of generation to determine if it is hazardous, and if so, to then determine the appropriate LDR standard." [May 12, 1997; 62 *FR* 26007]

Q *D002 corrosive wastewater is treated in a wastewater treatment system that produces a nonhazardous sludge. What LDR treatment standards apply to the sludge?*

A The POG of the sludge is where it is removed from the wastewater treatment system. [RO 50940] Because the sludge is not hazardous at this point, it is not subject to LDR treatment standards, including any standards for underlying hazardous constituents. [RO 14216] (Note that this example appears to follow the same logic as in the previous example.)

Q *Ground water contaminated with unused methanol (U154) is mixed with other wastewaters such that the mixture is not ignitable. The mixture is then treated in a wastewater treatment system that produces a sludge. What LDR treatment standards apply to the sludge?*

A The sludge is in a different treatability group than the influent wastewater and is, therefore, a newly generated waste. U154 methanol was listed solely for ignitability, and because the influent wastewater mixture is no longer ignitable, the U154 code no longer applies to the treated wastewater or the sludge. Because the sludge does not exhibit any characteristic, it is not a hazardous waste and is not subject to LDR treatment standards. [RO 14207]

14.1.5.3 Intermediate-step treatment residues

The next (and last) major discussion that addressed POG issues under the LDR program also came on May 11, 1999. [64 *FR* 25411–2] One of the components of this rule addressed "intermediate-step treatment residues":

"For treatment residuals that appear only at intermediate steps of a treatment train, there is no obligation to…determine whether the residual is itself characteristic. Intermediate-step treatment residuals are not newly generated hazardous wastes for LDR purposes…even when an intermediate treatment residual is sent offsite for further treatment (such as incinerator ash going offsite for stabilization and landfilling)…."

After considering the following example, it appears that this new interpretation doesn't work in the real world:

- A generator feeds a D018 characteristic benzene waste (nonwastewater) with lead contamination below the universal treatment standards to an onsite incinerator, which produces an ash (nonwastewater) residue. All of the benzene has been destroyed, but the lead has been concentrated. The facility owner/operator wants to send the ash to an offsite landfill.

- *For waste coding and manifesting purposes,* the owner/operator analyzes the ash (which, as a treatment residue, is a newly generated waste) and finds that it is characteristic for lead (i.e., it is D008). He/she manifests the ash as a hazardous waste to the offsite landfill for stabilization and land disposal.

- *For LDR purposes,* the landfill operator cannot dispose a characteristic waste unless it meets the associated treatment standard. Therefore, he/she stabilizes the ash prior to disposal. However, the quoted language above says that the incinerator operator does not have to analyze the ash if it is an "intermediate-step treatment residue." This implies that the ash would not have to be treated for lead prior to disposal. However, if the ash isn't treated for lead, it isn't an "intermediate-step treatment residue" anymore and EPA's logic falls apart.

14.1.6 POG for spills

If a spill occurs, the owner must make an "immediate response" and begin cleanup activities. The agency has said: "EPA has not established a definition of what constitutes an immediate response to a spill situation. The time frames and extent of immediate response must be judged by persons responding to discharges on an individual basis." [RO 12748]

But what is the POG for cleaning up spills? And what is the regulatory status of spill cleanup residues? These questions are answered in the following two subsections.

14.1.6.1 POG for spills of characteristic wastes/products

As shown in Figure 14-1, a tanker carrying gasoline is involved in an accident, spilling 500 gallons of gas onto the roadside. Most of the gas soaks into the soil, making recovery of the gasoline unlikely. Therefore, the contaminated soil will be excavated and disposed. When the contaminated soil is excavated and placed into containers for disposal, where is the POG? Some would argue that the gasoline becomes a waste at the moment it spills from the truck, before it hits the ground. This argument would say that because the gasoline exhibits a characteristic (i.e., ignitability and also toxicity for benzene), the excavated soil must also be a hazardous waste. This argument is not supported by EPA guidance.

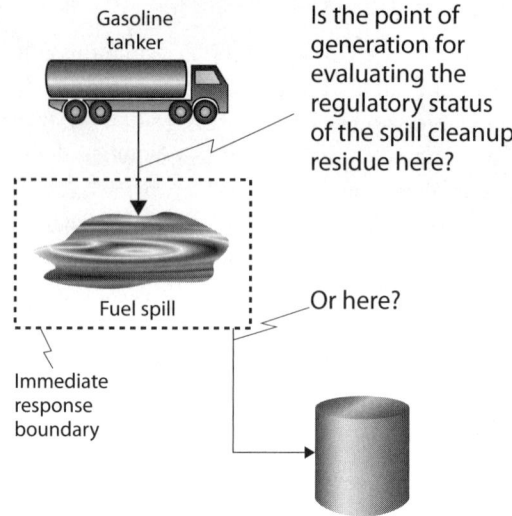

Figure 14-1: Point of Generation of Spill Cleanup Residues

For spills of characteristic wastes or products, the spill cleanup residues are evaluated based on their own properties—not the properties of the material spilled.

Source: McCoy and Associates, Inc.

For spills of characteristic wastes (D001–D043) or of characteristic products (e.g., gasoline, virgin acid) that become wastes as a result of the spill, it appears from the preponderance of EPA guidance that the POG of a solid waste during the spill cleanup is when the spill residue (e.g., contaminated soil, absorbent) is cleaned up. See for example, May 26, 1998; 63 *FR* 28617 and RO 13748, 14283, 14291, and 14588. Thus, the gasoline-contaminated soil would be evaluated based on its own properties, not the properties of the material that was spilled. The person cleaning up the spill would have to evaluate the soil to determine if it exhibits a characteristic. If it does, it is a hazardous waste; but if it doesn't exhibit any characteristic, it could be managed as nonhazardous waste. This concept is illustrated in the following example.

Q *A characteristic waste is spilled onto soil and the resulting contaminated soil does not exhibit any characteristic. What are the RCRA regulatory implications of this situation?*

A Here's EPA's answer: "It is conceivable that soil could be contaminated with a characteristic hazardous waste and yet not display a hazardous characteristic at the point of generation due to dilution in the soil matrix or to breakdown or alteration of the constituents in the soil environment. If this soil does not exhibit a characteristic when it is generated (excavated), then LDR requirements do not apply. However, treatment of the soil may still be required under cleanup authorities.... [A]ny deliberate mixing of prohibited hazardous waste with soil in order to change its treatment classification (i.e., from waste to contaminated soil) is illegal." [RO 14547]

14.1.6.2 POG for spills of listed wastes/products

For spills of F- or K-listed wastes, the POG of spill cleanup residues is when the spill residue (e.g., contaminated soil, absorbent) is cleaned up—just as for characteristic wastes. However, for spills of these wastes, the contained-in policy is applicable, meaning that we have to go back to what was spilled to determine if the spill cleanup residues are still listed. If the spill cleanup residues contain F- or K-listed wastes, they must be managed as if they are those listed wastes (see Section 5.3).

For spills of P- or U-listed chemicals, the POG of spill cleanup residues is also when the spill residue (e.g., contaminated soil, absorbent) is cleaned up. However, any spill cleanup residues must carry the same listed waste codes as the chemical that was spilled. [§261.33(d)]

A few of the F- and K-listed wastes and P- and U-listed chemicals were listed solely because they are ignitable, corrosive, or reactive. These ICR-only listed wastes are shown in Table 5-1 of Chapter 5. When one of these ICR-only listed wastes is spilled, the residue from cleaning up the spill is hazardous only if it exhibits a characteristic (see Section 5.3.1.1).

14.2 Unknown wastes

Have you ever come across one or more drums of material at your facility, and you truly have no idea what's in them nor can you determine where they came from? Or maybe someone did some "midnight dumping" of wastes onto your property. What do you do with this unknown waste material?

In some cases, particularly if individuals are dumping wastes illegally onto your property, calling the state or local hazardous materials response team may be the best answer. These groups will often take responsibility as the generator of the waste, and the proper characterization will be up to them. In other situations, your facility either is the rightful generator or wants to take on that responsibility. Then what do you do?

EPA considers hazardous wastes that have been dumped illegally to be a discharge and thus subject to the immediate response regulations at §§264.1(g)(8), 265.1(c)(11), and 270.1(c)(3). As discussed more fully in Section 7.5.4, these regulations exempt management activities performed to provide an immediate response for discharges of hazardous waste from the RCRA substantive management and permitting requirements. To qualify for the exemption, the treatment or containment activities must be for the initial, immediate response to the discharge. Once the immediate threat passes, all applicable RCRA standards apply, including the generator accumulation provisions of §262.34. The exemption applies only to treatment and storage activities; it does not relieve anyone of complying with RCRA requirements for the disposal of hazardous waste. [RO 14031]

If the materials are determined to be solid wastes (i.e., you want to get rid of them), you must make a hazardous waste determination just like for any other solid waste. You can do this by testing or knowledge. Determining whether an unknown material exhibits any of the hazardous waste characteristics or is a listed hazardous waste is discussed in the next two subsections.

©2015 McCoy and Associates, Inc.   **McCoy's RCRA Unraveled**

14.2.1 Determining if unknown wastes exhibit a characteristic

It is usually difficult to determine if an unknown waste is characteristically hazardous based solely on visual inspection and knowledge of how the characteristics are defined. So, a sample of the unknown will usually have to be sent for analysis. We have heard many different versions of what to ask the lab to test for, but the following is a consensus supplemented with limited EPA guidance:

- *Ignitability*—This should include a flash point test if the unknown is a liquid as determined by the discussions in Section 2.2.1.1; the flash point method should be one of the ones specified in §261.21(a)(1). Oxidation potential per §261.21(a)(4) is sometimes screened by using potassium iodide (KI) paper, which is especially suited to detect strong oxidizers. Samples that screen positive can be tested using SW–846 Method 1040—Test Method for Oxidizing Solids if confirmation testing is deemed necessary.

- *Corrosivity*—This should include a Method 9040C pH meter test if the unknown is aqueous as determined by the discussions in Section 2.3.1.

  Sometimes generators make pH determinations using other analytical methods. For example, pH paper can be used to make rough acid-base measurements; however, such a corrosivity determination would fall under the category of generator knowledge because the prescribed method (Method 9040C) was not used.

- *Reactivity*—Talk to the lab about their recommendations on testing for this characteristic. Some folks still ask (and some states require) that the rescinded HCN/H_2S release threshold tests be run to test for §261.23(a)(5) reactivity (see Section 2.4.2.1). Others just request total and amenable cyanides/sulfides analyses per Methods 9010C, 9012B, 9030B, 9031, and/or 9034.

- *Toxicity*—Have the lab run the TCLP for all 40 constituents. According to EPA/530/R-94/024 (cited in Section 18.2), once the TCLP is run, the extract obtained may be analyzed for the 40 constituents listed in §261.24 by any method, as long as that method has documented quality control and is sensitive enough to meet the regulatory limits. The agency noted that the following EPA test methods could be used in this step:

  – Methods 3010 and 6010 for arsenic, barium, cadmium, chromium, lead, silver, and selenium;

  – Method 7470 for mercury;

  – Methods 3510 and 8081 for pesticides;

  – Method 8151 for herbicides;

  – Method 8260 for volatile organics; and

  – Methods 3510 and 8270 for semivolatile organics.

  As an option to running the TCLP (which is quite expensive), some facilities will ask the lab to run totals for the 40 constituents in §261.24. Based on those results, the TCLP may then be requested only for those constituents with the potential to exceed regulatory levels. See Section 2.6.3.1 for a discussion of this option.

- *PCBs*—Although PCBs are regulated by TSCA, not by RCRA, unknowns should be tested for these constituents because 1) of how ubiquitous they are in oil-based substances, and 2) the regulatory requirements associated with their management depend on the PCBs concentration in the as-generated waste before treatment.

- *Asbestos*—Regulated by CAA, TSCA, and OSHA, not RCRA, asbestos is usually only tested where friable asbestos-containing material (ACM) may be present. ACM was commonly used in construction for thermal insulation systems, fireproofing, and acoustical insulation.

14.2.2 Determining if unknown wastes are listed

Hazardous waste listings are based on the sources of, or the processes that generated, the wastes rather than the concentrations of hazardous constituents; therefore, analytical testing alone, without information on a waste's source, will not produce information that will conclusively indicate whether a given waste is a listed hazardous waste. [RO 13181]

In some situations, it is hard for facility owners/operators to determine the source of the material or contamination, and so it is difficult to tell if a material is a listed waste. If the owner/operator has made a good-faith effort to find out if a material is a listed waste but cannot make such a determination because documentation on the source or process is unavailable or inconclusive, EPA allows the owner/operator to assume that the material is not a listed hazardous waste. (See, for example, December 21, 1988; 53 *FR* 51444, March 8, 1990; 55 *FR* 8758, and April 29, 1996; 61 *FR* 18805.) Consequently, if the material does not exhibit a hazardous waste characteristic, the RCRA hazardous waste regulations do not apply. This has been EPA's long-standing policy, which was reaffirmed in RO 14291.

Q At an industrial site, unknown solids are discovered in unmarked drums. Samples of the materials are analyzed indicating that they contain 2,3,4,5- and 2,3,4,6-tetrachlorophenol (constituents that, in part, form the basis for some of the dioxin hazardous waste listings—see F020, F023, F027, and F028). However, no information is available on the original processes that generated the materials. Are the solids considered a hazardous waste?

A If the waste cannot be traced back to an original process that would generate a waste meeting one of the dioxin waste listing descriptions, it is exempt from regulation as long as it does not exhibit any of the hazardous waste characteristics. Because the solids are not ignitable, corrosive, or reactive and neither chemical is included on the §261.24 table, the solids would not be hazardous wastes under the federal regulations. [RO 13586]

14.2.3 Accumulation time for unknown wastes

When materials of unknown origin and content are discovered that are subsequently going to be discarded (i.e., they are solid wastes), they must be characterized before disposal. If sampling and analysis are used for this purpose, and the results come back stating that the waste is indeed a RCRA hazardous waste, when does the generator accumulation time (i.e., the 90- or 180-day clock) specified in §262.34 begin?

Section 262.34 allows large quantity generators to accumulate hazardous waste onsite for 90 days or less (180 days or less for small quantity generators) without a RCRA permit or interim status if certain conditions are met. These conditions include:

- The date that the accumulation period begins must be clearly marked on each container, and
- Each container and tank must be labeled clearly with the words "Hazardous Waste."

According to EPA, if a container holding unknown waste (which is later discovered via analysis to be a RCRA hazardous waste) is not labeled as "Hazardous Waste" or marked with the accumulation start date *as soon as the waste is put in the container*, the generator has not met the requirements for the 90- or 180-day accumulation permit exemption. Consequently, he/she is operating a hazardous waste storage facility that needs a RCRA permit.

In order to qualify for the 90- or 180-day accumulation exemption, all containers of unknown waste should be labeled as hazardous and dated when the waste is first generated, even if the contents are not yet known to be hazardous. The accumulation period does *not* begin after the waste is analyzed and you get the results back from the lab saying the waste is hazardous. [EPA/233/B-00/001, RO 11424] One way to make it clear that the designation of the unknown waste may change is to add the words "Pending Analysis" after "Hazardous Waste."

Some states have a different interpretation than the federal guidance noted above, so you should check with your state on accumulation time for unknowns.

14.3 Spills and spill residues

Spills happen despite the best planning, employee training, and plant operations. In many situations, spill residues are hazardous waste that must be managed in accordance with the RCRA regulations. This section offers EPA guidance on several topics associated with spills: 1) do we have to clean up spills?, 2) spill response, 3) spills of characteristic wastes/products, 4) spills of listed wastes/products, 5) spills of oil or used oil, 6) used absorbents, 7) recycling spilled products, 8) corrective action requirements, 9) LDR implications, 10) spill reporting, and 11) spill response training.

14.3.1 Do we have to clean up spills?

EPA requires all spills of hazardous waste to be addressed promptly to minimize hazards to human health and the environment. If cleanup does not start promptly, EPA will consider the spill to be a hazardous waste disposal site subject to RCRA permitting requirements and/or enforcement action. [RO 12748, 13296, 14547, 14650] Where do the regulations say that we have to clean up spills? It's hard to find—we have to come at this from a little different direction.

If we spill a material and don't clean it up, that material is considered to be abandoned. [RO 14650] We noted in Section 1.1.4.1 that abandoned materials, as defined in §261.2(b), are one of the most common types of solid wastes. So, the abandoned spill material would be a solid waste, and, if that material meets the definition of a hazardous waste, "a spill of hazardous material to soil or ground water is normally a simple act of disposal." [June 1, 1990; 55 *FR* 22671] To keep from complying with the numerous, expensive requirements associated with operating a hazardous waste disposal area, we must clean up spills.

14.3.1.1 Spills of hazardous waste versus spills of products

If what we spill already meets the definition of a hazardous waste (i.e., a characteristic or listed waste), it makes sense that we have to clean it up and likely manage the spill residues as hazardous per Table 14-1. But what if we spill a product? Do we have to worry about that spill under RCRA? We do. The reason is that spilled product is not being used for its intended purpose; if we leave it in the soil, for example, it's abandoned as noted above. Therefore, it is a solid (and potentially hazardous) waste.

If we clean up the spilled product and can still use it for its intended purpose, it never became a solid waste and therefore avoids regulation as a hazardous waste. If the spilled product that we clean up is sent for reclamation, the cleanup residue can be considered an off-specification commercial chemical product and is not solid (or hazardous) waste via Table 1 in §261.2(c)—see Section 14.3.7 and RO 14650. Conversely, if we clean up a spilled

Table 14-1: Regulatory Status of Spill Cleanup Residues

Spilled material[1]	Disposition of spill cleanup residue		
	Abandoned (not cleaned up)	Cleaned up and disposed	Cleaned up and reused or reclaimed
Characteristic waste (D001–D043)	HW	Solid waste—determine if characteristic	Not applicable
Characteristic product (e.g., gasoline, acid)	HW	Solid waste—determine if characteristic	Off-spec product
Listed waste (F-, K-, P- or U-waste)	HW	HW[2]	Not applicable
Listed product (P- or U-chemical)	HW	HW[2]	Off-spec product

HW = hazardous waste.
[1] Any deliberate mixing of hazardous waste with soil to circumvent RCRA is illegal.
[2] If the spilled waste/product was listed solely because it is ignitable, corrosive, or reactive, the residue will be hazardous only if it exhibits a characteristic (see Section 5.3.1.1).

Source: McCoy and Associates, Inc.

product and decide to dispose the cleanup residue, it is a solid waste like any other, and a hazardous waste determination must be made.

Bottom line: Spilled material that is a hazardous waste *or becomes a hazardous waste as a result of the spill* must be cleaned up and managed under Subtitle C.

14.3.2 Spill response

If a spill occurs, we must make an "immediate response" and begin cleanup activities. Here's EPA's guidance on the topic:

"EPA has not established a definition of what constitutes an immediate response to a spill situation. The time frames and extent of immediate response must be judged by persons responding to discharges on an individual basis. Extended responses which are not judged to be immediate in nature may result in: 1) a modification to the facility's contingency plan, 2) an enforcement action for an inadequate contingency plan or permit violation, or 3) enforcement action for illegal disposal." [RO 12748]

EPA has defined "immediate" for the purpose of cleaning up incidental and infrequent wood preservative drippage (F032) at wood preserving facilities:

"[T]he agency intends, absent extenuating circumstances, that owners/operators respond to storage yard drippage that occurs when a facility is in operation within one consecutive working day. A facility is considered in operation on any day in which it is treating wood. For facilities which are not in operation during a storage yard drippage event, the agency expects the facility to clean up drippage within 72 hours of occurrence. EPA recognizes that the term 'immediate' must take into account the nature of the incident as well as facility-specific factors." [December 24, 1992; 57 *FR* 61494]

14.3.2.1 Generator requirements

Large quantity generators that accumulate hazardous waste onsite for 90 days or less without a permit must develop a contingency plan for their hazardous waste activities. [§262.34(a)(4)] This plan must address immediate actions that facility personnel will take in response to releases of hazardous waste. Small quantity generators don't have to have a formal contingency plan, but they must comply with the emergency response requirements in §262.34(d)(5).

14.3.2.2 TSD facility requirements

Sections 264.51 and 265.51 require owners/operators of TSD facilities to have contingency plans in place. As with large quantity generators, the plans must describe immediate actions that facility personnel will take in response to any unplanned release of hazardous waste or hazardous waste constituents. The contingency plan will be incorporated into the facility's permit.

14.3.2.3 Transporter requirements

Under §263.30, transporters are required to take immediate action to protect human health and the environment if a spill of hazardous waste occurs. These actions may include notifying the local authorities, diking the discharge area, etc. Furthermore, transporters must clean up any hazardous wastes so that the discharge no longer presents a hazard to human health and the environment. [§263.31] DOT also has regulations with which transporters must comply when responding to a spill of hazardous waste.

An authorized federal, state, or local official on the scene may waive the EPA ID number and manifest requirements for generators and transporters involved in immediate hazardous waste removal actions following a spill or leak during transportation. [§263.30(b)] Alternatively, EPA regions or authorized states can issue provisional EPA ID numbers to persons responding to a spill. This supposedly can be done over the phone. [RO 12016]

The immediate response spill containment or treatment activities are exempt from RCRA permitting or substantive RCRA standards. [November 19, 1980; 45 *FR* 76629] When the emergency is over, all applicable RCRA hazardous waste management provisions once again apply to the generator's/transporter's activities. For example, if the transporter has generated hazardous waste, he/she must comply with Part 262 requirements when the immediate actions are over.

14.3.2.4 Cleanup levels

The RCRA regulations do not address hazardous waste spills to any significant degree, and they certainly do not identify cleanup levels that must be achieved. All the regs say is that large quantity generators and TSD facilities are held to the standard of minimizing hazards to human health or the environment. [§§264/265.51] Additionally, transporters must clean up any spilled hazardous wastes so that the discharge no longer presents a hazard to human health and the environment. [§263.31]

EPA's philosophy is that different cleanup levels will be appropriate in different situations. As a result, cleanup levels are most appropriately set on a situation-specific basis. However, some states may require specific cleanup standards to be achieved. [RO 11848]

14.3.2.5 Exemption for immediate response treatment and containment activities

As noted in Section 14.3.1, EPA considers that an unremediated hazardous waste spill is essentially unpermitted disposal. However, the agency recognizes that requiring RCRA permits for spill cleanups would hinder prompt response actions. As a result, EPA allows immediate spill response actions to be performed without a RCRA permit. [§270.1(c)(3)—see Section 7.5.4] Treatment and containment activities conducted in immediate response to hazardous waste discharges and spills (including the use of emergency structures, such as tanks used for emergency secondary containment) are also exempt from Part 264 or 265 standards (except for the preparedness and prevention and contingency plan requirements). [§§264.1(g)(8) and 265.1(c)(11), RO 12298, 12748, 14031]

According to EPA:

"To qualify for the exemption, a unit must be intended exclusively for immediate response to discharges of hazardous wastes, such as burst pipes, ruptured containers or tanks, breached dikes, and the like. Structures used for responding to discharge events which occur periodically or repeatedly, or in which containment or treatment extends beyond the immediate response period, do not qualify for the exemption." [RO 12298]

For example, an owner/operator responding to a discharge might excavate soil contaminated with spilled hazardous waste and store it temporarily in containers prior to the removal of the material offsite. The container storage area would not be subject to Part 264 or 265 standards. [RO 13296] In another example, a tank is used to contain spilled residue from a truck loading/unloading area. In this case, the exemption from Part 264/265, Subpart J tank standards would apply to the tank only if the owner/operator could demonstrate that spills were "extremely rare and unpredictable events." [RO 12298] Furthermore, EPA stresses that "[t]reatment and containment activities conducted after the initial response period are subject to [Part 264 or 265] standards. A facility may qualify for an emergency permit under §270.61 for such treatment and containment activities occurring after the immediate response period." [RO 12748; see also RO 14031]

As another example, a tank car of liquid hazardous waste ruptures at a generator's site, and the waste spills on the ground. The facility immediately builds an emergency dike to contain the spilled waste and then pumps the liquid into drums. The drums of hazardous waste remain onsite for several weeks before being shipped offsite to an incinerator. EPA's evaluation of this scenario indicates that the construction and operation of the dike are not subject to RCRA standards or permitting, since it was an immediate response. However, once the liquid was contained, the immediate response was over, and the subsequent

McCoy's RCRA Unraveled ©2015 McCoy and Associates, Inc.

storage in drums is subject to 90/180/270-day accumulation provisions specified in §262.34. If the spilled waste and/or contaminated soil in the dike are treated as part of the immediate cleanup, such treatment would not be subject to RCRA regulation. However, if the treatment occurred beyond the immediate response (a judgment call between the facility and the state), an emergency permit would be required unless the generator treated the waste in a way that is exempt from permitting (e.g., in a 90-day unit). [November 19, 1980; 45 *FR* 76629]

Any tank that is not eligible for the §264.1(g)(8) or 265.1(c)(11) immediate response exemption is subject to the hazardous waste tank standards. Consequently, if the spilled material is accumulated in the tank for less than 90 days, only the generator standards under §262.34 would apply. However, if the accumulation time exceeds 90 days, the facility storage standards under Part 264 or 265 would apply. Any tank that will be used on a predictable basis for hazardous waste storage at a TSD facility must also be addressed in the facility's closure plan. [RO 12291]

14.3.2.6 Point of generation issues

The point of generation of spill cleanup residues was previously discussed in Section 14.1.6.

14.3.3 Spills of characteristic wastes/products

If a characteristic waste is released into the environment, the spill must be cleaned up. Additionally, the responsible party needs to determine (by testing or knowledge) if the excavated contaminated media exhibit a hazardous waste characteristic. If they do, they must be managed as hazardous waste until they no longer exhibit that characteristic. An example would be a spill of chromium-containing wastewater onto soil, resulting in contaminated soil that fails the TCLP for chromium. Such soil would have to be excavated and managed as D007 hazardous waste.

What about a leak or spill of a characteristic product that becomes a solid and hazardous waste as a result of the spill? For example, a tanker carrying gasoline is involved in an accident, spilling 500 gallons of gas onto the roadside. Most of the gas soaks into the soil, causing it to exhibit the toxicity characteristic for benzene. Recovering the gasoline is unlikely; therefore, the contaminated soil must be managed as a RCRA hazardous waste when disposed. [RO 12489]

EPA's latest discussion of how the contained-in policy relates to characteristic wastes corroborates the two paragraphs above. In the case of media that exhibit a characteristic of hazardous waste, the media are considered to contain hazardous waste as long as they exhibit a characteristic. Once the characteristic is eliminated (e.g., through treatment), the media are no longer considered to contain hazardous waste. [RO 14291]

If a product that exhibits a characteristic can be captured and recycled to a process or to a reclamation unit, the captured material is usually not a solid waste at all. [See Table 1 in §261.2(c) and Table 14-1 in this chapter.] It is a product, although it might be an off-specification product that needs to be reclaimed before use. Either way, the product is not subject to RCRA if it is put to beneficial use.

For example, a bulk oil storage terminal has a spill of characteristic product. The facility is able to capture the spilled material and return it to a refinery or other legitimate fuel production operation. In this situation, the spilled product is not a solid waste when reused. However, "mere assertion of an intent to recycle a commercial product spill does not convert the spill into a nonwaste. There must be objective indicia that recycling is reasonable, and that it will occur in a timely manner." [July 28, 1994; 59 *FR* 38539] The objective indicia referenced above are detailed in Section 14.3.7.

Q A product, which contains two active ingredients, toluene and benzene, spills on the ground. How is the spill residue regulated under RCRA?

> **A** If a product with more than one active ingredient is spilled, it will not be classified as a P- or U-listed spill residue per §261.33(d) (see Section 3.5.3.7). Therefore, the spill residue will be hazardous only by characteristic. If the soil mixed with the spilled product exhibits a characteristic, then the spill residue is a RCRA hazardous waste. If the soil mixture does not exhibit a characteristic of hazardous waste, RCRA is not applicable. (CERCLA reporting may be required if the reportable quantity is exceeded, since benzene and toluene are both hazardous substances—see Section 14.3.10.2.) [RO 12489]

14.3.4 Spills of listed wastes/products

Spills of listed hazardous waste/product can be broken into two major categories: 1) contamination resulting from spills of F- and K-wastes, and 2) contamination resulting from spills of commercial chemical products on the P- and U-lists.

14.3.4.1 Spills of F- and K-wastes

Spills of F- and K-wastes into the environment are regulated per the contained-in policy. Environmental media that contain an F- or K-waste must be managed as if they are that F- or K-waste. The media are subject to hazardous waste management until they are decontaminated such that they no longer contain the listed waste. (Refer to the contained-in policy discussion in Section 5.3.)

14.3.4.1.1 *Leachate from spill-contaminated soil*

Rainwater that percolates through soils contaminated with spills of several different listed hazardous wastes may qualify as multisource leachate (F039). EPA has determined that "water that has percolated through soils contaminated with more than one listed hazardous waste for which Part 268 treatment standards are in effect is normally F039." [RO 13509]

14.3.4.2 Spills of commercial chemical products (P- and U-wastes)

Any residue or contaminated soil, water, or debris resulting from cleanup of spills of any commercial chemical product, manufacturing chemical intermediate, or off-specification products listed in §261.33(e) and (f) (i.e., the P- and U-lists) is a RCRA hazardous waste when it is discarded or intended to be discarded. [§261.33(d)] In other words, we really don't need to apply the contained-in policy to residues resulting from spills of products on the P- and U-lists; they are P- and U-wastes by regulation.

The §261.33(d) language applies to all P- and U-spill residues that are discarded or intended to be discarded, regardless of where the spill occurs (e.g., onto land, into water, within a wholly contained building). [RO 13335] On the other hand, if a spill residue can be returned to a process or otherwise put to beneficial use, it would not be a listed hazardous waste—it would be a product. However, the burden of proof is on the generator to show that legitimate recycling will take place, as discussed in more detail in Section 14.3.7. [RO 13743]

If P- or U-spill residues are decontaminated such that they no longer meet the hazardous waste listing (i.e., they no longer contain the listed waste), they would no longer be regulated as a listed hazardous waste. [RO 13732] Another possibility is that a P- or U-listed product is spilled onto environmental media such as soil, but the resulting contaminant concentrations are below state-determined remedial levels. If the state is authorized for the base RCRA program, the state can make a no-longer-contains determination for the contaminated soil, and it would not be subject to RCRA hazardous waste management requirements before or during any excavations that might occur. [RO 13568]

Examples of spills of commercial chemical products are given below.

> **Q** *A thermometer in use at a weather observation station is blown off its hook and breaks, leaking mercury onto the ground. Is the contaminated soil a hazardous waste?*

A Unused mercury is a listed waste (U151) when discarded. However, the P- and U-lists of discarded commercial chemical products and spill residues apply only to unused materials. Since the thermometer (and contained mercury) in this case has been used, the U151 listing does not apply. However, if the soil exhibits the toxicity characteristic for mercury, the soil would need to be managed as a D009 hazardous waste. [RO 13372] (You may recall from our discussion in Section 3.5.3.1.2 that thermometers are manufactured articles, and the contained mercury in even an *unused* thermometer would also not be considered U151.)

Q *Overapplication of the pesticide heptachlor at a residence caused a swimming pool and the surrounding soil to become contaminated. When disposed, are the pool water and soil P059?*

A No. Under §261.33, listed commercial chemical products are not hazardous wastes when they are applied to the land if that is their normal manner of use. As a result, soil and/or surface water that are contaminated as a result of overapplication of these products during their normal manner of use are not considered hazardous wastes (unless the materials exhibit a hazardous waste characteristic). However, if a listed commercial chemical product is *spilled* during its normal manner of use, this constitutes disposal and the hazardous waste regulations will apply to cleanup residues. Since the contamination of the pool water and soil resulted from the normal use of the pesticide as a product, instead of from its disposal, they would not be considered hazardous unless characteristic. [RO 11291, 12357]

Q *A warehouse was used to store pesticides that are on the P-list. Wipe tests of the floor and walls in the building confirm the presence of these chemicals. If the residues resulted from spills of the P-chemicals, must they be managed as hazardous wastes?*

A The spilled materials appear to have been abandoned by accumulating them in the warehouse rather than disposing them elsewhere. They may also have been abandoned by essentially being disposed within the building itself. This makes them solid wastes, and if they were listed P- or U-chemicals, the residues would also be hazardous wastes. If the spills occurred after November 19, 1980 (the effective date of the RCRA regulations), "treatment and containment of spills (except in immediate response to spills) must comply with applicable Part 265 requirements. Since it appears that this is not an immediate response situation, the facility would be subject to an enforcement action for treating, storing, or disposing of hazardous waste without interim status or a permit, and could be required to take appropriate action to clean up the residues." [RO 11161]

Q *A facility bakes used railroad ties to recover creosote, and then it uses the recovered creosote as a wood preservative without further processing. The reclaimed creosote is again a commercial chemical product (i.e., the label on a drum of this material would say "Creosote"). What would be the regulatory status of the recovered creosote if it were subsequently spilled?*

A This spilled material would be hazardous waste U051. [RO 13572]

14.3.4.3 Did we spill a listed waste or not?

Determining the correct hazardous waste classification (if any) for listed spill residues (F-, K-, P-, or U-listed) depends on the regulatory status of the material before it was spilled. For example, consider a spill residue consisting of 15% tetrachloroethylene and 85% clay/dirt. If the solvent was a discarded commercial chemical product, manufacturing chemical intermediate, or off-specification commercial chemical product before it was spilled, the spill residue would be classified as a U210 waste. However, if the tetrachloroethylene was a nondegreasing spent solvent before the spill occurred, the waste would be classified as F002 via the contained-in policy. [RO 12906, 14194]

Q *During cleanup of an industrial facility, the owner/operator discovers that some of the soil onsite is contaminated with toluene. The facility's environmental manager notes that toluene is listed in Appendix VIII to Part 261 of the RCRA regulations (which is a list of "hazardous constituents") and, therefore, recommends that the soil be managed as a RCRA hazardous waste. Is the environmental manager correct?*

A No. Soil contaminated with toluene is not automatically a hazardous waste based solely on the fact that the chemical is listed in Appendix VIII. The soil would only be considered a hazardous waste if 1) the contamination was caused by a spill of one of the wastes listed in §261.31, 261.32, or 261.33; or 2) the soil exhibits one of the hazardous waste characteristics. [RO 12171, 12392] Regulated entities are not meant to use Appendix VIII to determine whether a waste (or spill residue) is hazardous. That appendix is solely for use by EPA in evaluating whether to list a waste as hazardous. [RO 11051, 11144, 12014, 12296, 13290] Additional discussion of this topic is contained in Section 18.2.

As noted in Section 14.2.2, the mere presence of contaminants in soil does not automatically make the soil a hazardous waste. The generator or owner/operator must make a good-faith effort to find out if a material is (or has been contaminated with) a listed waste. If documentation on the source of contamination is unavailable or inconclusive, EPA allows the facility to assume that the material is not a listed hazardous waste. Consequently, if the material does not exhibit a hazardous waste characteristic, the RCRA hazardous waste regulations do not apply. This has been EPA's long-standing policy, which was reaffirmed in RO 14291.

14.3.5 Spills of oil or used oil

Spills of oil and used oil are common in industry. How are they regulated under RCRA?

14.3.5.1 Oil spills

RCRA Section 7003 gives EPA the authority to require cleanup in situations where human health and/or the environment are imminently and substantially endangered. Under the agency's interpretation of this statutory provision, it has the authority to use Section 7003 to compel cleanup of oil spills, regardless of whether or not the spilled material meets the definition of a characteristic waste (e.g., via the toxicity characteristic for benzene). See Case Study 14-2. Additional cleanup authorities are available under the Oil Pollution Act and the Clean Water Act. [RO 13513]

Case Study 14-2: Spill of Crude Oil

A leak occurs in a pipeline carrying crude oil from the oil field to a refinery. As a result of the leak, the soil underneath the pipeline becomes contaminated. Is the contaminated soil excluded from RCRA regulation under the oil and gas exclusion of §261.4(b)(5)?

The RCRA oil and gas exclusion states that "[d]rilling fluids, produced waters, and other wastes associated with the exploration, development, or production of crude oil, natural gas, or geothermal energy" are not hazardous wastes. When creating this exclusion, Congress clearly meant it to apply only to wastes "intrinsically derived from primary field operations." The exclusion was not designed to exclude oil and gas wastes associated with transportation and manufacturing operations. According to EPA, "[f]or the purpose of the RCRA exemption, nonexempt transportation-related wastes are those resulting from any mode of transportation, including pipelines…." Consequently, in this situation, if the wastes from the pipeline leak exhibit the toxicity characteristic for benzene or exhibit any other hazardous waste characteristic, they would be hazardous wastes subject to RCRA regulation. [RO 11610]

14.3.5.2 Used oil spills

As set forth in §279.22(d), used oil generators must take the following response actions when used oil from aboveground tanks and containers is spilled:

- Stop the release,
- Contain the released used oil,
- Clean up and properly manage the spilled used oil and other materials, and
- Repair or replace any leaking used oil storage tanks or containers before putting them back in service.

These actions must be taken only if a release *to the environment* occurs. EPA has stipulated that releases to the environment do *not* include "releases within contained areas such as concrete floors or impervious containment areas, unless the releases go beyond the contained areas." [September 10, 1992; 57 *FR* 41586] Consequently, a spill from an aboveground tank into the tank's containment structure would not be considered a release to the environment and, therefore, would not be subject to the §279.22(d) response action requirements. However, used oil handlers "have an obligation to clean up used oil spills or leaks onto containment areas before the used oil reaches the environment. Such cleanup operations prevent the potential contamination of unprotected soils near storage and work areas. If a release of used oil goes beyond a containment pad and into the environment, then the response-to-releases requirements in §279.22(d) apply." [RO 14339]

Releases of used oil from underground storage tanks are subject to the Part 280, Subpart F corrective action requirements. [§279.22]

14.3.6 Used absorbents

Absorbents and pads used to contain accidental spills are normally not wastes at all—they are products performing their intended function. When a hazardous waste or a product that becomes hazardous waste when spilled mixes with product absorbent, the mixture is not a solid waste until the absorbent will be discarded; at that time, the resulting contaminated pad could be hazardous via application of the contained-in policy. Therefore, the regulatory status of contaminated absorbent will be identical to that of contaminated environmental media discussed in Sections 14.3.3 and 14.3.4.

If a characteristic waste or a characteristic product is absorbed, the used absorbent (when disposed) would be hazardous only if it exhibits a characteristic. Conversely, absorbents that are contaminated with listed wastes or products will carry the same listed code as the spilled material. [RO 11798, 14025] If an F003 spill is picked up with an absorbent pad, it will be an F003 waste via the contained-in policy only if it exhibits a characteristic. (The same logic would apply to spills of any of the other 29 ICR-only listed wastes—see Section 5.3.1.1.)

Q: Absorbent pads are used to absorb a range of petroleum products (e.g., gasoline, kerosene, fuel oil, etc.) resulting from a variety of activities, including spill cleanups and the cleaning of product tanks and containers. These pads are processed to recover the petroleum products, which are then sent either for further processing (including reprocessing by a petroleum refinery) or for burning as fuels. What is the regulatory status of the contaminated absorbent pads?

A: Absorbent pads containing petroleum products that are sent for recovery are excluded from the definition of solid waste and are not subject to regulatory controls under the RCRA hazardous waste regulations. This is a result of §261.2(c)(3) as it pertains to off-specification commercial chemical products being reclaimed. [RO 14503]

Absorbents containing used oil that are to be burned for energy recovery are subject to the Part 279 used oil management standards regardless of whether or not free-flowing oil is visible. [§279.10(c)(2)] On the other hand, if absorbents containing used oil will *not* be burned for energy recovery, they are subject to the used oil standards only if free-flowing used oil is visible.

[§279.10(c)] If they will be disposed (not burned for energy recovery), a hazardous waste determination must be made for them just like any other solid waste. [RO 11798]

14.3.7 Recycling spilled products

If spilled commercial chemical product is recycled (reclaimed) instead of disposed, it will generally be exempt from classification as a solid waste and therefore avoid regulation as a hazardous waste. [Table 1 in §261.2(c)] However, in order for this exemption to apply, the generator must prove that the material will be legitimately recycled. "[A] generalized assertion that a material is being recycled does not necessarily satisfy this burden." [RO 11713] EPA considers five factors when determining if a recycling exemption is appropriate for spill cleanup residue [June 1, 1990; 55 *FR* 22671]:

1. Whether the generator has begun to recycle the spill,
2. The length of time the spill residue has existed,
3. The value of the spilled material,
4. Whether it is technically feasible or practical to recycle the spill, and
5. Whether there is any history of the company recycling this type of spill residue.

If the generator cannot make a strong case that the spilled material will be legitimately recycled, "the materials are solid wastes immediately upon being spilled because they have been abandoned." Consequently, if the spilled material/cleanup residues are hazardous, they must be managed in compliance with the RCRA regulations. [RO 13743]

For example, recovery of free-phase hydrocarbons from the water table as part of a ground water remediation operation would likely generate an off-spec product that could be sent to a refinery outside the purview of RCRA. However, "only recovered materials that are primarily oil, and that can be inserted into a refinery's recovered oil system without pretreatment (or can be inserted directly into the refining process itself), would be considered…eligible for…exclusion. The management of petroleum-contaminated ground water in separation and treatment units is clearly solid waste (and potentially hazardous waste) management, essentially wastewater treatment." [July 28, 1994; 59 *FR* 38540]

We have noted a number of times in this section that commercial chemical products that are reclaimed are not solid (or hazardous) wastes via Table 1 in §261.2(c). However, we should point out here the actual wording of the regs: "Commercial chemical products listed in 40 *CFR* 261.33." In other words, only those chemicals on the P- or U-lists appear to be exempt. EPA has clarified numerous times, however, that products not specifically listed in §261.33 (e.g., gasoline) are also included in the exemption. [April 11, 1985; 50 *FR* 14219, RO 11713, 11726, 13356, 13490]

14.3.7.1 Leaking petroleum underground storage tanks

Free product is often recovered during cleanups of contaminated soil and ground water from leaking underground storage tanks (USTs) containing petroleum products. (Such petroleum USTs are regulated under the RCRA Subtitle I underground storage tank program, the regulations for which are codified in Part 280.)

EPA has determined that the Subtitle C hazardous waste regulations do not apply to spilled petroleum products that are subsequently recovered as free product for use or reuse *in their normal manner*. [RO 14650] An example of recovering spilled gasoline for reuse is given in Case Study 14-3. The generator may be required to show that the recovered material is suitable for, and actually is used as, a fuel or a constituent in fuel. However, if the material is destined for disposal, it would be a solid (and potentially hazardous) waste. [RO 11713]

Petroleum-contaminated media and debris are exempt from regulation as hazardous waste if they 1) exhibit a characteristic of D018–D043 only, and 2) are subject to the Part 280 underground storage tank corrective action regulations. [§261.4(b)(10)]

> **Case Study 14-3: Recovered Gasoline**
>
> Free product is recovered during a leaking underground storage tank cleanup at an abandoned gas station. The contaminated gasoline free product (which exhibits one or more hazardous characteristics) is sent offsite to a commercial waste management facility for use in a fuels blending program. Is the recovered gasoline regulated as a hazardous waste? What if the free product is reused as gasoline by either being re-refined at a refinery for resale or added to a large bulk storage tank for resale without processing?
>
> Recovered commercial chemical products (e.g., gasoline) normally used as fuels are not regulated as RCRA Subtitle C hazardous wastes if they are used as an ingredient to make a fuel (e.g., fuel blending). The RCRA regulations also would not apply if the free product is sent to a refinery for re-refining or combined directly with other gasoline for resale (with or without processing). [RO 11138, 11713]

14.3.8 Corrective action requirements

EPA is authorized under RCRA Section 3004(u) to require corrective action at solid waste management units (SWMUs) located at permitted facilities. EPA has determined that the term "solid waste management units" includes areas that have become contaminated by routine and systematic releases of hazardous waste or hazardous constituents. An example of such a SWMU would be an area where preservatives from pressure-treated wood were allowed to drip onto the soil for a number of years.

Conversely, corrective actions will *not* be required at areas where one-time, accidental spills occur if they cannot be linked to a discernible SWMU. For instance, an accidental spill from a truck at a RCRA-permitted facility would not be subject to corrective action if cleaned up promptly. [RO 12969] See Section 17.2.1 for additional discussion.

14.3.9 Land disposal restrictions

Under the land disposal restrictions program, characteristic wastes must be treated prior to land disposal so that all underlying hazardous constituents "which can reasonably be expected to be present at the point of generation of the hazardous waste" meet the universal treatment standards. (An exception to this requirement exists for characteristic wastes that are managed in a wastewater treatment system regulated under the Clean Water Act or injected into a Class I injection well.) Underlying hazardous constituents and their applicable universal treatment standards are listed in §268.48—see Section 13.3 for details.

Identifying underlying hazardous constituents in industrial waste streams is fairly straightforward. However, for soil (or other environmental media) that has been contaminated via spills, it is often difficult or impossible to determine exactly what constituents are reasonably expected to be present due to scarcity of records or knowledgeable plant personnel. In these situations, EPA stated that it is appropriate to analyze the soil for the entire list of §268.48 constituents. Such constituents detected at levels above the universal treatment standards would then be the constituents reasonably expected to be present at the point of generation. [RO 13748]

Although analyzing a spill for all underlying hazardous constituents would provide a rigorous answer, in our experience this is rarely done. Why? The cost to analyze a sample for all 240+ constituents in §268.48 is about $5,000. More commonly, the generator uses professional judgment, perhaps supplemented with some analytical screening tests, to identify underlying hazardous constituents.

14.3.10 Spill reporting

The following subsections briefly summarize the spill reporting required by several federal environmental programs.

14.3.10.1 RCRA spill reporting

For nontransportation-related releases of hazardous waste to the environment, RCRA requires notification to EPA or authorized states and follow-up reporting if the release occurred from a tank system. [§§264/265.196(d)] (There are no comparable notification/reporting requirements if the release occurs from a container.) This notification requirement is waived if the spill is ≤1 lb and is immediately contained and cleaned up. Note that EPA has stated in guidance that releases from tanks that are contained within the secondary containment system need not be reported. [EPA/530/SW-87/012, cited in Section 10.1.2.4]

Large quantity generators and permitted and interim status TSD facilities must have contingency plans specifying the actions that will be taken in response to fires, explosions, and releases of hazardous wastes to the environment. The regulations in §§264/265.56 include reporting and notification requirements that could be triggered by a serious spill.

14.3.10.2 CERCLA spill reporting

EPA also requires facilities to immediately notify the National Response Center (NRC) when a CERCLA hazardous substance (including all hazardous wastes) is released to the environment in excess of its reportable quantity (RQ) within a 24-hr period. This is actually a CERCLA, not RCRA, reporting requirement. [40 *CFR* 302.6] (RQs for chemicals and hazardous wastes are contained within 40 *CFR* 302.4 and also in Appendix A to the Hazardous Materials Table in 49 *CFR* 172.101.) If such notification is made under this CERCLA provision, the RCRA notification requirement noted above for releases from tanks is satisfied (although the follow-up reporting must still be completed).

Note that it is a spill of a hazardous substance that triggers CERCLA reporting requirements. All hazardous wastes would qualify as hazardous substances; however, spills of nonhazardous wastes that contain hazardous substances could also trigger CERCLA reporting. For example, certain mining and mineral processing wastes are excluded from the definition of hazardous waste by the Bevill exclusion (see Section 4.15). However, if a release to the environment of one of these excluded wastes exceeds an RQ of a hazardous substance (e.g., arsenic or mercury contained within the waste) within a 24-hr period, CERCLA reporting requirements must be met. [EPA/530/R-99/022, cited in Section 4.15.1]

Although (to our knowledge) the federal regs don't define "immediately," some states have. For example, several states require notification to the NRC within 15 minutes after discovery of a release. Also, EPA has a CERCLA reporting enforcement policy that notes "ordinarily, delays in making the required notifications should not exceed 15 minutes after the person in charge has knowledge of the release." This document gives suggested fines for three levels of late release reporting: 1) >15 minutes but ≤1 hour, 2) >1 hour but ≤2 hours, and 3) >2 hours after the person in charge had knowledge that an RQ of a substance was released. [*Enforcement Response Policy for EPCRA Sections 304, 311 and 312 and CERCLA Section 103*, September 30, 1999, available online at http://www2.epa.gov/sites/production/files/documents/epcra304.pdf]

Q *A product, which contains two active ingredients, toluene and benzene, spills on the ground. How is the spill residue regulated under RCRA? Could there be a CERCLA reporting requirement?*

A If a product with more than one active ingredient is spilled, it will not be classified as a P- or U-listed spill residue per §261.33(d) (see Section 3.5.3.7). Therefore, the spill residue will be hazardous only by characteristic. If the soil mixed with the spilled product exhibits a characteristic, then the spill residue is a RCRA hazardous waste. If the soil mixture does not exhibit a characteristic of hazardous waste, RCRA is not applicable. However, CERCLA reporting may be required if an RQ is exceeded, since benzene and toluene are both hazardous substances. [RO 12489]

Q *Approximately 50 gallons of wastewater that contains 15 ppm of benzene spills on soil due to a pipeline leak in the wastewater treatment system. Does this spill have to be reported to the NRC?*

A The RQ for benzene is 10 lb, as is the RQ for D018. [40 *CFR* 302.4] Fifty gallons of D018 wastewater is about 417 lb, well over its 10-lb RQ. However, a facility is allowed to use the RQ for benzene (10 lb) in lieu of the RQ for D018 (also 10 lb) if facility personnel know the concentration of the benzene in the wastewater which was spilled (this is known as the "CWA mixture rule"). [April 4, 1985; 50 *FR* 13463, EPA/540/R-98/022, June 1998, available from http://www.epa.gov/superfund/contacts/sfhotlne/cerep.pdf] A benzene concentration of 15 ppm in 50 gallons of water results in 0.006 lb benzene. Since less than 10 lb of benzene was spilled, this release is not reportable.

Q *A facility spills approximately 100 lb of F001 to the environment. Does the spill have to be reported to the NRC?*

A If a release of F001 occurred and the concentrations of the constituents in the waste are unknown, the RQ for F001 in 40 *CFR* 302.4 (10 pounds) would apply, and the spill would have to be reported. By contrast, if the facility can determine that the F001 hazardous waste contains 50% 1,1,1-trichloroethane and 50% water, the CWA mixture rule can be applied. Since 1,1,1-trichloroethane has a 1,000-lb RQ, the spill is not reportable until 2,000 lb are released. [EPA/540/R-98/022, cited earlier]

Q *A petroleum naphtha solvent (flash point = 130°F, density = 6 lb/gal) is used as a raw material for making various chemicals. However, a 55-gallon drum of this material is accidentally spilled, releasing the contents to the environment. If 200 lb of the spilled material, which is not on the 40 CFR 302.4 list of designated hazardous substances, is captured for recycling or reuse, does this spill have to be reported to the NRC?*

A Yes. The mass of 55 gallons of this solvent is 330 lb. EPA guidance at 50 *FR* 13460 (April 4, 1985) and 51 *FR* 34539 (September 29, 1986) requires the release of a nondesignated ignitable, corrosive, and/or reactive substance to be reported only if the amount not recovered equals or exceeds an RQ. The RQ for an unlisted hazardous waste exhibiting the characteristic of ignitability is 100 pounds. Because 130 lb of the spilled material (which is ignitable) will be disposed, the RQ is exceeded and the spill must be reported to the NRC.

14.3.10.3 EPCRA spill reporting

Releases of a CERCLA hazardous substance or an "extremely hazardous substance" (EHS) in an RQ or more must be reported to state emergency response commissions and local emergency planning committees. (This is an EPCRA requirement codified in 40 *CFR* 355.40.) Onsite releases of EHSs that do not migrate offsite are exempt from this notification requirement as long as no offsite individuals are potentially exposed. The list of EHSs is contained within Appendices A and B to 40 *CFR* Part 355. EPA's List of Lists (available at http://www.epa.gov/emergencies/tools.htm#lol) also contains the RQs for all CERCLA hazardous substances and EHSs. A written follow-up report must subsequently be submitted to these authorities as soon as practicable after the release.

14.3.10.4 SPCC spill reporting

The CWA requires facilities to report certain oil spills. Discharges of oil to navigable waters or adjoining shorelines must be reported immediately to the NRC if they: 1) violate applicable water-quality standards, 2) cause a film or "sheen" on the surface of the water or adjoining shorelines, or 3) cause a sludge or emulsion to be deposited beneath the surface of the water or upon adjoining shorelines. [40 *CFR* 110.3, 110.6] Finally, the EPA regional administrator and appropriate state agency must be notified when there is a discharge of:

- More than 1,000 gallons of oil in a single discharge to navigable waters or adjoining shorelines, or

- More than 42 gallons of oil in each of two discharges to navigable waters or adjoining shorelines occurring within any 12-month period. [40 CFR 112.4]

When determining the applicability of this SPCC reporting requirement, the gallon amount specified (either 1,000 or 42) refers to the amount of oil that actually reaches navigable waters or adjoining shorelines, not the total amount of oil spilled. [*Spill Prevention, Control, and Countermeasure (SPCC) Regulation, 40 CFR Part 112-A Facility Owner/Operator's Guide to Oil Pollution Prevention*, EPA/540/K-09/001, June 2010, available at http://www.epa.gov/oem/docs/oil/spcc/spccbluebroch.pdf]

14.3.10.5 DOT spill reporting

Transportation-related releases must also be reported to the NRC if a hazardous substance (including all hazardous wastes) is released to the environment in excess of its RQ within a 24-hr period. Additional reporting may be required in accordance with §263.30(c).

Additionally, any unintentional release of any quantity of hazardous waste during transportation (including loading, unloading, and temporary storage) triggers a Hazardous Materials Incident Report on DOT Form F 5800.1 within 30 days of discovery of the incident. [49 CFR 171.16 (a) (2)]

14.3.10.6 Release "to the environment"

The reporting requirements discussed above are all predicated on the spill being released into the environment. What constitutes release "to the environment"? CERCLA Section 101(8) defines "environment" to include "surface water, ground water, drinking water supply, land surface or subsurface strata, or ambient air within the United States."

EPA has provided limited guidance on what constitutes a release "to the environment":

- "Examples of such [reportable] releases are spills from tanks or valves onto concrete pads or into lined ditches open to the outside air, releases from pipes into open lagoons or ponds, or any other discharges that are not wholly contained within buildings or structures. Such a release, if it occurs in a reportable quantity (e.g., evaporation of an RQ into the air from a dike or concrete pad), must be reported under CERCLA. On the other hand, hazardous substances may be spilled at a plant or installation but not enter the environment, e.g., when the substance spills onto the concrete floor of an enclosed manufacturing plant. Such a spill would need to be reported only if the substance were in some way to leave the building or structure in a reportable quantity.... For the purposes of CERCLA, 'ambient air' shall refer to the air that is not completely enclosed in a building or structure and that is over and around the grounds of a facility. A release into the air of a building or structure that does not reach the ambient air (either directly or via a ventilation system) is not a reportable event under CERCLA." [April 4, 1985; 50 *FR* 13462]

- "Hazardous substances discharged in buildings or vehicles with active vents or openings, however, may become releases into the environment. For example, a spill of a hazardous substance onto a concrete floor of a totally enclosed manufacturing facility could be released into the environment if part of that substance seeps into the ground through cracks in the concrete or volatilizes into the atmosphere via process vents." [EPA/540/R-98/022, cited previously]

- A release of hazardous waste within a manufacturing building that flowed through the plant sewer system to an onsite, but outdoor, wastewater treatment system containing an open-top tank constituted a release into the environment, requiring notification to the NRC. The agency's reasoning was that the spill was not contained within a building or wholly enclosed structure. [June 1986 CERCLA Hotline question and answer]

A court case added another twist to the definition of release "to the environment." In a February 2000 ruling by an EPA administrative law judge (*In the matter of United States Leather, Inc.*, Docket No. EPCRA-7-99-0048), the judge ruled that a spill of sulfuric acid onto a concrete road (in an amount in excess of its RQ) *did* constitute a release into the environment, even though the acid did not penetrate the concrete, volatilize into the air, or run off the road into any adjoining soil or surface water. The decision in this case was based on the judge's finding that the concrete roadway fits within the meaning of "land surface" and "land," as those terms are used in the CERCLA and EPCRA statutes that trigger reporting.

Note that releases from hazardous waste tank systems (discussed in Section 10.3.3.1.1) that are completely contained within the tanks' secondary containment systems do not require reporting under §§264/265.196(d) since such releases never enter the environment. [EPA/530/SW-87/012, cited in Section 10.1.2.4]

14.3.11 Spill response training

Spill response training requirements for different entities that handle hazardous wastes are noted below—see Section 6.4.1.3.2 for additional details.

14.3.11.1 Generator requirements

Large quantity generators who accumulate waste onsite for 90 days or less must comply with the personnel training requirements in §265.16. [§262.34(a)(4)] These training programs must "ensure that facility personnel are able to respond effectively to emergencies by familiarizing them with emergency procedures, emergency equipment, and emergency systems…." [§265.16(a)(3)]

Small quantity generators must comply with §262.34(d)(5)(iii), which requires the generator to make sure that all employees are "thoroughly familiar with proper waste handling and emergency procedures, relevant to their responsibilities during normal facility operation and emergencies." In addition, each facility must have a designated emergency response coordinator per §262.34(d)(5)(i). [RO 11779]

14.3.11.2 TSD facility requirements

Emergency response training for personnel at TSD facilities in §§264/265.16 is the same as that for large quantity generators noted above.

14.3.11.3 Transporter requirements

DOT's hazardous materials regulations require hazardous materials transporters to train, test, certify, and develop/maintain records of current training for each of their employees. [49 *CFR* 171.8 and 172.704] The training requirements in DOT's regs are specific regarding spill response.

14.4 Gases

Under RCRA, EPA has the authority to regulate any material that meets the definition of "solid waste." This term is defined in the RCRA law as "any garbage, refuse, sludge from a waste treatment plant, water supply treatment plant, or air pollution control facility and other discarded material, including solid, liquid, semisolid, or *contained gaseous material*…." [Emphasis added.] [RCRA Section 1004(27)]

Many questions have arisen over exactly what types of gaseous materials fall under the RCRA regulatory umbrella. EPA guidance on the regulatory status of gaseous materials and residues in different situations is summarized below. The discussion is broken into two broad areas: gaseous emissions from manufacturing operations and emissions from hazardous waste management activities. Emissions associated with specific pieces of equipment are examined. But first, we start with a chronological review of the agency's interpretation of how gases should be regulated.

14.4.1 EPA's stance on gases has changed over time

When EPA was working on the RCRA regulations in the late 1970s, their thinking on contained gases was directed at compressed gases in cylinders. For example, when the regs were issued in May 1980, the agency required regulation of ignitable compressed gases in §261.21(a)(3) if they were discarded. A handful of the P- and U-listed chemicals are gases at standard conditions (see Table 14-2). If a bottle or cylinder of one of these ignitable or listed gases was headed for disposal, it would be a RCRA-regulated hazardous waste. EPA clearly has the statutory authority for regulating containerized gases under RCRA.

In the early- to mid-1980s, people in the regulated community started asking questions as to how waste gases and vapors flowing through pipes and ductwork are regulated. EPA initially took the position that only "true gases" flowing through equipment were excluded from the definition of solid waste. To EPA, true gases referred to gases or vapors that are not capable of being condensed and that remain gases at standard temperature and pressure (e.g., hydrogen or methane). Under this interpretation, condensable gases (i.e., those that would be liquid at standard conditions) that are wastes would be RCRA solid wastes and, if hazardous, would be subject to the applicable RCRA regulations *even if they remained in the gaseous state*. EPA took this stance to prevent generators from evading hazardous waste regulation by heating process wastes to the gaseous state to keep them out of RCRA. [February 10, 1984; 49 *FR* 5314]

Due to adverse comments received on that 1984 *Federal Register* discussion, however, the agency changed its tune on waste condensable gases and vapors in 1989. "EPA now believes our authority to identify or list a waste as hazardous under RCRA is limited to containerized or condensed gases [i.e., Section 1004(27) of RCRA excludes all other gases from the definition of solid wastes and thus cannot be considered hazardous wastes]." The agency conceded that "it cannot use its permitting procedures to mandate the production process design of a manufacturing facility so that it generates a waste as a liquid instead of (for example) installing some internal heating mechanism that generates the same liquid waste in the gaseous state. RCRA jurisdiction does not provide this kind of control over manufacturing processes. Of course, thermal treatment *after* a material

Table 14-2: Listed Hazardous Wastes That Are Gases at Room Temperature (68°F)

Hazardous waste code	Chemical name
P031	Cyanogen
P033	Cyanogen chloride
P063	Hydrogen cyanide
P096	Phosphine
U001	Acetaldehyde
U043	Vinyl chloride
U045	Chloromethane
U122	Formaldehyde
U135	Hydrogen sulfide

Source: McCoy and Associates, Inc.

becomes a hazardous waste [i.e., heating it to a gaseous state] is fully regulated under RCRA." [December 11, 1989; 54 *FR* 50973]

That December 1989 preamble discussion reflects EPA's current policy. RCRA gives the agency the authority to regulate containerized gases (gases in bottles or cylinders) or condensed gases (i.e., liquids that result from changing the physical properties of a gas/vapor stream) when they will be discarded. Conversely, gases/vapors flowing through pipes and ductwork are not solid wastes subject to RCRA management, because a pipe or duct is not a container. Any liquid condensate from such a gas/vapor stream, however, may be subject to RCRA regulation if it meets the definition of a solid and hazardous waste.

EPA recently clarified that its 1989 statements and interpretations are still valid. In RO 14819, the agency noted that burning gaseous material, such as in fume incinerators (as well as other combustion units, including air pollution control devices that may combust gaseous material), does not involve treatment or other management of a solid waste. (See also December 23, 2011; 76 *FR* 80473, February 7, 2013; 78 *FR* 9128, and RO 14830.) In RO 14823, the agency also noted that it is not changing any of its previous statements and interpretations concerning landfill gas.

14.4.2 Gaseous emissions from manufacturing operations

In general, gaseous emissions from manufacturing operations are regulated under the Clean Air Act (CAA)—not RCRA. Similarly, the emission control devices or units themselves (e.g., scrubbers, carbon canisters, filters) will also be subject to CAA control. EPA guidance on selected applications and/or equipment associated with gaseous emissions from manufacturing processes is summarized below.

14.4.2.1 Gases vented from compressed gas cylinders

Due to unique ownership issues surrounding compressed gas containers or cylinders, EPA has provided guidance on who is responsible for managing compressed gas remaining in cylinders. Compressed gas cylinders, which are typically owned by the gas supplier, are usually returned to the supplier when they are empty or when the customer no longer needs them. The purpose of this shipment is for refilling and/or to return the supplier's property, not to discard the remaining contents of the cylinder. Therefore, the customer does not have any input on the final disposition of the residue in the cylinder, which occurs at the supplier's facility at the supplier's discretion.

Consequently, EPA has determined that returning a compressed gas cylinder to the supplier does not constitute generation of waste under RCRA. Neither the returned cylinder nor the residue it contains is a "solid waste," even if the cylinder is not empty (i.e., it is still pressurized). As such, the shipment of cylinders from the customer back to the supplier does not have to be manifested. However, DOT requirements will apply, and the cylinders may have to be transported as DOT hazardous materials. [August 18, 1982; 47 *FR* 36094, RO 14759, 14760, 14762]

If the gas supplier decides to discard the contents of returned cylinders, any liquid or physically solid waste removed from the cylinders is regulated if it is hazardous waste. If the supplier sends the cylinders offsite for treatment, storage, or disposal, they must be manifested if they are not empty and contain hazardous waste.

"However…the handling of gaseous residues removed from the cylinders and neutralization or scrubbing of gases prior to release are not subject to RCRA regulation [because they do not meet the definition of 'solid waste']. Any liquid or physically solid wastes derived from the treatment of hazardous compressed gas is still subject to RCRA regulations, if it is derived from listed waste or if the residue is hazardous under Part 261, Subpart C (characteristics)." [RO 12350]

Although the above quotation from EPA guidance might lead a facility to believe it is acceptable to vent unwanted gas from compressed gas cylinders, we have significant reservations about that practice. First, RO 11835 suggests that, if all of the materials generated by the venting (including the cylinders) will be discarded, then the practice might be considered treatment requiring a RCRA permit. Second, such venting may violate state air regulations or the conditions in a facility's air permit. These concerns argue against this practice, and we strongly recommend not venting gases from compressed gas cylinders to the atmosphere.

14.4.2.2 Activated-carbon/filtered control of manufacturing emissions

Activated-carbon units and filters are often used to control gaseous emissions from manufacturing operations. As noted above, these units are installed to comply with CAA requirements and are subject to the air regulations. The gas/vapors treated in carbon or filter units are not solid wastes (so they can't be hazardous wastes) because they are not containerized. Therefore, the units are not subject to RCRA standards and any organics that condense in the carbon/filter media do not derive from treatment of a hazardous waste (i.e., the derived-from rule is not applicable). [February 21, 1991; 56 *FR* 7200] However, this *Federal Register* preamble goes on to say that the nongas residue from carbon units (the spent carbon) could be hazardous waste if it is listed or if it exhibits a hazardous characteristic.

Determining if spent carbon or filters are characteristic is fairly straightforward. Characteristic spent carbon/filters sent for disposal must be managed as any other hazardous waste. However, characteristic carbon/filters sent for regeneration may not be solid or hazardous waste via Table 1 in §261.2(c). Why? Because spent carbon/filters from an air emission control device in manufacturing applications may meet the definition of a "sludge" under RCRA; characteristic sludges sent for reclamation are not solid or hazardous wastes per Table 1 in §261.2(c). From that perspective, a significant regulatory incentive exists to send all characteristic spent carbon/filters for regeneration. However, EPA guidance from the February 21, 1991 BIF rule preamble sheds significant uncertainty on this exclusion from the definition of solid waste:

"We considered whether other units truly are engaged in reclamation, or whether the regeneration of the carbon is just the concluding aspect of the waste treatment process that commenced with the use of activated carbon to adsorb waste contaminants, which are now destroyed in the 're-generation' process (just as rinsing out a container of hazardous waste is a stage in the storage process and does not constitute recycling of the container).... EPA does not believe that these [carbon regeneration units] are recycling units, but rather that regeneration is a continuation of the waste treatment process, that process consisting of removal of pollutants by adsorption followed by their destruction." [56 FR 7200]

Thus, entities wishing to manage characteristically hazardous spent carbon from pollution control applications outside of RCRA when the spent carbon will be sent for regeneration should check with their state or EPA regional RCRA authority to confirm that the §261.2(c) exclusion from the definition of solid waste is available.

Q *Vapor-recovery units (VRUs—carbon-containing filters) are used to capture offgases and liquid condensate from petroleum product storage tanks. The contaminated VRUs are removed and heated, releasing hydrocarbons that are returned to the petroleum refinery for reuse. The spent carbon is then tested for characteristics and land disposed. Are the VRUs subject to RCRA regulation when removed from service?*

A No. The captured hydrocarbons in the VRUs are not secondary materials; they are products rather than wastes and are not subject to RCRA regulation. The captured hydrocarbons would be considered commercial chemical products being reclaimed, which are excluded from the definition of solid waste per Table 1 in §261.2(c). [RO 14555]

Can spent activated carbon from manufacturing applications be listed hazardous waste? Typically, the answer is "No." When EPA wants to list a specific manufacturing process waste as hazardous, it assigns an F- or K-code. For spent carbon from air emission control applications, though, this isn't a very likely scenario. Could a listed code apply to the spent carbon per the derived-from rule if the carbon is located in an area where F- or K-wastes are being produced? No. The February 21, 1991 preamble language paraphrased above notes that the derived-from rule is not applicable in manufacturing settings. This position is confirmed in the following two examples.

Q *Activated-carbon canisters are used to collect solvent vapors (e.g., Freon 113, 1,1,1-trichloroethane, and methylene chloride) during their use as degreasing agents in a paint spray booth. (These chemicals are F001 listed wastes when they are used as degreasing solvents and are spent.) Are the spent canisters F001 when disposed or regenerated?*

A No. Solvent vapors are not solid wastes because they are uncontained. Furthermore, when the solvent vapor is adsorbed onto activated carbon, it would not be covered by the F001 listing or by the mixture rule. Instead, the spent canisters would be hazardous waste only if they exhibit any of the hazardous waste characteristics. [RO 11166]

Q *Carbon filters are used to capture volatilized air emissions from electroplating activities. When discarded, are these filters F006?*

A The filters would not be F006 because the air emissions are not wastewater. [RO 14108]

14.4.2.2.1 Spent carbon/filters from product handling areas

One situation exists, however, where spent carbon/filters from manufacturing operations could be a listed waste. That situation is where air emissions from product storage and packaging areas (what we call "product handling" areas) are controlled using activated-carbon or filter units. In such applications, the materials being handled are already commercial chemical products; that is, they already meet a P- or U-listing description when disposed because the products are chemicals on the P- or U-list. The products are simply being stored or packaged. In these cases, EPA has noted:

> "When these chemicals meet a listing description under §261.33, any discard of these materials (including these materials captured on filters or mixed with other wastes) are considered hazardous wastes and must be handled accordingly...." [RO 14095]

EPA has decided that waste materials generated from manufacturing operations prior to the point where a commercial chemical product is produced would be manufacturing process wastes that are not covered by the P- or U-listings. In other words, the original chemical product manufacturer would generate process wastes and not manage them as P- or U-wastes. However, once the commercial product is produced, product storage or packaging is simply product handling and does not produce manufacturing process wastes. Therefore, wastes such as spent carbon or filters that are generated at storage and packaging facilities that handle a P- or U-listed chemical would be listed hazardous wastes; they are listed material (including debris) because they "contain" the P- or U-listed chemical. This distinction between manufacturing process wastes and product handling wastes is discussed further in Section 3.5.3.8.1.

Q *A company packages finished phorate product. Phorate, which is a liquid at room temperature, appears on the P-list as P094. Phorate vapors are released to the air during packaging and are captured in carbon filters that are sent offsite for regeneration. What is the regulatory status of the spent carbon filters?*

A Because the phorate in the spent carbon will be destroyed during regeneration, EPA contends that the chemical is being discarded. Therefore, the spent carbon, which contains phorate, must be managed as hazardous waste P094. [RO 11248]

One piece of EPA guidance (from 1993) seems to conflict with the point that we just made. In this situation, a scrubber system is used to control air emissions from a tank that stores a P- or U-listed commercial chemical product. When asked about the status of the spent scrubber solution, EPA responded that it is not a listed hazardous waste because the gas/vapors contained in the solution are derived from a product, not a waste. Therefore, the derived-from rule is not applicable. Consequently, when the scrubber solution is discarded, it is hazardous only if it exhibits a characteristic. [RO 13626] Because the manufacturing process versus product handling policy document was issued in 1997, we believe that this 1993 guidance may not reflect current agency thinking.

14.4.2.3 Fume incinerators are regulated under the CAA—not RCRA

Fume incinerators are also installed as air pollution control devices under the CAA and are used to destroy gaseous emissions from various industrial processes. In line with the agency's position on activated-carbon units, EPA has determined that "in general, RCRA standards do not apply to fume incinerators because the input (an uncontainerized gas) is not a solid waste according to the definition set forth in §261.2." [December 11, 1989; 54 *FR* 50973; see also June 24, 1982; 47 FR 27530, December 23, 2011; 76 FR 80473, RO 14819] Therefore, burning gaseous emissions from industrial processes in fume incinerators is covered only by the CAA, not RCRA. This applies only if the fume incinerator is burning gases, not liquids (i.e., condensate). If a gas stream is condensed to a liquid, the liquid may be a RCRA hazardous waste due to the characteristics of ignitability or toxicity. If hazardous liquids were being fed into the fume incinerator, it would be subject to the RCRA Part 264 or 265, Subpart O incinerator standards. [OSWER Directive 9488.1986(06)] Case Study 14-4 takes an additional look at a fume incinerator.

14.4.3 Gaseous emissions from hazardous waste management activities

Gaseous emissions from hazardous waste management activities are often regulated under RCRA (sometimes in addition to the CAA) even though they are not solid wastes.

"EPA maintains separate and independent authority under RCRA to regulate certain types of uncontained gases whether or not they themselves are solid wastes (e.g., gases emitted from the management of hazardous waste).... As an example, EPA has authority to regulate emissions generated during treatment of hazardous waste, including volatilization and incineration of hazardous waste." [February 7, 2013; 78 FR 9128]

A few examples follow:

- Emissions from hazardous waste incinerators are regulated under Subpart O of Parts 264 and 265;
- Emissions from boilers and industrial furnaces that burn hazardous waste must be controlled under Part 266, Subpart H; and
- Organic emissions from certain process vents, equipment leaks, and units in hazardous waste service are regulated under Subparts AA, BB, and CC of Parts 264 and 265—see Chapter 15.

In addition to these examples, application of RCRA to emissions from specific hazardous waste management activities is discussed below.

Case Study 14-4: Fume Incinerator

A pesticide manufacturer uses a fume incinerator to burn emissions from the pesticide production process. Listed hazardous wastes are also burned in the fume incinerator as supplemental fuel, although the hazardous waste fuel has a high enough Btu value to constitute legitimate recycling. Does the fume incinerator have to be operated under a RCRA permit?

A fume incinerator used *only* to destroy gaseous emissions from an industrial process is not subject to RCRA regulation because the fume input (an uncontained gas) is not a solid waste; it is only subject to the CAA. However, in this situation, the pesticide manufacturer's fume incinerator burns listed hazardous waste in addition to gaseous emissions. Units burning listed wastes for energy recovery are exempt from the Part 264/265, Subpart O incinerator regulations only if the unit is a boiler or industrial furnace. There is no exemption for an incinerator burning hazardous wastes as fuel. Consequently, the unit must comply with Subpart O standards. [RO 12568]

14.4.3.1 Vent streams from hazardous waste management units

As noted previously, vent streams (i.e., gases and vapors) from hazardous waste management units are not classified as solid or hazardous wastes. Noncontainerized gases emitted from hazardous wastes are not themselves hazardous wastes because the RCRA statute implicitly excludes them. [EPA/450/3-89/021, see Section 15.1.1 for availability] Armed with that guiding principle, how are units that receive and treat such vent streams regulated, and what is the status of residues from such units?

14.4.3.1.1 *Boilers and process heaters used to destroy organic vapors*

Boilers and process heaters may be used to destroy organic emissions from hazardous waste treatment, storage, or disposal units. Since the organic vapors emitted from hazardous waste are not solid or hazardous waste, control devices installed specifically to comply with the Subpart CC organic vapor control requirements are not hazardous waste management units and do not require RCRA permits. Therefore, the standards in Part 264 or 265, Subpart O; Part 265, Subparts P and Q; and Part 266, Subpart H do not apply to such control devices. However, if a boiler or process heater that is used as a hazardous waste treatment unit (i.e., a RCRA BIF) is also used as a Subpart CC vapor control device, the unit must be operated in compliance with applicable Part 266, Subpart H requirements. [EPA/453/R-94/076b, see Section 15.3.1 for availability]

14.4.3.1.2 Activated-carbon control of organic vapors

Activated carbon is also used to control emissions from hazardous waste treatment, storage, or disposal activities. What is the status of the spent carbon? Because the gases/vapors entering the unit are not solid waste, the spent carbon is not necessarily a hazardous waste. If the gases/vapors entering the unit were emitted from the treatment, storage, or disposal of a characteristic hazardous waste, the spent carbon would be hazardous only if it exhibits a characteristic and is to be disposed. If the carbon was used for Subpart AA or CC compliance and will be disposed, the RCRA regulations specify the disposal options in §264.1033(n) or 265.1033(m). Characteristic carbon sent for regeneration may not be a solid or hazardous waste per Table 1 in §261.2(c). See the discussion of this topic and applicable EPA guidance in Section 14.4.2.2.

Conversely, spent carbon from a unit that was used to capture emissions from treating, storing, or disposing a listed hazardous waste is considered to be listed waste under the derived-from rule in §261.3(c)(2). [February 9, 1996; 61 FR 4910, EPA/453/R-94/076b] In general, the regeneration, reactivation, or disposal of such listed spent carbon is subject to RCRA because it is thermal treatment of a hazardous waste; the options for managing the spent carbon are described in §264.1033(n) or 265.1033(m). [February 21, 1991; 56 FR 7200] Note, however, that spent carbon that will be reclaimed under one of the exclusions promulgated in the October 30, 2008 definition of solid waste rule is not solid or hazardous waste. See Section 11.2.3 for details.

When an activated-carbon filtration unit is attached to the vent piping on a permitted hazardous waste storage tank to capture air emissions, this filtration unit is considered an appurtenance to the storage tank and, therefore, is not considered separately during permitting. It would be regulated under the storage permit issued for the tank. [RO 12139] On the other hand, if the activated-carbon unit is associated with a generator's 90-day tank, "the operation of a carbon adsorption system associated with a RCRA 90-day accumulation tank that is exempt from permitting requirements pursuant to §262.34(a)(1)(ii) does not require a RCRA permit under EPA's interpretation of the federal regulations." [RO 14781]

14.4.3.2 Gases vented from treating hazardous ground water

During air stripping of hazardous ground water, volatile organic contaminants are often released as gases/vapors. These volatile organic emissions are not solid wastes (i.e., they are not contained gaseous materials) and, therefore, cannot be regulated as hazardous waste. Even so, releases of hazardous constituents to the air from hazardous waste management or solid waste management units at RCRA-permitted or interim status facilities are subject to RCRA correction action. Therefore, EPA advises the use of a carbon unit on top of an air stripper to reduce or eliminate volatile organic emissions. [RO 12783, 13155]

14.4.3.3 Landfill gas condensate

Gas produced from biological degradation in landfills is sometimes collected. Condensation generated from such landfill gas is typically disposed. The regulatory status of landfill gas condensate depends on the origin of the waste contained in the landfill. Three situations may apply:

1. The landfill gas condensate may be derived from a fill that contains exclusively household waste. The RCRA household hazardous waste exclusion [§261.4(b)(1)] applies to household hazardous waste throughout its entire management cycle, from collection through final disposal (including treatment of residues). As a result, landfill gas condensate derived from a fill that contains only household hazardous waste is not a hazardous waste.

2. Condensate derived from processing landfill gas from a fill that contains municipal waste, nonhazardous industrial waste, or unlisted hazardous waste is a hazardous waste only if it exhibits one or more of the hazardous waste characteristics.

3. Condensate derived from a fill containing listed hazardous waste is always a hazardous waste via the derived-from rule (unless it is specifically delisted). [RO 12362]

14.4.3.4 Synthesis gas

Under §261.4(a)(16), EPA excludes gaseous waste-derived fuel (called synthesis gas or syngas) from the definition of solid waste (and therefore from the definition of hazardous waste) if it meets the specifications set forth in §261.38. The exclusion applies to syngas fuel (containing a mixture of hydrogen and carbon monoxide) that is produced from the thermal treatment of hazardous waste.

When this exclusion was proposed, several commenters questioned EPA's authority to require syngas fuels to meet specifications before they are excluded from RCRA since they are uncontained gases. The agency responded that it has the authority to regulate fuels produced from hazardous waste under RCRA Section 3004(q). The fact that syngas is a gas, instead of a liquid or solid, does not matter because it is still produced from hazardous wastes that are being thermally processed [i.e., materials that are recycled by being burned for energy recovery or used to produce a fuel that will be burned for energy recovery are defined as solid wastes under §261.2(c)(2)(i)(A) and (B)].

14.5 Point of generation for waste military munitions

At the onset of RCRA in 1980, many federal facilities operated on the assumption that the hazardous waste regulations did not apply to them. Citing sovereign immunity, they managed their waste independently of the RCRA program. Some facilities, recognizing the danger of waste chemicals to human health and the environment, voluntarily complied with RCRA; however, their compliance consistency was hit-or-miss at best. Recognizing the environmental costs for RCRA noncompliance at federal facilities, Congress enacted the Federal Facilities Compliance Act (FFCA) in 1992. The act established that federal facilities (which are those facilities within the Executive branch of the U.S. government) were not immune from federal, state, or local enforcement that results from noncompliance with the hazardous waste regulations. In short, federal facilities were required to adhere to RCRA regulations; otherwise, they could face civil or criminal penalties. Based on the FFCA and subsequent implementing regulations, military waste, including waste military munitions, falls within the realm of RCRA regulation.

Congress was especially concerned about waste military munitions. Throughout the early 1990s, the Department of Defense (DOD) relied on a patchwork of guidance, interpretations, and rules from states on when their munitions became a solid (and potentially hazardous) waste. Through the FFCA, Congress directed EPA to consult with DOD to finalize a nationally uniform rule identifying when conventional and chemical military munitions become a solid and hazardous waste under RCRA. On February 12, 1997, EPA published the military munitions rule, which became effective August 12, 1997. [62 *FR* 6622] In this section, we examine the point of generation of solid and hazardous wastes for waste military munitions.

14.5.1 What are "military munitions"?

The term "military munitions" (defined in §260.10) applies to all ammunition products and components produced or used by or for the U.S. Armed Services for national defense and security, including munitions under the control of the Coast Guard, National Guard, and the U.S. Department of Energy (DOE). The munitions rule also covers those parties under contract or acting as an agent for the above mentioned entities. [62 *FR* 6624] Military munitions include, but are not limited to: confined gaseous, liquid, and solid propellants, explosives, pyrotechnics, chemical warfare agents, chemical munitions, rockets, missiles, bombs, warheads, mortar rounds, ammunition, grenades, mines, torpedoes, demolition charges, and devices and components thereof. The term does not include wholly inert items and improvised explosive devices. Nuclear weapons managed under DOE's nuclear weapons program are also not recognized as military munitions unless such weapons have been sanitized to the degree specified by the Atomic Energy Act of 1954.

Most military munitions are products that will be or have been used for their intended purpose. Thus, they are not solid or hazardous waste. However, there are a number of situations where these munitions can become a discarded material and a solid waste, subjecting them to the RCRA program. The February 1997 rule amended §261.2(a)(2) to include a fourth type of discarded material: those military munitions identified as solid waste under the provisions of §266.202. That section in Part 266, Subpart M sets out when military munitions are and are not solid wastes. The rest of this section discusses when military munitions become solid wastes.

14.5.2 Determining if military munitions are solid wastes

Part 266, Subpart M characterizes the status of four activities involving military munitions:

1. Unused munitions that are designated for disposal;

2. Unused munitions that are disassembled, repaired, or otherwise recovered;

3. Munitions that are used for training, research and development, or evaluation; and

4. Range clearance activities.

14.5.2.1 Unused munitions that are designated for disposal

Part 266, Subpart M cites four situations where unused military munitions become a solid waste because they are designated for disposal. Those situations are shown in Figure 14-2 and noted below:

1. Munitions that have been or are being disposed, burned, or otherwise treated prior to disposal [§266.202(b)(1)];

2. Munitions removed from storage for the purposes of being disposed, burned, or otherwise treated prior to disposal [§266.202(b)(2)];

3. Munitions that are leaking or deteriorated [§266.202(b)(3)]; and

4. Munitions declared by an authorized military official to be a solid waste [§266.202(b)(4)].

14.5.2.1.1 *Munitions that are disposed, burned, or otherwise treated*

A military munition becomes discarded, and therefore a solid waste, when it has been abandoned by being disposed (e.g., buried or landfilled), burned, detonated, or otherwise treated prior to disposal. However, if burning or detonation is performed as part of a training exercise, a solid waste is not generated because the munition is a product being used for its intended purpose (see Section 14.5.2.3). Also, actions taken as part of an emergency response (e.g., stabilization, detonation) can be exempt from Subtitle C standards under the immediate response exemption (see Section 7.5.4). Unused munitions that were buried in the past are considered solid wastes and potentially hazardous waste and would be regulated as such if unearthed and further managed. [62 *FR* 6626]

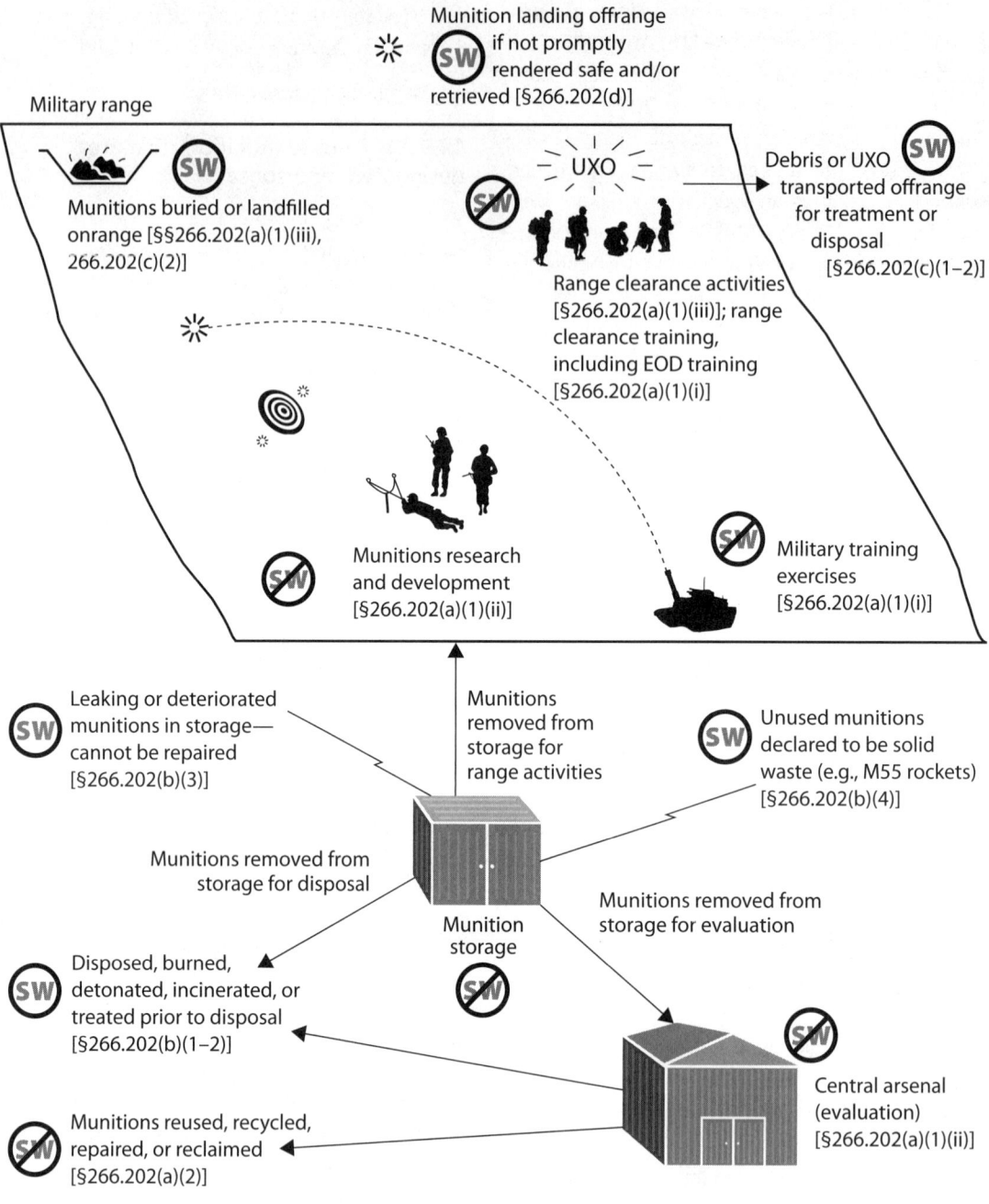

Figure 14-2: Point When Military Munitions Become Solid Wastes

EOD = explosive ordnance disposal; UXO = unexploded ordnance.
Source: McCoy and Associates, Inc.; adapted from §266.202.

14.5.2.1.2 Munitions removed from storage and then disposed, burned, or otherwise treated

EPA's approach for determining when unused munitions become a solid waste is the same approach the agency used for unused commercial chemical products held by manufacturers or their customers. That is, the unused product (in this case a munition) will not be a solid waste until there is an intent to discard the material. Unfortunately DOD's complex system of accounting and management controls and the numerous management options available for munitions (e.g., reconditioning, reuse, and sale) made it difficult to determine at what point there is an intent to discard the munition. For example, aged, damaged, or outdated munitions may be sent to a central arsenal for evaluation. Upon evaluation, the munitions may be put back into service, reconditioned, disposed, sold to other nations, etc. Just because a munition was slated for evaluation at a central arsenal did not mean there was an intent to discard. EPA wanted a simple enforceable approach to determining when a solid waste munition was generated, and the agency recognized that DOD has in place extensive storage standards that are protective of human health and the environment. So EPA took the approach that a munition becomes discarded (thus, a solid waste) when it is removed from military magazines or other storage for the purposes of disposal, burning, incineration, or other treatment prior to disposal. At that point, the munition will be a solid waste. This approach is known as the "magazine door" rule.

14.5.2.1.3 Munitions that are leaking or deteriorated

Military munitions in storage are not solid wastes. However, if a munition is found to be deteriorated or damaged to the point that it cannot be put into serviceable condition and cannot reasonably be recycled or used for other purposes, it is a solid waste. For example, a stabilizer is added to a military propellant during the manufacturing process. Over time the stabilizer in the propellant can deteriorate to the point the unstabilized propellant poses an autoignition hazard. The only options available for this unstable propellant are treatment or disposal. Under these circumstances, the munition in storage would be a solid waste. [62 *FR* 6627]

However, in preamble guidance, EPA noted that if a leaking chemical munition or agent container can be repaired or reclaimed, the munition would still be considered a product. [62 *FR* 6627]

14.5.2.1.4 Munitions determined by an authorized military official to be a solid waste

An authorized military official (usually the designated disposition authority or DDA) may declare a stockpiled military munition to be a solid waste subject to RCRA regulation. For example, in 1984 the Army determined that M55 rockets (which contain a chemical agent) are solid waste. DOD made this decision because 1) the rockets' delivery system no longer existed, 2) the rockets would not be used in military operations, and 3) they would not be sold or reclaimed.

14.5.2.2 Unused munitions that are disassembled, repaired, or otherwise recovered

Military munitions that are repaired, reused, recycled, reclaimed, disassembled, reconfigured, or otherwise subjected to materials recovery activities are not solid waste (see Figure 14-2), unless such activities result in use constituting disposal or burning for energy recovery. [§266.202(a)(2)] See Table 14-3 for examples of how this principal applies to munitions.

14.5.2.3 Munitions used for training, research and development, or evaluation

EPA has consistently held that products used for their intended purpose are outside the scope of RCRA, even if such use results in land application. For example, the proper application of a pesticide will usually result in soil contamination. However, as long as the product was used for its intended purpose, RCRA doesn't regulate the activity or the resulting contamination. [January 4, 1985; 50 *FR* 628,

Table 14-3: Examples of Military Munitions That Are Not Solid Wastes When Disassembled and/or Recycled

Military munition	RCRA-exempt disassembly/recycling activity
Mortar rounds, artillery ammunition, etc.	Recovery of explosive filler using a heated medium and then reusing or reformulating the filler for military or commercial explosives; the remaining inert metal parts may be reused as is or sent for scrap metal recycling
Large rocket motors, DOT Hazard Class 1.1	Removal of propellant using a dry machining process and then reusing the propellant as an ingredient in commercial blasting charge formulation; the remaining inert metal parts may be reused as is or sent for scrap metal recycling
Large rocket motors, DOT Hazard Class 1.3	Removal of propellant using a high-pressure water jet and then reclaiming ammonium perchlorate from the propellant for use as an ingredient in large rocket motor propellant formulation; the remaining inert metal parts may be reused as is or sent for scrap metal recycling
L8A1 smoke grenades	Removal of red phosphorus composition for reloading into new L8A3 grenades
White phosphorus munitions	Removal of white phosphorus composition and conversion to saleable phosphoric acid in a modified furnace

Source: McCoy and Associates, Inc.; adapted from November 8, 1995; 60 *FR* 56472.

RO 11291, 12357] By the same logic, EPA has consistently held that the use of munitions for their intended purpose (e.g., firing a round of ammunition, dropping a bomb from an aircraft, throwing a grenade) does not constitute "discard" and therefore is not a waste management activity. [62 *FR* 6628]

Part 266, Subpart M cites two situations where used military munitions are not a solid waste; rather, they are products used for their intended purpose. Those situations are:

1. Military training exercises [§266.202(a)(1)(i)]; and

2. Weapons testing, research and development, or evaluation [§266.202(a)(1)(ii)].

14.5.2.3.1 Military training exercises

Munitions used for the training of military personnel and explosive ordnance disposal (EOD) personnel are not solid waste under RCRA. Such training, which can include the destruction of unused artillery or mortar round propellant, is a legitimate use of a product, and it lies outside the realm of RCRA.

Q During a bombing run as part of a training exercise, four of the bombs hit the ground but fail to detonate. The pilot requests permission to take a strafing run at the bombs to detonate them. However, the flight operations officer thinks that such an activity would necessitate a RCRA Subpart X permit. Are the unexploded bombs an explosive waste subject to Subpart X permitting prior to detonation?

A No. Since it is normal that not all bombs detonate when dropped, any unexploded bombs are within the normal use pattern of training/target practice. There is no intent to discard the bombs. Subsequent detonation of the bombs in place is not subject to Subpart X permitting. The training mission (or further use of the bombs) can include the strafing run, any dismantling or deactivation of the bombs, or detonating them in place by other means. However, if at any point the bombs are collected and shipped to a place other than the training range (or another training range) to be open burned/open detonated, then that OB/OD site is subject to the permitting requirements of RCRA. [RO 11370]

14.5.2.3.2 Weapons testing, research and development, or evaluation

Munitions used during research, development, testing, and evaluation programs are not regulated under RCRA. This includes using munitions during the testing and evaluation of weapons and weapon systems. EPA views such operations as legitimate use of a material or product for its intended purposes.

A fired military munition would also not be a solid waste if shipped offrange for further evaluation, unless the evaluation is related to treatment or disposal. [62 FR 6628]

14.5.2.4 Range clearance operations

The recovery, collection, and onrange destruction of debris and unexploded ordnance is a common practice at military ranges. During these activities, debris and unexploded ordnance are collected and rendered safe by EOD specialists. EPA considers such onrange management activities to be a necessary part of the safe use of munitions for their intended purpose. As such, the agency excludes range clearance exercises at active or inactive ranges from RCRA regulation. [§266.202(a)(1)(iii)]

However, if used or fired munitions are removed from their landing spot and shipped offrange for treatment or disposal, debris or unexploded ordnance would be a solid waste. [§266.202(c)(1)–(2)] Recognizing the inherent danger of buried munitions, EPA clarified that the onrange disposal (e.g., recovery, collection, and subsequent burial) of unexploded ordnance and debris would generate a solid waste. [§266.202(a)(1)(iii), (c)(2)] In either case, EPA believes the munition is no longer being used for its intended purpose and there is clearly an intent to discard or abandon the munition.

On rare occasions, a fired munition will land offrange. What is the status of munitions that have been used for their intended purpose, but have landed offrange? EPA addressed these rare situations in §266.202(d). Munitions that land offrange and are not promptly collected or rendered safe are solid waste and, further, are potentially subject to RCRA's corrective action and imminent and substantial endangerment provisions. It is EPA's belief that failure to render safe or retrieve a munition that lands offrange is an intent to discard in the same manner that failure to clean up a spill is an intent to discard. In those situations where retrieval or remediation is infeasible (e.g., impact spot could not be located), the operator of the range must maintain a record of the event for as long as any threat remains. The record must include the type of munition and its location (to the extent the location is known).

Q: *What is the regulatory status of fired munitions and munitions fragments that are collected and sent offrange for metals recycling?*

A: Fired munitions and munitions fragments being sent offrange for reclamation are solid waste per §266.202(c)(1). Though not explicitly stated in Part 266, Subpart M, those items that meet the definition of scrap metal can be managed as RCRA-exempt scrap metal under §261.6(a)(3)(ii) if they meet the acceptance criteria of the scrap metal recycler. For example, expended small arms cartridge casings sometimes fail the TCLP for lead but would not be D008 if scrap metal recycled.

14.5.3 Military munitions that are solid waste may also be hazardous waste

Once a military munition is determined to be a solid waste, the generator must make a hazardous waste determination. A solid waste will be a hazardous waste if it is listed in Part 261, Subpart D or if it exhibits a characteristic of ignitability, corrosivity, reactivity, or toxicity from Part 261, Subpart C. These munitions will normally not be listed, but they could be reactive (D003) and/or toxic due to heavy metals and/or 2,4-dinitrotoluene. Thus, once military munitions become solid waste, they could be subject to all RCRA Subtitle C hazardous waste management regulations.

An example of the above is spent M5-HC smoke pots used during range training exercises. Testing shows that these spent units fail the TCLP for cadmium and lead, thereby requiring the D006 and D008 codes, respectively, when disposed. ["Hazardous Waste Study No. 37-7016-97/98," U.S. Army Environmental Center]

Air Emission Standards

How air emission standards apply to equipment/units in hazardous waste service

Leaking equipment, such as tanks, pumps, valves, open-ended lines, connectors, and pressure-relief devices, are collectively a significant source of organic emissions from petroleum refineries, chemical manufacturing facilities, and other industrial plants. As these components wear from normal use, leaks may develop releasing organic vapors and/or volatile liquids into the environment.

Air emissions from equipment in hazardous waste service are also a concern and subject to regulation. In general, the RCRA air emission standards are unit specific; that is, applicable standards are determined by the type of equipment or unit being used to manage hazardous wastes. Table 15-1 correlates the regulatory air emission standards with the various types of hazardous waste management equipment/units.

The work practices that reduce organic emissions from leaking equipment are collectively known as leak detection and repair (LDAR) programs. LDAR programs typically have their regulatory basis through both the Clean Air Act (CAA) and RCRA, depending on if the organic material is a hazardous waste. It is important to note that many of the work practices and recordkeeping provisions associated with the RCRA standards are similar to, if not the same as, those required by the CAA. In crafting the RCRA air emission standards, EPA decided that the best way to eliminate any regulatory overlap between the RCRA standards and the various CAA emission standards was to exempt from the RCRA Subpart AA, BB, and CC rules any process vents, equipment components, and waste management units that are using air emission controls in accordance with applicable CAA requirements.

Given that a facility may elect to comply with RCRA air emission standards through compliance with applicable CAA standards, failure to comply with CAA provisions could result in RCRA noncompliance. Moreover, the magnitude of a RCRA violation in such a situation could be more substantive—consider the situation where a pump barrier seal fails, releasing a volatile liquid hazardous waste onto the ground (but also some volatiles into the air). Fixing the broken barrier seal per CAA LDAR regulations and, therefore RCRA air emission regulations, does not necessarily mitigate the release/disposal of hazardous waste to the ground.

Chapter 16 discusses enforcement issues pertaining to an assortment of noncompliance scenarios, including noncompliance with the RCRA air emission standards. However, it seems appropriate to mention here that EPA identified cutting hazardous air pollutants, including excess emissions

Table 15-1: Air Emission Standards Applicable to Hazardous Waste Equipment/Units

Type of equipment or unit	Applicable regulations	Regulated hazardous waste	Text reference
Chemical/physical separation equipment, including: 　Distillation units 　Fractionation units 　Thin-film evaporation units 　Solvent extraction units 　Air stripping units 　Steam stripping units	Parts 264/265, Subpart AA	Containing ≥10 ppmw organics	Section 15.1
Equipment associated with pipe runs, including: 　Valves 　Pumps 　Compressors 　Pressure-relief devices 　Sampling systems 　Open-ended valves or lines 　Flanges and other connectors	Parts 264/265, Subpart BB	Containing ≥10% (by wt) organics	Section 15.2
Tanks	Parts 264/265, Subpart CC	Containing ≥500 ppmw VO	Section 15.3.4
Containers	Parts 264/265, Subpart CC	Containing ≥500 ppmw VO	Section 15.3.5
Surface impoundments	Parts 264/265, Subpart CC	Containing ≥500 ppmw VO	Section 15.3.6
Boilers and industrial furnaces	Part 266, Subpart H	All	Not covered
Incinerators	Parts 264/265, Subpart O	All	Not covered

VO = volatile organics.
Source: McCoy and Associates, Inc.

---

caused by facilities' failure to comply with LDAR regulations, as a national enforcement initiative for fiscal years 2014 through 2016. [*National Enforcement Initiatives for Fiscal Years 2014–2016*, available at http://www2.epa.gov/enforcement/national-enforcement-initiatives] To the extent that a facility elects to comply with the RCRA air emission standards by relying on compliance with the CAA, a CAA LDAR inspection is, essentially, a RCRA air emission standards inspection.

Before we jump into the details, be aware of an EPA Region 4 RCRA information resource site that contains training and guidance documents on the Subparts AA, BB, and CC air emission standards. [See http://www.trainex.org/web_courses/subpart_x/EPA CD Content/TopicSearch.htm.] As an example of what the site has to offer, one document, *The ABC's of Subparts AA, BB, and CC: A Practical Guide to Compliance with RCRA Air Emission Standards*, compiled by the Indiana Department of Environmental Management, discusses misconceptions and common violations of the RCRA air emission regulations. Another document, *General Recordkeeping and Reporting Guidance for Waste Management Units Requiring Air Emission Controls Under RCRA Air Standard Subpart CC*, is a 36-page guidance manual written by EPA Region 4. It summarizes the recordkeeping and reporting requirements for tanks, containers, closed-vent systems, control devices, etc. under Subpart CC.

15.1 Air emission standards for process vents—Subpart AA

When hazardous wastes containing organics are treated or recycled in chemical/physical separation equipment (such as solvent stills), high levels of air emissions may be produced. Consider the simple solvent still shown in Figure 15-1. Spent solvent is

Figure 15-1: Potential Air Emission Sources From a Solvent Still

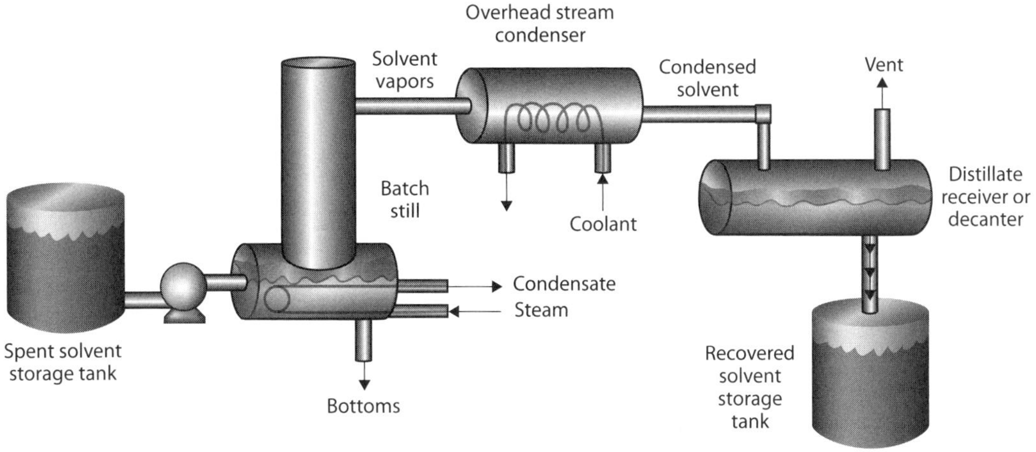

Source: Adapted from EPA/450/3-89/021.

heated in the batch still so that volatile organics are vaporized. The solvent vapors leave the still in what is typically called the "overhead stream" and pass into a condenser. Condensed liquids (recovered solvent) along with noncondensable gases flow into a distillate receiver, where the gases and recovered solvent separate into two phases. The liquids flow to a storage tank (product storage) and the noncondensed gases are simply vented to atmosphere through an open vent pipe. The purpose of Parts 264/265, Subpart AA is to limit the quantity of organic emissions from process vents such as the one shown on the distillate receiver in Figure 15-1.

15.1.1 Applicability of the Subpart AA regulations

Figure 15-2 identifies the conditions under which the Subpart AA air emission control standards for process vents will apply to a hazardous waste management operation. Important details referenced in this diagram are discussed below.

Perhaps the most useful reference available from EPA on Subpart AA applicability is EPA/450/3-89/021, *Technical Guidance Document for RCRA Air Emission Standards for Process Vents and Equipment Leaks*, July 1990. This document can be obtained online at http://nepis.epa.gov/EPA/html/Pubs/pubtitleOAR.html by downloading the report numbered 450389021.

15.1.1.1 Units handling hazardous waste

Note that the Subpart AA standards apply only to units handling hazardous waste. For example, if a steam stripper is being used to clean up contaminated ground water, but the ground water does not contain a listed waste and does not exhibit a characteristic of hazardous waste, Subpart AA control requirements would not apply. Subpart AA also does not apply to process operations associated with manufacturing (e.g., fractionation columns used to separate components during manufacture of a chemical product).

15.1.1.2 Only specific separation processes are subject to controls

Only certain types of hazardous waste process equipment having the potential to emit large quantities of volatile organics are regulated under Subpart AA. These types of processes are defined in §§264/265.1031 as follows:

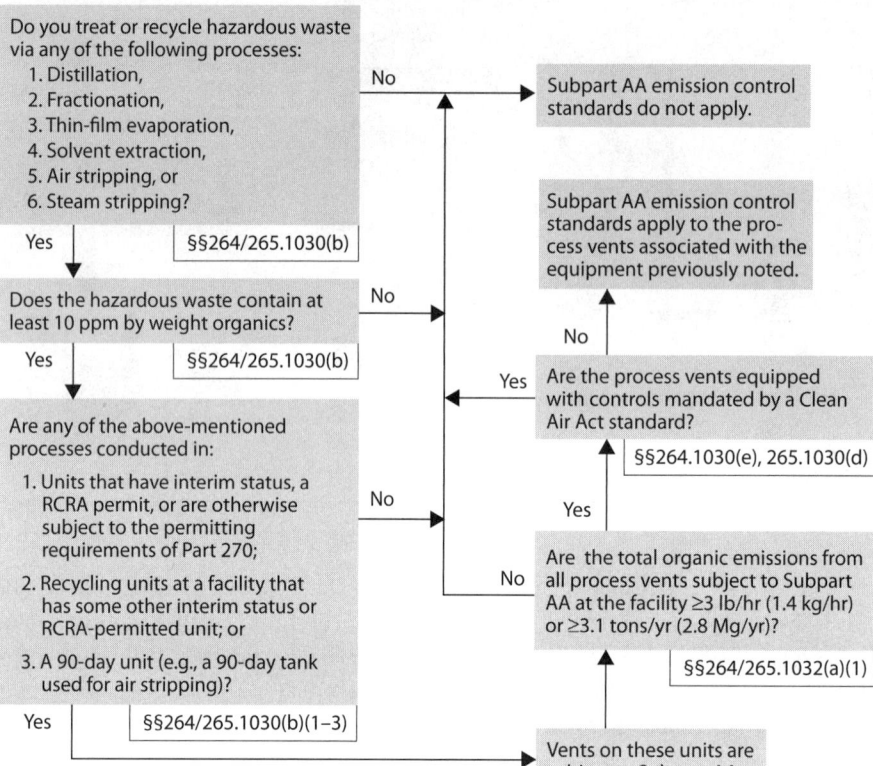

Figure 15-2: Determining If Subpart AA Process Vent Standards Apply

Source: McCoy and Associates, Inc.

1. *Distillation*—"an operation, either batch or continuous, separating one or more feed stream(s) into two or more exit streams, each exit stream having component concentrations different from those in the feed stream(s). The separation is achieved by the redistribution of the components between the liquid and vapor phase as they approach equilibrium within the distillation unit."

2. *Fractionation*—"a distillation operation or method used to separate a mixture of several volatile components of different boiling points in successive stages, each removing from the mixture some proportion of one of the components."

3. *Thin-film evaporation*—"a distillation operation that employs a heating surface consisting of a large diameter tube that may be either straight or tapered, horizontal or vertical. Liquid is spread on the tube wall by a rotating assembly of blades that maintain a close clearance from the wall or actually ride on the film of liquid on the wall."

4. *Solvent extraction*—"an operation or method of separation in which a solid or solution is contacted with a liquid solvent (the two being mutually insoluble) to preferentially dissolve and transfer one or more components into the solvent."

5. *Air stripping*—"a desorption operation employed to transfer one or more volatile components from a liquid mixture into a gas (air) either with

or without the application of heat to the liquid. Packed towers, spray towers, and bubble-cap, sieve, or valve-type plate towers are among the process configurations used for contacting the air and a liquid."

6. *Steam stripping*—"a distillation operation in which vaporization of the volatile constituents of a liquid mixture takes place by the introduction of steam directly into the charge."

These definitions follow conventional processing terminology and do not have any hidden meaning or nuances of particular importance.

EPA clarified in *Federal Register* preamble language that portable equipment would be subject to Subpart AA control standards as well as stationary units. [December 8, 1997; 62 *FR* 64639]

15.1.1.3 Organic concentration of 10 ppmw triggers requirements

The Subpart AA regulations apply only where organic concentrations of at least 10 ppmw are involved. [§§264/265.1030(b)] The details of determining whether a waste meets the 10-ppmw limit are rather complex and are codified with the testing procedures in §§264/265.1034(d). When direct measurement (as opposed to process knowledge) is used, all waste streams entering one of the processes described above must be identified. At least four grab samples of each of these waste streams must be taken when the organic concentrations are expected to be at a maximum.

For wastes generated onsite, the grab samples must be taken before the waste is exposed to the air. Each sample is analyzed using SW–846 Method 9060—Total Organic Carbon. The four (or more) results for each waste stream are averaged and then a time-weighted, annual-average total organic concentration for all waste streams processed in the unit is calculated. If this time-weighted annual average equals or exceeds 10 ppmw, the unit is potentially subject to Subpart AA controls.

If the time-weighted annual average is <10 ppmw (i.e., the unit is not subject to Subpart AA), wastes that are generated continuously must be reanalyzed at least annually or whenever the wastes change or the processes that generate or treat the wastes change. [§§264/265.1034(e)]

Up-to-date information on whether or not a process vent is subject to regulation must be recorded in the facility operating record. [§§264/265.1035(f)]

15.1.1.3.1 *Knowledge-based determinations*

Rather than using direct measurement to determine the organics concentration, an owner/operator can use his/her knowledge of the waste's composition to determine if the organic content is less than 10 ppmw. If a knowledge-based determination is made, §§264/265.1034(d)(2) specify that documentation of the waste determination is required.

The first example of using knowledge in §§264/265.1034(d)(2) discusses understanding of the process and allows an owner/operator to document that "no organic compounds are used" in the production process. In some instances, knowing that your production process does not use organics may be sufficient. However, using knowledge of the process should also entail a detailed look at all the chemical constituents formed in the production process.

Other examples in §§264/265.1034(d)(2) utilize knowledge of the process which stems from analytical testing. In the second example, an owner/operator may use information from waste generated by a process at another facility provided that the processes are identical and the other facility has previously demonstrated, by direct measurement, that the waste stream has a total organic content less than 10 ppmw. In the third example, prior speciated analytical results may be used for a waste stream for which an owner/operator can document that no process changes have occurred since that analysis that could affect the total organic concentration. In both the second and third examples, if analytical results were obtained for another regulatory purpose, or the analytical results may not have considered organics formed in the production process, direct measurement may be the most conservative approach.

15.1.1.4 Only certain categories of units are regulated

Not all units as described above are subject to Subpart AA. Only units that fit into three categories are regulated. First, units that are subject to the permitting standards of Part 270 (i.e., they either are in interim status or have a RCRA permit) are subject to Subpart AA. Such units are not very common because most of the processes described above are recycling processes, and recycling units are not subject to permitting per §261.6(c)(1). The following units that are exempt from RCRA permitting are also exempt from Subpart AA:

- Manufacturing process units (such as distillation columns that generate hazardous waste still bottoms; these units are not subject to Subpart AA because they are not managing hazardous wastes)—see exemption in §261.4(c).
- Totally enclosed treatment facilities—see exemption in §270.1(c)(2)(iv).
- Elementary neutralization units—see exemption in §270.1(c)(2)(v).
- Wastewater treatment units—see exemption in §270.1(c)(2)(v).
- Closed-loop recycling units—see exemption in §261.4(a)(8).

Second, if a recycling unit is located at a facility that has a RCRA permit (e.g., for hazardous waste storage tanks), the recycling unit is subject to Subpart AA. [§261.6(d)] On the other hand, if the recycling unit is located at a generator's facility that does not have or need a RCRA permit for any treatment, storage, or disposal activities, the recycling unit is not subject to Subpart AA.

Third, 90-day units (that are not recycling units) are subject to Subpart AA. This applicability provision is unlikely to be triggered because 90-day units are typically not used for distillation, fractionation, solvent extraction, thin-film evaporation, etc. One situation where Subpart AA applies is when a 90-day tank is used to air strip wastewater for organics removal and the treated water is disposed by underground injection.

The following examples should help clarify the applicability of Subpart AA. [EPA/450/3-89/021, RO 14461]

Q: *A large quantity generator stores spent solvent in a 90-day tank, processes it in a batch still, and then reuses the recovered solvent in a manufacturing process. No other hazardous waste management activities occur at the site. Is the batch still subject to Subpart AA?*

A: No. Recycling units at facilities that don't have interim status or a RCRA permit for some other hazardous waste management activity are not subject to Subpart AA.

Q: *In the preceding example, the spent solvent is stored in a tank for more than 90 days prior to distillation. Is the still subject to Subpart AA?*

A: Maybe. If the solvent recovery system meets the closed-loop recycling with reclamation criteria of §261.4(a)(8), the still is not processing solid or hazardous waste and Subpart AA does not apply. On the other hand, if §261.4(a)(8) does not apply, the still is subject to Subpart AA because storing spent solvents for more than 90 days requires a RCRA storage permit. Recycling units at facilities subject to RCRA permitting are subject to Subpart AA.

Q: *A generator processes spent solvent (D001) in a still without prior storage. Still bottoms are accumulated in containers for less than 90 days before being shipped offsite. Somewhere else on the site, hazardous wastewater is stored in a surface impoundment. Does Subpart AA apply to the still?*

A: Yes. Because a RCRA permit is required for the surface impoundment, Subpart AA applies to the still.

Q: *At an industrial facility, hazardous ground water is pumped to a feed tank and then to an air stripping column for organics removal. The ground water contains >10 ppmw organics. Treated ground water is piped directly to the facility's NPDES permitted outfall point. Does Subpart AA apply to the air stripper?*

A Probably not. The air stripper probably meets the definition of a tank and qualifies for the wastewater treatment unit exemption, as would the feed tank. [§§264.1(g)(6), 265.1(c)(10), 270.1(c)(2)(v)] Subpart AA does not apply to wastewater treatment units as defined in §260.10.

Q *In the previous example, treated ground water is reinjected rather than being discharged under an NPDES permit. Does Subpart AA apply to the air stripper?*

A Yes. Under these circumstances, the air stripper is not a wastewater treatment unit and may be subject to permitting unless it is operated as a 90-day tank. However, 90-day tanks used for air stripping (and that are not used for recycling) are subject to Subpart AA.

Q *An offsite solvent recycler stores spent solvents in tanks prior to processing them in a fractionation column. Does Subpart AA apply to the vents associated with the column?*

A Yes. The offsite facility needs a RCRA permit to store the solvent in tanks (i.e., the tanks are not 90-day units because the recycler is not the generator of the spent solvent). Fractionation units recycling hazardous wastes at facilities having a RCRA permit are subject to Subpart AA.

Q *A generator excavates hazardous soil containing >10 ppmw organics and plans to put the soil in a roll-off box fitted with air sparging equipment. Is this activity subject to Subpart AA?*

A Probably. This sounds like air stripping, even though it doesn't meet the definition in §264.1031. The roll-off box is probably a 90-day container (otherwise, the generator would need a treatment permit). Air stripping units that are 90-day containers are subject to Subpart AA.

15.1.1.5 Only certain process vents are regulated

Subpart AA controls emissions from "process vents," which are defined as "any open-ended pipe or stack that is vented to the atmosphere either directly, through a vacuum-producing system, or through a tank (e.g., distillate receiver, condenser, bottoms receiver, surge control tank, separator, or hot well) associated with hazardous waste distillation, fractionation, thin-film evaporation, solvent extraction, or air or steam stripping operations." [§§264/265.1031]

Particular attention needs to be directed to the term "associated with" in this definition. The preamble to the original Subpart AA rule [June 21, 1990; 55 *FR* 25461] stated:

> "Thus, the final vent standards apply to: 1) vents on distillation, fractionation, thin-film evaporation, solvent extraction, and air or steam stripping processes and vents on condensers serving these processes; and 2) vents on tanks (e.g., distillate receivers, bottoms receivers, surge control tanks, separator tanks, and hot wells associated with distillation, fractionation, thin-film evaporation, solvent extraction, and air or steam stripping processes) if emissions from these processes are vented through the tank. For example, *uncondensed overhead* emitted from a distillate receiver (which fits the definition of a tank) serving a hazardous waste distillation process unit is subject to these [Subpart AA] air controls. On the other hand, emissions from vents on tanks or containers that do not derive from a process unit specified above are not covered by these rules. For example, if the condensed (recovered) solvent is pumped to an intermediate holding tank following the distillate receiver mentioned in the above example, and the intermediate storage tank has a pressure-relief vent (e.g., a conservation vent) serving the tank, this vent will not be subject to the process vent standards." [Emphasis in original.]

Figure 15-3 illustrates these concepts. Four vents are located on the storage tanks and distillation system. Vents #1 and 4 are not associated with the distillation system and are not subject to Subpart AA. Uncondensed overhead vapors from the distillation process are not being released from these

Figure 15-3: Identifying Vents Subject to Subpart AA

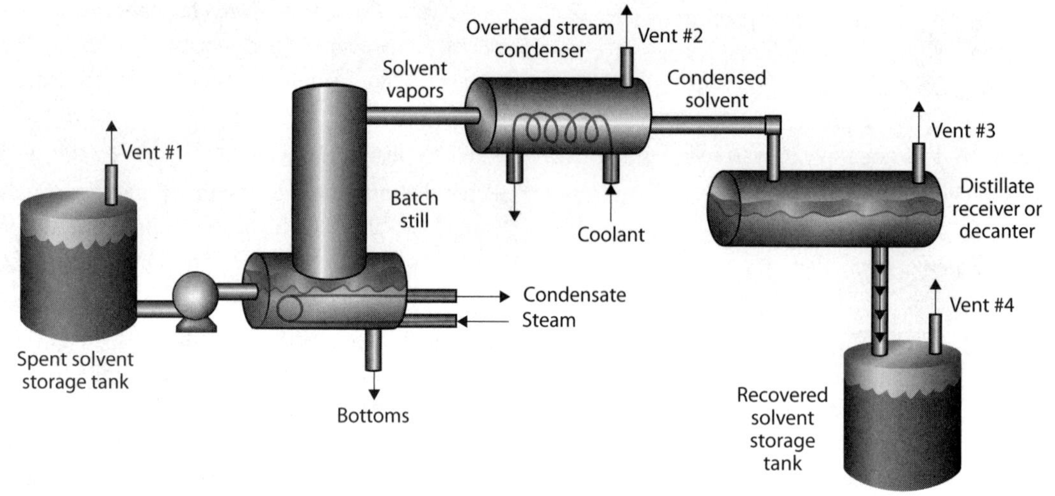

Source: Adapted from EPA/450/3-89/021.

vents. These vents are probably conservation vents designed to protect the mechanical integrity of the tanks. On the other hand, Vents #2 and 3 are releasing uncondensed overhead vapors from the distillation process and are subject to Subpart AA. [EPA/450/3-89/021]

15.1.1.6 Process vents in compliance with CAA requirements are exempt

To eliminate regulatory overlap between the CAA (which may also regulate emissions from vents) and RCRA, EPA codified an exemption at §§264.1030(e)/265.1030(d). In the *Federal Register* preamble establishing this exemption, the agency stated that if Subpart AA-regulated process vents are equipped with and operating air emission controls in accordance with CAA requirements in 40 *CFR* Part 60—Standards of Performance for New Stationary Sources, Part 61—National Emission Standards for Hazardous Air Pollutants, or Part 63—National Emission Standards for Hazardous Air Pollutants for Source Categories, this constitutes compliance with Subpart AA. Two caveats exist, however, for this CAA exemption:

1. "[T]oday's exemption is only available at a facility where each and every process vent that would otherwise be subject to Subpart AA is equipped with, and operating, air emission controls in compliance with an applicable CAA standard under Parts 60, 61, or 63…. [T]o comply with the requirements at paragraphs [§264.1030(e) or §265.1030(d)], the emissions from each Subpart AA process vent must be routed through an air emission control device; a vent that is in compliance with a CAA standard under an exemption from control device requirements is not in compliance with those provisions of Subpart AA." [December 8, 1997; 62 *FR* 64638–9]

2. "A process vent that is in compliance with a CAA standard under an exemption from control requirements (i.e., is not equipped with and operating a control device) does not meet the criteria established in the provisions paragraph §264.1030(e) or §265.1030(d) of Subpart AA. Therefore, a unit that does not use the required air emission controls but is in compliance with a NESHAP through an 'emission

averaging' or 'bubbling' provision does not qualify for the exemption. Similarly, if the Clean Air Act standard for the particular unit is no control (for example, because the Maximum Achievable Control Technology (MACT) floor for the source category is no control and the agency decided not to apply controls more stringent than the floor), the exemption from the RCRA standards under §264.1030(e) or §265.1030(d) of Subpart AA would not apply since the unit would not actually be controlled (i.e., equipped and operating air emission controls) under provisions of the MACT standard.

"To take the above example a step further, at a facility where all but one of the Subpart AA process vents are equipped with air emission controls for compliance under CAA rules and the one uncontrolled Subpart AA process vent is also in compliance with a CAA regulation but is not controlled for air emissions, the facility's Subpart AA process vents do not meet the applicability exemption criteria as stated in Subpart AA and thus are not exempt from the rule under §264.1030(e) or §265.1030(d). Despite this restriction, EPA considers this alternative to provide the facility owner or operator with a broader degree of compliance flexibility, and less extensive monitoring, recordkeeping, and reporting requirements under RCRA." [EPA Region 4, *CAA and RCRA Overlap Provisions in Subparts AA, BB, and CC of 40 CFR Parts 264 and 265*, October 2000, available at http://www.trainex.org/web_courses/subpart_x/TopicSearch pdf files/pdf docs ABC/Final Overlap Provisions.pdf]

15.1.2 Complying with Subpart AA

The preceding discussion serves to identify process vents that are subject to Subpart AA. Once affected vents are identified, the next step is to determine how to comply with Subpart AA control standards.

15.1.2.1 Subpart AA emission limits

Once an owner/operator has identified process vents subject to Subpart AA, he/she must determine the organic emission rate for each vent and then aggregate the emissions for the entire facility. Emission rates may be determined via engineering calculations or direct source tests. The regulations [§§264/265.1035(b)(2)(ii)] specify:

"[D]eterminations of vent emissions...must be made using operating parameter values (e.g., temperatures, flow rates, or vent stream organic compounds and concentrations) that represent the conditions that result in maximum organic emissions, such as when the waste management unit is operating at the highest load or capacity level reasonably expected to occur. If the owner or operator takes any action (e.g., managing a waste of different composition or increasing operating hours of affected waste management units) that would result in an increase in total organic emissions from affected process vents at the facility, then a new determination is required."

The aggregated facility emission rate is compared against both a short-term limit [3 lb/hr (1.4 kg/hr)] and a long-term limit [3.1 tons/year (2.8 Mg/yr)]. If the organic emissions from all affected vents at the facility never exceed either limit, no controls are required.

If the aggregated facility emission rate exceeds either limit, two options are available [§§264/265.1032(a)]:

1. Reduce facility-wide emission rates below both limits, or
2. Install control devices capable of reducing facility-wide, aggregated emissions by 95 weight percent [enclosed combustion devices such as fume incinerators, boilers, or process heaters may meet alternative performance standards per §§264/265.1033(c)].

Note that emission reductions have to be met on a facility-wide, aggregated basis. [EPA/530/K-02/003I, October 2001, available from http://www.epa.gov/wastes/inforesources/pubs/training/air.pdf] In other words, if a facility has 10 affected vents, nine of which have very low emissions and one that has high emissions, installing a control device on only the high-emission source might succeed in meeting the emission limits. In this case, no controls would be necessary for the other nine vents.

15.1.2.2 Subpart AA control devices

If emissions must be reduced, the reduction may be accomplished via changes in operating practices or via the use of add-on control devices. Add-on control devices would be connected to the process vent via a "closed-vent system," which is defined as "a system that is not open to the atmosphere and that is composed of piping, connections, and, if necessary, flow-inducing devices that transport a gas or vapor from a piece or pieces of equipment to a control device." [§§264/265.1031]

The regulations do not require the use of any specific type of control equipment; however, performance requirements are specified for vapor-recovery systems [condensers and carbon adsorbers—see §§264/265.1033(b)], enclosed combustion devices [§§264/265.1033(c)], and flares [§§264/265.1033(d)]. Other types of control devices may be used as long as they meet the emission limits cited previously.

Q *In order to control emissions from a distillation column, a facility wants to utilize a vapor-recovery system that includes the primary condenser attached to the column. How does the use of this condenser figure into the 95% recovery efficiency requirement?*

A According to EPA, "[v]apor-recovery systems whose primary function is the recovery of organics for commercial or industrial use or reuse (e.g., a primary condenser on a waste solvent distillation unit) are not considered a control device and should not be included in the 95-percent emission reduction determination." [June 21, 1990; 55 *FR* 25463]

EPA/450/3-89/021 gives detailed information on various control devices for complying with the Subpart AA regulations.

15.1.2.3 Inspection and monitoring

If control devices are used to comply with Subpart AA requirements, the devices must be routinely inspected and monitored in accordance with §§264/265.1033(f) and (i). In general, vent stream flow rates must be recorded at least every hour, process-specific parameters (e.g., combustion chamber temperatures in fume incinerators) must be monitored continuously, and the monitoring devices must be inspected at least once each operating day. If a problem is found, corrective measures must be implemented immediately.

Closed-vent systems conveying vapors to the control device are also subject to inspection per §§264/265.1033(k–m). These systems must either operate with no detectable emissions (<500 ppmv above background) or operate under negative pressure. After an initial inspection, annual reinspection is required; reinspection is also required whenever equipment components are changed or repaired. If a leak is detected, emissions must be controlled as soon as practicable, but within 15 calendar days at most.

15.1.2.4 Recordkeeping and reporting

According to EPA:

"The RCRA air emission standards are meant to be self-implementing. Consequently, EPA does not play an active role in ensuring that facilities are in compliance with the regulations on a day-to-day basis. Instead, the agency verifies that owners and operators are complying with the regulations through facility inspections and audits. Subpart AA [§§264/265.1035] requires owners and operators to keep detailed records in the facility's operating log to demonstrate compliance with the standards. Design documentation and monitoring, operating, and inspection information for each unit subject to the Subpart AA standards must be kept up-to-date and in the facility's operating log for at least three years.

"Subpart AA also requires that, every six months, owners and operators of permitted facilities report to the Regional Administrator any instances during that time period when a control device exceeded or operated outside of its design specifications (as indicated by monitoring devices) for a period of longer than 24 hours without being corrected (§264.1036). The report must indicate the duration of each exceedance and any corrective measures taken to remedy the situation. If during the six-month period no exceedances occur at the facility, the owner and operator need not submit a report to the Regional Administrator." [EPA/530/K-02/003I]

Note that the only difference between the Subpart AA regulations in Part 264 and Part 265 is that Part 264 includes this reporting requirement for permitted facilities. Interim status facilities, regulated under Part 265, are not required to submit the report.

15.2 Air emission controls for equipment leaks—Subpart BB

When organic-containing hazardous waste flows through pipe runs, certain types of equipment in the pipe runs can be a significant source of organic emissions. Figure 15-4 shows a hypothetical pipe run containing the types of equipment that Parts 264/265, Subpart BB regulations are designed to control. In this figure, hazardous waste from a process vessel is pumped first to a 90-day tank and then to a loading rack where tank trucks pick up the waste. Organic emissions (which EPA terms "fugitive emissions") may occur from leaking valve stems, pump shafts, pressure-relief devices, sampling systems, open-ended valves or lines, and flanges. If the pipe run meets certain applicability requirements, a leak detection and repair program will have to be established and these emissions will have to be controlled under Parts 264/265, Subpart BB.

15.2.1 Applicability of the Subpart BB regulations

Figure 15-5 identifies the conditions under which the Subpart BB fugitive emission control standards will apply to a hazardous waste management operation. Important details referenced in this diagram are discussed below.

As with Subpart AA, one of the most useful references available from EPA on Subpart BB applicability is EPA/450/3-89/021, *Technical Guidance Document for RCRA Air Emission Standards for Process Vents and Equipment Leaks*, July 1990. This document is available online at http://nepis.epa.gov/EPA/html/Pubs/pubtitleOAR.html by downloading the report numbered 450389021.

Additional guidance and experience on making Subpart BB equipment leak inspections are contained in

Figure 15-4: Potential Air Emission Sources in Pipe Runs

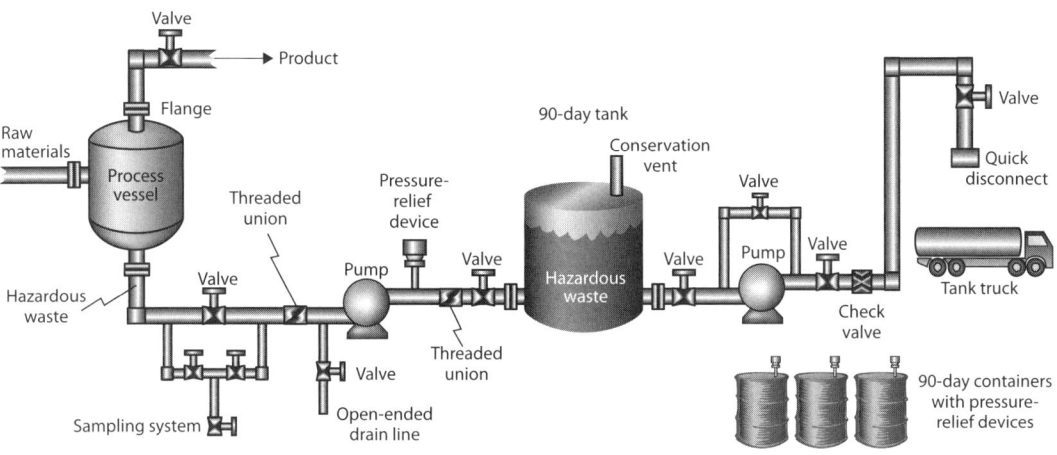

Source: McCoy and Associates, Inc.

Figure 15-5: Determining If Subpart BB Fugitive Emission Standards Apply

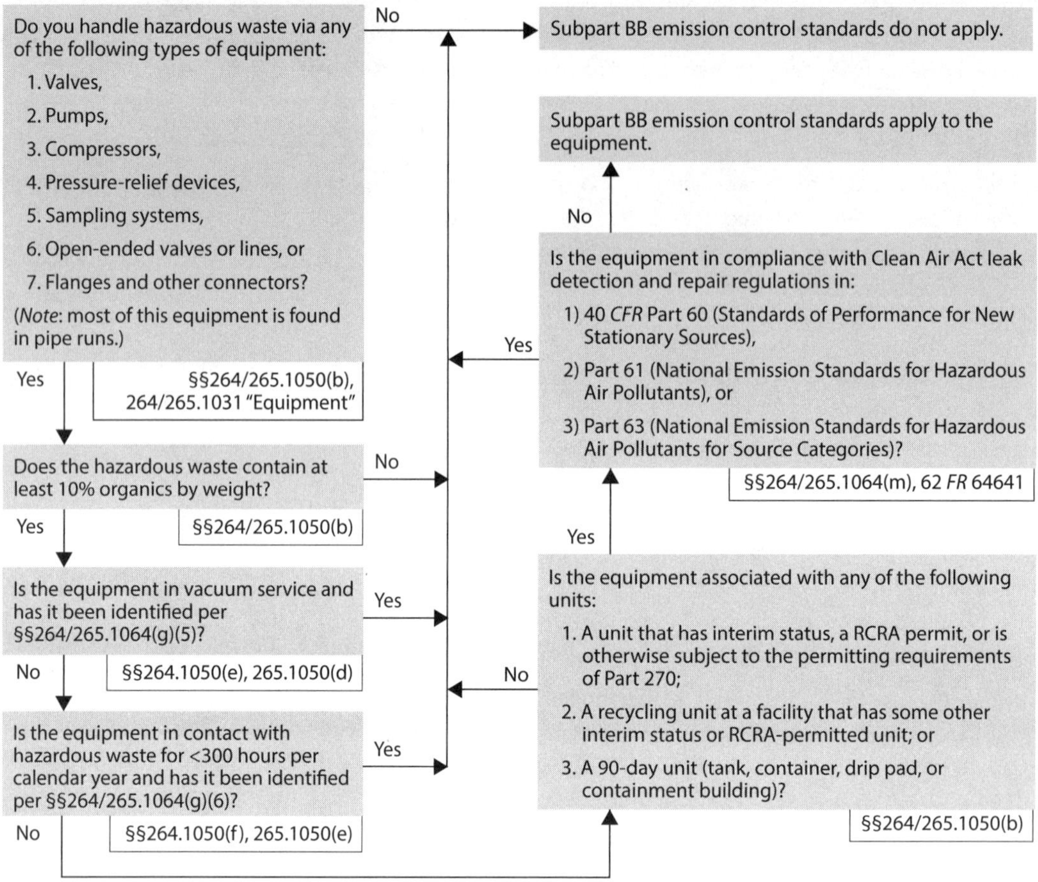

Do you handle hazardous waste via any of the following types of equipment:
1. Valves,
2. Pumps,
3. Compressors,
4. Pressure-relief devices,
5. Sampling systems,
6. Open-ended valves or lines, or
7. Flanges and other connectors?

(Note: most of this equipment is found in pipe runs.)

- **No** → Subpart BB emission control standards do not apply.
- **Yes** → §§264/265.1050(b), 264/265.1031 "Equipment"

Does the hazardous waste contain at least 10% organics by weight?
- **No** → Subpart BB emission control standards do not apply.
- **Yes** → §§264/265.1050(b)

Is the equipment in vacuum service and has it been identified per §§264/265.1064(g)(5)?
- **Yes** → (exempt path)
- **No** → §§264.1050(e), 265.1050(d)

Is the equipment in contact with hazardous waste for <300 hours per calendar year and has it been identified per §§264/265.1064(g)(6)?
- **Yes** → (exempt path)
- **No** → §§264.1050(f), 265.1050(e)

Is the equipment in compliance with Clean Air Act leak detection and repair regulations in:
1) 40 *CFR* Part 60 (Standards of Performance for New Stationary Sources),
2) Part 61 (National Emission Standards for Hazardous Air Pollutants), or
3) Part 63 (National Emission Standards for Hazardous Air Pollutants for Source Categories)?

- **Yes** → §§264/265.1064(m), 62 *FR* 64641
- **No** ↓

Is the equipment associated with any of the following units:
1. A unit that has interim status, a RCRA permit, or is otherwise subject to the permitting requirements of Part 270;
2. A recycling unit at a facility that has some other interim status or RCRA-permitted unit; or
3. A 90-day unit (tank, container, drip pad, or containment building)?

- **Yes** → Subpart BB emission control standards apply to the equipment. §§264/265.1050(b)
- **No** → Subpart BB emission control standards do not apply.

Source: McCoy and Associates, Inc.

Inspection Manual: Federal Equipment Leak Regulations for the Chemical Manufacturing Industry, EPA/305/B-98/011, December 1998, available from http://www.epa.gov/compliance/resources/publications/assistance/sectors/insmanvol1.pdf.

15.2.1.1 Equipment managing hazardous waste

The Subpart BB standards apply only to equipment used to manage hazardous waste. For example, Figure 15-4 shows a valve on the pipe run that transports product from the process vessel—this valve is not subject to any Subpart BB requirements, because it is not in contact with hazardous waste.

15.2.1.2 Only specific types of equipment are subject to controls

Subpart BB applies to fugitive emissions from certain types of fluid handling equipment. In an operating facility, most of this equipment is found in pipe runs that convey hazardous waste from one location to another. The regulations identify the types of equipment subject to controls but don't provide any practical information about the equipment. For this type of information, EPA's publication EPA/450/3-89/021, summarized below, is very helpful.

The following types of equipment are subject to Subpart BB standards:

1. *Valves* can emit organics to the atmosphere around the valve stem. Some valves are eligible for reduced controls under §§264/265.1057(f), because they have no external actuating mechanism in contact with the hazardous waste stream. These "leakless" or "sealless" valves come in three types: sealed bellows, diaphragm, and pinch valves.

2. *Pumps* having rotating shafts (e.g., centrifugal pumps, gear pumps) or reciprocating shafts (e.g., piston pumps) can emit hydrocarbon vapors where the shafts penetrate the pump housing around the shaft seal. Some specialized (and uncommon) pumps are sealless and designed not to leak (i.e., they have no externally actuated shaft penetrating the pump housing) and are eligible for reduced controls under §§264/265.1052(e). These pumps are generally classified as sealless centrifugal pumps, sealless gear pumps, and diaphragm pumps.

3. *Compressors* can emit hydrocarbons around the shaft seal. Even though the Subpart BB regulations include compressor standards, we don't think they are triggered very often because gases in piping systems are not considered to be solid or hazardous wastes. Therefore, compressors (which only handle gases) are rarely in hazardous waste service. (Refer to Section 14.4.1 for a discussion of how gases are regulated under RCRA.) Why did EPA include compressor standards in Subpart BB? The Subpart BB regs were copied from Clean Air Act regulations that apply to equipment regardless of whether it is in hazardous waste service or not.

4. *Pressure-relief devices* are typically either spring loaded or consist of a rupture disk. If a spring-loaded device is set to relieve pressure very close to the system operating pressure it will "simmer" and emit organics to the atmosphere. Similarly, if a spring-loaded device does not properly reseat, it will leak. Once a rupture disk releases pressure via failure, it will emit hydrocarbons until replaced. Pressure-relief devices in either gas or liquid service are subject to Subpart BB.

5. *Sampling systems* are emission sources when sample lines are purged prior to sampling and the purged material drains onto the ground or into the sewer where it can evaporate. The Subpart BB controls address the disposition of these purged liquids. In situ sampling systems (i.e., nonextractive samplers or in-line samplers) and sampling systems without purges are not subject to such controls. [§§264/265.1055(c)] (Note, however, that if the sampling system includes any open-ended valves, the next category applies.)

6. *Open-ended valves or lines* are, by definition, open to the atmosphere and can be a source of emissions. If the valve is faulty or not completely closed, significant emissions can result. Examples are purge valves, drain valves, and vent valves.

7. *Flanges and other connectors* can produce emissions if, for example, a flange gasket is leaking. What is a "connector"? Section 264.1031 defines a connector as "flanged, screwed, welded, or other joined fittings used to connect two pipelines or a pipeline and a piece of equipment." In guidance, EPA states: "For reporting and recordkeeping purposes, the definition of 'connector' includes only flanged fittings (e.g., those screwed, welded, or otherwise joined are not flanges)." Also the §264.1031 definition of "connector" states: "For purposes of reporting and recordkeeping, connector means flanged fittings that are not covered by insulation or other materials that prevent location of the fittings." Therefore, screwed

unions, quick-disconnect hose fittings, and strainer housings with flange-type tops are not "connectors" for the purposes of reporting and recordkeeping. For purposes of inspection, leak monitoring, and repair, screwed unions, quick-disconnect hose fittings, and strainer housings would be subject to Subpart BB. [RO 11802]

15.2.1.3 Organic concentration of 10 percent by weight triggers requirements

Per §§264/265.1050(b), the Subpart BB regulations apply only where organic concentrations of at least 10% by weight are involved; however, facilities must document per §§264/265.1064(k) that certain equipment is exempt from these requirements because the organics concentration within the equipment is <10%. The test methods used to determine if a waste meets the 10% limit are codified in §§264/265.1063(d). When direct measurement (as opposed to process knowledge) is used, samples are analyzed using any of the methods cited below:

- SW–846 Method 9060—Total Organic Carbon;
- ASTM D2267-88—Standard Test Method for Aromatics in Light Naphthas and Aviation Gasolines by Gas Chromatography;
- ASTM E169-87—Standard Practices for General Techniques of Ultraviolet-Visible Quantitative Analysis;
- ASTM E168-88—Standard Practices for General Techniques of Infrared Quantitative Analysis; or
- ASTM E260-85—Standard Practice for Packed Column Gas Chromatography.

Samples collected for testing must be representative of the highest total organic content in hazardous waste that is expected to contact the equipment.

Q *A two-phase liquid is stored in a hazardous waste tank. How does the facility owner sample the waste to determine whether it has a 10% organic content (which would subject all equipment associated with the tank to Subpart BB)?*

A *"If the organic content fluctuates or the equipment handles more than one waste stream, determination will be based on the maximum total organics content of a waste stream contained or contacted by the equipment.... [W]hen the waste is stratified, it is necessary to obtain and integrate subsamples from all layers of the waste material."* [EPA/450/3-89/021, cited previously]

Up-to-date information on the organic concentration of waste in contact with Subpart BB equipment must be recorded in the facility operating record. [§§264/265.1064(b)(1)(iv), (k)(3)]

15.2.1.3.1 Knowledge-based determinations

Rather than using the direct measurement options previously explained, an owner/operator may use knowledge to determine whether the equipment contains or contacts hazardous waste with an organic concentration ≥10% by weight. If a knowledge-based determination is made, §§264/265.1063(d)(3) specify that documentation of the waste determination is required.

The first example of using knowledge in §§264/265.1063(d)(3) discusses understanding of the process and allows an owner/operator to document that "no organic compounds are used" in the production process. In some instances, knowing that your production process does not use organics may be sufficient. However, using knowledge of the process should also entail a detailed look at all the chemical constituents formed in the production process.

Other examples in §§264/265.1063(d)(3) utilize knowledge of the process which stems from analytical testing. In the second example, an owner/operator may use information from waste generated by a process at another facility provided that the processes are identical and the other facility has previously demonstrated, by direct measurement, that the waste stream has a total organic content less than 10 percent. In the third example, prior speciated analytical results may be used for a waste stream for which an owner/operator can document that no process changes have occurred since that analysis that could affect the total organic concentration. In both the second and third examples, if analytical results were obtained for another regulatory purpose, or the analytical results may not have considered organics formed in the process, direct measurement may be the most conservative approach.

15.2.1.4 A 300-hr/year contact time is required

If a piece of equipment as described above contacts hazardous waste for less than 300 hours per calendar year, it is not subject to Subpart BB standards. [§§264.1050(f)/265.1050(e)]

> "[T]he amount of time that equipment contains hazardous waste, whether at operating capacity or as a residue, is considered time that the equipment 'contains or contacts' hazardous waste. Thus, if Subpart BB equipment contains Subpart BB-regulated hazardous waste residues for more than 300 hours during a calendar year, that equipment would not be exempt…. EPA purposefully worded the provision to say 'contains or contacts' because the emissions from the equipment are related to the organic hazardous waste that is in the equipment; even if the process or equipment is not in service, the organic hazardous waste in contact with the equipment has the potential to volatilize, and EPA considers it necessary to subject the equipment to the requirements of Subpart BB…. [F]or this regulatory requirement, instances during which equipment contains or contacts Subpart BB-regulated waste need not be consecutive; it is only required that the sum of all time that the equipment contains or contacts Subpart BB-regulated waste is less than 300 hours per calendar year." [December 8, 1997; 62 *FR* 64641]

If a facility exempts its equipment from Subpart BB requirements because it contains or contacts hazardous waste with an organic concentration of at least 10% for <300 hours/year, it must document such exemption per §§264/265.1064(g)(6).

15.2.1.5 Only equipment associated with certain categories of units is regulated

Not all equipment as described above is subject to Subpart BB. Only equipment associated with units that fit into three categories is regulated. [§§264/265.1050(b)]

First, equipment associated with units that are subject to the permitting standards of Part 270 (i.e., they either have a RCRA permit or are in interim status) is subject to Subpart BB. Equipment associated with the following types of units that are exempt from RCRA permitting is exempt from Subpart BB:

- Manufacturing process units (these units are not subject to Subpart BB because they are not managing hazardous wastes)—see exemption in §261.4(c).
- Totally enclosed treatment facilities—see exemption in §270.1(c)(2)(iv).
- Elementary neutralization units—see exemption in §270.1(c)(2)(v).
- Wastewater treatment units—see exemption in §270.1(c)(2)(v).
- Closed-loop recycling units—see exemption in §261.4(a)(8).

Second, if a recycling unit is located at a facility that has a RCRA permit (e.g., for hazardous waste storage tanks), the equipment connected to the recycling unit is subject to Subpart BB. [§261.6(d)] On the other hand, if the recycling unit is located at a generator's facility that does not have or need a RCRA permit for any treatment, storage, or disposal activities, equipment associated with the recycling unit is not subject to Subpart BB.

Third, equipment connected to 90-day units (that are not recycling units) is subject to Subpart BB.

Case Study 15-1 and the following examples illustrate the applicability of Subpart BB.

 A facility has a RCRA-permitted container storage area. Do any Subpart BB standards apply to the containers?

Case Study 15-1: Equipment Subject to Subpart BB

In Figure 15-6, liquids are periodically removed from a process reactor vessel. These liquids flow through a piping manifold that can divert the liquid to either of two destinations: 1) reuse/recycle within the plant, or 2) a 90-day or permitted hazardous waste storage tank for storage prior to disposal. The liquid contains greater than 10% organics, and liquid destined for waste disposal is in the piping for more than 300 hours per year. Is any of the equipment in the pipe run subject to Subpart BB, or is all of the equipment considered to be part of a manufacturing process unit that is exempt from RCRA regulation under §261.4(c)?

When the liquid is removed from the process reactor vessel, the owner/operator knows whether the liquid will be recycled or disposed. If the material is to be disposed, it is a hazardous waste and the point of generation is where the liquid leaves the bottom of the reactor vessel. "[A]ny process transfer equipment, even if normally used for production purposes, that is also used to transfer hazardous waste residue during equipment washout/cleanout procedures to a hazardous waste storage/treatment tank, would be considered part of a hazardous waste tank system and thus subject to standards for such." [RO 13790] The equipment in the pipe run from the bottom of the reactor vessel to the distribution manifold and then to the hazardous waste storage tank is considered to be equipment ancillary to the storage tank and is subject to Subpart BB. [RO 14469]

Figure 15-6: Applicability of Subpart BB to Equipment in Pipe Runs

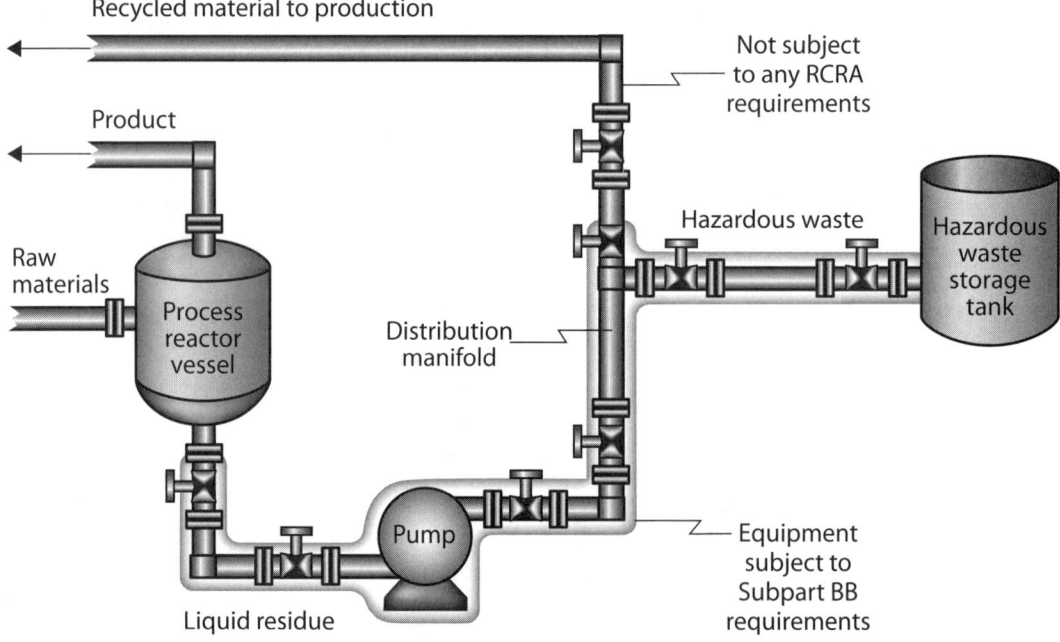

Source: McCoy and Associates, Inc.; adapted from RO 13790, 14469.

A Maybe. Most of the equipment subject to Subpart BB is not typically associated with containers. Although containers may have pressure-relief devices as shown in Figure 15-4, we have an email from EPA indicating that a pressure-relief device mounted in the vapor space of a 90-day container would not be subject to Subpart BB. Instead, such a pressure-relief device would be subject to Subpart CC requirements in §265.1087(c)(3)(iv), (d)(3)(iv). On the other hand, equipment in a pipe run that is used to fill 90-day containers would be subject to Subpart BB.

Q *A large quantity generator accumulates spent solvent in a 90-day tank prior to processing it in a batch still. The generator does not have a RCRA permit. Does Subpart BB apply to this operation?*

A Yes. Equipment in the pipe runs to and from the 90-day tank is subject to Subpart BB.

Q *A wood preserving facility operates a drip pad under the §262.34(a)(1)(iii) 90-day provisions. Does Subpart BB apply to this operation?*

A Maybe. Subpart BB applies to 90-day units regulated under §262.34(a). These §262.34(a) regulations apply to tanks, containers, Subpart W drip pads, and Subpart DD containment buildings. However, §§264/265.1050(b), which specify that Subpart BB applies to 90-day units, only refer to tanks and containers. If wood preserving wastes with ≥10% organics are contained in a waste collection system associated with the drip pad, and the system includes any of the equipment cited in Section 15.2.1.2, the facility owner/operator should seek a determination from EPA or the state as to whether Subpart BB applies.

Q *A state does not recognize EPA's closed-loop recycling exemption codified at §261.4(a)(8); therefore, under state regulation, material in the recycling system (including pipe runs) is a hazardous waste. Are the pipe runs subject to Subpart BB regulations?*

A EPA has no jurisdiction over state regulations that are broader in scope than the federal requirements. Although the state can choose to make the closed-loop unit subject to Subpart BB, EPA's air emission regulations do not apply. [RO 13487]

15.2.1.6 Equipment in compliance with CAA requirements is exempt

To eliminate regulatory overlap between the CAA (which also regulates fugitive emissions) and RCRA, EPA codified an exemption at §§264/265.1064(m). In the *Federal Register* preamble establishing this exemption, the agency stated that, if the relevant documentation requirements of 40 *CFR* Part 60—Standards of Performance for New Stationary Sources, Part 61—National Emission Standards for Hazardous Air Pollutants, or Part 63—National Emission Standards for Hazardous Air Pollutants for Source Categories are met, this constitutes compliance with Subpart BB.

> "[This provision allows owners and operators of] any equipment that contains or contacts hazardous waste that is subject to Subpart BB and also subject to regulations in 40 *CFR* Part 60, 61, or 63 to determine compliance with Subpart BB by documentation of compliance with the relevant provisions of the Clean Air Act rules.... Because compliance with Subpart BB is demonstrated through recordkeeping, this recordkeeping revision has the effect of exempting equipment that would otherwise be subject to Subpart BB from Subpart BB requirements, provided the equipment is operated, monitored, and repaired in accordance with an applicable CAA standard, and appropriate records are kept to that effect." [December 8, 1997; 62 *FR* 64641]

One caveat exists for this CAA exemption:

> "EPA does not consider it appropriate to allow the [40 *CFR* Part 63] Subpart RR drain system requirements to substitute for the more extensive open-ended valve and line requirements of Subpart BB, because application of the Subpart RR standards to Subpart BB equipment would not provide an equivalent level of organic emission control as would be achieved by compliance

15.2.2 Complying with Subpart BB

The preceding discussion serves to identify equipment that is subject to Subpart BB. Once such equipment is identified, the next step is to determine how to comply with Subpart BB fugitive emission standards. If a piece of equipment is subject to Subpart BB, fugitive emissions are controlled in several ways:

1. Design standards are imposed in limited circumstances. For example, pumps equipped with a dual mechanical seal system that includes a barrier-fluid system are subject to reduced monitoring requirements. [§§264/265.1052(d)]

2. Inspection requirements are frequently imposed. For example, many pumps must be inspected weekly for indications that liquids are dripping from the seal. [§§264/265.1052(a)(2)]

3. Monitoring requirements are often obligatory. For example, many valves must be monitored monthly to detect leaks. [§§264/265.1057(a)]

4. Unexpected leak detection requires action. In some cases, equipment doesn't have to be inspected or monitored on a routine basis; however, if someone notices that the equipment is leaking, it must be monitored and fixed. [§§264/265.1058(a)]

15.2.2.1 Light liquid service

Because "light" liquids are more likely to produce fugitive organic emissions, equipment handling these types of wastes is subject to more stringent design/inspection/monitoring standards. *"In light liquid service* means that the piece of equipment contains or contacts a waste stream where the vapor pressure of one or more of the organic components in the stream is greater than 0.3 kilopascals (kPa) at 20°C, the total concentration of the pure organic components having a vapor pressure greater than 0.3 kilopascals (kPa) at 20°C is equal to or greater than 20 percent by weight, and the fluid is a liquid at operating conditions." [§§264/265.1031] EPA noted in guidance that the definition applies only to the organic components of the waste stream—not to nonorganic constituents that meet the vapor pressure criteria (e.g., water). [December 8, 1997; 62 FR 64639]

If the equipment contains or contacts a hazardous waste stream that meets the definition above, it is "in light liquid service." If the waste stream doesn't meet all three criteria in the definition, the equipment is "in heavy liquid service."

To help people make the light vs. heavy liquid service determination, EPA included guidance in EPA/450/3-89/021, as referenced in Section 15.2.1. According to this document, each organic compound in the waste, its concentration, and its vapor pressure must be known. A complete analysis (e.g., gas chromatography/mass spectroscopy) or engineering estimate should be used to determine the waste stream composition. Vapor pressures for many constituents at 20°C are available in chemical literature. If the vapor pressure for a particular component is not available, the pure component can be tested using ASTM Method D2879 to determine the vapor pressure. Table 15-2 gives vapor pressures of common solvents for comparison purposes.

15.2.2.2 The LDAR program

In general, Subpart BB is a leak detection and repair (LDAR) program. Inspection and monitoring

Table 15-2: Vapor Pressures of Common Solvents

Solvent	Vapor pressure at 20°C, kPa
Acetone	24.6
Ethanol	5.9
Isopropyl alcohol	4.1
Methanol	12.7
Methyl ethyl ketone	9.4
Methyl isobutyl ketone	2.1
Methylene chloride	45.2
Mineral spirits	0.27
Tetrachloroethylene	1.7
Toluene	5.1
1,1,1-Trichloroethane	2.3
Trichloroethylene	7.8
Xylene(s)	1.3

Source: EPA/450/3-89/021.

requirements are imposed for each type of equipment meeting the Subpart BB applicability requirements discussed in Section 15.2.1. Such inspections and monitoring are designed to detect leaks; once a leak is detected, repair must be initiated within 5 days and completed as soon as practicable, not to exceed 15 days.

Table 15-3 summarizes the inspection and monitoring requirements for the various types of Subpart BB equipment. As noted in the table, the inspection/monitoring options for valves are the most complex. Where the table is not self-explanatory, references are given to section numbers in Parts 264/265 where the details may be found.

Note that closed-vent systems connected to a control device may be used to control emissions from leaking equipment. This would essentially consist of a hood over/around the equipment to capture any emissions. Sections 264/265.1060 impose standards on the closed-vent systems and control devices that are taken from Subpart AA.

15.2.2.2.1 Identification of Subpart BB-regulated equipment

Sections 264.1050(d)/265.1050(c) require that each piece of equipment subject to Subpart BB be marked in such a manner that it can be distinguished readily from other pieces of equipment at the plant that are not subject to LDAR requirements. The most common method of complying with this requirement is to tag each piece of Subpart BB-regulated equipment. However, EPA has clarified verbally that "the language does not necessarily require physical tagging or marking, and that a plant site plan or diagram would meet the intent of the §265.1050(c) marking requirements. We emphasize, however, that the authorized state agency may have a more stringent interpretation."

There is an almost identically worded requirement for marking equipment in the CAA regulations for equipment leaks (fugitive emission sources). [§61.242-1(d)] Written guidance from EPA Region IV on that requirement notes that an alternative marking system of assigning each flange or other connector an identification number corresponding to the nearest piece of physically tagged equipment and recording those numbers in a log would be acceptable. [This guidance is available at http://cfpub.epa.gov/adi/ by querying for NR89 in the Control Number field.]

15.2.2.3 Recordkeeping and reporting

EPA and state inspectors rely primarily on their review of a facility's records to determine if the facility is complying with Subpart BB. The extensive recordkeeping requirements for Subpart BB are codified at §§264/265.1064. Many facilities choose to comply with these requirements and their CAA LDAR requirements via the service of a contractor that conducts the inspections and monitoring and also maintains all of the associated documentation.

For permitted facilities only, §264.1065 imposes reporting requirements. These regulations require owners/operators to file a report every six months identifying leaking valves, pumps, and compressors that were not repaired within 15 days. Where control devices are used to manage fugitive emissions, instances of control device failures/excursions exceeding 24 hours in duration must also be reported. If all leaking valves, pumps, and compressors are repaired within 15 days and all control devices function properly, no semiannual report is required.

15.3 Air emission controls for tanks, containers, and surface impoundments—Subpart CC

When hazardous waste containing volatile organics is stored/treated in certain tanks, containers, or surface impoundments, significant air emissions can result. These emissions can be controlled in three basic ways: 1) change the waste generating process to reduce or eliminate the volatile organic content, 2) treat the waste to remove the volatile organics, or 3) manage the waste in units that are closed such that emissions either can't escape or are captured in some type of control system. The air emission standards for hazardous waste tanks,

Table 15-3: Inspection/Monitoring Requirements for Subpart BB-Regulated Equipment

Equipment type	Inspection/monitoring frequency	Delay of repair allowed?	Exceptions	Regulatory citation
Valves				
In gas/vapor or light liquid service	Monthly: Method 21 portable analyzer[1] Quarterly monitoring[3] Semiannual monitoring[5] Annual monitoring[7,8]	Technical infeasibility[2] Equipment isolation[4] Excessive emissions[6] Depleted valve supplies[9]	Sealless valves [§§264/265.1057(f)] Unsafe to monitor [§§264/265.1057(g)] Difficult to monitor [§§264/265.1057(h)]	§§264/265.1057
In heavy liquid service	5 days after potential leak detection[10]	Same as above		§§264/265.1058
Pumps				
In light liquid service	Monthly: Method 21 portable analyzer[1] Weekly: visual inspection for liquid drippage	Technical infeasibility[2] Equipment isolation[4] Dual mechanical seals[11]	Dual mechanical seals [§§264/265.1052(d)] Sealless pumps [§§264/265.1052(e)] Closed-vent systems [§§264/265.1052(f)]	§§264/265.1052
In heavy liquid service	5 days after potential leak detection[10]	Same as above		§§264/265.1058
Compressors[13]	Check barrier-fluid seal sensor daily	Technical infeasibility[2] Equipment isolation[4]	Closed-vent systems [§§264/265.1053(h)] No detectable emissions [§§264/265.1053(i)]	§§264/265.1053
Pressure-relief devices				
In gas/vapor service	5 days after pressure release[12]	Technical infeasibility[2] Equipment isolation[4]	Closed-vent systems [§§264/265.1054(c)]	§§264/265.1054
In light or heavy liquid service	5 days after potential leak detection[10]	Same as above		§§264/265.1058
Sampling systems	None[13]	NA[13]	In situ systems [§§264/265.1055(c)] Systems without purges [§§264/265.1055(c)]	§§264/265.1055
Open-ended valves or lines	None[13]	NA[13]	None	§§264/265.1056
Flanges and other connectors	5 days after potential leak detection[10]	Technical infeasibility[2] Equipment isolation[4]	Inaccessible or ceramic lined [§§264/265.1058(e)]	§§264/265.1058

[1] An instrument reading of 10,000 ppmv or more indicates a leak.
[2] Repairing equipment is technically infeasible without a hazardous waste unit shutdown. [§§264/265.1059(a)]
[3] Valves that don't leak in two consecutive months may be monitored quarterly. [§§264/265.1057(c)]
[4] Equipment is isolated from the hazardous waste management unit and doesn't contact hazardous waste with ≥10% by weight organics. [§§264/265.1059(b)]
[5] If no more than 2% of valves are leaking during two consecutive quarters, monitoring may be done every six months. [§§264/265.1062(b)(2)]
[6] Emissions due to repair will exceed emissions resulting from delay of repair. [§§264/265.1059(c)]
[7] If no more than 2% of valves are found to be leaking after five consecutive quarters, monitoring may be done annually. [§§264/265.1062(b)(3)]
[8] If no more than 2% of valves are found to be leaking, valves can be tested annually. [§§264/265.1061]
[9] Inadequate repair assemblies are available and next unit shutdown will be within 6 months. [§§264/265.1059(e)]
[10] Potential leak detection occurs if a leak is seen, heard, smelled, or otherwise detected. [§§264/265.1058(a)]
[11] Pumps equipped with dual mechanical seals and barrier fluid must be repaired within 6 months. [§§264/265.1059(d)]
[12] No later than five days after a pressure release, the device must be monitored to confirm no detectable emissions, as indicated by an instrument reading of less than 500 ppmv above background using Method 21. [§§264/265.1054(b)(2)]
[13] Compliance achieved via equipment design standards. See referenced regulatory citation for details.
Source: McCoy and Associates, Inc.; adapted from Subpart BB of 40 *CFR* Parts 264 and 265.

containers, and surface impoundments are codified in Parts 264/265, Subpart CC.

The Subpart CC regulations in Parts 264 and 265 differ only in three respects: 1) Part 264 includes reporting requirements (§264.1090) applicable to permitted facilities, 2) the definitions applicable to Subpart CC are found only in §265.1081, and 3) Part 265 includes an implementation schedule (§265.1082). Because the implementation schedule section only appears in Part 265, subsequent sections (which are identical between Parts 264 and 265) are out of sync numerically. For example, waste determination procedures are codified in Part 264 as §264.1083 but as §265.1084 in Part 265.

15.3.1 Applicability of the Subpart CC regulations

Figure 15-7 identifies the conditions under which the Subpart CC emission control standards will apply to a hazardous waste management operation. Important details referenced in this diagram are discussed below.

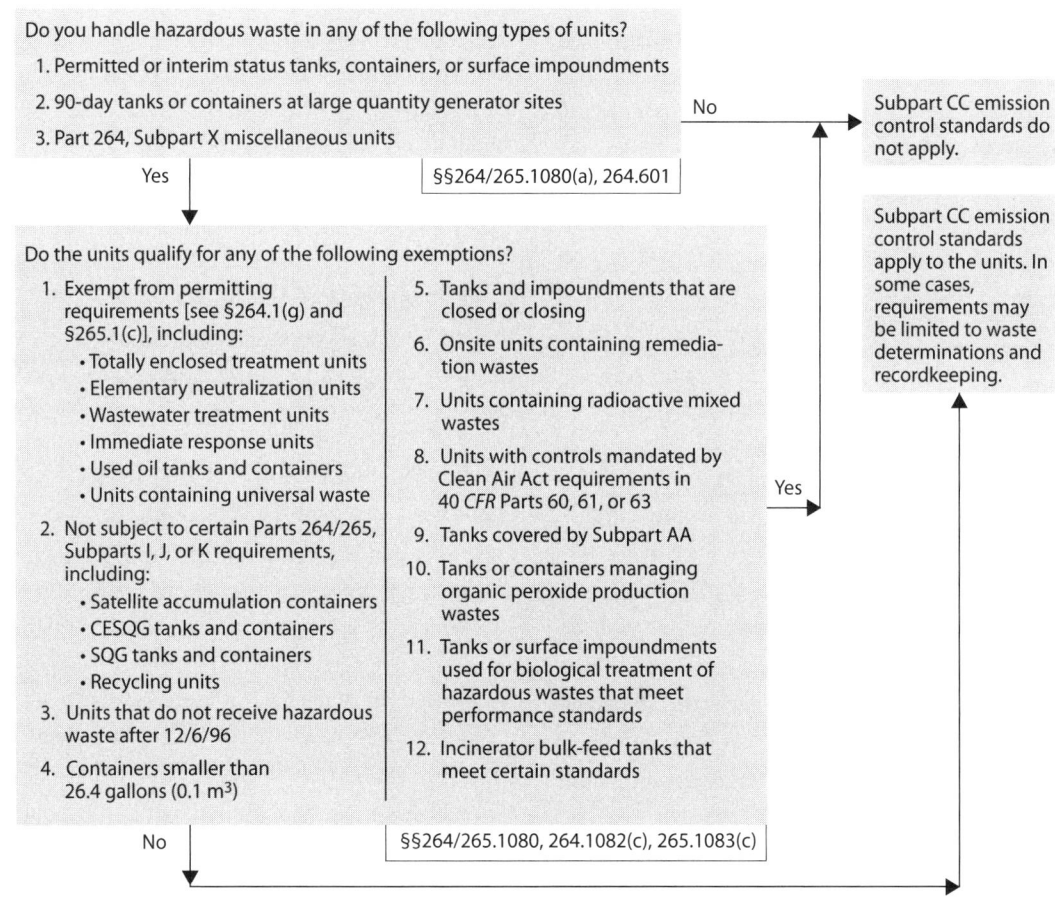

Figure 15-7: Determining If Subpart CC Emission Control Standards Apply

Source: McCoy and Associates, Inc.

One of the most useful references available from EPA on Subpart CC applicability is EPA/453/R-94/076b, *Background Information for Promulgated Organic Air Emission Standards for Tanks, Surface Impoundments, and Containers*, November 1994. This document is available online at http://nepis.epa.gov/EPA/html/Pubs/pubtitleOAR.html by downloading the report numbered 453R94076B.

15.3.1.1 Units managing hazardous waste

Note that the Subpart CC standards apply only to units managing hazardous waste. For example, if a generator produces a characteristic hazardous waste, decharacterizes it, and then manages the nonhazardous residue in a tank or container, Subpart CC requirements do not apply to that tank or container.

Q *An organic-containing hazardous waste flows into a plant sewer system. Eventually, the plant sewer discharges to a Subtitle D (nonhazardous) surface impoundment. Is the surface impoundment subject to Subpart CC?*

A If the waste entering the impoundment is not hazardous and hazardous wastes are not generated in the impoundment, it is not managing hazardous wastes; therefore, the impoundment is not subject to Subpart CC.

15.3.1.2 Only specific types of units are subject to controls

Subpart CC applies to three general types of waste management units: tanks, containers, and surface impoundments. Specifically, units potentially subject to these regulations are:

1. Tanks that are permitted or have interim status,
2. Tanks at large quantity generator facilities (i.e., 90-day tanks),
3. Containers in permitted or interim status storage areas,
4. Containers at large quantity generator facilities (i.e., 90-day containers),
5. All surface impoundments used to manage hazardous waste, and
6. Permitted Part 264, Subpart X miscellaneous units. The permitting authority will determine if a Subpart X unit most closely resembles a tank, container, or surface impoundment; the appropriate Subpart CC requirements will then apply.

Note that tanks or containers may be used to manage solids and still be subject to Subpart CC.

15.3.1.2.1 *Exemptions*

Many types of tanks, containers, and impoundments are exempt from Subpart CC requirements. In general, a unit must be subject to Part 264 or Part 265, Subpart J (for tanks), Subpart I (for containers), or Subpart K (for surface impoundments) before Subpart CC requirements will apply. Conversely, any units exempt from these regulatory subparts are also exempt from Subpart CC. In addition, several specific exemptions appear in §§264/265.1080, 264.1082, and 265.1083. The following units are not subject to Subpart CC:

1. Units exempt from permitting [§264.1(g)] or interim status [§265.1(c)] requirements, including:
   - Totally enclosed treatment units,
   - Elementary neutralization units,
   - Wastewater treatment units,
   - Immediate response units,
   - Used oil tanks and containers, and
   - Units containing universal waste.

2. Units that are not subject to certain Subpart I, J, or K requirements, including:
   - Satellite accumulation containers (although satellite accumulation containers are subject to some Subpart I requirements, §265.178, which imposes Subpart CC requirements, is not applicable) [§262.34(c)(1)(i)],
   - Units at conditionally exempt small quantity generator facilities [§261.5(b)],
   - Units at small quantity generator facilities [§262.34(d)(2–3)], and
   - Recycling units [§261.6(c)(1)].

3. Units that don't receive hazardous waste after December 6, 1996. [§§264/265.1080(b)(1)]

4. Containers smaller than 26.4 gallons (0.1 m³). [§§264/265.1080(b)(2)]

5. Tanks and impoundments that are closed or closing. [§§264/265.1080(b)(3–4)]

6. Onsite units containing remediation wastes. [§§264/265.1080(b)(5)] Note that this exemption applies only to remediation wastes generated during RCRA corrective action or CERCLA cleanups or cleanups required under similar state or federal authorities. Remediation wastes from voluntary cleanups, with no state involvement or oversight, would not qualify. Also, as soon as the waste is moved offsite, the exemption no longer applies. [*RCRA Subpart CC: Questions and Answers*, March 2002, available from http://www.trainex.org/web_courses/subpart_x/EPA CD Content/TopicSearch.htm]

7. Units containing radioactive mixed wastes. [§§264/265.1080(b)(6)]

8. Units with controls mandated by CAA requirements in 40 *CFR* Parts 60, 61, or 63. [§§264/265.1080(b)(7)] See Section 15.3.1.2.2 for more discussion on this exemption.

9. Tanks with vents subject to the Subpart AA controls (e.g., a tank with a process vent through which distillation equipment is vented). [§§264/265.1080(b)(8)]

10. Tanks or containers managing hazardous waste generated from the manufacture of organic peroxides. [§§264/265.1080(d)]

11. Tanks or surface impoundments used for biological treatment of hazardous wastes that meet performance standards of §264.1082(c)(2)(iv) or 265.1083(c)(2)(iv). [§§264.1082(c)(3)/265.1083(c)(3)]

12. Incinerator bulk-feed tanks. [§§264.1082(c)(5)/265.1083(c)(5)] Incinerator bulk-feed tanks that are located in an enclosure connected to a control device and that are in compliance with certain benzene NESHAP control requirements are not subject to Subpart CC.

The regulations also contain two site-specific exemptions from the Subpart CC requirements for the Stonewall pharmaceutical plant in Elkton, Virginia [§§264/265.1080(e)] and the OSi Specialties plant in Sistersville, West Virginia [§§264/265.1080(f–g)].

15.3.1.2.2 *Equipment in compliance with CAA requirements is exempt*

To eliminate regulatory overlap between the CAA and RCRA, EPA codified an exemption at §§264/265.1080(b)(7). Hazardous waste tanks, containers, and surface impoundments that are in compliance with an applicable CAA regulation under 40 *CFR* Part 60, 61, or 63 are not subject to Subpart CC. However, there are three limitations to the exemption for units complying with an applicable CAA regulation:

1. For a tank that uses an enclosure as a part of the air emission control system rather than the more conventional tank controls involving covers, the enclosure and control device used must comply with the technical requirements for enclosures and combustion devices in §§264.1084(i)/265.1085(i).

2. A unit that does not use the CAA-required air emission controls but is in compliance with a CAA regulation through an "emission averaging" or "bubbling" provision does not qualify for the exemption. EPA lacks assurance that emissions from the unit resulting from such averaging are controlled to the extent necessary to protect human health and the environment. [November 25, 1996; 61 *FR* 59939]

3. If the CAA standard for the particular unit requires no control, the exemption from the RCRA standards under §§264/265.1080(b)(7) would not apply since the unit would not actually be equipped with and operating air emission controls under provisions of the applicable CAA standard. EPA believes the best way to assure protectiveness under the Subpart CC rule

is to require controls on each particular unit. [CAA and RCRA Overlap Provisions in Subparts AA, BB, and CC of 40 CFR Parts 264 and 265, cited in Section 15.1.1.6]

15.3.1.2.3 Examples

Q *Is a sump subject to the requirements of Subpart CC?*

A Maybe. Section 10.2 discusses the regulatory status of sumps. Some sumps are clearly exempt from Subpart CC, such as temporary sumps (also known as "immediate response sumps" or "dry sumps"), elementary neutralization units, and sumps that meet the definition of a wastewater treatment unit. On the other hand, if a sump is used as a 90-day tank or is a primary containment sump, Subpart CC applies. [RO 12442]

"The tank control equipment requirements specified under the Subpart CC standards (e.g., a covered tank vented to a control device, external floating roof) do apply to sumps. However, the Subpart CC standards also require that the owner and operator use enclosed pipes or other closed systems to transfer hazardous waste to or from a tank required to use the Subpart CC air emission controls. In the case where the sump is used to transfer wastewater, for example, EPA considers the individual drain system requirements specified in the benzene waste operations NESHAP under 40 *CFR* 61.346(a)(1) or 40 *CFR* 61.346(b)(1) through (b)(3) to define a 'closed system' and to provide adequate emission control for a sump." [EPA/453/R-94/076b, cited previously]

Q *Is a roll-off box subject to Subpart CC regulations?*

A Maybe. A roll-off box is a container that is typically used to store solids. If the solids are remediation wastes from a RCRA corrective action or CERCLA cleanup, the roll-off box is exempt from Subpart CC while onsite at the generating facility. [§§264/265.1080(b)(5)] On the other hand, if the solids are process wastes or remediation wastes from a voluntary cleanup, Subpart CC would apply.

Q *Laboratory wastes are stored in 5-gallon plastic carboys. Does Subpart CC apply to these containers?*

A No. Any hazardous waste container smaller than 26.4 gallons in design capacity is exempt from Subpart CC. [§§264/265.1080(b)(2)]

Q *A small quantity generator accumulates wastes in 180/270-day tanks and containers, which are exempt from Subpart CC. However, a couple of times a year, the SQG generates more than 1,000 kg of hazardous waste in a month. As a large quantity generator in these months, do the tanks and containers now have to meet Subpart CC standards?*

A The general requirements applicable to episodic generators such as this are discussed in Section 6.1.2.1. Hazardous wastes generated during months that the facility is a LQG must be managed in accordance with Subpart CC, if all other applicability requirements are met. If wastes generated during the LQG months are kept segregated from the wastes generated during SQG months, the 180/270-day tanks and containers would not become subject to Subpart CC. However, if wastes are commingled from SQG and LQG months, the Subpart CC standards would apply to all tanks and containers.

Q *If I remove a hazardous waste tank, container, or surface impoundment from service, is this an acceptable means of complying with Subpart CC?*

A Yes and no. Subpart CC doesn't apply to units that are closed or closing. However, if you stored/treated wastes in units subject to Subpart CC after December 6, 1996 (and didn't have an implementation plan), you were out of compliance with Subpart CC and may be subject to enforcement penalties if this comes to the attention of the regulators. See §265.1082 for the dates associated with installing Subpart CC controls.

15.3.1.3 Volatile organic concentration of 500 ppmw triggers requirements

Subpart CC emission control requirements apply only where regulated tanks, containers, or surface impoundments manage hazardous wastes containing at least 500 parts per million by weight (ppmw) volatile organics at the "point of waste origination." How are volatile organics defined? What is the "point of waste origination"? These two concepts are discussed below.

15.3.1.3.1 *Volatile organics*

For the record, the term "volatile organic concentration" is defined as follows [§265.1081]:

"Volatile organic concentration or *VO concentration* means the fraction by weight of the volatile organic compounds contained in a hazardous waste expressed in terms of parts per million (ppmw) as determined by direct measurement or by knowledge of the waste in accordance with the requirements of §265.1084 of this subpart. For the purpose of determining the VO concentration of a hazardous waste, organic compounds with a Henry's law constant value of at least 0.1 mole-fraction-in-the-gas-phase/mole-fraction-in-the-liquid-phase (0.1 Y/X) (which can also be expressed as 1.8×10^{-6} atmospheres/gram-mole/m^3) at 25 degrees Celsius must be included. Appendix VI of this subpart presents a list of compounds known to have a Henry's law constant value less than the cutoff level."

The reference to §265.1084 in the above definition should be to §265.1084(a)(3)(iii) when VO concentration is determined by testing and §265.1084(a)(4) when VO concentration is determined via knowledge.

The bottom line is, when testing is used to determine VO concentration, this parameter is determined by the results of one or more analytical tests. EPA developed a test method (Method 25D in 40 *CFR* Part 60, Appendix A) specifically to determine VO concentration. This is a relatively aggressive test wherein nitrogen gas is bubbled through the waste sample for 30 minutes at 75°C. The concentrations of organics and chlorinated compounds in the purge gas are determined using two different types of detectors. The results for organics are reported as total methane, and chlorinated compounds are reported as total Cl. The two results are added to give VO concentration.

Method 25D was developed for use where no information is available (or needed) on the specific compounds present in the waste. If the specific chemicals present in the waste are known, EPA allows other test methods to be used—see §265.1084(a)(3)(iii)(A–B).

Having said all this, we rarely find that people analyze their hazardous wastes to determine VO concentrations. Why? As will be explained in a moment, a better way to comply with Subpart CC is to simply assume that all hazardous wastes will contain ≥500 ppmw VO and manage the wastes in units having controls that meet Subpart CC standards.

Knowledge of the waste(s) composition may also be used in lieu of testing; however, any such knowledge-based determinations must be fully documented per §265.1084(a)(4).

15.3.1.3.2 *Point of waste origination*

For the generator of a hazardous waste, the point of waste origination is the same as the point of generation—see Section 14.1. For offsite TSD facilities, the point of waste origination is where the facility accepts delivery or takes possession of the hazardous waste. Signing the manifest accompanying the waste is typically considered to be the time at which possession is established.

Q *A large facility stores/treats hazardous wastewater in several tanks before disposing the water in underground injection wells. It is very likely that some of the individual wastewater streams flowing into the storage/treatment tanks contain more than 500 ppm VO at the point of generation. Are these tanks subject to Subpart CC?*

A The answer to this question depends on the regulatory status of the tanks. Assuming that the facility is a large quantity generator, the tanks are most likely to be 90-day tanks or interim status/permitted tanks, both of which are subject to Subpart CC. (The tanks cannot be exempt wastewater treatment units because underground injection wells are not CWA systems—see Section 7.2.1.1.) If the tanks are elementary neutralization units or are equipped with emission controls in accordance with a CAA requirement in 40 CFR Parts 60, 61, or 63, they would not be subject to Subpart CC.

15.3.2 Basic intent of Subpart CC

Before describing the emission controls required by Subpart CC for hazardous waste tanks, containers, and impoundments, it's helpful to understand the basic intent of the program.

Take a look at the hypothetical facility depicted in Figure 15-8. The facility is a large quantity generator, and hazardous wastes are generated at three locations. The plant also has a RCRA permit to operate two waste storage/treatment tanks and a hazardous waste impoundment. Wastes at the point of waste origination (point of generation) are initially collected in 90-day containers. Based on a waste analysis program, the VO concentrations of wastes A, B, and C have been determined to be 400, 0, and 600 ppmw, respectively.

Because Subpart CC is triggered at a VO concentration of 500 ppmw, waste C and its associated container will be subject to emission controls. The containers used to collect wastes A and B are not subject to Subpart CC.

Wastes A, B, and C are pumped into Tank 1. Because Tank 1 receives a waste having ≥500 ppmw VO at the point of origination (waste C), it is subject to Subpart CC requirements. If wastes A, B, and C are generated in equal volumes, the average VO concentration in Tank 1 would be about 333 ppmw. This concentration is unimportant; what matters is that one of the wastes being managed in Tank 1 exceeded 500 ppmw VO at the point of origination. Now, Tank 1 and every other nonexempt tank, container, or surface impoundment downstream from Tank 1 will be subject to Subpart CC (with one exception as noted below), because they will also be receiving waste C.

Tank 2, which receives the wastes from Tank 1, is subject to Subpart CC controls because it is also receiving hazardous waste that had ≥500 ppmw VO

Figure 15-8: Basic Intent of Subpart CC

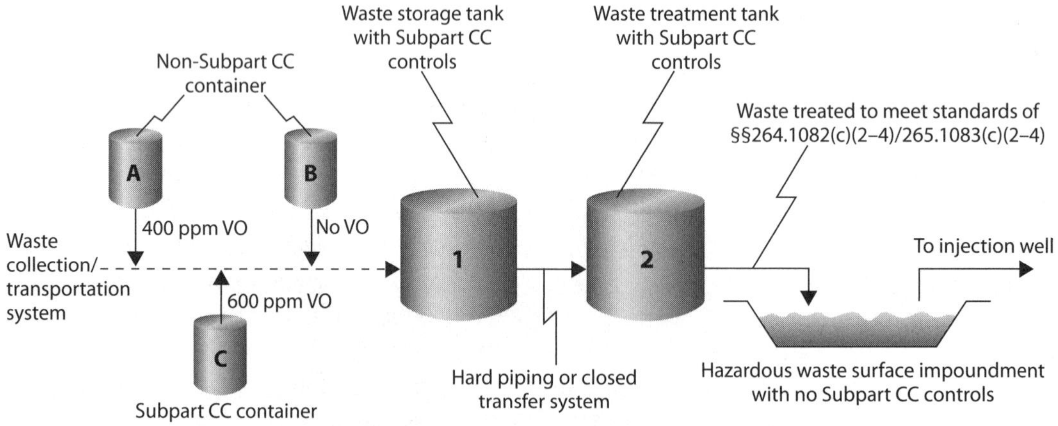

Source: McCoy and Associates, Inc.

at its point of origination. In this tank, however, waste treatment is occurring that removes or destroys VO. If the amount of treatment meets certain performance levels specified in §§264.1082(c)(2–4)/265.1083(c)(2–4), any downstream units receiving the treated waste are no longer subject to Subpart CC controls. Therefore, the hazardous waste surface impoundment receiving the waste from Tank 2 is not subject to Subpart CC standards. (This is a very important outcome because, as will be discussed in Section 15.3.6, controlling organic emissions from impoundments is very difficult and expensive.)

15.3.3 Compliance options for Subpart CC

If a tank, container, or surface impoundment is potentially subject to Subpart CC (i.e., it is not an exempt unit cited in Section 15.3.1.2.1), compliance can be achieved in six ways:

1. *Demonstrate <500 ppmw VO via sampling and analysis*—A waste determination can be made for each waste managed in the units via sampling and analysis. If all analyses show VO concentrations are below 500 ppmw, no controls are required as long as these determinations are documented annually. [§§264.1082(c)(1)/265.1083(c)(1)] If any single waste determination shows VO concentrations are ≥500 ppmw at the point of origination, Subpart CC emission controls will be required.

2. *Demonstrate <500 ppmw VO via knowledge-based determinations*—A waste determination can be made for each waste managed in the units via knowledge. If all knowledge-based determinations show VO concentrations are below 500 ppmw, no controls are required as long as these determinations are documented annually. [§§264.1082(c)(1)/265.1083(c)(1)] If any single waste determination shows VO concentrations are ≥500 ppmw at the point of origination, Subpart CC emission controls will be required.

3. *Make process changes*—Under limited circumstances, manufacturing processes might be changed such that hazardous wastes produced contain <500 ppmw VO.

4. *Render the waste nonhazardous*—Characteristic wastes and ICR-only listed wastes might be treated to render them nonhazardous prior to placing them in a tank, container, or surface impoundment.

5. *Treat wastes for VO reduction*—If wastes containing ≥500 ppmw VO are treated to meet specified performance levels, no Subpart CC controls will be required for units that subsequently manage the residues. [§§264.1082(c)(2–4)/265.1083(c)(2–4)]

6. *Meet emission control standards*—All units receiving hazardous wastes comply with Subpart CC emission control requirements. Under these circumstances, it doesn't matter what the VO concentration of the hazardous waste is; even if it is ≥500 ppmw, any required controls are in place.

The practical advantages/disadvantages of each of these options are discussed below.

15.3.3.1 Sampling and analysis

Sampling and analyzing every hazardous waste to determine if the VO concentration is <500 ppmw is probably the worst option. For every waste potentially subject to Subpart CC, at least four samples must be taken (within a one-hour period) in accordance with a written sampling plan. Analytical results from each sample are averaged and, if the *average* VO concentration is <500 ppmw, no controls will be required (i.e., the average of the four or more samples represents one waste determination for the waste).

If the waste has very consistent VO concentrations over time, four samples could suffice. However, if the waste composition varies significantly or changes in accordance with manufacturing process parameters or any other variable, enough samples must be taken to capture this variability. In addition, these samples must be collected over an averaging period that cannot exceed one year. A new determination must be made at the beginning of each averaging period (i.e., a new waste determination must be made at least once a year).

Alternatively, if four samples can be collected during the one-hour period when VO concentrations are at a maximum during the averaging period, the average of these four samples can be used to determine if controls will be required.

The details associated with making waste determinations are codified in §§264.1083/265.1084. Also, see RO 14594. We will not scrutinize the details here, because we rarely encounter anyone who chooses the waste sampling and analysis option.

15.3.3.2 Knowledge-based determinations

Rather than testing a waste, a regulated party can use his/her knowledge of the waste's properties/composition to determine VO concentrations. For example, if a process uses only inorganic chemicals as raw materials and organic chemicals are not generated in the process, this knowledge could be used to establish the fact that Subpart CC would not apply to any units managing the hazardous waste generated from the process. If a knowledge-based determination is made, §265.1084(a)(4) specifies the documentation that must be provided. In many cases, knowledge will be based on some type of analytical data.

TSD facilities frequently use knowledge-based determinations wherein the generator of a waste provides the TSD facility with VO data (as a component of the waste profile form).

Q *If the owner/operator of a TSD facility uses a generator's waste analysis information to demonstrate that the waste is not subject to Subpart CC, how often must the generator provide updated data?*

A "If the waste analysis information received from the generator with the first shipment of waste is representative of subsequent shipments to the TSDF, the TSDF can continue to rely on the original waste analysis information, within certain limits. Owners/operators are required to update the waste analysis information at least once every twelve months following the date of the original analysis [§265.1083(c)(1)].

"It is not the responsibility of the generator to supply the TSDF with waste analysis documentation, rather it is the TSDF's option to use this information to perform volatile organic concentration determinations. In all cases, it is the responsibility of the person with custody of the waste to obtain valid information to make compliance determinations. Therefore, the TSDF should only use shipping papers, waste certifications, or other generator-prepared information in which they have confidence." [RO 14121]

Again, unless a generator is working only with inorganic materials, it is unusual for parties to make VO determinations based on knowledge that is devoid of analytical data. Hence, this option is chosen infrequently.

15.3.3.3 Process changes

Occasionally, it might be feasible to change a manufacturing process such that any hazardous wastes produced contain less than 500 ppm VO. For example, a higher purity feedstock might be used that minimizes the production of high-VO waste products or by-products. In our experience, compliance with Subpart CC is rarely achieved via this option.

15.3.3.4 Rendering the waste nonhazardous

Recall that Subpart CC applies only to tanks, containers, and impoundments used to manage hazardous waste. If a waste is generated and rendered nonhazardous before it enters one of these units, Subpart CC doesn't apply. Therefore, certain characteristic wastes and ICR-only listed wastes might be treated to render them nonhazardous as a means of achieving Subpart CC compliance. According to EPA, "the Subpart CC rule no longer applies once these wastes are decharacterized, i.e., no longer exhibit a characteristic of a hazardous waste.... Also, since the rules do not prohibit any method which removes a hazardous characteristic, dilution can be used for this purpose...." [December 8, 1997; 62 *FR* 64644] However, any dilution to render the waste noncharacteristic would be treatment (see Chapter 7), and the LDR dilution prohibition (see Section 13.11) would also have to be addressed if the waste is to be land disposed.

15.3.3.5 Treating wastes for VO reduction

In §§264.1082(c)(2–4)/265.1083(c)(2–4), EPA provides a number of options for treating, removing, or destroying VO such that downstream units will not require Subpart CC controls. The conditions under which treated wastes may be placed in uncontrolled units are discussed below. Figure 15-9 illustrates what each of these options looks like in the left-hand column; the right-hand column identifies the analytical work required to establish compliance with each option.

***Option 1**—VO concentration of treated wastes is <500 ppmw*

Under Option 1, if a waste being treated all comes from one point of origination, the treated waste must have a VO concentration of <500 ppmw. Diagram 1a in Figure 15-9 shows the treatment process removing 50 kg VO from a waste stream in order to meet the 500-ppmw outlet limit. Note from the right-hand column that VO analyses will be required at the point of origination for the inlet stream and also on the outlet stream.

Where multiple hazardous wastes are mixed prior to treatment, a calculation is performed essentially requiring that each hazardous waste stream exceeding 500 ppmw VO at the point of origination be treated to lower the concentration of that stream to 500 ppmw. The calculation assumes that no treatment is conducted on waste streams having less than 500 ppmw at the point of origination. Diagram 1b in Figure 15-9 illustrates this scenario.

Option 1 is identified at §§264.1082(c)(2)(i)/265.1083(c)(2)(i); the applicable equation for calculating the treatment process VO outlet concentration is found at §265.1084(b)(4). Note that the calculation procedure discounts any effects of VO reduction due solely to dilution.

***Option 2**—95% of VO has been removed and treated VO is <100 ppmw*

The equations used to calculate VO removal efficiency in Option 2 (which the regulations term "organic reduction efficiency") are given in §265.1084(b)(5), and are essentially material balance equations based on VO levels at the point of origination for each waste stream and the outlet from the treatment process. This option is codified at §§264.1082(c)(2)(ii)/265.1083(c)(2)(ii) and is illustrated in Diagram 2 in Figure 15-9. To meet the performance specification, total VO on a mass basis entering the treatment system must be reduced by 95%. Additionally, the VO concentration in the treated waste must be <100 ppmw. The number of VO determinations required for this option are the same as for Option 1.

***Option 3**—Mass of VO removed is ≥ mass of VO that was greater than 500 ppmw at point of origination*

For Option 3, the mass of VO in each stream that exceeds 500 ppmw at the point of origination is calculated and totaled. This total is the "required organic mass removal rate" [equations are given in §265.1084(b)(7)]. The mass of VO removed in the treatment process, based on inlet/outlet measurements, is termed the "actual organic mass removal rate" [equations are given in §265.1084(b)(8)]. The actual VO mass removal rate must be greater than or equal to the required VO mass removal rate before the waste may be placed in an uncontrolled unit. This option is codified at §§264.1082(c)(2)(iii)/265.1083(c)(2)(iii) and illustrated in Diagram 3 in Figure 15-9. Note that this option requires a VO determination at the point of origination of each inlet stream.

***Option 4**—For biosystems, VO reduction efficiency is ≥95% and organic biodegradation efficiency is ≥95%*

Option 4 is applicable to wastes treated in biological treatment systems. Two criteria must be met. First, the VO removal efficiency (as described for Option 2 above) must be at least 95%. Second, the organic biodegradation efficiency must also be at least 95%. This parameter is determined using the equations in §265.1084(b)(6), which reference a procedure specified in 40 *CFR* Part 63, Appendix C.

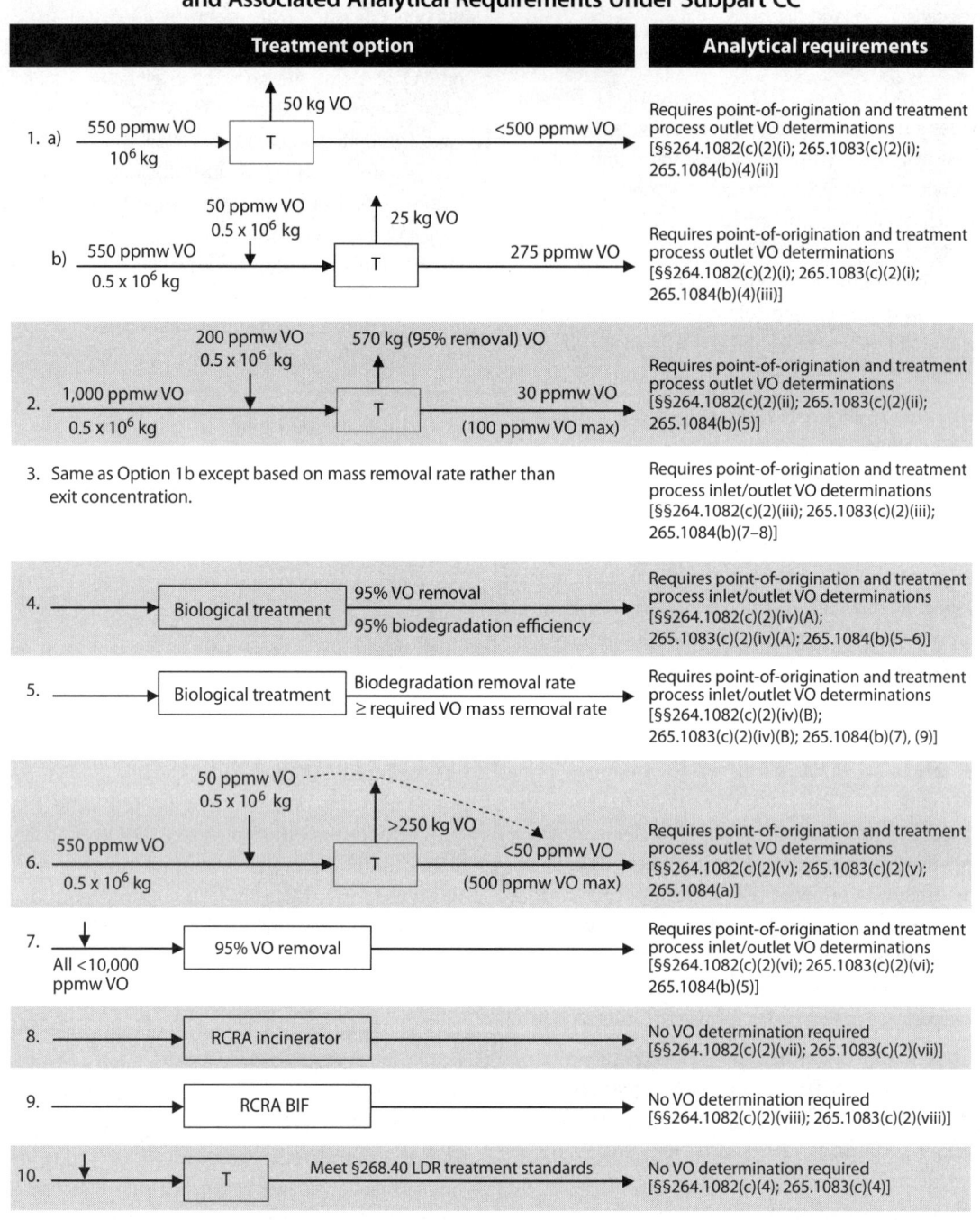

Figure 15-9: Waste Treatment Options and Associated Analytical Requirements Under Subpart CC

Source: McCoy and Associates, Inc.

This procedure involves complex calculations to determine whether organics are being biodegraded or destroyed in the biotreatment system. This option is codified at §§264.1082(c)(2)(iv)(A)/ 265.1083(c)(2)(iv)(A), and it is illustrated in Diagram 4 in Figure 15-9. The same VO determinations as for Option 3 are required.

Option 5—*For biosystems, the organic biodegradation rate is ≥ the required VO removal rate*

For this option, a required mass removal rate for VO is calculated based on the waste streams being treated that have ≥500 ppmw VO at the point of origination [using the equations in §265.1084(b)(7) as in Option 3]. If the organic biodegradation efficiency [using the equations in §265.1084(b)(6) as in Option 4] is multiplied by the mass of VO entering the biotreatment process, a biodegradation removal rate can be calculated per §265.1084(b)(9). If this value is greater than or equal to the required VO removal rate from Option 3, the treated waste may be placed in uncontrolled units. This option is codified at §§264.1082(c)(2)(iv)(B)/265.1083(c)(2)(iv)(B). Diagram 5 in Figure 15-9 illustrates the concept; note that extensive inlet stream VO analysis is required.

Option 6—*Treatment unit outlet VO concentration is < lowest point-of-origination VO concentration (500 ppmw max)*

Option 6 is actually an alternative to Option 1 that eliminates the need to calculate a required treatment process outlet VO concentration limit. "As an alternative to calculating the exit concentration limit for a treatment process [Option 1], the Subpart CC standards [§§264.1082(c)(2)(v)/265.1083(c)(2)(v)] allow the owner or operator to treat the mixed hazardous wastes to a volatile organic concentration level that is less than or equal to the lowest waste volatile organic concentration at the point of waste origination for all of the individual hazardous waste streams mixed together prior to entering the treatment process." [December 6, 1994; 59 *FR* 62915] Although the preamble guidance just quoted says "less than or equal to," the language of the regs themselves only says "less than." In no case can the VO concentration at the treatment process exit be over 500 ppmw. Refer to Diagram 6 in Figure 15-9; note that VO determinations are required at the point of origination for each inlet stream and on the outlet stream.

Option 7—*95% VO removal and all individual inlet streams contain <10,000 ppmw VO*

This option, which is codified in §§264.1082(c)(2)(vi)/ 265.1083(c)(2)(vi), requires the owner/operator to certify that the VO concentration of each hazardous waste stream at the point of origination is less than 10,000 ppmw and to achieve a VO removal efficiency of 95%. If these conditions are met, no further controls would be required on downstream units. This option also requires extensive inlet stream analysis, as indicated in Diagram 7 in Figure 15-9.

Option 8—*Burn the hazardous waste in a RCRA incinerator*

Options 8 and 9 are in §§264.1082(c)(2)(vii–viii)/ 265.1083(c)(2)(vii–viii). A significant advantage to treating the hazardous waste in a RCRA-permitted incinerator or RCRA-permitted boiler or industrial furnace (BIF) is that no VO determinations are required (assuming that the waste is stored/ transported in controlled units).

Option 9—*Burn the hazardous waste in a RCRA BIF*

See Option 8.

Option 10—*Treat the waste to meet LDR standards*

This option, which is codified in §§264.1082(c)(4)/ 265.1083(c)(4), really has no direct involvement with VO. Instead, the waste is simply treated such that the organic constituents meet LDR standards specified in §268.40. EPA's strategy for this option is "if hazardous wastes meet generally applicable LDR treatment standards for organics, their concentrations of organics are in virtually every case going to be less than warrants control under the Subpart CC rules (i.e., volatile organic concentrations will be less than 500 ppmw)." [December 8, 1997; 62 *FR* 64643]

This exemption does not apply to listed hazardous wastes having LDR treatment standards solely for

> **Case Study 15-2: Subpart CC LDR Exemption**
>
> The following examples/guidance are taken from the December 8, 1997 *Federal Register*. [62 FR 64644–5]
>
> I. A mixture of listed wastes F001 and F006 has been treated to meet LDR treatment standards and will be placed in a 90-day container. Does Subpart CC apply to the container?
>
> No. Even though F006 has no LDR treatment standards for organic constituents, F001 does. As long as the mixture meets the LDR treatment standards for the F001 organic constituents, Subpart CC does not apply to the container.
>
> II. As a result of mixing listed wastes, the mixture meets LDR treatment standards (via dilution). Does this qualify the mixture for the Subpart CC LDR exemption?
>
> No. The LDR standards need to be achieved by treatment that removes or destroys the organic hazardous constituents (or the wastes may meet the treatment standard as generated).
>
> III. A waste is hazardous for lead (D008). Because this is an inorganic constituent, is the waste ineligible for the Subpart CC LDR exemption?
>
> Characteristic hazardous wastes may also contain underlying hazardous constituents (UHCs) that are identified in the universal treatment standards table of §268.48. Because the treatment standards for UHCs are applicable to characteristic wastes, the Subpart CC LDR exemption is available in this instance if the treatment standards for all organic UHCs are met in the lead waste. The same principle applies to other characteristic wastes, such as ignitable, corrosive, and reactive wastes.
>
> IV. How would the Subpart CC LDR exemption apply to a mixture of F001 and D018 wastes?
>
> The waste must meet the treatment standards for F001, benzene (D018), and any organic UHCs in the waste. Alternatively, the D018 characteristic can be eliminated before the waste is managed in a tank, container, or surface impoundment, in which case only the F001 constituents would have to meet treatment standards.
>
> V. Hazardous soil meets the alternative LDR treatment standards for soil (§268.49). Is it eligible for the Subpart CC LDR exemption?
>
> No. "[T]he exemption from Subpart CC standards applies only to hazardous wastes that have been treated to meet the treatment standards set out in 40 *CFR* 268.40." Although the LDR exemption does not apply in this case, if the soil is a remediation waste from a RCRA corrective action, CERCLA cleanup, or cleanup resulting from similar state or federal authorities, an exemption from Subpart CC controls applies—see Section 15.3.1.2.1 and RO 14445.

inorganic constituents (e.g., F006). [December 8, 1997; 62 *FR* 64644] Examples of treating hazardous wastes to meet LDR standards, thus exempting subsequent management units from Subpart CC compliance, are given in Case Study 15-2.

In practice, we see little interest in any of the options that require VO analysis. Obviously, many parties burn their high-VO wastes in RCRA incinerators or BIFs or treat their wastes to meet LDR standards. However, this is not typically conducted for purposes of Subpart CC compliance; it is accomplished for other waste management reasons.

About the only time we can envision these Subpart CC treatment provisions to be of much use is in a situation similar to that shown in Figure 15-8. The surface impoundment in this figure is managing hazardous waste that contained ≥500 ppmw VO at the point of waste origination. Unless the waste is treated in Tank 2 to meet one of the options cited above, the surface impoundment will have to be equipped with Subpart CC emission controls. As we will see in Section 15.3.6, this would be a very undesirable outcome.

15.3.3.6 Meeting emission control standards for all potentially affected units

By far the most common method of complying with Subpart CC when placing hazardous wastes in nonexempt tanks, containers, or surface impoundments is to ensure that the units meet emission control standards. If this option is chosen, it doesn't matter what the VO concentration of the hazardous waste is—the necessary controls will be in place. By choosing this compliance option, all of the complexities of sampling, analyses, waste determinations, averaging periods, etc. are avoided.

The central question involved with this option is: How difficult is it to meet emission control standards under Subpart CC? The answer is that, for hazardous waste tanks and containers, most facilities will meet the control standards in the course of their normal operations and little if any additional equipment or investment will be required. The story is different for hazardous waste surface impoundments.

The next three sections describe the Subpart CC emission control standards applicable to hazardous waste tanks, containers, and surface impoundments.

15.3.4 Control standards for tanks

As discussed in Section 15.3.1.2.1, a number of exemptions exist that may keep a tank with hazardous waste in it from having to comply with Subpart CC. If no exemptions apply, EPA divides Subpart CC-regulated tanks into two groups, Level 1 and Level 2. Differentiating between the two levels is based on the size of the tank, the maximum organic vapor pressure of the hazardous waste in the tank, and whether waste stabilization is occurring (see Table 15-4). Each level has different emission control requirements, as described in the tank standards in §§264.1084/265.1085.

15.3.4.1 Level 1 tanks

Level 1 tanks are used to manage hazardous wastes that have low vapor pressures. Since the wastes have a low potential for emitting organics to the atmosphere, the control requirement is very rudimentary: Level 1 tanks must have a fixed roof. (A fixed roof means a cover mounted in a stationary position that does not move with fluctuations in liquid level.) Figure 15-10 illustrates a typical fixed-roof tank.

Table 15-4: Matrix for Determining Subpart CC Level 1 or 2 Tank Controls

Tank design capacity[1]	Maximum organic vapor pressure of hazardous waste in tank	Stabilization occurring in tank?	Subpart CC level of control
<20,000 gallons (75 m^3)	≤11.1 psi (76.6 kPa)	Yes	Level 2
		No	Level 1
	>11.1 psi (76.6 kPa)	Yes	Level 2
		No	Level 2
≥20,000 gallons (75 m^3) and <40,000 gallons (151 m^3)	≤4.00 psi (27.6 kPa)	Yes	Level 2
		No	Level 1
	>4.00 psi (27.6 kPa)	Yes	Level 2
		No	Level 2
≥40,000 gallons (151 m^3)	≤0.75 psi (5.2 kPa)	Yes	Level 2
		No	Level 1
	>0.75 psi (5.2 kPa)	Yes	Level 2
		No	Level 2

[1] Capacities specified as gallons are approximate; the Subpart CC regulations specify capacities in m^3. Basis is design capacity, not fill level, of the tank.

Source: Adapted from EPA/530/F-98/011, July 1998, available at http://www.epa.gov/epawaste/hazard/tsd/td/ldu/pdf/subcc.pdf.

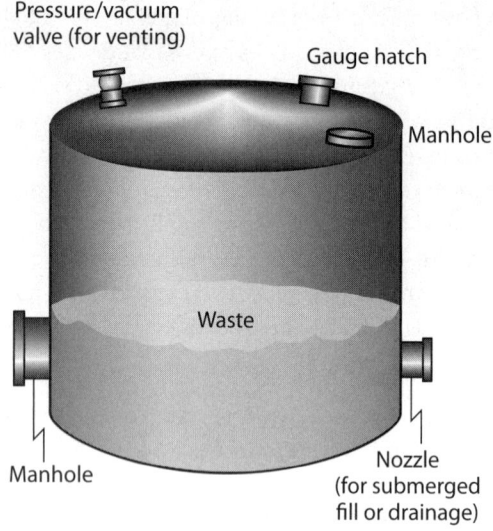

Figure 15-10: Typical Fixed-Roof Tank

Source: McCoy and Associates, Inc.

15.3.4.1.1 Prerequisites for Level 1 tanks

To qualify for Level 1 under Subpart CC, tanks must meet four prerequisites:

1. The organic vapor pressure of the waste cannot exceed certain values that are dependent on tank capacity [§§264.1084(b)(1)(i)/265.1085(b)(1)(i)],

2. Wastes in the tank cannot be heated such that the vapor pressures identified in Item 1 above are exceeded [§§264.1084(b)(1)(ii)/265.1085(b)(1)(ii)],

3. Wastes may not be stabilized in the tanks [§§264.1084(b)(1)(iii)/265.1085(b)(1)(iii)], and

4. When wastes are transferred from one Subpart CC Level 1 tank to another, hard piping or some other enclosed means of conveyance is required. The same requirement applies when wastes are transferred from a Subpart CC impoundment to a Subpart CC tank. [§§264.1084(j)/265.1085(j)]

Hard piping is not required when wastes are transferred from a tank to a container. [November 25, 1996; 61 FR 59946]

Table 15-4 identifies the applicable Subpart CC tank level based on the first three prerequisites summarized above. The first step in determining if a hazardous waste tank may comply with the Level 1 tank standards is to determine the maximum organic vapor pressure of the waste in the tank. Several analytical methods are cited in §§264.1083(c)/265.1084(c)(3)(ii), or knowledge may be used for determining vapor pressure in accordance with §§264.1083(c)/265.1084(c)(4). One industrial facility with which we are familiar conservatively calculates maximum organic vapor pressure as follows: the vapor pressure of the most volatile constituent in the waste is evaluated at the highest temperature to which the waste will ever be subjected. This value is used as the maximum organic vapor pressure for Subpart CC tank compliance. In *RCRA Subpart CC: Questions and Answers,* cited previously, EPA notes that the maximum organic vapor pressure should be based on a worst-case scenario, meaning that it needs to be based on the highest expected temperature of the waste and on any stirring being conducted in the tank.

Next, the maximum organic vapor pressure of the hazardous waste must be compared against the capacity of the tank to be used for its storage. If the maximum organic vapor pressure of the waste is no more than the values noted in Table 15-4 (and waste stabilization is not occurring in the tank), a Level 1 fixed-roof tank can be used for the hazardous waste storage tank. If the waste has a vapor pressure higher than that associated with the capacity of the tank to be used (and/or waste stabilization is occurring in the tank), the tank will have to meet Level 2 standards. Also, the owner/operator can choose to use Level 2 controls (which are somewhat more stringent) in lieu of simply using a fixed-roof tank.

Table 15-5 lists the vapor pressures and maximum Level 1 tank capacities for some common solvents.

Table 15-5: Vapor Pressures of Common Solvents and Allowable Level 1 Tank Capacity

Solvent	Vapor pressure at 20°C, psi	Maximum Level 1 tank capacity, gal[1]
Acetone	3.57	<40,000
Ethanol	0.85	<40,000
Isopropyl alcohol	0.60	No limit
Methanol	1.85	<40,000
Methyl ethyl ketone	1.37	<40,000
Methyl isobutyl ketone	0.31	No limit
Methylene chloride	6.56	<20,000
Mineral spirits	0.04	No limit
Tetrachloroethylene	0.25	No limit
Toluene	0.73	No limit
1,1,1-Trichloroethane	0.33	No limit
Trichloroethylene	1.13	<40,000

[1] Capacities specified as gallons are approximate; the Subpart CC regulations specify capacities in m^3. Basis is design capacity, not fill level, of the tank.

Source: McCoy and Associates, Inc.; adapted from EPA/450/3-89/021.

15.3.4.1.2 Specifications for the fixed roof

Specifications for the fixed-roof tank include:

- The roof may be an integral part of the tank, or it may be removable; however, it must be designed to "form a continuous barrier over the entire surface area of the hazardous waste in the tank." [§§264.1084(c)(2)(i)/265.1085(c)(2)(i)]

- No visible cracks, holes, gaps, or open spaces are allowed between roof sections or between the roof edge and the tank wall. [§§264.1084(c)(2)(ii)/265.1085(c)(2)(ii)]

- Openings in the roof (other than conservation vents or pressure/vacuum-relief valves) must be equipped with a closure device (e.g., a hatch cover or cap) or the opening must be vented to a control device. No performance standards are specified for the control device. Closure devices can be open during routine inspection, maintenance, or other activities that need to be conducted during normal operations (e.g., removing sludges). [§§264.1084(c)(2)(iii)/265.1085(c)(2)(iii)]

Any hatches or openings that are in contact with the waste placed in the tank (i.e., those below the normal high liquid level of the tank) are subject to Subpart BB requirements, if applicable. Conversely, those hatches or openings that are above the normal high liquid level are subject to Subpart CC. [EPA/453/R-94/076b, cited previously, *RCRA Subpart CC: Questions and Answers*, cited previously]

- The fixed roof and closure devices must be made of materials that will minimize emissions to the atmosphere and be compatible with the waste. [§§264.1084(c)(2)(iv)/265.1085(c)(2)(iv)]

- Pressure/vacuum-relief valves and conservation vents, which vent directly to atmosphere, are allowed, as long as they operate with no detectable emissions when closed. [§§264.1084(c)(3)(ii)/265.1085(c)(3)(ii)] Organic vapors are not continuously vented through such devices, but emissions occur only during those situations when tank operating conditions require vents to be open. An EPA Region IV document gives a procedure to follow to determine compliance with §§264.1084(c)(3)(ii)/265.1085(c)(3)(ii) for pressure/vacuum-relief valves and conservation vents, including determination of relieving set points. [*Guidance Document for RCRA Hazardous Waste Air Emission Standards Under 40 CFR Parts 264 and 265*, October 30, 2000, available from http://www.trainex.org/web_courses/subpart_x/EPA CD Content/TopicSearch.htm]

We have received *verbal* EPA headquarters clarification that conservation vents associated with Subpart CC-regulated tanks are *not* subject to Subpart BB requirements for pressure-relief devices in gas/vapor service. This clarification is confirmed in the Region IV guidance document referenced above. (Under Subpart BB, the device would have to be monitored within five days after a pressure release to confirm that there are no detectable emissions.)

- To prevent physical damage or permanent deformation to the tank and associated equipment, emergency venting through a safety device directly to atmosphere is allowed. [§§264.1084(c)(3)(iii)/ 265.1085(c)(3)(iii)] In the guidance document referenced above, Region IV noted that these safety devices (which are typically rupture discs, O-rings, or other one-time emergency-use devices) are subject to the Subpart BB standards in §§264/265.1054 for pressure-relief devices in gas/vapor service.

- The fixed roof and closure devices must be visually inspected on or before the date wastes are first placed in the tank and then at least annually thereafter. Any defects must be repaired, and inspection records must be maintained as noted below. [§§264.1084(c)(4)/265.1085(c)(4)] Guidance for conducting such inspections, including a checklist, is given in *General Inspection Guidance for Waste Management Units Requiring Air Emission Controls Under RCRA Air Standard Subpart CC*, November 2000, available at http://www.trainex.org/web_courses/subpart_x/EPA CD Content/TopicSearch.htm.

15.3.4.1.3 Recordkeeping and reporting

If a tank qualifies for Level 1 controls, documentation of the maximum organic vapor pressure of the hazardous waste must be maintained in the facility's records. [§§264.1089(b)(2)(i)/265.1090(b)(2)(i)]

Additionally, records of all inspections must be kept at the facility for a minimum of three years after the date of inspection.

15.3.4.2 Level 2 tanks

Level 2 tanks satisfy the emission control requirements for any hazardous waste subject to Subpart CC; they are the only available option if a waste/tank design does not meet the Level 1 tank prerequisites noted in Section 15.3.4.1.1. Five tank/control device options are available for Level 2 tank storage.

15.3.4.2.1 Compliance options

Option 1—Fixed roof with internal floating roof

Figure 15-11 illustrates a fixed-roof tank with an internal floating roof. In this design, the internal roof floats on the liquid surface, suppressing vaporization

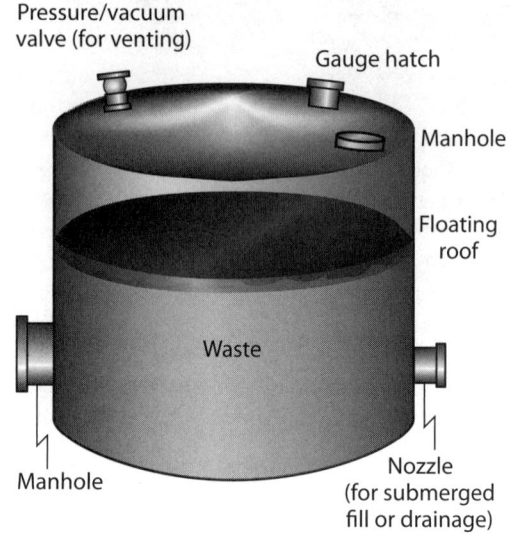

Figure 15-11: Fixed-Roof Tank With Internal Floating Roof

Source: McCoy and Associates, Inc.

of the volatile organics contained in the waste. Additionally, the seals between the edge of the floating roof and the tank wall minimize emissions. The regulations pertaining to this option are codified in §§264.1084(e)/265.1085(e). Detailed specifications for design, operation, and inspection are included in the regs and are not discussed further here.

Option 2—External floating roof

A tank with an external floating roof is illustrated in Figure 15-12. This type of tank is sometimes used at refineries, chemical plants, and fuel storage depots. Again, the floating roof suppresses vaporization of the volatile organic compounds contained in the waste. While an external floating roof is more complex than one might suspect (e.g., the floating roof needs two continuous seals between the roof and the tank wall, a drain system to remove precipitation, support legs, gauge ports, guide-pole wells, bleeder vents, etc.), these units are widely available.

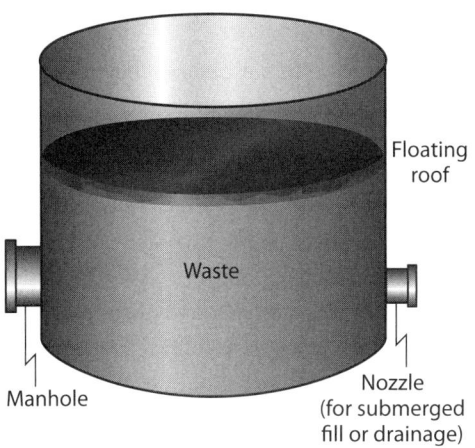

Figure 15-12: Tank With External Floating Roof

Source: McCoy and Associates, Inc.

The regulations pertaining to this option are codified in §§264.1084(f)/265.1085(f).

Option 3—*Fixed roof vented to a control device*

Figure 15-13 illustrates a fixed-roof tank that is vented to a control device. As the liquid level rises in the tank, vapors are routed via a closed-vent system to a control device that may be 1) a vapor-recovery system such as a condenser or carbon adsorption system, 2) an enclosed combustion device such as a fume incinerator or boiler, or 3) a flare. In general, the control device must reduce organics in the vapor stream by at least 95%. The regulations for this option are codified in §§264.1084(g)/265.1085(g), which describe the fixed roof, and §§264.1087/265.1088, which describe the closed-vent system and control device. This is a common compliance option for hazardous waste tanks at refineries and chemical plants, because such tanks are often vented to the flare.

The regs in §§264.1084(g)(1)(ii)/265.1085(g)(1)(ii) require each opening in the fixed roof that is not

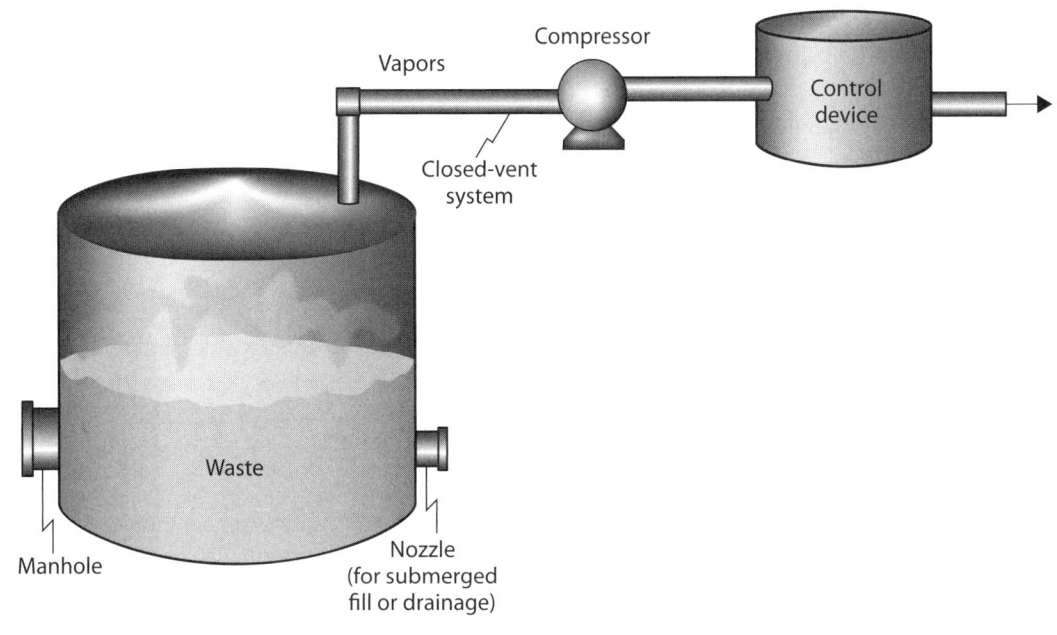

Figure 15-13: Fixed-Roof Tank With Closed-Vent System and Control Device

Source: McCoy and Associates, Inc.

vented to the control device to be provided with a closure device. If the pressure in the tank vapor headspace is greater than or equal to atmospheric when the control device is operating, each closure device must be designed to operate with no detectable organic emissions. In EPA/453/R-94/076b, EPA notes the following:

> "The final Subpart CC standards require the cover and all cover openings to be designed to operate with no detectable organic emissions when all cover openings are secured in the closed, sealed position. For a vertical wall tank, the no detectable emission requirement under the Subpart CC standards applies to the tank top cover or tank roof, to the junction between the cover and the tank walls, and to openings on that portion of the tank walls that do not directly contact the waste placed in the tank when the tank is filled to maximum capacity. EPA does not intend this requirement to apply to the seams and welds on the tank walls nor to piping connections through the tank walls or bottom through which waste is transferred to or from the tank."

Note that we don't think the compressor indicated in Figure 15-13 is subject to Subpart BB fugitive emission requirements, because it does not contain or contact hazardous waste. Instead, the compressor is part of the closed-vent system subject to §§264.1087(b)/265.1088(b) standards. These sections reference inspection/monitoring requirements in §§264.1033(l)/265.1033(k), which require a no detectable emissions condition (<500 ppmv organics above background using a portable VOC analyzer).

Option 4—*Pressure tank*

Pressure tanks are conceptually very simple. They are typically cylindrical or perhaps spherical in shape, and, as waste is pumped into the tank, the vapor headspace decreases and the pressure inside the tank increases. A pressure tank operates as a closed system, with no detectable emissions occurring during filling and emptying. [February 9, 1996; 61 *FR* 4908] The only release of pressure occurs if 1) the liquid level in the tank decreases, or 2) a pressure/vacuum-relief valve opens to prevent structural damage to the tank. The regulations pertaining to this option are codified in §§264.1084(h)/265.1085(h).

Option 5—*Enclosure vented to an enclosed combustion device*

Sometimes hazardous waste tanks must have open tops. For example, a commercial TSD facility may solidify wastes in a large, open-top, below-grade, concrete tank. Wastes are placed in the tank, solidification agents are added, and a backhoe or some other type of device is used to mix the tank contents. Under these circumstances, the tank is in a room (enclosure), with exhaust fans pulling emissions from the room to a control device such as a fume incinerator. Under the Level 2 standards [§§264.1084(i)/265.1085(i)], specifications are established for the enclosure and the control device.

15.3.4.2.2 Other requirements

When wastes are transferred from one Subpart CC Level 2 tank to another, hard piping or some other enclosed means of conveyance is required. The same requirement applies when wastes are transferred from a Subpart CC impoundment to a Subpart CC tank. [§§264.1084(j)/265.1085(j)] Hard piping is not required when wastes are transferred from a tank to a container. [November 25, 1996; 61 *FR* 59946]

15.3.4.2.3 Recordkeeping and reporting

The recordkeeping and reporting requirements associated with Level 2 tanks are included in §§264.1089–1090/265.1090 and are summarized by EPA as follows:

> "All tanks regulated by the Subpart CC standards must be regularly inspected. Inspection procedures and requirements vary by type of tank control used. Records of all inspections, regardless of the control level, must be kept at the facility for a minimum of 3 years after the date of inspection.

> "Owners or operators of tanks with internal or external floating roofs using Tank Level 2 controls are required to notify the Regional Administrator before conducting inspections. More detailed recordkeeping and inspection requirements are also required for floating-roof tanks and tanks or

enclosures that vent to a control device." [EPA/530/F-98/011]

Guidance for conducting such inspections, including a checklist, is given in *General Inspection Guidance for Waste Management Units Requiring Air Emission Controls Under RCRA Air Standard Subpart CC*, November 2000, availability as given in Section 15.3.4.1.2.

15.3.4.3 Tank examples

Q *A fixed-roof tank connected to a control device (Option 3 above) at a RCRA-permitted facility has two large doors that allow sludge to be pumped to the tank from tankers or emptied into the tank from containers. Subpart CC regulations specify that whenever hazardous waste is in the tank, the closure devices (doors) must be closed other than during inspections, maintenance, sampling, sludge cleanout, etc. [§264.1084(g)(2)(i)] Will a hard-piped system be required for transferring waste from the tankers/containers into the tank?*

A The preamble to a set of Subpart CC technical corrections [November 25, 1996; 61 *FR* 59946] specifies that a "transfer of hazardous waste between a tank and container is not required to be done in a closed system." However, exposure of the waste in the tank to the atmosphere must be limited by keeping the doors open only during periods of active waste addition. [RO 14288]

Q *If organic vapors from a Level 2 hazardous waste tank are routed to an industrial boiler that functions as a control device (Option 3 above), does the boiler have to be a RCRA BIF that requires a permit?*

A No. "The organic vapors emitted from hazardous waste are not hazardous wastes. Therefore, the control devices installed specifically to comply with Subpart CC organic vapor control requirements are not hazardous waste management units and are not required to be permitted under RCRA." [EPA/453/R-94/076b]

Q *Activated-carbon adsorption is used as the control device for Level 2 hazardous waste tanks (Option 3 above). When spent, how must the carbon be managed?*

A "Spent carbon, with adsorbed organics, used to control air emissions from hazardous waste treatment, storage, or disposal is not necessarily a hazardous waste. It is a hazardous waste if it exhibits a characteristic, or if it was used to capture emissions from treating listed hazardous waste." [EPA/453/R-94/076b, February 9, 1996; 61 *FR* 4910] The RCRA regs specifically identify the acceptable methods for managing spent carbon that is a hazardous waste. [§§264.1033(n)/265.1033(m), 264.1087(c)(3)(ii)/265.1088(c)(3)(ii)] Characteristic carbon sent for regeneration may not be a solid or hazardous waste per Table 1 in §261.2(c). See the discussion of this topic and applicable EPA guidance in Section 14.4.2.2.

Q *Does a Level 2 tank that routinely vents to an incinerator meet the definition of a "pressure tank" (Option 4 above)?*

A No. A tank that routinely vents to any device or unit (other than through emergency or safety devices) does not meet the definition of a pressure tank. All openings in a pressure tank must be equipped with closure devices that function with no detectable emissions, and the tank must operate as a closed system with no venting. [§§264.1084(h)/265.1085(h), RO 14461]

Q *An enclosure, inside of which is an open-top hazardous waste tank (Option 5 above), at a RCRA-permitted facility is connected through a closed-vent system to an incinerator that functions as a control device. Occasionally, the incinerator has to be shut down for maintenance (e.g., rebricking), which may take three weeks or longer. Can the tank continue to be operated while the incinerator is out of service for maintenance?*

A The regulations specify that periods of planned, routine maintenance of the control device shall not exceed 240 hours (10 days) per year. [§§264.1087(c)(2)(i)/265.1088(c)(2)(i)] If the incinerator maintenance activities can be completed within this time frame, no method of emission control is required. [RO 14461] Where the maintenance period exceeds 240 hours, three options appear to be available: 1) remove all hazardous waste from the tank during periods of incinerator maintenance, 2) install a back-up enclosed combustion device (e.g., a fume incinerator) to be used when the primary incinerator is out of service, or 3) seek a site-specific alternative from the regulators.

In one situation, the third option was chosen, and EPA specified that a back-up activated-carbon system could be used that is operated at all times when the incinerator is out of service and hazardous waste remains in the tank. EPA's justification for this approach was as follows: "The additional control of air emissions during the 240 hours of allowed maintenance (as compared to no control) would help offset any loss in emission control efficiency (enclosed combustion devices versus activated carbon) occurring during the ten or so additional days during which the incinerator is not operating due to planned maintenance. The closed-vent system and back-up carbon system must be designed and operated in accordance with the requirements of §264.1087." [RO 14288]

Q *A grinder is located immediately prior to slurry tanks in a incinerator feed preparation system. Is the grinder subject to Subpart CC?*

A Yes. The grinder would be considered a Subpart X miscellaneous unit, which would be subject to Subpart CC. In most instances, grinders meet the definition of a tank and could meet either Level 1 or Level 2 tank standards. [RO 14392]

Q *Can a single pressure-relief valve be installed on the pipe header that manifolds a series of Level 1 hazardous waste tanks together?*

A Yes. A single pressure-relief valve can be installed on a manifold vent header for Level 1 tanks. [§§264.1084(c)(2)(iii)/265.1085(c)(2)(iii)]

15.3.5 Control standards for containers

First, remember that a number of exemptions exist (see Section 15.3.1.2.1) that may keep a hazardous waste container from having to comply with Subpart CC. For example, satellite accumulation containers, containers at conditionally exempt small quantity generator and small quantity generator facilities, and containers with capacities ≤26.4 gallons (0.1 m^3) are all exempt from Subpart CC controls. Additionally, RCRA-empty containers (see Section 9.5) are not subject to these standards.

If no exemptions apply, EPA divides Subpart CC-regulated containers into three groups, Level 1, Level 2, and Level 3. Differentiating among the three levels is based on the size of the container, vapor pressure of the constituents in the hazardous waste, and whether waste stabilization is occurring. Table 15-6 is a matrix for determining the applicable Subpart CC container level. Each level has different emission control requirements, as described in the container standards in §§264.1086/265.1087.

Note that containers include transport vehicles, such as tank trucks and rail cars. Thus, the Subpart CC container standards apply to these units if they are in hazardous waste service.

Table 15-6: Matrix for Determining Subpart CC Level 1, 2, or 3 Container Controls

Container design capacity[1]	Container in light material service?[2]	Stabilization occurring in container?	Subpart CC level of control
≤26.4 gallons (0.1 m^3)	NA	NA	Container is exempt from Subpart CC
>26.4 gallons (0.1 m^3) and ≤121 gallons (0.46 m^3)	Yes	Yes	Level 3
	No	No	Level 1
	No	Yes	Level 3
	Yes	No	Level 1
>121 gallons (0.46 m^3)	Yes	Yes	Level 3
	No	No	Level 1
	No	Yes	Level 3
	Yes	No	Level 2

[1] Capacities specified as gallons are approximate; the Subpart CC regulations specify capacities in m^3. Basis is design capacity, not fill level, of the container.

[2] "*In light material service* means the container is used to manage a material for which both of the following conditions apply: The vapor pressure of one or more of the organic constituents in the material is greater than 0.3 kilopascals (kPa) at 20°C; and the total concentration of the pure organic constituents having a vapor pressure greater than 0.3 kPa at 20°C is equal to or greater than 20 percent by weight." [§265.1081]

Source: Adapted from EPA/530/F-98/011.

15.3.5.1 Level 1 containers

As noted in Table 15-6, Level 1 containers consist of two types of hazardous waste units. First, they are small-capacity containers (26–121 gallons) in which waste stabilization is not occurring; thus, all 55-gallon drums are Level 1 containers for the purposes of Subpart CC (as long as no waste stabilization is being conducted).

Additionally, larger capacity containers (>121 gallons) used to manage hazardous wastes having low vapor pressures (i.e., the containers are not "in light material service") are also Level 1 containers for Subpart CC purposes, as long as no waste stabilization is being conducted. Typical examples of such Level 1 containers are roll-off boxes containing hazardous soil.

15.3.5.1.1 Compliance options

Since these units have a low potential for emitting organics to the atmosphere, the control requirement for Level 1 containers is very simple: keep the containers closed. Three container design/closure options are available for Level 1 containers.

Option 1—*DOT-compliant shipping container*

This option requires the container to be designed, constructed, and closed in accordance with all DOT regulations in 49 *CFR* Part 178 or 179 for shipping containers. The standards applicable to this option [§§264.1086(f)/265.1087(f)] note that no exceptions under 49 *CFR* Part 178 or 179 regulations are allowed for DOT containers except for lab packs meeting the exceptions for combination packagings in 49 *CFR* 173.12(b).

People often ask if they can use 55-gallon drums that formerly held product to store hazardous waste and ship it offsite. The DOT regs at 49 *CFR* 173.12(c) allow this practice on a one-time basis and exempt such reuse from the drum reconditioning and reuse provisions in 49 *CFR* 173.28. However, if these containers are subject to Level 1 control, the DOT-compliant shipping container option cannot be used to comply with Subpart CC air emission standards. Why? Because the Subpart CC regulations do not allow any exceptions to the DOT standards except for lab packs. Thus, facilities can store hazardous waste in reused containers if they meet Part 264/265, Subpart I standards, they can ship these reused containers holding hazardous waste based on 49 *CFR* 173.12(c), but they cannot take advantage of the DOT-compliant shipping container option for complying with Subpart CC for Level 1 containers—one of the other two options discussed below will have to be used. In guidance, EPA noted the following:

> "EPA does not consider it appropriate that a container which is a 'non-reusable container (NRC)' or 'single-trip container (STC)' according to DOT requirements, be repeatedly used while at the facility site (i.e., emptied and refilled) for the handling of hazardous waste subject to Subpart CC rules. Before a DOT container can be reused, even within the boundaries of a facility site, it must comply with the DOT reconditioning and reuse provisions of the hazardous materials regulations in 49 *CFR* 173.28." [November 25, 1996; 61 *FR* 59947]

Option 2—*Covered container with no visible gaps or other openings*

Level 1 containers complying with Subpart CC using this option must have a tight-fitting cover such that, when closed, there are no visible holes, gaps, spaces, or other openings. The regulations say that the cover must form a continuous barrier over the container openings. The cover may be a separate cover installed on the container (e.g., a lid on a drum or a secured tarp on a roll-off box) or an integral part of the container (e.g., screw-type caps on portable tanks or bulk cargo containers).

This is actually the easiest of the Level 1 container compliance options. Since Part 264/265, Subpart I container standards have always required that we keep hazardous waste containers closed at all times except when adding or removing waste, Subpart CC has not added any additional burden for containers meeting Level 1 criteria.

***Option 3**—Open-top container with vapor barrier*

This option allows the container to be open, but an organic vapor-suppressing barrier (e.g., foam) must be applied over the hazardous waste surface. We have never met anyone who is complying with Subpart CC standards for Level 1 containers using this option.

15.3.5.1.2 Container loading/opening

There are no restrictions on how hazardous waste should be loaded into a Level 1 container. Once hazardous waste is in the container, it must be kept closed at all times except during waste addition or removal, waste sampling or depth measurement, maintenance of equipment inside the container, or other activities that need to be conducted during normal operations. [§§264.1086(c)(3)/265.1087(c)(3)]

The regulations noted above limit how long a container lid or other closure device can be open. The cover/closure devices must be in place if 1) waste will not be added or removed for 15 minutes, or 2) the person performing the loading or unloading leaves the immediate vicinity of the container, whichever occurs first.

Pressure-relief devices are allowed on Level 1 containers to prevent unsafe operating conditions, as long as they operate with no detectable organic emissions (<500 ppmv above background) when closed. [§§264.1086(c)(3)(iv)/265.1087(c)(3)(iv)]

15.3.5.1.3 Inspection and monitoring

Level 1 container inspection and monitoring requirements are found at §§264.1086(c)(4)/265.1087(c)(4) and are summarized as follows:

"Owners and operators of containers using either [Level 1 or Level 2] controls in accordance with the provisions of the rule are required to visually inspect the container and its cover and closure devices to check for defects at the time the owner or operator first manages a hazardous waste in the container or accepts possession of the container at the facility, with the exception of those containers emptied within 24 hours of being received. Also, in the case when a container used for managing hazardous waste remains at the facility for a period of 1 year or more, the container and its cover and closure devices are to be visually inspected to check for defects at least once every 12 months." [November 25, 1996; 61 *FR* 59948]

Most facilities make it a standard practice to inspect a container every time hazardous waste is added or removed to ensure that it is closed and the closure devices are in good condition.

Any defects found during the inspections must be repaired within 5 days after detection, and an initial attempt at repair must be made within 24 hours of detection.

15.3.5.1.4 Recordkeeping and reporting

The only Subpart CC recordkeeping/reporting requirement associated with a Level 1 container is data showing that larger capacity containers (>121 gallons) qualify for Level 1 (i.e., the container is not in light material service)—see §§264.1086(c)(5)/265.1087(c)(5). This recordkeeping requirement applies only if the container is not a DOT-compliant shipping container.

15.3.5.2 Level 2 containers

As noted in Table 15-6, Level 2 containers consist of larger capacity containers (>121 gallons) that are "in light material service" with no waste stabilization being conducted.

15.3.5.2.1 Compliance options

Since these units have a higher potential for emitting organics to the atmosphere, the control requirement for Level 2 containers is more complicated. Three container design/closure options are available for Level 2 containers.

***Option 1**—DOT-compliant shipping container*

This option (which is the same as for Level 1 containers) requires the container to be designed, constructed, and closed in accordance with all DOT regulations in 49 *CFR* Part 178 or 179 for shipping containers. The standards applicable to this option [§§264.1086(f)/265.1087(f)] note that no exceptions under 49 *CFR* Part 178 or 179 regulations are allowed for DOT containers except for lab packs meeting the exceptions for combination packagings in 49 *CFR* 173.12(b).

***Option 2**—Container with no detectable emissions*

Subpart CC compliance for a Level 2 container can also be achieved by filling it with hazardous waste, closing it, and then sniffing it with a portable VOC analyzer in accordance with Method 21 in Appendix A of 40 *CFR* Part 60. If no detectable organic emissions are found (<500 ppmv above background), the container meets the air emission control standards. The portable analyzer should be used to check every closure device and pressure-relief valve. [§§264.1086(g)/265.1087(g)] Although there is no requirement for periodic Method 21 leak monitoring of containers, we have typically seen that, if the same container is repeatedly filled with hazardous waste and then emptied, the Method 21 test is conducted each time it is full.

***Option 3**—Container that is vapor tight*

The third option for a Level 2 container is to conduct an annual pressure tightness test per Method 27 in Appendix A of 40 *CFR* Part 60, as modified by §§264.1086(h)/265.1087(h).

15.3.5.2.2 Container loading/opening

The regulations require hazardous waste to be transferred into or out of a Level 2 container in a manner that will minimize hazardous waste exposure to the atmosphere. Examples of such loading are:

- Submerged-fill lines for loading liquids into containers,
- Vapor-balancing or vapor-recovery systems to collect and control vapors displaced from the container during loading, and
- Fitted openings in the top of containers that allow waste loading and subsequent purging (e.g., with nitrogen gas) of the transfer line before disconnecting it from containers.

Once hazardous waste is in the container, it must be kept closed at all times except during waste addition or removal, waste sampling or depth measurement, maintenance of equipment inside the container, or other activities that need to be conducted during normal operations. [§§264.1086(d)(3)/265.1087(d)(3)]

The regulations noted above limit how long a container lid or other closure device can be open. The cover/closure devices must be in place if 1) waste will not be added or removed for 15 minutes, or 2) the person performing the loading or unloading leaves the immediate vicinity of the container, whichever occurs first.

Pressure-relief devices are allowed on Level 2 containers to prevent unsafe operating conditions, as long as they operate with no detectable organic emissions (<500 ppmv above background) when closed. [§§264.1086(d)(3)(iv)/265.1087(d)(3)(iv)]

15.3.5.2.3 Inspection and monitoring

The inspection and monitoring requirements for Level 2 containers under Subpart CC are the same as for Level 1 containers noted in Section 15.3.5.1.3.

15.3.5.2.4 Recordkeeping and reporting

There are no recordkeeping/reporting requirements associated with Level 2 containers. However, if someone is using Option 3 (vapor-tight container) to comply with Subpart CC, we recommend that they document the annual pressure tightness test.

15.3.5.3 Level 3 containers

If a facility wants to stabilize a hazardous waste in a container that has a capacity >26.4 gallons, any time the container is open/vented, it must either be 1) vented directly through a closed-vent system to a control device, or 2) located in an enclosure which is vented to a control device. Under the Level 3 standards [§§264.1086(e)(1–2)/265.1087(e)(1–2)], specifications are established for the control device and the enclosure.

Per §265.1081, waste stabilization includes mixing hazardous waste with binders or other materials and curing the resulting hazardous waste and binder mixture. Waste stabilization doesn't include addition of absorbent materials to the surface of a waste to absorb free liquid if no mixing, agitation, or curing occurs. Stabilization is generally not occurring if absorbent is added to the surface of the waste (if no mixing occurs) at the end of a work day or at the completion of a waste transfer (see Section 7.5.3).

However, the addition of absorbent to waste material while it is being transferred from one container into another will likely involve mixing of the absorbent into the waste; therefore, this activity is considered waste stabilization and requires Level 3 container controls. [December 8, 1997; 62 FR 64651]

EPA Region 7 has prepared a document entitled *RCRA Subpart CC Guidance Module for Container Level 3 Controls.* This guidance is available at http://www.trainex.org/web_courses/subpart_x/EPA CD Content/TopicSearch.htm.

15.3.5.3.1 Container loading/opening

The Subpart CC regulations require hazardous waste to be loaded into a Level 3 container in the same manner as discussed above (Section 15.3.5.2.2) for Level 2 units.

Safety devices, such as pressure-relief valves, may be installed and operated as necessary to prevent unsafe conditions on any container, enclosure, closed-vent system, or control device used to comply with the Level 3 container requirements. [§§264.1086(e)(3)/265.1087(e)(3)]

15.3.5.3.2 Inspection and monitoring

Any closed-vent system and control device associated with a Level 3 container under Subpart CC must be inspected and monitored per §§264.1086(e)(4)/265.1087(e)(4).

15.3.5.3.3 Recordkeeping and reporting

Recordkeeping requirements associated with Level 3 containers consist of [§§264.1089(d)/265.1090(d)]:

1. Closed-vent system and control device documentation per §§264.1089(e)/265.1090(e), and

2. Records that show that the enclosure meets the criteria in Procedure T under 40 CFR 52.741, Appendix B.

15.3.5.4 Container examples

The following examples illustrate container management under Subpart CC.

Q *How can roll-off boxes containing hazardous waste meet Subpart CC provisions?*

A If the container is *not* in light material service (i.e., the waste has a fairly low vapor pressure), then the roll-off box is a Level 1 container for Subpart CC purposes. This would be typical for hazardous soil in a roll off. We're not DOT experts, but our understanding is that a roll-off box is a bulk packaging that is authorized for the transportation of low-hazard solid materials per 49 CFR 173.240(c). That section of DOT's hazardous materials regulations identifies "sift-proof non-DOT specification portable tanks and closed bulk bins" as authorized for such transportation, and it is our understanding that DOT considers a tarped roll-off box to meet this definition. However, because they are non-DOT specification units, there are no specs in 49 CFR Parts 178 or 179 for the construction of these boxes. EPA's guidance in this matter is clear:

> "A container not subject to 49 CFR Part 178 or 179 is thus not eligible to comply with the Subpart CC rule through the requirements of [the DOT-compliant container option]." [December 8, 1997; 62 FR 64653]

Thus, the box will have to meet Option 2 or 3 for Level 1 containers noted in Section 15.3.5.1.1 above. [RO 14127] This would likely mean placing a tarp over the box as allowed in Option 2. "If a roll-off container is not in light material service, then use of a tarp with no visible holes or gaps or open spaces…is an example of a suitable Level 1 control device. However, use of tarps in this instance…requires closure suitable to weather conditions, including exposure to wind, moisture and sunlight." [RO 14826]

If the roll-off box is in light material service, then it is a Level 2 container. Again, Subpart CC cannot be met using the DOT-compliant shipping container option, so Options 2 or 3 in Section 15.3.5.2.1 will have to be employed. Unless the roll-off box is supplied with metal lids and gasketed covers, the Method 21 sniff test in Option 2 will be hard to meet. "The use of a tarp [on a roll-off box] would not be an acceptable Level 2 control device." [RO 14826]

It may not be possible to pass a pressure tightness test with a roll-off box. Case Study 15-3 includes additional guidance on this topic.

 Do Subpart CC container standards apply to vacuum trucks?

Case Study 15-3: Subpart CC Issues for Solvent-Contaminated Debris in a Roll-Off Box

How do the Subpart CC air emission standards apply to hazardous waste managed in roll-off boxes? Consider the following scenario: rags and paint filters contaminated with spent solvents are to be discarded. These solid wastes, which are determined by the generator to be hazardous, are managed in a roll-off box. Because the facility is a large quantity generator, the roll-off is a 90-day container subject to §262.34(a)(1)(i). This section of the regs requires 90-day containers to comply with the Subpart CC air emission standards. What are the Subpart CC implications for such a container of hazardous waste?

Determining if the box is Level 1 or 2

Compliance standards for a roll-off box managing hazardous waste depend on whether the box meets Subpart CC Level 1 or 2 criteria. The roll-off box will be a Level 2 container if it is "in light material service." That term is defined in §265.1081 as a container managing material in which the total concentration of pure organic constituents with a vapor pressure greater than 0.3 kPa at 20°C is greater than 20% by weight. To calculate the 20%-by-weight factor, the weight of all of the organic constituents with pure vapor pressures greater than 0.3 kPa at 20°C should be divided by the total weight of the waste (i.e., the rags and paint filters). If a container is not "in light material service" and waste is not being stabilized in the container, the Level 1 container standards apply.

The determination of "in light material service" is made for the waste managed in the container—not necessarily the waste at its point of generation. Thus, the determination would be made on the waste "as it is in the container." [RO 14568]

Subpart CC compliance for a Level 1 roll-off box

Based on the above discussion, the generator should be able to determine whether the box is "in light material service." If the roll-off box is not in light material service, it will be a Level 1 container. The Subpart CC regulations at §265.1087(c)(1)(ii) allow such a container to be equipped with a cover that forms a continuous barrier over the container openings such that when the cover is secured in the closed position, there are no visible holes, gaps, or other open spaces into the interior of the container. An example would be a suitably secured tarp on the box. An integral sliding top on the box, supported by roller bearings and running within a fixed track such that it achieves a no-visible-gap closure, would also be acceptable, providing it meets the standards in §265.1087(c)(1–2). [RO 14568]

Subpart CC compliance for a Level 2 roll-off box

Although not discussed in RO 14568, if the roll-off box is in light material service, it must meet the requirements for a Level 2 container; Subpart CC standards for a Level 2 container, as detailed in Section 15.3.5.2.1, are more stringent. Since many roll-off boxes are not constructed to DOT specifications, and the other two options are difficult to meet for such containers, compliance with Subpart CC for a roll-off box in Level 2 service will be hard to achieve. "The use of a tarp [on a roll-off box] would not be an acceptable Level 2 control device." [RO 14826] Other containers that meet DOT specs may be a better management option.

A No. Although Subpart CC applies to containers, including transport vehicles, EPA is not sure that it is possible to control the organic emissions from a vacuum truck and has therefore concluded that vacuum trucks are not regulated under these standards. [EPA/453/R-94/076b] The agency also notes "[w]ith regard to vacuum trucks, EPA has always intended the Subpart CC final rules to allow containers to vent emissions directly to the atmosphere during filling operations. This would include use of a vacuum system to fill a tank truck (i.e., a container under RCRA).... EPA intended to allow venting during waste transfer operations, either through the opening through which the waste is transferred, or through a second opening that would serve as a vent.... The fact that EPA is not requiring control of vacuum trucks is also discussed in [EPA/453/R-94/076b] where it is clear that EPA is fully aware that a practical means of controlling the exhaust from the vacuum pump on a vacuum truck has not been demonstrated." [December 8, 1997; 62 FR 64651]

15.3.6 Control standards for surface impoundments

Few facilities knowingly operate hazardous waste surface impoundments these days, but there are still a few out there. If the impoundment is receiving hazardous waste that at the point of waste origination contained ≥500 ppmw VO, it is subject to Subpart CC emission controls (unless exempted per one of the items in Section 15.3.1.2.1 or unless the waste has been treated to reduce VO as explained in Section 15.3.3.5).

The two compliance options under Subpart CC for surface impoundments will be difficult and expensive. They include [§§264.1085/265.1086]:

1. A floating membrane cover, providing a continuous barrier (no visible cracks, holes, or gaps) over the entire surface area of the liquid in the impoundment. The cover has to be high-density polyethylene with a 2.5-mil minimum thickness or an alternative material providing equivalent protection. [§§264.1085(c)/265.1086(c)]

2. A nonfloating cover that is vented through a closed-vent system to a control device. This would typically consist of an air-supported structure (again with no visible cracks, holes, or gaps), with the air being vented to a control device. It could also be a rigid cover. [§§264.1085(d)/265.1086(d)]

Either of the above options would be very expensive to implement for a large hazardous waste surface impoundment. Thus, owners/operators of these units typically choose to treat the hazardous waste for VO reduction before it enters the unit. As noted in Section 15.3.3.5, treating wastes for VO reduction obviates the need for VO emission controls on the surface impoundment. [§§264.1082(c)(2–4)/265.1083(c)(2–4)]

Where a cover is being used, facilities are required to visually inspect the impoundment cover and closure devices to check for defects at least once every 12 months. Any defects found during the inspections must be repaired within 45 days after detection, and an initial attempt at repair must be made within 5 days of detection. A repair may be delayed beyond 45 days if it requires emptying the impoundment or removing it from service and no alternative capacity is available at the facility.

Documentation describing the surface impoundment cover must be maintained in the facility's files; this includes a certification that the cover meets the specifications in §§264.1085/265.1086. Additionally, records of all annual inspections must be maintained at the facility for at least three years from the date of inspection. Finally, if the cover consists of an air-supported or rigid cover vented to a control device, closed-vent system and control device documentation per §§264.1089(e)/265.1090(e) is required.

15.3.7 Recordkeeping and reporting

Specific recordkeeping/reporting requirements for tanks, containers, and surface impoundments have been discussed in the previous sections. The only difference between the Subpart CC air emission standards for tanks, containers, and surface impoundments in Part 264 versus Part 265 is that there are no reporting requirements in Part 265. Thus, there are no specific reporting requirements for interim status or large quantity generator facilities under Subpart CC, although there are a number of records that must be maintained at the facility, as discussed previously. Permitted facilities must report the following to EPA [§264.1090]:

1. Each occurrence when hazardous waste is placed in a tank, container, or surface impoundment exempted from using air emission controls per §264.1082(c) that is no longer complying with the exemption. The report must be submitted within 15 days of the time the owner/operator becomes aware of the noncompliance.

2. A Level 1 hazardous waste tank that is no longer complying with the Level 1 requirements. The report must be submitted within 15 days of the time the owner/operator becomes aware of the noncompliance.

3. A control device that is not complying with standards for a continuous period of at least 24 hours in any 6-month period or a flare that operates with visible emissions for at least 5 minutes in any 2-hour period. The report is required semiannually, unless no noncompliance as noted above was observed.

Inspections and Enforcement

Avoid RCRA violations with regular training and vigilant compliance procedures

Despite the best efforts of environmental personnel, noncompliance with federal and/or state hazardous waste regulations periodically occurs. This chapter identifies common RCRA violations that have resulted in notices of violations (NOVs) and enforcement actions following EPA and/or state inspections of facilities subject to RCRA. We present this information in the hopes that it will help our readers avoid RCRA compliance problems. This chapter also includes a section (16.6) on how to prepare for an inspection and a section (16.7) on "cradle-to-grave" liability.

16.1 Common RCRA violations

Although we haven't found any nationwide statistics on the most common RCRA violations, the state of Colorado provided us with a summary of its findings. Table 16-1 identifies the most common RCRA violations in Colorado, for both large and small quantity generators of hazardous waste, from 2003 through 2012. We believe that these lists are representative of what happens in other states as well. Each RCRA violation is discussed in more detail in the subsections that follow.

Table 16-1: Most Common RCRA Violations in Colorado

No.	Large quantity generators	Small quantity generators
1	Training violations	Training violations
2	Contingency plan violations	Labeling violations
3	Labeling violations	Inadequate emergency response provisions
4	Failure to make hazardous waste determinations	Failure to make hazardous waste determinations
5	Open containers	Open containers
6	Land disposal restrictions violations	Used oil violations
7	Satellite accumulation violations	Satellite accumulation violations
8	Improper tank management	Recordkeeping problems
9	Preparedness and prevention violations	Land disposal restrictions violations
10	Used oil violations	Failure to perform weekly container inspections

Source: Colorado Department of Public Health and the Environment (2003–2012).

16.1.1 Training violations

We believe facilities typically receive NOVs for either 1) not providing adequate training to their personnel, or 2) more commonly, not having documentation showing that their people received training (even if they actually did). The RCRA training requirements are codified in §264.16 (for permitted TSD facilities), §265.16 (for interim status TSD facilities and large quantity generators), and §262.34(d)(5)(iii) (for small quantity generators). Conditionally exempt small quantity generators have no RCRA training obligations.

Our discussion of training issues is contained in Section 6.4 of this book. As noted in that section, the regulations are not very specific with respect to training program content. However, they are very specific with respect to the training records that must be maintained at a facility. Therefore, most RCRA training violations stem from failure to comply with the recordkeeping requirements, rather than the content of the training program itself.

We get a call about once a month from a customer who is expecting a state or EPA inspection and can't find the certificate of attendance that he or she received at one of our RCRA training seminars. These people know that, no matter what else may be inspected, training records are a "for-sure" item (so we email them a copy).

To minimize compliance problems associated with training, establish a bulletproof training records retention program to meet the requirements of §§264/265.16(d–e). This will go a long way towards eliminating the number one source of NOVs for noncompliance with the RCRA regs. See Section 6.4.1.5 for help in this area.

We are aware of at least one case where, based on collected training logs and employment records, EPA inspectors found that no RCRA-trained personnel were employed during certain time periods when the facility was shipping hazardous waste offsite. To remedy such situations, some facilities have successfully coordinated RCRA training and records retention with their human resources department. In doing so, the human resources staff and environmental department personnel can work towards ensuring no gaps in trained personnel result as people transfer within or leave an organization.

Another common trigger for NOVs related to training is when an EPA or state inspector observes noncompliance with some other aspect of the RCRA regs. In this case, they will often check to see if the responsible personnel were trained in that area. For example, if hazardous waste containers are left open or are not properly labeled, in addition to issuing an NOV for those violations, inspectors may also want to review the training program for facility personnel responsible for those activities. If the training program does not adequately address container management and labeling, this could be the basis for another NOV. To help prevent such enforcement issues, review the top ten RCRA violations for your generator class in Table 16-1, and make sure these activities are covered adequately in the training programs for the appropriate personnel.

16.1.2 Contingency plan violations

The requirement for large quantity generators to have a contingency plan is in §262.34(a)(4), which references compliance with Part 265, Subpart D. Small quantity generators are subject to the less-stringent provisions in §262.34(d)(5), which are discussed in Section 16.1.11. TSD facilities are also required to have contingency plans. [See Part 264/265, Subpart D.] One of the most common reasons for NOVs associated with contingency plans at large quantity generator and TSD facilities is that the emergency coordinators' information (names, phone numbers, addresses, etc.) in the contingency plan is out of date. [§§264/265.54(d)] Keeping this information up to date is a challenge, because it changes periodically, but it's important and something inspectors check carefully. Incomplete contingency plans also regularly result in NOVs.

Another difficulty with contingency plans is that they must be reviewed/coordinated with local authorities (police, fire departments, hospitals, and emergency response teams) and sometimes this doesn't happen. Sections 264/265.52(c) require the contingency plan to describe arrangements and coordination between the facility and these local authorities. This review might have been done initially, but any significant changes to a facility's operations that cause the contingency plan to be revised should also be brought to the attention of the local authorities, and their review of the revised contingency plan should be conducted when the revised plan is sent to them per §§264/265.53(b). [RO 14832] In the rare event that the local authorities will not accept the contingency plan (or revisions to it) from your facility, document your attempt to provide them these materials. You may also want to seek guidance from your state or EPA regional office on how to proceed.

An area that inspectors are starting to review more carefully is the evacuation routes/procedures in the contingency plan. Inadequate discussions and drawings of evacuation routes/procedures and/or inadequate evacuation drills may result in an NOV.

Finally, there has been a trend in recent years towards developing an integrated facility contingency plan, covering all regulatory (environmental, health, and safety) requirements. These so-called "one plans" can be a good idea, but they are sometimes written by non-RCRA personnel who don't fully understand the RCRA contingency plan requirements. Thus, a state RCRA or EPA inspection of the resulting document may uncover noncompliance with the Part 264/265, Subpart D or §262.34(d)(5) provisions.

Section 6.3.8 summarizes other contingency plan requirements under RCRA for both large and small quantity generators.

16.1.3 Labeling violations

By far the most common labeling violations are hazardous waste accumulation tanks or containers at generator facilities not being labeled with the words "Hazardous Waste" or containers not being labeled with the 90- or 180-day accumulation start date. These requirements are found in §262.34(a)(2–3) of the RCRA regulations [small quantity generators get there by reference in §262.34(d)(4)].

It's not too difficult to comply with the RCRA labeling requirements for an accumulation tank; just put a big "Hazardous Waste" sign or label on the tank and track the start and end of each accumulation period in a logbook. As noted in Section 6.3.6.3.2, however, EPA's guidance says that accumulation start dates should be marked on tanks as well as containers.

The vast majority of NOVs in this area are for improper/missing container labeling. Make sure your hazardous waste training and management procedures get your people in the mindset that 90- or 180-day containers of hazardous waste must be labeled "Hazardous Waste" and have the 90- or 180-day accumulation start date on them. To further the labeling mindset among employees and avoid labeling problems, some facilities have successfully instilled RCRA labeling awareness in their maintenance department. In doing so, when maintenance personnel repair, replace, wash, or paint a hazardous waste tank or container, proper RCRA labeling becomes a shared responsibility. Section 9.3 summarizes the federal RCRA labeling requirements for hazardous waste containers.

One of the problems that crops up is that many 90- or 180-day accumulation containers often begin their lives as satellite accumulation containers. The labeling requirements for satellite containers are different from those for 90- or 180-day containers. For example, satellite containers do not have to be labeled "Hazardous Waste"; they can be labeled with other words that identify the contents of the containers (but they must have a label on them). Additionally, these containers are usually dated when the 55-gallon limit is exceeded (for nonacute hazardous waste). That date of excess satellite accumulation is not necessarily the same date as the start of the 90- or 180-day accumulation period, depending on how long it takes (up to 3 days) for personnel to move it into the 90- or 180-day accumulation area. Therefore, the container may have to be relabeled when it is moved into the 90- or 180-day accumulation area. A "Hazardous Waste" label is required, as well as the 90- or 180-day accumulation start date marking, when that container enters the 90- or 180-day accumulation area. However, some facilities use the date of excess satellite accumulation as the start of the 90- or 180-day clock.

16.1.4 Failure to make hazardous waste determinations

As stated early and often in this book, anyone who generates a solid waste is required by RCRA to determine if it is a hazardous waste (see §262.11). Yet when state or EPA personnel are inspecting a site, they often find drums or buckets of "stuff" in production areas, wastewater treatment plants, warehouses, maintenance areas, behind buildings, and along fence lines that nobody at the facility seems to know anything about. These containers often have no labels on them or the labels that are on them do not reflect the container contents. Such "orphan" materials are usually unneeded or out-of-date products (e.g., paint, cleaners) or wastes that someone at the plant (including contractors) wants to get rid of, but they haven't followed plant procedures for characterizing and storing the waste materials. Usually in these cases, the inspector will try to determine whether the material is a solid waste; if it is, and at the discretion of the inspector, such a facility can be cited for not making a hazardous waste determination for the solid waste. Numerous enforcement cases and significant monetary penalties have resulted from these situations, as noted in Section 16.2.2.

In a case with which we are familiar, a state inspector observed a facility managing a solid waste stream as nonhazardous. The inspector asked to see the waste analysis data supporting the nonhazardous waste determination, but the facility responded that they had made a knowledge-based determination that the material was not hazardous. When the inspector then asked to see the information used to make the knowledge-based determination, the facility couldn't produce anything. As a result, the inspector cited the facility for not making a hazardous waste determination for the solid waste stream. (The facility subsequently had the material analyzed and it was indeed nonhazardous, but the NOV for not making a hazardous waste determination was still valid.) Thus, whenever knowledge-based determinations are made, they should be documented (see Section 6.7.3.2).

Ensure that procedures have been established for your facility that detail what should happen to materials that are not needed anymore—who should be contacted, how should they be characterized if they are solid waste, to where should they be moved, who should label them, etc. You should also periodically train your people and remind them of the potentially costly problem they could create by moving a drum of stuff out of their area to make it someone else's problem—it could wind up being everyone's problem when an inspector spots it.

Finally, all hazardous waste determinations for each solid waste generated at a facility should be documented. Whether they are made via testing or knowledge, Section 6.7.3 details the hazardous waste determination recordkeeping requirements.

16.1.5 Open containers

Even at large facilities with good environmental programs in place, it is not uncommon to find open hazardous waste containers during times when no waste is being added or removed. It just seems to be human nature. This is a violation. [§§264/265.173(a) as referenced from §262.34(a)(1)(i) for large quantity generators and §262.34(d)(2) for small quantity generators] Satellite accumulation containers must also be closed per §262.34(c)(1)(i), which references §265.173(a). Train your facility personnel not to leave hazardous waste containers open and continually reinforce this point with signs or other means. Section 9.2 includes a discussion of what EPA means by a "closed" container.

16.1.6 Land disposal restrictions violations

The land disposal restrictions (LDR) in Part 268 of the RCRA regulations are so complex, that a myriad of noncompliance issues could crop up. From a hazardous waste generator's perspective, however, the most common source of violations is LDR paperwork. In many cases, the only way that regulators can determine if a generator is in compliance with LDR requirements is by examining copies of the paperwork (notifications and certifications) required by the regulations. If a generator's LDR paperwork is in order, the likelihood that the facility will encounter LDR enforcement problems is greatly reduced.

Typical LDR paperwork problems to watch out for include:

1. The LDR form is unavailable for a manifested waste shipment as required in §268.7(a)(8).

2. The LDR form does not accurately reflect the shipped waste. For example, all of the appropriate hazardous waste codes have not been included—see Section 13.6.

3. The LDR form does not list applicable underlying hazardous constituents for characteristic wastes (this is especially common for small facilities that have the TSD facility complete the LDR notification). Even if a contractor supplies and completes an LDR form for a generator, the accuracy and completeness of the form are the generator's responsibility.

4. For hazardous debris, the LDR notification does not identify the contaminants subject to treatment.

5. The wrong LDR form is used, or where multiple forms (i.e., notifications or certifications) are needed for a single shipment, a required form is omitted.

6. The one-time notice required by §268.7(a)(7) for wastes that are excluded from regulation subsequent to their point of generation is missing from the facility's records.

A discussion of how to comply with the LDR paperwork requirements is included in Section 13.12.

16.1.7 Satellite accumulation violations

There are several potential sources of violations associated with the satellite accumulation regulations in §262.34(c). Perhaps the most common NOVs are for open satellite containers. Section 9.2.1 covers recent EPA guidance on keeping satellite units closed. Personnel training on this topic is vital for compliance.

Another likely cause of noncompliance is accumulating more than 55 gallons of nonacute hazardous waste in a satellite area (e.g., having more than one 55-gallon drum in a SAA). RO 14826 is clear that "[w]hile most facilities will have a SAA with only one container (usually a 55-gallon container), some facilities may generate multiple waste streams in a SAA and require more than one container, although they cannot accumulate more than 55 gallons total regardless of the number of containers."

The dating requirement associated with satellite containers is often misunderstood and misapplied. Not dating the container when "in excess of" 55 gallons of nonacute hazardous waste is accumulated or not moving the excess amount to another location (e.g., a 90- or 180-day accumulation area) within 3 days are common problems. See Section 6.2.4 for EPA guidance in this area.

Finally, defending your claims to your state or EPA inspectors that your satellite containers are "at or near any point of generation" and "under the control of the operator of the process generating the waste" can be a constant struggle. Interpretation of these terms can be a state-specific or even inspector-specific determination; EPA guidance on these requirements is in Section 6.2.2.

16.1.8 Improper tank management

Tanks used for hazardous waste management are subject to Part 264/265, Subpart J standards. Large quantity generators operating 90-day tanks get there by reference from §262.34(a)(1)(ii); small quantity generators' 180-day tanks are subject only to the less-stringent provisions of §265.201 per reference from §262.34(d)(3).

A number of requirements in Subpart J could get a facility in trouble, but the most likely are 1) inadequate or missing secondary containment for the tank and/or ancillary equipment, 2) failure to conduct and/or document daily inspections of the tank and ancillary equipment, and 3) noncompliance with the Subpart CC air emission standards. The primary considerations for each of these three potential problem areas are summarized in Sections 10.3, 10.5, and 15.3.4, respectively.

16.1.9 Preparedness and prevention violations

Large and small quantity generators and interim status TSD facilities are subject to the preparedness and prevention provisions of Part 265, Subpart C. Large quantity generators get there via reference from §262.34(a)(4)—small quantity generators from §262.34(d)(4). This subpart requires the following:

1. An accessible alarm/communication system capable of providing emergency instruction to facility personnel;

2. Accessible telephones or two-way radios capable of summoning assistance from local authorities (e.g., police, fire departments, and emergency response teams);

3. Fire extinguishers, water hose stations, automatic sprinklers, and other fire control equipment, spill control equipment, and decontamination equipment;

4. Testing/maintenance of the above equipment/systems to ensure they are operational;

5. Adequate aisle space to allow emergency/spill response;

6. Operation/maintenance procedures that minimize the possibility of fire, explosion, or spills; and

7. Coordination with local authorities on how emergencies will be addressed by both facility and offsite emergency personnel.

Inspections by state or EPA personnel typically uncover missing, inoperable, deteriorated, patched, jury-rigged, or otherwise inadequate equipment; inadequate aisle space; fire hazards; and/or no or out-of-date arrangements between the facility and local emergency responders. Because personnel safety may be compromised if these requirements are not met, preparedness and prevention issues should be a high priority at hazardous waste facilities. Agreements between the facility and emergency responders are discussed further in Section 6.7.7.

16.1.10 Used oil violations

Under the federal regulations, the used oil generator requirements in Part 279 boil down to storage in tanks or containers, marking such units with the words "Used Oil," and cleaning up any releases that may occur. (Of course, there are lots of details, as noted in Section 12.2.1.) Of these three basic requirements, the one that gets complied with the least is marking used oil storage containers with the words "Used Oil." It's a pretty simple requirement, but one that gets forgotten frequently. Every bucket and other small container holding used oil needs to be so labeled. Train your facility personnel on this issue and remind them of it frequently. Also, not promptly cleaning up small leaks from used oil collection tanks is a source of NOVs.

16.1.11 Inadequate emergency response provisions

Small quantity generators must comply with the emergency response provisions of §262.34(d)(5) in addition to the Part 265, Subpart C preparedness and prevention requirements discussed above. For these generators, the requirements in §262.34(d)(5) take the place of having a formal contingency plan. We believe that the requirements in §262.34(d)(5)(i–ii) are the ones that result in the most NOVs. Specifically, facilities often fail to 1) designate an emergency coordinator at the facility or have one on call, and/or 2) post the information required in §262.34(d)(5)(ii) next to plant telephones. These potential problems can be addressed fairly easily and inexpensively. (A sample emergency contacts telephone posting form is available from McCoy and Associates' website at http://www.mccoyseminars.com/forms.cfm.)

16.1.12 Recordkeeping problems

The RCRA regulations require numerous records to be maintained. Although this issue only made it into the top ten violations for small quantity generators, we suspect that large quantity generators are issued NOVs regularly for incomplete or missing RCRA records. The recordkeeping provisions are in §§262.40–262.43 for large quantity generators and §262.44 for small quantity generators. Conditionally exempt small quantity generators have almost no recordkeeping requirements. A summary of the RCRA recordkeeping requirements for the three generator classes is included as Table 6-6 in Section 6.7. You may want to review that section carefully to verify that the required records are being maintained by your facility and the record retention periods are being satisfied.

The recordkeeping and reporting requirements for TSD facilities are codified in Parts 264/265, Subpart E.

16.1.13 Failure to perform weekly container inspections

All facilities that accumulate hazardous waste in containers (except for conditionally exempt small quantity generators) must inspect their container accumulation areas weekly. [§§264/265.174 as referenced from §262.34(a)(1)(i) for large quantity generators and §262.34(d)(2) for small quantity generators] Inspections should be focused on proper labeling and closure and on any container leakage, deterioration, corrosion, or bulging. Satellite accumulation containers do not need to be inspected weekly, unless it is a state program requirement. [RO 14418, 14703]

Container accumulation area inspection logs are virtually certain to be scrutinized during a state or EPA inspection. Having these logs fully completed per the suggestions in Section 9.6 will help minimize NOVs in this area. A sample container inspection checklist is available from McCoy and Associates' website at http://www.mccoyseminars.com/forms.cfm.

16.2 RCRA enforcement

This section consists of two parts: a short summary of EPA's RCRA civil penalty policy (which the agency uses to calculate monetary penalties for noncompliance with the RCRA regulations), and a sampling of RCRA enforcement cases.

16.2.1 RCRA civil penalty policy

To impose consistent penalties for RCRA violations, EPA developed a civil penalty policy. The policy in use today was issued June 2003, and it is applicable to all administrative enforcement actions brought under RCRA. (Judicial cases do not necessarily use this policy.) There are four steps involved in calculating a penalty. Each step is briefly described below. (The policy may be downloaded from http://www2.epa.gov/sites/production/files/documents/rcpp2003-fnl.pdf.)

16.2.1.1 Step 1—Determine gravity-based penalty

The first step in developing a monetary penalty for RCRA noncompliance involves calculation of a gravity-based penalty—this portion is related to how serious the violation is. The gravity-based penalty is based on the potential for harm and extent of nonconformity with the regulations. The potential for harm is assigned one of three values: major, moderate, or minor. The major category applies where there is a substantial risk of exposure of human or environmental receptors to hazardous wastes or a substantial adverse impact on the RCRA regulatory program. For example, a large release of hazardous waste to soil and/or ground water would be considered major. Additionally, not using the exact wording specified in financial assurance documents could render them useless; this would also be a major violation. The moderate potential-for-harm category reflects significant risk of exposure or adverse regulatory impact. Inspecting container accumulation areas once or twice per month instead of once a week would be an example of moderate potential for harm. The minor potential-for-harm category reflects relatively low risk of exposure/impact; typing names in the manifest certification block instead of hand-written signatures would be minor potential for harm.

The extent of deviation from the RCRA regulations is also assessed as being major, moderate, or minor. For example, a permitted facility not having a closure plan would be a major deviation from the regs, having an incomplete closure plan would be a moderate deviation, and having a small omission in a decontamination step in the plan would be a minor deviation.

Once the potential for harm and extent of deviation from the regulations are known, the matrix shown in Table 16-2 is used to assess a gravity-based penalty *for Day 1 of the violation*. Selection of the exact penalty amount within the specified ranges in the matrix is at the discretion of enforcement personnel.

EPA increased the maximum civil penalty that may be assessed under RCRA authority to $37,500. [December 11, 2008; 73 *FR* 75340] Table 16-3 summarizes the maximum penalty amounts.

RCRA authorizes separate penalties for each distinct violation. Where multiple violations are alleged in a complaint, EPA may conduct a penalty calculation for each. In practice, however, the agency sometimes incorporates similar violations into one count for penalty calculation purposes.

16.2.1.2 Step 2—Calculate multiday penalty

Under RCRA, each day that a violation goes uncorrected can constitute a separate violation. Therefore, Step 2 involves calculating the additional penalty if a violation has gone on for more than one day. As with the gravity-based penalty, the potential for harm and extent of deviation from the regulations are assessed, and the matrix in Table 16-4 is used to establish a daily penalty. The dollar amount selected at the discretion of enforcement personnel is then multiplied by the number of days of violation minus one (i.e., Days 2 through the end of the violation) to determine the multiday component of the penalty.

Table 16-2: Matrix for Assessing Gravity-Based Penalty

Potential for harm	Extent of deviation from the RCRA regulations		
	Major	Moderate	Minor
Major	$37,500–$28,326	$28,325–$21,244	$21,243–$15,579
Moderate	$15,578–$11,330	$11,329–$7,082	$7,081–$4,249
Minor	$4,248–$2,124	$2,123–$708	$707–$142

Source: Amendments to EPA's Civil Penalty Policies to Implement the 2008 Civil Monetary Penalty Inflation Adjustment Rule, December 29, 2008.

Table 16-3: RCRA Statutory Maximum Civil Penalties

RCRA statute section number	Civil penalty description	Maximum civil penalty
3008(a)(3), 3008(g)	Violation of the RCRA Subtitle C hazardous waste program	$37,500 per day
3008(c)	Continued violation of a compliance order	$37,500 per day
3008(h)(2)	Violation of a corrective action compliance order	$37,500 per day
3013(e)	Failure to comply with an order to monitor for the presence or release of a hazardous waste	$7,500 per day
7003(b)	Failure to comply with an imminent and substantial endangerment order	$7,500 per day
9006(a)(3)	Violation of an underground storage tank compliance order	$37,500 per day
9006(d)(1)	Failure to notify state/local agency of the existence of an underground storage tank subject to RCRA Subtitle I	$16,000 per tank
9006(d)(2)	Violation of the RCRA Subtitle I underground storage tank requirements	$16,000 per tank per day

Source: Adapted from 40 *CFR* 19.4.

Table 16-4: Matrix for Assessing Daily Penalty for Multiday Violations

Potential for harm	Extent of deviation from the RCRA regulations		
	Major	Moderate	Minor
Major	$7,500–$1,417	$5,667–$1,060	$4,250–$780
Moderate	$3,114–$566	$2,268–$351	$1,417–$214
Minor	$851–$143	$423–$143	$143

Source: Revised Penalty Matrix for RCRA Civil Penalty Policy, July 17, 2009.

Violations that occurred for long periods of time are typically assessed multiday penalties for Days 2 through 180 only. However, at the discretion of enforcement personnel, multiday penalties for Days 181+ may be added.

Despite the matrix given in Table 16-4, EPA noted in its 2003 civil penalty policy that:

"While this policy provides general guidance on the use of multiday penalties, nothing in this policy precludes or should be construed to preclude the assessment of penalties of up to $27,500 [$37,500 starting January 12, 2009] for each day after the first day of any given violation. Particularly in circumstances where significant harm has in fact occurred and immediate compliance is required to avert a continuing threat to human health or the environment, it may be appropriate to demand the statutory maximum."

16.2.1.3 Step 3—Determine any adjustment factors

The multiday component of the penalty from Step 2 is added to the gravity-based penalty for Day 1 (from Step 1), and that gives a total proposed penalty amount. The following factors may be considered to adjust the proposed penalty:

- Good-faith efforts to bring the facility into compliance—The degree of cooperation and promptness in reporting the violation and/or instituting measures to remedy it will be evaluated by enforcement personnel. This factor may adjust the penalty up or down.

- Degree of willfulness and/or negligence—How much control the violator had over events leading up to the violation will be assessed, as will the possibility that the violator could reasonably have foreseen and taken precautions to prevent the violation. This factor may also adjust the penalty up or down.

- History of noncompliance—The similarity of the present violation to past RCRA noncompliance at the facility will be evaluated. This factor can adjust the penalty up only.

- Ability to pay—The financial capabilities of the violator will be considered when determining the penalty amount and/or payment schedule. For example, more harm to the environment may occur if the penalty bankrupts a company, rendering it unable to afford its environmental obligations (such as cleaning up a hazardous waste spill). However, EPA reserves the right to seek penalties that might put a company out of business. This factor can adjust the penalty down only.

- Implementation of supplemental environmental projects (SEPs)—If the violator agrees to implement beneficial SEPs at the facility in addition to bringing the facility into RCRA compliance, the penalty may be reduced. A SEP involves actions a violator agrees to undertake to protect human health and the environment in exchange for a monetary penalty reduction.

- Risk and/or cost of litigation—EPA may reduce a penalty if the violator will settle without going to litigation. However, if lengthy negotiations cause the violation to continue longer than anticipated, the penalty may be increased.

The adjustments discussed above are based on percentages. Typically, an adjustment percentage will be 25% per factor or less. The adjustment percentages from more than one factor are cumulative.

16.2.1.4 Step 4—Adjust for economic benefits of noncompliance

To keep violators from benefitting financially from not complying with the RCRA regulations, EPA's civil penalty policy requires that violators be forced to relinquish any such benefit. The objective is to eliminate any economic incentives for noncompliance because allowing violators to benefit from noncompliance places them at an unfair competitive advantage. Therefore, any calculated economic benefit is added to the gravity-based penalty determined in Step 1. (The adjustment factors discussed in Section 16.2.1.3 do not apply to the economic benefit component of the penalty.)

Typical areas of noncompliance that may financially benefit facilities are:

- Not making hazardous waste determinations,
- No or inadequate contingency plan,
- No or inadequate training program for hazardous waste facility personnel,
- No or inadequate ground water monitoring,
- No or inadequate financial assurance,
- Not providing closure or post-closure care,
- Improper land disposal of hazardous waste,
- Inadequate or delayed cleanup of hazardous waste discharges,
- No or inadequate Part B permit submittals, and
- Not meeting minimum technological requirements for land disposal units.

Both delayed and avoided costs are considered when determining the economic benefits of noncompliance. Delayed costs are capital expenditures that have been deferred as a result of the facility's failure to comply with regulatory requirements. Examples include failure to install ground water monitoring wells, failure to submit a Part B permit application, or failure to develop a waste analysis plan. Operating and maintenance costs avoided because of the facility's failure to comply with a RCRA requirement (e.g., failure to perform ground water monitoring, sampling, and analyses) would be considered avoided costs.

The agency uses a complex computer model called BEN to calculate economic benefits from delaying and/or avoiding pollution control expenditures and/or from any illegal competitive advantage obtained because of a violation. BEN addresses capital and operating costs and accounts for the time value of money with inflation adjustments and a discount/compound rate tailored to the time period from the violation up through the present. The BEN model is available for download at http://www2.epa.gov/enforcement/penalty-and-financial-models.

16.2.2 RCRA enforcement cases

Application of the RCRA civil penalty policy discussed in the previous section can be quite subjective. Different people may use the policy in similar situations and yet calculate significantly different penalties. Also, there are usually settlement negotiations conducted between the violator and EPA, which may result in a penalty that is significantly less than that initially proposed by the agency.

To illustrate application of the penalty policy, selected RCRA enforcement cases are presented in Table 16-5. These cases give you an idea of the consequences of RCRA noncompliance; they were taken from 1) annual accomplishment reports issued by EPA's Office of Enforcement and Compliance Assurance (OECA), 2) recent enforcement cases listed on OECA's website at http://www2.epa.gov/enforcement/cases-and-settlements, and 3) EPA regional websites.

16.2.2.1 EPA's right to overfile

As a final note related to recent RCRA enforcement cases, EPA's ability to overfile (bring action against a facility after a RCRA-authorized state has already settled a RCRA violation that, in the opinion of the EPA regional enforcement program, conflicts with state authorization agreements) is inconsistent throughout the country. In the states governed by the U.S. Court of Appeals for the Eighth Circuit (Arkansas, Iowa, Minnesota, Missouri, Nebraska, North Dakota,

Table 16-5: Selected RCRA Enforcement Cases

Case type	EPA region	Facility type	Alleged RCRA violation(s)	Penalty[1]
Civil	I	Hardware manufacturing	■ Failed to provide adequate hazardous waste training to facility personnel ■ Failed to properly label and close hazardous waste containers ■ Failed to provide adequate aisle space between containers ■ Failed to update and submit a revised contingency plan to local authorities	■ $39,705
Civil	I	Real estate developer	■ Failed to make hazardous waste determination for electroplating waste sludge ■ Improperly disposed of 212 tons of hazardous waste	■ $227,500
Civil	I	Chemical manufacturing	■ Diluted hazardous waste as a substitute for adequate treatment	■ $32,000
Civil	I	University	■ Failed to make hazardous waste determinations ■ Improper storage of hazardous waste ■ Improper storage of universal waste	■ $60,000
Civil	I	University	■ Improper storage of waste picric acid	■ $126,600
Civil	I	Commercial hazardous waste storage and transfer facility	■ Failed to adequately characterize waste ■ Failed to properly maintain hazardous waste tanks ■ Failed to provide adequate secondary containment ■ Improper storage of incompatible wastes	■ $650,000 ■ $1.06 million SEP
Civil	II	Recycler	■ Failed to notify and receive consent from EPA before exporting non-working computer monitors	■ $199,900
Civil	II	University	■ Failed to make hazardous waste determinations ■ Improper storage of hazardous waste, including open containers ■ Stored hazardous waste without a permit	■ $145,000
Civil	II	Petroleum product distribution terminal	■ Treated, stored, and disposed hazardous waste without a permit	■ $8.2 million ■ $3 million SEP
Civil	II	Chemical manufacturing	■ Failed to record hazardous waste tank inspections ■ Failed to conduct Subpart CC inspections ■ Failed to record maximum organic vapor pressures	■ $50,000
Civil	II	Hazardous waste TSD facility	■ Unpermitted storage of hazardous waste outside of facility ■ Storage of hazardous waste without adequate secondary containment ■ Improperly shipped non-empty hazardous waste containers back to generator	■ $250,000

Table 16-5: Selected RCRA Enforcement Cases

Case type	EPA region	Facility type	Alleged RCRA violation(s)	Penalty[1]
Civil	II	Public health agency (two facilities)	■ Failed to make hazardous waste determinations ■ Failed to properly label and manage hazardous waste containers ■ Failed to comply with preparedness and prevention requirements	■ $68,000
Civil	III	Aerosol packaging and filling	■ Failed to properly inspect, label, and close hazardous waste containers ■ Failed to provide adequate hazardous waste training to facility personnel; failed to maintain adequate training records, including job title and the name of the employee filling each job ■ Failed to obtain TSD facility signature on four hazardous waste manifests within 60 days of shipment ■ Failed to maintain adequate contingency plan	■ $25,000
Civil	III	Air National Guard base	■ Stored hazardous waste without a permit ■ Failed to provide adequate hazardous waste training to facility personnel ■ Failed to properly label hazardous waste containers ■ Failed to maintain proper records	■ $75,000
Civil	III	Specialty steel manufacturing	■ Stored hazardous waste without a permit ■ Failed to keep hazardous waste containers closed ■ Failed to perform weekly inspections of hazardous waste containers ■ Failed to provide hazardous waste training to facility personnel ■ Failed to properly label hazardous waste containers ■ Failed to comply with universal waste container labeling/marking requirements ■ Improperly disposed universal waste lamps ■ Failed to properly complete hazardous waste manifests ■ Failed to comply with preparedness and prevention requirements	■ $150,000
Civil	III	Wood treatment	■ Failed to properly label, date, and close hazardous waste containers ■ Failed to clean, maintain, and enclose a drip pad	■ $43,600
Civil	III	Chemical manufacturing	■ Failed to make a hazardous waste determination for a benzene-containing waste stream ■ Failed to comply with waste-specific LDR prohibitions for chlorotoluene production wastes ■ Disposed hazardous waste without a RCRA permit	■ $95,000 ■ Remedial cost (≥$25,000)

Table 16-5: Selected RCRA Enforcement Cases

Case type	EPA region	Facility type	Alleged RCRA violation(s)	Penalty[1]
Civil	III	Defense and aerospace	■ Operated a hazardous waste storage facility without a permit or interim status ■ Failed to maintain adequate contingency plan ■ Failed to meet new tank system design and installation requirements ■ Failed to conduct and record hazardous waste tank inspections ■ Failed to meet air emission standards marking requirements ■ Failed to meet air emission monitoring requirements for pumps in light liquid service ■ Failed to meet air emission standards test methods and procedures ■ Failed to meet air emission standards recordkeeping requirements	■ $325,000
Criminal	III	Chemical distribution	■ Illegally stored hazardous waste without a permit ■ Illegally exported hazardous waste without consent of the receiving country ■ Illegally shipped hazardous waste without manifests to unpermitted facilities	■ 6-mo in-home confinement for company executive ■ $200,000 fine ■ $1.8 million restitution
Criminal	III	Wood and metal components manufacturing	■ Stored hazardous waste without a permit	■ $200,000 fine ■ $250,000 SEP ■ 3 years probation for company employee
Civil	IV	Protective coatings	■ Failed to make hazardous waste determination ■ Failed to perform weekly inspections of hazardous waste containers ■ Failed to develop personnel training program ■ Improper management of hazardous waste containers ■ Failed to update and submit revised contingency plan to local authorities and first responders	■ $55,000
Civil	IV	Polyvinyl chloride manufacturing	■ Failed to comply with land disposal restrictions for wastes entering a surface impoundment ■ Disposed hazardous waste without a RCRA permit	■ $610,000 (multimedia settlement) ■ $2,900,000 SEP[2]
Civil	IV	Fertilizer manufacturer	■ Failed to make hazardous waste determinations ■ Treated, stored, and disposed hazardous waste without a permit ■ Failed to comply with LDR requirements ■ Failed to comply with recordkeeping requirements	■ $701,500

©2015 McCoy and Associates, Inc. McCoy's RCRA Unraveled

Table 16-5: Selected RCRA Enforcement Cases

Case type	EPA region	Facility type	Alleged RCRA violation(s)	Penalty[1]
Civil	V	Waste management	■ Failed to comply with preparedness and prevention requirements ■ Failed to properly manage hazardous waste tanks and secondary containment systems ■ Failed to comply with monitoring, inspection, and recordkeeping requirements	■ $20,000 ■ $69,771 SEP
Civil	V	Chemical manufacturing	■ Failed to immediately notify the National Response Center for spills of hazardous substances	■ $20,000
Civil	V	Chemical manufacturing	■ Failed to comply with secondary containment requirements for hazardous waste tanks ■ Failed to meet hazardous waste tank labeling, certification, and inspection requirements ■ Stored hazardous waste without a permit ■ Failed to maintain proper shipping records	■ $101,900
Civil	V	Paint manufacturing	■ Failed to comply with Subpart BB organic air emissions standards	■ $100,115
Civil	V	Die casting and plating	■ Failed to keep hazardous waste containers closed ■ Failed to perform weekly inspections of hazardous waste containers ■ Failed to provide tank overfill prevention controls and secondary containment systems ■ Failed to provide adequate aisle space between containers ■ Stored and disposed hazardous waste without a permit	■ $50,000 ■ $250,000 in ground water contamination cleanup
Civil	V	Furniture manufacturing	■ Failed to keep copies of manifests ■ Failed to provide hazardous waste training to facility personnel ■ Failed to maintain adequate contingency plan and an emergency coordinator familiar with all aspects of the contingency plan	■ $27,300
Criminal	V	Uranium processing	■ Stored radioactive mixed waste without a permit	■ $11.8 million ■ Five years probation
Criminal	V	Electroplating	■ Illegally stored hazardous waste	■ $250,000 fine ■ $22,800 restitution ■ $77,200 SEP
Civil	VI	Oil refinery	■ Operated hazardous waste surface impoundments without a permit	■ $734,008
Civil	VI	Chemical manufacturing	■ Sent hazardous spent sulfuric acid to a neighboring fertilizer plant (use constituting disposal) outside of the RCRA program	■ $1.485 million
Criminal	VI	Waste transportation	■ Falsified manifest information by shipping drums of hazardous waste as nonhazardous material ■ Illegally shipped hazardous waste to landfills not permitted to accept hazardous waste	■ 5-mo prison term ■ 5-mo in-home confinement and 2 years supervised release ■ $10,000 fine

McCoy's RCRA Unraveled ©2015 McCoy and Associates, Inc.

Table 16-5: Selected RCRA Enforcement Cases

Case type	EPA region	Facility type	Alleged RCRA violation(s)	Penalty[1]
Criminal	VI	Environmental consulting	■ Illegally imported and stored hazardous waste	■ 15-mo prison term for company employee ■ $1,500 fine
Civil	VII	University	■ Failed to make hazardous waste determinations ■ Failed to properly mark and close hazardous waste containers ■ Improperly treated hazardous waste by allowing solvents and solvent-soaked rags to evaporate prior to disposal as solid waste	■ $39,431
Civil	VII	Hospital	■ Improperly treated hazardous waste by allowing paint, solvents, and solvent-soaked rags to evaporate prior to disposal as solid waste ■ Failed to conduct and record hazardous waste container inspections ■ Failed to properly mark hazardous waste containers ■ Failed to properly label used oil containers ■ Failed to properly mark universal waste batteries	■ $83,488
Civil	VII	Veterinary pharmaceuticals	■ Failed to make hazardous waste determinations ■ Stored hazardous waste without a permit ■ Failed to comply with generator requirements ■ Shipped hazardous waste without a manifest to an unpermitted facility ■ Improper management of universal waste lamps ■ Improper management of used oil	■ $68,475 ■ $300,000 SEP
Civil	VII	Hardware manufacturing	■ Stored hazardous waste without a permit ■ Failed to make hazardous waste determination ■ Failed to manage used oil and universal waste in accordance with applicable regulations	■ $31,379 ■ $91,809 SEP
Civil	VII	Aircraft component manufacturing	■ Shipped hazardous waste without a manifest ■ Failed to make hazardous waste determinations ■ Failed to follow RCRA training, preparedness and prevention, and contingency plan regulations ■ Failed to properly mark and close hazardous waste containers	■ $132,500
Civil	VII	Hazardous waste TSD facility	■ Failed to make hazardous waste determinations ■ Failed to properly mark and close hazardous waste containers ■ Failed to comply with incompatible waste provisions for containers ■ Failed to meet air emission standards for hazardous waste tanks ■ Failed to ensure the integrity of a secondary containment structure	■ $150,000
Civil	VIII	Oil refinery	■ Stored hazardous waste in a wastewater pond	■ $900,000

Table 16-5: Selected RCRA Enforcement Cases

Case type	EPA region	Facility type	Alleged RCRA violation(s)	Penalty[1]
Criminal	VIII	Phosphorus manufacturing	■ Illegally stored hazardous waste	■ $16.2 million fine ■ $1.8 million restitution
Civil	IX	Engine manufacturing and repair	■ Failed to make a hazardous waste determination ■ Treated, stored, and disposed hazardous waste without a permit ■ Operated a hazardous waste boiler without a permit ■ Shipped hazardous waste without a manifest ■ Accumulated waste acids and waste cyanides without adequate segregation ■ Failed to follow RCRA training, preparedness and prevention, and contingency plan regulations ■ Failed to inspect, label, close, and provide aisle space for hazardous waste containers ■ Failed to inspect containment building ■ Failed to inspect, meet performance standards for, and provide secondary containment for hazardous waste tanks	■ $5,000,000 ■ $1,000,000 SEP
Civil	IX	Metal finishing	■ Failed to notify EPA/state and obtain an EPA ID number for hazardous waste activities ■ Failed to make hazardous waste determinations ■ Improperly stored hazardous waste without a permit ■ Failed to provide hazardous waste training to facility personnel ■ Failed to comply with biennial reporting requirements	■ $150,000
Civil	IX	Raceway	■ Failed to make hazardous waste determinations ■ Shipped hazardous waste without a manifest ■ Failed to comply with biennial reporting requirements	■ $31,851
Civil	IX	Aircraft maintenance	■ Failed to keep hazardous waste containers closed ■ Failed to properly label hazardous waste containers ■ Stored hazardous waste for greater than 90 days without a permit	■ $850,000
Civil	IX	Commercial hazardous waste landfill	■ Failed to follow proper lab quality control procedures ■ Land disposed hazardous waste that didn't meet treatment standards	■ $400,000 ■ $600,000 additional compliance costs
Civil	IX	Wafer fabrication	■ Failed to properly label and close hazardous waste containers ■ Failed to maintain adequate contingency plan ■ Failed to adequately train personnel ■ Failed to provide leak detection for hazardous waste tank ■ Failed to control hazardous waste tank emissions ■ Failed to monitor ancillary equipment ■ Failed to maintain adequate contingency plan	■ $62,500

Table 16-5: Selected RCRA Enforcement Cases

Case type	EPA region	Facility type	Alleged RCRA violation(s)	Penalty[1]
Civil	IX	Medical device manufacturing	■ Failed to manage hazardous waste in a closed container in good condition ■ Failed to adequately train personnel ■ Failed to maintain adequate contingency plan	■ $31,500
Civil	IX	Pipe fabrication	■ Failed to contain release of hazardous waste	■ $158,000
Criminal	IX	Individual	■ Illegally stored hazardous waste without a permit which resulted in explosion and fire	■ 5-year prison term ■ $800,000
Civil	X	Wood treatment	■ Failed to maintain adequate financial assurance even though it was a provision of an existing consent order	■ $7,880
Civil	X	Scrap equipment yard	■ Failed to properly label used oil containers ■ Failed to respond to releases of used oil	■ $32,882
Civil	X	Chemical manufacturing	■ Failed to immediately notify the National Response Center for spills of hazardous substances	■ $17,000 ■ $72,000 SEPs
Civil	X	Auto salvage	■ Stored hazardous waste in used oil drums ■ Provided used oil containing potentially hazardous waste to private entities to burn in space heaters	■ $69,000
Civil	X	Defense facility	■ Failed to make hazardous waste determination ■ Failed to have an adequate hazardous waste training plan ■ Failed to properly label hazardous waste containers ■ Failed to perform weekly inspections of hazardous waste containers ■ Improper management of hazardous fluorescent lamps	■ $45,700
Criminal	X	Paint manufacturing	■ Illegally transported and disposed hazardous waste	■ 30 days in-home confinement and 6 mo supervised release for company executive ■ $50,000 fine ■ $40,000 restitution

[1] SEP means supplemental environmental project. A SEP involves actions a violator agrees to undertake to protect human health or the environment in exchange for a monetary penalty reduction.
[2] SEP requires facility to strip vinyl chloride out of process wastewaters prior to entering an earthen surface impoundment.
Source: EPA's Office of Enforcement and Compliance Assurance and EPA regional enforcement programs.

and South Dakota), EPA's ability to overfile has been successfully challenged. [*Harmon Industries, Inc. v. EPA*, No. 98-3775 (1999)]

In the rest of the country, the agency apparently does have overfiling authority. Typically, EPA will overfile when, in EPA's opinion, an authorized state fails to take timely and appropriate action. Customarily, EPA will alert the state that the proposed resolution of an alleged violation is inadequate before the state executes a settlement with a facility. If the authorized state fails to address the concern, the likelihood of EPA overfiling increases. U.S. circuit courts for the Ninth and Tenth Circuits have both decided that EPA retains secondary enforcement authority and may bring an enforcement action against a party, even if an authorized state has already settled with that party. [*United States v. Allen Elias*, No. 00-30145 (2001) and *United States v. Power Engineering Company et al.*, No. 01-1217 (2002), respectively] In the other circuits, we are unaware of any ongoing cases addressing this issue; thus, the agency would appear to have overfiling authority in these areas of the country until this authority is successfully challenged.

Although the case in the Tenth Circuit was appealed to the U.S. Supreme Court, the high court decided not to review it. Therefore, no nationally consistent legal position currently exists on this issue.

16.3 EPA's national RCRA enforcement initiatives

Every year or two, EPA puts in writing what its enforcement priorities will be for the immediate future. The agency evaluates significant environmental risks and noncompliance patterns associated with various industrial sectors, specific regulatory requirements, and geographic areas.

Based on the above assessment, EPA has established national enforcement initiatives for fiscal years 2014–2016. [http://www2.epa.gov/enforcement/national-enforcement-initiatives] Of the six national initiatives, three could potentially affect enforcement at facilities subject to RCRA:

1. *Reducing pollution from mineral processing operations*—EPA believes that mining and mineral processing facilities generate more toxic and hazardous waste than any other industrial sector. Based on numerous inspections of these facilities, EPA has "found significant non-compliance with hazardous waste and other environmental laws." Therefore, EPA will continue its enforcement initiative to bring these facilities into compliance.

2. *Assuring energy extraction sector compliance with environmental laws*—EPA is continuing its initiative to assure that energy extraction activities are complying with federal requirements to prevent pollution of the nation's air, water, and land. This initiative is being concentrated in particular areas of the country where energy extraction activities are concentrated.

3. *Cutting toxic air pollution that affects communities' health*—Included in this initiative is a focus on excess emissions caused by facilities' failure to comply with EPA's leak detection and repair requirements.

16.4 ECHO—EPA's enforcement database

OECA updates a website called Enforcement and Compliance History Online (ECHO) at http://echo.epa.gov. The ECHO website provides compliance and enforcement information for more than 800,000 regulated facilities nationwide, and yours is very likely included. This information, previously available primarily through Freedom of Information Act (FOIA) requests, has been consolidated into a database with a web interface.

EPA worked with state governments in developing the ECHO database. Data on ECHO are pulled from several sources including the Air Facility System, Permit Compliance System, RCRAInfo, Integrated Compliance Information System, and National Compliance Database.

The web interface allows you to search the database by any combination of the following criteria:

- Regulatory program (CAA, CWA, RCRA, or SDWA),
- Facility name and type,
- Geographic location,
- Inspection/enforcement history,
- Compliance status,
- TRI chemical releases, and
- Minority population within three miles of the facility.

Detailed facility reports show inspections, violations, enforcement actions, and penalties for the past three years. The site also offers an online means to report data errors. Because this information is now easily available to the public, those in the regulated community would be wise to periodically check reports on their facilities and report any errors.

16.5 EPA's self-disclosure policy

Under EPA's self-disclosure policy, the agency can substantially reduce civil penalties for companies that voluntarily disclose and promptly correct violations that are identified through self-auditing. [*Incentives for Self-Policing: Discovery, Disclosure, Correction and Prevention of Violations*, April 11, 2000; 65 *FR* 19617] If a facility discloses (in writing) a specific violation to EPA within 21 days of discovery of its occurrence, the "gravity" (punitive) component of any resulting penalty may be fully or substantially eliminated; however, EPA reserves the right to collect penalties that offset any economic benefit that results from noncompliance. Several specified conditions must be met for facilities to take advantage of the policy, and there are exceptions in cases involving serious harm to public health or the environment.

As an example of the significant incentives this policy can provide, EPA recently waived 100% of $1.6 million in gravity-based penalties associated with a large retail company's violations of the CWA, EPCRA, and RCRA, because the company satisfied all of the self-disclosure policy conditions. (A gravity-based penalty of $78,625 was assessed against the company for other violations for which the company failed to satisfy the self-disclosure policy).

Information on the program and interpretive guidance are available at http://www.epa.gov/compliance/incentives/auditing/auditpolicy.html. Many states have voluntary disclosure policies as well, so check with your state.

16.6 Preparing for a RCRA Inspection

This section contains recommendations on how to prepare for a RCRA inspection. We begin by discussing the regulators' legal authority to inspect facilities.

16.6.1 State's/EPA's right to inspect

RCRA Section 3007(a) provides state and EPA inspectors the authority to conduct inspections:

"For the purposes of developing or assisting in the development of any regulation or enforcing the provisions of this chapter, such officers, employees or representatives are authorized—

(1) to enter at reasonable times any establishment or other place where hazardous wastes are or have been generated, stored, treated, disposed of, or transported from;

(2) to inspect and obtain samples from any person of any such wastes and samples of any containers or labeling for such wastes.

"Each such inspection shall be commenced and completed with reasonable promptness. If the officer, employee or representative obtains any samples, prior to leaving the premises, he shall give to the owner, operator, or agent in charge a receipt describing the sample obtained and if requested a portion of each such sample equal in volume or weight to the portion retained. If any analysis is made of such samples, a copy of the results of such analysis shall be furnished promptly to the owner, operator, or agent in charge."

Additionally, Section 3007 requires facilities to "furnish information relating to such wastes and permit such person at all reasonable times to have access to, and to copy all records relating to such wastes."

EPA believes that the above inspection authority extends to any solid waste that the state/EPA reasonably believes may meet the statutory definition of a hazardous waste. Thus, non-Subtitle C units at a facility may be subject to inspection where there is some basis for concluding that they may provide information relating to hazardous wastes. [RO 14042]

16.6.2 Types of inspections

Multimedia inspections are possible (including evaluation of CWA NPDES or POTW permit requirements and CAA Title V permit requirements in addition to RCRA requirements), but we will focus on the RCRA issues.

There are many types of inspections, as noted in the Table 16-6; however, the compliance evaluation inspection is the primary mechanism used by state/EPA personnel for ensuring facility compliance with the many RCRA requirements and for detecting and verifying RCRA violations.

16.6.3 State/EPA inspections involve three stages

Facility personnel involvement in a state/EPA inspection can be divided into three distinct stages:

1. Preparing for the inspection,
2. The inspection itself, and
3. Inspection follow-up.

Each of these stages is discussed in the next subsections.

16.6.3.1 Preparing for the inspection

"The battle is won or lost before it is ever fought." [Sun Tsu, *The Art of War*] The time spent in preparation for an inspection will contribute significantly to how well the inspection itself goes.

A single point of contact (POC) for all RCRA inspections and an alternate POC should be designated. All RCRA inspections should then be coordinated through these POCs.

Most states give three to five days advance notice of an inspection. If you have notice of a state/EPA inspection, that information should be communicated

Table 16-6: Types of RCRA Inspections

Type of inspection	Description
Compliance evaluation inspection (most common)	Routine inspections of hazardous waste generators and TSD facilities to evaluate compliance with RCRA requirements[1] ■ Onsite examination of generation, treatment, accumulation/storage, and disposal areas ■ Records review
Case development inspection	Inspections conducted when RCRA violations are suspected or revealed during a compliance evaluation inspection; conducted to gather data in support of an enforcement action
Comprehensive ground water monitoring evaluation	Inspections conducted to ensure ground water monitoring systems are functioning properly (typically only conducted at TSD facilities operating land disposal units)
Compliance sampling inspection	Inspections conducted to collect samples for laboratory analysis; may be conducted in conjunction with a compliance evaluation inspection or any other type of inspection
Laboratory audit	Inspections of laboratories to evaluate proper sample handling and analysis protocols

[1]The inspection and sampling authority of RCRA Section 3007 is available for determining and assuring compliance with any Subtitle C requirement. [RO 14042]

Source: Adapted from OSWER Directive 9938.02(b), available from http://www.understandrcra.com/RCCD/RCRAInspectionManual.pdf.

to 1) the plant manager/superintendent, 2) onsite and corporate EH&S personnel, 3) onsite and corporate legal counsel, 4) operating unit line managers who are responsible for the process equipment near where HW tanks or satellite containers are located, and 5) plant security. Then, onsite EH&S personnel will prepare diligently for the upcoming RCRA inspection.

Keep in mind, however, that the language in RCRA Section 3007 "to enter at reasonable times any establishment" gives state/EPA personnel the right to conduct surprise or "no-knock" inspections, where they show up unannounced, show proper identification/credentials, and ask to be admitted. Therefore, "preparing for a RCRA inspection" should be a mindset and level of awareness for all RCRA facility personnel that an inspection could occur at any time. An extremely high level of RCRA compliance at all times is critical.

The primary elements involved in preparing for a RCRA inspection are detailed in the next subsections.

16.6.3.1.1 Review past inspection reports/compliance history

A review of previous inspection reports, letters from the state/EPA, administrative actions, etc. should be undertaken. Any noncompliance issues raised during previous inspections/actions will be areas that the current inspectors will almost certainly be evaluating/auditing. Also, any citizen complaints concerning alleged violations at the facility will almost certainly be evaluated/audited, so legal counsel should be consulted to develop facility responses.

16.6.3.1.2 Review hazardous waste management equipment/areas

Hazardous waste treatment, accumulation/storage, and disposal areas are almost certain to be inspected. A pre-inspection review by facility personnel should include:

- Satellite and 90/180-day accumulation areas at generator facilities. These areas should be reviewed to ensure they meet all §262.34 requirements (see Sections 6.2, 6.3, and 16.1). Any rusted or deformed containers, poorly stacked drums, etc. are potential signs of a deficient waste management system.

- Permitted container storage areas, tanks, containment buildings, incinerators, boilers/industrial furnaces, surface impoundments, waste piles, land treatment units, landfills, drip pads, and miscellaneous units at TSD facilities. These units should be reviewed to ensure they meet all permit conditions.

- Permit-exempt units (i.e., elementary neutralization units, totally enclosed treatment facilities, immediate response units, wastewater treatment units, and recycling units) at both generator and TSD facilities. These units should be reviewed to ensure they meet the unit-specific definitions in the RCRA regs and are indeed exempt from substantive RCRA standards (see Chapter 7). For example, if a claim is made by a facility that an in-ground basin is an exempt wastewater treatment unit, the facility should be able to show that it passes the "parking lot test" (see Sections 7.2.1.3 and 10.1.3.2).

16.6.3.1.3 Review process areas

EPA's guidance notes that a compliance evaluation inspection will also include an examination of hazardous waste *generation* areas. Thus, a pre-inspection review by facility personnel should include a review of:

- Process areas and point of generation (POG) issues/determinations for all waste streams generated within each process. This review includes inputs/outputs from ancillary equipment, such as filters, heat exchangers, condensers, scrubbers, baghouses, strippers, flare knockout pots, etc. This review may uncover additional hazardous waste streams that are not being handled under RCRA.

- Management practices for all such wastes from their POG to accumulation/storage/treatment areas, including diversions, by-passes, and overflows. For example, are wastes discharged to a concrete slab and from there into a floor drain? Do wastes drain into a pit, sump, pond, or lift station before they are transferred to accumulation/storage/treatment areas; if so, what is the regulatory status of that pit, sump, pond, or lift station? Is there potential for mislabeling, misplacing, or mishandling wastes? Are wastes properly tracked to enable proper identification at accumulation/storage/treatment areas?

- The management of equipment cleanings (e.g., flushes, rinsewaters, sludges/solids), catalyst or carbon change outs, other maintenance wastes, and truck/railcar washings. This includes discharges to sewers and vacuum trucks.

- Manufacturing process units that are out of service [see §261.4(c)].

- The co-products and by-products produced in the process and their management (particularly by-products).

- Management of off-spec products. How are they stored/inventoried, and are they managed as products or wastes?

- Any secondary (non-virgin) materials used as inputs to a process. This includes materials that are by-products from another process at the facility or an offsite facility.

- Any release or spill points, including any process or waste units that are releasing material to the environment. This could include dripping/leaking equipment with no secondary containment. Puddles or stained areas around units, erosion, and dead vegetation are signs of past or current releases that may not have been remediated properly or immediately.

- Management of any contaminated rainwater generated during heavy rain events.

- The onsite sewer systems, wastewater treatment and water cooling systems, sludge treatment equipment, and what happens when such treatment equipment "goes down." Any out-of-service units in the wastewater treatment and water cooling systems should be evaluated. Vacuum truck discharges and cleanouts should also be evaluated.

- Maintenance areas and the wastes generated from such maintenance.

- Onsite laboratories and how lab wastes are managed.

Plant environmental personnel should perform these process reviews *before* the inspectors visit and make sure all waste streams have been characterized and are being managed accordingly. One of the most common violations found during inspections is that a waste is generated during a process upset or other uncommon situation, is not characterized per §262.11, and is managed in a way the facility environmental staff is unaware of. This could result in significant RCRA violations, including management of a hazardous waste in an unpermitted unit (e.g., a surface impoundment).

16.6.3.1.4 Review product and raw material warehouses/storage areas

Because of the abandoned material definition in §261.2(b)(3), state/EPA inspectors have the right to audit product and raw material warehouses, lockers, and other areas where products and raw materials are stored, looking for products/raw materials that are being accumulated in lieu of being burned, incinerated, or otherwise disposed. This would include products/raw materials that will not be used at the plant but are being stored because it is less expensive than sending them offsite for proper treatment/disposal.

16.6.3.1.5 Review contractor areas and waste generation activities

Contractor laydown and material storage areas can be a source of RCRA violations. For example, maintenance wastes (rags, burned-out bulbs, etc.) and unusable products are sometimes stored in these areas without consideration of facility procedures or the RCRA regs. In addition, contractors may generate wastes in other areas of the facility during the course of their work and fail to manage

16.6.3.1.6 Review shared services with other companies

Numerous facilities have multiple companies operating at the same location, resulting in shared services (e.g., raw material/intermediate/product storage areas, wastewater treatment systems, utility systems, maintenance areas, 90-day areas, etc.). Contract agreements or other documents should be in place specifying who has generator responsibilities under RCRA.

16.6.3.1.7 Review applicable RCRA records

Depending on the type of inspection, an extensive review of the facility's RCRA-required records will likely be conducted. The records/documents that will most likely be reviewed at generator facilities during an inspection are listed in Table 16-7.

Section 262.40 (generator recordkeeping) does not specify that all records be kept onsite. For example, copies of manifests and biennial reports can be kept at corporate headquarters. [RO 12199] However, RCRA Section 3007 notes that facilities must "permit such person at all reasonable times to have access to, and to copy all records relating to such wastes." So, reasonable access to all waste management records must be provided to inspection personnel. To be proactive and to show inspectors that the facility is cooperative and environmentally responsible, some of the records/documents noted in Table 16-7 could be placed into a notebook or on a

Table 16-7: RCRA Inspection Document Request List

Record/document	Place record/document in notebook/CD before inspector(s) arrive?
Facility map or sketch showing location of all waste management units, including satellite accumulation areas	Yes
Brief description, including simplified process flow diagrams, of facility operations/process areas and wastewater treatment system(s)	Yes
Current copy of state-required Notice of Registration, if applicable	Yes
Air, RCRA, and wastewater permits, if applicable	No[1]
List of all container accumulation/storage areas and satellite accumulation areas, including container area identity, location, wastes handled, and number and types of containers	Yes
List of tanks, if any, that manage hazardous waste, including tank identity, location, date/year put into service, whether or not tank has secondary containment, secondary containment volume calculations, information on impervious liner, overfill protection and release detection systems, wastes handled, and tank system integrity assessment records	Yes
Waste characterization documentation, including analytical data and SDSs, and waste profiles for waste shipped offsite[2]	No[1]
Hazardous waste manifests[2]	No[1]
Exception reports or letters to the state for any manifests not received back from designated TSD facilities within regulatory time frames[2]	No[1]
Manifest discrepancy reports sent from designated TSD facilities[2]	No[1]
LDR notifications/certifications and all supporting data, including analytical data and process knowledge, for LDR determinations[2]	No[1]
Waste analysis plan, if applicable	Yes
Universal waste shipment records, if applicable	No[1]
Container accumulation/storage area and tank inspection logs[2]	No[1]

Table 16-7: RCRA Inspection Document Request List

Record/document	Place record/document in notebook/CD before inspector(s) arrive?
Employee training records (initial and annual refresher, if required) for hazardous waste handlers, including job titles and descriptions, name of each employee, training plan, and documentation of the type and amount of training each employee has received	No[1]
Preparedness and prevention plan, including a list and description of all emergency response equipment onsite with its location	Yes
Inspection schedule and logs for all safety and emergency response equipment	No[1]
Arrangements/agreements with local police departments, fire departments, hospitals, and local/state emergency response agencies that may be called upon to provide emergency services	Yes
Current contingency plan and evidence it has been submitted to all local police departments, fire departments, hospitals, and local/state emergency response agencies that may be called upon to provide emergency services	Yes
Summary reports and documentation of all incidents that required implementation of the contingency plan[2]	No[1]
Reports of any releases to the environment (reportable and nonreportable)	No[1]
Source reduction and waste minimization plan (pollution prevention plan) and annual progress reports	No[1]
Annual waste summaries[2], if required, and last biennial report	Yes
Hazardous waste fee payment documentation	No[1]
Documentation to show compliance with the organic air emission regulations of Part 264/265, Subparts AA, BB, and CC ■ For Subpart BB, information to determine whether equipment is in light or heavy liquid service, a list of all pumps, valves, connectors, etc. subject to monitoring and inspection, documentation of monitoring, and leak repair information ■ For Subpart CC, information to determine whether containers are subject to Level 1, 2, or 3 controls and to determine whether tanks are subject to Level 1 or 2 controls, and container/tank inspection logs	No[1]
Used oil and used oil filter management records, if required	No[1]
Notifications for hazardous waste intended to be exported, if applicable[2]	No[1]
Notification forms for recycled hazardous wastes and records documenting that recyclable materials are being legitimately recycled	No[1]
List of primary offsite disposal/treatment/recycling facilities utilized	No[1]
Deed recordation documents for closed solid waste land disposal units	No[1]
Information on RCRA ground water monitoring status, if required: ■ Site map locating each monitoring well and associated waste management unit ■ Changes to the RCRA monitoring well system since the last ground water monitoring inspection ■ Sampling and analysis plan	No[1]

LDR = land disposal restrictions; SDS = safety data sheet; TSD = treatment, storage, and disposal.

[1] Although these records/documents do not have to be placed in a notebook/CD to be made available to the inspector(s), the location of all of these records/documents should be confirmed so that they can be produced quickly if the inspector(s) requests them.

[2] For last three years.

Source: McCoy and Associates, Inc.; adapted from information obtained from states of Colorado and Texas.

CD that will be made available to inspection personnel at the time of inspection.

For RCRA-permitted facilities, waste analysis plans will likely be the subject of much review and discussion between plant and inspection personnel. Additional permit conditions will likely be reviewed, but it is difficult to know in advance exactly what the inspectors will want to review. The inspector(s) will likely read your permit beforehand, however, and so if you haven't read your permit in a while, think about reading through it before the inspection to see if there are any areas you are unaware of or not in compliance with.

Section 4.4 of OSWER Directive 9938.02(b), cited in Table 16-6, includes listings of records to be maintained by generators, transporters, TSD facilities, and used oil processors, burners, and marketers. Submittals that must be made to EPA (which should be available for inspection) are also listed.

In addition, spill/incident records should be available for inspection. These include all spills, not just those required to be reported to the National Response Center or to state/local entities. These records should include a description of all spill response activities.

During the pre-inspection records review, facility personnel should look for the following:

- That all of the required records have been completed and maintained in paper or electronic form,
- That all of the required records were completed on time per RCRA regulations,
- That all of the required records were distributed to entities specified in the regs, and
- That specific records are consistent with each other (e.g., manifests and accompanying LDR notifications).

16.6.3.1.8 *Prepare facility personnel*

Inspectors will often want to interview specific personnel who are responsible for certain waste management activities, but they also like to talk with random people found to be working in waste management capacities. These conversations are a way for inspectors to identify inconsistencies in explanations of procedures or operations that could indicate 1) possible noncompliance, requiring further investigation; and 2) inadequacies in the RCRA facility personnel training program. Facility personnel should be coached on what type of responses they should and should not give if state/EPA inspectors ask them a question during the site inspection. For example, employees should never answer a question if they are not absolutely sure of the answer. Also, employees should always answer truthfully to questions from an inspector, but they should not volunteer other information. The plant manager and/or legal counsel should be encouraged to have these discussions with facility personnel.

When EH&S personnel get advanced notice of an inspection, it is worth sending an email to the proper operating unit line managers, giving them a quick refresher on 1) specific waste units in certain areas (e.g., hazardous waste tank secondary containment should be clean and dry, etc.), 2) general hazardous waste management (proper labeling, lids on containers, info legible on the labels, etc.), and 3) emphasizing good housekeeping. EH&S personnel should also assure operating unit managers it is acceptable to have satellite accumulation containers in the operating units as long as they are managed correctly; otherwise, their first response to the email may be to send all satellite drums to the 90-day facility, causing that area to become congested.

16.6.3.2 The inspection itself

Usually, but not always, the inspection itself consists of three stages:

1. Pre-inspection (opening) conference,
2. Records/physical inspection, and
3. Post-inspection (closing) conference.

Each of these is covered in the next subsections. But first, we start with the ramifications of denying access to state or EPA inspectors.

16.6.3.2.1 Denying access

As noted previously, RCRA Section 3007 gives inspectors the legal right "to enter at reasonable times any establishment." If facility personnel deny inspectors entry to the facility or any portion of the facility, inspectors have the legal right to obtain a search warrant to obtain entry. According to OSWER Directive 9938.02(b), cited previously, the following are legally indefensible actions that deny inspectors access:

- Refusing to allow inspectors to bring in necessary equipment, such as cameras;
- Refusing access to certain records/documents;
- Refusing entry due to a strike and/or plant shutdown; and
- Refusing entry due to inspectors' refusal to sign a waiver or other legal document restricting the owner's/operator's liabilities or obligations.

Conversely, inspectors may be denied access to a facility or any portion of the facility if they do not have the safety equipment required by a facility (e.g., per OSHA or NIOSH requirements). In such a case, it will generally be possible for inspectors to obtain access at another time by satisfying the safety equipment requirements. National security concerns have also been used to deny inspectors access to portions of a site.

If a search warrant is presented, the POC should immediately notify the plant manager/superintendent and onsite and corporate legal counsel.

16.6.3.2.2 Pre-inspection (opening) conference

The first stage of a RCRA inspection usually consists of an opening conference. The POC should ensure that the following items are covered during the opening conference:

- Introductions—The POC will introduce the primary plant contacts, and each inspector will be asked to introduce themself and provide their background/affiliation.

- Scope and objectives of the inspection—The inspectors should clarify whether the inspection is a 1) routine, periodic assessment of RCRA compliance; 2) review of facility activities or status with respect to an ongoing enforcement action; 3) review of facility compliance with conditions and/or deadlines delineated in the facility's RCRA permit; and/or 4) response to information received concerning alleged violations at the facility (e.g., a citizen's complaint). The duration of the inspection (hours vs. days) should also be established at this time.

- Identification of the areas the inspectors intend to inspect—After the inspectors identify these areas (and in what order they will be inspected), the POC should assign appropriate plant personnel to accompany them. The POC should ensure that operations/production personnel have cleared process areas for inspection; for example, any significant maintenance work/turnaround activities could limit inspection access. Remediation units, such as corrective action units and/or ground water monitoring wells may also be reviewed if the facility has a RCRA permit. Inspectors should be accompanied at all times and should be discouraged from wandering around a facility looking at areas or equipment beyond the scope of the inspection as identified at the opening conference. Onsite transportation should be arranged by facility personnel.

- Sampling—Whether sampling will be required during the inspection should be discussed; if samples will be required, where will they be taken? Generally, inspection personnel bring their own sampling equipment. If not, plant personnel should arrange for sampling equipment. RCRA Section 3007 gives the facility the right to ask for sample splits, and this should be done.

- Health and safety considerations—Inspection personnel must be made aware of the potential hazards they may encounter during field inspections and must have received appropriate training (either provided by the plant or by their own agencies) to enter designated areas safely. Required personal protective equipment and other safety equipment must also be used and either provided by the plant or brought with inspection personnel. Evacuation routes/staging areas should be reviewed. EPA inspection personnel will generally be trained in accordance with EPA Order 1440.2 (Appendix B of EPA's *NPDES Compliance Inspection Manual*, see http://www.epa.gov/compliance/resources/publications/monitoring/cwa/inspections/npdesinspect/npdesinspect.pdf), but state inspection personnel may or may not have commensurate training.

- Records—The records that the inspectors want to review should be identified.

- Ground rules should be established for 1) photographs (make sure camera passes are available) and whether copies of photographs taken by inspection personnel can be obtained by the facility after the inspection, 2) copying of records, 3) taking/splitting samples and receipt by the facility of subsequent analytical results obtained by the agency, 4) treatment of confidential business information (see Section 16.6.3.2.3 for additional information on CBI), 5) whether copies of forms/checklists completed by inspection personnel can be made and left at the facility at the closing conference, 6) inspectors must be accompanied by facility personnel at all times, 7) inspectors must keep their state/EPA identification in sight at all times, and 8) how lunch will be handled.

- Process overviews from process engineers and operators (optional, depending on whether the inspectors will review process areas)—Inspectors may ask for copies of the process drawings.

- A closing (post-inspection) conference should be scheduled.

A pre-inspection conference checklist that can be used, covering the items noted above, is given in Table 16-8.

Note that sometimes inspectors, immediately upon entry to a facility, will proceed with a visual inspection of certain operations or units, looking for a suspected violation, before facility personnel have time to stop, correct, or conceal any possible violation. In such cases, a pre-inspection conference may not be held or could be held after this initial inspection.

16.6.3.2.3 Confidential business information

During the opening conference, the facility should inform inspectors of any information/processes that are confidential business information (CBI). If such a claim is made, inspectors should provide appropriate forms to complete to document all CBI. If inspectors do not have such forms, all information/processes that are CBI should otherwise be documented and agreed upon during the opening conference.

16.6.3.2.4 Records/physical inspection

It is important that an atmosphere of respect and cooperation be established between facility personnel and state/EPA inspectors. Even hostile inspectors should be treated with respect and courtesy. Furthermore, we highly recommend that if an inspector points out a problem, noncompliance, or "area of concern," the facility POC should either correct it right there in the inspector's presence or attempt to correct it by the post-inspection conference.

Records inspection—Focus will be on reviewing RCRA-required records (see Section 16.6.3.1.7 for the types of records that will likely be reviewed and should be made available). Inspectors will check for 1) the presence of required records and plans, 2) the dates of the documents to ensure they are kept up-to-date and/or are maintained for the required period, and 3) any suspected falsification of data. Inspectors expect that there will be some records retrieval that must be done, so getting applicable files (that are not available in the notebook/CD) should not be a problem. If advanced notice of the inspection is available, a conference room should be reserved close to where the majority of RCRA compliance records are maintained. That gives the inspector(s) room to review records in a neutral, reasonably quiet environment. If inspectors want copies of

Table 16-8: Pre-Inspection Conference Checklist

Item	Responsibility	Completed
Introductions of plant and inspection personnel	Both	
Scope and objectives of the inspection	Inspection personnel	
Identification of the plant areas the inspectors intend to inspect (and in what order), plant personnel that will accompany them, and what transportation needs to be provided	Both	
Whether sampling will be conducted during the inspection and, if so, where? Review required sampling equipment	Both	
Health and safety considerations, including safety training, PPE, evacuation routes/staging areas, etc.	Plant personnel	
Identification of records that inspection personnel want to review	Inspection personnel	
Inspection ground rules, including		
■ Taking photographs (plant to make camera passes available) and whether copies of photographs taken by inspection personnel can be obtained by the facility after the inspection	Both	
■ Copying records	Plant personnel	
■ Taking/splitting samples and receipt by the facility of subsequent analytical results obtained by the agency	Both	
■ Treatment of confidential business information	Both	
■ Whether copies of forms/checklists completed by inspection personnel can be made and left at the facility	Both	
■ Inspectors must be accompanied by facility personnel at all times	Plant personnel	
■ Inspectors must keep their state/EPA/other identification in sight at all times	Inspection personnel	
■ How lunch will be handled, if applicable	Both	
Process overviews from process engineers and operators (optional, depending on whether the inspectors will review process areas)	Plant personnel	
Schedule a post-inspection conference	Both	

PPE = personal protective equipment.

Source: McCoy and Associates, Inc., available at http://www.mccoyseminars.com/forms.cfm.

certain records, plant personnel should make the specified copies—not the inspectors.

Physical inspection—Focus may be:

- "End of pipe" waste management operations—For example, inspectors may want to discuss and review waste generation and management practices (e.g., where wastes are generated and how they get to SAAs, 90/180-day areas, and other waste management areas).

- Process review and point of generation (POG) issues/determinations—Inspectors may want to determine the POG and waste classification for all waste streams generated within each process. Based on drawings and physical inspection of each process, inspectors may focus on the POG and subsequent management of every waste stream, air emission, and wastewater stream. This type of inspection includes how equipment cleanings and other maintenance wastes are generated/managed. During review of processes, inspectors may ask about any samples or other measurements that have been collected for generated waste streams.

- Surface impoundments—Incorrect waste determinations (e.g., not enough samples) may lead to disposal of hazardous waste into impoundments that are not RCRA-permitted. Equipment cleanouts or untreated waste streams from process upsets (as opposed to normal nonhazardous flows) that are discharged to wastewater treatment systems that include ponds can cause this situation. In addition to the issue of hazardous waste flowing into impoundments that are not RCRA-permitted, LDR violations can also occur since ponds are land disposal units.

- Sampling of materials—Inspectors may believe sampling is required:
    - To verify whether the waste is hazardous or nonhazardous (e.g., unknown wastes, suspected waste misclassification);
    - To demonstrate whether a facility is mislabeling or mishandling wastes;
    - To demonstrate compliance with permit conditions, including waste analysis plan requirements;
    - To identify any improper onsite disposal in piles, ponds, or landfills;
    - To identify any unpermitted or questionable discharges or suspected contamination of storm-water run-off;
    - To demonstrate that a release has occurred or is occurring (e.g., stains or discolorations on soil/gravel in process or waste management areas); or
    - To verify that the correct wastes are being managed in the facility's various waste management units.

Plant personnel accompanying state/EPA inspectors should document in a logbook or equivalent:

- Date, time, and names/titles of all inspection personnel;
- What exactly was inspected/reviewed/copied;
- The questions the inspectors ask and what information is provided to inspectors;
- The date, time, and location of all photographs taken; and
- The exact location of any samples taken and split samples that will be maintained by the facility.

16.6.3.2.5 Post-inspection (closing) conference

The post-inspection conference is the last stage of the inspection. The POC should ensure that the following items are covered during the closing conference:

- A review of inspectors' preliminary inspection findings, including problems and/or areas of potential noncompliance;
- Clarification of any misunderstandings that inspection personnel may have, based on the first item above;
- Inform inspectors if any problems or areas of potential noncompliance have been corrected;
- Any help/suggestions inspectors might have to 1) respond to or correct any observed noncompliance, and 2) improve the facility's waste minimization efforts;
- Whether copies of forms/checklists completed by inspection personnel can be made and left at the facility;
- Any additional questions that inspectors have, including questions about potential violations uncovered during the inspection;
- Receipts describing all samples obtained and split samples should be provided by inspectors; and
- Follow-up procedures, including 1) how results of the inspection will be used and what further communications the state or EPA region may have with the facility; 2) what analytical procedures agency personnel will conduct on samples and a request for copies of all analytical results thus obtained; 3) whether copies of photographs taken will be sent from inspection personnel to the facility after the inspection; and 4) what other data/information required by inspection personnel should subsequently be sent to inspectors, such as answers to questions raised that could not be answered during the inspection and

copies of records that could not be produced during the inspection.

A post-inspection conference checklist that can be used, covering the items noted above, is given in Table 16-9.

16.6.3.3 Inspection follow-up

If the inspector(s) identified any problems and/or areas of potential noncompliance, the facility should investigate them and bring the facility into compliance as soon as possible. This should be documented.

It may take a significant amount of time for the regulatory agency to send a written report after the inspection. If the facility has heard nothing from the agency within 60 days after the inspection, the POC should contact the lead inspector and inquire about 1) the status of the inspection report, 2) whether any violations were recorded, and 3) whether any enforcement action will be initiated. The POC should always try to obtain a copy of the inspection report that the lead inspector filed with his/her agency. Even if the inspection report will not be released by the agency, the POC should try to determine if any violations will be recorded. Once the facility has addressed and corrected any such violations, the facility should notify the agency.

If a violation is discovered during the inspection, an enforcement action may be initiated by the

Table 16-9: Post-Inspection Conference Checklist

Item	Responsibility	Completed
Have inspector(s) review preliminary inspection findings, including problems and/or areas of potential noncompliance	Inspection personnel	
Plant personnel should clarify any misunderstandings that inspection personnel may have, based on first item above, and should inform inspectors if any problems or areas of potential noncompliance have been corrected	Plant personnel	
Have inspector(s) comment on any help/suggestions relative to ■ Plant responding to or correcting any observed noncompliance ■ Improving the facility's waste minimization efforts	 Inspection personnel Inspection personnel	
Ask whether copies of forms/checklists completed by inspection personnel can be made and left at the facility	Plant personnel	
Ask whether there are any additional questions that inspectors have, including questions about potential violations uncovered during the inspection	Plant personnel	
Request inspectors to provide ■ A receipt describing all samples obtained ■ Split samples for all samples	 Both Both	
Review follow-up procedures, including ■ How results of the inspection will be used and what further communications the state or EPA region may have with the facility ■ What analytical procedures agency personnel will conduct on samples and a request for copies of all analytical results thus obtained ■ A request for copies of all photographs taken by inspection personnel ■ What other data/information required by inspection personnel will be sent at a later time, such as answers to questions raised that could not be answered during the inspection and copies of records that could not be produced during the inspection	 Inspection personnel Both Plant personnel Plant personnel	

Source: McCoy and Associates, Inc., available at http://www.mccoyseminars.com/forms.cfm.

agency. The purpose of the enforcement action is to compel the facility to return to compliance with RCRA requirements and possibly to pay a monetary penalty for the noncompliance. Enforcement actions may result in one of the following:

- Administrative action (e.g., warning letter, notice of violation [NOV], administrative order, administrative penalty, permit action),
- Civil court action, or
- Criminal court action.

In all cases, the facility should work with its onsite and corporate legal counsel to respond to any of the above. The agency may conduct a follow-up inspection to ensure compliance has been achieved.

16.6.4 Resources for RCRA inspection checklists

Plant personnel can use a variety of inspection checklists to perform their pre-inspection reviews. Some of the checklists available online are noted below.

1. EPA inspection checklists:
    - *Protocol for Conducting Environmental Compliance Audits for Hazardous Waste Generators Under RCRA*, EPA/305/B-01/003, June 2001, available from http://cfpub.epa.gov/compliance/resources/policies/incentives/auditing/.
    - *Protocol for Conducting Environmental Compliance Audits of Treatment, Storage and Disposal Facilities under RCRA*, EPA/305/B-98/006, December 1998, available from http://cfpub.epa.gov/compliance/resources/policies/incentives/auditing/.

2. State inspection checklists—Numerous states have inspection checklists available on their websites for small and large quantity hazardous waste generators, small and large quantity universal waste handlers, etc. The user can search on specific state websites in the hazardous or universal waste area or use web browser search engines to find and download these documents.

16.7 "Cradle-to-grave" RCRA liability—What does that mean?

The term "cradle to grave" is not in the RCRA statute, and it is not in the federal RCRA regulations. It first showed up during the legislative development of the RCRA law—see House of Representatives Report No. 1491, 94th Cong., 2d Session (1976). The first time EPA appears to have used the term was when issuing the RCRA rules in 1980:

- "Subtitle C of the Act fosters these objectives by providing for the identification of hazardous wastes, the establishment of a 'cradle-to-grave' hazardous waste tracking system, and the development of standards and permit requirements for the treatment, storage, and disposal of hazardous waste." [February 26, 1980; 45 *FR* 12724–5]

- "Subtitle C of RCRA establishes a federal program to provide comprehensive regulation of hazardous waste. When fully implemented, this program will provide 'cradle-to-grave' regulation of hazardous waste." [May 19, 1980; 45 *FR* 33066]

So the question that comes up is this—Does the term "cradle to grave" mean:

1. That a generator retains RCRA liability for its hazardous waste from the point of generation (cradle) through the point of land disposal (grave); or

2. That once a generator receives a signed manifest from a TSD facility that is permitted to receive the waste, its RCRA liability ends.

16.7.1 RCRA statutory liability for generators

The RCRA statute includes provisions that establish both civil and criminal penalties for noncompliance. The law also gives EPA the authority to restrain any hazardous waste management activities that present an "imminent and substantial endangerment" to human health or the environment.

16.7.1.1 Civil penalties

RCRA Section 3002(a) sets out standards applicable to generators of hazardous waste. Requirements included are for:

1. Recordkeeping;
2. Container labeling;
3. Use of appropriate containers;
4. Furnishing information on the general chemical composition of hazardous waste;
5. Use of the manifest and any other reasonable means necessary to assure that all such hazardous waste generated is designated for treatment, storage, or disposal in, and arrives at, treatment, storage, or disposal facilities; and
6. Submission of reports.

RCRA Section 3002(b) requires generators to certify on the manifest that they have a waste minimization program and that they have selected the practicable method of treatment, storage, or disposal currently available to them that minimizes the present and future threat to human health and the environment.

RCRA elsewhere prescribes that any violation of the RCRA hazardous waste laws and regulations may result in a maximum civil penalty of $25,000 per day (now up to $37,500 per day). [Section 3008(a)(3) and (g)] Thus, generators of hazardous waste can be held liable for civil penalties 1) for failure to complete a manifest, 2) for failure to use a hazardous waste transporter with an EPA ID number, or 3) if the wastes do not arrive at a RCRA-permitted facility.

16.7.1.2 Criminal penalties

RCRA Section 3008(d) provides that criminal penalties can be imposed upon any person who knowingly:

1. Transports or causes to be transported any hazardous waste to a facility that does not have a RCRA permit.
2. Treats, stores, or disposes any hazardous waste (A) without a RCRA permit, (B) in violation of any material condition or requirement of such permit, or (C) in violation of any material condition or requirement of any applicable interim status regulations or standards.
3. Omits material information or makes any false material statement or representation on a document used for RCRA compliance.
4. Generates, stores, treats, transports, disposes, exports, or otherwise handles any hazardous waste and who destroys, alters, conceals, or fails to file any record, application, manifest, report, or other document required to be maintained or filed for purposes of compliance.
5. Transports a hazardous waste without a manifest, or causes a hazardous waste to be transported without a manifest.
6. Exports a hazardous waste (A) without the consent of the receiving country, or (B) where there exists an international agreement between the United States and the government of the receiving country establishing notice, export, and enforcement procedures for the transportation, treatment, storage, and disposal of hazardous wastes, in a manner which is not in conformance with such agreement.
7. Stores, treats, transports, or causes to be transported, disposes of, or otherwise handles any used oil not identified or listed as a hazardous waste—(A) in violation of any material condition or requirement of a permit, or (B) in violation of any material condition or requirement of any applicable regulations or standards.

16.7.1.3 Imminent and substantial endangerment

Section 7003 of RCRA provides EPA with authority to issue administrative orders and institute civil actions requiring abatement of conditions that may "present an imminent and substantial endangerment to health or the environment." That statutory section gives EPA the authority to address imminent hazards and creates liability for those persons who are contributing or have contributed to the handling, storage, treatment, transportation, or disposal of solid and hazardous waste.

This statutory provision is particularly applicable at operations in the petroleum industry where CERCLA's petroleum exclusion is applicable.

16.7.2 RCRA regulatory liability for generators

The federal RCRA regulations delineate numerous requirements for hazardous waste generators in 40 *CFR* Part 262, which implement the statutory language in RCRA Section 3002. The regs also note that a hazardous waste generator "is subject to the compliance requirements and penalties prescribed in Section 3008 of [RCRA] if he does not comply with the requirements of this part." [§262.10(g)]

The generator regs go on to say at §262.10(h) that a TSD facility "who initiates a shipment of hazardous waste…must comply with the generator standards established in this part." Thus, a TSD facility (e.g., a RCRA-permitted incinerator) that treats a generator's hazardous waste and then manifests hazardous ash to a RCRA-permitted landfill has to take on the generator responsibilities for the ash.

Finally, the generator regs at §262.12(c) require that the generator use a transporter with an EPA ID number and use a TSD facility with an EPA ID number.

16.7.3 EPA guidance on generator liability

EPA guidance regarding RCRA liability for generators follows:

- "The Agency believes that one way to address this problem is to limit the activities of the waste hauler to the transportation of wastes and to make the generator responsible for determining where the waste will be disposed of and ensuring that it gets there. Therefore, this section has been revised to require the generator to designate one facility on the manifest. Further, the generator must ascertain that the designated facility is permitted to accept his particular waste." [February 26, 1980; 45 *FR* 12728]

- "The Agency policy is that if the generator, in good faith, follows all the generator requirements in the regulations implementing RCRA at 40 *CFR* 262, through and including the receipt of a signed copy of his manifest from the designated facility or the filing of an exception report, and complies with the other recordkeeping requirements in Subpart D, the Agency would not take action under 40 *CFR* 262 against the generator for mishandling of hazardous waste by the designated permitted facility. Part of the generator's responsibility, of course, is to ascertain that the designated facility is permitted to handle his particular waste…. However, notwithstanding the fact that the generator has complied with the requirements in 40 *CFR* 262, the Agency may still take legal action, in appropriate cases, pursuant to the emergency provisions in any appropriate environmental statutes including Section 7003 of RCRA…." [RO 11022]

- "[I]t has consistently been EPA's position that the disposal facility remains responsible for ensuring that restricted wastes are not disposed except in full compliance with all applicable treatment standards. See 51 *Fed. Reg.* 40597 (Nov. 7, 1986). A rule of strict liability applies under RCRA, so that a disposal facility can be liable for improper disposal of untreated waste even if it does so in the good faith belief that the treatment standard does not apply. As noted above, this is no different from the regime under which disposal facilities operate generally as to other RCRA requirements." [RO 13630]

- "Though the generator is responsible for properly identifying and classifying the waste, the TSDF will be held liable by enforcement authorities if it violates its permit conditions and any other applicable regulations." [EPA/530/R-94/024, available at http://www.epa.gov/osw/hazard/tsd/ldr/wap330.pdf]

- "Under the full Subtitle C program, only the waste handler that violates a hazardous waste regulation is 'liable' (i.e., subject to enforcement) for that violation. Generators of hazardous waste are not responsible for mismanagement by subsequent waste handlers." [RO 14088]

16.7.4 What about CERCLA liability

RCRA regulates how wastes should be managed to avoid potential threats to human health and the environment. CERCLA, on the other hand, comes into play when mismanagement occurs or has occurred; that is, CERCLA applies when there has been a release to the environment of hazardous substances (including hazardous wastes) that may present an imminent and substantial danger to public health. [§300.3(a)(2)]

While RCRA liability is limited to that discussed above, CERCLA liability is much more far-reaching. EPA's thoughts on the use of CERCLA statutory provisions for assigning liability for mismanagement of hazardous waste include:

- "As you know, the generator retains potential liability under Superfund for future mismanagement of hazardous waste even after it has left his site and is out of his possession." [RO 11589]

- "Generators are responsible for subsequent mismanagement under CERCLA, however. The [universal waste (UW)] rule does not change CERCLA liability. Since UW are still hazardous wastes, persons who generate UW remain liable under CERCLA for remediation of any releases of UW." [RO 14088]

- "Please be aware that generators may be held liable under the Comprehensive Environmental Response, Compensation and Liability Act (CERCLA) for any environmental damages caused by the release of a hazardous material into the environment." [RO 14462]

16.7.5 Ways to minimize RCRA/CERCLA liability for hazardous waste shipments

To minimize the potential RCRA and/or CERCLA liability as discussed above, consider implementing the following three practices:

1. Audit offsite TSD facilities to which your hazardous wastes will be shipped. This can be accomplished using personnel from your own company or using third-party audits (e.g., CHWMEG, Inc.).

2. Request certificates of destruction and/or certificates of disposal from offsite TSD facilities.

3. If you ship your hazardous waste to a TSD facility but it (or treatment residue) is then shipped to another TSD facility, request copies of the manifest that is prepared at the first facility for the subsequent transportation of the waste to the ultimate treatment and disposal facility. [RO 11589]

16.7.6 Conclusion

The term "cradle to grave" has been interpreted in different ways by different entities. Hazardous waste generators are subject to minimal liability under RCRA Subtitle C if they 1) diligently follow Part 262 requirements while the waste is onsite, 2) ship the waste using reputable hazardous waste transporters, 3) use only audited TSD facilities that are permitted to accept the waste, and 4) receive a signed copy of the manifest from the designated TSD facility for every shipment. However, any mismanagement of the hazardous waste after it has been received by the TSD facility may still subject the generator to liability under CERCLA or RCRA Section 7003.

Corrective Action

17

Investigating and cleaning up releases from SWMUs

Prior to 1984, EPA had limited authority to require cleanup or other remediation at hazardous waste treatment, storage, and disposal (TSD) facilities. The regulations at that time only required facility owners and operators to address releases of hazardous wastes and constituents to ground water from "regulated units." ("Regulated units" are defined in §264.90 as surface impoundments, waste piles, land treatment units, or landfills that received hazardous waste after July 26, 1982.) However, the regulations did not include any provisions requiring cleanup (i.e., corrective action) of releases from other types of units. For example, the 1976 RCRA statute didn't give the agency any authority over ground water contamination caused by an old (pre-RCRA) landfill or surface impoundment.

The Hazardous and Solid Waste Amendments (HSWA), passed by Congress in 1984, substantially expanded EPA's authority to require corrective action at permitted and interim status TSD facilities. The primary source of that authority is RCRA Section 3004(u), as contained in HSWA:

"Continuing releases at permitted facilities—Standards promulgated under this section shall require, and a permit issued after November 8, 1984 by the Administrator or a state shall require, corrective action for all *releases* of *hazardous waste or constituents* from any *solid waste management unit* at a treatment, storage, or disposal *facility seeking a permit* under this subchapter, *regardless of the time at which waste was placed in such unit*. Permits issued under Section 3005 shall contain schedules of compliance for such corrective action (where such corrective action cannot be completed prior to issuance of the permit) and assurances of financial responsibility for completing such corrective action." [Emphasis added.]

The overall effect of this statutory section is that permitted and interim status TSD facilities can be required to clean up contamination resulting from the release of hazardous constituents from nonhazardous waste management units, regardless of when the release occurred. The solid waste management units do not have to be active to be subject to these requirements, and cleanup is not limited to the owner's/operator's property unless access is denied by an offsite property owner. [See RCRA Section 3004(v), as contained in HSWA.]

One purpose of the corrective action program is to ensure that RCRA-permitted facilities do not become Superfund sites. Congress reasoned that, if releases of hazardous constituents from waste management units at active RCRA-permitted facilities are not cleaned up and the facilities are later abandoned or sold for nonindustrial use, they could end up on the National Priorities List (i.e., the list of Superfund sites). Congress believed that the current facility owners/operators should be responsible for cleaning up releases at RCRA-permitted facilities, so that these sites don't eventually end up in the Superfund program.

This chapter summarizes the regulations and guidance EPA and authorized states have relied on since 1984 to implement corrective action at TSD facilities around the United States. EPA estimates that about 3,779 facilities are in need of corrective action. [http://www.epa.gov/epawaste/hazard/correctiveaction/facility/index.htm]

Section 17.1 summarizes the origins of the corrective action program and discusses how we got to where we are today. Sections 17.2–17.7 define the primary terms (e.g., solid waste management unit) and triggers that determine the scope and applicability of this program and contrast corrective action with Superfund cleanups. Sections 17.8–17.10 review the steps required to implement a full corrective action at a TSD facility and discuss strategies EPA uses to measure corrective action progress and completion, both at an individual facility and nationwide. The remediation waste management units EPA has established to provide increased cleanup flexibility at facilities undergoing corrective action are reviewed in Sections 17.11–17.14, while the use of a streamlined permit for remediation waste management sites is discussed in Section 17.15. (Strategies for managing hazardous soil and debris generated during remediations at these sites are discussed in Sections 13.7 and 13.8, respectively.) Finally, Sections 17.16 and 19.1.4 list guidance documents and tabulate the primary remedial options available to remediation project personnel.

17.1 Few regulations codified

The corrective action program is an important part of RCRA, but the federal regulations contain few specific corrective action requirements. Prior to HSWA, EPA established a program (in Part 264, Subpart F) for monitoring and remediating releases to ground water from permitted hazardous waste surface impoundments, waste piles, land treatment units, and landfills. These regulations were codified in 1982 and are not the subject of this chapter.

In 1985, EPA added §264.101 to Subpart F. This section basically reiterates the language in RCRA Section 3004(u); that is, it requires corrective action for releases from solid waste management units (SWMUs) at facilities seeking a RCRA permit. In 1987, EPA amended §264.101 to implement RCRA Section 3004(v); that is, the regs were revised to indicate that a facility owner/operator must implement corrective action beyond its facility boundary, where necessary, unless it cannot get permission from the neighboring property owner. At the same time, EPA amended the Part B permit application requirements, adding a provision that requires permit applicants to submit information on the SWMUs at their facility. [§270.14(d)]

EPA proposed a major rule on July 27, 1990 [55 FR 30798] that would have codified detailed procedures and technical requirements (in a new Part 264, Subpart S) for implementing corrective action under Section 3004(u). The 1990 proposal was very controversial, however, and most of it was never finalized. Even so, soon after it was published, EPA and states authorized to administer the corrective action program began using the proposed rule and associated preamble as the primary guidance for the corrective action program. One part of the 1990 Part 264, Subpart S proposal was finalized in 1993; it deals with corrective action management units (CAMUs) and will be discussed in Section 17.11.

On May 1, 1996 [61 FR 19432], EPA published an advanced notice of proposed rulemaking (ANPRM) outlining its strategy for codifying corrective action regulations. The ANPRM summarized the agency's experience with corrective action up to that point, identified several problems with the way the program was being implemented, and requested comment on a number of issues. EPA now considers the 1996 ANPRM the primary guidance for implementing corrective action. [RO 14021]

The agency announced its decision to withdraw most provisions of the 1990 Subpart S proposal on October 7, 1999. [64 FR 54604] The regulations in Subpart S that had already been finalized (i.e., the CAMU regulations) were not withdrawn. At the same time, the agency indicated that it has no plans to issue additional corrective action regulations. As a result, the statutory corrective action requirements of Section 3004(u) continue to be implemented based on policy documents and guidance, such guidance primarily being the 1990 Subpart S proposal and the 1996 ANPRM. These two documents are the basis for much of this chapter. We turn now to EPA's definitions of the terms Congress used in RCRA Section 3004(u), as enacted by HSWA.

17.2 Solid waste management units

Facilities subject to corrective action are required to investigate and possibly clean up releases of hazardous wastes or constituents from "solid waste management units." According to EPA, a SWMU is "any discernible unit at which solid wastes have been placed at any time, irrespective of whether the unit was intended for the management of solid or hazardous waste. Such units include any area at a facility at which solid wastes have been routinely and systematically released." [55 FR 30808]

A discernable unit in this context includes all of the types of units typically used for waste management, such as landfills, surface impoundments, land treatment units, waste piles, tanks, sumps, container storage areas, incinerators, and injection wells. Note that wastewater treatment units and waste recycling units are generally exempt from the hazardous waste management requirements of RCRA but are considered SWMUs subject to corrective action. [55 FR 30808] Industrial sewers that collect wastes from manufacturing processes are SWMUs, as are open ditches that convey wastewater.

Old, inactive waste disposal areas, such as old disposal trenches, pits, and open burn areas, would qualify as SWMUs, regardless of when the disposal occurred. Maintenance areas have also been identified as SWMUs at permitted facilities. Additional units/areas that have been identified as SWMUs are listed in Table 17-1.

Table 17-1: Examples of Units/Areas Identified as SWMUs

- Coal pile surface water run-off collection ponds
- Spent battery accumulation areas
- Asbestos storage/disposal areas
- Septic tanks
- Aboveground used oil storage areas
- Ordnance storage/disposal areas
- Dust collectors associated with paint, wood, and metal shops
- Fluorescent light bulb disposal areas
- Pesticide rinsate tanks and rinse areas
- Sandblasting waste storage/disposal areas
- Underground used oil storage tanks
- Abandoned underground piping such as fuel lines from old underground tank farms
- Waste tire storage areas
- Perchloroethylene tanks associated with dry cleaners
- Storage lockers for small waste containers
- Popping furnaces
- Spill areas associated with products and wastes
- Firefighter burn pads
- Salvage yards
- Drainage ditches
- Wash racks
- Low-level radioactive waste ponds
- Rifle bore cleaning areas
- Fuel oil tanks which have releases from them

Source: McCoy and Associates, Inc.

Q *Should a scrap metal storage area be considered a SWMU?*

A Even though scrap metal destined for recycling is not regulated as a hazardous waste under RCRA, it is considered a solid waste [see Table 1 in §261.2(c)]. Therefore, a scrap metal storage area could be a SWMU. [RO 12415]

Q *If a storm water retention pond contains sediments that fail the TCLP, would it be considered a SWMU?*

A The term "SWMU" includes any unit at a facility from which hazardous constituents might migrate, irrespective of whether the unit was intended for the management of solid or hazardous waste. Therefore, storm water retention ponds containing sediments that fail the TCLP would generally be considered SWMUs. [RO 14253]

Q *Military firing ranges and impact areas are often hazardous due to the presence of unexploded ordnance. Are these areas solid waste management units?*

A No. These areas should not be considered SWMUs. Unexploded ordnance and fragments of exploded ordnance fired during target practice are not discarded materials—they are materials being used for their intended purpose. Hence, these materials are not solid wastes. [55 FR 30809]

17.2.1 Routine and systematic releases

The definition of SWMU includes "any area at a facility at which solid wastes have been routinely and systematically released." One example of such a SWMU would be a "kickback drippage" area at a wood preserving facility, where pressure-treated wood is stored in a manner that allows preservative fluids to routinely and systematically drip on the soil. Another example might be an area at a facility where rail cars are loaded and unloaded. If a hose used to load/unload the cars was disconnected and dropped on the ground repeatedly, releasing a small amount of material that, over time, resulted in contaminated soil, this area would be considered a SWMU. Still another example would be an outdoor area at a facility used for solvent washing of large parts. If the solvents were allowed to continually drain on the soil, that area could be considered a SWMU.

At one point in time, EPA issued guidance indicating that routine, systematic, and *deliberate* releases are subject to corrective action. The agency later decided that areas which have become contaminated by routine and systematic releases of hazardous wastes or constituents are SWMUs. "It is not necessary to establish that such releases were deliberate in nature." [RO 12969]

The "routine and systematic" language is important to remember during the SWMU identification process. Facilities have been able to use this language to their advantage in removing potential SWMUs from the state's preliminary list.

Q *Are one-time spills or leakage from product storage or production processes subject to corrective action?*

A "A one-time spill of hazardous wastes (such as from a vehicle traveling across the facility) would not be considered a solid waste management unit. If the spill were not cleaned up, however, such a spill would be illegal disposal, and therefore subject to enforcement action under Section 3008(a) or Section 7003 of RCRA. Similarly, leakage from a chemical product storage tank would generally not constitute a solid waste management unit; such 'passive' leakage would not constitute a routine and systematic release since it is not the result of a systematic human activity. Likewise, releases from production processes, and contamination resulting from such releases, will generally not be considered solid waste management units, unless [EPA or the state] finds that the releases have been routine and systematic in nature. (Such releases could, however, be addressed as illegal disposal....)" [55 FR 30809]

17.2.2 Manufacturing and product storage areas

The last question and answer raises another question: "Can manufacturing process units or product storage tanks be considered SWMUs subject to corrective action?" EPA has given conflicting answers to this question.

In a 1988 guidance document, the agency indicated product or process units would not be SWMUs. But, areas contaminated by routine and systematic discharges from product or process units would be SWMUs. [RO 13125] In 1991, EPA again held the position that a product tank would not be considered a SWMU because it was used exclusively to store product. However, the agency did not render a definitive opinion as to whether the area surrounding and underneath the leaking tank should be considered a SWMU, choosing to rely instead on other authorities [i.e., the "omnibus" provisions in §270.32(b)(2)] to address the releases at the site. [RO 13441]

In 1998, EPA took a more aggressive posture as to whether manufacturing process units could be considered solid waste management units. The agency noted in RO 14309 that manufacturing process units often hold materials that can be classified as solid wastes and potentially hazardous wastes (e.g., precipitated residues). Even though these materials are exempt from hazardous waste regulation under §261.4(c), they are still considered solid wastes, thereby rendering the manufacturing process units as solid waste management units. However, EPA may exercise differing statutory authority to require cleanup at the facility.

We have two things to say about this issue: 1) It would be helpful if EPA issued additional guidance to clarify whether the 1998 interpretation reflects the agency's current stance on manufacturing process units as SWMUs; and 2) such manufacturing units are potentially subject to the corrective action program as SWMUs, but only RCRA-permitted and interim status facilities are subject to corrective action.

17.2.3 EPA discourages arguing about "SWMUs"

The definition of "SWMU" is often a point of disagreement when corrective action permits or orders are issued. Facility owners/operators often argue that the RCRA corrective action program should be focused on waste management units and that nonwaste management related releases (e.g., spills of raw materials or products) should be addressed by other cleanup programs. On the other hand, EPA believes that corrective action can be used to address all unacceptable risks to human health or the environment at RCRA-permitted facilities. Citing their authority to require cleanup of non-SWMU related releases under other statutory authorities, EPA asserted that "extended debate or litigation over a particular SWMU designation will in many cases be unproductive for all parties and, as a general principle, EPA discourages debate on these issues…." [May 1, 1996; 61 *FR* 19443]

17.2.4 Areas of concern

When EPA or a state issues a corrective action permit or order, they sometimes require facilities to investigate releases in an "area of concern." This term has no specific definition; it is basically a catch-all term used to require facilities to investigate potential releases, regardless of whether they are associated with a specific SWMU. For example, when an overseeing agency believes one-time spills of hazardous wastes or constituents have not been adequately cleaned up, these releases are often addressed as areas of concern. Depending on the extent of contamination associated with such releases, the area may subsequently be designated a SWMU.

17.3 Hazardous waste and constituents

Under the corrective action program, RCRA Section 3004(u) requires facilities to be concerned with releases of "hazardous waste or constituents" from SWMUs. The hazardous constituents do not necessarily have to be derived from hazardous waste; they may be derived from nonhazardous wastes. In 1990 [55 *FR* 30874], EPA proposed the following definitions for these terms:

"*Hazardous constituent* means any constituent identified in Appendix VIII of 40 *CFR* Part 261, or any constituent identified in Appendix IX of 40 *CFR* Part 264."

"*Hazardous waste* means a solid waste, or combination of solid wastes, which because of its quantity, concentration, or physical, chemical, or infectious characteristics may cause, or significantly contribute to, an increase in mortality or an increase in serious irreversible, or incapacitating reversible, illness; or pose a substantial present or potential hazard to human health or the environment when improperly treated, stored, transported, or disposed of, or otherwise managed. The term hazardous waste includes hazardous constituent as defined above."

A few notes on these definitions are in order.

1. The proposed definitions were never finalized but are the only guidance for these terms available.
2. The definition of "hazardous constituent" given above references Appendix VIII of Part 261 and Appendix IX of Part 264. Appendix IX is basically a subset of Appendix VIII; that is, it consists of those Appendix VIII constituents for which it is feasible to analyze in ground water samples. However, Appendix IX also includes constituents not found in Appendix VIII that are commonly addressed at Superfund ground water cleanups.
3. The definition of "hazardous waste" given above is the same as the statutory definition found in Section 1004(5) of the RCRA statute but is broader than the regulatory definition in §261.3.

Finally, EPA indicated that investigation of releases from SWMUs should focus on the hazardous wastes and constituents likely to have been released at a particular site, based on the available information. "Only where very little is known of waste characteristics, and where there is a potential for a wide spectrum of wastes to have been released, would the owner/operator be required to perform extensive or routine analysis for a broader spectrum of wastes." [55 *FR* 30809–10]

17.4 Releases

A "release," as included in RCRA Section 3004(u), is any "spilling, leaking, pumping, pouring, emitting, emptying, discharging, injecting, escaping, leaching, dumping, or disposing into the…air, surface water, ground water, and soils…." [July 15, 1985; 50 *FR* 28713] This is essentially the same as the CERCLA definition of release. EPA also considers abandoned or discarded containers that have hazardous waste or constituents inside a "release." [55 *FR* 30808] In some situations, EPA will impose RCRA corrective action requirements on releases that were permitted under some other environmental program, such as the Clean Water Act.

17.5 Facility

For purposes of Section 3004(u) corrective action, a "facility" is all contiguous property under the control of the owner or operator seeking a RCRA permit. [§260.10] Two parcels of land under common ownership but separated by a road or public highway are considered to be contiguous and constitute a single facility. The term "contiguous property" is also significant when applied to a facility that is owned by one entity and operated by another. In 1990, EPA gave the following example: If a 100-acre parcel of land is owned by a company that leases five acres of it to another company, and the second company stores hazardous waste on the five leased acres, the facility for purposes of corrective action would be the entire 100-acre parcel. [55 *FR* 30808] In 1996, the agency reconsidered this approach, noting that there are differing views regarding the policy merits of the foregoing interpretation and inviting further public comment. [61 *FR* 19442]

Another situation in which the definition of "facility" is important is when two adjacent properties are owned by subsidiaries of the same corporate parent. For example, in some parts of the country it is common for a petroleum refinery and a chemical plant to be co-located (i.e., they share a common boundary), as shown in Figure 17-1. The refinery is owned by XYZ Refining Company, and the chemical plant is owned by XYZ Chemical Company. The chemical plant is seeking a RCRA permit for a hazardous waste management unit. Ordinarily, these would be considered two separate facilities; for example, they have two separate EPA ID numbers. However, because both companies are subsidiaries of the same parent, XYZ Corporation, then, even though only the chemical plant is seeking a RCRA permit, both properties constitute one facility for purposes of corrective action. [55 FR 30808]

Another common scenario involves two geographically separated parcels of land owned by the same person. If the properties are connected by ditches, bridges, sewer systems, or other links under the control of the facility owner/operator, they can be considered a single facility. In one case, evaporation ponds three miles from a refinery were considered part of the facility, because they were linked to the refinery by a drainage ditch controlled (although not owned) by the refinery owner. [61 FR 19442]

17.6 Applicability issues

The corrective action requirements apply at hazardous waste TSD facilities. These include permitted facilities and facilities that have, have had, or should have had interim status. In general, corrective action requirements do not apply to generators who aren't subject to RCRA permitting requirements.

Two types of permits will make a facility subject to corrective action: 1) operating permits (sometimes referred to as "Part B" permits) for treatment, storage, or disposal of hazardous waste; and 2) post-closure permits. Post-closure permits are required for any surface impoundment, waste pile, land treatment unit, or landfill that received hazardous waste after July 26, 1982, or which ceased the receipt of hazardous waste prior to July 26, 1982 but did not certify closure until after January 26, 1983. However, a post-closure permit is not required if the unit is "clean closed." EPA's position is that clean closure levels should be equivalent to corrective action cleanup levels, and they should be risk-based. [RO 11959]

The following types of RCRA permits do not usually trigger corrective action [61 FR 19441]:

- Permits for land treatment demonstrations;
- Emergency permits;
- Ocean disposal permits; and
- Research, development, and demonstration permits.

The aforementioned permits are relatively rare. EPA's rationale for not requiring corrective action at facilities that receive these types of permits is discussed in the July 27, 1990 Subpart S proposal. [55 FR 30806]

Figure 17-1: Two Adjacent Plants May Be One "Facility"

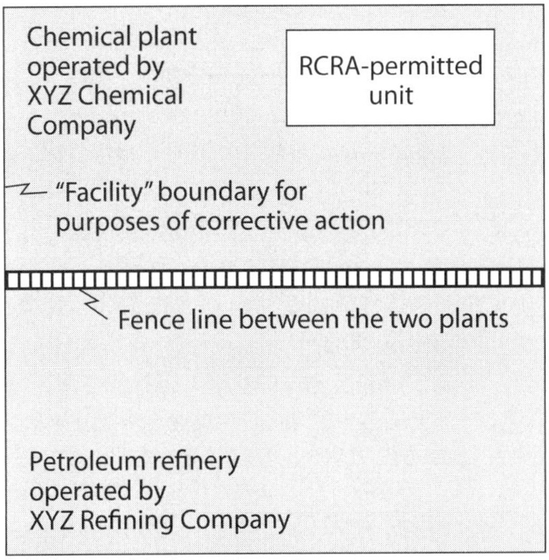

Source: McCoy and Associates, Inc.

Another type of permit that does not trigger facility-wide corrective action is a remedial action plan (RAP). A RAP is a permit (an enforceable document) that EPA or a state can issue to authorize treatment, storage, or disposal of remediation waste at a remediation waste management site. The regulations in Subpart H of Part 270 describe what a RAP must include and how such a plan can be obtained. RAPs and remediation waste management sites are discussed more fully in Section 17.15.

One final note on the applicability of corrective action at permitted facilities. Facilities that obtain a traditional RCRA Part B permit solely for the management of remediation waste are not subject to corrective action. See §264.101(d) and November 30, 1998; 63 FR 65883.

17.6.1 Corrective action at interim status facilities

Interim status TSD facilities are facilities that:

1. Were "in existence" on the effective date of statutory or regulatory requirements that rendered the facility subject to RCRA permitting,
2. Notified EPA or the state of their hazardous waste management activities and obtained an EPA ID number, and
3. Submitted a Part A permit application.

"In existence" as used above means the facility was actually treating, storing, or disposing hazardous waste or was under construction for such purposes. Refer to the definition of "existing hazardous waste management facility" in §260.10 for details on the meaning of "under construction."

In general, facilities that have qualified for interim status under Part 270, Subpart G are seeking a RCRA permit and, therefore, are subject to corrective action. Certainly, this is the case for operating facilities, but what about facilities that qualified for interim status in the past but have subsequently decided that they would prefer not to have a RCRA permit? This makes no difference to EPA—the agency's position is that once a facility has interim status, that facility is subject to corrective action. If a facility will not cooperate, and EPA believes there has been a release of hazardous waste or constituents, the agency can order the facility to undertake corrective action. [RO 12516] This authority is found in RCRA Section 3008(h), which is entitled "Interim status corrective action orders." This section allows agency action outside of the permitting process.

Q *In the past, some facility operators have submitted Part A permit applications (one of the requirements to qualify for interim status) but never actually treated, stored, or disposed hazardous waste. Would such a facility be subject to corrective action?*

A No. If a facility never did treat, store, or dispose hazardous waste, EPA does not consider that facility to have attained interim status, even though a Part A application was submitted (i.e., a "protective filing"). Facilities that have never engaged in treatment, storage, or disposal of hazardous waste are not subject to the corrective action provisions of RCRA Section 3004(u) or 3008(h). [September 25, 1985; 50 FR 38948, RO 12590]

17.6.2 Additional applicability examples

Q *A facility owner is preparing a RCRA permit application for the thermal treatment of hazardous waste generated onsite. Before treating this waste, the facility accumulates it onsite for less than 90 days in containers and tanks in compliance with §262.34. Does the facility operator have to include the 90-day accumulation units in its permit application?*

A If a generator accumulation unit meets the definition of a SWMU, the owner or operator is required to identify the unit in Part B of the permit application. [RO 14710]

Q *An interim status facility completes closure of its hazardous waste management units and ceases operation. Can the facility avoid corrective action?*

A No. The corrective action order provisions of Section 3008(h) can be used to require corrective action after closure. [RO 12444]

Q *An interim status facility completes closure of its hazardous waste management units and then resumes operation as a generator (i.e., the facility no longer accumulates wastes for more than 90 days). Can the facility avoid corrective action?*

A No. Section 3008(h) can be used to require corrective action at the facility because it operated under interim status at one time. However, the likelihood that the facility will actually have to begin the corrective action process may be reduced, because large quantity generators conducting 90-day accumulation will have a lower priority at the jurisdictional regulatory agency than permitted or interim status facilities with respect to corrective action.

Q *A facility that treats hazardous waste never achieved interim status (i.e., it did not submit a Section 3010 notification and/or a Part A permit application). Is this facility subject to corrective action?*

A Yes. The administrative order provisions of Section 3008(h) can be used to force facilities that should have obtained or that lost interim status to conduct corrective actions. [RO 11425, 12516]

Q *Can a facility owner/operator be required to implement corrective action beyond the facility boundary?*

A Yes. Unless the owner/operator demonstrates that, despite its best efforts, it was unable to obtain the necessary permission from the affected property owner to undertake such actions, corrective action beyond the facility boundary is required. [§264.101(c)]

Q *A facility owner finds that his/her property was contaminated due to actions of the prior owner. Is the current owner subject to corrective action requirements for problems caused by the previous owner?*

A Yes. Corrective action requirements apply regardless of when waste was placed in a solid waste management unit. "Accordingly, the owner or operator of a solid waste management unit containing only waste placed there by a previous owner would be fully responsible for corrective action for any release from such unit. This interpretation would not, of course, preclude such owner or operator from bringing any action otherwise allowed by law against the previous owner seeking remuneration for the costs of corrective action." [July 15, 1985; 50 *FR* 28714]

Q *Will closed municipal landfills suspected of containing hazardous wastes be subject to corrective action under RCRA?*

A Closed municipal landfills which are suspected of containing hazardous wastes would not be subject to the corrective action provisions under RCRA unless: 1) the municipal landfill is part of a facility and another part of the facility requires a RCRA permit, or 2) the facility has interim status or was required to have interim status. On the other hand, the municipal landfill is still subject to the provisions of CERCLA. [RO 11089]

17.7 RCRA corrective action vs. Superfund

Most facilities that are subject to corrective action under RCRA are potentially subject to cleanup under other state or federal authorities including CERCLA. CERCLA is an acronym for the Comprehensive Environmental Response, Compensation, and Liability Act, also known as Superfund. The RCRA corrective action program and CERCLA have the same fundamental goal—to clean up contaminated sites. Therefore, EPA believes both programs should arrive at similar remedial solutions.

"Generally, cleanup of any given site or area at a facility under RCRA corrective action or CERCLA will substantively satisfy the requirements of both programs. We believe that, as a general matter, RCRA and CERCLA program implementers can defer cleanup activities from part or all of a site to one program with the expectation that no further cleanup will be required under the other program. For example, when investigations or studies have been completed under one program, there should be no need to review or repeat those investigations or studies under another program. Similarly, a remedy that is acceptable to one program can be presumed to meet the standards of the other. The same principle should apply to authorized state corrective action programs and state CERCLA analogous programs." [61 *FR* 19441]

EPA emphasized the concept of parity between RCRA and CERCLA cleanups by establishing a policy for deleting active RCRA-permitted facilities from the Superfund's National Priorities List and deferring their cleanup to the RCRA corrective action program. [March 20, 1995; 60 *FR* 14641, November 24, 1997; 62 *FR* 62523] For additional information on the concept of parity, see RO 11959.

17.8 Corrective action process

Based on the complexity, agency oversight, and expense of the steps required to implement a full corrective action program (as detailed in this section), it is easy for some facilities to get bogged down in the process and difficult for them to actually clean up their sites. Through guidance, EPA is trying to encourage states to work with facilities to emphasize results rather than process:

"The purpose of the corrective action program is to stabilize releases and clean up RCRA facilities in a timely manner, not to ensure compliance with or fulfillment of a standardized process. Program implementors and facility owners/operators should focus on environmental results rather than process steps and ensure that each corrective action related activity at any given facility directly supports cleanup goals at that site." [61 *FR* 19441]

EPA's latest efforts in this area appear promising. Led by personnel at Regions 3 and 7, the agency is attempting to improve the efficiency of corrective action through the "Lean" process improvement system. "Lean" refers to a collection of principles and methods that focus on the identification and elimination of non-value-added activities involved in producing a product or delivering a service. More about EPA's Lean efforts is available at http://www.epa.gov/lean/government/index.htm.

The first corrective action process selected for efficiency improvement is the RCRA facility investigation (RFI), which characterizes the nature and extent of contamination at a facility (see Section 17.8.2). Applying Lean to remove various redundant steps within the RFI process and front-loading goals and expectations through a "corrective action framework" (CAF) offers the potential for significant time savings. A CAF guide and template are available at http://www.epa.gov/waste/hazard/correctiveaction/lean_effort.htm. Currently, this Lean RFI process is being piloted in all EPA regions.

The corrective action process involves five major elements or phases common to most environmental restoration projects: 1) an initial site assessment, 2) site characterization, 3) interim measures, 4) evaluation and selection of remedial alternatives, and 5) implementation of selected remedies. "These activities are not always undertaken as a linear progression towards final facility cleanup, but can be implemented flexibly to most effectively meet site-specific corrective action needs." [*Introduction to RCRA Corrective Action*, EPA/530/K-02/017I, October 2001, available from EPA at http://www.epa.gov/wastes/inforesources/pubs/training/cact.pdf] A process flow chart of the RCRA corrective action process is presented in Figure 17-2, and major elements are discussed in the following subsections.

17.8.1 RCRA facility assessment

The corrective action process typically starts when a facility applies for a RCRA permit by submitting its Part B permit application. The application must identify and describe any SWMUs at the facility. According to §270.14(d), the permit application must include the following information for each SWMU at the facility:

- Location of the unit on a topographic map of the facility;

- Designation of the type of unit;

Figure 17-2: RCRA Corrective Action Process

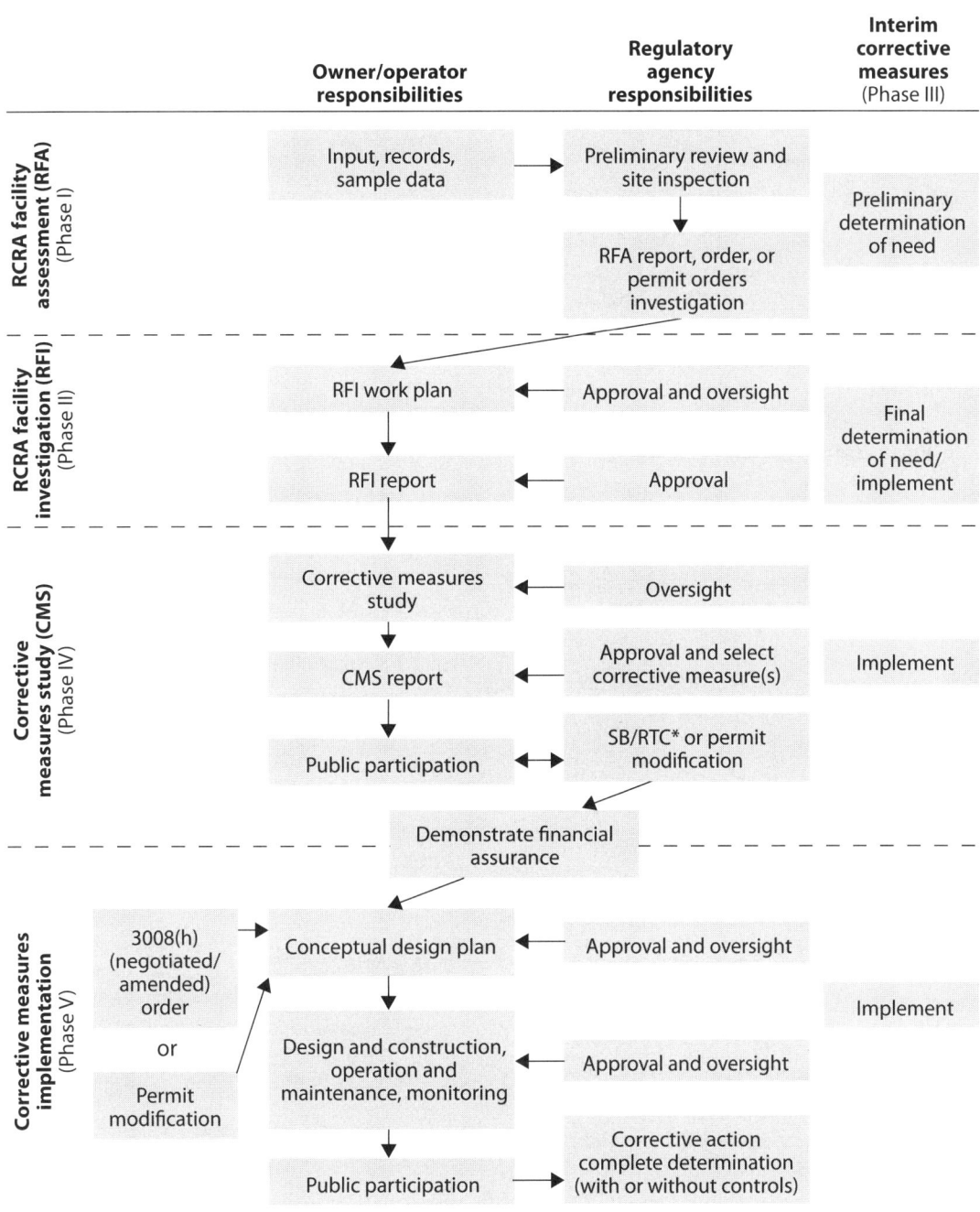

* The statement of basis/response to comments (SB/RTC) or permit modification documents the selected corrective measure(s).
Source: Adapted from OSWER Directive 9902.3-2A.

- General dimensions and structural description, with drawing if available;
- Dates the unit was operated; and
- Specification of all wastes that have been managed at the unit, to the extent available.

In addition, the facility owner/operator must submit all available information pertaining to releases from their SWMUs and may be required to sample and analyze ground water, soil, and other environmental media at the facility.

The information submitted by the permit applicant is then used to prepare the RCRA facility assessment (RFA). The purpose of the RFA is to identify all SWMUs at the facility and determine if there have been releases or suspected releases from them that require further investigation. An RFA is analogous to the preliminary assessment/site investigation conducted at Superfund sites.

Ordinarily, the RFA will involve a visual inspection of the facility and a review of the existing information on environmental conditions at the site. Existing plant records (e.g., inspection reports, permit applications, citizen complaints), interviews with personnel familiar with the facility, and historical monitoring data and documents (e.g., old photographs of the site, documentation of fish kills, odors, and/or worker illnesses) are reviewed for this purpose.

Typically, not much field work is conducted during the RFA, but the regulators may request additional information and/or sampling and analytical work. This information may be used to focus subsequent, more detailed investigation on areas, releases, and exposure pathways that pose the most significant risks to human health or the environment, or to eliminate areas from consideration during the next step in the corrective action process.

The RFA is usually conducted by the regulatory agency that is writing the facility's permit (i.e., the state or EPA if the state is not authorized to administer corrective action). However, in the 1996 ANPRM, EPA indicated that facility owners/operators may conduct their own RFA:

"Where RFAs have not yet been completed, facility owners/operators may choose to conduct their own site assessment and submit the report to EPA for review. If EPA believes the site assessment is adequate, the site assessment may be approved and adopted as the RFA for the facility…. Facility owners/operators who choose to conduct or update their own RFAs should ensure that they address all solid waste management units and other areas of concern at the facility." [61 FR 19443–4]

If the facility is operating under interim status and is not seeking a RCRA permit, the RFA may take place before the facility closes. [EPA/530/K-02/017I]

Q *If a release of hazardous waste or constituents is identified after a RCRA Part B permit has been issued, can EPA reopen the permit and modify it to include additional investigation and/or corrective measures? Does the "permit as a shield" protect a facility from such action until the permit comes up for renewal?*

A EPA has the authority to reopen the permit and impose corrective action requirements if new information about a release comes to light after the original permit was issued. The "permit as a shield" would not protect a facility from such an action. [RO 13134]

The most comprehensive guidance available on RFAs is *RCRA Facility Assessment Guidance*, EPA/530/86/053, October 1986. This guidance document is available at http://www.epa.gov/osw/hazard/correctiveaction/resources/guidance/sitechar/rfaguid.pdf.

If, based on the RFA, it seems likely that there has been a release from one or more SWMUs at a facility, the implementing agency will order an investigation. At that point, the facility owner or operator would begin the RCRA facility investigation.

17.8.2 RCRA facility investigation

A RCRA facility investigation (RFI) is a thorough, detailed site investigation conducted by the facility owner/operator and is comparable to the remedial investigation in the Superfund program. The purpose of an RFI is to characterize the nature, extent, direction, rate, movement, and concentration of releases from SWMUs at a facility. The scope and complexity of an RFI will depend on the nature and extent of the contamination, whether releases have migrated beyond the facility boundary, the amount of existing information at the site, and other pertinent factors. In the 1990 Subpart S proposal, EPA provided the following list of specific information that may be required in an RFI [55 *FR* 30810]:

- Characterization of the environmental setting;
- Characterization of SWMUs;
- Description of the human and environmental receptors that are, have been, or may potentially be exposed to contaminants released at the site;
- Information that will assist the regulatory agency in assessing the risks to human and environmental receptors at the site;
- Extrapolation of future contaminant movement;
- Results of laboratory, bench-scale, or pilot-scale tests conducted to determine the potential effectiveness of treatment or other technologies that may be used as remedies at the site; and
- Statistical analyses to aid in the interpretation of data obtained in the investigation.

Not all of the information listed above will be required in every RFI. The specific information required for a particular facility will typically be identified in an RFI work plan prepared by the facility owner/operator. EPA or the state must approve the plan before the RFI moves forward. The approved plan typically becomes an enforceable part of the schedule of compliance in the facility's permit. Requirements for an RFI may also be specified in a corrective action order.

The permit or order may also include "action levels" for specific constituents in specific media under investigation. If the RFI indicates that the action levels have been exceeded, interim measures and/or a corrective measures study may be required for a SWMU or group of SWMUs. On the other hand, where action levels have not been exceeded, the agency may determine that no further action is required under corrective action for a SWMU or group of SWMUs. Action levels are discussed in greater detail in Section 17.8.2.1. At the conclusion of an RFI, the facility will submit a final RFI report to EPA or the state. A summary of this report will be distributed to the interested public.

As noted in Figure 17-2, significant agency oversight and approval are required during the RFI. As a result, this phase of the corrective action process often takes many years. We know of situations where a facility has sent their RFI work plan to the state for approval, and nothing was heard from the state for three or four years. Sometimes, just when some progress is being made, there will be turnover at the state agency, and the review process starts all over again. In the 1996 ANPRM, EPA suggested ways to focus and accelerate the RFI process, including the use of conceptual site models, innovative site characterization technologies, and tailored data quality objectives. [61 *FR* 19444]

The most comprehensive guidance on the RFI process is *Interim Final RCRA Facility Investigation (RFI) Guidance*, EPA/530/SW-89/031, May 1989. This four-volume document can be accessed at http://www.epa.gov/epawaste/hazard/correctiveaction/resources/guidance/sitechar/index.htm.

17.8.2.1 Action levels

Action levels are contaminant concentrations used to indicate whether cleanup of a release from a SWMU or group of SWMUs will be required. Contamination that is present below action levels would not generally be subject to cleanup or further investigation. Action levels are media-specific; that is, the action level for a given contaminant will vary depending on whether the contaminant is present in soil, ground water, surface water, or air. EPA recommends that action levels be health risk-based and/or environmental risk-based where environmental receptors are of concern. Action levels are often established at the more protective end of the risk range (e.g., one-in-one-million excess cancer risk to the maximally exposed individual over a lifetime of exposure) using conservative exposure and land-use assumptions. Maximum contaminant levels (MCLs) developed under the Safe Drinking Water Act are sometimes used as action levels for ground water. However, action levels based on less conservative assumptions may be appropriate based on site-specific conditions. For example, if the current and anticipated future uses of a site are industrial, action levels based on industrial exposure scenarios would be appropriate. [61 *FR* 19446]

Action levels can be developed on a facility-specific basis or can be taken from standardized lists. Some states and EPA regions have developed standardized risk-based action levels for RCRA corrective action facilities and other cleanup sites (e.g., sites subject to cleanup under state Superfund-like programs). One widely distributed list of action levels was contained in Appendix A of the preamble to the 1990 Subpart S proposal. [55 *FR* 30865] EPA subsequently indicated that some of the action levels in the 1990 proposal may no longer be appropriate for use—either because they are overly conservative or because they are based on out-of-date toxicity data. [61 *FR* 19446]

EPA's soil screening guidance presents a framework for developing risk-based soil screening levels. These screening levels (i.e., contaminant concentration levels) can be used to eliminate areas, pathways, and/or chemicals of concern from further consideration during site investigations. Three guidance documents are available: *Soil Screening Guidance: Fact Sheet*, EPA/540/F-95/041, July 1996; *Soil Screening Guidance: User's Guide*, EPA/540/R-96/018, July 1996; and *Soil Screening Guidance: Technical Background Document*, EPA/540/R-95/128, July 1996. These guidance documents are available at http://www.epa.gov/superfund/health/conmedia/soil/index.htm.

17.8.2.2 Determination of no further action

Based on the results of the RFI, a facility may ask the implementing agency to determine that no further action is required for a SWMU or group of SWMUs. This request must be accompanied by documentation that shows there are no releases of hazardous waste or constituents from a SWMU or group of SWMUs at the facility that pose a threat to human health or the environment. This determination will be straightforward where the facility owner/operator can show that no release has occurred; however, such a determination may also be appropriate in situations where releases that exceed action levels have occurred. For example, when concentrations of hazardous constituents exceed action levels, but the contamination is in a highly saline aquifer that is not a potential source of drinking water, no further action may be necessary. Other examples would be where contamination in ground water can be shown to originate from a source outside the facility or where the risks posed by the contaminants are insignificant compared to the risks posed by naturally occurring contaminants at the facility. [55 *FR* 30813]

A determination of no further action requires a Class 3 permit modification at a permitted facility. A Class 3 permit modification is the most difficult permit modification to get approved and requires an opportunity for pubic involvement. Procedures for obtaining a Class 3 permit modification are specified in §270.42(c). In guidance, EPA indicated that it may be appropriate in some cases for EPA or a state to make a completion determination (e.g., a determination that no further action is required) for a portion of a facility. This might enable the facility owner to sell that portion of the facility. [February 25, 2003; 68 *FR* 8762]

17.8.3 Interim corrective measures

Since 1990, EPA has emphasized the importance of interim actions and site stabilization in the corrective action program. Interim measures may be undertaken during the site investigation process and well before a final remedy is selected. Typically, interim measures are used to control ongoing releases or abate obvious risks to human health or the environment. For example, a contaminated drinking water supply might necessitate an interim action to provide an alternative source of drinking water. The discovery of damaged or leaking drums that contain hazardous waste might be addressed by overpacking and/or removing the drums from the facility. Interim measures should be employed as early in the corrective action process as possible and should be compatible with, or a component of, the final remedy. [61 FR 19446; see also RO 11648]

17.8.4 Corrective measures study

Contamination at most RCRA facilities can be addressed using a number of remedial alternatives, each of which have certain advantages and disadvantages. Some remedial options will be very protective of human health and the environment but very expensive to implement. For example, excavating all of the contaminated soil at a facility and sending it offsite to a hazardous waste incinerator may be very protective of the health of those living near the facility, but it will likely be very costly. Other approaches will be less expensive but also less protective. For example, capping contamination in place or relying on natural attenuation are less expensive alternatives but may also be less effective in protecting human health and the environment.

Before choosing a cleanup approach, the regulators and facility owner/operator typically evaluate a range of remedial alternatives for each SWMU or group of SWMUs that require cleanup. This evaluation process is referred to as the corrective measures study (CMS), and it is comparable to the feasibility study conducted under the Superfund program. The CMS is normally conducted by the facility owner/operator, with oversight from the regulatory agency. The CMS report, which is the deliverable from this stage, proposes appropriate corrective measures for each SWMU or group of SWMUs. Included in the report are measures to control the source of contamination and actions required to abate problems caused by migration of contaminants from the source. The recommendations in the CMS report should include proposed media cleanup levels, points of compliance, and compliance time frames.

At some facilities, the CMS will be a complex, time-consuming, and costly process. However, EPA believes that at many facilities a more streamlined, highly focused CMS will be sufficient to identify an acceptable, environmentally protective remedy. In the 1996 ANPRM [61 FR 19447], EPA discussed opportunities to streamline the CMS. One way to do this is to integrate the CMS with the site investigation. This may enable the owner/operator to focus the data collection effort on collection of information needed to support an appropriate remedy. For example, in a situation where the contamination involves a large mixed-fill landfill, the remedial alternatives will likely involve physical and institutional controls. If these alternatives are identified early in the RFI, the owner/operator may be able to focus on collection of data necessary to support development of effective controls.

In "A Study of the Implementation of the RCRA Corrective Action Program," April 9, 2002 (available as report number 531R02001 from http://nepis.epa.gov/EPA/html/Pubs/pubtitleOSWER.html), EPA indicated that a corrective measures study should focus on "realistic remedies" and encouraged the use of presumptive remedies. Presumptive remedies are preferred cleanup technologies for common categories of sites (e.g., wood treatment sites, or sites with volatile organic contaminants in soil). EPA began developing presumptive remedy guidance under the Superfund program to expedite the remedy selection process at those sites. EPA's policies and procedures related to presumptive remedies, along with a sampling of these remedies developed for Superfund sites, may be accessed at http://www.epa.gov/superfund/policy/remedy/presump/pol.htm.

17.8.4.1 Performance-based approach

At facilities that will use a "performance-based" approach to cleanup, the CMS does not have to be submitted to an overseeing agency for review and approval. Instead, EPA or the state would oversee the facility investigation and work with the facility owner/operator to develop remedial goals. After the remedial goals undergo public review and comment and are approved by the agency, the facility owner/operator would design and implement a remedy that will achieve the remedial goals. Obviously, EPA or the state would monitor compliance with the remedial goals. If the goals are not achieved, additional remediation will likely be required. Many states attribute the success of their corrective action programs, in part, to use of a performance-based approach to achieve goals to which all parties have agreed rather than requiring agency review and approval of an expensive and time-consuming CMS. [61 FR 19447]

17.8.5 Remedy selection

As indicated above, the purpose of a CMS is to identify and evaluate various remedial alternatives. At the conclusion of a CMS, the facility owner/operator recommends remedies or remedial goals (if a performance-based approach is being used). However, the regulators actually select the remedy or remedies for a facility.

In the 1996 ANPRM [61 FR 19448], EPA identified its remedial expectations as follows:

- Treatment should be used to address the principal threats posed by a site whenever practicable and cost-effective. Principal threats for which treatment is most likely to be appropriate include contamination that is highly toxic, highly mobile, or that cannot be reliably contained and that would present a significant threat to human health or the environment if exposure occurs.

- Engineering controls, such as containment, should be used for wastes and contaminated media that can be reliably contained, pose relatively low long-term threats, or for which treatment is impracticable.

- Institutional controls, such as water- and land-use restrictions, should be used primarily to supplement engineering controls. In general, institutional controls should not be the sole remedial action. Additionally, institutional controls should not be considered "no further action" remedies: "Institutional controls, while not actively cleaning up the contamination at the site, can control exposure and, therefore, are considered to be limited action alternatives." ["Summary of Key Existing EPA CERCLA Policies for Groundwater Restoration," June 26, 2009, available at http://www.epa.gov/superfund/health/conmedia/gwdocs/index.htm] Guidance on institutional controls is available at http://www.epa.gov/epawaste/hazard/correctiveaction/resources/guidance/index.htm.

- A combination of methods (e.g., treatment, containment, and institutional controls) may be necessary to protect human health and the environment.

- Innovative remediation technologies should be used if they will perform better or cost less than conventional technologies.

- EPA expects to return ground water to its maximum beneficial use whenever practicable, within a time frame that is reasonable. Often, this means cleaning up contaminated ground water to drinking water standards.

- Contaminated soil should be remediated as necessary to prevent or limit direct exposure of human and environmental receptors and prevent the transfer of contaminants to air, ground water, or surface water (e.g., by airborne emissions, leaching, or run-off).

EPA believes that final remedies selected for RCRA corrective action facilities should achieve the following three performance standards [55 FR 30823]:

1. *Protect human health and the environment*—This is EPA's basic mandate under RCRA. Current and reasonably anticipated future land uses should be considered. This standard requires remedies to include measures that may not be directly related to media cleanup or source control. For example, at some facilities, it may be necessary to provide an alternative drinking water supply or to construct barriers to prevent people from entering contaminated areas.

2. *Achieve media cleanup objectives*—The cleanup objectives should include proposed media cleanup levels, points of compliance, and compliance time frames. These concepts are discussed in greater detail below.

3. *Remediate the source of releases*—The source includes both the location of the original release and any location to which significant masses of contaminants have migrated. This will typically involve removal or treatment of the source; however, in some situations, containment technologies and/or institutional controls may be sufficient.

17.8.5.1 Media cleanup levels

Media cleanup levels are contaminant concentrations that may remain in the soil, ground water, surface water, and air at a facility when the cleanup is complete. The concentrations should be risk-based and should take into account site-specific conditions. At some facilities, generic cleanup levels, such as SDWA maximum contaminant levels or state cleanup standards, will be used. At other facilities, a site-specific risk assessment will be required to develop cleanup levels. Both approaches require a site-specific risk-based decision. When generic cleanup levels are used, the assumptions used to develop the cleanup levels should be consistent with conditions at the facility. Site-specific risk assessments conducted at RCRA facilities often rely on the extensive guidance EPA has developed for risk assessments conducted under CERCLA (available at http://www.epa.gov/oswer/riskassessment/risk_superfund.htm).

Cleanup levels for carcinogens should reduce the excess cancer risk to an individual over a lifetime of exposure to between one in one million and one in ten thousand (10^{-6} to 10^{-4} risk range). EPA's preference is to set cleanup levels at the more protective end of the risk range (i.e., based on 10^{-6} excess cancer risk); however, cleanup levels anywhere within this range may be acceptable. For noncarcinogens, EPA recommends that cleanup levels be set at concentrations that will not result in deleterious effects to humans over a lifetime of exposure. This means the hazard index should be less than or equal to one.

Media cleanup levels must also be protective of environmental receptors (i.e., plants and animals). EPA's latest guidance on this topic is also available at http://www.epa.gov/oswer/riskassessment/risk_superfund.htm. This site provides guidance to site managers who are responsible for designing and conducting technically defensible ecological risk assessments.

Cleanup levels may be based on future land use. For example, contaminated soil at an industrial site may not have to be cleaned up to residential standards if the site will never be used for residential purposes. At some facilities, deed restrictions are used to ensure that the property will not be converted to residential use in the future.

17.8.5.2 Points of compliance

The point of compliance (POC) is the location or locations at which media cleanup levels are achieved. These are typically established as follows [61 *FR* 19450]:

- For air releases, the POC is the location of the person most exposed or some point closer to the source of the release.

- For surface water, the POC is the point at which releases could enter the surface water. A POC may also be established for sediments at facilities where sediments may be affected.

- For soil, the POC is any point where direct contact to the soil may occur.

- For ground water, the POC is throughout the plume of contaminated ground water or, when waste is left in place, at the boundary of the waste management area encompassing the original source.

The POCs listed above are typically used at corrective action facilities, but these are not hard-and-fast rules. For example, when determining the POC for ground water at a facility, the potential future uses of the ground water and the practicality of ground water remediation should be considered. Two useful guidance documents for evaluating ground water POCs are: *Handbook of Groundwater Protection and Cleanup Policies for RCRA Corrective Action*, EPA/530/R-04/030, April 2004 (available online at http://www.epa.gov/epawaste/hazard/correctiveaction/resources/guidance/gw/gwhandbk/index.htm), and *Guidance for Evaluating the Technical Impracticability of Ground Water Restoration*, EPA/540/R-93/080, September 1993 (available from http://www.epa.gov/superfund/health/conmedia/gwdocs/techimp.htm).

17.8.5.3 Compliance time frames

When EPA or the state selects a remedy at a corrective action facility, the agency will issue a permit modification or a corrective action order that specifies the remedy and establishes a remedy implementation schedule. EPA's preference is for expeditious stabilization of releases, followed by timely completion of corrective actions and full restoration of contaminated ground water. However, the agency recognizes that uncertainties associated with remediation, especially ground water remediation, make it difficult to specify when a remedy will be completed.

17.8.5.4 Natural attenuation

Natural attenuation is an approach to remediation in which natural processes, such as biodegradation, dispersion, dilution, and/or adsorption, are used to achieve remedial goals. At some facilities, natural attenuation can be used to remediate contaminated ground water. Natural attenuation remedies are *not* "no further action" remedies. Considerable design, construction, operation, maintenance, and monitoring costs may be involved in such a cleanup option.

Remedies involving natural attenuation should include: a thorough site characterization, source control or removal where appropriate, documentation of the existence of attenuation processes at a facility and of their ability to achieve remedial objectives, a long-term monitoring plan, and, in some cases, a fall-back plan in case natural attenuation appears to be ineffective. [61 *FR* 19451]

17.8.5.5 Summary of the remedy selection process

Remedy selection is the responsibility of the regulatory agency that is overseeing corrective action at a facility. However, in actual practice, selecting a remedy is a negotiated process. A facility owner/operator may propose a remedy that is affordable, but not optimum in terms of protecting human health and the environment over the long term. The regulators may counter with a more expensive and more protective remedy. If the facility owner/operator cannot afford the more expensive remedy, it probably will not be implemented. Whatever remedy the two parties ultimately agree upon is subject to review and comment by interested members of the public before it becomes an enforceable requirement of the facility's permit.

17.8.6 Financial assurance

After a remedy or remedies have been selected for a facility, owners/operators are required to provide assurances of financial responsibility. [§264.101(b–c)] In other words, the facility owner/operator must be able to show that he/she can pay for the cleanup. RCRA permits include financial assurance provisions. Financial assurance is also typically included in corrective action orders. On October 24, 1986 [51 *FR* 37854], EPA proposed detailed regulations governing financial assurance for corrective action. Under the proposed rule, the mechanisms that could be used to provide financial assurance included trust funds, surety bonds guaranteeing performance, letters of credit, financial tests, and corporate guarantees. These are similar to the mechanisms used to provide financial assurance for closure and post-closure care of RCRA-permitted facilities. EPA subsequently indicated that insurance would also be an acceptable mechanism. Financial assurance is typically required at the time of remedy selection. The 1986 proposal also discussed remediation cost-estimating procedures for determining the amount of financial assurance required.

In 1990, EPA proposed to require a demonstration of financial responsibility for corrective action within

120 days of the permit modification used to select a remedy. [55 *FR* 30855] Neither of the proposed rules dealing with financial assurance for corrective action were finalized. The only codified regulations are in §264.101(b) and (c). Therefore, EPA and states have considerable discretion in this area. A useful guidance document is *Transmittal of Interim Guidance on Financial Responsibility for Facilities Subject to RCRA Corrective Action*, September 30, 2003, available at http://www2.epa.gov/enforcement/interim-guidance-financial-responsibility-facilities-subject-rcra-corrective-action.

17.8.7 Corrective measures implementation

The final step in the corrective action process is referred to as corrective measures implementation. This involves design and construction, operation and maintenance, and monitoring of the selected remedy, which is comparable to the remedial design/remedial action phase of a Superfund cleanup. For example, at this stage in the process, hazardous waste in a SWMU that is the source of a release to ground water may be excavated and shipped offsite for treatment and disposal. At the same time, a ground water pump-and-treat system may be installed. The remedial measures selected for a facility may be prescribed in the facility's permit or corrective action order. In other cases, a performance-based approach, in which the facility owner/operator has considerable flexibility in deciding how to meet remedial goals, may be used. In either case, periodic reports on the progress of the cleanup will be required.

Information on 30 specific corrective actions is contained in *Treatment Experiences at RCRA Corrective Actions*, EPA/542/F-00/020, December 2000, available at http://nepis.epa.gov/EPA/html/Pubs/pubtitleOSWER.html by downloading the report numbered 542F00020.

17.9 Completion determinations

On February 25, 2003 [68 *FR* 8757], EPA finalized "Guidance on Completion of Corrective Action Activities at RCRA Facilities." The standard for any corrective action to be complete is that protection of human health and the environment has been achieved. In the guidance, EPA describes two types of completion:

1. *Corrective action complete without controls*—This type of determination is used in cases where either 1) no corrective action was required; or 2) corrective action was necessary, the remedy was implemented successfully, and no further activity or controls are necessary to protect human health and the environment. Once such a determination is received from a state or EPA region, the facility will likely be eligible for release from financial assurance requirements. Furthermore, those portions of a facility receiving this determination would generally return to unrestricted use.

2. *Corrective action complete with controls*—This type of determination is used to designate a situation where 1) a full set of corrective measures has been defined; 2) the facility has completed construction and installation of all remedial measures; 3) site-specific media cleanup objectives have been met; and 4) all that remains are operation, maintenance, and monitoring actions and/or compliance with and maintenance of any institutional controls. This determination recognizes that protection of human health and the environment has been achieved as long as ongoing operation, maintenance, and monitoring requirements are met. The guidance also states that "enforceable mechanisms" (e.g., permits, orders, etc.) should be used to assure continued compliance.

EPA's guidance indicates that, in some situations, it may be appropriate to subdivide a facility for purposes of corrective action. This would allow the implementing agency (state or EPA region) to make a corrective action complete determination for a part of the facility. If corrective action is complete and there are no controls required for a portion of a facility, presumably that portion of the facility could be sold.

In the guidance, EPA indicated that "it is important to provide meaningful opportunities for public participation as part of a completion determination." [68 FR 8763] At permitted facilities, a determination that corrective action is complete will usually require a Class 3 permit modification. At nonpermitted facilities where corrective action is complete and all other RCRA obligations at the facility have been satisfied, EPA or the authorized state may acknowledge completion of corrective action by terminating interim status through final administrative disposition of the facility's permit application [see §270.73(a)]. At nonpermitted facilities, where corrective action is complete but controls are required, interim status would not be terminated.

17.10 Environmental indicators

In 1999, EPA developed specific goals for the RCRA corrective action program. The goals were developed in response to the Government Performance and Results Act of 1993 (GPRA). The GPRA requires federal agencies to develop plans for what they intend to accomplish, measure how well they are doing, make appropriate decisions based on the information they have gathered, and communicate information about their performance to Congress and to the public.

By the year 2020, EPA and the authorized states plan to have largely completed the work of implementing final remedies at all facilities requiring corrective action. While working toward the 2020 goal, EPA decided to ensure that sites presenting the greatest risk to human health and the environment were dealt with first. Accordingly, program goals focused on two "environmental indicators" (EIs) designed to stabilize the program's most threatening sites:

- The human exposures EI ensures that people near a particular site are not exposed to unacceptable levels of contaminants.

- The ground water EI ensures that contaminated ground water does not spread and further contaminate ground water resources.

In guidance, EPA defined those EIs as follows:

"Current human exposures under control" means "that there are no 'unacceptable' human exposures to 'contamination' (i.e., contaminants in concentrations in excess of appropriate risk-based levels) that can be reasonably expected under current land- and ground water-use conditions (for all 'contamination' subject to RCRA corrective action at or from the identified facility (i.e., site-wide))."

"Migration of contaminated ground water under control" means "that the migration of 'contaminated' ground water has stabilized, and that monitoring will be conducted to confirm that contaminated ground water remains within the original 'area of contaminated ground water' (for all ground water 'contamination' subject to RCRA corrective action at or from the identified facility (i.e., site-wide))." [RO 14335]

These EIs were supposed to "aid facility decision makers by clearly showing where risk reduction is necessary, thereby helping regulators and facility owners/operators reach agreement earlier on stabilization measures or cleanup remedies that must be implemented." [EPA/530/K-02/017I] However, EPA cautioned that EIs "were primarily designed to stabilize environmental problems, and are not final cleanup determinations." [RO 14672]

The agency's latest GPRA goals for corrective action were to document by 2014 that: 1) 51% of the 3,779 corrective action facilities have put a final remedy in place, 2) 87% of facilities have current human exposures under control, and 3) 78% of those facilities have migration of contaminated ground water under control.

EIs are mostly an internal EPA reporting and accountability issue and have little to do with facility owners/operators. Related guidance and information on EPA's progress towards those EI goals are available at http://www.epa.gov/osw/hazard/correctiveaction/eis/.

17.11 Corrective action management units

A corrective action management unit (CAMU) is an area or unit within a facility that is used for treatment, storage, or disposal of remediation wastes. The basic purpose of these units is to encourage aggressive remediation of contaminated sites by providing a location where cleanup wastes can be managed without triggering the land disposal restrictions (LDR) or minimum technological requirements (MTRs). The CAMU provisions are codified at §§264.550–552 and 264.555.

A 40-page packet of EPA guidance on CAMUs is available at http://www.epa.gov/epawaste/hazard/correctiveaction/resources/guidance/remwaste/refrnces/02camutu.pdf.

17.11.1 Background

The RCRA LDR program and MTRs were established to minimize the threat to human health and the environment associated with the disposal of hazardous wastes. When EPA issued these standards, it was primarily thinking of protecting the environment from the disposal of manufacturing process wastes. However, when these rules are applied to hazardous wastes generated during cleanup of contaminated sites, they tend to be a disincentive to aggressive cleanup for the following reasons:

- First, hazardous waste (including contaminated soil) that is excavated during a site cleanup and that will subsequently be placed in a land disposal unit, such as a landfill, is subject to the LDR program. Sometimes, this means the waste must be incinerated to meet the low LDR concentration-based standards. Burning large volumes of waste and contaminated soil is very expensive. As a result, facility owners/operators may try to avoid or delay remediation projects that involve excavation of large quantities of hazardous soil.

- Second, land disposal units that receive hazardous waste are subject to the MTRs. This is generally interpreted to mean two liners, a leachate collection system, and ground water monitoring. Such units are more expensive to build and operate than disposal units that are not subject to MTRs. Therefore, disposal of waste in MTRs-compliant units is more expensive than disposal in non-MTRs units. As a result, facility owners/operators may try to avoid or delay remediation projects that involve excavation of large quantities of hazardous waste or soil.

In an effort to eliminate these obstacles to site remediation, EPA promulgated the CAMU regulations on February 16, 1993. [58 *FR* 8658] This rule gave facilities a way of managing remediation wastes without incurring the substantial costs associated with the LDR program and MTRs. Shortly thereafter, the CAMU rule was challenged in court by the Environmental Defense Fund. This organization considered the rule to be insufficiently protective of human health and the environment. (See *Environmental Defense Fund v. EPA*, Docket No. 93-1316, U.S. Court of Appeals for the District of Columbia Circuit.)

As a result of the litigation, significant revisions to the CAMU regulations were issued on January 22, 2002. [67 *FR* 2962] These revised, more-stringent requirements apply to CAMUs that are not grandfathered under the old 1993 CAMU rules. Grandfathered units are identified in §264.550(b) as CAMUs that were approved before April 22, 2002, or as units for which substantially complete permit applications had been submitted by November 20, 2000. Grandfathered CAMUs are regulated under §264.551; however, the rules for grandfathered CAMUs are not of widespread interest. Conversely, if you wanted to permit a CAMU today, the rules for new CAMUs (i.e., units that are not grandfathered) are contained in §264.552 and are discussed in the following subsections.

17.11.2 CAMU-eligible wastes

CAMUs can only be used to manage "CAMU-eligible wastes," which are defined as "all solid and hazardous wastes, and all media (including ground water, surface water, soils, and sediments) and debris, that are managed for implementing cleanup. As-generated wastes (either hazardous or nonhazardous) from ongoing industrial operations at a site are not CAMU-eligible wastes." [§264.552(a)(1)]

Containers and tanks that are excavated during cleanup (and materials they hold) are CAMU-eligible. Soil that is contaminated by releases (e.g., leachate) from operating hazardous waste units is CAMU-eligible when managed for implementing cleanup. In addition, "soil or other materials contaminated by product spills or releases from ongoing industrial processes are not considered as-generated wastes and, as such, are CAMU-eligible when managed for implementing cleanup." [67 FR 2967] Wastes from closed or closing land-based units are CAMU-eligible; however, wastes removed from nonpermanent units (containers, tanks, waste piles) are not. [67 FR 2968]

Liquids may not be placed in CAMUs unless it will facilitate the remedy selected for the waste. [§264.552(a)(3)] EPA can prohibit the placement of waste in a CAMU if the agency believes the cleanup waste resulted from mismanaging as-generated wastes. For example, if a facility disposes hazardous waste that does not meet the applicable LDR treatment standard and subsequently has to exhume the waste, EPA could prevent the exhumed waste from being managed in a CAMU. The idea is to prevent parties from being rewarded for past noncompliance.

17.11.3 CAMU designations

A CAMU must be located within the contiguous property under the control of the owner/operator where the wastes to be managed in the CAMU originated. It may contain both contaminated and uncontaminated areas. The areal extent of a CAMU will be established through discussions/consultations between the owner/operator and the regulators. More than one CAMU may be designated at a facility.

An owner/operator can attain the benefits of a CAMU by voluntarily entering into the corrective action process, or by obtaining a remedial action plan (RAP—see Part 270, Subpart H and Section 17.15).

However, an owner/operator cannot unilaterally establish a CAMU—only EPA or an authorized state may do so. A Class 3 permit modification or RAP may be used to establish the CAMU and will also allow for public participation. Section 3008(h) orders can also be used to designate CAMUs. EPA expects that CAMUs may also be used as applicable or relevant and appropriate requirements (ARARs) for the remediation of CERCLA sites. [November 30, 1998; 63 FR 65880]

An example of how a CAMU could be useful during a corrective action cleanup is given in Case Study 17-1.

Case Study 17-1: Corrective Action Using a CAMU

An example of how a CAMU might be used at a facility undergoing corrective action is shown in Figure 17-3. Before remediation begins, the facility/regulators have identified four SWMUs, three of which are located in the floodplain of a river. The remedial goal is to treat the waste from each of the SWMUs and move the wastes from the SWMUs located in the floodplain to a more protective location. Corrective action implementation for these units includes four steps:

1. EPA or the state designates SWMU 4 as a CAMU.

2. The CAMU-eligible wastes from the four SWMUs are removed and treated in a temporary onsite treatment unit.

3. SWMU 4 is retrofitted with a composite liner and leachate collection system.

4. The remediation wastes can be placed in the CAMU after they have been treated to meet the standards in §264.552(e)(4). Other design, operation, closure, and post-closure requirements for the CAMU would be specified according to the criteria in §264.552. [*Handbook From the RCRA Corrective Action Workshop on Results-Based Project Management*, EPA, February 2000]

Figure 17-3: How a CAMU May Be Used at a Corrective Action Site

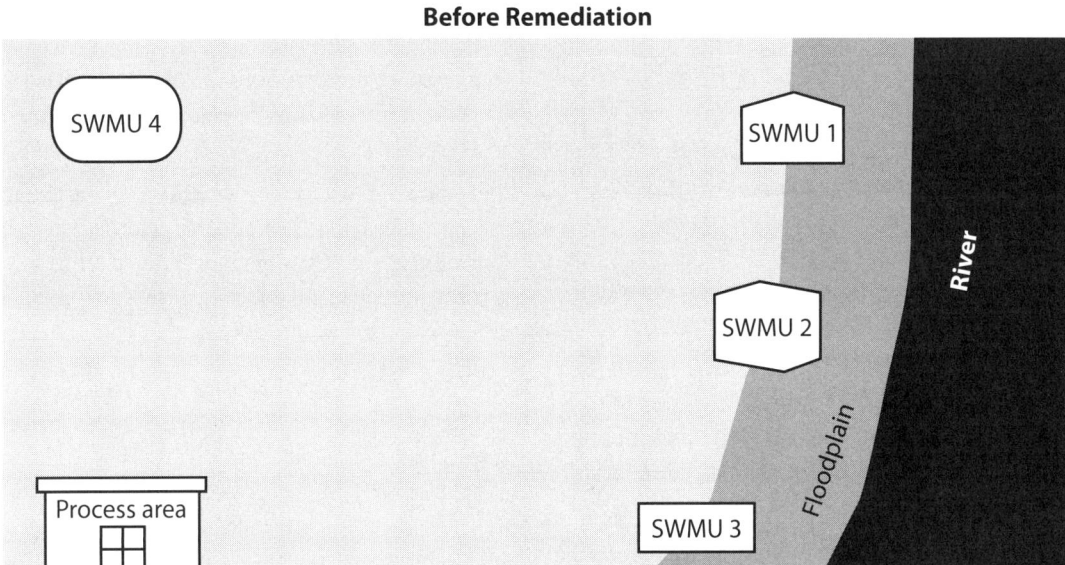

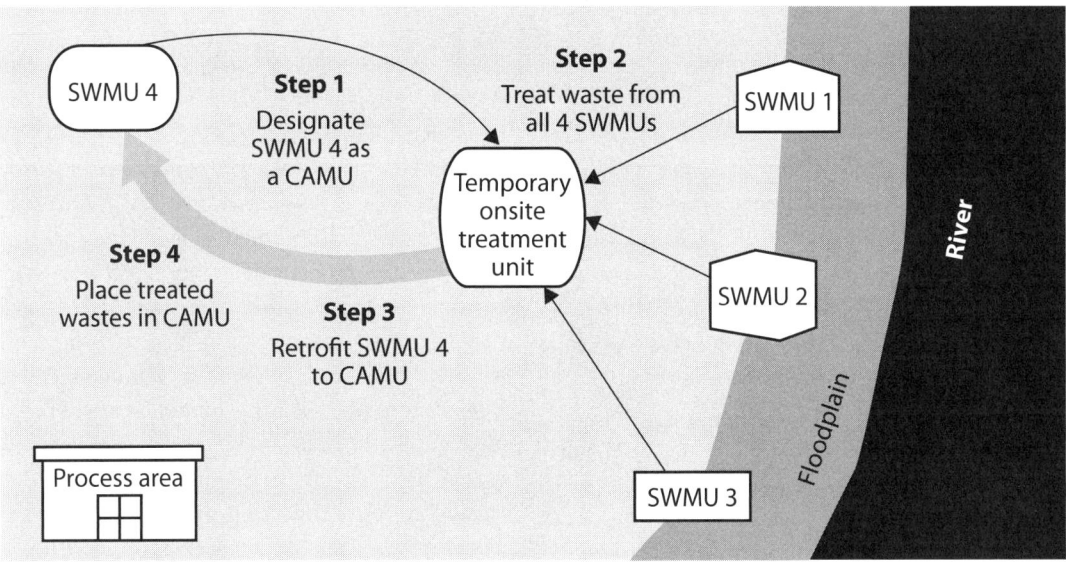

Source: Adapted from EPA.

17.11.4 Liners and caps

As stated previously, CAMUs are not subject to RCRA's minimum technological requirements. However, CAMUs (other than grandfathered CAMUs) used as disposal units (i.e., the CAMUs are the final resting place of the wastes) must have liners. The liner standards are similar to those for Subtitle D landfills—a minimum composite liner consisting of a 30-mil synthetic membrane (60-mil for HDPE) on top of 2 feet of compacted soil is required. A leachate collection system capable of maintaining less than 30 cm of leachate on the liner is also required. Alternative liner designs may be approved on a case-by-case basis. [§264.552(e)(3)]

CAMUs used for disposal must be closed with a cap that meets essentially the same performance standards as currently required for closed hazardous waste landfills.

Ground water monitoring is required whenever the CAMU contains a disposal unit. EPA must be notified of any ground water contamination originating from the CAMU, and corrective action requirements would apply. Treatment- and storage-only CAMUs do not need ground water monitoring.

17.11.5 Treatment requirements

The LDR program in Part 268 does not apply to wastes placed into or within a CAMU. [§264.552(a)(4)] However, before wastes may be disposed in a CAMU (other than a grandfathered CAMU), "principal hazardous constituents" (PHCs) must be treated to meet standards that are similar to the LDR alternative soil standards in §268.49—see Section 13.7. (Wastes may be stored in the CAMU without being treated first.)

PHCs include constituents that meet two criteria. First, they must otherwise be subject to treatment under the LDR program for as-generated wastes. Second, PHCs must pose higher risks than site-specific cleanup levels. In general, these are constituents that, if carcinogens, exceed a 10^{-3} risk level, or if noncarcinogens, exceed the reference dose by an order of magnitude.

Once identified, PHCs in CAMU-eligible wastes must be treated to reduce their concentrations by 90% but do not have to be treated to below 10 times the UTS (universal treatment standards) levels in §268.48. PHCs on debris must be treated to this same standard or to the LDR alternative debris standards in §268.45—see Section 13.8.

Case-by-case "adjustment factors" may be applied that could increase or decrease the amount of treatment required. One adjustment factor is "views of the affected local community." The treatment requirements for CAMU-eligible wastes are codified in §264.552(e)(4).

17.11.6 Treatment/storage CAMUs

CAMUs used solely to treat or store wastes are not subject to the requirements for liners, caps, and treatment discussed in the preceding sections. Instead, they are subject to essentially the same performance standards as staging piles—see Section 17.13. However, if wastes remain in these units for more than 2.5 years, liners and ground water monitoring must be provided. [67 FR 2995–6]

17.11.7 Offsite disposal of CAMU-eligible wastes

Under §264.555, CAMU-eligible wastes may be disposed at an offsite hazardous waste landfill that meets Part 264, Subpart N landfill standards (e.g., the landfill must have a double synthetic liner and leachate collection system). PHCs in the cleanup waste must meet treatment standards established by regulators in the originating state. The same treatment standards that are available for onsite CAMU disposal apply; that is, PHC concentrations must be reduced by 90% but treatment to less than 10 times UTS is not required. Some of the adjustment factors that can be used to increase or decrease the amount of treatment required when wastes are disposed in an onsite CAMU are not available when the wastes are disposed offsite.

The receiving landfill must have a RCRA permit (i.e., interim status landfills may not receive CAMU-eligible wastes). The landfill's permit must specifically allow receipt of CAMU-eligible wastes; Class 2 permit modifications could be used for this purpose. Once the permit modification is approved, the landfill could potentially accept any CAMU-eligible waste approved for offsite disposal by regulators overseeing remediation sites.

Before an offsite landfill can dispose waste from a specific remediation site, three criteria must be met: 1) the public must be notified of the landfill facility's intent to receive waste from a particular cleanup, 2) the public has 15 days to provide comments to the landfill's regulators, and 3) the landfill's regulators have up to 60 days to approve the disposal. Disposal cannot occur without notification from EPA or the state that they do not object to placement of the CAMU-eligible wastes.

17.11.8 State authorization issues

In states that are not authorized to administer the corrective action program, EPA administers corrective action and implements the CAMU regulations described in this section. States that are authorized to administer corrective action were not required to adopt the CAMU regulations promulgated in 1993, because they were less stringent than the pre-existing rules for remediation waste management. However, states that are authorized to administer the CAMU regs were also required to adopt the more stringent CAMU regulations promulgated by EPA on January 22, 2002. In order to prevent a slowdown in the state approval of CAMUs, states that were CAMU-authorized in 2002 were granted "interim authorization by rule" to administer the new standards.

17.12 Temporary units

Temporary units (TUs) are tanks or container storage areas used to store or treat hazardous remediation wastes during site cleanups. A TU must be located within the contiguous property under the control of the owner/operator where the wastes to be managed originated. The TU regulations are in §264.553.

EPA and states may authorize the use of TUs and replace the otherwise applicable Parts 264 and 265 design, operating, or closure standards for hazardous waste tanks or containers with alternative standards that can be determined on a site-specific basis. The alternative standards must protect human health and the environment. The idea is to provide more flexibility to regulators overseeing remediation projects. The regulators may authorize the use of TUs in a Class 2 permit modification, a RCRA Section 3008(h) order, a RAP (see Section 17.15), or a CERCLA record of decision (ROD).

TUs may also be used for activities that are not part of a selected remedy, interim measure, or site stabilization activity. For example, prior to selecting a remedy, large quantities of investigation-derived wastes may be generated during a facility's RFI. A TU could be used to store or treat such waste. However, the owner/operator would have to seek a Class 2 permit modification or similar type of approval to use the TU.

Wastes may be stored/treated in TUs for only one year; however, a one-year extension is available. At the end of the unit's approved life, management of remediation waste in the unit must cease and the unit must be closed. Alternatively, the unit could be retrofitted to meet Part 264/265 standards, as addressed through a permit modification.

A 40-page packet of EPA guidance on TUs is available at http://www.epa.gov/epawaste/hazard/correctiveaction/resources/guidance/remwaste/refrnces/02camutu.pdf.

17.13 Staging piles

EPA developed the concept of a staging pile to solve a particular problem. To explain the problem, we begin with some definitions from §260.10:

- "Remediation waste" means all solid and hazardous wastes, and all media (including ground water, surface water, soils, and sediments) and debris, that are managed for implementing cleanup."

- A waste "pile" is defined as "any noncontainerized accumulation of solid, nonflowing hazardous waste that is used for treatment or storage and that is not a containment building."

Waste piles are considered to be land disposal units and, therefore, must be equipped with liners, leachate collection systems, and ground water monitoring systems (i.e., MTRs), as specified in Parts 264/265, Subpart L. Furthermore, hazardous waste must meet LDR treatment standards prior to placement in a waste pile. Sometimes people excavate large quantities of hazardous remediation waste during a cleanup project. The problem is that if hazardous remediation waste is placed in a pile, in general, the pile is a waste pile subject to the aforementioned requirements. This would make it very difficult to excavate remediation waste and store or treat it in a pile. One option would be to use a CAMU for this purpose, but obtaining approval for a CAMU can be very difficult. Another option would be to locate the pile inside the "area of contamination" (AOC). The area of contamination policy is discussed in Section 17.14. However, this may not be practical in some cases where it is necessary to pile the remediation waste in an area outside the area of contamination.

EPA's solution to the problem described above was to develop a new type of remediation waste management unit called a "staging pile." A staging pile is defined as "an accumulation of solid, nonflowing remediation waste that is not a containment building and that is used only during remedial operations for temporary storage at a facility." [§260.10] Keep in mind that a staging pile cannot be used to manage hazardous wastes from ongoing industrial processes.

Remediation waste may be placed in a staging pile without triggering the LDR program or MTRs for hazardous waste piles. These units must be located within the contiguous property under the control of the owner/operator where the wastes to be temporarily stored originated. The staging pile regulations are in §264.554.

These piles are intended to "facilitate short-term storage of remediation wastes so that sufficient volumes can be accumulated for shipment to an offsite treatment facility or for efficient onsite treatment."

[November 30, 1998; 63 FR 65909] Staging piles can be used to store contaminated soil temporarily while decisions on the final remedy for site cleanup are being finalized. In addition, staging piles may be used for physical operations intended to prepare wastes for subsequent treatment (e.g., mixing, sizing, blending, and other similar physical operations). [§264.554(a)(1)]

A two-year time limit (with a possible 180-day extension) applies to staging piles from the time the owner/operator first places remediation waste in the pile. If the remediation waste is expected to need storage for longer than 2 years, a CAMU is more appropriate. Storage of hazardous waste in approved staging piles is exempt from the LDR storage prohibition of §268.50.

Staging piles provide an alternative to seeking a CAMU and may be used in situations where the AOC concept does not apply. For example, where a site has noncontiguous areas of contaminated soil, a staging pile in one of those areas may be used to temporarily store waste from the other contaminated areas prior to further management. [63 FR 65920]

Staging piles are not self-implementing; that is, the owner/operator of a facility undergoing remedial operations must get EPA's or the state's approval to use such a pile. A Class 2 permit modification or RAP (see Section 17.15) is required for a permitted TSD facility, while an interim status facility or generator must have the staging pile designated in a RAP, closure plan, or order. Such approval will also include design standards and operating practices to meet the following performance standard: the staging pile must prevent or minimize releases and cross-media transfers of hazardous wastes and constituents into the environment (e.g., through the use of liners, covers, run-off/run-on controls, ground water and/or air monitoring, and inspections). [§264.554(d)(1)(ii)]

Special requirements apply for ignitable, reactive, or incompatible wastes placed in staging piles. EPA expects that staging piles may be used as

ARARs for the remediation of CERCLA sites. [63 *FR* 65932]

Staging piles must be closed within 180 days after their operating term expires. There are two sets of closure standards: 1) if the pile is located in an uncontaminated area, it must be clean closed; 2) if the staging pile is located in a previously contaminated area, the final cleanup of the contaminated subsoil may be coordinated with the overall site remedy. [§264.554(j–k)]

An example of using a staging pile during facility cleanup is given in Case Study 17-2.

17.14 The area of contamination policy

This policy is best introduced by EPA guidance:

"In what is typically referred to as the area of contamination (AOC) policy, EPA interprets RCRA to allow certain discrete areas of generally dispersed contamination to be considered RCRA units (usually landfills). Because an AOC is equated to a RCRA land-based unit, consolidation and in situ treatment of hazardous waste within the AOC do not create a new point of hazardous waste generation for purposes of RCRA. This interpretation allows wastes to be consolidated or treated in situ within an AOC without triggering land disposal restrictions or minimum technology requirements. The AOC interpretation may be applied to any hazardous remediation waste (including nonmedia wastes) that is in or on the land. Note that the AOC policy only covers consolidation and other in situ waste management techniques carried out within an AOC." [RO 14291]

Under the AOC policy, the site owner/operator may consolidate contaminated soil within the AOC. Normally, excavation of contaminated soil is considered the point of generation, but under the AOC policy, consolidation is not considered to be removal from the land (i.e., generation). Thus, contaminated soil can be consolidated within the AOC and a hazardous waste determination can be made after such consolidation. [RO 14283, 14338] "In an area of generally dispersed soil contamination, soil may be consolidated or managed within the area of contamination to facilitate sampling, for example, to ensure that soil samples are representative or to separate soil from nonsoil materials." [May 26, 1998; 63 *FR* 28619]

Case Study 17-2: Corrective Action Using a Staging Pile

An example of how a staging pile may be used at a facility involved with corrective action is shown in Figure 17-4. Before remediation, the facility/regulators have identified three SWMUs, each containing old, weathered sludges. Soil contamination is associated with two of the SWMUs and the production building. The remedial goal is to treat the wastes from all three SWMUs and the contaminated soil onsite and then ship the treated wastes and soil offsite for disposal at a Subtitle D (nonhazardous waste) landfill. The remediation utilizes five steps:

1. EPA or the state authorizes construction of a staging pile within the area of contaminated soil.

2. Remediation wastes from throughout the site are consolidated into the staging pile.

3. A soil washing unit is constructed using a tank that meets the requirements for a generator 90-day unit (i.e., §262.34).

4. Remediation waste from the staging pile is treated in batches in the 90-day unit. Care must be taken to ensure that the 90-day accumulation provisions are not violated.

5. Following treatment, the remediation waste meets the LDR treatment standards and is determined to no longer contain hazardous waste. The treated material is sent offsite as nonhazardous waste for disposal in a Subtitle D facility. [*Handbook From the RCRA Corrective Action Workshop on Results-Based Project Management*, EPA, February 2000]

Figure 17-4: How a Staging Pile May Be Used at a Corrective Action Site

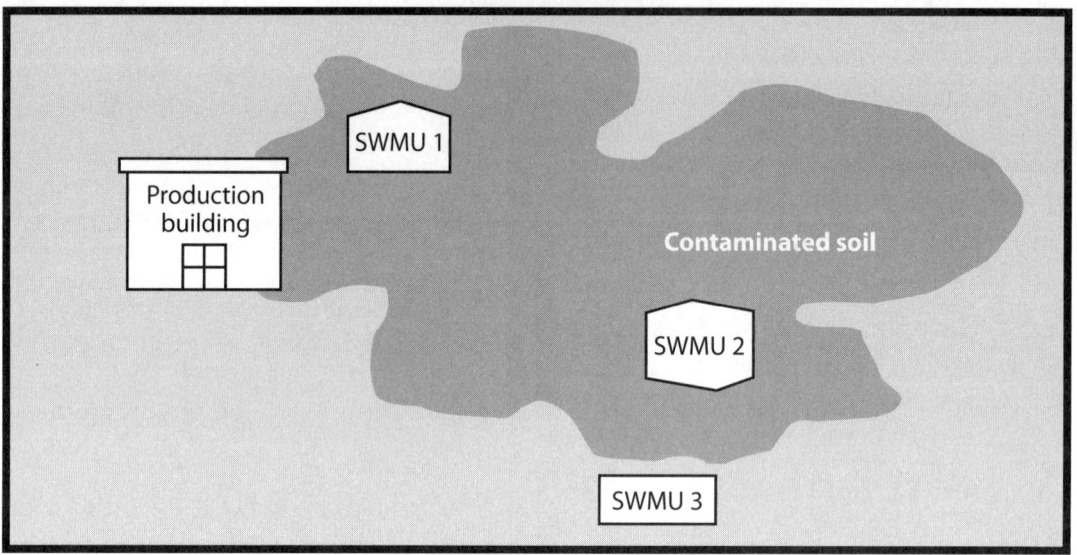

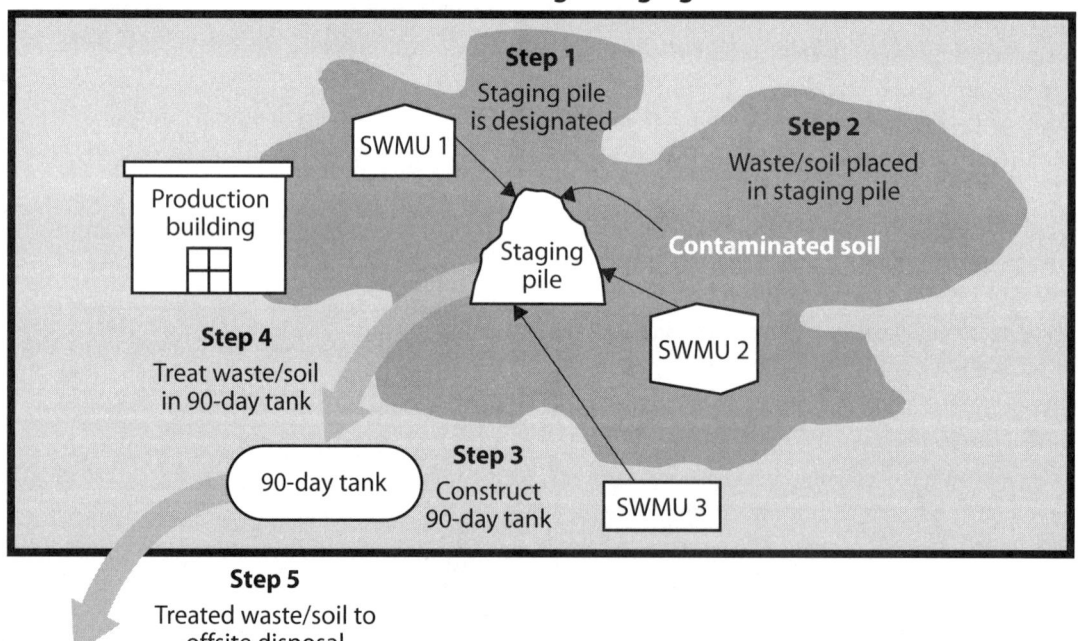

Source: Adapted from EPA.

Hazardous remediation waste is generated when it is removed from the AOC. Thus, in cases where a site owner/operator wants to consolidate remediation waste from separate, noncontiguous contaminated areas, the AOC concept cannot be used, but a staging pile can, as described in Section 17.13.

The AOC policy was developed in the context of the CERCLA program. [December 21, 1988; 53 *FR* 51444, March 8, 1990; 55 *FR* 8758] However, it also applies to RCRA corrective action sites, cleanups under state law, and voluntary cleanups. At the federal level, advance approval is not required for people to take advantage of the AOC policy. [RO 11954] However, some states have more-stringent standards that require consultation and/or prior approval before use of an AOC. EPA encourages people to consult with the appropriate agency to ensure that the policy is implemented correctly. A 42-page packet of EPA guidance on the AOC policy is available at http://www.epa.gov/epawaste/hazard/correctiveaction/resources/guidance/remwaste/refrnces/01aoc.pdf.

How the AOC policy may be used to facilitate cleanup is illustrated in Case Study 17-3 and the following.

Q *During a site investigation, a soil boring is taken to obtain a sample. Excess soil not needed for the sample contains hazardous waste (e.g., it's contaminated with F001) and is left on the ground near the borehole in the AOC. Is this allowable?*

A Based on *Management of Investigation-Derived Wastes During Site Inspections* (OERR Directive 9345.3-02, May 1991, available as report number 540G91009 from http://nepis.epa.gov/EPA/html/Pubs/pubtitleOSWER.html), this would be allowable and, in fact, is recommended by EPA. However, this is CERCLA guidance and should be used with caution when managing hazardous soil at sites not subject to CERCLA. During site investigations conducted under RCRA corrective action, we recommend that facility owners/operators develop an investigation-derived waste management plan and seek approval of the plan from the agency overseeing the site investigation.

Note that EPA's example, summarized in Case Study 17-3, allows final disposal of hazardous remediation waste in an AOC. Other agency guidance concurs, noting that allowable hazardous remediation waste management activities in an AOC include storage, in situ treatment, and disposal. [EPA/530/K-02/017I, OSWER Directive 9347.3-05FS]

17.15 Remedial action plans

On November 30, 1998, EPA published a final rule referred to as the "hazardous waste identification rule for contaminated media," often called the HWIR-media rule. [63 *FR* 65874] The goal of this rule was to make it easier to clean up sites contaminated with hazardous waste. One of the difficulties

---

Case Study 17-3: Corrective Action Using the AOC Policy

An example of how the AOC policy may be used at a facility conducting corrective action is shown in Figure 17-5. Before remediation, the facility/regulators have identified three SWMUs containing old, weathered sludges and associated contaminated soil. Contaminated soil is also located outside the production building. All three SWMUs are located inside the contaminated area. The goal is to consolidate the wastes from the SWMUs and the contaminated soil in a lined, land-based unit under a cap. The remedy implementation for these units consists of four steps:

1. An AOC is designated by the facility, working in conjunction with the regulators.
2. A lined, land-based unit is constructed within the AOC.
3. Sludges from all three SWMUs and contaminated soil are consolidated in the lined unit.
4. The unit is capped. [*Handbook From the RCRA Corrective Action Workshop on Results-Based Project Management*, EPA, February 2000]

Figure 17-5: How an AOC May Be Used at a Corrective Action Site

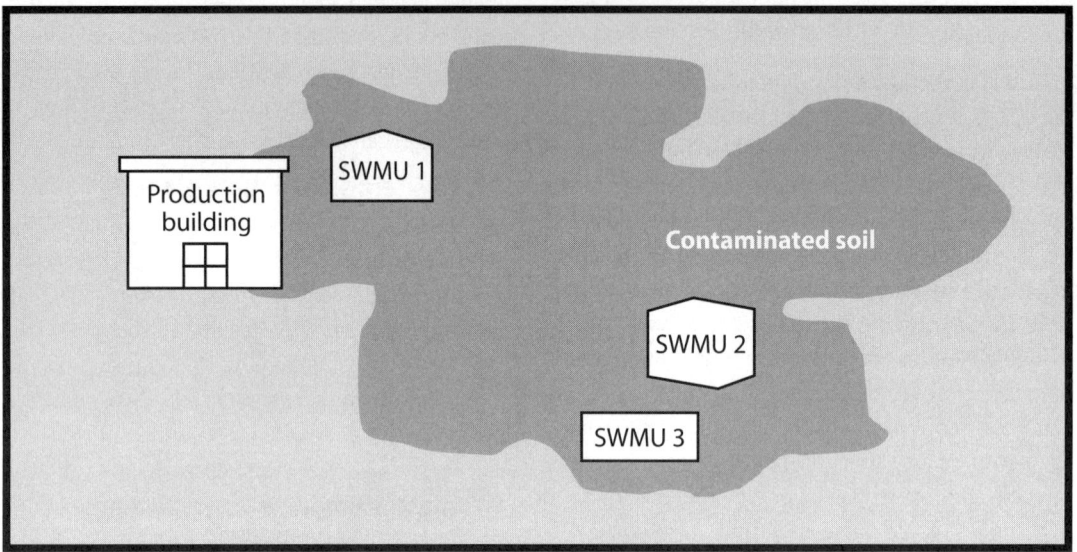

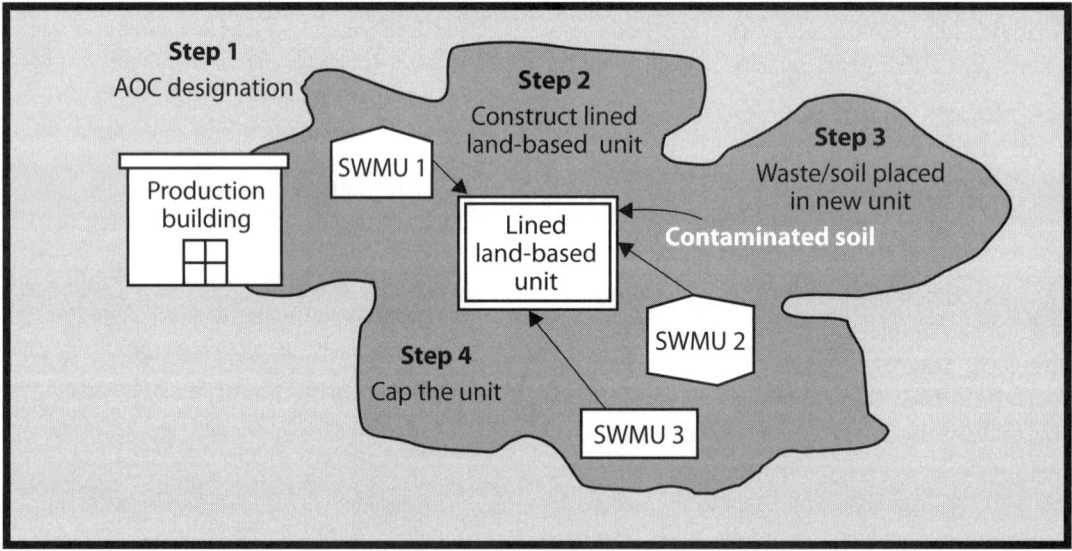

Source: Adapted from EPA.

people encounter in cleaning up these sites is the RCRA permit program. If cleanup of a site will require onsite treatment, storage, or disposal of hazardous waste, a RCRA permit may be required. In general, these permits are difficult and time-consuming to obtain. Furthermore, as discussed above, a facility seeking a RCRA permit is subject to facility-wide corrective action. EPA realized that some people who may have been inclined to voluntarily clean up a contaminated area decided not to initiate the cleanup because it would require a RCRA permit and associated corrective action.

To solve this problem, EPA established requirements in the HWIR-media rule for a new type of RCRA permit that is supposed to be easier to get and does not trigger facility-wide corrective action. The new permit is called a "remedial action plan" (RAP). A RAP is a permit (an enforceable document) that EPA or a state can issue to authorize treatment, storage, or disposal of "remediation waste" at a "remediation waste management site." [§270.80(a)] These terms are defined below.

"Remediation waste" means "all solid and hazardous wastes, and all media (including ground water, surface water, soils, and sediments) and debris, that are managed for implementing cleanup." [§260.10] Remediation waste can be generated from all types of cleanups—not just from RCRA corrective actions—and may include waste generated outside of a facility's boundaries; thus, material removed to offsite locations may continue to meet the definition of remediation waste. Remediation waste does not include hazardous process waste generated from ongoing manufacturing operations. However, it does include wastes generated from treating remediation wastes (e.g., carbon canisters and sludges produced from treating extracted ground water or soil vapors). [63 *FR* 65881]

"Remediation waste management site" means "a facility where an owner or operator is or will be treating, storing, or disposing of hazardous remediation wastes. A remediation waste management site is not a facility that is subject to corrective action under 40 CFR 264.101, but is subject to corrective action requirements if the site is located in such a facility." [§260.10] The requirements in Subparts B, C, and D in Part 264 do not apply to remediation waste management sites, unless the site is located at a facility that is subject to traditional RCRA permitting for management of nonremediation wastes. [§264.1(j)]

The regulations in Subpart H of Part 270 describe what a RAP must include and how such a plan can be obtained. Although the agency has streamlined the information submittal and approval process for obtaining a RAP, public involvement is still required. Facilities operating under interim status may use a RAP without losing that status.

The substantive, unit-specific requirements for hazardous waste management units in Part 264 must be referenced in a RAP and complied with. For example, hazardous remediation waste storage tanks must have secondary containment. However, combustion units cannot be included in such plans. As noted in previous sections, site-specific requirements for CAMUs, temporary units, and/or staging piles may be specified in a RAP.

No RAP is needed where a RCRA permit would not otherwise be required; for example, a RAP is not required for the treatment or accumulation of remediation waste in a 90-day tank or container, a wastewater treatment unit, or at a CERCLA cleanup.

After receipt of the RAP application, EPA or the authorized state will review the application, solicit public comment, and then either approve or deny the RAP. According to EPA, "[t]he most critical part of the [agency's] determination is whether or not operation according to the RAP will ensure compliance with applicable Part 264, 266, and 268 requirements." [63 *FR* 65894] Like RCRA Part B permits, RAPs have a maximum 10-year life, and those that specify land disposal must be reviewed every 5 years.

17.16 Useful references

As mentioned in Section 17.1, the RCRA corrective action program is implemented primarily through guidance and policy documents rather than rigid regulations. The most useful references we've found on the corrective action process (many of which we previously referenced in our discussions above) are summarized below, with their availability:

- *RCRA Corrective Action Plan,* OSWER Directive 9902.3-2A, May 31, 1994. This document provides a detailed discussion of the corrective action process. It includes model work plans and guidance on developing site-specific scopes of work. The document is available from EPA at http://www.epa.gov/epawaste/hazard/correctiveaction/resources/guidance/index.htm.

- *RCRA Corrective Action Inspection Guidance Manual,* EPA/520/B-95/001, May 1995. The manual may be ordered from NTIS [(703) 605-6050] as PB95-269189.

- *Soil Screening Guidance: Fact Sheet,* EPA/540/F-95/041, July 1996; *Soil Screening Guidance: User's Guide,* EPA/540/R-96/018, July 1996; and *Soil Screening Guidance: Technical Background Document,* EPA/540/R-95/128, July 1996. These guidance documents are available at http://www.epa.gov/superfund/health/conmedia/soil/index.htm.

- *Best Management Practices for Soils Treatment Technologies,* EPA/530/R-97/007, May 1997. This document discusses protective design/performance steps to prevent or decrease the likelihood of cross-media transfers of contaminants during soil remediation activities and is available from EPA at http://www.epa.gov/epawaste/hazard/correctiveaction/resources/guidance/index.htm.

- *Management of Remediation Waste Under RCRA,* EPA/530/F-98/026, October 1998. [RO 14291]

- *Transmittal of the National Enforcement Strategy for RCRA Corrective Action,* April 27, 2010, available at http://www2.epa.gov/enforcement/guidance-national-enforcement-strategy-rcra-corrective-action.

- *Results-Based Approaches and Tailored Oversight Guidance for Facilities Subject to Corrective Action,* EPA/530/R-03/012, September 2003, available at http://www.epa.gov/epawaste/hazard/correctiveaction/resources/guidance/index.htm.

- *Region 6 Corrective Action Strategy,* November 2008, available at http://www.epa.gov/earth1r6/6pd/rcra_c/pd-o/riskman.htm.

- *Handbook of Groundwater Protection and Cleanup Policies for RCRA Corrective Action,* EPA/530/R-04/030, April 2004, available at http://www.epa.gov/epawaste/hazard/correctiveaction/resources/guidance/gw/gwhandbk/index.htm.

- *Transmittal of Guidance on Enforcement Approaches for Expediting RCRA Corrective Action,* January 2, 2001, available from EPA at http://www2.epa.gov/enforcement/guidance-enforcement-approaches-expediting-rcra-corrective-action.

Waste Characterization/Sampling

Making hazardous waste determinations

Although the preceding chapters have provided hundreds of pages of guidance on complying with the RCRA regulations, the question often remains: "How do I know whether the stuff in this drum is a hazardous waste?" Hopefully, this chapter will shed a little more light on this difficult issue.

We begin in Section 18.1 by reviewing the regulatory basics of making a hazardous waste determination. Some practical tips for making an assessment based on analytical data are provided in Section 18.2; guidance covering waste analysis plans is included. If you don't use testing but want to make a knowledge-based hazardous waste determination, what constitutes "acceptable knowledge"? This ill-defined topic is addressed in Section 18.3. (You'll note in both of these sections, though, that a combination of analytical data and knowledge is often employed.)

In Section 18.4, we discuss waste characterization and compliance issues related to the land disposal restrictions.

Sampling issues are tackled in Section 18.5. How do you take a representative sample of material in a container or tank, sludge in a lagoon, or a 10-ft filter cartridge? This is one of the most complex topics in this book, and we cover sampling plans, data quality objectives, statistical methods, and how to obtain "representative samples."

18.1 Waste characterization

Before we begin our discussions, note that EPA offers a "Hotline" service for people with questions about RCRA-related waste characterization issues. The Methods Information Communication Exchange (MICE) will accept your questions, especially in regard to what analytical methods to use, at (703) 818-3238. Or, you can email the MICE at mice@techlawinc.com. Check them out—they do a good job, respond quickly, and provide this service for free.

The first thing to remember with waste characterization is: Each individual generator of a solid waste is responsible for evaluating his/her own waste and making a hazardous waste determination. "A generator's failure to properly analyze, label, and accumulate waste does not exempt the waste from regulation." [RO 11424] The RCRA regulations place the burden on the generator to determine whether a solid waste is hazardous—EPA or your state will not make the determination for you. [December 18, 1978; 43 *FR* 58969, RO 11158, 11223, 11501, 11599, 11806]

As noted in Section 2.1 of this book, whenever facilities are trying to make a hazardous waste determination for a particular material, they should ask themselves the following four questions *in order*:

1. Is it a solid waste?
2. Is it exempt?
3. Is it listed?
4. Is it characteristic?

This four-question hierarchy comes right out of §262.11, which is the waste characterization section in the federal RCRA regulations. This section requires facilities to make a hazardous waste determination for each solid waste they generate. Each of these four questions is evaluated in more detail below.

Note that this discussion covers the federal waste characterization process only—states may have more stringent requirements. For example, some states have fewer exemptions and/or expanded hazardous waste listings and characteristics.

18.1.1 Is it a solid waste?

The answer to the first question is typically based on knowledge. As discussed in Chapter 1 of this book, if the material you're managing is being abandoned or is inherently waste-like, the material is a solid waste. If it is being recycled, Chapters 1 and 11 note that the material *may* be a solid waste. As noted in those two chapters, there are a few exclusions from the definition of solid waste for certain materials being recycled in certain ways. You claim these exclusions based on knowledge of the recycling process and the regulations. Finally, if your material is a waste military munition, it is likely to be a solid waste as discussed in Section 14.5.

18.1.2 Is it exempt?

If you are managing a solid waste, the second question to ask [from §262.11(a)] is whether the material is excluded or exempt from the RCRA regulations under §261.4. Most of these exclusions and exemptions are discussed in Chapter 4, and they are largely based on application of knowledge enhanced by EPA guidance.

One exclusion to the RCRA regulations that is discussed in Section 4.10 is the household hazardous waste exclusion in §261.4(b)(1). This exclusion allows solid wastes generated in households or other residential buildings (e.g., single and multiple residences, military base housing units, bunkhouses, crew quarters) to be disposed as ordinary household trash, whether or not they are listed or characteristic hazardous waste. No hazardous waste determination needs to be made for these residential wastes. [RO 11958]

For sources that are not exempt from the RCRA regulations, a hazardous waste determination must be made for all solid waste generated, even if the facility is a conditionally exempt small quantity generator (CESQG) subject to the reduced RCRA requirements in §261.5. CESQGs (e.g., retail stores that generate less than 100 kilograms of hazardous waste per month) are still required to make §262.11 determinations. [RO 11958, 14030]

18.1.3 Is it listed?

A solid waste that is not exempt from the RCRA regulations will have to be evaluated as to its hazardousness per the third and fourth questions. The regs require that generators first determine whether the solid waste is a listed waste. [§262.11(b)] This third step is accomplished by determining if the solid waste meets any of the listing descriptions in §§261.31–261.33 (see Chapter 3).

The first list of hazardous wastes is the F-list, found in §261.31. This list identifies hazardous wastes from nonspecific sources and includes, among other things, spent solvents, heat treating and electroplating wastes, and dioxin wastes.

The K-list (in §261.32) is the second list of hazardous wastes. It identifies hazardous wastes from specific manufacturing processes. Wastes from specific sources within various industries such as petroleum refining and chemical manufacturing are included in this group of listed wastes.

The third and fourth lists of hazardous wastes are the P- and U-lists in §261.33. These lists identify unused commercial chemical products (and their spill residues) that are hazardous wastes when discarded.

Finally, some states regulate what we call "state-listed" wastes. These are wastes that are not on any of EPA's F-, K-, P-, or U-lists, but they are considered listed hazardous wastes within that state's borders. For example, some states regulate PCBs or used oil as listed wastes and have special state waste codes for those wastes.

Determining whether a solid waste is a listed waste is primarily a knowledge-based evaluation. Analytical testing alone, without information on a waste's source, will not produce information that will conclusively indicate whether a given waste is a listed hazardous waste. [RO 14291] We'll talk more about this in Section 18.3.2.

A mixture of a listed waste and other solid waste, or a residue from treating a listed waste, is likely still a listed waste, regardless of the concentrations of hazardous constituents in the mixture or treatment residue. However, there are some exceptions to these mixture and derived-from rules, as detailed in Sections 5.1 and 5.2.

Also, soil, ground water, and other environmental media, which are not solid wastes, may become contaminated with a listed waste. In this situation, EPA's contained-in policy says media that contain a listed hazardous waste must be managed as if they are that listed waste. Again, there are some exceptions, as discussed in Section 5.3.1.

Similarly, debris (e.g., from building demolition activities) can also be contaminated with listed wastes. The contained-in policy has been codified for contaminated debris: debris that contains a listed hazardous waste must be managed as if it is that listed hazardous waste unless or until it no longer contains the hazardous waste. Reference Section 5.3.2 of this book and §261.3(f).

18.1.4 Is it characteristic?

Finally, whether or not the solid waste is listed per the third question above, the generator must ask this fourth question. This is necessary, in part, to comply with the land disposal restrictions program in Part 268. [§262.11(c)]

There are four characteristics that a waste may exhibit: ignitability, corrosivity, reactivity, and/or toxicity. Each of these characteristics is explained in detail in Chapter 2 of this book. Note that even if it is known that a waste exhibits one characteristic, it still must be evaluated for the other three. [§261.20(b), RO 13570]

In addition to process wastes, contaminated environmental media (e.g., soil or ground water) and debris may also exhibit hazardous waste characteristics and thus be subject to RCRA.

The exciting part about determining if a solid waste exhibits a characteristic, however, is that §262.11(c) gives generators the option of using testing *or* knowledge for this purpose. That is, the determination may be made by 1) analyzing a representative sample of the waste, *or* 2) using knowledge of the waste itself and/or the process and materials that generated the waste.

We summarize EPA's guidance on making analysis-based determinations in Section 18.2 and knowledge-based determinations in Section 18.3. Before moving to those sections, though, we offer the following assistance on exactly who at a facility can make a hazardous waste determination, and how often a waste stream should be recharacterized.

18.1.5 Who makes the determination?

Section 262.11 states "A person who generates a solid waste…must determine if that waste is a hazardous waste…." A "person" is defined as "an individual, trust, firm, joint stock company, federal agency, corporation (including a government corporation), partnership, association, state, municipality, commission, political subdivision of a state, or any interstate body." [§260.10]

According to EPA, a "person" is not limited to a specific individual. Therefore, any individual who is part of the "person" may make a hazardous waste determination. The hazardous waste determination is not limited to the individual who actually produces the solid waste. For example, Environmental, Health, and Safety (EH&S) personnel may make a hazardous waste determination for a waste produced by an individual operator, technician, or researcher, as long as the EH&S personnel and the producer are part of the same "person" (e.g., academic institution).

It is the "person's" responsibility to ensure that the individuals within the organization who are making the hazardous waste determination obtain all the necessary information from whichever individuals within the organization have that data. For example, a hazardous waste determination in a laboratory setting would ideally be a collaborative effort between the individual researcher who produces the waste and EH&S personnel who may make the hazardous waste determination. That is, EH&S personnel making a hazardous waste determination should receive sufficiently accurate and detailed information about each waste and the process that generated it from the individual technician or researcher to ensure accurate waste identification. [RO 14618]

18.1.6 How often should I recharacterize?

In 1978, EPA proposed an annual reassessment of all solid wastes. Such a characterization frequency was *never* finalized:

"The deletion of the retesting requirements that were contained in the proposed regulations does not, of course, relieve a generator of solid waste from his continuing responsibility to know whether his wastes are hazardous. If there is a significant change in the materials, processes, or operation which indicate the waste has become hazardous, the generator must repeat the determination. EPA recognizes the potential burden that this places on certain manufacturers whose products, processes, and wastes change frequently, for example chemical specialty producers or other batch-type producers. This burden is created, however, by [the RCRA statute] which demands special attention and care for the handling and management of hazardous wastes. Those persons whose wastes are sometimes hazardous and sometimes nonhazardous have the same obligation as any other generator to ensure that all their hazardous wastes are managed in accordance with the requirements of Subtitle C and regulations implementing those statutory provisions." [February 26, 1980; 45 *FR* 12727]

Although we have not seen concrete guidance from EPA, our general observations on this issue are that generators will have to recharacterize a particular solid waste stream when:

- The raw materials, process, or operation that produces the waste changes;
- The waste is sent to a different TSD facility for the first time;
- New analytical data are required by the TSD facility with which you are doing business to recertify the waste profile (this is often every one or two years);
- Questions arise during transportation or receipt of the waste at the designated facility (e.g., the waste received at the TSD facility does not match the waste profile or manifest); or
- Questions are raised through internal or external audits as to the regulatory status of the waste.

For complying with the treatment standards in the land disposal restrictions (LDR) program, EPA *has* given us guidance:

"For each waste stream, the waste constituents regulated under the land disposal restrictions rule must be comprehensively analyzed. Although the frequency of testing will depend to some extent upon the variability of the waste stream, the agency recommends that a comprehensive analysis of each waste stream be performed at least annually by the generator or treater. When the comprehensive analysis is performed, however, it must contain data on all the applicable constituents in [the §268.40 treatment standards table] so that the [TSD facility] owner/operator will be able to determine whether the waste meets all applicable treatment standards." [November 7, 1986; 51 FR 40598]

Although this guidance obviously recommends an annual assessment, keep in mind that it is for wastes that are subject to the LDR program. This guidance does not tell you specifically how often you should make a hazardous waste determination for each solid waste you generate at your facility.

Still, generating facilities that are proactive in terms of environmental compliance have told us they test everything once a year. If something changed in your waste, making hazardous what used to be nonhazardous, and you don't know about it, you're out of compliance and subject to enforcement. The onus is on the generator to be right in his/her hazardous waste determinations, and a good way of staying right is to test or otherwise evaluate your waste streams every year.

Permitted TSD facilities must have a waste analysis plan (WAP), which is approved at the time the permit is issued. Under the federal RCRA regs, generators are required to have a WAP only if they are treating hazardous waste in a 90- or 180-day accumulation unit for the purpose of meeting LDR treatment standards. One of the WAP components is the frequency with which the facility will analyze or reanalyze all wastes and treatment residues. [§§264/265.13(b)(4)]

18.2 Analysis-based determinations

To determine if a solid waste is a listed hazardous waste, you must know the source of the waste and/or the process that produced it. Thus, as noted in Section 18.1.3 above, making a listed waste determination is primarily accomplished using knowledge. Determining if a solid waste exhibits a characteristic, however, can be accomplished using testing or knowledge. This section discusses characteristic hazardous waste determinations based on analytical testing. Alternatively, making a knowledge-based determination for both listed and characteristic wastes is discussed in Section 18.3.

Although the agency allows generators the option of using knowledge to make these assessments, "[w]henever feasible, the preferred method to [make hazardous waste determinations] is to conduct sampling and laboratory analysis because it is more accurate and defensible than other options." [*Waste Analysis at Facilities That Generate, Treat, Store, and Dispose of Hazardous Wastes*, EPA/530/R-94/024, April 1994, available at http://www.epa.gov/waste/hazard/tsd/ldr/wap330.pdf]

One caution before we get into some analytical details: Just because you determine through analysis that Part 261, Appendix VIII constituents are present in a waste does *not* mean you should assume the waste is hazardous. Although this appendix is entitled "Hazardous Constituents," *generators are not supposed to use Appendix VIII or §261.11 to identify hazardous wastes*. Rather, EPA lists solid wastes with significant concentrations of Appendix VIII constituents as hazardous if *the agency* determines (per the §261.11 criteria) that the wastes pose a potential and substantial threat to human health and the environment. The presence of one or more of these hazardous constituents within a waste does not by itself determine that the waste is hazardous. [RO 11144, 12014, 12296, 13290]

For example, soil or ground water that is found through analysis to be contaminated with one of the Appendix VIII constituents (e.g., toluene) is not automatically a hazardous waste. Only if you can determine using knowledge the source of the constituent (e.g., from a spill of pure toluene or spent solvent toluene) can you correctly assign a listed hazardous waste code. Otherwise, the soil or ground water will be hazardous only if it exhibits a characteristic. [RO 11051, 12171, 12392, 14291] See Sections 14.2 and 14.3.4.3.

Using analysis to make hazardous waste evaluations for characteristic wastes is discussed in Section 18.2.2. Details on developing a waste analysis plan for making such determinations are given in Section 18.2.3. Procedures for obtaining a "representative" sample or samples and for statistically evaluating multiple analytical results are discussed in Section 18.5. Before discussing the specifics of making an analysis-based determination for the characteristics, we want to briefly discuss some general analytical issues.

18.2.1 General analytical considerations

Dating back to the beginning of the regulatory program in 1980, EPA has been fairly prescriptive in specifying test methods to show compliance with RCRA. Most of the methods codified within the RCRA regulations are EPA methods contained in the agency's methods compendium, entitled *Test Methods for Evaluating Solid Waste, Physical/Chemical Methods*, more commonly known as SW–846.

18.2.1.1 SW–846

SW–846 is accessible online at http://www.epa.gov/epawaste/hazard/testmethods/sw846/online/index.htm. The first time you look at the compendium, you may find that it appears to be a huge compilation of test methods, with no apparent logic as to when a specific method should be used to analyze for a specific constituent (called an analyte). There are generally several alternative test methods for the same class of analytes, any of which could be selected to measure target constituents in the context of a particular RCRA compliance situation. SW–846 is designed to address all testing a facility might need to conduct to comply with RCRA, but a method may be indicated as being approved for determination of analytes that might not need to be evaluated in a particular situation.

An SW–846 user selects the list of target analytes based on the regulatory purposes for which the analysis is being performed. [RO 13274] For example, if the reason for testing is to determine if the waste exhibits the toxicity characteristic, then the facility needs to analyze the Method 1311 extract only for the 40 chemicals listed in Table 1 in §261.24.

The variety of waste matrices encountered in the RCRA waste management and cleanup programs is too diverse to expect prescriptive, one-size-fits-all sampling and analytical methods to work in all situations. Therefore, the RCRA regs include references to only 27 SW–846 test methods that are required to be used in specific applications. "In other situations, SW–846 functions as a guidance document setting forth acceptable, although not required, methods to be implemented by the user, as appropriate, in responding to RCRA-related sampling and analysis requirements." [July 25, 1995; 60 *FR* 37975] EPA allows the use of non-SW–846 methods, provided the method meets acceptable quality assurance/quality control (QA/QC) standards for the particular application. "If you can demonstrate that your method of sampling and data interpretation is scientifically and statistically correct, then you may use that procedure in place of a SW–846 method." [RO 13557] EPA reiterated this in a formal policy statement at 78 *FR* 63188 (October 23, 2013). Guidance for acceptable QA/QC standards are contained in Chapter One of SW–846. [August 31, 1993; 58 *FR* 46044, June 13, 1997; 62 *FR* 32457]

Methods in SW–846 do not need to be implemented exactly as written, and performance data presented in these methods should not be used as regulatory default or absolute QC requirements. [June 13, 1997; 62 *FR* 32457, RO 13707] "The agency wishes to stress that flexibility in the use of equipment, glassware, and procedures is allowed pursuant to the SW–846 Disclaimer and Sections 2.1.1 and 2.1.2 of Chapter Two." [January 13, 1995; 60 *FR* 3091]

If an SW–846 method cannot "get the right answer" due to analytical difficulties with the matrix or some other reason, modification of the method or selection of a different method is required. "Having run a method 'as written' is no excuse for reporting faulty data…. Especially when unusual or complex matrices are involved, SW–846 methods must still undergo a 'demonstration of applicability' to establish adequate analytical performance in the context of that application." [*The Relationship Between SW–846, PBMS, and Innovative Analytical Technologies*, EPA/542/R-01/015, October 2001, available online at http://www.clu-in.org/download/char/sw-846.pdf]

The June 13, 1997 *Federal Register* explained that non-SW–846 equipment or procedures may be used "provided that method performance appropriate for the intended RCRA application is documented. Such performance includes consideration of precision, accuracy (or bias), recovery, representativeness, comparability, and sensitivity (detection, quantification, or reporting limits) relative to the data quality objectives for the intended use of the analytical results." [62 *FR* 32457]

Some waste matrices are considered difficult to analyze when certain RCRA analytical methods are used (e.g., analysis of total cyanides in electroplating wastes that have been treated with polysulfides). RO 14317 gives EPA's position on the use of analytical methods when low percentages of target analytes are recovered: "Inadequate recovery of target analytes from the RCRA-regulated waste matrices of concern demonstrates that the analytical conditions selected are inappropriate for the intended application." This guidance cites acceptable recoveries of: 1) 70% or greater for organics extracted from standard matrices (e.g., ground water, aqueous leachate, soil); 2) 50% or greater for organics extracted from difficult matrices (e.g., sludge, ash, stabilized waste); 3) 80% or greater for volatile organics; and 4) 80–120% for inorganic analytes in most matrices.

18.2.1.2 Performance-based measurement

The agency is moving away from the use of prescriptive testing to a performance-based measurement system (PBMS). A PBMS approach conveys what needs to be accomplished, but not specifically how to do it. The use of PBMS for RCRA was announced on May 8, 1998 [63 *FR* 25430], proposed on October 30, 2002 [67 *FR* 66252], and finalized on June 14, 2005. [70 *FR* 34538]

Performance-based monitoring objectives include [67 *FR* 66256, 70 *FR* 34540, EPA/542/R-01/015, RO 14590]:

- Allowing more flexibility in method choice for RCRA-required testing,

- Stimulating the development of innovative and/or more cost-effective analytical techniques,

- Focusing on measurement objectives rather than on specific measurement techniques (e.g., allowing the use of less expensive semiquantitative results in lieu of rigorous quantitative data), and

- Promoting more timely releases of new SW–846 methods by decoupling the methods from their specified use in the RCRA regs.

Using a PBMS methodology, any analytical method (whether or not it currently is published in SW–846) may be used to generate data as long as it can be demonstrated to measure:

- The constituent of concern,

- In the waste matrix of concern,

- At the concentration level of concern, and

- At the degree of accuracy required for the site-specific application. [EPA/542/R-01/015]

In a July 22, 2009 *Federal Register* preamble, EPA identified four steps needed to ensure continued progress towards a PBMS approach [74 *FR* 36198]:

1. Emphasizing flexibility in choosing sampling/analytical approaches to meet measurement-based regulatory requirements,

2. Developing method validation processes that confirm quality requirements (e.g., accuracy) are achieved,

3. Increasing stakeholder collaboration/participation in development of these validation processes, and

4. Providing timely agency assessment/review of new or modified measurement technologies and procedures that are alternatives to traditional analytical methods.

18.2.1.3 Method-defined parameters

Under the June 14, 2005 rule, an SW–846 method is required only in those situations where it is the only method capable of measuring the property (i.e., a method-defined parameter). In other words, a method-defined parameter is a RCRA regulatory parameter (i.e., a property or characteristic) that is defined by the outcome of a specific test method. When testing for these parameters, the method must be followed exactly as written.

A total of 27 method-defined parameters remain in the RCRA regs. [70 *FR* 34547] For example, SW–846 methods continue to be required for determining compliance with 1) the corrosivity characteristic [Methods 9040C and 1110A], 2) the toxicity characteristic (Method 1311—the TCLP), and 3) the ban on free liquids in wastes to be landfilled (Method 9095B—the paint filter test).

If the method is not performed exactly as written (e.g., if a different leaching solution or leaching time period is used when conducting the TCLP), the result for the measured parameter cannot be used to interpret compliance with the corresponding regulation. [EPA/542/R-01/015]

18.2.2 Analyzing for characteristics

The SW–846 test methods used to make an analysis-based determination for the four hazardous waste characteristics are discussed in detail in Sections 2.2–2.6 of this book. However, according to January 13, 1995, July 25, 1995, and June 14, 2005 preamble language [60 *FR* 3089, 60 *FR* 37975, and 70 *FR* 34550, respectively], only the methods specified in §261.22 for the corrosivity characteristic (Methods 9040C and 1110A) and in §261.24 for the toxicity characteristic (Method 1311—the TCLP) are required.

Making an analysis-based determination of whether a solid waste exhibits a characteristic is based on testing a "representative" sample of the solid waste. [§§261.20–24] Procedures for obtaining representative samples are discussed in Section 18.5. Tips for determining whether a solid waste exhibits a characteristic based on analysis follow.

18.2.2.1 Ignitability

Solid wastes that are liquids with a flash point <140°F are D001 ignitable wastes under §261.21(a)(1). If testing is used to determine whether a solid waste is ignitable, the flash point can be determined using one of two ASTM methods standardized into SW–846 [the Pensky-Martens closed-cup tester (Method 1010A) or the Setaflash closed-cup tester (Method 1020B)]. The choice of the method should be based on the waste being tested. [RO 11594] For example, the Setaflash method is applicable only to liquids with viscosities below 150 stokes at 25°C. The Pensky-Martens test can be used for liquids that contain nonfilterable, suspended solids, while the Setaflash method should not be used for such wastes.

When testing is used to determine the flash point of a waste, a closed-cup tester should be used. Note that flash point can also be obtained through open-cup testing, but those results are only valid if the measured flash point is <140°F (i.e., if the waste is determined to be hazardous). "Ordinarily, open-cup tests...will produce higher flash points than the closed-cup tests required by EPA." [RO 12296]

Some confusion has arisen over using one of the two approved flash point tests for wastes that are essentially all solids (e.g., solvent-contaminated rags or gasoline-contaminated soil) or for semisolid wastes containing both a liquid and solid phase. According to EPA, "[n]either test, however, is approved by ASTM for use in evaluating the flash point of solids or sludges." [OSWER Directive 9443.00-1A] "If your samples contain filterable solids, they are not amenable to the Pensky-Martens flash point test. Flash point testing is only appropriate for liquid samples. It should not be applied to solids." [RO 13759] See also RO 12909, 13550, and 14669.

Based on this guidance, wastes that are semisolids (e.g., sludges) or essentially all solids should be evaluated using the Method 9095B paint filter test. If no free liquid comes through the filter, the waste is not a liquid and will not be D001 via the §261.21(a)(1) criteria. [RO 11619, 11787, 13328] However, the waste may still be an ignitable solid or DOT oxidizer, which are also D001 hazardous wastes. See §261.21(a)(2) and (4), respectively. If the paint filter test produces a free-liquid phase from wastes that are semisolid or mostly solid, EPA's guidance is that you should test each phase separately: the liquid by flash point testing and the solids by Method 1030. [RO 13759] The purpose and limitations of Method 1030 are discussed in Section 2.2.2.

18.2.2.2 Corrosivity

Aqueous wastes with a pH ≤2.0 or ≥12.5 as measured by a pH meter using Method 9040C are D002 characteristic hazardous wastes. [§261.22(a)(1)] If the pH of the waste is above 12.0, pH measurement for the corrosivity characteristic must be taken when the sample is at 25±1°C. Measurement of pH for wastes with pH levels less than 12.0 may be made at other sample temperatures. [April 4, 1995; 60 *FR* 17003]

Liquids that corrode carbon steel at >0.25 inches per year are also corrosive hazardous wastes. [§261.22(a)(2)] The National Association of Corrosion Engineers Standard TM–01–69 must be used to measure the corrosion rate. This method has been standardized in SW–846 as Method 1110A, and EPA requires generators to use this standardized version when running this test. [June 14, 2005; 70 *FR* 34549]

Method 9045 in SW–846 can be run to determine the pH of solids. The test is applicable to solids (e.g., soil), sludges, or nonaqueous liquids that contain <20% water by volume. Basically, water is added to the solids or soil, and the pH of the resulting solution is measured. Although this test is available to generators, EPA has clarified that "Method 9045 is not [to be] used for corrosivity characteristic determinations." [April 4, 1995; 60 *FR* 17003] That is EPA's position at the federal level; thus, materials that are normally considered to be corrosive, such as lye or solid acids, are not D002 when they become wastes. [RO 13533] However, some states have chosen to have a more stringent corrosivity definition and require use of this test for potentially corrosive solids.

18.2.2.3 Reactivity

The regulatory language defining the reactivity characteristic in §261.23(a) does not specify any test methods for determining if a waste exhibits this characteristic. The agency has found no appropriate test methods for this purpose, noting that existing methods: 1) are not general enough to measure the various reactivity criteria, 2) do not take enough factors (e.g., mass, surface area) into account to reflect the reactivity of the whole waste (as opposed to the reactivity of the sample itself), and 3) require subjective interpretation of the results instead of providing pass/fail results. [May 19, 1980; 45 *FR* 33110]

Previous agency guidance indicating that solid wastes releasing more than 250 mg of hydrogen cyanide gas/kg of waste or more than 500 mg of hydrogen sulfide gas/kg of waste should be regulated as reactive hazardous wastes has been rescinded. EPA has withdrawn the guidance, along with test methods for determining the amount of gases released, from Chapter 7 of SW–846. EPA discovered that "critical errors" were made in developing these release thresholds and test methods. Therefore, the agency no longer supports the use of these thresholds for making reactive waste determinations. [June 14, 2005; 70 *FR* 34548, RO 14177] However, you need to check to see if your state RCRA program still requires use of these release thresholds and test methods to determine reactivity of cyanide- and sulfide-bearing wastes.

18.2.2.4 Toxicity

As noted in Section 18.2.1.3, the toxicity characteristic is a method-defined parameter, and the method that must be used to determine toxicity is the TCLP (Method 1311 in SW–846). When testing for toxicity, once the TCLP is run, the extract obtained is analyzed for the 40 constituents listed in §261.24. The agency noted that the following EPA test methods could be used in this step [EPA/530/R-94/024]:

- Methods 3010 and 6010 for arsenic, barium, cadmium, chromium, lead, silver, and selenium;
- Method 7470 for mercury;

- Methods 3510 and 8081 for pesticides;
- Method 8151 for herbicides;
- Method 8260 for volatile organics; and
- Methods 3510 and 8270 for semivolatile organics.

However, SW–846 methods do not have to be used to analyze the extract for these constituents. Any method can be used as long as that method has documented quality control and is sensitive enough to meet the regulatory limits. [RO 11568, 11579, 11649]

Sometimes, used antifreeze, upon TCLP testing, is found to contain high levels of arsenic and/or selenium; however, knowledge indicates that neither arsenic nor selenium are present in the materials of construction of the radiator and neither are contained in any process fluids which could leak into the antifreeze. An EPA-funded study found that arsenic and/or selenium may appear as false positives during the analytical process. Although the study was not able to definitively determine the cause of the analytical discrepancy, it did indicate that false positives were not observed when using Methods 6020 or 7060, instead of the more common Method 6010, to test for heavy metals in TCLP leachates from used antifreeze. Note that Method 7060 was deleted from SW–846 in 2007. A complete discussion of these results is contained in *Waste Analyses Project for Auto Dealerships—Waste Antifreeze Summary*, September 2006, available online at http://www.understandrcra.com/rccd/WasteAntifreeze.pdf.

In some situations (particularly for oily wastes and organic liquids), dilution of TCLP liquid extracts for subsequent analysis has resulted in detection limits above regulatory levels. EPA's guidance in these cases is that it is not possible to determine conclusively whether the waste is hazardous or not. "[I]f no other information is available to assist a generator to make a hazardousness determination in light of the inconclusive TCLP results [i.e., if a knowledge-based determination cannot be made], it would generally be prudent for the generator to manage that waste as a hazardous waste." [RO 11579]

Other suggestions for measuring analytes in the liquid phase produced when running the TCLP on oily matrices include [RO 11627, 11649, 13485]:

- For volatile organics, the traditional purge and trap GC/MS method (Method 8240) does not always provide detection limits that are sufficiently low. As an alternative, EPA recommends modifying an existing headspace screening method (Method 3810) to include isotope dilution. This modified method includes the addition of several standard isotopes that correspond to each of the target analytes.

- For semivolatile organics, the existing SW–846 methods are adequate for analyzing most samples, but oily matrices require dilutions that sometimes yield unacceptable detection limits. To improve the detection levels, EPA recommends a specific ion monitoring (SIM) option on the GC/MS. Instead of scanning the sample for a full spectrum of semivolatile compounds, the agency found that analytes with lower concentration could be easily detected using SIM.

Extraction using a zero-headspace extraction (ZHE) vessel is specified in the TCLP when testing for a volatile constituent in §261.24. However, if the analysis of an extract obtained using a simple bottle extractor results in a concentration of the volatile compound in excess of its regulatory level, the test does not need to be rerun using a ZHE vessel. Results from a bottle extractor *cannot* be used, however, to show that the concentration of a volatile constituent is below regulatory levels. [November 20, 1997; 62 *FR* 62084]

Although the TCLP specifies the use of a minimum 100-g sample size, using a sample size <100 g is highly recommended for testing radioactive mixed wastes with concentrations of radionuclides that may present serious radiation exposure hazards. [November 20, 1997; 62 *FR* 62084] No matter what sample size is used, however, the sample must be representative.

18.2.2.5 Knowledge often included

There often is an element of knowledge associated with making an analysis-based determination of whether a solid waste exhibits a characteristic. Consider the accidental discharge of gasoline from a tanker. The contaminated soil is excavated for offsite disposal. Is it characteristic? You might use knowledge to determine the soil is not ignitable (if there is no free liquid), not corrosive, and not reactive. Regarding toxicity, you know that the primary toxic constituents in gasoline are the BTEX chemicals (benzene, toluene, ethylbenzene, and xylenes) and that, of those, only benzene is regulated under §261.24. Thus, any toxicity testing would likely be limited to a relatively inexpensive TCLP test for benzene (and perhaps lead if that constituent is thought to be present). The important point to remember is that you can use knowledge to identify the constituents and characteristics that couldn't be there, simplifying and minimizing the cost of any analytical work. [RO 11603, 14695]

Such generator knowledge is often used in making characteristic waste determinations in manufacturing plants. For example, a facility may know a particular process does not use any pesticides or herbicides, so these chemicals don't need to be analyzed for (via the TCLP) in process residues. Other groups of chemicals can sometimes be categorically eliminated from analysis using process knowledge, such as metals, halogenated organics, etc. Process knowledge regarding possible chemical reactions can also be used to identify constituents for which analysis may be required. [November 20, 1997; 62 *FR* 62082–5, RO 12830, 13406, 14695]

If the characterization is for a waste such as sludge in a lagoon, then the selection of constituents for testing will be dependent on the historical introduction of materials to the unit, including process wastewaters and surface run-off. Changes in influent composition over time as process changes occur must be evaluated. [RO 13406]

18.2.3 Waste analysis plans

The RCRA regulations require certain facilities managing hazardous waste to establish a waste analysis plan (WAP). A WAP documents the procedures used to obtain a "representative sample" of the waste and to conduct a detailed physical/chemical analysis of that sample. Procedures for obtaining a "representative" sample or samples and for statistically evaluating multiple analytical results are discussed in Section 18.5.

Under the federal RCRA regs, generators are required to have a WAP in only one situation: when they are treating hazardous waste in a 90- or 180-day accumulation unit for the purpose of meeting LDR treatment standards. All permitted and interim status TSD facilities must have a WAP that complies with §§264/265.13 requirements. Available guidance follows for generators and TSD facilities preparing WAPs.

18.2.3.1 WAPs for generators

If generators treat hazardous waste in accumulation units for the purpose of meeting an LDR treatment standard, they must develop and follow a written WAP. [§268.7(a)(5)] The WAP should include six elements [EPA/530/R-94/024]:

1. Description of the facility-specific processes, wastes, activities, and waste management units covered by the WAP;

2. Identification of the waste constituents and other parameters to be evaluated for the waste streams, both before and after treatment, and the rationale for the selection of these parameters;

3. Identification of how the generator will obtain a representative sample of the waste, both before and after treatment;

4. Analytical methods to be used to evaluate the waste streams;

5. Testing frequency for the waste, both before and after treatment; and

6. Special procedural requirements—in this case, procedures the generator will follow to treat the waste and meet the applicable LDR standards.

Although there is no required format for the WAP, incorporating these six elements should satisfy regulatory requirements. EPA/530/R-94/024 contains significant guidance and details on preparing WAPs, including a sample WAP for a generator treating to meet LDR treatment standards. The WAP must be maintained at the generator's facility and be available for inspection.

The above requirements apply to large quantity generators who use 90-day tanks, containers, or containment buildings to treat their hazardous waste. Such entities get to the WAP requirements in §268.7(a)(5) by reference from §262.34(a)(4). Small quantity generators can only use tanks or containers as 180-day units. The only way a small quantity generator could use a containment building for waste treatment would be by complying with the large quantity generator standards specified in §262.34(a) for the management of wastes in containment buildings (including a 90-day onsite accumulation limit). [EPA/530/K-05/008, RO 13696] These small quantity generators are subject to §268.7(a)(5) by reference in §262.34(d)(4).

As exemplified below, a WAP is not required for generators who are treating hazardous wastes in units other than 90- or 180-day accumulation units.

Q *Is a WAP required for a generator who treats D002 spent acid in an elementary neutralization unit (e.g., a tank)?*

A No. Generators who treat hazardous wastes in exempt units, such as elementary neutralization units, are not required to prepare a WAP. However, many such generators may elect to develop a WAP as a practical measure. [EPA/530/R-94/024, page 4-22]

EPA encourages generators not required to have a WAP to prepare one anyway. For example, if a number of samples of a waste are required to characterize the material before disposal, a WAP (although not required) is a good practice. This subject is discussed more thoroughly in Section 18.5.

18.2.3.2 WAPs for TSD facilities

Permitted and interim status TSD facilities are subject to detailed waste analysis requirements in §§264/265.13. Those sections require that a TSD facility obtain a detailed physical/chemical analysis of a representative sample of a hazardous waste before it can treat, store, or dispose the waste.

Sources of this information include data provided by the generator of the waste (if received from offsite), published or historical data, and analyses performed by the TSD facility itself. Often, a combination of these information sources will be required to fully characterize the waste. The bottom line is that the WAP must identify the physical/chemical data necessary to properly treat, store, or dispose a waste at the TSD facility. [OSWER Directive 9523.00-10, RO 12212]

If the TSD facility accepts hazardous waste from conditionally exempt small quantity generators, it does not have to address such wastes in the WAP. [RO 12376]

For permitted TSD facilities, the WAP is part of their permit. Interim status TSD facilities are also required to have a WAP. The WAP for these facilities consists of the same six elements as discussed above for generators [EPA/530/R-94/024]:

1. Description of the facility-specific processes, wastes, activities, and waste management units covered by the WAP;

2. Identification of the waste constituents and other parameters to be evaluated for received wastes and treated wastes (if any), and the rationale for the selection of these parameters;

3. Identification of how the facility will obtain a representative sample of received wastes and treated wastes (if any);

4. Analytical methods that will be used to evaluate the waste streams;

5. Testing frequency for received wastes and treated wastes (if any); and

6. Special procedural requirements (e.g., additional waste analysis requirements for specific waste management methods [§§264/265.13(b)(6)] and procedures for: receiving

wastes generated offsite; selecting the appropriate LDR treatment standards and treating wastes to meet these standards; obtaining Subpart CC air emission control compliance data; characterizing incompatible, ignitable, and reactive wastes; etc.).

TSD facility WAPs may be pretty simple or quite complex, depending on the hazardous waste operations conducted at the facility. For example, if the only RCRA-permitted operation at a plant is storage of a relatively homogeneous liquid hazardous waste in a tank, the WAP may be quite simple, identifying analytical parameters such as ignitability and reactivity to illustrate that the waste can be safely stored before shipment to an offsite facility for further storage, treatment, and/or disposal. The plant would also be concerned with waste/tank materials compatibility, as well as the compatibility of different batches of wastes that are stored together; waste analysis data would be required to assure that compatibility exists. An example of the waste analysis data that must be known to properly manage a waste received at an offsite solvent recovery plant is given in Table 18-1.

For a hazardous waste incinerator or landfill, however, the WAP will be extensive and detailed. Sufficient waste analysis data will be required to show that the waste can be properly treated/disposed, considering all LDR implications. EPA provides significant waste analysis guidance for hazardous waste incinerators and also for boilers and industrial furnaces that burn hazardous waste in *Waste Analysis Guidance for Facilities That Burn Hazardous Wastes—Draft*, EPA/530/R-94/019, October 1994 [available from http://nepis.epa.gov/EPA/html/Pubs/pubtitleOSWER.html by downloading the report numbered 530R94019]. EPA/530/R-94/024 also contains significant guidance and details on preparing WAPs for TSD facilities, including three sample WAPs for facilities performing treatment, incineration, and landfilling.

18.2.3.2.1 Characterization of each movement of waste

One of the most important aspects of waste analysis at TSD facilities is the need to characterize each movement of waste to ensure that the information provided on the waste manifest correctly identifies the waste. [§§264/265.13(a)(4), (c)] What amount of inspection, sampling, and analysis must be conducted to achieve this regulatory requirement? Each facility's WAP should answer this question, and the answer depends on a variety of site-specific factors. We offer the following general guidance.

Commonly, TSD facilities visually inspect every bulk shipment and every container and its contents for the physical state of the waste, phase separation, texture, color, and odor. They also typically

---

Table 18-1: Example Analytical Data for a Waste Received at an Offsite Solvent Recovery Plant

Waste No. 346A (EPA Code F002): 1,1,1-trichloroethane from degreasing of cutting tools

Characteristics	Value	Comments
1,1,1-trichloroethane (%)	83±5	More than 40% required to be recoverable
Metal contaminants (mg/L)		Metal contaminants partition to bottoms sludge, which is sent to TSD facility
Chromium	210±30	
Copper	650±80	
Lead	40±10	
Nickel	800±100	
Silver	25±5	
Viscosity (cp)	0.85	Measured to ensure pumpability
Flash point (°F)	149	Measured to ensure handling safety
Specific gravity	1.30	Measured to assess separability of phases

Source: Adapted from OSWER Directive 9523.00-10.

sample/analyze every bulk shipment of waste received. If it is a drummed shipment of a new waste, the facility often analyzes samples from each of the first ten drums (to establish a consistency basis), and then some fraction of the drums after that (if the shipment consists of the same waste stream). According to EPA/530/R-94/019, ASTM Method D140 (available from http://www.astm.org) can be used to estimate the number of containers of a waste stream to be sampled. That method suggests that the number of drums to be randomly selected for sampling should equal the cube root of the total number in the shipment.

Analytical parameters evaluated for bulk loads and selected drums include flash point; cyanide, halogen, and heavy metal concentration; pH; solid vs. liquid concentration; specific gravity; and total organic carbon. The analytical parameters selected for inspection/analysis are called the "fingerprint" parameters and are dependent on the type of waste and the treatment process to be utilized. This fingerprint or spot check evaluates consistency between the received waste and the accompanying manifest. [RO 12943] If something looks suspicious during this initial inspection/analysis, more samples of the bulk load or more drums (or every drum) will be evaluated.

According to EPA, operator knowledge is not an appropriate substitute for fingerprint inspection/analysis, except in the case when the TSD facility accepts manifested waste from a site owned by the same company. Further, if a TSD facility relies on waste analysis data provided by the generator, it is still the receiving facility's responsibility to accurately identify the waste per §§264/265.13(a)(4), (c). The TSD facility will generally be liable in enforcement situations if it violates its permit conditions or any applicable RCRA regulations. [EPA/530/R-94/024]

If a TSD facility receives a shipment of several sealed drums of mixed wastes (i.e., wastes that are both radioactive and hazardous), a representative sample from only one drum may be adequate if the owner/operator has knowledge that the chemical composition of the waste is identical in every drum. [November 20, 1997; 62 *FR* 62086]

18.3 Knowledge-based determinations

To determine if a solid waste is a listed hazardous waste, you must know the source of the waste. Thus, as noted in Section 18.1.3 above, making a listed waste determination is primarily achieved using knowledge. Assessing whether a waste exhibits a characteristic, however, can be accomplished using testing or knowledge. This section discusses listed and characteristic hazardous waste determinations based on knowledge. Analysis-based determinations for characteristic wastes are discussed in Section 18.2.2.

Here is a typical quote from EPA guidance regarding knowledge-based determinations:

"It is important to keep in mind that EPA does not require testing to determine whether a waste is hazardous; the generator may use other information (such as knowledge of the process by which the waste was generated) in making that determination." [RO 11649]

However, the agency also says that generators are "required to be correct in their determination" [RO 11599], and "if subsequent testing by EPA or others demonstrates that the waste was hazardous, an incorrect determination made based on knowledge would leave a waste generator…vulnerable to enforcement action." [RO 11958] "[G]enerators and subsequent handlers would be in violation of RCRA if they managed hazardous waste, erroneously classified as nonhazardous, outside of the RCRA hazardous waste system." [November 20, 1997; 62 *FR* 62083] "You do not have to run all or any of the constituents listed in the TCLP test if you believe, and can demonstrate through process knowledge, that your waste is nonhazardous. However, keep in mind, if enforcement takes a sample and finds that your waste is indeed hazardous, then enforcement actions would take place." [RO 14695]

The violations alluded to in the preceding paragraph could include noncompliance with the §262.11 waste characterization provisions and also the hazardous waste storage, transportation, and treatment/disposal requirements (including LDR regulations).

Generators must have "sufficient information to make an accurate determination." [RO 13570] If you determine using knowledge that your waste is nonhazardous, "this belief [must] be based on an objective review of the materials and processes involved in the generation of the waste." [February 26, 1980; 45 *FR* 12727] Objective means based on observable phenomena—presented factually. You can't rely on a subjective determination (defined in our dictionary as "taking place within an individual's mind such as to be unaffected by the external world"). In other words, it can't be based on "Well, I don't think it's hazardous." Or, "The previous environmental guy told me it wasn't hazardous." Inspectors reviewing your hazardous waste determinations will be looking for facts, not opinions.

> "The agency has…rejected the suggestion that a 'good faith' mistake provision should be included in the regulation to excuse inadvertent mistakes in the determination of whether a waste is hazardous. The determination is the crucial, first step in the regulatory system, and the generator must undertake this responsibility seriously. The declaration [that a waste is or is not hazardous without testing] provided for in the regulation must be based on factors which are subject to objective review. A deliberate or negligent oversight, for example overlooking the presence of hazardous substances in the feedstock, would not support the declaration." [February 26, 1980; 45 *FR* 12727]

Bottom line: If you want to make a knowledge-based determination that a waste is not hazardous, it's important to be right.

Given the importance of this issue, facilities wishing to minimize the costs associated with testing wastes often assume a questionable waste is hazardous and handle it accordingly. [November 20, 1997; 62 *FR* 62083] "If a person believes his waste to be hazardous, he may also simply declare it to be so without any references to [the regs] or to scientific literature." [December 18, 1978; 43 *FR* 58969]

Using generator knowledge to determine if wastes are listed or characteristic is discussed in Sections 18.3.2 and 18.3.3, respectively. The unique aspects of making a knowledge-based determination for radioactive mixed wastes are summarized in Section 18.3.4. Before discussing the specifics of making a knowledge-based determination, we want to briefly address one key question: what knowledge would be acceptable to the agency if you don't have analytical data?

18.3.1 What is acceptable knowledge?

The regulations in §262.11(c)(2) are vague as to what information EPA considers adequate for making a knowledge-based assessment: "knowledge of the hazardous characteristic of the waste in light of the materials or the processes used." Referring again to EPA's guidance in EPA/530/R-94/024, the agency considers acceptable knowledge to include:

1. Process knowledge,
2. Waste analysis data obtained from other facilities, and
3. Old analytical data.

Each of these types of acceptable knowledge is discussed in the following subsections.

18.3.1.1 Process knowledge

Most people have a good idea of what process knowledge is (e.g., knowledge of the process itself and the inputs, reactions, and operating status of the process). But, EPA has given us some definitions for this term in guidance.

> "Process knowledge refers to detailed information on processes that generate wastes subject to characterization, or to detailed information (e.g., waste analysis data or studies) on wastes generated from processes similar to that which generated the original waste. Process knowledge includes, for example, waste analysis data obtained

by TSDFs from the specific generators that sent the waste offsite, and waste analysis data obtained by generators or TSDFs from other generators, TSDFs or areas within a facility that test chemically identical wastes." [November 20, 1997; 62 *FR* 62081]

EPA provides a similar definition in EPA/530/R-94/024.

We interpret the first part of the above definition ("information on processes that generate wastes") to include data such as:

- Material balances, including an analysis of constituents of concern in all raw materials, intermediate products, by-products, and final products of contributing processes and chemical reactions occurring in these processes;

- Engineering production data, treatment schematics, and other plans;

- Safety data sheets (discussed more below);

- Process kinetic information and process rates; and

- Other engineering calculations (e.g., does the process concentrate or dilute the waste?).

Although not specific to waste characterization, the Subpart CC air emission standards include a good summary of process knowledge [§265.1084(a)(4), December 6, 1994; 59 *FR* 62916]:

- Material balances for the source or process generating the waste, including raw materials or intermediate products fed to a process;

- Constituent-specific chemical test data for the waste from previous testing at the facility that are still applicable to the current waste;

- Previous test data from other locations using substantially similar processes and/or managing the same type of waste streams; and

- Other knowledge based on information in manifests, shipping papers, or waste certification notices.

18.3.1.1.1 Safety data sheets and other manufacturers' data

Safety data sheets (SDSs) associated with products are often a significant component of process knowledge. These could be products being used in a process or products being disposed. SDSs are developed by product manufacturers and give lots of useful information about the ingredients and physical/chemical properties of the materials. Some SDSs even contain a section on RCRA classification for the unadulterated product. [RO 14686]

SDSs are commonly used to determine whether a waste could be one of the F-, P-, or U-listed wastes. Also, flash point; pH; stability; OSHA classification; reactivity; fire and explosion; handling, storage, and transportation; and incompatibility data on these sheets can be used in making knowledge-based decisions about whether the waste exhibits the characteristics of ignitability, corrosivity, or reactivity. For example, information under the reactivity data section of an SDS that notes "Oxidation; ignition sources" may indicate that, when the product becomes a solid waste, it exhibits the reactivity characteristic via §261.23(a)(6). [RO 11398] "Using the SDS of a product is generally an acceptable means to determine whether or not any of the product's constituents or properties would make it a characteristic or listed waste, when discarded." [RO 14790]

For the toxicity characteristic, chemical composition data on the SDS are helpful. However, be advised that SDSs are not always an appropriate reference for determining if a material exhibits the toxicity characteristic. The OSHA regs, as modified to reflect the provisions of the United Nations Globally Harmonized System of Classification and Labeling of Chemicals, generally require manufacturers to identify constituents present in the material at concentrations ≥1% (10,000 ppm) for noncarcinogens or ≥0.1% (1,000 ppm) for carcinogenic constituents. [Appendix A to 29 *CFR* 1910.1200] Therefore, the product might contain toxicity characteristic constituents above RCRA regulatory levels even though they are not identified on the SDS.

Another concern with the use of SDSs in making knowledge-based hazardous waste determinations is that they quickly get out-of-date. Thus, knowledge-based evaluations made using old data sheets may be incorrect.

Even EPA admits, "[a]lthough an SDS may often be adequate, other information can also be relevant." [RO 14790] The concerns raised in the paragraphs above are the reasons the state of Indiana notes in its hazardous waste determination guidance: "Therefore, SDSs should be viewed in a supporting fashion and not as the sole means of providing generator knowledge."

In addition to SDSs, other data from manufacturers are sometimes used by generators to make hazardous waste determinations. For example, manufacturers of low-mercury fluorescent light bulbs have conducted TCLP testing on their bulbs to show that they do not exhibit the toxicity characteristic when disposed. Generators often (with state agency concurrence) use such manufacturers' information to make nonhazardous waste determinations for these materials. In another situation, the American Dental Association has conducted research showing that scrap dental amalgam doesn't exhibit the toxicity characteristic, and EPA noted that generators could cite that research as "applying knowledge of his waste in determining the regulatory status." [RO 11457] However, confirmatory TCLP testing in both situations noted above would be prudent.

18.3.1.2 Analytical data from other facilities

The second type of "acceptable knowledge," per EPA, is analytical data obtained from other facilities. We discussed this a little already when we reviewed some of EPA's guidance on "process knowledge." That term includes test data from other locations that use substantially similar processes and/or are managing substantially similar waste streams. Analytical data from other facilities could be used in a couple of situations:

1. There is another plant that conducts substantially similar processes and/or is managing substantially similar waste streams to yours. They have recently completed a detailed waste characterization of some of their waste streams, and you want to use their data to make knowledge-based hazardous waste determinations for your wastes. EPA allows test data on wastes known to be very similar to yours to be used in lieu of testing. [RO 13506] Just make sure their feedstocks and processes really are similar to yours and that the data are current, or the wastes won't necessarily be the same.

2. You work at a TSD facility and are relying on waste analysis data from offsite generators that are sending you their wastes. You can use these data to determine if the wastes are hazardous. Although TSD facilities may rely generally on information provided to them by generators, they are required to conduct periodic analyses for purposes of complying with the LDR program (all in accordance with their WAP). [RO 11545] Additionally, TSD facilities should become thoroughly familiar with the generator's processes (e.g., by visiting the site or obtaining split samples for analysis) to verify the integrity of the data. "[A]n offsite TSDF is not relieved of its responsibility to obtain accurate waste analysis data despite the submission of erroneous information provided to the TSDF by the generator." [EPA/530/R-94/024]

18.3.1.3 Old analytical data

The third type of "acceptable knowledge," according to EPA, is old analytical data. We think of old analytical data as information from two or three years ago. But, believe it or not, EPA is talking about pre-RCRA data.

Normally, data that is 30-plus years old won't be useful to you today in making hazardous waste determinations. However, as noted in Section 18.3.1.1 when we discussed "process knowledge," test data for the waste from previous testing at the facility that are still applicable to the current waste may be used as knowledge. "This information then becomes the basis for future generator knowledge about the waste. If the waste proves nonhazardous [from previous representative sampling and analysis], as long as the process or type of material…doesn't change, further testing should be unnecessary as documented generator knowledge has proven it does not pose a hazard." [RO 11829]

One caveat is that the ability of analytical equipment to detect low concentrations of contaminants has improved over the years. Constituents that were determined by previous analytical work to be nondetectable may be detectable using more sophisticated equipment and techniques available today. [EPA/530/R-94/024]

18.3.1.4 "Acceptable knowledge" looks like waste analysis data

Although EPA says in the regs and guidance that you can use your knowledge to make a hazardous waste determination, when the agency defines what it means by "acceptable knowledge," in many cases it looks like they mean waste analysis data. That's what EPA prefers. "Compliance is best ensured through sampling and analysis." [EPA/530/R-02/003] The agency clearly believes that sampling and analysis are more accurate, reliable, and defensible. Still, there are times when it makes sense to use knowledge, as detailed in the following sections.

18.3.2 Knowledge for listings

Listed wastes are hazardous simply because they meet the description of a waste on one of the four lists in Subpart D of Part 261. In general, it doesn't matter what the concentrations of hazardous constituents are in a listed waste; what matters for purposes of the hazardous waste determination is the source of the waste (i.e., the process that generated it). [RO 13181] (Of course, the hazardous constituent concentrations are important for determining LDR compliance, as discussed in Section 18.4.) You can't really test a waste to determine if it's listed—

what would you test for? Heat-exchanger bundle cleaning sludge from exchangers at refineries is listed waste K050 because it meets the listing description in §261.32, not because it contains certain levels of toxic constituents. "A generator who produces such residues should know, without any sampling or analysis, that these wastes are listed RCRA hazardous wastes by examining the…hazardous waste description in the hazardous waste lists." [November 20, 1997; 62 *FR* 62082]

In fact, even if none of the constituents for which the waste was listed (see Part 261, Appendix VII) can be found in a waste, the waste is still a listed hazardous waste if it meets a listing description in §§261.31–261.33. [RO 13586, 14103, 14482, 14691, 14699]

As such, "application of acceptable knowledge is appropriate because the physical/chemical makeup of the waste is generally well known and consistent from facility to facility." [EPA/530/R-94/024] Therefore, analytical testing alone, without using knowledge of a waste's source, will not produce information that will conclusively indicate whether a given waste is a listed hazardous waste. [RO 12171, 12392, 14291] Some tips on making a listed waste determination using knowledge are given below.

18.3.2.1 F- and K-wastes

The F- and K-wastes are found in §§261.31 and 261.32, respectively, of the federal RCRA regulations. They are manufacturing process wastes. Solid wastes are identified as F- or K-wastes by "comparing the specific process that generated the waste to those processes listed in [the regs] (rather than conducting a chemical/physical analysis of the waste)." [EPA/530/R-94/024]

18.3.2.1.1 *F-wastes*

The concept behind the F-listed wastes is that they are manufacturing process wastes produced by a wide variety of industrial operations; that is, they are process wastes generated from nonspecific industries/sources. For example, any type of facility that uses any of the solvents in the F001–F005 listings for their solvent properties will likely produce an F001–F005 listed hazardous waste when the solvent becomes spent.

Lots of issues crop up when generators are trying to determine whether one or more of the F-listings apply to their waste streams. For example, when determining if an F001–F005 listed waste code applies to a spent solvent mixture, the generator needs to know the *before-use* concentrations of the solvents in the mixture. This information usually comes from the solvent manufacturer in its SDS. Our discussions in Sections 3.2–3.4 of this book should help when making knowledge-based determinations for the F-listed hazardous wastes.

In addition, EPA prepared a listing background document for every F-waste that is quite specific about the particular waste streams the agency wanted to capture via the listing. EPA has tried to be very specific in both the language of the listing description in the regs and also in the listing background document as to "where in the [manufacturing] process wastes are generated so as to enable generators to determine more easily if their wastes are listed." [May 19, 1980; 45 *FR* 33114] Copies of these listing background documents can be obtained online from http://www.regulations.gov. Search for RCRA-2004-0016 for F-listed wastes. Alternatively, you can request this information by emailing rcra-docket@epa.gov or phoning the EPA Docket Center at (202) 566-0270.

18.3.2.1.2 *K-wastes*

K-wastes are manufacturing process wastes from specific industries/sources. This list is subdivided into groups of wastes generated by various industries (e.g., the inorganic pigment manufacturing industry, the organic chemical manufacturing industry, etc.) In order to use the K-list, a generator first determines if his/her facility operations fit within any of the industry categories (e.g., wood preservation or inorganic pigment manufacturing). Once the industry category is identified, the generator checks to see if his/her wastes meet any of the K-waste listing descriptions in that category.

The generator needn't care at all about the K-wastes associated with the other industries. For example, a generator in the petroleum refining industry may generate any of the nine K-wastes included in that industry category; but they will not generate any of the K-wastes in any of the other industry categories. Of course, if a facility fits into several industry categories (e.g., it is a petroleum refinery and an organic chemical manufacturing facility), it could generate the K-wastes associated with each industry category.

The listing descriptions associated with the K-wastes are generally very specific and clear. When questions are put to EPA as to the applicability of a particular K-waste code, the agency notes in general that "[o]ur interpretations on the applicability of RCRA [K] codes are based on the consideration of 1) the descriptive regulatory language, 2) the regulatory intent of the original listing, and 3) facts specific to the waste stream at issue." [RO 13679]

When people aren't sure if they generate a particular K-waste, we usually suggest that they obtain and read the listing background document for that waste. Copies of these listing background documents can be obtained online from http://www.regulations.gov. Search for RCRA-2004-0017 for K-listed wastes. Alternatively, you can request this information by emailing rcra-docket@epa.gov or phoning the EPA Docket Center at (202) 566-0270.

18.3.2.2 *P- and U-wastes*

The last two lists of hazardous wastes, the P- and U-lists, are both contained in §261.33. These two lists identify pure unused commercial chemical products, off-specification products, and cleanup residues from spilling such products that, when discarded, pose a threat to human health or the environment, usually due to their toxicity. Unused product formulations that contain a P- or U-listed chemical as the sole active ingredient are also assigned the corresponding P- or U-code when these products are to be discarded. SDSs are usually very helpful when making knowledge-based determinations for the P- and U-listed hazardous wastes. Still, considerable confusion exists about when to apply the P- and U-codes. A complete discussion of this topic is presented in Section 3.5.

18.3.2.3 Analysis can be included

Sometimes, analytical data may be used along with process knowledge to make a listed waste determination. For example, analyses of soil or ground water samples may indicate that the media is contaminated with a hazardous constituent (e.g., toluene). This information can be used, along with knowledge of the source of the toluene, to determine that the soil or ground water contains listed hazardous waste. However, you can correctly assign a listed hazardous waste code only if you can determine through knowledge the source of the contaminant (e.g., from a spill of pure toluene or spent solvent toluene). Otherwise, the soil or ground water will be hazardous only if it exhibits a characteristic. [RO 12392]

EPA's guidance for CERCLA sites is consistent:

"[A]t many CERCLA sites, no information exists on the source of the wastes nor are references available citing the date of disposal. The lead agency should use available site information, manifests, storage records, and vouchers in an effort to ascertain the source of these contaminants. When this documentation is not available, the lead agency may assume that the wastes are not listed RCRA hazardous wastes...." [December 21, 1988; 53 *FR* 51444; see also March 8, 1990; 55 *FR* 8758]

18.3.2.3.1 *Wipe/chip sampling for contaminated debris*

Another situation where analytical data can be combined with process knowledge is when making a listed waste determination for contaminated debris. The physical nature of debris (e.g., a pump, piping, pieces of a building, a concrete pad) makes it difficult to obtain a representative sample. However, surface wipe or solvent wash samples can be collected for nonporous debris, or chip sampling can be conducted for porous debris, followed by laboratory analysis of the samples to determine if hazardous contaminants are present on the debris. Once the contaminants, if any, are known, knowledge may be used to determine where those contaminants came from and whether the debris is hazardous waste because it contains a listed waste [see §268.2(h)]. [EPA/530/R-94/024]

The MICE service, referenced in Section 18.1, had this to say about the use of wipe sampling for RCRA waste characterization (in a January 2008 email):

"While wipe sampling has been described in the open literature and is cited in some 40 *CFR* sections, one of the biggest problems is that it is very difficult to interpret the results. By its very nature, the analysis of whatever material is used to wipe a surface (often filter paper wet with a solvent) yields the mass of the analyte(s), for example nanograms of dioxin or micrograms of another analyte. However, there is no straightforward way in which to convert that mass into a concentration per unit area, nor any good way in which to compare results from different wipes, except to say that one wipe picked up more material than another. If the surfaces that are wiped have different characteristics, for example a smooth metal surface versus a rough concrete block, there is no way to judge the efficiency of the wiping process itself. Therefore, use caution in interpreting the results of wipe samples."

18.3.3 Knowledge for characteristics

Knowledge of a waste and its constituents can be used to determine whether a waste exhibits a characteristic. When EPA proposed the RCRA regulations in 1978, the agency wanted to give generators the option of making knowledge-based determinations. EPA stated:

"A person who has knowledge of the raw materials input into his process and knows these materials to be present in the waste may utilize this information to determine whether the waste would match the characteristics set forth in [the regs] without testing. This can be accomplished by using the manufacturer's specifications and data or by consulting scientific literature and comparing the physical and chemical properties of the raw materials in the waste to the characteristics…which make a waste hazardous." [December 18, 1978; 43 *FR* 58969]

Some tips on making a characteristic waste determination using knowledge are given below.

18.3.3.1 Ignitability

As noted in the following examples, knowledge is often used to make determinations of whether a waste exhibits the ignitability characteristic.

Q: *Gasoline-contaminated soil passes the paint filter test (i.e., no liquid passes through the filter). Can the generator determine if the soil exhibits the ignitability characteristic without running a flash point test?*

A: Yes. The soil is not a liquid because it passed the paint filter test. Therefore, it cannot be ignitable per the federal regulations at §261.21(a)(1), regardless of whether it flashes at <140°F. The generator can also use knowledge to determine whether the soil could cause fire through friction, moisture absorption, or spontaneous ignition [i.e., the criteria in §261.21(a)(2)]. Since this is not likely, the generator can determine using knowledge that the soil is not a D001 ignitable hazardous waste.

Q: *Wastewater from a manufacturing plant contains 30 ppm hexane, 20 ppm toluene, and 40 ppm cyclohexanone. Can the generator determine if the wastewater exhibits the ignitability characteristic without running a flash point test?*

A: Probably. The generator may use knowledge to determine that, even though there are constituents in the wastewater that are ignitable in their pure form, the concentrations of these chemicals in the plant wastewater stream are not high enough to cause it to have a flash point <140°F. If the generator is unsure, he or she should test a sample of the wastewater to verify that the flash point is >140°F.

Q: *A facility is getting ready to discard unused paint. The environmental manager finds an old SDS associated with a similar paint product that shows a flash point of <140°F. Should the manager apply the D001 code to the paint?*

A: Maybe. SDSs are useful tools in making knowledge-based determinations for products to be discarded. But, the SDS used should be up-to-date and applicable to the specific product intended for disposal. In this case, using the old, nonapplicable SDS to code the discarded paint as D001 could be overregulating the material since it may actually have a flash point >140°F.

Q: *A waste consists of 77% water, 13% alcohol, and 10% nonalcoholic organic liquid. Can the generator determine if the waste exhibits the ignitability characteristic without running a flash point test?*

A: Yes. The alcohol exclusion in §261.21(a)(1) applies to solutions containing <24% alcohol by volume. EPA clarified that the presence of a nonalcoholic constituent will not require the waste to be regulated as D001, even if the mixture has a flash point <140°F. [RO 13548]

Q: *A sludge is generated from treating wastewaters from an electroplating operation. The sludge (which we know meets the F006 listing) contains 75% liquid. Can the generator determine if the waste is also D001 without running a flash point test?*

A: Maybe. If the liquid is all water, the generator can document that the sludge cannot exhibit the ignitability characteristic with such high water content. On the other hand, if information about the manufacturing process indicates that a flammable organic solvent may appear in the waste due to solvent cleaning operations, knowledge alone without testing may be inadequate. [*Petitions to Delist Hazardous Waste*, EPA/530/R-93/007, March 1993, available at http://nepis.epa.gov/EPA/html/Pubs/pubtitleOSWER.html by downloading the report numbered 530R93007]

18.3.3.2 Corrosivity

As with ignitability, knowledge is often used to determine if a solid waste exhibits the characteristic of corrosivity.

Q *Sometimes, generators make rough pH determinations using pH paper. Is this a knowledge-based determination?*

A Yes. Because the prescribed method for measuring pH (a pH meter using Method 9040C) was not used, such a corrosivity determination would fall under the category of generator knowledge.

Q *A facility produces a filter cake that is essentially dry. Is the cake, which contains sodium hydroxide precipitate, a corrosive hazardous waste when disposed?*

A No. There are no corrosive solids at the federal level. Method 9040C specifies that it should only be used when the aqueous phase constitutes at least 20% of the total volume of the waste. Therefore, any waste for which this method is applicable must contain at least 20% free water by volume. [RO 11738] Since the cake is not aqueous, it cannot be corrosive by the §261.22(a)(1) criteria. Also, since the waste is not a liquid (based on the paint filter test), it cannot be corrosive under §261.22(a)(2). Some states, however, have more stringent requirements. For example, in some states, a waste that might have a pH ≤2.0 or ≥12.5 if mixed with water must be evaluated for corrosivity.

Q *An old sodium hydroxide product (containing 50% NaOH in aqueous solution) will be disposed. The SDS shows a pH of 14. Can a chemist use knowledge to assume that the product is D002 when disposed?*

A Yes. It is clear without the need for pH testing that the product will exhibit the corrosivity characteristic.

Q *Can a generator determine based on knowledge whether his or her waste is corrosive under §261.22(a)(2) because it corrodes carbon steel, or does Method 1110A have to be used to measure the corrosion rate?*

A EPA's guidance suggests that knowledge can often be used to determine if a waste will corrode carbon steel, thereby making it D002. "If a given waste has a pH of 5 and is known not to contain any oxidizing materials, then the corrosivity toward steel test may not be warranted. The generator of the waste should decide which tests need to be performed on a case-by-case basis. Available engineering data are usually quite useful in determining which tests are most appropriate for a particular waste." [OSWER Directive 9443.01(80)]

18.3.3.3 Reactivity

The regulatory language in §261.23(a) does not specify any test methods for determining if a waste exhibits the reactivity characteristic. The agency has found no appropriate test methods for this purpose; thus, all D003 determinations are knowledge-based. A generator compares his/her knowledge of the solid waste to the narrative description of reactivity in §261.23(a).

According to EPA: "Most generators of reactive wastes are aware that their wastes possess this property and require special handling. This is because such wastes are dangerous to the generators' own operations and are rarely generated from unreactive feedstocks. Consequently, the prose definition [found in §261.23(a)] should provide generators with sufficient guidance to enable them to determine whether their wastes are reactive." [May 19, 1980; 45 *FR* 33110]

One of the reactivity criteria is specific to cyanide- and sulfide-bearing wastes. Section 261.23(a)(5) identifies a reactive waste as "a cyanide- or sulfide-bearing waste which…can generate toxic gases, vapors, or fumes *in a quantity sufficient to present a danger to human health or the environment*." [Emphasis added.] Since EPA has rescinded its previous guidance on what constitutes such a quantity, generators must rely on knowledge of their wastes to make a proper D003 classification. [June 14, 2005; 70 *FR* 34548, RO 14177] In old guidance documents (none of which are in EPA's RCRA Online database), the agency noted that, if the answer to any of the four questions below is "Yes," the waste should be considered a D003 reactive hazardous waste:

1. Has the waste ever caused injury to a worker because of hydrogen cyanide (HCN) or hydrogen sulfide (H_2S) generation?
2. Have the OSHA workplace air concentration limits for either HCN or H_2S been exceeded in areas where the waste is generated, stored, or otherwise handled?
3. Have air concentrations of HCN or H_2S above a few parts per million been encountered in areas where the waste is generated, stored, or otherwise handled?
4. Would a chemist with knowledge of the waste believe that one or more of the above might occur if the waste was subject to acidic conditions?

[Four EPA Headquarters memos, the most recent of which is a letter from William Collins, Jr. to Donald Searles, U.S. Attorney for the Eastern District of California, December 17, 1996]

Other guidance for making knowledge-based determinations for this characteristic is summarized below:

- EPA has concluded that waste ammunition up to and including 0.50 caliber is not reactive and that the disposal of such ammunition is not subject to hazardous waste requirements (unless it exhibits some other characteristic). [RO 12339]
- The U.S. Army Toxic and Hazardous Materials Agency concluded that explosives-contaminated soil/sediment containing 12% explosives or less will not propagate a detonation or explode when heated under confinement. [*Testing to Determine Relationship Between Explosive-Contaminated Sludge Components and Reactivity*, AMXTH-TE-CR-86096, January 1987] Appendix A of SW–846 Method 8330B says the same thing. The U.S. Army Environmental Center considers all soils containing more than 10% by weight secondary explosives to be susceptible to initiation and propagation. [*Handbook: Approaches for the Remediation of Federal Facility Sites Contaminated With Explosive or Radioactive Wastes*, EPA/625/R-93/013, September 1993, available at http://nepis.epa.gov/EPA/html/Pubs/pubtitleORD.html by downloading the report numbered 625R93013]
- EPA has not been able to categorically determine whether or not various types of waste aerosol cans that may have contained a wide range of products are reactive. As a result, it is the responsibility of the generator to determine if a waste aerosol can is hazardous in accordance with §262.11. [RO 11806]

18.3.3.4 Toxicity

The toxicity characteristic is one of the method-defined parameters (i.e., the TCLP defines the toxicity characteristic); as such, the TCLP must be followed exactly as written. (See Section 18.2.1.3.) However, knowledge is often used in part or in whole when making toxicity characteristic determinations, as noted in the three items below.

1. The generator can use material balances and/or knowledge of the raw materials and the processes that generated the waste to make a toxicity determination. If a generator has good knowledge that a certain waste would pass or fail the TCLP, no testing is necessary. An example is the U.S. Army managing waste munitions as toxic for lead and/or 2,4-dinitrotoluene; the Army didn't want to perform the TCLP because of the inherent safety problems associated with the method's particle-size reduction criteria. Thus, the Army could simply declare the waste to be hazardous for toxicity and manage it as such. [RO 11608] Where no information or knowledge is available to assist a generator, or when TCLP results are not available or are inconclusive, the agency noted that it might be prudent for the generator to manage the waste as hazardous. [RO 11579, 11592] Sometimes,

a generator will use a combination of testing and knowledge; that is, if the generator has knowledge that certain §261.24 constituents cannot be present in the waste, the TCLP will have to be performed only for the other constituents that may be in the waste. [EPA/530/R-93/007, RO 11603, 14695]

2. An alternative test method can be used that may be more appropriate for the waste matrix. For example, the oily waste extraction procedure (OWEP—Method 1330 in SW–846) can be used to evaluate metal concentrations in oily sludges, slop oil emulsions, and other oily wastes (although this method is typically used only in support of delisting petitions). This method was developed for wastes containing oil or grease in concentrations of 1% or greater. [RO 11522, 12450] This would be considered a knowledge-based determination because the specified method (the TCLP) is not being used.

3. A generator can use total waste analyses to determine that a waste does not exhibit the toxicity characteristic. Although based on analytical results, this procedure is considered knowledge-based because the prescribed method for measuring toxicity (Method 1311) is not used. [RO 14533] Using total waste analyses in lieu of TCLP results is described in detail in Section 2.6.3.1. A total waste analysis is also a convenient and cost-effective screening tool for determining if a TCLP test is needed. [RO 14695]

Examples of making knowledge-based toxicity characteristic determinations are given below.

Q: *A facility uses mineral spirits (which do not produce F-listed spent solvents) in a solvent application. Can facility personnel use knowledge to determine if the resulting spent solvent exhibits a characteristic?*

A: Yes. Facility personnel can use knowledge (probably from an SDS) of the mineral spirits' flash point and chemical composition *plus* knowledge of how the solvent was used to conclude that the spent solvent will or will not exhibit any characteristics. [RO 11829, 13570]

Q: *A paint removal product is sprayed onto a surface that previously was coated with lead-based paint, and the product is allowed to cure. During the curing process, the paint adheres to the product, facilitating its removal by conventional coating removal techniques. Can the generator use the product manufacturer's data, which shows that paint debris removed using its product passes the TCLP, to determine that his/her paint chips will not be hazardous?*

A: Each generator is responsible for making a toxicity determination for its paint removal debris. Although the generator may use the manufacturer's data in making a knowledge-based assessment, the generator must be right. If the generator assumes (based on manufacturer's data) that the removed paint is nonhazardous, and subsequent testing by the state or EPA demonstrates that the chips are hazardous, an incorrect determination made based on knowledge would leave the generator in violation of RCRA. [RO 11958, 14069] Therefore, confirmatory TCLP testing for lead would be prudent.

Q: *Lead-based paint is identified (e.g., by historical knowledge, x-ray fluorescence detector, or laboratory analysis) on a building to be demolished. The building consists of wood, brick, cement foundations/floors, steel members and stairs, plaster, dry wall, etc. If the entire building is demolished, can knowledge be used to make a hazardous waste determination for the resulting demolition debris?*

A: According to EPA information, wood (e.g., building walls) painted with lead-based paint before 1950 may contain 1,400–20,000 ppm lead. [*Construction and Demolition Waste Landfills*, February 1995, available at http://www.epa.gov/wastes/hazard/generation/sqg/const/cdrpt.pdf] Nevertheless, EPA noted in preamble language that a "representative sample of demolition debris subjected to the TCLP is not likely to exceed the TC regulatory limit for lead because of the small amount of paint in relation to the overall waste stream...." [December 18, 1998; 63 *FR* 70206]

A U.S. Army Environmental Hygiene Agency report noted that whole-building demolition debris (e.g., from World War II-era structures) can be statistically characterized as nonhazardous waste based on the following assumptions:

- Components such as asbestos and PCBs (e.g., from light ballasts and roofing tars) are not present or are removed and disposed separately.
- Metal components such as ductwork, furnaces/boilers, piping, and siding are removed to the extent feasible as scrap metal for reuse/recycling.
- All remaining material comprises a single waste stream at the point of generation (when the building is demolished). This debris is handled as a single, discrete waste stream and disposed all together.

Conversely, debris that is generated during renovation, maintenance, or abatement activities, such as paint chips, blast grit/media, or personal protective equipment, is more likely to be characterized as "hazardous" due to the concentrated mass of lead-based paint. For these types of wastes, TCLP testing would likely be required, but hazardous waste generation can be minimized through waste segregation techniques. [*Lead-Based Paint Contaminated Debris*—Waste Characterization Study No. 37-26-JK44-92, May 1992–May 1993, available from http://www.understandrcra.com/RCCD/LeadContaminatedDebris.pdf and *Waste Characterization of Lead Paint-Containing Wastes*, June 2009, available from http://www.understandrcra.com/RCCD/LeadPaintWaste.pdf]

Note that when building debris is deemed to be nonhazardous based on knowledge, some limited sampling and TCLP testing for lead may be warranted to minimize liability.

18.3.3.5 Analysis may be included

A generator will often use a combination of testing and knowledge in evaluating whether a waste exhibits a characteristic. For example, the paint filter test may be used to determine if a waste is a liquid. Based on that information, a knowledge-based ignitability and/or corrosivity determination can sometimes be made.

If the generator has knowledge that certain §261.24 constituents could not be present in a waste, the TCLP will have to be performed only for the other constituents that might be in the waste. [EPA/530/R-93/007, RO 11603, 14695]

Process, screening, or other data that can help make an assessment regarding possible characteristics are sometimes available for a waste. Here's EPA's discussion for making knowledge-based determinations at CERCLA sites:

"EPA wants to make clear, however, that a decision that a waste is not characteristic in the absence of testing may not be arbitrary, but must be based on site-specific information and data collected on the constituents and their concentrations during investigations of the site. Based on site data, it will be very clear in some cases that a waste cannot be characteristic; for example, if a waste does not contain a constituent regulated as...toxic, a decision that the waste does not exhibit this characteristic can reliably be made without testing for...toxicity. EPA does not expect to undertake testing when it can otherwise be determined with reasonable certainty whether or not the waste will exhibit a characteristic." [March 8, 1990; 55 *FR* 8762]

Another situation where analysis can be combined with knowledge is when making a characteristic determination for contaminated debris. The physical nature of debris (e.g., a pump, piping, pieces of a building, a concrete pad) doesn't lend itself to sampling and analysis. However, surface samples (e.g., of lead-based paint) can be collected and analyzed. Once the contaminants and their concentrations are known, knowledge (e.g., weight averaging) may be used to determine whether the debris is characteristic. [EPA/530/R-94/024]

18.3.4 Use knowledge for mixed wastes

In EPA/530/R-94/024, EPA identified three specific situations where it was more appropriate to use knowledge than sampling and analysis. We've discussed the first two: making listed waste determinations, and determining if contaminated debris is hazardous. Here comes the third: use knowledge if sampling and analyzing your waste would pose unacceptable health or safety risks. The primary example of such a situation is when facilities make hazardous waste determinations for radioactive waste. Mixed waste (i.e., waste that contains both hazardous waste and source, special nuclear, or by-product material subject to the Atomic Energy Act) is subject to both RCRA and Atomic Energy Act regulations.

EPA and the Nuclear Regulatory Commission (NRC) published joint guidance on this topic on November 20, 1997. [62 *FR* 62079] One of the predominant themes in this document is that "testing of mixed waste may be avoided. This is important when handling mixed waste, since each sampling, workup, or analytical event may involve an incremental exposure to radiation. This guidance encourages mixed waste handlers to use waste knowledge, such as process knowledge, where possible in making RCRA hazardous waste determinations involving mixed waste." [62 *FR* 62081]

The EPA/NRC guidance provided the following examples of knowledge-based hazardous waste determinations:

- Maintenance personnel using trichloroethylene (TCE) to remove paint from a radioactively contaminated surface can determine that the resulting waste is F002, after considering the listing description and, if starting with a solvent blend, the percent TCE by volume in the solvent blend before use.

- A facility may use a surrogate material (i.e., a chemically identical material with significantly less or no radioactivity) to determine the RCRA status of a waste. For example, a nonradioactive lead-bearing material may be tested to determine if it exhibits the toxicity characteristic for lead (D008) in lieu of running a TCLP analysis on the actual radioactive waste. Use of a surrogate is acceptable as long as it "faithfully represents" the properties of the actual waste.

- Personnel can use old records or knowledge of the processes that generated the waste to determine if old, "legacy" wastes are hazardous. However, EPA and the NRC recognize "that certain…waste streams, such as wastes from remediation activities or wastes produced many years ago, may have to be identified using laboratory analysis, because of a lack of waste or process information on these waste streams." [62 *FR* 62083]

- A generator can use total waste analysis to determine if a waste exhibits the toxicity characteristic (as discussed in Section 18.3.3.4). The grinding or milling step when running the TCLP on radioactive wastes raises significant radiation exposure concerns. Additionally, total analyses may result in the use of smaller sample sizes and generation of less laboratory wastes.

Q *Americium-beryllium sealed sources are neutron emitting sources that are used in the agriculture, oil-well logging, oil refining, and construction industries. According to one report, several thousand sealed sources may enter the radioactive waste stream annually. Such waste sources are not listed and do not exhibit the ignitability, corrosivity, or reactivity characteristic. However, they often contain lead- or silver-based solder and so could exhibit the toxicity characteristic. What is an acceptable method of characterizing these wastes for toxicity without conducting the TCLP on one of the radioactive sealed sources?*

A "Given the safety concerns with mixed wastes, a combination testing/mass balance approach may be appropriate to characterize solder from sealed sources. Either information on the composition of the solder, or TCLP testing (on a non-radioactive sample) would be a starting point. Then, based upon the percentage of the whole material that is solder, a 'theoretical' TCLP concentration may be determined, using an assumption of no contribution of TCLP constituents from the non-solder portion of the waste." [RO 13545]

18.4 LDR program characterization

Characterizing a waste as to its hazardousness was discussed in Sections 18.1–18.3. In this section, we examine the means that a generator or TSD facility may use to determine if a waste meets its LDR treatment standard prior to land disposal.

The basic intent of characterization for the LDR program is shown in Figure 18-1. Per §268.7(a), hazardous waste generators must determine whether their wastes meet LDR treatment standards at the point of generation or must be treated before they can be land disposed. The federal regs give generators the option of making this determination by testing or knowledge. (An April 4, 2006 final rule [71 *FR* 16872] allowed generators to choose not to determine if their hazardous waste requires treatment prior to land disposal. If the generator chooses this approach, he/she must manifest the waste to a RCRA-permitted hazardous waste treatment facility that will have the responsibility for determining if treatment is required.)

Treatment facilities that receive and treat hazardous wastes to meet the LDR standards must test the wastes and residues from treating the wastes in accordance with their WAP to ensure they meet the applicable standards. [§268.7(b)] Facilities that recycle hazardous wastes (e.g., solvent recyclers) are a subset of treatment facilities and must meet the same LDR testing requirements.

Finally, land disposal facilities that dispose hazardous wastes and residues from treating such wastes must spot test them in accordance with their WAP to ensure they meet LDR standards before they are land disposed. [§268.7(c)]

Additional characterization requirements under the LDR program are discussed below for each of these three entities. But first, we look at the overall strategy for showing that an LDR treatment standard has been met.

18.4.1 Meeting LDR standards

We'll give you the bottom line first: "While a statistical evaluation is used to determine if a waste is hazardous, all parts of the waste must be treated to meet the applicable [LDR treatment] standards, not just a representative sample. Thus, if results show that 'hot spots' remain, this is presumptive evidence that treatment was not effective and there is noncompliance with the LDR treatment requirements…. [I]n application of the land disposal treatment

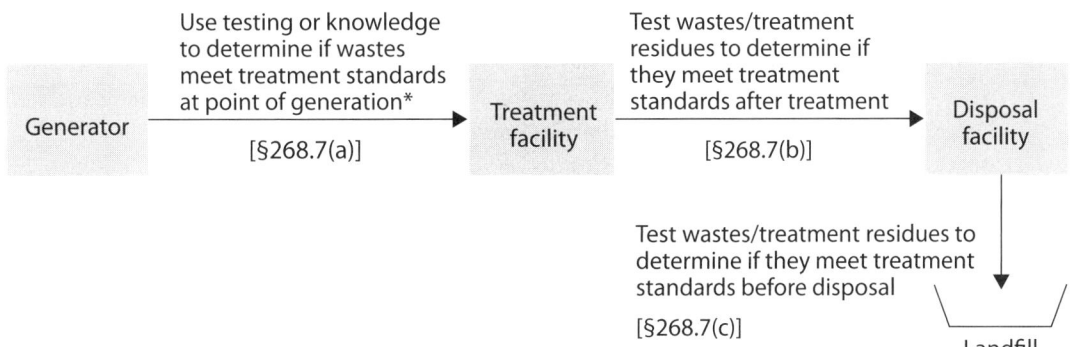

Figure 18-1: Basic Intent of Characterization for the LDR Program

*Generators may choose not to make this determination and instead allow the treatment facility to make it. [71 *FR* 16872]
Source: McCoy and Associates, Inc.

standards, all portions of the waste must meet the applicable treatment standards, i.e., no portion may exceed the regulatory limit." [May 26, 1998; 63 FR 28567] See also May 12, 1997; 62 FR 26047 and *Guidance on Demonstrating Compliance With the Land Disposal Restrictions (LDR) Alternative Soil Treatment Standards*, EPA/530/R-02/003, July 2002 (available from http://www.epa.gov/epawaste/hazard/tsd/ldr/soil_f4.pdf). This approach gives regulators the ability to take just one sample for compliance determinations.

In the process of developing the LDR standards, the agency analyzed two types of samples: grab samples and composite samples. Grab samples are discrete, one-time samples taken from any part of the waste at a specific point in time. EPA set treatment standards for the nonwastewater forms of all hazardous wastes and the D004–D011 wastewaters based on analyses of grab samples. ["Wastewaters" are defined in §268.2(f) as "wastes that contain less than 1% by weight total organic carbon (TOC) and less than 1% by weight total suspended solids (TSS)." "Nonwastewaters" are defined in §268.2(d) as all waste forms that are not wastewaters.]

A composite sample is a combination of samples collected at various locations for a given waste, or samples collected over time from the waste stream. Treatment standards for hazardous wastewaters (other than D004–D011 wastewaters) were set by the agency using composite samples.

The RCRA regulations require that compliance with LDR treatment standards be based on the same type of sampling that was used to initially establish the standard. Thus, for all nonwastewaters and D004–D011 wastewaters, compliance with LDR standards is based on analysis of grab samples. On the other hand, compliance with LDR standards for all wastewaters (other than D004–D011 wastewaters) is based on analysis of composite samples. [§268.40(b), Footnotes 3 and 5 to the §268.40 table of treatment standards, June 23, 1989; 54 FR 26606, June 1, 1990; 55 FR 22539] In addition, §268.40(b) requires compliance for wastewaters to be based on "maximums for any one day," implying that the hazardous constituent concentrations in a daily composite sample (which may be made up of several discrete grab samples) cannot exceed the concentration-based standards in the §268.40 table of treatment standards.

18.4.2 Generator LDR testing/knowledge

As noted above, a generator of hazardous waste must determine if the waste meets its applicable treatment standard or requires treatment before it can be disposed. The regulations allow the generator to use testing or knowledge (or both) for this determination. If the generator claims the waste meets the treatment standard, he/she must certify this per §268.7(a)(3). "Where this determination is based solely on the generator's knowledge of the waste, the agency is requiring that the generator maintain in the facility operating record all supporting data used to make this certification." [November 7, 1986; 51 FR 40597]

Most hazardous wastes do not meet their LDR treatment standard at the point of generation, so the generator may not need to certify anything when shipping the waste offsite. Instead, as discussed in Section 13.12 of this book, the generator simply sends a notification to the TSD facility, noting that it is shipping a waste subject to the LDR program that requires treatment to meet the appropriate standards before land disposal.

An April 4, 2006 final rule [71 FR 16872] allows generators to choose not to determine if their hazardous waste requires treatment prior to land disposal. If the generator chooses this approach, he/she must manifest the waste to a RCRA-permitted hazardous waste treatment facility that will have the responsibility for determining if treatment is required. In this case, the LDR notification sent with the waste will include only 1) the waste code(s), 2) the manifest tracking number of the first shipment, and 3) the following statement: "This hazardous waste may or may not be subject to the

LDR treatment standards. The treatment facility must make the determination."

To save money on sampling/analytical costs, generators often want to obtain LDR compliance data for a waste at the same time they are determining whether the waste is hazardous. This is particularly true when characterizing a site that may contain hazardous soil. [EPA/530/R-02/003] EPA has clarified that generators can determine if their wastes must be treated to meet LDR treatment standards (before they can be land disposed) at the same time as they make hazardous waste determinations. [§268.7(a)(1), April 4, 2006; 71 *FR* 16872] For example, a generator could make a single request to a laboratory to perform analytical testing to determine whether a sample of its waste 1) is hazardous, and 2) meets LDR treatment standards for primary waste codes and any underlying hazardous constituents. However, keep in mind that making a hazardous waste determination is different from determining compliance with LDR treatment standards. The former can be based on statistics—the latter cannot.

Generators who treat their hazardous wastes onsite to meet LDR standards are subject to treatment facility LDR testing requirements and to the WAP requirements in §268.7(a)(5). Generators that dispose hazardous wastes onsite are also subject to disposal facility LDR testing requirements.

18.4.2.1 Identifying underlying hazardous constituents

The LDR treatment standards for certain characteristic wastes include treatment requirements for underlying hazardous constituents (UHCs). Section 13.3 of this book gives a discussion of what UHCs are and when we have to worry about them. To recap, UHCs are any constituents in the universal treatment standards (UTS) table in §268.48 that the generator reasonably expects to be present in his/her characteristic waste at its point of generation at concentrations greater than the corresponding UTS concentrations.

How are generators supposed to determine whether they have UHCs in their wastes? They can use testing or knowledge—or a combination of the two. EPA notes that generators may rely on knowledge of the raw materials used, the process, and the potential reaction products. [May 24, 1993; 58 *FR* 29872, EPA/530/R-94/024] However, RO 13748 states that, where no institutional knowledge exists concerning waste contaminants (e.g., for soil contaminated by a previous party), analysis should be conducted for the entire list of §268.48 constituents.

In other guidance [RO 14325], EPA states that the generator "may use testing or knowledge to ascertain if [UHCs] are present at levels above UTS. Where this determination is based on testing, any of the constituents not shown to be below UTS and not otherwise known to be below UTS should be listed as a 'constituent of concern' [on LDR paperwork]." If a one-time analysis for all of the §268.48 constituents is conducted, subsequent analysis may be limited to only those UHCs identified in the initial sampling and analysis. [May 24, 1993; 58 *FR* 29872, EPA/530/R-94/024]

The problem with this approach is the extremely high analytical cost associated with analyzing a waste for all §268.48 constituents (anecdotally reported to us as being in the range of $3,000 to $5,000 per sample). In our experience and in discussions with regulators, we find that most generators rely on analytical results that have been collected for other purposes (e.g., for waste identification) and on process knowledge to identify UHCs in their wastes.

For example, a facility may know a particular process does not use any pesticides or herbicides, metals, halogenated organics, etc.; thus, these are constituents for which residues don't need to be analyzed when determining UHCs. [November 20, 1997; 62 *FR* 62082–5, RO 12830, 13406] For nonhalogenated organics, although the facility can determine all such substances that are used as raw materials or feedstocks in the process, it may be unsure of particular constituents that may be present in residues as reaction by-products. These will be constituents for which the facility may have to test. This may be a one-time analysis, or there may be a need to periodically retest the residue stream (e.g., if the process or material inputs change).

EPA and state regulators don't tend to second-guess generators' conclusions regarding UHC identification, unless they (the regulators) have specific knowledge that the generator has overlooked.

The last sentence of §268.48(a) notes that compliance with the universal treatment standards will be established based on grab sampling, unless otherwise noted in the §268.48 table. Footnote 2 of that table specifies composite sampling for wastewater forms of wastes/residues. We have never seen any guidance on this subject, but our understanding is that compliance with treatment standards for D004–D011 wastewaters that contain UHCs would also be based on grab samples—to be consistent with §268.40(b).

18.4.3 Treatment facility LDR testing

Treatment facilities receive hazardous wastes from offsite generators and treat/recycle the wastes to meet appropriate LDR treatment standards. Residues from the treatment/recycling process must be tested prior to land disposal according to the treatment facility's WAP to determine if treatment has achieved the required levels. If the standards have been achieved, the residues may be shipped to a land disposal facility or disposed onsite if the facility is permitted for that activity. In either case, the treatment facility must certify and document that the waste meets LDR requirements. "For instance, if the waste analysis plan calls for testing of each batch of waste from an incineration process, these data must be submitted to the land disposal facility along with the certification statement." [November 7, 1986; 51 *FR* 40597]

"Treatment facilities must periodically test the treated waste residue from prohibited wastes to determine whether it meets the…treatment standards and may not rely on materials and process knowledge to make this determination." [November 20, 1997; 62 *FR* 62083] However, testing is not required for residues from wastes that must be treated by specified methods (i.e., those wastes that have treatment standards that are five-letter codes in the §268.40 table of treatment standards).

Treatment facilities have to be careful when using statistical sampling and analysis, which may be allowed by their WAPs, to establish compliance with LDR standards, because compliance with the treatment standards is not based on statistics. All portions of a waste must meet the treatment standard in order to be land disposed. That means every sample must meet the standard. See Section 18.4.1. Furthermore, according to EPA:

> "[A] waste analysis plan cannot immunize land disposal of prohibited wastes [those that do not meet treatment standards], although such plans may be written to authorize types of sampling and monitoring different from those used to develop the treatment standards. If a waste analysis plan were to authorize a different mode of sampling or monitoring, there would need to be a demonstration that the plan (and the specific deviating feature) is adequate to assure compliance with Part 268…. This might require, for example, a demonstration of statistical equivalence between a composite sampling protocol and one based on grab sampling, or a demonstration of why monitoring for a subset of pollutants would assure compliance of those not monitored. (*EPA repeats that enforcement of the rule is based on the treatment standard, not the facility's waste analysis plan, so that enforcement officials would normally take grab samples and analyze for all constituents regulated by the applicable treatment standards.*)" [Emphasis added.] [June 23, 1989; 54 *FR* 26606]

EPA/530/R-94/019 makes similar statements. Thus, a treatment facility that analyzes composite samples of nonwastewater treatment residues to show compliance with LDR standards (per its WAP) is at risk. If enforcement personnel take a grab sample that shows the treatment standard is not being met, the facility is in violation of the LDR regulations, regardless of whether the composite samples show compliance with the standard. "[A] facility remains strictly liable for meeting the treatment standards, so that if it disposes a waste that does not meet a treatment standard, it is in violation of the land disposal restrictions regulations." [June 23, 1989; 54 *FR* 26606]

Q *A RCRA-permitted incinerator receives hazardous wastes carrying numerous waste codes from several different generators. Underlying hazardous constituents (UHCs) will also require treatment in some wastes that carry characteristic codes. How does the facility keep track of the waste codes going into the incinerator so it can assign codes (per the derived-from rule) to the ash? Is the ash tested only for the waste codes and UHCs that were fed into the unit, or is the ash tested for everything?*

A Here are the answers we received from one hazardous waste incinerator: The ash and filter cake (from the incinerator stack scrubber system) generated are assigned all of the hazardous waste codes that the facility is permitted to accept and incinerate. The ash is tested for everything (total organics and metals by TCLP) according to the facility's WAP. In particular, each ash bin is tested for all LDR organics; if it fails, the ash is fed back into the incinerator. If it passes, TCLP is run for all LDR metals. If the ash passes the TCLP, it is placed into a hazardous waste landfill. If it fails TCLP, it is stabilized. The stabilized material is evaluated again via TCLP. When it passes, it is landfilled.

18.4.4 Disposal facility LDR testing

The disposal facility's WAP must address the procedures for testing incoming wastes/treatment residues to ensure they conform to the certification made by the generator or treatment facility that the wastes/residues meet LDR standards. "The main objective of corroborative testing is to provide an independent verification that a waste meets the LDR treatment standard." [November 20, 1997; 62 *FR* 62088]

"For each waste stream, the waste constituents regulated under the land disposal restrictions rule must be comprehensively analyzed. Although the frequency of testing will depend to some extent upon the variability of the waste stream, the agency recommends that a comprehensive analysis of each waste stream be performed at least annually by the generator or treater. When the comprehensive analysis is performed, however, it must contain data on all the applicable constituents in [the §268.40 treatment standards table] so that the [disposal facility] owner/operator will be able to determine whether the waste meets all applicable treatment standards. If the owner/operator of the land disposal facility does not receive this information in writing from the generator or treatment facility, he must perform the analysis to determine whether the waste meets the treatment standards according to the waste analysis plan." [November 7, 1986; 51 *FR* 40598; see also RO 12943]

The agency noted in 1990 that "treatment and disposal facilities may generally rely on information provided to them by generators or treaters of the waste. However, treatment and disposal facilities must conduct periodic detailed physical and chemical analysis on their waste streams to assure that the appropriate Part 268 treatment standards are being met." [June 1, 1990; 55 *FR* 22669]

While screening of each incoming shipment will usually be limited to relatively simple and rapid tests, the disposal facility has a responsibility to identify any wastes that exceed treatment standards. Some flexibility is allowed under §§264/265.13(c) as to the extent of analysis necessary for each shipment. [RO 12943]

However, EPA provided the following warning:

"[A] disposal facility might violate the land disposal restrictions while at the same time comply with the provisions of its waste analysis plan.... In any case, enforcement of the land disposal restrictions is based on grab samples (except as described [for wastewaters]) and analysis of all constituents regulated by the applicable treatment standards, not on the facility's waste analysis plan." [June 1, 1990; 55 *FR* 22539]

Generators and treatment facilities that conduct onsite disposal of hazardous wastes meeting LDR standards are also subject to disposal facility LDR testing requirements.

18.4.5 Documentation for LDR program characterization

The LDR program is quite specific as to what documentation (notifications/certifications) is required for generators and TSD facilities. Generators are allowed to use testing or knowledge to determine if their hazardous wastes meet LDR treatment standards; TSD facilities must test to meet this requirement. Required LDR program documentation can be found in:

- The Generator Paperwork Requirements Table in §268.7(a)(4) and the generator documentation requirements section [§268.7(a)(8)], the Treatment Facility Paperwork Requirements Table in §268.7(b)(3), and the disposal facility paperwork requirements in §268.7(c);

- Section 13.12 of this book;

- Appendix C to EPA/530/R-94/024; and

- The waste analysis plan at TSD facilities and at generator sites that are required to have such a document.

18.5 Sampling issues

When making an analysis-based hazardous waste determination, one or more samples of the waste are taken for subsequent testing. Numerous questions arise as to where the samples should be taken, how many samples should be collected, etc. This section tries to address these issues.

18.5.1 "Representative sample"

The RCRA regulations (at §§261.20–24) require the use of a "representative sample" when testing to determine if a waste exhibits a hazardous characteristic. This term is defined in §260.10 as "a sample of a universe or whole (e.g., waste pile, lagoon, ground water) which can be expected to exhibit the average properties of the universe or whole."

EPA acknowledges that "the representativeness of the sample is critical to the accuracy of the waste characterization" [RO 11624] and has identified a few sampling methods that can be used to obtain a representative sample; these methods are listed in Appendix I to Part 261 and are referenced in the regs at §261.20(c). Samples taken using the ASTM or EPA sampling protocols specified in this appendix will be considered by EPA to be representative of the waste if it is an extremely viscous liquid; crushed or powdered material; soil-, rock-, or fly ash-like material; containerized liquid waste; or liquid waste in a pit, pond, or lagoon. However, be careful about using these methods "blindly"; if the sample obtained is not representative of the waste, you may still be held to the representative-sample standard.

Note that the comment below §261.20(c) allows you to use an alternative sampling method (one not specified in Part 261, Appendix I) without demonstrating equivalency to your state or EPA region. No matter what method is employed, however, all samples obtained for determining characteristics must meet the definition of a representative sample as given above. [May 19, 1980; 45 *FR* 33108]

The definition of representative sample allows a hazardous waste determination to be based on the *average* properties of the solid waste—not on the lowest or highest contaminant concentration. In the past, whenever anyone in the regulated community asked EPA what it meant by a "representative sample," the agency always referred them to Chapter 9 of SW–846. [See, for example, RO 11603 and 11907.] In Chapter 9 (the current version of which is dated September 1986), the agency states that the regulated community may make hazardous waste determinations based on a 90% probability that the actual average value of a contaminant of interest is less than the corresponding regulatory threshold. This degree of certainty is obtained by calculating an 80% confidence interval and requires statistical manipulation of the sample average value. This concept is discussed in detail in Section 18.5.3.1.

However, EPA issued additional sampling guidance in 2002. [*RCRA Waste Sampling—Draft Technical Guidance,* EPA/530/D-02/002, August 2002, available from http://www.epa.gov/osw/hazard/testmethods/sw846/pdfs/rwsdtg.pdf] Although this guidance notes that "[r]epresentative samples are obtained by controlling (at acceptable levels) random variability [represented by population variance] and systematic error (or bias) in sampling and analysis," this document seemed to signal a change in EPA's philosophy regarding the concept of a "representative sample" (i.e., the use of average properties to make a hazardous waste determination):

"An enforcement official, when conducting a compliance sampling inspection to evaluate a waste handler's compliance with a 'do not exceed' standard, [may] take only one sample. Such a sample may be purposively selected based on professional judgment. This is because all the enforcement official needs to observe—for example to determine that a waste is hazardous—is a single exceedance of the standard." [EPA/530/D-02/002, p. 11]

This quotation suggests that, even though all your monthly or quarterly *averages* of a toxicity constituent concentration are below the regulatory threshold (i.e., what you thought was a representative sample), if an inspector takes a single grab sample that results in a level that is above the regulatory level, you've got serious trouble. This "zero exceedances allowed" philosophy was originally discussed in the February 8, 1990 *Federal Register*:

"While the regulated community is concerned with proving the negative or absence of any hazardous constituents or characteristics, regulatory agencies are often concerned with demonstrating the opposite—that the waste concentration of a specific analyte in a waste exceeds a regulatory level, or that the waste exhibits a hazardous waste characteristic. Sampling strategies for these situations…often do not require a precise determination of the actual magnitude of the property…. Depending on the degree to which the property of interest is exceeded, testing of samples which represent all aspects of the waste or other material may not be necessary to prove that the waste is subject to regulation." [55 *FR* 4441–2]

This discussion was contained in the preamble to proposed sampling/analytical changes, some of which were never finalized; therefore, it is not clear if this is EPA's official position or not. However, the agency's view in the 2002 draft sampling guidance (which references the above February 1990 preamble language) seemed to be: a waste is hazardous (partially at least) if analysis of a single grab sample shows it to be hazardous, even if the statistical approach in Chapter 9 of SW–846 shows the waste to be nonhazardous. EPA took the same approach in a draft revision to *Waste Analysis at Facilities that Generate, Treat, Store, and Dispose of Hazardous Wastes*, EPA/530/R-12/001, January 2013, available at http://www.epa.gov/epawaste/hazard/tsd/permit/tsd-regs/tsdf-wap-guide2.pdf. This document was released for public comment in 2013 and again in 2014. It will probably be finalized in late 2014 or early 2015.

This "zero exceedances allowed" approach is a very stringent one, and the regulated community has questioned whether EPA is trying to effect a significant change to the RCRA program using the 2002 and 2013 draft guidance. Industry harshly criticized the agency for this position and approach in comments submitted on the draft documents. Industry commented that such an approach, in which a waste essentially cannot be proven to be nonhazardous no matter how many samples are analyzed or what results are obtained, while a single biased sample exceeding a numeric value is sufficient to show that the waste is hazardous, is contrary to EPA's own regulations, unsupported by technical basis, and unwarranted by the risk that would be posed.

Partially as a result of those comments, EPA has not finalized the 2002 draft guidance (as of October 2014), and the agency noted that "[u]ntil we make a final decision on the guidance, Chapter Nine [of SW–846] is the applicable guidance." [RO 14743] See also EPA's sampling technical guidance website at http://www.epa.gov/osw/hazard/testmethods/sw846/samp_guid.htm, which says basically the same thing. Thus, until the agency finalizes the draft sampling technical guidance, the SW–846, Chapter 9 methodologies (discussed in Section 18.5.3.1) are still applicable. But, you need to know what, if any, policy your state/EPA regional regulators have regarding this approach.

EPA's intent is to eventually eliminate Chapter 9 of SW–846 and use instead the finalized sampling guidance document. Thus, these two documents, Chapter 9 of SW–846 and *RCRA Waste Sampling—Draft Technical Guidance* (EPA/530/D-02/002), are the basis for much of the discussion that follows.

18.5.1.1 How do I get a "representative sample"?

How do you get a representative sample as defined in the previous section? Although not based on EPA guidance, ASTM D5956, *Standard Guide for Sampling Strategies for Heterogeneous Wastes* (available from http://www.astm.org), notes that a representative sample can be a single sample, a set of samples, or one or more composite samples.

18.5.1.1.1 A single sample

Particularly if the volume of waste is small, uniform, and/or consistent over time, a single sample should suffice as representative. For example, analysis of a single glass thief sampler or COLIWASA (an acronym for <u>co</u>mposite <u>l</u>iquid <u>wa</u>ste <u>sa</u>mpler) sample of a drummed liquid or sludgy waste (over the full depth of the material) should be representative of the contents of the drum; for solids, an auger or coring type sampler will typically be required (again, over the full depth). [Appendix I of Part 261; Environmental Response Team SOP 2009, November 1994, available from http://www.ert.org/products/2009.pdf; EPA/530/D-02/002] If the drum contains physically solid material, the core will subsequently be ground (if necessary) to pass through a 9.5-mm standard sieve, and a TCLP test can then be performed on a well mixed subsample of the solids. [*Characterizing Heterogeneous Wastes: Methods and Recommendations*, EPA/600/R-92/033, February 1992, available from http://nepis.epa.gov/EPA/html/Pubs/pubtitleORD.html by downloading the report numbered 600R92033]

18.5.1.1.2 A set of samples

What about situations where the volume of waste is large, or the waste is heterogeneous and/or inconsistent over time? In these situations, EPA is generally looking for a "representative selection" of samples to characterize the waste, not just a single sample. [RO 11907] Both Chapter 9 of SW–846 and EPA/530/D-02/002 recommend strategies for making hazardous waste determinations (and give several examples) based on taking numerous samples and statistically massaging the resulting analytical data. That is the second approach to developing a representative sample. Detailed examples of this approach are given in Section 18.5.2.2.

18.5.1.1.3 One or more composite samples

The third approach to developing a representative sample is to gather a number of samples of a waste or contaminated medium either over time (temporally) or physical space (spatially) and combine them into one or more samples. If the field and laboratory subsampling are done correctly, the combined sample "can be expected to exhibit the average properties of the universe or whole" and, therefore, is a representative sample by definition.

Obviously, there are significant economic incentives for choosing this third approach—sample transportation and analytical costs go way down. However, some state or federal regulators may not allow composite samples; to some, compositing equals dilution of hot spots. One reason that regulators sometimes struggle with the use of composite samples is due to a misunderstanding of what they represent. The term composite is used generically and not always in the proper context.

ASTM D5956, *Standard Guide for Sampling Strategies for Heterogeneous Wastes* and D6051, *Standard Guide for Composite Sampling and Field Subsampling for Environmental Waste Management Activities* (both available from http://www.astm.org) define a composite sample as "a combination of two or more samples." However, it is more complicated than that.

A true composite sample is a combination of two or more representative samples, each one taken from and representing an individual "decision unit." A decision unit is a volume or mass of material (such as waste or contaminated soil) about which a decision will be made. [EPA/530/D-02/002] (For example, a decision unit could be 100 drums of waste, a pile of dirt, or a roll-off box of sludge from a wastewater treatment plant.) When the composite is formed by combining representative samples from several decision units, concentration information regarding each specific decision unit is lost. The goal of composite sampling is to reduce analytical costs at the expense of loss of information regarding specific decision units.

When measuring contaminant concentration using a composite sample, only three conclusions can be drawn: 1) all of the decision units must be below the regulatory limit; 2) one or more of the decision units *may* be above the limit, but we do not know which; or 3) one or more of the decision units *must* be above the limit, but we do not know which. In the latter two cases, the individual representative samples used to form the composite must be analyzed to determine which, if any, of the decision units have a concentration over the limit (see Section 18.5.2.4.2). Not knowing the contaminant concentrations of individual decision units due to the mixing of material from different units is a concern and must be considered.

There is a difference between composite and *MULTI INCREMENT*® samples. (*MULTI INCREMENT*® is a registered trademark of EnviroStat, Inc.) A Multi Increment sample is made up of portions of material from within the same decision unit and is used to estimate the average properties of the material (e.g., contaminant concentration) within the unit. A Multi Increment sample typically consists of, at a minimum, 40 small discrete samples, called increments, representing all temporal or spatial parts of the waste or contaminated medium in the decision unit. Thus, a Multi Increment sample is a representative sample. There is no loss of information from this type of sample since all the material is from within the same decision unit. Any "dilution" concerns with losing information are invalid since all the material is from the same decision unit. With Multi Increment samples, the goal is to represent a decision unit as defensibly and cost-effectively as possible. If material from several Multi Increment samples from different decision units is combined, the result would be a composite sample with the concerns noted above.

For example, contaminated soil from a recent spill is excavated and placed into 5 roll-off boxes. Most regulators and disposal facilities will consider each box a separate decision unit (see Section 18.5.2.4.1). A true composite sample would consist of taking a representative sample from each of the 5 roll offs and combining them into one composite sample. How would you get a representative sample from each roll off? You would take multiple discrete samples (increments) from different locations within each box—several samples from the top, several from the middle, and several from the bottom. [Such samples are very difficult to obtain once the box is full; it is much easier to sample solids while they are in motion (i.e., when filling the box).] When combined, these 40 or more increments, which include a little bit of soil from multiple locations within the box, make up a single Multi Increment sample. Because each Multi Increment sample represents the soil from all areas of the box, it is a representative sample.

In our example of contaminated soil in 5 roll-off boxes (each of which is a discrete decision unit), a composite sample would consist of combining the Multi Increment sample from each box. This composite sample would then be analyzed, and the resulting analytical results would be used to make inferences about each of the 5 decision units (boxes), although concentration information regarding the 5 specific decision units is lost. Following this approach, a nonhazardous waste determination could be made for all 5 boxes, even though one definitely contained hazardous soil. You can see the problems a regulator or disposal facility might have with this approach: the concentrations of toxic constituents in significantly contaminated soil in one decision unit (roll off) could be diluted with lightly contaminated soil from the other 4 units. Since regulators and nonhazardous waste disposal facilities will often not accept the results of a composite sample, we will likely have to make a hazardous waste determination for each box. In that case, the Multi Increment sample from each box (the representative sample) will be analyzed, and a hazardous waste determination will be made for each box (decision unit).

The Multi Increment sample approach is being used successfully by various regulated entities around the country and is discussed in detail in the following training program:

> Sampling for Defensible Environmental Decisions
> EnviroStat, Inc.
> PO Box 636
> Ft. Collins, CO 80522
> (970) 689-5700
> http://www.envirostat.org/
> Presenter: Chuck Ramsey

Examples of the Multi Increment sample approach to developing a representative sample are given in Section 18.5.2.3.

18.5.1.2 Representative samples for LDR compliance

Although a "representative sample" can be used to show a waste is or is not hazardous, that approach cannot be used to show compliance with LDR treatment standards. As noted in Section 18.4.1, all portions of the waste must meet the treatment standard.

18.5.2 Sampling and analysis plan/DQOs

EPA recommends that you have a sampling and analysis plan in place when taking a representative sample(s). You may not be required to have one if you're just a generator, but it is a good idea anyway. [August 31, 1993; 58 FR 46044] A sampling and analysis plan will document the steps you will take to produce characterization data that are "scientifically valid, defensible, and of known precision and accuracy." [Chapter 1 of SW–846] A sampling and analysis plan helps site personnel collect data of the right type, quality, and quantity to support waste characterizations. The two primary components of a sampling and analysis plan are a field sampling plan and a quality assurance project plan.

> "[T]he primary objectives of a sampling plan for a solid waste are twofold: namely, to collect samples that will allow measurements of the chemical properties of the waste that are both accurate and precise. If the chemical measurements are sufficiently accurate and precise, they will be considered reliable estimates of the chemical properties of the waste." [Chapter 9 of SW–846]

In terms of sampling, precision is a measurement of the closeness of agreement among repeated measurements. Precision in sampling can be improved by increasing the number of samples, increasing the volume (mass) of each sample, or by employing a Multi Increment sampling strategy (discussed in Section 18.5.2.3).

Accuracy reflects the closeness or correctness of a measured sample value to the true value. Accuracy cannot be measured—it can only be surmised from taking representative samples of a solid waste and analyzing them. In more recent guidance, EPA tends to use the term "bias" instead of accuracy—bias being the systematic or consistent over- or underestimation of the true value. Bias includes [EPA/530/D-02/002]:

1. Sampling bias—This bias consists of three subcategories: 1) improper selection and use of sampling and subsampling devices (minimized by ensuring all of the material of interest is accessible by the sampling tool), 2) improper selection of the sample collection strategy (minimized by ensuring the sampling protocol is impartial so there is an equal chance for each part of the waste to be included in the sample—that is, achieving some form of randomness), and 3) the loss or addition of contaminants during sampling and sample handling (minimized by using sampling devices made of materials that do not sorb or leach constituents of concern and by proper decontamination and sample handling procedures).

2. Analytical bias (e.g., systematic error caused by instrument contamination, improper calibration, matrix effects, or sample shipping and handling problems).

3. Statistical bias (e.g., incorrect assumptions are made about the sample distribution).

Here's an example of sampling bias: if large pieces of waste material (e.g., slag) are not collected and included with a sample, a negative bias of contaminant concentrations may result (analytical levels may be lower than what is actually representative). Small particle size can also bias a sample; some contaminants adsorb more readily onto small particles, so a small-particle-size sample may result in a positive bias (analytical levels may be higher than what is representative). [*Superfund Program Representative Sampling Guidance, Volume 4: Waste*, EPA/540/R-95/141, December 1995 (available from http://www.ert.org/products/SF-WASTE.pdf)]

Numerous references are available to help you develop a sampling plan. The ones we've run across and their availability are given in Table 18-2. Although these documents are too voluminous and complex to summarize here, they contain details on developing sampling plans, and they also provide lengthy discussions of the data quality objective (DQO) process. DQOs include both regulatory and scientific objectives and specify a systematic planning process for developing the sampling design, data analysis, and quality assurance/control procedures to support the waste characterization.

In addition to the references in Table 18-2, considerable guidance on developing DQOs is available at http://vsp.pnnl.gov/dqo/. Another free tool that can be used in conjunction with DQOs is Visual Sample Plan (VSP) software. The software can be downloaded from http://vsp.pnnl.gov/, and is a tool to help you develop a sampling plan for characterizing surface soil, building surfaces, water bodies, or other similar decision units. Statistical methods for analyzing laboratory results are included.

The agency's general DQO process is summarized in Figure 18-2. Although the sampling plan and associated DQOs can be quite complex for evaluating contaminated soil and ground water at a large facility, they can be fairly simple for determining if a waste is characteristically hazardous.

For example, a maintenance facility uses plastic blasting media to strip paint off of equipment. The spent blasting media/paint fines are collected in a roll-off box for disposal. The facility used knowledge to determine that the spent blasting media/paint fines are not listed and do not exhibit the characteristic of ignitability, corrosivity, or reactivity. However, sampling and analysis will be used to determine if the waste exhibits the toxicity characteristic due to cadmium and/or chromium, which are known to be in the paint. The DQO process associated with making a hazardous waste determination for the material filling the roll-off box is also shown in Figure 18-2.

Guidance that addresses a number of specific questions that often arise when preparing a sampling plan is reviewed below.

18.5.2.1 Where should I take a sample?

In general, sample location will be defined by the output of Step 4 of the DQO process. Inputs to that step may include the point at which a solid

Table 18-2: References Useful When Preparing a Sampling Plan

Reference	Availability
Methods for Evaluating the Attainment of Cleanup Standards, Volume 1: Soils and Solid Media, EPA/230/02-89/042, February 1989	http://clu-in.org/download/stats/vol1soils.pdf
Soil Sampling Quality Assurance User's Guide, EPA/600/8-89/046, March 1989	http://clu-in.org/download/char/soilsamp.pdf
A Rationale for the Assessment of Errors in the Sampling of Soils, EPA/600/4-90/013, May 1990	http://clu-in.org/download/stats/rationale.pdf
Preparation of Soil Sampling Protocols: Sampling Techniques and Strategies, EPA/600/R-92/128, July 1992	http://www.epa.gov/oust/cat/mason.pdf
Chapters 1 and 9 of SW–846, July 1992	http://www.epa.gov/epawaste/hazard/testmethods/sw846/online/index.htm
RCRA Ground-Water Monitoring: Draft Technical Guidance, EPA/530/R-93/001, November 1992	http://www.epa.gov/epawaste/hazard/correctiveaction/resources/guidance/sitechar/gwmonitr/rcra_gw.pdf
Waste Analysis at Facilities That Generate, Treat, Store, and Dispose of Hazardous Wastes, EPA/530/R-94/024, April 1994	http://www.epa.gov/osw/hazard/tsd/ldr/wap330.pdf
Superfund Program Representative Sampling Guidance, Volume 1: Soil, OSWER Directive 9360.4-10, EPA/540/R-95/141, December 1995	http://www.ert.org/products/SOIL-PT1.pdf
Superfund Program Representative Sampling Guidance, Volume 4: Waste, OSWER Directive 9360.4-14, EPA/540/R-95/141, December 1995	http://www.ert.org/products/SF-WASTE.pdf
Superfund Program Representative Sampling Guidance, Volume 5: Water and Sediment, OSWER Directive 9360.4-16, EPA/540/R-95/141, December 1995	http://www.ert.org/products/SF-WATR1.pdf
Soil Screening Guidance: User's Guide, EPA/540/R-96/018, July 1996	http://www.epa.gov/superfund/health/conmedia/soil/index.htm
Soil Screening Guidance: Technical Background Document, EPA/540/R-95/128, July 1996	http://www.epa.gov/superfund/health/conmedia/soil/index.htm
Environmental Quality Technical Project Planning Process, EM 200-1-2 (U.S. Army Corps of Engineers), August 1998	https://www.clu-in.org/conf/tio/triad_012302/tpp.pdf
Guidance on Demonstrating Compliance With the LDR Alternative Soil Treatment Standards, EPA/530/R-02/003, July 2002	http://www.epa.gov/epawaste/hazard/tsd/ldr/soil_f4.pdf
RCRA Waste Sampling—Draft Technical Guidance, EPA/530/D-02/002, August 2002	http://www.epa.gov/osw/hazard/testmethods/sw846/samp_guid.htm
Guidance on Environmental Data Verification and Data Validation, EPA QA/G-8, November 2002	http://www.epa.gov/quality/qa_docs.html
Guidance for Quality Assurance Project Plans, EPA QA/G-5, December 2002	http://www.epa.gov/quality/qa_docs.html
Guidance on Choosing a Sampling Design for Environmental Data Collection, EPA QA/G-5S, December 2002	http://www.epa.gov/quality/qa_docs.html
Guidance for Geospatial Data Quality Assurance Project Plans, EPA QA/G-5G, March 2003	http://www.epa.gov/quality/qa_docs.html
Guidance for Obtaining Representative Laboratory Analytical Subsamples from Particulate Laboratory Samples, EPA/600/R-03/027, November 2003	http://www.clu-in.org/download/char/epa_subsampling_guidance.pdf
Systematic Planning: A Case Study for Hazardous Waste Site Investigations, EPA QA/CS-1, February 2006	http://www.epa.gov/quality/qa_docs.html
Data Quality Assessment: A Reviewer's Guide, EPA QA/G-9R, February 2006	http://www.epa.gov/quality/qa_docs.html
Guidance on Systematic Planning Using the Data Quality Objectives Process, EPA QA/G-4, February 2006	http://www.epa.gov/quality/qa_docs.html
Data Quality Assessment: Statistical Methods for Practitioners, EPA QA/G-9S, February 2006	http://www.epa.gov/quality/qa_docs.html

Source: McCoy and Associates, Inc.

Figure 18-2a: The Data Quality Objectives Process

General DQO process

1. STATE THE PROBLEM
Summarize the contamination problem that will require new environmental data, and identify the resources available to resolve the problem; develop conceptual site model.

↓

2. IDENTIFY THE DECISION
Identify the decision that requires new environmental data to address the contamination problem.

↓

3. IDENTIFY INPUTS TO THE DECISION
Identify available information (including waste generation/process knowledge) and additional information needed to support the decision and specify which inputs require new environmental measurements.

↓

4. DEFINE THE STUDY BOUNDARIES
Specify the spatial and temporal aspects of the waste or environmental media that the data must represent to support the decision.

↓

5. DEVELOP A DECISION RULE
Develop a logical "if . . . then . . ." statement that defines the conditions that would cause the decisionmaker to choose among alternative actions.

↓

Continue with Figure 18-2b.

Example DQO process for characterizing spent blasting media/paint fines filling a roll-off box

Determine if the spent blasting media/paint fines should be classified as a hazardous waste (requiring Subtitle C management) or as nonhazardous industrial waste (that can be land disposed in a Subtitle D industrial landfill).

↓

Determine if the spent blasting media/paint fines are characteristically hazardous under the RCRA regulations.

↓

- The decision will be based on the quantity of material generated over a one-month period, not to exceed the volume of a 10-yd³ roll-off box.
- Cadmium and chromium are constituents of concern in the solid waste, based on process knowledge.
- Analyze samples for total cadmium and chromium (SW-846 Methods 3050/6010); if total analyses indicate TC levels could be exceeded, run TCLP (SW-846 Method 1311).
- Use minimum sample mass of 100 g.

↓

- Samples will be obtained from material discharging into the roll-off box from the blast booth.
- Decision boundary is the length of time over which the decision applies (i.e., if a process change occurs in the blast media or application, the boundary would change).

↓

- TC regulatory levels for cadmium and chromium in §261.24 cannot be equaled or exceeded; otherwise the solid waste is a hazardous waste.
- Although not required by regulation, the planning team chose 90% as the level of confidence to use in deciding that the solid waste is nonhazardous, consistent with SW-846, Chapter 9.
- The decision rule is "If the upper 90th percentile TCLP concentrations for cadmium and chromium in the solid waste are less than their respective regulatory levels of 1.0 and 5.0 mg/L, then the waste will be classified as nonhazardous; otherwise, the solid waste will be considered a hazardous waste."

↓

Continue with Figure 18-2b.

©2015 McCoy and Associates, Inc. **McCoy's RCRA Unraveled**

Figure 18-2b: The Data Quality Objectives Process (Continued)

General DQO process—Continued

From Figure 18-2a.

6. SPECIFY LIMITS ON DECISION ERRORS
Specify the decisionmaker's acceptable limits on decision errors, which are used to establish performance goals for limiting uncertainty in the data.

7. OPTIMIZE THE DESIGN FOR OBTAINING DATA
Identify the most resource-effective sampling and analysis design for generating data that are expected to satisfy the DQOs.

Example DQO process for characterizing spent blasting media/paint fines filling a roll-off box—Continued

From Figure 18-2a.

- The baseline hypothesis is that the spent blasting media/paint fines are hazardous; that is, the true proportion of samples with concentrations of cadmium or chromium less than their regulatory levels is less than 0.9.
- Type I error (false rejection) is concluding that the true proportion of the waste that is nonhazardous is >0.9 when it actually is <0.9 (i.e., concluding the waste is not hazardous when, in fact, it is). Because a false rejection would lead the facility to ship untreated hazardous waste to a nonhazardous waste landfill, the planning team set the acceptable probability of making a Type I error at 10% (i.e., they are willing to accept a 10% chance of concluding the waste is nonhazardous when at least a portion of it is).
- Type II error (false acceptance) is concluding that the true proportion of the waste that is nonhazardous is <0.9 when it actually is >0.9 (i.e., concluding the waste is hazardous when, in fact, it isn't). Because a false acceptance is less of a concern than a false rejection, the planning team set the acceptable probability of making a Type II error at 25%.
- Based on the first six steps of the DQO process, the planning team selected a systematic sampling strategy with total metals analysis for cadmium and chromium with the condition that, if any total sample analysis result indicated the maximum theoretical leachate concentration exceeded the TC regulatory level, then the TCLP would be conducted for that sample.
- Because the roll-off box is filled about once every 30 days, one random sample was obtained each day for 30 days as the waste discharged into the box. A flat 12-inch polyethylene pan with vertical sides was used to collect each primary field sample. Since each sample so collected was about 2 kg, the field team used "fractional shoveling" to reduce the sample mass to about 300 g (the sample size requested by the analytical lab).

DQO = data quality objective; TC = toxicity characteristic; TCLP = toxicity characteristic leaching procedure.
Adapted from EPA QA/G-4 and EPA/530/D-02/002.

waste is generated. EPA addressed this issue on March 29, 1990:

"The current rules require that the determination of whether a waste is hazardous be made at the point of generation (i.e., when the waste becomes a solid waste).... EPA is retaining the existing approach of requiring sampling at the point of generation." [55 FR 11830; see also EPA/530/D-02/002]

So, where is the point of generation of a waste? We covered this issue in some detail in Section 14.1. Recapping the highlights, a waste is generated upon removal from a manufacturing process or waste treatment unit. When wastes are generated in enclosed systems, EPA stated that sample taps can typically be installed for waste characterization purposes, or, in extreme cases, wastes can be sampled when they leave closed systems. [July 8, 1987; 52 FR 25766] However, wastes must be characterized as to their hazardousness before they are commingled (mixed) with other waste streams. [RO 11631]

Examples of the point of generation for various solid wastes (such that a hazardous waste determination needs to be made) include:

- For wastewater generated in a pulp and paper mill bleach plant, at the outlet of the plant prior to mixing with other wastewater streams [RO 11631];

- For solvents used in parts washers, when the parts washer apparatus is removed from a drum of used solvent [RO 12790];

- For baghouse dust generated from a manufacturing operation, when the material is removed from the baghouse (although the agency has allowed facilities to use the point where the dust exits an ash storage silo if that silo is directly connected to the baghouse by piping) [RO 11921, 12824, 14200];

- For waste generated by paint removal operations, once the paint has been removed from the surface of the structure [RO 14069];

- For P- or U-listed chemicals, when they are discarded or intended to be discarded [RO 12012];

- For still bottoms resulting from distillation of spent solvent, when they are removed from the distillation unit [RO 11420, 12865, 13280];

- For waste acid to be stored in containers or tanks, when the acid is poured into or enters the containers/tanks [May 19, 1980; 45 FR 33096]; and

- For waste filter cake being placed into a roll-off box, on the conveyor belt or as the cake enters the roll off. [EPA/530/D-02/002]

For contaminated soil or ground water, the point of generation is when the soil is excavated or the ground water is pumped out of the ground. [May 26, 1998; 63 FR 28617, RO 13748, 14283]

When sampling a large, heterogeneous, and/or inconsistent waste stream, the sample location will also depend on the collection strategy chosen. Different sample collection strategies are discussed briefly below.

18.5.2.1.1 *Probability sampling*

Probability sampling refers to all parts of the waste or contaminated media having a known probability of being included in the sample. All probability sampling makes use of randomness in some way, which allows statistical statements to be made about the quality of conclusions that are derived from the samples and subsequent analytical work.

The three primary sample collection strategies that use probability are summarized below. Table 18-3 summarizes the advantages and limitations of each. The following discussion is taken from Chapter 9 of SW–846, EPA/230/2-89/042, EPA/530/D-02/002, EPA QA/G-4, and EPA QA/G-5S. Document availabilities are given in Table 18-2.

1. *Simple random sampling*—This simplest type of probability sampling assumes every part of the waste or media has an equal chance of being selected for sampling; further, each sample point is selected independently from all others. This strategy is often used if little is known about any heterogeneity or stratification in a waste or media, or if a batch of waste is randomly heterogeneous with regard to its physical/chemical properties and

Table 18-3: Guidance for Selecting Probability Sampling Strategies

Sampling strategy	Usefulness	Advantages	Limitations
Simple random sampling	■ Useful when the population of interest is relatively homogeneous (i.e., there are no major patterns or "hot spots" expected) ■ Useful when little information is available about a waste	■ Provides statistically unbiased estimates of the mean, proportions, and variability ■ Easy to understand and implement	■ Least preferred if patterns or trends are known to exist and are identifiable ■ Localized clustering of sample points can occur by random chance
Stratified random sampling	■ Most useful for estimating a parameter (e.g., the mean) for wastes exhibiting high heterogeneity (e.g., distinct portions or components of the waste have high and low constituent concentrations or characteristics) ■ Useful when the pattern of contamination (e.g., any cyclical patterns) is known	■ Ensures more uniform coverage of the entire target population ■ Achieves greater precision in estimates of the mean and variance ■ Reduces costs over simple random and systematic sampling strategies because fewer samples may be required ■ Enables computation of reliable estimates for population subgroups of special interest	■ Requires some prior knowledge of the waste or media to define strata and to obtain a more precise estimate of the mean ■ Statistical procedures for calculating the number of samples, the mean, and the variance are more complicated than for simple random sampling
Systematic sampling	■ Useful for estimating spatial patterns or trends over time and can be used to identify "hot spots" ■ Useful when little information is available about a waste	■ Preferred over simple random sampling when sample locations are random within each systematic grid or interval ■ Practical and easy method for designating sample locations ■ Ensures uniform coverage of site, unit, or process ■ May be lower cost than simple random sampling because it is easier to implement	■ May be misleading if the sampling interval is aligned with the pattern of contamination (e.g., any cyclical patterns), which could happen inadvertently if there is inadequate prior knowledge of the pattern of contamination ■ Not truly random, but can be modified through use of the "random within grids" design

Source: Adapted from EPA/530/D-02/002.

that randomness remains constant from batch to batch. The waste or media is divided into equal-sized or equal-massed grids or units, and a series of sequential numbers is assigned to each. (The total number of possible sampling locations should be much greater than the number of samples to be collected.) Then, grid or unit numbers are selected randomly (using a random number table or generator), and a sample is collected from the corresponding location/unit (see Figure 18-3a). Note that, if sampling over time instead of location, the time interval is divided into equal-sized intervals, assigned a series of sequential numbers, and time intervals for sampling are selected randomly.

2. *Stratified random sampling*—If the waste or media contains significant, nonoverlapping, and identifiable portions in terms of its chemical or physical properties, the waste or media is said to be stratified. A sampling strategy is employed to

Figure 18-3: Spatial Probability Sampling Strategies
(Two-Dimensional Plan Views)

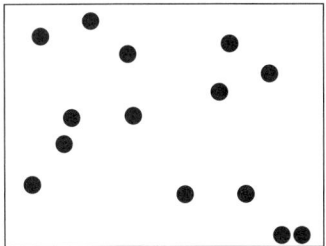

a) Simple random sampling

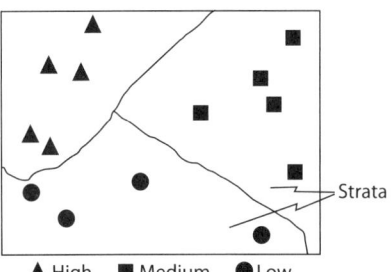

▲ High   ■ Medium   ● Low
b) Stratified random sampling

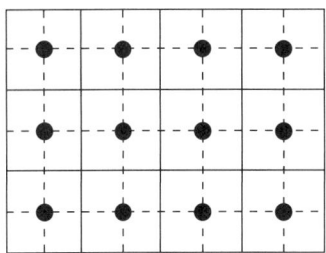

c) Systematic grid sampling

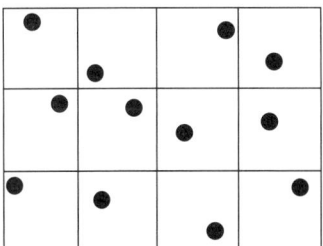

d) Systematic sampling using "random within grids"

Source: Adapted from EPA/530/D-02/002 and EPA QA/G-5S.

segregate the known areas of nonrandom properties into nonoverlapping groups called strata. Each stratum should be relatively homogenous (i.e., the variability within each stratum should be much less than that observed over the entire waste or media). Sampling depth, terrain characteristics, concentration level, particle size, previous cleanup efforts, and the presence of unusual contaminants can be used as the basis for designating strata. After segregation, which may occur spatially or temporally, the simple random sampling technique discussed above is applied separately to each stratum, as shown in Figure 18-3b. Obviously, this sampling strategy requires knowledge of the waste and professional judgment. If we have knowledge that spatial or temporal patterns occur, stratified sampling can be a cost-effective approach for sampling heterogeneous wastes. A stratified strategy for sampling for energetic material residues on a military firing range is discussed in Method 8330B. Another example of this sampling strategy is given in Case Study 18-1.

3. *Systematic sampling*—Using this strategy, the waste or media is divided into equal-sized or equal-massed spatial grids (e.g., square, rectangular, or triangular grids in two or three dimensions) or units, and a random starting location within one of the grids or units is selected. Samples are subsequently taken in the same spot in every grid or unit (reference Figure 18-3c). This achieves an equally spaced, uniform spread of sampling points compared to simple random sampling. Of course, systematic sampling can be conducted at fixed time intervals for a waste being continuously generated. This strategy is effective for media or wastes when we have no knowledge of patterns or strata. For example, waste or media in a roll-off box could be divided into a two-dimensional (plan view) grid of squares, and a core sample could be obtained from the middle of each square. Figure 18-3d illustrates that randomness can be improved by

> **Case Study 18-1: Stratified Random Sampling of Spent Fluorescent Light Tubes**
>
> A facility is relamping a warehouse and is not planning to manage the spent fluorescent tubes under the universal waste program (i.e., it wants to send them to a nearby nonhazardous waste landfill if they are nonhazardous). Can analysis of a single spent fluorescent tube be used to determine the regulatory status of all spent bulbs generated at the facility?
>
> No. "Based on one tube, we have no way to assess the variability between fluorescent lamps (new versus old, different manufacturers, different wattage, etc.). A representative selection of lamps randomly chosen should be analyzed to make this determination." [RO 11907] If we have no acceptable knowledge to extrapolate the test result for one lamp to all the others, analyzing one tube isn't sufficient.
>
> However, what if we could separate (stratify) the spent bulbs into discrete populations using knowledge? For example, if we know using records that the lamps in the warehouse came from only two manufacturers and we know the dates of manufacture, we could divide all the bulbs into two or more populations. Then, running the TCLP on a few bulbs randomly selected from each population would likely satisfy the representative sample criterion.

conducting random sampling within the grids. (Using random sampling within the grids is similar to stratified random sampling discussed just above; however, instead of using knowledge to identify the strata, they are identified arbitrarily. Also, only one sample is taken per stratum.)

18.5.2.1.2 Authoritative sampling

Authoritative sampling uses the judgment and knowledge of the samplers to determine where samples should be taken, as opposed to probability sampling. Because this strategy is not based on probability, it is not possible to make statistical inferences (e.g., mean or variance) for the property of interest. "Also, due to its subjective nature, the use of authoritative sampling by the regulated community to demonstrate compliance with regulatory standards generally is not advisable except in those cases in which a small volume of waste is in question or where the concentration is either well above or well below the regulatory threshold." [EPA/530/D-02/002]

Thus, authoritative sampling is normally *not* used for making hazardous waste determinations. Instead, it can be used to stratify areas that are contaminated and areas that are clean to facilitate subsequent hazardous waste determinations.

Two types of authoritative sampling are in use, as discussed below. The discussion is taken from Chapter 9 of SW–846, EPA/230/2-89/042, EPA/530/D-02/002, EPA QA/G-4, and EPA QA/G-5S. Document availabilities are given in Table 18-2. The advantages and limitations of each sampling strategy are summarized in Table 18-4.

1. *Judgmental sampling*—This sampling strategy uses process or site knowledge to choose one or more sampling locations that represent the average or typical properties of the waste or contaminated media. No randomization is included. Confirming the existence of contamination at specific locations (e.g., from a spill or leak) is a typical application of this sampling approach. For example, when investigating subsurface soil contamination, the investigators can obtain a few soil borings from the areas suspected of having the highest contaminant concentrations. If the mean contaminant concentration for any individual boring exceeds the applicable screening or regulatory threshold value, additional (probability-based) investigation should be conducted. Samples are often collected in areas of stained soil; this is also a judgmental sampling strategy. Case Study 18-2 illustrates the use of judgmental sampling.

Table 18-4: Guidance for Selecting Authoritative Sampling Strategies

Sampling strategy	Usefulness	Advantages	Limitations
Judgmental sampling	■ To generate rough estimates of the average concentration or typical property ■ To obtain preliminary information about a waste stream or site to facilitate planning or to gain familiarity with the waste matrix for analytical purposes ■ To assess the usefulness of samples drawn from a small portion of the waste or site ■ To screen samples in the field to identify "hot spot" samples for subsequent analysis in a laboratory	■ Can be very efficient with sufficient knowledge of the site or waste generation process ■ Easy to do and explain	■ The utility of the sampling strategy is highly dependent on expert knowledge of the waste ■ Not very useful if the waste is highly heterogeneous and/or the existence of any strata is not known ■ It is difficult to prove prejudice was not used in selecting sampling locations ■ Nonprobability-based so inference to the general population is difficult ■ Cannot determine reliable estimates of variability
Biased sampling	■ To estimate "worst-case" or "best-case" conditions (e.g., to estimate the maximum contaminant concentration in an area of contaminated soil or ground water)	■ Same as for judgmental sampling	■ Same as for judgmental sampling

Source: Adapted from EPA/530/D-02/002.

2. *Biased sampling*—When trying to establish worst- or best-case estimates of a property of a waste or contaminated media—as opposed to the average or typical property—this strategy is useful. Samples are taken where one expects very high or very low concentrations. A sample taken at the source of a release could serve as an estimate of the worst-case constituent concentration likely to be found in the affected soil or ground water. Process knowledge is used to identify samples that will likely have the highest and lowest contaminant levels, defining the extent of the problem/cleanup.

18.5.2.1.3 *Sampling heterogeneous wastes*

Few wastes are truly homogeneous—there are just different degrees of heterogeneity. Problematic heterogeneous wastes include contaminated soil, waste equipment (e.g., spent filters, carbon canisters), demolition debris, waste construction material, nonreusable containers (e.g., drums, paint cans), post-consumer wastes (e.g., batteries and automobiles), and other wastes from laboratory and manufacturing operations that are often physically solids. Because of their relatively large particle size and varied composition, heterogeneous wastes are more difficult to characterize than more uniform materials such as liquids and sludges.

The primary obstacle to characterizing heterogeneous wastes and contaminated soil is obtaining representative samples. Laboratories are usually not set up to reduce the size of large items to the small quantities needed to perform analytical tests. EPA noted on December 18, 1998 [63 *FR* 70196] that the heterogeneity of debris, difficulties encountered during the laboratory sample size-reduction step, and variable weathering of the debris matrix all make it very difficult to generate reproducible TCLP data on certain debris, such as debris coated with lead-based paint.

The other major problem is determining the level of heterogeneity of the contaminant(s) of interest. For example, contaminants can have different concentrations in different locations.

Wastes are generally heterogeneous in two ways [*Preparation of Soil Sampling Protocols: Sampling Techniques and Strategies,* EPA/600/R-92/128, July 1992, available from http://www.epa.gov/swerust1/cat/mason.pdf]:

> **Case Study 18-2: Judgmental Sampling of a Smelter Slag Pile**
>
> A large pile of smelter slag is to be sampled to determine if it exhibits a characteristic. However, slag particle sizes range from dust to 1,300-lb casts. How do plant personnel obtain a representative sample?
>
> Nonrandom sampling based on the judgment and experience of the sampler may be acceptable (i.e., judgmental sampling). For example, collecting samples only of material less than 1/2 inch in diameter is acceptable and provides a reasonable approximation to a random sample if the sampling team can show that large particles should have similar composition to smaller ones because they resulted from the same process. Thus, this nonrandom sampling can be expected to be representative of the pile if the sampling team can prove that the small particles are representative of the large ones (i.e., the chemical composition of the waste is independent of particle size). However, properties of the waste such as its leaching potential may still be dependent on particle size. [February 8, 1990; 55 *FR* 4443]

1. *Compositional heterogeneity* occurs due to differences in contaminant concentrations among the different size (or type) of particles in a waste. An example of this type of heterogeneity is the differences in lead concentration among the different materials associated with building siding painted with lead-based paint (LBP). Obviously, most or all of the lead is contained in the paint film compared to the wood or metal of the walls. Or, in a large filter cartridge contaminated with benzene, compositional heterogeneity occurs due to the different benzene concentration associated with the end caps, the screening, and the filter media itself. Similarly, the difference in chromium concentration among varying particles sizes of chromium-contaminated soil exemplifies compositional heterogeneity. The chromium concentration associated with the gravel, the fines, and the pieces of chromium within the soil are all different.

2. *Distributional heterogeneity* derives from the nonrandom distribution of the different size (or type) of particles in a waste. Continuing the above examples, different areas of the building siding may have been painted with more coats or greater thicknesses of LBP; additionally, the areas around windows, doors, and rooflines may have different proportions of the wood/metal to paint. For the filter cartridge, the benzene concentration in the cap, screening, and filter media may be higher at the process fluid inlet end of the cartridge compared to the outlet end. A nonrandom distribution of the gravel, the fines, and/or the pieces of chromium within the soil (e.g., more pieces of chromium in one area of the contaminated soil then the others) would make it hard to develop a representative sample.

How do we overcome these two sources of heterogeneity? We have to take enough samples (increments) and enough mass to get all of the different sizes (or types) of particles in their correct proportions. [January 9, 1992; 57 *FR* 990] For example, when characterizing demolition debris, some of which is coated with lead-based paint, EPA noted "a representative sample for a TCLP analysis would represent both painted and unpainted components in the proportion that they are present in the debris." [December 18, 1998; 63 *FR* 70206]

For the siding contaminated with LBP, we *could* obtain random samples of the siding from all of the different areas, homogenize them and then subsample for laboratory analysis via the TCLP. A less costly scenario would be to use a mass balance approach. Using this technique, we calculate or measure the total amount of lead in the very thin paint layer per square foot of building siding, and divide that by the weight of the building wall per square foot. That will give a total lead concentration in mg/kg. If that value is <100 mg/kg, the building siding cannot be hazardous for lead (see Section 18.3.3.4). If the value is >100 mg/kg, it's time to run the TCLP on a representative sample. Case Study 18-3 looks at the characterization of LBP-contaminated windows that will be removed from a building and disposed.

For the filter cartridge discussed previously in this section, we're trying to determine if it is characteristically hazardous for benzene. Most of the benzene is probably in the filter media, but the end caps and screening have some concentration associated with them as well. The TCLP specifies a minimum 100-g sample size; our spent cartridge weighs 50 lb. Using the same methodology as noted in Case Study 18-3 for the LBP-contaminated windows, the weight of the end caps, the screening, and the filter media are determined using manufacturer's information or by disassembling a filter. Then, using scissors, tin snips, a hack saw, etc., we take numerous random samples from different areas of each of the three types of material. If the end caps, screening, and filter media make up 10%, 20%, and 70% of the total weight, respectively, we send a Multi Increment sample to the lab for TCLP testing containing those three material types in that weight ratio.

In the case of chromium-contaminated soil, using a strategy such as systematic grid sampling or systematic sampling using "random within grids" with large enough sample sizes should achieve a representative sample.

The following sampling techniques should be considered when sampling contaminated debris [EPA/530/D-02/002]:

- Coring and cutting pieces followed by crushing and grinding (either in the field or laboratory),
- Drilling a number of holes and sending the cuttings for analysis, or
- Grinding an entire piece of debris via a tub grinder followed by conventional sampling techniques.

Once we have a suitable Multi Increment sample, we can subsample in the field or laboratory. A document providing an excellent discussion of subsampling heterogeneous materials to obtain a representative sample is *Guidance for Obtaining Representative Laboratory Analytical Subsamples from Particulate Laboratory Samples*, availability as given in Table 18-2.

Additional suggestions for sampling heterogeneous wastes are contained in Chapter 5 of *Characterizing Heterogeneous Wastes: Methods and Recommendations*, EPA/600/R-92/033, February 1992 (available from http://nepis.epa.gov/EPA/html/Pubs/pubtitleORD.html by downloading the report numbered 600R92033) and ASTM D5956, *Standard Guide for Sampling Strategies for Heterogeneous Wastes* (available from http://www.astm.org).

18.5.2.2 How many samples do I need?

If you have waste in a large roll-off box, tank, waste pile, or lagoon, is it possible to take just one sample and call it representative? What if you are trying to make a hazardous waste determination for contaminated soil or ground water? Can one sample of the dirt or ground water be expected to exhibit the average properties of all the media? Doubtful, especially when the one sample is a single grab sample (rather than a Multi Increment sample made by combining multiple individual increments—see Section 18.5.2.3).

As noted above, when the waste stream is large, heterogeneous, and/or inconsistent over time, multiple samples will likely be required to characterize the waste. So, how many samples do you need to take?

EPA's guidance notes that the number of samples required for making hazardous waste determinations will usually be based on the sample collection strategy and statistical approach identified in the sampling plan. Variables that may affect how many samples will be required include properties of the waste matrix, degree of confidence required, access to sampling points, and resource

Case Study 18-3: Hazardous Waste Characterization for Heterogeneous Materials

Sampling very heterogeneous materials can be quite a challenge. As with all samples, it is necessary to ensure that some of every material in the decision unit is collected in the sample *in the same proportions as in the decision unit.* [January 9, 1992; 57 *FR* 990] With very heterogeneous materials, it is sometimes easier to sample materials of like properties individually and then combine them into the final sample.

If the solid waste is windows consisting of a frame painted with lead-based paint and leaded glass, it is easier to sample the frame and glass separately and then combine them to make up the field sample. The first step would be to separate the glass from the frame. The next step would be to weigh the individual phases. If a frame weighs 4 kg and the glass weighs 1 kg, the Multi Increment sample collected must have a frame-to-glass weight ratio of 4:1. For a 500-gram field sample, there would need to be 400 grams of frame and 100 grams of glass in the sample.

The frames are sampled by taking slices or complete cores of the frames at multiple locations. If the frames are uniform and the paint seems uniform, then fewer slices/cores are needed. If the frame or paint is not uniform (e.g., frames are different thicknesses, different color paints), then enough slices/cores need to be taken (in the correct proportion) to represent all of the different parts of all the frames in the Multi Increment sample. The glass can be sampled by collecting many small pieces to obtain the desired mass. If the TCLP will be performed, there are certain size requirements that must be followed (the Multi Increment sample will be ground to pass through a 9.5-mm standard sieve). The laboratory must perform adequate subsampling to represent the field sample.

As an example, the table below gives results for making a hazardous waste determination using two different strategies: 1) grab sampling and analysis of four different window frames and associated glass, and 2) the strategy that incorporates increments from all four frames and associated glass into a single Multi Increment sample.

	Lead in frame (mg/L TCLP)	Lead in glass (mg/L TCLP)	Lead in window[1] (mg/L TCLP)	TC limit for lead (mg/L TCLP)	Conclusion
Grab sample 1	3.0	4.0	3.2	5.0	Window is non-hazardous
Grab sample 2	6.0	7.0	6.2	5.0	Window is hazardous
Grab sample 3	6.0	ND	4.8	5.0	Window is non-hazardous
Grab sample 4	4.0	10.0	5.2	5.0	Window is hazardous
Multi Increment sample[2]	NA	NA	4.8	5.0	All windows are non-hazardous

ND = nondetect; TC = toxicity characteristic.
[1] Lead in window is based on weighting the TCLP result for lead in the frame at 80% and the TCLP result for lead in the glass at 20%.
[2] The Multi Increment sample was prepared using portions of material from all four windows, in the frame-to-glass weight ratio of 4:1.

When grab sample results are used, even though one phase may be above the limit (see Grab sample 3), if the average is below the limit (and if the window is integral and disposed as a unit), then it is nonhazardous. Conversely, even though one phase may be below the limit (see Grab sample 4), if the average is above the limit, the entire window must be disposed as hazardous waste unless the generator physically segregates it into separate phases.

The benefit of the Multi Increment sample is that all of the windows (frames plus glass) are included in the sample, making it a much more representative sample of the decision unit (all of the windows that are being disposed). Additionally, only one TCLP test needs to be conducted, saving on analytical costs. Of course, you have to be willing to base your characterization of the windows on this one result.

constraints. [See Sections 5.4 and 5.5 of EPA/530/D-02/002 and EPA QA/G-5S.] Thus, best professional judgment (and, possibly, the services of an environmental consulting firm) will come into play. EPA has free software available to estimate the appropriate number of samples under certain conditions. Called DEFT (for Decision Error Feasibility Trials), the software and user's guide can be downloaded from http://www.epa.gov/quality/qa_docs.html.

According to EPA/530/R-94/019, ASTM D140 (*Standard Practice for Sampling Bituminous Materials*, available from http://www.astm.org) can be used to estimate the number of containers of the same waste stream that must be sampled. That method suggests that the number of drums to be randomly selected for sampling equals the cube root of the total number in the shipment. Always round up to the next whole number. For example, the cube root of 28 is about 3.04, so if you have 28 containers of the same waste, you need to sample four containers.

EPA has provided examples of how to make hazardous waste determinations in certain waste management scenarios. In those examples, the agency shows how to calculate how many samples are needed and from where they would be collected. Selected examples from EPA guidance are presented in Case Studies 18-4 and 18-5.

Note that it may be prudent to collect more samples than the number calculated by statistical equations. This will provide spare samples in case of sample contamination, unexpected problems, data outliers, or poor preliminary estimates of the mean and standard deviation. Also, the equations used to calculate the number of samples required do not take into consideration any quality control samples (e.g., duplicates, blanks, split samples) needed to support the project.

18.5.2.3   How should I take a sample?

As discussed in Section 18.5.1.1.3, a Multi Increment sample is usually the best way to obtain a representative sample of a decision unit. (The difference between a Multi Increment and composite sample is discussed in that section also.) Multi Increment samples are formed by physically combining and mixing multiple aliquots (from different times or locations of the decision unit) into a single sample. Multi Increment sampling can produce more precise estimates of the average (mean) constituent concentration(s) in a waste stream or contaminated medium with fewer analytical samples. This is because a Multi Increment sample contains the same proportion of contaminant particles of different sizes, composition, and configuration as exists within the decision unit selected for sampling. Such samples emulate putting the whole population in a test beaker. Since Multi Increment samples average out the high and low contaminant concentrations, they make more representative samples, as that term is defined. Table 18-6 summarizes the advantages and limitations of this type of sampling.

In many cases, Multi Increment sample means are actually higher than discrete "grab" sample means. This results from the fact that, in most contaminant distributions, the mode concentration (that value occurring most frequently) is lower than the mean. Multi Increment samples find the "hot spots" that grab samples normally miss, also increasing the mean contaminant concentration. Thus, Multi Increment samples are actually more conservative and protective.

According to EPA, Multi Increment sampling is appropriate when determining the general characteristics or the representativeness of certain wastes when we are considering methods of treatment or disposal (e.g., wastes in a pile or surface impoundment). However, Multi Increment sampling should be performed only on like waste streams—not on dissimilar wastes or on wastes from different sources or decision units (e.g., drums with unknown contents or dissimilar materials). Taking Multi Increment samples

Case Study 18-4: Simple Random Sampling of Lagoon Sludge

Wastewater from a manufacturing process is stored/treated in a lagoon. The wastewater entering the lagoon is nonhazardous, but it contains barium. The barium accumulates in the sludge on the bottom of the lagoon. A preliminary study of barium levels in the sludge was conducted two years ago. Two samples of sludge were taken from the upper end of the lagoon; barium concentrations in the TCLP leachate were 86 and 90 mg/L. Two samples were taken from the lower end of the lagoon, and the barium concentrations in the TCLP leachate from those samples were 98 and 104 mg/L. Do we have enough samples to determine if the sludge in the lagoon is hazardous?

1. Although the four old samples cannot be considered to completely characterize the barium concentration in the lagoon sludge today, they are a good starting point. Preliminary estimates of the mean ($\bar{x}$) and variance (s^2) are calculated from the old data as:

$$\bar{x} = \frac{\sum_{i=1}^{n} x_i}{n} = \frac{86 + 90 + 98 + 104}{4} = 94.50 \text{ mg/L}$$

and

$$s^2 = \frac{\sum_{i=1}^{n} x_i^2 - \left(\sum_{i=1}^{n} x_i\right)^2 / n}{n-1}$$

$$= \frac{35{,}916.00 - 35{,}721.00}{3} = 65.00$$

2. Based on these preliminary estimates of $\bar{x}$ and s^2, as well as the knowledge that the regulatory threshold (RT) for barium is 100 mg/L via the TCLP, a preliminary value for the required number of samples (n_1) can be calculated, assuming the data are in (or have been transformed to) a normal distribution:

$$n_1 = \frac{t_{.20}^2 \, s^2}{(RT - \bar{x})^2} = \frac{(1.638^2)(65.00)}{5.50^2} = 5.77$$

where $t_{.20}$ is the student's "t" value selected from Table 18-5.

3. As indicated above, the minimum number of sludge samples (n_1) to be collected in order to characterize the barium concentration in the lagoon sludge is six (we always have to round up). Nine samples (three extra samples for protection against sample contamination, unexpected problems, or poor preliminary estimates of $\bar{x}$ and s^2) are randomly collected from the lagoon, as shown in Figure 18-4. All samples consist of the greatest volume of sludge that can be practically collected. Six are randomly selected, and the three extra samples are suitably processed and stored for possible later analysis. Once the six new samples are analyzed, we will have to recalculate the minimum number of samples required to characterize the sludge, based on the new values of $\bar{x}$ and s^2. Thus, this is often an iterative process. [Chapter 9 of SW–846]

---

of waste in containers, tanks, or other units can be accomplished, but only after screening to prevent mixing incompatible wastes. [EPA/540/R-95/141]

The general strategy for Multi Increment sampling is to combine a number of individual samples together (either randomly or systematically as discussed below) to form a single sample. Individual samples to be combined should generally be of the same mass or volume. The Multi Increment sample is thoroughly mixed and homogenized, and then one or more subsamples are taken for subsequent analysis for the constituents of concern. Mixing of individual samples and subsequent subsampling in the field are discussed in detail in EPA/530/D-02/002, EPA/600/R-92/128, EPA/600/R-03/027, and ASTM D6051. Using the Multi Increment sampling concept to obtain a

Table 18-5: Tabulated Values of Student's "t" Value for Evaluating Solid Wastes

Degrees of freedom (n-1)[1]	Student's "t" value for an 80% CI[2]	Student's "t" value for a 90% CI[3]	Student's "t" value for a 95% CI[4]
1	3.078	6.314	12.706
2	1.886	2.920	4.303
3	1.638	2.353	3.182
4	1.533	2.132	2.776
5	1.476	2.015	2.571
6	1.440	1.943	2.447
7	1.415	1.895	2.365
8	1.397	1.860	2.306
9	1.383	1.833	2.262
10	1.372	1.812	2.228
11	1.363	1.796	2.201
12	1.356	1.782	2.179
13	1.350	1.771	2.160
14	1.345	1.761	2.145
15	1.341	1.753	2.131
16	1.337	1.746	2.120
17	1.333	1.740	2.110
18	1.330	1.734	2.101
19	1.328	1.729	2.093
20	1.325	1.725	2.086
21	1.323	1.721	2.080
22	1.321	1.717	2.074
23	1.319	1.714	2.069
24	1.318	1.711	2.064
25	1.316	1.708	2.060
30	1.310	1.697	2.042
40	1.303	1.684	2.021
60	1.296	1.671	2.000
120	1.289	1.658	1.980
∞[5]	1.282	1.645	1.960

CI = confidence interval.

[1] Degrees of freedom are equal to the number of samples collected from a solid waste less one (for simple random sampling).

[2] Tabulated "t" values are for a two-tailed confidence interval and a probability of 80% (the same values are applicable to a one-tailed confidence interval and a probability of 90%).

[3] Tabulated "t" values are for a two-tailed confidence interval and a probability of 90% (the same values are applicable to a one-tailed confidence interval and a probability of 95%).

[4] Tabulated "t" values are for a two-tailed confidence interval and a probability of 95% (the same values are applicable to a one-tailed confidence interval and a probability of 97.5%).

[5] The last row of the table (for infinite degrees of freedom) gives the "t" values for a standard normal distribution.

Source: Adapted from EPA QA/G-9S.

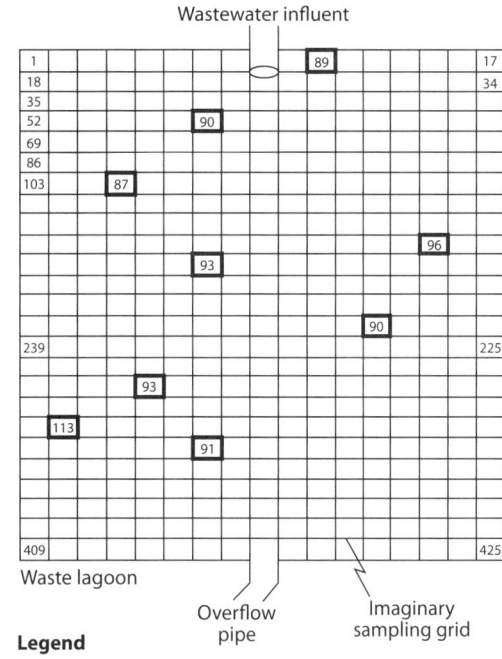

Figure 18-4: Simple Random Sampling of a Waste Lagoon Containing Barium-Contaminated Sludge

Legend
1-425 Units in sampling grid
90 Barium concentrations (mg/L) via the TCLP associated with nine samples of sludge

Adapted from Chapter 9 of SW–846.

representative sample of energetic materials dispersed on a military training range is discussed in Appendix A to SW–846 Method 8330B.

The subsampling can be done in the field, to reduce sample shipping costs, or in the analytical lab. Although economically advantageous to subsample in the field, gravity separation of the resulting subsample during shipment to the lab may occur. Therefore, EPA recommends that, if practical, mixing and subsampling operations should be conducted at the lab.

Caution: Mixing and subsampling procedures inherent in Multi Increment sampling will almost certainly cause losses of volatile organic compounds

Case Study 18-5: Simple Random Sampling of Material Filling a Roll-Off Box

A maintenance facility uses plastic blasting media to strip paint off of equipment. The spent blasting media/paint fines are collected in a roll-off box for disposal. The facility used knowledge to determine that the spent blasting media/paint fines are not listed and do not exhibit the characteristic of ignitability, corrosivity, or reactivity. However, sampling and analysis will be used to determine if the waste exhibits the toxicity characteristic due to cadmium and/or chromium, which are known to be in the paint.

In developing the sampling plan, the facility decided on dual, tough acceptance criteria: first, a 90% confidence that at least 90% of the solid waste would result in TCLP values for cadmium and chromium that are less than their corresponding toxicity characteristic values, and second, that no exceedance of the cadmium and chromium regulatory values will be allowed in any of the samples actually taken. If these criteria are met, the solid waste will be characterized as nonhazardous.

1. Per EPA/530/D-02/002, the equation used to calculate the minimum number of samples required given these statistical criteria is:

$$n = \frac{\log \alpha}{\log \rho}$$

where α is the probability of concluding that the waste is nonhazardous when, in fact, it is hazardous, and ρ is the minimum proportion of the waste that is required to be below toxicity characteristic levels. In this example, α is 0.10 because the acceptable probability of concluding that the waste is nonhazardous when at least a portion of the waste is hazardous is 10%. Conversely, ρ is 0.9 in this scenario because 90% of the samples from the solid waste must meet the regulatory levels for cadmium and chromium. Doing the math gives a calculated value for n of 21.8. We always have to round up, in this case to 22 samples.

2. Because it takes about a month to fill a 10-yd³ roll-off box, the facility decides to take a random sample of the spent blasting media/paint fines discharged to the box each day; thus, 30 samples were taken. Since only 22 samples were needed, 22 of the 30 samples were randomly selected for laboratory analysis. (The waste would then be regarded as nonhazardous if all 22 samples are below their respective TCLP limits for both cadmium and chromium.)

(VOCs). Therefore, combining individual increments in the field will likely be inappropriate when analyzing samples for such compounds. [RO 13406] Specialized procedures may be used for Multi Increment sampling when determining VOC concentrations, such as that discussed in *Standard Guide for Sampling Waste and Soils for Volatile Organic Compounds*, ASTM D4547 (available from http://www.astm.org).

Properly combined samples can be used to estimate the mean and variance of the constituent concentrations in a waste stream. Because the Multi Increment sampling process is a physical way of averaging out variabilities in concentrations among locations or time intervals, the resulting concentration data should be more normally distributed than data from individual samples—avoiding the need to transform the data.

In general, the number of Multi Increment samples required to calculate a mean constituent concentration can be calculated using the same equations as if individual samples will be used, with adjustments as discussed in Section 5.4 of EPA/530/D-02/002. To establish the sampling uncertainty for estimating mean concentrations of contaminants, replicate Multi Increment samples must be taken. For example, Method 8330B recommends that triplicate Multi Increment samples should be collected for

Table 18-6: Advantages and Limitations Associated With Multi Increment Sampling

Advantages

- Reduces the cost of estimating constituent mean concentrations, especially in situations where analytical costs greatly exceed sampling costs or when analytical capacity is limited
- Improves the precision of constituent mean concentration estimates (reduces between-sample variances); a set of n Multi Increment samples, each made up of m individual grab samples (for a total of $n \times m$ grab samples), from a heterogeneous waste provides a more precise estimate of the mean than n grab samples
- Increases the number of increments, thereby reducing grouping and segregation errors
- Can be used as a screening tool to identify if additional, separate analyses need to be performed (e.g., during the early stages of a study to locate areas requiring increased evaluation due to potentially high levels of contaminants)
- Can be used when the contaminant distribution is nonrandom and the majority of individual analytical results are nondetects for the contaminant of interest

Limitations

- May degrade the integrity of the individual samples due to physical mixing in some situations (e.g., chemical precipitation or volatilization of constituents can occur)[1]
- Cannot be used if RCRA regs require use of grab samples (see Section 18.4.1)
- Should not be used if specific data points are required to generate percentiles[2]
- Should not be used if individual samples are incompatible and may react when mixed, or when discrete properties (e.g., flash point or pH) may change upon mixing
- Should not be used if testing of individual samples is required later, or when the constituent concentrations need to be evaluated to check for a spatial or temporal correlation

[1]In the case of samples containing volatile constituents, combining individual sample extracts at the laboratory may be an alternative to field compositing.
[2]Multi Increment samples reflect a physical rather than mathematical mechanism for averaging. Therefore, such sampling should generally be avoided if statistical parameters other than the mean (e.g., percentiles) are being evaluated.

Source: McCoy and Associates, Inc.; adapted from EPA/530/D-02/002.

---

each type of activity under investigation. To avoid collecting co-located samples and to be random, each replicate of Multi Increment samples should be collected starting at different corners of the decision unit or different random start locations within the same starting corner. If replicate samples are not included in a sampling plan, sampling error cannot be estimated.

For physically solid waste, it is always best to take samples while the solids are in motion (e.g., filling a container, creating a pile), because you will have access to the entire population of material—don't wait until the container is filled or the pile is formed.

Q *How would Multi Increment sampling be used to determine if an ash stream exhibits the toxicity characteristic?*

A EPA provided an example of an acceptable strategy to characterize the ash stream as follows. Collect 14 Multi Increment samples, each consisting of 8 grab samples (1 taken every hour during an 8-hour shift). If the plant operates 24 hours/day, make a second 8-hour Multi Increment sample during a different shift (resulting in 14 Multi Increment samples taken over a period of 7 days). If the plant operates only 8 hours/day, it will take 2 weeks to obtain the 14 Multi Increment samples. Each Multi Increment sample should be developed by thoroughly mixing each of the 8 grab samples (a cement or other mechanical mixer can be used). Then obtain a subsample from each Multi Increment sample by taking a full core or slice through the mixed Multi Increment sample. Send each subsample to a lab for TCLP analysis. [*Guidance for the Sampling and Analysis of Municipal Waste Combustion Ash for the Toxicity Characteristic*, EPA/530/R-95/036, June 1995, available from http://nepis.epa.gov/EPA/html/Pubs/pubtitleOSWER.html by downloading the report numbered 530R95036]

18.5.2.3.1 Containers

Containerized wastes usually vary more in a vertical than horizontal direction due to settling of solids, variations in densities of liquid phases, and periodic variations in the composition of the waste entering the container. [Chapter 9 of SW–846, EPA/530/R-93/007] As mentioned previously, analysis of a single COLIWASA or full-depth core sample of a drummed waste should be representative of the contents of the drum.

However, a single sample of waste in larger-capacity containers may not be representative. Characterizing contaminated soil already loaded into a roll-off box, for example, is a difficult job. Four or five samples taken from the top 4 inches are not representative of all of the soil in the box. Either core samples will have to be taken at different levels and different locations within the box, or the box will have to be emptied and filled again (or loaded into a different box) while taking samples as the dirt is reloaded. Those samples should then be combined into a Multi Increment sample.

Q *Sludge from a manufacturing process is collected in roll-off boxes. Can one composite sample from four or five roll-off boxes be considered representative of all the waste in those boxes?*

A No. Assuming that the roll-off boxes are filled gradually over a given time period, grab samples should be taken from each of the periodic loads transferred to each roll-off box. All grab samples representing wastes transferred to a single roll-off box should then be combined to form a single Multi Increment sample. This sample should then be subjected to the appropriate testing requirements, so a hazardous waste determination can be made for each box. [RO 13343]

The use of Multi Increment samples from containers is appropriate to account for spatial variations within a single batch of waste (e.g., a rail car load of sludge from a wastewater treatment unit), but this technique should not be used to reflect average constituent concentrations for wastes originating from different sources. [EPA/530/R-94/019, EPA/540/R-95/141]

For example, a facility wants to make a hazardous waste determination for a continuously generated industrial process waste that it already stored in 100 drums. It would seem reasonable for facility personnel to take a representative sample (representing different depths of the material in each drum) from each of 10–20 drums (if solid, personnel may have to take the samples while recontainerizing the waste since it is hard to core solid material), and combine these into one composite sample, a subsample of which is sent for analysis (plus a duplicate for QC purposes). Although this seems to be reasonable and is consistent with EPA's guidance noted in the previous paragraph, our experience is that regulators often consider each container to be a discrete waste management or decision unit and want generators to make a hazardous waste determination for each one, versus evaluating the average properties of the waste contained in all of the containers as a whole. See Section 18.5.2.4.1.

In another example, a facility asked EPA for guidance on making hazardous waste determinations for waste nitrocellulose filter fabric, which it bales for disposal. The agency responded "First you must define the disposal unit. Is it each bale, analogous to individual drums, or is it the total accumulation of bales, analogous to a warehouse of drums. Because of the physical form of this waste, composite samples will be difficult to test, so individual grab samples should be used. If some of the waste proves to be characteristically hazardous, you will need to consult a statistician on an appropriate design to assure the level of confidence you need to make a decision on the hazard posed by the entire waste. As an initial step, use judgmental sampling of the different types of filter waste. Are they characteristically hazardous? If you can't identify individual components (filter types) of the waste which are hazardous, that is good assurance that the entire waste is nonhazardous and that further sampling and testing is not [necessary]." [RO 14259]

18.5.2.3.2 Tanks

The goal of tank sampling is to acquire a sufficient number of samples from different locations and depths within the tank to provide analytical data that are representative of the entire tank contents. If the tank is open-top, a representative set of samples may be taken using a three-dimensional simple random sampling strategy, as illustrated in Case Study 18-6.

Often, the tank is closed, restricting sampling access to inspection ports. In such a case, professional judgment will come into play as to whether taking a set of samples at different depths from one or two inspection ports will provide representative samples of the entire tank contents. However, in tanks (as well as containers), wastes have a much greater tendency to be nonrandomly heterogeneous in a vertical rather than a horizontal direction due to settling of solids, variations in densities of liquid phases, and periodic variations in the composition of the waste entering the tank. [Chapter 9 of SW–846, EPA/530/R-93/007]

If the waste in the tank is known to consist of two or more discrete strata (e.g., liquid over a sludge), a more precise representation of the tank contents can be obtained by using a stratified random sampling strategy (i.e., by sampling each stratum using two- or three-dimensional simple random sampling).

Case Study 18-6: Simple Random Sampling (With Compositing) of Tanks

A 290,000-gal manufacturing process waste storage tank has received manufacturing residues over a period of several months. Because the tank is approaching capacity, the facility wants to unload the contents into tank trucks and send them offsite for disposal. However, a young environmental engineer at the plant insists that the contents are a solid waste and need to be characterized as to their hazardousness. Because the tank is open top, the entire horizontal surface area is accessible for sampling. The tank diameter is 50 ft with a 20-ft height. How should the tank be sampled for this purpose?

Although no data on tank waste composition existed prior to this sampling effort, taking 15 vertically composited samples was thought to be a reasonable approach. The plant decided to select sampling points randomly along the circumference (157 ft) and radius (25 ft). A random number table was used for this purpose. The samples were taken using a weighted bottle. The bottle was lowered into the liquid at the randomly selected circumference and radius coordinates, and the liquid was sampled at those coordinates 10 times (each sample taken from a different 2-ft interval of tank depth), thereby sampling the entire tank depth of 20 ft. Each of the 10 samples taken from that coordinate was poured into a large sample container, which was used for compositing. The process was repeated for the other 14 randomly selected sampling points, resulting in 15 vertically composited samples.

Subsamples (aliquots) of the 15 composite samples were then further composited into 5 samples that were actually sent to the lab. Following laboratory analysis, the equation below was used to determine if enough samples were analyzed to make a statistically sound hazardous waste characterization.

$$n_1 = \frac{t^2_{.20} \; s^2}{(RT - \bar{x})^2}$$

If the number of samples analyzed was not sufficient, then the 15 vertically composited samples would be recomposited to a lesser degree or analyzed individually. [Chapter 9 of SW–846]

Useful guidance on tank sampling is found in Chapter 9 of SW–846 and the Environmental Response Team's SOP 2010, November 1994 (available online at http://www.ert.org/products/2010.pdf).

18.5.2.3.3 Surface impoundments/lagoons

One of EPA's national environmental enforcement initiatives is addressing the illegal disposal of hazardous wastes in unlined surface impoundments. These are units with no RCRA permit and without double liners required for hazardous waste units per Part 264/265, Subpart K.

A common situation at an industrial facility is process wastewater entering a nonhazardous, unlined lagoon. The facility has to demonstrate that the process wastewater is not hazardous (e.g., for benzene), because hazardous wastewater cannot enter an impoundment unless the unit has a RCRA permit. How does a facility make such a demonstration? Process wastewater is typically sampled using an automatic (systematic) sampler, in which samples are collected over a day (24-hour composites). These composites are then averaged over one month (or one quarter). One of the individual, 24-hour composite samples could be over the regulatory limit (e.g., 0.5 mg/L for benzene), but the average of the month (or quarter) is less than the limit. In this case, will EPA or the state be concerned about the one day over the limit?

We have seen little on-point guidance from EPA on this issue, but we have to go back to the regulatory definition of "representative sample." This definition requires generators to determine the *average* properties of the waste. Chapter 9 of SW–846 notes that:

"Another technique for increasing sampling precision is to maximize the physical size (weight or volume) of the samples that are collected. That has the effect of minimizing between-sample variation and, consequently, decreasing [the standard deviation of the mean, $s_{\bar{x}}$]. Increasing the number or size of samples taken from a population, in addition to increasing sampling precision, has the secondary effect of increasing sampling accuracy."

Chapter 9 also discusses the concept of a representative database as opposed to individual values. Once this database of samples is collected, SW–846 goes on to show how to use statistics to determine if the wastewater entering the lagoon is hazardous or not (see our discussion of this statistical approach in Section 18.5.3.1). In one of its examples in Chapter 9, EPA noted that one of the samples could have a constituent concentration over the regulatory limit and yet the solid waste as a whole was "definitively concluded" to be nonhazardous.

"The concept is that while each individual sample or even composite may not be representative of the [waste], the sum total of samples taken will represent the average property of the waste even though the entire waste stream was not sampled." [February 8, 1990; 55 *FR* 4443]

This may be helpful to the facility in supporting the notion that the individual values themselves are not representative and cannot be used for compliance determinations. It is the average properties of the wastewater over some time frame that is mutually agreed upon with the regulators (e.g., a month or quarter), that determines the regulatory status of the wastewater and, therefore, the lagoon.

That's the theory—the reality may be different. Some at EPA take the position that if the wastewater entering the impoundment is a hazardous waste *at any point in time*, the unit may become subject to RCRA requirements. Thus, discussions with state or EPA regulatory personnel should be held to determine an adequate sampling plan for the impoundment, including allowed averaging period, minimum number of days to sample per averaging period, minimum number of samples/times per day, and permissible averaging. The use of an automatic wastewater sampler vs. grab samples taken by personnel should be spelled out in the plan.

One thing about which facility personnel should be vigilant is identifying all inlets and inputs to the impoundment. For example, inlet pipes can sometimes be submerged; plant personnel may use flexible hoses for temporary discharges into a lagoon. Of particular concern are intermittent discharges into impoundments (e.g., process equipment cleanouts, untreated wastewater streams resulting from wastewater treatment unit malfunctions, spills routed directly to the impoundment, vac truck discharges, untreated waste streams from process upsets). Such discharges can occur without environmental personnel knowledge, and [according to EPA's National Enforcement Investigations Center (NEIC)], most surface impoundments that are RCRA-regulated became so because of intermittent discharges—not normal flows—into the unit. For intermittent discharges, EPA recommends that each different discharge be sampled separately using grab samples.

Another scenario that can result in an enforcement situation is generation of hazardous sludge in an impoundment receiving nonhazardous influent wastewater. Although organics may volatilize or biodegrade in an impoundment, metals will not. Thus, sludge that builds up over time in an impoundment may contain toxic concentrations of metals or other toxic constituents in §261.24. Even though the influent is nonhazardous, an impoundment storing characteristic sludge is a hazardous waste unit that requires a RCRA permit and must be in compliance with Part 264/265, Subpart K, etc. To prevent this abysmal situation, sludge in a lagoon should be sampled periodically. One approach, as noted in Case Study 18-4, is to use a simple random sampling strategy to collect a handful of samples (in that example—nine) across the surface area of the impoundment and send them all to a lab for analytical testing. As an alternative to that approach, those samples, plus several more to make sure the facility is collecting samples from all parts of the population (note the large areas that were not sampled in Figure 18-4), could have been combined into one Multi Increment sample. The average properties of the sludge could have been determined just as well or better by analyzing that one sample at 1/9th the analytical cost. As with the influent wastewater, discussions with state or EPA regulatory personnel should be held to determine an adequate sampling plan for the sludge in the unit. For example, the sampling should take into consideration any contaminant concentration variability across the impoundment and/or with sludge depth. [RO 13406]

18.5.2.3.4 Piles

Physically solid material is sometimes stored in piles prior to disposal. EPA provided guidance on how to collect representative samples from waste piles in the guidance manual, *Petitions to Delist Hazardous Waste*, EPA/530/R-93/007, March 1993, available at http://nepis.epa.gov/EPA/html/Pubs/pubtitleOSWER.html by downloading the report numbered 530R93007. RO 11201 implies that the sampling strategy given in that document can be used in other situations (i.e., not involving delistings) when a representative sample of solids in a pile needs to be obtained.

In summary, the pile should be divided into four equal quadrants. Each quadrant is further subdivided into a 10 x 10 grid, and a number of these 100 subdivisions are randomly selected for sampling. Samples are taken using a full-vertical core at each sampling point. The collected samples from each quadrant are combined to form a representative sample of each quadrant. In this guidance manual, EPA recommends that the four quadrant samples be analyzed separately, but the Multi Increment sampling approach would recommend that they be combined into a single Multi Increment sample representing the entire pile (which is the decision unit).

18.5.2.4 "Hot spots"

"Hot spots" are areas within a waste stream or contaminated media that have elevated contaminant concentrations. How do we address them when we are developing our sampling plan or trying to make a hazardous waste determination?

Because hazardous waste determinations are based on obtaining a "representative sample," which requires determination of "the average properties of the universe or whole," hot spots should have no more influence on our determination than areas with little or no contamination. We are not asked by the RCRA regulations to determine the highest contaminant concentrations or the lowest, but the *average*.

There can be "hot spots" within a solid waste just as there can be "low spots." But if, using a well thought out and defensible sampling plan, the waste can be shown *on average* to be nonhazardous, it can be managed as such. That's what the regs imply and that's what the SW–846, Chapter 9 examples show. However, EPA provided the following sobering guidance in its 2002 draft sampling guidance:

> "If the objective is to determine if a solid waste is a hazardous waste or to measure attainment of [an LDR] treatment standard for a hazardous waste, then any obvious 'hot spots' or high concentration wastes should be characterized separately from low concentration wastes to minimize mixing of hazardous waste with nonhazardous wastes and to prevent impermissible dilution." [EPA/530/D-02/002, p. 57]

This quote seems to say that hot spots have to be segregated and managed as hazardous waste, even if the waste on average is nonhazardous.

Thus, we have two sets of EPA guidance at odds with each other. As stated previously, our recommendation is to work with your state before characterizing a large volume of solid waste. If the state accepts the SW–846 approach for taking a "representative sample," a statistical nonhazardous waste demonstration will be allowed (even though there may be a hot spot or two over the limit). State involvement/support can be very helpful when you send the material to a nonhazardous waste receiving facility if its fingerprint analysis indicates the waste in one or two individual containers is hazardous.

Developing a sampling plan to find hot spots in soil and solid media is discussed in detail in Chapter 9 of *Methods for Evaluating the Attainment of Cleanup Standards, Volume 1: Soils and Solid Media*, EPA/230/02-89/042, February 1989 (http://clu-in.org/download/stats/vol1soils.pdf).

18.5.2.4.1 *Hot spots in containerized waste*

Sometimes, a large amount of the same waste or contaminated media is containerized. Can the waste or media be evaluated statistically as a whole? For example, can every tenth drum of 750 drums of the same waste be sampled and the analytical results evaluated to make a statistical hazardous waste determination for the entire amount? If the statistics show that the waste as a whole is nonhazardous, does that mean that we wouldn't have to manage the individual drums that do test hazardous as hazardous waste?

Probably not. Once a waste is containerized, we find that states/EPA often want generators to make a hazardous waste determination for *each* container, versus statistically evaluating all of the waste as a whole. If statistics can be used, then a small number of individual drums containing hazardous waste could go to a nonhazardous landfill—this just doesn't seem right to EPA:

> "[I]f a decision, based on the data collected, results in a large volume of waste being classified as nonhazardous, when in fact a portion of the waste exhibits a hazardous waste characteristic (e.g., due to the presence of a 'hot spot'), then the waste generator could potentially be found in violation of RCRA. To limit risk of managing hazardous waste with nonhazardous waste, the waste handler should consider dividing the waste stream into smaller decision units—such as the volume of waste that would be placed into an individual container to be shipped for disposal—and make a separate waste classification decision regarding each decision unit." [EPA/530/D-02/002, p. 38]

On the frequently asked questions (FAQs) page of EPA's sampling website (http://www.epa.gov/epawaste/hazard/testmethods/faq/faqs_sampl.htm), the agency notes under its composite sampling discussion that "[i]f some of the drums might be classified as hazardous when characterized individually, then we recommend you make a waste classification decision on each drum to avoid the possibility of mixing hazardous waste with nonhazardous waste."

Nonhazardous waste disposal facilities will have the same concerns; they may want to see data for each container showing that the waste is nonhazardous. If there are many containers, the nonhazardous disposal facility may be satisfied if you have data for several (but not all) of the containers, if all the data you do have shows the waste is clearly nonhazardous.

18.5.2.4.2 Compositing to identify hot spots

Sample compositing is usually conducted to reduce analytical costs. One perceived limitation of compositing is that one or more individual samples comprising the composite could be "hot" (exceed the regulatory threshold—RT) but remain undetected due to the averaging effect inherent in compositing. However, if the sampling objective is to determine if one or more individual samples is hot, composite sampling can still be used.

The procedure used to identify potential hot spots using composite samples is as follows. For a contaminant concentration x_i measured from a composite made up of n individual samples, if $x_i < RT/n$, no individual sample concentration can be greater than the regulatory threshold. But, if $x_i > RT/n$, then at least one of the n individual sample concentrations may be greater than the regulatory threshold. Thus, no more than $(n)(x_i)/RT$ individual sample concentrations can be greater than the regulatory threshold. Note that we round *down* when performing this calculation.

If we detect a possible hot spot using this process, we will likely have to go back and analyze the individual samples used to form the composite showing potential exceedances of the regulatory threshold. Therefore, a prudent practice is to save splits of each sample used to form a composite for possible future analysis, where handling considerations (e.g., holding times) permit. See Case Study 18-7.

18.5.2.5 Field sampling techniques

Selecting the proper sampling devices can significantly reduce sampling errors. A considerable number of resources are available to help you choose the correct equipment, depending on the type of waste and unit to be sampled. The following documents will quickly point the user to the proper sampling equipment and decontamination procedures:

- 40 *CFR* Part 261, Appendix I;
- Chapter 7 of EPA/530/D-02/002, http://www.epa.gov/epawaste/hazard/testmethods/sw846/samp_guid.htm;
- Chapter 2 of EPA/530/R-94/024, http://www.epa.gov/osw/hazard/tsd/ldr/wap330.pdf;
- Chapter 9 of SW–846, http://www.epa.gov/epawaste/hazard/testmethods/sw846/online/index.htm;
- *Standard Guide for Selection of Sampling Equipment for Waste and Contaminated Media Data Collection Activities*, ASTM D6232, http://www.astm.org;
- *Superfund Program Representative Sampling Guidance, Volume 4: Waste*, OSWER Directive 9360.4-14, EPA/540/R-95/141, December 1995, http://www.ert.org/products/SF-WASTE.pdf;
- *USACE Sample Collection and Preparation Strategies for VOCs in Solids*, October 1998, http://clu-in.org/download/stats/sampling.pdf; and
- *Field Analytical and Site Characterization Technologies*, EPA/542/R-97/011, November 1997, http://www.epa.gov/swerust1/cat/fasc.pdf.

18.5.2.6 Sample holding times

Sample holding times are maximum time periods specified in various analytical test methods; if a sample is stored before analysis for longer than the specified holding times, the subsequent analytical results may be suspect due to constituent volatilization or degradation while awaiting testing.

Case Study 18-7: Using Composite Sampling to Locate a Hot Spot

A facility has 20 drums of lead-containing smelter slag that is entirely solid phase. The waste drums are placed on pallets—4 drums per pallet for a total of 5 pallets of waste—and the facility must evaluate the waste as a possible D008 hazardous waste. Personnel take an individual sample from each of the 20 drums, but then composite the 4 samples from each pallet of drums, resulting in 5 composite samples.

To minimize analytical costs, the facility first performs a total lead analysis on each composited sample, resulting in the following: 6, 9, 18, 20, and 45 mg/kg total lead. From Section 2.6.3.1 of this book, if we multiply the regulatory threshold for lead (5.0 mg/L via the TCLP) by 20, the total lead regulatory threshold is 100 mg/kg.

If the lead concentration in a composite sample is less than RT/n (i.e., 100/4 = 25 mg/kg lead in this example), no individual sample concentration comprising the composite can have a lead concentration greater than the regulatory threshold. That is the case for each of the pallets of drums except for the fifth one. For the composite sample that contained 45 mg/kg total lead, one or more of the individual samples may be over 100 mg/kg.

Using the mathematical fact that no more than $(n)(x_i)/RT$ individual sample concentrations can be greater than the regulatory threshold (and remembering that we round *down* when performing this calculation), $(4 \times 45)/100 = 1.8$ rounded down equals 1 individual sample *possibly* exceeds the 100-mg/kg threshold. However, we do not know which (if any) of the 4 drums on the fifth pallet exceeds this value; so, we have to analyze a split sample (or new sample) from each of the 4 drums to determine which, if any, of the 4 drums exceeds 100 mg/kg total lead. If we find such a "hot" drum, we can either declare it to be D008 or have a TCLP test performed on a representative sample to find out if lead truly leaches out at greater than 5.0 mg/L. [EPA/530/D-02/002]

Holding times is a topic that has been discussed at length in various waste analysis plan guidance documents (such as those listed in Table 18-2). Guidance for specific applications is given below.

EPA states that holding time for a given sample begins at the time the sample is taken, and ends at the completion of its analysis. [RO 11306, 13589] For example, if one is to analyze a sample of ground water for semivolatile organics using Methods 3510 and 8270, the water must be extracted within 7 days from the time the sample was taken (the holding time specified in Method 3510) and then the organic extract analyzed within 40 days from the date that the water was extracted (the holding time specified for Method 8270 samples).

As long as the holding time for each sequential step in a determination is not exceeded, the holding time criteria are not exceeded, and the analytical results should be valid. Conversely, "[r]esults of samples not analyzed within the specified holding time will be considered minimum values. That is, the actual concentration will be assumed for regulatory purposes to be equal to or greater than the concentration determined after the holding time has expired." [February 8, 1990; 55 *FR* 4443]

When conducting a TCLP analysis, the holding times in Method 1311 are dependent on the type of constituent being evaluated. For volatiles (e.g., benzene), the holding time is 14 days from sample collection to completion of TCLP extraction and another 14 days from TCLP extraction to completion of extract analysis. Holding times for metal constituents (except for mercury) are 180 days from sample collection to completion of TCLP extraction and another 180 days from TCLP extraction to completion of extract analysis.

However, if any of the specified TCLP holding times are exceeded, the analytical results will be considered minimum concentrations. That is, if a sample exceeds a holding time and analysis demonstrates that concentrations are above the regulatory threshold for one or more constituents, then these concentrations can be treated as minimum values and the waste is hazardous for the toxicity characteristic—no further testing is required. If, on the other hand, a sample exceeds a specified holding time and analysis demonstrates that concentrations are below the regulatory threshold for one or more constituents, further testing is necessary to demonstrate that the waste is nonhazardous. [RO 13612]

Sometimes, data outliers result from analysis of samples whose holding times were exceeded. Therefore, if the holding time of a sample has expired, the facility should generally resample before analysis. [EPA/530/R-94/019]

18.5.2.7 Minimizing sampling error

A well-written sampling plan will address the variability and bias in the sampling and analysis steps. Error can be introduced primarily in two areas: 1) field sample collection and handling, and 2) laboratory sample handling and analysis. Normally, errors in the field during sample collection are much greater than laboratory error. The following resources provide good recommendations for identifying and minimizing sampling errors:

- Chapter 6 of EPA/530/D-02/002 (available online at http://www.epa.gov/epawaste/hazard/testmethods/sw846/samp_guid.htm);

- *A Rationale for the Assessment of Errors in the Sampling of Soils*, EPA/600/4-90/013, May 1990 (available from http://clu-in.org/download/stats/rationale.pdf);

- *Preparation of Soil Sampling Protocols: Sampling Techniques and Strategies*, EPA/600/R-92/128, July 1992 (available from http://www.epa.gov/swerust1/cat/mason.pdf);

- *Superfund Program Representative Sampling Guidance, Volume 4: Waste*, OSWER Directive 9360.4-14, EPA/540/R-95/141, December 1995 (available at http://www.ert.org/products/SF-WASTE.pdf); and

- *Correct Sampling Using the Theories of Pierre Gy*, March 1999 (available from http://www.epa.gov/nerlesd1/factsheets/csutpg.pdf).

Although typically of less concern, analytical error in the laboratory can also result in inaccurate data. When contracting with an analytical laboratory, a number of areas should be evaluated and specified, particularly homogenization, particle-size reduction, and subsampling that minimizes compositional and distributional heterogeneity. A number of improper laboratory practices to watch out for are listed in Section 4.2 of EPA QA/G-8 (available from http://www.epa.gov/quality/qa_docs.html).

Proper laboratory subsampling is very important to produce high-quality data. EPA appears to be concerned about this issue and is beginning to emphasize this component of sampling as evidenced by the discussion of subsampling concerns in an October 2006 revision to SW–846 Method 8330B.

18.5.2.8 Choosing a sampling strategy

The sampling strategy selected will depend on the DQO process. Based on the pros and cons of the different sampling strategies discussed in the preceding sections of this book, EPA has identified the most appropriate sampling strategy for a number of different situations, as summarized in Table 18-7.

Whichever sampling strategy you choose, we strongly recommend that you consult with your state or EPA region before you embark on a large-scale sampling project. Using the DQO process, discussed in Section 18.5.2, develop a sampling and analysis plan stating 1) how you will sample the waste, 2) how you will have the samples analyzed, 3) how you will evaluate the analytical data, and 4) what actions will be taken based on the results of the evaluation. Not only is this what EPA recommends (see November 20, 1997; 62 *FR* 62082, October 30, 2002; 67 *FR* 66257, RO 11624, 11829, 13406), but our experience indicates that, if your regulators review and approve your sampling and analysis plan, they will be

Table 18-7: Choosing the Appropriate Sampling Strategy

Primary objective	Secondary objective	Constraints	Appropriate sampling strategy
Obtain representative sample from continuous or batch process waste	Obtain samples that represent temporal variability	Adequate budget	Temporal Multi Increment sampling
Estimate contaminant mean concentration	Produce information on spatial or temporal patterns	Adequate budget	Systematic sampling
Estimate contaminant mean concentration	Produce an equally precise estimate of the mean with fewer analyses	Budget constraints	Multi Increment sampling
Estimate contaminant mean concentration or proportion/percentile	Increase the precision of the estimate with the same number of samples or achieve the same precision with fewer samples	Spatial or temporal information on contaminant patterns is available	Stratified sampling
Identify hot spots	Classify all units/containers at reduced cost	Need split samples for possible retesting	Systematic or Multi Increment sampling
Identify when/where contamination is present	Identify hot spots	Adequate budget	Systematic sampling
Perform screening phase of a small-scale investigation	Assess whether further sampling is warranted	Budget/schedule constraints	Judgmental sampling

Source: Adapted from EPA QA/G-5S, EPA/530/R-93/007, and EPA/230/2-89/042.

more willing to go along with the conclusions (e.g., hazardous vs. nonhazardous) that you draw from the data you obtain.

18.5.2.8.1 Preliminary information

Preliminary information may improve the effectiveness of any sampling strategy. A preliminary sampling investigation is important if the objective is to fully characterize a waste stream using a probability-based sampling strategy. The mean and variance calculated from preliminary data can provide a starting point for calculation of the number of samples required in the full-scale evaluation. According to Chapter 9 of SW–846 and EPA/530/D-02/002, the preliminary data can come from:

1. A pilot study during which a relatively small number of samples (e.g., four or five) provide a suitable preliminary estimate of the contaminant mean and variance,
2. Process engineering data, or
3. Data from other waste streams or similar waste streams from other facilities.

18.5.3 Statistical analyses

So, you developed a sampling plan, and you sent a number of representative samples to the lab for analysis. When you get the results back from the lab, it may seem obvious from those results that your waste is hazardous, or that it's nonhazardous. Sometimes, however, it won't be obvious. For example, what if some of the sample results indicate the waste is hazardous and other results indicate it's not hazardous? Chapter 9 in SW–846 implies that, in these situations, generators should use statistics to determine if their wastes are hazardous. Here are some quotes from EPA guidance on this issue:

"[A] statistical evaluation is used to determine if a waste is hazardous…." [May 26, 1998; 63 *FR* 28567]

"A statistical approach can characterize concentrations of constituents in…wastes generated onsite. It is appropriate for 'consistent' feed streams (for example, hazardous waste generated by a specific onsite production process…), for which there is reasonable expectation that each constituent will be normally distributed about a mean…. It should also be understood that, with the use of a statistical approach, there is a finite probability that a facility can be found to be out of compliance based on sampling and analysis. If such a circumstance occurs, use of a statistical sampling and analysis strategy is not a shield against enforcement action and the adequacy of the analysis may be considered in penalty calculations." [EPA/530/R-94/019]

EPA has developed numerous guidance documents for data analysis, primary among them being Chapter 9 of SW–846, EPA/530/D-02/002, EPA QA/G-8, EPA QA/G-9R, and EPA QA/G-9S, with document availabilities as given in Table 18-2.

Let's start with a simple example: we're trying to figure out if a waste stream from one of our manufacturing processes is hazardous. We use knowledge to determine that the waste is not listed and that it doesn't exhibit the characteristics of ignitability, corrosivity, or reactivity. Regarding the characteristic of toxicity, our process knowledge indicates that it could only be toxic for benzene. We ask a technician to go out each day for four days and get a sample of the waste. These four samples are sent to the lab for TCLP analysis, and the lab sends us a letter back reporting benzene concentrations in the four samples of 0.1, 0.1, 0.3, and 0.7 mg/L via the TCLP (the characteristic level for benzene is 0.5 mg/L). Is our process waste hazardous or not?

Our first thought is to somehow throw out that 0.7 value ("it must be bad data, or caused by laboratory error, or it's an outlier"). But, do we really have any good reason to believe the 0.7-mg/L value isn't just as valid as any of the other results? Probably not.

If we have to include the 0.7-mg/L value, let's average the results. After all, the definition of "representative sample" in the regs tells us to determine "the average properties of the universe or whole." The average of those four data points is 0.3 mg/L. That's less than the regulatory level, so can we conclude that our process waste is nonhazardous? Maybe, but there's more to it than just simple averaging.

As discussed in the following subsections, when making hazardous waste determinations, EPA's guidance tells us that we must not only find the average properties of the waste samples (e.g., the average concentration of contaminants such as 0.3 mg/L benzene in the preceding example), but we also have to determine the statistical level of confidence that we have in that average value.

18.5.3.1 An 80% confidence interval of the mean

One approach for determining how much confidence we have in an average contaminant concentration based on analysis of samples is to calculate a range of concentrations that brackets the true mean contaminant concentration with a known level of confidence. When we send several samples to a lab and get the analytical results back, we can easily calculate a sample mean concentration ($\bar{x}$) and standard deviation (s, a measure of the extent to which individual sample concentrations are dispersed around the mean) for each contaminant of concern. However, it is not the variation of individual sample concentrations that is of interest. Rather, it is the variation that characterizes the sample mean itself. That measure of dispersion is called the standard deviation of the mean (also the standard error of the mean or standard error) and is designated as $s_{\bar{x}}$. These two values, $\bar{x}$ and $s_{\bar{x}}$, are used to estimate the concentration range (called an interval) within which the true mean concentration (μ) of the contaminant probably occurs, under the assumption that the individual concentrations exhibit a normal (bell-shaped) distribution.

To calculate the standard error of the mean, sample variance (s^2) and standard deviation (s) are required. These parameters are easily calculated from the analytical data and give us a measure of the scatter or dispersion of the data around the mean.

For the purposes of evaluating solid wastes, EPA prescribes in Chapter 9 of SW–846 that, for each contaminant of concern, a confidence interval (CI) be determined within which there is an 80% probability that μ occurs. An 80% confidence interval is just a range of concentrations that gives us an 80% probability that μ lies within that range; thus, there is a 20% chance that μ lies outside that range. Figure 18-5 is a representation of the 80% CI for our benzene example.

Once that CI is calculated based on the analytical results, the upper limit of the 80% CI is compared with the appropriate regulatory threshold. If the upper limit is less than the regulatory threshold, the contaminant is not considered to be present in the waste at a hazardous level; otherwise, the opposite conclusion is drawn.

Looking again at our 80% CI in Figure 18-5, because the upper limit of the 80% CI is greater than 0.5 mg/L, we would conclude, tentatively at least, that our process waste is hazardous for benzene. That is, although the average value for benzene calculated from our four samples (0.3 mg/L) is less than the regulatory level (0.5 mg/L), the scatter of the data is so large that we cannot conclude with 80% confidence that the true mean benzene concentration in the waste is less than 0.5 mg/L.

Even if the upper limit of the 80% CI is only slightly less than the regulatory threshold (i.e., the solid waste is just barely deemed nonhazardous), there is only a 10% (not 20%) chance that the true mean value of the contaminant is greater than the regulatory threshold. This is because values of a normally distributed contaminant that are outside the limits of an 80% CI are equally distributed between the lower and upper portions of the concentration range. The two-sided 80% CI is thus equivalent to a one-sided 90% CI. "Consequently, the CI employed to evaluate solid wastes is, for all practical purposes, a 90% interval." [Chapter 9 of SW–846]

In the February 8, 1990 *Federal Register* [55 FR 4443], EPA discussed this statistical approach (i.e., a 90% probability that the true mean is below the appropriate regulatory threshold) as follows:

> "The concept is that while each individual sample or even composite may not be representative of the [waste], the sum total of samples taken will represent the average property of the waste even though the entire waste stream was not sampled."

The basic equations and step-by-step procedures for calculating 80% CIs and determining if contaminants in a solid waste are present above characteristic levels are presented in Chapter 9 of SW–846. Spreadsheets (e.g., in Excel) can be developed to calculate 80% CIs; this practice is common in industry. This same approach (calculation of an 80% CI of the mean) is required in EPA/530/R-95/036.

In Case Study 18-4, a simple random sampling strategy was used to collect nine sludge samples from the lagoon and each sample was analyzed.

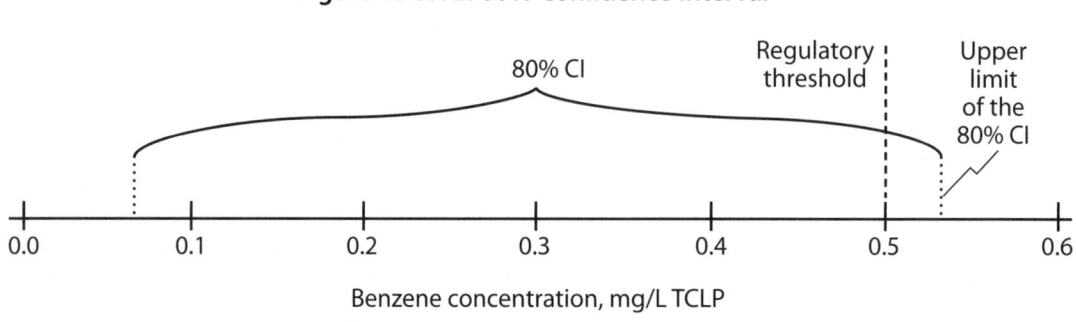

Figure 18-5: An 80% Confidence Interval

Source: McCoy and Associates, Inc.; adapted from Chapter 9 of SW–846.

Case Study 18-8 continues this example and illustrates the calculation of an 80% CI based on these samples.

As an alternative to the approach of analyzing multiple grab samples as noted in Case Study 18-8, those samples could have been combined into one Multi Increment sample (either in the field or at the lab), and the average properties of the sludge could have been determined by analyzing that one sample. There are significant advantages to this alternative approach. First, analytical costs are much lower. Second, the sludge can be more completely characterized if more than nine samples are taken to form the Multi Increment sample (note the large areas that were not sampled in Figure 18-4).

More recent EPA guidance discussing confidence limits does *not* specify use of an 80% CI for making hazardous waste determinations for solid wastes. Instead, the statistical criteria used are chosen based on the DQO process developed during preparation of the sampling plan. [EPA/530/D-02/002, EPA QA/G-9R] In these more recent guidance documents, the statistics EPA uses in its examples are much more stringent and the agency often uses a simple exceedance rule to classify a waste as hazardous; that is, if a single sample results in an exceedance of the corresponding regulatory threshold, the waste is hazardous. Note: if such a simple exceedance rule applies, no statistical test is capable of proving that a waste is nonhazardous, because without testing every possible sample that might ever be taken, it is not possible to be completely sure that no sample can exceed the regulatory limit.

18.5.3.2 Using a proportion or percentile

Here's an example of another statistical approach. Assume it isn't possible to know, based on sampling, what the concentration of a chemical of interest is in every bit of a waste stream that you are evaluating. However, we can estimate, using statistical methods, what proportion or percentage of the waste complies or does not comply with a particular regulatory standard. Such a percentage is usually called a percentile.

Just as we are able to calculate the confidence interval for the mean in Section 18.5.3.1, we can calculate a tolerance interval (TI) for a percentile. If the upper limit of the TI on the percentile is below the regulatory threshold, there is statistical evidence that the specified proportion of the waste or media meets the standard. If the upper limit on the percentile exceeds the standard (but all sample analytical results are below the standard), the waste or media could still be judged in compliance with the standard—but not with the specified degree of confidence.

Similar to calculating confidence limits for a mean, construction of a TI for a proportion or percentile is based on the assumption that the individual concentrations exhibit a normal distribution. Also, EPA guidance suggests that at least 4 samples/analytical results be used to determine the TI. [EPA/530/D-02/002]

For example, we may decide through the DQO process that we want to have a 90% confidence that 95% of the samples of a waste contain a toxic constituent at concentrations below the applicable §261.24 standard. The proportion or percentile is 95% with a 90% TI, as shown in Figure 18-6.

Making a hazardous waste determination using a percentile is exemplified in Case Study 18-9.

18.5.3.3 Nonnormal distributions

Both of the statistical evaluations of analytical data discussed above are based on the assumption that the data fit a normal distribution (i.e., a plot of concentration vs. frequency of measurement results in a bell-shaped curve). Unfortunately, data from waste characterizations are often highly skewed. If this occurs, statistical tools are available to transform the data into a roughly normal scale that is amenable to the above evaluations. Although mentioned in Chapter 9 of SW–846,

Case Study 18-8: Simple Random Sampling of Lagoon Sludge

We use statistical analysis to determine if the barium level in lagoon sludge is present at hazardous levels. This is a continuation of Case Study 18-4.

1. We had previously collected nine samples of the sludge. Six samples of sludge (n_1) are designated for immediate analysis and generate the following concentrations of barium from the TCLP test: 89, 90, 87, 96, 93, and 113 mg/L. Although the value of 113 mg/L appears unusual as compared with the other data, there is no obvious indication that the data are not normally distributed.

2. Values for $\bar{x}$ and s^2 and associated values for the standard deviation (s) and $s_{\bar{x}}$ are calculated using the analytical results as:

$$\bar{x} = \frac{\sum_{i=1}^{n} x_i}{n} = \frac{89 + 90 + \ldots + 113}{6} = 94.67 \text{ mg/L}$$

$$s^2 = \frac{\sum_{i=1}^{n} x_i^2 - \left(\sum_{i=1}^{n} x_i\right)^2 / n}{n-1} = \frac{54{,}224.00 - 53{,}770.67}{5}$$
$$= 90.67$$

$$s = \sqrt{s^2} = 9.52$$

and

$$s_{\bar{x}} = s/\sqrt{n} = 9.52/\sqrt{6} = 3.89$$

3. The value for $\bar{x}$ (94.67 mg/L) is less than the regulatory threshold (RT) of 100 mg/L. In addition, $\bar{x}$ is greater (only slightly) than s^2 (90.67), and as previously indicated, the raw data are not characterized by obvious abnormality. Consequently, the study is continued with the following calculations performed:

$$CI = \bar{x} \pm t_{.20} s_{\bar{x}} = 94.67 \pm (1.476)(3.89)$$
$$= 94.67 \pm 5.74 \text{ mg/L}$$

where the student's "t" value is selected from Table 18-5. Because the upper limit of the CI (100.41 mg/L) is greater than the applicable RT (100 mg/L), we cannot conclude with 90% confidence that barium is not present in the sludge at a hazardous concentration (i.e., we cannot conclude that the sludge is nonhazardous).

4. However, before this conclusion can be confirmed, the number of samples needed to characterize the barium concentration has to be re-estimated using $\bar{x}$ and s^2 from the six samples to make sure we had enough samples to make a statistically valid hazardous waste determination:

$$n_2 = \frac{t^2_{.20} s^2}{(RT - \bar{x})^2} = \frac{(1.476^2)(90.67)}{5.33^2} = 6.95$$

The value for n_2 (we always have to round up—in this case to 7) indicates that, based on the distribution of barium concentrations in the lagoon sludge, an additional ($n_2 - n_1 = 1$) sludge sample from the lagoon needs to be included in our analysis.

5. Fortunately, three extra samples were collected from the lagoon. All extra samples are analyzed, generating the following levels of barium in the TCLP test: 93, 90, and 91 mg/L. Consequently, $\bar{x}$, s^2, s, and $s_{\bar{x}}$ are recalculated using the analytical results from all nine samples as:

$$\bar{x} = \frac{\sum_{i=1}^{n} x_i}{n} = \frac{89 + 90 + \ldots + 91}{9} = 93.56 \text{ mg/L}$$

$$s^2 = \frac{\sum_{i=1}^{n} x_i^2 - \left(\sum_{i=1}^{n} x_i\right)^2 / n}{n-1} = \frac{79{,}254.00 - 78{,}773.78}{8}$$
$$= 60.03$$

$$s = \sqrt{s^2} = 7.75$$

and

$$s_{\bar{x}} = s/\sqrt{n} = 7.75/\sqrt{9} = 2.58$$

The value for $\bar{x}$ (93.56 mg/L) is again less than the RT (100 mg/L), and there is no indication that the nine data points, considered collectively, are abnormally distributed (in particular, $\bar{x}$ is now substantially greater than s^2). Consequently, the 80% CI is determined to be:

$$CI = \bar{x} \pm t_{.20} s_{\bar{x}} = 93.56 \pm (1.397)(2.58)$$
$$= 93.56 \pm 3.60 \text{ mg/L}$$

The upper limit of the CI (97.16 mg/L) is now less than the RT of 100 mg/L. Therefore, it can be "definitively concluded" (EPA's language) that barium is not present in the sludge at a hazardous level. [Chapter 9 of SW–846]

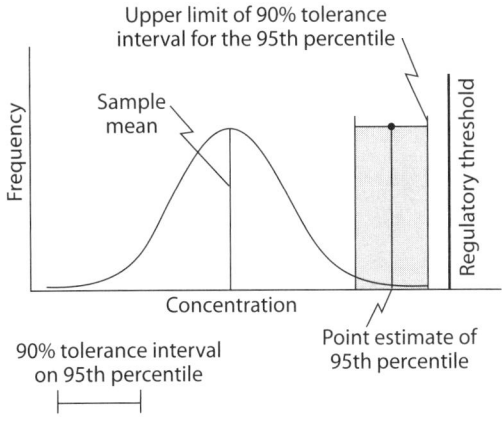

Figure 18-6: A Tolerance Interval for a Percentile

Adapted from EPA/530/D-02/002.

EPA no longer recommends either the arcsine or square-root transformations. [RO 13557]

Data can be checked for normality using graphical methods, such as histograms and normal probability plots, or by using numerical tests. Such numerical tests are described in EPA/530/D-02/002 and EPA QA/G-9S. EPA's most recent guidance recommends the Shapiro-Wilk test for determining normality. This test is appropriate if the number of samples is no more than 50. [EPA/530/D-02/002, RO 13557]

Lognormally distributed data (where a relatively small number of samples include some relatively large analytical values) can be transformed into a normal distribution by taking the logarithms of the data. [EPA QA/G-9S] However, care must be used when such a transformation is conducted. If the results of a test for normality on a log-transformed data set indicate that the original data set has a lognormal distribution, and an upper confidence limit on the mean is to be calculated, then a special approach is typically required. A thorough discussion of lognormal distributions is found in *The Lognormal Distribution in Environmental Applications*, EPA/600/S-97/006, December 1997, available at http://www.epa.gov/osp/hstl/tsc/Singh1997.pdf.

Most environmental data, although often highly skewed, rarely follow a lognormal pattern.

As noted in Section 18.5.2.3, the Multi Increment sampling process is a physical way of averaging out variabilities in concentrations among locations or time intervals. Therefore, concentration data resulting from Multi Increment samples should be more normally distributed than data from individual samples—avoiding the need to transform the data.

18.5.3.4 Handling nondetects

The concentrations of analytes in samples are often below the detection limit of the analytical procedure. These results are generally reported as "nondetects," along with the corresponding detection limit. In fact, the concentration of the chemical of interest lies somewhere between zero and the detection limit. EPA's general recommendations on handling nondetects depend on their frequency of occurrence in the sample analysis results and are summarized in Table 18-9.

Multi Increment sampling, as discussed in Section 18.5.2.3, physically averages out concentration variabilities. Thus, fewer nondetects are expected using this approach, versus discrete grab samples.

18.5.3.5 Outliers

An analytical result that is significantly different from other values in a sample data set is often called an "outlier." There are a number of causes of outliers [EPA/530/D-02/002]:

- Inconsistent sampling,
- Sample contamination,
- Analytical equipment malfunction,
- Inconsistent analytical procedures at the laboratory,
- Laboratory contamination,
- Errors in recording data values (e.g., slipping a decimal point) or codes at the laboratory, and
- From a true (although extreme) concentration of a chemical of concern (e.g., from a hot spot).

Case Study 18-9: Using a Percentile to Characterize a Solid Waste

A secondary lead smelter produces a slag that periodically exhibits the toxicity characteristic for lead. The facility needs to determine if a batch of slag is hazardous or not. The owner decides that he/she needs at least 90% confidence that 95% of the slag will be below the regulatory standard of 5.0 mg/L via the TCLP.

1. Ten samples of the slag are obtained using simple random sampling, and they are tested using the TCLP followed by inductively coupled plasma/atomic emission spectrometry of the extract (SW–846 Method 6010). The ten samples produce the following lead values: 2.5, 2.9, 3.2, 3.5, 3.5, 3.7, 3.9, 4.0, 4.7, and 5.3. Based on a normal probability plot, the data were judged to exhibit a normal distribution. Therefore, we proceed with the calculation on the original (untransformed) data.

2. Values for $\bar{x}$, s^2, and s are calculated using these ten samples as:

$$\bar{x} = \frac{\sum_{i=1}^{n} x_i}{n}$$

$$= \frac{2.5 + 2.9 + 3.2 + 3.5 + 3.5 + 3.7 + 3.9 + 4.0 + 4.7 + 5.3}{10}$$

$$= 3.7 \text{ mg/L}$$

$$s^2 = \frac{\sum_{i=1}^{n} x_i^2 - \left(\sum_{i=1}^{n} x_i\right)^2 / n}{n-1} = \frac{144.48 - 138.38}{9} = 0.677$$

$$s = \sqrt{s^2} = 0.823$$

3. A one-sided upper limit for a TI over a percentile is computed as follows [EPA/530/D-02/002]:

$$UL_{1-\alpha}(x_\rho) = \bar{x} + s \cdot \kappa_{1-\alpha,\rho}$$

Where $\kappa_{1-\alpha,\rho}$ is a statistical constant based on the percentile (ρ) being estimated, the desired confidence level ($1-\alpha$), and the number of samples. The values for $\kappa_{1-\alpha,\rho}$ are taken from Table 18-8; the constant for n = 10 samples needed to construct a 90% upper confidence limit for the 95th percentile is 2.568. Thus, the upper limit on the one-sided 90% TI in this example is:

$$UL_{0.90}(x_{0.95}) = 3.7 + (0.823)(2.568) = 5.8 \text{ mg/L}$$

4. Because the upper limit of the TI is greater than the regulatory threshold for lead of 5.0 mg/L, the owner cannot conclude with 90% confidence that 95% of the slag will be below the regulatory level. Therefore, the owner classifies the slag as hazardous waste. [EPA/530/D-02/002]

Data resulting from the first six causes of outliers noted above are no good and should be removed or corrected before statistical analysis. However, outliers caused by the last reason represent valid data and should not be excluded. For example, unusually high concentrations of a certain constituent may be real but infrequent, such as might be found in lognormally distributed data—it would not be appropriate to exclude such "outliers."

Not removing true outliers and removing false outliers both lead to a distortion of the waste characterization. How do you tell the true outliers from the false? Use statistical outlier tests and/or conduct limited retesting.

Outlier tests can be simple (e.g., constructing a probability plot of the data) or complex (e.g., using software). Statistical outlier tests are described in EPA/530/D-02/002 and EPA QA/G-9S.

EPA recommends the following five-step process for dealing with outliers [EPA QA/G-9S]:

1. Identify extreme values that may be potential outliers,
2. Apply statistical tests,
3. Review statistical outliers and decide on their disposition,
4. Conduct data analyses with and without outliers, and

Table 18-8: Tabulated Values of κ Constant for the 95th Percentile*

Number of samples	κ values			
	1−α = 0.80	1−α = 0.90	1−α = 0.95	1−α = 0.99
2	6.464	13.090	26.26	131.426
3	3.604	5.311	7.656	17.370
4	2.968	3.957	5.144	9.083
5	2.683	3.400	4.203	6.578
6	2.517	3.092	3.708	5.406
7	2.407	2.894	3.399	4.728
8	2.328	2.754	3.187	4.285
9	2.268	2.650	3.031	3.972
10	2.220	2.568	2.911	3.738
11	2.182	2.503	2.815	3.556
12	2.149	2.448	2.736	3.410
13	2.122	2.402	2.671	3.290
14	2.098	2.363	2.614	3.189
15	2.078	2.329	2.566	3.102
16	2.059	2.299	2.524	3.028
17	2.043	2.272	2.486	2.963
18	2.029	2.249	2.453	2.905
19	2.016	2.227	2.423	2.854
20	2.004	2.208	2.396	2.808
21	1.993	2.190	2.371	2.766
22	1.983	2.174	2.349	2.729
23	1.973	2.159	2.328	2.694
24	1.965	2.145	2.309	2.662
25	1.957	2.132	2.292	2.633
30	1.924	2.080	2.220	2.515
40	1.880	2.010	2.125	2.364
60	1.832	1.933	2.022	2.202
100	1.786	1.861	1.927	2.056

*Tabulated κ values are for the 95th percentile, the desired confidence level (1−α), and the number of samples available. κ values for other percentiles are available in EPA/530/D-02/002.

Source: Adapted from EPA/530/D-02/002.

Table 18-9: Guidance for Handling Nondetects*

Percent of data reported as non-detect	Recommended treatment of data
<15	Replace nondetects with detection limit divided by 2
15 to 50	Cohen's adjustment, trimmed mean, regression order statistics, or other treatment discussed in EPA QA/G-9S and EPA/530/D-02/002
>50	Regression order statistics, Helsel's method, or test for proportions, as discussed in EPA QA/G-9S and EPA/530/D-02/002

*EPA notes in the references for this table that "[a]lthough these guidelines are usually adequate, they should be implemented cautiously." [EPA QA/G-9S]

Source: Adapted from EPA QA/G-9S and EPA/530/D-02/002.

5. Document the entire outlier evaluation process.

Note that in Step 3, the tests alone cannot determine whether a statistical outlier should be discarded or corrected; that decision should be based on judgment or scientific grounds. If the data point is found to be an outlier, you may 1) correct the data point, 2) discard it, or 3) use it in analyses. For example, outliers resulting from laboratory errors in recording data values or codes can be corrected, whereas data points from a malfunctioning analytical instrument should be discarded. However, if scientific reasoning cannot explain an outlier, it should not be discarded from the data set—it may represent a legitimate extreme value—unless judgment by involved professionals allows it to be discarded.

For example, in a situation with which we are familiar, several years of semiannual ground water monitoring at a well produced results for a particular constituent that were consistently between 40–50 mg/L. Then one result came back from the lab at 440 mg/L. The lab could find no explanation; its records supported the 440-mg/L result, and the lab stood by its reported result. The high result was dutifully reported to regulatory authorities, with a notation that a decimal error was strongly suspected (e.g., a dilution of 10 times was recorded but not performed in the lab). No regulatory consequences resulted. Subsequently, 10 additional years of monitoring continued to produce results between 40–50 mg/L for the constituent. There are little scientific grounds for discarding the outlying result, but, could best professional judgment of the environmental professionals involved have allowed the result to be discarded?

Outliers caused by an error in sampling can be corrected through immediate resampling and reanalysis of the waste. Outliers caused by errors in analysis often can be corrected through reanalysis of the sample (if possible). [EPA/530/R-95/036]

Remediation and Demolition

Managing cleanup and demolition wastes

Many organizations across the United States are in the process of cleaning up sites that were contaminated before RCRA started in 1980. Although remediation and contaminated soil issues have been touched upon in earlier chapters, Section 19.1 pulls together all of the RCRA factors to be considered when embarking on a site cleanup.

Additionally, the demolition of obsolete equipment/buildings is a significant undertaking at many government and industrial sites. Can demolition debris be hazardous? How do I write a demolition work plan? Could the demolition of a contaminated building be considered hazardous waste treatment? These questions and more are answered in Section 19.2.

19.1 Remediation

Many government and industrial properties include areas of soil contamination. Much of this contamination occurred before 1980, which is when the RCRA regulations went into effect. Additionally, ground water aquifers underlying these sites may also be contaminated. Many of these facilities are now in the process of remediating this contaminated soil and/or ground water, either on a voluntary basis or under a RCRA or CERCLA order. This section looks at the RCRA issues associated with these remediation activities.

The major components of a remediation project are:

- Site investigation,
- Remedial alternatives analysis,
- Decision as to the selected remediation,
- Implementation of the selected remediation, and
- Project closure.

As noted in Table 19-1, the above components may have different names, depending upon the authority (if any) under which the remediation is being conducted.

This section provides a look at two of the remediation project components noted above: 1) site investigation (Section 19.1.1), and 2) the RCRA implications of implementing the selected remediation (Sections 19.1.2–19.1.4). Information on the other three components can be found in Chapter 17 for corrective action projects.

In addition to the RCRA implications of conducting a site remediation, there may be CERCLA and CAA requirements. The interface between RCRA and CERCLA during a site remediation is discussed in

Table 19-1: Major Components of a Remediation Project

Component	RCRA corrective action[1]	Superfund remedial action	Voluntary cleanup
Site investigation	RCRA facility investigation	Remedial investigation	Site investigation
Remedial alternatives analysis	Corrective measures study	Feasibility study	Remedial alternatives analysis
Selected remedy decision	Statement of basis	Record of decision	Selected remedy decision
Selected remedy implementation	Corrective measures implementation	Remedial design/remedial action, operation, and maintenance	Selected remedy implementation
Project closure	Corrective action completion determination	Completion of operation and maintenance and of ground or surface water restoration	Project closure

[1] The components in this column are discussed in detail in Chapter 17.
Source: McCoy and Associates, Inc.

Section 19.1.5. The possibility of triggering the Part 63, Subpart GGGGG MACT standards (under the CAA) during a remediation is summarized in Section 19.1.6.

19.1.1 Site investigation

The site investigation focuses on evaluating whether any contamination exists at a site and, if so, characterizing the nature and extent of that contamination. It may include the use of field instrumentation and/or laboratory analysis of samples of soil, soil gas, ground water, surface water, and sediments.

Early on, EPA recognized the need for site investigators to implement a systematic process to produce consistent and accurate data during an investigation:

"EPA has established policy which states that before information or data are collected on Agency-funded or regulated environmental programs and projects, a systematic planning process must occur during which performance or acceptance criteria are developed for the collection, evaluation, or use of these data." [EPA QA/G-4, available at http://www.epa.gov/quality/qa_docs.html]

To achieve that goal, the agency has developed the Data Quality Objectives (DQO) process. This process focuses on data collection planning when data will be used to determine whether or not site contamination exceeds regulatory exposure thresholds.

19.1.1.1 Systematic planning using the DQO process

The DQO process produces quantitative and/or qualitative statements (called the DQOs) that express the project-specific decision goals. The DQOs then are used to guide the design of sampling and analysis plans that will cost-effectively produce the required data. [EPA/542/R-01/014, October 2001, available at http://nepis.epa.gov/EPA/html/Pubs/pubtitleOSWER.html]

The DQO process is summarized in Section 18.5.2. Details of this process are available in EPA QA/CS-1, EPA QA/G-4, and EPA QA/G-4D, all available at http://www.epa.gov/quality/qa_docs.html.

19.1.1.2 Other methods of site investigation

Other methods of site investigation may also be employed. For example, the U.S. Army Corps of Engineers has established the Technical Project Planning (TPP) process to improve planning activities associated with hazardous, toxic, and radioactive waste site cleanup. The TPP process is a systematic planning process that involves four different phases:

- Phase I: Identify current project,
- Phase II: Determine data needs,
- Phase III: Develop data collection options, and
- Phase IV: Finalize data collection program.

More information on the TPP process can be found at http://www.usace.army.mil/missions/environmental/technicalprojectplanning.aspx.

Another site investigation approach is ASTM E1903, *Standard Practice for Environmental Site Assessments: Phase II Environmental Site Assessment Process* (available from http://www.astm.org). This process is primarily used by the business community to investigate a property before a real estate transaction.

EPA's latest efforts in the site investigation area seem to be centered around the "Triad Approach." As its name implies, this methodology consists of a three-pronged approach:

- Systematic planning,
- Dynamic work strategies, and
- Real-time measurement systems.

More information on the Triad Approach can be found at http://www.triadcentral.org/, and Table 19-2 contains a list of recent EPA guidance documents describing this methodology.

In addition, several states have developed their own processes for site investigation. Three such state processes are:

- New York: *Technical Guidance for Site Investigation and Remediation*, available at http://www.dec.ny.gov/docs/remediation_hudson_pdf/der10.pdf.
- New Jersey: *Technical Guidance for Site Investigation of Soil, Remedial Investigation of Soil, and Remedial Action Verification Sampling for Soil*, available at http://www.state.nj.us/dep/srp/guidance/srra/soil_inv_si_ri_ra.pdf.

Table 19-2: References Describing the Triad Approach to Site Investigation

Reference[1]
Using the Triad Approach to Improve the Cost-Effectiveness of Hazardous Waste Site Cleanups, EPA/542/R-01/016, October 2001
Technical and Regulatory Guidance for the Triad Approach: A New Paradigm for Environmental Project Management, EPA/542/B-04/002, December 2003
Improving Sampling, Analysis, and Data Management for Site Investigation and Cleanup, EPA/542/F-04/001a, April 2004
Case Study of the Triad Approach: Expedited Characterization of Petroleum Constituents and PCBs Using Test Kits and a Mobile Chromatography Laboratory at the Former Cos Cob Power Plant Site, EPA/542/R-04/008, June 2004
Use of Dynamic Work Strategies Under a Triad Approach for Site Assessment and Cleanup, EPA/542/F-05/008, September 2005
Grand Plaza Site Investigation Using the Triad Approach and Evaluation of Vapor Intrusion, EPA/540/R-07/002, September 2006
The Role of a Conceptual Site Model for Expedited Site Characterization Using the Triad Approach at the Poudre River Site, Fort Collins, Colorado, EPA/542/R-06/007, November 2006
Management and Interpretation of Data Under a Triad Approach, EPA/542/F-07/001, May 2007
Considerations for Applying the Triad Approach—Hartford Area Hydrocarbon Plume Site, Hartford, Illinois, EPA/542/R-06/008, August 2007
Demonstrations of Method Applicability Under a Triad Approach for Site Assessment and Cleanup, EPA/542/F-08/006, August 2008
Triad Issue Paper: Using Geophysical Tools to Develop the Conceptual Site Model, EPA/542/F-08/007, December 2008
Best Management Practices: Use of Systematic Project Planning Under a Triad Approach for Site Assessment and Cleanup, EPA/542/F-10/010, September 2010

[1] All of these references are available at http://nepis.epa.gov/EPA/html/Pubs/pubtitleOSWER.html.

Source: McCoy and Associates, Inc.

- New Hampshire: *Site Investigation Report Check List*, available at http://des.nh.gov/organization/divisions/waste/hwrb/sss/hwrp/guidance_documents.htm.

19.1.2 Point of generation for remediation wastes

Where is the point of generation (POG) for remediation wastes such as contaminated soil and ground water? The answer to this question is quite complicated, and it depends on three issues: 1) retroactivity of RCRA, 2) "active management," and 3) movement of wastes within a unit or area of contamination (AOC). These three topics are discussed in the next three subsections.

19.1.2.1 Retroactivity of RCRA

Do hazardous waste characteristics and/or listings apply today to wastes that were legally disposed before the characteristics and/or listings went into effect? If we disposed some waste in the 1970s that now exhibits a characteristic or meets a listing description, is that disposal site a hazardous waste landfill? Do we have to go back today and excavate that waste and manage it as hazardous? In other words, is RCRA retroactive?

According to EPA, hazardous waste characteristics and listings do apply retroactively to wastes disposed prior to the effective date of the characteristics and listings. The agency first stated its position as follows: "once a particular waste is listed, all wastes meeting that description are hazardous wastes *no matter when disposed*." [Emphasis added.] [August 17, 1988; 53 *FR* 31147; see also May 17, 1988; 53 *FR* 17586] As such, wastes that were not hazardous at the time of disposal, but which are subsequently identified or listed as hazardous wastes, become hazardous when the new characteristic or listing goes into effect. For example, spent solvents disposed in 1977 may meet the same listing description as spent solvents disposed in 2014. Having said that, such previously disposed waste *is not subject* to hazardous waste management regulations unless and until it is "actively managed."

19.1.2.2 "Active management"

EPA's retroactive application of new characteristics or listings does not mean that all newly identified or listed wastes must be removed from their historical disposal site for proper treatment. RCRA does not require such retroactive waste management; if no regulatory program (e.g., the RCRA corrective action program—see Chapter 17) is making you clean up the previously disposed waste, RCRA regulations do not apply to it. Nor does EPA impose any retroactive penalties for prior disposal of the waste, since it was legally disposed at that time. The RCRA regulations kick in only when the wastes are actively managed. Consequently, compliance with the RCRA Subtitle C regulations is not required at the disposal site (whether it is still operating or not) unless the previously disposed wastes are actively managed. [March 8, 1990; 55 *FR* 8762–3]

EPA has defined "active management" as "physically disturbing the accumulated wastes within a management unit or disposing additional hazardous wastes into existing waste management units containing previously disposed wastes." [September 1, 1989; 54 *FR* 36597, August 18, 1992; 57 *FR* 37298] The agency noted that "disturbing" a waste includes removing, excavating, mixing with other wastes, or other onsite treatment (including ex situ treatment). [RO 11954, 12995, 13057]

Thus, cleanups involving removal of wastes or contaminated media from a site (e.g., excavating solids or pumping ground water) are deemed to be active management. In these situations, all wastes that exhibit a characteristic or meet a listing description [including environmental media that "contain" listed wastes (see Section 5.3)] must be managed as hazardous wastes, even if they were disposed before the applicable characteristic or listing effective date. Furthermore, any mixtures of those wastes with other solid wastes and wastes derived from the treatment, storage, or disposal of those wastes may be regulated as hazardous via application of the mixture and derived-from rules (see Sections 5.1 and 5.2, respectively). Finally, all such wastes that are hazardous because they are being actively managed are subject to the LDR program. [August 17, 1988; 53 FR 31148]

The second part of the active management definition notes that continued use of an existing unit, after the effective date of an applicable hazardous waste identification or listing, for treatment, storage, or disposal of the newly identified or listed waste (or any other hazardous waste) will subject the unit and its contents to RCRA Subtitle C regulation. Conversely, if only nonhazardous waste is added to a waste management unit in which wastes were previously disposed before they were regulated as hazardous, this activity would not constitute active management. Additional retroactivity and active management examples are explored in Case Study 19-1.

19.1.2.2.1 Exception for plant construction activities

During excavation of soil as part of a plant construction activity, contaminated soil is sometimes unearthed. If the soil exhibits a characteristic or is contaminated with a listed hazardous waste, the soil is a hazardous waste when excavated. This has caused considerable concern since, in many cases (e.g., when a trench is being dug to lay a pipeline or conduit), the dirt is simply piled up next to the trench before it is put back in the hole. Does that make the hazardous soil subject to

---

Case Study 19-1: RCRA Retroactivity and Active Management

I. Since a manufacturing facility ceased operations in the 1960s, its onsite landfill has been inactive. In the 1980s and 1990s, several wastes that were previously disposed in the landfill were listed as RCRA hazardous wastes. Does the facility owner have to remove the wastes?

No. According to EPA, the wastes in the landfill are considered hazardous wastes even though they were disposed before they were subject to RCRA regulation. However, if the wastes are not actively managed, the RCRA Subtitle C requirements do not apply. If active management occurs (e.g., removal of the wastes for disposal), the hazardous waste regulations would be applicable. [May 17, 1988; 53 FR 17586]

II. Prior to 1980, a facility sandblasted structures that were painted with a lead primer. The sandblast residue was allowed to simply fall to the ground. As a result, several areas of the plant are likely to contain lead-contaminated soil. When, if ever, is the soil considered a hazardous waste (assuming that it would fail the TCLP)?

The soil becomes a hazardous waste if it is actively managed (e.g., excavated). The RCRA corrective action program may ultimately require that the soil be cleaned up, but contaminated soil that is left in place is not subject to any hazardous waste management requirements. Additionally, covering such toxic soils with sod, mulch, or gravel would not constitute generation of a hazardous waste, and such measures would not trigger Subtitle C obligations. [RO 11436, 11898]

III. Leaks from a storage tank allowed unused benzene to seep into ground water during the 1970s. Today, the ground water is being pumped and treated in a surface impoundment. Is the surface impoundment receiving hazardous waste?

Yes. Pumping the ground water constitutes active management. Since the ground water contains discarded commercial chemical product benzene (U019), it must be managed as if it is U019. Therefore, the surface impoundment is a RCRA-regulated hazardous waste unit. "When a site is 'reactivated,' (cleaned up, waste removed, etc.), the facility must comply with the hazardous waste regulations." [RO 12090]

generator standards, which would also make the piles of dirt Subtitle C waste piles subject to Part 264/265, Subpart L standards?

EPA has addressed this issue in guidance by stating that, as long as the hazardous dirt remains in the area of contamination (i.e., the contaminated soil is not moved to an uncontaminated area of the plant or shipped offsite) and is subsequently put back in the ground from which it was excavated, such operations do not produce a hazardous waste or subject it to hazardous waste regulation. Therefore, it would not have to be counted, the piles are not regulated waste piles, and the dirt is not subject to land disposal standards. [RO 11671] Although this guidance did not specifically say that the hazardous soil was never actively managed, the agency noted in EPA/530/K-05/011 (available at http://www.epa.gov/wastes/inforesources/pubs/training/gen05.pdf) that "excavation of contaminated soil during routine construction operations, such as pipeline installation, may not be considered active management if the soil is redeposited into the same excavated area. Site-specific situations should be discussed with the implementing agency." Thus, EPA has decided to give facilities conducting construction operations relief from managing contaminated soil as hazardous if the soil is returned to the trench. Note that states may not include this federal guidance as part of their state program, so use this with caution.

19.1.2.2.2 Leachate derived from previously disposed wastes

The derived-from rule states that residues derived from the treatment, storage, or disposal of a listed hazardous waste remain listed. [§261.3(c)(2)(i)—see Section 5.2.2] Based on this rule, hazardous waste listings apply to leachate derived from the disposal of listed hazardous wastes. This holds true for leachate derived from wastes disposed *before* an applicable listing effective date, even if the landfill ceased disposal of the waste before it became hazardous. If the leachate is actively managed, it is subject to hazardous waste regulation. [May 17, 1988; 53 *FR* 17586] The point of generation of the hazardous waste is when the leachate is first collected or otherwise actively managed. [August 6, 1998; 63 *FR* 42191] This interpretation was upheld by the U.S. Court of Appeals for the District of Columbia in 1989 (*Chemical Waste Management, Inc. v. EPA*, 869 F.2d 1526).

Although leachate resulting from the previous disposal of wastes that today meet a listing description is hazardous when actively managed, EPA noted that such active management of the leachate does *not*, by itself, subject landfills holding such wastes to RCRA hazardous waste regulation. Collection of hazardous leachate from otherwise inactive units does not activate the unit in terms of Subtitle C management. [August 17, 1988; 53 *FR* 31149, August 6, 1998; 63 *FR* 42191]

In many situations, active management of leachate is exempt from RCRA Subtitle C regulation because the leachate is managed under the CWA. The leachate is either discharged to a sewer line running to a publicly owned treatment works (POTW) or to a navigable water of the United States under an NPDES permit. [Such wastes are excluded from RCRA regulation at the point of discharge under §§261.4(a)(1) and 261.4(a)(2), respectively.] Management of leachate in wastewater treatment tanks prior to discharge under the CWA is also exempt from RCRA regulation under the wastewater treatment unit exemption of §§264.1(g)(6) and 265.1(c)(10)—see Section 7.2.

19.1.2.2.3 Wastes in surface impoundments

Wastes (particularly sludges) can remain in surface impoundments for a long time. Similar to the situation for landfills discussed above, wastes that were disposed in the impoundments before they were listed or identified as hazardous wastes are hazardous today, but they do not become subject to RCRA regulation unless the wastes are actively managed. The regulatory status of the wastes and impoundments depends on several factors, as discussed in EPA preamble language and correspondence. [September 27, 1990; 55 FR 39410, November 2, 1990; 55 FR 46383, RO 11826, 13510] Five different scenarios are envisioned for wastes that were deposited or generated in an impoundment *prior* to the effective date of the listing or identification:

1. If the wastes are removed from the unit before the effective date, the wastes and the impoundment are not subject to the hazardous waste regulations (as long as the unit does not receive or generate any hazardous waste after the effective date).

2. If the wastes remain in the unit (which is considered the final disposal site for the wastes) after the effective date, and the unit does not receive or generate any hazardous waste after the effective date, the hazardous wastes in the impoundment are not being actively managed. Therefore, neither the wastes nor the impoundment become subject to RCRA Subtitle C requirements.

3. If the wastes remain in the unit (which is *not* considered the final disposal site for the wastes) after the effective date, and the unit does not receive or generate any hazardous waste after the effective date, the hazardous wastes in the impoundment are being stored (i.e., actively managed). Therefore, the wastes and the impoundment become subject to RCRA Subtitle C requirements on the effective date. This scenario is based on the facility removing some or all of the waste from the unit on a *periodic* basis on or after the effective date.

4. If the wastes are removed from the unit after the effective date in a one-time removal as part of a closure, the wastes are subject to hazardous waste management requirements (i.e., they are being actively managed), but the impoundment is not a Subtitle C unit (as long as the unit does not receive or generate any hazardous waste after the effective date). EPA does not consider one-time removal of waste from a unit on or after the effective date, in and of itself, to make the impoundment a storage unit subject to Subtitle C.

5. If the wastes are scoured from the unit after the effective date (due to nonhazardous wastewater influent) and the unit's effluent is therefore listed or exhibits a characteristic, the unit generating the hazardous wastewater and any surface impoundment receiving that hazardous effluent would be subject to Subtitle C requirements.

19.1.2.2.4 *Corrective action and CERCLA provisions still apply*

As mentioned above, if units contain characteristic or listed wastes but are not being used for active waste management after the identification or listing effective date, they would not be subject to regulation under Part 264 or 265. However, inactive units that are located at facilities otherwise subject to Subtitle C interim status or permitting requirements are solid waste management units (SWMUs) subject to corrective active requirements under RCRA Section 3004(u) or 3008(h). In addition, the CERCLA cleanup authorities may also apply. [September 27, 1990; 55 FR 39410]

19.1.2.3 Movement of wastes within a unit or AOC

With the advent of the LDR program, the POG of remediation wastes became more complicated. When wastes are moved within a RCRA unit (e.g., a landfill), does this constitute waste generation, subjecting the moved waste to LDR treatment standards? EPA's position on this issue was explained on March 8, 1990 [55 FR 8759]: "EPA believes…that movement of waste within a unit does not constitute 'land disposal' for purposes of application of the RCRA LDRs." Although not specifically stated, the implication here is that moving waste within a unit is not considered to be an activity that generates a waste. However, moving a waste outside a unit and placing it in another unit can constitute waste generation: "EPA has consistently interpreted the [term] 'placement'…to mean the placement of hazardous wastes into [a land disposal unit], not the movement of waste within a unit." [55 FR 8759]

In remediation settings, EPA considers broad areas of generally dispersed contamination to be essentially the same as a RCRA unit (e.g., a landfill). Under the area of contamination (AOC) policy, wastes can be moved within the AOC without triggering LDR requirements or generating new wastes. (See Section 19.1.4.1.1.) When wastes are removed from the AOC, waste generation has occurred, and if the waste is placed in a land disposal unit, the LDR requirements will apply. [RO 14291]

19.1.2.4 Summary of POG for remediation wastes

In summary, the POG of a remediation waste is when it is actively managed, unless the waste is being moved within a unit or AOC. In this latter case, the POG would be when the waste leaves the unit or AOC.

The following examples illustrate how EPA has interpreted POG issues for remediation wastes.

Q *Soil that was contaminated by a previous site owner is being cleaned up by the current owner/operator. What is the POG of the soil?*

A In remediation activities, the POG is the point when/where the contaminated soil is excavated. [May 26, 1998; 63 FR 28617, RO 13748, 14283]

Q *Does simply moving contaminated soil constitute waste generation?*

A No. "In what is typically referred to as the area of contamination (AOC) policy, EPA interprets RCRA to allow certain discrete areas of generally dispersed contamination to be considered to be RCRA units (usually landfills). Because an AOC is equated to a RCRA land-based unit, consolidation and in situ treatment of hazardous waste within the AOC do not create a new point of hazardous waste generation for purposes of RCRA…. The AOC interpretation may be applied to any hazardous remediation waste (including nonmedia waste) that is in or on the land." [RO 14291] "Contaminated soil may be consolidated within an area of contamination before it is removed from the land (i.e., generated); the determination as to whether the soil exhibits a characteristic of hazardous waste or contains listed hazardous waste may be made after such consolidation." [RO 14283; see also RO 14338]

19.1.3 Are remediation wastes hazardous?

Once a facility knows that remediation wastes are generated (i.e., they are solid wastes) via application of EPA's guidance discussed in Section 19.1.2, a hazardous waste determination for those wastes must be made. Figure 19-1 is a logic diagram for determining the regulatory status of remediation wastes, and the evaluation steps in this figure are further discussed in the subsections that follow. Basically, this figure walks the reader through the four questions that must be evaluated when making a hazardous waste determination. Reference Sections 2.1 and 18.1 for more information on the four questions.

Once it is known that remediation waste is a solid waste, the second question to ask [from §262.11(a)] is whether the material is excluded or exempt from the RCRA regulations under §261.4. Most of these exclusions and exemptions are discussed in Chapter 4 (see specifically Table 4-1), and they are largely based on application of knowledge enhanced by EPA guidance. Some in the regulated community are surprised that there are very few exclusions and exemptions for remediation waste. Probably the most useful remediation waste exclusion is for petroleum-contaminated media (soil or ground water) generated during cleanup of petroleum underground storage tanks. [§261.4(b)(10)]

If generated and not exempt, a determination of whether the remediation waste meets a listing description or exhibits a characteristic will have to be made. Although it is fairly straightforward to determine if remediation wastes exhibit a characteristic,

Figure 19-1: RCRA Regulatory Status of Remediation Waste

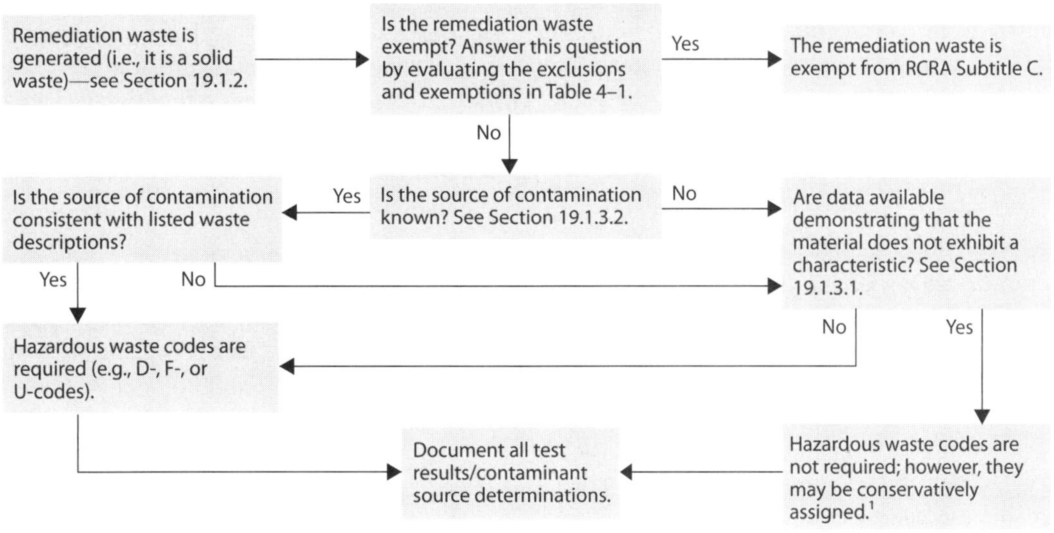

```
Remediation waste is          Is the remediation waste
generated (i.e., it is a      exempt? Answer this question   Yes    The remediation waste is
solid waste)—see              by evaluating the exclusions  ────►   exempt from RCRA Subtitle C.
Section 19.1.2.        ────►  and exemptions in Table 4–1.
                                         │ No
                                         ▼
Is the source of contamination  Yes   Is the source of contamination   No   Are data available
consistent with listed waste   ◄────  known? See Section 19.1.3.2.   ────►  demonstrating that the
descriptions?                                                                material does not exhibit a
  Yes │    No                                                                characteristic? See Section
       │    └──────────────────────────────────────────────────────►         19.1.3.1.
       ▼                                                                        No │    Yes
Hazardous waste codes are                                                          │      │
required (e.g., D-, F-, or  ◄──────────────────────────────────────────────────────┘      │
U-codes).                                                                                 ▼
       │                         Document all test                           Hazardous waste codes are
       └──────────────────────►  results/contaminant           ◄──────────   not required; however, they
                                 source determinations.                      may be conservatively
                                                                             assigned.¹
```

¹ For example, if contamination includes solvent chemicals, F001–F005 codes may be conservatively assigned even if source is unknown. Discussions with site or corporate legal counsel should be held during this step.

Source: McCoy and Associates, Inc.

---

this evaluation is discussed in Section 19.1.3.1. The more complicated issue of whether remediation wastes meet a listing description is tackled in Section 19.1.3.2.

19.1.3.1 Are remediation wastes characteristic?

Characteristic determinations for remediation wastes are not particularly difficult. Knowledge can usually be used to determine that remediation waste is not ignitable, corrosive, or reactive. An exception is for explosives-contaminated soil/sediment at a DOD facility; if the soil/sediment contains >10 percent explosives, it may propagate a detonation or explode when heated under confinement and so may carry the D003 code, as discussed in Section 2.4.3.2.

Here is what EPA has said regarding the use of knowledge to determine if remediation waste is characteristically toxic:

"EPA wants to make clear, however, that a decision that a waste is not characteristic in the absence of testing may not be arbitrary, but must be based on site-specific information and data collected on the constituents and their concentrations during investigations of the site. Based on site data, it will be very clear in some cases that a waste cannot be characteristic; for example, if a waste does not contain a constituent regulated as...toxic, a decision that the waste does not exhibit this characteristic can reliably be made without testing for...toxicity. EPA does not expect to undertake testing when it can otherwise be determined with reasonable certainty whether or not the waste will exhibit a characteristic." [March 8, 1990; 55 *FR* 8762]

Based on this language, if the facility has documentation showing that certain §261.24 constituents could not be present in a specific remediation waste, the TCLP will have to be performed only for the other toxicity constituents that cannot be eliminated through knowledge. [EPA/530/R-93/007, RO 11603, 14695] As an alternative to running the TCLP (which is quite expensive), some facilities may ask the lab to run totals analysis for those constituents. Based on the results, the TCLP may then be requested only for those constituents with the potential to exceed regulatory levels. See Section 2.6.3.1 for a discussion of this option.

Additional discussion on making a determination as to whether a specific remediation waste exhibits a hazardous characteristic is given in Sections 18.2.2 and 18.3.3.

19.1.3.2 Are remediation wastes listed?

Hazardous waste listings are based on the source of, or the process that generated, the waste rather than the concentrations of hazardous constituents in the waste; therefore, analytical testing alone, without information on a waste's source, will not produce information that will conclusively indicate whether a given waste meets a listing description. [RO 13181]

In some situations, it is hard for facility owners/operators to determine the source of the waste or contamination, and so it is difficult to tell if a waste is listed. If the owner/operator has made a good-faith effort to find out if a waste is a listed waste but cannot make such a determination because documentation on the source or process is unavailable or inconclusive, EPA allows the owner/operator to assume that the waste is not a listed hazardous waste. (See, for example, December 21, 1988; 53 *FR* 51444, March 8, 1990; 55 *FR* 8758, and April 29, 1996; 61 *FR* 18805.) Consequently, if the waste does not exhibit a characteristic, the RCRA hazardous waste regulations do not apply. This has been EPA's long-standing policy, which was reaffirmed in RO 14291.

This policy also applies to contaminated soil. According to EPA, the mere presence of contaminants in soil does not automatically make the soil a hazardous waste. The origin of the contaminants must be known in order for the soil to require management as a listed hazardous waste via the contained-in policy. "If the exact origin of the [contaminants] is not known, the soils cannot be considered RCRA hazardous wastes unless they exhibit one or more of the characteristics of hazardous waste...." [RO 12171; see also RO 12392, 14291]

Q: *Soil is found to be contaminated with 1,1,1-trichloroethane. The facility owner/operator is unable to determine the source of the contamination after making a good-faith effort. If the soil is excavated and shipped offsite for disposal, how is it regulated under RCRA?*

A: If the material contaminating the soil was a discarded commercial chemical product, manufacturing chemical intermediate, or off-specification product, the soil would be classified as a U226 waste via application of the contained-in policy. Conversely, if the 1,1,1-trichloroethane was used for its solvent properties before the spill occurred, the soil would be classified as F001 or F002. However, the facility owner/operator cannot determine the source of the contamination after making a good-faith effort; therefore, the only way the soil can be hazardous is if it exhibits a characteristic. The soil is not ignitable, corrosive, or reactive, and 1,1,1-trichloroethane is not a toxicity characteristic constituent. Therefore, the soil is not a hazardous waste under the federal regulations. [RO 12171]

What constitutes a "good-faith effort" to determine the source of the contamination? "The agency believes that by using available site- and waste-specific information such as manifests, vouchers, bills of lading, sales and inventory records, storage records, sampling and analysis reports, accident reports, site investigation reports, spill reports, inspection reports and logs, and enforcement orders and permits, facility owner/operators would typically be able to make these determinations. However...if information is not available or inconclusive, facility owner/operators may generally assume that the material[s] contaminating the media were not hazardous wastes." [April 29, 1996; 61 *FR* 18805]

Even though EPA's guidance summarized above suggests not putting any listed waste codes on remediation waste unless the contaminant source is known, facility personnel need to be very careful about this. We know of enforcement actions where site personnel knew of solvent chemical contamination in soil and ground water but did not assign listed codes during remediation because they could not pinpoint the exact source of contamination (e.g., whether the contamination was caused from spent solvents, new solvents, or nonsolvent chemical use such as freeze protection). Federal enforcement personnel claimed that since certain F-listed solvents were known to have been used extensively at the site for cleaning and degreasing, site personnel should have known that some or most of the contamination came from spent solvent usage, necessitating F-codes on remediation wastes. Thus, site environmental personnel should review listed waste determinations for remediation waste with site or corporate legal counsel.

Based on the concerns noted in the previous paragraph, listed codes are sometimes conservatively assigned even though the exact source of contamination is unknown. This consideration is built into the Figure 19-1 logic diagram for determining the regulatory status of remediation waste. Case Study 19-2 provides an example of assigning listed codes to remediation wastes contaminated with F-listed solvent chemicals.

Additional discussion on determining whether a specific remediation waste meets a hazardous listing description is given in Section 18.3.2.

19.1.3.2.1 *Soil contaminated with pesticides*

Soil at industrial facilities that are undergoing cleanups is occasionally found to be contaminated with pesticides. Questions arise over whether the contaminated soil is a hazardous waste. EPA's interpretation of this issue for the pesticide chlordane is as follows. According to §261.33(f), chlordane is a listed commercial chemical product that becomes hazardous waste U036 when it is discarded or intended to be discarded in its unused form. Therefore, if it is known that unused chlordane was dumped or spilled on the ground, the contaminated soil would be U036 when excavated. However, EPA "did not intend to cover those cases when the chemical is released into the environment as a result of use…. In addition, §261.2(c)(1)(ii) specifically states that commercial chemical products listed in §261.33 are not solid wastes (and, thus, not hazardous wastes) if they are applied to the land and that is their ordinary manner of use." Therefore, soil contaminated with chlordane as a result of normal application would only be regulated as hazardous waste (if it is excavated) if it exhibits one or more of the characteristics. Chlordane-contaminated soil could exhibit the toxicity characteristic as D020. [RO 11182]

This interpretation applies to the application of any product for its intended purpose, which incidentally contaminates soil or other environmental media. For example, it also applies to the practice of spraying buildings with pesticide to kill bugs. Any soil contaminated during this normal pesticide use would not be hazardous unless the soil exhibits a characteristic and is actively managed. A similar example is given in Case Study 19-3.

Section 262.70 exempts farmers from the hazardous waste regulations if they dispose waste pesticides on their own farms (and meet certain other conditions). However, if the soil where the waste pesticides were disposed is later excavated for offsite disposal, the disposal of the contaminated soil is not covered under the §262.70 exemption. If the soil exhibits a characteristic at the point of generation (excavation), it is subject to the RCRA hazardous waste regulations, including compliance with LDR treatment standards before it is land disposed. [RO 14588]

Case Study 19-2: Assigning Listed Waste Codes to Remediation Waste

A DOE facility is remediating several areas at the site by pumping ground water through granular activated carbon (GAC). When spent, the GAC is evaluated for disposal. Samples of GAC are sent to a lab for total analyses of numerous metal and organic constituents. If there are any "hits" of project-determined contaminants, two determinations are made: 1) if any of the constituents found indicate potential listed wastes, a due diligence process is initiated to determine the source of the contamination (e.g., from historical discard of listed spent solvents); and 2) the GAC is evaluated to determine if it exhibits a characteristic.

GAC from four areas at the site is being evaluated to determine possible hazardous waste codes:

- Area 1—Ground water is contaminated with 1,1,1-trichloroethane (TCA). Due diligence indicates that the TCA in that area was an ingredient in cutting oil used in a specialty machine shop. No F001–F005 waste codes should be assigned because the TCA was used as an ingredient in a product. EPA notes in RO 13257 that "The 1,1,1-trichloroethane, in this circumstance is being used as a cooling ingredient in the formulation of product cutting oil. The metal working facility is using the cutting oil to coat and lubricate metals during their drilling operation. When the cutting oil can no longer be used, it meets the definition of a spent material in 40 *CFR* 261.1(c)(1). However, even though the cutting oil meets the definition of a spent material, it does not meet the spent solvent listing because the cutting oil formulation was not used as a solvent as described by the December 31, 1985 *Federal Register*. Likewise, the 1,1,1-trichloroethane is an ingredient in the cutting oil and this is not a use covered by the F001–F005 spent solvent listings found in 40 *CFR* Section 261.31. Therefore, the spent cutting oil in this circumstance does not meet the spent solvent hazardous waste listings found in 40 *CFR* Section 261.31." Also see RO 11212. The GAC should also be evaluated for characteristics.

- Area 2—Ground water is contaminated with TCA. Due diligence indicates that there are two possible sources of TCA in that area: 1) use as an ingredient in cutting oil used in a specialty machine shop, and 2) use as a degreaser in an engine repair area. The GAC may be contaminated with TCA from degreasing operations, which would meet the listing description for F001. But, it may also be contaminated with TCA from nonsolvent sources. EPA's guidance notes the following: "If the exact origin of the toxicants is not known, the [media] cannot be considered RCRA hazardous wastes unless they exhibit one or more of the characteristics of hazardous waste...." [RO 12171] Also see RO 12392 and 14291, which contain similar guidance. Although the exact source of the TCA in the ground water (and on the resulting GAC) cannot be determined, this may be a situation where a hazardous waste code (i.e., F001) is conservatively assigned. The GAC should also be evaluated for characteristics.

- Area 3—Ground water is contaminated with TCA. Due diligence indicates that the only source of TCA in that area was use as a degreaser in an engine repair area. The GAC is contaminated with TCA from degreasing operations, which meets the listing description for F001. The GAC should also be evaluated for characteristics.

- Area 4—Ground water is contaminated with TCA. Due diligence cannot determine the source of the contamination. EPA's guidance is clear that the source of the contamination must be known in order for listed waste codes to apply. [RO 12171, 12392, 13181, 14291] Although the source of the TCA in the ground water (and on the resulting GAC) is unknown, this may be a situation where hazardous waste codes are conservatively assigned, since it is known that TCA was used as a degreaser at other areas of the site. The GAC should also be evaluated for characteristics.

> **Case Study 19-3: Soil Contaminated With 2,4,5-T**
>
> At a county public-works yard, the pesticide 2,4,5-trichlorophenoxyacetic acid (2,4,5-T) was sprayed to control bugs. Soil contamination occurred at the yard due to both use of the pesticide and spills of unused product on the ground. The soil is being excavated for offsite disposal. Since the F027 listing includes discarded, unused formulations of 2,4,5-T, is the contaminated soil a listed hazardous waste?
>
> Whether the soil contains a listed waste depends on how the 2,4,5-T got into the soil. If the soil was contaminated with *unused* 2,4,5-T that had been discarded, it would be considered an F027 hazardous waste. On the other hand, if the soil was contaminated as a result of the use of 2,4,5-T, it would not be a hazardous waste unless it exhibits a hazardous waste characteristic. [RO 12903] As a practical matter, if records or employee recollections indicate that spills of the unused pesticide occurred at the site, excavated soil would have to be managed as F027 hazardous waste.

19.1.3.2.2 Remediation wastes contaminated with ICR-only listed wastes

As noted in Section 5.1.2, there are 29 wastes that were listed solely because they exhibit the characteristic of ignitability, corrosivity, and/or reactivity (i.e., ICR-only listed waste). Even though a facility may have documentation that shows that certain remediation wastes are contaminated with one or more ICR-only listed waste, the remediation wastes will not be hazardous if, at their POG, they do not exhibit a characteristic. This is discussed in more detail in Section 5.3.1.1.

19.1.4 Hazardous remediation waste management options

If remediation wastes are determined to be hazardous, there are numerous options available under the RCRA regulations for managing them. The primary RCRA remediation options are summarized in Table 19-3. The allowable waste management activities and other information are given for each of the eight management options discussed in the table, as well as the regulatory citations and guidance documents relied upon in producing the table.

Some of the management options identified in Table 19-3 may be implemented onsite, while others will be conducted offsite. A discussion of these on- and offsite remediation waste management options is given in the sections that follow.

19.1.4.1 Onsite waste management options

If hazardous remediation wastes are generated, they may be managed onsite in the areas/units that are discussed in the following eight subsections.

19.1.4.1.1 *Areas of contamination*

The area of contamination (AOC) policy is discussed as it may be applied during corrective action at a RCRA-permitted site in Section 17.14. This section provides a general discussion of the AOC concept, which may be applied to any hazardous remediation waste (including nonmedia waste) that is in or on the land.

Defined as "a discrete area of generally dispersed contamination," an AOC is equated to a RCRA land-based unit. The AOC policy was developed in the context of the CERCLA program. [December 21, 1988; 53 *FR* 51444, March 8, 1990; 55 *FR* 8758] However, it also applies to RCRA corrective action sites, cleanups under state law, and voluntary cleanups. [RO 11954]

According to OSWER Directive 9347.3-05FS (available at http://www.epa.gov/superfund/policy/remedy/sfremedy/arars/rcra.htm), an AOC is delineated by the areal extent (or boundary) of contiguous contamination. Such contamination must be continuous, but may contain varying types and concentrations of hazardous substances. One or more AOCs may be delineated at a site; examples include:

Table 19-3: McCoy's Summary of RCRA Remediation Options

RCRA Unit or Policy	Allowable Management Activity	Potential Application	Location Requirements	Type of Approval Required	Source of Design and Operating Standards
Contained-in Policy	Hazardous waste determinations for contaminated media and debris.	For low levels of contamination, can be used to obtain a no-longer-contains determination for media and debris contaminated with listed hazardous waste.	Not restricted by location.	No-longer-contains determinations are made by the state or other authorized agency after submission of a petition. The agency may use any format or mechanism to document contained-in determinations. [RO 11948]	Not applicable.
Area of Contamination (AOC) Policy	Storage, in situ treatment, or disposal of hazardous waste, including remediation waste.[2]	In situ treatment, disposal (capping in place), and waste consolidation allowed within the AOC [OSWER Directive 9347.3-05FS]; however, treatment is subject to permitting. [RO 11826, 11692] Not limited to media. No inherent time limit for operation.	Broad areas of generally dispersed contamination that are not necessarily located at a RCRA TSD facility. [RO 11671]	Federal approval not required; states may require prior approval for AOC designations. [RO 11970]	If approved by the authorized agency, the approval document (e.g., a RAP) may specify design and operating conditions.
90-day Unit	Store and/or treat hazardous waste (in tanks, containers, or containment buildings)	Any type of hazardous waste (including remediation waste[2]) may be stored and/or treated in tanks, containers, or containment buildings for up to 90 days (180 days for small quantity generators). The primary disadvantage is the short time allowed for storage/treatment.	Must be located at the site where the wastes are generated. Must be at a facility in compliance with Part 265, Subparts C and D.	No RCRA permit is required at the federal level. Most, but not all, states recognize the §262.34 exemption.	Containers must comply with Part 265, Subpart I standards; tanks must comply with Subpart J standards; and containment buildings must comply with Subpart DD standards.
Temporary Unit	Treat or store remediation waste[2] in tanks or containers.	Wastes may be stored/treated for up to 1 year (a 1-year extension may be available). Not limited to media. Well suited for liquids and flowing sludges. Substitutes for staging pile where treatment will be conducted.	Must be within the contiguous property where the waste originates.	1. Class 2 permit modification at permitted facilities, 2. Revised Part A application at interim status facilities, 3. 3008(h) order, or 4. RAP or RAP modification.	Design and operating standards established on a case-by-case basis per §264.553(c).
Staging Pile	Store solid, nonflowing remediation waste[2] in piles. Mixing, sizing, blending, or similar pretreatment is allowed.	Wastes may be stored for up to 2 years (a 180-day extension may be available). Not limited to media (debris and nonflowing sludges are allowed). [63 FR 65912] Substitutes for AOC where wastes from several noncontiguous areas are to be consolidated.	Must be within the contiguous property where the waste originates.	1. An agency-initiated permit modification per §270.41, 2. Class 2 permit modification at permitted facilities, 3. Closure plan at interim status or permitted facilities, 4. 3008(h) order at interim status facilities, or 5. RAP or RAP modification.	Design and operating standards established on a case-by-case basis per §264.554(d).
Corrective Action Management Unit (CAMU)	Treat, store, or dispose remediation waste.[2]	Differs from AOC in that wastes may be consolidated from noncontiguous areas; wastes may be treated ex situ and placed in the CAMU; may be located in uncontaminated areas. "CAMU-eligible wastes" [see §264.552(a)(1)] may be managed in offsite hazardous waste landfills per §264.555.	Must be within the contiguous property where the waste originates.	1. An agency-initiated permit modification per §270.41, 2. Class 3 permit modification at permitted facilities, 3. 3008(h) orders at interim status facilities, or 4. RAP or RAP modification.	Areal configuration, design, operating, treatment, and closure standards established on a case-by-case basis per §264.552(e).
Conventional RCRA Treatment, Storage, or Disposal (TSD) Unit	Treat, store, or dispose any type of hazardous waste.	Generally designed for management of as-generated process wastes. These units are generally ill suited for the management of remediation wastes because of their regulatory complexity.	Conventional RCRA units are located at interim status or permitted facilities.	Existing units operate under interim status standards of Part 265. Facilities with permits operate under standards of Part 264. Permitting procedures of §§270.3–270.66 apply and are very specific, complex, and time-consuming.	Design and operating standards are specified in detail in Parts 264 and 265.
Remediation Waste Management Sites Approved via Remedial Action Plans (RAPs)	Treat, store, or dispose remediation waste.[2]	A RAP is a type of permit specifically created for managing remediation wastes at remediation waste management sites. It cannot be used to permit treatment, storage, or disposal of as-generated process wastes. [63 FR 65886] A RAP also cannot be used to permit combustion units. [§270.85(b)] However, RAPs may be utilized at facilities that are already permitted. [§270.85(c)]	Must be located: 1) at the AOC where the waste to be managed under the RAP originated, 2) in areas in close proximity to the AOC, or 3) alternative, more protective locations (including offsite locations) approved per §270.230.	A streamlined permit application and approval process have been developed for remediation waste management sites using RAPs in Part 270, Subpart H. RAP approvals are supposed to be much easier and less costly to obtain than conventional RCRA permits.	Design and operating standards for remediation waste management sites will be developed on a case-by-case basis per §270.135.

©2015 McCoy and Associates, Inc.

[1] "Placement" occurs when wastes are moved from one AOC to another (e.g., for consolidation) or when waste is actively managed (e.g., treated ex situ) within or outside the AOC and returned to the same or different AOC. [OSWER Directive 9347.3-05FS] If wastes are stored in drums that remain in the AOC and if the drums are not placed into a separate storage or treatment area, the wastes may be redeposited in the AOC without triggering "placement." [OSWER Directive 9345.3-03FS]

Corrective Action Triggered?	Ground Water Monitoring Required?	RCRA Closure Required?	Land Disposal Requirements (LDR) Applicable?	Minimum Technological Requirements (MTR) Applicable?	Regulatory Citation(s) or Guidance Documents
Not applicable.	Not applicable.	Not applicable.	If the waste was hazardous at the point of generation, LDRs may still apply after a no-longer-contains determination is made. [RO 11948]	Not applicable.	RO 11195, 11434, 14291 61 *FR* 18795 57 *FR* 21450 §261.3(f)(2) §268.7(d-e)
If a RCRA permit is not required for the activity in the AOC (e.g., for a treatment process), activities within the AOC will not trigger corrective action.	Not unless "placement"[1] has occurred such that a RCRA unit is created (e.g., a landfill) requiring ground water monitoring. [RO 13413]	Not unless "placement"[1] has occurred such that a RCRA unit is created (e.g., a landfill) requiring RCRA closure. [RO 13413]	As long as activities conducted in AOC do not constitute "placement,"[1] LDRs do not apply.	As long as activities conducted in AOC do not constitute "placement,"[1] MTRs do not apply.	RO 14291, 11954 55 *FR* 8758 63 *FR* 28619
Corrective action is not triggered unless facility is otherwise subject to RCRA permitting requirements or (for tanks and containment buildings only) if the unit cannot be clean closed and becomes subject to closure requirements for landfills.	Ground water monitoring not required.	Tanks and containment buildings are subject to closure requirements that generally involve decontamination.	LDRs do not apply because these are not land disposal units.	MTR standards (generally interpreted to be liner and leachate collection standards) do not apply.	RO 11425 §262.34
No, if approved via a RAP. Interim status and permitted facilities are already subject to corrective action.	No, temporary units are not land-based units requiring ground water monitoring.	Closure standards are established on a case-by-case basis per §264.553(d).	LDRs do not apply to wastes placed in temporary units. Wastes removed from temporary units are subject to LDRs if land disposed. [RO 14291]	No, design standards are established on a case-by-case basis.	58 *FR* 8673 §264.553
No, if approved via a RAP. Interim status and permitted facilities are already subject to corrective action.	Ground water monitoring may be required on a case-by-case basis to prevent cross media transfer of contaminants. [63 *FR* 65915]	Closure standards are specified in §264.554(j–k) and include clean closure requirements if the staging pile is located in a previously uncontaminated area.	LDRs do not apply to wastes placed in staging piles. Wastes removed from staging piles are subject to LDRs if land disposed. [§264.554(g)] Wastes in staging piles are also exempt from the LDR 1-year storage prohibition. [§268.50(g)]	MTRs do not apply to staging piles. [§264.554(g)]	63 *FR* 65909 67 *FR* 2997 §264.554
No, if approved via a RAP. Interim status and permitted facilities are already subject to corrective action.	Ground water monitoring is required per §264.552(e)(5), unless the CAMU is used for storage and/or treatment only. See also §264.552(g).	Closure standards are established on a case-by-case basis per §264.552(e)(6).	LDRs do not apply to wastes placed into or within a CAMU per §264.552(a)(4).	MTRs do not apply to CAMUs. [§264.552(a)(5)]	RO 11692, 14291 63 *FR* 65920 58 *FR* 8658 67 *FR* 2962 §264.552
Interim status and permitted facilities are subject to corrective action. However, "any facility that meets the definition of a 'remediation waste management site' regardless of whether its hazardous waste management activities are authorized by a RAP or a traditional RCRA permit, will not be subject to the facility-wide corrective action requirement." [63 *FR* 65883]	Ground water monitoring is required at facilities operating land-based units (landfills, surface impoundments, waste piles, and land treatment units).	Closure standards are specified for all conventional RCRA units. Post-closure standards may be applied to land-based units.	LDRs apply to any hazardous waste placed in a land-based unit (landfills, surface impoundments, waste piles, land treatment units, underground mines, and injection wells).	MTRs (generally interpreted to be liner and leachate collection standards) apply to land-based units.	Part 265, Subparts I–W Part 264, Subparts I–X
Remediation waste management sites, which are permitted by RAPs, are not subject to corrective action unless they are located at facilities that are already subject to corrective action. [§260.10 definition of "remediation waste management site"]	If a new land-based unit is created at a remediation waste management site, ground water monitoring (Part 264, Subpart F) will be invoked. Ground water monitoring may not be appropriate where old or existing land-based units are being addressed as part of a cleanup. [63 *FR* 65906]	Closure standards (Part 264, Subpart G) and unit-specific closure requirements apply to new units permitted under a RAP, but not to AOCs or to old units not already subject to Subtitle C. [63 *FR* 65906]	LDR standards (e.g., soil treatment standards) apply at remediation waste management sites. [§270.110(f)(3)]	MTRs (generally interpreted to be liner and leachate collection standards) will apply on a case-by-case basis. [§270.135(b)(1)]	63 *FR* 65884 63 *FR* 65905 Part 270, Subpart H §264.1(j)

[2]"Remediation waste" means all solid and hazardous wastes, and all media (including ground water, surface water, soils and sediments) and debris that are managed for implementing cleanup. [§260.10] "First, it should be noted that remediation waste includes only waste managed because of cleanup, and does not include wastes generated from ongoing hazardous waste operations, which are commonly referred to as 'newly generated,' 'as-generated,' or 'process' wastes. When managed as part of a legitimate cleanup action, any (non-'as-generated') hazardous wastes (for example, media, debris, sludges, or other wastes) are all remediation waste. Second, remediation waste includes both hazardous and nonhazardous solid wastes managed as a result of cleanup, including any wastes generated from treating remediation wastes (for example, carbon canisters and sludges generated from ground water pump-and-treat or soil vapor extraction systems)." [63 *FR* 65881]

- A waste source (e.g., waste pit, landfill, waste pile) and the surrounding contaminated soil.

- A waste source, and the sediments in a stream contaminated by the source, where the contamination is continuous from the source to the sediments.

- Several lagoons separated only by dikes, where the dikes are contaminated and the lagoons share a common liner.

EPA's key concept associated with its AOC policy is this: consolidation and in situ treatment of hazardous waste within the AOC do *not* create a new point of hazardous waste generation for purposes of RCRA. [RO 14291] With that concept in mind, the following guidance summarizes the usefulness of the AOC policy during site remediation:

- If little information concerning site contamination is available, or if no visual contamination exists, site managers may use their best professional judgment to delineate AOCs. [EPA/540/G-91/009, May 1991, available at http://nepis.epa.gov/EPA/html/Pubs/pubtitleOSWER.html]

- Normally, excavation of contaminated soil is considered the point of generation; but under the AOC policy, consolidation is not considered to be removal from the land (i.e., generation). Thus, contaminated soil can be consolidated within the AOC and a hazardous waste determination can be made after such consolidation. [RO 14283, 14338]

- Remediation wastes can be consolidated or treated in situ within an AOC without triggering land disposal restrictions (LDR) or minimum technology requirements (i.e., liner and leachate collection standards). [RO 14291]

- Hazardous remediation waste that is moved outside the boundaries of the AOC is generated and RCRA requirements apply. Thus, in cases where a site owner/operator wants to consolidate remediation waste from separate, noncontiguous contaminated areas, the AOC concept cannot be used. [November 30, 1998; 63 *FR* 65920, RO 11954]

- Soil-like investigation-derived waste (IDW) generated during a site investigation may be left in the AOC without generating hazardous waste. For example, soil cuttings from a soil boring could be deposited in a small area immediately adjacent to the borehole if the surface soil is similar to soil from the borehole. [EPA/540/G-91/009, cited previously] Nonsoil IDW wastes (e.g., disposable bailers, PPE, purge water) cannot be managed under the AOC policy. For example, if ground water is purged prior to taking a sample, the purge water should be containerized and managed as solid and potentially hazardous waste.

- Soil may be managed within the AOC to facilitate sampling, for example, to ensure that soil samples are representative or to separate soil from nonsoil materials. [May 26, 1998; 63 *FR* 28619]

- The AOC policy may be used when managing contaminated soil during construction activities (e.g., digging a trench to install a pipeline or electrical conduit). As long as the contaminated soil remains in the AOC (i.e., the contaminated soil is not moved to an uncontaminated area of the plant or shipped offsite) and is subsequently put back in the ground from which it was excavated, such operations do not produce a hazardous waste or subject it to hazardous waste regulation. Therefore, the contaminated soil would not have to be counted, the piles next to the excavation are not regulated waste piles, and the dirt is not subject to LDR standards. [RO 11671, EPA/530/K-05/011, September 2005, available at http://www.epa.gov/wastes/inforesources/pubs/training/gen05.pdf]

At the federal level, advance approval is not required for facilities to take advantage of the AOC policy. However, EPA encourages them to consult with the appropriate state agency to ensure they implement the AOC concept appropriately. [RO 11954] A 42-page packet of EPA guidance on the AOC policy is available at http://www.epa.gov/epawaste/hazard/correctiveaction/resources/guidance/remwaste/refrnces/01aoc.pdf.

19.1.4.1.2 *Satellite accumulation units*

Although SAAs are not included in our table of RCRA remediation options (Table 19-3), we see no reason why remediation wastes (such as contaminated PPE, purge water, waste samples) could not be put into a satellite accumulation container. Note that the satellite accumulation regulations did not envision facilities generating one-time wastes (such as soil borings) and essentially abandoning the wastes in an SAA. We are not aware of any specific EPA guidance on this issue; however, we suspect that some state agencies may not allow SAA containers to be used during remediation waste management.

The useful aspects of designating drums containing remediation wastes as satellite accumulation containers are that 1) there is no clock running, and 2) the drums are subject only to minimal labeling and RCRA container standards as noted in §262.34(c)(1). If alternatively, the generator designates drums holding remediation wastes as 90/180-day accumulation containers, then there would be a clock running and these units would be subject to additional labeling and RCRA container standards as noted in the following section.

19.1.4.1.3 *90/180-day units*

The first nonearthen units listed in Table 19-3 that can be used for onsite management of hazardous remediation wastes after they are generated are 90-day units. Of course, small quantity generators may use 180-day units for this purpose. Typically, such 90/180-day units would be tanks or containers, and these units would be subject to all of the RCRA requirements when in remediation waste service just as they are when in process waste service. In addition to the summary given in Table 19-3, Sections 6.3 and 7.3 detail the §262.34 regulatory requirements for these units when used for accumulation and treatment, respectively.

Examples of the use of 90/180-day units for managing hazardous remediation waste include:

- Blending hazardous contaminated soil and/or coal tar wastes from manufactured gas plant sites with coal or other fuels to produce a burnable, nonhazardous fuel [RO 11675, 11739, 14024, 14338, 14344];
- Accumulating F001 soil in roll-off boxes; and
- Treating hazardous soil during site remediation. [RO 14112, 14291, 14471]

There is one exemption from RCRA requirements for 90-day tanks and containers used to manage remediation wastes. Onsite units used to manage remediation wastes are exempt from Subpart CC air emission standards. [§265.1080(b)(5)] Note that this exemption applies only to remediation wastes generated during RCRA corrective action or CERCLA cleanups or cleanups required under similar state or federal authorities. Remediation wastes from voluntary cleanups, with no state involvement or oversight, would not qualify. Also, as soon as the waste is moved offsite, the exemption no longer applies. [*RCRA Subpart CC: Questions and Answers*, March 2002, available at http://www.trainex.org/web_courses/subpart_x/ EPA CD Content/TopicSearch.htm]

19.1.4.1.4 *Temporary units*

Temporary units are noted in Table 19-3 as tanks or containers that may be used to treat or store hazardous remediation wastes. A summary of the regulatory provisions applicable to these units is given in Table 19-3. Since these units are primarily used during corrective action at RCRA-permitted facilities, more detailed regulatory requirements are discussed in Section 17.12 of our chapter on corrective action.

19.1.4.1.5 *Staging piles*

A staging pile is an accumulation of solid, non-flowing remediation waste that is used only during remedial operations for temporary storage at a facility. [§260.10] As noted in Table 19-3, such piles may also be used to pretreat (e.g., mix, size, blend) hazardous remediation wastes. A summary of the regulatory provisions applicable to these units is given in Table 19-3. Since these units are primarily used during corrective action at RCRA-permitted facilities, more detailed regulatory requirements are discussed in Section 17.13 of our chapter on corrective action.

19.1.4.1.6 Corrective action management units (CAMUs)

A CAMU is an area at a facility used to manage remediation wastes during implementation of corrective action (a more exact definition is given in §270.2). Hazardous remediation wastes can be stored, treated, and even disposed in a CAMU. A summary of the regulatory provisions applicable to these units is given in Table 19-3. Since these units are primarily used during corrective action at RCRA-permitted facilities, more detailed regulatory requirements are discussed in Section 17.11 of our chapter on corrective action.

19.1.4.1.7 Remediation waste management sites

On November 30, 1998, EPA published a final rule referred to as the "hazardous waste identification rule for contaminated media," often called the HWIR-media rule. [63 FR 65874] In that rule, EPA emphasized that, to stimulate cleanup, remediation waste management activities should be regulated differently from as-generated process waste management. One of the provisions of the HWIR-media rule was the establishment of a new unit called a "remediation waste management site," defined in §260.10 as:

"[A] facility where an owner or operator is or will be treating, storing or disposing of hazardous remediation wastes. A remediation waste management site is not a facility that is subject to corrective action under 40 CFR 264.101, but is subject to corrective action requirements if the site is located in such a facility."

This concept allows the identification of an area at a facility that will be used solely to treat, store, and/or dispose hazardous remediation waste. To limit application of this new type of unit to cleanup activities, EPA defined "remediation waste" in §260.10 as:

"[A]ll solid and hazardous wastes, and all media (including ground water, surface water, soils, and sediments) and debris, that are managed for implementing cleanup."

While temporary units, staging piles, and CAMUs (as discussed in the preceding sections) are primarily used during corrective action at a RCRA-permitted site, a remediation waste management site (RWMS) may be used to manage hazardous remediation wastes generated during a wide range of cleanups conducted under many different types of cleanup authorities—not just RCRA corrective action. For example, a RWMS could be used during a voluntary cleanup at a site with no RCRA Part B permit. [63 FR 65881–2]

There are three primary benefits to using a RWMS: 1) use of a simplified permit to establish the site, 2) exclusion from facility-wide corrective action, and 3) compliance with performance standards instead of the stringent Part 264 requirements. These three benefits and other provisions of the RWMS option are discussed below.

A streamlined RCRA permit, called a remedial action plan (RAP), is the enforceable document that EPA or a state can issue to authorize treatment, storage, or disposal of remediation waste at a RWMS. [§270.80(a)] RAPs are more fully discussed in Section 17.15. Note that a RAP would not be required if the treatment and storage of hazardous remediation waste can be conducted in a 90/180-day accumulation unit (see Sections 7.3 and 6.3, respectively) and if no land disposal of hazardous remediation waste will occur at the site.

Neither a RAP nor a RWMS will trigger facility-wide corrective action at the facility. [§§260.10 (paragraph (3) of the "facility" definition), 264.1(j), 264.101(d)] EPA believes that applying the corrective action program to facilities not already subject to these requirements is a disincentive to voluntarily initiated cleanup actions. However, if the facility is already subject to the corrective action program because it has a conventional RCRA Part B permit or is operating under interim status, then corrective action requirements will apply to the entire facility, including the RWMS. A RWMS could be part of an operating (or closing) RCRA hazardous waste management facility that is already subject to the corrective action requirements; in those cases, identifying an area of the facility as a RWMS would not have any effect on the corrective action requirements for that site or the rest of the facility. [63 *FR* 65884] One final note on the applicability of corrective action—facilities that obtain a traditional RCRA Part B permit solely for the management of remediation waste are also not subject to corrective action. [§264.101(d), November 30, 1998; 63 *FR* 65883]

Performance standards for RWMSs are specified in §264.1(j) that replace the detailed requirements in Part 264, Subparts B, C, and D (general facility standards, preparedness and prevention, and contingency plans and emergency procedures, respectively).

To maximize flexibility, there is no limitation that remediation waste managed in a RWMS must originate from within the facility boundary. When EPA codified these provisions in the HWIR-media rule at §270.80(a), the agency said a RAP may be issued for a RWMS that will manage remediation wastes that originate from 1) onsite areas of contamination, 2) offsite areas "in close proximity to the [onsite] contaminated area," or 3) remote offsite locations per §270.230. These three locations are discussed in more detail below.

1. When used for onsite management of remediation waste, the RWMS can, but does not have to, be located near or in contaminated areas. For example, it might make more sense to locate the RWMS in another, noncontaminated area of the facility where utilities (e.g., electricity, steam, roadways, wastewater treatment plant, etc.) are available. Additionally, it may be more environmentally beneficial to locate the RWMS away from the area of contamination if the contaminated area is located in a potable well field, over a sole-source aquifer, or in a floodplain. [63 *FR* 65903]

2. The "close proximity" language in §270.80(a) allows remediation wastes that originate from outside the facility fence line, but in close proximity to an onsite contaminated area, to be managed in the onsite RWMS. For example, wastes that have migrated beyond the facility boundary may be excavated and managed in the onsite RWMS. [63 *FR* 65881]

3. Sometimes, it makes sense to locate the RWMS at an offsite location that is remote from the contaminated areas from which remediation wastes are generated. For example, it would likely be more cost-effective to establish a centralized RWMS to manage waste generated from the remediation of historical soil contamination at multiple remote pumping stations along a pipeline rather than to carry out remedial treatment at each station. Thus, the RCRA regulations allow designation (via a RAP) of a remote offsite location as a RWMS under §270.230. Keep in mind, however, that if owners/operators manage hazardous remediation waste during cleanup at their facility and ship that waste to an offsite RWMS, they must comply with all hazardous waste generator requirements, such as manifesting and transportation requirements. [63 *FR* 65904]

A RWMS may be used only to manage wastes generated because of cleanup; such a unit cannot be used to manage wastes generated from ongoing maintenance or manufacturing operations. [63 *FR* 65881] Both hazardous and nonhazardous solid wastes generated as a result of cleanup may be managed in a RWMS, including any wastes generated from treating remediation wastes (e.g., carbon canisters and sludges generated from ground water pump-and-treat or soil vapor extraction systems). [63 *FR* 65881]

19.1.4.1.8 *RCRA-permitted units*

Although generally used for the treatment, storage, and/or disposal of process wastes, RCRA-permitted units, such as a hazardous waste container storage area, tank, landfill, or even surface impoundment, can be used for the onsite management of hazardous remediation wastes. The requirements for obtaining a RCRA Part B permit for a unit that will manage hazardous remediation wastes are the same as those for permitting a process waste management unit. As noted in Table 19-3, these requirements are expensive, time-consuming, and trigger other RCRA provisions (e.g., the corrective action program).

There is one exemption from RCRA requirements for permitted tanks, containers, and surface impoundments used to manage remediation wastes. Onsite units used to manage remediation wastes are exempt from Subpart CC air emission standards. [§264.1080(b)(5)] Note that this exemption applies only to remediation wastes generated during RCRA corrective action or CERCLA cleanups or cleanups required under similar state or federal authorities. Remediation wastes from voluntary cleanups, with no state involvement or oversight, would not qualify. Also, as soon as the waste is moved offsite, the exemption no longer applies. [*RCRA Subpart CC: Questions and Answers*, cited previously]

19.1.4.1.9 Is soil consolidation considered dilution?

Sometimes, people ask if dilution is occurring if, during a site remediation, contaminated soil from various onsite locations is consolidated. The only guidance we have found that addresses this issue is as follows:

> "If mixing occurs through the normal consolidation of contaminated soil from various portions of a site that typically occurs during the course of remedial activities or in the course of normal earth-moving and grading activities, then the agency does not consider this to be intentional mixing of soil with nonhazardous soil for the purposes of evading LDR treatment standards. Therefore, it is not viewed as a form of impermissible dilution...."

> "Some situations may require soil mixing, as part of a pretreatment process, to facilitate and ensure proper operation of the final treatment technology to meet the LDR treatment standards. If the mixing or other pretreatment is necessary to facilitate proper treatment in meeting the LDR standards, then dilution is permissible. For example, addition of less-contaminated soil may be needed to adjust the contaminated soil Btu value, water content, or other properties to facilitate treatment. These adjustments would be for meeting the energy or other technical requirements of the treatment unit to ensure its proper operation. The agency views this type of pretreatment step as allowable, provided the added reagents or other materials produce chemical or physical changes and do not 1) merely dilute the hazardous constituents into a larger volume of waste so as to lower the constituent concentration, or 2) release excessive amounts of hazardous constituents to the air." [EPA/530/R-02/003; see also RO 14338]

19.1.4.2 Offsite waste management options

In addition to the onsite management options discussed above, hazardous remediation waste may be managed at offsite facilities, as discussed in the following three subsections.

19.1.4.2.1 Remediation wastes to Subtitle C TSD facility

The most likely option for the offsite management of hazardous remediation wastes is simply to send them to a commercial TSD facility. Such facilities are RCRA-permitted to accept and treat, store, and/or dispose hazardous wastes, including hazardous remediation wastes. If this option is chosen, the generating facility would manage the hazardous remediation wastes just like any other hazardous waste at its facility, and it would comply with all hazardous waste generator requirements, such as 90/180-day accumulation and manifesting requirements.

19.1.4.2.2 CAMU-eligible wastes to offsite Subtitle C landfill

As discussed in Sections 17.11 and 19.1.4.1.6, a CAMU may be used to store, treat, and even dispose hazardous remediation waste during implementation of corrective action at a facility (see §264.552). However, the requirements (and accompanying expense) to get a CAMU designated, built, and in operation are significant.

Instead of complying with these stringent requirements, the same remediation wastes that are eligible for management in an onsite CAMU can alternatively be shipped to an offsite Subtitle C hazardous waste landfill. These "CAMU-eligible wastes" are defined in §264.552(a)(1), and the regulatory requirements associated with this alternative are specified in §264.555. One noteworthy provision in §264.555(a) allows CAMU-eligible wastes that meet certain conditions to be land disposed without meeting the land disposal treatment standards in Part 268. These requirements are discussed in Section 17.11.7 of our chapter on corrective action.

19.1.4.2.3 *Hazardous remediation waste sent offsite triggers LDR program*

Remediation waste will be subject to the LDR program only if 1) it is generated (i.e., excavated), 2) it is hazardous (i.e., it exhibits a characteristic or contains a listed waste), and 3) it will be placed in a land disposal unit. For example, if hazardous soil is excavated and will be disposed in an offsite landfill, it will be subject to the LDR program just like any other hazardous waste, unless the hazardous soil is managed under §264.555, discussed above. Note that EPA has developed less-stringent LDR treatment standards for hazardous soil; these standards are discussed in detail in Section 13.7.

19.1.5 RCRA/CERCLA remediation interface

In general, RCRA applies to sites that are currently active; that is, a financially viable owner/operator exists, and the site is a hazardous waste generator, is operating under interim status, or has a RCRA permit. When cleanup of these active sites is required, the RCRA corrective action program is implemented if the site is operating under interim status or has a RCRA permit (see Chapter 17).

The Comprehensive Environmental Response, Compensation and Liability Act (CERCLA—which established the Superfund to help pay for cleanup) typically applies to sites that are not active RCRA sites. The sites may be abandoned (i.e., no owner/operator can be identified) or the owner/operator may be insolvent or unwilling to undertake cleanup actions. Although the Superfund provides funds for cleanup of sites where potentially responsible parties (PRPs) cannot be identified, EPA often seeks to recover funds from PRPs when they can be identified.

When inactive sites are to be cleaned up, activities are undertaken in accordance with CERCLA's National Contingency Plan (NCP). The NCP is codified at 40 *CFR* Part 300 and must be complied with if Superfund monies are to be used or if cost recovery for cleanup activities is to be sought from PRPs.

19.1.5.1 How RCRA sites become subject to CERCLA

Under certain circumstances, sites may become subject to both RCRA and CERCLA cleanup authorities. These situations are discussed in the next few sections.

19.1.5.1.1 *Imminent and substantial endangerment*

In some cases, an imminent hazard may exist that requires a very prompt response. Both RCRA and CERCLA provide authority for such responses:

- RCRA Section 7003(a) provides EPA with the authority to issue an administrative order (or file suit) where "any solid waste or hazardous waste may present an imminent and substantial endangerment to health or the environment."

- CERCLA Section 106(a) authorizes EPA to act where "there may be an imminent and substantial endangerment to the public health or welfare or the environment because of an actual or threatened release of a hazardous substance from a facility."

The choice of whether to use RCRA or CERCLA authorities (or both) for immediate responses often depends on the fact that hazardous wastes are a subset of hazardous substances. CERCLA hazardous substances are identified at §302.4 and include many CWA and CAA compounds that are not RCRA hazardous wastes.

If a facility has a release of a hazardous substance, but not a hazardous waste, a response under CERCLA authority would be most appropriate. If a facility has a release of hazardous wastes, which are a subset of CERCLA hazardous substances, a response under RCRA authority would be most appropriate.

19.1.5.1.2 *RCRA-permitted facilities*

Under the NCP, a projected Hazard Ranking Score (HRS) was developed for most RCRA-permitted sites, and any that could potentially become National Priority List (NPL) sites requiring CERCLA cleanup were identified. Although cleanup at a RCRA-permitted site would normally be conducted under RCRA corrective action provisions (see Chapter 17), EPA has stated that owners/operators who subsequently declare bankruptcy or who are unwilling to undertake corrective action become subject to CERCLA cleanup authority if warranted by the site's HRS score. [OSWER Directive 9932.1, available at http://nepis.epa.gov/EPA/html/Pubs/pubtitleOther.html by downloading the document numbered OSWERDIR99321] The policy of making owners/operators of RCRA-permitted facilities who are unwilling or financially unable to undertake corrective action subject to CERCLA is discussed in the August 9, 1988 *Federal Register*. [53 *FR* 30002]

Some RCRA-permitted facilities are already listed on the NPL. EPA's general policy is to delete RCRA sites that are on the NPL and defer their cleanup to the RCRA corrective action program. Criteria for deleting such RCRA sites from the NPL were issued on March 20, 1995 [60 *FR* 14641] and November 24, 1997. [62 *FR* 62523]

As noted in the previous paragraph, EPA's general policy is for RCRA facilities subject to both CERCLA and RCRA to be cleaned up under RCRA. However, in some cases, it may be more appropriate for the federal CERCLA program or a state/tribal "Superfund-like" cleanup program to take the lead. In these situations, the facility RCRA permit should defer corrective action at the facility to CERCLA or a state/tribal cleanup program. For example, where program priorities differ and a cleanup under CERCLA has already been completed or is underway at a RCRA-permitted facility, corrective action conditions in the RCRA permit could state that the existence of a CERCLA action makes separate RCRA action unnecessary. In this case, there would be no need for the corrective action program to revisit the remedy at some later point in time. [RO 11959]

While deferral from RCRA to CERCLA or vice versa is typically the most efficient and desirable way to address overlapping cleanup requirements, full deferral may not be appropriate at all facilities, and coordination between programs will be required. RO 11959 provides guidance for coordination between programs in these cases. For example, CERCLA or RCRA decision documents should be written so that cleanup responsibilities are divided and timing sequences are established.

Some of the most significant RCRA/CERCLA integration issues are associated with coordination of requirements for closure of RCRA-regulated units. [Such units are surface impoundments, waste piles, land treatment units, or landfills that receive (or have received) hazardous waste after July 26, 1982.] There are a couple of options for tackling this integration issue:

1. Under a rule issued by EPA on October 22, 1998 [63 *FR* 56710], EPA or an authorized state can defer cleanup, ground water monitoring, and financial responsibility requirements for a closed or closing RCRA-permitted unit that is located among SWMUs until such time as those requirements are identified for the surrounding SWMUs during the corrective action process. This type of deferral would usually require modification of the facility's closure plan for that unit. [63 *FR* 56724–5]

2. A cleanup plan for a CERCLA operable unit that physically encompasses a RCRA-regulated unit could be structured to provide for concurrent compliance with CERCLA and the RCRA closure

and post-closure requirements. In this example, the RCRA permit could be modified to cite the ongoing CERCLA cleanup and incorporate the CERCLA requirements by reference. [RO 11959]

19.1.5.1.3 *Federal facilities*

Federal facilities are subject to both RCRA and CERCLA because the Superfund Amendments and Reauthorization Act of 1986 (SARA) stated that federal facilities are subject to CERCLA. Similarly, the Federal Facilities Compliance Act of 1992 specified that federal facilities are subject to RCRA. Hence these facilities become subject to whichever RCRA/CERCLA provisions the regulatory agencies decide provide the greatest authority.

Many federal facilities are very large in size, and there are often several federal, state, and local regulatory agencies and other interested parties involved in their cleanup. Most federal sites have compliance agreements with the pertinent regulatory agencies; the agreements identify when and where CERCLA and/or RCRA authorities will apply. For example, it is common for the CERCLA process to be used at large federal facilities for cleanup of operable units. Although these facilities typically have RCRA permits, the corrective action requirements for SWMUs and closure plans for RCRA-permitted units are addressed through the CERCLA process.

Numerous federal facilities are listed on the NPL, and EPA recognizes that CERCLA Section 120 imposes special requirements on federal facilities—these requirements must be accommodated in any RCRA/CERCLA coordination approach. One of the methods that the agency has developed to implement this coordinated approach is identification of a single lead regulator to oversee the cleanup of federal facility sites on the NPL. This lead-regulator policy is contained in "Lead Regulator Policy for Cleanup Activities at Federal Facilities on the National Priorities List," November 6, 1997, available at http://www2.epa.gov/fedfac/lead-regulator-policy-cleanup-activities-federal-facilities-national-priorities-list.

19.1.5.2 How CERCLA sites become subject to RCRA

At a site being remediated under CERCLA authority, the cleanup provisions of the NCP require implementation of remedial actions when extensive cleanup activities, such as ground water remediation, are involved. Remedial actions involve a number of distinct steps. First, a remedial investigation (RI) is conducted to characterize the problem and determine the nature and extent of contamination. This is followed by a feasibility study (FS), which evaluates various methods to address site contamination. Formalized screening methods that consider the effectiveness, implementability, and cost of the various alternatives are applied to identify a limited number of options for detailed study.

Using criteria specified in §300.430(e)(9)(iii), the lead agency will identify the preferred remedial alternative for the site and seek public comment. When the decision is finalized, it is documented in the Record of Decision.

Finally, the remedial action undergoes detailed design, construction, and operation. For complex sites, the entire cleanup process may take a number of years, and in some cases decades, to complete.

19.1.5.2.1 *ARARs drive the cleanup process*

CERCLA Section 121 establishes cleanup standards for remedial actions. Included in that statutory section is the provision that the parties involved in the cleanup 1) evaluate standards, requirements, criteria, and limitations under state and federal environmental laws; and 2) ensure that the remedial action results in "a level or standard of control for such hazardous substance or pollutant or contaminant which at least attains such legally *applicable or relevant and appropriate* standard, *requirement*, criteria, or limitation." [Emphasis added.] The CERCLA regulations at §300.400(g) note how lead agencies are to identify these applicable or relevant and appropriate requirements (ARARs). In order to determine if RCRA regulatory provisions apply during

CERCLA cleanups, it is necessary to first see how EPA defines these terms:

"*Applicable requirements* means those cleanup standards, standards of control, and other substantive requirements, criteria, or limitations promulgated under federal environmental or state environmental or facility siting laws that specifically address a hazardous substance, pollutant, contaminant, remedial action, location, or other circumstance found at a CERCLA site. Only those state standards that are identified by a state in a timely manner and that are more stringent than federal requirements may be applicable." [§300.5] Applicable requirements are legally binding requirements that must either be met or waived.

"*Relevant and appropriate requirements* means those cleanup standards, standards of control, and other substantive requirements, criteria, or limitations promulgated under federal environmental or state environmental or facility siting laws that, while not 'applicable' to a hazardous substance, pollutant, contaminant, remedial action, location, or other circumstance at a CERCLA site, address problems or situations sufficiently similar to those encountered at the CERCLA site that their use is well suited to the particular site. Only those state standards that are identified in a timely manner and are more stringent than federal requirements may be relevant and appropriate." [§300.5] In other words, relevant requirements address a similar situation or problem; appropriate requirements are well suited to a particular site.

Evaluating whether a requirement is "applicable" is a legal and jurisdictional determination. The jurisdictional requirements for RCRA Subtitle C applicability are discussed in Section 2.3 of EPA/540/G-89/006 (available from http://nepis.epa.gov/EPA/html/Pubs/pubtitleOSWER.html). In general, RCRA Subtitle C requirements for the treatment, storage, or disposal of hazardous waste will be applicable at a CERCLA site if a combination of the following conditions are met:

- The waste exhibits a characteristic or is listed under RCRA; and
- The waste was 1) treated, stored, or disposed after the effective date of the RCRA requirements under consideration; or 2) the activity at the CERCLA site constitutes treatment, storage, or disposal as defined by RCRA.

Under the second scenario, for example, if the lead agency determines that RCRA characteristic or listed hazardous waste is present at the site (even if the waste was disposed before the effective date of the requirement) and the proposed CERCLA action involves treatment, storage, or disposal as defined under RCRA, then RCRA requirements related to those actions would be applicable.

If a requirement is applicable, all substantive parts must be followed. An applicable requirement is one with which a private party would have to comply by law if the same action was being undertaken apart from CERCLA authority. All jurisdictional prerequisites of the requirement must be met in order for the requirement to be applicable.

A requirement that is "relevant and appropriate" may "miss" on one or more jurisdictional prerequisites for applicability but still make sense, given the circumstances of the site. The determination of relevant and appropriate relies on professional judgment, and is based on environmental and technical factors at the site. There is more flexibility in the relevance and appropriateness determination; a requirement may be "relevant" in that it covers situations similar to that at the site but may not be "appropriate" to apply for various reasons and, therefore, not be well suited to the site. In some situations, only portions of a requirement or regulation may be judged relevant and appropriate. [OSWER Directive 9234.2-01/FS-A, available at http://www.epa.gov/superfund/policy/remedy/sfremedy/arars/overview.htm]

For example, if a hazardous substance at the site is identical to a RCRA listed hazardous waste, but its source is unknown, RCRA requirements will not be applicable but may be relevant and appropriate if the action taken is regulated by RCRA (e.g., storage). Significant guidance is available in EPA/540/G-89/006 for evaluating whether RCRA requirements are relevant and appropriate in such a scenario.

Once a requirement is deemed relevant and appropriate, it must be attained (or waived) during remedial actions that are carried out onsite. Relevant and appropriate requirements do not have to be met when wastes are sent offsite. If a requirement is not both relevant and appropriate, it is not an ARAR. [EPA/542/R-92/005, available from http://nepis.epa.gov/EPA/html/Pubs/pubtitleOSWER.html]

There are three types of ARARs that must be examined at a specific CERCLA site:

1. Chemical-specific ARARs—These are usually health- or risk-based numerical values that establish the acceptable amount or concentration of a chemical that may be found in, or discharged to, the ambient environment. For example, the RCRA toxicity characteristic levels codified in §261.24 are a factor of 100 times either SDWA MCL standards or health-based data. Thus, the RCRA ARAR for lead is 0.05 mg/L, which assumes a ground water exposure scenario. Chemical-specific ARARs are given in Exhibit 1-1 of EPA/540/G-89/006.

2. Location-specific ARARs—These are restrictions placed on the concentration of hazardous substances or the conduct of activities solely because they are in specific locations. Examples of special locations include floodplains, wetlands, historical places, and sensitive ecosystems or habitats. RCRA ARARs that involve location-specific standards are codified at §264.18: 1) new TSD facilities cannot be located within 200 feet of a fault that has had displacement in Holocene time; 2) facilities in 100-year floodplains must be designed to prevent washout in a 100-year flood; and 3) bulk liquid wastes cannot be placed in salt domes, salt beds, or underground mines or caves. Location-specific ARARs for other statutes are given in Exhibit 1-2 of EPA/540/G-89/006.

3. Action-specific ARARs—These are usually technology- or activity-based requirements that are selected to accomplish a remedy. Action-specific requirements indicate how a selected alternative must be achieved. RCRA corrective action requirements are the most similar to cleanup activities typically taken at CERCLA sites. Examples of other RCRA requirements that are action-specific ARARs include: capping; closing land disposal units; container storage; constructing new land disposal units; dike and impoundment design, construction, and operation; ground water protection and monitoring; incineration; land treatment; placement of wastes in a land disposal unit (the land disposal restrictions); and tank storage. Action-specific ARARs are given in Exhibit 1-3 of EPA/540/G-89/006.

Detailed procedures that should be used to identify RCRA-based ARARs are given in EPA/540/G-89/006. They must be determined on a site-specific basis, considering the specific chemicals at the site, special features of the site location, and the actions that are being considered as remedies.

Six statutory waivers from the requirement to attain ARARs are codified at §300.430(f)(l)(ii)(C) and are discussed in some detail in EPA/540/G-89/006.

Q *RCRA hazardous waste is going to be left (disposed) at a CERCLA site. Are RCRA closure requirements ARARs?*

A As noted earlier in this section, the basic prerequisites for applicability of closure requirements at a CERCLA site are: 1) the waste must be hazardous; and 2) the unit (or AOC) must have received waste after the RCRA requirements became effective, either because of the original date of disposal or because the CERCLA action constitutes disposal. If closure requirements under Subtitle C of RCRA are applicable, the disposal area must be closed in compliance with one of the closure options available in the RCRA regulations. For wastes left at the site, the landfill closure option will be applicable, which requires the unit to be capped with an impermeable cover and includes 30 years of long-term cover maintenance, operation of a leachate collection and removal system, ground water monitoring, etc. However, if Subtitle C closure requirements are not applicable based on the above prerequisites but are determined to be relevant and appropriate, then a "hybrid closure" that includes other types of closure designs may be used. The hybrid closure option would arise from a determination that only some elements of the above-noted RCRA closure requirements are relevant and appropriate. [December 21, 1988; 53 FR 51445–6, OSWER Directive 9234.2-01/FS-A, OSWER Directive 9234.2-04/FS, available at http://www.epa.gov/superfund/policy/remedy/sfremedy/arars/rcra.htm]

19.1.5.2.2 Substantive vs. administrative requirements

In the definitions for *applicable requirements* and *relevant and appropriate requirements* given in the previous section, only "substantive," as opposed to "administrative," requirements qualify as ARARs. In EPA/540/G-89/006, EPA has defined these two types of requirements as follows:

"Substantive requirements are those requirements that pertain directly to actions or conditions in the environment. Examples of substantive requirements include quantitative health- or risk-based restrictions upon exposure to types of hazardous substances (e.g., MCLs establishing drinking water standards for particular contaminants), technology-based requirements for actions taken upon hazardous substances (e.g., incinerator standards requiring particular destruction and removal efficiency), and restrictions upon activities in certain special locations (e.g., standards prohibiting certain types of facilities in floodplains)."

"Administrative requirements are those mechanisms that facilitate the implementation of the substantive requirements of a statute or regulation. Administrative requirements include the approval of, or consultation with administrative bodies, consultation, issuance of permits, documentation, reporting, recordkeeping, and enforcement. In general, administrative requirements prescribe methods and procedures by which substantive requirements are made effective for purposes of a particular environmental or public health program. For example, the requirement of the Fish and Wildlife Coordination Act to consult with the U.S. Fish and Wildlife Service, Department of the Interior, and the appropriate State agency before controlling or modifying any stream or other water body is administrative."

Cleanup actions that take place "on-site" must meet the substantive but not the administrative requirements. ("On-site" is defined in §300.5, to mean "the areal extent of contamination and all suitable areas in very close proximity to the contamination necessary for implementation of the response action.") Response actions carried out offsite are subject to all applicable provisions, including administrative requirements and any specified procedures for obtaining permits. [EPA/542/R-92/005]

Although administrative requirements are not ARARs at CERCLA sites, keep in mind that the CERCLA program has its own set of administrative procedures, which assure proper implementation of cleanup activities.

Based on the above definitions, the following general principles may be used in determining potential RCRA ARARs [March 8, 1990; 55 FR 8756–7, EPA/540/G-89/006, EPA/542/R-92/005]:

- Substantive requirements usually specify a concentration level or standard of control, and they could also provide performance criteria or location restrictions. For example, performance standards for incinerators and minimum technological requirements for double liners and leachate collection systems for landfills would be substantive requirements.

- LDR treatment standards would be substantive requirements; detailed guidance on whether LDR standards are ARARs is found in OSWER

Directives 9347.3-05FS and 9347.3-07FS, both available at http://www.epa.gov/superfund/policy/remedy/sfremedy/arars/rcra.htm. A closer look at making this determination is provided in Case Study 19-4.

- Monitoring requirements are considered substantive for the purpose of ascertaining whether the levels and limitations set in the decision document have been attained.

- RCRA permits are not required for CERCLA actions taken entirely onsite.

- Administrative RCRA requirements, such as reporting and recordkeeping requirements, are not applicable or relevant and appropriate for onsite activities at CERCLA sites.

- Consultations with administrative bodies are considered administrative requirements.

Based on the above principles, we have put together Table 19-4 to distinguish between substantive and administrative requirements for roll-off containers storing known hazardous wastes at a CERCLA site.

Case Study 19-4: Determining If LDR Standards Are ARARs at a CERCLA Site

A number of drums containing liquid wastes are discovered during a site investigation at a CERCLA site. No written documentation or specific knowledge of the wastes' source is available to identify with certainty the origins of the wastes. Laboratory analyses indicate that the wastes do not exhibit a characteristic, but they contain high concentrations of solvent chemicals indicative of industrial spent solvent streams. The CERCLA site manager determines that incineration would be technically suitable, with subsequent onsite land disposal of incinerator ash. Are LDR treatment standards ARARs for the incinerator ash?

In order for LDR standards to be applicable, the CERCLA hazardous substance has to be a RCRA hazardous waste. Since the waste does not exhibit a characteristic and the source is unknown, no hazardous waste codes can be assigned; thus, LDR treatment standards are not applicable. These standards are relevant and appropriate, however, if two primary considerations are satisfied:

- There is a close match between the CERCLA and LDR objectives—The objectives of the LDR program (see Chapter 13) are to reduce the toxicity and/or mobility of a hazardous waste, based on treatment using best demonstrated available technology prior to its land disposal. Maximum destruction of the drum contents is established as the remedial action objective, which is consistent with LDR program objectives. Thus, LDR standards appear to be well-suited to this situation.

- There is a close match between the constituents/matrix of the CERCLA waste and the constituents/matrix of the relevant RCRA waste codes—In this scenario, there is a general similarity between the drummed liquids and the spent solvents listed in the F001–F005 waste codes. Thus, the CERCLA waste is "sufficiently similar" to a listed RCRA waste code or family of waste codes such that the LDR standard for that waste code or family of waste codes is appropriate for the CERCLA waste.

Because the two considerations discussed above are satisfied, LDR standards are determined to be relevant and appropriate requirements for this CERCLA action involving treatment and onsite placement of the waste. [OSWER Directives 9347.3-05FS and 9347.3-07FS] Note that if the wastes were instead sent offsite for disposal, LDR standards would not have to be met because the wastes aren't hazardous at the point of generation.

Table 19-4: Substantive vs. Administrative Requirements for Roll-Off Containers Storing Known Hazardous Wastes at a CERCLA Site

Requirement	Substantive (required)	Administrative (not required)
Ensure containers are in good shape/not leaking or deteriorated [§§264.171/265.171]	X[1]	
Ensure materials of container construction are compatible with waste [§§264.172/265.172]	X[1]	
Ensure containers are kept closed at all times except when adding/removing waste [§§264.173/265.173]	X[1]	
Inspect containers at least weekly [§§264.174/265.174]	X[1]	
Provide containers with secondary containment [§264.175]	X[1]	
Comply with special requirements for D001/D003 wastes [§§264.176/265.176]	X[1]	
Comply with special requirements for incompatible wastes [§§264.177/265.177]	X[1]	
Close container storage area after use [§264.178]	X[1]	
Comply with one-year storage prohibition [§268.50]	X[1]	
Comply with Part 264/265, Subpart D contingency plan requirements		X[2]
Keep track of 90-day clock for containers		X[3]
Biennial reporting		X[4]

[1] Per EPA/540/G-89/006.
[2] Per OSWER Directive 9234.2-04/FS.
[3] Per EPA/540/R-98/020, available at http://www.epa.gov/superfund/contacts/sfhotline/arar.pdf.
[4] Per RO 14842.

Source: McCoy and Associates, Inc.

Q *If wastes from noncontiguous facilities are combined onto one site for treatment, is this viewed as an offsite activity, triggering RCRA administrative requirements such as permitting?*

A No. CERCLA Section 104(d)(4) authorizes EPA to treat two or more noncontiguous facilities as one site for purposes of response, if such facilities are reasonably related on the basis of geography or their potential threat to public health, welfare, or the environment. Because the combined remedial action constitutes onsite action, compliance with permitting or other RCRA administrative requirements would not be required. [OSWER Directive 9234.2-01/FS-A]

19.1.5.2.3 "To be considered" guidance

ARARs are legally binding laws and regulations. Other information that should be considered for CERCLA remedial actions consists of federal and state environmental and public health criteria, advisories, guidance, and proposed standards. These criteria and guidelines are called "to be considered" (TBC) materials and are meant to complement the use of ARARs—not to compete with or replace them. Because TBC guidelines are not ARARs, their identification and use are not mandatory. However, if no ARARs address a particular situation, TBC documents may be used as a basis for CERCLA actions. Exhibit 1-10 in EPA/540/G-89/006 identifies numerous TBC documents that should be considered, including the following types of RCRA documents:

- EPA RCRA design guidelines,
- Permitting guidance manuals,
- Technical resource documents, and
- Test methods for evaluating solid wastes, including SW–846.

19.1.6 Applicability of the G5 MACT During Site Remediation

In addition to RCRA, site remediation may also be subject to the maximum achievable control technology (MACT) standards in 40 *CFR* Part 63, Subpart GGGGG. For these CAA regulations to apply, all three of the following conditions must be met [§63.7881(a)1–3]:

1. The facility is, or will be, a major source of hazardous air pollutants (HAPs) during the site remediation.

2. The site is cleaning up "remediation material," which means a material that contains one or more volatile organic HAPs (or VOHAPs) listed in Table 1 of Subpart GGGGG and is either a material found in naturally occurring media or is found in substantially intact containers, tanks, storage piles, or other storage units. [§63.7957]

3. There are other nonremediation stationary sources at the facility that emit HAPs and meet the definition for another air toxics source category.

Note that the term VOHAP, as included in the second applicability condition above, has no RCRA counterpart. Therefore, no direct relationship exists between a CAA VOHAP and a RCRA hazardous waste. In other words, site remediation activities regulated under RCRA may or may not be regulated under the Subpart GGGGG MACT.

If all three of the applicability conditions noted above are met, and no exemptions from the standards apply, Subpart GGGGG MACT controls are required during a site remediation. The following is a synopsis of these CAA requirements:

- Process vents on certain remediation equipment must comply with §63.7885,
- Remediation material management units (e.g., tanks, containers, surface impoundments, oil/water separators) must comply with §63.7886,
- Equipment leaks on certain remediation equipment must comply with §63.7887,
- Remediation equipment must comply with numerous general requirements in §63.7935,
- Reports must be submitted as described in §§63.7950–63.7951, and
- Records must be kept per §63.7953.

19.2 Demolition

Much of the equipment and some of the structures built during the building boom of the 1950s, 60s, and 70s are now reaching the end of their useful life and are being decommissioned and demolished. This section looks at the RCRA issues associated with demolition activities.

19.2.1 Building/structure assessment and pre-demolition activities

Before beginning a demolition project, the building/structure owner/operator should develop a work plan. This plan guides personnel through the various steps of assessing and demolishing the building/structure and managing all generated materials and wastes. Sample demolition work plans can be found at:

- http://www.osha.gov/doc/outreachtraining/htmlfiles/demolit.html; and
- http://www.epa.gov/region1/ge/thesite/geplantarea/reports/20s30s40s/30s/258051.pdf.

Included in the demolition work plan will be a hazardous materials assessment. This assessment can be performed by facility personnel or by hiring an independent hazard assessment expert. It's referred to as a hazardous "materials" assessment—not a hazardous waste assessment—because waste is not generated until 1) components/equipment are removed from the building/structure for recycling or disposal prior to demolition, 2) residues are generated from cleaning/decontaminating the building/structure prior to demolition, and/or 3) debris is generated when the building/structure is actually demolished. (These point of generation concepts are discussed in more detail in Section 19.2.2.)

Once hazardous materials are identified during the assessment, they are usually removed from the building/structure prior to demolition. The pre-demolition removal of hazardous materials accomplishes two goals: 1) ensuring that these hazardous materials (many of which will be hazardous wastes when discarded) will be managed appropriately under the RCRA Subtitle C regulations instead of diluting them within the much larger volume of potentially nonhazardous demolition debris; and 2) increasing the likelihood that the large volume of demolition debris generated will be nonhazardous, allowing management in a Subtitle D landfill.

Table 19-5: Pre-Demolition Building/Structure Hazardous Materials Assessment

Assessment item	Text reference
Building/structure	
■ Assessment of any containers: 1) if product, remove for use; 2) if waste (e.g., used rags, used oil, used aerosol cans, unused but unusable product, drums of unknown material/waste), make a hazardous waste determination and manage accordingly; 3) if RCRA-empty, manage as nonhazardous waste or scrap metal	Sections 9.5, 14.2, 18.2, 18.3
■ Assessment of any noncontainerized debris (e.g., pallets, cardboard, tires, trash); make a hazardous waste determination and manage accordingly	Sections 18.2, 18.3
■ Assessment for lead-based paint on building/structure components	Sections 2.6.6, 19.2.1.1
■ Assessment for mercury-containing lamps, thermostats, switches, and other types of equipment that contain mercury	Sections 8.1, 8.3, 19.2.1.2
■ Assessment for chlorofluorocarbons (CFCs) in air conditioning, heat pump, or other refrigeration equipment	Sections 2.6.4, 19.2.1.3
■ Assessment for PCBs in transformers, capacitors, light ballasts, paint, caulking, roofing materials, and other building/structure materials	Sections 2.6.4, 2.8.4.1, 19.2.1.4
■ Assessment for asbestos in vinyl floor tiles, roofing felt, ceiling texture, cement stucco and gypsum plaster, roofing and siding shingles, insulation/fire-proofing, and other building/structure materials	Section 19.2.1.5
Process equipment within building/structure	Sections 4.17, 19.2.1.6
Utilities within building/structure	Sections 12.1.1, 19.2.1.7
Hazardous waste tanks/container accumulation areas within building/structure	Section 19.2.1.8
Spills/stains on walls, concrete	Section 19.2.1.9
Assessment of soil/ground water contamination (either before or after demolition)	Section 19.2.1.10

Source: McCoy and Associates, Inc.

The items to be covered in the pre-demolition building/structure hazardous materials assessment are included in Table 19-5; recommended pre-demolition material removal activities are also included in this table. Table 19-5 also includes references to sections in this book that will assist the reader in accomplishing these assessment and removal activities.

19.2.1.1 Lead-based paint assessment/removal

Painted surfaces of buildings/structures, such as structural steel, drywall, ceilings, and exterior surfaces, could contain lead-based paint (LBP). Most older buildings have multiple layers of paint and all layers should be considered. Thus, a lead survey before demolition is advised.

A logic diagram that may be used for managing buildings/structures contaminated with LBP is given in Figure 19-2. The first step is a LBP assessment; portable x-ray fluorescence (XRF) instruments are often used for the detection of lead in paint or coatings on buildings/structures.

If damaged or deteriorating lead-based coatings are found on the building/structure, Step 2 directs the owner/operator to remove or stabilize such materials prior to demolition. Otherwise, they will likely detach during demolition, causing lead contamination of the air and/or surrounding soil. Loose or peeling

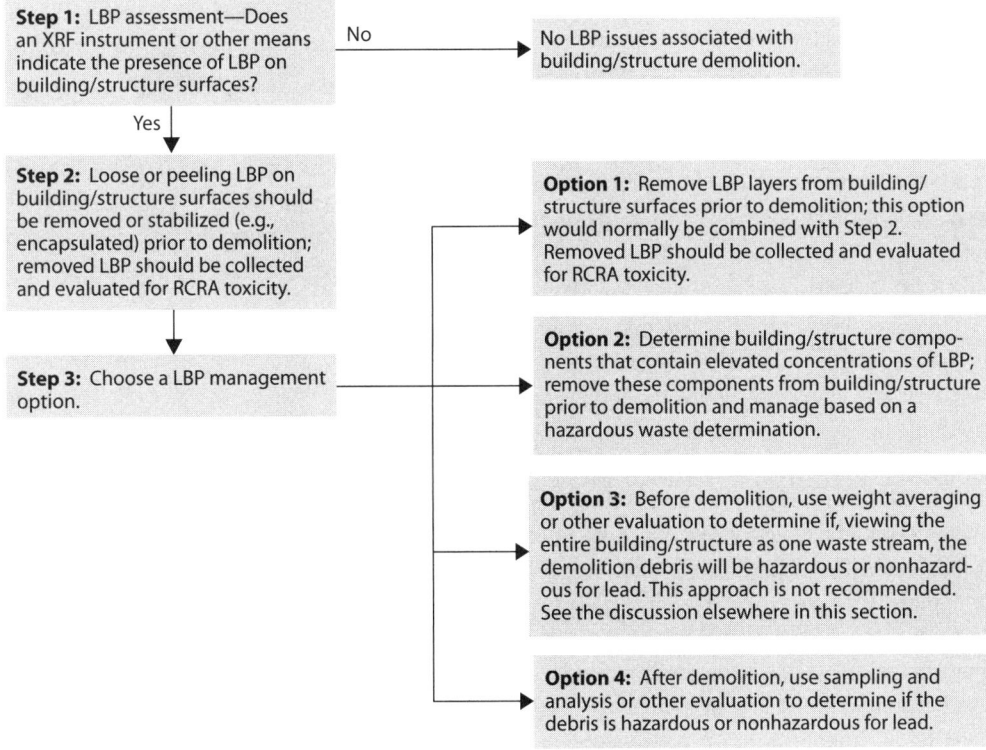

Figure 19-2: Managing Lead-Based Paint During Building/Structure Demolition

LBP = lead-based paint; XRF = x-ray fluorescence.
Source: McCoy and Associates, Inc.

LBP may be scraped, brushed, or even vacuumed to remove surface deposits before demolition. Whichever method is chosen, the removed LBP should be collected and evaluated for toxicity (D008).

Once loose or peeling LBP on building/structure surfaces is removed or stabilized, Step 3 is deciding on a LBP management option. Four options from Figure 19-2 are discussed below:

- Option 1: Removing LBP layers from building/structure surfaces prior to demolition—This is usually very expensive, but the advantage with this approach is that the debris generated from demolition will likely not be hazardous. There are a number of commercially available methods that can be used to remove LBP from a standing building/structure. Of course, any LBP dust/chips/debris generated during these activities would be subject to a hazardous waste determination. See Section 2.6.6 for a discussion of the RCRA issues associated with LBP removal.

- Option 2: Before demolition, those building/structure components that contain elevated concentrations of LBP are removed—Using this approach, it is likely that at least some of these removed building components will require management as D008 hazardous waste.

- Option 3: Before demolition, evaluate whether the building/structure taken as a whole will generate hazardous or nonhazardous demolition debris—Taken as a whole, the debris resulting from demolishing an entire building is not likely to be hazardous for lead. Although the concentration of lead in LBP can be quite high, the relatively thin layers of paint typically result in a mass of lead that is small compared to the total mass of the building/structure to be demolished. If the building as a whole is determined to be nonhazardous, the generator might decide to dispose of all of the demolition debris in a nonhazardous waste landfill. There are at least two potential problems with this approach. First, a significant amount of lead-contaminated material would be placed in the nonhazardous waste landfill. This may not be protective of human health and the environment and could conceivably result in CERCLA liability. Second, even though the building as a whole is determined to be nonhazardous, many regulators would consider each container of demolition waste to be a separate decision unit for purposes of making a hazardous waste determination. If the waste in an individual container is determined to be hazardous but is not managed as such, a regulatory agency could pursue enforcement action under RCRA.

- Option 4: After demolition, evaluate whether the entire debris pile is hazardous or nonhazardous—This is a tough option because it is quite difficult to obtain a representative sample (for subsequent analysis) or otherwise make a hazardous waste determination for the very heterogeneous debris generated from building/structure demolition. See Section 19.2.3 for more on this topic.

All of the options discussed above require a sampling plan and lead concentration determinations. The following information may be helpful:

- The concentration of lead in paint in ppm, percent weight, or mg/cm^2 can be determined using XRF analyzers. Alternatively, laboratory analysis of representative samples may be used.

- ASTM E1908, *Standard Guide for Sample Selection of Debris Waste from a Building Renovation or Lead Abatement Project for Toxicity Characteristic Leaching Procedure Testing for Leachable Lead* (available from http://www.astm.org). In general terms, this standard requires building components to be drilled to collect samples in proportion to the volume of those components in the entire building. The component samples are assembled, and the resulting assembled sample is analyzed according to the TCLP protocol.

- *Guidance for the Management and Disposal of Lead-Contaminated Materials Generated in the Lead Abatement, Renovation, and Demolition Industries*, Connecticut Department of Environmental Protection, updated May 18, 2007, available from http://www.ct.gov/deep/lib/deep/waste_management_and_disposal/construction_renovation_demolition/lead_debris_disposal_guidance_5-18-07.pdf.

- *Suggested Sampling Plans for Building Debris Disposal*, Washington Department of Ecology, available at http://www.ecy.wa.gov/programs/hwtr/dangermat/samplePlans.html.

- Significant information on EPA's regulations and guidance associated with LBP can be found at http://www2.epa.gov/lead/lead-laws-and-regulations.

19.2.1.2 Mercury-containing equipment assessment/removal

During a pre-demolition building/structure assessment for mercury-containing equipment, thermostats and lamps (e.g., fluorescent, mercury vapor, metal halide, high-pressure sodium) are of primary concern. Additionally, flow meters, pumps, thermometers, switches, and relays can all contain elemental mercury. Once identified, these items should all be removed from the building/structure prior to demolition and either 1) reused as products for their intended purpose (e.g., reusing lamps in another building), or 2) managed as universal waste or hazardous waste (D009).

EPA information on these materials is available at:

- Mercury-Containing Equipment, http://www.epa.gov/osw/hazard/wastetypes/universal/mce.htm;

- Recycling Mercury-Containing Light Bulbs (Lamps), http://www.epa.gov/osw/hazard/wastetypes/universal/lamps/index.htm; and

- Mercury Laws and Regulations, http://www.epa.gov/mercury/regs.htm.

19.2.1.3 CFCs assessment/removal

Although primarily a CAA consideration, CFCs can become subject to RCRA if they are a contained gas (see Section 14.4). Additionally, spent CFCs can exhibit the toxicity characteristic due to the small amounts of carbon tetrachloride (D019) or chloroform (D022) contaminants often found in them. An exclusion in §261.4(b)(12) allows nonhazardous management of these spent CFCs that exhibit the toxicity characteristic, but it applies only if the CFCs will be reclaimed for reuse (see Section 2.6.4).

During the building/structure assessment for CFCs, all air conditioning, heat pump, or other refrigeration equipment that contains CFC refrigerant should be identified. CAA-regulated refrigerants may not be released during salvaging or dismantling activities; they must be properly recovered using approved equipment operated by qualified technicians. Significant information is available from EPA on the management of CFCs from stationary refrigeration and air conditioning equipment at http://www.epa.gov/ozone/title6/608/index.html.

19.2.1.4 PCBs assessment/removal

PCBs are not hazardous waste subject to the federal RCRA regulations, although certain forms of PCB waste can be captured under state hazardous waste programs (e.g., New York). Because PCBs are ubiquitous in older buildings, a PCBs assessment should be conducted prior to demolition. Although not subject to RCRA, EPA regulates PCBs under TSCA, as codified at 40 *CFR* Part 761. These regulations, as well as final rules, proposed rules, and notices in the *Federal Register*; and interpretative guidance related to PCBs can be found at http://www.epa.gov/epawaste/hazard/tsd/pcbs/pubs/laws.htm.

Although the manufacture of PCBs was banned in 1979 by TSCA, PCBs can still be found in buildings/structures slated for demolition. The PCB items most commonly associated with old buildings/structures are transformers, capacitors, light ballasts, and hydraulic-based equipment. Finally, paint, caulking, roofing materials, and other building/structure materials may contain PCBs as discussed in the following EPA guidance documents:

- PCBs in light ballasts, http://www.epa.gov/wastes/hazard/tsd/pcbs/pubs/ballasts.htm and http://www.epa.gov/epawaste/hazard/tsd/pcbs/pubs/ballastchart.pdf.

- PCBs in caulk, http://www.epa.gov/pcbsincaulk/; and

- PCBs in caulk and other building materials, http://www.epa.gov/epawaste/hazard/tsd/pcbs/pubs/caulk/caulkremoval.htm.

Once identified, these PCB-containing materials should be removed from the building/structure prior to demolition and managed per the TSCA regulations noted above.

19.2.1.5 Asbestos assessment/removal

Asbestos is not a hazardous waste subject to the federal RCRA regulations, although certain forms of asbestos waste can be captured under state hazardous waste programs (e.g., California). Because this material is ubiquitous in older buildings and industrial structures, an asbestos survey should be conducted prior to demolition. Although not subject to RCRA, EPA regulates asbestos under 1) the CAA (through the asbestos NESHAP regulations codified at 40 *CFR* Part 61, Subpart M); and 2) TSCA (through the asbestos worker protection rule codified at 40 *CFR* Part 763, Subpart G). An excellent summary of EPA's asbestos regulations is contained in "Notification of Rules and Regulations Regarding the Demolition of Asbestos-Containing Structures," June 8, 2012, available at http://www2.epa.gov/sites/production/files/documents/nps5da.pdf.

OSHA, of course, regulates worker exposure to asbestos; those rules are codified at 29 *CFR* 1910.1001, 1915.1001, and 1926.1101. These OSHA standards, as well as final rules, proposed rules, and notices in the *Federal Register*; directives (instructions for compliance officers); and standard interpretations (official letters of interpretation of the standards) related to asbestos can be found at http://www.osha.gov/SLTC/asbestos/standards.html.

Once identified, asbestos-containing materials should be removed from the building/structure prior to demolition and managed per the TSCA/OSHA regulations noted above.

19.2.1.6 Process equipment within building/structure

Process equipment located within a building/structure slated for demolition will normally be decontaminated and removed prior to destruction of the building/structure. Because the process streams contained within this equipment were not wastes during operation, the residues remaining within this equipment are subject to §261.4(c). That section says that these residues are exempt from RCRA regulation until one of two things occurs:

1. The residues are removed from the process unit in which they were generated; or

2. The residues remain in the inactive unit for more than 90 days.

See Section 4.17 for a more complete discussion of the §261.4(c) exemption.

Once these residues potentially become subject to RCRA for one of the two reasons noted above, a hazardous waste determination must be made. If the residue is hazardous, it must be managed as such. After it has been decontaminated, the process equipment should be removed from the building/structure and 1) reused somewhere else as nonwaste equipment, 2) sent for scrap metal recycling if metallic, or 3) disposed.

19.2.1.7 Utilities within building/structure

When conducting a building/structure assessment, utility systems may present RCRA issues. Utility systems that should be evaluated for removal before demolition include:

- Hydraulic systems—Hydraulic systems, such as those used in production assembly lines and elevators, contain oil. This oil will need to be drained before the equipment, tanks, and piping are disassembled and removed. Providing the oil does not contain PCBs at a TSCA-regulated level, the drained hydraulic oil can be managed as used oil, as discussed in Section 12.1.1. Once decontaminated, the hydraulic equipment should be removed from the building/structure and 1) reused somewhere else as nonwaste equipment, 2) sent for scrap metal recycling if metallic, or 3) disposed.

- Drain lines and traps—An assessment of drain lines and traps within the building/structure prior to demolition will identify potential RCRA issues. Old lead pipes should be identified and removed; they can usually be managed as scrap metal. Mercury often accumulates in low spots or traps in drain lines (especially from laboratory sinks); these areas should be investigated, and any piping/traps containing elevated levels of mercury should be removed and managed as potential D009 wastes. Testing or knowledge should be used to determine if traps exhibit the toxicity characteristic; if so, they should be removed and managed as hazardous waste prior to demolition. Liquids, such as water, should be removed from drain lines prior to demolition and containerized for subsequent characterization and disposal.

- Sumps and floor drains/trenches—Building sumps and floor drains/trenches should be visually inspected for standing liquids and/or solids buildup. Any liquids should be drained and solids removed and—after a hazardous waste determination—managed appropriately.

- Tritium exit signs—Many self-luminous exit signs contain significant amounts of radioactive tritium. Self-luminous exit signs should have a permanent label affixed to the sign that identifies it as containing radioactive material. In addition, the label should include the name of the manufacturer, the product model number, the serial number, and the quantity of tritium contained. To avoid release of radioactive tritium, these signs should be removed before building/structure demolition. However, it is illegal to abandon or dispose these signs; instead, they should be returned to the manufacturer or sent to a facility licensed by the Nuclear Regulatory Commission (NRC) to accept them. The NRC's fact sheet on tritium exit signs is available at http://www.nrc.gov/reading-rm/doc-collections/fact-sheets/fs-tritium.html. EPA provides additional information on these signs at http://www.epa.gov/radtown/exit-signs.html.

- Uninterruptible power supply (UPS) and other battery systems—Lead-acid batteries are typically used in UPS systems, but other types such as lithium and nickel-cadmium may also be used. When removed prior to demolition, batteries from UPS and other battery systems may be reused as products or sent for recycling or disposal as universal waste or potentially hazardous waste.

- Fire extinguishers/fire suppression systems—Prior to demolition, fire extinguishers/fire suppression systems should be removed. Both portable and installed fire suppression systems should be evaluated to determine if they contain halons. Halon fire extinguishers/fire suppression systems should not be discharged, as intentionally releasing these substances is prohibited under federal regulations. Halons and halon-containing equipment must be properly disposed in accordance with EPA standards. If halon-containing equipment is sent offsite for disposal, it must be sent to a manufacturer, fire equipment dealer, or recycler operating in accordance with National Fire Protection Association (NFPA 10 and NFPA 12A) standards. EPA's halon-handling website address is http://www.epa.gov/ozone/title6/608/halons/.

19.2.1.8 Hazardous waste tanks/container accumulation areas within building/structure

Hazardous waste tanks/container accumulation areas located within a building/structure slated for demolition will have to be closed as follows:

- Permitted hazardous waste tanks and container storage areas will have to be decontaminated and closed per the specific closure plan, as required in §264.112. In addition to meeting the general closure standards of §§264.111 and 264.114, RCRA-permitted tanks must be closed according to §264.197 and permitted container storage areas must be closed according to §264.178.

- 90-day accumulation tanks must be closed in compliance with §§265.111, 265.114, and 265.197(a–b).

- 90-day container accumulation areas must be closed in compliance with §§265.111 and 265.114. No unit-specific closure standards apply to 90-day containers. [RO 14321]

- 180-day accumulation tanks must comply with the tank closure requirement in §265.201(f).

- 180-day container accumulation areas are not subject to any closure requirements in the federal RCRA regulations. If no state provisions are applicable, they should be closed using best professional judgment. No unit-specific closure standards apply to 180-day containers. [RO 14321]

- Satellite accumulation areas are not subject to any closure requirements in the federal RCRA regulations. If no state provisions are applicable, they should be closed using best professional judgment.

In addition to the specific closure requirements noted above, any spills/stains on walls or concrete associated with these units should be addressed per Section 19.2.1.9 below.

19.2.1.9 Spills/stains on walls, concrete

Spills of products, process streams, maintenance materials, and wastes are a fact of life. Despite good housekeeping practices, spills may result in stains on building/structure walls or concrete.

As noted in Section 14.3.1, materials that are spilled and not cleaned up are considered to be abandoned and, therefore, solid waste per §261.2(b). If the solid waste meets the definition of hazardous waste, then the abandoned material could be hazardous waste. It is of particular importance to decontaminate buildings/structures before demolition so that hazardous demolition debris is not generated (see Section 19.2.3).

Sometimes the following question is raised: Is cleaning potentially hazardous spills/stains from a building/structure prior to demolition considered treatment? EPA answered this question as follows: "[A]n intact building would not yet be a solid waste, and therefore, extraction of contaminants would not involve hazardous waste treatment." [RO 11841]

19.2.1.10 Assessment of contaminated soil/ground water adjacent to or underlying building/structure

During the development of the demolition work plan, some thought should be given to assessing the potential for soil and/or ground water contamination adjacent to or underlying the building/structure. And if soil and/or ground water contamination is found, will it need to be remediated following demolition?

The pre-demolition evaluation of potential contamination will be based on 1) knowledge of spills or other releases from the building/structure during its operation; and 2) visual and/or olfactory evidence of potential contamination, as supplemented with field instruments and/or samples and laboratory analytical results. Additionally, visual and/or olfactory evidence of potential contamination may also be discovered during or after the demolition activities.

If soil and/or ground water contamination adjacent to or underlying the building/structure is discovered before, during, or after demolition, the site owner/operator will need to decide if remediation is required once demolition is complete. This determination is usually based on site-specific health risk-based action levels (e.g., contaminant concentrations that will result in one-in-one-million excess cancer risk to the maximally exposed population). If a decision is made to remediate contaminated soil and/or ground water, the RCRA implications of implementing the remediation are discussed in Sections 19.1.2–19.1.4.

Other factors that may trigger remediation of soil and/or ground water contamination adjacent to or underlying a demolished building/structure include:

- The soil and/or ground water contamination may require remediation as part of the formal closing of a hazardous waste management unit (e.g., hazardous waste tank or container accumulation area) that was located in or adjacent to the building/structure.

- Post-1980 spills or other releases of materials (products, process streams, or wastes) to the soil (which possibly migrated to ground water) that were not properly cleaned up could be a RCRA enforcement issue if not remediated. After RCRA started in 1980, if we spill a material and don't clean it up, that material is considered to be abandoned. [RO 14650] Abandoned materials, as defined in §261.2(b), are one of the most common types of solid waste. So, the abandoned spill material would be a solid waste, and, if that material meets the definition of a hazardous waste, the soil and/or ground water contamination could be considered an unpermitted hazardous waste disposal area.

19.2.2 Point of generation for demolition wastes

A hazardous waste determination is supposed to be made at the point of generation (POG) of a solid waste. There are three primary POGs for demolition wastes:

1. Components/equipment are removed from the building/structure for reuse, recycling, or disposal prior to demolition;
2. Residues are generated from cleaning/decontaminating the building/structure prior to demolition; and
3. Debris is generated when the building/structure is demolished.

19.2.2.1 POG for components/equipment removal

As noted in Section 19.2.1, components/equipment that may be hazardous should be removed prior to demolishing the building/structure. For example, lead pipes and flashing, fluorescent lamps and PCB ballasts, mercury thermostats, batteries, fire extinguishers, exit signs, etc. should be removed and reused, recycled, or disposed.

If reused (e.g., a used battery removed from a building slated for demolition could be reused as a battery in a different building or application without reclamation), the components/equipment never became a waste subject to RCRA. [January 4, 1985; 50 *FR* 624, RO 11258, 11541, 11868, 14281, 14677] In this scenario, there is no POG of demolition wastes.

If recycled (e.g., a battery sent for reclamation), then the components/equipment would be considered spent material. Because of the asterisk at the intersection of "Spent materials" and "Reclamation" in Table 1 in §261.2(c), components/equipment removed from a building/structure prior to demolition that are sent for reclamation are solid wastes, requiring a hazardous waste determination. See Section 1.2.5.1 for a more thorough discussion of this topic. In this alternative, the POG for these demolition wastes would be when a decision is made to send them for recycling; usually, this decision will have been made when the demolition work plan was finalized, and so the POG would typically be at the point the components/equipment are removed from the building/structure. (Note that spent materials sent for reclamation are not solid wastes if one of the reclamation exclusions discussed in Sections 11.2.3.1 and 11.2.3.2 applies.)

If disposed, components/equipment removed from a building/structure prior to demolition are solid wastes, requiring a hazardous waste determination. In this alternative, the POG for these demolition wastes would be when a decision is made to send them for disposal; usually, this decision will have been made when the demolition work plan was finalized, and so the POG would typically be at the point the components/equipment are removed from the building/structure.

19.2.2.2 POG for residues from cleaning/decontaminating

As noted in Section 19.2.1.9, spills/stains on walls and concrete should be removed before demolition, particularly if these spills/stains would otherwise result in the generation of hazardous demolition debris (reference Section 19.2.3). In addition to residues resulting from cleaning spills/stains, other residues can also be generated from the decontamination of the building/structure. Examples include depainting of building/structure components coated with LBP and the removal of asbestos-containing materials.

Where is the POG of these residues generated from cleaning/decontaminating the building/structure? "EPA considers the actual removal of the contaminants to be the point of waste generation and consequently, the point at which the RCRA regulations become applicable." [RO 11841]

19.2.2.3 POG for demolition debris

The POG for debris from a demolition project is when the debris is generated (e.g., when a building/structure is knocked down), not when a decision is made to demolish the building/structure.

> "[W]e believe that an intact, standing building continues to perform the essential functions of a building and so need not, and should not be considered to be 'discarded' under §261.2(a)(2)(i) until it is actually destroyed." [RO 11841]

19.2.3 Are demolition wastes hazardous?

Once a facility knows that demolition wastes are generated (i.e., they are solid wastes) via application of EPA's guidance discussed in Section 19.2.2, a hazardous waste determination for those wastes must be made. In EPA/530/K-04/005 (available at http://www.epa.gov/wastes/inforesources/pubs/infocus/rif-cd.pdf), EPA notes that "[m]ost [construction and demolition] debris is nonhazardous and is not regulated by EPA." However, that statement does not relieve generators of demolition wastes from making a hazardous waste determination. This determination is made by asking the four questions—reference Sections 2.1 and 18.1 for more information on the four questions that must be evaluated when making a hazardous waste determination.

Once it is known that demolition waste is a solid waste, the second question to ask [from §262.11(a)] is whether the material is excluded or exempt from the RCRA regulations. Most of these exclusions and exemptions are discussed in Chapter 4 (see specifically Table 4-1), and they are largely based on application of knowledge enhanced by EPA guidance. Probably the most useful demolition waste exemptions are for scrap metal [§261.6(a)(3)(ii)] and universal waste. [§261.9] For example, some components/equipment will be exempt as:

- Scrap metal—Including lead pipes, I-beams, radiators, and flashing; or

- Universal waste—Including fluorescent lamps, mercury thermostats and other mercury-containing equipment, and batteries.

If demolition waste meets the definition of scrap metal in §261.1(c)(6) or universal waste in §273.9, a hazardous waste determination is not required (if the waste will be managed according to the exemption-specific conditions). Instead, the reader can skip to Sections 19.2.4.1.1 and 19.2.4.2.1 for on- and offsite management of scrap metal or Sections 19.2.4.1.4 and 19.2.4.2.2 for on- and offsite management of universal waste.

If generated and not exempt, a determination of whether the demolition waste meets a listing description or exhibits a characteristic will have to be made. One of the primary challenges in determining if demolition wastes exhibit a characteristic is in obtaining a representative sample from this highly heterogeneous waste stream. The issues associated with evaluating demolition wastes for characteristics are discussed in Section 19.2.3.1. The complicated issue of whether demolition wastes meet a listing description is tackled in Section 19.2.3.2.

19.2.3.1 Are demolition wastes characteristic?

As noted in Section 19.2.2, there are three primary waste streams generated during a building/structure demolition project:

1. Components/equipment removed from the building/structure for recycling or disposal prior to demolition,

2. Residues generated from cleaning/decontaminating the building/structure prior to demolition, and

3. Debris generated when the building/structure is demolished.

Making a characteristic waste determination for each of these three waste streams is discussed in the three subsections below. Additional discussion on making a determination as to whether a specific demolition waste exhibits a hazardous characteristic is given in Sections 18.2.2 and 18.3.3.

19.2.3.1.1 Components/equipment removed from the building/structure

As noted in Section 19.2.2.1, potentially hazardous components/equipment that are removed prior to demolishing the building/structure are solid wastes if they will be recycled (e.g., reclaimed) or disposed.

Removed components/equipment that are solid waste and for which no exemption applies must be evaluated to determine if they exhibit a characteristic. In many cases, manufacturers' information or other knowledge can be used for this purpose. This knowledge can usually be used to determine that these demolition wastes are not ignitable, corrosive, or reactive. Regarding toxicity, if the facility has knowledge/documentation showing that certain §261.24 constituents could not be present in a specific demolition waste, the TCLP will have to be performed only for the other toxicity constituents that cannot be eliminated through knowledge. [EPA/530/R-93/007, RO 11603, 14695] As an alternative to running the TCLP, some facilities may ask the lab to run totals analysis for those constituents. Based on the results, the TCLP may then be requested only for those constituents with the potential to exceed regulatory levels. See Section 2.6.3.1 for a discussion of this option.

Based on the above evaluations, many removed components will not be characteristically hazardous. Note, however, that some nonhazardous components will require management per other regulatory programs, such as:

- PCB-containing equipment per TSCA (reference Section 19.2.1.4),

- Tritium-containing exit signs per NRC (reference Section 19.2.1.7), and

- Halon-based fire suppression systems per NFPA.

19.2.3.1.2 Residues generated from cleaning/decontaminating the building/structure

Residues generated from cleaning/decontaminating the building/structure include concrete chips/dust, wood chips/dust, paint chips/dust, used cleaning rags, used washwater, spent solvents, and asbestos-containing material. Knowledge can usually be used to determine if these demolition wastes are ignitable, corrosive, or reactive. Knowledge or TCLP testing will be required to make a toxicity determination. As an alternative to running the TCLP, some facilities may ask the lab to run totals analysis for the contaminants in §261.24. Based on those results, the TCLP may then be requested only for those constituents with the potential to exceed regulatory levels. See Section 2.6.3.1 for a discussion of this option.

Based on the above evaluations, many cleaning/decontamination residues will not be characteristically hazardous. Note, however, that some nonhazardous residues will require management per other regulatory programs, such as:

- Asbestos-containing material per CAA, TSCA, and OSHA (reference Section 19.2.1.5), and

- Potentially infectious substances (e.g., molds, fungi, or viruses) per OSHA.

19.2.3.1.3 Building/structure demolition debris

Before a hazardous waste determination is made for building/structure demolition debris, any metallic components, including structural steel, metal decking/siding/roofing, stairways, handrails, piping and other equipment, etc., should be segregated for scrap metal recycling. Nonmetallic building/structure demolition debris should be evaluated to determine if it exhibits a characteristic as follows.

Knowledge can usually be used to determine that building/structure demolition debris is not ignitable, corrosive, or reactive. If the facility has knowledge/documentation showing that certain §261.24 constituents could not be present in a specific demolition waste, the TCLP will have to be performed only for the other toxicity constituents that cannot be eliminated through knowledge. [EPA/530/R-93/007, RO 11603, 14695]

Oftentimes it is challenging to obtain representative samples of demolition debris to make a toxicity determination. For example, EPA noted at 63 *FR* 70196 [December 18, 1998] that the heterogeneity of debris and difficulties encountered during the laboratory sample size-reduction step both make it very difficult to generate reproducible TCLP data on debris. Although the agency noted at 57 *FR* 990 [January 9, 1992] that facility personnel should ensure that some of every material in the decision unit (amount of debris you are trying to characterize) is collected in the sample in the same proportions as in the decision unit, that is easier said than done. See Section 18.3.3.4 for examples of characterizing building demolition debris for toxicity.

As an alternative to running the TCLP, some facilities may ask the lab to run totals analysis for the constituents in §261.24. Based on those results, the TCLP may then be requested only for the constituents with the potential to exceed regulatory levels. See Section 2.6.3.1 for a discussion of this option.

Probably the most likely characteristic that building/structure debris will exhibit is toxicity for lead due to the presence of LBP. The options for making toxicity determinations for such debris were previously discussed in Section 19.2.1.1.

19.2.3.2 Are demolition wastes listed?

Demolition waste, particularly debris, must be managed as listed waste if it contains listed waste. This conclusion results from the definition of "Hazardous debris" in §268.2(h) and application of the contained-in policy, which is discussed in Section 5.3.2. Therefore, demolition waste that contains an F-, K-, P-, or U-waste must be managed as that listed hazardous waste.

How do owners/operators determine if contaminated demolition waste (e.g., demolished concrete slab with dark stains) is listed? Hazardous waste listings are based on the sources of, or the processes that generated, the wastes rather than the concentrations of hazardous constituents. Therefore, analytical testing alone, without information on a waste's source, will not produce information that will conclusively indicate whether a given demolition waste is a listed hazardous waste. [RO 13181]

In some situations, it is hard for facility owners/operators to determine the source of the contamination, and so it is difficult to tell if a demolition waste is a listed waste. If the owner/operator has made a good-faith effort to find out if the contamination is a listed waste but cannot make such a determination because documentation on the source or process is unavailable or inconclusive, EPA allows the owner/operator to assume that the material is not a listed hazardous waste. (See, for example, December 21, 1988; 53 *FR* 51444, March 8, 1990; 55 *FR* 8758, and April 29, 1996; 61 *FR* 18805.) Consequently, if the material does not exhibit a hazardous waste characteristic, the RCRA hazardous waste regulations do not apply. This has been EPA's long-standing policy, which was reaffirmed in RO 14291.

What constitutes a "good-faith effort" to determine the source of the contamination? "The agency believes that by using available site- and waste-specific information such as manifests, vouchers, bills of lading, sales and inventory records, storage records, sampling and analysis reports, accident reports, site investigation reports, spill reports, inspection reports and logs, and enforcement orders and permits, facility owner/operators would typically be able to make these determinations. However…if information is not available or inconclusive, facility owner/operators may generally assume that the material[s] contaminating the [debris] were not hazardous wastes." [April 29, 1996; 61 FR 18805]

The preceding discussion makes it clear why spills/stains should be cleaned from building/structure walls or floors prior to demolition. If the source of these spills/stains is listed hazardous waste and the spills/stains are not removed prior to demolition, the resulting demolition debris will be listed. Listed debris is very expensive to manage.

Additional discussion on using knowledge to make a determination as to whether a specific demolition waste is listed is given in Section 18.3.2.

19.2.4 Hazardous demolition waste management options

If demolition wastes are determined to be hazardous, there are several options available under the RCRA regulations for managing them. Some of the management options may be implemented onsite, while others will be conducted offsite. A discussion of these on- and offsite demolition waste management options is given in the sections that follow.

19.2.4.1 Onsite waste management options

If hazardous demolition wastes are generated, they may be managed onsite in the units that are discussed in the following five subsections.

19.2.4.1.1 *Scrap metal containers/yards*

During the demolition of a building/structure, metallic wastes will be generated, including structural steel, metal decking/siding/roofing, stairways, handrails, piping and other equipment, etc. As noted in Section 4.21, §261.6(a)(3)(ii) exempts such scrap metal—even if hazardous (e.g., coated with LBP)—from RCRA regulation if sent for recycling/reclamation. Therefore, it is unnecessary to make a hazardous waste determination (i.e., whether it exhibits a characteristic or is contaminated with listed hazardous waste) for material that meets the definition of scrap metal and will be recycled. [RO 11782, 11806, 11835, 14184]

Thus, bins filled with scrap metal are not subject to "Hazardous Waste" labeling, 90/180-day or speculative accumulation clocks, Subpart I container standards, etc. Similarly, scrap yards are not subject to RCRA signage requirements, liner standards, the speculative accumulation clock, etc.

Conversely, scrap metal that is to be disposed is not exempt; it is a solid waste, and the generator must determine whether it is hazardous and manage it in one of the units noted below if it is.

19.2.4.1.2 *Satellite accumulation units*

We see no reason why demolition wastes (such as contaminated PPE, LBP debris) could not be managed in an SAA. Note that the SAA regulations did not envision facilities generating one-time wastes (such as concrete chips from cleaning a floor stain) and essentially abandoning the wastes in an SAA. We are not aware of any specific EPA guidance on this issue; however, we suspect that some state agencies may not allow satellite accumulation containers to be used for demolition waste management.

The useful aspect of designating drums containing demolition wastes as satellite accumulation containers is that 1) there is no clock running, and 2) the drums are subject only to minimal labeling and RCRA container standards as noted in §262.34(c)(1). If alternatively, the generator designates drums holding demolition wastes as 90/180-day accumulation containers, then there would be a clock running and these units would be subject to additional labeling and RCRA container standards as noted in the following section.

19.2.4.1.3 *90/180-day units*

Typically, such 90/180-day units would be tanks or containers, and these units would be subject to all of the RCRA requirements when in demolition waste service just as they are when in process waste service. Sections 6.3 and 7.3 detail the §262.34 regulatory requirements for these units when used for accumulation and treatment, respectively.

Containment buildings may be used as 90-day accumulation/treatment units [see §262.34(a)(1)(iv)]. These units give generators a way to accumulate/treat hazardous debris without having to obtain a RCRA permit.

Examples of the use of 90/180-day units for managing hazardous demolition waste are given in Case Study 19-5. Two additional examples are:

- Accumulating hazardous building components coated with LBP in roll-off boxes; and

- Treating hazardous debris in containment buildings. [August 18, 1992; 57 *FR* 37242, RO 13553, 13696]

19.2.4.1.4 Universal waste units

Some hazardous wastes generated during building/structure demolition meet the definition of

---

Case Study 19-5: Demolishing a Building Contaminated With Listed Waste

A maintenance building is scheduled for demolition. During the pre-demolition hazardous materials assessment, stains are observed on a portion of the concrete floor. Due diligence results in documentation showing that the stains are from old spills of hazardous waste F001 that could not be completely removed. What are the building owner's options regarding the demolition and disposal of the stained concrete floor?

Option 1—The building owner demolishes the building, including the stained concrete floor (see Option 1 in Figure 19-3). The concrete pieces that are visually stained are segregated for management as F001 hazardous waste because they contain F001 hazardous waste via the contained-in policy. The building owner has two primary options for management of this hazardous waste:

- The F001 hazardous debris can be sent offsite to a Subtitle C TSD facility (see Section 19.2.4.2.3).

- The F001 debris can be treated without a RCRA permit in a 90-day accumulation unit (e.g., containment building) to meet the §268.45 LDR treatment standards for hazardous debris. As noted in §§261.3(f)(1) and 268.45(c), if the owner treats the debris to meet the performance standards for an extraction or destruction technology in Table 1 of §268.45, the treated debris can then be managed as nonhazardous (see Section 13.8.2). Any residues generated during this treatment, however, would require management as F001.

Option 2—Prior to demolishing the building, the owner meets with EPA or the state and receives a no-longer-contains determination from the agency for the contaminated concrete (see Option 2 in Figure 19-3). This option is codified at §261.3(f)(2). If the owner successfully receives such a determination from EPA or the state, the concrete pieces from subsequent demolition are considered nonhazardous and can be placed directly into a Subtitle D landfill.

Option 3—The building owner decontaminates the concrete floor before the building is demolished (see Option 3 in Figure 19-3). For example, the building owner could use facility or contractor personnel to grind the stained portion of the floor to remove the stains. As noted in Section 19.2.1.9, such grinding would not be considered treatment of hazardous waste, but it would generate hazardous waste (i.e., the concrete chips/dust resulting from the grinding would require management as F001). Because the concrete floor is not yet waste (as noted in Section 19.2.2.3), the cleaned concrete after grinding does not have to meet any particular RCRA standards. For example, no LDR treatment standards, including the standards for hazardous debris in §268.45, apply. However, if the LDR standards of §268.45 are met, it would seem reasonable (although we have seen nothing in writing from EPA on this subject) that the subsequently demolished concrete floor would not be hazardous at its point of generation. If nonhazardous debris is generated, it can be managed in a Subtitle D landfill.

Figure 19-3: Options for Demolishing a Building Contaminated With Listed Waste

Option 1 — Demolish building without no-longer-contains determination

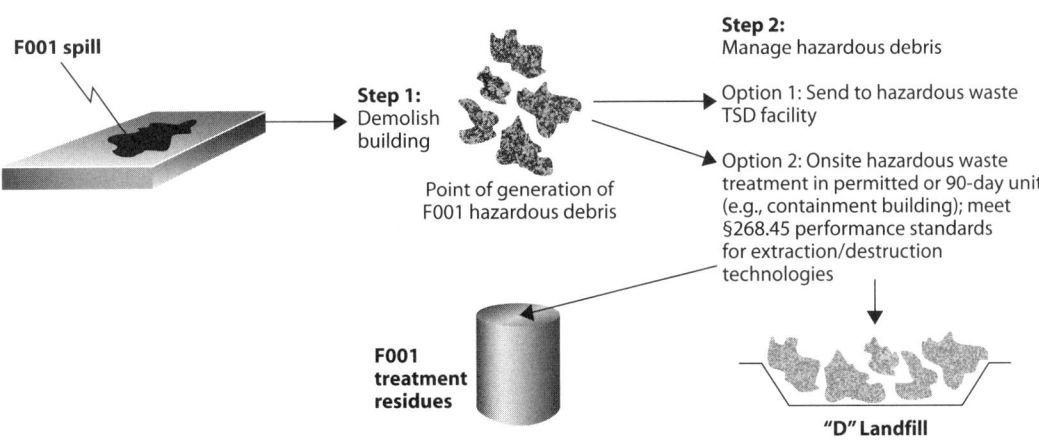

Option 2 — Demolish building with no-longer-contains determination

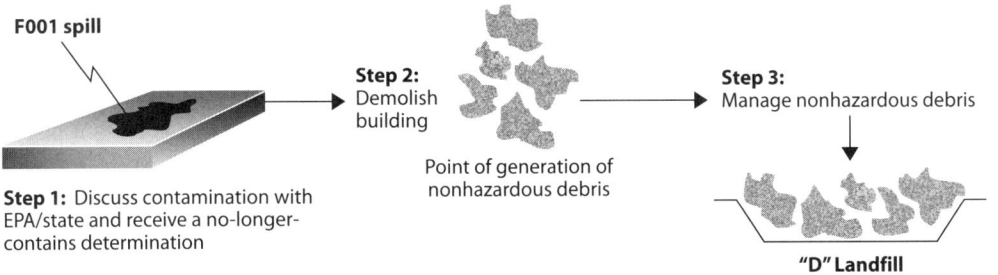

Option 3 — Decontaminate building before demolition

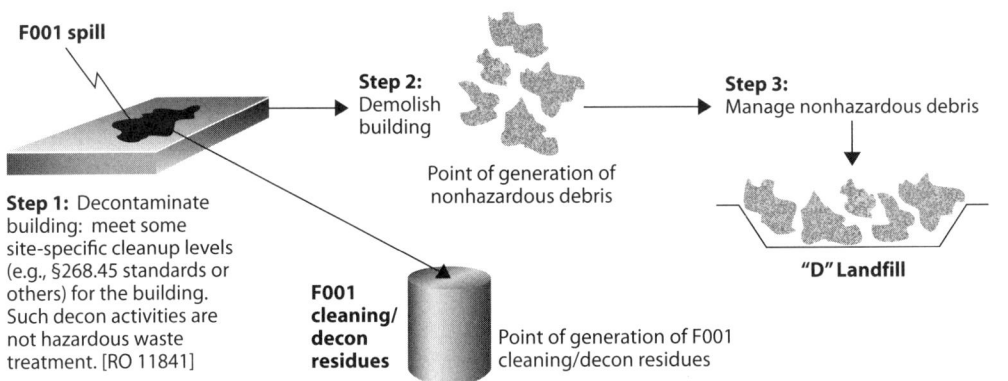

Source: McCoy and Associates, Inc.

universal wastes and so may be managed under the less-stringent universal waste program. For example, fluorescent lamps, UPS batteries, and mercury thermostats and switches removed from a building slated for demolition may all be managed in universal waste containers in accordance with Part 273, Subpart B or C, depending on whether the facility is a small or large quantity handler of universal waste. Section 8.3 provides a summary of these standards.

19.2.4.1.5 RCRA-permitted units

Although generally used for the treatment, storage, and/or disposal of process wastes, RCRA-permitted units, such as a hazardous waste container storage area, tank, landfill, or even surface impoundment, can be used for the onsite management of hazardous demolition wastes. The requirements for obtaining a RCRA Part B permit for a unit that will manage hazardous demolition wastes are the same as those for permitting a process waste management unit. These requirements are expensive, time-consuming, and trigger other RCRA provisions (e.g., the corrective action program).

19.2.4.2 Offsite waste management options

In addition to the onsite management options discussed above, hazardous demolition waste may be managed at offsite facilities, as discussed in the following four subsections.

19.2.4.2.1 *Metallic demolition wastes to scrap metal recycling facility*

During the demolition of a building/structure, metallic wastes will be generated, including structural steel, metal decking/siding/roofing, stairways, handrails, piping and other equipment, etc. As noted in Section 4.21, these metallic wastes—even if hazardous (e.g., coated with LBP)—may be sent to an offsite scrap metal recycling facility without complying with the RCRA Subtitle C program. Thus, offsite shipments of scrap metal are not subject to hazardous waste manifest or LDR paperwork requirements.

19.2.4.2.2 *Universal wastes to destination facilities*

All universal wastes generated during the demolition of a building/structure (see Section 19.2.4.1.4) may be sent offsite to a universal waste handler for accumulation or to a destination facility for storage, treatment, recycling, or disposal. Sections 8.3 and 8.4.2 provide a summary of the RCRA requirements applicable to these facilities. Offsite shipments of universal waste are not subject to hazardous waste manifest or LDR paperwork requirements.

19.2.4.2.3 *Nonmetallic demolition wastes to Subtitle C TSD facility*

Once the scrap metal has been segregated from the nonmetallic hazardous debris, the most likely option for the offsite management of hazardous demolition wastes is simply to send them to a commercial TSD facility. Several TSD facilities are RCRA-permitted to accept and treat, store, and/or dispose hazardous demolition wastes. If this option is chosen, the generating facility would manage the hazardous demolition wastes just like any other hazardous waste at its facility, and it would comply with all hazardous waste generator requirements, such as 90/180-day accumulation and manifesting requirements.

19.2.4.2.4 *Hazardous demolition waste sent offsite triggers LDR program*

Demolition waste will be subject to the LDR program only if 1) it is generated (i.e., components are removed from the building/structure, residues are produced from cleaning/decontaminating the building/structure, and/or the building/structure is demolished), 2) it is hazardous (i.e., it exhibits a characteristic or contains a listed waste), and 3) it will be placed in a land disposal unit. For example, if hazardous debris is generated and will be disposed in an offsite landfill, it will be subject to the LDR program just like any other hazardous waste. However, EPA has developed less-stringent LDR treatment standards for hazardous debris; these standards are discussed in detail in Section 13.8.

Appendix 1—List of Case Studies

Case Study 1-1: Co-Product Fuel Determination · · 18

Case Study 1-2: Mercury and Thermometers · · · · 21

Case Study 1-3: Zinc Fertilizer Production From Waste Materials · · · · · · · · · · · · 26

Case Study 2-1: Total Waste Analyses for Toxicity Determination · · · · · · · · · · · · · · · · 69

Case Study 2-2: Toxicity Characteristic Determination for Leather Shoes and Gloves · · · · 71

Case Study 2-3: Making a Hazardous Waste Determination for Lamps · · · · · · · · · · · · 83

Case Study 3-1: "Before Use" · · · · · · · · · · · · 96

Case Study 3-2: F003 Spent Solvent Mixtures · · · 98

Case Study 4-1: Domestic Sewage Exclusion Applicability · · · · · · · · · · · · · · · · · · 143

Case Study 4-2: Wastewater Diverted Into a Basin for Fire Training · · · · · · · · · · · · · · 149

Case Study 4-3: Is Lead Shielding Subject to RCRA? · · · · · · · · · · · · · · · · · · · 156

Case Study 4-4: Does the LDR Storage Prohibition Apply to Decay-in-Storage? · · · · · · · 158

Case Study 4-5: Equipment Cleaning Wastes · · · 186

Case Study 4-6: Natural Gas Compressor Stations · 193

Case Study 4-7: Scrubber Waste Generated at Geothermal Power Plants · · · · · · · · · · · 194

Case Study 4-8: De Minimis Wastewater Exemptions for Spent Solvents · · · · · · · · · · 210

Case Study 4-9: De Minimis Wastewater Exemptions for Commercial Products · · · · · · · 212

Case Study 4-10: De Minimis Wastewater Exemptions for Laboratory Wastes · · · · · · · · 214

Case Study 4-11: Is It a Manufacturing Process Unit or RCRA Tank? · · · · · · · · · · · · · 218

Case Study 6-1: 90-Day Accumulation/Treatment Provisions · · · · · · · · · · · · · · · · · · 305

Case Study 6-2: 90/180/270-Day Accumulation Quiz · · · · · · · · · · · · · · · · 317

Case Study 6-3: Definition of "Annual Review" · · 325

Case Study 6-4: Episodic Generators · · · · · · · 329

Case Study 6-5: Transportation of Explosives That Pose a Safety Threat · · · · · · · · · · · · 338

Case Study 6-6: Two Transporters Jointly Moving a Shipment · · · · · · · · · · · · · · · 342

Case Study 6-7: Transfers From Rail Cars to Tank Trucks · · · · · · · · · · · · · · · · · 343

Case Study 6-8: Shipments Between TSD Facilities · · · · · · · · · · · · · · · · · · · 345

Case Study 6-9: Consolidating Several Shipments of Imported Hazardous Waste · · · · · · · · 357

Case Study 7-1: Incidental Mixing vs. Purposeful Dilution · · · · · · · · · · · · · · · · · · · 383

Case Study 7-2: Sludge Dryers Are WWTUs · · · · 388

Case Study 7-3: Receiving Offsite Shipments of Hazardous Wastewater Into a WWTU · · · · · 395

Case Study 7-4: Treatment of Partially Full Cans of Unwanted Paint · · · · · · · · · · · · · · 402

Case Study 7-5: Metals Recovery in Industrial Furnaces · · · · · · · · · · · · · · · · · · · 407

Case Study 7-6: Liquids-in-Landfills Ban · · · · · · 414

Case Study 7-7: Explosives, Explosive Devices, and Other Shock-Sensitive Materials · · · · · · · 419

Case Study 8-1: Decharacterized Universal Waste Disposed in Municipal Landfills · · · · · · · · · 431

Case Study 9-1: Use of Commonly Employed Practices · · · · · · · · · · · · · · · · · · · 450

Case Study 9-2: EPA Region I Position on Tank Truck Washwater · · · · · · · · · · · · · · · 455

Case Study 9-3: Container Management Examples · · · · · · · · · · · · · · · · · · · 457

Case Study 10-1: Storage Prior to Recycling— Two Scenarios · · · · · · · · · · · · · · · · · 465

Case Study 10-2: Is It a Manufacturing Unit or RCRA Tank? · · · · · · · · · · · · · · · · 469

Case Study 11-1: Is Spent Solvent Reuse for Drum Washing Legitimate? · · · · · · · · · · 523

Case Study 11-2: Secondary Material Used as a Fertilizer Ingredient · · · · · · · · · · · · 532

Case Study 11-3: Putting the Five Steps Together · · 546

Case Study 12-1: Generator Requirements When Processor Disposes Used Oil Instead of Recycling It as Originally Planned · · · · · · · 552

Case Study 12-2: Mixing Hazardous Waste and Used Oil Is Treatment · · · · · · · · · · · 554

Case Study 12-3: Presumption May Be Rebutted Regardless of Halogen Level · · · · · · · · · · 558

Case Study 12-4: Used Oil Storage Tank Bottoms Burned for Energy Recovery · · · · · · · · · · 561

©2015 McCoy and Associates, Inc. **McCoy's RCRA Unraveled**

Appendix 1 List of Case Studies

Case Study 12-5: Servicing a Customer's Fleet of Vehicles · · · · · · · · · · · 575

Case Study 12-6: Used Oil Activities Conducted by Contractors · · · · · · · · · · · 576

Case Study 12-7: Burning Used Oil in a County Maintenance Facility Space Heater · · · · · · · 583

Case Study 13-1: When Does "Placement" Occur for Remediation Wastes? · · · · · · · · · · 591

Case Study 13-2: How LDR Treatment Standards Were Developed · · · · · · · · · · 595

Case Study 13-3: How Is "Point of Generation" Defined Under the LDR Program? · · · · · · · 597

Case Study 13-4: Treatment Standards for D001 Wastes · · · · · · · · · · · · · 607

Case Study 13-5: Treatment Standards for D002 Wastes · · · · · · · · · · · · · 611

Case Study 13-6: Treatment Standards for TC Wastes · · · · · · · · · · · · · · 613

Case Study 13-7: LDR Applicability to Partially Full Cans of Unwanted Paint · · · · · · · · 616

Case Study 13-8: Treatment Standards for Spent Solvent Wastes · · · · · · · · · · 617

Case Study 13-9: Alternative Treatment Standards for Soil · · · · · · · · · · · · 623

Case Study 13-10: Alternative Treatment Standards for Characteristic Soil · · · · · · · · 625

Case Study 13-11: Alternative Treatment Standards for TC Debris · · · · · · · · · · 630

Case Study 13-12: Alternative Treatment Standards for Listed Debris · · · · · · · · · 631

Case Study 13-13: Dilution Prohibition · · · · · · · 642

Case Study 13-14: Decharacterized Universal Waste Disposed in Municipal Landfills · · · · · · 658

Case Study 14-1: Point of Generation for Off-Shore Oil Platform Wastes · · · · · · · · · 665

Case Study 14-2: Spill of Crude Oil · · · · · · · 682

Case Study 14-3: Recovered Gasoline · · · · · · · 685

Case Study 14-4: Fume Incinerator · · · · · · · · 694

Case Study 15-1: Equipment Subject to Subpart BB · · · · · · · · · · · · · · · 718

Case Study 15-2: Subpart CC LDR Exemption · · 734

Case Study 15-3: Subpart CC Issues for Solvent-Contaminated Debris in a Roll-Off Box · · · 747

Case Study 17-1: Corrective Action Using a CAMU · · · · · · · · · · · · · · · · 806

Case Study 17-2: Corrective Action Using a Staging Pile · · · · · · · · · · · · · · 811

Case Study 17-3: Corrective Action Using the AOC Policy · · · · · · · · · · · · · 813

Case Study 18-1: Stratified Random Sampling of Spent Fluorescent Light Tubes · · · · · · · · 860

Case Study 18-2: Judgmental Sampling of a Smelter Slag Pile · · · · · · · · · · · · 862

Case Study 18-3: Hazardous Waste Characterization for Heterogeneous Materials · · 864

Case Study 18-4: Simple Random Sampling of Lagoon Sludge · · · · · · · · · · · · · 866

Case Study 18-5: Simple Random Sampling of Material Filling a Roll-Off Box · · · · · · · · 868

Case Study 18-6: Simple Random Sampling (With Compositing) of Tanks · · · · · · · · · 871

Case Study 18-7: Using Composite Sampling to Locate a Hot Spot · · · · · · · · · · · · 876

Case Study 18-8: Simple Random Sampling of Lagoon Sludge · · · · · · · · · · · · · 882

Case Study 18-9: Using a Percentile to Characterize a Solid Waste · · · · · · · · · · 884

Case Study 19-1: RCRA Retroactivity and Active Management · · · · · · · · · · · · 891

Case Study 19-2: Assigning Listed Waste Codes to Remediation Waste · · · · · · · · · · · 898

Case Study 19-3: Soil Contaminated With 2,4,5-T · · · · · · · · · · · · · · · · 899

Case Study 19-4: Determining If LDR Standards Are ARARs at a CERCLA Site · · · · · · · · 913

Case Study 19-5: Demolishing a Building Contaminated With Listed Waste · · · · · · · · 928

Appendix 2—List of Figures

Figure 1-1: Regulatory Status of Discarded Materials From a Manufacturing Plant · · · · · · · · 3

Figure 1-2: Relationships Among Discarded Materials, Solid Wastes, and Hazardous Wastes · · · 6

Figure 2-1: Regulatory Scenario for the Toxicity Characteristic · 63

Figure 3-1: How Solvent Mixtures Are Regulated · · 99

Figure 3-2: Hypothetical Electroplating Process · · 108

Figure 3-3: Sequential Treatment of Electroplating Wastewaters · · · · · · · · · · · · · · · · 112

Figure 3-4: Commercial Chemical Products Used in Manufacturing Compared to Product Handling · 132

Figure 4-1: The Domestic Sewage Exclusion · · · 141

Figure 4-2: The NPDES Discharge Exclusion · · · 148

Figure 4-3: Irrigation Return Flows Exclusion · · · 150

Figure 4-4: Typical Kraft Pulping Process · · · · · 163

Figure 4-5: Oil-Bearing Secondary Materials and Recovered Oil Exclusions · · · · · · · · · · · · · · 164

Figure 4-6: Oil-Bearing Secondary Materials Generated/Recycled at a Refinery · · · · · · · · · · · 166

Figure 4-7: Spent Caustic From a Petroleum Refinery · 170

Figure 4-8: Bevill Exclusion at a Coal-Fired Power Plant · 176

Figure 4-9: Regulatory Status of Fossil Fuel Combustion Wastes · 177

Figure 4-10: Regulatory Status of Oil and Gas E&P Wastes · 192

Figure 4-11: Extraction, Beneficiation, and Mineral Processing Wastes From Copper Production · 199

Figure 4-12: Regulatory Status of Mining and Mineral Processing Wastes · · · · · · · · · · · · · 205

Figure 4-13: Identification of Manufacturing Process Units vs. RCRA Tanks · · · · · · · · · · · · · 218

Figure 4-14: Applicability of §261.4(c) Exemption to Pipelines Occasionally Used to Transfer Hazardous Waste · 219

Figure 4-15: Analytical Sample Activities Exempt From RCRA Requirements · · · · · · · · · · 225

Figure 5-1: Regulatory Status of Mixtures Containing Bevill Wastes · · · · · · · · · · · · · · · · · 253

Figure 5-2: Regulatory Status of Mixtures Containing Oil and Gas Wastes · · · · · · · · · · · · 254

Figure 6-1: RCRA Training Recordkeeping Requirements for LQGs/TSD Facilities · · · · · · · 329

Figure 6-2: Uniform Hazardous Waste Manifest · · 336

Figure 7-1: WWTU Exemption Applicability at Common Wastewater Treatment Facilities · · · · 394

Figure 7-2: Scope and Applicability of the Permit Exemption for Recycling · · · · · · · · · · · · · · 404

Figure 7-3: Applicability of Elementary Neutralization Unit Exemption · · · · · · · · · · · · · 409

Figure 8-1: Management and Regulation of Universal Wastes · 427

Figure 8-2: Authorization Status of the State Universal Waste Programs · · · · · · · · · · · · · · 437

Figure 10-1: Regulation of Spent Solvent Management · 470

Figure 10-2: Aboveground Tank With External Liner Secondary Containment · · · · · · · · · · · · · 481

Figure 10-3: Secondary Containment Vault for Underground Tanks · · · · · · · · · · · · · · · · · · 481

Figure 11-1: Recycling Determination Process · · 495

Figure 11-2: Materials Reclaimed "Under the Control of the Generator" · · · · · · · · · · · · · · 501

Figure 11-3: The Transfer-Based Reclamation Exclusion · 504

Figure 11-4: Exclusion for Use as an Ingredient or Feedstock · 513

Figure 11-5: Recycling That Includes Both Reclamation and Use as an Ingredient · · · · · · · 517

Figure 11-6: Reclamation Voids the Use as an Effective Substitute Recycling Exclusion · · · · 521

Figure 11-7: Closed-Loop Recycling With No Reclamation · 524

Figure 11-8: Closed-Loop Recycling With Reclamation · 527

Figure 11-9: Use of Reclaimed Materials in Fertilizer Production · · · · · · · · · · · · · · · · · · 532

Figure 12-1: Applicability of the Used Oil Management Standards · · · · · · · · · · · · · · · · · 549

Figure 12-2: Used-Oil-Containing Material Eligible for Reduced Regulation · · · · · · · · · · · · 550

Figure 12-3: Regulation of PCB-Containing Used Oil · 563

©2015 McCoy and Associates, Inc. **McCoy's RCRA Unraveled**

Figure 12-4: Used Oil Inserted Into Crude Oil Pipelines · · · · · · · · · · · · · · · · 565

Figure 12-5: Used Oil Inserted Into Petroleum Refineries · · · · · · · · · · · · · · · · 568

Figure 12-6: Regulatory Requirements for Used Oil Generators · · · · · · · · · · · · 571

Figure 12-7: Regulatory Requirements for Used Oil Burners · · · · · · · · · · · · · · · 579

Figure 12-8: Regulatory Requirements for Used Oil Fuel Marketers · · · · · · · · · · 586

Figure 13-1: How LDR Treatment Standards Were Developed · · · · · · · · · · · · · · · 596

Figure 13-2: Hypothetical Wastewater Treatment System at a Chemical Plant · · · · · · · · 599

Figure 13-3: Key to Understanding the §268.40 Table of Treatment Standards · · · · · · · · · · 603

Figure 13-4: Using and Interpreting the §268.40 Table of Treatment Standards · · · · · · · · · · 606

Figure 13-5: Treatment of D001 High-TOC Wastes in a CWA System Without a Land Disposal Unit · · · · · · · · · · · · · · · · · 607

Figure 13-6: Treatment of D001 Wastes in a CWA System With a Land Disposal Unit · · · · · 608

Figure 13-7: Disposal of D001 Wastes in a Class I Injection Well · · · · · · · · · · · · · · · · · 609

Figure 13-8: Treatment of D002 Wastes in a CWA System · · · · · · · · · · · · · · · · · · 612

Figure 13-9: Treatment of D018 Wastewater in a CWA System · · · · · · · · · · · · · · · · 613

Figure 13-10: Treatment of a Lead-Containing Wastewater · · · · · · · · · · · · · · · · · · 614

Figure 13-11: Cradle-to-Grave Path for Recycling a Spent Solvent · · · · · · · · · · · · · · · · 617

Figure 13-12: Alternative Treatment Standards for Soil · 622

Figure 13-13: Contaminated Concrete Slab to Be Treated Via the Alternative Debris Standards · · · · · · · · · · · · · · · · · · 629

Figure 13-14: Aggregation for Centralized Treatment · · · · · · · · · · · · · · · · · · · 638

Figure 13-15: Dilution as a Consequence of Treatment · · · · · · · · · · · · · · · · · · · 639

Figure 13-16: Determining When Dilution Is Permissible · · · · · · · · · · · · · · · · · · 641

Figure 13-17: Basic Intent of the LDR Paperwork Requirements · · · · · · · · · · · · · · · · 642

Figure 13-18: Generators' LDR Paperwork Requirements for Wastes Shipped Offsite · · · · · 644

Figure 13-19: LDR Paperwork Requirements for Wastes Treated or Disposed Onsite · · · · · · 645

Figure 13-20: Treatment Facilities' LDR Paperwork Requirements · · · · · · · · · · · · · 646

Figure 13-21: Section 268.7(a)(7) One-Time Notice to File · · · · · · · · · · · · · · · · · 648

Figure 13-22: Master Form for LDR Notifications/Certifications That Are Not Available From Offsite Facilities · · · · · · · · · · · · · · · · 650

Figure 14-1: Point of Generation of Spill Cleanup Residues · · · · · · · · · · · · · · · · 672

Figure 14-2: Point When Military Munitions Become Solid Wastes · · · · · · · · · · · · · · 698

Figure 15-1: Potential Air Emission Sources From a Solvent Still · · · · · · · · · · · · · · · 705

Figure 15-2: Determining If Subpart AA Process Vent Standards Apply · · · · · · · · · · · · · 706

Figure 15-3: Identifying Vents Subject to Subpart AA · · · · · · · · · · · · · · · · · · · 710

Figure 15-4: Potential Air Emission Sources in Pipe Runs · · · · · · · · · · · · · · · · · · 713

Figure 15-5: Determining If Subpart BB Fugitive Emission Standards Apply · · · · · · · · · · · · 714

Figure 15-6: Applicability of Subpart BB to Equipment in Pipe Runs · · · · · · · · · · · · · 718

Figure 15-7: Determining If Subpart CC Emission Control Standards Apply · · · · · · · · 723

Figure 15-8: Basic Intent of Subpart CC · · · · · · 728

Figure 15-9: Waste Treatment Options and Associated Analytical Requirements Under Subpart CC · · · · · · · · · · · · · · · · · · 732

Figure 15-10: Typical Fixed-Roof Tank · · · · · · 736

Figure 15-11: Fixed-Roof Tank With Internal Floating Roof · · · · · · · · · · · · · · · · · · 738

Figure 15-12: Tank With External Floating Roof · · 739

Figure 15-13: Fixed-Roof Tank With Closed-Vent System and Control Device · · · · · · · · · · · 739

Figure 17-1: Two Adjacent Plants May Be One "Facility" · 791

Figure 17-2: RCRA Corrective Action Process · · · 795

Figure 17-3: How a CAMU May Be Used at a Corrective Action Site · · · · · · · · · · · · · · 807

Figure 17-4: How a Staging Pile May Be Used at a Corrective Action Site · · · · · · · · · · · · 812

Figure 17-5: How an AOC May Be Used at a Corrective Action Site · · · · · · · · · · · · 814

Figure 18-1: Basic Intent of Characterization for the LDR Program · · · · · · · · · · · · · · · 843

Figure 18-2: The Data Quality Objectives Process · · 855

Figure 18-3: Spatial Probability Sampling Strategies · · · · · · · · · · · · · · · · · · 859

Figure 18-4: Simple Random Sampling of a Waste Lagoon Containing Barium-Contaminated Sludge · 867

Figure 18-5: An 80% Confidence Interval · · · · · 880

Figure 18-6: A Tolerance Interval for a Percentile · · 883

Figure 19-1: RCRA Regulatory Status of Remediation Waste · · · · · · · · · · · · · · · · 895

Figure 19-2: Managing Lead-Based Paint During Building/Structure Demolition · · · · · · · · · · 917

Figure 19-3: Options for Demolishing a Building Contaminated With Listed Waste · · · · · · · · · 929

Appendix 3—List of Tables

Table 1-1: Determining If Recycled Materials Are Solid Wastes · · · · · · · · · · · · · · · · 11

Table 1-2: Questions Used to Determine If a Material Is a Co-Product Fuel · · · · · · · · · 17

Table 1-3: EPA Examples of Co-Products vs. By-Products · · · · · · · · · · · · · · · · · · · 19

Table 2-1: Hazardous Waste Categories · · · · · · 39

Table 2-2: Closed-Cup Flash Points for Common Materials · · · · · · · · · · · · · · · · · · 42

Table 2-3: RCRA Definitions for "Aqueous," "Liquid," and "Wastewater" · · · · · · · · · · · · 44

Table 2-4: Examples of Materials That EPA Has Determined Are Reactive · · · · · · · · · · 55

Table 3-1: Examples of F001, F002, F004, and/or F005 Spent Solvent Mixture Rule · · · · · 97

Table 3-2: Examples of F003 Spent Solvent Mixture Rule · 100

Table 3-3: Regulatory Status of Solvent-Contaminated Wipes Under EPA's July 31, 2013 Rule · 105

Table 3-4: Electroplating Operations That Produce F006 · · · · · · · · · · · · · · · · · · 113

Table 3-5: Manufacturing Operations That Do Not Produce F006 · · · · · · · · · · · · · 115

Table 4-1: Exclusions and Exemptions Available in the Federal RCRA Regulations · · · · · · · · · 138

Table 4-2: Low-Volume Wastes That Are "Uniquely Associated" With Fossil Fuel Combustion · 179

Table 4-3: Examples of Exploration, Development, or Production Wastes Excluded From Hazardous Waste Regulation · · · · · · · · · · · · · · · · · · 188

Table 4-4: Examples of Wastes That Do Not Qualify for the Oil, Gas, and Geothermal Energy Exclusion · · · · · · · · · · · · · · · · · · · 190

Table 4-5: Wastes That Are "Uniquely Associated" With Mining and Mineral Processing · · · · · · · 202

Table 4-6: Sample Quantity and Time Limits for Conducting Treatability Studies · · · · · · · · 230

Table 4-7: Scope of the Precious Metals Exemption · 234

Table 4-8: RCRA Requirements for Managing Spent Lead-Acid Batteries · · · · · · · · · · · · · 237

Table 4-9: Examples of Materials That Qualify for the §261.6(a)(3)(ii) Scrap Metal Exemption When Sent for Recycling/Reclamation · · · · · · 242

Table 5-1: Wastes Listed Solely Because They Exhibit a Characteristic · · · · · · · · · · · · · · · 249

Table 6-1: Hazardous Wastes That Do Not Have to Be Counted · · · · · · · · · · · · · · · · · 290

Table 6-2: Federal Requirements for Satellite Accumulation Units vs. 90-Day Containers · · · · 295

Table 6-3: Summary of Generator Waste Accumulation Provisions · · · · · · · · · · · · · · 301

Table 6-4: Elements of a Waste Minimization Program · 313

Table 6-5: Example Personnel Training Matrix for Large Quantity Generators · · · · · · · · · · 326

Table 6-6: Primary Recordkeeping Requirements for the Three Generator Classes · · · · · · · · · 361

Table 6-7: Miscellaneous Recordkeeping Requirements for the Three Generator Classes · · 367

Table 6-8: Activities That Trigger RCRA Subtitle C Notification or Renotification · · · · · · · · · · · 371

Table 6-9: Issuing EPA ID Numbers for Different Situations · 372

Table 7-1: Treatment Status of Miscellaneous Hazardous Waste Activities · · · · · · · · · · · · 385

Table 8-1: Types of Universal Wastes in the Federal Program · · · · · · · · · · · · · · · · · · · 424

Table 8-2: Requirements for Small and Large Quantity Handlers of Universal Wastes · · · · · · 429

Table 8-3: DOT Shipping Descriptions for Batteries Commonly Managed as Universal Waste · 434

Table 9-1: Design and Management Standards for Hazardous Waste Containers · · · · · · · · · 443

Table 10-1: Treatment Exemptions Applicability to Tanks · 467

Table 10-2: Design and Management Standards for Hazardous Waste Tanks · · · · · · · · · · · · 475

Table 10-3: Specific Design Requirements for Tank Secondary Containment Systems · · · · · · 482

Table 10-4: Inspection Requirements for Hazardous Waste Tanks · · · · · · · · · · · · · · 490

Table 11-1: Determining If Recycled Materials Are Solid Wastes · · · · · · · · · · · · · · · · · · 497

Table 11-2: Questions That May Be Used to Distinguish Between Legitimate and Sham Recycling · 543

©2015 McCoy and Associates, Inc.   **McCoy's RCRA Unraveled**

Appendix 3 — List of Tables

Table 12-1: Examples of Materials That, When Used, Qualify as Used Oil · · · · · · · · 551

Table 12-2: Examples of Materials That, When Used, Do Not Qualify as Used Oil · · · · · 551

Table 12-3: The Used Oil Fuel Specification · · · 580

Table 13-1: Alternative Treatment Standards for Gasoline-Contaminated Soil · · · · · · · · · · · 623

Table 13-2: Generator LDR Paperwork Requirements Table · · · · · · · · · · · · · · · · 647

Table 13-3: Treatment Facility LDR Paperwork Requirements Table · · · · · · · · · · · · · · · · 647

Table 14-1: Regulatory Status of Spill Cleanup Residues · 676

Table 14-2: Listed Hazardous Wastes That Are Gases at Room Temperature (68°F) · · · · · · · 690

Table 14-3: Examples of Military Munitions That Are Not Solid Wastes When Disassembled and/or Recycled · · · · · · · · · · · · · · · · · 700

Table 15-1: Air Emission Standards Applicable to Hazardous Waste Equipment/Units · · · · · · 704

Table 15-2: Vapor Pressures of Common Solvents · 720

Table 15-3: Inspection/Monitoring Requirements for Subpart BB-Regulated Equipment · · · · · 722

Table 15-4: Matrix for Determining Subpart CC Level 1 or 2 Tank Controls · · · · · · · · · · · 735

Table 15-5: Vapor Pressures of Common Solvents and Allowable Level 1 Tank Capacity · · · · · · 737

Table 15-6: Matrix for Determining Subpart CC Level 1, 2, or 3 Container Controls · · · · · · · 742

Table 16-1: Most Common RCRA Violations in Colorado · · · · · · · · · · · · · · · · · · · 751

Table 16-2: Matrix for Assessing Gravity-Based Penalty · 758

Table 16-3: RCRA Statutory Maximum Civil Penalties · 758

Table 16-4: Matrix for Assessing Daily Penalty for Multiday Violations · · · · · · · · · · · · · 759

Table 16-5: Selected RCRA Enforcement Cases · · 761

Table 16-6: Types of RCRA Inspections · · · · · · 770

Table 16-7: RCRA Inspection Document Request List · 773

Table 16-8: Pre-Inspection Conference Checklist · · 778

Table 16-9: Post-Inspection Conference Checklist · · 780

Table 17-1: Examples of Units/Areas Identified as SWMUs · 787

Table 18-1: Example Analytical Data for a Waste Received at an Offsite Solvent Recovery Plant · · 829

Table 18-2: References Useful When Preparing a Sampling Plan · · · · · · · · · · · · · · · · · 854

Table 18-3: Guidance for Selecting Probability Sampling Strategies · · · · · · · · · · · · · · · · 858

Table 18-4: Guidance for Selecting Authoritative Sampling Strategies · · · · · · · · · · · · · · · · 861

Table 18-5: Tabulated Values of Student's "t" Value for Evaluating Solid Wastes · · · · · · · · · 867

Table 18-6: Advantages and Limitations Associated With Multi Increment Sampling · · · 869

Table 18-7: Choosing the Appropriate Sampling Strategy · 878

Table 18-8: Tabulated Values of κ Constant for the 95th Percentile · · · · · · · · · · · · · · · · 885

Table 18-9: Guidance for Handling Nondetects · · 885

Table 19-1: Major Components of a Remediation Project · 888

Table 19-2: References Describing the Triad Approach to Site Investigation · · · · · · · · · · 889

Table 19-3: McCoy's Summary of RCRA Remediation Options · · · · · · · · · · · · · · · 900

Table 19-4: Substantive vs. Administrative Requirements for Roll-Off Containers Storing Known Hazardous Wastes at a CERCLA Site · · · 914

Table 19-5: Pre-Demolition Building/Structure Hazardous Materials Assessment · · · · · · · · · 916

Index

A

ABANDONED MATERIALS
- Are solid wastes...7, 22
- Materials that are burned or incinerated...7
- Materials that are disposed...7
- Materials that are stored in lieu of being abandoned...8

ABSORBENT ADDITION...414
- Permitting exemption...414, 460, 467
- To containers...414, 460, 746

ABSORBENTS
- Absorbent addition exemption...414, 460, 467
- Adding to bulk hazardous wastes...414
- Adding to containerized hazardous wastes...414, 460, 746
- Contained-in policy...251, 683
- Contaminated with petroleum products...29, 683
- Contaminated with spent solvents...102, 250, 683
- Contaminated with spilled material...683
- Contaminated with used oil...560, 584, 683
- Mixture rule...250
- Use constituting disposal...534
- Used to contain spills...102, 250, 683

ACADEMIC LABS
- Subpart K...377

ACCUMULATION
- Aerosol cans...61
- Fluorescent light tubes...81
- Universal wastes...432

90-DAY ACCUMULATION...300
- 90-day clock...304, 445, 488, 675
- Accumulation time extensions...307
- Air emission standards...305, 314, 398, 708, 717, 724
- Allowable units...300, 398
- Closure standards...314, 475, 922
- Containers...303, 398, 439, 445
- Containment buildings...303, 398, 632
- Corrective action...314, 793
- Drip pads...303, 398, 719
- Financial assurance...315
- Manufacturing process units...221
- New clock for returned shipments...307, 352
- Permitting exemption...397, 467
- Point of waste generation...668
- Required provisions...300
- Still bottoms...305
- Summary of requirements...300
- Tanks...302, 398, 463, 465, 469, 478, 488
- Training requirements...314, 322, 752
- Treatment allowed...303, 397, 402, 616, 679, 709
- Unknown wastes...304, 675
- Used during demolition activities...927
- Used during remediation activities...903
- Used to mix used oil and hazardous waste...554

180-DAY ACCUMULATION...300
- 180-day clock...304, 445, 488, 675
- Accumulation time extensions...307
- Allowable units...300, 398
- Closure standards...315, 475, 922
- Containers...303, 398, 439, 445
- Corrective action...314, 793
- Financial assurance...315
- New clock for returned shipments...307, 352
- Permitting exemption...397, 467
- Required provisions...301
- Summary of requirements...300
- Tanks...302, 398, 463, 465
- Training requirements...314, 328, 752
- Treatment allowed...303, 397, 402, 616
- Unknown wastes...304, 675
- Used during demolition activities...927
- Used during remediation activities...903
- Used to mix used oil and hazardous waste...554

ACETALDEHYDE
- Gas at standard temperature...453

ACETONE
- Burning for energy recovery...20
- Solvent use...87, 89, 91, 95, 101, 263, 529

ACETYLENE
- Ignitable compressed gas...49

ACRYLONITRILE
- De minimis wastewater exemption...212
- Destroyed by fire...264

ACTIVATED CARBON
- Carbon regeneration...256, 407
- For Subpart CC compliance...741-742
- From aerosol can puncturing...60
- From treating hazardous wastes...263, 695
- From treating manufacturing gases/vapors...94, 124, 691
- Spent material or sludge?...14
- Vapor-recovery units...30, 692

ACTIVE MANAGEMENT...890-892

Index

A

ACUTE HAZARDOUS WASTES
- Acute hazardous wastes are counted separately...293
- Definition...284
- Rendering containers empty...121, 126, 451
- Satellite accumulation...293-294

ADHESIVES
- Mixed with spent solvent...251
- Waste adhesive is not spent solvent...93

AEROSOL CANS...58
- Accumulation...61
- As state universal wastes...426
- Burning the removed propellant...20, 60
- Disposal...59
- Empty cans may be reactive...58, 381, 449, 451, 839
- Liquid contents...60
- Pesticides...59
- Propellants...60
- Puncturing...59, 245
- Puncturing may be treatment...60, 381
- Recycling...58, 243, 245, 649
- Regulatory status of cans...58, 242-243, 245
- Scrap metal...58, 243, 245, 649
- State vs. federal determination...61

AGRICULTURAL WASTE
- Exclusion...174

AIR EMISSION STANDARDS...703-750
- 90-day accumulation units...305, 314, 398, 708, 717, 724
- Satellite accumulation units exempt...299, 724
- Subpart AA standards...314, 398, 410, 443, 515, 704
- Subpart BB standards...314, 398, 405, 410, 443, 515, 713
- Subpart CC standards...314, 398, 410, 443, 721

AIR STRIPPERS
- As 90- or 180-day accumulation units...401
- As air emission control devices...706, 708
- As tanks...465
- As wastewater treatment units...387
- Ground water treatment prior to spray irrigation...591

ALCOHOLS
- Alcohol-content exclusion...46
- In wastewater treatment units...387
- Spent solvents...46

ALDRIN
- Used to make lab standards...130

ALUMINUM...47

ALUMINUM ALKYLS
- Reactivity...54

ALUMINUM CHAFF
- Reactivity...55

ALUMINUM PHOSPHIDE
- Use as fumigant...124

AMMONIUM NITRATE...581

AMMONIUM SULFATE
- Lead-acid batteries...236

AMMUNITION
- Reactivity...57, 839

ANALYTICAL COSTS...41

ANALYTICAL METHODS
- See SW–846

ANALYTICAL REQUIREMENTS
- Difficult-to-analyze matrices...823
- Land disposal restrictions...600, 603, 843

ANALYTICAL SAMPLE EXEMPTION...224
- Applies to product samples?...228
- Disposal of excess sample...227
- DOT requirements still apply...224
- Importing samples...226
- Sample analysis...227
- Sample shipment...226
- Sample storage...226, 468
- Scope...224
- Waste counting...291
- Who is the generator?...225

ANCILLARY EQUIPMENT
- Are containers ancillary equipment?...389, 410, 442, 474
- Definition...389, 473
- Drip pads...473
- Leak testing...487
- Secondary containment...469, 483
- Tank systems...409, 465, 473, 479, 483

ANTIFREEZE
- Point of generation...667
- Test methods...826
- Toxicity...73

APPENDIX I TO PART 261...848

APPENDIX VII TO PART 261...85, 834

APPENDIX VIII TO PART 261...17, 64, 75, 121, 154, 682, 790, 821

APPENDIX IX TO PART 264...790

AQUEOUS...44, 46, 51, 53

AREA OF CONCERN...789

AREA OF CONTAMINATION (AOC)...811
- Approval required?...813, 902
- Containers within...591
- For corrective actions...811
- Investigation-derived wastes...813, 902
- Land disposal restrictions...591-592, 902
- Piling contaminated soil within...810
- Plant construction activities...891, 902
- Point of generation issues...811, 891, 894, 902
- Remediation activities...899

ASBESTOS
- Building/structure assessment...920

ASBESTOS WASTES
- In unknown wastes...674

Index

ASPHALT
 Use constituting disposal...27, 546

ATOMIC ENERGY ACT
 Interface with RCRA...151, 634

B

BAGHOUSE DUST
 Copper recovery...531
 Lead...407
 Metals recovery...526
 Point of generation...667, 857
 Sulfur/chlorides recovery...407
 Zinc recovery...516, 518

BATTERIES
 Battery casings are containers...441
 Cadmium...76, 424, 426, 438, 604, 612, 634
 DOT requirements...434
 Landfilling...431, 442, 658
 LDR storage prohibition...637
 Lead-acid batteries...32, 236, 401, 424, 430, 438, 441, 465, 635, 637
 Lithium-sulfur dioxide...55, 431, 658
 Mercury-Containing and Rechargeable Battery Management Act...438
 Nickel-cadmium...426, 438
 Not scrap metal...22, 241
 Radioactive...604, 634-635
 Rechargeable batteries...438
 Reclamation...31, 236
 Recycling...236, 438
 Removed before demolition...921
 Spent materials...12, 236, 426
 Treatment of battery waste...401, 431, 465, 658
 Universal wastes...236, 424, 430-431, 658
 Waste counting...238, 287
 When they are debris...440, 626-627
 Zinc...612

BENEFICIATION WASTES
 Bevill wastes...196, 198
 Commercial chemical products...201
 Exclusion...196, 198
 Mixture rule for Bevill wastes...252
 Regulatory determination...197
 Residues from treating excluded wastes...204
 Spent solvents...201
 Uniquely associated wastes...201
 Wastes that are excluded...198

BENZENE
 Benzene NESHAP rule...211, 725-726
 Burning for energy recovery...28
 De minimis wastewater exemption...209
 Gasoline tank bottoms...221
 Solvent use...89
 Spilled...679, 686, 891
 Stripped from wastewater...385, 387
 Treatment...385

BERYLLIUM
 Wastes from machining beryllium metal...123
 Wastes in glove boxes...123

BEVILL AMENDMENT
 Bevill wastes...175, 252, 266
 Bevill-exempt devices...116, 652
 Derived-from rule for Bevill wastes...266
 LDR paperwork...181, 208, 649, 652
 Mixture rule for Bevill wastes...252

BIENNIAL REPORT...366
 Episodic generators...285, 321, 368
 Generator certification...369
 Large quantity generators...365-366
 Mixed waste...160, 368
 State-specific requirements...369
 Waste codes on...368
 Waste minimization...312, 369
 Weight designation...369

BLASTING CAPS
 Reactivity...55

BLENDERS
 As 90- or 180-day accumulation units...401

BOILERS AND INDUSTRIAL FURNACES
 BIF rule...31, 270, 420, 554, 692
 Boiler chemical cleaning wastes...180
 Burning propellant from aerosol cans...60
 Burning small quantities of hazardous waste in onsite units...419
 Combustion (CMBST)...605, 609, 615
 For Subpart CC compliance...733, 741
 Gaseous emissions...694
 Generation of Bevill wastes...175
 LDR paperwork...652
 Metals recovery...407
 Recycling exemption eligibility...406, 498, 528, 534
 Used oil combustion...555, 578, 581-582, 584-585
 Waste analysis plans...829

BOMB SQUADS...419

BURNING FOR ENERGY RECOVERY...28
 Aerosol can propellant...20, 60
 Burning small quantities of hazardous waste in onsite units...419
 By-products...18, 535
 Commercial chemical products...19, 28, 61, 136, 510, 518, 522
 Definition...10, 498
 Fuel blending...30, 383, 401, 408, 464, 510, 513, 518, 522, 560, 652, 664
 Household hazardous wastes...174
 Materials whose ordinary manner of use is as fuels...28
 Natural gas pipeline condensate...29, 196
 Pesticides...20, 510

Index

Recycling exemption is voided...528, 534
Sham burning...30
Sludge...28
Solid wastes...7, 10, 28, 530
Spent materials...28
Start-up fuel for an incinerator...28
Still bottoms...29
Used oil...29, 555, 561, 578, 581-582, 584-585

BUTANE...61

BUTYL ALCOHOL
Isomer issues...120
Solvent use...89

BY-PRODUCTS...15
Burning for energy recovery...18, 535
Definition...10, 498
May be characteristic or listed in Table 1...11, 498
Not the same as co-products...10, 15-16
Reclamation...32, 509, 516
Still bottoms...16

C

CADMIUM
Batteries...76, 424, 426, 438, 604, 612, 634
Sandblast residue...519
Toxicity...519

CALCIUM CARBIDE SLAGS...47

CAPPING WASTES IN PLACE...591

CARBAMATE WASTES
Biological treatment sludge...270
De minimis wastewater exemptions...215
Exception to derived-from rule...270
Production wastes...215

CARBON DIOXIDE
Geologic sequestration exclusion...246

CARBON DISULFIDE
Contaminating equipment...134
De minimis wastewater exemption...209
LDR paperwork...655
Solvent use...89

CARBON TETRACHLORIDE
Contaminated soil...277
De minimis wastewater exemption...209
Recycling...519
Solvent use...88, 101

CATALYSTS
De minimis wastewater exemption...211
Reclamation...31
Sole active ingredient...130

CATHODE RAY TUBES (CRTs)
As state universal wastes...426
Cullet...79
Exempt from RCRA if recycled...78
Exported...79
Regulatory status...78
Sent to a reseller are not solid wastes...79
Toxicity...77
Used as a fluxing agent...79

CEMENT KILNS
Bevill wastes...116, 252, 652

CEMENT/AGGREGATE
Cement kilns...116, 252, 652
Exception to derived-from rule...270
Recycling exemption eligibility...528, 533
Use constituting disposal...25-26, 533, 652

CERCLA ISSUES
Applicable or relevant and appropriate requirements (ARARs)...806, 811, 909
Assigning EPA ID number to site...372
CERCLA reporting requirements...84, 458, 486, 529, 572, 584, 680, 686, 913
Ground water reinjection...592
Inactive units...893, 907
Investigation-derived wastes...813
Landfills...75, 458, 893, 911
Leaching from lead-based paint waste...75, 381
Liability...41, 75, 458, 666, 784
NPL sites...908-909
Presumptive remedies...799
RCRA corrective action parity...793, 908
RCRA/CERCLA interface...907
Risk assessments...801

CESIUM
In cooling water...154

CHARACTERISTIC WASTES...39-84
Adding characteristic codes to listed wastes...619
Aerosol cans...58, 381, 449, 451, 839
Cathode ray tubes (CRTs)...77
Characteristic determinations for demolition wastes...924
Characteristic determinations for remediation wastes...895
Characteristic determinations for unknown wastes...673-674
Contained-in policy...275
Corrosive wastes...51, 825, 838
Debris...630
Demolition wastes...924
Derived-from rule...262
Determination using analysis...674, 824, 841, 895, 924
Determination using knowledge...826, 836, 895, 924
Fluorescent light tubes...80
Ignitable wastes...42, 824, 837
Land disposal restrictions...255, 274, 602, 624, 672
LDR paperwork requirements...654
Mixed with used oil...553
Mixture rule...249-250, 383
Point of generation...662, 672
Reactive wastes...54, 825, 838
Remediation wastes...895
Spilled...679
Toxic wastes...62, 825, 839

Index

CHLORAL
    And hydrate are U034...120

CHLORAMBUCIL
    Container residues...127

CHLORATES
    Ignitability...49

CHLORDANE
    Pesticide use...27, 124
    Sole active ingredient in pesticide...129

CHLORINATED FLUOROCARBONS (CFCs)
    See Also FREON
    Blowing agents...96
    Building/structure assessment...919
    De minimis wastewater exemption...209
    Exemption for reclaimed CFC refrigerants...70, 72, 125
    In used oil...558, 577
    LDR treatment standards...617
    May be D019 or D022 wastes...70, 72
    Refrigerant use...88, 124
    Solvent use...71, 88, 692

CHLOROBENZENE
    De minimis wastewater exemption...209
    Recycling...519
    Solvent use...88

CHLOROFORM...214

CHLOROMETHANE
    Gas at standard temperature...453

CHROMIC ACID...15, 513

CHROMIUM
    Electroplating operations...33
    Sandblast residue...519
    Toxicity...64, 68, 519
    Trivalent vs. hexavalent chromium...64, 70-72

CIRCUIT BOARDS
    See PRINTED CIRCUIT BOARDS

CLEAN AIR ACT
    Benzene NESHAP rule...211
    Controls emissions from manufacturing operations...4
    Interface with RCRA...3, 5, 60, 210, 691, 693, 710, 719, 725
    Nonhazardous secondary materials that are not solid waste when combusted...582
    Nonhazardous secondary materials that are solid waste when combusted...584

CLEAN WATER ACT
    CWA-equivalent treatment...386, 604-605
    Interface with RCRA...4, 141, 150, 208, 278, 385, 892
    Land disposal restrictions...146-147, 591, 597, 599, 604, 607, 611, 613-615, 649
    NPDES program...4, 147, 385, 467
    Pretreatment program...4, 142, 385
    Releases permitted under may be subject to corrective action...790
    Reporting for discharges of oil...687
    SPCC program...573-574, 581
    Used oil management...561
    Wastewater treatment unit exemption...386

CLEANUP RESIDUES...678
    Absorbents...250, 683
    Characteristic waste determination...679
    Cleanup levels...678
    Contained-in policy...680
    From spills of P- and U-chemicals...127, 277, 680
    From spills of product...127, 277, 677
    Land disposal restrictions...591
    Listed waste determination...681
    Manifesting...677
    Spent solvents...102, 209
    Storage in tanks...220, 418, 678
    Universal wastes...429, 433

CLOSURE
    90- or 180-day accumulation units...314, 475, 922
    Containers...443, 921
    Corrective action...792
    Mixed waste units...160
    Permitted units...921
    Staging piles...811
    Tanks...475, 921
    Temporary units...809

CO-PRODUCTS...16
    Exempt from RCRA regulations...15
    Fuel determination...18
    Intentionally produced...15
    Not the same as by-products...10, 15
    Produced to specifications...15
    Ready market...15
    Used as is without additional processing...15

COMMERCIAL CHEMICAL PRODUCTS...2, 16, 122
    Are not solid wastes when reclaimed...20, 126, 134-135, 684, 699
    Burning for energy recovery...19, 28, 61, 136, 510, 518, 522
    Cleanup residues...127, 277, 677
    Contained-in policy...134, 277, 280
    Container residues...126, 451, 454
    De minimis wastewater exemption...133, 211
    Definition...10, 498
    Discarded...2, 87, 119, 835
    Fertilizer...26, 533
    Free product recovered from leaking tanks...20, 511, 684
    From a manufacturing plant...2
    Gases/vapors...453, 690
    Hazard codes...120
    Inactive ingredients...128
    Isomer issues...120
    Manufactured articles...18, 121, 125
    Manufacturing chemical intermediates...125
    Manufacturing process wastes...131, 693
    May include products not on the P- or U-lists...17, 20, 498
    Military munitions...699
    Mixed wastes...154

Index

Mixed with used oil...19, 135
Must be unused...123
Obsolete chemicals...5, 8, 19
Off-specification products...29, 94, 125, 259, 518, 522
Oil and natural gas E&P wastes...187, 195
Point of generation...666, 857
Product handling wastes...131
Product samples...228
Pure, unused chemicals...122
Recycling...19-20, 32, 126, 134-135, 684
Returned to the manufacturer...20, 126, 134
Shelf life exceeded...19
Sole active ingredient...128, 686
Spilled...2, 127, 223, 417, 676, 679-680
Technical grades...128
Two different meanings...17, 510
Unreacted reagents...13, 133
Use constituting disposal...19, 27, 135
Used for their intended purpose...2, 123, 700
Used to make lab standards...130, 227
Waste counting...287

COMPARABLE FUELS
Exclusion...18, 169
Synthesis gas...696

COMPRESSED GAS CYLINDERS...49, 690
Returned to supplier...453, 691

COMPRESSOR STATIONS...193

COMPUTERS
Sent to a reseller are not solid wastes...79

CONDITIONALLY EXEMPT SMALL QUANTITY GENERATORS
180-day clock?...305
Accumulation quantity limits...319
Definition...284, 316
Derived-from rule...266, 275, 318
Electronic wastes...77

EPA ID number exemption...319, 377
Generation rate limits...319
Hazardous waste determinations...284, 316, 362, 818
Laboratories...127, 227, 232
Land disposal restrictions exemption...320, 587, 589
Light bulbs...81
Manifesting exemption...320, 335, 362
Mixed waste...155, 321
Mixture rule...173, 259, 288, 318
Not the same as households...171, 173
Offsite transportation...320
Recordkeeping requirements...321, 362, 757
Satellite accumulation...299, 320
Small quantity burner exemption...420
Subpart CC exemption...724
Summary of requirements...316
Training...320, 322, 330
Universal waste management...321, 429
Used oil exemption...318, 321, 553, 557
Waste accepted at TSD facilities...828
Waste counting...288, 318
Waste treatment requirements...319, 410

CONFIDENTIAL BUSINESS INFORMATION
Reviewed during inspections...777

CONTAINED-IN POLICY...275
Absorbents...250, 683
Applies to media contaminated with P- and U-wastes...277
Applies to process wastes...281
Applies to spills...680
Characteristic wastes...276
Contained-in determinations...276
Contaminated debris...271, 279, 440, 633
Contaminated environmental media...276

Contaminated ground water...277
Contaminated rainwater...278
Contaminated sediments...150, 278
Contaminated soil...277, 624
F001 wastes...281
F003 wastes...101, 250, 278, 683
F004 wastes...227, 280
F007 wastes...277
F020 wastes...280
ICR-only listed wastes...278
Laboratory wastes...280, 440
Land disposal restrictions...278, 598
Listed wastes...276
Maintenance wastes...280
Manufactured articles...280
P- and U-wastes...134, 277, 280
P022 wastes...134
Personal protective equipment...280, 633
Rags contaminated with spent solvents...103
Risk assessments...276
Spent solvents...102
U051 wastes...277
U096 wastes...251
U211 wastes...277
U220 wastes...280

CONTAINER RESIDUES
In vials and syringes...127, 451
Mixture rule...258
P- and U-wastes...126, 451, 454
Status of removed residues...440, 452, 454
Warfarin...452
Waste counting...288, 458

CONTAINERS...439-462
90- or 180-day clock...306, 445
See Also EMPTY CONTAINERS
Absorbent addition...414, 459, 746
Absorbent addition exemption eligibility...415, 460
Are portable tanks containers?...395, 440, 464

Are they ancillary equipment?...389, 410, 442, 474
Are they debris?...441, 627-628
As elementary neutralization units...408
Batteries...441
Closure...443, 921
Consumer goods...446
Counting cleanout residues...288, 458
Fiber drums...460
Identified on manifests...346
Inspections...443, 458, 757
Intact containers...441, 627-628
Integrity assessments...443
Lab packs...416, 442, 459-460, 655
Labeling...445-446, 753
Laboratory equipment...440
Made of lead...28, 73, 156, 449
Materials of construction...443
Mixed wastes...159, 161
Must be closed...442, 754
Notices of violation...753-754, 757
On pallets...446, 448
Pumps...440, 442
Reuse...448
Sampling...870, 874
Satellite accumulation units...294, 459
Secondary containment...443, 447-448
Special requirements for ignitable and reactive wastes...443
Special requirements for incompatible wastes...443, 460
Storage within an AOC...591
Subpart BB air emission controls...443, 717
Subpart CC air emission controls...305, 443, 724, 742
Transportation...441, 446, 451, 460
Used for 180-day accumulation...303, 398, 439, 445

Used for 90-day accumulation...303, 398, 439, 445
Used for immediate waste transfer...292
Used oil storage...570, 578, 584
Waste counting of contents...287, 441

CONTAINMENT BUILDINGS
90-day clock...307
Can they serve as secondary containment?...465, 483
Used for 90-day accumulation...303, 398, 632
Used for treating contaminated debris...632

CONTAMINATED DEBRIS...626
Batteries...441, 626-627
Characteristic...630
Contained-in policy...271, 279, 440, 633
Containers...441, 627-628
Exception to derived-from rule...270, 629-630
Exemption for petroleum contamination...27, 70, 72, 628, 684
Filters...30, 96, 111, 131-133, 280, 287, 441, 691, 863
From product handling...131, 280
Hazardous waste determination...836, 841
Laboratory equipment...440
Land disposal restrictions...270, 279, 441, 626, 930
Lead-based paint chips/wood...406, 840, 864
Listed...631, 926
Macroencapsulation...633, 635
Mercury-contaminated items...424, 430, 441, 627, 630, 633
P- and U-wastes...128, 134, 633, 693
Personal protective equipment...633
Pumps...440, 442
Refractory brick...633
Sampling...861
Tanks...628

Treated in 90-day accumulation units...401, 632
Treatment by immobilization technologies...630-632
Treatment in containment buildings...632
Treatment residues...630
Treatment subject to permitting...631

CONTAMINATED GROUND WATER
Assessment during building/structure demolition...922
CERCLA issues...592
Contained-in policy...277
Corrective action...798, 801
De minimis wastewater exemption...212
Exemption for petroleum contamination...27, 70, 72, 628, 684
F039 wastes...275, 619
From pre-RCRA spills...891
Gases/vapors emitted during treatment...695, 708
Gasoline...511
Hazardous waste determination...836
Is treatment subject to permitting?...419
Land disposal restrictions...278
Landfill leachate...275, 619
P- and U-wastes...127
Point of generation...857, 890
Relationship to NPDES discharges...149
Storage/treatment in tanks...465
Treated in 90-day accumulation units...401
Underground injected...592
Used for spray irrigation...591

CONTAMINATED SEDIMENTS
Contained-in policy...150, 278
Corrective action...801
Relationship to NPDES discharges...150

CONTAMINATED SOIL
Assessment during building/structure demolition...922

Commercial chemical products...128
Contained-in policy...277, 624
Corrective action...798, 801
D007 wastes...68, 625
D008 wastes...625
D009 wastes...635
D018 wastes...623
Derived-from rule...626
Dilution prohibition...638, 906
Excavation activities...277, 592, 638, 811, 891
Exemption for petroleum contamination...27, 70, 72, 628, 684
Explosives...57, 839, 895
F003 wastes...101
From plant construction activities...891
From spills...419, 679, 681
Gasoline...623
Hazardous waste determination...827, 836
Land disposal restrictions...278, 592, 621, 655, 659, 685, 907
LDR paperwork requirements...655
Lead contaminated...625
Listed wastes...624
Listing determinations from unknown contamination...896
P- and U-wastes...127
PCB contaminated...625
Pesticides...27, 124, 681, 897
Piled in a CAMU...810
Piled in an AOC...810
Point of generation...857, 890, 894
Radioactive mercury...635
Reactivity...57, 839
Recycling...27, 407, 546
Residues from treatment...626
Soil screening guidance...798
Solvents...102
Stored in a waste pile or tank?...472
Treatability variances...621
Treated in 90-day accumulation units...401, 709
Used to make asphalt...27, 546

CONTINGENCY PLAN
During immediate responses...417
Large quantity generators...301, 310, 338, 365, 677, 752
Notices of violation...752
OSHA emergency action plans...310
Satellite accumulation...299, 310
TSD facilities...677

CONTRACTORS...331
Are cogenerators of the wastes they produce...305, 331, 341, 428, 576
Equipment maintenance contractors...333
Household contractors...332
Manifesting and recordkeeping requirements...332, 341
Painters...332
Removing lead-based paint...381
Removing shipboard wastes...333
Sample collectors...332
Solvent recyclers...333
Storage of wastes generated by contractors...332
Training...323, 332
Universal wastes...428
Who counts the waste?...332

COPPER
Plating...25
Recovered from a sludge...510
Recovered from baghouse dust...531
Recovered from F006 wastes...513
Use as ingredient...517-518

COPPER SULFATE...25

CORRECTIVE ACTION...785-816, 900-901
90- or 180-day accumulation units...314, 793
Action levels...797
Applicability...791
Area of concern...789
Area of contamination (AOC)...811
Beyond the facility boundary...786, 793

Compliance time frame...802
Contaminated ground water...798, 801
Contaminated sediment...801
Contaminated soil...798, 801
Corrective action management units...805
Corrective measures implementation...803
Corrective measures study...799
Determination of completion...803
Elementary neutralization units...411
Environmental indicators...804
Facility...790
Financial assurance...802
Five-phase process...794
For spills...529, 685, 788
Ground water reinjection...592
Guidance documents...815
Hazardous waste and constituents...789
Inactive units...893
Institutional controls...800
Interim measures...798
Interim status facilities...792
Investigation-derived wastes...813
Landfills...793, 893
Maintenance wastes...787
Manufacturing process units...223, 789
Media cleanup levels...801
Mixed wastes...158
Natural attenuation...802
No further action required...798
Parity with CERCLA cleanups...794, 908
Permit triggers program...379, 791, 908
Points of compliance...801
RCRA facility assessment...794
RCRA facility investigation...796
Recycling units...787
Regulated units...785
Regulations...786
Releases...790
Releases from underground used oil storage tanks...683

Remedial action plans (RAPs)...792, 806, 809-810, 813
Remedy selection...800
Scrap metal management area...23, 788
Sewer line leaks...145, 787
Solid waste management units...787
Staging piles...809, 811
Temporary units...809
TSD facilities...791
Wastewater treatment units...787
Wood preservative wastes...788

CORRECTIVE ACTION MANAGEMENT UNITS...805
CAMU-eligible wastes...805
Grandfathered units...805
Land disposal restrictions...589, 591, 805
Liners and caps...808
Minimum technological requirements...805
New units...805
Offsite disposal of CAMU-eligible wastes...808, 906
Piling contaminated soil within...810
Regulatory agency must designate...806
Treatment requirements...808
Used during remediation activities...904

CORROSIVITY...51
Aqueous...44, 51
Aqueous liquids...52-53
Aqueous nonliquids...53
Aqueous wastes with low or high pH...52, 825, 838
Corrosive solids...54, 825, 838
Determination using analysis...52, 674, 825
Determination using knowledge...52, 838
Lead-acid batteries...236, 401, 465
Liquid...44, 51
Liquids that corrode carbon steel...53, 825, 838
Nonaqueous liquids...53-54
State vs. federal determination...52

Suspensions, colloids, and gels...53
Unknown wastes...674

CRADLE-TO-GRAVE LIABILITY...781

CREOSOTE
Burning for energy recovery...20, 136
Contained-in policy...277
Creosote-treated railroad ties...121, 252
Mixture rule...252
Spilled...681

CRESOLS
De minimis wastewater exemption...209
Solvent use...89
Use as reactant...93

CRESYLIC ACID
De minimis wastewater exemption...209
Solvent use...89

CUMENE HYDROPEROXIDE
Mixture rule...255
Spilled...251

CYANIDES
Dilution prohibition...639
F007 wastes...116, 252
F008 wastes...116
F009 wastes...117
Reactivity...54, 56, 116, 674, 825, 838
Stabilization...639
Treatment...385

CYANOGEN
Gas at standard temperature...453

CYANOGEN CHLORIDE
Gas at standard temperature...453

CYCLOHEXANE
Ignitability...45

CYCLOHEXANONE
LDR paperwork...655
Solvent use...89

CYCLOPHOSPHAMIDE
And hydrate are U058...120
Container residues...127

CYLINDERS
See COMPRESSED GAS CYLINDERS

D

D001 WASTES...42
Alcohol-content exclusion...46
Derived-from rule...262
Dilution prohibition...642
Fuel blending...652
In recycling units...708
In thermal treatment units...399
In wastewater treatment units...387, 389
Land disposal restrictions...157, 604-605, 615, 640, 652
Mixed wastes...157
Mixed with used oil...465, 553
Mixture rule...251, 383
Overlap with D003 reactives...47
Point of generation...664, 667, 671
Products that contain solvents...93
Rags...49, 107
Setback requirements for containers...443
Special management procedures for tanks...475, 482

D002 WASTES...51
Are no longer corrosive at the TSD facility...41, 351
Declared hazardous wastes...41
Dilution prohibition...642
In elementary neutralization units...408, 410, 828
In manufacturing process units...221
In wastewater treatment units...387, 389
Land disposal restrictions...255, 411, 610
Lead-acid batteries...236, 401, 465
Mixture with oil and natural gas E&P wastes...253
Point of generation...380, 666, 670-671, 857
Sham recycling...542

Index

D003 WASTES...54
  Aerosol cans...58, 381, 449, 451, 839
  Immediate response exemption...57, 419
  Land disposal restrictions...604, 610, 640
  Mixed with used oil...554
  Open burning...590
  Overlap with D001 ignitable solids...47
  Oxygen breathing apparatus canisters...457
  Setback requirements for containers...443
  Special management procedures for tanks...475, 482

D004–D011 WASTES
  Land disposal restrictions...611, 640

D005 WASTES
  Land disposal restrictions...594

D006 WASTES
  Land disposal restrictions...604, 612, 634

D007 WASTES
  Contaminated soil...68, 625
  Land disposal restrictions...615
  Mixture with Bevill wastes...253
  Spilled...679
  Trivalent vs. hexavalent chromium...64, 71-72

D008 WASTES
  Contaminated soil...625
  Decharacterized and sent to Subtitle D facilities...653
  Dilution prohibition...642
  Foundry sands...74
  Land disposal restrictions...73, 604, 614-615, 625, 634, 640, 653
  Lead shot from a shooting range...243, 245
  Lead-acid batteries...236, 401, 465
  Lead-lined pipe and shielded cable...73
  Mixed wastes...157
  Mixture with Bevill wastes...253
  Point of generation...664, 671
  Radioactive lead solids...604, 634-635, 842
  Total analyses in lieu of TCLP...69

D009 WASTES
  Contaminated debris...627
  Contaminated soil...635
  Land disposal restrictions...604, 612, 634
  Mixed wastes...157, 635

D011 WASTES
  Land disposal restrictions...605, 634
  Photoprocessing operations...33, 407
  Point of generation...666

D012–D017 WASTES
  Land disposal restrictions...605, 611

D018 WASTES
  Contaminated soil...623
  Exemption for USTs holding petroleum...27, 70, 72, 628, 684
  Incinerated...671
  Land disposal restrictions...613, 623
  Mixture with oil and natural gas E&P wastes...254
  Spilled...679

D019 WASTES
  Used CFC refrigerants...70, 72

D020 WASTES...897

D021 WASTES
  In PCBs...72

D022 WASTES...214
  Used CFC refrigerants...70

DAUNOMYCIN
  Container residues...127

DDT
  In wool blankets...125

DE MINIMIS WASTEWATER EXEMPTIONS...208
  Applicability of land disposal restrictions program...216, 649, 652
  Applicability to produced sludges...215
  Carbamate production wastes...215
  Commercial chemical products...133, 211
  De minimis losses...212
  F- and K-wastes...211
  Ground water...212
  Laboratory wastes...213
  P- and U-wastes...133, 211
  Refinery wastes...210
  Rinsate from empty containers...456
  Spent solvents...95, 209
  Spills...211
  Storage/treatment in tanks...468
  Waste counting...289

DEBRIS
  See CONTAMINATED DEBRIS

DECLARED HAZARDOUS WASTES...40, 831
  D002 wastes...41
  F001 wastes...92
  May save analytical costs...41
  Waste munitions...67

DELISTING PETITIONS...75, 85, 265
  Electroplating sludges...113
  Land disposal restrictions...598
  Using the OWEP instead of the TCLP...67

DEMOLITION ACTIVITIES...915
  90- or 180-day accumulation units...927
  Asbestos assessment...920
  Building/structure assessment...915
  CFCs assessment...919
  Contaminated soil/ground water assessment...922
  Land disposal restrictions...930
  Lead-based paint assessment...917
  Mercury-containing equipment assessment...919
  Offsite management...930
  Onsite management...927
  PCBs assessment...919

Point of waste generation...923
Pre-demolition activities...915
Process equipment assessment...920
RCRA-permitted units...930
Satellite accumulation units...927
Universal wastes...928
Utilities assessment...920

DERIVED-FROM RULE...260
Background...261
Bevill wastes...266
Cement produced from hazardous waste fuels...270
Characteristic wastes...262
Contaminated debris...270, 629-630
Contaminated soil...626
D001 wastes...262
Exception for debris...270, 630
Exceptions for certain filtrate and supernatant...111, 117, 267
Exceptions to...265
Exemption for de minimis losses of listed wastes...215
F001 wastes...264
F002 wastes...265
F003 wastes...100, 263
F005 wastes...93, 264-265
F006 wastes...264
F020 wastes...263
For conditionally exempt small quantity generators...266, 275, 318
HTMR residues from treating F006, K061, and K062...116, 269, 273
ICR-only listed wastes...262-263
Incinerator residues...123, 263
Ion-exchange resins...264
K045 wastes...263
K062 wastes...111
Landfill gas condensate...265, 696
Landfill leachate...274, 892
LDR ramifications...274, 595, 597
Listed wastes...262-263, 663
Oil and natural gas E&P wastes...266

POTW sludge...146
Precious metals...233
Precipitation run-off...275
Refinery wastes...266
Residues from exempt waste...265
Residues from reclamation...103, 263-264, 271, 406, 529
Spent solvents...29
Spiking feed during incinerator trial burn...123, 264
Status of reclaimed materials...101, 272
U009 wastes...264
Waste coding for treatment residues...262
Waste counting...292
Wastewater treatment sludges...263, 390
Wastewater treatment units...271, 390

DESIGNATED FACILITY
On manifest...307, 344, 352

DESTINATION FACILITIES
Universal wastes...427, 431, 435, 658

DICHLOROBENZENE
De minimis wastewater exemption...209
Solvent use...88
Use as drain cleaner...124

DIELDRIN
Used to make lab standards...130

DIETHYLHEXYL PHTHALATE (DEHP)
DEHP-containing light ballasts...83
Manufacturing use...131

DILUTION PROHIBITION...637
Adding reagents...639
Addition of iron fines...74
Aggregation for centralized treatment...638
Attaches at point of generation...597
Contaminated soil...638, 906
Cyanide wastes...639
D001 wastes...642
D002 wastes...642

D008 wastes...642
F003 wastes...642
F005 wastes...642
For incinerators...640
In sewer lines...383
Mixture rule...247, 254
Prior to CWA-permitted discharge...604, 608
Prior to underground injection...604, 642

2,4-DINITROTOLUENE
Toxicity...64, 67, 839

DIOXIN WASTES
Container secondary containment...448
Description...86
Empty containers...451
From the production of pentachlorophenol...16
Inherently waste-like...8, 531
Small quantity burner exemption...420

DISCARDED MATERIALS...1
Garbage and refuse...1, 3
Gaseous emissions from manufacturing operations...5
Incidental residues...2, 5
Includes recycled materials...5
Materials thrown away, abandoned, or destroyed...2, 4
Obsolete chemicals...5, 8, 19
Process wastes...5
Products that are spilled or discarded...2, 87, 119, 835
Sludge...1, 3
Spent materials...2, 5
Wastewater...4

DOMESTIC SEWAGE EXCLUSION...4, 140
Applicability to F006 wastes...112, 146
Applicability to federally owned treatment works...146
Applicability to sewer line leaks...145
Applicability to sewer sludge...145
Domestic sewage...142

Index

Notification for discharging hazardous waste to a POTW...144
Point of applicability...143
POTW...142
Storage/treatment in tanks...466

DOT REGULATORY INTERFACE
Analytical sample exemption...224
Aqueous...46
Batteries...434
Combining wastes...302, 382
Combustible liquids...46
Compressed gas cylinders...49, 453, 690-691
Container issues...441, 446, 460
Empty containers...456
Flammable liquids...42
Ignitable compressed gases...49, 690
Ignitable solids...47
Manifesting...334
Onsite shipments...338
Oxidizers...49
Reuse of containers...448
Spill response...677
Spontaneous combustion...47
Treatability study exemption...229
Universal wastes...429, 433
Used oil...581
"Offeror" on the manifest...349, 352

DRIP PADS
90-day clock...307
Are they ancillary equipment?...473
Are they tanks?...473
Immediate responses...419
Subject to Subpart BB air emission standards?...719
Used for 90-day accumulation...303, 398, 719

DRY CLEANING WASTES
Are F002 wastes...92
Spent filters...287, 406
Spent solvents...92, 265, 289, 406
Tetrachloroethylene...265, 287, 387
Waste counting...287, 289

DUST SUPPRESSANTS
Land disposal restrictions...610
Use constituting disposal...19, 135
Used oil...578

E

EARTHEN PITS...590

ELECTRIC ARC FURNACE DUST
Recycling...407, 517, 525, 533
Waste coding...620

ELECTRONIC WASTES...75
Printed circuit boards...80
Regulatory status...76
Sent to a reseller are not solid wastes...76
Toxicity...76

ELECTROPLATING PROCESS...108
Anodizing...118
Baths...109
Bright dripping...109
Chemical conversion coating...114, 117
Chemical etching and milling...109, 117
Chromating...114
Cleaning and stripping...110
Coloring of metals...114
Copper plating...25
Dragout...109, 116, 252
Electroless plating baths...116
Electrowinning process baths...116
Galvanizing...15, 113, 513
Immersion plating...114
Phosphating...114, 118
Photoresist stripping...114
Rinsewaters...109, 112, 116, 264
Zinc plating...15, 113, 513

ELECTROPLATING WASTES
See Also F006 WASTES, F007 WASTES, F008 WASTES, F009 WASTES, and F019 WASTES
Ion-exchange resins...15, 111
Mixture rule...252
Precious metals...109-110, 232
Smelting is reclamation...33, 115, 273, 407, 513

ELEMENTARY NEUTRALIZATION UNITS...408
At generator's facilities...410
Can be containers...408
Can be sumps...409, 478-479
Can be tanks...408, 465
Corrective action...411
D002 wastes...408, 410, 828
Definition...408
Land disposal restrictions...411
May become solid waste management units...411
Permitting exemption...408, 442, 467, 478-479
Subpart AA exemption...708
Subpart BB exemption...717
Subpart CC exemption...724
Waste analysis plan not required...828
Waste counting...291

EMPTY CONTAINERS...448
Aerosol cans...59, 381, 449, 451, 839
Alternative cleaning methods...452
Beneficial use of residues...126, 454
CERCLA liability for residues...458
Containers that held acute hazardous wastes...121, 126, 451
Containers that held compressed gas...453
Containers that held nonacute hazardous waste...449
De minimis wastewater exemption...456
Dioxin wastes...451
Discharge of residues to CWA systems...455
DOT definition of...456
Emptied using commonly employed practices...449
F004 wastes...227, 280, 450
Filter cartridges are not containers...450
Laboratory wastes...440, 450

Manifesting nonempty shipments...351, 456
One inch of residue...450
Paint cans...457
Pressurized rail cars...453
Status of removed residues...440, 452, 454
Triple rinsing...451
Used oil placed into...556
Waste counting...458
Weight-limit alternative to one inch of residue...450

EMPTY TANKS...306, 488, 528

ENFORCEMENT ISSUES...751-784
Common RCRA violations...751
Economic benefits from noncompliance...760
EPA enforcement database...768
EPA's right to overfile...760
Gravity-based penalty...757
Multiday penalty...758
National RCRA enforcement priorities...768
RCRA civil penalty policy...757
Recent RCRA enforcement cases...760
Regulatory liability for generators...783
Self-disclosure policy...769
Supplemental environmental projects...759
Types of enforcement actions...780
Types of inspections...770

EPA ID NUMBERS
Emergency EPA ID numbers...375
Exemptions...319, 377, 417
For separate facilities...167, 372, 653, 791
Generators...144, 234, 237, 317, 332, 340, 354, 370, 377, 417, 446, 677
Importers...372
Interim status facilities...792
Multiple generators on a single contiguous property...371
Obtaining an EPA ID number...144, 234-235, 237, 317, 356, 370, 395, 405, 429, 792

On container labels...446
On manifests...332, 340, 354-356, 377, 417, 677
One EPA ID number for one site...370
Recyclers...405, 653
Renotification...371, 376
Ships...358, 372, 375
Superfund sites...373
Temporary EPA ID numbers...375
Transporters...234, 236, 337, 341, 355, 370, 372, 377, 417, 677
TSD facilities...234, 352, 370, 395, 653
Universal waste handlers...371, 429
Used oil burners...371, 581, 584-585
Used oil generators...574, 576
Used oil marketers...371, 585
Used oil processors...371, 578
Used oil transporters...574

EPCRA ISSUES
TRI reporting...458

EPINEPHRINE
Container residues...127, 451
Isomer issues...120

EPISODIC GENERATORS...285
Biennial report...285, 368
Recordkeeping requirements...285
Renotification...377
Subpart CC standards...726
Training...286, 329

ESTERS
Isomer issues...120

ETHANOL
Solvent use...169

2-ETHOXYETHANOL
De minimis wastewater exemption...209
Solvent use...89
Use as ingredient...94

ETHYL ACETATE
Solvent mixtures...98
Solvent use...89, 95

ETHYL BENZENE
Solvent use...89

ETHYL ETHER
Immediate response...419
Solvent use...89

ETHYLENE GLYCOL
In antifreeze...74

EVAPORATION PONDS...4

EVAPORATORS
As 90- or 180-day accumulation units...399
As totally enclosed treatment facilities...413
As wastewater treatment units...388

EXCLUSIONS AND EXEMPTIONS...137-246
90- or 180-day accumulation units...397, 467
Absorbent addition...414, 460, 467
Agricultural waste/manure returned to the soil as fertilizer...174
Alcohols in ignitable wastes...46
Analytical sample exemption...224, 291, 468
Burning on-spec used oil...581
Cathode ray tubes (CRTs) that are recycled...78
Characteristic by-products that are reclaimed...32
Characteristic sludges that are reclaimed...32
Closed-loop recycling with no reclamation...524
Closed-loop recycling with reclamation...90, 405, 471, 526
Commercial chemical products that are reclaimed...20
Comparable fuels...169
Conditionally exempt small quantity generator waste...77, 81, 127, 171, 173, 227, 232, 266, 275, 284, 288, 299, 305, 322, 330, 335, 362, 377, 410, 429, 553, 557, 587, 589, 724, 757, 818, 828
De minimis wastewater exemptions...95, 133, 208, 289, 456, 468
Documentation requirements...140, 542, 544

Domestic sewage exclusion...4, 112, 140, 466
Elementary neutralization units...291, 408, 442, 465, 467, 478-479, 828
EPA ID numbers...377
Fossil fuel combustion wastes...175
General...40, 818
Geologic sequestration of carbon dioxide...246
Hazardous wastes exempt from Subtitle C regulation...140
Household hazardous wastes...61, 69, 77-78, 81, 92, 171, 196, 236, 259, 265, 332, 557, 589, 696, 818
Immediate responses...57, 417, 467, 469, 477, 479, 678
In situ mining...161
Irrigation return flows...150
Lead-acid batteries...236
Manufactured gas plant wastes...67
Materials that are not solid wastes...6, 137
Mining overburden returned to the mine site...175
Mining wastes...196, 252, 266
Mixed wastes...467
NPDES discharge exclusion...4, 147, 467
Oil and natural gas E&P wastes...48, 182, 256, 266, 468, 665, 682
Oil-bearing secondary materials...164
PCB dielectric fluids...72
Petroleum underground storage tanks...27, 70, 72, 628, 684
Precious metals recycling...33, 232, 273, 407, 542
Pulping liquors...162
Radioactive wastes...153
Reclaimed CFC refrigerants...70, 73, 125
Recovered oil from associated chemical plant...168
Recovered oil from petroleum operations...166
Recycled CRTs...78
Recycled materials...467
Recycling units/processes...401, 515, 529, 589
Residues from empty containers...440, 452, 454
Scrap metal...22, 58, 72-73, 240-241, 628
Secondary materials from mineral processing...36
Small quantity burner exemption...419
Solid wastes that are not hazardous wastes...7, 137
Solvent-contaminated wipes...104
Spent caustic solutions from petroleum refining...169
Substitution for a commercial product...521
Sulfuric acid recycling...520, 526
Totally enclosed treatment facilities...411, 467
Treatability study exemption...228, 468
Trivalent chromium wastes...71-72
Use or reuse as ingredient or feedstock...512
Used oil filters...559
Wastes in active manufacturing process units...216, 289, 468
Wastewater treatment units...384, 442, 464, 467, 473, 478-479, 668
Zinc fertilizers...26, 531

EXPLOSIVES
Burned in incinerators...263
Contaminated soil...57, 839
Immediate responses...419
Is treatment subject to permitting?...419
Reactivity...57, 419

EXPORTING HAZARDOUS WASTE
Assigning EPA ID number...372
Cathode ray tubes (CRTs)...79
For excluded reclamation...35
LDR paperwork...654
Manifesting...356
Precious metals...235
Primary exporter...357-358, 433
Recordkeeping requirements...366, 433
To Organization for Economic Cooperation and Development (OECD) countries...240
Treatability study samples...231
Universal wastes...432

F

F001 WASTES...88
90-day accumulation...305
Carbon tetrachloride...88, 101
Chlorinated fluorocarbons (CFCs)...71, 88
Contained-in policy...281
Derived-from rule...264
Difference between F001 and F002...88
Hazard codes...88
Land disposal restrictions...597, 617
Methylene chloride...88, 692
Mixture rule...251, 692
Mixture with Bevill wastes...253
Solvent mixtures...96
Still bottoms...101
Tetrachloroethylene...88
1,1,1-Trichloroethane...88, 617, 692
Trichloroethylene...88, 102, 620
Used oil presumed to be...103, 555

F002 WASTES...88
Chlorobenzene...88
Derived-from rule...265
Dichlorobenzene...88
Difference between F001 and F002...88
Dry cleaning wastes...92
Hazard codes...88
Methylene chloride...88, 90, 95
Recycling...94
Solvent mixtures...96
Tetrachloroethylene...88, 92
1,1,2-Trichloro-1,2,2-trifluoroethane...88, 92
1,1,1-Trichloroethane...88, 126

Index

1,1,2-Trichloroethane...88
Trichloroethylene...88, 842
Trichlorofluoromethane...88
Used oil presumed to be...103, 555
Waste counting...287

F003 WASTES...89
  Acetone...87, 89, 91, 95, 101, 263, 529
  Applying the D001 code...100, 620
  Butyl alcohol...89
  Contained-in policy...101, 250, 278, 683
  Contaminated soil...101
  Cyclohexanone...89
  Derived-from rule...100, 263
  Dilution prohibition...642
  Ethyl acetate...89, 95, 98
  Ethyl benzene...89
  Ethyl ether...89
  Hazard codes...89
  How nonignitable F003 spent solvents are regulated...89, 91
  Land disposal restrictions...255, 598
  Methanol...89, 91, 101, 169, 250-251, 272, 617
  Methyl isobutyl ketone...89
  Mixture rule...100, 250-251
  Mixture with Bevill wastes...253
  Rags...107
  Recycling...515, 664
  Solvent mixtures...97
  Spilled...102, 250
  Still bottoms...103, 263
  Xylene...89, 91, 95, 98, 515, 617, 664

F004 WASTES...89
  Contained-in policy...228, 280
  Cresols...89
  Cresylic acid...89
  Empty containers...227, 280, 450
  Hazard codes...89
  Nitrobenzene...89
  Solvent mixtures...96

F005 WASTES...89
  Applying the D001 code...620
  Benzene...89
  Carbon disulfide...89
  Derived-from rule...93, 264-265
  Dilution prohibition...642
  2-Ethoxyethanol...89
  Hazard codes...89
  Isobutanol...89
  Methyl ethyl ketone...89, 617
  2-Nitropropane...89
  Pyridine...89
  Solvent mixtures...96
  Toluene...89-91, 93-94, 96, 98, 154, 265, 617
  Waste counting...292

F006 WASTES...108
  Air emissions...693
  Baths and rinsewaters...109
  Cement/aggregate production...26
  Code only applies to sludge...111
  Confusion with K062...110
  Derived from treating characteristic wastewaters...263, 410
  Derived-from rule...264
  Dragout is not spent...109
  Exception to derived-from rule...111, 116-117, 267, 269
  Filters...111
  Filtrate and supernatant...111, 267
  From anodizing...109
  From chemical etching and milling...109
  From cleaning and stripping...109-110
  From common and precious metals electroplating...33, 109
  From gold plating...109
  From printed circuit board manufacturing...114
  From sequential treatment...112
  From surface impoundments...112
  From wastewater mixtures...113, 258
  Generated offsite...112, 263
  HTMR residues...116, 269, 273
  Ion-exchange resins...15, 111
  Land disposal restrictions...598, 659
  Methods to avoid F006 generation...113
  Mixture rule...257
  Paint stripping does not produce F006...110
  Point of generation...670
  Processes that do not generate F006...108, 113
  Recycling...115, 264, 407, 513
  Rinsewaters...109, 112, 116, 264
  Sent to POTW...112, 146
  Stabilization is treatment...385, 401
  Stabilization of cyanides...639
  Storage...114
  Treatment...385
  Waste counting...292

F007 WASTES...116
  Contained-in policy...277
  Cyanides...116, 252
  Dragout is not spent...116, 252
  Reactivity...116

F008 WASTES...116
  Cyanides...116
  Reactivity...116

F009 WASTES...117
  Cyanides...117

F019 WASTES...117
  Chemical conversion coating...114
  Exception to derived-from rule...117, 267
  Exemption for zinc phosphating in motor vehicle manufacturing...118
  Exemption for zirconium phosphating...118
  Filtrate and supernatant...117, 267
  From sequential treatment...117
  HTMR residues...273
  Mixture rule...257

F020–F023, F026–F028 WASTES
  See DIOXIN WASTES

F020 WASTES
  Contained-in policy...280

Index

Derived-from rule...263

F021 WASTES...16

F024 WASTES
Description...86
Synthesis gas production...535

F025 WASTES
Description...86

F027 WASTES...899
Sole active ingredient...131

F032 WASTES
Description...86

F034 WASTES
Description...86

F035 WASTES
Description...86

F037 WASTES...620
Description...86
Mixture rule...257
Oil-bearing secondary materials exclusion...164

F038 WASTES
Description...86
Mixture rule...257
Oil-bearing secondary materials exclusion...164

F039 WASTES
Contaminated ground water...275, 619
Description...86
From rainwater percolating through contaminated soil...680
Land disposal restrictions...619, 639

FACILITY
Corrective action...790

FEDERAL FACILITIES COMPLIANCE ACT
DOE facilities storing mixed waste...158
Domestic sewage exclusion extended to FOTWs...146
Extended storage of mixed waste...634
Federal facilities are subject to RCRA...909
Management of military munitions...696
Shipboard wastes...333

FEDERAL INSECTICIDE, FUNGICIDE, AND RODENTICIDE ACT
Interface with RCRA...2, 59, 252, 452

FEDERALLY OWNED TREATMENT WORKS
Domestic sewage exclusion...146

FERRIC CHLORIDE
Corrosivity...53
Production...519
Wastewater conditioning...27, 53, 522, 534

FERTILIZER
Exclusion for agricultural waste/manure returned to the soil as fertilizer...174
Irrigation return flows exclusion...150
Land disposal restrictions...533, 610
Made from K062 wastes...25
Made from sulfuric acid...24-25, 610
Product exemption...26, 533
Recycling exemption eligibility...528
Use constituting disposal...13, 24-25, 531-532, 610, 654
Zinc...26, 531

FILTER PRESSES
As recycling units...529
As totally enclosed treatment facilities...412
As wastewater treatment units...388

FILTERS AND FILTER MEDIA
Contaminated with P- and U-chemicals...131-133, 280
F006 wastes...111
Filter cartridges are not containers...287, 450
Filtrate from F006/F019 sludges...111, 117, 267
From dry cleaning...287, 406
From treating manufacturing gases/vapors...691
From venting aerosol can propellant...60
Fuel filters...245, 559
Hazardous waste determination...96, 863
Point of generation of filter cake...857
Used oil filters...241, 243-244, 441, 558, 584
Vapor-recovery units...30, 692
Waste counting...287

FINANCIAL ASSURANCE
90- or 180-day accumulation units...315
Corrective action...802

FIRE EXTINGUISHERS
Halon systems...921
Removed before demolition...921

FIRE RESIDUES...264

FLASH POINT TESTING...43, 824, 837

FLUORESCENT LIGHT TUBES...80
Accumulation...81
Ballasts...83
CERCLA reporting requirements...84
Classified as spent materials...80
Crushing...81, 432
Hazardous waste determination...82, 833
Managed as hazardous wastes...81
Managed as universal waste...425, 432
Transportation...82

FORMALDEHYDE
Gas at standard temperature...453
Off-spec product...126
Use as preservative...124, 130
Use as sanitizer...130

FOSSIL FUEL COMBUSTION WASTES
Beneficial uses...181
Bevill wastes...175, 252
Boiler chemical cleaning wastes...180
Commercial chemical products...178
Derived-from rule for Bevill wastes...266
Equipment cleaning wastes...178

Index 955

Exclusion...175
Four large-volume wastes...175
Importing...181
Other low-volume wastes...178
Regulatory determination...175, 177-178
Spent solvents...178
Uniquely associated wastes...178
Wastes from other sources...177
Wastes that are excluded...178
Wastes that are not excluded...178

FOUNDRIES
Cleaning foundry sands...406
Mixing iron fines with foundry sands...74

FREE PRODUCT...20, 511
From underground storage tanks...684

FREON
Ignitability...45
Use as ingredient...94
Use as solvent...71, 692
Used CFC refrigerants...70, 72

FUEL BLENDING...30
D001 wastes...652
In accumulation units...401
Is treatment...383, 401, 408, 464
LDR paperwork...652
Off-spec gasoline...510
Off-spec jet fuel...518, 522
Point of generation...664
Reclamation...513
With used oil...560, 570, 576

FUME INCINERATORS
To control air emissions...693

G

GASES/VAPORS...689
Commercial chemical products...453, 690
Condensed gases...690
Contained/containerized gases...49, 60, 690
Control devices do not require RCRA permits...694
Discarded materials...5
Emitted during treatment...695, 708
Flowing through pipes/ductwork are not solid wastes...690
From hazardous waste management activities...5, 694
From manufacturing operations...5, 30, 691-692
Ignitable compressed gases...49, 690
Landfill gas...695
P- and U-wastes...453, 690
Subpart AA standards...694
Subpart BB standards...694
Subpart CC standards...694
Synthesis gas...406, 535, 696
Treated by activated carbon...94, 124, 691, 695
Treated in boilers and process heaters...694
Treated in fume incinerators...693
True gases...690
Vented from compressed gas cylinders...453, 691

GASOLINE
Gasoline tank bottoms...221
Gasoline-contaminated soil...623
Off-spec gasoline burned for energy recovery...29, 510-511
Recovered from leaking tanks...685

GENERATOR ISSUES...283-378
90- or 180-day accumulation...300, 307
90- or 180-day storage...352
Cogenerators...305, 331, 341, 372, 428, 570, 576
Contractors...305, 331, 341, 428, 576
Episodic generators...285, 329, 377, 726
Four options to store...293
Generator status can change from month to month...284
Hazardous waste determinations...817
How often should I recharacterize?...820
Land disposal restrictions...588, 844
LDR paperwork...362, 640, 643
Manifesting...340, 348
Obtaining an EPA ID number...144, 234-235, 237, 317, 356, 370, 395, 429
P- and U-chemicals...134
Preparing for a RCRA inspection...769
Recordkeeping...360, 757
Satellite accumulation...293
Shipments to recyclers...403
Three classes of generators...284
Training...314, 322, 328, 752
Used oil generators...569-570
Waste analysis plans...364, 400, 654, 827
Waste counting...283
Who makes a hazardous waste determination?...819

GREASE
Might be managed as used oil...551

GROUND WATER
See CONTAMINATED GROUND WATER

H

HALOGEN-ACID FURNACES
Burning inherently waste-like materials...8, 531

HAZARD CODES
F001 wastes...88
F002 wastes...88
F003 wastes...89
F004 wastes...89
F005 wastes...89
For ICR-only listed wastes...249
P- and U-wastes...120

HAZARDOUS AND SOLID WASTE AMENDMENTS OF 1984 (HSWA)
Corrective action program...23, 685, 785, 893
Domestic sewage exclusion notification...144

©2015 McCoy and Associates, Inc. **McCoy's RCRA Unraveled**

Index

Hazardous waste used as fuel...696
Land disposal restrictions program...157, 587
Liquids-in-landfills ban...414
Revision of toxicity characteristic...64

HAZARDOUS WASTE DETERMINATION...817-886
Acute hazardous wastes...284
Adding characteristic codes to listed wastes...619
Ask the four questions...40, 817
Characteristic wastes...39-84, 819, 824, 836, 895, 924
Debris...836, 840
Declared hazardous wastes...40-41, 92, 831
Fluorescent light tubes...82, 833
For conditionally exempt small quantity generators...284, 316, 362, 818
Ground water...836
Hazardous waste categories...39
How often?...820
Listed wastes...85, 102, 674, 818, 834, 836, 896, 926
Mass balance approach...842, 862
Mixed wastes...842
Nonacute hazardous wastes...284
Notices of violation...754
Safety data sheets...832, 837-838
Soil...827, 836
Solid wastes...6, 40, 596, 818
Spill residues...679, 681
Unknown wastes...673
Using analysis...821, 841
Using knowledge...826, 830
Who makes the determination?...819

HAZARDOUS WASTE IDENTIFICATION RULE (HWIR)...813

HEADWORKS EXEMPTIONS
See DE MINIMIS WASTEWATER EXEMPTIONS

HEPTACHLOR
Impurity in pesticide...129

HERBICIDES...19, 135
Toxicity...63

HEXACHLOROBENZENE
Toxicity...64

HEXACHLOROETHANE
Incinerator trial burn...123, 264

HIGH-TEMPERATURE METALS RECOVERY (HTMR)...115, 273
Exception to derived-from rule for HTMR residues...116, 269, 273

HOUSEHOLD HAZARDOUS WASTES
Aerosol cans...61, 172
Burned in resource recovery facilities...174
Cathode ray tubes (CRTs)...78
Dry cleaning wastes...92, 171
Electronic wastes...77, 172
Exemption for...171, 265, 696, 818
Importing...174
Land disposal restrictions exemption...589
Lead-acid batteries...172, 236
Lead-based paint...69, 172, 332
Light bulbs...81, 172
Mixed with used oil...557
Mixture rule...173, 259
Natural gas regulators...172, 196
Treatment...173

HYDROCHLORIC ACID
Co-product...16
Produced in halogen-acid furnaces...8
Recycled...522
Reused without reclamation...522
Use constituting disposal...530

HYDROCHLOROFLUOROCARBONS (HCFCs)...71
Solvent use...88

HYDROFLUORIC ACID
Reused without reclamation...522

HYDROFLUOROCARBONS (HFCs)...71
Solvent use...88

HYDROGEN CYANIDE
Gas at standard temperature...453

HYDROGEN SULFIDE...193
Gas at standard temperature...453

I

ICR-ONLY LISTED WASTES
Are nonhazardous if not characteristic...87, 89, 899
Cleanout residues...220
Contained-in policy...278
Derived-from rule...262-263
Hazard codes...249
Interstate transportation...87
Land disposal restrictions...255, 274, 598, 618
Manifesting exemption...338
Mixed with used oil...555
Mixture rule...121, 249-250, 383
Remediation wastes...899

IGNITABILITY...42
Alcohol-content exclusion...46
Aqueous...44, 46
Determination using analysis...43, 674, 824
Determination using knowledge...45, 837
Flash point testing...43, 824, 837
Ignitable compressed gases...49, 690
Ignitable solids...45-46, 48, 107, 824, 837
Liquid...43-44
Liquids with low flash points...42, 107, 824
Overlap with D003 reactives...47
Oxidizers...49, 263, 674
Solvent-contaminated rags...49, 107
Unknown wastes...674

IMMEDIATE RESPONSES...417
90- or 180-day accumulation units...401

Drip pads...419
EPA ID number exemption...377, 417
For explosives...419
For reactive wastes...57, 419
For spills...417, 677
For unknown wastes...673
Laboratory wastes...419, 667
Manifesting exemption...338, 417
May trigger contingency plan...417
Permitting exemption...57, 417, 467, 469, 477, 479, 678
Subpart CC exemption...724
Sumps...418
Tanks...418, 469, 477, 678

IMMINENT AND SUBSTANTIAL ENDANGERMENT
Bevill-excluded wastes...201
Both RCRA and CERCLA provide response authority...907
Clandestine drug laboratories...373
Oil spills...682
Part of cradle-to-grave liability...782
Waste military munitions...701

IMMOBILIZATION TECHNOLOGIES...630-632

IMPORTING HAZARDOUS WASTE
Analytical samples...226
Assigning EPA ID number...372
Fossil fuel combustion wastes...181
Household hazardous wastes...174
Land disposal restrictions...598
Manifesting...356
Precious metals...235
Treatability study samples...231
Waste counting...289

IN SITU MINING
Exclusion...161

IN SITU TREATMENT
Land disposal restrictions...591-592

INCIDENTAL RESIDUES
Discarded materials...2, 5

INCINERATOR RESIDUES
Can be treated in a totally enclosed treatment facility?...412
Change-in-treatability-group principle...669
Derived-from rule...123, 263
From burning D018...671
From burning explosives...263
From burning ignitable wastes...262
From burning residues from empty containers...452
From disassembly...280
From trial burns...123, 264
Gaseous emissions...694
Mixture rule...263
New point of generation?...664
Refractory bricks...514
Sampling...869
Spent fluidized-bed media...281

INCINERATORS
As totally enclosed treatment facilities...413
Burning lab packs...460
Combustion (CMBST)...605, 609, 615
For Subpart CC compliance...733, 741
Impermissible dilution...640
Recycling exemption eligibility...498, 528
Start-up fuel...28
Trial burns...123, 264
Used oil combustion...578, 582
Waste analysis plans...829

INCOMPATIBLE WASTES
Container requirements...443, 460

INHERENTLY WASTE-LIKE MATERIALS
Are solid wastes...8, 531

INJECTION WELLS
Dilution prohibition...604, 642
Discarded wastewater...4
Land disposal restrictions...587, 589, 592, 604, 609
Used for geologic sequestration of carbon dioxide...246

INK
Mixed with spent solvent...251
Mixing tubs...90
Waste ink is not spent solvent...93
Waste ink recycling...102

INSECTICIDES
Toxicity...63

INSPECTIONS
Applicable records...773, 777
Confidential business information...777
Containers...443, 458, 757
Denying access to state/EPA inspectors...776
Enforcement actions resulting from...780
Hazardous waste management equipment/areas...771, 778
Inspection checklists...781
Post-inspection conference...779
Pre-inspection conference...776
Preparing for...769
Process areas...771, 778
Recordkeeping requirements...364
Sampling during inspections...779
Satellite accumulation containers...299, 459
Secondary containment...485, 489
State/EPA right to inspect...360, 366, 769
Subpart AA air emission controls...712
Subpart BB air emission controls...720
Subpart CC air emission controls...738, 744-746, 748
Surface impoundments...778
Tanks...475, 489, 738, 740
Types...770
Video surveillance...489

INTERIM STATUS
Corrective action...792

Index

INVESTIGATION-DERIVED WASTES
  Land disposal restrictions...591
  Left in areas of contamination (AOCs)...813, 902

ION-EXCHANGE RESINS
  Derived-from rule...264
  Electroplating operations...15, 111
  F006 wastes...15, 111
  Mixed wastes...154
  Photoprocessing operations...33
  Sludge...15, 33, 498

IRON AND STEEL INDUSTRY WASTES
  See Also STEEL PRODUCTION WASTES
  HTMR residues...273

IRON SPONGE
  Ignitability...48
  Uniquely associated waste...186

IRRIGATION RETURN FLOWS
  Exclusion...150

ISOBUTANOL
  De minimis wastewater exemption...209
  Isomer issues...120
  Solvent use...89

ISOPROPYL ALCOHOL
  Rags contaminated with...107
  Reused without reclamation...522
  Solvent use...218, 469

K

K016 WASTES
  Recycling...519

K019 WASTES
  Synthesis gas production...535

K020 WASTES
  Synthesis gas production...535

K021 WASTES
  Land disposal restrictions...595

K045 WASTES
  Derived-from rule...263

K047 WASTES
  Recycling...520

K048 WASTES
  Cement production...534
  Oil-bearing secondary materials exclusion...164

K049 WASTES
  Mixture rule...252
  Oil-bearing secondary materials exclusion...164

K050 WASTES...217
  De minimis wastewater exemption...211
  Oil-bearing secondary materials exclusion...164

K051 WASTES
  Mixture rule...252, 257
  Oil-bearing secondary materials exclusion...164

K052 WASTES
  Cement production...534
  Oil-bearing secondary materials exclusion...164

K061 WASTES
  Cement/aggregate production...25
  Exception to derived-from rule...269
  HTMR residues...116, 269, 273
  Recycling...407, 517, 525, 533
  Waste coding...620

K062 WASTES
  Confusion with F006...110
  Derived-from rule...111
  Exception to derived-from rule...269
  Fertilizer production...25
  HTMR residues...116, 269
  Recycling...493, 519, 521-522, 541

K069 WASTES
  Land disposal restrictions...619

K085 WASTES...16
  Recycling...519

K086 WASTES...90

K088 WASTES
  Vitrification...518

K116 WASTES...595

K156 WASTES...215

K157 WASTES...215

K169 WASTES
  De minimis wastewater exemption...211
  Oil-bearing secondary materials exclusion...164

K170 WASTES
  De minimis wastewater exemption...211
  Oil-bearing secondary materials exclusion...164

K171 WASTES
  De minimis wastewater exemption...211
  Exception to derived-from rule...270
  Not excluded when reclaimed...35

K172 WASTES
  De minimis wastewater exemption...211
  Exception to derived-from rule...270
  Not excluded when reclaimed...35

K174 WASTES
  Conditional listings...85

L

LAB PACKS
  Absorbent addition...416, 459
  Containers...459
  Incineration...460
  Land disposal restrictions...655
  Landfilling...442, 459

LABELING
  Containers...445-446, 753
  Tanks...753
  Used oil containers...570, 756
  Violations...753

LABORATORY WASTES
  Academic labs...377
  Alternative laboratory waste management program...371
  Analytical sample exemption...227

Clandestine drug laboratories...373, 667
Conditionally exempt small quantity generators...127
Contained-in policy...280, 440
Containers...440
Contractors...332
De minimis wastewater exemption...213
Disposed or recycled analytical supplies...227
Does repackaging require permitting?...416
Empty containers...440, 450
Immediate responses...419, 667
Lab packs...416, 442, 459-460, 655
LDR paperwork...652
P- and U-chemicals used to make lab standards...130, 227
Solvents used to make lab standards...130
Spent solvents...91, 227
Treatability study exemption...231
Vials and syringes...127, 451
Waste counting...291

LAMPS...80
Universal wastes...425, 430

LAND DISPOSAL PROGRAM FLEXIBILITY ACT OF 1996...604

LAND DISPOSAL RESTRICTIONS...587-660
Abandoned lead-lined pipe and shielded cable...73
Aggregation for centralized treatment...638
Alternative treatment methods...610, 635, 658
Analytical requirements...600, 603, 843
Applying characteristic codes to listed wastes...619
Area of contamination (AOC)...591-592, 902
As ARARs...912
Best demonstrated available technology (BDAT)...593
Bevill-exempt devices...181, 208, 649, 652
Capacity variances...593
Capping wastes in place...591

Change-in-treatability-group principle...597, 607, 669
Characteristic wastes...255, 274, 602, 624, 672
Combustion (CMBST)...605, 609, 615
Concentration-based treatment standards...593, 602, 605, 616, 846
Conditionally exempt small quantity generators...320, 587, 589
Constituents of concern...593, 600, 602, 618, 655, 845
Contained-in policy...278, 598
Containers versus debris...441, 627-628
Contaminated debris...270, 279, 441, 626, 930
Contaminated ground water...278
Contaminated soil...278, 592, 621, 655, 659, 685, 907
Corrective action management units...589, 591, 805
D001 wastes...157, 604-605, 615, 640, 652
D002 wastes...255, 411, 610
D003 wastes...604, 610, 640
D004–D011 wastes...611, 640
D005 wastes...594
D006 wastes...604, 612, 634
D007 wastes...615
D008 wastes...73, 604, 614-615, 625, 634, 640, 653
D009 wastes...604, 612, 634
D011 wastes...605, 634
D012–D017 wastes...605, 611
D018 wastes...613, 623
De minimis losses of listed wastes...216, 649, 652
Deactivation (DEACT)...599, 605, 608, 611
Delisted wastes...598
Demolition wastes...930
Derived-from rule...274, 595, 597
Dilution prohibition...74, 247, 254, 383, 597, 604, 608, 637-639
Discharges to POTWs...146, 591, 604, 607, 611, 649
Disposal facility requirements...588-589, 847
Dust suppressants...610

Exemptions...591
Exported wastes...654
F001 wastes...597, 617
F003 wastes...255, 598
F006 wastes...598, 659
F039 wastes...619, 639
Fertilizers must meet treatment standards...533, 610
For spills...685
Generator requirements...588, 844
Hazardous constituents...593, 600, 602, 618, 655, 845
How often should I recharacterize?...820
HSWA requirements...587
ICR-only listed wastes...255, 274, 598, 618
Imported wastes...598
In situ treatment...591-592
Injection wells...587, 589, 592, 604, 609
Intermediate-step treatment residues...671
Investigation-derived wastes...591
K021 wastes...595
K069 wastes...619
Lab packs...655
Land treatment units...589-590
Landfills...431, 589, 658
Listed wastes...256, 274, 615, 624
Mixed wastes...157, 160, 604, 634
Mixture rule...254, 618
Must determine if a waste is characteristic...40, 619
No-migration units...592, 657
Nonwastewater treatment standards...594, 603, 844
Notices of violation...754
NPDES discharges...591, 599, 604, 607
One-time notice of §268.7(a)(7)...147, 187, 215, 255, 396, 591, 607, 611, 643, 649
Open burning/open detonation...590
Paint...615-617

Paperwork requirements...161, 216, 238, 362, 429, 435, 587, 617, 640, 643, 649, 654, 754, 848
Permit as a shield...589
Placement...591, 894
Point of generation...254, 274, 594, 597, 669, 843
Polychlorinated biphenyls (PCBs)...625
Precious metals...589, 653
Program objectives...592, 843
Prohibited wastes...157, 602
Recycling...588-590, 617
Regulated parties...587
Remediation wastes...591, 621, 907
Residues from elementary neutralization units...411
Residues from wastewater treatment units...396
Restricted wastes...602
Sampling...844, 846
Satellite accumulation...299
Sludge...607, 610-611, 614-615
Specified-method treatment standards...593, 595, 597, 603, 605, 615, 626, 631, 846
Spent solvents...590, 617, 655
Spray irrigation...148, 591
Staging pile...589, 591, 810
State-regulated wastes...589
Still bottoms...590
Storage prohibition...299, 590, 597, 634, 636
Subcategories of wastes...597, 602, 612
Subpart CC compliance...733
Surface impoundments...146, 589
Tank-based wastewater treatment systems...607
Transporters...587
Treatability groups...594-595, 597
Treatability variances...592, 621, 624, 659
Treatment before discharge to an FOTW...146
Treatment facility requirements...846

Treatment in CWA/CWA-equivalent systems...591, 599, 604, 607, 611, 613, 615, 649
Treatment standards...254, 593, 602, 843
TSD facilities...588-589, 846-847
U096 wastes...255, 618
Underground injection...587, 589, 592, 604, 609
Underground mines...589
Underlying hazardous constituents (UHCs)...147, 598, 605, 611, 613-614, 618, 630, 636, 652, 685, 734, 845
Universal treatment standards...600, 618
Universal wastes...273, 431, 435, 627, 657-658
Used oil...554, 656
Vaults...589
Waste analysis plans...161, 364, 400, 654, 827-828, 846-847
Waste characterization...843
Waste coding...619
Waste piles...589
Wastewater treatment standards...594, 603, 844

LAND FARM
Land disposal restrictions...589-590

LAND TREATMENT UNITS
Land disposal restrictions...589-590
Spray irrigation...592

LANDFILL GAS CONDENSATE
Derived-from rule...265, 696
From landfill gas...695

LANDFILL LEACHATE
Contaminating ground water...275, 619
Created from spills...680
Derived from pre-RCRA wastes...892
Derived-from rule...274, 892

LANDFILLS
Active management...890, 892
Batteries...431, 442, 658
CERCLA issues...75, 893, 911
Corrective action...793, 893

Lab packs...442, 459
Land disposal restrictions...431, 589, 658
Landfill gas...695
Leachate collection sumps...478-479
Liquids-in-landfills ban...414, 442, 459
Offsite disposal of CAMU-eligible wastes...808, 906
Spray irrigation of cap...148
Vaults are regulated as landfills...473

LARGE QUANTITY GENERATORS
90-day accumulation...300, 397, 465
Accumulation time extensions...307
Air emission standards...305, 314, 398, 708, 717, 724
Allowable units for 90-day accumulation...300, 398
Biennial report...365
Closure standards...314, 475
Container standards...303, 398, 439, 445
Definition...284
Exception reports...353, 362
Preparedness and prevention/contingency plans...301, 308, 310, 338, 364-365, 677, 752, 756
Quiz...317
Recordkeeping requirements...360, 757
Required provisions...300
Spill reporting...686
Spill response...677
Spill response training...689
Summary of requirements...300
Tank standards...302, 398, 463, 465, 469, 478, 488
Training...314, 322, 752
Treatment in 90-day units...303, 397, 402, 616, 709
Waste minimization program...348

LARGE QUANTITY HANDLERS
Universal wastes...427

LAUNDRIES...143

LEAD
- Agents that mask the TCLP...74
- Baghouse dust...407
- Batteries...32, 236, 401, 424, 430, 438, 441, 465, 635, 637
- Building/structure assessment...917
- Containers...28, 73, 449
- Contaminated soil...625
- In laboratory wastes...228
- Lead shot from a shooting range...242-243, 245
- Lead-based paint...69, 74, 332, 381, 384, 401, 406, 515, 840, 862, 864, 917
- Lead-lined pipe and shielded cable...73
- Recovered from a sludge...510
- Recovered from wastes...515
- Shielding...156, 635
- Toxicity...67, 519, 839-840, 864

LEAK DETECTION SYSTEMS...475, 484

LIGHT BULBS...80
- See Also LAMPS and FLUORESCENT LIGHT TUBES

LIQUID...43-44, 46, 51, 53

LIQUID-LIQUID EXTRACTION
- Using listed solvents...95

LISTED WASTES...85
- Adding characteristic codes to listed wastes...619
- Conditional listings...85
- Contained-in policy...275
- Contaminated soil...624
- Debris...631, 926
- Delisting petitions...75, 85, 265
- Demolition wastes...926
- Derived-from rule...262-263, 663
- Determination using analysis...836
- Determination using knowledge...85, 102, 674, 834, 896
- Discarded chemicals...87, 119, 835
- F-wastes...86, 680, 834
- From manufacturing...131
- From nonspecific sources...86
- From product handling...131, 280, 693
- From specific sources...86
- K-wastes...86, 680, 835
- Land disposal restrictions...256, 274, 615, 624
- Listing determinations for unknown wastes...102, 675, 896
- Mixed with used oil...554
- Mixture rule...249, 251, 383
- P- and U-wastes...87, 118, 835
- Point of generation...662, 673
- Remediation wastes...896
- Spilled...680

LISTING BACKGROUND DOCUMENTS...86, 835

LITHIUM-CONTAINING MATERIALS
- Batteries...55, 431, 658
- Reactivity...54

M

MAGNESIUM...47

MAINTENANCE WASTES
- Cogenerators...333
- Contained-in policy...280
- Corrective action...787
- From pipelines...372
- Shipboard wastes...375

MANGANESE
- Batteries...438

MANIFESTING...334
- Background...334
- Combining wastes...343, 382
- Conditionally exempt small quantity generators...320, 335, 362
- Container shipments...460
- Container types...346
- Continuation sheets...353
- Designated facilities...307, 344, 352
- Discrepancies...350-351
- DOT shipping descriptions...334, 344
- Electronic manifest distribution...359
- Electronic manifest systems...355, 359, 362
- Emergency response information...339
- Exception reports...353, 362
- Exemptions from...226, 229, 242, 335, 356, 417, 429
- Explosives...338
- Exports...356
- Filling out the manifest...339, 351
- Generator certification...348
- Generators receiving rejected wastes...352
- Generator's name and EPA ID number...332, 340, 355-356, 377, 417, 677
- Hazardous wastewater/sludge shipments to offsite wastewater treatment units...395
- ICR-only listed waste exemption...338
- Immediate response exemption...338, 417
- Imports...356
- Large quantity generators...354
- Making changes...353
- Management method codes...353
- Manifest retention...354, 360
- Mixed wastes...156, 160
- Nonempty shipments back to a generator...351, 456
- Offeror certification...349
- Onsite shipment exemption...337
- PCB information...348
- Rail shipments...355
- Rejected loads...351
- Shipments to recyclers...403
- Small quantity generators...335, 354
- Special handling instructions and additional information...348, 436
- Spill response...677
- State-listed wastes...347
- Training required to sign...348
- Transfer facilities...341
- Transfers from rail cars to tank trucks...343
- Transporters...341

Index

TSD facilities...345, 353
Uniform hazardous waste manifest...335-336
Units of measurement...347
Universal wastes...355, 429, 433, 436
Waste codes on...347, 460, 668
Wastes generated by a contractor...332, 341
Weight designation...287, 346, 441

MANUFACTURED ARTICLES
Are not P- or U-wastes...18, 21, 121, 125
Contained-in policy...280
Releases from...681

MANUFACTURED GAS PLANT WASTES
Exemption from toxicity...67
Treated in 90-day accumulation units...401

MANUFACTURING CHEMICAL INTERMEDIATES...125
De minimis wastewater exemption...212

MANUFACTURING PROCESS UNITS
Cleaning or decommissioning...217, 920
Counting cleanout residues...288
Examples of...217, 277
May become 90-day accumulation units...221
Pipelines...217, 718
Point of waste generation...218, 381, 469, 471, 663, 857
Recycling exemption...471
Residues subject to regulation if remain in unit more than 90 days...221
Residues subject to regulation when removed from the unit...220
Ships...220
Subject to corrective action?...223, 789
Surface impoundments don't qualify...217
Temporary removal from service...222
Waste counting...289

Wastes are not P- or U-wastes...131
Wastes in units are not subject to regulation...216, 289, 468

MARINE PROTECTION, RESEARCH, AND SANCTUARIES ACT
Interface with RCRA...150

MELPHALAN
Container residues...127

MERCURY
Are mercury-containing items containers?...440
Batteries...438
Mercury Export Ban Act...637
Mercury-Containing and Rechargeable Battery Management Act...438
Mercury-containing equipment...21, 23, 73, 424-425, 430, 441, 627, 630, 633
Not oil and natural gas E&P wastes...193
Not scrap metal...22-23
Recycling...273, 407, 514, 637
Spilled...681
Thermometers...21, 121, 125, 135
Universal wastes...21, 23, 73, 424-425, 430, 441

MERCURY-CONTAINING EQUIPMENT
Building/structure assessment...919
Universal wastes...21, 23, 73, 424-425, 430, 441

METAL SMELTERS
Closed-loop recycling with no reclamation...526
F006 wastes...33, 115, 273, 513
Lead-acid batteries...239
Lead/copper recovery...515
Reclamation units...31, 407

METHANOL
Burning for energy recovery...20, 136
LDR paperwork...655
Oil and natural gas E&P wastes...185
Solvent use...89, 91, 101, 169, 250-251, 272, 617

Spilled...251
Use as reactant...93
Used to make lab standards...91, 130

METHYL BROMIDE
Use as fumigant...124

METHYL ETHYL KETONE
De minimis wastewater exemption...209
Solvent use...89, 617

METHYL ISOBUTYL KETONE
Solvent use...89

METHYLENE CHLORIDE
Blowing agents...96
De minimis wastewater exemption...209-210
Product packaging...133
Product to be disposed...222
Reaction and synthesis media...90
Solvent use...88, 692
Use as ingredient...94, 130
Use in liquid-liquid extraction...95

MILITARY MUNITIONS
Are not solid wastes when reclaimed...699
Declared to be hazardous...67
Definition...697
Disassembled or recycled...700
Lead shot from a shooting range...245
May be solid wastes...697
Point of generation...696
Products not wastes...699
Range clearance operations...701
Rule required by Federal Facilities Compliance Act...696
Used for their intended purpose...700, 788

MINERAL PROCESSING WASTES
Bevill wastes...196, 204, 252
Closed-loop recycling with no reclamation...524
Exclusion...196, 204
Reclamation...36, 207
Regulatory determination...204
Wastes that are excluded...204

Wastes that are not excluded...204

MINING WASTES
Bevill wastes...196-197, 252, 266
Commercial chemical products...201
Exclusion...196-197
Exclusion for mining overburden returned to the mine site...175
In situ mining exclusion...161
Regulatory determination...197
Residues from treating excluded wastes...204
Spent solvents...201
Uniquely associated wastes...201
Wastes that are excluded...197

MISCELLANEOUS UNITS
Are they tanks?...471
Subpart CC air emission controls...724, 742

MITOMYCIN C
Container residues...127

MIXED WASTES...151
At commercial facilities...152, 159, 634, 636
At DOE facilities...152, 160, 634
Batteries...635
Closure of units...160
Conditionally exempt small quantity generators...155, 321
Containers...159, 161
Correction action...158
Decay-in-storage...156, 158
Discharged to a POTW...146
Dual regulatory scheme...158
Exclusion...151, 467
Hazardous waste determination using knowledge...842
Ion-exchange resins...154
Land disposal restrictions...157, 160, 604, 634
Lead...157
Low-level mixed wastes...153, 158
Manifesting...156, 160

Mercury...157, 635
Naturally occurring and/or accelerator-produced radioactive material (NARM)...151, 153
Operator training...159
Scintillation fluids...152-154, 157
Source, special nuclear, and by-product materials...152
Subpart CC exemption...725
Tanks...159
Treated in 90-day accumulation units...401
Treatment...159
Waste accepted at TSD facilities...830

MIXTURE RULE...247
Absorbents...250
Background...248
Bevill wastes...252
Characteristic wastes...249-250, 383
Cleaning out tanks and containers...288
Conditionally exempt small quantity generator wastes...173, 259, 288, 318
Container residues...258
D001 wastes...251, 383
Dilution prohibition...254
Does not apply to certain mixtures...250
Domestic sewage...146
Electroplating wastes...252
Exemption for de minimis losses of listed wastes...208
F001 wastes...251, 692
F003 wastes...100, 250-251
F006 sludge mixed with other sludge...113, 258
F006 wastes...257-258
F019 wastes...257
F037 wastes...257
F038 wastes...257
Has mixing occurred?...256
Household hazardous wastes...173, 259
ICR-only listed wastes...121, 249-250, 383
Incinerator residues...263
Is mixing permissible under the LDR program?...248

Is mixing treatment?...248, 256, 383
K049 wastes...252
K051 wastes...252, 257
LDR ramifications...254, 618
Listed wastes...249, 251, 383
Mixtures of excluded hazardous secondary materials and hazardous waste...260
Mixtures of product with wastewaters...259
Oil and natural gas E&P wastes...253
Point of generation of waste mixture...662
Precipitation run-off...257
Refinery wastes...252, 257
Sewer line leaks...145
Spent solvents...94
Spiking feed during incinerator trial burn...123
Still bottoms...263
Tank bottoms...488
Used oil...259, 553
Waste coding for resulting mixture...250

MOBILE EQUIPMENT
As recycling units...305, 333, 405
Totally enclosed treatment facility exemption...413

N

NATURAL GAS PIPELINE CONDENSATE...195
Burning for energy recovery...29, 196

NATURAL GAS REGULATORS
Not oil and natural gas E&P wastes...196
Scrap metal?...23, 73, 243

NATURALLY OCCURRING AND/OR ACCELERATOR-PRODUCED RADIOACTIVE MATERIAL (NARM)...151, 153

NATURALLY OCCURRING RADIOACTIVE MATERIAL (NORM)...151

NICKEL
Batteries...426, 438
Electroplating operations...33

Index

Recovered from F006 wastes...513

NICOTINE
Isomer issues...120
Unused patches...125, 129

NITRATES
Ignitability...49

NITRIC ACID
Is it ignitable?...50

NITROBENZENE
De minimis wastewater exemption...209
Solvent use...89

NITROGLYCERINE
Hazard code of reactivity...120

2-NITROPROPANE
Solvent use...89

NO SMOKING SIGNS
Signage...321

NOTICES OF VIOLATION (NOVs)...751-784
Common RCRA violations...751
Container/tank labeling...753
Contingency plan...752
Inadequate emergency response provisions...756
LDR paperwork...754
Making hazardous waste determinations...754
Open containers...754
Preparedness and prevention...756
Recordkeeping...757
Satellite accumulation...755
Tank management...755
Training...752
Used oil management...756
Weekly container inspections...757

NOTIFICATIONS
Episodic generators...377
For discharging hazardous waste to a POTW...144
For obtaining an EPA ID number...144, 234-235, 237, 317, 356, 370, 395, 405, 429, 792
Renotification for updating an EPA ID number...376

Small quantity burner exemption...421

NPDES DISCHARGE EXCLUSION...4, 147
Applicability to surface impoundments...149
Applies to point-source discharges only...147
Land disposal exemption...591, 604
Status of dredged sediments...150
Storage/treatment in tanks...467

O

OCCUPATIONAL SAFETY AND HEALTH ADMINISTRATION
Basis for safety data sheets...67, 832
Emergency action plans...308, 310
Interface with RCRA...327, 330, 430, 432
Mercury emissions from crushing bulbs...82

OFF-SHORE PLATFORM WASTES
Point of generation...665

OFF-SPECIFICATION PRODUCTS...94, 125, 259, 518, 522
Fuels...29

OIL AND NATURAL GAS E&P WASTES...182
Commercial chemical products...186
Compressor-station wastes...193
Crude oil processes...184, 191, 194
Derived-from rule...266
Development...183
Equipment cleaning wastes...186
Exclusion...48, 182, 256, 266, 468, 665, 682
Exploration...183
Gas plants...186
Gas-plant cooling-tower wastes...196
Geothermal energy processes...185

Geothermal energy scrubber wastes...194
Iron sponge...48, 186
Manufacturing wastes...194
Methanol...185
Mixture rule...253, 256
Natural gas pipeline condensate...195
Natural gas processes...184, 191
Natural gas regulators...196
Primary field operations...185
Production...183
Regulatory determination...182
Residues from treating exempt wastes...266
Service company wastes...195
Spent solvents...186
Transportation wastes...191
Underground natural gas storage fields...196
Uniquely associated wastes...185
Use constituting disposal...530
Used oil recycling...564
Wastes that are excluded...187
Wastes that are not excluded...187
Workover wastes...184

OIL-BEARING SECONDARY MATERIALS
Exclusion...164

ONSITE
Manifesting exemption...337
Offsite facilities need separate EPA ID numbers...167, 374, 653, 791

OPEN BURNING/OPEN DETONATION
Land disposal restrictions...590
Reactive wastes...590
Treatment in 90-day units...304

ORGANIZATION FOR ECONOMIC COOPERATION AND DEVELOPMENT (OECD)
Exporting spent lead-acid batteries...240

OSMIUM TETROXIDE
Sole active ingredient...129

Index

OXIDIZERS...49, 263
  Organic peroxides...50
  Screening using potassium iodide paper...674

OXYGEN BREATHING APPARATUS CANISTERS...457

P

P- AND U-WASTES...118
  CAS number issues...119
  Cleanup residues...127, 277, 680
  Contained-in policy...134, 277, 280
  Container residues...126, 451, 454
  Contaminated debris...128, 134, 633, 693
  Contaminated ground water...127
  Contaminated soil...127
  De minimis wastewater exemption...133, 211
  Gases/vapors...453, 690
  Generator issues...134
  Hazard codes...120
  Inactive ingredients...128
  Isomer issues...120
  Manufactured articles...18, 21, 121, 125
  Manufacturing chemical intermediates...125
  Manufacturing process wastes...131, 693
  Mixed wastes...154
  Mixed with used oil...19, 135
  Must be unused...123
  Off-specification products...29, 94, 125, 259, 518, 522
  Point of generation...666, 857
  Product handling wastes...131, 280, 693
  Pure, unused chemicals...122
  Recycling...20, 126, 134-135, 684
  Regulatory thresholds...119, 121
  Returned to the manufacturer...20, 126, 134
  Sole active ingredient...128, 686
  Spilled...127, 417, 676, 679-680
  Technical grades...128
  Tool for identifying...121
  Unreacted reagents...13, 133
  Use the List of Lists...119
  Used for their intended purpose...123
  Used to make lab standards...130
  Waste counting...287

P001 WASTES
  Container residues...452
  Product handling wastes...132

P004 WASTES
  Sole active ingredient...130

P015 WASTES
  Beryllium wastes in glove boxes...123
  Wastes from machining beryllium metal...123

P022 WASTES
  Contained-in policy...134

P031 WASTES
  Empty containers...453

P033 WASTES
  Empty containers...453

P037 WASTES
  Sole active ingredient...130

P042 WASTES...451
  Container residues...127
  Isomer issues...120

P059 WASTES
  Sole active ingredient...129

P063 WASTES
  Empty containers...453

P075 WASTES
  Isomer issues...120
  Unused patches...125, 129

P087 WASTES
  Sole active ingredient...129

P094 WASTES...693

P096 WASTES
  Empty containers...453

P120 WASTES
  Sole active ingredient...130

PAINT
  As state universal wastes...426
  Blending into fuel...640
  Empty cans...457
  Hazardous waste determination...91, 93, 616, 837
  Land disposal restrictions...615-617
  Lead-based paint...69, 74, 332, 381, 384, 401, 406, 515, 840, 862, 864, 917
  Mixed with spent solvent...251
  Off-spec paint reclaimed...22
  Painting contractors...332
  Point of generation for paint removal...75, 381, 667, 857
  Sandblast residue...14, 25, 519, 534, 853, 891
  Stripped/cleaned using solvents...91, 110, 130, 617, 664
  Thinned with solvent...94
  Treatment to render nonhazardous...402, 616
  Waste paint is not spent solvent...93-94, 402, 616
  Waste paint is not U-listed...121, 402, 616

PAINT FILTER TEST...43, 51, 442, 448, 824, 837
  Absorbent addition...415

PARKING LOT TEST...388, 472, 477-478

PARTS WASHERS...252, 662, 857
  Point of waste generation...14, 90
  Waste counting...289

PENTACHLOROPHENOL
  Dioxin wastes...16
  Sole active ingredient...131

PERCHLOROETHYLENE
  See TETRACHLOROETHYLENE

PERMANGANATES
  Ignitability...49-50

PERMIT AS A SHIELD
  Corrective action requirements...796
  Land disposal restrictions requirements...589

Index

P

PERMITTING
  Absorbent addition exemption...414, 460, 467
  Closure...921
  Containment buildings...632
  Corrective action...379, 791, 908
  Emergency permits...417
  Exemption for 90- or 180-day accumulation units...300, 397
  Exemption for burning small quantities of hazardous waste in onsite units...419
  Exemption for elementary neutralization units...408, 442, 478-479
  Exemption for immediate responses...417
  Exemption for recycling...401, 515
  Exemption for totally enclosed treatment facilities...411
  Exemption for wastewater treatment units...384, 442, 464, 473, 478-479, 668
  Exemptions...227, 229, 379, 466
  For treating contaminated debris...631
  For treating contaminated ground water...419
  Permit modifications for corrective action...798, 804, 806, 809-810
  Remedial action plans (RAPs)...792, 806, 809-810, 813, 904

PEROXIDES...263
  Ignitability...49-50

PERSONAL PROTECTIVE EQUIPMENT...227
  Contained-in policy...280, 633
  Contaminated with P- and U-chemicals...132

PESTICIDES
  Burning for energy recovery...20, 510
  Contaminated soil...27, 124, 681, 897
  Contamination resulting from use...2, 27, 123, 897
  In aerosol cans...59
  Irrigation return flows exclusion...150
  Land disposal restrictions...610
  Mixed with used oil...556
  Rinsed out of containers...450, 452
  Spilled...128, 681
  Universal wastes...425, 430
  Use constituting disposal...24, 534

PETROLEUM PRODUCTS
  Captured in absorbent pads...29, 683
  Captured in vapor-recovery units...30, 692
  Exemption for USTs holding petroleum...27, 70, 72, 628, 684
  Mixtures of product and wastewater...259
  Off-specification variants used as fuel...20, 29, 510-511, 513
  Use constituting disposal...27, 546
  Used oil recycled into petroleum industry...564

pH DETERMINATIONS...52, 825, 838

PHARMACEUTICAL WASTES
  Discarded chemicals...129
  Methanol...272
  P001 listing...452
  P042 listing...120
  QC residues...129
  Returned to the manufacturer...20, 135
  U248 listing...452
  Used vials and syringes...127, 451-452
  Waste counting...287, 442

PHASE SEPARATION
  Pretreatment...609, 667

PHORATE...693

PHOSPHINE
  Gas at standard temperature...453

PHOSPHORIC ACID
  Reused without reclamation...522

PHOTOPROCESSING OPERATIONS
  D011 wastes...33, 407
  Ion-exchange resins...33
  Recovery of silver...31, 33, 233, 273, 407, 666
  Spent fixer solution...14, 33, 407, 666
  Wastewater treatment...14

PICKLE LIQUOR
  See SPENT PICKLE LIQUOR

PICRIC ACID
  Immediate response...419
  Reactivity...57

PILOT PLANTS...214

PIPES, SEWERS, AND OTHER CONVEYANCES
  Ancillary equipment...389, 409, 465, 469, 473, 479, 483, 487
  As elementary neutralization units...409
  As manufacturing process units...217
  As totally enclosed treatment facilities...413-414
  As wastewater treatment units...387, 389
  Associated with closed-loop recycling...528
  Cleaned out before demolition...921
  Controlling fugitive emissions from...714
  Pipes are designed to convey, not treat...383
  Sewer line leaks...787
  Subject to Subpart BB standards...217

PLACEMENT
  Point of waste generation...894

POINT OF GENERATION...661
  Antifreeze...667
  Baghouse dust...667, 857
  Boiler chemical cleaning wastes...180, 666
  Change-in-treatability-group principle...669
  Characteristic wastes...662, 672

Commercial chemical products...666, 857
Contaminated ground water...857, 890
Contaminated soil...857, 890, 894
D001 wastes...664, 667, 671
D002 wastes...380, 666, 670-671, 857
D008 wastes...664, 671
D011 wastes...666
De minimis amounts of listed wastes...215
Demolition wastes...923
F006 wastes...670
For 90-day accumulation units...668
For areas of contamination (AOC)...811, 891, 894, 902
For filter cake...857
For land disposal purposes...254, 274, 594, 597, 669, 843
For manufacturing process units...218, 380, 469, 471, 663, 857
For recycling operations...406, 516, 531, 666
For tank systems...473, 662
For waste coding...663
For waste counting...668
For wastes removed from storage...664, 669
For wastewater treatment units...668
Incinerator residues...664
Intermediate-step treatment residues...668, 671
Listed wastes...662, 673
May extend beyond initial point of generation...662
Mixtures...662
Off-shore oil platform wastes...665
P- and U-wastes...666, 857
Paint wastes...75, 381, 667, 857
Parts washers...14, 662
Remediation wastes...890
Sampling...853
Scrap metal...23, 242
Shipboard wastes...220, 333, 372, 375
Sludge in a surface impoundment...666
Solids removed from ignitable wastes...667
Spills...672
Still bottoms...304, 664, 667, 857
Treatment residues have a new point of generation...406, 663
TSD facilities are generators of treatment residues...664
U154 wastes...671
Waste military munitions...696
Wastewater...668
When is a hazardous waste generated?...380, 662
Where is a hazardous waste generated?...380, 662
Where is a solid waste generated?...663, 857

POLYCHLORINATED BIPHENYLS (PCBs)
Building/structure assessment...919
Contaminated soil...625
Exemption for PCB dielectric fluids...72
In unknown wastes...674
In used oil...557, 562, 580-581, 585
Land disposal restrictions...625
May be D021 wastes...72
PCB information on manifests...348
PCB-containing light ballasts...83
Rinsing a PCB container...452
Solvents used to remove PCBs from transformers...92
Testing for PCBs...563

POST-CLOSURE PERMITS...791
No training required...330

POTASSIUM...47

POTASSIUM HYDROXIDE
Corrosivity...54
Used potassium hydroxide used as ingredient in fertilizer...13

POTW
See PUBLICLY OWNED TREATMENT WORKS

PRECIOUS METALS
Derived-from rule...233
Electroplating wastes...109-110, 232
Exemption...33, 232
Exemption records...367
Exporting...235
Importing...235
In printed circuit boards...232, 244
Land disposal restrictions...589, 653
Recycling...33, 232-233, 273, 407, 542
Sham recycling...235, 542

PRECIPITATION RUN-OFF...257, 275, 278, 447

PREPAREDNESS AND PREVENTION
During immediate responses...417
Large quantity generators...301, 308, 338, 364, 756
Minimum requirements...308, 756
Notices of violation...756
OSHA emergency action plans...308
Recordkeeping requirements...364
Small quantity generators...301, 308, 338, 364-365, 756

PRINTED CIRCUIT BOARDS
Exemption for shredded boards...80, 244
F006 wastes...114
Off-spec boards reclaimed...22, 80, 244
Precious metals...232, 244
Production by-products...518
Scrap metal...72, 80, 242-243

PRIVATELY OWNED TREATMENT WORKS...142

PROCESSED CRT GLASS
Cullet...79
Used as a fluxing agent...79

PRODUCT HANDLING...131, 280
Air emissions...693

Index

PRODUCTS
See COMMERCIAL CHEMICAL PRODUCTS

PROPANE...61
Ignitable compressed gas...49

PUBLICLY OWNED TREATMENT WORKS (POTWs)...142
Discharges to sewer line...4
Domestic sewage exclusion...4, 112, 466
Land disposal exemption...146, 591, 604, 607, 611, 649
LDR paperwork requirements...649
Pretreatment program...385
Sludge...145

PULP AND PAPER WASTES
Pulping liquors exclusion...162

PULPING LIQUORS
Exclusion...162

PUMPS
Are containers?...440, 442
Controlling fugitive emissions from...715

PYRIDINE
De minimis wastewater exemption...209
Sole active ingredient...129
Solvent use...89
Toxicity...64

Q

QUANTITATION LIMIT...64

R

RADIOACTIVE LEAD SOLIDS
D008 wastes...604, 634-635, 842
Determination using knowledge...842
Gloves...635
LDR treatment subcategory...634
Not debris...626
Not exempt in CWA systems...604
Radioactive lead-acid batteries...635

Submarine reactor compartments...635
Tanks and containers...635

RADIOACTIVE MIXED WASTES
See MIXED WASTES

RADIOACTIVE WASTES
Batteries...604, 634-635
Decay-in-storage...158
Disposed in lead containers...28, 73
High-level wastes...152, 604, 634
Land disposal restrictions...604, 634
Low-level wastes...153
Mercury...635
Naturally occurring and/or accelerator-produced radioactive material (NARM)...151, 153
Transuranic wastes...152
Tritium exit signs...921

RAGS
Contaminated with F003 spent solvents...107
Contaminated with F005 spent solvents...264
Contaminated with isopropyl alcohol...107
Disposed...104
Ignitable...49, 107
Laundered and reused...104
Managed under a tolling agreement...104
Solvent contaminated...49, 103-104
Used oil contaminated...107, 560

RAILROAD TIES
Creosote-treated railroad ties...121, 252

RAINWATER...257, 275
Contained-in policy...278
Percolating through contaminated soil...680

REACTIVITY...54
Aerosol cans...58, 381, 451
Aluminum chaff...55
Blasting caps...55
Cyanide and sulfide wastes...54, 56, 116, 674, 825, 838

Determination using analysis...56, 674, 825
Determination using knowledge...55-56, 838
Explosives...57, 419
Explosives-contaminated soil...57, 839, 895
F007 wastes...116
F008 wastes...116
Immediate response exemption...57
Lithium/sulfur dioxide batteries...55
No test methods...55
Overlap with D001 ignitable solids...47
Small arms ammunition...57, 839
Unknown wastes...674
Waste in lagoons...58

RECLAIMED SOLVENTS...101
Point of generation...664, 857
Reused as solvent...101, 272, 292, 305, 333, 384, 407, 465, 513, 515, 523, 525, 529
Used as fuel...101, 272, 292

RECLAMATION...31, 512
See Also RECYCLING
Batteries...31, 236
By-products...32, 509, 516
Cleaning foundry sands...406
Commercial chemical products...20, 126, 134-135, 684, 699
Definition...10, 498, 512
Derived-from rule...263, 406
Dewatering...513
F006 wastes...115, 264, 513
Fuel blending...513
Incidental processing...31, 512
Metal smelters...31, 33, 115, 273, 407, 513, 515, 526
Military munitions...699
Mineral processing wastes...36
Silver wastes...31, 33, 273, 407
Sludge...32, 510, 516, 518
Solid wastes...10
Spent catalysts...31
Spent materials...32, 426

RECORDKEEPING REQUIREMENTS...360

Biennial report...160, 365-366
Conditionally exempt small quantity generators...321, 362, 757
Contingency plan...365
Documenting an exemption...542, 544
Emergency responder agreements...364
Episodic generators...285
Exception reports...362
For proving a recycling exemption...35, 542, 544
Generators...360, 757
Hazardous waste characterizations...362
Hazardous waste exports...366, 433
Inspection of applicable records...773, 777
Inspection records...364
Knowledge-based HW determinations...363
LDR paperwork...161, 238, 362, 429, 435, 587, 617, 640, 643, 649, 654, 754, 848
Manifests...354, 360
Minimum requirements...360, 757
Miscellaneous records...366
Notices of violation...757
Small quantity burner exemption...421
Subpart AA air emission controls...712
Subpart BB air emission controls...721
Subpart CC air emission controls...738, 740, 744-746, 748
Test results and waste analyses...363
Training records...328, 364, 752
Treatability study exemption...232
Universal waste shipments...429
Used oil burners...584
Used oil generators...576
Used oil marketers...585
Used oil processors...578
Used oil transporters...576
Waste analysis plans...364, 827
Wastes generated by a contractor...332

RECOVERED OIL
Exclusion...166, 168

RECYCLERS
EPA ID number...405, 653

RECYCLING...493-546
Activities subject to RCRA control...9, 11, 493, 497
Aerosol cans...59, 243, 245, 649
Batteries...236, 438
Burning for energy recovery voids exemption...528, 534
Categorizing...10, 498
Closed-loop recycling with no reclamation...162, 524
Closed-loop recycling with reclamation...90, 162, 405, 471, 526
Commercial chemical products...20, 32, 126, 134-135, 684
Contaminated soil...27, 407, 546
D001 wastes...708
Definition...493
Derived-from rule...103, 264, 271, 406, 529
Discarded materials...5
Documentation requirements...35, 542, 544
Economics...540, 544
Excluded materials...33, 103, 264, 271, 529
F002 wastes...94
F003 wastes...515, 664
F006 wastes...115, 264, 407, 513
Fossil fuel combustion wastes...181
Four use/reuse recycling exemptions...511
In a continuous industrial process is exempt from RCRA...32
Incidental processing...31, 512
Is viewed prospectively...23, 32, 242, 509
K061 wastes...407, 517, 525, 533
K062 wastes...493, 519, 521-522, 541

Land disposal restrictions...588-590, 617
LDR paperwork requirements...653
May occur at two locations...273, 406, 516
Military munitions...700
Partial recycling by generator...406, 515
Permitting exemption...401, 467
Point of waste generation...406, 516, 531, 666
Precious metals...33, 232-233, 273, 407, 542
Qualifying criteria for use/reuse exemptions...529
Reclamation under the control of the generator...34, 260, 371, 499
Scrap metal...245, 927, 930
Secondary materials...263, 471, 493-546
Sham recycling...30, 235, 545
Solid wastes...9
Spent solvents...101, 272, 292, 305, 333, 384, 407, 465, 513, 515, 525, 529, 664, 708, 857
Stabilized waste...515
Storage prior to recycling...403, 471
Substitution for a commercial product...521
Sulfuric acid...520, 526
Transfer-based reclamation...34, 260, 371, 504
Transfer-based reclamation in a foreign country...35, 508
Units are permit exempt...402, 515, 589
Units exempt from Subpart AA standards...708
Units exempt from Subpart BB standards...717
Units exempt from Subpart CC standards...724
Units subject to air emission standards...405, 515, 708, 717
Units subject to corrective action?...787
Use constituting disposal voids exemption...528, 531

Use or reuse as ingredient or feedstock...512
Used oil...552
Waste counting...291
Wastewater...272, 515, 529

REFINERY HAZARDOUS WASTE FUELS
Exception to derived-from rule...269

REFINERY WASTES
De minimis wastewater exemption...210
Derived-from rule...266
Mixture rule...252, 257
Oil-bearing secondary materials exclusion...164
Oil-bearing wastewater...259, 562
Recovered oil exclusion...166
Skinner memo...267
Spent catalyst...35
Spent caustic solutions exclusion...169

REFRACTORY BRICK...633

RELEASE TO THE ENVIRONMENT...486, 683, 688

RELEASES
See Also SPILLS
Universal wastes...429, 433
Used oil...572, 584, 683

REMEDIAL ACTION PLANS (RAPs)
Do not trigger corrective action...792, 904
Used to designate a CAMU...806
Used to designate a remediation waste management site...815, 904
Used to designate a staging pile...810
Used to designate a temporary unit...809

REMEDIATION ACTIVITIES...887
90- or 180-day accumulation units...401, 903
Active management...891
Area of contamination policy...813, 899
Contaminated soil...277, 592, 638, 811, 891
Corrective action...785-816, 900-901
Corrective action management units...904
Good-faith effort to determine contaminant source...896
Investigation-derived wastes...591, 813
Land disposal restrictions...591, 621, 907
Offsite management...906
Onsite management...899
Point of waste generation...890
RCRA-permitted units...905-906
Remedial action plans (RAPs)...792, 813, 904
Remediation waste...809, 815
Remediation waste management sites...815, 904
Satellite accumulation units...300, 903
Staging piles...903
Subpart CC exemption...725
Temporary units...903

REMEDIATION WASTE MANAGEMENT SITES
Used during corrective action...815
Used during remediation activities...904

REPORTABLE QUANTITY (RQ)
Lamp ballasts and bulbs...84
Spill reporting...686
Spills of products...680, 686
Spills of used oil...572

REPORTING
CERCLA reporting...84, 458, 486, 529, 572, 584, 680, 686, 913
Spill reporting...475, 486, 572, 574

REPRESENTATIVE SAMPLE...848
A composite sample...850
A set of samples...850
A single sample...850
Based on average properties...848
Data analysis...879
Definition...848
For fluorescent lamps...83, 425
For LDR purposes...852

Methods in Appendix I to Part 261...848

RESOURCE CONSERVATION AND RECOVERY ACT (RCRA)
Ban on liquids in landfills...414
Civil penalties...782
Conflict with Atomic Energy Act...154
Control of hazardous waste injection...592
Cradle-to-grave liability...781
Criminal penalties...782
Definition of mixed waste...152-153
Definition of solid waste...1, 147, 689
Domestic sewage exclusion...141
Hazardous waste activity notification...290, 370
Imminent and substantial endangerment...201, 373, 682, 701, 782, 907
Maximum civil penalties...758
Small quantity burner exemption...420
State/EPA right to inspect...360, 366, 769
States may have more stringent regulations...509
Subtitle C...1-2
Subtitle D...2
Waste minimization...311

RETROACTIVITY OF HAZARDOUS WASTE LISTINGS...890

RINSEWATERS
Containing spent solvents...95
From electroplating operations...109, 112, 116, 264
From rinsing containers...258, 450, 455
From rinsing tanks...468

RISK ASSESSMENTS
CERCLA...801
Contained-in policy...276

S

SAFE DRINKING WATER ACT
Class VI wells for geologic sequestration of carbon dioxide...246

Interface with RCRA...4, 62, 209, 798

SAFETY DATA SHEET (SDS)
  For determining if a material is a co-product fuel...17
  For determining if wastes are hazardous...832
  For determining legitimate recycling...543
  For ignitability determinations...47
  For lithium-sulfur dioxide batteries...658
  For toxicity determinations...67
  For used oil contamination rebuttals...557
  Keep copies as hazardous waste characterization records...362-363

SALTS
  Isomer issues...120

SAMPLING...848
  Accuracy...852
  At point of generation...853
  Authoritative sampling...860
  Bias...853
  Choosing a sampling strategy...877
  Composite samples...603, 844, 846, 851, 865, 875
  Containers...870, 874
  Data quality objectives...852
  Debris...861
  Difficult-to-analyze matrices...823
  Field techniques...875
  Grab samples...603, 844, 846, 865
  Handling nondetects...883
  Heterogeneous wastes...861
  Holding times...66, 875
  Hot spots...843, 873, 875
  Hotline...817
  Incinerator residues...869
  Laboratory subsampling...877
  Land disposal restrictions...844, 846
  Methods in Appendix I to Part 261...848
  Minimizing error...877
  MULTI INCREMENT® samples...851, 865
  Nonnormal distributions...881
  Number of samples required...863
  One sample?...850
  Outliers...877, 883
  Precision...852
  Probability sampling...857
  Representative sample...848
  Sampling during inspections...779
  Sampling plan...852
  Statistical analyses...878
  Surface impoundments...866, 872, 882
  Tanks...871
  Waste piles...862, 873
  Wipe sampling...836

SANDBLAST RESIDUE...14, 25, 519, 534, 853, 891

SANITARY WASTES...144

SATELLITE ACCUMULATION...293
  Acute hazardous wastes...294
  Areas...294
  At or near the point of generation...296
  Centralized locations...296
  Comparison with 90-day containers...294
  Conditionally exempt small quantity generators...299, 320
  Contingency plan...299, 310
  Dating requirements...297, 753
  Designating areas...294
  For universal wastes?...432
  Inspections...299, 459
  Labeling requirement...446
  Laboratory settings...297
  Land disposal restrictions...299, 636
  Marking requirement...295
  Must be a container...294
  Notices of violation...755
  Physical limitations...296
  Quantity limits...294, 298
  Roll-off boxes for satellite accumulation units?...299
  Subpart CC exemption...299, 724
  Time limits...299
  Training for operators...299, 324
  Treatment allowed?...300
  Two units in series?...296
  Under the control of the operator...296
  Units must be closed...296, 442, 444
  Used during demolition activities...927
  Used during remediation activities...300, 903
  Waste coding...296
  Waste counting...295

SCINTILLATION FLUIDS...91, 152-154, 157

SCRAP METAL...22
  Aerosol cans...58, 242-243, 245, 649
  Coated with oil...243, 561
  Corrective action...23, 788
  Definition...10, 240, 498
  Does not include batteries...22, 241
  Does not include drosses, slags, or sludges...22, 241
  Does not include liquid mercury...22, 241
  Dust created during recycling...256, 526
  Examples...242
  Excluded scrap metal...23, 241
  Exempt from RCRA if recycled...22, 58, 71, 73, 80, 240, 242, 628, 701, 927, 930
  Exemption applies at point of generation...23, 242
  Lead shot from a shooting range...242-243
  Mixed with something else...243
  Painted with lead-based paint...73
  Printed circuit boards...72, 80, 242-243
  Some scrap metal is not exempt...242
  Unnecessary to make hazardous waste determination...241
  Used oil filters...241-244, 441, 559

SECONDARY CONTAINMENT
- Containment buildings...465, 483
- Design requirements...447, 480
- Double-walled tanks...480, 482
- External liners...480, 482-483
- For ancillary equipment...469, 483
- For containers...443, 447-448
- For dioxin wastes...448
- For tanks...465, 475, 478-479
- For underground storage tanks...483, 487
- For used oil containers...573-574, 578, 584
- For used oil tanks...573-574, 578, 584
- Inspections...485, 489
- Installation deadlines...486
- Must be impermeable...483
- Required capacity...447-448, 480
- Sumps...478-479
- Variances...487
- Vaults...480, 482-483

SECONDARY MATERIALS
- Classifying...9, 497
- Definition...9, 494
- From mineral processing...36
- Is it a solid waste?...494, 516
- Is recycling subject to RCRA?...263, 471, 494
- Is wastewater conditioning subject to RCRA?...534
- Recycling...493-546

SEDIMENTS
- See CONTAMINATED SEDIMENTS

SHIPBOARD WASTES...333, 375
- Point of generation...220, 333

SIGNAGE
- No smoking signs...321

SILVER
- Photoprocessing operations...14, 33, 233, 273, 407, 666
- Precious metals recycling...233

SLUDGE...14, 33
- Baghouse dust...407, 516, 518, 526, 531, 667, 857
- Burning for energy recovery...28
- Carbamate biological treatment sludge...270
- Definition...3, 9, 14, 498
- Discarded materials...1, 3
- Dryers...388
- Electric arc furnace dust...25
- F006 wastes...111, 258
- From de minimis losses...215
- From wastewater treatment...263, 385, 388, 392, 401, 466
- Ion-exchange resins...15, 33, 498
- Land disposal restrictions...607, 610-611, 614-615
- May be characteristic or listed in Table 1...11, 498
- Mixture rule...257
- Reclamation...32, 510, 516, 518
- Refinery wastes...257
- Removed from a POTW...145
- Spent carbon...14
- Sulfuric acid...25
- Use constituting disposal...25, 531, 533
- Versus a spent material...14, 33

SMALL QUANTITY GENERATORS
- 180-day accumulation...300, 315, 397, 465
- Accumulation time extensions...307
- Allowable units for 180-day accumulation...300, 398
- Closure standards...315, 475
- Container standards...303, 398, 439
- Definition...284
- Exception reports...354, 362
- Manifesting exemption...335
- Preparedness and prevention...301, 308, 311, 338, 364-365, 756
- Quiz...317
- Recordkeeping requirements...360, 757
- Required provisions...301
- Spill response...677
- Spill response training...689
- Subpart CC exemption...724
- Summary of requirements...300
- Tank standards...302, 398, 463, 465
- Training...314, 328, 752
- Treatment in 180-day units...303, 397
- Waste minimization program...349

SMALL QUANTITY HANDLERS
- Universal wastes...427

SMELTING
- Closed-loop recycling with no reclamation...526
- F006 wastes...33, 115, 273, 407, 513
- Lead-acid batteries...239
- Lead/copper recovery...515
- Reclamation...31, 407, 513

SODIUM...47

SODIUM HYDROXIDE
- Corrosivity...52, 54
- Recycling...522

SODIUM SULFITE
- Production...520

SODIUM-CONTAINING MATERIALS
- Reactivity...54

SOIL
- See CONTAMINATED SOIL

SOLDER DROSS
- Not scrap metal...23

SOLE ACTIVE INGREDIENT
- In commercial chemical products...128
- In P- and U-wastes...128

SOLID WASTE MANAGEMENT UNITS...787
- 90- or 180-day accumulation units...314, 793
- Area of concern...789
- Definition...787
- Elementary neutralization units...411
- Inactive units...893
- Maintenance areas...787
- Manufacturing process units...223, 789
- Military firing range...788

RCRA facility investigation...796
Recycling units...787
Releases must be routine and systematic...788
Scrap metal management area...23, 788
Sewer line leaks...145, 787
Spill locations...529, 685, 788
Storm water retention pond...788
Wastewater treatment units...396, 787
Wood preserving areas...788

SOLID WASTES...1-38
Abandoned materials...7, 22
By-products...10
Commercial chemical products...10
Defined to capture recycling activities...7, 9
Definition...1
Determination using knowledge...818
Documentation for claims that materials are not solid wastes...542, 544
Inherently waste-like materials...8, 531
Management of nonhazardous wastes...2, 41
Materials that are burned or incinerated...7, 10, 28, 530
Materials that are disposed...7
Materials that are not solid wastes...6
Materials that are reclaimed...10
Materials that are speculatively accumulated...10, 530
Materials that are used constituting disposal...530
Obsolete chemicals...8
Recycled materials...9
Relationship to hazardous wastes...1, 40
Scrap metal...10
Sludge...9
Solid wastes that are not hazardous wastes...7
Spent materials...9
Subset of discarded materials...1, 6
Waste military munitions...697

SPECULATIVE ACCUMULATION...36
Definition...10, 498
Lead-acid batteries...238
Solid wastes...10, 530
When materials that are not recycled become wastes...37

SPENT CARBON
See ACTIVATED CARBON

SPENT MATERIALS...12
Batteries...12, 236, 426
Burning for energy recovery...28
Definition...9, 498
Discarded materials...2, 5
Do not include leftover fuels...13
Do not include unreacted raw materials...13, 133
Fluorescent light tubes...80
Reclamation...32, 426
Sandblast grit...14, 25, 519, 534, 857, 891
Spent carbon...14, 124, 256, 263, 407, 691, 695, 741
Spent photoprocessing fixer solution...14
Spent pickle liquor...25, 27
Spent refrigerants...70, 124
Spent solvents...14, 522
Sulfuric acid...24
Thermostats...12
Use constituting disposal...24-25
Versus a sludge...14, 33
Wastewater...12

SPENT PICKLE LIQUOR
Corrosivity...52
Exemption for sludge...268
Fertilizer production...25
K062 wastes...110
Recycling...493, 519, 521-522, 541
Used to condition wastewater...27

SPENT SOLVENTS...88
Absorbents contaminated with spent solvents...102, 250, 683
Alcohols...46
Before use...96-98, 835
Contained-in policy...102

Contaminated soil...102
Criteria that must be met...89
De minimis wastewater exemptions...95, 209
Derived-from rule...29
Description...88, 835
Difference between F001 and F002...88
Dry cleaning wastes...92, 265, 289, 406
Extractants...91, 95
Laboratory activities...91, 227
Land disposal restrictions...590, 617, 655
LDR paperwork requirements...655
Manufacturing operations...90
Mixed with adhesives...251
Mixed with used oil...103, 401, 465, 554-555
Mixture rule...94
Mixtures...96
Not used oil...90, 551
Oil and natural gas E&P wastes...186, 194
Painting operations...91, 94, 130, 218, 469
Process wastes are not F001–F005...92, 95
Products containing solvents are not F001–F005...92, 402, 616
Rags contaminated with spent solvents...49, 103-104
Reaction and synthesis media...90, 93, 96
Reclaimed solvents...101, 272, 292, 305, 333, 384, 407, 465, 513, 515, 523, 525, 529, 664, 708, 857
Recycling...101, 272, 292, 305, 333, 384, 407, 465, 513, 515, 523, 525, 529, 664, 708, 857
Solvent dragout is not spent...95
Solvents used as ingredients...93
Solvents used as reactants...92
Solvents used to make lab standards...91, 130
Spent materials...14, 522
Spills...102, 209
Still bottoms...29, 101, 103, 263, 333, 376, 406, 529, 708

Uncontained vapors are not solid wastes...692
Unused solvents may be P- or U-wastes...94
Use as reactants or ingredients are not F001–F005...92, 94
Waste counting...291

SPILL PREVENTION, CONTROL, AND COUNTERMEASURES PROGRAM...260, 547, 571, 573-574, 581
Contingency plan...310, 365

SPILL RESIDUES
See CLEANUP RESIDUES

SPILLS...675
Absorbents used to contain spills...102, 250, 683
Assigning EPA ID number...372
Characteristic wastes...679
Cleanup does not require a permit...417, 678, 922
Cleanup levels...678
Commercial chemical products...2, 127, 223, 417, 676, 679-680
Corrective action...529, 685, 788
De minimis wastewater exemption...211
Detection...484
Do we have to clean up spills?...676
F- and K-wastes...680
Free product recovery...20, 511, 684
From manufacturing process units...223, 788
Hazardous waste...417, 676
Immediate response...417, 677
Land disposal restrictions...685
Large quantity generators...677
May become solid waste management units...529, 685, 788
Oil...682
P- and U-wastes...127, 679-680
Pesticides...128, 681, 899
Point of generation...672
Pre-RCRA spills...891
Recycling...684

Release to the environment...486, 683, 688
Small quantity generators...677
Spent solvents...102, 209, 250
Spill reporting...475, 486, 572-573, 685
Spill response...429, 433, 475, 484, 486, 572-573, 584, 677
Spill response training...689
To soil...419, 679, 681
Transporters...677
TSD facilities...677
Unremediated spills are solid wastes...676
Used oil...572, 584, 683

STABILIZATION
Absorbent addition...417
Cyanides...639
F006 wastes...385, 401
Paint...402
Recycling stabilized waste...515

STAGING PILES...809
Alternative to AOC policy...810, 813
Alternative to CAMUs...810
Closure...811
Corrective action...809, 811
Land disposal...589, 591, 810
Regulatory agency must designate...810
Staging piles or waste piles?...810
Used during remediation activities...903

STATE-LISTED WASTES...88, 819
Land disposal restrictions...589
Manifesting...344, 347
Universal wastes...426

STATE-SPECIFIC REQUIREMENTS
Biennial report...369

STEAM STRIPPERS
As air emission control devices...707
As wastewater treatment units...387

STEEL PRODUCTION WASTES
Can be treated in a totally enclosed treatment facility?...412-413
Electric arc furnace dust...517, 525, 533, 620
Pickling wastes...25, 27, 52, 110, 493, 519, 521-522, 541
Treatment?...381

STILL BOTTOMS
90-day accumulation...305
Burning for energy recovery...29
By-products...16
Land disposal restrictions...590
Mixture rule...263
Point of generation...304, 664, 667, 857
Processed to recover products...16
Spent solvents...29, 101, 103, 263, 333, 376, 406, 529, 708
Waste counting...291

STORAGE
Analytical samples...226, 468
Four storage options...293
Point of generation issues...664, 669
Prior to recycling...403, 465, 471
Treatability study samples...230, 468
Used oil...570

STORAGE PROHIBITION...636
1-year time limit...636
Attaches at point of generation...597
Batteries...637
Exempt wastes...636
Land disposal restrictions...299, 590, 597, 634, 636
Mixed wastes...634
Need to treat UHCs...636

STREPTOZOTOCIN
Container residues...127

SUBPART AA STANDARDS...704
90-day accumulation units...305, 314, 398, 708
Applicability...705
Control devices...712

Emission control requirements...711
Exemptions...390, 410, 708, 710
Inspection and monitoring...712
Organic concentration determinations...707
Process vents covered...709
Recordkeeping and reporting...712
Recycling units...405, 515, 708
TSD facilities...708

SUBPART BB STANDARDS...713
90-day accumulation units...305, 314, 398, 717
Applicability...219, 713
Containers...443, 717
Drip pads...719
Exemptions...390, 410, 717, 719
Inspection and monitoring...720
Leak detection and repair requirements...720
Organic concentration determinations...716
Recordkeeping and reporting...721
Recycling units...405, 515, 717
Tanks...469, 719
TSD facilities...717

SUBPART CC STANDARDS...721
90-day accumulation units...305, 314, 398, 724
Activated carbon...741-742
Applicability...723
Basic intent...728
Compliance options...729
Containers...305, 443, 724, 742
Do not apply if waste is nonhazardous...730
Episodic generators...726
Exemptions...390, 410, 724
Inspection and monitoring...738, 744-746, 748
Land disposal restrictions...733
Miscellaneous units...724, 742
Point of waste origination...727

Recordkeeping and reporting...738, 740, 744-746, 748
Sampling and analysis...729
Sumps...726
Surface impoundments...724, 748
Tanks...475, 724, 735
TSD facilities...724, 730
VO concentration determinations...729
VO removal or destruction...731
Volatile organics...727

SUBSTANTIVE RCRA REGULATIONS
Waste counting...286, 289

SUBTITLE D FACILITIES
LDR paperwork...653

SULFIDES
Reactivity...54, 56, 674, 825, 838

SULFURIC ACID
Corrosivity...52
Fertilizer production...24-25, 532, 610
Production...535
Recycling...520, 526
Reused without reclamation...522

SUMPS
Are surface impoundments if they fail the parking lot test...388, 478
Are they tanks?...388, 473, 477, 479
As 90-day accumulation tanks...478, 726
As elementary neutralization units...409, 478-479
As immediate response units...418
As totally enclosed treatment facilities...413
As wastewater treatment units...388, 478-479
Cleaned out before demolition...921
Definition...477
Is treatment allowed in sumps?...479
Landfill leachate collection sumps...478-479

Parking lot test...477-478
Primary containment sumps...478
Secondary containment sumps...478-479
Subpart CC air emission controls...726
Temporary sumps...477, 479

SURFACE IMPOUNDMENTS
Active management...892
Are they tanks?...472
Cannot be manufacturing process units...217
Definition...472
Earthen pits...590
Evaporation ponds...4
F006 wastes...112
Inspections...778
Land disposal restrictions...146, 589
Management of hazardous wastewater...4, 520
Management of reactive wastes...58, 590
Parking lot test...388, 472, 477-478
Point of generation of the sludge...666
Recycling unit exemption...406
Relationship to NPDES outfall points...149
Sampling...866, 872, 882
Storage of solid-waste-exempt material allowed?...162, 516
Storm water retention ponds...788
Subpart CC air emission controls...724, 748
Used as a source of fire training water...149
Used oil storage...574

SW–846...822
Method 1010—Pensky-Martens closed-cup tester...43, 824
Method 1020—Setaflash closed-cup tester...43, 824
Method 1030—Ignitability of solids...45, 48, 824
Method 1040—Test Method for Oxidizing Solids...49, 674

Index

Method 1050—Substances likely to spontaneously combust...47

Method 1110—Corrosivity toward steel...51, 53, 824-825, 838

Method 1311—Toxicity Characteristic Leaching Procedure...43, 65, 824-825

Method 1330—Oily waste extraction procedure...67, 75, 840

Method 3010—Acid digestion of aqueous samples and extracts for total metals...674

Method 3510—Separatory funnel liquid-liquid extraction...674

Method 6010—Inductively coupled plasma-atomic emission spectrometry...74, 674

Method 6020A—Inductively coupled plasma-mass spectrometry...674

Method 7060—Graphite furnace atomic absorption spectrophotometry...74

Method 7470—Mercury in liquid waste...674

Method 8081—Organochlorine pesticides by gas chromatography...674

Method 8082—Polychlorinated bephenyls (PCBs) by gas chromatography...563

Method 8095—Explosives by gas chromatography...55

Method 8151—Chlorinated herbicides by GC using methylation...674

Method 8260—Volatile organic compounds by gas chromatography/mass spectrometry...674

Method 8270—Semivolatile organic compounds by gas chromatography/mass spectrometry...674

Method 8330—Nitroaromatics, nitramines, and nitrate esters by high performance liquid chromatography (HPLC)...867, 877

Method 9000—Determination of water in waste materials by Karl Fisher titration...51

Method 9001—Determination of water in waste materials by quantitative calcium hydride reaction...51

Method 9040C—pH measurement...51-52, 674, 824-825, 838

Method 9045—Soil and waste pH...54, 825

Method 9060—Total organic carbon...707, 716

Method 9095B—Paint filter test...43, 51, 415, 442, 448, 824, 837

Methods don't have to be used exactly as written...822

Representative sample...83, 425, 848, 879

Required use of methods...66, 822, 824

Testing for reactive cyanides and sulfides...56

T

2,4,5-T...899

TANK SYSTEM...387
  Definition...464

TANKS
  90- or 180-day clock...306, 488
  Ancillary equipment...409, 465, 473, 479, 483
  Are portable tanks containers?...395, 440, 464
  Are they surface impoundments?...472
  As elementary neutralization units...408, 465
  As immediate response units...220, 418, 469, 477, 678
  As totally enclosed treatment facilities...412
  As wastewater treatment units...385, 464, 467-468, 473
  Associated with closed-loop recycling...471, 528-529
  At transfer facilities...466
  Closure...475, 921
  Counting cleanout residues...288
  Definition...387, 464
  Double-walled tanks...482, 485
  Drip pads are not tanks...473
  Empty for purposes of stopping the 90-day clock...306, 488
  Empty tanks...306, 488, 528
  Free product recovered from leaking tanks...684
  Holding spill cleanup residues...220, 418, 678
  Inspections...475, 489, 738, 740
  Integrity assessments...475-476
  Labeling...753
  Leak detection systems...475, 484
  Leak testing...475, 487
  Lift stations...464
  Mixed wastes...159
  New vs. existing...474
  Notices of violation...753, 755
  Operating procedures/equipment...475
  Parking lot test...388, 472, 477-478
  Permitting exemptions...384, 466
  Point of generation of hazardous waste...473
  Portable tanks...395, 440, 464
  Requirements for new tanks...475
  Residues...488
  Sampling...871
  Secondary containment...448, 465, 475, 478-479
  Special requirements for ignitable and reactive wastes...475, 482
  Special requirements for incompatible wastes...475
  Spill reporting...475, 486
  Spill response...475, 484, 486
  Status of removed residues...468
  Subpart BB air emission controls...469, 719
  Subpart CC air emission controls...475, 724, 735
  Sumps...473, 477, 479
  Temporary tanks...469
  Underground storage tanks may be Subpart J tanks...468

Used for 180-day accumulation...303, 398, 463, 465
Used for 90-day accumulation...302, 398, 463, 465, 469, 478, 488
Used oil storage...570, 578, 584
Used to mix used oil and hazardous waste...465, 554
Waste analysis and trial storage/treatment tests...475
Waste piles or tanks?...471
When they are debris...628

TECHNICAL GRADE...128

TEMPORARY UNITS
Closure...809
Corrective action...809
Used during remediation activities...903

TEST METHODS
See SW–846

TETRACHLOROETHYLENE
De minimis wastewater exemption...209
Dry cleaning wastes...265, 287, 387
Production...519
Solvent use...88, 92
Waste counting...287

THERMOMETERS...21
Manufactured articles...121, 125
Universal wastes...21, 424

THERMOSTATS
Spent materials...12

TITANIUM SWARF...46

TOLLING AGREEMENTS...335

TOLUENE...96, 98, 131
Contained in filters...280
De minimis wastewater exemption...209
Product packaging...133
Reaction and synthesis media...93
Scintillation fluids...153-154
Solvent mixtures...98
Solvent use...89-91, 93-94, 265, 617
Spilled...679, 682, 686

Use as carrier...96
Use as ingredient...94, 130, 686
Use as reactant...93

TOLUENEDIAMINE
And isomers are U221...120

TOTAL ANALYSIS
In lieu of TCLP...67

TOTALLY ENCLOSED TREATMENT FACILITIES...411
Can be tanks...412
Definition...411
Evaporators...413
Filter presses...412
Incinerators...413
Permitting exemption...411, 467
Subpart AA exemption...708
Subpart BB exemption...717
Subpart CC exemption...724
Waste counting...292

TOXIC SUBSTANCES CONTROL ACT (TSCA)
Management of PCB dielectric fluids...72
PCB-containing light ballasts...83
PCBs in used oil...563, 580-581, 585
Rinsing a PCB container...452
Solvents used to remove PCBs from transformers...92

TOXICITY...62
Antifreeze...73
Chromium level based on total chromium...64
CRTs...77
Determination using analysis...65, 674, 825, 895
Determination using knowledge...67, 839, 895
Determination using total analyses...67, 840, 842
Dual-phase wastes...69
Electronic wastes...76
Exemptions...69
Land disposal restrictions...611
Liquids...68
Nonaqueous wastes...69
Old EP toxicity test...64

Solids...68
State vs. federal determination...62
Unknown wastes...674

TOXICITY CHARACTERISTIC LEACHING PROCEDURE (TCLP)
Alternative analytical methods...67, 826
Cost...65
Development...64
Exemption for MGP wastes...67
Masking using iron additives...74
Method description...43, 65, 824-825
Problems with oily wastes/ organic liquids...66-67, 840
Reproducibility...65
Sample holding times...66, 876
Total analysis in lieu of TCLP...67, 840, 842
Use in delisting petitions...75
Used for evaluating unknowns...674

TRAINING...322
Annual review...324
Basic training requirements...324
Conditionally exempt small quantity generators...320, 322, 330
Contractors...323, 332
Emergency response...325
Episodic generators...286, 329
For mixed waste storage...159
For signing manifests...348
Hazardous waste management procedures...324
Large quantity generators...314, 322, 752
Notices of violation...752
Overlap between RCRA and OSHA...327
Recordkeeping requirements...328, 364, 752
Requirements for instructors...328
Satellite accumulation units...299, 324
Small quantity generators...314, 328, 752
Spill response...689

Index

Three training types...327
TSD facilities...330
Universal waste handlers...330, 429
When must training be completed?...324
Who must be trained?...323

TRANSPORTERS
Combining wastes...302, 382
Definition...341
EPA ID number...234, 236, 337, 341, 355, 370, 372, 377, 417, 677
Land disposal restrictions exemption...587
Lead-acid batteries...236
Manifesting...341
May be subject to generator requirements...343, 382, 588
Of containers...441, 446, 451, 460
Of mixed waste...160
Oil and natural gas E&P wastes...191
Pressurized rail cars...453
Remanifesting wastes...302
Spill reporting...688
Spill response...677
Spill response training...689
Tank storage allowed at transfer facilities?...466
Transfer facilities...320, 341, 405, 433, 466
Universal wastes...427, 433
Used oil...574, 581, 583

TREATABILITY STUDY EXEMPTION...224
Disposal of excess sample/treatment residues...231
DOT requirements still apply...229
Exporting samples...231
Importing samples...231
Mobile treatment units...231
Reporting and recordkeeping...232
Sample quantity and time limits...229
Sample shipment...230
Sample storage...230, 468
Sample treatment...230
Scope...229

What is a treatability study?...228

TREATABILITY VARIANCES...592, 621, 624, 659

TREATMENT
Adding absorbents to wastes...414
Battery waste...401, 431, 465, 658
Before wastes are placed in CAMUs...808
Bulking and containerization...382
Burning small quantities of hazardous waste in onsite units...419
Conditionally exempt small quantity generators...319, 410
Contaminated debris...401, 631
Contaminated ground water...401, 465
Contaminated soil...401, 626, 709
Counting treatment residues...293
Definition...380
For Subpart CC compliance...731
Fuel blending...401, 408, 464
Household hazardous wastes...173
Immediate responses...417
In 90- or 180-day accumulation units...303, 397, 402, 616, 679, 709
In containment buildings...632
In elementary neutralization units...408
In satellite accumulation units...300
In sequential units...304
In situ treatment...592
In sumps...479
In totally enclosed treatment facilities...411
In wastewater treatment units...384
Miscellaneous processes...401
Mixed wastes...159
Mixing hazardous waste with used oil...554

Mixture rule...248, 256, 383
Point of generation issues...380, 663
Puncturing aerosol cans...59-60, 381
Recycling...384, 401
Residues...390, 410, 663
Thermal treatment...304, 399
Treatability study exemption...229
Treatment residues have a new point of generation...406, 663
Truck washing...457
Universal wastes...431-432, 658

1,1,2-TRICHLORO-1,2,2-TRIFLUOROETHANE
Solvent use...88, 92

1,1,1-TRICHLOROETHANE
Contaminated soil...896
De minimis wastewater exemption...209-210
Off-spec product...126
Solvent use...88, 94, 126, 617, 692
Use as ingredient...94

1,1,2-TRICHLOROETHANE
Solvent use...88

TRICHLOROETHYLENE
Contaminated soil...102
De minimis wastewater exemption...209
Off-spec product...126
Solvent use...88, 102, 620, 842
Treatment...385

TRICHLOROFLUOROMETHANE
Solvent use...88

TRITIUM...154

TSD FACILITIES
Are generators of treatment residues...664
Can also use 90- or 180-day accumulation...302
Certification of waste receipt...353
Container standards...439
Contingency plans...677
Converting to 90- or 180-day units...315

Corrective action...791
EPA ID number...234, 352, 370, 395, 653, 792
Generators of F006 wastes...112, 263
Land disposal restrictions...588-589, 846-847
LDR paperwork...642-643, 653
Manifesting...307, 344, 352-353
Preparing for a RCRA inspection...769
Receiving D002 wastes...41, 351
Receiving demolition wastes...930
Receiving remediation wastes...906
Recycled material handling...509
Spill reporting...686
Spill response...677
Spill response training...689
Subpart AA air emission standards...708
Subpart BB air emission standards...717
Subpart CC air emission standards...724, 730
Tank standards...463, 480
Training...330
Treatment residues have a new point of generation...664
Using generators' data...833
Waste analysis plans...828, 846

U

U001 WASTES
  Empty containers...453

U009 WASTES...212
  Derived-from rule...264

U010 WASTES
  Container residues...127

U019 WASTES
  Burning for energy recovery...28
  Spills...621, 891

U028 WASTES
  Product handling wastes...131

U031 WASTES
  Isomer issues...120

U034 WASTES
  Isomer issues...120

U035 WASTES
  Container residues...127

U036 WASTES...897
  Contamination resulting from use...27, 124
  Sole active ingredient...129

U043 WASTES
  Empty containers...453

U044 WASTES...214

U045 WASTES
  Empty containers...453

U051 WASTES...136
  Burning for energy recovery...20, 136
  Contained-in policy...277
  Creosote-treated railroad ties...121
  Mixture rule...252
  Spilled...681

U058 WASTES
  Container residues...127
  Isomer issues...120

U059 WASTES
  Container residues...127

U080 WASTES...222
  Product handling wastes...133
  Sole active ingredient...130

U096 WASTES
  Contained-in policy...251
  Land disposal restrictions...255, 618

U122 WASTES
  Empty containers...453
  Land disposal restrictions...626
  Off-spec formaldehyde...126
  Sole active ingredient...130

U131 WASTES
  Incinerator trial burn...123, 264

U135 WASTES
  Empty containers...453

U140 WASTES
  Isomer issues...120

U150 WASTES
  Container residues...127

U151 WASTES
  Spilled...681
  Vial of mercury...121

U154 WASTES
  Burning for energy recovery...136
  Mixture rule...251
  Point of generation of residues...671

U196 WASTES
  Sole active ingredient...129

U202 WASTES
  Use constituting disposal...19, 135

U206 WASTES
  Container residues...127

U211 WASTES
  Contained-in policy...277
  Contaminated soil...277

U220 WASTES...131
  Contained-in policy...280
  Product handling wastes...133
  Sole active ingredient...130

U221 WASTES
  Isomer issues...120

U226 WASTES
  Off-spec 1,1,1-trichloroethane...94, 126

U228 WASTES...102
  Off-spec trichloroethylene...126

U237 WASTES
  Container residues...127

U239 WASTES
  Burning for energy recovery...29

U248 WASTES
  Container residues...452
  Listings are concentration-based...123

U249 WASTES
  Listings are concentration-based...123

Index

UNDERGROUND INJECTION
   Discarded wastewater...4
   Land disposal restrictions...587, 589, 592, 604, 609

UNDERGROUND MINES
   Land disposal restrictions...589

UNDERGROUND STORAGE TANKS
   Cleanups...684
   Exemption for petroleum contamination...27, 70, 72, 628, 684
   May be Subpart J tanks...468
   Secondary containment...483, 487
   Used oil storage...573, 581

UNDERLYING HAZARDOUS CONSTITUENTS (UHCs)...598, 605
   Determination of...845
   For Subpart CC compliance...734
   From spills...685
   In D001 wastes...605
   In D002 wastes...611
   In debris...630
   In hazardous waste fuel...652
   In listed wastes...618
   In toxic wastes...613-614
   In wastes discharged to sewers...147
   Subject to storage prohibition...636

UNIVERSAL TREATMENT STANDARDS...600, 618

UNIVERSAL WASTES
   Accumulation...432
   Accumulation time limit...429
   Applicability determinations...428
   Batteries...236, 424, 430-431, 658
   Conditionally exempt small quantity generators...321, 429
   Contractors...428
   Counting...428
   Definition...424
   Destination facilities...427, 431, 435, 658, 930
   Exports...432
   Fluorescent light tubes...425, 432
   Generated during demolition activities...928
   Labeling...429
   Lamps...425, 430
   Land disposal restrictions...273, 431, 435, 627, 657-658
   Large quantity handlers...371, 427
   Manifesting...355, 429, 433, 436, 930
   Mercury-containing equipment...21, 23, 73, 424-425, 430, 441
   No LDR notifications required...429, 930
   Overlap with OSHA training requirements...330
   Pesticides...425, 430
   Program structure...426
   Releases...429, 433
   Small quantity handlers...427
   State authorization...436
   State universal wastes...426
   Thermometers...21
   Training...330, 429
   Transporters...427, 433
   Treatment...431-432, 658
   Waste tracking...429

UNKNOWN WASTES...673
   90- or 180-day accumulation...304, 675
   90- or 180-day clock...304, 675
   Characteristic wastes...673-674
   Containing asbestos...674
   Containing polychlorinated biphenyls (PCBs)...674
   Contaminated soil...896
   Immediate response...673
   Listed wastes...674, 896
   Pesticide-contaminated soil...897

URACIL MUSTARD
   Container residues...127

USE CONSTITUTING DISPOSAL...24
   Absorbent production...534
   Animal feed production...28, 533
   Asphalt production...27
   Cement/aggregate production...25-26, 533, 652
   Commercial chemical products...19, 27, 135
   Definition...10, 498
   Dust suppressants...19, 135
   Excludes wastewater conditioning...27, 534
   Fertilizer production...13, 24-25, 531-532, 610, 654
   Land disposal restrictions...654
   Oil and natural gas E&P wastes...196, 530
   Pesticide production...24, 534
   Petroleum-contaminated soil...27, 546
   Recycling exemption is voided...528, 531
   Sludge...25, 531, 533
   Solid wastes...10, 530
   Spent materials...24-25

USED OIL...547-586
   Absorbents containing used oil...560, 584, 684
   Burner requirements...371, 583
   Burning for disposal...578, 582
   Burning for energy recovery...29, 555, 561, 578, 581-582, 584-585
   CFCs...558, 577
   Cogenerator situations...333, 570, 575-576
   Conditionally exempt small quantity generator exemption...318, 321, 553, 557
   Consolidation...548
   Cutting oil...94, 557, 561, 577
   Definition...548
   Disposal...552, 578
   Does not include spent solvents...90, 551
   Dust suppressants...578
   Examples...548
   Filters...241-244, 441, 558, 585
   Fuel specification...557, 578, 581-582

Generator requirements...569-570
Grease...551
Household do-it-yourselfers...557, 569, 583
Land disposal restrictions...554, 656
Marketer requirements...371, 585
Materials that are not used oil...551
Mixed with chlorinated pesticides...556
Mixed with commercial chemical products...19, 135
Mixed with fuel...560, 570, 576
Mixed with hazardous waste...259, 465, 553, 578
Mixed with solid and nonhazardous wastes...560, 562, 584
Mixed with spent solvents...103, 401, 465, 554-555
Mixture rule...259, 553
Notices of violation...572, 756
Off-spec used oil...582
On-spec used oil...581
PCBs...557, 562, 580-581, 585
Placed into empty containers...556
Presumed to be recycled...552
Processing exemptions...577
Processor requirements...371, 578
Rags containing used oil...107, 560
Rebuttable presumption...103, 555, 578, 584
Recovered from contaminated materials...561
Recycled into petroleum industry...564
Releases...572-573, 584, 683
Residues from managing used oil...561
Scrap metal contaminated with used oil...243, 561
Space heater exemption...583
Spill reporting...572
Storage requirements...570
Storage tank bottoms...561
Stored in containers...570, 578, 584

Stored in tanks...570, 578, 584
Synthetic oil...548
Transportation requirements...371, 574, 581
Transporter exemptions...333, 574, 583
Used oil processing...576
Wastewater containing used oil...167, 562, 574, 577

V

VANADIUM PENTOXIDE
In catalysts...130

VAULTS...482
Are landfills—not tanks...473
Land disposal restrictions...589
Secondary containment...480, 483

VINYL CHLORIDE
Gas at standard temperature...453
Residues remaining in rail cars...453

VITRIFICATION
Is recycling subject to RCRA?...518

W

WARFARIN...281
Container residues...452
Discarded product...123
Product packaging...132

WASHWATER...220
Contaminated with P- and U-chemicals...132
Pesticide operations...123

WASTE ANALYSIS PLANS
For burning hazardous wastes...829
For generators treating to meet an LDR standard...364, 400, 654, 827
For treatment of mixed waste...161
For TSD facilities...828, 846-847
Not required for treatment in certain units...828

WASTE CODING
Adding characteristic codes to listed wastes...619
Derived-from rule...262
Land disposal restrictions...619
Mixture rule...250
Must codes appear on 90- or 180-day containers?...317, 446
On manifests...347, 460, 668
Point of generation...663
Required for wastes in satellite accumulation...296

WASTE COUNTING...283
Acute hazardous wastes...293
Analytical samples...291
Batteries...238, 287
Conditionally exempt small quantity generators...288, 318
Container cleanout residues...288, 458
Containerized waste...287, 441
Contractors...332
Converting from volume to mass...284
De minimis losses...289
Derived-from rule...292
Don't double count wastes...293
Dry cleaning wastes...287, 289
Episodic generators...285
Generators...283
Imported wastes...289
Manufacturing process units...288-289
Mixed wastes...160
P- and U-wastes...287
Point of waste generation...668
Residues from empty containers...458
Spent filter cartridges...287
Spent solvents...291
Tank cleanout residues...288
Treatment residues...293
Universal wastes...428
Wastes in satellite accumulation units...295
Wastes managed in elementary neutralization units...291

Index

Wastes managed in recycling units...291
Wastes managed in totally enclosed treatment units...292
Wastes that must be counted...286
Wastes that need not be counted...289, 429

WASTE MINIMIZATION
Biennial report...312, 369
Generator certification...312, 348, 351-352
Generator's waste minimization program...311
Large quantity generators...312, 348
RCRA statutory authority...311
Small quantity generators...312, 348

WASTE PILES
Definition...472
Land disposal...589
Sampling...862, 873
Storage of solid-waste-exempt material allowed?...516
Waste piles or staging piles?...810
Waste piles or tanks?...471

WASTEWATER
Conditioning...27, 53, 522, 534
Containing spent solvents...95
Containing used oil...167, 562, 574, 577
CWA regulation...385
Discarded materials...4
Point of generation...668
Recycling...272, 515, 529
Spent materials...12
Treatment sludge...263, 385, 388, 392, 401

Wastewater treatment unit exemption...44, 387, 442

WASTEWATER TREATMENT UNITS
Can be sumps...478-479
Corrective action...396, 787
CWA-equivalent treatment...386, 604-605
Definition...385
Derived-from rule...271, 390
Designated facility status on manifests...344
Equipment eligible for exemption...387, 708
Evaporators...388
Filter presses...388
Land disposal restrictions...396, 607
Leaking...396
May become solid waste management units...396, 787
Must be dedicated for use with onsite wastewater treatment systems...391
Must be tanks...385
Permitting exemption...384, 442, 464, 467, 473, 478-479, 668
Point of waste generation...668
Subpart AA exemption...708
Subpart BB exemption...717
Subpart CC exemption...724
Zero-discharge facilities...386, 604-605

WATERS OF THE UNITED STATES...148

WET AIR OXIDATION
Totally enclosed treatment facility exemption eligibility...413

WIPES
Solvent contaminated...104

WOOD PRESERVATIVE WASTES
Containing tri-, tetra-, or pentachlorophenol...131
Corrective action...788
Immediate responses...419
On drip pads...473, 719

X

XYLENE...98, 617
Burning for energy recovery...29
Scintillation fluids...153
Solvent mixtures...98
Solvent use...89, 91, 95, 515, 617, 664
Use as ingredient...121
Use as reactant...93

Z

ZERO-DISCHARGE FACILITIES
As wastewater treatment units...386, 604-605

ZINC
Batteries...438, 612
Dewatering...513
Fertilizer production...26, 531
Galvanizing...15, 113, 513
Recovered from baghouse dust...516, 518
Recovered from K061 wastes...407, 517

ZINC PHOSPHIDE
Discarded product...123

ZIRCONIUM...157

Acronyms

AA	Atomic absorption
AEA	Atomic Energy Act of 1954
AEC	Atomic Energy Commission
ANPRM	Advance notice of proposed rulemaking
AOC	Area of contamination
API	American Petroleum Institute
ARAR	Applicable or relevant and appropriate requirement
ASTM	American Society for Testing and Materials
BATF	Bureau of Alcohol, Tobacco, and Firearms
BDAT	Best demonstrated available technology
BIF	Boiler and industrial furnace
C&D	Construction and demolition
CAA	Clean Air Act
CAMU	Corrective action management unit
CAS	Chemical Abstracts Service
CCP	Commercial chemical product
CDX	Central data exchange
CERCLA	Comprehensive Environmental Response, Compensation, and Liability Act of 1980
CERCLIS	Comprehensive Environmental Response, Compensation, and Liability Information System
CESQG	Conditionally exempt small quantity generator
CFC	Chlorofluorocarbon
CFR	*Code of Federal Regulations*
CI	Confidence interval
CMS	Corrective measures study
CN	Course notes
COLIWASA	Composite liquid waste sampler
CPU	Central processing unit
CRT	Cathode ray tube
CWA	Clean Water Act
DAF	Dilution and attenuation factor
DEA	Drug Enforcement Administration
DEFT	Decision error feasibility trial
DEHP	Diethylhexyl phthalate
DET	Determination of equivalent technology
DIY	Do-it-yourselfer
DOD	Department of Defense
DOE	Department of Energy
DOT	Department of Transportation
DQO	Data quality objective
DSW	Definition of solid waste
DTC	Drum-top crusher
E&P	Oil & gas exploration and production
EAF	Electric arc furnace
ECHO	Enforcement and Compliance History Online
EH&S	Environmental, health, and safety
EHS	Extremely hazardous substance
EI	Environmental indicator
ENU	Elementary neutralization unit
EOD	Explosive ordnance disposal
EP	Extraction procedure
EPA	Environmental Protection Agency
EPCRA	Emergency Planning and Community Right-to-Know Act
FAQ	Frequently asked question
FFC	Fossil fuel combustion
FFCA	Federal Facilities Compliance Act of 1992
FGD	Flue gas desulfurization
FIFRA	Federal Insecticide, Fungicide, and Rodenticide Act
FOIA	Freedom of Information Act
FOTW	Federally owned treatment works
FR	*Federal Register*
FRH	Flameless ration heater
GC	Gas chromatography
GC/ECD	Gas chromatography with an electron capture detector
GPRA	Government Performance and Results Act of 1993
HCFC	Hydrochlorofluorocarbon
HFC	Hydrofluorocarbon
HID	High-intensity discharge
HLMW	High-level mixed waste
HMR	Hazardous materials regulations
HON	Hazardous organic NESHAP
HRS	Hazard ranking score
HSWA	Hazardous and Solid Waste Amendments of 1984
HTMR	High-temperature metals recovery
HWIR	Hazardous waste identification rule
ICP	Inductively-coupled plasma
ICR	Ignitable, corrosive, and reactive wastes, or Information collection request

Acronyms

ICRT	Ignitable, corrosive, reactive, and TC wastes	OECD	Organization for Economic Cooperation and Development
ID	Identification	OSC	On-scene coordinator
IDW	Investigation-derived wastes	OSWER	Office of Solid Waste and Emergency Response
IPA	Isopropyl alcohol	OWEP	Oily waste extraction procedure
LBP	Lead-based paint	PBMS	Performance-based measurement system
LDAR	Leak detection and repair	PCB	Polychlorinated biphenyl
LDR	Land disposal restrictions	PCE	Perchloroethylene
LLMW	Low-level mixed waste	PE	Professional engineer
LLRWDF	Low-level radioactive waste disposal facility	PEL	Permissible exposure level
LQG	Large quantity generator	PHC	Principal hazardous constituent
MACT	Maximum achievable control technology	POC	Point of compliance or Point of contact
MCE	Mercury-containing equipment	POG	Point of generation
MCL	Maximum contaminant level	POTW	Publicly owned treatment works
MEK	Methyl ethyl ketone	PPE	Personal protective equipment
mg/kg	Milligrams per kilogram	ppm	Parts per million
mg/L	Milligrams per liter	ppmv	Parts per million by volume
MGP	Manufactured gas plant	ppmw	Parts per million by weight
MICE	EPA's Methods Information Communication Exchange	PRP	Potentially responsible party
MPU	Manufacturing process unit	QA	Quality assurance
MS	Mass spectrometry	QC	Quality control
MSDS	Material safety data sheet	RAP	Remedial action plan
MTR	Minimum technological requirements	RCRA	Resource Conservation and Recovery Act
MTU	Mobile treatment unit	RD&D	Research, demonstration, and development
NAAQS	National ambient air quality standards	RFA	RCRA facility assessment
NAICS	North American industrial classification system	RFI	RCRA facility investigation
NARM	Naturally occurring and/or accelerator-produced radioactive material	RIA	Regulatory impact analysis
NCP	National contingency plan	RO	EPA's RCRA Online database
NEIC	National Enforcement Investigation Center	ROD	Record of decision
NESHAP	National emission standards for hazardous air pollutants	RQ	Reportable quantity
NIPDWS	National interim primary drinking water standards	RR	*McCoy's RCRA Reference*
NORM	Naturally occurring radioactive material	RT	Regulatory threshold
NOV	Notice of violation	RU	*McCoy's RCRA Unraveled*
NPDES	National pollutant discharge elimination system	RWMS	Remediation waste management site
NPL	National priorities list	SAA	Satellite accumulation area
NPRM	Notice of proposed rulemaking	SCR	Selective catalytic reduction
NRC	Nuclear Regulatory Commission, National Response Center, or Non-reusable container	SDS	Safety data sheet
		SDWA	Safe Drinking Water Act
		SEP	Supplemental environmental project
		SIC	Standard industrial classification
		SIM	Specific ion monitoring
		SLAB	Spent lead-acid battery
NSPS	New source performance standards	SPCC	Spill prevention, control, and countermeasures
NWW	Nonwastewater	SQG	Small quantity generator
OBA	Oxygen breathing apparatus	SSL	Soil screening level
OECA	Office of Enforcement and Compliance Assurance	STC	Single-trip container
		SWDA	Solid Waste Disposal Act of 1965

SWMU	Solid waste management unit	UHC	Underlying hazardous constituent
TC	Toxicity characteristic	UIC	Underground injection control
TCE	Trichloroethylene	UST	Underground storage tank
TCLP	Toxicity characteristic leaching procedure	UTS	Universal treatment standards
TENORM	Technologically enhanced naturally occurring radioactive material	UXO	Unexploded ordnance
		VO	Volatile organics
TETF	Totally enclosed treatment facility	VOC	Volatile organic compound
TI	Tolerance interval	VRU	Vapor-recovery unit
TIM	Technical information memorandum	VSP	Visual sample plan
TOC	Total organic carbon	VTD	Vacuum thermal desorption
TRI	Toxic Release Inventory	WAP	Waste analysis plan
TRU	Transuranic	WIPP	Waste Isolation Pilot Plant
TSCA	Toxic Substances Control Act	WW	Wastewater
TSD	Treatment, storage, and disposal	WWTU	Wastewater treatment unit
TSS	Total suspended solids	XML	Extensible markup language
TU	Temporary unit	ZHE	Zero-headspace extraction

Acronyms